SYMBOLS

n	An integer (1, 2, 3, . . .)
n	Principal quantum number
n	Turns per unit length
N	Number of turns
N	Normal (perpendicular) force
p	Object distance
p	Pressure
P	Power
PE	Potential Energy
q	Image distance
Q	Electric charge
Q	Heat
r	Angle of reflection, refraction
r	Internal resistance
r, R	Radius
R	Range
R	Rate of flow
R	Electrical resistance
s	Displacement from equilibrium position
s	Distance; displacement
S	Shear modulus
t	Time
T	Period
T	Temperature
T	Tension
v	Speed
v	Velocity
V	Potential difference
V	Volume
w	Energy density
w	Weight
W	Work
X_C	Capacitive reactance
X_L	Inductive reactance
Y	Young's Modulus
Z	Atomic number
Z	Impedance
α (alpha)	Angular acceleration
α (alpha)	Temperature coefficient of resistivity
β (beta)	Sound intensity level
γ (gamma)	Electromagnetic photon
γ (gamma)	Surface tension
Δ (delta)	"Change in"
η (eta)	Viscosity
θ (theta)	An angle
λ (lambda)	Wavelength
μ (mu)	Coefficient of friction
μ (mu)	Magnetic permeability
ρ (rho)	Resistivity
σ (sigma)	Stefan-Boltzmann constant
Σ (sigma)	"Sum of"
τ (tau)	Torque
ϕ (phi)	Phase angle
Φ (phi)	Magnetic flux
ω (omega)	Angular velocity

POWERS OF TEN

10^{-10}	$= 0.000,000,000,1$	10^{0}	$= 1$
10^{-9}	$= 0.000,000,001$	10^{1}	$= 10$
10^{-8}	$= 0.000,000,01$	10^{2}	$= 100$
10^{-7}	$= 0.000,000,1$	10^{3}	$= 1000$
10^{-6}	$= 0.000,001$	10^{4}	$= 10,000$
10^{-5}	$= 0.000,01$	10^{5}	$= 100,000$
10^{-4}	$= 0.000,1$	10^{6}	$= 1,000,000$
10^{-3}	$= 0.001$	10^{7}	$= 10,000,000$
10^{-2}	$= 0.01$	10^{8}	$= 100,000,000$
10^{-1}	$= 0.1$	10^{9}	$= 1,000,000,000$
10^{0}	$= 1$	10^{10}	$= 10,000,000,000$

MULTIPLIERS FOR SI UNITS

a	atto-	10^{-18}	da	deka-	10^{1}
f	femto-	10^{-15}	h	hecto-	10^{2}
p	pico-	10^{-12}	k	kilo-	10^{3}
n	nano-	10^{-9}	M	mega-	10^{6}
μ	micro-	10^{-6}	G	giga-	10^{9}
m	milli-	10^{-3}	T	tera-	10^{12}
c	centi-	10^{-2}	P	peta-	10^{15}
d	deci-	10^{-1}	E	exa-	10^{18}

MODERN TECHNICAL PHYSICS

FIFTH EDITION

MODERN TECHNICAL PHYSICS

ARTHUR BEISER

FIFTH EDITION

ADDISON-WESLEY PUBLISHING COMPANY, INC.

Reading, Massachusetts • Menlo Park, California
Don Mills, Ontario • Wokingham, U.K. • Amsterdam • Sydney
Singapore • Tokyo • Madrid • Bogota • Santiago • San Juan

Sponsoring Editor: Andrew Crowley
Production Coordinator: Kristina M. Montague
Copy Editor: Linda Thompson
Book Designer: Janet Bollow
Cover Designer: Michael Rogondino
Artists: Georg Klatt; Ben Turner Graphics
Cover Photograph: © Peter Runyon, The Image Bank

Photo credits appear on page 905

Library of Congress Cataloging in Publication Data

Beiser, Arthur.
  Modern technical physics.

  Includes index.
  1. Physics. I. Title.
QC23.B4143 1987   530   86-17615
ISBN 0-8053-0684-6

ABCDEFGHIJ–DO– 89876

PREFACE

SCOPE OF THE BOOK

Physics, whose concern is the fundamental structure, properties, and behavior of matter, forms the core of modern technology. Students who plan careers in technology, whatever the branch, must have some competence in physics, and the purpose of this book is to help provide it.

Modern Technical Physics is a straightforward presentation of the laws of motion, the properties of matter in bulk, vibrations and waves, heat and thermodynamics, electromagnetism, optics, and, finally, atomic and nuclear physics. The mathematical level has been kept to elementary algebra and simple trigonometry, which are reviewed in two appendixes to the extent required by the book.

STUDENT AIDS

A variety of aids are provided to help the reader master the text. These aids, which make up a built-in study guide, are as follows:

1. *Chapter Objectives.* A list of exactly what each chapter should enable the student to do is given at the beginning of the chapter.

2. *Text Notes.* Significant ideas are flagged by brief marginal notes next to their mention in the text. Basic equations are named and frequently stated in words as well as in symbols to make clear what they signify.

3. *Examples.* Several hundred worked examples are included in the text to show how physical ideas are put into practice.

4. *Important Terms and Formulas.* The meanings of important terms are given at the end of each chapter together with a list of formulas needed to solve problems based on the chapter material. These lists act as chapter summaries.

5. *Multiple Choice Questions.* A typical chapter has 30 multiple choice questions (with answers) that serve as a quick check on understanding. Correct answers provide reinforcement and encouragement, and incorrect ones identify areas of weakness.

6. *Exercises.* A typical chapter has 50 or 60 exercises (some have over 100) arranged in groups that correspond to sections in the text. The exercises in each group are further divided into three categories. Those with italic numbers (*1, 2, 3, . . .*) are questions that test understanding of basic principles and simple problems that usually need only one or two steps to solve. Those with roman numbers (5, 6, 7, . . .) range from fairly easy to modestly challenging; most students should be able to cope with them. Exercises given boldface numbers (**10, 11, 12,** . . .) are more difficult.

7. *Outline Solutions.* Outline solutions for all the odd-numbered exercises are provided at the back of the book. Each outline solution shows how the problem can be attacked and gives the answer, but some of the intermediate steps are omitted. In this way the reader who is puzzled by a certain type of problem can find out how to proceed and still have something left to do on his or her own. The outline solutions supplement the worked examples in the body of the text to furnish more than a thousand model answers to typical questions and problems. Mastering the solved exercises should bring the unsolved even-numbered ones within the competence of the reader.

8. *Math Appendixes.* Only elementary algebra and simple trigonometry are employed in the book, and these are reviewed in two appendixes. Powers-of-ten notation for small and large numbers is carefully explained and the use of electronic calculators for such purposes as finding powers and roots is described.

9. *Glossary.* A comprehensive glossary of the basic terms of physics appears at the back of the book.

THIS EDITION

In preparing the fifth edition of *Modern Technical Physics,* a great many small changes were made throughout the text to iron out rough spots and to clarify or amplify arguments. SI units were given increased emphasis. In addition, a number of sections either were rewritten or are entirely new. Although none of the alterations is very large, the total amounts to a major improvement. Since classroom experience with the previous edition showed that its organization and level of difficulty were satisfactory, they were not altered. To help make the reader's task easier, several dozen more figures and photographs were added, for a total of over 800.

The end-of-chapter exercises were given a complete overhaul. As described earlier, they are now arranged in groups according to the text section they correspond to and are further divided into three categories according to difficulty. This scheme should help both students and teachers. Many of the old exercises and multiple choice questions were replaced and others added; the total is now 2500, one-third more than in the previous edition.

ACKNOWLEDGMENTS

In preparing this edition of *Modern Technical Physics*, I have had the benefit of comments by David Mills, College of the Redwoods, California; James B. Shaw, Jr., Williamsport Area Community College, Pennsylvania; and John Thornton, Stark Technical College, Ohio. Their generous help was of great value and is much appreciated.

Readers who have suggestions for improving the book are cordially invited to send them to me in care of the publisher.

Arthur Beiser

CONTENTS

MODERN TECHNICAL PHYSICS

FIFTH EDITION

1

DESCRIBING MOTION

Everything in the universe is moving, from the electrons in an atom to the stars in the sky. Since the goal of physics is to understand the nature and behavior of the universe, physicists are concerned with moving things in nearly all their work. The same laws of motion that hold for electrons and stars hold in the more familiar world of technology, and these laws are a good starting point in the study of applied physics. The subject of this chapter and the next is motion along a straight line, the simplest kind of all. We shall find that introducing the quantities speed and acceleration to supplement the basic ones of time and distance enables us to deal with even seemingly complicated motions in a straightforward way.

1–1 UNITS

The raw material of physics consists of measurements. The conclusions physicists draw from their measurements are used by them and other scientists to investigate nature further, and by engineers to design such things as bridges, airplanes, and computers.

CHAPTER OBJECTIVES

Completing this chapter should enable you to:

1. Properly use units in making calculations.

2. Change the units in which a quantity is expressed from those of one system to those of another system.

3. Distinguish between instantaneous speed and average speed.

4. Determine the instantaneous speed of a moving object from a graph of distance versus time.

5. Use the formula $\bar{v} = s/t$ to solve problems that involve distance, time, and average speed.

6. Find the acceleration of an object whose speed is changing.

7. Use the formula $a = (v_f - v_0)/t$ to solve problems that involve acceleration, initial and final speeds, and time.

8. Use the formula $s = v_0 t + \frac{1}{2} at^2$ to solve problems that involve distance, initial speed, acceleration, and time.

9. Use the formula $v_f^2 = v_0^2 + 2as$ to solve problems that involve acceleration, distance, and initial and final speeds.

10. Analyze situations in which the acceleration of an object is opposite in direction to its initial speed.

11. Apply the formulas that describe accelerated motion to objects that fall from rest.

12. Solve problems that concern objects thrown upward or downward with initial speeds.

Before we actually take up the study of physics, then, we must review how physical quantities are expressed and treated mathematically.

The process of measurement is basically one of comparison. A certain standard quantity of some kind, called a *unit,* is first established, and other quantities of the same kind are compared with it. When we say a ladder is 2.6 m long, we mean that its length is 2.6 times a certain distance called the meter whose magnitude is fixed by international agreement. The result of every measurement must therefore have two parts, a number to answer the question "How many?" and a unit to answer the question "Of what?"

Measurement requires units

Nearly all quantities in the physical world can be expressed in terms of only four fundamental units, those of length, time, mass, and electric current. Thus every unit of area (the square meter, for instance) is the product of a length unit and a length unit; every unit of speed (the kilometer per hour, for instance) is a length unit divided by a time unit; every unit of force (the pound, for instance) is the product of a mass unit and a length unit divided by the square of a time unit; and every unit of electric charge (the coulomb, for instance) is the product of a unit of electric current and a time unit.

Four basic units are sufficient for most quantities

TABLE 1–1
Basic SI units

Quantity	Symbol	Unit	Unit Symbol
Length	L	meter	m
Mass	m	kilogram	kg
Time	t	second	s
Electric current	I	ampere	A
Temperature	T	kelvin	K
Luminous intensity	I_L	candela	cd

SI units are used in science

A *system of units* is a set of specified units of length, time, mass, and electric current from which all other units are to be derived. All units are arbitrary. Thousands of different ones have been adopted at various times and places in the course of history, a number of which survive. The most widely used is the *metric system,* which was introduced in France nearly two centuries ago. The virtues of the metric system soon led to its universal adoption by scientists, more slowly to its spread throughout the world for engineering and everyday life. The Système Internationale d'Unités (SI) is the current version of the metric system and is the one emphasized in this book. The fundamental SI units are listed in Table 1–1.

The British system is rarely used outside the U.S.

Units of the *British system,* which are seldom found elsewhere (even in Great Britain) these days, are still widely used in the United States. Although they are on the way out here as well, for the time being American students of technology need to be acquainted with them. In the British system the units of length and mass are respectively the foot and the slug (the pound is a unit of force). Units of time and electrical units are the same in both systems.

Calculations are easy with SI units

Because a basic unit in a system may not always be convenient in size for a given measurement, other units have come into use within each system. Thus long distances are usually given in miles rather than in feet, or in kilometers rather than in meters. The great advantage of the SI system is that it is wholly decimalized (Fig. 1–1), which makes calculations easy, whereas the British system is quite irregular in this respect. (1 km = 1000 m, for example, but 1 mi = 5280 ft.) The chief units in each system, together with their equivalents in the other system, are given inside the back cover of the book. The conventional prefixes used with SI units of all kinds are listed in Table 1–2; powers-of-ten notation is reviewed in Appendix B.

Standard units

It is essential that standard units be absolutely constant in magnitude. Whenever possible in recent years they have been redefined in terms of quantities in nature that are regarded as invariant under all circumstances and not subject to change in time.

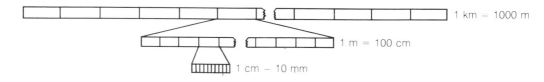

1 km – 1000 m

1 m = 100 cm

1 cm – 10 mm

FIG. 1–1 There are 1000 m in a kilometer, 100 cm in a meter, and 10 mm in a centimeter.

Prefix	Power of Ten	Abbre- viation	Pronunciation	Example
atto-	10^{-18}	a	at' toe	1 aC $=$ 1 attocoulomb $= 10^{-18}$ C
femto-	10^{-15}	f	fem' toe	1 fm $=$ 1 femtometer $= 10^{-15}$ m
pico-	10^{-12}	p	pee' koe	1 pf $=$ 1 picofarad $= 10^{-12}$ f
nano-	10^{-9}	n	nan' oe	1 ns $=$ 1 nanosecond $= 10^{-9}$ s
micro-	10^{-6}	μ	my' kroe	1 μA $=$ 1 microampere $= 10^{-6}$ A
milli-	10^{-3}	m	mil' i	1 mg $=$ 1 milligram $= 10^{-3}$ g
centi-	10^{-2}	c	sen' ti	1 cl $=$ 1 centiliter $= 10^{-2}$ l
kilo-	10^{3}	k	kil' oe	1 kN $=$ 1 kilonewton $= 10^{3}$ N
mega-	10^{6}	M	meg' a	1 MW $=$ 1 megawatt $= 10^{6}$ W
giga-	10^{9}	G	ji' ga	1 GeV $=$ 1 gigaelectronvolt $= 10^{9}$ eV
tera-	10^{12}	T	ter' a	1 Tm $=$ 1 terameter $= 10^{12}$ m
peta-	10^{15}	P	pe' ta	1 Ps $=$ 1 petasecond $= 10^{15}$ s
exa-	10^{18}	E	ex' a	1 EJ $=$ 1 exajoule $= 10^{18}$ J

TABLE 1–2
Subdivisions and multiples of SI units are widely used, and each is designated by a prefix according to the corresponding power of ten

The second, for instance, is now defined in terms of the microwave radiation given off under certain circumstances by the ^{137}Cs atom: 1 s is equal to the time needed for 9,192,631,770 cycles of this radiation to be emitted. The meter in turn is defined in terms of the second and the speed of light: 1 m is the distance traveled in 1/299,792,458 s by light waves in a vacuum. No longer must the entire world rely upon the measurement between scratches on a certain platinum-iridium bar kept at Sèvres, France.

Often a quantity expressed in terms of a certain unit must instead be expressed in terms of another unit of the same kind. For instance, we might find from a European map that Amsterdam is 648 km from Berlin and want to know what this distance is in miles. To carry out such a conversion we must keep in mind two rules:

1. Units are treated in an equation in exactly the same way as any algebraic quantity, and may be multiplied and divided by one another;
2. Multiplying or dividing a quantity by 1 does not affect its value.

How to convert units

To convert 648 km to its equivalent in miles we note from the table inside the back cover that

$$1 \text{ km} = 0.621 \text{ mi}$$

Therefore

$$0.621 \frac{\text{mi}}{\text{km}} = 1$$

and multiplying or dividing any quantity by 0.621 mi/km does not affect its value but only changes the units in which it is given. Hence we have

$$s = (648 \text{ km}) \left(0.621 \frac{\text{mi}}{\text{km}} \right) = 402 \text{ mi}$$

since km/km $=$ 1 and so drops out. (When multiplied out completely, 648 $\times$ 0.621 $=$ 402.408. However, as discussed in Appendix B–4, we can keep only as many significant figures as there are in the least accurately known quantity used in a calculation. Since this means three significant figures here, the result must be expressed as 402 mi.)

Example   The piston displacement of the Volvo MD21A diesel engine is 2.11 liters, where 1 liter (L) = 1000 cm^3. Express this volume in cubic inches.

Solution   The displacement of the engine is V = 2110 cm^3. Since 1 cm = 0.394 in., 1 cm^3 = (0.394 in.)^3 = 0.0612 in.^{3}, and

$$V = (2110 \, \text{cm}^3) \left(0.0612 \, \frac{\text{in.}^3}{\text{cm}^3} \right) = 129 \, \text{in.}^3$$ ■

Example   An American driving in Europe finds that her car averages 9.5 km/L of gasoline, where 1 gal = 3.785 L. Express the fuel economy of the car in miles/gallon.

Solution   Here two units are to be converted, which we can do in a single step:

$$\left(9.5 \, \frac{\text{km}}{\text{L}} \right) \left(0.621 \, \frac{\text{mi}}{\text{km}} \right) \left(3.785 \, \frac{\text{L}}{\text{gal}} \right) = 22 \, \text{mi/gal}$$ ■

Example   (a) Express a length of 47.2 mm in meters. (b) Express an electric current of 6.83 × 10^{-5} A in microamperes.

Solution   (a) We follow the same procedure as in any other conversion of units. Since 1 mm = 10^{-3} m = 0.001 m,

$$s = (47.2 \, \text{mm}) \left(10^{-3} \, \frac{\text{m}}{\text{mm}} \right) = 47.2 \times 10^{-3} \, \text{m} = 0.0472 \, \text{m}$$

(b) What we must find is how many microamperes there are in the given current I. Since 1 μA = 10^{-6} A, we therefore divide I by 10^{-6} A/μA. As discussed in Appendix B–2, $10^n / 10^m = 10^{n-m}$, so

$$I = \frac{6.83 \times 10^{-5} \, \text{A}}{10^{-6} \, \text{A/μA}} = 6.83 \times 10^{-5-(-6)} \, \text{μA} = 6.83 \times 10^{-5+6} \, \text{μA}$$

$$= 6.83 \times 10^1 \, \text{μA} = 68.3 \, \text{μA}$$ ■

1–2   SPEED

Average speed

As we all know, the *speed* of a moving object is the rate at which it covers distance. It is important to distinguish between average speed and instantaneous speed. The *average speed* $\bar{v}$ of something that travels the distance s in the time interval t is

$$\bar{v} = \frac{s}{t} \qquad\qquad \textit{Average speed} \quad (1\text{–}1)$$

$$\text{Average speed} = \frac{\text{distance traveled}}{\text{time interval}}$$

Thus a car that has gone 180 km in 3 h had an average speed of

$$\bar{v} = \frac{s}{t} = \frac{180 \, \text{km}}{3 \, \text{h}} = 60 \, \text{km/h}$$

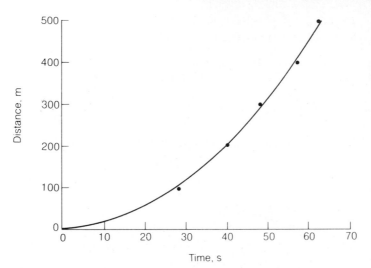

FIG. 1–2(a) A graph of the data in Table 1–3. Each point represents the result of one measurement. A smooth curve does not quite go through all the points, but we may reasonably attribute the discrepancies to experimental error.

The average speed of the car is only part of the story of its journey, however, because knowing $\bar{v}$ does not tell us whether the car had the same speed for the entire 3 h or sometimes went faster than 60 km/h and sometimes slower.

Let us suppose we are in a car starting from rest and that we note the time at which its odometer indicates it has covered 100 m, 200 m, 300 m, and so on. The data might appear as in Table 1–3. When we plot these data on a graph, as in Fig. 1–2(a), we find that the line joining the various points is not a straight line but shows a definite upward curve. In each successive equal time interval (as marked off at the bottom of the graph), the car covers a greater distance than before—it is going faster and faster.

Total distance, m	0	100	200	300	400	500	**TABLE 1–3**
Elapsed time, s	0	28	40	49	57	63	

Even though the car's speed is changing, at every moment it has a certain definite value (which is what is indicated by its speedometer). To find this *instantaneous speed* v at a particular time t, we draw a straight line tangent to the distance-time curve at that value of t. The length of the line does not matter. Then we determine v from the tangent line from the formula

Instantaneous speed

$$v = \frac{\Delta s}{\Delta t} \qquad\qquad \textit{Instantaneous speed} \quad (1\text{–}2)$$

where Δs is the distance interval between the ends of the tangent and Δt is the time

FIG. 1–2(b) The procedure for finding the instantaneous speed at $t = 40$ s from the data of Table 1–3. Because the graph of distance versus time is not a straight line, the instantaneous speed is changing continuously.

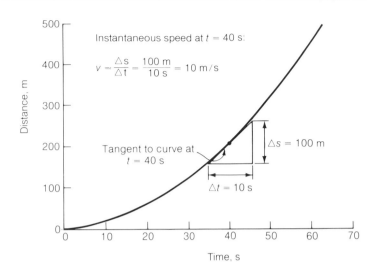

interval between them. (Δ is the Greek capital letter *delta*.) The instantaneous speed of the car at $t = 40$ s is, from Fig. 1–2(b),

$$v = \frac{\Delta s}{\Delta t} = \frac{100 \text{ m}}{10 \text{ s}} = 10 \text{ m/s}$$

Table 1–4 shows the instantaneous speeds of the car at 10-s intervals as determined from the graph.

TABLE 1–4

Elapsed time, s	0	10	20	30	40	50	60
Instantaneous speed, m/s	0	2.5	5.0	7.5	10	12.5	15

Constant speed

When the instantaneous speed of an object does not change, it is moving at *constant speed*. Figure 1–3 is a distance-time graph of a car that has a constant speed of 7.5 m/s; the curve is, of course, a straight line. For the case of constant speed, Eq. (1–2) provides two useful formulas. The first gives the distance covered in a given period of time:

$$s = vt \qquad (v = \text{constant}) \tag{1–3}$$
$$\text{Distance} = \text{speed} \times \text{time}$$

The second formula gives the time needed to cover a given distance:

$$t = \frac{s}{v} \qquad (v = \text{constant}) \tag{1–4}$$
$$\text{Time} = \frac{\text{distance}}{\text{speed}}$$

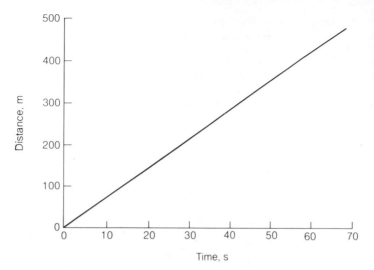

FIG. 1–3 The distance-time graph of a car traveling at a constant speed of 7.5 m/s is a straight line.

Thus if the speed of something is constant, we can predict exactly how far it will go in a given period of time; or, given the distance, we can determine the time required.

Example Echoes return in 2.5 s to a person standing in front of a cliff. How far away is the cliff? The speed of sound in air at sea level is 343 m/s when the temperature is 20°C.

Solution The total distance a sound travels in 2.5 s is

$$s = vt = (343 \text{ m/s})(2.5 \text{ s}) = 858 \text{ m}$$

The sound must travel to the cliff and back, and so the cliff is half this distance away, which is $\frac{1}{2} \times 858$ m $= 429$ m. ■

Example If the highway speed limit were increased from 55 mi/h to 65 mi/h, how much time would be saved on a 100-mi trip made at the maximum permitted speed?

Solution At $v_1 = 55$ mi/h the time needed for the trip is $t_1 = s/v_1$ and at $v_2 = 65$ mi/h the time needed is $t_2 = s/v_2$. Hence the time saved is

$$\Delta t = t_1 - t_2 = \frac{s}{v_1} - \frac{s}{v_2} = \frac{100 \text{ mi}}{55 \text{ mi/h}} - \frac{100 \text{ mi}}{65 \text{ mi/h}}$$

$$= (1.82 - 1.54)\text{h} = 0.28 \text{ h}$$

Since 1 h = 60 min, this result can also be expressed as

$$\Delta t = (0.28 \text{ h})(60 \text{ min/h}) = 17 \text{ min}$$ ■

Example An airplane takes off at 9:00 A.M. and flies in a straight path at 300 km/h until 1:00 P.M. At 1:00 P.M. its speed is increased to 400 km/h and it maintains this speed in the same direction until it lands at 3:30 P.M. What is the average speed of the airplane for the entire flight?

Solution   To calculate the average speed of the airplane we must know the total distance s it has covered and the total time t involved. In the first part of the flight, which lasted 4.00 h, the airplane covered

$$s_1 = v_1 t_1 = (300 \text{ km/h})(4.00 \text{ h}) = 1200 \text{ km}$$

and in the second part, which lasted 2.50 h, it covered

$$s_2 = v_2 t_2 = (400 \text{ km/h})(2.50 \text{ h}) = 1000 \text{ km}$$

The total distance and total time of the flight are

$$s = s_1 + s_2 = (1200 + 1000) \text{ km} = 2200 \text{ km}$$

$$t = t_1 + t_2 = (4.00 + 2.50) \text{ h} = 6.50 \text{ h}$$

and the average speed is therefore

$$v = \frac{s}{t} = \frac{2200 \text{ km}}{6.50 \text{ h}} = 338 \text{ km/h} \qquad\blacksquare$$

1–3   ACCELERATION

Acceleration

In the real world few objects move at constant speed for very long. Something whose speed is increasing or decreasing is said to be *accelerated*. The same term is used for a changing direction, even with the speed staying the same (Fig. 1–4). In this chapter we are only concerned with straight-line motion; accelerations that involve changes in direction are discussed later.

Force and acceleration

What is it that causes something to be accelerated? The answer is a force, which for the moment we can think of as a push or a pull. If no force acts on it, an object at rest remains at rest and an object in motion continues in motion along a straight line at constant speed. The connection between force and acceleration is the subject of Chapter 2.

Acceleration is rate of change of speed

Just as speed is the rate of change of distance with time, acceleration is the rate of change of speed with time. If an object's speed is v_0 to begin with and changes to v_f during a time interval t, its acceleration a is given by the formula

$$a = \frac{v_f - v_0}{t} \qquad\qquad \textit{Acceleration} \quad (1\text{–}5)$$

$$\text{Acceleration} = \frac{\text{change in speed}}{\text{time interval}}$$

Positive and negative acceleration

When the final speed v_f is greater than the initial speed v_0, the acceleration is positive, which signifies that the object is going faster and faster. When the final speed is less than the initial speed, the acceleration is negative, which signifies that the object is going slower and slower.

Constant acceleration

Let us return to the data on the car of Tables 1–3 and 1–4 and plot a graph of its instantaneous speed v versus time t, as in Fig. 1–5. All the points lie on a straight line, which means that v is directly proportional to t. Although the car's speed is not constant,

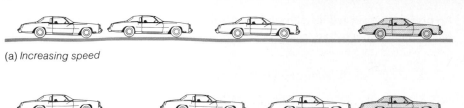

(a) *Increasing speed*

(b) *Decreasing speed*

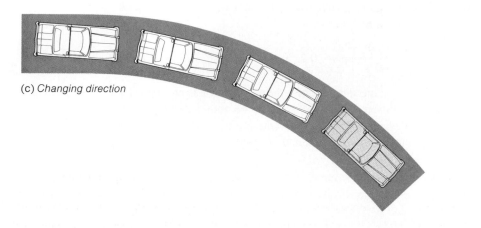

(c) *Changing direction*

FIG. 1–4 The successive positions of three accelerated cars after equal periods of time. (a) The car is going faster and faster, so it travels a longer distance in each period of time. (b) The car is going slower and slower, so it travels a shorter distance in each period of time. (c) The car's speed is constant, but its direction changes.

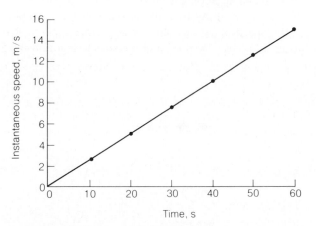

FIG. 1–5 A graph of instantaneous speed versus time for the data of Table 1–4. Although the car's speed is not constant, it varies in a uniform way with time. This is an example of constant acceleration.

it varies in a uniform way with time; as time goes on, the speed increases exactly in proportion. Therefore the car's acceleration is constant.

From its definition, acceleration is expressed in terms of

Dimensions of acceleration

$$\frac{\text{Speed}}{\text{time}} = \frac{\text{distance/time}}{\text{time}} = \frac{\text{distance}}{\text{time} \times \text{time}} = \frac{\text{distance}}{\text{time}^2}$$

An object whose speed increases by 10 m/s in each second would accordingly have its acceleration expressed as

$$a = 10\frac{\text{m/s}}{\text{s}} = 10\,\text{m/s}^2$$

The initial speed of the car whose motion we have been considering is $v_0 = 0$ and after, say, $t = 20$ s it is $v_f = 5.0$ m/s. Hence the car's acceleration is

$$a = \frac{v_f - v_0}{t} = \frac{(5.0 - 0)\ \text{m/s}}{20\ \text{s}} = 0.25\ \text{m/s}^2$$

If we make the same calculation at the later time $t = 40$ s, we have, since v_f is now 10 m/s according to Table 1–4,

$$a = \frac{(10 - 0)\ \text{m/s}}{40\ \text{s}} = 0.25\ \text{m/s}^2$$

The value of a is the same because the acceleration is constant. If the acceleration were not constant, different values of a would be obtained at different times.

Not all accelerations are constant, of course, but a great many real motions are best understood by idealizing them in terms of constant accelerations.

Example　Discuss the motion of the car whose speed-time graph is shown in Fig. 1–6.

Solution　A horizontal line on a v-t graph means that v does not change, so the acceleration is 0 at first. A line sloping upward means a positive acceleration (v increasing), and a line sloping downward means a negative acceleration (v decreasing). Hence the car begins to move at the constant speed of 10 m/s, then is accelerated at 2 m/s^2 to the speed of 20 m/s, travels at 20 m/s for 5 s, and finally undergoes a negative acceleration of -4 m/s^2 until it comes to rest at $t = 25$ s. ∎

FIG. 1–6　Instantaneous speed versus time for a certain car.

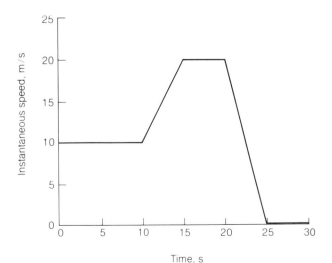

1-4 SPEED AND ACCELERATION

In the event an object starts to accelerate from some initial speed v_0, its change in speed at during the time interval t in which the acceleration a (assumed constant) occurs is added to v_0 (Fig. 1-7). Hence the final speed v_f at a time t after the acceleration begins is the initial speed v_0 plus the change in speed at:

Final speed of accelerated object

$$v_f = v_0 + at \qquad\qquad \textit{Final speed} \quad (1-6)$$

Final speed = initial speed + speed change

The same formula can be obtained by solving Eq. (1-5) for v_f.

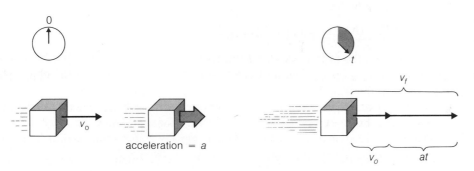

FIG. 1-7 The final speed v_f of an accelerated body is equal to its initial speed v_0 plus the change at that occurred during the interval t.

acceleration = a

Example A car accelerates at the constant rate of 5 ft/s^2 from an initial speed of 20 mi/h. Find its speed in mi/h after 10 s.

Solution After 10 s the increase at in the car's speed is

$$at = \left(5\frac{\text{ft}}{\text{s}^2}\right)(10\,\text{s}) = 50\,\text{ft/s}$$

Since 1 ft/s = 0.682 mi/h,

$$at = \left(50\frac{\text{ft}}{\text{s}}\right)\left(0.682\frac{\text{mi/h}}{\text{ft/s}}\right) = 34\,\text{mi/h}$$

and the final speed v_f of the car is

$$v_f = v_0 + at = (20 + 34)\,\text{mi/h} = 54\,\text{mi/h} \qquad\blacksquare$$

Equations (1-5) and (1-6) can be used regardless of the algebraic signs of v_0 and a. In the case of a car, a negative speed customarily implies motion opposite to a specified direction, which we might refer to as backward motion (though the car may actually be facing either way). A negative acceleration means that the car is slowing if v_0 is positive or increasing in backward speed if v_0 is negative.

If v_0 is negative and a is positive, the car is traveling backward at a decreasing speed. The importance of keeping track of the algebraic signs of the various quantities is obvious.

Example   A car has an initial speed of 20 m/s and an acceleration of -1 m/s^2. Find its speed after 10 s and after 50 s.

Solution   The speed after 10 s is

$$v_f = v_0 + at = 20 \text{ m/s} - (1 \text{ m/s}^2)(10\text{s}) = (20 - 10) \text{ m/s} = 10 \text{ m/s}$$

which is half what it was initially. After 50s, the speed of the car, if a stays the same, is

$$v_f = v_0 + at = 20 \text{ m/s} - (1 \text{ m/s}^2)(50 \text{ s}) = (20 - 50) \text{ m/s} = -30 \text{ m/s}$$

which is in the opposite direction and greater than the original speed v_0.   ■

1–5   DISTANCE AND ACCELERATION

As we have seen, the final speed of an object that has been accelerated for the time t is given by $v_f = v_0 + at$. The next question to ask is, how far does the object go during the time interval t?

Average speed during constant acceleration

We know that, in general, $s = \bar{v}t$, so if we can determine the average speed during the time interval t we can also find s, the distance through which the body moves. Because the acceleration a is constant, v is changing at a uniform rate, and the average speed can be found in the same way as the average of two numbers:

$$\bar{v} = \frac{v_0 + v_f}{2} \quad (a = \text{constant}) \qquad \textit{Average speed} \quad (1–7)$$

$$\text{Average speed} = \frac{\text{initial speed } + \text{ final speed}}{2}$$

If a is not constant, this formula does not hold.

Distance covered by accelerated object

Here the initial speed is v_0 and the final speed is $v_0 + at$; hence

$$\bar{v} = \frac{v_0 + v_0 + at}{2} = v_0 + \tfrac{1}{2} at$$

The distance traveled is accordingly

$$s = \bar{v}t = (v_0 + \tfrac{1}{2} at) \times t$$

which yields the very useful formula

$$s = v_0 t + \tfrac{1}{2} at^2 \qquad \textit{Distance under constant acceleration} \quad (1–8)$$

Figure 1–8 illustrates this result. When the initial speed is $v_0 = 0$, we have simply

$$s = \tfrac{1}{2} at^2 \quad (v_0 = 0) \qquad\qquad\qquad (1–9)$$

Example   A car has an initial speed of 20 m/s and an acceleration of -1 m/s^2. Find its displacement after the first 10 s and after the first 50 s from the moment the acceleration begins.

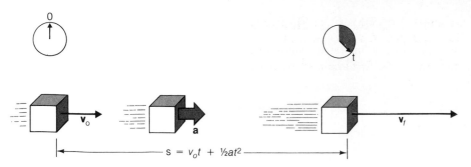

FIG. 1–8 A body that has the initial speed v_0 at $t = 0$ and undergoes the constant acceleration a travels a distance of $s = v_0t + \frac{1}{2}at^2$ in the time t.

FIG. 1–9 Position and speed, at various times, of a car that has a constant acceleration of -1 m/s^2 and whose initial speed is 20 m/s. At $t = 20$ s the car has come to a stop, and then it begins to move in the negative (backward) direction. At $t = 40$ s the car is back where it started, but moving in the opposite direction to its initial one. At $t = 50$ s the car is 250 m behind its starting point.

Solution In the first 10 s the car travels

$$s = v_0t + \frac{1}{2}at^2 = (20 \text{ m/s})(10\text{s}) - \frac{1}{2}(1 \text{ m/s}^2)(10 \text{ s})^2$$
$$= (200 - 50)\text{m} = 150 \text{ m}$$

and its displacement after 50 s is

$$s = v_0t + \frac{1}{2}at^2 = (20\text{ m/s})(50\text{ s}) - \frac{1}{2}(1 \text{ m/s}^2)(50\text{ s})^2$$
$$= (1000 - 1250) \text{ m} = -250\text{ m}$$

This result means that the car is 250 m *behind* its starting point after 50 s have elapsed (Fig. 1–9). ■

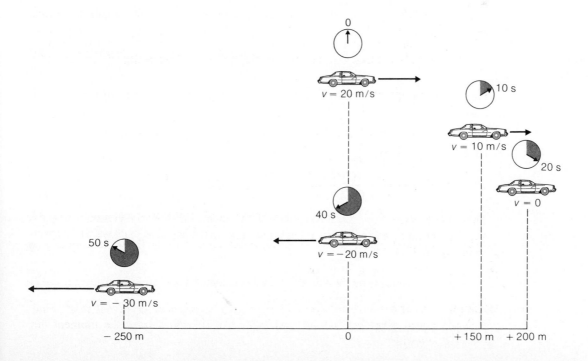

1–6 DISTANCE, SPEED, AND ACCELERATION

Additional formulas for accelerated motion

It is not hard to find relationships among s, v_0, v_f, and a that do not directly involve the time t. The first step is to rewrite the defining formula for acceleration, which is

$$a = \frac{v_f - v_0}{t}$$

in the equivalent form

$$t = \frac{v_f - v_0}{a}$$

This formula gives the time during which the speed of the body changed from v_0 to its final value of v_f. Now we substitute this expression for t into the formula for the distance traveled:

$$s = v_0 t + \tfrac{1}{2} a t^2$$

The result is

$$
\begin{aligned}
s &= v_0 \frac{(v_f - v_0)}{a} + \tfrac{1}{2} a \frac{(v_f - v_0)^2}{a^2} \\
&= \frac{v_0 v_f}{a} - \frac{v_0^2}{a} + \frac{v_f^2}{2a} - \frac{v_0 v_f}{a} + \frac{v_0^2}{2a} \\
&= \frac{v_f^2 - v_0^2}{2a}
\end{aligned}
$$

Distance under constant acceleration (1–10)

Also, by multiplying both sides of the last equation by $2a$ and rearranging the terms, we arrive at the formula

$$v_f^2 = v_0^2 + 2as \qquad\qquad \textit{Speed under constant acceleration}\quad (1–11)$$

Example How far will the car of the previous example have gone when it comes to a stop?

Solution When the car is at rest, $v_f = 0$. Since $v_0 = 20$ m/s and $a = -1$ m/s^2, we have from Eq. (1–10)

$$s = \frac{v_f^2 - v_0^2}{2a} = \frac{0 - (20 \text{ m/s})^2}{(2)(-1 \text{ m/s}^2)} = \frac{400 \text{ m}^2/\text{s}^2}{2 \text{ m/s}^2} = 200 \text{ m}$$

The car comes to a stop after it has gone 200 m. Then, since the acceleration is negative and is assumed here to continue, it begins to move in the negative (backward) direction, as in Fig. 1–9. ■

Another useful formula that involves distance and speed follows from Eq. (1–7). Multiplying both sides by the time t gives

$$\bar{v} t = \left(\frac{v_0 + v_f}{2} \right) t$$

Since $\bar{v}t$ is the distance s traveled in the time t by an object whose acceleration is constant,

$$s = \left(\frac{v_0 + v_f}{2} \right) t \qquad\qquad \textit{Distance under constant acceleration} \quad (1\text{--}12)$$

This formula differs from Eq. (1–10) in that it contains the time t but not the acceleration a. Thus Eq. (1–12) is handy when the value of a is neither known nor has to be determined.

1–7 HOW TO SOLVE PROBLEMS

Most people who study physics find the subject matter interesting as well as useful. When the time comes to put theory into practice, however, there is often a feeling of alarm. This is a natural response, but an orderly approach permits even a newcomer to physics to solve apparently complicated problems. Such an approach can be summarized as follows:

1. Draw a neat, reasonably accurate sketch where appropriate. **Sketch**
2. Write down all the known quantities. If necessary, convert them to the proper units. **Known quantities**
Put these quantities on the sketch.
3. Write down the unknown quantity (or quantities) and show it on the sketch as well. **Unknown quantities**
4. Find the basic equation that connects known and unknown quantities. **Basic equation**
5. Solve the basic equation for the unknown quantity. The appendix may help: All the **Working equation**
mathematics needed for this book is reviewed there.
6. Substitute the known quantities in the resulting formula, including all the units. If **Substitution**
more information was given than needed to work out the problem, which sometimes
happens, one or more of the known quantities will not have to be used.
7. Carry out the indicated operations to get the answer. Be sure the units are the same **Calculation**
on both sides of the equals sign and that the correct number of significant figures
appears in the answer.
8. Ask yourself if the answer is reasonable. Often a scale drawing can be made in Step **Check**
1 and the answer checked by a measurement on the drawing. If the units of the answer
are strange to you, you might express the answer in more familiar units to see if it
makes sense.

Of course, not all these steps may be required in a given case. The sketch can be left out when the problem is as simple as finding average speed from the distance covered in a stated time, for instance. Often the basic equation of Step 4 is already in the right form and Step 5 is not needed. On the other hand, in a complex situation, more than one basic equation may have to be used, and the various quantities must then be combined in another step before solving for the unknown quantity. Despite the occasional exception, however, experience shows it is wise to follow the above procedure as closely as the problem permits.

All problems that involve motion under constant acceleration can be solved using one or more of the formulas listed in Table 1–5 for Step 4. In Step 5 these formulas

TABLE 1–5
Formulas for motion under
constant acceleration

$$v_f = v_0 + at$$

$$s = v_0 t + \tfrac{1}{2} at^2$$

$$s = \left(\frac{v_0 + v_f}{2} \right) t$$

$$v_f^2 = v_0^2 + 2as$$

may have to be rewritten to suit a particular combination of known and unknown quantities. For instance, the last formula listed can be solved for four different quantities:

$$v_f = \sqrt{v_0^2 + 2as}$$

$$v_0 = \sqrt{v_f^2 - 2as}$$

$$a = \frac{v_f^2 - v_0^2}{2s}$$

$$s = \frac{v_f^2 - v_0^2}{2a}$$

Example   The driver of a train traveling at 30 m/s applies the brakes when he passes an amber signal. The next signal is 1.5 km down the track and the train reaches it 75 s later. Assuming a uniform acceleration, find the speed of the train at the second signal.

Solution   Step 1. *Sketch*. See Fig. 1–10.
        Step 2. *Known Quantities*.

Initial speed $= v_0 = 30$ m/s

Distance of acceleration $= s = 1.5$ km $= 1500$ m

Time of acceleration $= t = 75$ s

FIG. 1–10

Step 3. *Unknown Quantity.*

Final speed $= v_f = ?$

Step 4. *Basic Equation.* We do not need to know the actual value of the train's acceleration to find v_f if we use the third formula in Table 1–5, which is

$$s = \left(\frac{v_0 + v_f}{2}\right)t$$

Step 5. *Working Equation.* We solve for v_f as follows:

$$s = \left(\frac{v_0 + v_f}{2}\right)t$$

$$2s = (v_0 + v_f)t$$

$$\frac{2s}{t} = v_0 + v_f$$

$$v_f = \frac{2s}{t} - v_0$$

Step 6. *Substitution.*

$$v_f = \frac{2s}{t} - v_0 = \frac{(2)(1500\,\text{m})}{75\,\text{s}} - 30\,\text{m/s}$$

Step 7. *Calculation.*

$$v_f = \frac{(2)(1500\,\text{m})}{75\,\text{s}} - 30\,\text{m/s} = 40\,\text{m/s} - 30\,\text{m/s}$$

$$= 10\,\text{m/s}$$

Step 8. *Check.* The answer seems reasonable. One way to check further is to work out the train's acceleration. From the first formula in Table 1–5 we have

$$v_f = v_0 + at$$

$$a = \frac{v_f - v_0}{t} = \frac{(10 - 30)\,\text{m/s}}{75\,\text{s}} = -0.27\,\text{m/s}^2$$

The train must slow down by 0.27 m/s in each second, which is plausible. With this value of a we obtain, as before,

$$v_f = v_0 + at = 30\,\text{m/s} + (-0.27\,\text{m/s}^2)(75\,\text{s})$$

$$= 30\,\text{m/s} - 20\,\text{m/s} = 10\,\text{m/s}$$

An elaborate check like this one is not always necessary, but it is a good idea if we are not sure of our algebra in Step 5 or of our arithmetic in Step 7. ■

1–8 ACCELERATION OF GRAVITY

Experiment and observation are the foundations of science

Drop a stone, and it falls. Does the stone fall at a constant speed, or is it accelerated? Does the motion of the stone depend upon its weight, or its size, or its color? More than two thousand years ago questions such as these were answered by Greek philosophers, notably Aristotle, on the basis of "logical reasoning" only. To them it seemed reasonable that heavy things should fall faster than light things, for example. Almost nobody felt it necessary to perform experiments to seek information on the physical universe until Galileo (1564–1642) revolutionized science by doing just that: performing experiments. Modern science owes its success in understanding and utilizing natural phenomena to its reliance upon experiment and observation.

The acceleration of gravity near the earth's surface is approximately the same everywhere

What Galileo found, as the result of careful measurements made on balls rolling down inclined planes, was that *all freely falling objects have the same acceleration* at the same place near the earth's surface. This acceleration, which is called the accel-

FIG. 1–11 If air resistance is neglected, all freely falling objects near the earth's surface have the same acceleration of $g = 9.8$ m/s². The longer an object falls, the greater its speed.

FIG. 1–12 All objects that fall in a vacuum near the earth's surface have the same downward acceleration. In air, however, a stone falls faster than a feather because air resistance affects it less.

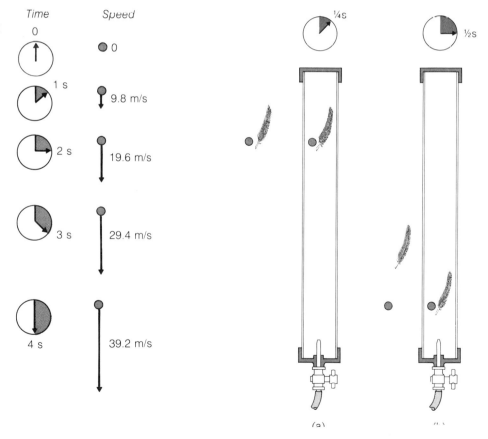

eration of gravity (symbol g), has the value

$$g = 9.81 \text{ m/s}^2 = 32.2 \text{ ft/s}^2 \qquad \textit{Acceleration of gravity}$$

to three significant figures. In most problems in elementary physics it is sufficient to let $g = 9.8 \text{ m/s}^2$ or $g = 32 \text{ ft/s}^2$.

An object falling from rest thus has a speed of 9.8 m/s (32 ft/s) after the first second, a speed of 19.6 m/s (64 ft/s) after the next second, and so on. The longer the time during which a stone falls after being dropped, the greater its speed when it hits the ground (Fig. 1–11). But the stone's *acceleration* is always the same.

Another aspect of Galileo's work deserves comment. His conclusion that all things fall with the same constant acceleration is an *idealization* of reality. The actual accelerations with which objects fall depend upon many factors: the location on the earth, the size and shape of the object, and the density and state of the atmosphere. For example, a bullet falls faster than a feather does in air because of the effects of buoyancy and air resistance. Galileo saw that the basic phenomenon was a constant acceleration downward, with other factors acting merely to cause deviations from the constant value. In a vacuum the bullet and feather fall with exactly the same acceleration (Fig. 1–12).

The drag force due to air resistance on an object of a given size and shape depends upon the speed of the object—the faster it goes, the more the drag. In the case of a falling object, the drag force increases as the speed increases until finally it cannot go any faster. The object then continues to fall at a constant *terminal speed* (Fig. 1–13). The terminal speed of a person in free fall is about 54 m/s (120 mi/h), whereas it is only about 6.3 m/s (14 mi/h) with an open parachute. In the absence of air resistance, raindrops would reach the ground at speeds high enough to be dangerous.

Terminal speed of falling object

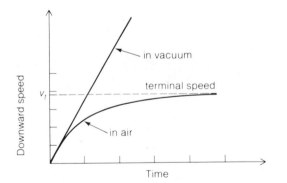

FIG. 1–13 This graph compares how the speed of a falling body varies with time when it is in a vacuum and when it is in air. In a vacuum the speed is given by $v = gt$ and increases without limit until the body strikes the ground. In air the terminal speed of v_t is eventually reached and the body continues to fall at this speed. The terminal speed of a person falling from an airplane is considerably reduced by a parachute, which enables a safe landing. (Photo: Werner H. Muller, Peter Arnold, Inc.©)

Why clouds stay aloft   A cloud consists of tiny water droplets or ice crystals whose terminal speeds are so small, often 1 cm/s or less, that very little updraft is needed to keep them suspended indefinitely. However, when a cloud is rapidly cooled, the water droplets or ice crystals grow in size and weight until their terminal speeds become too great for updrafts to keep them aloft. The result is a fall of raindrops or snowflakes from the cloud.

1–9   FREE FALL

We can apply the formulas derived earlier for motion under constant acceleration to objects in free fall. It is important to keep in mind that the direction of the acceleration of gravity g is always downward, no matter whether we are dealing with a dropped object or with one that is initially thrown upward.

Example   A stone is dropped from the top of New York's Empire State Building, which is 450 m high. Neglecting air resistance, how long does it take the stone to reach the ground? What is its speed when it strikes the ground? (See Fig. 1–14.)

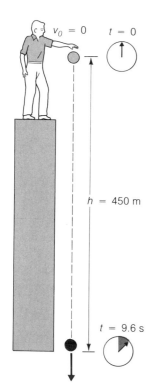

FIG. 1–14

Solution   In a problem like this one where the motion is entirely downward it is easiest to consider down as the positive direction, so that $h = 450$ m and $g = 9.8$ m/s^2. The general formula for the distance traveled in the time t by an accelerated object is

$$s = v_0 t + \tfrac{1}{2} a t^2$$

Here $v_0 = 0$, since the stone is simply dropped with no initial speed, and the acceleration is $a = g$. Hence we have

$$h = \tfrac{1}{2} g t^2$$

where h represents vertical distance from the starting point. First we solve this formula for t, which yields

$$t = \sqrt{\frac{2h}{g}}$$

and then we substitute $h = 450$ m and $g = 9.8$ m/s^2 to obtain

$$t = \sqrt{\frac{2h}{g}} = \sqrt{\frac{(2)(450 \text{ m})}{9.8 \text{ m/s}^2}} = \sqrt{92} \text{ s} = 9.6 \text{ s}$$

Knowing the duration of the fall makes it simple to compute the stone's final speed:

$$v_f = v_0 + at = 0 + gt = (9.8 \text{ m/s}^2)(9.6 \text{ s}) = 94 \text{ m/s} \qquad \blacksquare$$

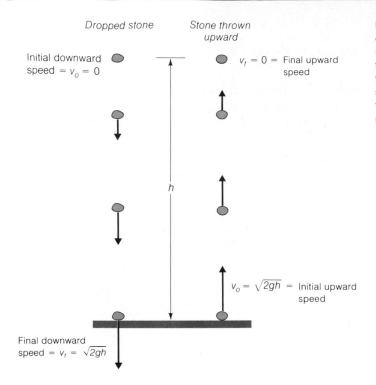

FIG. 1–15 A stone dropped from a height h reaches the ground with the speed $\sqrt{2gh}$. In order to reach the height h, a stone thrown upward from the ground must have the minimum speed $\sqrt{2gh}$.

The speed that a dropped object has when it reaches the ground is the same as the speed with which it must be thrown upward from the ground to rise to the same height (Fig. 1–15). To prove this statement, we refer to the formula $v_f^2 = v_0^2 + 2as$ and replace the s with h to give

The speed needed to reach a certain height equals the speed after falling from that height

$$v_f^2 = v_0^2 + 2ah$$

When a stone is dropped, its acceleration a equals the acceleration of gravity g, the initial speed v_0 is zero, and so

$$v_f^2 = 0 + 2gh \qquad v_f = \sqrt{2gh}$$

When the stone is thrown upward, on the other hand, $a = -g$ (since the downward acceleration is opposite in direction to the upward initial speed), and at the top of its path $v = 0$. Hence

$$0 = v_0^2 - 2gh \qquad v_0 = \sqrt{2gh}$$

The speed is the same in both cases.

Example　A stone is thrown upward with an initial speed of 48 ft/s (Fig. 1–16). (a) What will its maximum height be? (b) When will it return to the ground? (c) Where will it be after 1.0 s? (d) Where will it be after 2.0 s?

Solution　(a) We make use of Eq. (1–10) to find the highest point the stone will reach, reckoning up as positive (+) and down as negative (−). Here $v_0 = +48$ ft/s and $a = -32$ ft/s^2; at the top of the stone's path, $s = h$ and $v_f = 0$. Hence

$$h = \frac{v_f^2 - v_0^2}{2a} = \frac{0 - (48\,\text{ft/s})^2}{(2)(32\,\text{ft/s}^2)} = \frac{2304\,(\text{ft/s})^2}{64\,\text{ft/s}^2} = 36\,\text{ft}$$

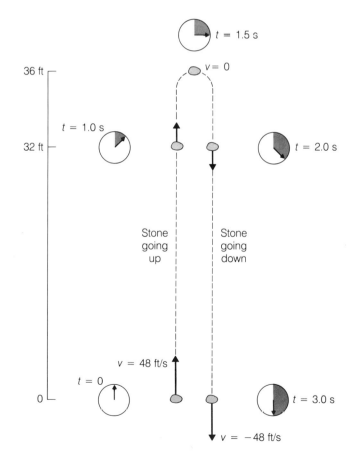

FIG. 1–16　The path of a stone thrown upward with an initial speed of 48 ft/s. Air resistance is neglected.

(b) When will the stone strike the ground? An object takes exactly as long to fall from a certain height h as it does to rise that high (provided that h is its maximum height, as it is here), just as an object's final speed when dropped from a height h is the same as the initial upward speed needed for it to get that high. From Eq. (1–9),

$$h = \tfrac{1}{2} gt^2$$

which we solve for t to find that

$$t = \sqrt{\frac{2h}{g}} = \sqrt{\frac{(2)(36 \text{ ft})}{32 \text{ ft/s}^2}} = 1.5 \text{ s}$$

Because the stone takes as long to rise as to fall, the total time it is in the air is twice 1.5 s or 3.0 s.

(c) To find the height of the stone a given time after it was thrown upward, we make use of Eq. (1–8),

$$s = v_0 t + \tfrac{1}{2} at^2$$

Here $s = h_1$, $v_0 = 48$ ft/s, $a = -32$ ft/s^2, and $t = 1.0$ s, so that

$$h_1 = (48 \text{ ft/s})(1.0 \text{ s}) - \tfrac{1}{2}(32 \text{ ft/s}^2)(1.0 \text{ s})^2 = 32 \text{ ft}$$

(d) When we substitute $t = 2.0$ s in the above formula, the height we find is once more

$$h_2 = (48 \text{ ft/s})(2.0 \text{ s}) - \tfrac{1}{2}(32 \text{ ft/s}^2)(2.0 \text{ s})^2 = 32 \text{ ft}$$

What this result means is that at 1.0 s the stone is at a height of 32 ft on its way up; then it goes on further to its maximum height of 36 ft, which it reaches at 1.5 s; and at 2.0 s it is again 32 ft above the ground but now on its way down. ∎

IMPORTANT TERMS

To measure a quantity means to compare it with a standard quantity of the same kind called a **unit.** Almost all physical quantities can be expressed in terms of four fundamental measurements: length, mass, time, and electric current. In a **system of units,** a set of units of length, mass, time, and electric current is specified from which all other units in the system are derived. The **SI** (metric) system is used in everyday life in much of the world and universally by scientists; it is gradually replacing the **British system** in the United States.

The **average speed** of a moving object is the distance

it covers in a time interval divided by the time interval. The object's **instantaneous speed** at a certain moment is the rate at which it is covering distance at that moment.

The **acceleration** of an object is the rate at which its speed changes with time. Changes in direction are also considered as accelerations.

The **acceleration of gravity** is the acceleration of a freely falling body near the earth's surface. The symbol of the acceleration of gravity is g, and its value is 9.8 m/s^2 (32 ft/s^2).

IMPORTANT FORMULAS

Speed: $v = \dfrac{s}{t}$

Constant speed: $s = vt$

$\qquad\qquad\qquad t = \dfrac{s}{v}$

Acceleration: $a = \dfrac{v_f - v_0}{t}$

Final speed under constant acceleration:

$v_f = v_0 + at$

$v_f^2 = v_0^2 + 2as$

Distance under constant acceleration:

$s = v_0 t + \frac{1}{2} at^2$

$s = \left(\dfrac{v_0 + v_f}{2}\right) t$

Free fall from rest: $h = \frac{1}{2} gt^2$

$\qquad\qquad\qquad v_f = \sqrt{2gh}$

[Air resistance is assumed negligible in the following exercises and problems.]

MULTIPLE CHOICE

1. The prefix micro represents
   a. $\frac{1}{10}$.
   b. $\frac{1}{100}$.
   c. $\frac{1}{1,000}$.
   d. $\frac{1}{1,000,000}$.

2. The British unit of length closest to the meter is the
   a. inch.
   b. foot.
   c. yard.
   d. mile.

3. Of the following, the shortest is
   a. 1 mm.
   b. 0.01 in.
   c. 0.001 ft.
   d. 0.00001 km.

4. Of the following, the longest is
   a. 10^4 in.
   b. 10^4 m.
   c. 10^3 ft.
   d. 0.1 mi.

5. The number of cubic centimeters in a cubic foot is approximately
   a. 1.7×10^3.
   b. 1.7×10^4.
   c. 2.8×10^4.
   d. 1.7×10^5.

6. The number of seconds in a month is approximately
   a. 2.6×10^6.
   b. 2.6×10^7.
   c. 2.6×10^8.
   d. 2.6×10^9.

7. A height of 5 ft 8 in. is equivalent to
   a. 173 cm.
   b. 177 cm.
   c. 207 cm.
   d. 223 cm.

8. An example of an object whose motion is *not* accelerated is a car that
   a. turns a corner at the constant speed of 10 km/h.
   b. descends a hill at the constant speed of 30 km/h.
   c. descends a hill at a speed that increases from 20 km/h to 40 km/h uniformly.
   d. climbs a hill, goes over its crest, and descends on the other side, all at the constant speed of 30 km/h.

9. On a distance-time graph, a horizontal straight line corresponds to motion at
   a. zero speed.
   b. constant speed.
   c. increasing speed.
   d. decreasing speed.

10. On a distance-time graph, a straight line sloping upward to the right corresponds to motion at
   a. zero speed.
   b. constant speed.
   c. increasing speed.
   d. decreasing speed.

11. On a speed-time graph, the motion of a car traveling along a straight road with the uniform acceleration of 2 m/s^2 would appear as a
   a. horizontal straight line.
   b. straight line sloping upward to the right.
   c. straight line sloping downward to the right.
   d. curved line whose downward slope to the right increases with time.

12. The acceleration of a stone thrown upward is
   a. greater than that of a stone thrown downward.
   b. the same as that of a stone thrown downward.
   c. smaller than that of a stone thrown downward.
   d. zero until it reaches the highest point in its motion.

13. A stone is thrown upward from a roof at the same time

as another, identical stone is dropped from there. The two stones
    a. reach the ground at the same time.
    b. have the same speed when they reach the ground.
    c. have the same acceleration when they reach the ground.
    d. none of the above.

14. Two balls are thrown vertically upward, one with an initial speed twice that of the other. The ball with the greater initial speed will reach a height

    a. $\sqrt{2}$ that of the other.
    b. twice that of the other.
    c. 4 times that of the other.
    d. 8 times that of the other.

15. A bicycle travels 12 km in 40 min. Its average speed is
    a. 0.3 km/h.   b. 8 km/h.
    c. 18 km/h.   d. 48 km/h.

16. A car that travels at 40 mi/h for 2 h, at 50 mi/h for 1 h, and at 20 mi/h for $\frac{1}{2}$ h has an average speed of
    a. 31.4 mi/h.   b. 40 mi/h.
    c. 45 mi/h.   d. 55 mi/h.

17. A pitcher takes 0.1 s to throw a baseball, which leaves his hand with a speed of 30 m/s. The ball's acceleration was
    a. 3 m/s^2.   b. 30 m/s^2.
    c. 300 m/s^2.   d. 3000 m/s^2.

18. How long does a car with an acceleration of 2 m/s^2 take to go from 10 m/s to 30 m/s?
    a. 10 s   b. 20 s
    c. 40 s   d. 400 s

19. A car undergoes a constant acceleration of 6 m/s^2 starting from rest. In the first second it travels
    a. 3 m.   b. 6 m.
    c. 18 m.   d. 36 m.

20. An airplane requires 20 s and 400 m of runway to become airborne, starting from rest. Its speed when it leaves the ground is
    a. 20 m/s.   b. 32 m/s.
    c. 40 m/s.   d. 80 m/s.

21. A car has an initial speed of 50 ft/s and an acceleration of 4 ft/s^2. In the first 10 s after the acceleration begins, the car travels
    a. 200 ft.   b. 500 ft.
    c. 700 ft.   d. 900 ft.

22. A car has an initial speed of 50 ft/s and an acceleration of -4 ft/s^2. In the first 10 s after the acceleration begins, the car travels

    a. 100 ft.   b. 300 ft.
    c. 500 ft.   d. 700 ft.

23. How far does the car of Question 22 go before coming to a stop?
    a. 312.5 ft.   b. 400 ft.
    c. 625 ft.   d. 1250 ft.

24. A spacecraft takes 200 km after being launched to reach the escape speed from the earth of 11.2 km/s. Its average acceleration in terms of the acceleration of gravity g is
    a. 8g.   b. 16g.
    c. 32g.   d. 64g.

25. A police car leaves in pursuit of a holdup car $\frac{1}{2}$ h after the latter has left the scene of the crime at 60 mi/h. How fast must the police car go if it is to catch up with the holdup car in 1 h?
    a. 90 mi/h   b. 100 mi/h
    c. 110 mi/h   d. 120 mi/h

26. A wheel falls from an airplane flying horizontally at an altitude of 490 m. If there were no air resistance, the wheel would strike the ground in
    a. 10 s.   b. 50 s.
    c. 80 s.   d. 100 s.

27. The wheel of Question 26 will strike the ground with a speed of
    a. 49 m/s.   b. 98 m/s.
    c. 490 m/s.   d. 9604 m/s.

28. A stone is dropped from a cliff. After it has fallen 30 m its speed is
    a. 17 m/s.   b. 24 m/s.
    c. 44 m/s.   d. 588 m/s.

29. A ball thrown vertically upward at 25 ft/s continues to rise for
    a. 2.5 s.   b. 5 s.
    c. 7.5 s.   d. 10 s.

30. In Question 29, how much time will elapse before the ball strikes the ground?
    a. 2.5 s   b. 5 s
    c. 7.5 s   d. 10 s

The exercises at the end of each chapter are given in order by section. The typeface of the exercise number indicates its degree of difficulty:

      Italic = question or easy problem
     Standard = moderately difficult problem
    Boldface = complicated or difficult problem

EXERCISES

1–1   Units

1. The tallest tree in the world is a Sequoia in California that is 368 ft high. How high is this in meters? In kilometers?

2. A moderately large tree gives off 300 gal of water daily to the atmosphere. How many cubic meters is this?

3. The speedometer of a European car is calibrated in kilometers per hour. What is the car's speed in miles per hour when the speedometer reads 50?

4. In 1968 the horse Dr. Fager ran a mile in 1 min 32.2 s. Find his average speed in meters per second.

5. An acre contains 4840 yd^2, where 1 yd = 3 ft. How many square meters is this? How many acres are there in a square kilometer?

6. Find the volume in cubic meters of a swimming pool whose dimensions are 50 ft $\times$ 20 ft $\times$ 6 ft.

7. A "board foot" is a unit of lumber measure that corresponds to the volume of a piece of wood 1 ft square and 1 in. thick. How many cubic inches are there in a board foot? How many cubic feet? How many cubic centimeters?

8. Water emerges from the nozzle of a fountain in Arizona at 75 km/h and reaches a height of 170 m. Express the water speed in meters per second and in miles per hour and the height in feet.

1–2   Speed

9. A pitcher throws a baseball at 90 mi/h. How much time does it take the ball to reach the batter 60 ft away?

10. In 1977 Steve Weldon ate 91.44 m of spaghetti in 28.73 s. At the same speed, how long would it take Mr. Weldon to eat 15 m of spaghetti?

11. The starter of a race stands at one end of a line of runners. What is the difference in time between the arrival of the sound of his pistol at the nearest runner and at the most distant runner 10 m farther away? The speed of sound in air is 343 m/s. (In a sprint, 0.01 s can mean the difference between winning and coming in second.)

12. The speed of light is 3 $\times$ 10^8 m/s. How long does it take light to reach the earth from the sun, which is 1.5 $\times$ 10^{11} m away?

13. The average lifetime of many unstable elementary particles is about 10^{-23} s, and they move at nearly the speed of light (3 $\times$ 10^8 m/s) after being created in high-energy collisions between other elementary particles. How far does such a particle travel on the average before decaying?

14. A car travels 540 km in 4.5 h. How far will it go in 8 h at the same average speed? How long will it take to go 200 km at this speed?

15. A snake is slithering toward you at 1.5 m/s. If you start walking when it is 5 m away, how fast must you go in order that the snake not overtake you when you have gone 100 m?

16. The figure below is a graph that shows the speed of an object plotted against time. Find the total distance the object travels during the period covered by the graph.

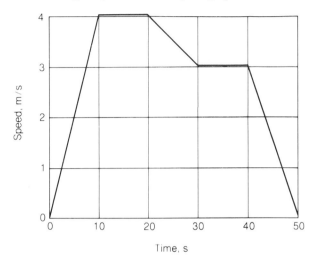

17. An airplane takes off at 9:00 A.M. and flies at 300 km/h until 1:00 P.M. At 1:00 P.M. its speed is increased to 400 km/h and it maintains this speed until it lands at 3:30 P.M. What is the airplane's average speed for the entire flight?

18. A car travels at 100 km/h for 2 h, at 60 km/h for the next 2 h, and finally at 80 km/h for 1 h. Find its average speed for the entire 5 h.

19. A woman jogs halfway to her destination at 8 km/h and walks the rest of the way at 6 km/h. What is her average speed for the entire trip?

20. A car covers one-quarter of the distance to the next town at 40 km/h, another quarter at 50 km/h, and the rest at 90 km/h. Find the car's average speed for the entire distance.

21. A man drives one-third of the distance home at 25 mi/h. How fast must he drive the rest of the way in order to average 35 mi/h for the entire trip?

22. A boat moving downstream (with the current) in a river takes 4 min 7 s to cover a measured mile between marks on shore and 7 min 30 s to cover the same distance moving upstream (against the current). (a) Find the boat's speed relative to the water under the assumption it is constant. (b) Find the speed of the current in the river.

1–3 Acceleration

1–4 Speed and Acceleration

23. A red ball is thrown horizontally at a certain speed and a green ball is thrown upward at the same speed. Compare their accelerations.

24. The acceleration of a certain moving object is constant in magnitude and direction. Must the path of the object be a straight line? If not, give an example.

25. Can a rapidly moving object have the same acceleration as a slowly moving one?

26. The figure shows distance-time graphs for nine cars.
   a. Which cars are or have been moving in the forward direction?
   b. Which cars are or have been moving in the backward direction?
   c. Which car has the highest constant speed?
   d. Which car has the highest constant speed in the forward direction?

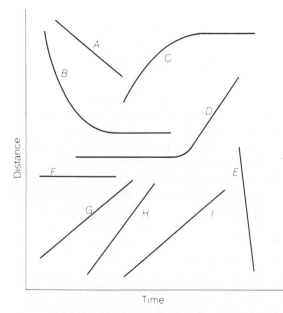

   e. Which car has the highest constant speed in the backward direction?
   f. Which cars have the same speed?
   g. Which car has not moved at all?
   h. Which car has accelerated from rest to a constant speed?
   i. Which car has been brought to a stop from an initial speed in the forward direction?
   j. Which car has been brought to a stop from an initial speed in the backward direction?

27. (a) A car's speed increases from 8 m/s to 20 m/s in 10 s. Find its acceleration. (b) The car's speed then decreases from 20 m/s to 10 m/s in 5 s. Find its acceleration now.

28. The tires of a certain car begin to lose their grip on the pavement at an acceleration of 15 ft/s^2. If the car has this acceleration, how many seconds does it require to reach a speed of 75 ft/s starting from 30 ft/s?

29. A car starts from rest and reaches a speed of 22 m/s in 20 s. (a) What was its acceleration? (b) How long would the car take to go from 22 m/s to 30 m/s with the same acceleration?

30. The brakes of a car moving at 14 m/s are suddenly applied and the car comes to a stop in 4 s. (a) What was its acceleration? (b) How long would the car take to come to a stop starting from 20 m/s with the same acceleration? (c) How long would the car take to slow down from 20 m/s to 10 m/s with the same acceleration?

31. A passenger in an airplane flying from New York to Los Angeles notes the time at which he passes over various cities and towns. With the help of a map he determines the distances between these landmarks, and compiles the table shown below. Plot the distance covered by the airplane versus time from these data, and describe the airplane's motion with the help of the graph.

Time (P.M.):	4:00	5:12	5:41	6:14
Distance (km):	0	660	926	1267

Time (P.M.):	6:39	7:54	9:18	10:00
Distance (km):	1525	2300	3028	3392

32. A European train passes successive kilometer posts at the times given below. Plot the data on a graph and determine whether the train's speed is constant over the entire distance or not. If it is not constant, plot the train's speed in each time interval versus time and find the acceleration. What are the train's initial and final speeds?

Distance (km):	0	1	2	3	4
Time (s):	0	64	114	156	193

Distance (km):	5	6	7	8
Time (s):	227	259	292	324

33. The odometer of a car is checked at 1-min intervals and the readings below are obtained. Calculate the speed of the car in each time interval and plot the results on a graph. Describe the motion of the car with the help of this graph.

Time (min):	0	1	2	3
Distance (mi):	42.20	42.74	43.64	44.90

Time (min):	4	5	6	7
Distance (mi):	46.34	47.78	49.22	50.66

Time (min):	8	9	10
Distance (mi):	51.74	52.10	52.10

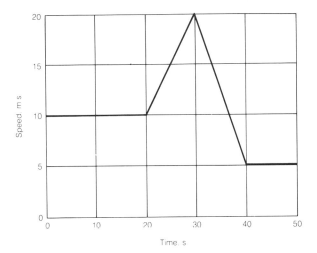

34. The above speed-time graph represents the motion of a certain car. (a) Find the car's acceleration at $t = 15, 25, 35,$ and 45 s. (b) Find the distance covered by the car from $t = 0$ to $t = 30$ s and from $t = 30$ s to $t = 50$ s.

1–5 Distance and Acceleration

35. A Porsche reaches a speed of 26 mi/h from a standing start in 15.5 s. What distance does it cover while doing so?

36. A Ferrari covers 100 m from a standing start in 6 s at constant acceleration. Find its final speed.

37. A DC-8 airplane has a takeoff speed of 80 m/s, which it reaches 35 s after starting from rest. (a) How much time does the airplane spend in going from 0 to 20 m/s? What distance does it cover in doing so? (b) How much time does the airplane spend in going from 60 to 80 m/s? What distance does it cover in doing so? (c) What is the minimum length of the runway?

38. The engineer of a train traveling at 80 km/h applies the brakes when he passes an amber signal. The next signal is one km down the track and the train reaches it 75 s later. Assuming a uniform deceleration, find the speed of the train at the second signal.

39. A truck accelerates from rest at 1 ft/s^2 for 30 s, continues at constant speed for 2 min, and then comes to a stop in 15 s. What distance did it cover?

40. A sprinter accelerates from rest until he reaches a speed of 12 m/s and then continues running at this speed. If he takes 11 s to cover 100 m, what was his acceleration and how long did it last?

41. A car is stationary in front of a red traffic light. As the light turns green a truck goes past at a constant speed of 15 m/s. At the same moment, the car begins to accelerate at 1.25 m/s^2; when it reaches 25 m/s, the car continues at that speed. When does the car pass the truck? How far will they have gone from the traffic light at that time?

1–6 Distance, Speed, and Acceleration

42. A golf cart has an acceleration of 1.2 ft/s^2. What is its speed after it has covered 30 ft starting from rest?

43. A spacecraft has an acceleration of magnitude 5g. What distance is needed for it to attain a speed of 10 km/s?

44. The brakes of a certain car produce an acceleration of -5 m/s^2. (a) If the car is moving at 20 m/s when the brakes are applied, how far does it go in the first second afterward? (b) How far does the car go in the course of being slowed down from 20 m/s to 10 m/s?

45. A car is approaching a traffic light at 20 m/s when it turns red. The driver takes 1 s to react and put his foot on the brake pedal, after which the car decelerates at 2 m/s^2. Find the distance needed for the car to come to a stop after the light turns red.

46. An express train passes a certain station at 20 m/s. The next station is 2 km away and the train reaches it 1 min later. (a) Did the train's speed change? (b) If it did, what was its speed at the second station, assuming a constant acceleration?

47. A bus travels 400 m between two stops. It starts from rest and accelerates at 1.5 m/s^2 until it reaches a speed of 9 m/s. The bus continues at this speed and then decelerates at 2 m/s^2 until it comes to a halt. Find the total time required for the journey.

1–8 Acceleration of Gravity

48. A movie is shown that appears to be of a ball falling through the air. Is there any way to determine from what appears on the screen if the movie is actually of a ball being thrown upward but the film is being run backward in the projector?

49. A hunter aims a rifle directly at a squirrel on a branch of a tree. The squirrel sees the flash of the rifle's firing. Should the squirrel stay where it is or drop from the branch in free fall at the instant the rifle is fired?

50. A person at the masthead of a sailboat moving at constant speed drops a wrench. The person is 20 m above the boat's deck at the time, and the stern of the boat is 20 m aft of the mast. Is there a minimum speed the sailboat can have such that the wrench does not land on the deck? If so, what is this speed?

51. Is it true that an object dropped from rest falls three times farther in the second second after being released than it does in the first second?

1–9   Free Fall

52. How fast must a ball be thrown upward to reach a height of 12 m?

53. Divers in Acapulco, Mexico, leap from a point 36 m above the sea. What is their speed when they enter the water?

54. A stone is dropped from a cliff 490 m above its base. How long does the stone take to fall?

55. A ball dropped from the roof of a building takes 4 s to reach the street. How high is the building?

56. A bullet is fired vertically upward and returns to the ground in 20 s. Find the height it reaches.

57. Find the initial and final speeds of a ball thrown vertically upward that returns to the thrower 3 s later.

58. A stone is thrown vertically upward at 9.8 m/s. When will it reach the ground?

59. A ball is thrown vertically downward at 10 m/s. What is its speed 1 s later? 2 s later?

60. A ball is thrown vertically upward at 10 m/s. What is its speed and direction 1 s later? 2 s later?

61. A lead pellet is propelled vertically upward by an air rifle with an initial speed of 16 m/s. Find its maximum height.

62. From Fig. 1–13 we can see that the air resistance experienced by a falling object is not an important factor until a speed of about half its terminal speed is reached. The terminal speed of a baseball is 40 m/s. How much time is needed for a dropped baseball to reach a speed of half this? How far does it fall in this time?

63. The acceleration of gravity at the surface of Mars is 3.7 m/s². (a) If a stone thrown upward on Mars reaches a height of 15 m, find its initial speed. (b) What is the total time of flight?

64. The acceleration at the surface of Venus is 8.9 m/s². Would a ball thrown upward on Venus return to the ground sooner or later than a ball thrown upward with the same speed on the earth?

65. In a novel, a pirate forces a captured sailor to "walk the plank," from which he falls into the sea 22 ft below. The author claims the sailor's entire life flashes before him during his fall. (a) How much time is available for this? (b) With what speed will the sailor strike the water?

66. A stone is thrown vertically upward with a speed of 60 ft/s from a bridge. On the way down, the stone just misses the bridge and falls into the water 5.0 s after having been thrown upward. Find the height of the bridge above the water.

67. A British parachutist bails out at an altitude of 150 m and accidentally drops his monocle. If he descends at the constant speed of 6 m/s, how much time separates the arrival of the monocle on the ground from the arrival of the parachutist himself?

68. A Russian balloonist floating at an altitude of 150 m accidentally drops his samovar and starts to ascend at the constant speed of 1.2 m/s. How high will the balloon be when the samovar reaches the ground?

69. An orangutan throws a coconut vertically upward at the foot of a cliff 40 m high while his mate simultaneously drops another coconut from the top of the cliff. The two coconuts collide at an altitude of 20 m. What was the initial speed of the coconut that was thrown upward?

70. When a flea jumps, it accelerates through about 0.8 mm (a little less than the length of its legs) and is able to reach a height of as much as 10 cm. (a) Find the flea's acceleration (assumed constant) and its speed at takeoff when this occurs. (b) When a person jumps, the acceleration distance is about 50 cm. If the acceleration of a person were the same as that of a flea, find the speed at takeoff and the height that would be reached.

71. A helicopter is climbing at 8 m/s when it drops a pump near a leaking boat. The pump reaches the water 4 s afterward. How high was the helicopter when the pump was dropped? When the pump reached the water?

72. A person in an elevator drops an apple from a height 2 m above the elevator's floor and, with a stopwatch, times the fall of the apple to the floor. What is found (a) when the elevator is ascending with an acceleration of 1 m/s²; (b) when it is descending with an acceleration of 1 m/s²; (c) when it is ascending at the constant speed of 3 m/s; (d) when it is descending at the constant speed of 3 m/s; and (e) when the cable has broken and it is descending in free fall?

73. An elevator has a maximum acceleration of ± 1.5 m/s² and a maximum speed of 6 m/s. Find the shortest period of time required for it to take a passenger to the tenth floor of a building from street level, a height of 50 m, with the elevator coming to a stop at this floor.

74. A rocket is launched upward with an acceleration of 100 m/s^2. Eight seconds later the acceleration stops when the fuel is exhausted. Find (a) the highest speed the rocket attains; (b) the maximum altitude; (c) the total time of flight; and (d) the speed with which the rocket strikes the ground.

75. A girl throws a ball vertically upward at 10 m/s from the roof of a building 20 m high. (a) How long will it take the ball to reach the ground? (b) What will its speed be when it strikes the ground?

76. A girl throws a ball vertically downward at 10 m/s from the roof of a building 20 m high. (a) How long will it take the ball to reach the ground? (b) What will its speed be when it strikes the ground?

ANSWERS TO MULTIPLE CHOICE

1. d.	**7.** a	**13.** c	**19.** a	**25.** a
2. c	**8.** b	**14.** c	**20.** c	**26.** a
3. b	**9.** a	**15.** c	**21.** c	**27.** b
4. b	**10.** b	**16.** b	**22.** b	**28.** b
5. c	**11.** b	**17.** c	**23.** a	**29.** a
6. a	**12.** b	**18.** a	**24.** c	**30.** b

2

FORCE AND MOTION

Thus far we have only discussed how motion is described mathematically. But why does anything move in the first place? Why do some things move faster than others? Why are some accelerated and others not? How can a body traveling in one direction be accelerated in the opposite direction? All these are reasonable questions and ones we must be able to answer if we are to understand the factors at work in the world around us that produce the physical phenomena we observe and if we are to harness these factors to meet our needs. Almost three centuries ago Isaac Newton (1642–1727) formulated three principles based upon observations he and others had made which summarize so much of the behavior of moving bodies that they have become known as the laws of motion. These laws form the subject of this chapter.

2–1 FIRST LAW OF MOTION

In everyday life it is a familiar observation that stationary objects tend to remain stationary and moving objects tend to continue moving. A certain amount of effort is

CHAPTER OBJECTIVES

Completing this chapter should enable you to:

1. Use the first law of motion to find out whether or not something is being acted upon by a net force.

2. Use the second law of motion, $F = ma$, to solve problems that involve force, mass, and acceleration.

3. Distinguish between mass and weight.

4. Understand the relationship between the slug and the pound.

5. Use the formula $w = mg$ to find the weight of an object of given mass and to find the mass of an object of given weight.

6. Use the third law of motion to relate action and reaction forces.

7. Calculate the minimum force needed to move an object against the influence of friction.

8. Calculate the frictional force acting on a moving object.

9. Analyze the motion of an object subject to friction.

needed to start a cart moving on a level road, and once in motion a certain amount of effort is also required to bring it to a stop. Of course, in the case of a cart friction is a factor in its reluctance to begin to move, but friction is only part of the story. Even without friction, a cart at rest on a level road will remain at rest unless something pushes it. And without friction, a cart moving along a level road at a certain speed will continue to move at that speed indefinitely.

Newton's first law of motion is a statement of the above behavior:

An object at rest will remain at rest and an object in motion will continue in motion in a straight line at constant speed in the absence of any interaction with the rest of the universe.

First law of motion

We might object that, although everyday experience does indicate that things in motion *tend* to remain in motion along a straight line at constant speed, sooner or later they usually come to a stop and often deviate from a straight path as well. Even heavenly bodies such as the sun, moon, and planets are neither at rest nor pursue straight paths.

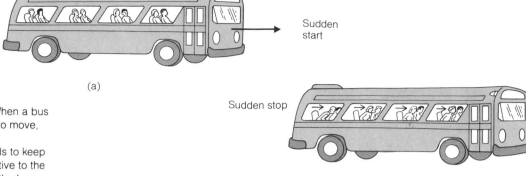

(a)

(b)

FIG. 2-1 (a) When a bus suddenly starts to move, the inertia of the passengers tends to keep them at rest relative to the earth. (b) When the bus comes to a sudden stop, their inertia tends to keep them moving. Both effects illustrate the first law of motion.

But these observations do not mean that the first law of motion is wrong; they merely emphasize how difficult it is to avoid interactions between something and its environment. A cart rolling along a smooth, perfectly level road will not continue forever owing to friction and air resistance, but we are free to imagine what would happen if the air were to disappear and the friction were to vanish.

All matter exhibits inertia

The reluctance of an object to change its state of rest or uniform motion is a property of matter known as *inertia*. When a bus suddenly starts to move, its passengers seem to be pushed backward (Fig. 2–1). What is actually happening is that inertia tends to keep their bodies in place relative to the earth while the bus carries their feet forward. When a bus suddenly stops, on the other hand, the passengers seem to be pushed forward. What is actually happening is that inertia tends to keep their bodies moving while the bus has come to a halt.

2-2 MASS

Mass is a measure of inertia

A quantitative measure of the inertia of an object at rest is its *mass*. The greater the resistance something offers to being set in motion, the greater its mass. The inertia of a lead ball is more than that of an aluminum ball of the same size, as we can tell by kicking them in turn, so the mass of the lead ball is more than that of the aluminum one (Fig. 2–2).

We can arrive at a precise definition of mass by using a simple experiment to compare the inertias of two objects. What we do is put a small spring between them, push them together so the spring is compressed, and tie a string between them to hold the assembly in place (Fig. 2–3). Now we cut the string. The compressed spring pushes the objects apart, and A flies off to the left at the speed v_A while B flies off to the right at the speed v_B. Object A has a lower speed that object B, and we interpret this difference to mean that A exhibits more inertia than B.

We repeat the experiment a number of times using springs of different stiffness, so that the recoil speeds are different in each case. What we find each time is that A

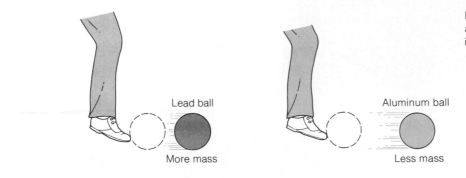

FIG. 2-2 The mass of an object determines its inertia.

Lead ball

Aluminum ball

More mass

Less mass

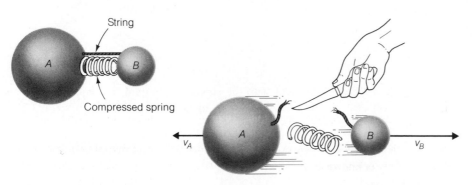

FIG. 2-3 An experiment to compare the masses of two bodies. When the string is cut, the body with the greater mass moves off with the smaller speed.

String

Compressed spring

moves more slowly than B, and that, regardless of the exact values of v_A and v_B, their *ratio*

$$\frac{v_B}{v_A}$$

is always the same.

The fact that the speed ratio v_B/v_A is constant gives us a way to specify exactly what we mean by mass. If we denote the masses of A and B by the symbols m_A and m_B respectively, we define the ratio of these masses to be

$$\frac{m_A}{m_B} = \frac{v_B}{v_A} \tag{2-1}$$

Operational definition of mass

The object with the greater mass has the lower speed, and vice versa. The above procedure provides a definite, experimental method for finding the ratio between the masses of any two objects—it is an *operational definition*.

The next step is to select a certain object to serve as a standard unit of mass. By international agreement this object is a platinum cylinder at Sèvres, France, called the *standard kilogram*. The mass of any other object in the world can be determined by a recoil experiment with the standard kilogram. (There are easier ways to measure mass, needless to say, but we are interested in basic principles for the present.) A liter (1000 cm^3) of water, which is equal to 1.057 qt, has a mass of 1 kg.

Standard kilogram

FIG. 2-4

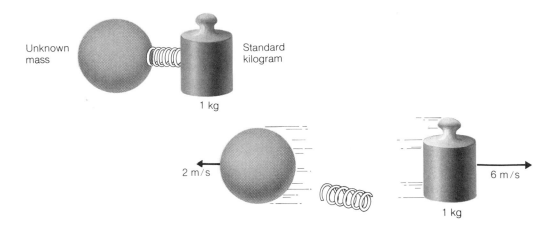

Unknown mass

Standard kilogram

1 kg

2 m/s

6 m/s

1 kg

Example   In a recoil experiment with the standard kilogram, an object of unknown mass moves off at 2 m/s and the standard kilogram moves off at 6 m/s (Fig. 2-4). What is the mass of the object?

Solution   If we call the unknown mass A and the standard kilogram B, then $m_B = 1$ kg, $v_A = 2$ m/s, and $v_B = 6$ m/s. We find that

$$m_A = m_B \left(\frac{v_B}{v_A}\right) = (1\,\text{kg})\left(\frac{6\,\text{m/s}}{2\,\text{m/s}}\right) = 3\,\text{kg} \qquad \blacksquare$$

Why is it necessary to define mass in this seemingly roundabout way? After all, everybody knows that the mass of an object refers to the amount of matter it contains. The trouble is that "amount of matter" is a vague concept: it could refer to an object's volume, to the number of atoms it contains, or to yet other properties. It has proved most fruitful to choose the inertia of an object as a measure of the quantity of matter it contains and to define the object's mass in terms of this inertia as shown in an appropriate experiment.

In the British system of units the *slug* rather than the kilogram is the standard unit of mass. The slug and its relation to the pound (which is a unit of force, not of mass) will be discussed later in this chapter.

2-3   FORCE

Interactions produce changes in speed and direction

According to the first law of motion, the direction and speed of an object do not change as long as it is isolated from the rest of the universe. When the object interacts with something else, however, its state of motion may be altered. We can interact with a football by kicking it, and the result is a change in the football's speed from $v_1 = 0$ to some value v_2 (Fig. 2-5). Or the interaction can take the form of catching a moving football, in which case again the football's speed changes. An interaction can also

FIG. 2–5 Three examples of how an interaction may give rise to an acceleration of an object.

change the direction in which an object moves, as for instance when a football bounces off a tree.

Some interactions are more effective than others in influencing the motion of an object. A swift kick affects a football more than a gentle tap does. The concept of *force* can be used to put the matter on a precise basis. In general, **Force**

A force is any influence that can change the speed or direction of motion of an object.

This definition is in accord with the notion of a force as a "push" or a "pull," but it goes further since no direct contact is necessarily implied. No hand reaches up from

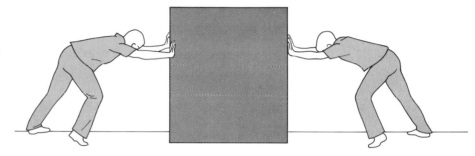

the earth to pull a dropped stone downward, yet the increasing downward speed of the falling stone means that a force is acting upon it (Fig. 2–6).

It is entirely possible for two or more forces to act upon an object without affecting its state of motion; the forces may be such as to cancel one another out. What is required for a change of motion is a *net force,* often called an *unbalanced force.* When an object is acted upon by several forces that cancel out, the forces are said to be *balanced forces* and the object is then in *equilibrium* (Fig. 2–7). But each of the forces acting by itself is capable of accelerating the object.

We can therefore restate the first law of motion in terms of net force:

Restatement of first law of motion

In the absence of a net force acting on it, an object at rest will remain at rest and an object in motion will continue in motion at constant speed in a straight line.

2–4   SECOND LAW OF MOTION

In the *second law of motion* we have a quantitative definition of force:

Second law of motion

The net force acting upon an object is equal to the product of the mass and the acceleration of the object; the direction of the force is the same as that of the object's acceleration.

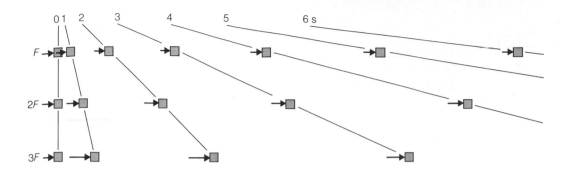

In equation form,

$$F = ma$$

Force = mass × acceleration

Second law of motion   (2–2)

The second law of motion provides us with a way to analyze and compare forces in terms of the accelerations they give rise to. Thus a force which causes something to have twice the acceleration another force produces must be twice as great as the other one (Fig. 2–8). An object moving to the right but going slower and slower is accelerated to the left. Hence there is a force toward the left acting on it (Fig. 2–9). The first law of motion is clearly a special case of the second: When the net force on an object is zero, its acceleration is also zero.

The second law is in accord with the definition of mass given earlier since the smaller the mass of an object acted upon by a given force, the greater its acceleration and hence final speed if it starts from rest (Fig. 2–10).

The relationship between force and acceleration means that the less the acceleration, the less the force. A person dropping to the ground from a height can therefore reduce the force of his or her impact by coming to a stop gradually instead of suddenly, which the person can accomplish by bending his or her knees upon striking the ground instead of landing with stiff legs. In this way t is greater, and even though $v_f - v_0$ and m remain the same, F will be correspondingly smaller.

It is convenient to have a special unit for force. In the SI system of units the appropriate unit is the *newton* (abbreviated N):

FIG. 2–8 The acceleration of an object is proportional to the net force applied to it. Successive positions of a block are shown at 1-s intervals while forces of F, $2F$, and $3F$ are applied.

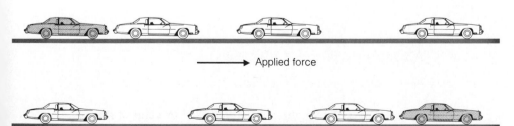

→ Applied force

← Applied force

FIG. 2–9 A force and the acceleration it produces are always in the same direction.

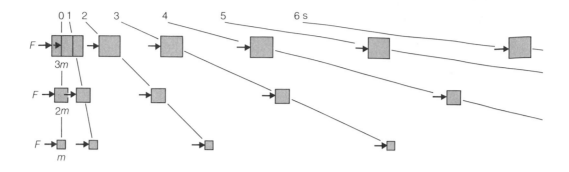

The newton is the SI unit of force

A newton is that force which, when applied to a 1-kg mass, gives it an acceleration of 1 m/s^2.

The newton is not a fundamental unit like the meter, second, and kilogram, and in some calculations it may have to be replaced by its equivalent in terms of the latter. This equivalent may be found as follows:

$$F = ma$$

$$1\,\text{N} = (1\,\text{kg})\left(1\,\frac{\text{m}}{\text{s}^2}\right) = 1\,\text{kg} \cdot \text{m/s}^2$$

A newton is equivalent to 0.225 lb, a little less than $\frac{1}{4}$ lb. A pound is nearly $4\frac{1}{2}$ N.

FIG. 2–10 When the same force is applied to objects of different masses, the resulting accelerations are inversely proportional to the masses. Successive positions of blocks of mass m, $2m$, and $3m$ are shown at 1-s intervals while identical forces of F are applied.

Let us examine a few problems in order to become familiar with the application of the second law of motion. These problems are, of course, artificial and oversimplified, but they do illustrate the power of the second law in situations involving force and motion.

Example A 60-g tennis ball approaches a racket at 30 m/s, is in contact with the racket's strings for 5 ms (1 ms = 1 millisecond = 10^{-3} s), and then rebounds at 30 m/s (Fig. 2–11). What was the average force the racket exerted on the ball?

FIG. 2–11

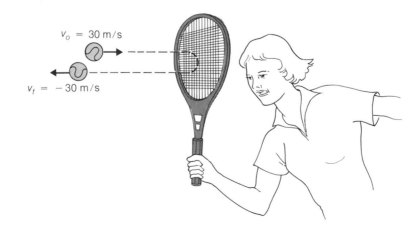

Solution   The tennis ball experienced a change in speed of

$$\Delta v = v_f - v_0 = (-30 \text{ m/s}) - (30 \text{ m/s}) = -60 \text{ m/s}$$

so its acceleration was

$$a = \frac{\Delta v}{\Delta t} = \frac{-60 \text{ m/s}}{5 \times 10^{-3} \text{ s}} = -1.2 \times 10^4 \text{ m/s}^2$$

The corresponding force is, since 60 g = 0.060 kg,

$$F = ma = (-0.060 \text{ kg})(1.2 \times 10^4 \text{ m/s}^2) = -720 \text{ N}$$

The minus sign means that the force was in the opposite direction to that of the ball when it approached the racket. The British equivalent of 720 N is 162 lb.   ■

Example   During performances of the Bouglione Circus in 1976, John Tailor was fired from a compressed-air cannon whose barrel was 20 m long. Mr. Tailor emerged from the cannon (twice daily, three times on Saturdays and Sundays) at 40 m/s. If Mr. Tailor's mass was 70 kg, find the average force on him during the firing of the cannon (Fig. 2–12).

Solution   We start by finding Mr. Tailor's acceleration with the help of Eq. (1–11),

$$v_f^2 = v_0^2 + 2as$$

Here $v_0 = 0$, $v_f = 40$ m/s, and $s = 20$ m, so

$$v_f^2 = 0 + 2as$$

$$a = \frac{v_f^2}{2s} = \frac{(40 \text{ m/s})^2}{(2)(20 \text{ m})} = 40 \text{ m/s}^2$$

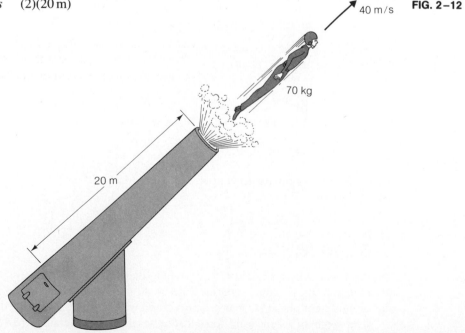

40 m/s   **FIG. 2–12**

70 kg

20 m

FIG. 2–13 Successive distances and speeds of a 4-kg mass acted upon by a 10-N force.

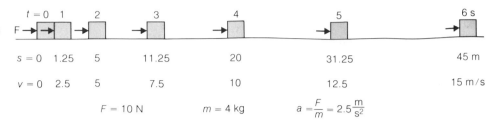

The corresponding average force is

$$F = ma = (70 \text{ kg})(40 \text{ m/s}^2) = 2800 \text{ N}$$

which is about 630 lb. ∎

Example A force of 10 N is applied to a 4.0-kg block that is at rest on a perfectly smooth, level surface. Find the speed of the block and how far it has gone after 6.0 s.

Solution We start from the second law of motion,

$$F = ma$$

We know what F and m are, so we find the acceleration a of the block as follows:

$$a = \frac{F}{m} = \frac{10 \text{ N}}{4.0 \text{ kg}} = \frac{10 \text{ kg} \cdot \text{m/s}^2}{4.0 \text{ kg}} = 2.5 \text{ m/s}^2$$

The direction of the acceleration is the same as that of the force.

To find the speed of the block after $t = 6.0$ s, we use the formula $v = v_0 + at$ and obtain

$$v = v_0 + at = 0 + (2.5 \text{ m/s}^2)(6.0 \text{ s}) = 15 \text{ m/s}$$

For the distance the block travels in $t = 6.0$ s at an acceleration of $a = 2.5 \text{ m/s}^2$ we require Eq. (1–8),

$$s = v_0 t + \tfrac{1}{2} at^2$$

The block started from rest, so $v_0 = 0$ and we have

$$s = \tfrac{1}{2} at^2 = \tfrac{1}{2}(2.5 \text{ m/s}^2)(6.0 \text{ s})^2 = 45 \text{ m}$$

After 6.0 s a 4.0-kg mass acted upon by a 10-N force will have gone 45 m and have a speed of 15 m/s (Fig. 2–13). ∎

2–5   WEIGHT

Weight is a force

The force with which an object is attracted to the earth is called its *weight*. Weight is different from mass, which is a measure of the inertia an object exhibits. Although they are different physical quantities, mass and weight are closely related.

The weight of a stone is the force that causes it to be accelerated when it is dropped.

All objects in free fall near the earth's surface have a downward acceleration of $g = 9.8$ m/s^2 $= 32$ ft/s^2, the acceleration of gravity. If the stone's mass is m, then the downward force on it, which is its weight w, can be found from the second law of motion, $F = ma$, by letting $F = w$ and $a = g$. Evidently

$$w = mg \qquad\qquad \textit{Weight} \quad (2-3)$$

Weight = mass × acceleration of gravity

The weight of any object is equal to its mass multiplied by the acceleration of gravity. Since g is a constant near the earth's surface, the weight w of an object is always directly proportional to its mass m: A large mass is heavier than a small one.

Weight is proportional to mass

The mass of an object is a more basic property than its weight, because its mass when at rest is the same everywhere in the universe whereas the gravitational force on it depends upon its position relative to the earth or to some other astronomical body. A 100-kg person weighs 980 N (221 lb) on the earth, but he or she would weigh 2587 N (582 lb) on Jupiter, 372 N (84 lb) on Mars, 162 N (36 lb) on the moon, and 0 in space far from the sun and other stars.

The mass of something is the same everywhere, unlike its weight

Example A loaded elevator whose total mass is 800 kg is suspended by a cable whose maximum permissible tension is 20,000 N. What is the greatest upward acceleration possible for the elevator under these circumstances? What is the maximum possible downward acceleration?

Solution When the elevator is at rest, or moving at constant speed, the tension in the cable is just its weight of

$$mg = (800 \text{ kg})(9.8 \text{ m/s}^2) = 7840 \text{ N}$$

To accelerate the elevator upward, an additional tension F is required in order to provide a net upward force (Fig. 2–14). Since the total tension cannot exceed 20,000 N, the greatest accelerating force available is

$$F = F_{max} = mg = 20,000 \text{ N} - 7840 \text{ N} = 12,160 \text{ N}$$

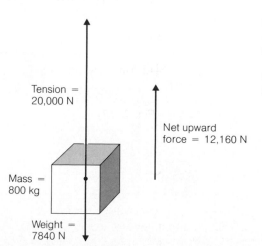

Tension = 20,000 N

Net upward force = 12,160 N

Mass = 800 kg

Weight = 7840 N

FIG. 2–14 The net upward force on an elevator of mass 800 kg is 12,160 N when the tension in its supporting cable is 20,000 N.

The elevator's acceleration when this net force is applied is

$$a = \frac{F}{m} = \frac{12,160\,\text{N}}{800\,\text{kg}} = 15.2\,\text{m/s}^2$$

For the elevator to exceed the downward acceleration of gravity $g = 9.8\,\text{m/s}^2$, a downward force besides the weight of the elevator is needed. Since this cannot be provided by a supporting cable, the maximum downward acceleration is $9.8\,\text{m/s}^2$. ■

Example   Figure 2–15 shows a 12-kg block, A, which hangs from a string that passes over a pulley and is connected at its other end to a 30-kg block, B, which rests on a frictionless table. Find the accelerations of the two blocks under the assumption that the string is massless and the pulley is massless and frictionless. What is the tension in the string?

Solution   Because the blocks are joined by the string, their accelerations have the same magnitude a even though different in direction. The net force on B is equal to the tension T in the string; so, from the second law of motion, taking the left as the $+$ direction so that a will come out positive,

$$F_B = T = m_B a$$

In the case of A, the net force is the difference between its weight $m_A g$, which acts downward, and the tension T in the string, which acts upward. Considering the downward direction as positive, so that the two accelerations will have the same sign,

$$F_A = m_A g - T = m_A a$$

Adding these two equations together eliminates T, whose value we do not know at this point, and permits us to find a:

$$F_A + F_B = m_A g - T + T = m_A a + m_B a$$

$$m_A g = (m_A + m_B)a$$

$$a = \frac{m_A g}{m_A + m_B} = \frac{(12\,\text{kg})(9.8\,\text{m/s}^2)}{12\,\text{kg} + 30\,\text{kg}} = 2.8\,\text{m/s}^2$$

FIG. 2–15 Both blocks have the same acceleration.

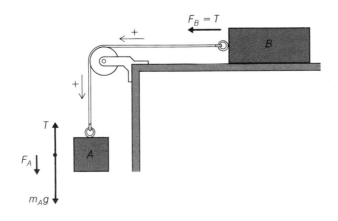

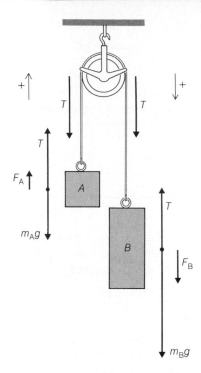

FIG. 2-16 The force on each block is the difference between its weight and the tension in the string.

The tension in the string is

$$T = m_B a = (30 \text{ kg})(2.8 \text{ m/s}^2) = 84 \text{ N}$$

If B were fixed in place, the tension would equal the weight of A, or $m_A g = 117.6 \text{ N}$. Here the tension is less because B moves in response to the pull of A's weight, but it is not zero because of B's inertia. ∎

Example Figure 2-16 shows the same two blocks, A and B, suspended by a string on either side of a massless, frictionless pulley. Find the accelerations of the two blocks and the tension in the string.

Solution Here A moves upward and B moves downward, both with accelerations having the same magnitude a. Applying the second law of motion to the two blocks gives

$$F_A = T - m_A g = m_A a$$
$$F_B = m_B g - T = m_B a$$

where we have considered up as $+$ for A and down as $+$ for B. To eliminate T we add these equations, and then solve for a:

$$F_A + F_B = T - m_A g + m_B g - T = m_A a + m_B a$$

$$(m_B - m_A)g = (m_A + m_B)a$$

$$a = \frac{(m_B - m_A)g}{m_A + m_B} = \frac{(30\,\text{kg} - 12\,\text{kg})(9.8\,\text{m/s}^2)}{12\,\text{kg} + 30\,\text{kg}} = 4.2\,\text{m/s}^2$$

The tension in the string may be found from the first equation above:

$$T = m_A a + m_A g = m_A(a + g) = (12\ \text{kg})(4.2 + 9.8)\text{m/s}^2 = 168\ \text{N} \qquad \blacksquare$$

2-6   BRITISH UNITS OF MASS AND FORCE

The slug and pound are the British units of mass and force

In the British system the unit of mass is the *slug* and the unit of force is the *pound* (lb). Something whose mass is 1 slug experiences an acceleration of 1 ft/s^2 when a net force of 1 lb acts on it. Thus

$$1\ \text{lb} = 1\,\frac{\text{slug} \cdot \text{ft}}{\text{s}^2}$$

The slug is an unfamiliar unit because in everyday life weights rather than masses are specified in the British system: We go shopping for 10 lb of apples, not $\frac{1}{3}$ slug of apples. In the metric system, on the other hand, masses are normally specified: European grocery scales are calibrated in kilograms, not in newtons.

Relation between slugs and pounds

In order to convert a weight in pounds to a mass in slugs we make use of Eq. (2-3) to obtain

$$m(\text{slugs}) = \frac{w\ (\text{lb})}{g\ (\text{ft/s}^2)}$$

Since g, the acceleration of gravity at the earth's surface, has the value 32 ft/s^2,

$$m\ (\text{slugs}) = \frac{w\ (\text{lb})}{32\ \text{ft/s}^2}$$

The *weight* of a 1-slug mass is 32 lb, and the *mass* of a 1-lb weight is $\frac{1}{32}$ slug (Table 2-1).

Relation between kilograms and pounds

One kilogram corresponds to 2.21 pounds in the sense that the weight of 1 kilogram is 2.21 pounds.

One pound corresponds to 0.454 kilogram in the sense that the mass of 1 pound is 0.454 kilogram.

Example   A net force of 15 lb acts on an 8.0-lb object. Find the object's acceleration.

Solution   First we calculate the mass of the object, which is

$$m = \frac{w}{g} = \frac{8\ \text{lb}}{32\ \text{ft/s}^2} = 0.25\ \text{slug}$$

From the second law of motion $F = ma$ we find that the acceleration is

$$a = \frac{F}{m} = \frac{15\,\text{lb}}{0.25\,\text{slug}} = 60\,\text{ft/s}^2$$

We can work out problems like this in a single step by using w/g in place of m in the second law, so that $F = (w/g)a$ and

$$a = \frac{F}{w/g} = \frac{Fg}{w} = \frac{(15\,\text{lb})(32\,\text{ft/s}^2)}{8.0\,\text{lb}} = 60\,\text{ft/s}^2 \qquad \blacksquare$$

Example A 3000-lb car has an initial speed of 10 mi/h. How much force is required to accelerate the car to a speed of 50 mi/h in 9.0 s?

Solution We begin by converting the speeds from mi/h to ft/s. From the table inside the back cover of this book, 1 mi/h = 1.47 ft/s. To two significant figures,

$$v_0 = (10\,\text{mi/h})\left(1.47\,\frac{\text{ft/s}}{\text{mi/h}}\right) = 15\,\text{ft/s}$$

$$v_f = (50\,\text{mi/h})\left(1.47\,\frac{\text{ft/s}}{\text{mi/h}}\right) = 73\,\text{ft/s}$$

The car's acceleration is

$$a = \frac{v_f - v_0}{t} = \frac{(73 - 15)\text{ft/s}}{9.0\,\text{s}} = 6.4\,\text{ft/s}^2$$

The force that must act on the car is therefore

$$F = ma = \left(\frac{w}{g}\right)a = \left(\frac{3000\,\text{lb}}{32\,\text{ft/s}^2}\right)(6.4\,\text{ft/s}^2) = 600\,\text{lb} \qquad \blacksquare$$

TABLE 2–1
Units of mass and weight

System of units	Unit of mass	Unit of weight	Acceleration of gravity g	To find mass m given weight w	To find weight w given mass m
SI	Kilogram (kg)	Newton (N)	9.8 m/s^2	$m\,(\text{kg}) = \dfrac{w\,(\text{N})}{9.8\,\text{m/s}^2}$	$w\,(\text{N}) = m\,(\text{kg}) \times 9.8\,\text{m/s}^2$
British	Slug	Pound (lb)	32 ft/s^2	$m\,(\text{slugs}) = \dfrac{w\,(\text{lb})}{32\,\text{ft/s}^2}$	$w\,(\text{lb}) = m\,(\text{slugs}) \times 32\,\text{ft/s}^2$

Conversion of units:	1 slug = 14.6 kg	1 N = 0.225 lb
	1 kg = 0.0685 slug	1 lb = 4.45 N

2–7   THIRD LAW OF MOTION

The third of Newton's laws of motion states that

Third law of motion

When an object exerts a force on another object, the second object exerts on the first a force of the same magnitude but in the opposite direction.

If we call one of the interacting objects A and the other B, then according to the third law of motion

$$F_{AB} = -F_{BA} \qquad\qquad \textit{Third law of motion} \quad (2\text{–}4)$$

Here F_{AB} is the force A exerts on B and F_{BA} is the force B exerts on A (Fig. 2–17).

Action and reaction forces

Forces never occur all by themselves. When we think just one force is at work in a situation, a closer look always shows another force present as well. The key idea here is that the forces do not act on the same object. The third law of motion always applies to two different forces on two different things—the *action force* that one object exerts on another, and the equal but opposite *reaction force* that the second exerts on the first.

Every action force is accompanied by a reaction force

There is no such thing as a single force in the universe. Four examples of action-reaction pairs of forces are shown in Fig. 2–18. (a) We push against a wall; the wall pushes back on us. (b) We throw a ball; as we are pushing it into the air, it is pushing back on our hand. (c) We fire a rifle, and the expanding gases in its barrel push the bullet out; the gases push back on the rifle, which gives rise to the recoil force we feel with our shoulder. (d) An apple falls because of the downward gravitational pull of the earth; there is an equal upward pull by the apple on the earth which we cannot detect because the earth is so much more massive than the apple, but it nevertheless exists.

Let us apply the third law to a few more situations in order to appreciate its significance. A 1-kg book lies stationary on a table, pressing down on the table with a force of 9.8 N (Fig. 2–19). The table pushes upward on the book with the reaction force of 9.8 N. Why doesn't the book fly upward into the air? The answer is that the upward force of 9.8 N on the book merely balances its weight of 9.8 N, which acts

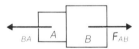

FIG. 2–17 According to the third law of motion, $F_{AB} = -F_{BA}$.

FIG. 2–18 Every action force in the universe is accompanied by an equal and opposite reaction force. The two forces act on different bodies.

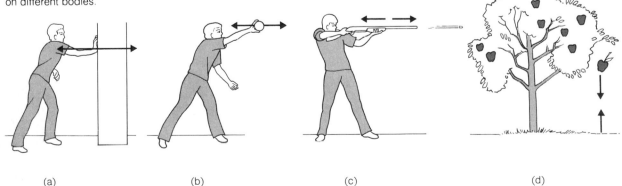

(a)   (b)   (c)   (d) downward. If the table were not there to cancel out the latter 9.8-N force, the book would, of course, be accelerated downward.

Another illustration of the third law of motion is the operation of walking. We push backward with one foot, and the earth pushes forward on us. The forward reaction force exerted by the earth causes us to move forward, and at the same time, the backward force of our foot causes the earth to move backward (Fig. 2–20). Owing to the earth's enormously larger mass, its motion cannot be detected practically, but it is there. Why is it that there is no reaction force on us, responding to the earth's push on our foot, to keep us from moving? The explanation is that every action-reaction pair of forces acts on *different* bodies. We push on the earth, the earth pushes back on us. If there are no *additional* forces present to impede our motion (for instance, pressure by a wall directly in front of us), we proceed to undergo a forward acceleration.

Suppose we find ourselves on a frozen lake with a perfectly smooth surface. Now we cannot walk because the absence of friction prevents us from exerting a backward force on the ice that would produce a forward force on us. But what we can do is exert a force on some object we may have with us, say a rock. We throw the rock forward by applying a force to it; at the same time the rock is pressing back on us with the identical force but in the opposite direction, and as a result we find ourselves moving backward (Fig. 2–21).

Walking and reaction forces

Throwing an object produces a backward reaction force

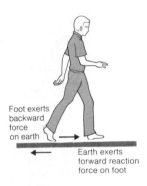

FIG. 2–20 When we push backward on the earth with one foot, the opposite reaction force of the earth pushes forward on us. The latter force causes us to move forward.

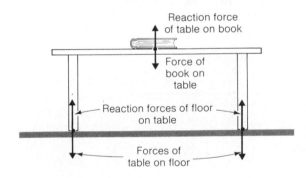

FIG. 2–19 The action-reaction forces between a book and a table and between the table and the floor.

FIG. 2–21 If we throw a rock while standing on a frozen lake, the reaction force pushes us backward.

We shall learn in Chapter 6 how these notions are expressed in the *principle of conservation of momentum,* one of the most useful formulations of the laws of motion.

2−8   FRICTION

Frictional forces impede motion

Frictional forces are of many kinds, but they all act to impede motion. The effects of friction must be distinguished from those of inertia. The term *inertia* refers to the fact that bodies maintain their original states of rest or of motion in the absence of net forces on them; but even the smallest force is sufficient to accelerate a body despite its inertia. The term *friction,* on the other hand, refers to actual forces that come into being when two surfaces are in contact that act to oppose motion between them, and may even prevent it.

Friction can be useful

Often friction is desirable: The fastening action of nails, screws, and bolts and the resistive action of brakes depend upon it. Walking would be impossible without friction. In many situations, however, friction merely reduces efficiency, and great efforts are made in industry to minimize it through the use of lubricants—notably grease and oil—and special devices—notably the wheel. About half the power of a typical automobile engine is wasted in overcoming friction in the engine itself and in its drive train.

To clearly understand the characteristic properties of frictional forces, let us consider what happens when we move a box across a level floor (Fig. 2−22). As we begin to push, the box remains in place because the floor exerts a force on the bottom of the

FIG. 2−22   (a) As force is applied to a box on a level floor, the frictional resistive force increases to a certain maximum, decreases somewhat as the box begins to move, and then remains constant. (b) The net force on the box equals the applied force minus the frictional force. The frictional force is always equal to or less than the applied force.

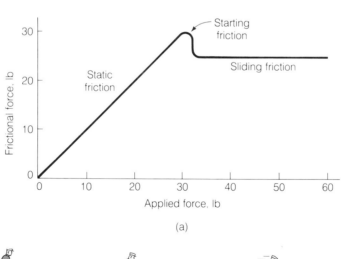

(a)

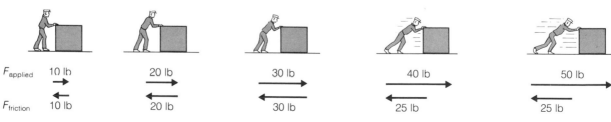

(b)

box which opposes the force we apply. This opposition force is friction, and it arises from the nature of the contact between the floor and the box. As we push harder, the frictional force also increases to match our efforts, until finally we are able to exceed the frictional force and begin to move the box. The frictional force is parallel to the contacting surfaces of the floor and the box.

As indicated in Fig. 2–22(a) the force between two stationary surfaces in contact that resists any motion between them is called *static friction*. In a given situation, static friction has a certain maximum value called *starting friction*. When the applied force exceeds starting friction, one surface begins to move relative to the other. The *sliding friction* that occurs after that is somewhat less than the starting friction, so less force is needed to keep the box of Fig. 2–22 in motion than the force needed to set it moving in the first place.

Static, starting, and sliding friction

Because the maximum static friction is always greater than sliding friction, a car's tires provide more traction when they are not slipping. This holds true whether the intention is to accelerate the car forward (spinning the wheels on snow or ice is not very effective) or to slow it down (locking the wheels by applying too much force to the brakes reduces the retarding force on the car).

Sliding friction is less than static friction

Friction has two chief causes. One is the mechanical interlocking of irregularities in the two surfaces, as in Fig. 2–23(a), that prevents one surface from sliding evenly past the other. The second effect is the tendency for materials in very close contact to stick together because of attractive forces between their respective atoms; the origin of such forces is discussed in Chapter 29. The tiny "welds" that occur where projections on the surfaces are pressed tightly together must be broken in order to move one surface over the other, and as the motion continues, new welds form and must be broken in turn.

Origin of friction

Lubricants reduce friction by separating two contacting surfaces with an intermediate layer of a softer material. Instead of rubbing against each other, the surfaces rub against the lubricant (Fig. 2–23(b)). Depending upon the specific use, the most suitable lubricant may be a gas, a liquid, or a solid. Most lubricants are oils derived from petroleum; grease consists of oil to which a thickening agent has been added to prevent the oil from running out from between the surfaces involved. The joints of the limbs of the human body are lubricated by a substance called *synovial fluid*, which resembles blood plasma.

Lubricants

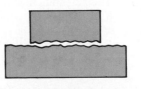

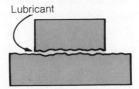

FIG. 2–23 A lubricant reduces friction by providing a soft material able to flow readily to separate surfaces in contact.

Lubricant

FIG. 2–24 The flattening of a wheel and the indentation of the surface it presses on both contribute to rolling friction. (Photo: © Roy King.)

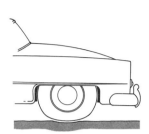

The wheel reduces friction

There is no relative motion between the rim of a wheel and a smooth surface over which it rolls if wheel and surface are both rigid, and there is accordingly no frictional resistance to overcome. By reducing friction so drastically, the wheel makes it possible to transport loads from one place to another without the enormous forces that dragging them would require. Most aspects of our present technological civilization rely upon the wheel in one way or another.

Rolling friction

Neither wheels nor the surfaces on which they travel can ever be perfectly rigid. Figure 2–24 shows the flattening of the wheel and the indentation of the surface which both contribute to *rolling friction.* Because the wheel and surface must be constantly deformed as the wheel rolls, a force is needed to keep the wheel rolling. However, this force is usually many times smaller than that needed to overcome sliding friction. Balls and rollers are widely used to reduce the friction between a rotating shaft and the bearings that hold it in place by replacing sliding friction with rolling friction. Tire deformation accounts for most of the frictional force on a car at speeds of up to 50 to 80 km/h, when air resistance starts to dominate.

2–9   COEFFICIENT OF FRICTION

It is a matter of experience that the frictional force exerted by one surface upon another depends upon two factors:

1. The kinds of surfaces that are in contact.
2. The perpendicular force with which either surface is pressed against the other. This perpendicular force is usually referred to as the *normal force,* symbol N.

The greater the normal force, the greater the friction, regardless of area in contact

The more tightly two objects are pressed together by a normal force, the greater the friction between them. For this reason an empty box is easier to push across a floor than a similar box loaded with something heavy (Fig. 2–25). Equally familiar is the effect of the nature of the contacting surfaces. For instance, it takes more than three times as much force to push a wooden box across a wooden floor as it does to push a steel box of the same weight across a steel floor. Interestingly enough, the area in contact between the two surfaces is not important: It is just as hard to push a small 50-kg box over a given floor as it is to push a large 50-kg box of the same material over the same floor.

To a good degree of approximation the following formula relates the normal force N pressing one surface against another with the frictional force F_f that results:

$$F_f = \mu N \qquad\qquad \textit{Frictional force} \quad (2-5)$$

Frictional force = coefficient of friction $\times$ normal force.

The quantity μ (Greek letter *mu*) is called the *coefficient of friction* and is a constant for a given pair of surfaces. Table 2-2 is a list of coefficients of static and sliding friction for a number of sets of surfaces.

Coefficient of friction

Materials in Contact	Coefficient of Static Friction, μ_s	Coefficient of Sliding Friction, μ
Wood on wood	0.5	0.3
Wood on stone	0.5	0.4
Steel on steel (smooth)	0.15	0.09
Metal on metal (lubricated)	0.03	0.03
Leather on wood	0.5	0.4
Rubber tire on dry concrete	1.0	0.7
Rubber tire on wet concrete	0.7	0.5
Glass on glass	0.94	0.40
Steel on Teflon	0.04	0.04

TABLE 2-2
Approximate coefficients of static and sliding friction for various materials in contact

When an object resting on a surface is pushed, the force of static friction increases with the applied force until the limiting value of $\mu_s N$ is reached. That is, when no motion occurs, Eq. (2-5) gives the starting frictional force, not the actual frictional force: The actual frictional force has the same magnitude as the applied force, but is in the opposite direction. When the applied force is greater than the starting frictional force of $\mu_s N$, motion begins and the coefficient of sliding friction must then be used. In this case Eq. (2-5) gives the actual frictional force, which no longer depends on the applied force and is the same over a fairly wide range of speeds.

Sliding friction is independent of speed

On a level surface, normal force equals weight in the absence of applied forces with vertical components

When an object on a horizontal surface is being pushed or pulled by a horizontal force, the normal force N is just the object's weight $w = mg$. In this case

$$F_f = \mu N = \mu w = \mu mg \qquad (2\text{--}6)$$

Example   A 100-kg wooden crate is at rest on a level stone floor. (a) What is the minimum horizontal force needed to start the crate moving? (b) What is the minimum horizontal force needed to keep the crate in motion at constant speed? (c) What will happen if a horizontal force of 500 N is applied to the crate?

Solution   (a) The starting frictional force is given by Eq. (2–6) using the coefficient of static friction for these surfaces from Table 2–2:

$$F_f(\text{starting}) = \mu_s mg = (0.5)(100\text{kg})(9.8 \text{ m/s}^2) = 490 \text{ N}$$

The crate will begin to move only if a force of more than 490 N is applied to it.

(b) When the crate is moving, the coefficient of sliding friction must be used to find the frictional force, which is

$$F_f(\text{sliding}) = \mu mg = (0.4)(100 \text{ kg})(9.8 \text{ m/s}^2) = 392 \text{ N}$$

Although 490 N at least of force must be exerted to set the crate in motion, only 392 N of force is needed to continue the motion at constant speed.

(c) Since the applied force of $F_A = 500$ N exceeds the starting force of 490 N, the crate will begin to move (Fig. 2–26). Since the applied force also exceeds the force of sliding friction $F_f = 392$ N, the crate will be accelerated. To find the acceleration, we begin by calculating the net force on the moving crate, which is

$$F = F_A - F_f = 500 \text{ N} - 392 \text{ N} = 108 \text{ N}$$

The crate's acceleration is, from the second law of motion $F = ma$,

$$a = \frac{F}{m} = \frac{108 \text{ N}}{100 \text{ kg}} = 1.08 \text{ m/s}^2$$

■

Example   The coefficient of sliding friction between a rubber tire and a dry concrete road is 0.7. What is the distance in which a car will skid to a stop on such a road if its brakes are locked when it is moving at 50 mi/h (80 km/h)?

FIG. 2–26

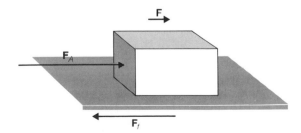

Solution   The first step is to calculate the acceleration of the car under these circumstances. The normal force N is the car's weight of mg, and the friction force is

$$F = \mu N = \mu mg$$

From the second law of motion, $F = ma$, and

$$\mu mg = ma$$
$$a = \mu g = (0.7)(32\,\text{ft/s}^2) = 22.4\,\text{ft/s}^2$$

The car's weight does not matter. We can find the distance s from $v_f^2 = v_0^2 + 2as$. Here $v_f = 0$ and

$$v_0 = (50\,\text{mi/h})\left(1.467\,\frac{\text{ft/s}}{\text{mi/h}}\right) = 73.4\,\text{ft/s}$$

Therefore, since $a = -22.4\,\text{ft/s}^2$ (the minus sign is needed because the acceleration is negative),

$$s = -\frac{v_0^2}{2a} = -\frac{(73.4\,\text{ft/s}^2)}{(2)(-22.4\,\text{ft/s}^2)} = 120\,\text{ft}$$

Which is 37 m.

■

IMPORTANT TERMS

The **inertia** of an object refers to the apparent resistance it offers to changes in its state of motion. The property of matter that manifests itself as inertia is called **mass.** The unit of mass in the SI system is the **kilogram,** in the British system the **slug.**

A **force** is any influence that can cause an object to be accelerated. The unit of force in the SI system is the **newton,** in the British system the **pound.**

The **weight** of an object is the gravitational force exerted on it by the earth. The weight of an object is proportional to its mass.

Newton's **first law of motion** states that, in the absence of a net force acting on it, an object at rest will remain at rest and an object in motion will continue in motion at constant velocity. The **second law of motion** states that a net force acting on an object causes it to have an acceleration proportional to the magnitude of the force and inversely proportional to the object's mass; the acceleration is in the same direction as the force. The **third law of motion** states that when an object exerts a force on another object, the second object exerts on the first a force of the same magnitude but in the opposite direction.

The term **friction** refers to the resistive forces that arise to oppose the motion of an object past another with which

it is in contact. **Sliding friction** is the friction resistance an object in motion experiences, whereas **static friction** is the frictional resistance a stationary object must overcome in order to be set in motion.

The **coefficient of friction** is the constant of proportionality for a given pair of contacting surfaces that relates the frictional force between them to the normal force with which one presses against the other; usually the coefficient of static friction is greater than that of sliding friction.

IMPORTANT FORMULAS

Second law of motion: $F = ma$

Weight: $w = mg$

Third law of motion: $F_{AB} = -F_{BA}$

Frictional force: $F_f = \mu N$

MULTIPLE CHOICE

1. When an object undergoes an acceleration,
   a. its mass increases.
   b. its speed increases.
   c. it falls toward the earth.
   d. a force acts upon it.

2. A sheet of paper can be withdrawn from under a bottle of milk without toppling it if the paper is jerked out quickly. This is an example of
    a. inertia.
    b. weight.
    c. acceleration.
    d. the third law of motion.

3. The weight of an object
    a. is the quantity of matter it contains.
    b. refers to its inertia.
    c. is basically the same quantity as its mass but expressed in different units.
    d. is the force with which it is attracted to the earth.

4. The acceleration of gravity on the surface of Mars is $3.7 \, \text{m/s}^2$. Compared with his or her mass and weight on the earth, an astronaut on Mars has
    a. less mass and less weight.
    b. less mass and more weight.
    c. the same mass and less weight.
    d. less mass and the same weight.

5. An automobile that is towing a trailer is accelerating on a level road. The force that the automobile exerts on the trailer is
    a. equal to the force the trailer exerts on the automobile.
    b. greater than the force the trailer exerts on the automobile.
    c. equal to the force the trailer exerts on the road.
    d. equal to the force the road exerts on the trailer.

6. When a horse pulls a wagon, the force that causes the horse to move forward is the force
    a. he exerts on the wagon.
    b. the wagon exerts on him.
    c. he exerts on the ground.
    d. the ground exerts on him.

7. The action and reaction forces referred to in Newton's third law of motion
    a. act upon the same object.
    b. act upon different objects.
    c. need not be equal in magnitude but must have the same line of action.
    d. must be equal in magnitude but need not have the same line of action.

8. A woman whose weight is 500 N is standing on a floor. The force the floor exerts on her is
    a. less than 500 N.
    b. equal to 500 N.
    c. more than 500 N.
    d. any of the above, depending on her location on the earth.

9. A jumper of weight w presses down on the ground with a force of magnitude F and becomes airborne as a result. The magnitude of the force the ground exerted on the jumper must have been
    a. equal to w and less than F.
    b. equal to w and equal to F.
    c. more than w and equal to F.
    d. more than w and more than F.

10. When an object exerts a normal force on a surface, the force is
    a. equal to the object's weight.
    b. frictional in origin.
    c. parallel to the surface.
    d. perpendicular to the surface.

11. To set an object in motion on a surface usually requires
    a. less force than to keep it in motion.
    b. the same force as that needed to keep it in motion.
    c. more force than to keep it in motion.
    d. only as much force as is needed to overcome inertia.

12. The frictional force between two surfaces in contact does not depend on
    a. the normal force pressing one against the other.
    b. the areas of the surfaces.
    c. whether the surfaces are stationary or in relative motion.
    d. whether a lubricant is used or not.

13. When a 1-N force acts on a 1-kg object that is able to move freely, the object receives
    a. a speed of 1 m/s.
    b. an acceleration of $0.102 \, \text{m/s}^2$.
    c. an acceleration of $1 \, \text{m/s}^2$.
    d. an acceleration of $9.8 \, \text{m/s}^2$.

14. When a 1-N force acts on a 1-N object that is able to move freely, the object receives
    a. a speed of 1 m/s.
    b. an acceleration of $0.102 \, \text{m/s}^2$.
    c. an acceleration of $1 \, \text{m/s}^2$.
    d. an acceleration of $9.8 \, \text{m/s}^2$.

15. When a 1-lb force acts on a 1-slug object that is able to move freely, the object receives
    a. a speed of 1 ft/s.
    b. an acceleration of $0.031 \, \text{ft/s}^2$.
    c. an acceleration of $1 \, \text{ft/s}^2$.
    d. an acceleration of $32 \, \text{ft/s}^2$.

16. When a 1-lb force acts on a 1-lb object that is able to move freely, the object receives
    a. a speed of 1 ft/s.

b. an acceleration of 0.031 ft/s^2.

c. an acceleration of 1 ft/s^2.

d. an acceleration of 32 ft/s^2.

17. A net force of 1 N acts on a 2-kg object, initially at rest, for 2 s. The distance the object moves during that time is

a. 0.5 m.

b. 1 m.

c. 2 m.

d. 4 m.

18. A 500-g ball lying on the ground is kicked with a force of 250 N. If the kick lasts 0.02 s, the ball flies off with a speed of

a. 0.01 m/s.

b. 0.1 m/s.

c. 2.5 m/s.

d. 10 m/s.

19. A 2400-kg car accelerates from 10 m/s to 30 m/s in 6 s. The force on the car is

a. 816 N.

b. 8000 N.

c. 12,000 N.

d. 78,400 N.

20. A certain force gives a 5-kg object an acceleration of 2.0 m/s^2. The same force would give a 20-kg object an acceleration of

a. 0.5 m/s^2.

b. 2.0 m/s^2.

c. 4.9 m/s^2.

d. 8.0 m/s^2.

21. A force of 10 N gives an object an acceleration of 5 m/s^2. What force would be needed to give it an acceleration of 1 m/s^2?

a. 1 N

b. 2 N

c. 5 N

d. 50 N

22. The weight of 600 g of salami is

a. 0.061 N.

b. 5.9 N.

c. 61 N.

d. 5880 N.

23. The mass of 8 lb of pineapples is

a. 0.25 slug.

b. 0.5 slug.

c. 4 slugs.

d. 256 slugs.

24. A force of 300 lb slows down a wagon from 60 ft/s to 20 ft/s in 10 s. The wagon's weight is

a. 75 lb.

b. 300 lb.

c. 1200 lb.

d. 2400 lb

25. An upward force of 600 N acts on a 50-kg dumbwaiter. The dumbwaiter's acceleration is

a. 0.82 m/s^2.

b. 2.2 m/s^2.

c. 11 m/s^2.

d. 12 m/s^2.

26. An 80-kg firefighter slides down a brass pole 8 m high in a fire station while exerting a frictional force of 600 N on the pole. He reaches the bottom of the pole in a time of

a. 1.46 s.

b. 1.87 s.

c. 2.64 s.

d. 6.96 s.

27. The coefficient of static friction between two wooden surfaces is

a. 0.5.

b. 0.5 N.

c. 0.5 kg/N.

d. 0.5 N/kg.

28. A brick has the dimensions 8 cm × 16 cm × 32 cm. The force of starting friction between the brick and a wooden floor is

a. greatest when the brick rests on the 8 cm × 16 cm face.

b. greatest when the brick rests on the 8 cm × 32 cm face.

c. greatest when the brick rests on the 16 cm × 32 cm face.

d. the same regardless of which face it rests on.

29. The coefficient of static friction for steel on ice is 0.1. The force needed to set a 70-kg skater in motion is approximately

a. 0.1 N.

b. 0.7 N.

c. 7 N.

d. 70 N.

30. A horizontal force of 150 N is applied to a 51-kg carton on a level floor. The coefficient of static friction is 0.5 and that of sliding friction is 0.4. The frictional force acting on the carton is

a. 150 N.

b. 200 N.

c. 250 N.

d. 500 N.

31. A toboggan reaches the foot of a hill at a speed of 4 m/s and coasts on level snow for 15 m before coming to a stop. The coefficient of sliding friction is

a. 0.004.

b. 0.05.

c. 0.16.

d. 0.27.

32. A car is moving at 20 m/s on a road such that the coefficient of friction between its tires and the road is 0.8. The minimum distance in which the car can be stopped is

a. 25.5 m.

b. 32.7 m.

c. 51 m.

d. undeterminable unless the car's mass is known.

33. A force of 40 lb is needed to set a 100-lb steel box moving across a wooden floor. The coefficient of static friction is

a. 0.08.

b. 0.25.

c. 0.4.

d. 2.5.

34. The coefficients of static and sliding friction for wood on wood are respectively 0.5 and 0.3. If a 100-lb wooden box is pushed across a horizontal wooden floor with just enough force to overcome the force of static friction, its acceleration is

a. 0.5 ft/s^2.

b. 6.4 ft/s^2.

c. 9.6 ft/s^2.

d. 16 ft/s^2.

EXERCISES

2–4   Second Law of Motion

1. When a body is accelerated, a force is invariably acting upon it. Does this mean that, when a force is applied to a body, it is invariably accelerated?

2. Compare the tension in the coupling between the first two cars in a train with the tension in the coupling between the last two cars when the train's speed is increasing.

3. It is less dangerous to jump from a high wall onto loose earth than onto a concrete pavement. Why?

4. Measurements are made of distance versus time for three moving objects. The distances are found to be directly proportional to t, t^2, and t^3, respectively. What can you say about the net force acting on each of the objects?

5. A 25-g snail goes from rest to a speed of 3 mm/s in 5 s. How much horizontal force does the ground exert on it?

6. A 430-g soccer ball lying on the ground is kicked and flies off at 25 m/s. If the duration of the impact was 0.01 s, what was the average force on the ball?

7. An empty truck whose mass is 2000 kg has a maximum acceleration of 1 m/s^2. What is its maximum acceleration when it is carrying a 1000-kg load?

8. A car whose mass (including the driver) is 1600 kg has a maximum acceleration of 1.2 m/s^2. If three 80-kg passengers are also in the car, find its maximum acceleration then.

9. A net horizontal force of 4000 N is applied to a 1400-kg car. What will the car's speed be after 10 s if it started from rest?

10. A bicycle and its rider together have a mass of 80 kg. If the bicycle's speed is 6 m/s, how much force is needed to bring it to a stop in 4 s?

11. A 12,000-kg airplane launched by a catapult from an aircraft carrier is accelerated from 0 to 200 km/h in 3 s. (a) How many times the acceleration of gravity is the airplane's acceleration? (b) What is the average force the catapult exerts on the airplane?

12. A force of 20 N gives an object an acceleration of 5 m/s^2. (a) What force would be needed to give the same object an acceleration of 1 m/s^2? (b) What force would be needed to give an acceleration of 10 m/s^2?

13. A 2000-kg truck is braked to a stop in 15 m from an initial speed of 12 m/s. How much force was required?

14. A 100-kg motorcycle carrying a 70-kg rider comes to a stop in 40 m from a speed of 50 km/h when its brakes are applied. Find the force exerted by the brakes and the force experienced by the rider.

15. A billy goat running at 5 m/s butts his head against a tree. If his head has a mass of 4 kg and it comes to a stop in 5 mm, find the average force of the impact. What do you think will happen to the goat?

16. A car moving at 10 m/s (22.4 mi/h) strikes a stone wall. (a) The car is very rigid and the 80-kg driver comes to a stop in a distance of 0.2 m. What is his average acceleration and how does it compare with the acceleration of gravity *g*? How much force acted upon him? Express this force in both newtons and pounds. (b) The car is so constructed that its front end gradually collapses upon impact, and the driver comes to a stop in a distance of 1 m. Answer the same questions for this situation.

17. A driver shifts into neutral when her 1200-kg car is moving at 80 km/h and finds that its speed has dropped to 65 km/h 10 s later. What was the average drag force acting on the car?

18. The driver of the car of Exercise 17 presses down on the gas pedal and thereby applies a force of 1300 N to the car. How long does it take the car to go from 65 km/h back to 80 km/h?

2–5   Weight

19. When a force equal to its weight is applied to a body free to move, what is its acceleration?

20. A person in an elevator suspends a 1-kg mass from a spring balance. What is the nature of the elevator's motion when the balance reads 9.0 N? 9.8 N? 10.0 N?

21. Can anything ever have a downward acceleration greater than *g*? If so, how can this be accomplished?

22. A balloon whose total mass is *M* is falling with the constant acceleration *a*. How much ballast should be thrown overboard to make the balloon float at constant altitude?

23. A force of 20 N acts upon a body whose weight is 8 N. (a) What is the mass of the body? (b) What is its acceleration?

24. A force of 20 N acts upon a body whose mass is 4 kg. (a) What is the weight of the body? (b) What is its acceleration?

25. A mass of 8 kg and another of 12 kg are suspended by a string on either side of a frictionless pulley. Find the acceleration of each mass.

26. A 100-kg man slides down a rope at constant speed. (a) What is the minimum breaking strength the rope must have? (b) If the rope has precisely this strength, will it support the man if he tries to climb back up?

27. How much force is needed to give a 5-kg box an upward acceleration of 2 m/s^2?

28. How much force must you supply to give a 1-kg object an upward acceleration of $2g$? A downward acceleration of $2g$?

29. A 1000-kg elevator has a downward acceleration of 1 m/s^2. What is the tension in its supporting cable?

30. A 1000-kg elevator has an upward acceleration of 1 m/s^2. What is the tension in its supporting cable?

31. A parachutist whose total mass is 100 kg is falling at 50 m/s when her parachute opens. Her speed drops to 7 m/s in a vertical distance of 40 m. What total force did her harness have to withstand? How many times her weight is this force?

32. A person stands on a scale in an elevator. When the elevator is at rest, the scale reads 700 N. When the elevator starts to move, the scale reads 600 N. (a) Is the elevator going up or down? (b) Does it have a constant speed? If so, what is this speed? (c) Does it have a constant acceleration? If so, what is this acceleration?

33. A 1200-kg elevator is supported by a cable that can safely withstand a tension of no more than 15 kN. (a) What is the maximum upward acceleration the elevator can have? (b) The maximum downward acceleration?

34. A 60-kg person stands on a scale in an elevator. How many newtons does the scale read (a) when the elevator is ascending with an acceleration of 1 m/s^2; (b) when it is descending with an acceleration of 1 m/s^2; (c) when it is ascending at the constant speed of 3 m/s; (d) when it is descending at the constant speed of 3 m/s; (e) when the cable has broken and the elevator is descending in free fall?

35. A 40-kg kangaroo exerts a constant force on the ground in the first 60 cm of her jump, and rises 2.0 m higher. When she carries a baby kangaroo in her pouch, she can rise only 1.8 m higher. What is the mass of the baby kangaroo?

36. In the arrangement shown (next column), find the mass M that will result in the 2-kg block remaining at rest. Assume that the pulleys are massless and frictionless.

2–6   British Units of Mass and Force

37. A force of 100 lb acts upon a body whose mass is 1.5 slugs. (a) What is the weight of the body? (b) What is its acceleration?

38. A force of 100 lb acts upon a body whose weight is 96 lb. (a) What is the mass of the body? (b) What is its acceleration?

39. How much force is needed to give a 6-lb box an upward acceleration of 4 m/s^2?

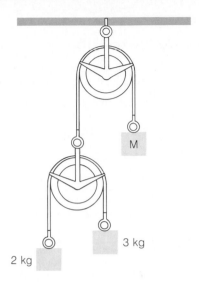

40. Find the force needed to accelerate a 2200-lb car from a standing start to a speed of 25 mi/h in 15 s.

41. A weight of 9 lb and another of 6 lb are suspended by a string on either side of a frictionless pulley. Find the acceleration of each weight.

42. A 2400-lb elevator is supported by a cable that can safely withstand a tension of no more than 3000 lb. (a) What is the maximum upward acceleration the elevator can have? (b) The maximum downward acceleration?

2–7   Third Law of Motion

43. Can we conclude from the third law of motion that a single force cannot act upon a body?

44. Since the opposite forces of the third law of motion are equal in magnitude, how can anything ever be accelerated?

45. An engineer designs a propeller-driven spacecraft. Because there is no air in space, he incorporates a supply of oxygen as well as a supply of fuel for the motor. What do you think of the idea?

46. A car with its engine running and in gear goes up a hill and then down on the other side. What forces cause it to move upward? Are they the same as the forces that cause it to move downward afterward?

47. A bird is in a closed box that rests on a scale. Does the reading on the scale depend on whether the bird is standing on the bottom of the box or is flying around inside it? What would happen if the bird were in a cage instead?

2–9   Coefficient of Friction

48. Ships are often built on ways that slope down to a nearby body of water. Normally a ship is launched before most of its interior and superstructure have been installed, and is completed when afloat. Is this done because the additional weight would cause the ship to slide down the ways prematurely?

49. A person is pushing a box across a floor. Can the frictional force on the box ever exceed its own weight? Can the frictional force ever exceed the applied force?

50. A woman prevents a 2-kg brick from falling by pressing it against a vertical wall. The coefficient of static friction is 0.6. What force must she use? Is this more or less than the weight of the brick?

51. An eraser is pressed against a vertical blackboard with a horizontal force of 10 N. The coefficient of friction between eraser and blackboard is approximately 0.2. Find the force parallel to the blackboard required to move the eraser.

52. A 100-kg wooden crate rests on a level wooden floor. What is the minimum force required to move it at constant speed across the floor?

53. A 240-lb wooden crate rests on a level wooden floor. What is the minimum force required to move it at constant speed across the floor?

54. A tennis ball rolling along a floor is found to be slowing down with an acceleration of $-1.6 \, \text{ft/s}^2$. Find the coefficient of rolling friction.

55. A driver sees a cow in the road ahead and applies the brakes so hard that they lock and the car skids to a stop in 80 ft. If the road is concrete and it is a dry day, what was the car's speed in mi/h when the brakes were applied?

56. A wooden block whose initial speed is 9 ft/s slides on a smooth floor for 6 ft before it comes to a stop. (a) Find the coefficient of friction. (b) If the block weighs 5 lb, how much force would be needed to keep it moving at constant speed across the same floor?

57. A tennis ball whose initial speed is 2 m/s rolls along a floor for 5 m before coming to a stop. Find the coefficient of rolling friction.

58. A 1250-kg car reaches a speed of 2 m/s after being pushed by two people for 12 m starting from rest. If the coefficient of rolling friction is 0.007, find the force the people exerted on the car.

59. A very strong 100-kg man is having a tug-of-war with a 1500-kg elephant. The coefficient of static friction for both on the ground is 1.0. (a) Find the force needed to move the elephant. (b) What would happen if the man could actually exert such a force?

60. Find the maximum possible accelerations the engines of the following cars can provide on a level road when the coefficient of friction is μ. In each case the front wheels of the car support 60% of its weight.
  a. a car with rear-wheel drive
  b. a car with front-wheel drive
  c. a car with four-wheel drive

61. The coefficient of friction between a tire and a certain road is 0.7. (a) What is the maximum possible acceleration of a car on this road? (b) What is the minimum distance in which the car can be stopped when it is moving at 100 km/h?

62. A truck moving at 15 m/s is carrying a 2000-kg steel girder that rests on the bed of the truck without any fastenings. The coefficient of static friction between the girder and the truck bed is 0.5. (a) Find the minimum distance in which the truck can come to a stop without having the girder move forward. (b) If the girder had a mass of 3000 kg, would this distance be different?

63. A railway boxcar is set in motion along a track at the same initial velocity as a truck on a parallel road. If the only horizontal forces acting on both vehicles are due to rolling friction with the respective coefficients of 0.0045 and 0.04, which vehicle will come to a stop first? How many times farther will the other vehicle travel?

64. If μ is the coefficient of static friction between the cart A and block B in the figure below, what acceleration must the cart have if B is not to fall?

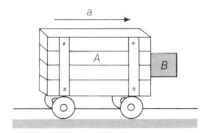

65. A 5-kg block resting on a horizontal surface is attached to a 5-kg block that hangs freely by a string passing over a pulley, an arrangement similar to that shown in Fig. 2–15. (a) If there is no friction between the first block and the surface, what is the block's acceleration? (b) If the coefficient of friction between the first block and the surface is 0.2, what is the block's acceleration?

66. A 10-kg block resting on a horizontal surface is attached to a 5-kg block that hangs freely by a string passing over a pulley, an arrangement similar to that shown in Fig. 2–15. (a) If there is no friction between the first block and the surface, what is the block's acceleration? What is the block's acceleration if the coefficient of friction between the first block and the surface is (b) 0.4? (c) 0.5? (d) 0.6?

67. Two wooden blocks rest on a wooden table, as in the figure. The coefficient of sliding friction is 0.3. Find the minimum force needed to move the 3-kg block at constant speed in each case.

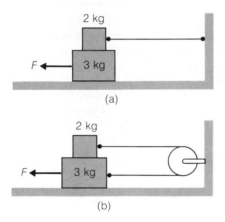

(a)

(b)

ANSWERS TO MULTIPLE CHOICE

1. d	**8.** b	**15.** c	**22.** b	**29.** d
2. a	**9.** c	**16.** d	**23.** a	**30.** a
3. d	**10.** d	**17.** b	**24.** d	**31.** b
4. c	**11.** c	**18.** d	**25.** b	**32.** a
5. a	**12.** b	**19.** b	**26.** a	**33.** c
6. d	**13.** c	**20.** a	**27.** a	**34.** b
7. b	**14.** d	**21.** b	**28.** d	

3

VECTORS

Some physical quantities are completely specified by just a number and a unit. It is quite sufficient to say that a person's mass is 70 kg, that the area of a farm is 160 acres, or that the frequency of a sound wave is 660 cycles per second. These are examples of *scalar quantities*. In other cases more needs to be said to specify a quantity. A person's mass is the same whether he stands on his feet or on his head, but the motion of a car headed north at 50 mi/h is not at all the same as the motion of a car going in a circle at 50 mi/h. A quantity whose direction is significant is called a *vector quantity*. Vector quantities occur often in physics, and their arithmetic is different from the ordinary arithmetic of scalar quantities.

3–1 VECTOR QUANTITIES

Displacement is a vector quantity

The simplest example of a vector quantity is *displacement*, which is a change in position. If we are told that a certain airplane leaves New York City and flies for 640 km

CHAPTER OBJECTIVES

Completing this chapter should enable you to:

1. Determine whether a quantity is vector or scalar in nature.

2. Draw a vector on a suitable scale that shows the magnitude and direction of a vector quantity.

3. Add vectors on a diagram to find their resultant.

4. Use the Pythagorean theorem to find the length of an unknown side of a right triangle.

5. Use the sine, cosine, and tangent functions of an angle in solving vector problems.

6. Subtract one vector from another of the same kind.

7. Resolve a vector into components in two perpendicular directions.

8. Find the magnitude and direction of a vector from its components.

9. Add vectors using their components.

10. Resolve the forces acting on an object on an inclined plane into components parallel and perpendicular to the plane, and then analyze the object's motion.

11. Analyze the motion of an object that has both horizontal and vertical components of motion.

12. Find the horizontal range of a projectile given its initial speed and direction.

(a scalar statement), we know nothing about its actual path or final destination; it could land anywhere within a circle of radius 640 km whose center is New York. If we are told instead that the airplane flies for 640 km to the west (a vector statement), we know where it is headed and can identify its destination as Cleveland (Fig. 3–1).

Two other vector quantities of special significance are *force* and *velocity*. The velocity of a body incorporates both its speed (a scalar quantity) and its direction: The speed of a car might be 60 km/h, whereas its velocity is 60 km/h to the north. This may seem a minor distinction, but it is not. Given its speed and direction, we can calculate where a moving body will be after a certain time interval; but if we do not know the direction, we cannot determine the path of the body at all.

Force and velocity are also vector quantities

The most straightforward way to represent a vector quantity is to draw a straight line with an arrowhead at one end to indicate the direction of the quantity. The length of the line is proportional to the magnitude of the quantity. A line of this kind is called a *vector*.

A vector is a directed line segment that represents a vector quantity

FIG. 3–1 An airplane that leaves New York and flies for a distance of 640 km may land anywhere within the circle shown. If the airplane undergoes a displacement of 640 km to the west, it lands in Cleveland. Distance is a scalar quantity; displacement is a vector quantity.

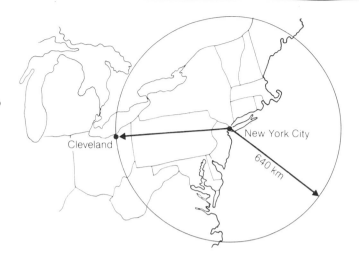

FIG. 3–2 The arrow is a vector that represents a 640-km westward displacement.

Symbols for vector and scalar quantities

Figure 3–2 shows how a 640-km westward displacement might be represented by a vector. The compass rose establishes the orientation of the displacement, and the distance scale establishes the relationship between length in the diagram and the actual length.

Vector quantities besides displacements can also be represented by vectors. The length of the vector in each case is proportional to the magnitude of the quantity it represents, and its direction is the direction of the quantity. Thus a velocity whose magnitude is 25 m/s is represented on a scale of 1 mm = 1 m/s by an arrow 25 mm long, and a force whose magnitude is 1500 N is represented on a scale of 1 mm = 100 N by an arrow 15 mm long (Fig. 3–3). The directions of the arrows indicate the directions of the quantities.

The symbols of vector quantities are customarily printed in boldface type (**F** for force), while italics are used both for scalar quantities (*m* for mass) and for the magnitudes of vectors. Thus we might denote a 240-km westward displacement by the symbol **A** and its magnitude of 240 km by the symbol *A*. In handwriting, vector quantities are indicated by placing arrows over their symbols—for instance, $\vec{A}$.

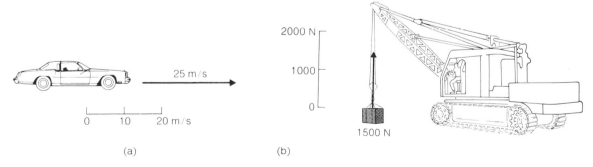

(a)   (b)

FIG. 3–3 (a) A vector that represents a velocity of 25 m/s. (b) A vector that represents a force of 1500 N.

3–2 VECTOR ADDITION

Ordinary arithmetic is used to add two or more scalar quantities of the same kind together. Ten kg of potatoes plus 4 kg of potatoes equals 14 kg of potatoes. The same kind of arithmetic is used for vector quantities of the same kind when the directions are the same. Thus a total upward force of 100 N is required to lift a 100-N weight, regardless of whether the force is applied by one, two, or ten people. It is the total force on an object, not the individual forces, that affects the motion of the object.

Adding scalar quantities

What do we do when the directions are different? A man who walks 10 km north and then 4 km west does *not* undergo a net displacement of 14 km from his starting point, even though he has walked a total of 14 km. A vector diagram provides a convenient method of determining his actual displacement. The procedure is to make a scale drawing of the successive displacements **A** and **B,** as in Fig. 3–4, and to join the starting point and the end point with a single vector **R**. The net displacement is **R,** whose length corresponds to 10.8 km and whose direction corresponds to 22° west of north.

The general rule for adding vectors is illustrated in Fig. 3–5. To add the vector **B** to the vector **A,** we follow these two steps:

1. Shift B parallel to itself until its tail is at the head of A. In its new position B must still have its original length and direction.

How to add two vectors

2. Draw a new vector R (called the *resultant*) from the tail of A to the head of B. The vector R is equal to A + B.

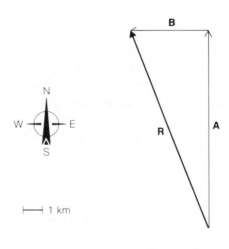

FIG. 3–4 The successive displacements of 10 km north, **A**, and 4 km west, **B**, result in a net displacement **R** of 10.8 km in a direction 22° west of north.

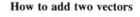

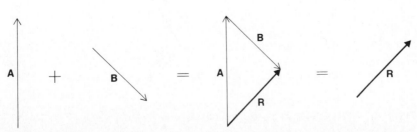

FIG. 3–5 The addition of two vectors to obtain the resultant **R**.

The sum of two or more vectors is called their resultant

To find the magnitude R of the resultant $\mathbf{R}$, all we need is to measure the length of $\mathbf{R}$ on the diagram and compare this length with the scale. The direction of $\mathbf{R}$ with respect to $\mathbf{A}$ or $\mathbf{B}$ may be found with a protractor.

The same procedure may be used with any number of vectors: Place the tail of each vector at the head of the previous one, keeping their lengths and original directions unchanged, and draw a vector $\mathbf{R}$ from the tail of the first vector to the head of the last. $\mathbf{R}$ is the sum required. Figure 3–6 shows how four vectors, whose initial magnitudes and directions differ, are added together. We note that the order in which the vectors are added does not affect the result; that is,

Vectors can be added in any order

$$\mathbf{A} + \mathbf{B} = \mathbf{B} + \mathbf{A} \qquad\qquad\qquad \textit{Vector addition} \quad (3–1)$$

Example   A boat heads east at 10.0 km/h in a river that flows south at 3.0 km/h. What is the boat's velocity relative to the earth?

Solution   We add the vector that represents the river current to the vector that represents the boat's velocity relative to the river, as in Fig. 3–7. This procedure leads to a single vector, which indicates that the boat's velocity relative to the earth is 10.4 km/h in a direction about 17° south of east.

For rough calculations it is sufficient to determine the magnitude and direction of a resultant directly from a vector diagram with ruler and protractor. More accurate results may be obtained with the help of trigonometry. In the case of this problem, the

FIG. 3–6   The addition of four vectors to obtain the resultant $\mathbf{R}$. The order in which the vectors are added does not affect the magnitude or direction of $\mathbf{R}$.

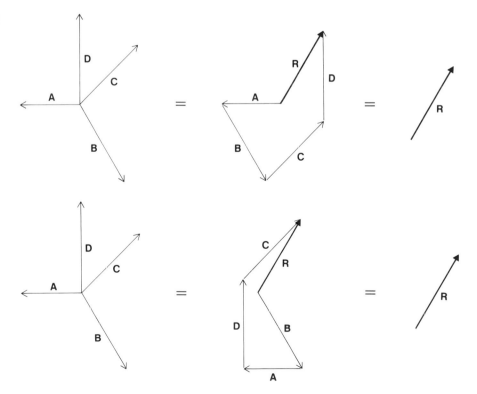

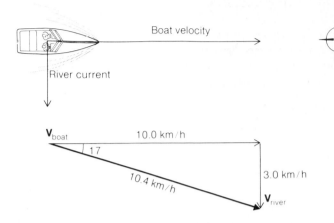

FIG. 3–7 A boat headed east at 10.0 km/h in a river that flows south at 3.0 km/h moves relative to the earth at 10.4 km/h in a direction 17° south of east.

fact that $\mathbf{v}_{boat}$ is perpendicular to $\mathbf{v}_{river}$ means that we can use the Pythagorean theorem to find the magnitude v of the resultant velocity $\mathbf{v}$. According to this theorem, the square of the hypotenuse (the long side) of a right triangle equals the sum of the squares of the other two sides. Hence

$$v^2 = v_{boat}^2 + v_{river}^2$$
$$v = \sqrt{v_{boat}^2 + v_{river}^2}$$
$$= \sqrt{(10.0\,\text{km/h})^2 + (3.0\,\text{km/h})^2} = \sqrt{109\,(\text{km/h})^2} = 10.4\,\text{km/h}$$

The angle θ between $\mathbf{v}$ and $\mathbf{v}_{boat}$ is specified by

$$\tan\theta = \frac{v_{river}}{v_{boat}} = \frac{3.0\,\text{km/h}}{10.0\,\text{km/h}} = 0.30$$

When we examine a table of trigonometric functions, we find that the angle whose tangent is closest to 0.30 is 17°, since $\tan 17° = 0.306$. Hence θ is slightly less than 17°. To the nearest degree,

$$\theta = 17°$$

Alternatively, we can use an electronic calculator to find θ by entering 0.30 and pressing the $\tan^{-1}$ (sometimes arctan or inv tan) key. Elementary trigonometry is reviewed in Appendixes A–7 and A–8 and summarized in Fig. 3–8. ■

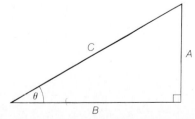

$$\sin\theta = \frac{\text{opposite side}}{\text{hypotenuse}} = \frac{A}{C}$$

$$\cos\theta = \frac{\text{adjacent side}}{\text{hypotenuse}} = \frac{B}{C}$$

$$\tan\theta = \frac{\text{opposite side}}{\text{adjacent side}} = \frac{A}{B}$$

Pythagorean theorem: $A^2 + B^2 = C^2$

FIG. 3–8 Basic trigonometric formulas.

3–3   VECTOR SUBTRACTION

The negative of a vector points in the opposite direction

It is sometimes necessary to subtract one vector from another. To subtract **A** from **B**, for example, we first form the negative of **A**, denoted − **A,** a vector that has the same length as **A** but points in the opposite direction (Fig. 3–9). We then add − **A** to **B** in the usual manner. This procedure may be summarized as

$$\mathbf{B} - \mathbf{A} = \mathbf{B} + (-\mathbf{A}) \qquad\qquad \textit{Vector subtraction} \quad (3-2)$$

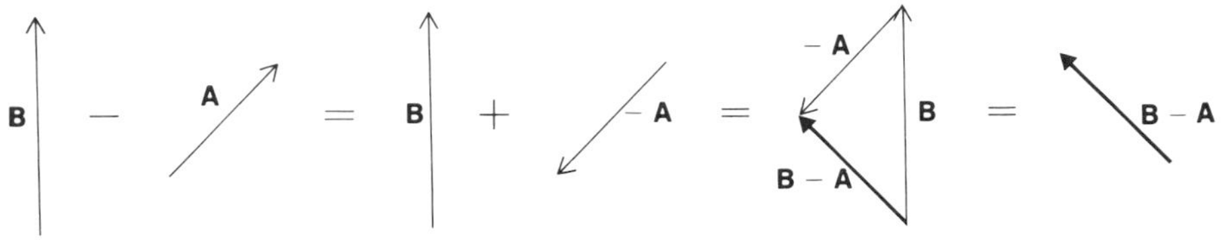

FIG. 3–9 Vector subtraction.

3–4   RESOLVING A VECTOR

Components of a vector

Just as we can add together two or more vectors to give a single resultant vector, so we can break up a vector into two or more others. The process of replacing one vector by two or more others is called *resolving* the vector, and the new vectors are called the *components* of the original one (Fig. 3–10). Often the best way to analyze a physical problem is to resolve a vector into components. Almost always the components of a vector are chosen to be perpendicular to one another.

Horizontal and vertical components

Figure 3–11 shows a boy pulling a wagon with a rope at the angle θ above the ground. Only part of the force he exerts affects the horizontal motion, since the wagon moves horizontally while the force **F** is not a horizontal one. We can resolve **F** into two components, **F**$_x$ and **F**$_y$, where

$$\mathbf{F}_x = \text{horizontal component of } \mathbf{F}$$

$$\mathbf{F}_y = \text{vertical component of } \mathbf{F}$$

It is a good habit to distinguish between the original vector and its components when all are on the same diagram by drawing two short lines across the original vector, as in Fig. 3–11. This is a reminder that the original vector is no longer to be considered since it has been replaced by its components.

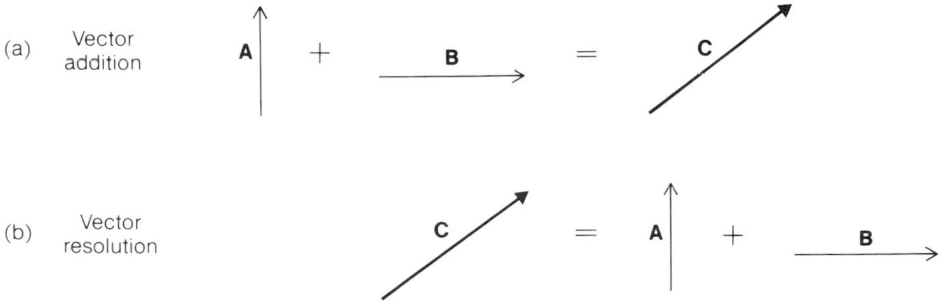

FIG. 3–10 (a) In vector addition, two or more vectors are combined to form a single vector. (b) In vector resolution, a single vector is replaced by two or more others whose sum is the same as the original vector. Here **A** and **B** are the components of the vector **C**.

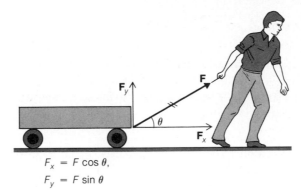

FIG. 3–11 The resolution of a force vector into horizontal and vertical components. The original vector is indicated by two short lines across it.

$F_x = F \cos \theta,$
$F_y = F \sin \theta$

The magnitudes of $\mathbf{F}_x$ and $\mathbf{F}_y$ are

$$F_x = F \cos \theta \tag{3–3}$$

$$F_y = F \sin \theta \tag{3–4}$$

The horizontal component $\mathbf{F}_x$ is responsible for the wagon's motion, while the vertical component $\mathbf{F}_y$ merely pulls upward on it. $\mathbf{F}_x$ is the projection of $\mathbf{F}$ in the horizontal direction, and $\mathbf{F}_y$ is the projection of $\mathbf{F}$ in the vertical direction.

Example   If the force the boy exerts on the wagon in Fig. 3–11 is 60 N and $\theta = 24°$, find F_x and F_y. Does their sum equal 60 N? Should it?

Solution   We have (Fig. 3–12)

$$F_x = F \cos \theta = (60 \, \text{N})(\cos 24°) = 54.8 \, \text{N}$$

$$F_y = F \sin \theta = (60 \, \text{N})(\sin 24°) = 24.4 \, \text{N}$$

The algebraic sum of the magnitudes F_x and F_y is 54.8 N + 24.4 N = 79.2 N, although the boy has exerted a force of only 60 N. Where is the mistake? The answer is that there is no mistake: $\mathbf{F}_x$ and $\mathbf{F}_y$ are vectors whose directions are different, so they can *only* be added vectorially. The algebraic sum of the magnitudes F_x and F_y has no meaning. If we add $\mathbf{F}_x$ and $\mathbf{F}_y$ vectorially, with the help of the Pythagorean theorem we find that

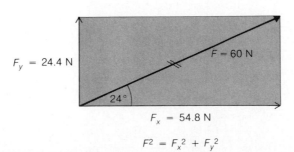

FIG. 3–12

$F_y = 24.4 \, \text{N}$

$F = 60 \, \text{N}$

$24°$

$F_x = 54.8 \, \text{N}$

$$F^2 = F_x^2 + F_y^2$$

$$F^2 = F_x^2 + F_y^2$$
$$F = \sqrt{F_x^2 + F_y^2} = \sqrt{(54.8\,\text{N})^2 + (24.4\,\text{N})^2} = \sqrt{3598\,\text{N}^2} = 60\,\text{N}$$

which is equal to the force the boy exerts. ■

Example   A woman on the ground sees an airplane climbing at an angle of 35° above the horizontal. She gets into her car and by driving at 70 mi/h is able to stay directly below the airplane. What is the airplane's speed?

Solution   The car's velocity must be equal to the horizontal component $\mathbf{v}_x$ of the airplane's velocity $\mathbf{v}$ (Fig. 3–13). Since

$$v_x = v \cos \theta$$

we have for the airplane's speed

$$v = \frac{v_x}{\cos \theta} = \frac{70\,\text{mi/h}}{\cos 35°} = 85\,\text{mi/h}$$ ■

Example   A rope whose working strength is 2000 N is used to tow a 1000-kg car up a 10° incline, as in Fig. 3–14. Find the maximum acceleration that can be given to the car.

Solution   The component of the weight mg of the car that is parallel to the incline is $mg \sin \theta$. If T is the maximum tension in the rope, the maximum net force along the incline that can be applied to the car is

$$F = T - mg \sin \theta$$

From the second law of motion,

$$T - mg \sin \theta = ma$$
$$a = \frac{T}{m} - g \sin \theta = \frac{2000\,\text{N}}{1000\,\text{kg}} - (9.8\,\text{m/s}^2)(\sin 10°)$$
$$= 2.0\,\text{m/s}^2 - 1.7\,\text{m/s}^2 = 0.3\,\text{m/s}^2$$

On a level road, $\theta = 0$ and $\sin \theta = 0$, and the maximum acceleration would be $2.0\,\text{m/s}^2$. ■

FIG. 3–13

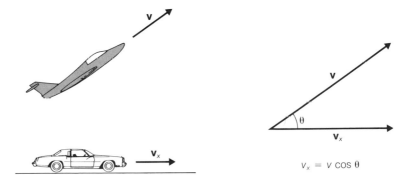

$$v_x = v \cos \theta$$

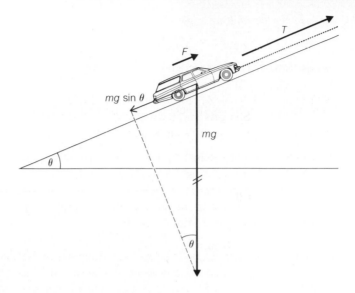

FIG. 3-14

Example A person whose shoes have leather heels is walking on a wooden floor. (a) Find the maximum angle the forward-swinging leg may make with the vertical in order that the heel not slip on the floor. (b) How is this angle affected if the floor is wet, which reduces the coefficient of static friction?

Solution (a) The geometry of the situation is shown in Fig. 3-15. The force **F** that the leg exerts on the floor can be resolved into a normal component **N** and a component **F**$_x$ parallel to the floor, where

$$N = F \cos \theta \qquad F_x = F \sin \theta$$

The condition for the heel not to slip is that the frictional force **F**$_f$ have the same magnitude as **F**$_x$. Since

$$F_f = \mu N = \mu F \cos \theta$$

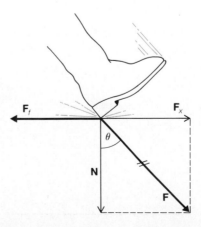

FIG. 3-15 If slipping is not to occur, the angle θ must be smaller than a certain critical angle that depends on the coefficient of static friction between the heel and the floor.

we have for the limiting angle

$$F_f = F_x$$
$$\mu\, F \cos\theta = F \sin\theta$$
$$\tan\theta = \mu$$

From Table 2–2 the coefficient of static friction for leather on wood is 0.5, so

$$\tan\theta = 0.5 \quad \text{and} \quad \theta = 27°$$

If θ is equal to or less than 27°, the heel will not slip.

   (b) Reducing μ reduces θ, so smaller steps have to be taken on a slippery surface. Slipping is a sudden process because, as it starts, the smaller coefficient of sliding friction applies, and the frictional force drops sharply. This is the reason it is difficult to keep from falling once slipping begins.   ■

Example   A wooden chute is being built along which wooden crates of merchandise are to be slid down into the basement of a store. (a) What angle with the horizontal should the chute make if the crates are to slide down at constant speed? (b) With what force must a 200-lb crate be pushed in order to start it sliding down the chute if the angle of the chute is that found in (a)?

Solution   (a) The procedure here is first to resolve the weight **w** of the crate, which is a force of magnitude w that acts downward, into a component **F** parallel to the plane and a component **N** perpendicular to the plane. With the help of Fig. 3–16 we find that

$$F = w \sin\theta \qquad N = w \cos\theta$$

When the crate slides down at constant speed, there is no net force acting on it, according to Newton's first law of motion. Hence the downward force along the chute must exactly balance the force of sliding friction, which means that

$$F = \mu N$$
$$w \sin\theta = \mu\, w \cos\theta$$
$$\mu = \frac{\sin\theta}{\cos\theta} = \tan\theta$$

From Table 2–2 the value of μ for wood on wood is 0.3, and so $\theta = 17°$.

FIG. 3–16 The weight **w** of a block on an inclined plane can be resolved into forces parallel and perpendicular to the plane. At an angle θ such that $\tan\theta - \mu_s$, the block slides down the plane at constant speed.

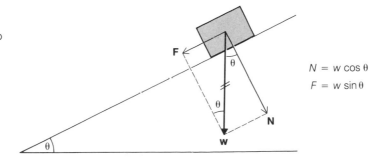

$N = w \cos\theta$
$F = w \sin\theta$

(b) We note that the coefficient of static friction here is $\mu_s = 0.5$. Hence the force of static friction to be overcome is

$$F_f = \mu_s N = \mu_s w \cos \theta$$

This is greater than the force of sliding friction, and so a crate will not begin to move without a push. The force component along the plane due to the crate's own weight is $F = w \sin \theta$. Here F is less than F_f, and so, if we call F' the outside force parallel to the plane required to move the crate,

Outside force + forward component of weight = backward friction force

$$F' + w \sin \theta = \mu_s w \cos \theta$$
$$F' = w(\mu_s \cos \theta - \sin \theta)$$
$$= 200 \, \text{lb} \, (0.5 \cos 17° - \sin 17°) = 37 \, \text{lb} \qquad \blacksquare$$

3–5 VECTOR ADDITION BY COMPONENTS

As we have seen, it is easy to add two or more vectors together graphically by drawing them head-to-tail and joining the tail of the first vector to the head of the last to form the resultant **R**. There are two ways to solve a problem of this kind by trigonometry. One is the direct method of adding two of the initial vectors together, then adding a third to the sum of the first two, a fourth to the sum of the first three, and so on. This usually means a lot of work and presents many chances for error.

A better way to add several vectors is to work in terms of their components. The procedure is as follows for vectors that lie in the same plane:

1. Resolve the initial vectors into their components in the x and y directions.
2. Add the components in the x direction to give $\mathbf{R}_x$ and add the components in the y direction to give $\mathbf{R}_y$. That is,

Component method of vector addition

$$\mathbf{R}_x = x\text{-component of } \mathbf{R}$$
$$= \mathbf{A}_x + \mathbf{B}_x + \mathbf{C}_x + \ldots = \text{sum of } x\text{-components} \qquad (3\text{–}5)$$
$$\mathbf{R}_y = y\text{-component of } \mathbf{R}$$
$$= \mathbf{A}_y + \mathbf{B}_y + \mathbf{C}_y + \ldots = \text{sum of } y\text{-components} \qquad (3\text{–}6)$$

3. Find the magnitude and direction of the resultant **R** from the components $\mathbf{R}_x$ and $\mathbf{R}_y$. From the Pythagorean theorem,

$$R = \sqrt{R_x^2 + R_y^2} \qquad (3\text{–}7)$$

The direction of **R** can be found from the values of the components by trigonometry; the best way to do this depends upon the problem at hand. Figure 3–17 shows the connection between head-to-tail vector addition and the component method for the case of two vectors.

Example   A sailboat is headed due north at a forward speed of 6.0 knots (kn). The pressure of the wind on its sails causes the boat to move sideways to the east at 0.5 kn. A tidal current is flowing to the southwest at 3.0 kn. What is the velocity of the sailboat

FIG. 3–17 Vector
addition by components.

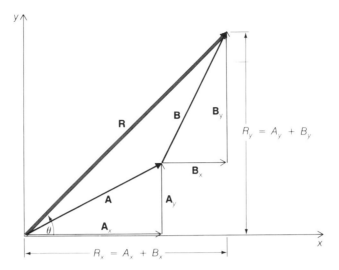

relative to the earth's surface? [A *knot* is a unit of speed equal to one nautical mile per hour. The nautical mile is widely used in air and sea navigation because it is the same in length as one minute (1′) of latitude, where 60′ = 1°. Since 1 nautical mile = 1.852 km = 6076 ft, 1 kn = 1.852 km/h = 1.151 mi/h.]

Solution   For convenience, we shall call north the $+y$-direction, south the y-direction, east the $+x$-direction, and west the $-x$-direction. With the help of Fig. 3–18 we see that the magnitudes of the components of the three velocity vectors are

$$A_x = 0 \qquad\qquad C_x = -(3.0\text{ kn})(\cos 45°)$$
$$A_y = 6.0\,\text{kn} \qquad\quad = -2.1\,\text{kn}$$
$$B_x = 0.5\,\text{kn} \qquad C_y = -(3.0\text{ kn})(\sin 45°)$$
$$B_y = 0 \qquad\qquad\quad = -2.1\,\text{kn}$$

We then add together the components in each direction:

$$R_x = A_x + B_x + C_x \qquad\qquad R_x = A_y + B_y + C_y$$
$$= 0 + 0.5\,\text{kn} - 2.1\,\text{kn} \qquad\quad = 6.0\,\text{kn} + 0 - 2.1\,\text{kn}$$
$$= -1.6\,\text{kn} \qquad\qquad\qquad\qquad = 3.9\,\text{kn}$$

The magnitude of the resultant velocity is

$$R = \sqrt{R_x^2 + R_y^2} = \sqrt{1.6^2 + 3.9^2}\,\text{kn} = \sqrt{17.8}\,\text{kn} = 4.2\,\text{kn}$$

Thus the speed of the boat relative to the earth is 4.2 kn. Its direction, as shown in the last part of the diagram, is west of north at an angle θ with north. We can find the value of θ by first finding the value of $\tan \theta$:

$$\tan \theta = \frac{R_x}{R_y} = \frac{1.6}{3.9} = 0.410 \qquad \text{so} \qquad \theta = 22°$$ ∎

FIG. 3–18

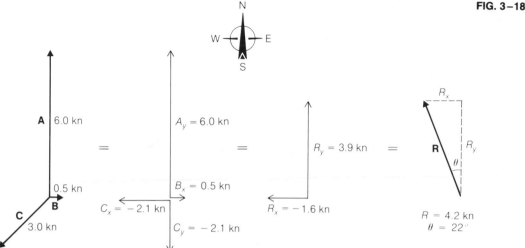

3–6   MOTION IN A VERTICAL PLANE

An object that moves through space usually has a curved path rather than a perfectly straight one. Our strategy in attacking problems of this kind is to resolve the object's acceleration **a** (assumed constant) and initial velocity $\mathbf{v}_0$ into their horizontal components $\mathbf{a}_x$ and $\mathbf{v}_{0x}$ and vertical components $\mathbf{a}_y$ and $\mathbf{v}_{0y}$. With the help of the formulas of Chapter 1 we then examine separately the object's motion in each of these directions. Finally we can combine $\mathbf{v}_x$ and $\mathbf{v}_y$ to find **v,** and $\mathbf{s}_x$ and $\mathbf{s}_y$ to find **s,** at any time t after the start of the motion by vector addition.

Suppose that we drop a ball A from the edge of a table while rolling an identical ball B off to the side (Fig. 3–19). At the moment the balls leave the table, A has zero velocity while B has the horizontal velocity $\mathbf{v}_0$. The velocity components of the balls therefore have the magnitudes

$$v_{Ax} = 0 \qquad v_{Bx} = v_0 \qquad v_{Ay} = 0 \qquad v_{By} = 0$$

Both balls reach the floor at the same time, even though B has traveled some distance s away from the table. The reason for this behavior is that the acceleration of gravity is the same for all bodies near the earth regardless of their state of motion; both A and B started out with no vertical velocity, both underwent the same downward acceleration, and so both took the same period of time to fall.

Vertical motion is independent of horizontal motion

An object in free fall descends the distance

$$h = \tfrac{1}{2} g t^2$$

in the time t when it starts with no vertical component of velocity. Hence both balls require the time

$$t = \sqrt{\frac{2h}{g}}$$

FIG. 3–19 **FIG. 3–19** Ball *A* is dropped from the edge of a table while ball *B* is simultaneously rolled off the edge with the initial horizontal speed v_0.

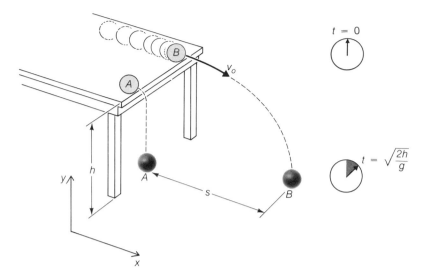

to reach the floor. If the table is 1.0 m high, then

$$t = \sqrt{\frac{2 \times 1.0\,\text{m}}{9.8\,\text{m/s}^2}} = 0.45\,\text{s}$$

While it is falling, ball *B* is also moving horizontally with the speed v_0. When it strikes the floor it will have traveled the horizontal distance

$$s = v_0 t$$

Let us say that $v_0 = 5.0$ m/s. Therefore

$$s = (5.0 \text{ m/s})(0.45 \text{ s}) = 2.3 \text{ m}$$

Example   Find the speeds with which balls *A* and *B* strike the floor.

Solution   The final velocity of *A* has only the single component

$$v_{Ay} = gt$$

and so

$$v_A = v_{Ay} = (9.8 \text{ m/s}^2)(0.45 \text{ s}) = 4.4 \text{ m/s}$$

The final velocity of *B*, however, has both horizontal and vertical components, namely

$$v_{Bx} = v_0 \qquad v_{By} = gt$$

From Fig. 3–20 we see that since $\mathbf{v}_{Bx}$ is perpendicular to $\mathbf{v}_{By}$, their vector sum $\mathbf{v}_B$ has the magnitude

$$v_B = \sqrt{v_{Bx}^2 + v_{By}^2} = \sqrt{v_0^2 + (gt)^2}$$

Since $v_0 = 5.0$ m/s and $t = 0.45$s,

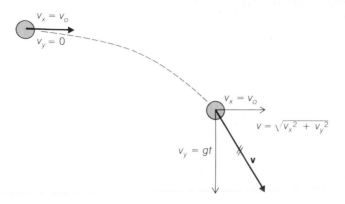

FIG. 3–20 The horizontal and vertical components of the velocity of ball B must be added vectorially to determine the magnitude of its velocity.

$$v_B = \sqrt{(5.0\,\text{m/s})^2 + (9.8\,\text{m/s}^2 \times 0.45\,\text{s})^2} = 6.7\,\text{m/s}$$

It is worth noting that the *vector sum* of $\mathbf{v}_{Bx}$ and $\mathbf{v}_{By}$ has the magnitude 6.7 m/s, whereas the *algebraic sum* of v_{Bx} and v_{By} is 9.4 m/s. The latter figure is, of course, completely meaningless, since velocity is a vector quantity and velocity addition must obey the rules of vector addition. ∎

Velocity addition must be done by vector methods

3–7 PROJECTILE FLIGHT

A more general case of motion in a vertical plane is exemplified by the flight of a projectile, for instance a rocket. Let us ignore the curvature of the earth and the frictional resistance of the atmosphere to the passage of the rocket, and assume that the rocket uses up its fuel at a distance from its launching point that is small compared with the total distance it travels. If the initial velocity $\mathbf{v}_0$ of the rocket makes an angle of θ with level ground, we can resolve $\mathbf{v}_0$ into components whose magnitudes are

$$v_{0x} = v_0 \cos\theta \qquad \textit{Initial velocity components} \quad (3\text{–}8)$$

$$v_{0y} = v_0 \sin\theta \qquad\qquad\qquad\qquad\qquad\qquad (3\text{–}9)$$

We can once again use the formulas for straight-line motion to examine the horizontal and vertical aspects of the rocket's flight separately, since these are independent of each other. If there are no horizontal forces on the rocket, its horizontal velocity component v_x remains constant during the flight. The vertical component v_y, however, gradually drops to zero due to the downward acceleration of gravity, and then becomes more and more negative (meaning that the rocket falls faster and faster) until the ground is reached (Fig. 3–21). Thus at the time t the rocket's velocity components become

Only the vertical component of projectile's velocity changes in flight

$$v_x = v_{0x} \qquad\qquad \textit{Velocity components at later time} \quad (3\text{–}10)$$

$$v_y = v_{0y} - gt \qquad\qquad\qquad\qquad\qquad\qquad (3\text{–}11)$$

If the rocket starts from the point $x = 0$, $y = 0$, then at the time t its horizontal and vertical coordinates of position will be

FIG. 3–21 In rocket flight, the horizontal component of a rocket's velocity is constant in the absence of air resistance.

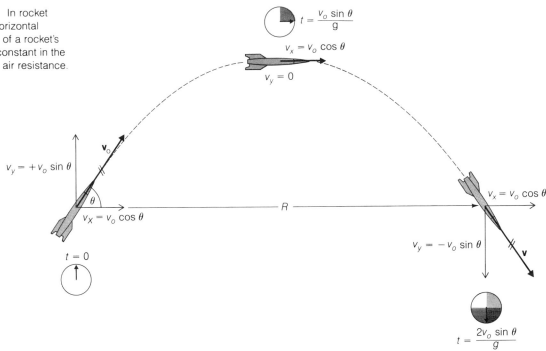

$$x = v_{0x}t \qquad\qquad\qquad \textit{Position of projectile} \quad (3\text{–}12)$$

$$y = v_{0y}t - \tfrac{1}{2}gt^2 \qquad\qquad\qquad\qquad\qquad\qquad (3\text{–}13)$$

Maximum height of projectile path

We can use the above formulas to find the maximum height H reached by a rocket and its range R, which is how far from its launching point it strikes the ground. At the top of its path, the rocket has no vertical component of velocity, so if T is the time needed to reach this height,

$$v_y = 0 = v_{0y} - gt = v_0 \sin\theta - gT$$

$$T = \frac{v_0}{g}\sin\theta$$

The maximum height is found by substituting this value of T in Eq. (3–13), with $y = H$:

$$H = v_{0y}T - \tfrac{1}{2}gT^2$$

$$= (v_0\sin\theta)\left(\frac{v_0}{g}\sin\theta\right) - \tfrac{1}{2}g\left(\frac{v_0}{g}\sin\theta\right)^2$$

$$H = \frac{v_0^2}{2g}\sin^2\theta \qquad\qquad\qquad \textit{Maximum height} \quad (3\text{–}14)$$

We note that $\sin^2\theta = (\sin\theta)^2$.

To find the range R, we note that the total time of flight equals $2T$, since the rocket needs the same amount of time T to return to the ground as the time T it took to reach the top of its path. (We are assuming that the rocket takes off and lands at the same height.) Hence with $x = R$ we have

Range of projectile

$$R = v_{0x}t = (v_0 \cos \theta)(2T) = (v_0 \cos \theta)\left(\frac{2v_0 \sin \theta}{g}\right)$$

$$= \frac{2v_0^2}{g} \sin \theta \cos \theta$$

This formula can be simplified by making use of the trigonometric identity

$$\sin \theta \cos \theta = \tfrac{1}{2} \sin 2\theta$$

The rocket's range may therefore be written

$$R = \frac{v_0^2}{g} \sin 2\theta \qquad\qquad \textit{Range of projectile} \quad (3–15)$$

This formula gives the range of a rocket (or any other projectile, for that matter, provided it obeys the restrictions given earlier) in terms of its initial speed v_0 and the angle θ at which it is launched. We see that R is a maximum when $\sin 2\theta = 1$, since 1 is the highest value the sine function can have. Since $\sin 90° = 1$, the maximum range occurs when the initial angle θ is 45°. Any other angle, greater or smaller, will result in a shorter range (Fig. 3–22). If θ_1 is an angle that leads to a range R, the other angle θ_2 for the same range is given by the formula

Maximum range occurs for $\theta = 45°$

$$\theta_2 = 90° - \theta_1 \qquad\qquad \textit{Angles for same range} \quad (3–16)$$

Example   An arrow leaves the bow of an archer at a speed of 30 m/s and an angle of 30° above the horizontal. (a) How far away will the arrow reach the ground? (b) What will its maximum altitude be? (c) What is the maximum range of an arrow with this speed? (d) At what angles could the archer point the arrow if it is to reach a target 70 m away? Neglect the height of the bow above the ground.

Solution   (a) From Eq. (3–15) we find that, since $2\theta = 60°$ here, the range of the arrow is

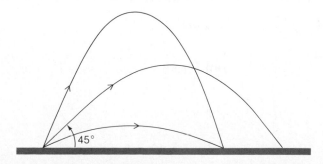

FIG. 3–22   In the absence of air resistance, the maximum range of a projectile occurs when it is fired at an angle of 45°.

$$R = \frac{v_0^2}{g} \sin 2\theta = \frac{(30 \text{ m/s})^2}{9.8 \text{ m/s}^2} \sin 60° = 80 \text{ m}$$

(b) From Eq. (3–14) the maximum altitude the arrow reaches is

$$H = \frac{v_0^2}{2g} \sin^2 \theta = \frac{(30 \text{ m/s})^2}{(2)(9.8 \text{ m/s}^2)} \sin^2 30° = 11.5 \text{ m}$$

(c) The maximum range corresponds to $\sin 2\theta = 1$ in Eq. (3–15), so

$$R_{max} = \frac{v_0^2}{g} = \frac{(30 \text{ m/s})^2}{9.8 \text{ m/s}^2} = 92 \text{ m}$$

(d) To find the smaller angle that leads to a given range R, we solve Eq. (3–15) for $\sin 2\theta$, which gives us

$$\sin 2\theta_1 = \frac{Rg}{v_0^2} = \frac{(70 \text{ m})(9.8 \text{ m/s}^2)}{(30 \text{ m/s})^2} = 0.762$$

Thus

$$2\theta_1 = \sin^{-1} 0.762 = 50° \quad \text{and} \quad \theta_1 = 25°$$

The other angle for the same range is (Fig. 3–23)

$$\theta_2 = 90° - \theta_1 = 90° - 25° = 65° \qquad ■$$

FIG. 3–23

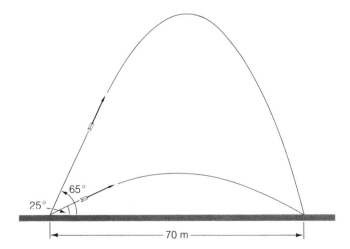

IMPORTANT TERMS

A **scalar quantity** is one that has magnitude only. A **vector quantity** is one that has both magnitude and direction. Thus time is a scalar quantity, and force is a vector quantity.

A **vector** is an arrowed line whose length is proportional to the magnitude of some vector quantity and whose direction is that of the quantity. A **vector diagram** is a scale drawing of the various forces, velocities, or other vector quantities involved in the motion of a body.

In the graphical method of **vector addition,** the tail of each successive vector is placed at the head of the previous one, with their lengths and original directions kept unchanged. The **resultant** is a vector drawn from the tail of the first vector to the head of the last.

A vector can be **resolved** into two or more other vectors called the **components** of the original vector. Usually the components of a vector are chosen to be in mutually perpendicular directions.

IMPORTANT FORMULAS

Vector addition: $\mathbf{A} + \mathbf{B} = \mathbf{B} + \mathbf{A}$

Vector subtraction: $\mathbf{B} - \mathbf{A} = \mathbf{B} + (-\mathbf{A})$

Components of a vector: $\mathbf{A} = \mathbf{A}_x + \mathbf{A}_y$
$$A_x = A \cos \theta$$
$$A_y = A \sin \theta$$

Vector addition by components:
$$R = \sqrt{R_x^2 + R_y^2}$$

where $R_x = A_x + B_x + C_x + \ldots$
$R_y = A_y + B_y + C_y \ldots$

Angle θ between $\mathbf{R}$ and $\mathbf{R}_x$: $\tan \theta = \dfrac{R_y}{R_x}$

Range of projectile: $R = \dfrac{v_0^2}{g} \sin 2\theta$

Angles for same range: $\theta_2 = 90° - \theta_1$

MULTIPLE CHOICE

1. Of the following units, which could be associated with a vector quantity?
   a. meters/minute
   b. quarts/second
   c. hours
   d. cubic feet

2. Which of the following statements is incorrect?
   a. All vector quantities have directions.
   b. All vector quantities have magnitudes.
   c. All scalar quantities have directions.
   d. All scalar quantities have magnitudes.

3. The minimum number of unequal forces whose vector sum can equal zero is
   a. 1.
   b. 2.
   c. 3.
   d. 4.

4. An engine block is supported by a rope hoist attached to an overhead beam. When the block is pulled to one side by a horizontal force exerted by another rope, the tension in the rope hoist
   a. is less than before.
   b. is unchanged.
   c. is greater than before.
   d. may be any of the above, depending on the magnitude of the horizontal force.

5. Which of the following pairs of displacements cannot be added to give a resultant displacement of 2 m?
   a. 1 m and 1 m
   b. 1 m and 2 m
   c. 1 m and 3 m
   d. 1 m and 4 m

6. Which of the following sets of forces cannot have a vector sum of zero?
   a. 10, 10, and 10 lb
   b. 10, 10, and 20 lb
   c. 10, 20, and 20 lb
   d. 10, 20, and 40 lb

7. Which of the following sets of displacements might be able to return a car to its starting point?
   a. 2, 8, 10, and 25 km
   b. 5, 20, 35, and 65 km
   c. 60, 120, 180, and 240 km
   d. 100, 100, 100, and 400 km

8. A displacement of 9 m and another of 6 m can be added to give a resultant displacement of
   a. 0 m.
   b. 1.5 m.
   c. 4 m.
   d. 16 m.

9. An airplane whose airspeed is 200 km/h is flying in a wind of 80 km/h. The speed of the airplane relative to the ground must be between
   a. 80 and 200 km/h.
   b. 80 and 280 km/h.
   c. 120 and 200 km/h.
   d. 120 and 280 km/h.

10. The magnitude of the resultant of two forces is a minimum when the angle between them is
   a. 0.
   b. 45°.
   c. 90°.
   d. 180°.

11. A man walks 8 km north and then 5 km in a direction 60° east of north. His resultant displacement from his starting point is

a. 11 km.   b. 12 km.
c. 13 km.   d. 14 km.

12. Two forces of 10 N each act on an object. The angle between the forces is 120°. The magnitude of their resultant is
a. 10 N.   b. 14 N.
c. 17 N.   d. 20 N.

13. The resultant of a 4-N force acting upward and a 3-N force acting horizontally is
a. 1 N.   b. 5 N.
c. 7 N.   d. 12 N.

14. The angle between the resultant of Question 13 and the vertical is approximately
a. 37°.   b. 45°.
c. 53°.   d. 60°.

15. An airplane travels 100 miles to the north and then 200 miles to the east. The displacement of the airplane from its starting point is approximately
a. 100 mi.   b. 200 mi.
c. 220 mi.   d. 300 mi.

16. At what angle east of north should the airplane of Question 15 have headed in order to reach its destination in a straight flight?
a. 22°   b. 45°
c. 50°   d. 63°

17. Three forces act on an object: 300 N to the north, 200 N to the east, and 100 N to the west. The magnitude of their resultant is
a. 200 N.   b. 316 N.
c. 400 N.   d. 424 N.

18. The direction of the resultant of the above forces is
a. north.   b. 18° east of north.
c. 18° north of east. d. 72° west of north.

19. A car of weight w is on a road that is at an angle of θ with the horizontal. The component of the car's weight parallel to the road is
a. $w \sin \theta$.   b. $w \cos \theta$.
c. $w/\sin \theta$.   d. $w/\cos \theta$.

20. A horizontal and a vertical force combine to give a resultant force of 50 N that acts in a direction 30° above the horizontal. The magnitude of the horizontal force is
a. 16.7 N.   b. 25 N.
c. 28.9 N.   d. 43.3 N.

21. A vector **A** lies in a plane and has the components A_x and A_y. The magnitude of A_x of $\mathbf{A}_x$ is equal to
a. $A - A_y$.   b. $\sqrt{A} - \sqrt{A_y}$.
c. $\sqrt{A - A_y}$.   d. $\sqrt{A^2 - A_y^2}$.

22. An escalator has a velocity of 3.0 m/s at an angle of 60° above the horizontal. The vertical component of its velocity is
a. 1.5 m/s.   b. 1.8 m/s.
c. 2.6 m/s.   d. 3.5 m/s.

23. Ball A is thrown horizontally and ball B is dropped from the same height at the same moment.
a. Ball A reaches the ground first.
b. Ball B reaches the ground first.
c. Ball A has the greater speed when it reaches the ground.
d. Ball B has the greater speed when it reaches the ground.

24. A ball is thrown horizontally from a moving car. While it is in flight (if air resistance is neglected) it is *not* true that
a. its speed changes.
b. its acceleration changes.
c. its direction of motion relative to the car changes.
d. its direction of motion relative to the road changes.

25. A ball rolls off the edge of a horizontal roof at 10 m/s. Two seconds later the speed of the ball will be
a. 10 m/s.   b. 19.6 m/s.
c. 29.6 m/s.   d. 22 m/s.

26. A 200-g ball is dropped from a cliff. A wind blowing on the ball produces a 1-N horizontal force on it. The resultant acceleration of the ball is
a. 8.4 m/s².   b. 9.8 m/s².
c. 11 m/s².   d. 14.8 m/s².

27. A ball is thrown at a 30° angle above the horizontal with a velocity of 10 ft/s. After $\frac{1}{2}$ s the horizontal component of its velocity will be
a. 9 ft/s.   b. 10 ft/s.
c. 12 ft/s.   d. 19 ft/s.

28. An archer shoots an arrow at an angle of 40° above the horizontal. The arrow would travel farther if the angle were
a. 30°.   b. 45°.
c. 50°.   d. 75°.

EXERCISES

3–1   Vector Quantities

3–2   Vector Addition

3–3   Vector Subtraction

1. What kind of quantity is the magnitude of a vector quantity? What kind of quantity is the resultant of two vector quantities of the same kind?

2. A bird lands on an overhead power cable, which sags slightly under its weight. Is it possible to prevent any such sagging by applying enough tension to the cable?

3. The resultant of three vectors is zero. Must they all lie in a plane?

4. The resultant of two forces **X** and **Y** is **R**. **X** and **R** are shown at left in the figure. Which of the other vectors is **Y**?

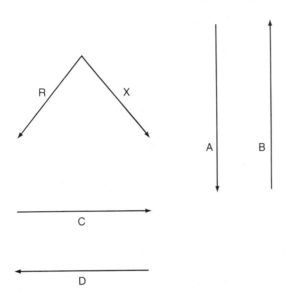

5. Three forces, each of 10 N, act on the same object. What is the maximum total force they can exert on the object? The minimum total force?

6. A 100-N sack of potatoes is suspended by a rope. A person pushes sideways on the sack with a force of 40 N. What is the tension in the rope?

7. A person walks 70 m to an elevator and then ascends 40 m. Find the magnitude and direction of the person's displacement from the starting place.

8. A driver becomes lost and travels 12 km west, 5 km south, and then 8 km east. Find the magnitude and direction of the car's displacement from the starting place.

9. Two cars leave a crossroads at the same time, one headed north at 50 km/h and the other headed east at 70 km/h. How far apart are they after 0.5 h? After 2 h?

10. The resultant of two perpendicular forces has a magnitude of 40 N. If the magnitude of one of the forces is 25 N, what is the magnitude of the other force?

11. The ketch *Minots Light* is heading northwest at 7 kn through a tidal stream that is flowing southwest at 3 kn. Find the magnitude and direction of its velocity relative to the earth's surface.

12. A sphere that weighs 100 N rests in the angle formed by two boards 90° apart, as shown. The boards are very smooth, so the only forces they exert on the sphere are perpendicular to their surfaces. Find the magnitude of each of these forces.

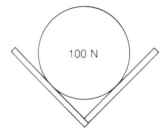

3–4   Resolving a Vector

13. A woman is rowing at 8 km/h in a river 1.5 km wide in which the current is 5 km/h. In what direction should she head in order to get across the river in the shortest possible time? How much time will she take?

14. A horizontal and a vertical force combine to give a resultant force of 10 N that acts in a direction 40° above the horizontal. Find the magnitudes of the horizontal and vertical forces.

15. A sailboat cannot sail directly to windward but must "tack" back and forth at a certain angle with respect to the direction from which the wind is blowing. Which sailboat has the greater component of velocity to windward, the *Alpha* whose velocity is 5 km/h at an angle of 40° off the wind, or the *Beta* whose velocity is 6 km/h at an angle of 50° off the wind?

16. The ship *Salmonella* is heading due east at 18 km/h in the presence of a north wind of 12 km/h. What is the horizontal component of velocity of the smoke relative to the ship as it leaves the funnel? Relative to the earth?

17. A woman pushes a 50-N lawn mower with a force of 25 N. If the handle of the lawn mower is 45° above the horizontal, how much downward force is being exerted on the ground by the lawn mower?

18. A weight of 80 N is at rest on an inclined plane that makes an angle of 40° with the horizontal. Find the components of this weight parallel and perpendicular to the plane.

19. A man on the ground observes an airplane climbing at an angle of 37° above the horizontal. He gets in his car and

by driving at 70 km/h is able to stay directly below the airplane. What is the airplane's speed?

20. An airplane whose speed is 150 km/h climbs from a runway at an angle of 20° above the horizontal. What is its altitude 1 min after takeoff? How many kilometers does it travel in a horizontal direction in this period of time?

21. An airplane is heading southeast when it takes off at an angle of 25° above the horizontal at 200 km/h. (a) What is the vertical component of its velocity? (b) What is the horizontal component of its velocity? (c) What is the component of the velocity of the plane toward the south?

22. The shadow of an airplane taking off moves along the runway at 170 km/h. If the sun is directly overhead and the airplane's air speed is 200 km/h, find the angle at which it is climbing.

23. On a windless day, raindrops that fall on the side windows of a car moving at 10 m/s are found to make an angle of 50° with the vertical. Find the speed of the raindrops relative to the ground.

24. A boat moving at 15 km/h is crossing a river 2 km wide in which the current is flowing at 5 km/h. (a) If the boat heads directly for the other shore, how long will the trip take? (b) In what direction should the boat head if it is to reach a point on the other shore directly opposite the starting point? (c) How long will the crossing take in case (b)?

25. Two tugboats are towing a ship. Each exerts a horizontal force of 5.0 tons, and the angle between the two ropes is 30°. What is the resultant force exerted on the ship?

26. A string with a weight at its lower end is suspended inside a car. When the car starts to move, the string makes an angle of 10° with the vertical. What is the car's acceleration? What happens when the car later travels at constant velocity?

27. A 10-kg crate and a 100-kg crate are both sliding without friction down a plane inclined at 20° with the horizontal. What is the acceleration of each crate?

28. A sprinter presses on the ground with a force equal to three times his own weight at a 50° angle with the horizontal at the start of a race. What is his forward acceleration?

29. A sled slides down a snow-covered hill at constant speed. If the hillside is 10° above the horizontal, what is the coefficient of sliding friction between the runners of the sled and the snow?

30. A cyclist finds that she is able to coast at constant speed along a road that slopes downward at an angle of 1° with the horizontal. If she and her bicycle together have a mass of 70 kg, find the force required to propel them at constant speed along a level road.

31. A horse pulls a 300-kg sled at constant speed over level snow by a rope that is 35° above the horizontal. If the coefficient of friction is 0.10, find the force the horse exerts on the rope.

32. A 500-lb crate is being slid down a ramp that makes an angle of 20° with the horizontal. The coefficient of friction is 0.3. How much force parallel to the plane must be applied to the crate if it is to slide down at constant speed? In which direction must the force be applied?

33. A block slides down an inclined plane 9 m long that makes an angle of 38° with the horizontal. The coefficient of sliding friction is 0.25. If the block starts from rest, find the time required for it to reach the foot of the plane.

34. If the block of Exercise 33 has a mass of 50 kg, find the minimum force required to move it upward along the plane. What should the direction of this force be?

35. A skier starts from rest and slides 50 m down a slope that makes an angle of 40° with the horizontal. She then continues sliding on level snow. (a) If the coefficient of friction between skis and snow is 0.10 and air resistance is neglected, what is the speed of the skier at the foot of the slope? (b) How far away from the foot of the slope does she come to a stop?

36. A block takes twice as long to slide down an inclined plane that makes an angle of 35° with the horizontal as it does to fall freely through the same vertical distance. What is the coefficient of friction?

37. A ship whose mass is 2×10^7 kg rests on launching ways that slope down to the water at an angle of 6°. The ways are greased to reduce the coefficient of sliding friction to 0.11. Will the ship slide down the ways into the water by itself? If not, find the force needed to winch the ship down into the water.

3–5   Vector Addition by Components

38. Two billiard balls are rolling on a flat table. One has the velocity components $v_x = 1$ m/s, $v_y = 2$ m/s. The other has the velocity components $v_x = 2$ m/s, $v_y = 3$ m/s. If both balls started from the same point, what is the angle between their paths?

39. The following forces act on an object resting on a level, frictionless surface: 10 N to the north, 20 N to the east, 10 N at an angle 40° south of east, and 20 N at an angle 50° west of south. Find the magnitude and direction of the resultant force acting on the object.

40. Find the magnitude and direction of the resultant of a 40-N force that acts at an angle of 63° clockwise from the +y-axis, a 15-N force that acts at an angle of 120° clockwise

from the $+y$-axis, and a 30-N force that acts at an angle of $310°$ clockwise from the $+y$-axis.

41. An airplane flies 200 km east from city A to city B, then 200 km south from city B to city C, and finally 100 km northwest to city D. How far is it from city A to city D? In what direction must the airplane head to return directly to city A from city D?

42. In going from one city to another, a car whose driver tends to get lost goes 30 km north, 50 km west, and 20 km in a direction $30°$ south of east. How far apart are the cities? In what direction should the car have headed to travel directly from the first city to the second?

43. The yacht *Quicksilver* is headed north at 7.0 knots and the yacht *Lianda* is headed east at 8.0 knots. Find (a) the velocity of *Quicksilver* relative to *Lianda*, and (b) the velocity of *Lianda* relative to *Quicksilver*.

44. A car is headed east at 50 km/h and a truck is headed northwest at 70 km/h. Find (a) the velocity of the car relative to the truck, and (b) the velocity of the truck relative to the car.

45. The vector **A** has a magnitude of 10 cm and points $37°$ clockwise from the $+y$-direction. The vector **B** has a magnitude of 10 cm and points $37°$ clockwise from the $+x$-direction. Find the magnitude and direction of **A** + **B**, **A** − **B**, and **B** − **A**.

46. The vector **A** has a magnitude of 20 cm and points $20°$ clockwise from the $+y$-direction. The vector **B** has a magnitude of 10 cm and points $60°$ clockwise from the $+y$-direction. Find the magnitude and direction **A** + **B**, **A** − **B**, and **B** − **A**.

47. A person walks 4 km in one direction and 2 km in another direction. If he ends up 5 km from his starting point, what is the angle between the two directions?

48. A person walks 4 km in one direction and 2 km in another direction. If she ends up 3 km from her starting point, what is the angle between the two directions?

3–6 Motion in a Vertical Plane

49. A rifle is aimed directly at the bull's-eye of a target 50 m away. If the bullet's speed is 350 m/s, how far below the bull's-eye does the bullet strike the target?

50. An airplane is flying at an altitude of 8000 m at a speed of 900 km/h. At what distance ahead of a target must it drop a bomb? If the airplane turns around at the moment the bomb is released, how far from the target will it be at the time the bomb strikes the ground?

51. An airplane is flying at an altitude of 500 m over the ocean at a speed of 60 m/s. A ship is moving along a path directly under that of the airplane but in the opposite direction at 10 m/s. At what horizontal distance from the ship should the pilot of the airplane drop a parcel meant to land on the ship's deck?

52. A rescue line is to be thrown horizontally from the bridge of a ship 30 m above sea level to a lifeboat 30 m away. What speed should the line have?

53. An airplane is in level flight at a speed of 400 km/h and an altitude of 1000 m when it drops a bomb. Find the bomb's speed when it strikes the ground.

54. A ball is rolled off the edge of a table with a horizontal velocity of 3 ft/s. What will be the magnitude and direction of the ball's velocity 0.1 s later?

55. A ball is thrown horizontally from the roof of a building 60 ft high at 80 ft/s. What will be the magnitude and direction of the ball's velocity when it strikes the ground?

56. A ball is thrown horizontally toward the north from a rooftop at 8 m/s. A 10-m/s wind is blowing from the east. (a) What is the speed of the ball relative to the ground after 2 s? (b) What angle does its velocity make relative to the vertical at this time? (c) What angle does its velocity make relative to due north at this time?

3–7 Projectile Flight

57. Does the speed of a projectile sent off at a $45°$ angle of elevation vary in its path? If so, where is the speed greatest and where is it least?

58. What effect does doubling the initial speed of a projectile have on its range?

59. Find the minimum initial speed of a champagne cork that travels a horizontal distance of 11 m.

60. In April 1959, Miss Victoria Zacchini was fired 155 ft from a cannon in Madison Square Garden, New York City. What was the minimum muzzle speed of the cannon in mi/h?

61. A football leaves the toe of a punter at an angle of $50°$ above the horizontal. What was its minimum initial speed if it travels 40 yd?

62. Find the range of an arrow that leaves a bow at 50 m/s at an angle of $50°$ above the horizontal. *Note:* $\sin(90° + \theta) = \cos\theta$.

63. A blunderbuss can fire a slug 100 m vertically upward. (a) What is its maximum horizontal range? (b) With what speeds will the slug strike the ground when fired upward and when fired so as to have maximum range?

64. A person can throw a ball a maximum distance of L. If the ball is thrown upward with the same initial speed, how high will it go?

65. What percentage increase in initial speed is required to increase the range of a javelin by 20%?

66. A golf ball leaves a tee at 60 m/s and strikes the ground 200 m away. At what two angles with the horizontal could it have begun its flight? Find the time of flight and maximum altitude in each case.

67. A shell is fired at a velocity of 300 m/s at an angle of 30° above the horizontal. (a) How far does it go? What are its time of flight and maximum altitude? (b) At what other angle could the shell have been fired to have the same range?

What would its time of flight and maximum altitude have been in this case?

ANSWERS TO MULTIPLE CHOICE

1. a	**8.** c	**15.** c	**22.** c
2. c	**9.** d	**16.** d	**23.** c
3. c	**10.** d	**17.** b	**24.** b
4. c	**11.** a	**18.** b	**25.** d
5. d	**12.** a	**19.** a	**26.** c
6. d	**13.** b	**20.** d	**27.** a
7. c	**14.** a	**21.** d	**28.** b

4

EQUILIBRIUM

Thus far our main concern has been with moving objects, whose study is known as dynamics. We now take up statics, which is the study of objects at rest. Something at rest that has no tendency to move or to rotate is said to be in equilibrium. If only a single force acts, equilibrium is obviously impossible, but two or more forces can balance one another out to leave no net force. Even then equilibrium may not occur, because if the lines of action of the forces do not meet at a common point, the object they act on will start to spin. In this chapter we will learn about the conditions for equilibrium and how they can be used to solve a variety of problems important in technology.

4–1 TRANSLATIONAL EQUILIBRIUM

Translational equilibrium

An object that has no net force acting on it is said to be in *translational equilibrium*. The important point is that the object has no linear acceleration. According to the first

CHAPTER OBJECTIVES

Completing this chapter should enable you to:

1. Draw a free-body diagram that shows the forces acting on an object.

2. Resolve the forces acting on an object into their horizontal and vertical components.

3. Write the equations that express the translational equilibrium of an object.

4. Establish by calculation whether an object is in translational equilibrium.

5. Find the force needed to balance out the other forces acting on an object so that it will then be in translational equilibrium.

6. Distinguish between concurrent and nonconcurrent forces.

7. Determine the torque exerted by a force about a given pivot point.

8. Write the equations that express the rotational equilibrium of an object.

9. Establish by calculation whether an object is in rotational equilibrium.

10. Find the center of gravity of an object both by calculation and by experiment.

11. Determine whether an object is in stable, unstable, or neutral equilibrium.

law of motion, such an object need not be at rest but may instead be moving along a straight path at constant speed.

The condition for translational equilibrium may be expressed in the form

$$\Sigma \mathbf{F} = 0 \qquad\qquad \textit{Translational equilibrium} \quad (4–1)$$

where the symbol Σ (Greek capital letter *sigma*) means "sum of" and $\mathbf{F}$ refers to the various forces acting on a specific object. This is simply the mathematical way of stating that, at equilibrium, the forces are such as to add vectorially to zero.

In many equilibrium situations all the various forces lie in the same plane. When this is the case, we can establish a set of x-y coordinate axes wherever convenient in the plane and then resolve each force $\mathbf{F}$ into the components $\mathbf{F}_x$ and $\mathbf{F}_y$. Thus we can replace Eq. (4–1), which is a vector equation, with the two scalar equations

Force components in each direction must total zero for equilibrium

$$\text{Sum of } x \text{ force components} = \Sigma F_x = 0 \qquad \textit{Translational equilibrium} \quad (4–2)$$
$$\text{Sum of } y \text{ force components} = \Sigma F_y = 0 \qquad\qquad\qquad\qquad (4–3)$$

It is usually much easier to calculate the components of each force acting and then make use of Eqs. (4−2) and (4−3) than it is to work directly with the forces involved.

There are five steps to follow in working out a problem that concerns the translational equilibrium of an object:

Procedure for evaluating translational equilibrium

1. Draw a sketch of the forces that act on the object. (This is called a *free-body diagram*.) Do not show the forces that the object exerts on anything else, since such forces do not affect the equilibrium of the object itself.

2. Choose a convenient set of coordinate axes and resolve the various forces acting on the object into components along these axes. Be sure to use + and − signs consistently. If the y-axis is vertical and the x-axis is horizontal, for instance, an upward force would be considered as +, a downward force as −, a force to the right as +, and a force to the left as −.

3. Set the sum of the force components along each axis equal to 0, as specified in Eqs. (4−2) and (4−3). This is the condition for equilibrium.

4. Solve the resulting equations for the unknown quantity or quantities.

5. Substitute the numerical values of the known quantities to find the answer.

Several examples will make clear the above procedure. In the first, shown in Fig. 4−1, we have the simple case of a box of weight w being supported by a single weightless rope. We note that the only force a rope can exert is a pull along its length in a direction away from the point of attachment at each end. If the weight of the rope is negligible, the two pulls are of equal magnitude. The box will be in equilibrium when all the forces acting on it cancel one another out, which here means that

$$\Sigma F_y = 0$$

since there are no forces in the x-direction. The tension T in the rope acts upward (the +y-direction) on the box and the box's weight w acts downward (the −y-direction) on it; hence

$$\Sigma F_y = T - w = 0$$

and

$$T = w \tag{4−4}$$

The tension in the rope must equal the weight being supported.

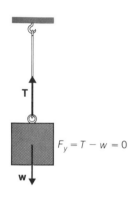

FIG. 4−1 A suspended object is in equilibrium when the tension in the rope is equal in magnitude to the weight of the object.

$F_y = T - w = 0$

Resolving rope tension into components

In Fig. 4−2 the same box is suspended from two ropes, A and B, which are at the angles θ and ϕ, respectively, with the horizontal. We begin by resolving the tension in each rope into components in the x- and y-directions, so that we have T_{Ay} and T_{By} upward, T_{Ax} to the left, and T_{Bx} to the right. From Fig. 4−2(b) we have

$T_{Ax} = -T_A \cos \theta$	$T_{Bx} = T_B \cos \phi$	*Horizontal components*
$T_{Ay} = T_A \sin \theta$	$T_{By} = T_B \sin \phi$	*Vertical components*

We may now ignore the actual tensions T_A and T_B, and treat their components as individual forces acting at the same point as the weight w, as in Fig. 4−2(c). For the forces in the horizontal and vertical directions to cancel separately,

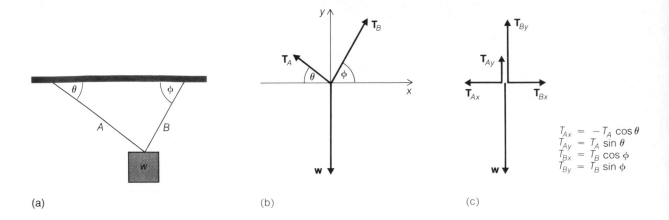

(a)   (b)   (c)

$$T_{Ax} = -T_A \cos \theta$$
$$T_{Ay} = T_A \sin \theta$$
$$T_{Bx} = T_B \cos \phi$$
$$T_{By} = T_B \sin \phi$$

$$\Sigma F_x = T_{Bx} + T_{Ax} = 0 \qquad \text{Horizontal direction} \quad (4\text{–}5)$$
$$\Sigma F_y = T_{Ay} + T_{By} - w = 0 \qquad \text{Vertical direction} \quad (4\text{–}6)$$

FIG. 4–2 (a) A box of weight w is suspended by two ropes. (b) A free-body diagram of the forces acting on the box. (c) At equilibrium the sum ΣF_x of the horizontal force components and the sum ΣF_y of the vertical force components each equal zero.

Example   In the above situation $w = 100$ N, $\theta = 37°$, and $\phi = 60°$. Find the tension in each rope.

Solution   From Eq. (4–5) we find that

$$T_{Bx} + T_{Ax} = 0$$
$$T_B \cos \phi - T_A \cos \theta = 0$$
$$T_B = T_A \frac{\cos \theta}{\cos \phi} = T_A \frac{\cos 37°}{\cos 60°} = 1.6 T_A$$

From Eq. (4–6) we find that

$$T_{Ay} + T_{By} - w = 0$$
$$T_A \sin \theta + T_B \sin \phi - 100\,\text{N} = 0$$

Substituting $1.6 T_A$ for T_B in the last equation, we obtain

$$T_A \sin \theta + 1.6 T_A \sin \phi - 100\,\text{N} = 0$$
$$T_A (\sin 37° + 1.6 \sin 60°) = 100\,\text{N}$$
$$T_A = 50\,\text{N}$$

The tension in rope A is 50 N. Since $T_B = 1.6 T_A$ here,

$$T_B = (1.6)(50\,\text{N}) = 80\,\text{N}$$

The tension in rope B is 80 N. The algebraic sum of the tensions in the two ropes is 130 N, which is more than the weight being supported, but the vector sum of $\mathbf{T}_A$ and $\mathbf{T}_B$ is 100 N, acting upward, which balances the downward force $\mathbf{w} = 100$ N.   ■

Example   A 150-lb box is suspended from the end of a horizontal strut, as in Fig.

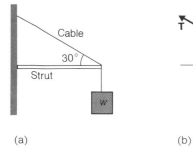

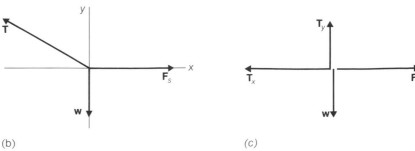

(a)   (b)   (c)

FIG. 4–3 (a) A box of weight w is suspended from the end of a horizontal strut held in place by a cable attached to the wall. (b) A free-body diagram of the forces acting on the end of the strut. (c) At equilibrium, $\Sigma F_x = 0$ and $\Sigma F_y = 0$.

4–3. Find the tension in the cable supporting the strut under the assumption that the strut's weight is negligible.

Solution It is easiest to consider the equilibrium of the end of the strut. The three forces that act on the end of the strut are the tension **T** in the cable, the outward force $\mathbf{F}_s$ exerted by the strut itself, and the weight **w** of the box. The horizontal and vertical components of the tension T are, from Fig. 4–3,

$$T_x = -T \cos 30°$$
$$T_y = T \sin 30°$$

The end of the strut is in equilibrium when

$$\Sigma F_x = T_x + F_s = 0$$
$$\Sigma F_y = T_y - w = 0$$

All we need is the second of these equations to find T:

$$T_y - w = 0$$
$$T \sin 30° = w$$
$$T = \frac{w}{\sin 30°} = \frac{150\,\text{lb}}{0.500} = 300\,\text{lb}$$ ∎

Example Find the inward force the strut of the previous example exerts on the wall.

Solution Since this force is horizontal, we need the equation that expresses the equilibrium of the strut in the x-direction, which is

$$\Sigma F_x = T_x + F_s = 0$$

Since $T = 300$ lb and $T_x = -T \cos 30°$,

$$\Sigma F_x = -T \cos 30° + F_s = 0$$
$$F_s = T \cos 30° = (300\,\text{lb})(\cos 30°) = 260\,\text{lb}$$ ∎

Example A 200-kg motor is suspended from the ceiling by a cable 6.0 m long, as in Fig. 4–4(a). If the coefficient of friction between his shoes and the floor is $\mu_s = 0.6$, the maximum horizontal force a 70-kg person can exert before he starts to slip is

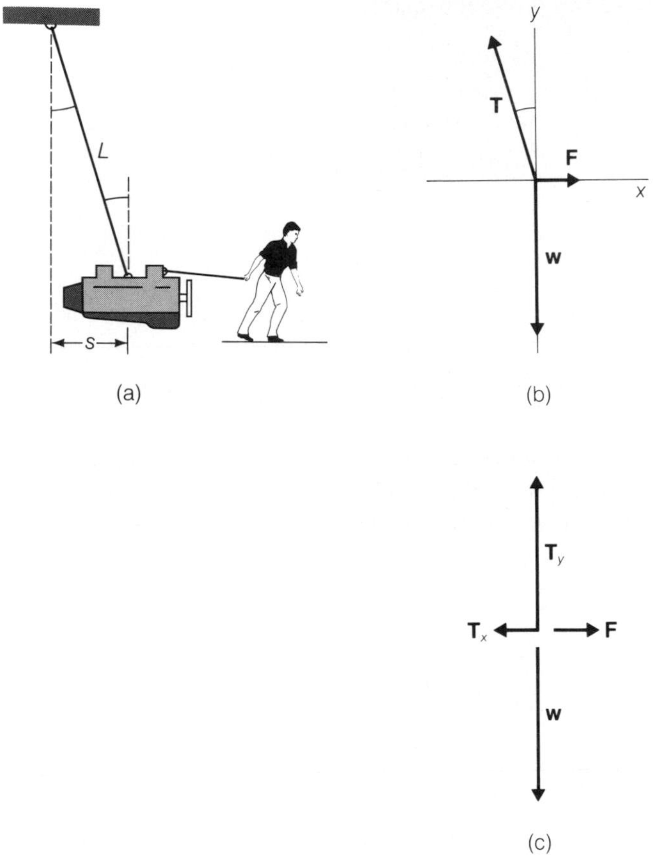

FIG. 4–4

(a)

(b)

(c)

$F = \mu_s\, mg = 412$ N. Find the horizontal displacement of the motor when a horizontal force of 412 N is applied to it.

Solution   Since we know that L in Fig. 4–4(a) is 6.0 m, we must find the angle θ in order to determine the displacement s. We begin by resolving the tension **T** in the cable into its horizontal and vertical components, which are

$$T_x = -T \sin \theta \qquad T_y = T \cos \theta$$

Next we substitute into Eqs. (4–2) and (4–3) to obtain

$$\Sigma F_x = T_x + F = -T \sin \theta + F = 0$$
$$\Sigma F_y = T_y - w = T \cos \theta - w = 0$$

We now solve the first equation for $\sin \theta$ and the second for $\cos \theta$:

$$T \sin \theta = F \qquad T \cos \theta = w$$
$$\sin \theta = \frac{F}{T} \qquad \cos \theta = \frac{w}{T}$$

Since $\sin \theta / \cos \theta = \tan \theta$,

$$\tan \theta = \frac{\sin \theta}{\cos \theta} = \frac{F/T}{w/T} = \frac{F}{w} = \frac{F}{mg}$$

$$= \frac{412\,\text{N}}{(200\,\text{kg})(9.8\,\text{m/s}^2)} = 0.210$$

and

$$\theta = \tan^{-1} 0.210 = 12°$$

It is not necessary to know the tension T in order to find θ. From Fig. 4−4(a) we have for the horizontal displacement s of the motor

$$s = L \sin \theta = (6.0\text{ m})(\sin 12°) = 1.2\text{ m} \qquad \blacksquare$$

Example   A boom hinged at the base of a vertical mast is used to lift a 2000-lb load, as in Fig. 4−5. Find the tension in the cable from the top of the mast to the end of the boom. The weight of the boom itself can be neglected.

Solution   We shall consider the equilibrium of the end of the boom. As shown in Fig. 4−5, three forces act here: the weight **w** of the boat, the tension **T** in the cable, and the outward reaction force **R** of the boom. Because **R** is along the same line as the boom,

$$\theta_3 = \theta_1 = 40°$$

The sum of the acute angles of a right triangle is always 90° (see Appendix A−8), hence

$$\theta_2 + \theta_4 = 90°$$

$$\theta_4 = 90° - \theta_2 = 90° - 60° = 30°$$

The x- and y-components of **T** and **R** are as follows:

$$T_x = -T \cos \theta_4 \qquad R_x = R \cos \theta_3$$

$$T_y = T \sin \theta_4 \qquad R_y = R \sin \theta_3$$

Hence the conditions for equilibrium in the x- and y-directions are

FIG. 4−5 The vectors in (b) are not drawn to scale.

$$\Sigma F_x = -T \cos \theta_4 + R \cos \theta_3 = 0$$

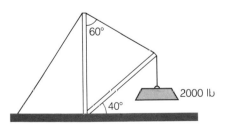

(a)

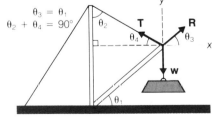

(b)

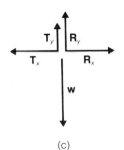

(c)

$$\Sigma F_y = T \sin \theta_4 + R \sin \theta_3 - w = 0$$

We are not interested in the value of R. To eliminate R we begin by solving the first equilibrium equation for it:

$$R \cos \theta_3 = T \cos \theta_4$$

$$R = T \left(\frac{\cos \theta_4}{\cos \theta_3} \right)$$

Substituting this value of R in the second equilibrium equation gives

$$T \sin \theta_4 + T \left(\frac{\cos \theta_4}{\cos \theta_3} \right) \sin \theta_3 - w = 0$$

Since $\sin \theta_3 / \cos \theta_3 = \tan \theta_3$,

$$T \sin \theta_4 + T \cos \theta_4 \tan \theta_3 - w = 0$$

$$T (\sin \theta_4 + \cos \theta_4 \tan \theta_3) = w$$

$$T = \frac{w}{\sin \theta_4 + \cos \theta_4 \tan \theta_3}$$

$$= \frac{2000 \, \text{lb}}{\sin 30° + (\cos 30°)(\tan 40°)}$$

$$= \frac{2000 \, \text{lb}}{1.227} = 1630 \, \text{lb}$$

Another way to eliminate R is to align the x- and y-axes so that the x-axis is along the boom. Then we need consider only the equilibrium of the boom end in the new y-direction, which will involve only components of $\mathbf{w}$ and $\mathbf{T}$. The advantage of proceeding as we did here is that, although there is a little more algebra, the geometry is less confusing when the x-axis is horizontal and the y-axis is vertical.   ■

4–2   TORQUE

When the lines of action of the various forces that act on an object intersect at a common point, they do not tend to set the object in rotation. Such forces are said to be *concurrent*.

Concurrent forces

If the lines of action of the various forces do *not* intersect, the forces are *noncurrent*, and the body may be set into rotation even though the vector sum of the forces may equal zero. In Fig. 4–6(a) the three applied forces are concurrent and, if their

Noncurrent forces

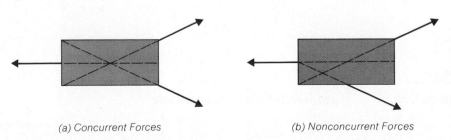

(a) Concurrent Forces   *(b) Nonconcurrent Forces*

FIG. 4–6   (a) The lines of action of concurrent forces intersect at a common point. (b) When the lines of action do not intersect at a common point, the forces are nonconcurrent and the object cannot be in rotational equilibrium.

FIG. 4–7 A seesaw is
balanced when $w_1L_1 = w_2L_2$.

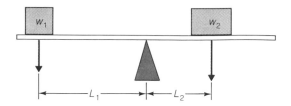

vector sum is zero, the object is in equilibrium. In Fig. 4–6(b) the same forces are nonconcurrent, and their combined effect is to set the object spinning counterclockwise. If we want the term equilibrium to mean the absence of a rotational acceleration as well as the absence of a linear acceleration, we must supplement $\Sigma \mathbf{F} = 0$ with another condition that the forces on a body must obey if it is to be in equilibrium.

Condition for balance

A hint as to the nature of this additional condition may be obtained by watching children on a seesaw at a playground (Fig. 4–7). A small child can balance a large child merely by sitting farther from the pivot. Two children of the same weight will not balance unless they sit the same distance from the pivot, though the exact distance does not matter. Evidently both the magnitudes of the forces (here the weights of the children) and their lines of action determine whether or not the object is in equilibrium. If we were to try various combinations of weights and distances from the pivot, we would find that the seesaw is balanced when the product $w_1 L_1$ of the weight w_1 and distance from the pivot L_1 of one child is equal to the product $w_2 L_2$ of the other child's weight and distance from the pivot.

Moment arm and torque

To make our discussion perfectly general, let us consider a force $\mathbf{F}$ acting upon an object free to rotate about some pivot point O (Fig. 4–8). The perpendicular distance L from O to the line of action of the force is called the *moment arm* of the force about O. The product of the magnitude F of the force and its moment arm L is known as the *torque* of the force about O. The symbol for torque is τ (Greek letter *tau*), so that

$$\tau = FL$$
$$\text{Torque} = \text{force} \times \text{moment arm}$$

Torque (4–7)

FIG. 4–8 A measure of
the turning effect of a
force about a pivot point
O is its torque τ, which is
equal to the product FL of
the magnitude F of the
force and the moment arm
L. If the line of action of
the force passes through
the pivot point, the
moment arm is $L = 0$ and
$\tau = 0$ also.

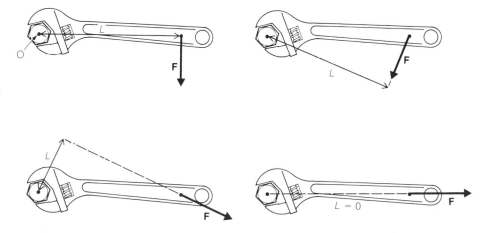

In the SI system torque is expressed in newton · meters (N · m); in the British system, **Units of torque**
in lb · ft.

The greater the torque applied to an object, the greater the tendency of the object
to be set into rotation.

Example   The cylinder-head bolts on the diesel engine of a truck are supposed to be
tightened to a torque of 70 N · m. If the center of the handle on the wrench used is 35
cm from the axis of the bolt, what force perpendicular to the handle should be applied?

Solution   The moment arm here is $L = 35$ cm $= 0.35$ m. Since $\tau = FL = 70$ N · m,
the required force is

$$F = \frac{\tau}{L} = \frac{70\,\text{N} \cdot \text{m}}{0.35\,\text{m}} = 200\,\text{N}$$ ∎

4–3   ROTATIONAL EQUILIBRIUM

By convention a torque that tends to produce a counterclockwise rotation is considered **Positive and negative**
positive and a torque that tends to produce a clockwise rotation is considered negative **torques**
(Fig. 4–9). Thus the condition for an object to be in rotational equilibrium is that the
sum of the torques acting upon it about any point, using the above convention for plus
and minus signs, be zero:

$$\Sigma\tau = 0 \hspace{3cm} \textit{Rotational equilibrium} \quad (4\text{–}8)$$

Of course, if the various forces that act do not all lie in the same plane, it is necessary
that the sum of the torques in each of three mutually perpendicular planes be zero.

It is possible to prove that if the sum of the torques on an object in translational **Torques can be**
equilibrium is zero about any point, it is also zero about all other points. Hence the **calculated about any**
location of the point about which torques are calculated in an equilibrium problem is **point**
completely arbitrary; *any* point will do. (Of course, *all* torques must then be calculated
about this point.) Let us verify this statement with an example.

Example   Figure 4–10 shows a rod 6.0 ft long that has weights of 10 lb and 30 lb at
its ends. We assume that the weight of the rod is negligible. At what point should the
rod be picked up if it is to have no tendency to rotate? In other words, where is the
balance point of the rod?

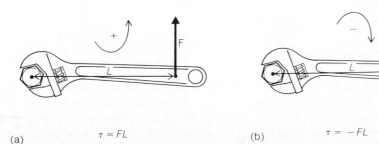

(a) $\tau = FL$ (b) $\tau = -FL$

FIG. 4–9   (a) A torque
that tends to produce a
counterclockwise rotation
is considered positive. (b)
A torque that tends to
produce a clockwise
rotation is considered
negative.

FIG. 4–10 Torques are computed about the unknown balance point of the rod in solution 1.

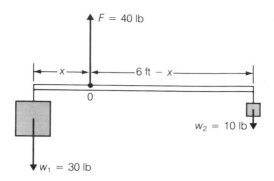

Solution 1   We first compute torques about the unknown balance point. If x is the distance of the 30-lb weight from this point, the 10-lb weight is $(6.0 \text{ ft} - x)$ from it on the other side. The torques these weights exert are

$$\tau_1 = w_1 L_1 = +30x \text{ lb}$$
$$\tau_2 = w_2 L_2 = -10(6.0 \text{ ft} - x) \text{ lb}$$

and equilibrium will result when

$$\Sigma \tau = \tau_1 + \tau_2 = 30x \text{ lb} - 10(6.0 \text{ ft} - x) \text{ lb} = 0$$
$$30x = 10(6.0 \text{ ft} - x) = 60 \text{ ft} - 10x$$
$$40x = 60 \text{ ft}$$
$$x = \frac{60 \text{ ft}}{40} = 1.5 \text{ ft}$$

When the rod is picked up 1.5 ft from the 30-lb weight, the two weights exert opposite torques of the same magnitude $(45 \text{ lb} \cdot \text{ft})$ about this point, so the rod is in balance.

Solution 2   Let us solve the same problem by calculating torques about the middle of the rod, as shown in Fig. 4–11. Here y represents the distance between the balance point and the middle of the rod. We have three torques to take into account now:

$$\tau_1 = w_1 L_1 = +(30 \text{ lb})(3.0 \text{ ft}) = +90 \text{ lb} \cdot \text{ft}$$
$$\tau_2 = FL_2 = -40y \text{ lb}$$
$$\tau_3 = w_2 L_3 = -(10 \text{ lb})(3.0 \text{ ft}) = -30 \text{ lb} \cdot \text{ft}$$

Equilibrium will result when

$$\Sigma \tau = \tau_1 + \tau_2 + \tau_3 = 90 \text{ lb} \cdot \text{ft} - 40y \text{ lb} - 30 \text{ lb} \cdot \text{ft} = 0$$
$$40y = 90 \text{ ft} - 30 \text{ ft} = 60 \text{ ft}$$
$$y = \frac{60 \text{ ft}}{40} = 1.5 \text{ ft}$$

The location of the balance point is the same regardless of the particular point about which torques are calculated. It is usually wise to calculate torques about the point of application of one of the forces that act on an object, since this makes it unnecessary to consider the torque produced by that force and thereby simplifies the arithmetic. ■

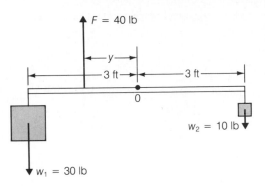

FIG. 4–11 Torques are computed about the center of the rod in solution 2.

$F = 40$ lb

y

3 ft

3 ft

0

$w_2 = 10$ lb

$w_1 = 30$ lb

Stable equilibrium

Unstable equilibrium

Neutral equilibrium

FIG. 4–12 These cones are all in equilibrium, but only one of them is in a stable position.

Not all equilibrium situations are necessarily stable. For instance, a cone balanced on its apex is in equilibrium, but it will fall over when disturbed even slightly (Fig. 4–12). This is an example of an *unstable equilibrium*. The same cone on its base will return to its original position if tipped over a little; hence it is in *stable equilibrium* on its base. There is a third possibility as well, illustrated by a cone lying on its side. If such a cone is displaced, it remains in equilibrium in its new position with no tendency either to move further or to return to where it was before. A cone on its side is said to be in *neutral equilibrium*.

Types of equilibrium

4–4 CENTER OF GRAVITY

The *center of gravity* (CG) of an object is that point from which it can be suspended in any orientation without tending to rotate (Fig. 4–13). Each of the particles of the object has a certain weight, and therefore exerts a torque about whatever point the object is suspended from. There is only a single point in an object about which all these torques cancel out no matter how it is oriented; this is its center of gravity. For equilibrium purposes we can therefore regard the entire weight of an object as a downward force acting from its center of gravity.

We can now recognize the distinctions between the different kinds of equilibrium shown in Fig. 4–12. The left-hand cone is in stable equilibrium because its center of gravity has to be raised to change its orientation; the center cone is in unstable equilibrium because any change in its orientation lowers its center of gravity; and the right-

Center of gravity

FIG. 4–13 A body suspended from its center of gravity is in equilibrium in any orientation.

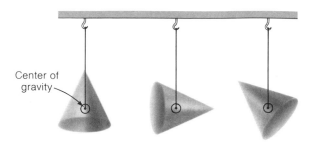

Center of gravity

hand cone is in neutral equilibrium because the height of its center of gravity does not change when it is rolled along on its side.

An object's weight acts from its center of gravity

If the rod of Figs. 4–10 and 4–11 had the weight w instead of being weightless, we could take into account its effect on the location of the balance point by including the torque due to a force of magnitude w acting downward at the center of the rod. The center of gravity of a uniform object of regular shape is located at its geometrical center. The center of gravity of an irregular object need not even be located within the object itself: The center of gravity of a seated person, for example, is a few inches in front of the person's abdomen.

Cantilever

Example A beam that projects beyond its supports is called a *cantilever;* a diving board is an example. Find the forces exerted by the two supports of the 4-m, 50-kg uniform diving board shown in Fig. 4–14 when a 60-kg woman stands at its end.

Solution The weight of the diving board is $w_1 = m_1 g_1 = 490$ N and that of the woman is $w_2 = m_2 g = 588$ N. It is obvious that the force F_1 exerted by the left-hand support must be downward. (If we nevertheless were to consider F_1 as upward, the result would be a negative value for F_1, signifying the opposite direction.) We can calculate F_1 without knowing F_2 by computing torques about the point of application of F_2. The center of gravity of the board is at its middle, which is 1.2 m from the pivot point. Hence the three torques that act about this point are

$$\tau_1 = +F_1 x_1 = +(F_1)(0.8 \text{ m}) = 0.8\,F_1 \text{ m}$$
$$\tau_2 = -w_1 x_2 = -(490 \text{ N})(1.2 \text{ m}) = -588 \text{ N} \cdot \text{m}$$
$$\tau_3 = -w_2 x_3 = -(588 \text{ N})(3.2 \text{ m}) = -1882 \text{ N} \cdot \text{m}$$

and so

$$\Sigma\tau = \tau_1 + \tau_2 + \tau_3 = 0.8\,F_1 \text{ m} - 588 \text{ N} \cdot \text{m} - 1882 \text{ N} \cdot \text{m} = 0$$
$$0.8 F_1 = 588 \text{ N} + 1882 \text{ N} = 2470 \text{ N}$$
$$F_1 - \frac{2470 \text{ N}}{0.8} = 3088 \text{ N}$$

We can find F_2 by considering the translational equilibrium of the loaded board:

$$F_2 = F_1 + w_1 + w_2 = 3088 \text{ N} + 490 \text{ N} + 588 \text{ N} = 4166 \text{ N} \qquad \blacksquare$$

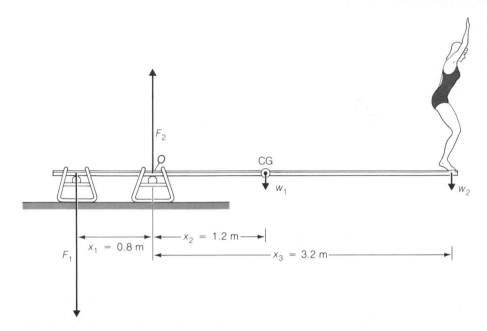

FIG. 4–14 The pivot point about which torques are calculated is marked O. (Photo: © Roy King.)

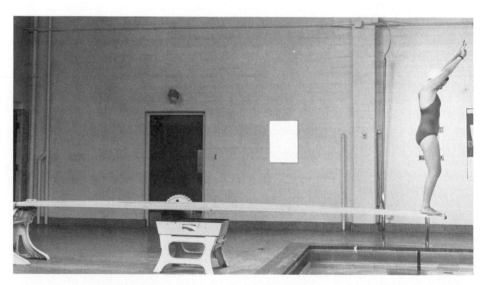

Example A horizontal beam 12.0 ft long is supported at one end by a vertical post and at the other by a cable that makes an angle of 40° with the beam, as in Fig. 4–15. A load of 3000 lb is suspended from the outer end of the beam, which itself weighs 430 lb. Find the tension in the cable.

Solution Four forces act on the beam in this problem: the load $\mathbf{w}_1$, the beam's own weight $\mathbf{w}_2$, which may be regarded as acting at its center, the tension $\mathbf{T}$ in the cable, and the force $\mathbf{F}$ exerted by the post on the inner end of the beam. By calculating torques

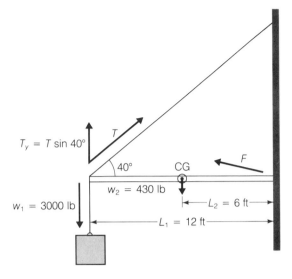

about the inner end of the beam we eliminate F, a convenient simplification. The torques τ_1 exerted by the load, τ_2 by the beam's weight, and τ_3 by the cable on the beam about its inner end are respectively

$$\tau_1 = w_1 L_1 = +(3000 \text{ lb})(12 \text{ ft}) = 36{,}000 \text{ lb} \cdot \text{ft}$$
$$\tau_2 = w_2 L_2 = +(430 \text{ lb})(6 \text{ ft}) = 2580 \text{ lb} \cdot \text{ft}$$
$$\tau_3 = T_y L_1 = -(T \sin \theta) L_1 = -(12T)(\sin 40°)\text{ft} = -7.71 \, T \text{ ft}$$

(In finding τ_3 the vertical component $\mathbf{T}_y$ of the tension in the rope was used as the force; the length L_1 of the beam is the moment arm because the beam is horizontal. The sign of τ_3 is negative, since it acts clockwise; the other torques are positive, since they act counterclockwise.) The condition for the equilibrium of the beam is

$$\Sigma\tau = \tau_1 + \tau_2 + \tau_3 = 36{,}000 \text{ lb} \cdot \text{ft} + 2580 \text{ lb} \cdot \text{ft} - 7.71 \, T \text{ ft} = 0$$

Solving for the tension T gives

$$7.71 \, T = 36{,}000 \text{ lb} + 2580 \text{ lb} = 38{,}580 \text{ lb}$$
$$T = \frac{38{,}580 \text{ lb}}{7.71} = 5004 \text{ lb}$$

Since we are entitled to only three significant figures here, the answer is better expressed as $T = 5000$ lb. ■

Example   A ladder 4.0 m long is leaning against a frictionless wall with its lower end 1.6 m away from the wall, as in Fig. 4–16. If the ladder weighs 150 N, how much force does it exert on the wall and on the ground?

Solution   Because the wall is frictionless, the only force it can exert on the ladder is the horizontal reaction force $\mathbf{F}_1$. The other forces on the ladder are its weight $\mathbf{w}$, which acts downward from its midpoint, and the reaction force $\mathbf{F}_2$ exerted by the ground.

Since $\mathbf{F}_{2y}$, the upward component of $\mathbf{F}_2$, and $\mathbf{w}$ are the only vertical forces and $\mathbf{w}$ is in the $-y$-direction,

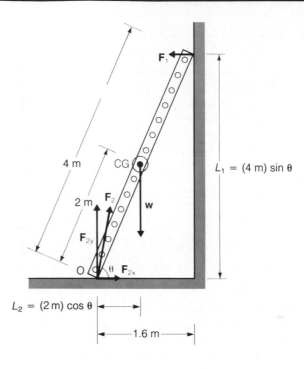

FIG. 4-16 To simplify the calculation, the pivot point about which torques are calculated is the bottom of the ladder.

$$\Sigma F_y = F_{2y} - w = 0$$
$$F_{2y} = w = 150\,\text{N}$$

Similarly the horizontal force $\mathbf{F}_1$ acting to the left must have the same magnitude as F_{2x} acting to the right:

$$\Sigma F_x = F_{2x} - F_1 = 0$$
$$F_{2x} = F_1$$

Before going further we need the value of the angle θ. From Fig. 4-16,

$$\cos\theta = \frac{1.6\,\text{m}}{4\,\text{m}} = 0.400$$
$$\theta = \cos^{-1} 0.400 = 66°$$

Let us find F_1 first. The easiest way is to calculate torques about the lower end of the ladder, since then we will have only the two torques produced by $\mathbf{w}$ and by $\mathbf{F}_1$ to consider. We have

$$\tau_1 = F_1 L_1 = (F_1)(4\,\text{m})\sin 66° = 3.65 F_1\,\text{m}$$
$$\tau_2 = -wL_2 = -(150\,\text{N})(2\,\text{m})\cos 66° = -122\,\text{N}\cdot\text{m}$$

The sum of these torques must be zero for the ladder to be in equilibrium, so

$$\tau = \tau_1 + \tau_2 = 0$$
$$\tau_1 = -\tau_2$$
$$3.65 F_1\,\text{m} = -(-122\,\text{N}\cdot\text{m}) = 122\,\text{N}\cdot\text{m}$$

$$F_1 = \frac{122\,\text{N}\cdot\text{m}}{3.65\,\text{m}} = 33.4\,\text{N}$$

The force the wall exerts on the ladder is 33.4 N, and this is equal in magnitude to the force the ladder exerts on the wall.

Now we must find F_2. Since $F_{2x} = F_1$ and $F_{2y} = w$, the total force the ground exerts on the ladder is

$$F_2 = \sqrt{F_{2x}^2 + F_{2y}^2} = \sqrt{F_1^2 + w^2}$$
$$= \sqrt{(33.4\,\text{N})^2 + (150\,\text{N})^2} = 154\,\text{N}$$

The force the ladder exerts on the ground has the same magnitude. ∎

4−5 FINDING THE CENTER OF GRAVITY OF AN IRREGULAR OBJECT

The center of gravity of an object of regular shape and uniform composition is always at its geometrical center. To obtain the center of gravity of an irregular object, we can use the experimental method shown in Fig. 4−17. An analytical procedure that follows from the definition of center of gravity can be applied to give a more accurate result. What is done is simply to break up the object into two or more separate ones whose centers of gravity are known, to consider each of these component objects as particles joined by weightless rods, and then to calculate the balance point of the resulting assembly. If the object is complex, for example the hull of a ship, a great many separate elements must be considered if the result is to be accurate, but often some degree of regularity is present to make the task easier.

How to calculate the location of a center of gravity

To see how such a calculation can be made, let us consider a system of three particles whose masses are m_1, m_2, and m_3 that are located x_1, x_2, and x_3 respectively from one end of a weightless rod, as in Fig. 4−18. The center of gravity of the system is located at some distance X from the end of the rod, which means that the torque exerted about this end by a single particle of mass $M = m_1 + m_2 + m_3$ located at X will be exactly the same as the torque exerted by the actual system of the three particles.

FIG. 4−17 To find the center of gravity of a flat body, suspend it and a plumb bob successively from two different points on its edge. The center of gravity is located at the intersection of the two lines of action of the plumb bob.

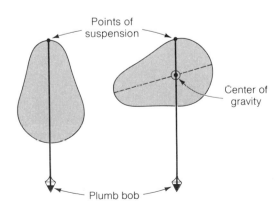

Points of suspension

Center of gravity

Plumb bob

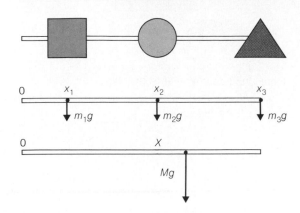

FIG. 4-18 The center of gravity of a complex object can be found by considering it as a system of particles and applying Eq. (4-9).

Hence

$$\Sigma\tau = m_1 g x_1 + m_2 g x_2 + m_3 g x_3 = MgX = (m_1 + m_2 + m_3)gX$$

so that

$$X = \frac{m_1 x_1 + m_2 x_2 + m_3 x_3}{m_1 + m_2 + m_3}$$

This formula can be generalized to a system of any number of particles (or objects that can be represented by particles located at their centers of gravity) in a straight line, giving

$$X = \frac{m_1 x_1 + m_2 x_2 + m_3 x_3 + \cdots}{m_1 + m_2 + m_3 + \cdots} \qquad \text{CG of composite object} \quad (4-9)$$

In the British system of units, the same formula can be used with weight w in place of mass m, since $w = mg$. Thus

$$X = \frac{w_1 x_1 + w_2 x_2 + w_3 x_3 + \cdots}{w_1 + w_2 + w_3 + \cdots} \qquad \text{CG of composite object} \quad (4-10)$$

Example The hand, forearm, and upper arm of a certain woman have the respective masses 0.4 kg, 1.2 kg, and 1.9 kg, and their centers of gravity are respectively 0.60 m, 0.40 m, and 0.15 m from her shoulder joint (Fig. 4-19). Find the distance of the center of gravity of her entire unbent arm from the shoulder joint.

Solution From Eq. (4-9)

$$X = \frac{(0.4 \text{ kg})(0.60 \text{ m}) + (1.2 \text{ kg})(0.40 \text{ m}) + (1.9 \text{ kg})(0.15 \text{ m})}{(0.4 + 1.2 + 1.9) \text{ kg}}$$

$$= 0.29 \text{ m}$$

The center of gravity of the entire arm is 0.29 m from the shoulder joint. ■

If the irregular object whose center of gravity is to be found lies in a plane rather than along a straight line (or is three-dimensional), the same procedure is applied along two (or three) coordinate axes. An example will make this clear.

FIG. 4–19

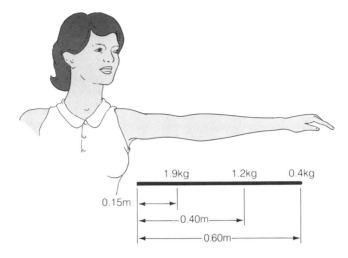

FIG. 4–20

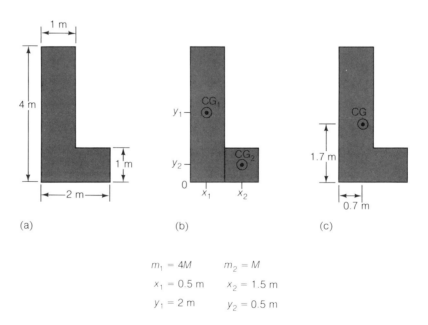

(a)   (b)   (c)

$$m_1 = 4M \qquad m_2 = M$$
$$x_1 = 0.5 \text{ m} \qquad x_2 = 1.5 \text{ m}$$
$$y_1 = 2 \text{ m} \qquad y_2 = 0.5 \text{ m}$$

Example   Find the center of gravity of the L-shaped steel plate shown in Fig. 4–20.

Solution   We imagine the plate to consist of two sections, one a rectangle 4 m long by 1 m wide and the other a square 1 m on a side. The centers of gravity of these sections are at their geometric centers, as shown in Fig. 4–20(b). Now we replace each section by a particle at its center of gravity. If the steel plate has a mass of M per square meter, particle 1 has a mass of $4M$ and particle 2 has a mass of M, since their areas are respectively 4 m^2 and 1 m^2. The x- and y-coordinates of the center of gravity of the entire plate are as follows:

$$X = \frac{m_1x_1 + m_2x_2}{m_1 + m_2} = \frac{(4M)(0.5 \text{ m}) + (M)(1.5 \text{ m})}{4M + M} = 0.7 \text{ m}$$

$$Y = \frac{m_1y_1 + m_2y_2}{m_1 + m_2} = \frac{(4M)(2 \text{ m}) + (M)(0.5 \text{ m})}{4M + M} = 1.7 \text{ m}$$

The location of this point is shown in Fig. 4–20(c). We did not need to know the value of M in this case since the steel plate is uniform. ■

4–6 FORCE AND CENTER OF GRAVITY

How an object responds to a net force acting on it depends on whether the line of action of the force passes through the object's center of gravity or not. If it does, the force has no effect on the object's rotational motion (Fig. 4–21). If the object is not spinning to begin with, such a force will simply accelerate it in the direction of the force according to $\mathbf{F} = m\mathbf{a}$. However, if the line of action of the force does not pass through the center of gravity, the object will start to spin.

Why some forces produce rotation

A wheel on a level surface has its center of gravity directly above its point of contact with the surface, as in Fig. 4–22(a). The line of action of the reaction force of the surface on the wheel, which is what supports the wheel, passes through the wheel's center of gravity, and, unless it is pushed, the wheel remains at rest. If the

Why wheels roll downhill

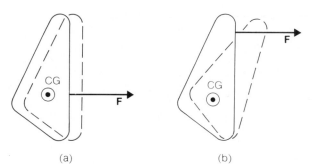

(a) (b)

FIG. 4–21 (a) A body remains in rotational equilibrium when the line of action of an applied force passes through its center of gravity. (b) When the line of action of an applied force does not pass through its center of gravity, the body is given a rotational as well as a translational acceleration.

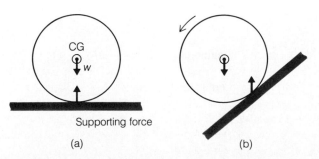

Supporting force

(a) (b)

FIG. 4–22 A wheel rolls downhill on an inclined surface because the line of action of the supporting force does not pass through its center of gravity, which leads to a net torque on the wheel.

surface is tilted, however, the line of action of the supporting force does not intersect the wheel's center of gravity, and the result is a torque on the wheel that starts it rolling (Fig. 4–22(b)).

We have all noticed that a car tilts backward when it is accelerated, and that it tilts forward when the brakes are applied. These effects occur because the forces applied to the car in each case are horizontal ones that act where the tires touch the road, and so are not in line with the car's center of gravity. The car rotates when the resulting torques act on it until the springs of the suspension system restore rotational equilibrium (Fig. 4–23).

An accelerated car tilts backward, a braked car tilts forward

Example   When the 4000-lb truck shown in Fig. 4–24(a) is at rest, its weight is equally distributed between its front and rear wheels. Find the downward force on each pair of wheels when the truck has an acceleration of 5 ft/s². Assume that the truck has rear-wheel drive. Would the results be different if it had front-wheel drive?

Solution   Figure 4–24(b) is a free-body diagram of the accelerated truck. The horizontal reaction force **F** exerted by the road on the truck's rear wheels has the magnitude

$$F = ma = \frac{w}{g}a = \frac{(4000\text{ lb})(5\text{ ft/s}^2)}{32\text{ ft/s}^2} = 625\text{ lb}$$

The sum $N_1 + N_2$ of the upward reaction forces of the road must equal the truck's weight w, so

$$N_1 + N_2 = w$$
$$N_2 = w - N_1$$

FIG. 4–23   A car tilts backward when accelerated (a) and forward when braked (b) because the lines of action of the applied forces in each case do not pass through the car's center of gravity. In (a) the car is assumed to have rear-wheel drive; if the car has front-wheel drive, the reaction forces act on its front tires, but the situation is otherwise the same.

The torques exerted by F, N_1, and N_2 about the center of gravity of the truck are respectively

$$\tau_1 = Fh = (625\text{ lb})(2.3\text{ ft}) = 1438\text{ lb}\cdot\text{ft}$$
$$\tau_2 = -N_1 L_1 = -4.2N_1\text{ ft}$$
$$\tau_3 = N_2 L_2 = (w - N_1)L_2 = wL_2 - N_1 L_2 = (4000\text{ lb})(4.2\text{ ft}) - (N_1)(4.2\text{ ft})$$
$$= 16{,}800\text{ lb}\cdot\text{ft} - 4.2N_1\text{ ft}$$

The condition for rotational equilibrium is

Reaction forces of road
on tires when accelerated

(a)

Reaction forces of road
on tires when braking

(b)

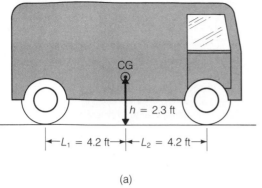

(a)

(b)

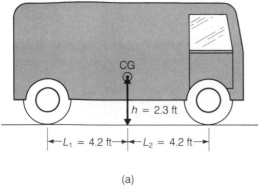

FIG. 4–24 The pivot point about which torques are calculated is the center of gravity of the truck.

$$\Sigma\tau = \tau_1 + \tau_2 + \tau_3$$
$$= 1438\,\text{lb} \cdot \text{ft} - 4.2N_1\,\text{ft} + 16,800\,\text{lb} \cdot \text{ft} - 4.2N_1\,\text{ft}$$
$$= 0$$

Solving for N_1 gives

$$4.2N_1\,\text{ft} + 4.2N_1\,\text{ft} = 1438\,\text{lb} \cdot \text{ft} + 16,800\,\text{lb} \cdot \text{ft}$$
$$8.4N_1 = 18,238\,\text{lb}$$
$$N_1 = \frac{18,238\,\text{lb}}{8.4} = 2171\,\text{lb}$$

which means that

$$N_2 = w - N_1 = 4000\,\text{lb} - 2171\,\text{lb} = 1829\,\text{lb}$$

The downward forces on the wheels are equal in magnitude to the respective reaction forces, so that the downward force on the rear wheels is 342 lb greater than that on the front wheels. This greater force causes the truck to tilt backward. When the truck is not accelerated, the downward forces are equal and the truck is level. The results are the same whether the truck has front- or rear-wheel drive because h, the height of the truck's center of gravity, is the same in both cases. ■

IMPORTANT TERMS

When the net force acting on an object is zero, the object is in **translational equilibrium.** When the net torque acting on it is zero, the object is in **rotational equilibrium.**

The **torque** of a force about a particular pivot point is the product of the magnitude of the force and the perpendicular distance from the line of action of the force to the pivot point. The latter distance is called the **moment arm** of the force. For an object to be in equilibrium there must be neither net force nor net torque acting upon it.

The **center of gravity** of an object is that point from which it can be suspended in any orientation without tending to rotate. The weight of an object can be considered as a downward force acting on its center of gravity.

IMPORTANT FORMULAS

Equilibrium of a particle: $\Sigma F_x = 0$
$\Sigma F_y = 0$

Torque: $\tau = FL$

Rotational equilibrium of object:
$\Sigma\tau = 0$ about any point

Center of gravity:

$$X = \frac{m_1 x_1 + m_2 x_2 + m_3 x_3 + \cdots}{m_1 + m_2 + m_3 + \cdots}$$

MULTIPLE CHOICE

1. Which of the following sets of horizontal forces could leave a body in equilibrium?

  a. 25, 50, and 100 lb   b. 5, 10, 20, and 50 lb
  c. 8, 16, and 32 lb   d. 20, 20, and 20 lb

2. Which of the following sets of horizontal forces could not leave an object in equilibrium?

  a. 6, 8, and 10 N
  b. 10, 10, and 10 N
  c. 10, 20, and 30 N
  d. 20, 40, and 80 N

3. In general, the number of scalar equations that must be satisfied if an object free to move in a plane is to be in equilibrium is

  a. 2.   b. 3.
  c. 4.   d. 6.

4. In general, the number of scalar equations that must be satisfied if an object free to move in three dimensions is to be in equilibrium is

  a. 2.   b. 3.
  c. 4.   d. 6.

5. An object in equilibrium may *not* have

  a. any forces acting upon it.
  b. any torques acting upon it.
  c. velocity.
  d. acceleration.

6. Two ropes are used to support a stationary weight W. The tensions in the rope must

  a. each be $W/2$.
  b. each be W.
  c. have a vector sum of magnitude W.
  d. have a vector sum of magnitude greater than W.

7. A weight is suspended from the middle of a rope whose ends are at the same level. In order for the rope to be perfectly horizontal, the forces applied to the ends of the rope

  a. must be equal to the weight.
  b. must be greater than the weight.
  c. might be so great as to break the rope.
  d. must be infinite.

8. A picture is supported by two strings that run from its upper corners to a nail on the wall. The tension in each string is T. The weight of the picture is

  a. less than T.   b. less than $2T$.
  c. between T and $2T$.   d. more than $2T$.

9. If the sum of the torques on an object is zero about a certain point,

  a. it is zero about no other point.
  b. it is zero about some other points.
  c. it is zero about all other points.
  d. the object must be in equilibrium.

10. In an equilibrium problem, the axis about which torques are computed

  a. must pass through one end of the object.
  b. must pass through the center of gravity of the object.
  c. must intersect the line of action of at least one force acting on the object.
  d. may be located anywhere.

11. The center of gravity of an object

  a. is always at its geometrical center.
  b. is always in the interior of the object.
  c. may be outside the object.
  d. is sometimes arbitrary.

12. A torque of 15 lb · ft is required to cut a thread on the end of a pipe. If the handles on the die are 9 in. from its axis on each side, the force each hand must apply is

  a. 0.83 lb.   b. 5 lb.
  c. 10 lb.   d. 22.5 lb.

13. A 70-kg person stands on the end of a 40-kg diving board whose support is 3 m away. The torque exerted on the support is

  a. 270 N · m.   b. 2058 N · m.
  c. 2646 N · m.   d. 3234 N · m.

14. A 10-lb box is suspended by a string from an overhead support. If a horizontal force of 5.8 lb is applied to the box, the string will make an angle with the vertical of

  a. 30°.   b. 45°.
  c. 60°.   d. 75°.

15. A 0.5-kg picture is supported by two strings that run from its upper corners to a nail on the wall. If each string makes a 40° angle with the vertical, the tension in each is

  a. 2.9 N.   b. 3.2 N.
  c. 3.8 N.   d. 4.9 N.

16. A 60-kg object is attached to one end of a steel tube 2.4 m long whose mass is 40 kg. The distance from the loaded end to the balance point is

  a. 48 cm.   b. 60 cm.
  c. 80 cm.   d. 160 cm.

17. A uniform wooden plank 2 m long that weighs 200 N is supported by a sawhorse 40 cm from one end. What downward force on the end nearest the sawhorse is needed to keep the plank level?

    a. 200 N   b. 300 N
    c. 500 N   d. 800 N

18. A weight of 1 kN is suspended from the end of a horizontal strut 1.5 m long that projects from a wall. A supporting cable 2.5 m long goes from the end of the strut to a point on the wall 2 m above the strut. The tension in the cable is

    a. 0.8 kN.   b. 1 kN.
    c. 1.25 kN.   d. 1.67 kN.

19. The inward force the strut of Question 18 exerts on the wall is

    a. 0.75 kN.   b. 0.94 kN.
    c. 1.25 kN.   d. 1.34 kN.

20. The entire 200-N weight of a door 2 m high and 0.8 m wide is supported by a hinge at a lower corner. The force the door exerts on the hinge at the top corner is

    a. 40 N.   b. 80 N.
    c. 200 N.   d. 400 N.

21. The force the above door exerts on the lower hinge is

    a. 200 N.   b. 204 N.
    c. 215 N.   d. 240 N.

EXERCISES

4–1 Translational Equilibrium

1. A ladder rests against a frictionless wall. (a) In what direction must the force the ladder exerts on the wall be? (b) How is the force the ladder exerts on the ground related to the weight of the ladder? Why?

2. Would you expect any difference in how soon you get tired between standing with your feet together and standing with them some distance apart? Why?

3. A horizontal beam 6 m long projects from the wall of a building. A guy wire that makes a 40° angle with the horizontal is attached to the outer end of the beam. When a weight of 100 N is attached to the end of the beam, what is the tension in the guy wire? (Neglect the weight of the beam.)

4. A 40,000-kg airplane is provided with 130 kN of thrust by its engines as it takes off at an angle of 25° above the horizontal. If the velocity of the airplane is constant, how much upward force on it is due to the lift produced by the flow of air over its wings?

5. A 50-kg object is suspended from two ropes that each make an angle of 30° with the vertical. What is the tension in each rope?

6. Find the maximum load that can be supported by the arrangement shown if the force on the strut is not to exceed 3000 N.

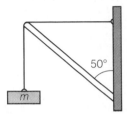

7. A 60-kg girl sitting in a canvas sling is being hoisted up a mast by a rope that passes through a pulley at the masthead and then down to a winch on the deck. She presses her feet against the mast to hold herself clear. How much force must her legs exert if her CG is 60 cm from the mast when she is 4 m below the masthead?

8. A 500-kg load of bricks is lifted by a crane alongside a building under construction. A horizontal rope is used to pull the load to where it is required, and the supporting cable is then at an angle of 10° from the vertical. What is the tension in the rope? In the supporting cable?

9. A horizontal force of 80 N acts upon a 5-kg object suspended by a string. What is the direction and magnitude of the force the string must provide to keep the object at rest?

10. A 600-lb compressor is suspended by two ropes, one that makes an angle of 20° with the vertical and the other that makes an angle of 30° with the vertical. Find the tension in each rope.

11. A 200-kg engine is suspended from the ceiling by a cable 6.0 m long. (a) If the coefficient of static friction between his shoes and the floor is 0.6, find the maximum horizontal force a 70-kg person can exert on the engine before beginning to slip. (b) Find the horizontal displacement of the engine when a horizontal force of this magnitude is applied to it. (*Hint:* Use the equilibrium conditions to find the angle between the cable and the vertical.)

12. The femur is the large bone in the thigh. When a fractured femur is set, the thigh muscles tend to pull the segments together so tightly that the healed bone would be shorter than its original length. To prevent this, a traction apparatus like the one shown (next page) can be attached to

the lower leg to exert a force that opposes the muscular contraction. The advantage of this particular arrangement is that the injured leg can be moved around a fair amount without much effect on the applied force. Find the magnitude of the resultant force on the femur and its direction relative to the horizontal for the situation shown. Assume a mass of 3.5 kg for the lower leg.

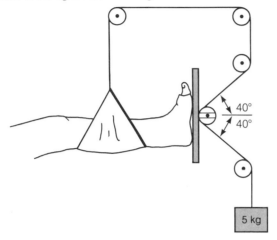

13. A bird sits on a telephone wire midway between two poles 20 m apart. The wire, assumed weightless, sags by 50 cm. If the tension in the wire is 90 N, find the mass of the bird.

14. A car is stuck in the mud. To get it out, the driver ties one end of a rope to the car and the other to a tree 30 m away. He then pulls sideways on the rope at its midpoint. If he exerts a force of 500 N, how much force is applied to the car when he has pulled the rope 1.5 m to one side?

15. Find the compressive force on the boom of the crane shown and the tension in the horizontal supporting cable. Assume both cables are fixed at the end of the boom.

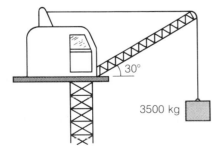

16. A 500-kg load is supported by a pair of hingeless legs as shown below. There is no friction between the legs and the floor. Find the tension in the rope.

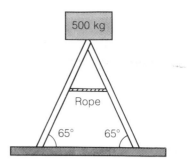

17. Find the compressive force on each of the legs supporting the 100-lb weight in the arrangement shown below.

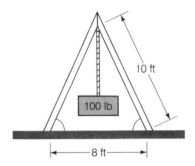

18. An object is attached to the midpoint of a string L long whose ends are tied to rings that fit loosely on a horizontal rod, as shown. If the coefficient of static friction between the rings and the rod is μ, find the maximum distance x between the rings before they begin to slip in terms of L and μ.

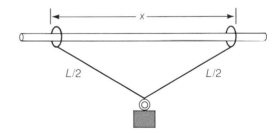

19. A person holds a 10-kg pail of water with his upper arm at his side and his forearm outstretched, as shown below. The palm of his hand is 35 cm from his elbow, his upper arm is 20 cm long, and his biceps muscle is attached to his forearm 5 cm from his elbow. The person's forearm (including his hand) has a mass of 3 kg and its CG is 16 cm from the elbow. Find the force the biceps muscle exerts to support the forearm and pail.

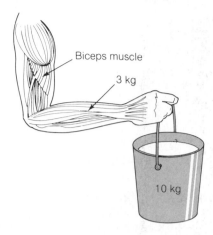

Biceps muscle

3 kg

10 kg

4–2 Torque

20. A person on a bicycle presses down on each pedal through the front half of its circle. In which position is the torque a maximum? In which position is it zero?

21. What determines the moment arm of a force?

22. A *couple* consists of two forces whose magnitudes are the same that act in opposite directions along parallel lines of action a certain distance *d* apart. If the magnitude of each force is *F,* find a formula for the torque exerted by a couple.

23. The torque required to fracture a person's tibia (the large bone in the lower leg) is about 100 N · m. If the distance between the pivoting heel of the safety binding on a ski and the toe release is 35 cm, what is the maximum horizontal force at which the release should be set to open?

4–3 Rotational Equilibrium

24. In an equilibrium problem, under what circumstances is it necessary to consider the torques exerted by the various forces?

25. About which point should the torques of the various forces that act on a body be calculated when this is necessary?

26. A weightless ladder 6 m long rests against a frictionless wall at an angle of 60° above the horizontal. A 75-kg person is 1.2 m from the top of the ladder. What horizontal force at the bottom of the ladder is needed to keep the ladder from slipping?

27. A 15-kg child and a 25-kg child sit at opposite ends of a 4-m seesaw pivoted at its center. Where should a third child whose mass is 20 kg sit in order to balance the seesaw?

28. A horizontal weightless boom is attached to a wall at its inner end and is supported at its outer end by a cable that

makes an angle of 60° with the boom. A load of 400 lb is attached to the outer end of the boom. Find the tension in the cable and the compression force in the boom.

29. One end of a 50-lb plank 16 ft long rests on the floor and the other end is held up by a carpenter so that the plank makes an angle of 25° with the floor. If the force the carpenter exerts is perpendicular to the plank, what is its magnitude?

30. When a person stands on one foot with his heel raised, the entire reaction force of the floor, which is equal to his weight *w,* acts upward on the ball of his foot, as in the figure. In order to raise his heel, he must apply the upward force F_1 via his Achilles tendon, so the downward force F_2 on his ankle is greater than his weight *w.* Find the values of F_1 and F_2 for a 75-kg man for whom $L_1 = 5$ cm and $L_2 = 15$ cm. If the distance L_1 were greater, would this mean an increase or a decrease in the muscular force F_1?

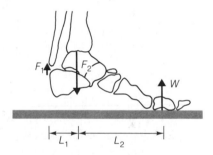

31. A fishing rod 3 m long is attached to a pivoting holder at its lower end. The rod is 50° above the horizontal. A fisherman grasps the rod 1 m from its lower end. What horizontal force must the fisherman exert to keep the rod at this angle when reeling in a 10-kg fish from directly below the rod's upper end?

4–4 Center of Gravity

32. A square table has legs at three of its corners. Will the table fall over? If not, what is the maximum mass that can be placed on the unsupported corner before it falls over?

33. The wheels of a certain truck are 1.8 m apart, and the truck falls over when tilted sideways at 30° from the horizontal. How high above the road is its CG?

34. A 60-kg woman stands upright with her feet 20 cm apart and her arms outstretched on both sides. In her right hand she holds a dumbbell 70 cm from the centerline of her body. Find the maximum mass the dumbbell can have if she is not to topple over.

35. The asymmetric dumbbell shown is balanced on a frictionless surface. Eventually the dumbbell falls over with the 2-kg ball moving to the right. What is the final position of the 1-kg ball? (*Hint:* Consider the motion of the CG of the dumbbell.)

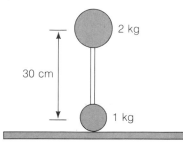

36. A crate whose total mass is 80 kg rests on a floor such that the coefficient of static friction is 0.4. The crate is 1.5 m high and 1.0 m square, and its CG is at its geometric center. (a) Find the minimum horizontal force needed to start the crate sliding across the floor. (b) What is the maximum height above the floor at which this force can be applied without tipping the crate over? (*Hint:* Assume the crate has just begun to tip over so only the far edge of its base is touching the floor and calculate moments about this edge.)

37. Three people are carrying a horizontal ladder 4 m long. One of them holds the front end of the ladder and the other two hold opposite sides of the ladder the same distance from its far end. What is the distance of the latter two people from the far end of the ladder if each person supports one-third of the ladder's weight?

38. A person holds a 10-kg board that is 2 m long in a horizontal position by supporting it with one hand underneath it 40 cm from one end while the other hand presses down at that end, as shown. Find the force each hand exerts. What percentage of the board's weight is each force?

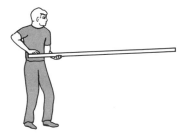

39. The front and rear axles of a 24-kN truck are 4.0 m apart. The CG of the truck is located 2.5 m behind its front axle. Find the weight supported by the front wheels of the truck.

40. A 25-N bag of cement is placed on a 4-m-long plank 1.8 m from one end. The plank itself weighs 8 N. Two men pick up the plank, one at each end. How much weight must each support?

41. The CG of a 150-kg polar bear standing on all fours is 100 cm from her feet and 80 cm from her hands. Find the force the ground exerts on each of her hands and feet.

42. A butcher with his upper arm vertically at his side and his forearm horizontal presses down on a scale with his hand, which is 40 cm from his elbow joint. His forearm and hand together weigh 30 N, and their CG is 20 cm from his elbow joint. The triceps muscle on the back of his upper arm is attached 2.5 cm behind the center of the elbow joint and it is exerting an upward force of 1 kN. What is the reading on the scale?

43. A 40-kg boom 3 m long is hinged to a vertical mast and held in position by a rope at its end that is attached to the mast 1 m above the hinge pin. If the boom is horizontal and is uniform, find the tension in the rope.

44. A 20-kg object is suspended from one end of a 30-kg wooden beam 3 m long. Where should the beam be picked up so that it remains horizontal?

45. The front wheels of a certain car are found to support 600 kg and the rear wheels 400 kg. The car's wheelbase (distance between axles) is 2.2 m. How far from the forward axle is the CG of the car?

46. The figure is a simplified diagram of a person bending over with the back horizontal. Typically a person's upper body (head, torso, and arms) comprises 65% of the total weight and has its CG 72% of the height above the feet. In the case of an 80-kg person 1.8 m tall, the upper body weight is 510 N and its CG is 36 cm above the base of the spine. This weight is supported by the back muscles, whose effect is equivalent to that of a single force **F** acting 42 cm above the base of the spine at an angle of 12°. (a) Find the magnitude of **F**. (b) Find the magnitude of the reaction force **R** that acts on the base of the spine. How many times greater than the weight of the upper body is this force? (Because bending over, particularly when picking something up, puts so much stress on the lower spine, this posture is best avoided. When an object is to be picked up, the back should be kept as vertical as possible and the actual lifting done by the leg muscles.)

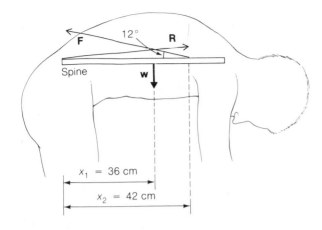

47. The arm of a certain person has a mass of 3 kg and its CG is 28 cm from the shoulder joint. When the arm is outstretched so that it is horizontal, the shoulder muscle supporting it has a line of action 15° above the horizontal. (a) If this muscle is attached to the upper arm 13 cm from the shoulder joint, find the force developed by the muscle. (b) Find the force when the outstretched hand 60 cm from the shoulder joint is used to support a 1-kg load.

48. A uniform 10-kg ladder 2.5 m long is placed against a frictionless wall with its base on the ground 80 cm from the wall. Find the magnitudes of the forces exerted on the wall and on the ground.

49. A uniform 15-kg ladder 3 m long rests against a frictionless wall at a point 2.4 m off the floor. Find the vertical and horizontal components of the force exerted by the ladder on the floor.

50. A uniform 12-kg ladder 2.8 m long rests against a vertical frictionless wall with its lower end 1.1 m from the wall. If the ladder is to stay in place, what must be the minimum coefficient of friction between the bottom of the ladder and the ground?

51. A door 2.4 m high and 0.8 m wide has hinges at the top and bottom of one edge. The entire 200-N weight of the door is supported by the upper hinge. Find the magnitude and direction of (a) the force the door exerts on the upper hinge; (b) the force the lower hinge exerts on the door.

52. A door 3 m high and 1.2 m wide has hinges on one edge that are 30 cm above the bottom and 30 cm below the top. The entire 500-N weight of the door is supported by the lower hinge. Find the magnitude and direction of (a) the force the door exerts on the lower hinge; (b) the force the upper hinge exerts on the door.

4–5 Finding the Center of Gravity of an Irregular Object

53. A 60-kg woman is sitting with her upper body and lower legs vertical and her thighs horizontal. The mass of her upper body is 40 kg and its CG is 32 cm above her hips; the mass of her thighs is 13 kg and their CG is 16 cm in front of her hips; and the mass of her lower legs and feet is 7 kg and their CG is 23 cm below her knees and 37 cm in front of her hips. Find the location of the CG of the seated woman.

54. Find the location of the CG of the flat object shown.

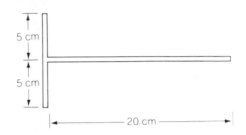

55. The metal plate shown has a circular hole 10 cm in radius. Find the location of the CG of the plate.

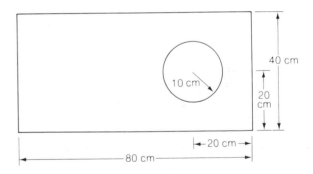

ANSWERS TO MULTIPLE CHOICE

1. d	**6.** c	**10.** d	**14.** a	**18.** c
2. d	**7.** d	**11.** c	**15.** b	**19.** a
3. b	**8.** b	**12.** c	**16.** a	**20.** a
4. d	**9.** c	**13.** c	**17.** b	**21.** b
5. d				

5

ENERGY

We all use the word energy, but how many of us know exactly what it means? We speak of the energy of a lightning bolt or of an ocean wave; we say that an active person is energetic; we hear a candy bar described as being full of energy; we read that most of the world's electricity will come from nuclear energy in the years to come. What do a lightning bolt, an ocean wave, an active person, a candy bar, and an atomic nucleus have in common?

In general terms, energy refers to an ability to accomplish change. All changes in the physical world involve energy, usually with energy being transformed from one sort into another. But "change" is not a very precise concept, and we must clarify our ideas before going further. What we shall do is first define a quantity called work, and then see how it permits us to discuss energy and its relation to change in the orderly manner of science.

CHAPTER OBJECTIVES

Completing this chapter should enable you to:

1. Find the work done by a force that moves an object through a certain displacement.

2. Take into account a difference in direction between an applied force and the displacement of the object it acts upon when calculating the work done.

3. Find the work done when an object is raised through a given height.

4. Determine the power output of an energy source.

5. Use the watt, ft · lb/s, and horsepower interchangeably to express power.

6. Relate the power developed by an energy source with its output force and the speed of the object being acted upon.

7. Distinguish among the kinetic, potential, and rest forms of energy.

8. Calculate the kinetic energy of a moving object.

9. Calculate the gravitational potential energy of an object relative to a given reference level.

10. Calculate the rest energy of an object of given mass.

11. Use the principle of conservation of energy to solve problems that involve moving objects.

5–1 WORK

All changes in the physical universe are the result of forces. Forces set things in motion, change their paths, and bring them to a stop; forces pull things together and push them apart. The physical quantity called *work* is a measure of the amount of change (in a general sense) a force gives rise to when it acts upon something.

Force and change

When we push against a brick wall, nothing happens. We have applied a force, but the wall has not yielded and shows no effects. On the other hand, when we apply exactly the same force to a ball, the ball flies through the air for some distance. In the second case something has happened because of our push, whereas in the first case there has been no result (Fig. 5–1).

What is the basic difference between the two situations? In the first case, where we pushed against a wall, the wall did not move. But in the second case, where we

FIG. 5–1 Work is done by a force when the object it acts upon is displaced while the force is applied.

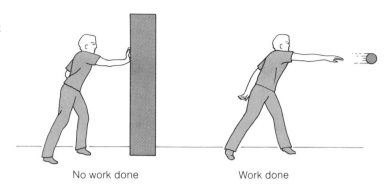

No work done   Work done

threw the ball, the ball *did* move while the force was being applied and before it left our hand. The displacement of the body while the force acted on it was what made the difference.

If we think carefully along these lines, we will see that whenever a force acts so as to produce motion in an object, the force acts while the object is being moved. In order to make this notion definite, we may make a preliminary definition of work as follows:

Preliminary definition of work

The work done by a force acting on an object that moves in the same direction as the force is equal to the magnitude of the force multiplied by the distance through which the force acts.

In equation form,

$$W = Fs \qquad (\textbf{F parallel to s}) \qquad\qquad Work \quad (5-1)$$

Work = force × distance

No work is done without both force and displacement

This definition is a great help in clarifying the effects of forces. Unless a force acts through a distance, no work is done no matter how great the force. And even if an object moves through a distance, no work is done unless a force is acting upon it or it exerts a force on something else (Fig. 5–2).

What we usually think of as work is in accord with the above statement. When something happens because a person applies a force, we say that he or she has done some work. Here the concept has simply been broadened to include inanimate forces. (We must still be careful, though; while we may get tired after pushing against a brick wall for an hour, we still have done no work on the wall if it remains in place.)

Units of work

In the SI system, work has a special unit, the *joule,* abbreviated J. One joule is equal to the work done by a force of 1 N acting through a distance of 1 m. That is,

$$1 \text{ joule} - 1 \text{ J} = 1 \text{ N} \cdot \text{m}$$

The joule is named after the English scientist James Joule and is pronounced "jool."

In the British system, the unit of work is the *foot-pound,* abbreviated ft · lb. One ft · lb is equal to the work done by a force of 1 lb acting through a distance of 1 ft.

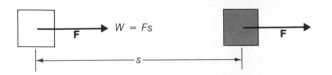

FIG. 5–2

Work = force × distance
through which force acts
(parallel case).

$W = Fs$

F F

When there is no
net applied force, no work
is done even though the
object may move.

$F = 0$ $W = 0$ $F = 0$

When the object acted upon
by a force remains at rest,
no work is done.

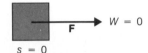

F $W = 0$

$s = 0$

Example A 100-lb wooden box is pushed across a horizontal floor with a force of
50 lb. The coefficient of sliding friction is $\mu = 0.4$. (a) Find the work done in pushing
the box 60 ft. (b) How much work went into overcoming friction and how much into
accelerating the box?

Solution (a) The work done by the applied force of 50 lb is

$$W = Fs = (50 \text{ lb})(60 \text{ ft}) = 3000 \text{ ft} \cdot \text{lb}$$

The weight of the box does not enter into the calculation, since only the force exerted
on the box determines the total amount of work done.

(b) We need to know the box's weight to find the frictional force opposing its
motion. Since the normal force here is the box's weight w, the frictional force is

$$F_f = \mu N = \mu w = (0.4)(100 \text{ lb}) = 40 \text{ lb}$$

The work done to overcome this force is

$$W_f = F_f s = (40 \text{ lb})(60 \text{ ft}) = 2400 \text{ ft} \cdot \text{lb}$$

Hence the work that went into accelerating the box was

$$3000 \text{ ft} \cdot \text{lb} - 2400 \text{ ft} \cdot \text{lb} = 600 \text{ ft} \cdot \text{lb} \qquad ■$$

The definition of work given above has an important qualification: The force **F**
must be in the same direction as the vector displacement **s**. If **F** and **s** are not parallel,
we must replace F in the formula $W = Fs$ by the magnitude of its component $\mathbf{F}_s$ in
the direction of the displacement s (Fig. 5–3). The magnitude of this component is

**Force and displacement
need not be parallel**

$$F_s = F \cos \theta$$

where θ is the angle between **F** and **s.** Hence the most general definition of work is

$$W = Fs \cos \theta \qquad\qquad\qquad Work \quad (5–2)$$

**General definition of
work**

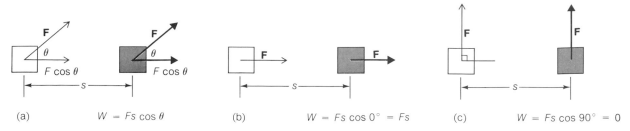

(a) $W = Fs \cos \theta$ (b) $W = Fs \cos 0° = Fs$ (c) $W = Fs \cos 90° = 0$

FIG. 5–3 The work done by a force depends upon the angle θ between the force and its displacement. (a) In general, $W = Fs \cos \theta$. (b) When **F** is parallel to **s**, $W = Fs$. (c) When **F** is perpendicular to **s**, $W = 0$ since **F** then has no component in the direction of **s** and therefore does not affect its motion.

FIG. 5–4

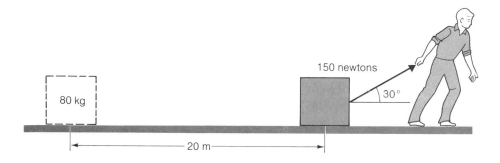

Work is a scalar quantity; there is no direction associated with work even though it depends upon two vector quantities, force and displacement.

No work is done when F is perpendicular to s

Equation (5–2) is always correct, since when the force and the displacement are parallel, $\theta = 0$ and $\cos 0 = 1$. When the force and the displacement are perpendicular, $\theta = 90°$ and $\cos 90° = 0$, so $W = 0$. In order for work to be done, the applied force must have a component in the direction of the displacement.

Example A person pulls an 80-kg crate for 20 m across a level floor using a rope that is 30° above the horizontal. The person exerts a force of 150 N on the rope (Fig. 5–4). How much work is performed?

Solution The work done is

$$W = F s \cos \theta = (150\,\text{N})(20\,\text{m})(\cos 30°)$$
$$= (150)(20)(0.866)\,\text{N} \cdot \text{m} = 2.6 \times 10^3 \text{J}$$

Evidently the 80-kg mass of the crate has no significance here—it is the force exerted on the crate that determines how much work is done. ■

5–2   WORK DONE AGAINST GRAVITY

It is easy to find the work done in lifting an object against gravity. The force of gravity on an object of mass m is the same as its weight $w = mg$. Hence in order to raise the

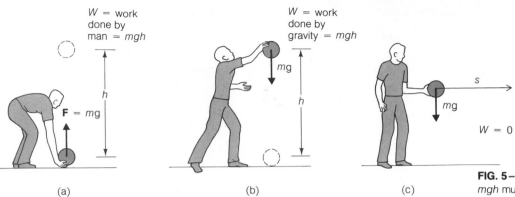

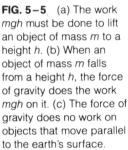

FIG. 5–5 (a) The work *mgh* must be done to lift an object of mass *m* to a height *h*. (b) When an object of mass *m* falls from a height *h*, the force of gravity does the work *mgh* on it. (c) The force of gravity does no work on objects that move parallel to the earth's surface.

object to a height *h* above its original position, a force of *mg* must be exerted on it. Since $F = mg$ and $s = h$ here, the work done is

$$W = Fs = mgh \tag{5–3}$$

To lift an object of mass *m* to a height *h* requires the performance of *mgh* of work (Fig. 5–5). In the British system, since $mg = w$, Eq. (5–3) becomes just $W = wh$.

Work done against gravity is independent of path

It is important to note that only the height *h* is involved in work done against the force of gravity. The particular route taken by an object being raised is not significant (Fig. 5–6); excluding any frictional effects, exactly as much work must be expended to climb a flight of stairs as to go up in an elevator to the same floor (though not by the person involved!).

Example How much work is done in lifting a 30-kg load of bricks to a height of 20 m on a building under construction?

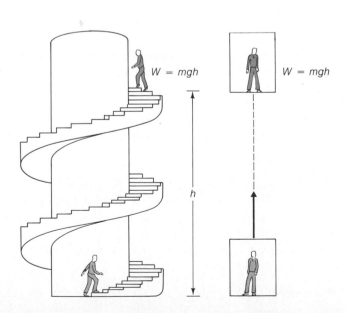

FIG. 5–6 In the absence of friction, the work done in lifting a mass *m* to a height *h* is *mgh* regardless of the exact path taken.

Solution   Here $m = 30$ kg and $h = 20$ m, so that

$$W = mgh = (30 \text{ kg})(9.8 \text{ m/s}^2)(20 \text{ m}) = 5.9 \times 10^3 \text{ J}$$ ∎

Example   Eating a banana enables a person to perform about 3×10^4 ft · lb of work, figuring his or her efficiency at 10%. To what height does eating a banana enable a 125-lb woman to climb?

Solution   Since $W = wh$ when British units are used,

$$h = \frac{W}{w} = \frac{3 \times 10^4 \text{ ft} \cdot \text{lb}}{125 \text{ lb}} = 240 \text{ ft}$$ ∎

5–3   POWER

Power is rate of doing work

Often the time needed to perform a task is just as significant as the actual amount of work required. Given enough time, even the feeblest motor can raise the Sphinx as high as we like. However, if we want to carry out a certain operation quickly, we try to obtain a motor whose output of work is rapid in terms of the total required. The rate at which work is done is therefore an important engineering quantity. This rate is called *power:* The faster some agency can do work, the more *powerful* it is.

If an amount of work W is performed in a time interval t, the power involved is

$$P = \frac{W}{t} \qquad\qquad\qquad \textit{Power} \quad (5–4)$$

$$\text{Power} = \frac{\text{work done}}{\text{time interval}}$$

SI units of power

In the SI system, where work is measured in joules and time in seconds, the unit of power is the *watt:*

1 watt = 1 W = 1 J/s

The watt is rather small for most industrial purposes; even an electric clock requires several watts of power. The larger *kilowatt* (kW) is accordingly in common use, where

1 kW = 1000 watts

The *kilowatt-hour* (kWh) is often used as a unit of work. Since 1 kW = 1000 W = 1000 J/s and 1 h = 3600 s,

1 kilowatt-hour = 1 kWh = 3.60×10^6 J

British units of power

In the British system, where energy is measured in foot-pounds and time in seconds, the unit of power is the *ft · lb/s*. The ft · lb/s is also inconveniently small and has been replaced by the larger *horsepower* (hp) in engineering practice. The horsepower was introduced two centuries ago by James Watt to compare the output of the steam engine he had perfected with a more familiar source of power. Today the horsepower is defined as

1 hp = 550 ft · lb/s

FIG. 5-7

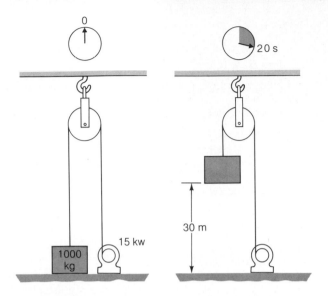

In the United States the watt and kilowatt are employed in connection with electric power, while mechanical power is customarily specified in horsepower. To convert power figures from one system to the other, we note that

$$1 \text{ hp} = 746 \text{ W} = 0.746 \text{ kW}$$
$$1 \text{ kW} = 1.34 \text{ hp} = 737 \text{ ft} \cdot \text{lb/s}$$

Example   An electric motor with an output of 15 kW provides power for the elevator of a 6-story building. If the total mass of the loaded elevator is 1000 kg, what is the minimum time needed for it to rise the 30 m from the ground floor to the top floor (Fig. 5–7)?

Solution   The work done in raising the elevator through a height h is

$$W = mgh$$

Since $P = W/t$, the time needed for the motor to raise the elevator by 30 m is

$$t = \frac{W}{P} = \frac{mgh}{P} = \frac{(1000 \text{ kg})(9.8 \text{ m/s}^2)(30 \text{ m})}{15 \times 10^3 \text{ W}} = 20 \text{ s} \quad \blacksquare$$

When a constant force performs work on an object moving at constant speed in the same direction as that of the force, the power delivered is equal to the product of the force F and the speed v of the object. We can see why when we express the power in terms of the work done by the force in the time t:

Force, speed, and power

$$P = \frac{W}{t} = \frac{Fs}{t}$$

But the speed of an object that travels the distance s in the time t is

$$v = \frac{s}{t}$$

Hence

$$P = Fv$$

Power = force × speed

In the general case where the angle between **F** and **v** is θ, from Eq. (5–2) we have

$$P = Fv \cos \theta \qquad\qquad\qquad\qquad\qquad\qquad Power \quad (5\text{–}5)$$

Example   A swimmer develops an average power of 200 W as she covers 100 m in 80 s. What is the resistive force exerted by the water on her?

Solution   The swimmer's speed is $v = 100$ m/80 s $= 1.25$ m/s. Since $P = Fv$ here,

$$F = \frac{P}{v} = \frac{200\,\text{W}}{1.25\,\text{m/s}} = 160\,\text{N}$$

which is about 36 lb.   ■

Human power output

A person in good physical condition is usually capable of a continuous power output of about 75 W, which is 0.1 hp. An athlete such as a runner or swimmer may have a power output several times greater during a distance event. What limits the power output of a trained athlete is not muscular development but the supply of oxygen via the bloodstream from the lungs to the muscles, where it is needed for the metabolic processes that extract work from nutrients. However, for a momentary effort such as that of a weight lifter or a jumper, an athlete's power output may exceed 5 kW.

Example   A trash compactor with a $\frac{1}{4}$-hp motor can apply a crushing force of 1 ton (2000 lb). How fast does its ram move, assuming 80% efficiency?

Solution   The power developed by the motor is

$$P = (\tfrac{1}{4}\text{hp})\left(550\frac{\text{ft}\cdot\text{lb/s}}{\text{hp}}\right) = 138\,\text{ft}\cdot\text{lb/s}$$

The efficiency of any machine is given by the ratio between the power it delivers and the power supplied to it:

$$\text{Eff} = \frac{\text{power output}}{\text{power input}} = \frac{P_{\text{out}}}{P_{\text{in}}}$$

We therefore have here

$$P_{\text{out}} = (\text{Eff})(P_{\text{in}}) = (0.80)(138\;\text{ft}\cdot\text{lb/s}) = 110\;\text{ft}\cdot\text{lb/s}$$

Since $P = Fv$, the ram's speed is

$$v = \frac{P}{F} = \frac{110\;\text{ft}\cdot\text{lb/s}}{2000\;\text{lb}} = 0.055\;\text{ft/s}$$

which can be expressed in inches per second as

$$v = \left(0.055\frac{\text{ft}}{\text{s}}\right)\left(12\frac{\text{in.}}{\text{ft}}\right) = 0.66\;\text{in./s} \qquad\qquad ■$$

5-4 ENERGY

From the straightforward notion of work we proceed to the complicated and many-sided concept of *energy:*

Energy is that property whose possession enables something to perform work.   **Energy and work**

When we say that something has energy, we mean it is able (directly or indirectly) to exert a force on something else and do work on it. On the other hand, when we do work on something, we have added to it an amount of energy equal to the work done. The units of energy are the same as those of work, the joule and the foot-pound.

There are three broad categories of energy:

1. *Kinetic energy,* which is the energy something possesses by virtue of its motion.   **Categories of energy**
2. *Potential energy,* which is the energy something possesses by virtue of its position.
3. *Rest energy,* which is the energy something possesses by virtue of its mass.

All modes of energy possession fit into one or another of these three categories.   **Heat is a form of energy**
For instance, it is convenient for many purposes to think of heat as a separate form of energy, but what this term actually refers to is the sum of the kinetic energies of the randomly moving atoms and molecules in a body of matter.

5-5 KINETIC ENERGY

When we perform work on a ball by throwing it, what becomes of this work?

Let us suppose we apply the uniform force $\mathbf{F}$ to the ball for a distance s before it leaves our hand, as in Fig. 5-8(a). The work done on the ball is therefore, since $\cos \theta = 1$,

$$W = Fs \qquad\qquad (5-6) \quad \textbf{Work done on ball}$$

The mass of the ball is m. As we throw it, its acceleration has the magnitude

$$a = \frac{F}{m} \qquad\qquad (5-7)$$

according to the second law of motion, $\mathbf{F} = m\mathbf{a}$ (Fig. 5-8(b)).

We know from the formula

$$v_f^2 = v_0^2 + 2as$$

that when an object starting from rest ($v_0 = 0$) undergoes an acceleration of magnitude a through a distance s, its final speed v is related to a and s by

$$v_f^2 = 2as \qquad\qquad (5-8) \quad \textbf{Final speed of ball}$$

This relationship is pictured in Fig. 5-8(c).

If we now substitute F/m for a in Eq. (5-8), we find that

$$v_f^2 = 2as = 2s\left(\frac{F}{m}\right)s$$

which we can rewrite as

$$Fs = \tfrac{1}{2} mv_f^2 \qquad\qquad (5-9)$$

How work done is related to final speed

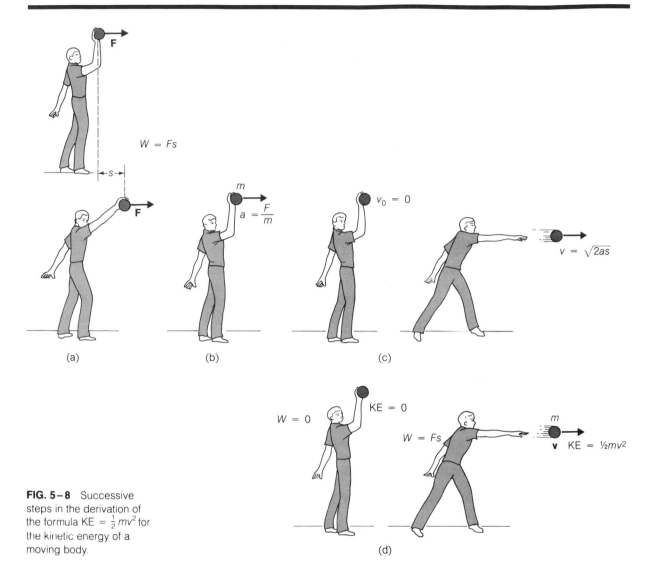

FIG. 5-8 Successive steps in the derivation of the formula KE $= \frac{1}{2} mv^2$ for the kinetic energy of a moving body.

The quantity on the left-hand side, *Fs,* is the work our hand has done in throwing the ball, as in Eq. (5-6). The quantity on the right-hand side, $\frac{1}{2} mv_f^2$, must therefore be the energy acquired by the ball as a result of the work we did on it. This energy is *kinetic energy,* energy of motion. That is, we interpret the preceding equation as follows (Fig. 5-8(d)):

$$Fs = \tfrac{1}{2} mv_f^2$$

Work done on ball = kinetic energy of ball

The symbol for kinetic energy is KE. The kinetic energy of an object of mass *m* and velocity *v* is therefore

All moving objects have kinetic energy

$$KE = \tfrac{1}{2} mv^2 \qquad\qquad\qquad\qquad\qquad\qquad\qquad \textit{Kinetic energy} \quad (5\text{--}10)$$

A moving object is able to perform an amount of work equal to $\frac{1}{2}mv^2$ in the course of being stopped.

Example A 1-lb hammer head strikes a nail at a speed of 12 ft/s and drives it $\frac{1}{4}$ in. into a wooden board (Fig. 5–9). What is the average force on the nail?

Solution We note that $\frac{1}{4}$ in. $= (\frac{1}{4})(\frac{1}{12})$ ft $= \frac{1}{48}$ ft. The initial kinetic energy of the hammer head is $\frac{1}{2}mv^2 = \frac{1}{2}(w/g)v^2$ and the work done on the nail is Fs. Hence $Fs = \frac{1}{2}(w/g)v^2$ and

$$F = \frac{wv^2}{2gs} = \frac{(1\,\text{lb})(12\,\text{ft/s})^2}{(2)(32\,\text{ft/s}^2)(\frac{1}{48})\,\text{ft}} = 108\,\text{lb}$$ ■

Example Find the kinetic energy of a 1200-kg car when it is moving at 25 km/h (16 mi/h) and when it is moving at 100 km/h (62 mi/h).

Solution Since

$$v_1 = (25\,\text{km/h})\left(0.278\frac{\text{m/s}}{\text{km/h}}\right) = 7.0\,\text{m/s}$$

$$v_2 = (100\,\text{km/h})\left(0.278\frac{\text{m/s}}{\text{km/h}}\right) = 27.8\,\text{m/s}$$

we have (Fig. 5–10)

$$KE_1 = \tfrac{1}{2}mv_1{}^2 = \tfrac{1}{2}(1200\,\text{kg})(7.0\,\text{m/s})^2 = 2.9 \times 10^4\,\text{J}$$
$$KE_2 = \tfrac{1}{2}mv_2{}^2 = \tfrac{1}{2}(1200\,\text{kg})(27.8\,\text{m/s})^2 = 46.4 \times 10^4\,\text{J}$$

At 100 km/h the car has 16 times as much kinetic energy as it does at 25 km/h. The fact that kinetic energy, and hence ability to do work (that is, damage), is proportional to the *square* of the speed is responsible for the severity of automobile accidents at high speeds.

■

FIG. 5–9

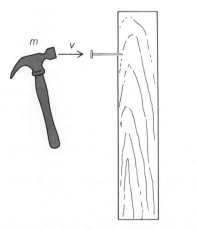

$\frac{1}{2}mv^2 = Fs$

FIG. 5–10

1200 kg

25 km/h

$KE_1 = 2.9 \times 10^4$ J

100 km/h

$KE_2 = 46.4 \times 10^4$ J

Example   What is the power output of the engine of a 1200-kg car if the car can go from 25 km/h to 100 km/h in 12 s?

Solution   The work needed to accelerate the car is, from the preceding problem,

$$W = KE_2 - KE_1 = 46.4 \times 10^4 \text{ J} - 2.9 \times 10^4 \text{ J} = 43.5 \times 10^4 \text{ J}$$

The power needed to provide this amount of work in 12 s is

$$P = \frac{W}{t} = \frac{43.5 \times 10^4 \text{ J}}{12 \text{ s}} = 3.63 \times 10^4 \text{ W} = 36.3 \text{ kW}$$

which is equivalent to

$$P = 36.3 \text{ kW} \times 1.34 \frac{\text{hp}}{\text{kW}} = 48.6 \text{ hp}$$

5–6   POTENTIAL ENERGY

A raised object has the potential of doing work

When we drop a stone from a height h, it falls faster and faster and finally strikes the ground. In striking the ground the stone does work; if it is sufficiently heavy and has fallen from a great enough height, the work done by the stone is obvious in the form of a hole (Fig. 5–11). Evidently the stone at its original location h above the ground had a capacity to do work, even though it was stationary at the time. The work the stone can perform in falling to the ground is called its *potential energy*, symbol PE.

We have already calculated that the work we must do to raise the stone to the height h is

$$W = mgh \tag{5–3}$$

where m is the stone's mass. This amount of work can also be done *by* the stone after dropping from the height h. Since the raised stone has the potentiality of doing the amount of work mgh, we define its *potential energy* as

$$PE = mgh \qquad\qquad \textit{Gravitational potential energy} \quad (5\text{–}11)$$

the product of its mass, the acceleration of gravity, and its height. In the British system of units, weights rather than masses are usually specified. Since $w = mg$, we may also write Eq. (5–11) as

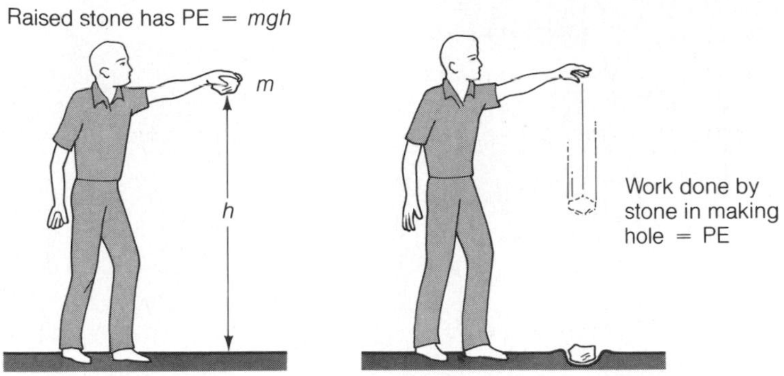

Raised stone has PE = *mgh*

FIG. 5-11

m

h

Work done by
stone in making
hole = PE

$$PE = wh \qquad\qquad (5\text{--}12)$$

which is more convenient in treating problems in this system of units.

The gravitational potential energy of an object depends upon the reference level from which its height h is measured. For example, the potential energy of a 1.0-kg book held 10 cm above a desk is

$$PE = mgh = (1.0\,\text{kg})(9.8\,\text{m/s}^2)(0.10\,\text{m}) = 0.98\,\text{J}$$

with respect to the desk (Fig. 5–12). However, if the book is 1.0 m above the floor of the room, its potential energy is

$$PE = mgh = (1.0\,\text{kg})(9.8\,\text{m/s}^2)(1.0\,\text{m}) = 9.8\,\text{J}$$

with respect to the floor. And the book may conceivably be 100 m above the ground so its potential energy is

$$PE = mgh = (1.0\,\text{kg})(9.8\,\text{m/s}^2)(100\,\text{m}) = 980\,\text{J}$$

with respect to the ground. The height h in the formula $PE = mgh$ means nothing unless the base height $h = 0$ is specified.

Potential energy is always a relative quantity. Just as the kinetic energy of a moving object depends upon with respect to what its speed is measured, so the potential energy of an object subject to a force depends upon the reference position chosen. Further, the potential energy is not a property of the object by itself but of the *system* of the object and the body that exerts the force on it. The potential energy of a stone held above the earth's surface is shared by both the stone and the earth. When the stone is released, both it and the earth move toward each other. However, the earth's motion is extremely small because of its immense mass relative to that of the stone. It is therefore appropriate to think of the entire potential energy of the system as belonging only to the stone, but in the case of two objects more nearly comparable in mass, it must be kept in mind that the potential energy belongs to the system, not just to one of them.

Potential energy depends upon reference level

Potential energy is a property of system of interacting objects

FIG. 5-12 The potential energy of an object depends upon the reference level from which its height h is measured.

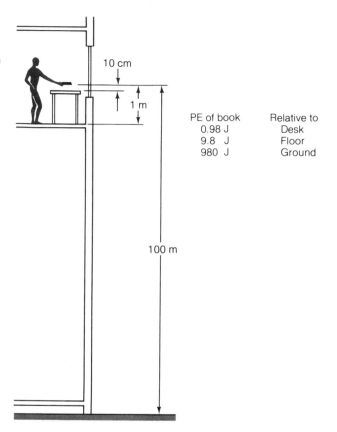

PE of book	Relative to
0.98 J	Desk
9.8 J	Floor
980 J	Ground

Example Compare the potential energy of a 1200-kg car at the top of a hill 30 m high (Fig. 5–13) with its kinetic energy when moving at 100 km/h.

Solution The car's potential energy relative to the bottom of the hill is

$$PE = mgh = (1200 \text{ kg})(9.8 \text{ m/s}^2)(30 \text{ m}) = 3.53 \times 10^5 \text{ J}$$

In an earlier example we found that the kinetic energy of the car at 100 km/h is 4.64 $\times$ 10⁵ J, which is greater than its potential energy at the top of the hill. This means that a crash at 100 km/h (62 mi/h) into a stationary obstacle will yield more work—that is, do more damage—than dropping the car 30 m (98 ft).

FIG. 5-13

1200 kg

$PE = 3.53 \times 10^5$ J

$h = 30$ m

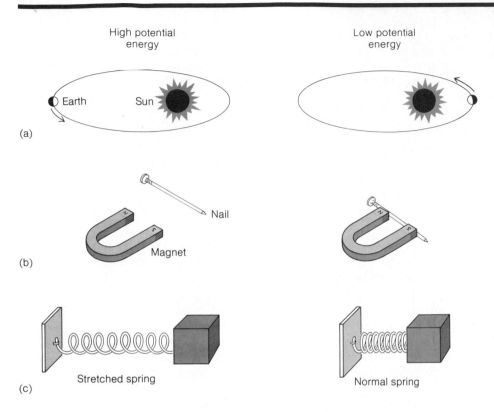

High potential energy   Low potential energy

Earth Sun

(a)

Nail

Magnet

(b)

Stretched spring   Normal spring

(c)

FIG. 5–14 Three examples of potential energy. In each case the potential energy is a property of the entire system.

We have spoken of one type of potential energy only—namely, the potential energy possessed by an object raised above some reference level in the earth's gravitational field. The concept of potential energy is a much more general one, however, for it refers to the energy something has because of its position regardless of the nature of the force acting on it. The earth itself, for instance, has potential energy with respect to the sun, since if its orbital motion were to stop it would fall toward the sun (Fig. 5–14). An iron nail has potential energy with respect to a nearby magnet, since it will fly to the magnet if released. An object at the end of a stretched spring has potential energy with respect to its position when the spring has its normal extension, since if it is let go the object will move as the spring contracts. In each of these cases the object in question is able to do work starting from its original position.

Potential energy is a general concept

5–7 REST ENERGY

Every body of matter possesses a certain inherent amount of energy called *rest energy* even if it is not moving (so that KE = 0) and is not being acted upon by a force (so that PE = 0). A body whose mass when it is at rest is m_0 has a rest energy E_0 of

$$E_0 = m_0 c^2 \qquad\qquad\qquad \textit{Rest energy} \quad (5\text{–}13)$$

All objects have energy by virtue of their mass alone

where c is the speed of light, 3×10^8 m/s. (The mass of a moving body increases with its speed, so that m is not always equal to m_0; the difference is only significant at

Rest energies are very large

speeds near that of light, however.) Equation (5–13) was discovered by Albert Einstein in the early years of this century and has been verified by many experiments since then.

Why are we not aware of rest energy as we are aware of kinetic and potential energies? After all, a 1-kg object—such as this book—contains the rest energy

$$m_0 c^2 = (1 \text{ kg})(3 \times 10^8 \text{ m/s})^2 = 9 \times 10^{16} \text{ J}$$

which is enough energy to send a payload of perhaps a million tons to the moon. How can so much energy be bottled up without revealing itself in some manner?

Rest energy is liberated in many familiar processes

In fact, all of us *are* familiar with processes in which rest energy is liberated, only we do not usually think of them in these terms. In every chemical reaction in which energy is given off, for instance a fire, a certain amount of matter is being converted into energy in the form of heat, which is molecular kinetic energy. But the amount of matter that vanishes in such reactions is so small that it escapes our notice. When 1 kg of dynamite explodes, 6×10^{-11} kg of matter is transformed into energy. The lost mass is so tiny—a fraction of the total mass involved—as to be impossible to detect directly, but it results in the giving off of

$$m_0 c^2 = (6 \times 10^{-11} \text{ kg})(3 \times 10^8 \text{ m/s})^2 = 5.4 \times 10^6 \text{ J}$$

of energy, which is very hard to avoid detecting.

The radiation of the sun and stars is powered by the conversion of matter into energy in reactions that involve atomic nuclei, as discussed in Chapter 30. The immense destructive force of nuclear weapons has the same source.

5–8   CONSERVATION OF ENERGY

Conservation principles

A *conservation principle* states that no matter what changes a system of some kind that is isolated from the rest of the universe undergoes, a certain quantity keeps the same value it had originally. For example, the law of conservation of mass in chemistry holds that the total mass of the products of a chemical reaction is the same as the total mass of the original substances. The increase in mass of a piece of iron when it rusts therefore indicates that the iron has combined with some other material, rather than having decomposed, as the early chemists believed. In fact, the gas oxygen was discovered in the course of seeking this other material.

Usefulness of conservation principles

Given one or more conservation principles that apply to a given system, we can immediately determine which kinds of events can take place in the system and which cannot. Thus when iron rusts, the gain in mass means that it has combined chemically with something else. In physics it is often possible to draw some conclusions about the behavior of the particles that make up a system without a detailed study, basing our analysis simply upon the conservation of some particular quantities.

The first conservation principle we shall study is that of *conservation of energy:*

Conservation of energy

The total amount of energy in a system isolated from the rest of the universe always remains constant, although energy transformations from one form to another, including rest energy, may occur within the system.

This principle is perhaps the most fundamental generalization in all of science, and no violation of it has ever been found.

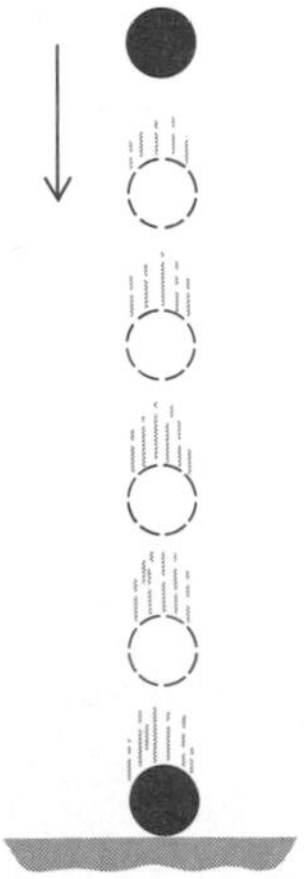

Height	1-kg ball	PE = mgh	KE = ½mv²	PE + KE
50 M	●	490 J	0 J	490 J
40	○	392	98	490
30	○	294	196	490
20	○	196	294	490
10	○	98	392	490
0	●	0	490	490

FIG. 5–15 The total energy of a falling ball remains constant as its potential energy is transformed into kinetic energy.

A falling ball provides a simple example of conservation of energy. As it falls, its initial potential energy is converted into kinetic energy, so that the total energy of the stone remains the same. The potential energy of a 1-kg ball 50 m above the ground is $mgh = 490$ J, and its total mechanical energy is 490 J until it interacts with the ground and transfers energy to it (Fig. 5–15).

Another example is the motion of a planet about the sun. Planetary orbits are elliptical, so that at different points in its orbit the planet is at different distances from the sun. When the planet is close to the sun, it has a low potential energy, just as a stone near the ground has a low potential energy; when the planet is far from the sun, it has a high potential energy (Fig. 5–14(a)). Since the sum of the planet's potential energy and kinetic energy must be constant, we conclude that the kinetic energy (and therefore speed) of the planet is at a maximum when it is nearest the sun and at a minimum when it is farthest from the sun. The earth's orbital speed varies between 29 and 30 km/s for this reason.

Energy conservation in planetary motion

Newton's laws of motion enable us—in theory—to solve all mechanical problems, that is, problems that involve forces and moving objects. However, these laws are actually useful only in the simplest cases because in order to apply them we must take into detailed account all the various forces acting on each object at every point in its path, which is usually a difficult and complicated procedure. The great advantage of the principle of conservation of energy is that it often permits us to draw definite conclusions about the relationship between the initial and final states of motion of some object or system of objects without having to investigate exactly what happens in between.

Example A skier is sliding downhill at a constant speed of 8 m/s when he reaches an icy patch on which his skis move with negligible friction (Fig. 5–16). If the icy patch is 10 m high, what is the skier's speed at its bottom?

Solution Because the path of the skier on the icy patch is complicated, to apply $\mathbf{F} = m\mathbf{a}$ here would be very difficult. Using conservation of energy, however, makes the

FIG. 5–16
(Photo: Margaret
Durrance/Photo
Researchers, Inc.)

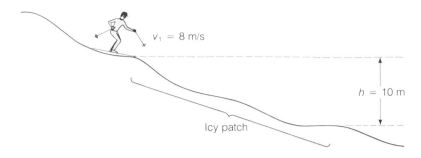

problem quite easy. At the top of the icy patch, the skier's initial kinetic energy is

$$KE_1 = \tfrac{1}{2} mv_1^2$$

where $v_1 = 8$ m/s, and his potential energy relative to the bottom of the patch is

$$PE = mgh$$

where $h = 10$ m. At the bottom the skier's kinetic energy is

$$KE_2 = KE_1 + PE$$

and so

$$\tfrac{1}{2} mv_2^2 = \tfrac{1}{2} mv_1^2 + mgh$$
$$v_2^2 = v_1^2 + 2gh$$
$$v_2 = \sqrt{v_1^2 + 2gh}$$
$$= \sqrt{(8\,\text{m/s})^2 + (2)(9.8\,\text{m/s}^2)(10\,\text{m})} = 16\,\text{m/s}$$

We did not have to know the skier's mass to find his final speed.   ■

As we have seen, it is possible to do work on something and thereby give it kinetic or potential energy which, in turn, is capable of reappearing as work. When we throw a stone upward into the air, the work we do appears first as kinetic energy; as the stone rises, the kinetic energy gradually becomes potential energy; at its highest point, the stone has only potential energy which, as the stone begins to fall, is converted back into kinetic energy; and, when the stone strikes the ground and makes a hole in it, the kinetic energy turns into work. But what happens to the energy used in opposing frictional forces?

To find the answer, all we need do is rub one piece of wood against another. After a short time, it is obvious that the contacting surfaces of the pieces of wood are warmer than before. This is a quite general observation: Work done against frictional forces produces a rise in the temperature of the objects involved. What is happening is that the energy that disappears from the motion of each object as a whole reappears as additional kinetic energy of random motion of the molecules of which the objects are composed. This additional molecular energy is manifested as a rise in temperature, as explained in Chapter 15.

Work done against friction becomes heat

This discussion can be summarized in the following equation:

Work done *on* object = change in object's KE

     + change in object's PE

     + work done *by* object

Another way to express energy conservation

The kinetic energy of the object may either increase or decrease; the same is true of its potential energy. It is the work done *by* the object that may be converted into heat.

Example A 25-kg box is pulled up a ramp 20 m long and 3 m high by a constant force of 120 N (Fig. 5–17). If the box starts from rest and has a speed of 2 m/s at the top, find the force of friction between box and ramp.

Solution According to conservation of energy, the total work W done by the applied force of 120 N must equal the work W_f done against friction (which becomes heat) plus the change ΔKE in the box's kinetic energy and the change ΔPE in its potential energy. Hence

$$W = W_f + \Delta KE + \Delta PE$$
$$W_f = W - \Delta KE - \Delta PE$$

The total work done is

$$W = Fs = (120 \text{ N})(20 \text{ m}) = 2400 \text{ J}$$

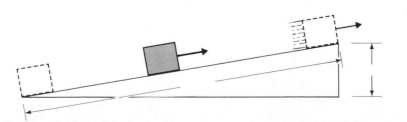

FIG. 5–17

Since the box starts from rest, its change in kinetic energy is equal to its kinetic energy at the top of the ramp, and so

$$\Delta KE = \tfrac{1}{2}mv^2 = \tfrac{1}{2}(25 \text{ kg})(2 \text{ m/s})^2 = 50 \text{ J}$$

The change in the box's potential energy is

$$\Delta PE = mgh = (25 \text{ kg})(9.8 \text{ m/s}^2)(3 \text{ m}) = 735 \text{ J}$$

Hence the work done against friction is

$$W_f = W - \Delta KE - \Delta PE = (2400 - 50 - 735) \text{ J} = 1615 \text{ J}$$

Because $W_f = F_f s$, the frictional force is

$$F_f = \frac{W_f}{s} = \frac{1615 \text{ J}}{20 \text{ m}} = 81 \text{ N}$$

IMPORTANT TERMS

Work is a measure of the change (in a general sense) a force gives rise to when it acts upon something. When an object undergoes a displacement while a force acts on it, the work done by the force is equal to the product of the displacement and the component of the force in the direction of the displacement. In the SI system the unit of work is the **joule** and in the British system it is the **foot-pound.**

The rate at which work is done is called **power.** The unit of power in the metric system is the **watt,** which is equal to 1 J/s, and in the British system it is the **foot-pound per second.** The **horsepower** is a unit of power equal to 550 ft · lb/s, which is 746 watts.

Energy is that which may be converted into work. When something possesses energy, it is capable of performing work or, in a general sense, of accomplishing a change in some aspect of the physical world. The units of energy are those of work.

The three broad categories of energy are **kinetic energy,** which is the energy something possesses by virtue of its motion; **potential energy,** which is the energy something possesses by virtue of its position in a force field; and **rest energy,** which is the energy something possesses by virtue of its mass.

The principle of **conservation of energy** states that the total amount of energy in a system isolated from the rest of the universe always remains constant, although energy transformations from one form to another, including rest energy, may occur within the system.

IMPORTANT FORMULAS

Work: $W = Fs \cos\theta$

Work in lifting object: $W = wh = mgh$

Power: $P = \dfrac{W}{t} = Fv \cos\theta$

Kinetic energy: $KE = \tfrac{1}{2}mv^2$

Gravitational potential energy: $PE = wh = mgh$

Rest energy: $E_0 = m_0c^2$

MULTIPLE CHOICE

1. According to the principle of conservation of energy (with energy interpreted as including rest energy), energy can be
  a. created but not destroyed.
  b. destroyed but not created.

c. both created and destroyed.

d. neither created nor destroyed.

2. A golf ball and a ping-pong ball are dropped in a vacuum chamber. When they have fallen halfway down, they have the same

a. speed.

b. potential energy.

c. kinetic energy.

d. rest energy.

3. In the formula $E = m_0c^2$, the symbol c represents

a. the speed of the body.

b. the speed of sound.

c. the speed of light.

d. the rest energy of 1 kg of matter.

4. Which of the following is not a unit of power?

a. joule-second

b. watt

c. newton-meter per second

d. horsepower

5. To keep a vehicle moving at the speed v requires a force F. The power needed is

a. Fv.

b. $\frac{1}{2}Fv^2$.

c. F/v.

d. F/v^2.

6. A 2-kg book is held 1 m above the floor for 50 s. The work done is

a. 0.

b. 10.2 J.

c. 100 J.

d. 980 J.

7. The work done in lifting 30 kg of bricks to a height of 20 m on a building under construction is

a. 61 J.

b. 600 J.

c. 2940 J.

d. 5880 J.

8. Six thousand foot-pounds of work is used to raise an object to a height of 100 ft. The weight of the object is

a. 1.875 lb.

b. 18.75 lb.

c. 60 lb.

d. 1920 lb.

9. A sedentary person requires about 6 million J of energy per day. This rate of energy consumption is equivalent to about

a. 70 W.

b. 335 W.

c. 600 W.

d. 250,000 W.

10. A 100-lb boy runs up a staircase to a floor 15 ft higher in 5.45 s. His power output is

a. 1/64 hp.

b. 1/2 hp.

c. 1 hp.

d. 275 hp.

11. An 800-kg white horse has a power output of 1 kW. The maximum force it can exert at a speed of 2 m/s is

a. 0.625 N.

b. 1.6 N.

c. 6.125 N.

d. 500 N.

12. A 1-kg mass has a potential energy of 1 J relative to the ground when it is at a height of

a. 0.102 m.

b. 1 m.

c. 9.8 m.

d. 32 m.

13. A 1-N weight has a potential energy of 1 J relative to the ground when it is at a height of

a. 0.102 m.

b. 1 m.

c. 9.8 m.

d. 32 m.

14. A 1-slug mass has a potential energy of 1 ft · lb relative to the ground when it is at a height of

a. 0.031 ft.

b. 1 ft.

c. 9.8 ft.

d. 32 ft.

15. A 1-lb weight has a potential energy of 1 ft · lb relative to the ground when it is at a height of

a. 0.031 ft.

b. 1 ft.

c. 9.8 ft.

d. 32 ft.

16. A total of 4900 J is expended in lifting a 50-kg mass. The mass is raised to a height of

a. 10 m.

b. 98 m.

c. 960 m.

d. 245,000 m.

17. A 16-slug mass is lifted to a height of 10 ft. Its potential energy is

a. 5 ft · lb.

b. 20 ft · lb.

c. 160 ft · lb.

d. 5120 ft · lb.

18. A 1-kg mass has a kinetic energy of 1 J when its speed is

a. 0.45 m/s.

b. 1 m/s.

c. 1.4 m/s.

d. 4.4 m/s.

19. A 1-N weight has a kinetic energy of 1 J when its speed is

a. 0.45 m/s.

b. 1 m/s.

c. 1.4 m/s.

d. 4.4 m/s.

20. A 1-slug mass has a kinetic energy of 1 ft · lb when its speed is

a. 0.25 ft/s.

b. 1 ft/s.

c. 1.4 ft/s.

d. 8 ft/s.

21. A 1-lb weight has a kinetic energy of 1 ft · lb when its speed is

a. 0.25 ft/s.

b. 1 ft/s.

c. 1.4 ft/s.

d. 8 ft/s.

22. Car A has a mass of 1000 kg and a speed of 60 km/h, and car B has a mass of 2000 kg and a speed of 30 km/h. The kinetic energy of car A is

a. half that of car B.

b. equal to that of car B.

   c. twice that of car *B*.

   d. four times that of car *B*.

23. A 1000-kg car whose speed is 80 km/h has a kinetic energy of

   a. 2.52×10^4 J.   b. 2.47×10^5 J.

   c. 2.42×10^6 J.   d. 3.20×10^6 J.

24. A 3200-lb car whose speed is 60 mi/h has a kinetic energy of

   a. 1.8×10^5 ft · lb.

   b. 3.9×10^5 ft · lb.

   c. 7.7×10^5 ft · lb.

   d. 1.4×10^7 ft · lb.

25. Burning a ton of coal liberates 3×10^{10} J of energy. The mass lost by a ton of coal when it burns is

   a. 0.33 μg.   b. 0.67 μg.

   c. 1 μg.   d. 100 kg.

26. The height above the ground of a child on a swing varies from 0.5 m at the lowest point to 2.0 m at the highest point. The maximum speed of the child is

   a. about 5.4 m/s.

   b. about 7.7 m/s.

   c. about 29.4 m/s.

   d. dependent on the child's mass.

27. The 2-kg head of an axe is moving at 60 m/s when it strikes a log. If the blade of the axe penetrates 2 cm into the log, the average force it exerts is

   a. 3 kN.   b. 90 kN.

   c. 72 kN.   d. 180 kN.

EXERCISES

5–1 Work

5–2 Work Done Against Gravity

1. Under what circumstances (if any) is no work done on a moving body even though a net force acts upon it?

2. The sun exerts a force of 4×10^{28} N on the earth, and the earth travels 9.4×10^{11} m in its annual orbit of the sun. How much work is done by the sun on the earth in the course of a year?

3. The work done to lift a 30-kg mass is 4000 J. If the mass is at rest before and after its elevation, how high does it go?

4. A uniform lamppost 30 ft long that weighs 500 lb is lying on the ground. How much work is needed to raise it to a vertical position?

5. (a) A force of 130 N is used to lift a 12-kg mass to a height of 8 m. How much work is done by the force? (b) A force of 130 N is used to push a 12-kg mass on a horizontal, frictionless surface for a distance of 8 m. How much work is done by the force?

6. A 20-kg wooden box is pushed a distance of 15 m on a horizontal stone floor by a force just sufficient to overcome the friction between box and floor. The coefficient of friction is 0.4. (a) What is the required force? (b) How much work does the force do?

7. A 50-kg box is pushed across a horizontal floor by a horizontal force of 180 N. The coefficient of sliding friction is μ = 0.3. (a) Find the work done in pushing the box 20 m. (b) How much work went into overcoming friction and how much into accelerating the box?

8. A woman drinks a bottle of beer and proposes to work off its 460-kJ energy content by exercising with a 20-kg barbell. If each lift of the barbell from chest height to over her head is through 60 cm and the efficiency of her body is 10% under these circumstances, how many times must she lift the barbell?

9. A horse is towing a barge with a rope that makes an angle of 20° with the canal. If the horse exerts a force of 400 N, how much work does it do in moving the barge 1 km?

10. A horizontal force of 5 N is used to push a box up a ramp 5 m long that is at an angle of 15° above the horizontal. How much work is done?

11. Two people set out to climb to the summit of a 3000-m mountain starting from sea level. One of them sets out along a slope that averages 30° above the horizontal, the other along a slope that averages 40° above the horizontal. Each person has a mass of 80 kg and carries a 10-kg knapsack. Find the work done by each of them.

12. A man pulls a 150-lb crate for 80 ft across a level floor using a rope that is 30° above the horizontal. If the coefficient of friction between crate and floor is 0.30 and the man uses just enough force to move the crate without accelerating it, how much work does he perform? (Assume the rope is attached to the center of gravity of the crate.)

5–3 Power

13. Object *A* is acted upon by a constant force **F** and object *B*, which is identical to *A*, is acted upon by a force that is initially **F** but delivers a constant power. Which object travels the greatest distance in a given period of time?

14. A 15-kW motor is used to hoist an 800-kg bucket of concrete to the twentieth floor of a building under construction, a height of 90 m. If no power is lost, how much time is required?

15. A weightlifter raises a 150-kg barbell from the floor to a height of 2.2 m in 0.8 s. What is his average power output during the lift?

16. In 1932 five members of the Polish Olympic ski team climbed from the 5th to the 102nd floor of the Empire State building, a distance of approximately 350 m, in 21 min. If one of these men had a mass of 70 kg, how much power did he develop during the ascent?

17. Find the amount of work in foot-pounds that a $\frac{1}{3}$-hp electric drill can do in 1 min.

18. What is the minimum time needed by a 120-lb woman to climb a mountain 6000 ft high if her average power output is 0.05 hp?

19. The anchor windlass of a boat must be able to raise a total load (anchor plus chain) of 800 kg at a speed of 0.5 m/s. What should the minimum rating of the motor be, in kilowatts?

20. A car needs 13 kW to move along a level road at 60 km/h. What is the sum of the air resistance and rolling friction on the car at that speed?

21. Each of the four engines of a DC-8 airplane develops 7500 hp when the cruising speed is 240 m/s. How much thrust does each engine produce under these circumstances?

22. A motorboat requires 160 hp to move at the constant speed of 8 m/s. How much resistive force does the water exert on it at that speed?

23. In 1970 approximately 2×10^{20} J of work were performed throughout the world by inanimate devices of all kinds, perhaps 15 times as much as the muscle power provided in that year. The work was used for heat, light, transport, manufacturing, and so forth. About 98% of the work was ultimately derived from the fossil fuels coal, natural gas, and oil, the rest mainly from water power with a small (0.25% of the total) contribution from nuclear power stations. (a) Express the power consumption in 1970 in watts. (b) Find the average power consumption per person in watts and in horsepower on the assumption that the world's population in 1970 was 3.5×10^9.

24. A 75-kg man carrying a 10-kg pack climbs a mountain 2800 m high in 10 h. (a) What is his average power output? (b) The efficiency with which his body utilizes food energy for climbing is 15%. How many joules of food energy were needed for the climb?

25. A crane whose motor has a power input of 5 kW raises a 1200-kg beam through a height of 30 m in 90 s. Find the efficiency.

26. A trash compactor with a $\frac{1}{4}$-hp motor can apply a crush-

ing force of 8000 N. How fast does its ram move, assuming 80% efficiency?

27. The bilge pump of a boat is able to raise 200 L of water per minute through a height of 1.2 m. If the pump is 60% efficient, how much power must be supplied to the pump? The mass of 1 L of water is 1 kg.

28. In an effort to lose weight, a person runs 5 km per day at an average speed of 4 m/s. While running, the person's body processes consume energy at a rate of 1.0 kW. Fat has an energy content of about 40 kJ/g. How many grams of fat are metabolized during each run?

29. A boy pulls a sled with a force of 10 lb for 100 ft. The rope attached to the sled is at an angle of 30° above the horizontal. (a) How much work is done? (b) If the boy moves the sled 100 ft in 45 s, find his power output in horsepower.

30. An escalator carries passengers from one floor of a building to another 35 ft higher. It is designed to have a capacity of 200 passengers per minute, assuming an average weight per passenger of 150 lb. Find the required horsepower of the motor if half the work it does is dissipated as heat.

31. A person's metabolic processes can usually operate at a power of 6 W/kg of body mass for several hours at a time. If a 60-kg woman carrying a 12-kg pack is walking uphill with an energy-conversion efficiency of 20%, at what rate, in meters per hour, does she ascend?

32. A 60-kg woman is riding a 10-kg bicycle at a constant speed of 20 km/h. Friction and air resistance total 25 N. Find her power output (a) when the road is horizontal and (b) when it goes up a 5° hill.

33. The power required to propel a 1200-kg car at 40 km/h on a level road is 30 kW. (a) How much resistance must the car overcome at this speed? (b) How much power is needed for the car to ascend an 8° hill at the same speed?

34. An escalator 14 m long is carrying a 70-kg person from one floor to another floor 8 m higher. The linear speed of the escalator is 1.0 m/s. (a) How much work does the escalator do in carrying the person to the top? What is its power output while doing so? (b) The person is walking up the escalator at 0.8 m/s. Answer the same questions for this situation. (c) The person is walking down the escalator at 0.8 m/s. Answer the same questions for this situation.

5–5 Kinetic Energy

35. Find the kinetic energy of a 20-g bullet moving at 500 m/s.

36. Find the speed of a 2-g insect whose kinetic energy is 0.01 J.

37. Find the kinetic energy of a 256-lb ostrich running at 50 ft/s.

38. Find the kinetic energy of a 3200-lb car moving at 40 mi/h.

39. Find the kinetic energy of a 70-kg runner who covers 400 m in 45 s at constant speed.

40. Is the work needed to bring a car's speed from 10 to 20 km/h less than, equal to, or more than the work needed to bring its speed from 90 to 100 km/h? If the amounts of work are different, what is the ratio between them?

41. A 2-kg ball is at rest when a horizontal force of 5 N is applied. In the absence of friction, what is the speed of the ball after it has gone 10 m?

42. A 1-kg trout is hooked by a person fishing and swims off at 2.5 m/s. The person stops the trout in 50 cm by braking the reel. How much tension is exerted on the line?

43. A certain 800-kg car has a motor whose power output is 30 kW. If the car is carrying two passengers whose total mass is 150 kg, how long will it need to accelerate from 70 to 110 km/h?

44. A 70-kg sprinter pushes on the starting blocks for 0.2 s and leaves them with a speed of 5 m/s. He then continues to accelerate for a further 5 s until his speed is 12 m/s. Find the reaction force exerted by the starting blocks on the sprinter and his average power output in each of the accelerations.

45. What is the power output of a 1200-kg car that can go from 25 km/h to 100 km/h in 12 s? Neglect air resistance and rolling friction.

46. Steam enters a 10-MW-output turbine at 800 m/s and emerges at 100 m/s. Assuming 90% mechanical efficiency, what mass of steam passes through the turbine per second?

5–6 Potential Energy

47. A 60-kg woman stands on a diving board 2 m above the surface of a lake at a place where it is 4 m deep. What is her potential energy with respect to the water surface and with respect to the lake bottom?

48. A 3-kg stone is dropped from a height of 100 m. Find its kinetic and potential energies when it is 50 m from the ground.

49. A bullet moving at 500 m/s has a kinetic energy of 2500 J and a potential energy of 0.5 J at a certain moment. What is the bullet's mass and how high above the ground is it at that moment?

50. A 120-lb woman jumps off a wall 4 ft high and lands with her knees stiff. If her body is compressed by 1 in. on impact, find the average force exerted on her by the ground. What would the force be if she had bent her knees on impact, so that she came to a stop in 10 in.?

51. In the operation of a certain pile driver, a 500-kg hammer is dropped from a height of 5 m above the head of a pile. If the pile is driven 20 cm into the ground with each impact of the hammer, what is the average force on the pile when struck?

52. A certain frog's hind legs produce a force of 2.5 times its weight through a vertical distance of 9 cm, at which point the frog becomes airborne. (a) What is the frog's speed at takeoff? (b) What height above the ground does the frog reach? Assume a vertical jump.

53. A ball is dropped from a height of 1 m and loses 10% of its kinetic energy when it bounces on the ground. To what height does it rise?

54. A method for storing large amounts of energy involves pumping water from a low reservoir to a high one. The pumping is done by turbines powered by electric motors; when the water is allowed to descend through the same turbines, the motors act as generators to supply electricity during times of high demand. A system of this kind in Wales has a difference in water level of 440 m and can provide 1320 MW. If the efficiency with which the energy of the falling water is converted into electricity is 85%, how much water must pass through the turbines per minute? Neglect changes in the water levels of the reservoirs.

55. A waterfall is 30 m high and 10^4 kg of water flows over it per second. (a) How much power does this flow represent? (b) If all this power could be converted to electricity, how many 100-W light bulbs could be supplied?

5–7 Rest Energy

56. The potential energy of a golf ball in a hole is negative relative to the ground. Under what circumstances (if any) is its kinetic energy negative? Its rest energy?

57. A 1-kg ball is thrown into the air. When it is 10 m above the ground, the ball's speed is 3 m/s. The ball then has kinetic energy, potential energy relative to the ground, and rest energy. In which form is the largest part of the ball's total energy? The smallest part?

58. The combustion of 1 kg of gasoline liberates 47.3 MJ of energy. How much mass is lost when 1 kg of gasoline is burned in the engine of a car? Would you expect to be able to measure the change in mass?

59. The lightest particle in an atom is an electron, whose

rest mass is 9.1×10^{-31} kg. Find the energy equivalent of this mass.

60. The sun's mass is 2×10^{30} kg and it is currently radiating energy at a rate of about 4×10^{26} W as its hydrogen is converted into helium. If the sun has been radiating energy at this rate during the 4.5 billion years the earth has been in existence, what fraction of its original mass has been lost? What does this suggest about the possibility that the solar radiation rate has indeed been approximately constant during most of this period of time, as geological evidence indicates?

61. The source of the sun's energy (and therefore, directly or indirectly, of nearly all energy available on earth) is the conversion of hydrogen to helium. As described later in the book, the nuclei of four hydrogen atoms, each of mass 1.673×10^{-27} kg, join together in a series of separate reactions to yield a helium nucleus of mass 6.646×10^{-27} kg. How much energy is liberated each time a helium nucleus is formed? How many helium nuclei are formed to produce the 10^7 J a moderately active person requires per day?

5–8   Conservation of Energy

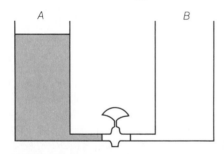

62. When the valve connecting the two tanks shown above is opened, water from A flows into B until their levels are the same. What percentage of the original PE of the water in tank A is changed into KE and eventually into heat in this process?

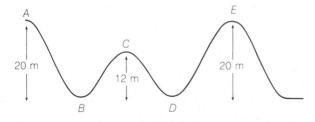

63. The car of a roller coaster starts from rest at A in the figure. Assuming no friction or air resistance, find the car's speed at B, C, D, and E.

64. A woman skis down a slope 300 ft high starting from rest. Her speed at the foot of the slope is 60 ft/s. What percentage of her initial potential energy was dissipated?

65. A force of 50 lb is used to lift a 40-lb object to a height of 20 ft. There is no friction present. (a) How much work is done by the force? (b) What is the change in the potential energy of the object? (c) What is the change in the kinetic energy of the object?

66. At her highest point, a 40-kg girl on a swing is 2 m from the ground, while at her lowest point she is 0.8 m from the ground. What is her maximum speed? On another swing a 50-kg boy undergoes exactly the same motion. What is his maximum speed?

67. Find the height of the bar a pole-vaulter can clear on the basis of the following assumptions: His running speed is 8 m/s; his center of gravity is initially 1.1 m above the ground and he pulls himself upward along the pole 0.6 m as the pole swings into a vertical position; and all his initial KE is converted into work done to raise his CG sufficiently to clear the bar.

68. A man uses a rope and system of pulleys to lift an 80-kg object to a height of 2 m. He exerts a force of 220 N on the rope and pulls a total of 8 m of rope through the pulleys in the course of raising the object, which is at rest afterward. (a) How much work does the man do? (b) What is the change in the object's PE? (c) If the answers to (a) and (b) are different, explain.

69. (a) A force of 8 N is used to push a 0.5-kg ball over a horizontal, frictionless table a distance of 3 m. If the ball starts from rest, what is its final kinetic energy? (b) The same force is used to lift the same ball a height of 3 m. If the ball starts from rest, what is its final kinetic energy?

70. A force of 200 N is used to lift a 15-kg object, initially at rest, to a height of 8 m. There is no friction present. What is the speed of the object at this height?

71. An 800-kg car coasts down a hill 40 m high with its engine off and its driver's foot pressing on the brake pedal. At the top of the hill the car's speed is 6 m/s and at the bottom it is 20 m/s. How much energy was converted to heat on the way down?

72. A 20-kg crate is dropped from a height of 15 m and reaches the ground in 1.9 s. If the force of air resistance on the crate is constant, find its magnitude. What percentage of the original potential energy of the crate is lost?

73. The bob of a pendulum 5 ft long is pulled aside so that the string is at an angle of 40° from the vertical. When the bob is released, with what speed will it pass through the bottom of its path?

74. A brick slides down a wooden plank 2 m long tilted so that one end is at a height of 1 m. The brick's speed at the bottom is 2.5 m/s. The plank is then sanded smooth and waxed so that the coefficient of friction is half what it was before, and the brick is slid down it again. What is the new speed of the brick at the bottom? Use energy considerations to find the answer.

ANSWERS TO MULTIPLE CHOICE

1. d	**7.** d	**13.** b	**19.** d	**25.** a
2. a	**8.** c	**14.** a	**20.** c	**26.** a
3. c	**9.** a	**15.** b	**21.** d	**27.** d
4. a	**10.** b	**16.** a	**22.** c	
5. a	**11.** d	**17.** d	**23.** b	
6. a	**12.** a	**18.** c	**24.** b	

6

MOMENTUM

So complex is the physical world that many different quantities turn out to be useful in describing its various aspects. We have already been introduced to length, time, mass, force, torque, work, and energy, and more are to come. There is nothing sacred about any of these quantities—it is entirely possible to do without one or another of them, but only at the cost of making physics a good deal more complicated than it already is. The idea behind the definition of each of the various physical quantities is to single out something that unifies a wide range of observations, so that it is then possible to boil down to a brief, clear statement a large number of separate discoveries about nature. In this chapter we shall learn how the concepts of linear momentum and impulse supplement those of work and energy to provide a convenient framework for analyzing the behavior of moving bodies.

6–1 LINEAR MOMENTUM

Linear momentum is a vector quantity

We all know that a baseball struck squarely by a bat is harder to stop than the same baseball thrown gently, and that the heavy iron ball used for the shotput is harder to

CHAPTER OBJECTIVES

Completing this chapter should enable you to:

1. Find the momentum of a moving object.

2. Find the impulse given to an object by a force that acts on it for a certain time.

3. Relate the impulse given to an object to the resulting change in its momentum.

4. Use the principle of conservation of momentum to analyze the motion of objects that push each other apart, as in the firing of a gun.

5. Explain how a rocket functions and why it needs nothing to push against.

6. Distinguish between elastic and inelastic collisions.

7. Analyze a completely inelastic collision to find the velocity of the resulting composite object and how much kinetic energy is lost.

8. Describe what happens when a moving object collides elastically with a stationary one whose mass is less than, equal to, or greater than the mass of the moving object.

9. State the condition for maximum energy transfer when a moving object collides head-on with a stationary one.

10. Relate the relative velocity with which two objects approach each other to the relative velocity with which they move apart after an elastic collision.

11. Do the same when the collision is inelastic by using the coefficient of restitution of the objects.

12. Find the coefficient of restitution for an object bouncing off a horizontal surface after having been dropped.

stop than a baseball whose velocity is the same (Fig. 6–1). These observations suggest that a measure of the tendency of a body to continue in motion at constant velocity is the product $m\mathbf{v}$ of its mass m and velocity $\mathbf{v}$.

The quantity $m\mathbf{v}$ is called the *linear momentum* of a moving body:

$$\text{Linear momentum} = m\mathbf{v} \qquad\qquad (6\text{–}1)$$

The symbol $\mathbf{p}$ is sometimes used to represent linear momentum. Linear momentum is a vector quantity whose direction is the direction of $\mathbf{v}$. The kinetic energy of a moving body, which also depends upon its mass and velocity since $\text{KE} = \frac{1}{2}mv^2$, is a scalar quantity with magnitude only.

Because $m\mathbf{v}$ describes the tendency of a moving body to pursue a straight path at constant speed, it is referred to as *linear* momentum. A different quantity, *angular momentum,* describes the tendency of a spinning body such as a top to continue to spin. When there is no question as to which is meant, linear momentum is usually referred to simply as momentum.

FIG. 6-1 The linear momentum *m***v** of a moving body is a measure of its tendency to continue in motion at constant velocity.

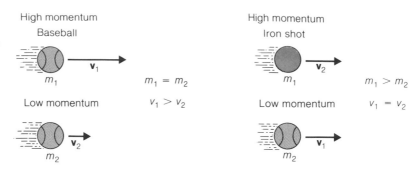

Momentum and kinetic energy refer to different properties of a moving object. An object's momentum determines its tendency to continue moving with constant speed and direction; its kinetic energy determines the amount of work it can do while being brought to a stop. (The kinetic energy also tells us how much work had to be done to bring the object from rest to the speed it has.) In the early days of physics, even after the time of Newton, the distinction between momentum and kinetic energy was as troublesome to scientists as it often is to students today.

6-2   IMPULSE

An impulse produces a change in momentum

To set something in motion from rest, a force must be applied for a period of time. We might expect that the greater force and the longer the time, the more momentum the object will have. This expectation is correct, and the product $\mathbf{F}\,\Delta t$ of a constant force $\mathbf{F}$ and the time interval Δt during which it acts is accordingly given the status of a physical quantity in its own right. This quantity is called *impulse:*

$$\text{Impulse} = \mathbf{F}\,\Delta t \tag{6-2}$$

Impulse, like momentum, is a vector quantity. Let us see how the momentum of an object is affected when it receives a certain impulse.

The second law of motion states that the force $\mathbf{F}$ applied to an object of constant mass m that undergoes the acceleration $\mathbf{a}$ is given by

$$\mathbf{F} = m\mathbf{a} \tag{6-3}$$

When a force $\mathbf{F}$ is applied at the time $t_1 = 0$ to an object whose initial velocity is $\mathbf{v}$, at the later time $t_2 = \Delta t$ its velocity will have changed to $\mathbf{v} + \Delta\mathbf{v}$ (Fig. 6-2). The object's acceleration in this time interval is

FIG. 6-2 (a) At the time $t_1 = 0$, a force $\mathbf{F}$ is applied to a body whose velocity is $\mathbf{v}$. (b) At the later time $t_2 = \Delta t$, the velocity of the body is $\mathbf{v} = \mathbf{v} + \Delta\mathbf{v}$.

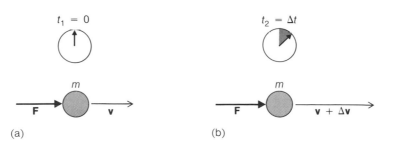

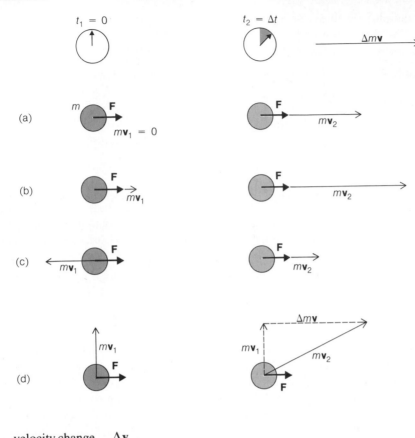

FIG. 6–3 Applying a constant force **F** to a mass m for a time Δt changes its momentum by $\Delta m\mathbf{v} = \mathbf{F}\,\Delta t$. At (a) the mass is initially at rest, at (b) its initial momentum is in the same direction as **F**, at (c) its initial momentum is in the opposite direction to **F**, and at (d) its initial momentum is perpendicular to **F**. Since momentum is a vector quantity, the momentum change $\Delta m\mathbf{v}$ must be added to the initial momentum $m\mathbf{v}$ by the process of vector addition.

$$\mathbf{a} = \frac{\text{velocity change}}{\text{time interval}} = \frac{\Delta \mathbf{v}}{\Delta t}$$

and so Eq. (6–3) becomes

$$\mathbf{F} = m\mathbf{a} = m\,\frac{\Delta \mathbf{v}}{\Delta t}$$

which we can rewrite as

$$\mathbf{F}\,\Delta t = m\,\Delta \mathbf{v}$$

Evidently the impulse provided by the force equals the momentum change of the object:

$$\mathbf{F}\,\Delta t = \Delta\,(m\mathbf{v}) \tag{6–4}$$

Impulse = momentum change

The impulse given to a body equals its momentum change

Figure 6–3 shows the effect of applying the constant force **F** for the time Δt to several objects with different momenta $m\mathbf{v}_1$. In each case, the final momentum $m\mathbf{v}_2$ is obtained by finding the vector sum $m\mathbf{v}_2 = m\mathbf{v}_1 + \mathbf{F}\,\Delta t$.

In the SI, the unit of impulse is the newton-second (N · s), and the unit of momentum is the (kg · m)/s; they are actually the same, of course, but it is often convenient to distinguish between them in this way. (The corresponding British units are the lb · s and (slug · ft)/s.)

Units of impulse and momentum

Example The head of a golf club is in contact with a 46-g golf ball for 0.50 ms (1 ms = 1 millisecond = 10^{-3} s), and as a result the ball flies off at 70 m/s. Find the average force that was acting on the ball during the impact.

Solution The ball starts from rest, hence its momentum change is

$$\Delta mv = (0.046 \text{ kg})(70 \text{ m/s}) = 3.22 \text{ kg} \cdot \text{m/s}$$

From Eq. (6–4) we therefore have

$$F = \frac{\Delta mv}{\Delta t} = \frac{3.22 \text{ kg} \cdot \text{m/s}}{5.0 \times 10^{-4} \text{ s}} = 6.44 \times 10^3 \text{ N}$$

whose British equivalent is 1450 lb. A force of the same magnitude but acting in the opposite direction (the *recoil force*) acts on the club's head during the impact, in accordance with the third law of motion. No golf club could withstand such a static load, but the impact is so brief that its only effect on the shaft is to temporarily bend it by a few cm. ∎

6–3 CONSERVATION OF MOMENTUM

Energy and work are scalar quantities; momentum and impulse are vector quantities

Energy and work are scalar quantities, having magnitude only. Despite the fundamental and all-inclusive character of the law of conservation of energy, it cannot by itself provide complete solutions to most problems that involve interacting bodies. A simple example is the firing of a rifle: Conservation of energy requires that the kinetic energies of the bullet and the recoiling rifle, plus the heat and sound energy that are liberated, must equal the chemical energy of the detonated explosive, but this does not tell us how the total energy is divided among the rifle, the bullet, and the atmosphere. Indeed, because energy is a scalar quantity, its conservation does not even imply that the bullet and rifle must move in opposite directions. To complete the solution of those problems in dynamics in which we lack a detailed knowledge of the active forces, an additional principle of a vector nature is required.

Let us consider a system of two or more particles instead of a single particle. If no forces from outside the system act upon its particles, the total linear momentum of the system, which is the sum

$$M\mathbf{V} = m_1\mathbf{v}_1 + m_2\mathbf{v}_2 + m_3\mathbf{v}_3 + \cdots \tag{6–5}$$

of the individual momenta of its particles, cannot change. With no force there is no impulse, hence no change in momentum. However, the *distribution* of the total momentum $M\mathbf{V}$ among the various particles in the system may change without $M\mathbf{V}$ changing in the absence of an external force.

Exchanging momenta among the members of a system does not change the total momentum of the system

Two particles A and B might collide and thereby exert forces upon each other. At every instant during their interaction these forces, $\mathbf{F}_{AB}$ acting on B and $\mathbf{F}_{BA}$ acting on A, obey Newton's third law of motion,

$$\mathbf{F}_{AB} = -\mathbf{F}_{BA}$$

This is shown in Fig. 6–4. Hence the impulses exchanged must be equal and opposite,

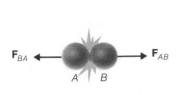

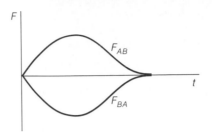

FIG. 6–4 Particles A and B collide and exert the forces F_{AB} and F_{BA} on each other. As shown in the graph, these forces are equal in magnitude and opposite in direction at all times. Their impulses are also equal and opposite, so the total momentum of A and B is left unchanged by the collision although it may be distributed differently between them.

$$\mathbf{F}_{AB}\,\Delta t = -\mathbf{F}_{BA}\,\Delta t$$

and the *total* momentum of the system of A and B together is the same after the collision as it was before.

Since we almost always can include the sources of all forces involved in a particular process within what we choose to be our "system," in such cases we have the condition that, no matter what interactions take place within the system, its total momentum never changes. Thus we have the principle of *conservation of linear momentum:*

Conservation of linear momentum

> **When the vector sum of the external forces acting upon a system of particles equals zero, the total linear momentum of the system remains constant no matter how the particles interact with one another.**

The total of the kinetic energies of the particles need not be the same before and after, however. The KE involved in an explosion is zero to start with, since it all comes from the chemical energy stored in the explosive material, but it is not zero afterward. In a collision, some or even all of the KE of the objects that collide may disappear into heat and sound energy. In both cases the total momentum does not change.

Let us consider a specific example easily treated with the help of conservation of momentum. Suppose we have an isolated particle of mass m, initially at rest, that suddenly explodes into two particles of masses m_1 and m_2, which fly apart (Fig. 6–5). The forces acting on the original particle that caused it to break up were internal ones, and no external force was present. Since m has the initial momentum of zero, the final momentum of m_1 and m_2, when added together, must also be zero. Hence

Analyzing an explosion

$$m\mathbf{v} = 0 = m_1\mathbf{v}_1 + m_2\mathbf{v}_2$$

and

$$m_1\mathbf{v}_1 = -m_2\mathbf{v}_2 \qquad \textit{Momentum conservation in explosion} \quad (6\text{–}6)$$

where $\mathbf{v}_1$ and $\mathbf{v}_2$ are the final velocities of the two fragments. We note immediately that these velocities must be in opposite directions along the same line. This problem could *not* be solved starting from $\mathbf{F} = m\mathbf{a}$, since we do not know exactly what the forces were that acted during the explosion.

In the British system, where the weight w of an object rather than its mass $m = w/g$ is usually given, we have $w_1\,\mathbf{v}_1/g = w_2\,\mathbf{v}_2/g$ and

$$w_1\,\mathbf{v}_1 = -\,w_2\mathbf{v}_2 \qquad\qquad (6\text{–}7)$$

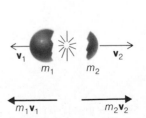

FIG. 6–5 The total momentum of a system of particles remains constant if no external forces act on the system.

FIG. 6-6 The forward momentum of the thrown camera is equal in magnitude to the backward momentum of the astronaut who threw it.

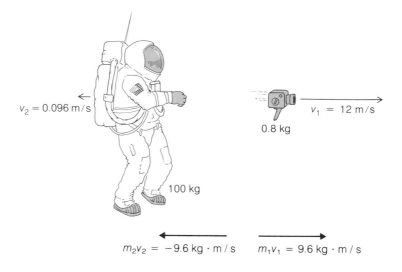

$v_2 = 0.096 \, \text{m/s}$

$v_1 = 12 \, \text{m/s}$

0.8 kg

100 kg

$m_2 v_2 = -9.6 \, \text{kg} \cdot \text{m/s}$ $m_1 v_1 = 9.6 \, \text{kg} \cdot \text{m/s}$

The g's have canceled out, so we need not find the masses of the objects involved if we know their weights.

Example   An astronaut in orbit outside an orbiting space station throws her 0.8-kg camera away in disgust when it jams (Fig. 6-6). If she and her space suit together have a mass of 100 kg and the speed of the camera is 12 m/s, how far away from the space station will she be in 1 h?

Solution   Here $m_1 = 0.8$ kg, $m_2 = 100$ kg, and $v_1 = 12$ m/s. From Eq. (6-6),

$$v_2 = -\frac{m_1}{m_2} v_1 = -\left(\frac{0.8 \, \text{kg}}{100 \, \text{kg}}\right)(12 \, \text{m/s}) = -0.096 \, \text{m/s}$$

After $t = 1$ h $= 3600$ s the astronaut will be

$$s = v_2 t = (0.096 \, \text{m/s})(3600 \, \text{s}) = 346 \, \text{m}$$

away from the space station. The total momentum of the system of astronaut plus camera remains zero until other forces act on them. ■

6-4   ROCKET PROPULSION

Rocket propulsion is based on momentum conservation

The principle underlying rocket flight is conservation of momentum. The total momentum of a rocket on its launching pad is zero. When it is fired, the exhaust gases shoot downward at high speed, and the rocket moves upward to balance the momentum of the gases (Fig. 6-7). Rockets do not operate by "pushing" against their launching pads, the air, or anything else; in fact, they perform best in space, where there is no

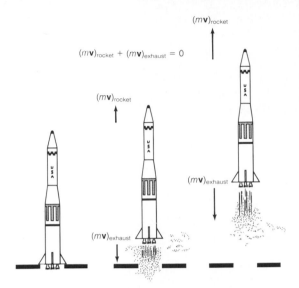

$(m\mathbf{v})_{\text{rocket}} + (m\mathbf{v})_{\text{exhaust}} = 0$

$(m\mathbf{v})_{\text{rocket}}$

$(m\mathbf{v})_{\text{rocket}}$

$(m\mathbf{v})_{\text{exhaust}}$

$(m\mathbf{v})_{\text{exhaust}}$

FIG. 6–7 Conservation of momentum in rocket flight. The downward momentum of the exhaust gases is exactly balanced by the upward momentum of the rocket itself.

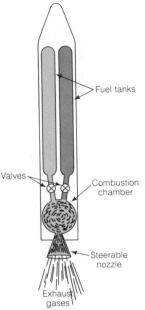

Fuel tanks

Valves

Combustion chamber

Steerable nozzle

Exhaust gases

FIG. 6–8 In a liquid-fueled rocket, the reacting substances are mixed and ignited in a combustion chamber and the resulting exhaust gases escape at high speed through a steerable nozzle. The liberated chemical energy becomes kinetic energy of the rocket and its exhaust. Liquid hydrogen and liquid oxygen are used to power many rockets, and their product is water vapor, H_2O. The combination of 1 kg of hydrogen and 8 kg of oxygen releases 245 MJ.

atmosphere to impede their motion. The energy of the rocket and its exhaust comes from chemical energy stored in the fuel (Fig. 6–8). The total momentum of the system of rocket plus exhaust, which is initially zero, does not remain constant after a launch from the earth because of the impulses provided by air resistance and the earth's gravitational pull.

The upward force exerted on a rocket by the expulsion of exhaust gases is called *thrust*. If a mass Δm of exhaust gas is ejected at the speed v in the time interval Δt, its momentum change is

$$\Delta(mv) = v\,\Delta m$$

From Eq. (6–4) $F\,\Delta t = \Delta(mv)$, so the associated force is

$$F = \frac{\Delta(mv)}{\Delta t} = v\,\frac{\Delta m}{\Delta t} \qquad\qquad \textit{Thrust} \quad (6-8)$$

The thrust of a rocket is the product of the exhaust speed and the rate at which fuel is consumed.

When the spacecraft is launched from the earth's surface, the thrust of its motor must exceed its initial weight $w_0 = m_0 g$ for it to rise from the ground. If the thrust remains constant, the acceleration of the spacecraft increases as its mass diminishes owing to the consumption of fuel. A final acceleration of several times g in magnitude is usual.

Example A certain rocket's motor consumes 100 lb of fuel per second, which it exhausts at 9000 ft/s. If the initial weight of the rocket is 10 tons of which 7 tons is fuel, find its initial and final accelerations.

Solution From Eq. (6–8) the thrust produced is, since $\Delta m = \Delta w / g$,

$$F = v \frac{\Delta m}{\Delta t} = \left(\frac{v}{g}\right)\frac{\Delta w}{\Delta t} = \left(\frac{9000\,\text{ft/s}}{32\,\text{ft/s}^2}\right)(100\,\text{lb/s}) = 28{,}125\,\text{lb}$$

Since the initial weight of the rocket is $w_0 = 10$ tons $= 20{,}000$ lb, the initial force available for its acceleration is $(F - w_0) = 8125$ lb, and so

$$F - w_0 = m_0 a_0 = \left(\frac{w_0}{g}\right) a_0$$

$$a_0 = \left(\frac{g}{w_0}\right)(F - w_0) = \left(\frac{32\,\text{ft/s}^2}{20{,}000\,\text{lb}}\right)(8125\,\text{lb}) = 13\,\text{ft/s}^2$$

The final weight of the rocket is $w = 3$ tons $= 6000$ lb, hence $(F - w) = 22{,}125$ lb is the net force and

$$a = \left(\frac{g}{w}\right)(F - w) = \left(\frac{32\,\text{ft/s}^2}{6000\,\text{lb}}\right)(22{,}125\,\text{lb}) = 118\,\text{ft/s}^2$$

which is $3.7g$. ∎

Factors that determine final speed of a rocket

Rocket propulsion is a gradual rather than an instantaneous process, with the fuel being burned and ejected as exhaust gases at a certain rate instead of in one lump. As a result, part of the momentum of the exhaust is wasted in pushing forward unburned fuel. When this factor is taken into account, the ratio V/v between the final speed of a rocket and the speed v of its exhaust gases turns out to depend on the ratio m/m_0 between its final and initial masses in the manner shown in Fig. 6–9. Thus the rocket considered in the above example would have a final speed of 10,800 ft/s, which is 7390 mi/h. This is very fast by ordinary standards, but is still only about 40% of the minimum speed needed to put a satellite in orbit around the earth (see Section 7–7).

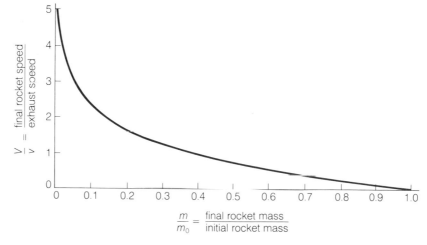

FIG. 6–9 How the ratio between the final speed of a rocket and the speed of its exhaust gases varies with the ratio between its final and initial masses. In order for a rocket's final speed V to exceed its exhaust speed v, at least 63% of its initial mass must be fuel. For V to be twice v, 86% of its initial mass must be fuel.

$$\frac{V}{v} = \frac{\text{final rocket speed}}{\text{exhaust speed}}$$

$$\frac{m}{m_0} = \frac{\text{final rocket mass}}{\text{initial rocket mass}}$$

To attain higher speeds than a single rocket is capable of, two or more rocket stages can be used. The first stage is a large rocket whose payload is another, smaller rocket. When the fuel of the first stage has been used up, its fuel tanks and engine are cast loose. Then the second stage is fired starting from a high initial speed instead of from rest and without the burden of the fuel tanks and engine of the first stage. This process can be repeated a number of times, depending upon the final speed required. The Saturn V launch vehicle that propelled the Apollo 11 spacecraft to the moon in July 1969 employed three stages, as shown in Fig. 6–10. At original ignition the entire assembly was 111 m long and had a mass of 2.9×10^6 kg (3240 tons).

Multistage rockets

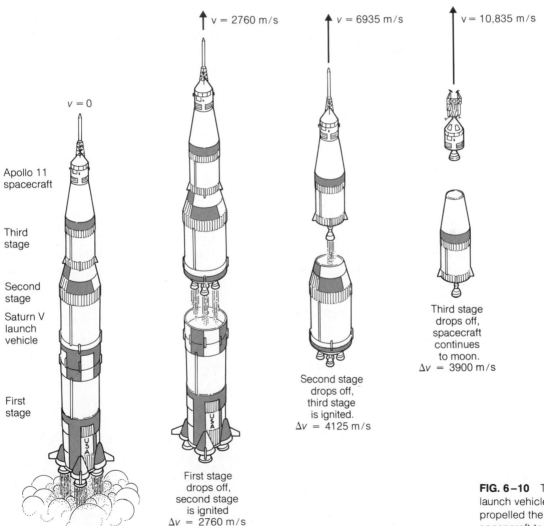

$v = 0$

Apollo 11 spacecraft

Third stage

Second stage

Saturn V launch vehicle

First stage

First stage is ignited.

$v = 2760$ m/s

First stage drops off, second stage is ignited
$\Delta v = 2760$ m/s

$v = 6935$ m/s

Second stage drops off, third stage is ignited.
$\Delta v = 4125$ m/s

$v = 10{,}835$ m/s

Third stage drops off, spacecraft continues to moon.
$\Delta v = 3900$ m/s

FIG. 6–10 The Saturn V launch vehicle that propelled the Apollo 11 spacecraft to the moon for the first manned landing used three rocket stages.

6−5 INELASTIC COLLISIONS

Momentum is redistributed in a collision

The law of conservation of momentum is essential in dealing with collisions between two or more objects. In such cases no external forces act on the participants, and therefore their total momentum before they collide equals their total momentum afterward. *The essential effect of the collision is to redistribute the total momentum of the objects.*

If we only know the masses and initial velocities of the objects involved in a collision, however, momentum conservation by itself does not tell us everything about their subsequent motion. An unknown amount of kinetic energy may be lost to heat, sound, or other forms of energy when the objects interact, and, though the need to conserve momentum does set limits to the result of the collision, it can go no further without additional information. These limiting cases, however, are worth examining.

Kinetic energy is conserved in an elastic collision

At one extreme are completely *elastic collisions* in which kinetic energy is conserved. During the actual collision, to be sure, some kinetic energy becomes elastic potential energy as the objects are deformed by the impact, but all of this energy is returned as the objects move apart.

Kinetic energy is not conserved in an inelastic collision; maximum kinetic energy loss occurs when the bodies stick together

At the other extreme are *completely inelastic collisions* in which the objects stick together permanently upon impact. The kinetic energy loss in a completely inelastic collision is the maximum possible consistent with momentum conservation. Such a collision can be analyzed on the basis of momentum conservation only.

Example A 5-kg lump of clay that is moving at 10 m/s to the left strikes a 6-kg lump of clay moving at 12 m/s to the right. The two lumps stick together after they collide. Find the final speed of the composite object and the kinetic energy dissipated in the collision.

Solution This is an example of a completely inelastic collision. If we call the mass of the final object M and its velocity V, conservation of linear momentum requires that

$$\text{Momentum afterward} = \text{momentum before}$$
$$MV = m_1 v_1 + m_2 v_2$$

Adopting the convention that motion to the right is $+$ and to the left is $-$, we have

$$m_1 = 5\,\text{kg} \qquad m_2 = 6\,\text{kg} \qquad M = m_1 + m_2 = 11\,\text{kg}$$
$$v_1 = -10\,\text{m/s} \quad v_2 = +12\,\text{m/s} \quad V = ?$$

Solving for V yields

$$V = \frac{m_1 v_1 + m_2 v_2}{M} = \frac{(5\,\text{kg})(-10\,\text{m/s}) + (6\,\text{kg})(12\,\text{m/s})}{11\,\text{kg}} = 2\,\text{m/s}$$

Since V is positive, the composite body moves off to the right (Fig. 6−11).

Energy and momentum are independent concepts. The lumps of clay before the collision have the kinetic energies

$$\text{KE}_1 = \tfrac{1}{2} m_1 v_1^2 = \tfrac{1}{2}(5\,\text{kg})(-10\,\text{m/s})^2 = 250\,\text{J}$$
$$\text{KE}_2 = \tfrac{1}{2} m_2 v_2^2 = \tfrac{1}{2}(6\,\text{kg})(12\,\text{m/s})^2 = 432\,\text{J}$$

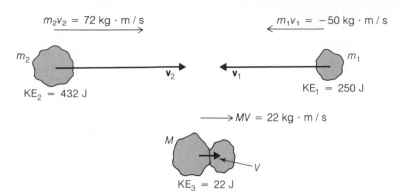

$m_2 v_2 = 72 \, \text{kg} \cdot \text{m} / \text{s}$

$m_1 v_1 = -50 \, \text{kg} \cdot \text{m} / \text{s}$

m_2

$\mathbf{v}_2$

m_1

$\mathbf{v}_1$

$KE_2 = 432 \, \text{J}$

$KE_1 = 250 \, \text{J}$

$MV = 22 \, \text{kg} \cdot \text{m} / \text{s}$

M

V

$KE_3 = 22 \, \text{J}$

FIG. 6–11 In a completely inelastic collision, the colliding bodies stick together. Kinetic energy is not conserved in such an event. Linear momentum is conserved in all collisions.

After the collision the new lump of clay has the kinetic energy

$$KE_3 = \tfrac{1}{2} MV^2 = \tfrac{1}{2} (11 \, \text{kg})(2 \, \text{m/s})^2 = 22 \, \text{J}$$

The total kinetic energy prior to the collision was $432 + 250$ or 682 J, while afterward it is only 22 J. The difference of 660 J was dissipated largely into heat energy in the collision, with some probably being lost to sound energy as well. ■

It is important to keep in mind the directional character of linear momentum. Sometimes a problem will involve bodies that move along the same straight line, as in the preceding example, but in general the bodies may move in two or three dimensions and we must be sure to take this into account by a vector calculation.

Momentum is a vector quantity and must be conserved in each direction

Example   A 60-kg man is sliding east on the frictionless surface of a frozen pond at a velocity of 0.50 m/s. He is struck by a 1.0-kg snowball whose velocity is 20 m/s toward the north. If the snowball sticks to the man, what is his final velocity?

Solution   Here linear momentum must be conserved separately in both the east-west and north-south directions, which we shall call the x- and y-axes respectively. Since

$$m_1 = 60 \, \text{kg} \qquad m_2 = 1.0 \, \text{kg} \qquad M = m_1 + m_2 = 61 \, \text{kg}$$
$$v_{1x} = 0.50 \, \text{m/s} \qquad v_{2x} = 0 \qquad V_x = ?$$
$$v_{1y} = 0 \qquad v_{2y} = 20 \, \text{m/s} \qquad V_y = ?$$

we have (see Fig. 6–12)

Momentum afterward = momentum before

x-direction: $MV_x = m_1 v_{1x} + m_2 v_{2x} = (60 \, \text{kg})(0.5 \, \text{m/s}) = 30 \, \text{kg} \cdot \text{m/s}$

y-direction: $MV_y = m_1 v_{1y} + m_2 v_{2y} = (1.0 \, \text{kg})(20 \, \text{m/s}) = 20 \, \text{kg} \cdot \text{m/s}$

Hence the magnitude MV of the momentum of man + snowball after the collision is

$$MV = \sqrt{(MV_x)^2 + (MV_y)^2} = \sqrt{(30 \, \text{kg} \cdot \text{m/s})^2 + (20 \, \text{kg} \cdot \text{m/s})^2} = 36 \, \text{kg} \cdot \text{m/s}$$

and the corresponding final speed is

$$V = \frac{MV}{M} = \frac{36 \, \text{kg} \cdot \text{m/s}}{61 \, \text{kg}} = 0.59 \, \text{m/s}$$

FIG. 6−12

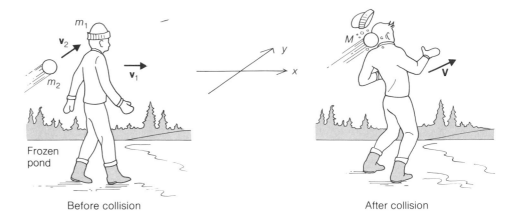

Before collision

After collision

FIG. 6−13

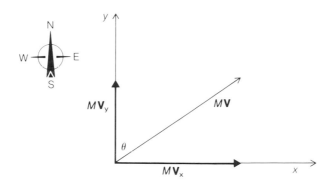

We can specify the direction in which the man + snowball combination moves after the collision in terms of the angle θ between the $+y$-direction (which is north) and MV:

$$\tan \theta = \frac{MV_x}{MV_y} = \frac{30 \, \text{kg} \cdot \text{m/s}}{20 \, \text{kg} \cdot \text{m/s}} = 1.5$$

$$\theta = \tan^{-1} 1.5 = 56°$$

Thus man + snowball move in a direction 56° to the east of north (Fig. 6−13). ∎

6−6   ELASTIC COLLISIONS

In a completely elastic collision, no kinetic energy is lost. Because many collisions are very nearly elastic, it is worth considering them in some detail.

Let us examine what happens when an object of mass m_1 and initial speed v_1 strikes another object of mass m_2 and initial speed v_2; both objects are moving along the same straight line before and after the collision, which is elastic. The speeds of the objects after the collision are respectively v_1' and v_2'. From conservation of energy we have

Kinetic energy before = kinetic energy after

$$\tfrac{1}{2} m_1 v_1^2 + \tfrac{1}{2} m_2 v_2^2 = \tfrac{1}{2} m_1 v_1'^2 + \tfrac{1}{2} m_2 v_2'^2$$
$$m_1(v_1^2 - v_1'^2) = m_2(v_2'^2 - v_2^2)$$

Since $(a + b)(a - b) = a^2 - b^2$ we can rewrite the last equation as

$$m_1(v_1 + v_1')(v_1 - v_1') = m_2(v_2' + v_2)(v_2' - v_2) \qquad (6-9)$$

From conservation of momentum we have

Momentum before = momentum after

$$m_1 v_1 + m_2 v_2 = m_1 v_1' + m_2 v_2'$$
$$m_1(v_1 - v_1') = m_2(v_2' - v_2) \qquad (6-10)$$

Now we divide Eq. (6–9) by Eq. (6–10):

$$\frac{m_1(v_1 + v_1')(v_1 - v_1')}{m_1(v_1 - v_1')} = \frac{m_2(v_2' + v_2)(v_2' - v_2)}{m_2(v_2' - v_2)}$$
$$v_1 + v_1' = v_2' + v_2$$
$$v_1 - v_2 = v_2' - v_1' = -(v_1' - v_2') \qquad (6-11)$$

Equation (6–11) means that the relative velocity $v_1 - v_2$ before the collision is the negative of the relative $v_1' - v_2'$ after the collision: The effect of the collision is to reverse the direction of the relative velocity without changing its magnitude. Thus the relative velocity with which the objects approach each other before the collision is equal to the relative velocity with which they move apart afterward.

Relative velocity of approach equals relative velocity of recession

Example   A ball rolling on a level table strikes head-on another identical ball that is at rest. What is the result of the collision?

Solution   Here $v_2 = 0$, so from Eq. (6–11)

$$v_1 = v_2' - v_1'$$

From conservation of momentum, since the balls have the same mass m,

$$m_1 v_1 + m_2 v_2 = m_1 v_1' + m_2 v_2'$$
$$m v_1 + 0 = m v_1' + m v_2'$$
$$v_1 = v_1' + v_2'$$

Setting equal these two formulas for v_1 gives

$$v_2' - v_1' = v_1' + v_2'$$
$$-2v_1' = 0$$
$$v_1' = 0$$

The first ball has no speed after the collision, which means it comes to a stop. For the final speed of the second ball, which was originally at rest, we can use either of the formulas for v_1 with $v_1' = 0$. Thus

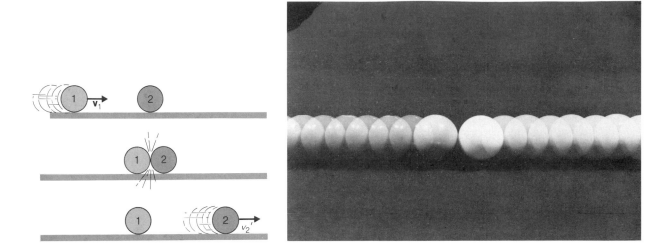

FIG. 6-14 A rolling ball makes a head-on collision with an identical stationary ball; the first ball stops and the second begins moving with the first's initial velocity. (Photo: © Roy King.)

$$v_1 = v_2' - v_1' = v_2' - 0 = v_2'$$

The second ball begins to move with the speed v_1. The two balls have thus exchanged their initial velocities (Fig. 6-14).   ∎

The results of the preceding problem help us to understand the operation of the toy shown in Fig. 6-15, which consists of a number of identical steel balls suspended by strings. When one ball at the left is pulled out and released, it swings to strike the row of stationary balls, and one ball at the right swings out in response. Why not two balls, or indeed all the others? The answer is that both momentum and kinetic energy can be conserved *only* if a single ball swings out on the right. If two balls were to swing out, then their joint mass is twice that of the ball on the left, and to conserve momentum their initial speeds must be half that of the ball on the left. But the combined kinetic energies of the two balls would then be half that of the ball on the left:

	One ball on left	Two balls on right
Momentum:	mv	$m\left(\dfrac{v}{2}\right) + m\left(\dfrac{v}{2}\right) = mv$
Kinetic energy:	$\frac{1}{2}\,mv^2$	$\frac{1}{2}\,m\left(\dfrac{v}{2}\right)^2 + \frac{1}{2}\,m\left(\dfrac{v}{2}\right)^2 = \frac{1}{4}\,mv^2$

Only if half the initial kinetic energy is lost can two balls swing out. The same reasoning shows why, if two balls are pulled out at the left and released, as in Fig. 6-15(b), two balls at the right will swing out after the collision, and so on.

What about the case when more than half the balls are pulled out to the left? In that event, the balls at the left cannot give up all their momentum and kinetic energy to the balls at the right, and so several of the pulled-out balls continue to move to the right after the collision.

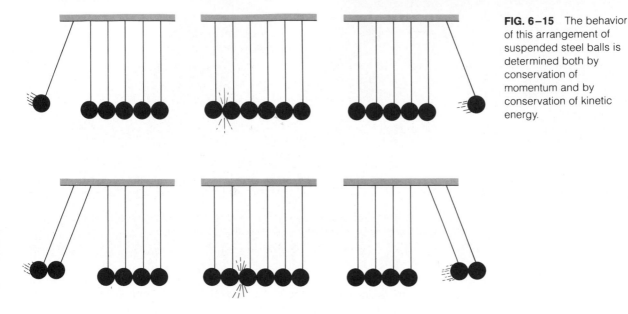

FIG. 6–15 The behavior of this arrangement of suspended steel balls is determined both by conservation of momentum and by conservation of kinetic energy.

6–7 ENERGY TRANSFER

A moving object strikes a stationary one, and as a result the second object is set in motion. What is the mass ratio between the two that will cause the struck object to have the highest possible speed after the impact? What mass ratio will lead to the greatest transfer of energy to the struck object? These are not questions of abstract interest only but are closely connected with a variety of actual problems that range from the design of golf clubs to the design of nuclear reactors.

For simplicity we will confine ourselves to head-on collisions in which both objects move along the same straight line. Let us consider an object of mass m_1 and initial speed v_1 that strikes a stationary object of mass m_2, after which their respective speeds are v_1' and v_2'. From conservation of momentum,

$$m_1 v_1 = m_1 v_1' + m_2 v_2'$$

$$m_1(v_1 - v_1') = m_2 v_2' \tag{6–12}$$

From Eq. (6–11), since $v_2 = 0$,

$$v_1 = v_2' - v_1' \tag{6–13}$$

Combining Eqs. (6–12) and (6–13) gives for the final speed of m_1

$$v_1' = \frac{m_1 - m_2}{m_1 + m_2} v_1 \tag{6–14}$$

and for the final speed v_2' of m_2

$$v_2' = \frac{2m_1}{m_1 + m_2} v_1 \tag{6–15}$$

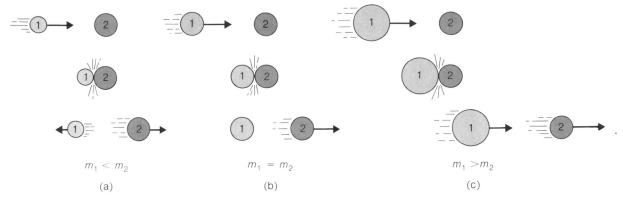

$m_1 < m_2$ $m_1 = m_2$ $m_1 > m_2$

(a)   (b)   (c)

FIG. 6–16 The result of an elastic head-on collision between a moving object of mass m_1 and a stationary object of mass m_2 depends upon the ratio of their masses.

How the effects of a collision depend on the relative masses of the colliding objects

Formulas (6–14) and (6–15) permit us to draw some general conclusions. If m_1 is less than m_2, v_1' is in the opposite direction to v_1: The lighter object rebounds from the heavier one (Fig. 6–16(a)). A ball striking a wall is an extreme example, where, since m_2 is virtually infinite compared with m_1, $v_1' = -v_1$. In the event that $m_1 = m_2$, $v_1' = 0$ and $v_2 = v_1$: The colliding object stops, while the struck one moves off with the same speed it had (Fig. 6–16(b)). When m_1 is greater than m_2, as in the case of a table-tennis serve, the colliding object continues on in the same direction after the impact but with reduced speed while the struck object moves ahead of it at a faster pace (Fig. 6–16(c)). When m_1 is much greater than m_2, the colliding object loses little speed while the struck one is given a speed nearly twice v_1.

The ratio between the kinetic energy KE_2' transferred to the initially stationary object and the kinetic energy KE_1 of the colliding object can be found with the help of Eq. (6–15):

$$\frac{KE_2'}{KE_1} = \frac{\frac{1}{2} m_2 v_2'^2}{\frac{1}{2} m_1 v_1^2} = \frac{4m_1 m_2}{(m_1 + m_2)^2} = \frac{4(m_2/m_1)}{(1 + m_2/m_1)^2} \tag{6–16}$$

Energy transfer is a maximum when the objects have the same mass

This formula is plotted in Fig. 6–17. Evidently the transfer of energy is a maximum for $m_1 = m_2$, when *all* the energy of m_1 is given to m_2. This is the situation illustrated in Fig. 6–14(b).

From Eq. (6–16) it would seem that the best mass for the head of a golf club would be the same as that of a golf ball, in order that all the energy of the club be given to the ball (assuming an elastic collision). In this case $v_2' = v_1$. However, according to Eq. (6–15), the greater the mass m_1 of the clubhead, the greater the ball's speed v_2' will exceed v_1, up to a limit of $2v_1$. The trouble with a heavy clubhead is twofold: It is hard to swing a heavy golf club as fast as a light one, and the more m_1 exceeds m_2, the smaller is the proportion of kinetic energy that is transferred to the ball—the extra effort does not provide a corresponding return. Experience has led golfers to use clubheads whose masses are typically about four times the 46-g mass of

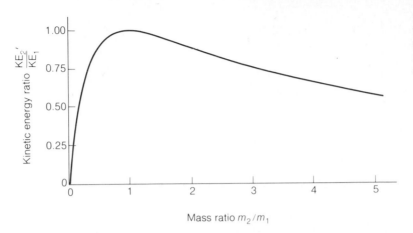

FIG. 6-17 Energy transfer in an elastic head-on collision between a moving object and a stationary one.

a golf ball where maximum distance is required, although this ratio does not seem to be very critical.

6-8 COEFFICIENT OF RESTITUTION

If a collision is not perfectly elastic, the loss of some of the original kinetic energy means that the relative speed of the objects involved after the collision will be less than their relative speed of approach. The ratio between the relative speeds is called the *coefficient of restitution,* symbol e:

Coefficient of restitution is a joint property of the colliding objects

$$e = \frac{v_2' - v_1'}{v_1 - v_2} \tag{6-17}$$

$$\text{Coefficient of restitution} = \frac{\text{relative speed after collision}}{\text{relative speed before collision}}$$

In a perfectly elastic collision, $e = 1$. In a perfectly inelastic collision, when the objects stick together and $v_2' = v_1'$, $e = 0$; the kinetic energy loss here is the maximum possible. A coefficient of restitution always applies jointly to the colliding objects, not to either one of them by itself.

When the coefficient of restitution is taken into account in a collision between a moving object and a stationary one,

$$v_2' = \frac{(1 + e)m_1}{m_1 + m_2} v_1 \tag{6-18}$$

$$\frac{KE_2'}{KE_1} = \frac{(1 + e)^2 m_1 m_2}{(m_1 + m_2)^2} = \frac{(1 + e)^2 (m_2/m_1)}{(1 + m_2/m_1)^2} \tag{6-19}$$

In the case of a golf club striking a golf ball, typical values might be $m_1 = 200$ g, $m_2 = 46$ g, and $e = 0.7$, from which we find that the ball moves off with a speed 38% greater than the speed of the clubhead and carries with it 44% of the clubhead's

FIG. 6-18

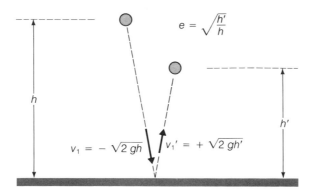

$$e = \sqrt{\frac{h'}{h}}$$

$$v_1 = -\sqrt{2gh}$$

$$v_1' = +\sqrt{2gh'}$$

original energy. If the collision were perfectly elastic, $e = 1$ and these figures would be 63% and 61% respectively.

Example   A tennis ball dropped on the floor from a height of 100 cm rebounds to a height of 60 cm. Find the coefficient of restitution.

Solution   Since the floor does not move, $v_2 = v_2' = 0$. Let us consider upward as $+$ and downward as $-$. The speed of the ball when dropped from the height $h = 100$ cm is $v_1 = -\sqrt{2gh}$. The ball's speed v_1' when it rebounds from the floor is given by $v_1' = +\sqrt{2gh'}$, where $h' = 60$ cm is the height it reaches. Hence, from Eq. (6–17),

$$e = \frac{v_2' - v_1'}{v_1 - v_2} = \frac{0 - v_1'}{v_1 - 0} = \frac{-\sqrt{2gh'}}{-\sqrt{2gh}} = \sqrt{\frac{h'}{h}}$$

This is a general conclusion for something dropped from a height h that rebounds to the new height h' from a level surface (Fig. 6–18). In the present case,

$$e = \sqrt{\frac{h'}{h}} = \sqrt{\frac{60\,\text{cm}}{100\,\text{cm}}} = \sqrt{0.6} = 0.77$$ ∎

Example   A 2-kg ball moving at 2 m/s to the right collides head-on with a 1-kg ball moving at 3 m/s to the left (Fig. 6–19). The coefficient of restitution is 0.7. Find the speeds and directions of the balls after the collision.

Solution   Let us call motion to the right $+$ and motion to the left $-$, so that

$$m_1 = 2\,\text{kg} \qquad m_2 = 1\,\text{kg}$$
$$v_1 = +2\,\text{m/s} \qquad v_2 = -3\,\text{m/s}$$
$$v_1' = ? \qquad v_2' = ?$$

From conservation of momentum,

Momentum before = momentum after
$$m_1v_1 + m_2v_2 = m_1v_1' + m_2v_2'$$

Since we have two unknown quantities, the final speeds v_1' and v_2', we need another equation relating them. From Eq. (6–17),

$$\text{Coefficient of restitution} = e = \frac{v_2' - v_1'}{v_1 - v_2}$$

Solving both equations for v_2' gives

$$v_2' = \frac{m_1 v_1 + m_2 v_2 - m_1 v_1'}{m_2}$$

$$v_2' = e(v_1 - v_2) + v_1'$$

Since both of these formulas equals v_2', they must be equal to each other. Hence

$$\frac{m_1 v_1 + m_2 v_2 - m_1 v_1'}{m_2} = e(v_1 - v_2) + v_1'$$

$$m_1 v_1 + m_2 v_2 - m_1 v_1' = e m_2 (v_1 - v_2) + m_2 v_1'$$

$$(m_1 + m_2) v_1' = m_1 v_1 + m_2 v_2 - e m_2 (v_1 - v_2)$$

$$v_1' = \frac{m_1 v_1 + m_2 v_2 - e m_2 (v_1 - v_2)}{m_1 + m_2} = -0.833 \text{ m/s}$$

With this value of v_1' we can find v_2' at once:

$$v_2' = e(v_1 - v_2) + v_1' = 2.667 \text{ m/s}$$

The first ball reverses its direction and moves off to the left at 0.833 m/s, which is slower than its original speed. The second ball also reverses its direction and moves off to the right at 2.667 m/s, also slower than its original speed. As we expect when e is less than 1, some of the initial KE of the system has been lost in the collision.

To check these results, we can calculate the total momentum of the system before and after the collision:

Before: $m_1 v_1 + m_2 v_2 = (2 \text{ kg})(2 \text{ m/s}) + (1 \text{ kg})(-3 \text{ m/s}) = +1 \text{ kg} \cdot \text{m/s}$

After: $m_1 v_1' + m_2 v_2' = (2 \text{ kg})(-0.833 \text{ m/s}) + (1 \text{ kg})(2.667 \text{ m/s}) = +1 \text{ m/s}$ ∎

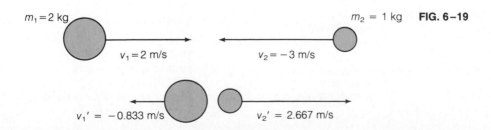

$m_1 = 2$ kg $v_1 = 2$ m/s $v_2 = -3$ m/s $m_2 = 1$ kg   **FIG. 6–19**

$v_1' = -0.833$ m/s $v_2' = 2.667$ m/s

IMPORTANT TERMS

The **linear momentum** of an object is the product of its mass and velocity. Linear momentum is a vector quantity having the direction of the object's velocity.

The **impulse** of a force is the product of the force and the time during which it acts. Impulse is a vector quantity having the direction of the force. When a force acts on an object that is free to move, its change in momentum equals the impulse given it by the force.

The law of **conservation of momentum** states that when the vector sum of the external forces acting upon a system of particles equals zero, the total linear momentum of the system remains constant.

The **thrust** of a rocket is the force that results from the expulsion of exhaust gases.

A **completely elastic collision** is one in which kinetic energy is conserved. A **completely inelastic collision** is one in which the objects stick together upon impact, which results in the maximum possible kinetic energy loss. Linear momentum is conserved in all collisions.

The **coefficient of restitution** is the ratio between the relative speeds of two colliding objects after and before they collide. It equals 1 for a completely elastic collision and 0 for a completely inelastic collision.

IMPORTANT FORMULAS

Linear momentum: $\mathbf{p} = m\mathbf{v}$

Impulse and momentum change: $\mathbf{F}\,\Delta t = \Delta(m\mathbf{v})$

Momentum conservation in collision:
$$m_1\mathbf{v}_1 + m_2\mathbf{v}_2 = m_1\mathbf{v}_1' + m_2\mathbf{v}_2'$$
Elastic collision: $v_1 - v_2 = -(v_1' - v_2')$

Coefficient of restitution: $e = \dfrac{v_2' - v_1'}{v_1 - v_2}$

MULTIPLE CHOICE

1. An object at rest may possess
   a. velocity.
   b. momentum.
   c. kinetic energy.
   d. potential energy.

2. An object in motion need not possess
   a. velocity.
   b. momentum.
   c. kinetic energy.
   d. potential energy.

3. An object that has momentum must also have
   a. acceleration.
   b. impulse.
   c. kinetic energy.
   d. potential energy.

4. Momentum is most closely related to
   a. kinetic energy.
   b. potential energy.
   c. impulse.
   d. power.

5. The impulse given to an object is equal to the consequent change in its
   a. velocity.
   b. momentum.
   c. kinetic energy.
   d. potential energy.

6. When the velocity of a moving object is doubled,
   a. its acceleration is doubled.
   b. its momentum is doubled.
   c. its kinetic energy is doubled.
   d. its potential energy is doubled.

7. If a shell fired from a cannon explodes in midair,
   a. its total momentum increases.
   b. its total momentum decreases.
   c. its total kinetic energy increases.
   d. its total kinetic energy decreases.

8. When two or more objects collide, it is always true that
   a. the momentum of each one remains unchanged.
   b. the kinetic energy of each one remains unchanged.
   c. the total momentum of all the objects remains unchanged.
   d. the total kinetic energy of all the objects remains unchanged.

9. An elastic collision conserves
   a. kinetic energy but not momentum.
   b. momentum but not kinetic energy.
   c. neither momentum nor kinetic energy.
   d. both momentum and kinetic energy.

10. A ball whose momentum is **p** strikes a wall and bounces off. The change in the ball's momentum is
   a. 0.
   b. **p**/2.
   c. **p**.
   d. 2**p**.

11. An iron sphere of mass 30 kg has the same diameter as an aluminum sphere of mass 10.5 kg. The spheres are simultaneously dropped from a cliff. When they are 10 m from the ground, they have identical
   a. accelerations.
   b. momenta.
   c. potential energies.
   d. kinetic energies.

12. The operation of a rocket is based upon
   a. pushing against its launching pad.
   b. pushing against the air.
   c. conservation of KE.
   d. conservation of linear momentum.

13. Object A strikes object B, which is initially at rest. The maximum transfer of energy from A to B occurs when m_A is
    a. less than m_B.
    b. equal to m_B.
    c. greater than m_B.
    d. any of the above, depending on the speed of A.

14. In the collision of the previous question, object B will acquire the most speed when m_A is
    a. less than m_B.
    b. equal to m_B.
    c. greater than m_B.
    d. any of the above, depending on the speed of A.

15. The average momentum of a 70-kg runner who covers 400 m in 50 s is
    a. 8.75 kg · m/s.
    b. 57 kg · m/s.
    c. 560 kg · m/s.
    d. 5488 kg · m/s.

16. A 1280-lb white horse is cantering at 20 ft/s. Its linear momentum is
    a. 400 slug · ft/s.
    b. 800 slug · ft/s.
    c. 8000 slug · ft/s.
    d. 25,600 slug · ft/s.

17. A 60-lb girl and a 50-lb boy face each other on frictionless roller skates. The girl pushes the boy, who moves away at a speed of 4 ft/s. The girl's speed is
    a. 2.1 ft/s.
    b. 3.3 ft/s.
    c. 4.0 ft/s.
    d. 4.8 ft/s.

18. An astronaut whose total mass is 100 kg ejects 1 gm of gas from his propulsion pistol at a speed of 50 m/s. His recoil speed is
    a. 0.5 mm/s.
    b. 5 mm/s.
    c. 5 cm/s.
    d. 50 cm/s.

19. A 200-g ball moving at 5 m/s strikes a wall perpendicularly and rebounds elastically at the same speed. The impulse given to the wall is
    a. 1 N · s.
    b. 2 N · s.
    c. 2.5 N · s.
    d. 10 N · s.

20. The additional force needed to maintain the speed of a pickup truck at a constant 20 m/s when rain begins to fall and to accumulate in its back at a rate of 3 kg/min is
    a. 0.
    b. 1 N.
    c. 10 N.
    d. 60 N.

21. A 1000-kg car moving north at 40 m/s collides with a 3000-kg truck moving south at 10 m/s. The two vehicles stick together and the wreckage starts to move
    a. north at 2.5 m/s.
    b. north at 10 m/s.
    c. south at 10 m/s.
    d. south at 17.5 m/s.

22. A 10-ton freight car moving at 5 ft/s collides with a stationary 15-ton freight car. The two cars couple together and move off at
    a. 0.5 ft/s.
    b. 2 ft/s.
    c. 3.3 ft/s.
    d. 5 ft/s.

23. A ball dropped from a height of 3 m bounces up to a height of 1 m. The coefficient of restitution is
    a. 0.33.
    b. 0.58.
    c. 1.7.
    d. 3.

EXERCISES

6–1 Linear Momentum

1. Is it possible for an object to have more kinetic energy but less momentum than another object? Less kinetic energy but more momentum?

2. When the momentum of an object is doubled in magnitude, what happens to its kinetic energy?

3. When the kinetic energy of an object is doubled in magnitude, what happens to its momentum?

4. The iron ball used in the shot put has a mass of 7.3 kg. What is the momentum of a shot whose kinetic energy is 700 J?

5. Find the momentum of a 3200-lb car moving at 30 mi/h.

6. Find the momentum of a 0.04-lb bullet whose kinetic energy is 1600 ft · lb.

7. Car A has a mass of 1000 kg and is moving at 60 km/h. Car B has a mass of 2000 kg and is moving at 30 km/h. Compare the kinetic energies and momenta of the two cars.

6–2 Impulse

8. A 1000-kg car strikes a tree at 30 km/h and comes to a stop in 0.15 s. Find its initial momentum and the average force on the car while it is being stopped.

9. A 160,000-kg DC-8 airplane is flying at 870 km/h. (a) Find its momentum. (b) If the thrust its engines develop is 340,000 N, how much time is needed for the airplane to reach this speed starting from rest? Neglect air resistance, changes in altitude, and the fuel consumed by the engines.

10. A $\frac{1}{3}$-lb baseball reaches the batter with a speed of 75 ft/s. After it has been struck, it leaves the bat at 100 ft/s in the opposite direction. If the ball was in contact with the bat for 0.001 s, find the average force exerted on it during this period.

11. A 170-g softball moving at 30 m/s is caught by a player. If the average force on the player's hands during the catch is 500 N, find the time in which the ball came to a stop.

12. Water emerging from a hose at a rate of 2 L/s and a speed of 8 m/s strikes a person. If the water loses all its momentum on impact, find the force on the person. (The mass of 1 L of water is 1 kg.)

13. A certain howitzer has a barrel 7.6 m long and fires a 65-kg projectile 20 cm in diameter at a muzzle speed of 400 m/s. Use three different ways, each based on an important physical principle, to find the average force on the projectile while it is in the howitzer barrel.

6–3 Conservation of Momentum

14. When a rocket explodes in midair, how are its total momentum and total kinetic energy affected?

15. How is the principle of conservation of linear momentum related to the definition of mass given in Chapter 3 and to Newton's first law of motion? In what way does this principle go beyond the definition of mass and the first law of motion?

16. (a) When an object at rest breaks up into two parts which fly off, must they move in exactly opposite directions? (b) When a moving object strikes a stationary one and the two do not stick together, must they move off in exactly opposite directions?

17. A railway car is at rest on a frictionless track. A man at one end of the car walks to the other end. (a) Does the car move while he is walking? (b) If so, in which direction? (c) What happens when the man comes to a stop?

18. An empty coal car coasts at a certain speed along a level railroad track without friction. (a) It begins to rain. What happens to the speed of the car? (b) The rain stops, and the collected water gradually leaks out. What happens to the speed of the car now?

19. A hunter has a rifle that can fire 60-g bullets with a speed of 900 m/s. A 40-kg leopard springs at him at 10 m/s. How many bullets must the hunter fire into the leopard in order to stop him in his tracks?

20. A 50-kg girl throws a 5-kg pumpkin at 10 m/s to a 50-kg boy, who catches it. If both are on a frictionless frozen lake, how fast does each of them move backward?

21. A 150-lb astronaut is drifting forward in an orbiting space shuttle at 1 in./s. To stop her, a fellow astronaut throws

her a 0.5-lb orange. If she comes to a stop when she catches the orange, what was its speed?

22. A hunter in a rowboat loses the oars and decides to set the boat in motion by firing his rifle astern five times. The total weight of the boat and its contents is 300 lb, the weight of each bullet is 0.04 lb, and the speed of the bullets is 2000 ft/s. What is the speed of the boat, assuming that water resistance is negligible? How far would it go in 1 h?

23. A 70-kg man and a 50-kg woman are in a 60-kg rubber dinghy. The man dives into the water with a horizontal speed of 3 m/s. If his swimming speed is 1 m/s, can he return to the dinghy? If not, can the woman change the dinghy's motion enough by diving off it at 3 m/s in the opposite direction? Could she then return to the dinghy herself if her swimming speed is also 1 m/s?

24. A spacecraft moving at 10 km/s breaks apart into two pieces of equal mass that continue moving in the original direction. If the speed of one of the pieces is 4 km/s, find the speed of the other one.

25. A spacecraft moving at 10 km/s breaks apart into two pieces of equal mass, one of which moves off at 4 km/s in a direction opposite to the original direction. Find the speed and direction of the other piece.

26. A certain cannon has a range of 2 km. One day a shell that it fires explodes at the top of its path into two equal fragments, one of which lands next to the cannon. How far away from the cannon does the other fragment land?

27. The cannon of Exercise 26 fires another shell that also explodes into two equal fragments at the top of its path. One of these fragments falls vertically downward. How far away from the cannon does the other fragment land?

28. A 500-kg cannon on a wheeled carriage is at the foot of a 25° slope with its barrel horizontal. The cannon fires a 30-kg shell at 300 m/s in a direction opposite to the slope. (a) What is the cannon's initial speed up the slope? (b) How far along the slope does the cannon travel before it comes to a stop and starts to roll back down? Disregard friction.

6–4 Rocket Propulsion

29. The motors of a spacecraft provide a total thrust of 1.8 MN at takeoff. If the exhaust speed is 2.5 km/s, find the rate at which fuel is being consumed.

30. The engine of a 2000-lb rocket standing vertically on the ground is being tested. If fuel is being burned at the rate of 10 lb/s with an exhaust speed of 10,000 ft/s, find the force needed to hold the rocket down.

31. A rocket fired from the earth's surface ejects 2% of its mass at a speed of 2 km/s in the first second of its flight. Find the initial acceleration of the rocket.

32. A rocket launched vertically from the earth has a mass of 2500 kg at takeoff. (a) If the rocket's initial acceleration is 0.5 g and it is using fuel at the rate of 15 kg/s, find the speed of the exhaust gases. (b) What is the acceleration of the rocket 1 min later? Ignore the decrease in g with altitude.

33. A certain rocket consumes 25 kg of fuel per second and its exhaust has a speed of 3000 m/s. If the initial mass of the rocket is 5000 kg of which 3500 kg is fuel, find its initial and final accelerations.

6–5 Inelastic Collisions

34. A 30-kg girl who is running at 3 m/s jumps on a stationary 10-kg sled on a frozen lake. How fast does the sled then move?

35. A neutron of mass 1.67×10^{-27} kg and speed 10^5 km/s collides with a stationary deuteron of mass 3.34×10^{-27} kg. The two particles stick together. What is the speed of the composite particle (called a *triton*)?

36. A 0.5-kg stone moving at 4 m/s overtakes a 4-kg lump of clay moving at 1 m/s. The stone becomes embedded in the clay. (a) What is the speed of the composite body after the collision? (b) How much KE is lost in the collision?

37. A 1000-kg car moving east at 80 km/h overtakes a 1500-kg car moving east at 40 km/h and collides with it. The two cars stick together. (a) What is the initial speed of the wreckage? (b) How much KE is lost in the collision?

38. A 1000-kg car moving east at 80 km/h collides head-on with a 1500-kg car moving west at 40 km/h, and the two cars stick together. (a) Which way does the wreckage move and with what initial speed? (b) How much KE is lost in the collision?

39. A 0.5-kg stone moving north at 4 m/s collides with a 4-kg lump of clay moving west at 1 m/s. The stone becomes embedded in the clay. What is the velocity (magnitude and direction) of the composite body after the collision?

40. A 2000-lb car traveling east at 30 mi/h collides with a 3000-lb car traveling north at 20 mi/h. The cars stick together after the collision. What is the velocity (magnitude and direction) of the wreckage?

41. A ballistic pendulum consists of a wooden block of mass M suspended by long strings from the ceiling. A bullet of mass m and speed v is fired horizontally into the block, which swings away until its height is the amount h above its

original height. (a) Find a formula that gives v in terms of g and the readily measurable quantities m, M, and h. (b) A 5-g bullet is fired into the 2-kg block of a ballistic pendulum, which rises by 10 cm. Find the bullet's speed.

42. As shown, a magnet A attached to a string R long is pulled out to one side so the string is horizontal. When the magnet is let go, it swings downward and strikes an iron cube B of the same mass that is resting on a frictionless surface. The two stick together and swing upward on the other side. What is the maximum value of θ, the angle between the string and the vertical?

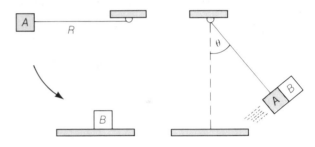

6–6 Elastic Collisions

43. The 176-g head of a golf club is moving at 45 m/s when it strikes a 46-g golf ball and sends it off at 65 m/s. Find the final speed of the clubhead after the impact, assuming that the mass of the club's shaft can be neglected.

44. A neutron of mass 1.67×10^{-27} kg and speed 10^5 m/s collides head-on with a stationary deuteron of mass 3.34×10^{-27} kg. The particles do not stick together, and the deuteron moves off at 6.67×10^4 m/s. What is the speed of the neutron? Is the collision elastic?

45. A billiard ball at rest is struck by another ball of the same mass whose speed is 5 m/s. After an elastic collision the incident ball goes off at an angle of 40° with respect to its original direction of motion and the struck ball goes off at an angle of 50° with respect to this direction. Find the final speeds of both balls.

46. A billiard ball at rest is struck by another billiard ball of the same mass whose speed is 4 m/s. After an elastic collision the incident ball goes off at an angle of 25° with respect to its original direction of motion. Find the angle the struck ball makes with this direction and the final speeds of both balls.

6–8 Coefficient of Restitution

47. A steel ball bearing is dropped on a steel plate from a height of 2 m. If the coefficient of restitution is 0.97, find the height to which the ball rebounds.

48. A rubber ball is dropped on the ground from a height of 150 cm and on its second rebound reaches a height of 50 cm. Find the coefficient of restitution.

49. A 5-kg ball moving at 6 m/s strikes a stationary 3-kg ball and continues in the same direction at 2 m/s. (a) Find the speed and direction of the 3-kg ball after the collision. (b) What is the coefficient of restitution?

50. A 3-kg ball moving at 6 m/s strikes a stationary 5-kg ball and rebounds in the opposite direction at 1 m/s. (a) Find the speed and direction of the 5-kg ball after the collision. (b) What is the coefficient of restitution?

51. A 12-lb ball moving at 20 ft/s strikes a stationary 30-lb ball that then moves off at 8 ft/s. (a) Find the speed and direction of the 12-lb ball after the collision. (b) What is the coefficient of restitution?

52. A 1-kg ball moving at 2 m/s in the $+x$-direction collides head-on with a 2-kg ball moving at 3 m/s in the $-x$-direction. The coefficient of restitution is 0.8. Find the velocities of the balls after the collision.

53. A 1-kg ball moving at 5 m/s in the $+x$-direction collides head-on with a 2-kg ball moving at 5 m/s in the $-x$-direction. The coefficient of restitution is 0.7. Find the velocities of the balls after the collision.

ANSWERS TO MULTIPLE CHOICE

1. d	**7.** c	**13.** b	**19.** b
2. d	**8.** c	**14.** c	**20.** b
3. c	**9.** d	**15.** c	**21.** a
4. c	**10.** d	**16.** b	**22.** b
5. b	**11.** a	**17.** b	**23.** b
6. b	**12.** d	**18.** a	

7

CIRCULAR MOTION

Nearly everything in the natural world travels in a curved path. Often these paths are either circles or are very close to being circles. For example, the orbits of the earth and the other planets about the sun are almost circular in shape, as is the orbit of the moon about the earth. On a smaller scale, a handy way to visualize an atom is to imagine it as having a central nucleus with electrons circling around. And, of course, circular motion is familiar in our own experience; it is hard to think of any important aspect of technology in which circular motion of some kind is not involved. We shall therefore find it both interesting and essential for our later work to consider this kind of motion in some detail.

7–1 CENTRIPETAL FORCE

An object traveling in a circle at a constant speed is said to undergo *uniform circular motion*.

CHAPTER OBJECTIVES

Completing this chapter should enable you to:

1. Explain the significance of centripetal force in motion along a curved path.

2. Relate the centripetal force on something in uniform circular motion to its mass, its speed, and the radius of its orbit.

3. Find the coefficient of friction needed for a car to make a turn on a level road without skidding.

4. Calculate the angle at which a highway curve should be banked for a given speed.

5. Describe the principles that underlie the operation of a centrifuge.

6. Discuss the motion of an object moving in a vertical circle and explain why its speed varies.

7. Understand the origin of the forces experienced by a flier whose airplane moves in a vertical circle.

8. Find the gravitational force one object exerts on another.

9. Account for the ability of a satellite to circle the earth in a stable orbit and find the speed needed for an orbit of given size.

Although the velocity of such an object has the same *magnitude* all along its path, the *direction* of the velocity changes constantly. A changing velocity means an acceleration, which in turn means that the object must be acted upon by a force. Since the object's path is a circle, the force on it must be directed toward the center of the circle (Fig. 7–1). This force is called *centripetal force*, literally "force seeking the center." Without it, circular motion cannot occur. In general,

Centripetal force is needed for motion along a curved path

Centripetal force = inward force on an object moving in a curved path

To verify the role of centripetal force in circular motion, we can whirl a ball at the end of a string (Fig. 7–2). As the ball swings around, we must continually exert an inward force on it by means of the string. If we let go of the string, the ball flies off tangent to its original circular path. With no centripetal force on it, the ball then proceeds along a straight path at constant velocity as the first law of motion predicts. (Actually, of course, the ball will fall to the ground eventually because of gravity, but this is another matter.)

FIG. 7–1   Even though the velocity **v** of a body traveling in a circle at constant speed has the same magnitude along its path, the direction of **v** changes constantly. The inward force that causes this change in direction is called *centripetal force*, **F$_c$**.

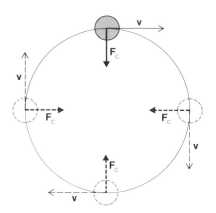

FIG. 7–2   When a ball is whirled at the end of a string, the tension in the string provides the centripetal force that keeps the ball moving in a circle. When the string is let go, the ball's momentum would carry it off in a straight path if gravity were absent.

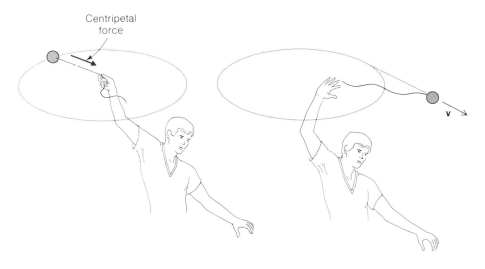

FIG. 7–3   The centripetal force exerted when a car rounds a curve on a level road is provided by friction between its tires and the road.

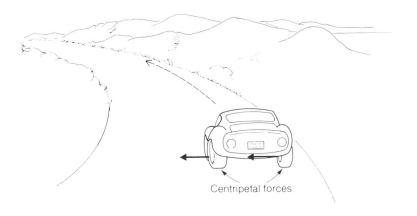

A centripetal force is acting whenever rotational motion occurs, since such a force is required to change the direction of motion of a particle from the straight line that it would normally follow to a curved path. Gravitation provides the centripetal forces that keep the planets moving around the sun and the moon around the earth. Friction between its tires and the road provides the centripetal force needed by a car in rounding a curve (Fig. 7–3). If the tires are worn and the road wet or icy, the frictional force is small and may not be enough to permit the car to turn.

7–2 CENTRIPETAL ACCELERATION

How large a centripetal force is needed to keep a given object moving in a circle with a certain speed? To find out, we must first find the acceleration of such an object. In the following derivation we shall consider the circular motion of a particle; the same arguments and conclusions hold for the circular motion of the center of gravity of an object of definite size.

In Fig. 7–4(a) a particle is shown traveling along a circular path of radius r at the constant speed v. At $t = 0$ the particle is at the point A, where its velocity is $\mathbf{v}_A$, and at $t = \Delta t$ the particle is at the point B, where its velocity is $\mathbf{v}_B$. The change $\Delta \mathbf{v}$ in the particle's velocity in the time interval Δt is $\Delta \mathbf{v} = \mathbf{v}_B - \mathbf{v}_A$, and its acceleration is

How formula for centripetal acceleration is derived

$$\mathbf{a} = \frac{\Delta \mathbf{v}}{\Delta t}$$

The vector triangle whose sides are $-\mathbf{v}_A$, $\mathbf{v}_B$, and $\Delta \mathbf{v}$ is similar to the space triangle whose sides are OA, OB, and s, as we can see from Fig. 7–4(c). Since v is the constant speed of the particle, the magnitudes of $-\mathbf{v}_A$ and $\mathbf{v}_B$ are both v. Also, OA and OB are radii of the circle, so their lengths are both r. Corresponding sides of similar triangles are proportional, hence

$$\frac{\Delta v}{v} = \frac{s}{r} \qquad \text{and} \qquad \Delta v = \frac{vs}{r}$$

The distance the particle actually covers in going from A to B is the arc joining these points, the length of which is $v\,\Delta t$. The distance s, however, is the chord joining A and B, as in Fig. 5–7(d). We are finding the *instantaneous* acceleration of the particle, and we are therefore concerned with the case where A and B are very close together, in which case the chord and arc are equal. Hence

$$s = v\,\Delta t$$

and we have

$$\Delta v = \frac{v^2\,\Delta t}{r}$$

The magnitude of the particle's acceleration is therefore

$$a_c = \frac{\Delta v}{\Delta t} = \frac{v^2}{r}$$

FIG. 7–4 (a) A and B are two successive positions, Δt apart, of a particle undergoing uniform circular motion at the speed v in a circle of radius r. (b) The velocity of the particle at A is $\mathbf{v}_A$ and at B is $\mathbf{v}_B$; the change in its velocity in going from A to B is $\Delta\mathbf{v} = \mathbf{v}_B - \mathbf{v}_A$. (c) The space and vector triangles are similar because both are isosceles with the long sides of each perpendicular to the corresponding long sides of the other. (d) The chord joining A and B is s, while the actual distance the particle traverses is $v\,\Delta t$. In calculating the instantaneous acceleration of the particle we are restricted to having A and B a very small distance apart, in which case the chord and arc have the same length. (e) The magnitude of the centripetal acceleration is $a_c = \Delta v/\Delta t = v^2/r$.

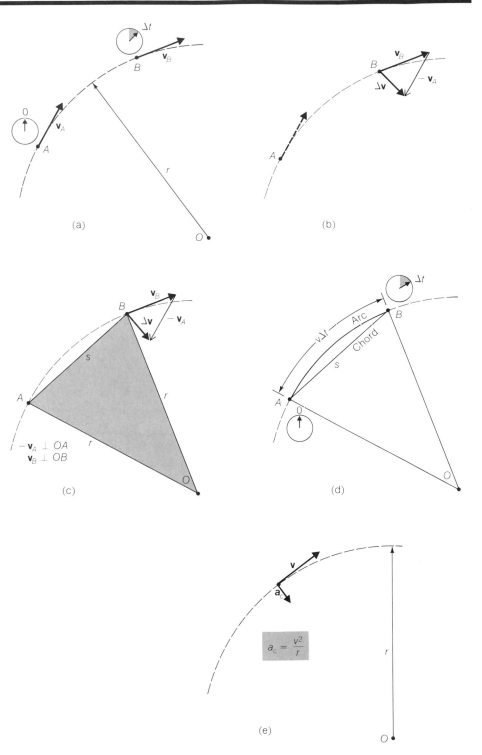

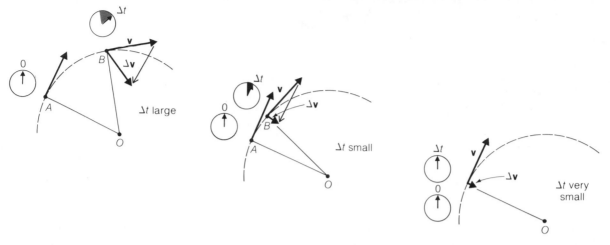

FIG. 7–5 The centripetal acceleration of a particle in uniform circular motion points toward the center of the circle.

In Fig. 7–4(c) $\Delta \mathbf{v}$ does not quite point toward O, the center of the particle's circular path. When A and B are very close together, however, $\Delta \mathbf{v}$ *does* point toward O (Fig. 7–5). Because the acceleration $\mathbf{a}_c$ is an instantaneous acceleration, we are solely concerned with the case when Δt and hence s are extremely small, and the direction of $\mathbf{a}_c$ is accordingly radially inward.

The direction of centripetal acceleration is toward center of circle

The inward, or *centripetal*, acceleration of a particle in uniform circular motion is proportional to the square of its speed and inversely proportional to the radius of its path:

$$a_c = \frac{v^2}{r} \qquad \textit{Centripetal acceleration} \quad (7-1)$$

The time needed by an object in uniform circular motion to make a complete revolution is called its *period*, symbol T. The distance the object travels in making a circle of radius r is $2\pi r$, the circle's circumference (Fig. 7–6). The speed of the object

Period of circular motion

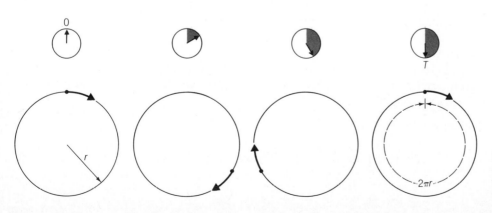

FIG. 7–6 A body in uniform circular motion whose period is T has a speed of $2\pi r/T$.

is the distance it covers (here $2\pi r$) divided by the time needed (here T), so

$$v = \frac{2\pi r}{T}$$ *Speed in orbit of period T*   (7–2)

This formula is often useful in working out problems in uniform circular motion.

Example   A yo-yo is whirled in a horizontal circle 80 cm in radius. If the yo-yo makes two revolutions per second, find its centripetal acceleration.

Solution   The yo-yo takes $\frac{1}{2}$ s for each orbit, so $T = 0.5$ s. The yo-yo's speed is given in terms of its orbit radius $r = 0.80$ m and period $T = 0.5$ s by Eq. (7–2). Using this formula for v in Eq. (7–1) gives for the centripetal acceleration

$$a_c = \frac{v^2}{r} = \left(\frac{2\pi r}{T}\right)^2 \left(\frac{1}{r}\right) = \frac{4\pi^2 r}{T^2} = \frac{(4\pi^2)(0.80\,\text{m})}{(0.5\,\text{s})^2} = 126\,\text{m/s}^2$$ ∎

7–3   MAGNITUDE OF CENTRIPETAL FORCE

From the second law of motion $\mathbf{F} = m\mathbf{a}$ we see that the centripetal force $\mathbf{F}_c$ that must be acting on an object of mass m in uniform circular motion is $\mathbf{F}_c = m\mathbf{a}_c$. Since the centripetal acceleration has the magnitude

$$a_c = \frac{v^2}{r}$$

the magnitude of the centripetal force is

$$F_c = \frac{mv^2}{r}$$ *Centripetal force*   (7–3)

The centripetal force that must be exerted to maintain an object in uniform circular motion increases with increasing mass and with increasing speed, with the force more sensitive to a change in speed since it is the square of the speed that is involved. An increase in the radius of the path, however, reduces the required centripetal force (Fig. 7–7). In the case of a car rounding a curve on a level road, the centripetal force is supplied by friction between the car's tires and the road. If the force needed to make a particular turn at a certain speed is too great for the road surface, the car skids.

Example   (a) Find the centripetal force needed by a 1200-kg car to make a turn of radius 40 m at a speed of 25 km/h (16 mi/h). (b) If the road is level, find the minimum coefficient of static friction between the car's tires and the road that will permit the turn to be made (Fig. 7–8).

Solution   (a) The car's speed is

$$v = (25\,\text{km/h})\left(0.278\,\frac{\text{m/s}}{\text{km/h}}\right) = 7.0\,\text{m/s}$$

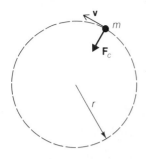

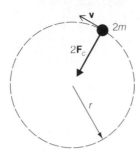

FIG. 7-7 Centripetal force.

The centripetal force on an object in uniform circular motion is equal in magnitude to mv^2/r.

Doubling the mass doubles the required centripetal force.

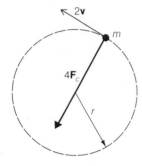

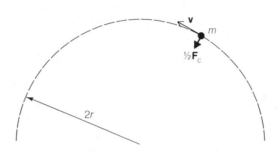

Doubling the speed, however, quadruples the required centripetal force.

Doubling the radius of the circle halves the required centripetal force.

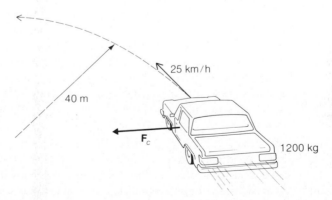

FIG. 7-8 A centripetal force of 1470 N is required by this car to make the turn shown. The corresponding coefficient of friction between its tires and the road is $\mu = 0.125$.

Accordingly the centripetal force is

$$F_c = \frac{mv^2}{r} = \frac{(1200\,\text{kg})(7.0\,\text{m/s})^2}{40\,\text{m}} = 1470\,\text{N}$$

(b) The frictional force is given by

$$F_f = \mu N$$

To find the coefficient of friction μ, we substitute 1470 N for F_f and the car's weight of $w = mg$ for the normal force N. Thus

$$\mu = \frac{F_f}{N} = \frac{F_c}{mg} = \frac{1470\,\text{N}}{(1200\,\text{kg})(9.8\,\text{m/s}^2)} = 0.125$$

which is available under most driving conditions.

It is worth noting that the coefficient of friction here does not depend on the car's mass, since

$$\mu = \frac{F_c}{mg} = \frac{mv^2/r}{mg} = \frac{v^2}{gr}$$

Because the required value of μ depends on v^2, high-speed turns on a level road can be dangerous. In the above case, increasing the speed to 60 km/h (37 mi/h) increases the needed μ to 0.72, which is too much for a wet road. The car would skid in such a situation. ∎

Why braking during a fast turn may be risky

Suppose a car is rounding a curve on a flat road at the highest speed possible without the tires slipping on the road. This means that the centripetal force on the car is the most that can be transferred to its tires from the road by static friction. If the driver now applies the brakes, an additional force is exerted on the tires at right angles to the centripetal force on them, and the resultant of these two forces is more than static friction between the tires and the road can provide. The tires therefore lose their grip on the road, the lower coefficient of sliding friction takes over, and the car skids toward the outside of the curve. (As Table 2–2 shows, the coefficient of static friction between a rubber tire and a concrete road is about 50% greater than the corresponding coefficient of sliding friction.) Braking a car during a fast turn is clearly hazardous, with the likelihood of a loss of control increasing as the braking force is increased. If a skid does occur, pressing harder on the brake pedal will not help since the tires are already slipping over the road. The only way to regain control is to ease up on the brakes and to momentarily straighten the wheels, which will allow the tires to recapture their grip on the road.

Banked turns

Example Usually the friction between its tires and the road is enough to provide a car with the centripetal force it needs to make a turn. However, if the car's speed is high or the road surface is slippery, the available frictional force may not be enough and the car will skid. To reduce the chance of skids, highway curves are often *banked*

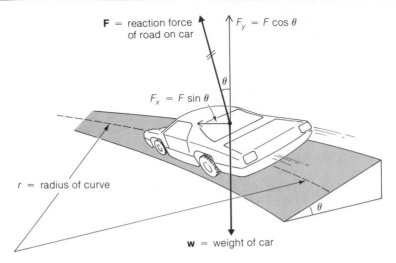

$\mathbf{F}$ = reaction force of road on car

$F_y = F \cos \theta$

$F_x = F \sin \theta$

r = radius of curve

$\mathbf{w}$ = weight of car

FIG. 7-9 When a car rounds a banked curve, the horizontal component F_x of the reaction force $\mathbf{F}$ of the road on the car provides it with the required centripetal force.

so that the roadbed tilts inward. The horizontal component of the reaction force of the road on the car (the action force is the car pressing on the road) then furnishes the required centripetal force. Find the proper banking angle for a car making a turn of radius r at the speed v.

Solution The reaction force $\mathbf{F}$ of the road on the car is perpendicular to the roadbed, as in Fig. 7-9, since friction is not involved here. This force can be resolved into two components, $\mathbf{F}_y$ which supports the weight $\mathbf{w}$ of the car, and $\mathbf{F}_x$ which is available to provide centripetal force. From the diagram,

$F_x = F \sin \theta$ = horizontal component of reaction force

$F_y = F \cos \theta$ = vertical component of reaction force

where θ is the angle between the roadbed and the horizontal. Since F_x furnishes the centripetal force F_c,

$$F_x = F_c$$
$$F \sin \theta = \frac{mv^2}{r}$$

The vertical component of the reaction force equals the car's weight, and so

$$F_y = mg$$
$$F \cos \theta = mg$$

We divide the first of these equations by the second to obtain

$$\frac{F \sin \theta}{F \cos \theta} = \frac{mv^2}{mgr}$$
$$\tan \theta = \frac{v^2}{gr} \qquad\qquad \textit{Banking angle} \quad (7\text{-}4)$$

Banking angle is independent of car's mass

The tangent of the proper banking angle θ varies directly with the square of the car's speed and inversely with the radius of the curve. The mass of the car does not matter. When a car goes around a curve at precisely the design speed, the reaction force of the road provides the centripetal force. If the car goes more slowly than this, friction tends to keep it from sliding down the inclined roadway; if the car goes faster, friction tends to keep it from skidding outward.   ■

7–4   THE CENTRIFUGE

A centrifuge speeds up the separation of substances of different density

A *centrifuge* is a device widely used to separate particles of some kind from a liquid in which they are suspended, for instance blood cells from plasma, or to separate liquids of different density from each other, for instance cream from milk. A simple type of centrifuge is shown in Fig. 7–10. As the centrifuge turns, the tubes swing upward, and the denser material migrates to the outer end of each tube.

No force pushes the denser material outward. Rather, each tube is pulled inward as the centrifuge turns, and the denser material responds less readily than the lighter material by virtue of its greater inertia and hence is left behind at the end of the tube. High speeds mean more effectiveness in separating substances having similar densities; some modern centrifuges operate at speeds exceeding 100,000 revolutions per minute.

Why the arms of a centrifuge rise with increasing speed

The faster the centrifuge turns, the greater the angle between its arms and the central shaft. Why? Let us look at Fig. 7–11, which shows a particle of mass m suspended by a massless string and whirled in a horizontal circle. At a given speed, the horizontal component T_x of the tension T in the string provides the centripetal force on the particle and its vertical component T_y is the force that supports the particle's weight w. When the particle's speed is increased, the tension T increases in magnitude so that T_x equals the new, larger centripetal force. But as T increases at a given angle θ, so does T_y. Now there is a net upward force on the particle, and it rises until T_y decreases to again equal w.

Conical pendulum

The arrangement of Fig. 7–11 is called a *conical pendulum* because the string traces out a cone in space as the particle moves in a circle. Reasoning similar to that in the case of a banked turn leads to the formula

$$\tan \theta = \frac{v^2}{gr}$$

FIG. 7–10 A simple centrifuge.

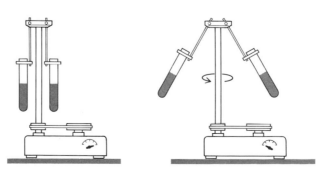

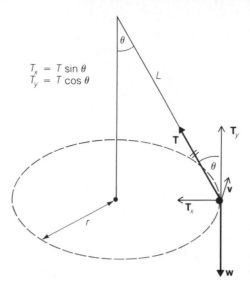

FIG. 7–11 A conical pendulum. When the particle's speed v increases, the angle θ also increases.

$$T_x = T \sin \theta$$
$$T_y = T \cos \theta$$

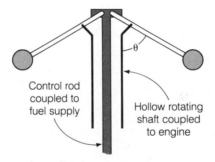

FIG. 7–12 A simple centrifugal governor. The control rod moves up when the shaft speed decreases and moves down when the shaft speed increases.

Control rod coupled to fuel supply

Hollow rotating shaft coupled to engine

for a conical pendulum. Though the string can approach close to the horizontal, it can never quite get there.

The *centrifugal governor* used to control the speeds of many engines is based on the above result. The hollow vertical shaft in Fig. 7–12 is driven by an engine of some kind at a certain speed. If the speed increases, the balls swing out farther; if the speed decreases, the balls swing in closer to the shaft. The rods to which the balls are attached can be coupled to the engine to reduce its fuel supply when θ increases and to increase its fuel supply when θ decreases. In this way the engine automatically maintains a steady speed despite changes in the power taken from it, which affect the amount of fuel it requires.

Centrifugal governor

7–5 MOTION IN A VERTICAL CIRCLE

Thus far we have been considering circular motion in a horizontal plane. Let us now look at what happens when an object moves in a vertical circle. If the object is whirling

FIG. 7–13. The speed of an object moving freely in a vertical circle is not constant. Shown are velocity vectors at four different positions on the circle.

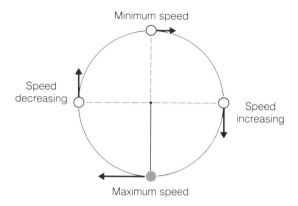

Minimum speed

Speed decreasing

Speed increasing

Maximum speed

FIG. 7–14

$$KE_1 = \tfrac{1}{2}mv_1^2$$

$$PE_1 = mgh = 2\,mgr$$

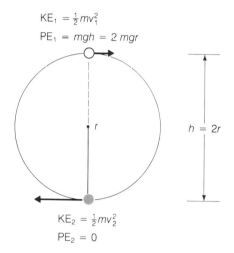

r

$h = 2r$

$$KE_2 = \tfrac{1}{2}mv_2^2$$

$$PE_2 = 0$$

Speed is highest at bottom of circle

freely, its total energy remains constant. All of this energy is kinetic energy at the bottom of the circle (Fig. 7–13). At the top of the circle, however, the total energy is divided between kinetic energy and potential energy, so the object's speed is least there. The motion, although circular, is not uniform: The object goes faster and faster as it moves downward and slower and slower as it moves upward.

Example   A yo-yo is being swung at constant energy at the end of an 80-cm string, as in Fig. 7–14. If its speed at the top of the circle is 3.5 m/s, find its speed at the bottom.

Solution   At the top of the circle the yo-yo has the potential energy

$$PE_1 = mgh = mg\,(2r) = 2mgr$$

and at the bottom $PE_2 = 0$. Hence

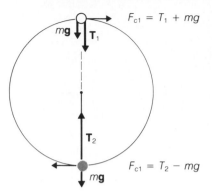

$F_{c1} = T_1 + mg$

mg T_1

T_2

$F_{c1} = T_2 - mg$

mg

FIG. 7–15 The centripetal force **F**$_c$ on a ball being whirled in a vertical circle is the vector sum of the tension **T** in the string and the component of the ball's weight $m\mathbf{g}$ toward the center of the circle.

Energy at bottom = energy at top

$$KE_2 + PE_2 = KE_1 + PE_1$$

$$\tfrac{1}{2}\,mv_2^2 + 0 = \tfrac{1}{2}\,mv_1^2 + 2mgr$$

$$v_2^2 = v_1^2 + 4gr$$

$$v_2 = \sqrt{v_1^2 + 4gr}$$

Here $v_1 = 3.5$ m/s and $r = 0.80$ m, so the speed of the yo-yo at the bottom of its path is

$$v_2 = \sqrt{(3.5\,\text{m/s})^2 + (4)(9.8\,\text{m/s}^2)(0.80\,\text{m})} = 6.6\,\text{m/s} \qquad \blacksquare$$

When a ball is whirled in a vertical circle at the end of a string, how does the tension in the string vary with the position of the ball? As we can see from Fig. 7–15, the centripetal force F_{c1} on the ball at the top of the circle is the sum of the string tension T and the ball's weight mg, so **How tension varies with position**

$$F_{c1} = T_1 + mg = \frac{mv_1^2}{r}$$

$$T_1 = \frac{mv_1^2}{r} - mg \qquad\qquad\qquad \textit{Top of circle} \quad (7\text{–}5)$$

At the bottom of the circle, $m\mathbf{g}$ is opposite in direction to $\mathbf{T}_2$, and so

$$F_{c2} = T_2 - mg = \frac{mv_2^2}{r}$$

$$T_2 = \frac{mv_2^2}{r} + mg \qquad\qquad\qquad \textit{Bottom of circle} \quad (7\text{–}6)$$

For a given speed v, the tension in the string is least at the top of the circle and most at the bottom.

Example An airplane pulls out of a dive in a circular arc whose radius is 3000 ft. If the speed of the airplane is a constant 600 ft/s (409 mi/h), find the force with which the 180-lb pilot presses down on his seat.

FIG. 7–16 Forces exerted by and on the pilot of an airplane making a vertical circle at constant speed.

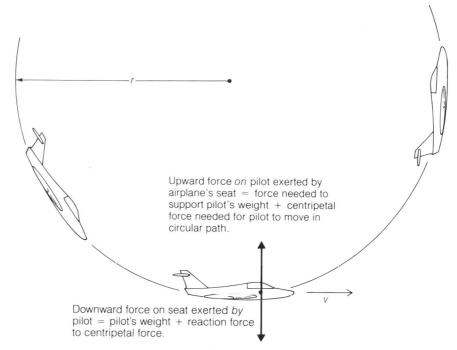

Upward force *on* pilot exerted by airplane's seat = force needed to support pilot's weight + centripetal force needed for pilot to move in circular path.

Downward force on seat exerted *by* pilot = pilot's weight + reaction force to centripetal force.

Solution   The downard force F the pilot exerts on his seat is the reaction to the upward force of the seat on him that supports his weight and keeps him in his circular path (Fig. 7–16). The force of the seat on the pilot thus plays the same role as the tension in the string does for an object being whirled in a vertical circle. Hence Eq. (7–6) applies here with T_2 being equal in magnitude to F, and we have

$$F = \frac{mv^2}{r} + mg = \left(\frac{w}{g}\right)\left(\frac{v^2}{r}\right) + w$$

$$= \frac{(180\,\text{lb})(600\,\text{ft/s})^2}{(32\,\text{ft/s}^2)(3000\,\text{ft})} + 180\,\text{lb}$$

$$= (675 + 180)\,\text{lb} = 855\,\text{lb}$$

The pilot presses down on his seat with a force of 855 lb, nearly five times his actual weight. Because there has been no compensating increase in his muscular strength, the pilot may be unable to move his arms and legs in order to control the airplane. A further complication is the tendency of the pilot's blood to leave his head because of inertia, leaving him with impaired vision ("blacked out") and perhaps unconscious. Special pressure suits have been devised that prevent disturbances in blood supply and in the positions of internal organs during severe accelerations in flight.

Accelerations are often measured in units of g   Because the forces exerted by and on a pilot or astronaut vary with his mass, it is often convenient to speak instead of his acceleration, by custom in units of g. The above pilot's acceleration is

$$a = \frac{F}{m} = \frac{F}{w/g} = \left(\frac{F}{w}\right)g = \left(\frac{855\,\text{lb}}{180\,\text{lb}}\right)g = 4.75g$$

The force on the pilot is $4.75mg = 4.75w$, which is 4.75 times his weight. ■

Anyone who has whirled a yo-yo knows that, if its speed at the top is less than a certain critical value v_0, the string goes slack and the upper part of the yo-yo's path is not circular. To find this critical speed we set $T_1 = 0$ in Eq. (7–5), with the result   **Critical speed at top of circle**

$$T_1 = \frac{mv_0^2}{r} - mg = 0$$

$$\frac{v_0^2}{r} - g = 0$$

$$v_0^2 = rg$$

$$v_0 = \sqrt{rg} \qquad\qquad \text{Critical speed} \quad (7\text{–}7)$$

For a yo-yo with an 80-cm string,

$$v_0 = \sqrt{(0.80\,\text{m})(9.8\,\text{m/s}^2)} = 2.8\,\text{m/s}$$

Example   A sled starts from rest and slides down the frictionless track shown in Fig. 7–17 to loop the loop without falling off. If $r = 10$ m, find the minimum value of h.

Solution   In the loop the force on the sled provided by the track takes the place of tension in a string. Equation (7–7) therefore applies here for the minimum speed of the sled at the top of the loop, so $v_0 = \sqrt{rg}$ and the sled's kinetic energy there is

$$\text{KE} = \tfrac{1}{2}mv_0^2 = \tfrac{1}{2}mrg$$

At the top of the track the sled is $h - 2r$ above the top of the loop and its potential energy there relative to the top of the loop is

$$\text{PE} = mg(h - 2r)$$

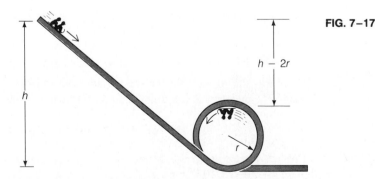

FIG. 7–17

Hence

$$PE = KE$$

$$mg(h - 2r) = \tfrac{1}{2}\,mrg$$

$$h - 2r = \frac{r}{2}$$

$$h = \frac{r}{2} + 2r = \tfrac{5}{2}\,r$$

Since $r = 10$ m here, $h = 25$ m.   ∎

7–6   GRAVITATION

The earth and the other planets follow approximately circular orbits around the sun. We conclude that the planets are being acted upon by centripetal forces that originate in the sun, since the sun is at the center of all the orbits. This much was generally understood by the middle of the seventeenth century, when Newton turned his mind to the question of exactly what the nature of the centripetal forces was.

Gravity holds the planets in orbits around the sun and the moon in an orbit around the earth

Newton proposed that the inward force exerted by the sun that is responsible for the planetary orbits is merely one example of a universal interaction, called *gravitation,* that occurs between all objects in the universe by virtue of their possession of mass. Another example of gravitation, according to Newton, is the attraction of the earth for nearby objects. Thus the centripetal acceleration of the moon and the downward acceleration g of objects dropped near the earth's surface have an identical cause, namely, the gravitational pull of the earth.

Newton was able to arrive at the form of the *law of universal gravitation* from an analysis of the motions of the planets about the sun:

Newton's law of gravitation

Every object in the universe attracts every other object with a force directly proportional to each of their masses and inversely proportional to the square of the distance separating them.

The law of gravitation is expressed in equation form as

$$F_{\text{grav}} = G\,\frac{m_A m_B}{r^2}$$ *Gravitational force*   (7–8)

where m_A and m_B are the masses of any two objects and r is the distance between them. The quantity G is a universal constant whose value is

$$G = 6.67 \times 10^{-11}\frac{\text{N} \cdot \text{m}^2}{\text{kg}^2}$$

$$= 3.44 \times 10^{-8}\frac{\text{lb} \cdot \text{ft}^2}{\text{slug}^2}$$ *Gravitational constant*

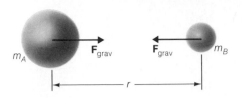

FIG. 7–18 The gravitational forces between two spherical objects.

The direction of the gravitational force is always along a line between the centers of mass of the two objects A and B. The force on A exerted by B is equal in magnitude to that on B exerted by A, but is in the opposite direction (Fig. 7–18).

Gravity acts along line between centers of mass of two objects

Example A grocer installs a 100-kg lead block under the pan of his scale. By how much does this increase the reading of the scale when 1 kg of onions are on the pan, if the centers of mass of the lead and of the onions are 0.3 m apart?

Solution The gravitational force of the lead on the onions is

$$F = G\frac{m_A m_B}{r^2} = \left(6.67 \times 10^{-11}\ \frac{\text{N} \cdot \text{m}^2}{\text{kg}^2}\right)\frac{(100\,\text{kg})(1\,\text{kg})}{(0.3\,\text{m})^2} = 7.4 \times 10^{-8}\ \text{N}$$

The increase in the scale reading is therefore

$$m = \frac{F}{g} = \frac{7.4 \times 10^{-8}\,\text{N}}{9.8\,\text{m/s}^2} = 7.6 \times 10^{-9}\,\text{kg} = 0.0000076\,\text{g}$$

so it is hardly worth the effort. Blowing gently on the onions will increase the reading over a million times more. ∎

The gravitational force of the earth on an object varies inversely with the square of the object's distance from the center of the earth. Hence a person's weight w at a distance r from the earth's center is

Inverse square variation of gravitational force

$$w = \left(\frac{r_e}{r}\right)^2 w_e \qquad (7\text{–}9)$$

where w_e is the person's weight on the earth's surface and r_e is the radius of the earth, which is

$$r_e = 6.37 \times 10^6\,\text{m} = 2.09 \times 10^7\,\text{ft} = 3960\,\text{mi} \qquad \textit{Earth's radius}$$

Example A woman weighs 128 lb on the earth's surface. (a) At what height above the surface would she weigh 80 lb? (b) What would her mass be there?

Solution (a) From Eq. (7–9) we have

$$w = \left(\frac{r_e}{r}\right)^2 w_e$$

$$\frac{r_e}{r} = \sqrt{\frac{w}{w_e}}$$

FIG. 7−19 How the weight of a 4-slug person varies with distance from the earth's center. The person's weight on the earth's surface is 128 lb.

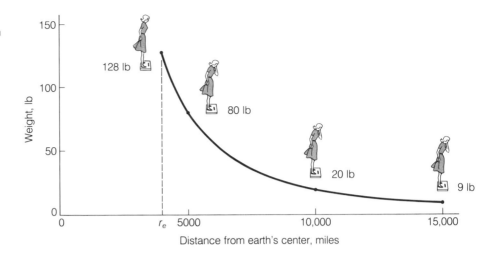

$$r = r_e \sqrt{\frac{w_e}{w}} = (3960\,\text{mi}) \sqrt{\frac{128\,\text{lb}}{80\,\text{lb}}} = 5009\,\text{mi}$$

Hence her height h above the earth's surface is

$$h = r - r_e = (5009 - 3960)\,\text{mi} = 1049\,\text{mi}$$

Figure 7−19 shows how the woman's weight varies with distance from the earth's center.

(b) The woman's mass of

$$m = \frac{w}{g} = \frac{128\,\text{lb}}{32\,\text{ft/s}^2} = 4.0\,\text{slugs}$$

is the same everywhere.   ■

7−7   EARTH SATELLITES

The first artificial satellite, Sputnik I, was launched by the Soviet Union in 1957. Since then thousands of others have been put into orbits around the earth, most of them by the United States and the Soviet Union. Men and women have been in orbit regularly since 1961, when a Russian cosmonaut circled the globe at a height of about 160 km (100 mi). The first American in orbit was John Glenn in the following year. More recently, the giant 75-ton U.S. space shuttles have voyaged into orbit and back many times.

A stable satellite orbit is possible at any distance from the earth

What keeps an artificial earth satellite from falling down? The answer, of course, is that it *is* falling down, but, like the moon, at just such a rate as to circle the earth in a stable orbit. Let us use what we know about gravitation and circular motion to investigate the orbits of earth satellites. In the following discussion the frictional resistance of the atmosphere, which ultimately brings down all artificial satellites, will be neglected.

Near the earth the gravitational force on an object of mass m is its weight

$$w = mg$$

where g is the acceleration of gravity at the location of the object. For uniform circular motion about the earth this force must provide the object with the centripetal force

$$F_c = \frac{mv^2}{r}$$

Hence the condition for a stable orbit is

$$w = F_c$$

$$mg = \frac{mv^2}{r}$$

$$v = \sqrt{rg} \qquad\qquad (7\text{--}10)$$

According to Eq. (7–9) the weight of an object the distance r from the earth's center is

$$w = \left(\frac{r_e}{r}\right)^2 w_e$$

Since $w = mg$ and m is the same everywhere,

$$mg = \left(\frac{r_e}{r}\right)^2 mg_e$$

$$g = \left(\frac{r_e}{r}\right)^2 g_e \qquad\qquad (7\text{--}11)$$

where g_e is the value of g at the earth's surface. Hence the satellite speed at the radius r is

$$v = \sqrt{rg} = \sqrt{\frac{r_e^2 g_e}{r}} \qquad\qquad \textit{Orbit of earth satellite} \quad (7\text{--}12)$$

The greater the orbital radius of a satellite, the smaller its speed.   **Distant satellites move slower than nearby ones**

For an orbit just above the earth's surface,

$$v_0 = \sqrt{r_e g_e} = \sqrt{(6.4 \times 10^6\,\text{m})(9.8\,\text{m/s}^2)} = 7.9 \times 10^3\,\text{m/s}$$

which is nearly 18,000 mi/h. Anything sent off tangent to the earth's surface at this speed will become a satellite of the earth. If it is sent off at a higher speed, its orbit will be elliptical rather than circular (Fig. 7–20). If the object's speed is great enough, it can escape permanently from the earth. The minimum *escape speed* is $\sqrt{2}\,v_0 = 1.41v_0$, which is 11.2×10^3 m/s, about 25,000 mi/h.   **Escape speed**

A satellite can be placed in a circular orbit of any radius if it is provided with a small rocket motor to give it the required impulse at the right altitude. About 3000

FIG. 7–20 Depending upon its speed, an object projected horizontally above the earth's surface may fall back to the earth, revolve around the earth in a circular orbit, revolve around the earth in an elliptical orbit, or escape permanently from the earth into space.

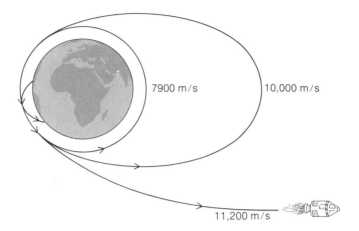

7900 m/s   10,000 m/s

11,200 m/s

satellites are now in orbits that range from 130 to 36,000 km above the earth. The closer satellites are "eyes in the sky" that survey the earth's surface, both for military purposes and to provide information on weather and earth resources such as mineral deposits, crops, and water. Several satellites at an altitude of 1230 km are used in the transit navigation system, which enables ships to find their positions with great accuracy day and night.

Geostationary orbits

The most distant satellites circle the equator exactly once a day, so they remain in place indefinitely over a particular location on the earth. A satellite in such a geostationary orbit can "see" about a third of the earth's surface. Nearly 200 of the satellites now in geostationary orbits are used to relay radio and television communications from one place to another, which is cheaper than using cables between them. A typical commercial communications satellite can relay 6000 telephone calls and two television programs at the same time; future ones will have still more capacity.

Example   Find the altitude of a geostationary orbit.

Solution   The speed of something in uniform circular motion with the period T is given by Eq. (7–2) as

$$v = \frac{2\pi r}{T}$$

where $T = 1$ day $= 86,400$ s. Another formula for the satellite speed is given by Eq. (7–12),

$$v = \sqrt{\frac{r_e^2 g_e}{r}}$$

Setting the two formulas equal enables us to eliminate v:

$$\frac{2\pi r}{T} = \sqrt{\frac{r_e^2 g_e}{r}}$$

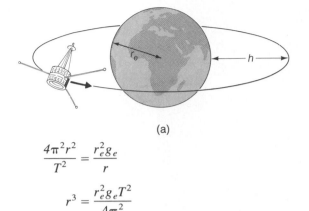

(a)

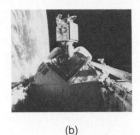

(b)

FIG. 7–21 (a) The altitude of a geostationary satellite is 35.9×10^6 m. (b) Communications satellite being launched from the cargo bay of the Space Shuttle.

$$\frac{4\pi^2 r^2}{T^2} = \frac{r_e^2 g_e}{r}$$

$$r^3 = \frac{r_e^2 g_e T^2}{4\pi^2}$$

Therefore

$$r = \sqrt[3]{\frac{r_e^2 g_e T^2}{4\pi^2}} = \sqrt[3]{\frac{(6.4 \times 10^6 \, \text{m})^2 (9.8 \, \text{m/s}^2)(8.64 \times 10^4 \, \text{s})^2}{4\pi^2}}$$

$$= \sqrt[3]{759 \times 10^{20}} \text{m} = \sqrt[3]{75.9 \times 10^{21}} \text{m} = 4.23 \times 10^7 \, \text{m}$$

(Appendix B–5 explains how to find roots such as this with a calculator.) The corresponding altitude h (Fig. 7–21) above the earth's surface is

$$h = r - r_e = 42.3 \times 10^6 \, \text{m} - 6.4 \times 10^6 \, \text{m} = 35.9 \times 10^6 \, \text{m}$$

which is about 22,000 miles. ∎

7–8 WEIGHTLESSNESS

Because an earth satellite is always falling toward the earth, an astronaut inside one feels "weightless." In reality, there *is* a gravitational force acting on him; what is missing to his senses is the upward reaction force provided by a stationary platform underneath him—the seat of a chair, the floor of a room, the ground itself. Instead of pushing back, the floor of the satellite falls just as fast as he does toward the earth.

It is useful to distinguish between the *actual weight* of an object, which is the gravitational force acting on it, and its *apparent weight,* which is the force it exerts on whatever it rests upon. We can think of the apparent weight of a person as the reading on a bathroom spring scale the person is standing on. An astronaut in an earth satellite has no apparent weight because he or she does not press down on the floor of the satellite.

A person jumping off a diving board is just as "weightless" in his descent as an astronaut, since nothing restricts his acceleration toward the earth either. But a person standing on the ground is acted upon by *both* the downward force of gravity and the upward reaction force of the ground: The latter force is what prevents him from simply dropping all the way down to the center of the earth. The human body (indeed, all

Actual and apparent weight

FIG. 7–22 An astronaut in orbit is weightless. (Photo: NASA)

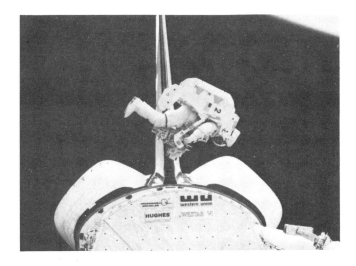

living organisms on the earth) evolved in the presence of both these forces, and various body functions, such as blood circulation, do not seem to take place efficiently in a "weightless" state. Future satellites and other spacecraft designed for long journeys may be set in rotation so that inertia will cause astronauts to press against the cabin sides, which will then press back (action-reaction again) and so bring about a situation corresponding to that on the earth's surface.

IMPORTANT TERMS

An object traveling in a circle at constant speed is said to be undergoing **uniform circular motion.**

The velocity of an object in uniform circular motion continually changes in direction although its magnitude remains constant. The acceleration that causes the object's velocity to change is called **centripetal acceleration,** and it points toward the center of the object's circular path.

The inward force that provides an object in uniform circular motion with its centripetal acceleration is called **centripetal force.**

Newton's **law of universal gravitation** states that every object in the universe attracts every other object with a force directly proportional to both their masses and inversely proportional to the square of the distance separating them.

IMPORTANT FORMULAS

Period of orbit: $T = \dfrac{2\pi r}{v}$

Centripetal acceleration: $a_c = \dfrac{v^2}{r}$

Centripetal force: $F_c = \dfrac{mv^2}{r}$

Law of gravitation: $F_{grav} = G\,\dfrac{m_A m_B}{r^2}$

Satellite orbit: $v = \sqrt{rg} = \sqrt{\dfrac{r_e^2 g_e}{r}}$

MULTIPLE CHOICE

1. In order to cause a moving object to pursue a circular path, it is necessary to apply
   a. inertial force.   b. gravitational force.
   c. frictional force.   d. centripetal force.

2. An object traveling in a circle at constant speed
   a. has a constant velocity.
   b. is not accelerated.
   c. has an inward radial acceleration.
   d. has an outward radial acceleration.

3. The acceleration of an object undergoing uniform circular motion is constant in
   a. magnitude only.
   b. direction only.

c. both magnitude and direction.

d. neither magnitude nor direction.

4. The centripetal force on a car rounding a curve on a level road is provided by

a. gravity.

b. friction between its tires and the road.

c. the torque applied to its steering wheel.

d. its brakes.

5. The centripetal force needed to keep the earth in orbit is provided by

a. inertia.

b. its rotation on its axis.

c. the gravitational pull of the sun.

d. the gravitational pull of the moon.

6. The radius of the path of an object in uniform circular motion is doubled. The centripetal force needed if its speed remains the same is

a. half as great as before.

b. the same as before.

c. twice as great as before.

d. four times as great as before.

7. If the earth were three times farther from the sun than it is now, the gravitational force exerted on it by the sun would be

a. three times as large as it is now.

b. nine times as large as it is now.

c. one-third as large as it is now.

d. one-ninth as large as it is now.

8. The moon's mass is 1.2% of the earth's mass. Relative to the gravitational force the earth exerts on the moon, the gravitational force the moon exerts on the earth

a. is smaller.

b. is the same.

c. is greater.

d. depends on the phase of the moon.

9. The speed needed to put a satellite in orbit does not depend upon

a. the radius of the orbit.

b. the shape of the orbit.

c. the value of g at the orbit.

d. the mass of the satellite.

10. An astronaut is "weightless" in a spacecraft

a. as it takes off from the earth.

b. in certain orbits only.

c. in all orbits.

d. only when the escape speed is exceeded.

11. A $\frac{1}{2}$-kg ball moves in a circle 0.4 m in radius at a speed of 4 m/s. Its centripetal acceleration is

a. 10 m/s^2.

b. 20 m/s^2.

c. 40 m/s^2.

d. 80 m/s^2.

12. The centripetal force on the ball of Question 11 is

a. 10 N.

b. 20 N.

c. 40 N.

d. 80 N.

13. A 3200-lb car moves in a circle 100 ft in radius at a speed of 20 ft/s. Its centripetal acceleration is

a. 0.2 ft/s^2.

b. 4 ft/s^2.

c. 20 ft/s^2.

d. 400 ft/s^2.

14. The centripetal force on the car of Question 13 is

a. 400 lb.

b. 640 lb.

c. 3200 lb.

d. 12,800 lb.

15. A toy cart at the end of a string 0.7 m long moves in a circle on a table. The cart has a mass of 2 kg and the string has a breaking strength of 40 N. The maximum speed of the cart is approximately

a. 1.9 m/s.

b. 3.7 m/s.

c. 11.7 m/s.

d. 16.7 m/s.

16. On a rainy day the coefficient of friction between a car's tires and a certain level road surface is reduced to half its usual value. The maximum safe speed for rounding the curve is

a. unchanged.

b. reduced to 25% of its usual value.

c. reduced to 50% of its usual value.

d. reduced to 71% of its usual value.

17. A car is traveling at 50 km/h on a road such that the coefficient of friction between its tires and the road is 0.5. The minimum turning radius of the car is

a. 9.9 m.

b. 39.4 m.

c. 947 m.

d. 5100 m.

18. A 2-kg stone at the end of a string 1 m long is whirled in a vertical circle. The tension in the string is 52 N when the stone is at the bottom of the circle. The stone's speed then is

a. 4 m/s.

b. 5 m/s.

c. 6 m/s.

d. 7 m/s.

19. A 3.2-lb stone at the end of a string 2 ft long is whirled in a vertical circle. The tension in the string is 5 lb when the stone is at the top of the circle. The stone's velocity then is

a. 1.8 ft/s.

b. 6 ft/s.

c. 10 ft/s.

d. 12.8 ft/s.

20. Mars is about 1.5 times as far from the sun as Earth and its mass is about 0.1 times Earth's mass. Relative to the gravitational force the sun exerts on Earth, the force it exerts on Mars is about

a. 0.0044 as much.

b. 0.0067 as much.

c. 0.044 as much.

d. 0.067 as much.

21. A woman has a mass of 60 kg at the earth's surface. At a height of one earth's radius above the surface her mass is
    a. 15 kg.
    b. 30 kg.
    c. 60 kg.
    d. 120 kg.

22. A man has a weight of 160 lb at the earth's surface. At a height of one earth's radius above the surface his weight is
    a. 40 lb.
    b. 80 lb.
    c. 160 lb.
    d. 320 lb.

23. Earth satellite A has an orbit four times greater in radius than satellite B. The orbital speed of A is
    a. $v_B/4$.
    b. $v_B/2$.
    c. $2v_B$.
    d. $4v_B$.

24. A 200-kg satellite circles the earth in an orbit 7×10^6 m in radius. At this altitude $g = 8.2$ m/s². The speed of the satellite is
    a. 38 m/s.
    b. 0.85 km/s.
    c. 7.6 km/s.
    d. 7.9 km/s.

EXERCISES

7–2 Centripetal Acceleration

1. Under what circumstances, if any, can an object move in a circular path without being accelerated?

2. A car makes a clockwise turn on a level road at too high a speed and overturns. Do its left or its right wheels leave the ground first?

3. Where should you stand on the earth's surface to experience the most centripetal acceleration? The least?

4. A car rounds a curve of radius 70 m at a speed of 12 m/s on a level road. Find its centripetal acceleration.

5. A phonograph record 12 in. in diameter rotates $33\frac{1}{3}$ times per minute. (a) What is the linear speed of a point on its rim in ft/s? (b) What is the centripetal acceleration of a point on its rim?

6. The minute hand of a large clock is 0.5 m long. (a) What is the linear speed of its tip in m/s? (b) What is the centripetal acceleration of the tip of the hand?

7. What is the minimum radius at which an airplane flying at 300 m/s can make a U-turn if its centripetal acceleration is not to exceed $4g$?

8. If the earth were to spin so fast that a person at the equator were weightless, what would the length of the day be?

9. An astronaut in training is seated at the end of a horizontal arm 7 m long. How many revolutions per second must the arm make for the astronaut to experience a horizontal acceleration of $4g$?

7–3 Magnitude of Centripetal Force

10. What is the centripetal force needed to keep a 3-kg mass moving in a circle of radius 0.5 m at a speed of 8 m/s?

11. A string 1 m long breaks when its tension is 100 N. What is the greatest speed at which it can be used to whirl a 1-kg stone? (Neglect the gravitational pull of the earth on the stone.)

12. What is the centripetal force needed to keep a 130-lb skater moving in a circle of radius 10 ft at a speed of 5 ft/s?

13. A 2000-kg car is rounding a curve of radius 200 m on a level road. The maximum frictional force the road can exert on the tires of the car is 4000 N. What is the highest speed at which the car can round the curve?

14. The maximum frictional force a road can exert on the tires of a certain 2500-lb car is 1500 lb. What is the highest speed in miles per hour at which the car can safely round a turn of radius 90 ft on a level road?

15. A car whose speed is 30 mi/h is traveling on a level road where the coefficient of static friction between its tires and the road is 0.7. Find the minimum turning radius of the car.

16. A box is resting on the flat floor in the rear of a station wagon moving at 15 m/s. What is the minimum radius of a turn the station wagon can make if the box is not to slip? Assume that $\mu_s = 0.4$.

17. A dime is placed 10 cm from the center of a record. The coefficient of friction between coin and record is 0.3. Will the coin remain where it is or will it fly off when the record turns at $33\frac{1}{3}$ rev/min? At 78 rev/min?

18. British railway engineers refer to the centripetal acceleration of a train rounding a curve in terms of "cant deficiency," the banking angle that would be needed for a passenger to feel no sideways force relative to the train. What is the centripetal acceleration that corresponds to the maximum permitted cant deficiency of 4.25°?

19. An airplane traveling at 500 km/h banks at an angle of 45° as it makes a turn. What is the radius of the turn in kilometers? Assume that the rudder is not used in making the turn.

20. A highway curve has a radius of 300 m. (a) At what angle should it be banked for a traffic speed of 100 km/h? (b) If the curve is not banked, what is the minimum coefficient of friction required between tires and road?

21. A curve in a road 8 m wide has a radius of 60 m. How much higher than its inner edge should the outer edge of the road be if it is to be banked properly for cars traveling at 30 km/h?

22. A car whose speed is 90 km/h rounds a curve 180 m in radius that is properly banked for a speed of 45 km/h. Find the minimum coefficient of friction between tires and road that will permit the car to make the turn.

7-4   The Centrifuge

23. A liquid sample in a centrifuge is being rotated in a circular path 6 cm in radius. If the centripetal acceleration of the sample is 10^5 g, at how many rpm is the centrifuge being spun?

24. A girl on a merry-go-round who is sitting 5 m from its axis is holding the string of a yo-yo. If the string makes an angle of 3° with the vertical, how long does it take the merry-go-round to make a complete rotation?

25. Show that $\cos \theta = g/4\pi^2 f^2 L$ for the conical pendulum of Fig 7-11, where f is the number of revolutions per second the particle makes.

26. (a) Find the tension T in the wire and the force F the horizontal beam exerts when the structure shown is at rest. What is the direction of **F**? (b) Find T and F when the structure is rotating twice per second. What is the direction of **F** now? Neglect the masses of the beam and of the wire.

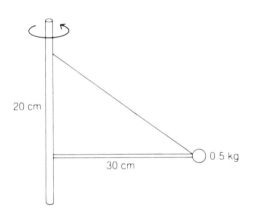

7-5   Motion in a Vertical Circle

27. A person swings an iron ball in a vertical circle at the end of a string. At what point in the circle is the string most likely to break? Why?

28. An airplane makes a vertical circle in which it is upside down at the top of the loop. Will the pilot fall out of his seat if he has no belt to hold him in place?

29. A string 0.8 m long is used to whirl a 2-kg stone in a vertical circle. What must be the speed of the stone at the top of the circle if the string is to be just taut? How does this speed compare with that required for a 1-kg stone in the same situation?

30. The 200-g head of a golf club moves at 45 m/s in a circular arc of 1 m radius. How much force must the player exert on the handle of the club to prevent it from flying out of his hands at the bottom of the swing? Assume that the shaft of the club has negligible mass.

31. A road has a hump 12 m in radius. What is the minimum speed at which a car will leave the road at the top of the hump?

32. A string 1 m long is used to whirl a $\frac{1}{2}$-kg stone in a vertical circle. What is the tension in the string when the stone is at the top of the circle moving at 5 m/s?

33. A physics instructor swings a pail of water in a vertical circle 4 ft in radius. What is the maximum time per revolution if the water is not to spill?

34. If the apparent weight of a person in an airplane is not to exceed 2 *mg*, find the maximum radius of a circular arc in which the airplane can pull out of a dive at a constant speed of 300 km/h.

35. A ball of mass m is swung in a vertical circle on a string just fast enough for the string to be barely taut at the top of the circle. Find the tension in the string at the bottom of the circle.

7-6   Gravitation

36. A track team on the moon could set new records for the high jump (if they did not need space suits, of course) because of the smaller gravitational pull there. Could sprinters also improve their times for the 100-m dash?

37. A 2-kg mass is 1 m away from a 5-kg mass. What is the gravitational force (a) that the 5-kg mass exerts upon the 2-kg mass; (b) that the 2-kg mass exerts upon the 5-kg mass? (c) If both masses are free to move, what are their respective accelerations in the absence of other forces?

38. Two identical lead spheres whose centers are 2 m apart attract each other with a force of 10^{-5} N. Find the mass of each sphere.

39. A bull and a cow elephant, each weighing 4000 lb, attract each other gravitationally with a force of 10^{-5} lb. How far apart are they?

40. A woman has a mass of 55 kg on the earth's surface. (a) What would she weigh at a distance from the earth's

center of three times the earth's radius? (b) What would her mass be there?

41. A man weighs 180 lb on the earth's surface. (a) What would he weigh at a distance from the earth's center of three times the earth's radius? (b) What would his mass be there?

42. The mass of the planet Saturn is 5.7×10^{26} kg and that of the sun is 2.0×10^{30} kg. The average distance between them is 1.4×10^{12} m. (a) What is the gravitational force the sun exerts on Saturn? (b) Assuming that Saturn has a circular orbit, what is its orbital speed?

43. The earth's average orbital radius is 1.5×10^{11} m and its average orbital speed is 3.0×10^{4} m/s. From these figures, together with the value of G, find the mass of the sun.

7–7   Earth Satellites

44. An earth satellite is placed in an orbit whose radius is half that of the moon's orbit. Is its time of revolution longer or shorter than that of the moon?

45. Two satellites are launched from a certain station with the same initial speeds relative to the earth's surface. One is launched toward the west, the other toward the east. Will there be any difference in their orbits? If so, what will the difference be and why?

46. For the moon to have the same orbit it has now, what would its speed have to be if the moon's mass were double its present mass?

47. The radius of the earth is 6.4×10^{6} m. What is the acceleration of a meteor when it is 8×10^{6} m from the center of the earth?

48. What is the acceleration of a meteor when it is one earth's radius above the surface of the earth? Two earth's radii?

49. A spacecraft with the kinetic energy KE circles the earth in an orbit just above its surface. How much kinetic energy would it need to escape permanently from the earth?

50. A satellite is to be put into orbit around the moon just above its surface. What should its speed be? Assume that the moon's radius is half that of the earth and that the acceleration of gravity at its surface is $g/6$.

51. Find the speed of an earth satellite whose orbit is 400 km above the earth's surface. What is the period of the orbit?

52. Find the altitude of an earth satellite whose period is 40 h.

53. The new Global Positioning System (GPS) developed by the United States, which is scheduled to become operational late in 1988, will employ 18 satellites that circle the earth in 12 h. Find the orbit radius that corresponds to this period.

ANSWERS TO MULTIPLE CHOICE

1. d	**7.** d	**13.** b	**19.** d
2. c	**8.** b	**14.** a	**20.** c
3. a	**9.** d	**15.** b	**21.** c
4. b	**10.** c	**16.** d	**22.** a
5. c	**11.** c	**17.** b	**23.** b
6. a	**12.** b	**18.** a	**24.** c

8

ROTATIONAL MOTION

Until now we have been considering only translational motion, motion in which something moves from one place to another. But rotational motion is also common. Wheels, pulleys, gears, propellers, drills, and phonograph records all rotate while carrying out their jobs. In the atomic world protons, neutrons, and electrons all rotate, and these rotations in part govern how they interact to form atoms and how atoms interact to form molecules, solids, and liquids. In this chapter our chief concern will be the rotational motion of a rigid body about a fixed axis. As we shall find, all the formulas that describe such motion are exact analogs of the formulas we have already used to describe translational motion.

8–1   ANGULAR MEASURE

We are accustomed to measuring angles in degrees, where 1° is defined as 1/360 of a full rotation. That is, a complete turn represents 360°.

CHAPTER OBJECTIVES

Completing this chapter should enable you to:

1. Use the radian as a unit of angle and convert radians to degrees and vice versa.

2. Find the arc length that corresponds to a given angle in a circle of known radius.

3. Find the angular speed and acceleration of a particle in circular motion and relate them to the particle's linear speed and acceleration.

4. Distinguish between the angular and centripetal accelerations of a particle moving in a circle.

5. Calculate the kinetic energy of a rotating body.

6. Determine the moment of inertia of an object about a certain axis of rotation.

7. Analyze the combined translational and rotational motion of a rolling object.

8. Find the angular acceleration of a rigid body acted on by a torque and calculate the angular speed of the body and the total angle through which it will have turned after a given time.

9. Use the formula $P = \tau\omega$ to solve problems that involve power, torque, and angular speed in rotational motion.

10. Describe how the principle of conservation of angular momentum is connected with the stability of rotating objects.

11. Relate the quantities and formulas that apply to linear motion to the corresponding ones that apply to angular motion.

A more suitable angular unit in technology is the *radian* (rad). The radian is defined with the help of a circle drawn with its center at the vertex of the angle in question. If the circle's radius is r and the arc cut by the angle is s as in Fig. 8–1, then the angle in radians is given by

The radian is a unit of angular measure

$$\theta = \frac{s}{r} = \frac{\text{arc length}}{\text{radius}} \qquad\qquad \textit{Radian measure} \quad (8–1)$$

That is, the angle θ between two radii of a circle, in radian measure, is the ratio of the arc s to the radius r. Evidently an angle of 1 rad has an arc length that is the same as the radius.

It is easy to find the conversion factor between degrees and radians and vice versa. We note that there are 360° in a complete circle, while the number of radians in a complete circle is

Conversion between degrees and radians

$$\theta = \frac{s}{r} = \frac{2\pi r}{r} = 2\pi$$

FIG. 8–1 The ratio of arc to radius gives the magnitude of an angle in radians.

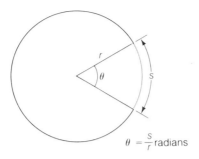

$$\theta = \frac{s}{r}\text{ radians}$$

FIG. 8–2 (a) 360° = 2π rad, so 1 rad = 57.30°. (b) A circle contains 2π rad, which is 6.28 rad.

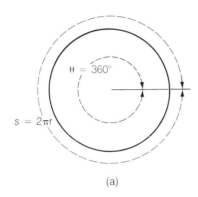

(a)

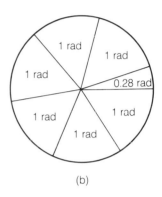

(b)

because the circumference of a circle of radius r is $2\pi r$ (Fig. 8–2). Hence

$$360° = 2\pi \text{ rad}$$

from which we find that

$$1° = 0.01745 \text{ rad} \quad \text{and} \quad 1 \text{ rad} = 57.30° \qquad \textit{Radians and degrees}$$

The conversions (2π rad/360°) and (360°/2π rad) are easier to remember.

The radian has no dimensions

The radian is an odd kind of unit because it has no dimensions—an angle expressed in radians is specified by a ratio of lengths, so it is really a pure number. In calculations the unit "rad" is always dropped at the end unless the result is an angular quantity.

Example A phonograph record 30 cm in diameter turns through an angle of 120°. How far does a point on its rim travel?

Solution First the angle is converted from degrees to radians:

$$\theta = (120°)\left(\frac{2\pi \text{ rad}}{360°}\right) = 2.09 \text{ rad}$$

The radius of the record is 15 cm, and so, from Eq. (8–1),

$$s = r\theta = (15 \text{ cm})(2.09 \text{ rad}) = 31.4 \text{ cm} \qquad ■$$

Example A television image with 525 horizontal lines is being displayed on a picture tube whose screen is 50 cm high. If a viewer's eyes can resolve detail to 0.0003 rad—

about 1′ (1 minute of arc), where 60′ = 1°—how far away should he be from the screen in order to just be able to see the separate lines?

Solution   The distance between adjacent lines is s = 0.5 m/525, hence

$$r = \frac{s}{\theta} = \frac{0.5\,\text{m}}{(525)(0.0003)} = 3.17\,\text{m}$$ ■

Sometimes it is useful to express angles in radian measure in terms of π itself. For example, an angle of 90° is $\frac{1}{4}$ of a complete circle, and so

Expressing angles in terms of π

$$90° = (\tfrac{1}{4}\text{circle})\left(2\pi\,\frac{\text{rad}}{\text{circle}}\right) = \pi/2\,\text{rad}$$

Of course, this has the same numerical value as

$$(90°)\left(\frac{2\pi\,\text{rad}}{360°}\right) = 1.571\,\text{rad}$$

since $\pi/2$ = 1.571. Table 8–1 gives some other examples of radian measure in terms of π and as decimals.

0° =	0 rad	
30° =	$\pi/6$ rad =	0.524 rad
45° =	$\pi/4$ rad =	0.785 rad
60° =	$\pi/3$ rad =	1.047 rad
90° =	$\pi/2$ rad =	1.571 rad
180° =	π rad =	3.142 rad
270° =	$3\pi/2$ rad =	4.712 rad
360° =	2π rad =	6.283 rad
720° =	4π rad =	12.566 rad

TABLE 8–1
Degrees to radians

8–2 ANGULAR SPEED

If a rotating body turns through the angle θ in the time t, its average *angular speed* ω (Greek letter *omega*) is

Angular speed

$$\omega = \frac{\theta}{t} \qquad\qquad\qquad \textit{Angular speed} \quad (8\text{–}2)$$

If θ is in radians and t in seconds, which are the usual units for these quantities, the unit of ω is the rad/s.

Two other common units of angular speed are the revolution per second (rps) and the revolution per minute (rpm), where

Revolutions per second and per minute

$$1\frac{\text{rev}}{\text{s}} = \left(1\frac{\text{rev}}{\text{s}}\right)\left(2\pi\frac{\text{rad}}{\text{rev}}\right) = 2\pi\,\text{rad/s} = 6.28\,\text{rad/s}$$

FIG. 8–3 The angular speed of a particle in uniform circular motion is $\omega = v/r$.

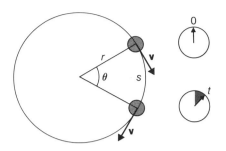

Axis of rotation

FIG. 8–4 The axis of rotation is that line of particles which does not move in a rotating body. (It may be a line in space.)

$$1\frac{\text{rev}}{\text{min}} = \left(1\frac{\text{rev}}{\text{min}}\right)\left(2\pi\frac{\text{rad}}{\text{rev}}\right)\left(\frac{1}{60\,\text{s/min}}\right) = \frac{\pi}{30}\,\text{rad/s} = 0.105\,\text{rad/s}$$

Let us consider a particle moving with the uniform speed v in a circle of radius r, as in Fig. 8–3. This particle travels the distance $s = vt$ in the time t. The angle through which it moves in that time is

$$\theta = \frac{s}{t} = \frac{vt}{r}$$

so that its angular speed is

$$\omega = \frac{\theta}{t} = \frac{vt}{rt}$$

or

Relation between linear speed and angular speed

$$\omega = \frac{v}{r} \qquad\qquad\qquad \textit{Angular speed} \quad (8\text{–}3)$$

$$\text{Angular speed} = \frac{\text{linear speed}}{\text{path radius}}$$

The above relationship can be written in another way:

$$v = \omega r \qquad\qquad\qquad\qquad\qquad\qquad (8\text{–}4)$$

$$\text{Linear speed} = \text{angular speed} \times \text{path radius}$$

The formulas of this section are valid only when ω is expressed in radian measure.

Axis of rotation

The *axis of rotation* of a rigid body turning in place is that line of particles which does not move (Fig. 8–4). Sometimes the axis of rotation is a line in space. All other particles of the body move in circles about the axis. Since $v = \omega r$, the farther a particle is from the axis, the greater its linear speed, although all the particles of the body (except those on the axis) have the same angular speed.

Example Find the linear speeds of points 2 cm and 15 cm from the axis of a phonograph record rotating at $33\frac{1}{3}$ rpm (Fig. 8–5).

Solution The angular speed of the record is

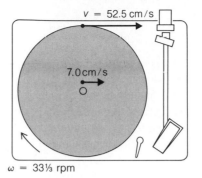

FIG. 8–5

$$\omega = (33\tfrac{1}{3}\,\text{rpm})\left(0.105\,\frac{\text{rad/s}}{\text{rpm}}\right) = 3.50\,\text{rad/s}$$

Hence a point 2 cm from the axis has a linear velocity of

$$v = \omega r = \left(3.5\,\frac{\text{rad}}{\text{s}}\right)(2\,\text{cm}) = 7.0\,\text{cm/s}$$

while a point on the record's rim, where $r = 15$ cm, has a velocity of

$$v = \left(3.5\,\frac{\text{rad}}{\text{s}}\right)(15\,\text{cm}) = 52.5\,\text{cm/s}$$ ∎

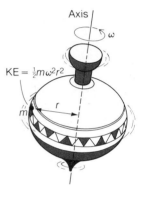

FIG. 8–6 The kinetic energy of each particle of a rotating body depends upon the square of its distance r from the axis of rotation.

8–3 ROTATIONAL KINETIC ENERGY

A rotating body has kinetic energy because the particles that make it up are moving, even though the body as a whole remains in place. The speed of a particle that is the distance r from the axis of a rigid body rotating with the angular speed ω is, as we know, $v = \omega r$ (Fig. 8–6). If the particle's mass is m, its kinetic energy is therefore

$$\text{KE} = \tfrac{1}{2}\,mv^2 = \tfrac{1}{2}\,m\omega^2 r^2 \tag{8–5}$$

Kinetic energy of particle in a rotating body

The body consists of many particles that need not have the same mass or be the same distance from the axis. However, all the particles have the same angular speed ω, which enables us to analyze the mechanics of rotating bodies in a very convenient manner. We start by noting that the total kinetic energy of all the particles may be written

$$\text{KE} = \Sigma\,(\tfrac{1}{2}mv^2) = \tfrac{1}{2}(\Sigma\,mr^2)\omega^2 \tag{8–6}$$

Kinetic energy of entire rotating body

where the symbol Σ means, as mentioned before, "sum of." Equation (8–6) states that the kinetic energy of a rotating rigid body is equal to one-half the sum of the mr^2 values of its particles multiplied by the square of its angular speed ω.

Moment of inertia is rotational analog of mass

The quantity

$$I = \Sigma\ mr^2 \qquad\qquad\qquad \textit{Moment of inertia} \quad (8-7)$$

is known as the *moment of inertia* of the body. It has the same value regardless of the body's state of motion. The farther a given particle is from the axis of rotation, the faster it moves and the more it contributes to the kinetic energy of the body. The moment of inertia of a body depends upon the way in which its mass is distributed relative to its axis of rotation; it is perfectly possible for one body to have a greater moment of inertia than another even though its mass may be much the smaller of the two.

Kinetic energy of rotating body in terms of moment of inertia

The kinetic energy of a body of moment of inertia I rotating with the angular speed ω is therefore

$$\mathrm{KE} = \tfrac{1}{2} I\omega^2 \qquad\qquad\qquad \textit{Rotational kinetic energy} \quad (8-8)$$

Evidently the rotational analog of mass is moment of inertia, just as the rotational analog of linear speed is angular speed. We shall find further support for the correspondence of mass and moment of inertia later in this chapter.

8−4   MOMENT OF INERTIA

How to calculate moments of inertia

A rigid body may be considered to be made up of a large number of separate particles whose masses are m_1, m_2, m_3, and so on (Fig. 8−7). To find the moment of inertia of such a body about a given axis, we multiply the mass of each of these particles by the square of its distance from the axis (r_1^2, r_2^2, r_3^2, and so on) and add all the mr^2 values. That is,

$$I = \Sigma\ mr^2 = m_1 r_1^2 + m_2 r_2^2 + m_3 r_3^2 + \cdots \qquad\qquad (8-9)$$

The unit of I is the kg · m^2 in the metric system and the slug · ft^2 in the British system.

Example   Find the moment of inertia of a thin ring of mass M and average radius R about an axis passing through its center and perpendicular to the plane in which it lies.

Solution   As in Fig. 8−8, we proceed by subdividing the ring into n segments, each of which is at a distance R from the axis. Hence

$$I = m_1 R^2 + m_2 R^2 + m_3 R^2 + \cdots + m_n R^2$$
$$= (m_1 + m_2 + m_3 + \cdots + m_n)R^2$$

But the sum of the masses of the segments is the same as the total mass M of the ring, and so

$$I = MR^2 \qquad\qquad\qquad\qquad\qquad \blacksquare$$

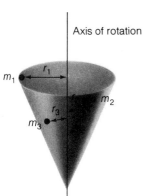

Axis of rotation

FIG. 8−7 A rigid body consists of a large number of particles each of which has a certain mass m and distance r from the axis.

When a body consists of a continuous distribution of matter, the more particles we imagine it to contain, the more accurate will be our value of its moment of inertia. While I can be calculated for a few simple bodies without difficulty by Eq. (8−9), in general either much labor or the use of advanced mathematics is required. Figure 8–9 gives the moments of inertia of several regularly shaped bodies in terms of the total mass M and dimensions of each. In Section 12–7 we shall learn of a simple experimental way to determine the moment of inertia of an object whose shape is too complex to permit I to be calculated readily.

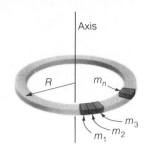

FIG. 8–8

Example   Rotating flywheels have been proposed for energy storage in electric power plants. The flywheels would be set in motion during off-peak periods by electric motors, which would act as generators to return the stored energy during times of heavy demand. (a) Find the kinetic energy of a 10^5-kg cylindrical flywheel whose radius is 2 m that rotates at 400 rad/s. (b) For how many hours could the flywheel supply energy at the rate of 1 MW?

Solution   (a) From Fig. 8–9 the moment of inertia of a solid cylinder is

$$I = \tfrac{1}{2} MR^2$$

The kinetic energy of the cylinder is therefore

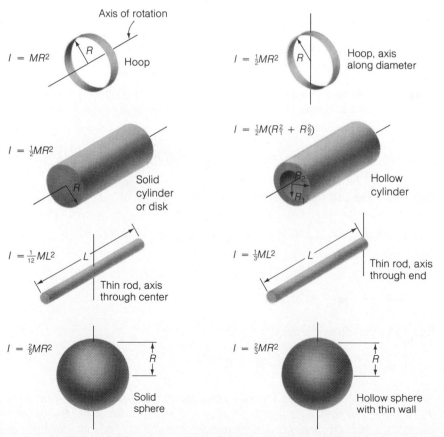

FIG. 8–9 Moments of inertia of various bodies each of mass M, about indicated axes.

$$KE = \tfrac{1}{2} I\omega^2 = \tfrac{1}{2}(\tfrac{1}{2}MR^2)\omega^2 = \tfrac{1}{4} MR^2\omega^2$$

$$= (\tfrac{1}{4})(10^5 \text{ kg})(2 \text{ m})^2(400 \text{ rad/s})^2 = 1.6 \times 10^{10} \text{ J}$$

(b) Since power = energy/time and 1 MW = 10^6 W, we have

$$t = \frac{KE}{P} = \frac{1.6 \times 10^{10} \text{ J}}{10^6 \text{ W}} = 1.6 \times 10^4 \text{ s}$$

There are 3600 s in an hour, and thus

$$t = \frac{1.6 \times 10^4 \text{ s}}{3600 \text{ s/h}} = 4.4 \text{ h}$$
∎

8–5   COMBINED TRANSLATION AND ROTATION

Total kinetic energy is sum of translational kinetic energy and rotational kinetic energy

When a rigid body is both moving through space and rotating, its total kinetic energy is the sum of its translational and rotational kinetic energies. The translational kinetic energy is calculated on the basis that the body is a particle whose linear speed is the same as that of the body's center of gravity; the rotational kinetic energy is calculated on the basis that the body is rotating about an axis that passes through the center of gravity. Thus

$$KE = KE_{\text{translation}} + KE_{\text{rotation}} = \tfrac{1}{2} mv^2 + \tfrac{1}{2} I\omega^2 \tag{8–10}$$

where m is the body's mass, v is the speed of its center of gravity, I is its moment of inertia about an axis through the center of gravity, and ω is its angular speed about that axis.

Example   A 16-lb wooden ball 1.0 ft in diameter is rolling at a speed of 10 ft/s. Find its total kinetic energy.

Solution   The ball's moment of inertia is, from Fig. 8–9,

$$I = \tfrac{2}{5}mR^2 = \tfrac{2}{5}\left(\frac{w}{g}\right) R^2 = \tfrac{2}{5}\left(\frac{16 \text{ lb}}{32 \text{ ft/s}^2}\right)(0.5 \text{ ft})^2 = 0.05 \text{ slug} \cdot \text{ft}^2$$

and its angular speed is

$$\omega = \frac{v}{r} = \frac{10 \text{ ft/s}}{0.5 \text{ ft}} = 20 \text{ rad/s}$$

The ball's total kinetic energy is

$$KE = \tfrac{1}{2} mv^2 + \tfrac{1}{2} I\omega^2 = \tfrac{1}{2}\frac{w}{g} v^2 + \tfrac{1}{2} I\omega^2$$

$$= \tfrac{1}{2}\left(\frac{16 \text{ lb}}{32 \text{ ft/s}^2}\right)(10 \text{ ft/s})^2 + \tfrac{1}{2}(0.05 \text{ slug} \cdot \text{ft}^2)(20 \text{ rad/s})^2$$

$$= 25 \text{ ft} \cdot \text{lb} + 10 \text{ ft} \cdot \text{lb} = 35 \text{ ft} \cdot \text{lb}$$
∎

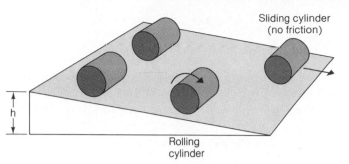

Sliding cylinder (no friction)

FIG. 8–10

h

Rolling cylinder

Example   Consider a cylinder of radius R and mass m that is at the top of an inclined plane (Fig. 8–10). Will it have a greater speed at the bottom if it slides down without friction or if it rolls down?

Solution   In the first case, we set the cylinder's initial potential energy of mgh equal to its final kinetic energy of $\frac{1}{2} mv^2$, and find that

$$PE = KE$$
$$mgh = \tfrac{1}{2} mv^2$$
$$v = \sqrt{2gh}$$

In the second case the cylinder has both translational and rotational kinetic energy at the bottom, so that

$$PE = KE$$
$$mgh = \tfrac{1}{2} mv^2 + \tfrac{1}{2} I\omega^2$$

The moment of inertia of the cylinder is $I = \frac{1}{2} mR^2$; if it rolls without slipping, its linear and angular speeds are related by the formula $\omega = v/R$. Hence

$$mgh = \tfrac{1}{2} mv^2 + \tfrac{1}{2}(\tfrac{1}{2} mR^2)\left(\frac{v^2}{R^2}\right) = \tfrac{1}{2} mv^2 + \tfrac{1}{4} mv^2 = \tfrac{3}{4} mv^2$$

$$v = \sqrt{\tfrac{4}{3} gh}$$

The cylinder moves more slowly when it rolls down the plane than when it slides without friction because some of the available energy is absorbed by its rotation. ■

8–6 ANGULAR ACCELERATION

A rotating body need not have a uniform angular speed ω, just as a moving particle need not have a uniform linear speed v. If the angular speed of a body changes by an amount $\Delta\omega$ in the time interval Δt, its average *angular acceleration* α (Greek letter *alpha*) is

Angular acceleration is analog of linear acceleration

$$\alpha = \frac{\Delta\omega}{\Delta t} \qquad\qquad \textit{Angular acceleration} \quad (8\text{--}11)$$

The unit of angular acceleration is the rad/s^2.

Tangential acceleration

A particle moving in a circle of radius r that experiences an angular acceleration α also experiences a linear acceleration a_T tangential to its path, since its orbital speed is changing (Fig. 8–11). From the definition $\omega = v/r$, if r is constant

$$\Delta\omega = \frac{\Delta v}{r} \qquad \text{and} \qquad \alpha = \frac{\Delta\omega}{\Delta t} = \frac{\Delta v}{r\Delta t}$$

Because the magnitude of the particle's linear acceleration along the direction of its velocity **v** is $a_T = \Delta v/\Delta t$, we see that

$$\alpha = \frac{a_T}{r} \qquad\qquad (8\text{--}12)$$

$$\text{Angular acceleration} = \frac{\text{tangential acceleration}}{\text{path radius}}$$

Conversely,

$$a_T = \alpha r \qquad\qquad (8\text{--}13)$$

Tangential acceleration = angular acceleration × path radius

We must be careful to distinguish between the *tangential acceleration* a_T of a particle, which represents a change in its speed, and its *centripetal acceleration* a_c, which represents a change in its direction of motion. A particle in circular motion with the speed v has, as we know, the centripetal acceleration

$$a_c = \frac{v^2}{r}$$

Tangential and centripetal accelerations are perpendicular

directed toward the center of its circular path. The accelerations $\mathbf{a}_T$ and $\mathbf{a}_c$ are therefore always perpendicular (Fig. 8–12). *All* particles in circular motion have centripetal accelerations. Only those particles whose speed changes in magnitude, however, have

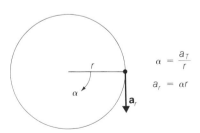

FIG. 8–11 The angular and tangential accelerations of a particle moving in a circle are proportional to each other.

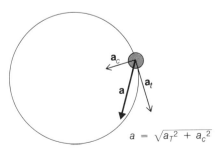

FIG. 8–12 The tangential and centripetal accelerations of a particle in circular motion are always perpendicular.

tangential accelerations. The centripetal acceleration of a particle moving in a circle of radius r can be expressed in terms of its angular speed ω as

$$a_c = \omega^2 r \qquad\qquad \textit{Centripetal acceleration} \quad (8-14)$$

because $v = \omega r$

Example   Find the total linear acceleration of a particle moving in a circle of radius 0.4 m, at the instant when the angular speed is 2 rad/s and the angular acceleration is 5 rad/s².

Solution   From Eqs. (8–13) and (8–14), for the particle's tangential and centripetal accelerations, we have

$$a_T = \alpha r = \left(5\frac{\text{rad}}{\text{s}^2}\right)(0.4\,\text{m}) = 2\,\text{m/s}^2$$

$$a_c = \omega^2 r = \left(2\frac{\text{rad}}{\text{s}}\right)^2 (0.4\,\text{m}) = 1.6\,\text{m/s}^2$$

We must add $\mathbf{a}_T$ and $\mathbf{a}_c$ vectorially, since they are vector quantities and are in different directions. As shown in Fig. 8–12, the magnitude a of the vector sum $\mathbf{a}$ and $\mathbf{a}_T$ and $\mathbf{a}_c$ is

$$a = \sqrt{a_T^2 + a_c^2} = \sqrt{2^2 + 1.6^2}\ \text{m/s}^2 = 2.6\,\text{m/s}^2 \qquad\blacksquare$$

Let us consider a rigid body whose angular acceleration has the constant value α. The body's angular speed is ω_0 at $t = 0$ and at the later time $t = t$ it has changed to

$$\omega_f = \omega_0 + \alpha t \qquad\qquad \textit{Angular speed under constant acceleration} \quad (8-15)$$

Angular displacement and speed under constant acceleration

In Chapter 1 we obtained the formula

$$s = v_0 t + \tfrac{1}{2} a t^2$$

for the displacement of an accelerated body. By means of the same reasoning, the angular displacement θ of a rigid body undergoing an angular acceleration turns out to be

$$\theta = \omega_0 t + \tfrac{1}{2} \alpha t^2 \qquad\qquad \textit{Angular displacement} \quad (8-16)$$

Similarly

$$s = \left(\frac{v_0 + v_f}{2}\right) t$$

becomes

$$\theta = \left(\frac{\omega_0 + \omega_f}{2}\right) t \qquad\qquad \textit{Angular displacement} \quad (8-17)$$

The angular analog of the linear formula

$$v_f^2 = v_0^2 + 2as$$

is the equally useful

$$\omega_f^2 = \omega_0^2 + 2\alpha\theta \qquad \textit{Angular speed under constant acceleration} \quad (8-18)$$

Example   The speed of a motor increases from 120 rad/s to 180 rad/s in 20 s. How many revolutions does it make in this period of time?

Solution   The angular acceleration of the motor is

$$\alpha = \frac{\Delta\omega}{\Delta t} = \frac{\omega_f - \omega_0}{\Delta t} = \frac{(180 - 120)\,\text{rad/s}}{20\,\text{s}} = 3.0\,\text{rad/s}^2$$

We can use either Eq. (8–16) or (8–18) to find θ, since we know ω_0, ω_f, α, and t. From Eq. (8–16) we have

$$\theta = \omega_0 t + \tfrac{1}{2}\alpha t^2 = \left(120\,\frac{\text{rad}}{\text{s}}\right)(20\,\text{s}) + \tfrac{1}{2}\left(3.0\,\frac{\text{rad}}{\text{s}^2}\right)(20\,\text{s})^2 = 3000\,\text{rad}$$

If we were to use Eq. (8–18), we would rewrite it as

$$\theta = \frac{\omega_f^2 - \omega_0^2}{2\alpha}$$

and get the same result. Since there are 2π rad in a revolution,

$$\theta = \frac{3000\,\text{rad}}{2\pi\,\text{rad/rev}} = 477\,\text{revolutions} \qquad\qquad ■$$

8–7   TORQUE AND ANGULAR ACCELERATION

According to Newton's second law of motion, a net force applied to a body causes it to be accelerated. What can cause a body capable of rotation to experience an angular acceleration?

To fix our ideas, let us look once more at a single particle of mass m restricted to motion in a circle of radius r (Fig. 8–13). A force $\mathbf{F}$ that acts upon the particle tangent to the particle's path gives it the acceleration a_T according to the formula

$$F = ma_T \qquad\qquad (8-19)$$

The tangential acceleration a_T here is equal to αr, and so

FIG. 8–13

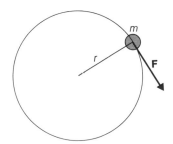

$$F = mr\alpha$$

Multiplying both sides of this equation by r gives

$$Fr = mr^2\alpha \tag{8–20}$$

We recognize Fr as the torque τ of the force F about the axis of the particle's rotation (see Section 4–2) and mr^2 as the particle's moment of inertia I. Equation (8–20) therefore states that

Torque produces angular acceleration

$$\tau = I\alpha \tag{8–21}$$

Torque = moment of inertia $\times$ angular acceleration

Although Eq. (8–21) was derived for the case of a single particle, it is also valid for any rotating body, provided that the torque and moment of inertia are both calculated about the same axis.

The formula $\tau = I\alpha$ is the fundamental law of motion for rotating bodies in the same sense that $F = ma$ is the fundamental law of motion for bodies moving through space. In rotational motion, torque plays the same role that force does in translational motion. The angular acceleration experienced by a body when the torque τ acts upon it is

Torque is rotational analog of force

$$\alpha = \frac{\tau}{I}$$

Thus α is directly proportional to the torque and inversely proportional to the body's moment of inertia.

Example A 2-kg grindstone 10 cm in radius is turning at 120 rad/s. The motor is switched off and a chisel is pressed against the grindstone with a force whose tangential component is 2 N (Fig. 8–14). How long will it take the grindstone to come to a stop?

Solution The grindstone is a solid cylinder and so its moment of inertia is

$$I = \tfrac{1}{2} MR^2 = \tfrac{1}{2}(2\,\text{kg})(0.1\,\text{m})^2 = 0.01\,\text{kg} \cdot \text{m}^2$$

The torque the chisel exerts on the grindstone is

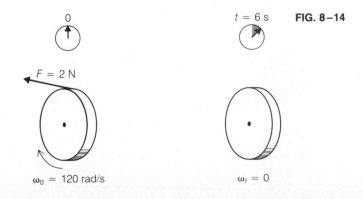

$t = 6\,\text{s}$ **FIG. 8–14**

$F = 2\,\text{N}$

$\omega_0 = 120\,\text{rad/s}$ $\omega_t = 0$

$$\tau = -Fr = -(2 \text{ N})(0.1 \text{ m}) = -0.2 \text{ N} \cdot \text{m}$$

where the minus sign is required because the torque acts to slow down the grindstone. The angular acceleration of the grindstone is

$$\alpha = \frac{\tau}{I} = \frac{-0.2 \text{ N} \cdot \text{m}}{0.01 \text{ kg} \cdot \text{m}^2} = -20 \text{ rad/s}^2$$

We now call upon Eq. (8–15), $\omega_f = \omega_0 + \alpha t$, and set $\omega_f = 0$ to correspond to the grindstone coming to a stop. The result is

$$\alpha t = -\omega_0$$

$$t = \frac{-\omega_0}{\alpha} = \frac{-120 \text{ rad/s}}{-20 \text{ rad/s}^2} = 6 \text{ s}$$ ∎

Center of gravity and line of action of applied force

When a net force acts on a body that is able to move freely, the body will experience both linear and angular accelerations unless the line of action of the force passes through the body's center of gravity (Fig. 8–15). In the latter case the body will be in rotational equilibrium, and if it was not rotating initially, it will keep its original orientation during its motion.

8–8 POWER

Mechanical energy is usually transmitted by rotary motion. The power output of almost every modern engine emerges via a rotating shaft, and more often than not this power is used in some form of rotation as well: The tires of a car, the propeller of a ship, the bit of a drill, the vanes of a centrifugal pump, and the rotor of a dynamo all function by turning.

Work and power in angular motion

The relationships among power, torque, and angular speed can be derived directly from the definitions of work and power. Let us consider a shaft of radius r on which a tangential force F acts, as in Fig. 8–16. After a time t the shaft has turned through the angle θ, the point at which the force is applied has moved through the distance s, and the force has done the amount of work

$$W = Fs = Fr\theta$$

FIG. 8–15 A force whose line of action passes through the center of gravity of a body cannot cause it to rotate.

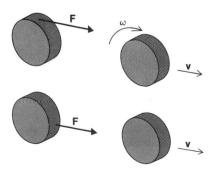

FIG. 8–16 The power transmitted by a rotating shaft is the product of the torque τ on the shaft and its angular speed ω, hence $P = \tau\omega$.

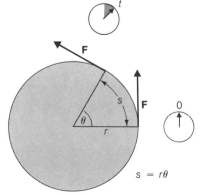

$$s = r\theta$$

Power is the rate at which work is done, and thus

$$P = \frac{W}{t} = \frac{Fr\theta}{t} = Fr\frac{\theta}{t}$$

We recognize that

$Fr = \tau =$ torque applied to shaft $\dfrac{\theta}{t} = \omega =$ angular speed of shaft

and therefore

$$W = \tau\theta \qquad\qquad\qquad\qquad\qquad Rotational\ work \quad (8-22)$$

Work = torque × angular displacement

$$P = \tau\omega \qquad\qquad\qquad\qquad\qquad Rotational\ power \quad (8-23)$$

Power = torque × angular speed

When energy is to be transmitted at a certain rate P by means of a rotating shaft, the higher the angular speed the lower the torque needed, and vice versa. Equation (8–23) holds only when the angular speed is constant and is the rotational analog of the formula

$$P = Fv$$

Power = force × speed

Example   A truck engine develops 230 hp at 5000 rpm. How much torque is exerted on the crankshaft?

Solution   We begin by converting 230 hp and 5000 rpm to their equivalents in lb · ft/s and rad/s:

$$P = (230\,\text{hp})\left(550\frac{\text{ft} \cdot \text{lb/s}}{\text{hp}}\right) = 1.27 \times 10^5\,\text{ft} \cdot \text{lb/s}$$

$$\omega = (5000\,\text{rpm})\left(0.1047\frac{\text{rad/s}}{\text{rpm}}\right) = 524\,\text{rad/s}$$

From Eq. (8–23) we obtain

$$\tau = \frac{P}{\omega} = \frac{1.27 \times 10^5\,\text{ft} \cdot \text{lb/s}}{524\,\text{rad/s}} = 242\,\text{lb} \cdot \text{ft} \qquad\qquad \blacksquare$$

Example   The alternator on the truck's engine produces 250 W of electric power when it turns at 60 rps (revolutions per second). The alternator's pulley has a radius of 4 cm. If the alternator is 95% efficient, find the difference between the tensions in the taut and slack parts of the V-belt connecting it to the engine.

Solution   The angular speed of the pulley is

$$\omega = \left(60\frac{\text{rev}}{\text{s}}\right)\left(2\pi\frac{\text{rad}}{\text{rev}}\right) = 377\,\text{rad/s}$$

Since Eff = $P_{\text{out}}/P_{\text{in}}$, the input power to the alternator pulley is

$$P_{\text{in}} = \frac{P_{\text{out}}}{\text{Eff}} = \frac{250\,\text{W}}{0.95} = 263\,\text{W}$$

FIG. 8–17 The torque on a V-belt pulley is due to the difference in the tensions in the taut and slack parts of the V-belt.

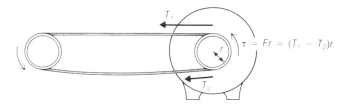

$$\tau = Fr = (T_1 - T_2)r$$

The torque on the alternator pulley is

$$\tau = \frac{P_{\text{in}}}{\omega}$$

and the net force on the pulley must therefore be

$$F = \frac{\tau}{r} = \frac{P_{\text{in}}/\omega}{r} = \frac{P_{\text{in}}}{\omega r} = \frac{263\,\text{W}}{(377\,\text{rad/s})(0.04\,\text{m})} = 17.4\,\text{N}$$

which is about 4 lb. This is the difference between the tensions in the two parts of the V-belt (Fig. 8–17). ■

8–9   ANGULAR MOMENTUM

We have all noticed the tendency of rotating objects to continue to spin unless they are slowed down by an outside agency. A top would spin indefinitely were it not for friction between its tip and the ground. The rotational quantity that corresponds to linear momentum is called *angular momentum*, and *conservation of angular momentum* is the formal way to describe the tendency of spinning objects to keep spinning.

The magnitude of the angular momentum L of a body depends upon its moment of inertia I and angular speed ω in the same way that its linear momentum depends upon its mass m and linear speed v:

$$L = I\omega \qquad\qquad \textit{Angular momentum} \quad (8\text{–}24)$$

Angular momentum = moment of inertia × angular speed

Angular momentum and angular velocity are vector quantities

Like linear momentum, angular momentum is a vector quantity. The direction associated with the angular momentum $\mathbf{L}$ of a spinning object is given by the right-hand rule shown in Fig. 8–18. This is also the direction associated with the object's *angular velocity* $\boldsymbol{\omega}$, which is the rotational analog of linear velocity v.

When there is no net torque on a rigid body, both its angular velocity and its angular momentum are constant. A deeper analysis shows that the angular momentum of a body does not change in the absence of a net torque on it even if it is *not* a rigid body, but is so altered during its motion that its moment of inertia changes. In a situation of this kind the angular velocity of the body also changes so that $\mathbf{L}$ stays the same. Thus we have the useful theorem of *conservation of angular momentum*:

Conservation of angular momentum

When the sum of the external torques acting upon a system of particles equals zero, the total angular momentum of the system remains constant.

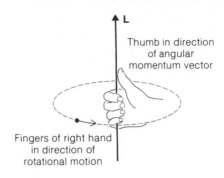

FIG. 8–18 Angular momentum is a vector quantity whose direction is given by the right-hand rule shown here.

A skater or ballet dancer doing a spin capitalizes upon conservation of angular momentum. In Fig. 8–19 a skater is shown starting her spin with her arms and one leg outstretched. By bringing her arms and extended leg inward, she reduces her moment of inertia considerably and as a result spins faster.

Example A skater has the moment of inertia 3 kg · m² when her arms are outstretched and 1 kg · m² when her arms are brought to her sides. She starts to spin at the rate of 1 revolution per second when her arms are outstretched, and then pulls her arms to her sides. (a) What is her final angular speed? (b) What are her initial and final kinetic energies?

Solution (a) Since the conversion factors would only cancel, it is simplest to use the rev/s as the unit of angular speed. Since $I_1 = 3$ kg · m², $I_2 = 1$ kg · m², and $\omega_1 = 1$ rev/s, from conservation of angular momentum we have

Angular momentum before = angular momentum afterward

$$I_1\omega_1 = I_2\omega_2$$

$$\omega_2 = \frac{I_1}{I_2}\omega_1 = \left(\frac{3\text{kg} \cdot \text{m}^2}{1\text{kg} \cdot \text{m}^2}\right)(1\text{ rev/s}) = 3\text{ rev/s}$$

The skater spins three times faster than she did with her arms outstretched.

I large, ω small   *I* small, ω large

FIG. 8–19 Angular momentum is conserved when a skater executes a spin. (Photo © Roy King.)

(b) For calculating kinetic energy we must use the rad/s as the unit of angular speed. Since 1 rev $= 2\pi$ rad, $\omega_1 = 2\pi$ rad/s and $\omega_2 = 6\pi$ rad/s. The skater's initial and final kinetic energies are therefore

$$KE_1 = \tfrac{1}{2}I_1\omega_1^2 = \tfrac{1}{2}\,(3\text{kg}\cdot\text{m}^2)(2\pi\,\text{rad/s})^2 = 390\text{J}$$

$$KE_2 = \tfrac{1}{2}I_2\omega_2^2 = \tfrac{1}{2}\,(1\text{kg}\cdot\text{m}^2)(6\pi\,\text{rad/s})^2 = 178\text{J}$$

The increase in kinetic energy is due to the work the skater had to do to pull in her arms. ■

Spin stabilization

Because angular momentum is a vector quantity, a torque must be applied to change the orientation of the axis of rotation of a spinning body as well as to change its angular speed. The greater the magnitude of **L,** the more torque is needed for it to deviate from its original direction. This is the principle behind the spin stabilization of projectiles, such as footballs and rockets. Such projectiles are set spinning about axes in their directions of motion so that they do not tumble and thereby offer excessive air resistance. A top is another illustration of the vector nature of angular momentum. A stationary top set on its tip falls over at once, but a rotating top stays upright until its angular momentum is dissipated by friction between its tip and the ground (Fig. 8–20).

The gyroscope

A *gyroscope* is a disk with a high moment of inertia whose axis is mounted on gimbals, as in Fig. 8–21. When it is set in motion, conservation of angular momentum leads to the disk's maintaining a fixed orientation in space regardless of how its support is moved. Thus a gyroscope can be used on an airplane as an artificial horizon when the true horizon is not visible, and as the basis of a compass that operates independently of the earth's magnetic field.

FIG. 8–20 The faster a top spins, the more stable it is. When its angular momentum has been lost through friction, the top falls over.

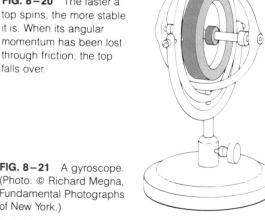

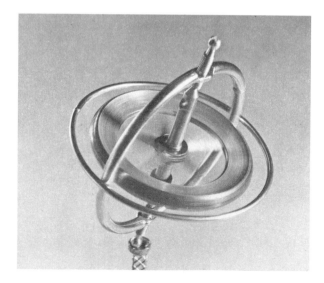

FIG. 8–21 A gyroscope. (Photo: © Richard Megna, Fundamental Photographs of New York.)

8–10   COMPARISON WITH LINEAR MOTION

As we have seen, there are many points of correspondence between angular and linear motion at constant speed and under constant acceleration. Table 8–2 lists the principal ones. Although the symbols are different, the formulas are the same because the basic concepts involved are so closely related.

Linear Quantity		Angular Quantity	
Distance	$s = v_0 t + \frac{1}{2} a t^2$	Angle	$\theta = \omega_0 t + \frac{1}{2} \alpha t^2$
	$s = \left(\dfrac{v_0 + v_f}{2}\right) t$		$\theta = \left(\dfrac{\omega_0 + \omega_f}{2}\right) t$
Speed	$v = v_0 + at$	Angular speed	$\omega = \omega_0 + \alpha t$
	$v^2 = v_0^2 + 2as$		$\omega^2 = \omega_0^2 + 2\alpha\theta$
Acceleration	$a = \Delta v/\Delta t$	Angular acceleration	$\alpha = \Delta\omega/\Delta t$
Mass	m	Moment of inertia	I
Force	$F = ma$	Torque	$\tau = I\alpha$
Momentum	$p = mv$	Angular momentum	$L = I\omega$
Work	$W = Fs$	Work	$W = \tau\theta$
Power	$P = Fv$	Power	$P = \tau\omega$
Kinetic energy	$KE = \frac{1}{2} mv^2$	Kinetic energy	$KE = \frac{1}{2} I\omega^2$

TABLE 8–2
Comparison of linear and angular quantities

IMPORTANT TERMS

The **radian** is a unit of angular measure equal to 57.30°. If a circle is drawn whose center is at the vertex of an angle, the angle in radian measure is equal to the ratio between the arc of the circle cut by the angle and the radius of the circle. A full circle contains 2π radians.

The **angular speed** ω of a rotating body is the angle through which it turns per unit time. The **angular acceleration** α of a rotating body is the rate of change of its angular speed with respect to time.

The **axis of rotation** of a rigid body turning in place is that line of particles which does not move.

All particles in circular motion experience centripetal accelerations, but only those particles whose angular speed changes have **tangential accelerations.**

The **moment of inertia** I of a body about a given axis is the rotational analog of mass in linear motion. Its value depends upon the way in which the mass of the body is distributed about the axis.

The **angular momentum** L of a rotating body is the product $I\omega$ of its moment of inertia and angular speed. The principle of **conservation of angular momentum** states that the total angular momentum of a system of particles remains constant when no net external torque acts upon it.

IMPORTANT FORMULAS

Radian measure: $\theta = \dfrac{s}{r}$

Angular speed: $\omega = \dfrac{\theta}{t}$

Linear speed: $v = \omega r$

Moment of inertia: $I = \Sigma\, mr^2$

Rotational kinetic energy: $KE = \frac{1}{2} I\omega^2$

Motion under constant angular acceleration:

$$\theta = \omega_0 t + \frac{1}{2} \alpha t^2$$
$$\omega_f = \omega_0 + \alpha t$$
$$\omega_f^2 = \omega_0^2 + 2\alpha\theta$$

Angular acceleration: $\alpha = \dfrac{\Delta\omega}{\Delta t} = \dfrac{a_T}{r}$

Torque: $\tau = I\alpha$

Work and power:

$W = \tau\theta$

$P = \tau\omega$

Angular momentum: $L = I\omega$

MULTIPLE CHOICE

1. In a rigid object undergoing uniform circular motion, a particle that is a distance R from the axis of rotation
   a. has an angular speed proportional to R.
   b. has an angular speed inversely proportional to R.
   c. has a linear speed proportional to R.
   d. has a linear speed inversely proportional to R.

2. The centripetal acceleration of a particle in circular motion
   a. is less than its tangential acceleration.
   b. is equal to its tangential acceleration.
   c. is more than its tangential acceleration.
   d. may be more or less than its tangential acceleration.

3. The rotational analog of force in linear motion is
   a. moment of inertia.
   b. angular momentum.
   c. torque.
   d. weight.

4. The rotational analog of mass in linear motion is
   a. moment of inertia.
   b. angular momentum.
   c. torque.
   d. angular speed.

5. A quantity not directly involved in the rotational motion of an object is
   a. mass.   b. moment of inertia.
   c. torque.   d. angular speed.

6. All rotating objects at sea level that have the same mass and angular speed also have the same
   a. angular momentum.
   b. moment of inertia.
   c. kinetic energy.
   d. potential energy.

7. The moment of inertia of an object does not depend upon
   a. its mass.
   b. its size and shape.
   c. its angular speed.
   d. the location of the axis of rotation.

8. A solid disk rolls from rest down an inclined plane. A hole is then cut in the center of the disk, and the resulting doughnut-shaped object is allowed to roll down the same inclined plane. Its speed at the bottom is
   a. less than that of the solid disk.
   b. the same as that of the solid disk.
   c. greater than that of the solid disk.
   d. any of the above, depending on the size of the hole.

9. A solid lead cylinder of radius R, a solid aluminum cylinder of radius R, a hollow lead cylinder of radius $R/2$, and a solid lead sphere of radius R all start rolling down an inclined plane at the same time. The one that reaches the bottom first is the
   a. solid lead cylinder.
   b. solid aluminum cylinder.
   c. hollow lead cylinder.
   d. solid lead sphere.

10. A solid iron sphere A rolls down an inclined plane, while an identical sphere B slides down the plane in a frictionless manner.
   a. Sphere A reaches the bottom first.
   b. Sphere B reaches the bottom first.
   c. They reach the bottom together.
   d. Which one reaches the bottom first depends on the angle of the plane.

11. A solid iron sphere A rolls down an inclined plane, while an identical sphere B slides down the plane in a frictionless manner. At the bottom the kinetic energy of sphere A is
   a. less than that of sphere B.
   b. equal to that of sphere B.
   c. more than that of sphere B.
   d. more or less than that of sphere B, depending on the angle of the plane.

12. Of the following properties of a yo-yo moving in a circle, the one that does not depend upon the radius of the circle is the yo-yo's
   a. angular speed.
   b. angular momentum.
   c. linear speed.
   d. centripetal acceleration.

13. A yo-yo being swung in a circle need not possess
   a. angular speed.
   b. angular momentum.
   c. angular acceleration.
   d. centripetal acceleration.

14. The total angular momentum of a system of particles
   a. remains constant under all circumstances.

b. changes when a net external force acts upon the system.

c. changes when a net external torque acts upon the system.

d. may or may not change under the influence of a net external torque, depending on the direction of the torque.

15. When a spinning skater pulls in his arms to turn faster,
   a. his angular momentum increases.
   b. his angular momentum decreases.
   c. his angular momentum remains the same.
   d. any of the above, depending on the circumstances.

16. A full circle contains
   a. $\pi/4$ radians.   b. $\pi/2$ radians.
   c. π radians.   d. 2π radians.

17. A radian is approximately equal to
   a. 1°.   b. 5°.
   c. 12°.   d. 60°.

18. An angle of $\pi/6$ rad is equivalent to
   a. 15°.   b. 30°.
   c. 60°.   d. 90°.

19. A wheel is 1 m in diameter. When it makes 30 rpm, the linear speed of a point on its circumference is
   a. $\pi/2$ m/s.   b. π m/s.
   c. 30π m/s.   d. 60π m/s.

20. An object undergoes a uniform angular acceleration. In the time t since the object started rotating from rest, the number of turns it makes is proportional to
   a. $\sqrt{t}$.   b. t.
   c. t^2.   d. t^3.

21. A wheel that starts from rest has an angular speed of 20 rad/s after being uniformly accelerated for 10 s. The total angle through which it has turned in these 10 s is
   a. 2π radians.   b. 40π radians.
   c. 100 radians.   d. 200 radians.

22. A flywheel rotating at 10 rps is brought to rest by a constant torque in 15 s. In coming to a stop the flywheel makes
   a. 75 rev.   b. 150 rev.
   c. 472 rev.   d. 600 rev.

23. A hoop rolls down an inclined plane. The fraction of its total kinetic energy that is associated with its rotation is
   a. $\frac{1}{4}$.   b. $\frac{1}{3}$.
   c. $\frac{1}{2}$.   d. $\frac{2}{3}$.

24. A torque of 30 N · m acts on a 5-kg wheel of moment of inertia 2 kg · m². If the wheel starts from rest, after 10 s it will have turned through

a. 750 rad.   b. 1500 rad.
c. 3000 rad.   d. 6000 rad.

25. The final KE of the wheel of Question 24 will be
   a. 9 kJ.   b. 22.5 kJ.
   c. 45 kJ.   d. 56.25 kJ.

26. A motor develops 56 kW when its crankshaft turns at 300 rad/s. The torque on the crankshaft
   a. is 30 N · m.
   b. is 187 N · m.
   c. is 16.8 kN · m.
   d. depends on the crankshaft diameter.

27. A tangential force of 10 N acts on the rim of a wheel 80 cm in diameter. The wheel takes 1.5 s to complete its first complete revolution starting from rest. The wheel's moment of inertia is
   a. 0.72 kg · m².   b. 0.96 kg · m².
   c. 1.8 kg · m².   d. 4.5 kg · m².

EXERCISES

8–1 Angular Measure

1. An applie pie is cut into nine equal pieces. What angle (in radians) is included between the sides of each piece?

2. How many radians does the second hand of a clock turn through in 30 s? In 90 s? In 105 s? Express the answers in terms of π.

3. The resolution of the human eye is about 1′, where $60' = 1°$. (a) Express an angle of 1′ in radians. (b) What is the length of the smallest detail that can be discerned when an object 25 cm away is being examined? (This is the distance of most distinct vision.)

4. In a good light, a golden eagle can detect a 16-in. hare at a distance of 1 mi. What angle does this represent, in radians?

8–2 Angular Speed

5. What is the angular speed in radians per second of the hour, minute, and second hands of a clock?

6. A car makes a U-turn in 5 s. What is its average angular speed?

7. In 1976, Kazuya Shiozaki made 49,299 turns in 5 h 37 min of skipping rope. Find the average angular speed of the rope.

8. The shaft of a motor rotates at the constant angular speed of 3000 rev/min. How many radians will it have turned through in 10 s?

9. The blades of a rotary lawnmower are 30 cm long and rotate at 315 rad/s. Find the linear speed of the blade tips and their angular speed in rpm.

10. A steel cylinder 4 cm in radius is to be machined in a lathe. At how many revolutions per second should it rotate in order that the linear speed of the cylinder's surface be 0.7 m/s?

11. In 1941 a wind-driven electric power station was set up at Grandpa's Knob, Vermont. The propeller was 40 m in diameter. When the propeller was turning at 30 rev/min, what was the linear speed of one of the blade tips in kilometers per hour?

12. The propeller of a boat rotates at 100 rad/s when the speed of the boat is 6 m/s. The diameter of the propeller is 40 cm. What is the speed of the tip of the propeller?

13. A barrel 80 cm in diameter is rolling with an angular speed of 5 rad/s. What is the instantaneous velocity of (a) its top, (b) its center, and (c) its bottom, all with respect to the ground?

14. A rotating platform is to be used to test aircraft equipment under accelerations of $6g$. If the equipment is 0.5 m from the axis, what angular speed is required?

15. A drill bit $\frac{1}{4}$ in. in diameter is rotating at 1200 rpm. (a) What is its angular speed? (b) What is the linear speed of a point on its circumference?

8–3   Rotational Kinetic Energy

8–4   Moment of Inertia

16. Many flywheels have most of their mass concentrated around their rims. What is the advantage of this?

17. A square and a rectangle of the same mass are cut from a sheet of metal. Which has the greater I about a perpendicular axis through its center?

18. The density of a body of matter is its mass divided by its volume. Aluminum has a density of 2.7×10^3 kg/m³ and iron has a density of 7.8×10^3 kg/m³. If an aluminum cylinder and an iron cylinder have the same length and the same mass, which has the greater moment of inertia?

19. The baton of a drum majorette is a 300-g uniform rod 80 cm long. What is its kinetic energy when it is twirled at an angular speed of 10 rad/s?

20. The baton of a drum majorette is a 0.7-lb uniform rod 2 ft long. What is its kinetic energy when it is twirled at an angular speed of 10 rad/s?

21. The *radius of gyration* of an object is the distance from

a specified axis of rotation to a point at which the object's entire mass may be considered to be concentrated from the point of view of rotational motion about that axis. Thus the moment of inertia of an object of mass M and radius of gyration k is $I = Mk^2$. (a) The radius of gyration about its center of a hollow sphere of radius R and mass M is $k = \sqrt{\frac{2}{3}}R$. Find its moment of inertia. (b) Find the radius of gyration of a solid sphere about its center.

22. The radius of gyration of an 80-kg flywheel 35 cm in radius is 30 cm. (a) Find its moment of inertia. (b) Find the radius of a solid disk with the same mass and moment of inertia. What does this suggest about the cross-sectional form of the flywheel?

23. A 7-kg bowling ball 30 cm in diameter rolls at a speed of 5 m/s. What is its total kinetic energy?

24. A uniform knitting needle 20 cm long is balanced on its point on a table. After a moment it falls over. Find the speed with which the end of the needle strikes the table.

25. A uniform thin rod of length L is pivoted about a horizontal axis at one end. If it is released from a horizontal position, what will its maximum angular speed be?

8–5   Combined Translation and Rotation

26. Which will roll to the bottom of an inclined plane first, a can that contains frozen orange juice or an identical can in which the juice has thawed and is liquid?

27. A hollow cylinder and a solid cylinder having the same mass and diameter are released from rest simultaneously at the top of an inclined plane. Which reaches the bottom first?

28. An aluminum cylinder of radius R, a lead cylinder of radius R, and a lead cylinder of radius $2R$ all roll from rest down an inclined plane. In what order will they reach the bottom?

29. Will a car coast down a hill faster if it has heavy tires or light tires?

30. A hollow cylinder of mass M and radius R rolls down an inclined plane of height h and reaches the bottom with a linear speed of $v = \sqrt{8gh/7}$. Find the moment of inertia of the cylinder.

31. A 3-kg hoop 1 m in diameter rolls down an inclined plane 10 m long that is at an angle of 20° with the horizontal. (a) What is the angular speed of the hoop at the bottom of the plane? (b) What is its linear speed? (c) What is its rotational kinetic energy? (d) What is its total kinetic energy?

32. A solid sphere 10 cm in radius starts from rest at the top of an inclined plane 10 m long and reaches the bottom in 7 s. What angle does the plane make with the horizontal?

8–6 Angular Acceleration

33. A truck undergoes an acceleration of 0.25 m/s^2. If its wheels are 1 m in diameter, what is their angular acceleration?

34. A wheel starts from rest under the influence of a constant torque and turns through 500 radians in 10 s. (a) What is its angular acceleration? (b) What is its angular speed at the end of these 10 s?

35. An engine idling at 10 rev/s is accelerated at 2.5 rev/s^2 to 20 rev/s. How many revolutions does it make during this acceleration?

36. A shaft turning at 200 rad/s is accelerated at 10 rad/s^2 until it reaches 300 rad/s. Find the angle through which the shaft turned during the acceleration.

37. A circular saw blade rotating at 15 rev/s is brought to a stop in 125 revolutions. How much time did this take?

38. A phonograph turntable slows down to a stop from an initial angular velocity of 3.5 rad/s in 20 s. (a) What is its acceleration? (b) How many turns does it make while slowing down?

8–7 Torque and Angular Acceleration

39. The lower the center of gravity of an object, the more stable it is. However, it is easier to balance a billiard cue vertically when its tip is on one's finger with its heavy handle up in the air rather than with its handle on the finger. Why?

40. A 60-lb cylindrical flywheel 1 ft in radius is acted upon by a torque of 5 lb · ft. (a) If it starts from rest, how much time is required to accelerate it to an angular speed of 30 rpm? (b) What is its kinetic energy at this angular speed?

41. A 200-kg cylindrical flywheel 0.3 m in radius is acted upon by a torque of 20 N · m. (a) If it starts from rest, how much time is required to accelerate it to an angular speed of 10 rad/s? (b) What is its kinetic energy at this angular speed?

42. A rope 1 m long is wound around the rim of a drum of radius 12 cm and moment of inertia $0.02 \text{ kg} \cdot \text{m}^2$. The rope is pulled with a force of 2.5 N. (a) Assuming that the drum is free to rotate without friction, what is its final angular speed? (b) What is its final kinetic energy? (c) How much work is done by the force?

43. A torque of 500 N · m is applied to a turbine rotating at 200 rad/s and after 40 s its speed has doubled. What is the turbine's moment of inertia?

44. A cylinder whose axis is fixed has a string wrapped around it which is pulled with a force equal to the cylinder's weight. Show that the acceleration of the string is equal to $2g$.

45. The uniform door to a bank vault is 2 m high and 1 m wide and is supported by frictionless hinges. To determine the door's mass, a person pushes against its free end with a constant force of 250 N always directed perpendicularly to its plane and finds that 1.5 s is needed to turn the door through 90°. What is the door's mass? (*Note:* The formula for the moment of inertia of a uniform rectangular plate about one of its edges is the same as one of the formulas given in Fig. 8–9.)

8–8 Power

46. A diesel engine can exert 140 N · m of torque when it is turning at 250 rad/s. Find the corresponding power output.

47. A 1-kW motor rotates at 125 rad/s. How much torque can it exert?

48. The cutting tool of a lathe exerts a torque of 15 lb · ft on a steel shaft. Find the power needed to turn the shaft at 8 rps.

49. A 1-hp motor rotates at 1200 rpm. How much torque in lb · ft can it exert?

50. A V-8 engine that develops 200 kW at 90 rev/s is coupled through a frictionless 5:1 reduction gear to a windlass drum 80 cm in diameter. (a) What is the heaviest mass the windlass can raise? (b) At what linear speed will it raise such a load?

51. An electric motor develops 5 kW at 2000 rev/min. The motor delivers its power through a spur gear 20 cm in diameter. If two teeth of the gear transmit torque to another gear at the same instant, find the force exerted by each gear tooth.

52. A high-speed elevator is being planned to lift a total load of 4 tons at 1200 ft/min. The winding drum is to be 6 ft in diameter. Neglecting losses, how much power is required from the driving motor, and at how many rpm should the power be developed?

53. A 30-rev/s motor operates a pump through a V-belt drive. The motor pulley is 20 cm in diameter and the tension in the belt is 180 N on one side and 70 N on the other. What is the power output of the motor?

8–9 Angular Momentum

54. All helicopters have two propellers. Some have both propellers on vertical axes but rotating in opposite directions, and others have one on a vertical axis and one on a horizontal axis perpendicular to the helicopter body at the tail. Why is a single propeller never used?

55. If the polar ice caps melt, how will the length of the day be affected?

56. (a) What is the angular momentum of a particle of mass m that moves in a circle of radius r at the speed v? (b) An earth satellite follows an elliptical orbit in which its maximum altitude above the earth's surface is 2000 km and its minimum altitude is 400 km. Use the result of (a) to find the ratio between the maximum and minimum speeds of the satellite.

57. The dimensions of all mechanical quantities can be expressed in terms of the fundamental dimensions of length L, time T, and mass M. For example, the dimensions of acceleration are LT^{-2}. Find the dimensions in terms of L, T, and M of force, torque, energy, power, linear momentum, angular momentum, and impulse.

58. An object of mass m on a frictionless table is attached to the end of a string that passes through a frictionless hole in the table, as shown below. The object is moving in a circle of radius R with the angular speed ω when the string is pulled through the hole until the object's orbital radius is $R/2$. (a) What is the new angular speed of the object? (b) How much work had to be done to pull in the string?

59. Two 0.4-kg balls are joined by a 1-m string and set whirling through the air at 5 rev/s about a vertical axis through the center of the string. After a while the string stretches to 1.2 m. (a) What is the new angular speed? (b) What are the initial and final kinetic energies of each ball? If these are different, account for the difference.

60. A woman is sitting on a piano stool with 3-kg dumbbells in her outstretched arms 60 cm from the axis of the stool. She is given a push and starts to turn at 1 rev/s. Her arms become tired, and she performs 70 J of work to pull the dumbbells to her lap, where they are 10 cm from the axis. Find the woman's moment of inertia about the axis of the stool. Neglect the contribution of her arms to the moment of inertia of the system.

61. A disk of moment of inertia 1 slug · ft^2 that is rotating at 100 rad/s is pressed against a similar disk that is initially at rest but is able to rotate freely. The two disks stick together and rotate as a unit. (a) What is the final angular speed of the combination? (b) If any kinetic energy was lost, where did it go? (c) The two disks are now separated. What are their new angular speeds? (d) If any kinetic energy was lost in the separation, where did it go?

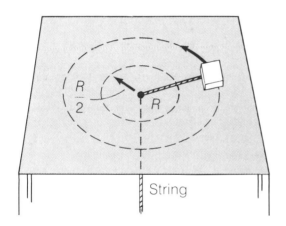

String

ANSWERS TO MULTIPLE CHOICE

1. c	**7.** c	**13.** c	**19.** a	**25.** b
2. d	**8.** a	**14.** c	**20.** c	**26.** b
3. c	**9.** d	**15.** c	**21.** c	**27.** a
4. a	**10.** b	**16.** d	**22.** a	
5. a	**11.** b	**17.** d	**23.** c	
6. d	**12.** a	**18.** b	**24.** a	

9

MECHANICAL ADVANTAGE

A machine is a device that transmits force or torque to accomplish a definite purpose. All machines, however complicated, are actually combinations of only three basic machines: the lever, the inclined plane, and the hydraulic press. Thus the train of gears that transmits power from the engine of a car to its wheels is a development of the lever, the screw jack that can raise one end of the car from the ground is a development of the inclined plane, and the brake system that permits a touch of the foot to stop the car is a development of the hydraulic press. In this chapter we shall look into the ideas underlying the operation of the first two kinds of basic machines and their common derivatives. The hydraulic press will be considered in Chapter 11.

9–1   MECHANICAL ADVANTAGE

Actual mechanical advantage

In a machine an applied force F_{in} is transformed into another force F_{out} that is different in magnitude or direction or both. The *actual mechanical advantage* (AMA) of a

CHAPTER OBJECTIVES

Completing this chapter should enable you to:

1. Distinguish between the ideal and actual mechanical advantages of a machine.

2. Calculate the ideal and actual mechanical advantages of a simple machine.

3. Relate the efficiency of a machine to its ideal and actual mechanical advantages.

4. Determine the required ratio between the moment arms of a lever for a particular application.

5. Design a block-and-tackle system to lift a given load.

6. Describe the operation of a chain hoist.

7. Determine the pulley diameter ratio or gear tooth ratio needed to obtain a given torque or angular speed from a given motor.

8. Find the mechanical advantages of an inclined plane, a cam, and a screw.

machine is the ratio between the output force F_{out} it exerts and the input force F_{in} provided to it:

$$\text{AMA} = \frac{F_{out}}{F_{in}} \qquad \textit{Actual mechanical advantage} \quad (9\text{–}1)$$

$$\text{Actual mechanical advantage} = \frac{\text{output force}}{\text{input force}}$$

A mechanical advantage greater than 1 signifies that the output force exceeds the input force, while a mechanical advantage less than 1 means that the output force is smaller. Usually the mechanical advantage is greater than 1, which makes it possible for a relatively small applied force to accomplish a task ordinarily beyond its capacity, but sometimes the reverse is true, as in the case of a pair of scissors where the range of motion is increased at the expense of a reduced force.

A distinction must be made between the *ideal mechanical advantage* (IMA) of a machine, which is the value of its mechanical advantage under ideal conditions, and

Ideal mechanical advantage

its actual mechanical advantage, which takes into account the effects of friction. Because friction can never be entirely eliminated, the actual mechanical advantage of every machine is less than its ideal mechanical advantage.

In order to obtain a general formula for ideal mechanical advantage that can be applied to any machine, let us consider the imaginary frictionless machine of Fig. 9–1. When we apply a force F_{in} to the rope at the right of the machine, a force of F_{out} is exerted on the load at the left. How the machine works does not matter, as long as there is no friction: It could be a system of pulleys, a system of gears, a hydraulic system, or something else.

How ideal mechanical advantage is defined

According to the law of conservation of energy, in the absence of friction the work W_{in} done *on* the machine by the input force must equal the work W_{out} done *by* the machine on the load. If the load moves through the distance s_{out} when the agent (our hand, perhaps) that applies the input force moves through the distance s_{in}, then since work = force × distance

$$W_{out} = W_{in}$$
$$F_{out}s_{out} = F_{in}s_{in}$$
$$\frac{F_{out}}{F_{in}} = \frac{s_{in}}{s_{out}} \qquad \textit{Ideal machine} \quad (9-2)$$

The ratio of the distances is the inverse of the ratio of the forces. The effect of friction is always to increase the input force F_{in} needed to produce a given output force F_{out}, but the distance ratio s_{in}/s_{out} does not change. Hence we can use s_{in}/s_{out} to define IMA:

$$\text{IMA} = \frac{s_{in}}{s_{out}} \qquad \textit{Ideal mechanical advantage} \quad (9-3)$$

$$\text{Ideal mechanical advantage} = \frac{\text{input distance}}{\text{output distance}}$$

Efficiency and mechanical advantage

In an actual machine, the presence of friction means that F_{in} must be greater than that specified in Eq. (9–2), and W_{in} must be greater than W_{out}. The lost work is work done to overcome friction and, as mentioned in Chapter 5, ends up as heat. Since the efficiency of any machine is

$$\text{Efficiency} = \frac{\text{work output}}{\text{work input}}$$

FIG. 9–1 An ideal machine operates without friction to transform an input force F_{in} to an output force F_{out} that may be greater or smaller than F_{in}. If F_{out} is greater than F_{in}, as shown here, the distance s_{out} through which the load is moved is less than the distance s_{in} through which the input force acts; the reverse is true if F_{out} is smaller than F_{in}.

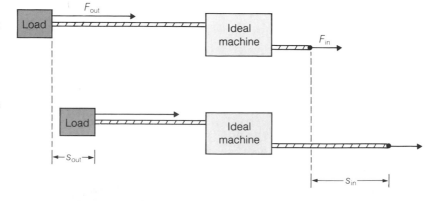

we have here

$$\text{Eff} = \frac{W_{\text{out}}}{W_{\text{in}}} = \frac{F_{\text{out}}s_{\text{out}}}{F_{\text{in}}s_{\text{in}}}$$

$$= \frac{\text{AMA}}{\text{IMA}} \qquad\qquad\qquad\qquad \textit{Efficiency} \quad (9\text{--}4)$$

The efficiency of a machine is equal to the ratio of its actual and ideal mechanical advantages. If we know the ideal mechanical advantage and the efficiency of a machine,

$$F_{\text{out}} = (\text{Eff})(\text{IMA})(F_{\text{in}}) \qquad\qquad\qquad\qquad (9\text{--}5)$$

In some machines, such as the lever, the efficiency may be very close to 100%, whereas in others, such as the screw jack, it may be 10% or less.

In most applications the higher the efficiency the better, which is why considerable attention is paid to proper bearings and adequate lubrication in industrial machinery. However, a low efficiency is occasionally necessary: For example, without a good deal of friction a screw jack under load would simply unwind the moment the applied force was released. Such a jack might have a theoretical mechanical advantage of 500, so even with an efficiency of only 10% the actual mechanical advantage would still be 50—enough for an applied force of 40 lb to lift a 1-ton load.

Example A woman uses a chain hoist to lift a 240-lb motor. She exerts a force of 27 lb on the chain and pulls a total of 24 ft of chain through the pulleys in order to raise the motor 2.5 ft. (a) Find the ideal mechanical advantage, actual mechanical advantage, and efficiency of the hoist. (b) How much force is needed to lift a 400-lb motor with the same hoist?

Solution (a) We do not have to know how a chain hoist operates in order to find these quantities (the chain hoist is described in Section 9–4). Here $F_{\text{out}} = 240$ lb, $F_{\text{in}} = 27$ lb, $s_{\text{out}} = 2.5$ ft, and $s_{\text{in}} = 24$ ft, hence

$$\text{IMA} = \frac{s_{\text{in}}}{s_{\text{out}}} = \frac{24\,\text{ft}}{2.5\,\text{ft}} = 9.6$$

$$\text{AMA} = \frac{F_{\text{out}}}{F_{\text{in}}} = \frac{240\,\text{lb}}{27\,\text{lb}} = 8.9$$

$$\text{Eff} = \frac{\text{AMA}}{\text{IMA}} = \frac{8.9}{9.6} = 0.93 = 93\%$$

(b) We must use the actual mechanical advantage to find F_{in}. From Eq. (9–1)

$$F_{\text{in}} = \frac{F_{\text{out}}}{\text{AMA}} = \frac{400\,\text{lb}}{8.9} = 45\,\text{lb} \qquad\qquad \blacksquare$$

Machines in combination

When two or more machines are coupled together so that the output of one is the input of the next, the overall mechanical advantage (ideal mechanical advantage or actual mechanical advantage) of the combination is the product of the mechanical advantages of each one:

$$MA = (MA_1)(MA_2)(MA_3)\ldots \qquad\qquad \textit{Overall MA} \quad (9-6)$$

The overall efficiency is similarly given by the product of the individual efficiencies:

$$Eff = (Eff_1)(Eff_2)(Eff_3)\ldots \qquad\qquad \textit{Overall efficiency} \quad (9-7)$$

Thus the overall ideal mechanical advantage of two machines connected with one feeding the other that each have an ideal mechanical advantage of 8 is (8)(8) = 64. If their efficiencies are each 75%, the overall efficiency is (0.75)(0.75) = 0.56 = 56%.

9–2   THE LEVER

A lever is a rigid rod pivoted at a point called the *fulcrum*. Figure 9–2 shows the most familiar type of lever, with a load at one end being raised by a force applied at the other.

From Fig. 9–2 we see that, because the angles through which the ends of the lever move are the same, the arcs s_{in} and s_{out} are respectively proportional to the lever arms L_{in} and L_{out}. The ideal mechanical advantage of the lever is therefore

$$IMA = \frac{s_{in}}{s_{out}} = \frac{L_{in}}{L_{out}} \qquad\qquad \textit{The lever} \quad (9-8)$$

The theoretical mechanical advantage of the lever is equal to the inverse ratio of the lever arms. A 4:1 ratio of lever arms, for instance, means a mechanical advantage of 4 when the input force is applied to the longer arm, a mechanical advantage of $\frac{1}{4}$ when it is applied to the shorter arm.

FIG. 9–2 The lever. The ratio of the input and output distances is equal to the ratio of the corresponding lever arms.

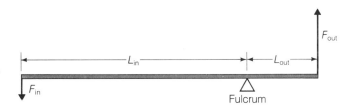

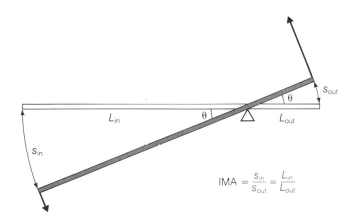

Example A steel rod 2.00 m long is to be used to pry up one end of a 500-kg crate. If a workman can exert a downward force of 350 N on one end of the rod, where should he place a block of wood to act as the fulcrum?

Solution The weight of the crate is $w = mg = 4900$ N. A force of half this is needed to lift one end, so the MA must be

$$\text{MA} = \frac{F_{\text{out}}}{F_{\text{in}}} = \frac{w/2}{F_{\text{in}}} = \frac{2450\,\text{N}}{350\,\text{N}} = 7$$

Hence the ratio of the lever arms should be 7:

$$L_{\text{in}} = 7L_{\text{out}}$$

Since the lever is 2.00 m long,

$$L_{\text{in}} + L_{\text{out}} = 2.00\,\text{m}$$
$$7L_{\text{out}} + L_{\text{out}} = 2.00\,\text{m}$$
$$L_{\text{out}} = \frac{2.00\,\text{m}}{8} = 0.25\,\text{m} = 25\,\text{cm}$$

The fulcrum should be placed 25 cm from the crate. ∎

The lever of Fig. 9-2, where the fulcrum is between the load and the applied force, is called a Class I lever. In a Class II lever, of which a wheelbarrow is an example, the load is between the fulcrum and the applied force, as in Fig. 9-3, while in a Class III lever, of which the human forearm is an example, the load and fulcrum are at the ends with the applied force between them. **The three classes of lever**

The various kinds of elementary lever are all handicapped by the limited angle through which they can operate. Certain developments of the lever, however, can readily be used on a continuous basis, and one or another of them is an important element in nearly every motor-driven machine. Perhaps the simplest is the *wheel and axle* (Fig. 9-4). A wheel of radius R is attached to an axle of smaller radius r. The input force acts tangentially on the wheel, and the output force is exerted by the rim of the axle; the center of the axle acts as the fulcrum. This arrangement is thus basically a Class II lever. Since the radii of the wheel and its axle are the lever arms of the input and output forces, **The wheel and axle is a development of the lever**

$$\text{IMA} = \frac{L_{\text{in}}}{L_{\text{out}}} = \frac{R}{r} \qquad\qquad \textit{Wheel and axle} \quad (9-9)$$

The larger the wheel relative to the axle, the greater the mechanical advantage. For this reason trucks and buses not equipped with power steering have larger steering wheels than those in ordinary cars, which permits the drivers of the larger vehicles to provide the greater steering torques required. Other examples of the wheel and axle are the screwdriver, the pedals and sprocket wheel of a bicycle, and the windlass (Fig. 9-5).

FIG. 9–3 The three classes of lever.

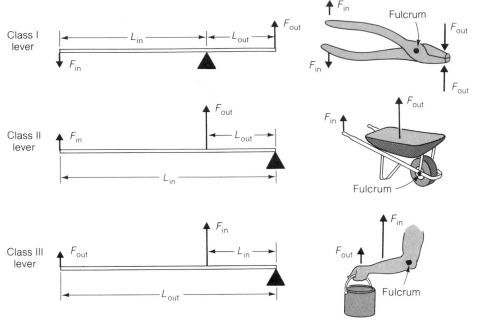

FIG. 9–4

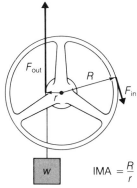

$$IMA = \frac{R}{r}$$

FIG. 9–5 Some examples of the wheel and axle. (a) Screwdriver. (b) Pedals and sprocket wheel of bicycle. (c) Windlass.

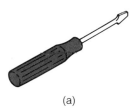

(a)

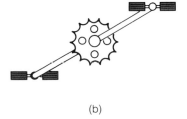

(b)

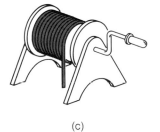

(c)

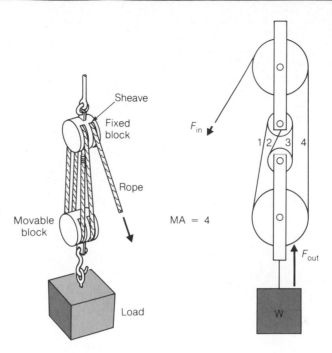

FIG. 9–6 The theoretical mechanical advantage of a block and tackle is equal to the number of strands of rope that support the movable block, in this case 4.

9–3 BLOCK AND TACKLE

A *block* consists of one or more pulleys (or *sheaves*) in a housing, as in Fig. 9–6. A *block and tackle* (or just *tackle*) consists of a pair of blocks with a rope threaded through them. The ideal mechanical advantage of this machine is easily found from Eq. (9–3). When the free end of the rope in Fig. 9–6 is pulled through a distance *d,* the movable block is raised through a height of $d/4$ since four strands of the rope must be shortened. Hence

$$\text{IMA} = \frac{s_{\text{in}}}{s_{\text{out}}} = \frac{d}{d/4} = 4$$

The ideal mechanical advantage of this block and tackle is 4.

   In general, the ideal mechanical advantage of a block and tackle is equal to the number of strands of rope that support the movable block and thereby the load. The strand to which the tension F_{in} is applied in Fig. 9–6 does not contribute to supporting the movable block; the upper pulley merely serves to change the direction of the applied force. If the entire pulley system is inverted, as in Fig. 9–7, the ideal mechanical advantage becomes 5 since the free end of the rope now exerts an upward force on the load.

Ideal mechanical advantage of block and tackle equals number of supporting strands

Example Figure 9–8 shows a pulley system that has a higher mechanical advantage than a block and tackle using the same number of pulleys. Find its ideal mechanical advantage.

FIG. 9–7 In this arrangement five strands of rope support the movable block, and so the theoretical mechanical advantage is 5.

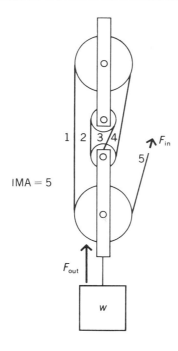

FIG. 9–8

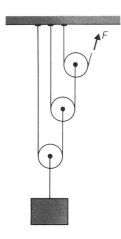

Solution In this arrangement, each pulley has an ideal mechanical advantage of 2 and its output force acts as the input force to the one below it. From Eq. (9–6) we have

$$IMA = (IMA_1)(IMA_2)(IMA_3) = (2)(2)(2) = 8$$ ■

9–4   CHAIN HOIST

The differential pulley of Fig. 9–9 is based upon both the wheel and axle and the block and tackle. Often called a *chain hoist* because it employs a chain that runs over notched pulleys to prevent slipping, this device is used to lift heavy objects in factories and machine shops.

How a chain hoist works

To operate the chain hoist, the free loop of the endless chain is pulled as shown. One strand of the chain around the movable pulley is shortened by the rotation of the larger upper pulley, while the other strand is lengthened by the rotation of the smaller upper pulley. Because of the difference in sizes of the upper pulleys, the net effect is to raise the movable block to which the load is attached. In order to lower the load, the other strand of the free loop of chain is pulled, and chain is now fed to the movable pulley from the larger upper pulley and taken up by the smaller one. The total length of the loop is thus increased, and the movable pulley falls.

The analysis of the chain hoist is straightforward. The circumference of the larger upper pulley is $2\pi R$. In a complete rotation of the pulley the input force acts through a distance of $s_{in} = 2\pi R$. Meanwhile the loop of chain around the movable pulley has been shortened by $2\pi R$ while being lengthened by $2\pi r$, the circumference of the smaller upper pulley, for a net decrease in length of $2\pi R - 2\pi r = 2\pi(R - r)$. Since this change is shared by both strands of the loop, the net effect is to raise the movable pulley by $s_{out} = \pi(R - r)$.

Hence

$$\text{IMA} = \frac{s_{in}}{s_{out}} = \frac{2\pi R}{\pi(R - r)} = \frac{2R}{R - r} \qquad \textit{Chain hoist} \quad (9-10)$$

The closer the upper pulleys are in size and the larger they are, the greater is the mechanical advantage of the chain hoist.

The presence of friction reduces the mechanical advantage of the chain hoist below its ideal mechanical advantage. Sometimes the friction actually plays a helpful role by preventing the load from descending when the chain is released; otherwise a ratchet must be used to control the chain.

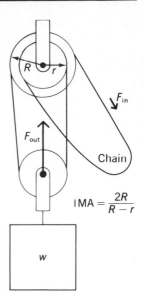

FIG. 9-9 The chain hoist is a combination of wheel and axle and block and tackle.

Example The upper pulleys of a chain hoist are 20 cm and 24 cm in diameter respectively. If the hoist is 80% efficient, find the load that can be lifted by an applied force of 200 N. How much chain must be pulled through the pulleys to raise the load by 1 m?

Solution (a) From Eq. (9-10) the ideal mechanical advantage of this hoist is

$$\text{IMA} = \frac{2R}{R - r} = \frac{(2)(12\,\text{cm})}{(12 - 10\,\text{cm})} = 12$$

Since it is 80% efficient, the actual mechanical advantage is

$$\text{AMA} = (\text{Eff})(\text{IMA}) = (0.80)(12) = 9.6$$

Hence an applied force of 200 N can lead to an output force of

$$F_{out} = (\text{AMA})(F_{in}) = (9.6)(200\,\text{N}) = 1920\,\text{N}$$

(b) The distance s_{in} through which the chain must be pulled is

$$s_{in} = (\text{IMA})(s_{out}) = (12)(1\,\text{m}) = 12\,\text{m}$$

Of course, the chain itself may be shorter than this, and merely travels through the hoist several times in the course of raising the load. ■

9-5 TORQUE TRANSMISSION

Torque can be transmitted from one rotating shaft to another in a number of ways. A common method makes use of pulleys joined by a belt (Fig. 9-10). Formerly the belts were always made of flat leather, but today fiber-reinforced rubber belts of V-shaped cross section are more often used. The advantage of these V-belts is that they cannot slide sideways on the pulleys while the large area in contact with the sides of the pulleys prevents slipping. Multiple pulleys may be used with several belts in especially heavy-duty applications.

V-belt drive

FIG. 9–10 A V-belt drive. Both pulleys rotate in the same direction. (Photo © Roy King.)

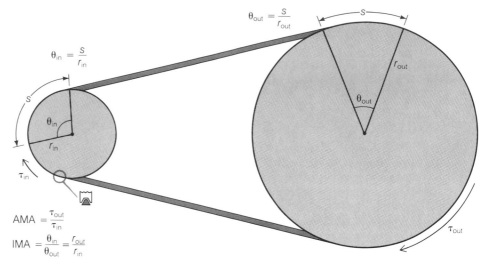

$$\theta_{in} = \frac{S}{r_{in}} \qquad \theta_{out} = \frac{S}{r_{out}}$$

$$AMA = \frac{\tau_{out}}{\tau_{in}}$$

$$IMA = \frac{\theta_{in}}{\theta_{out}} = \frac{r_{out}}{r_{in}}$$

Actual mechanical advantage and ideal mechanical advantage for a belt drive system

As we know, torque τ and angular displacement θ in rotational motion are the analogs of force F and displacement s in linear motion. Accordingly we have for the definitions of actual mechanical advantage and ideal mechanical advantage in a belt drive system

$$AMA = \frac{\tau_{out}}{\tau_{in}} \qquad\qquad\qquad\qquad\qquad\qquad \textit{Belt drive} \quad (9\text{--}11)$$

$$IMA = \frac{\theta_{in}}{\theta_{out}} \qquad\qquad\qquad\qquad\qquad\qquad \textit{Belt drive} \quad (9\text{--}12)$$

The output torque is given in terms of the efficiency and the ideal mechanical advantage by

$$\tau_{out} = (\text{Eff})(\text{IMA})(\tau_{in}) \qquad\qquad\qquad\qquad\qquad\qquad (9\text{--}13)$$

As in any machine, mechanical advantage in a belt drive is achieved at the expense of motion: The higher the output torque relative to the input torque, the more turns of the driving pulley are needed per turn of the driven one. The difference between the ideal mechanical advantage and actual mechanical advantage of a belt system is due to friction both in the belt itself and in the shaft bearings; the higher the tension in the belt, the more the friction from both sources, so the belt should only be tight enough to prevent it from slipping on the pulleys.

If the belt does not slip, the distance through which a point on the rim of the driving (input) pulley moves in any time interval is equal to the distance through which a point on the rim of the driven (output) pulley moves; both distances are marked s in Fig. 9–10. In this time the pulleys turn through the respective angles θ_{in} and θ_{out}. From Eq. (8–1)

$$\theta_{in} = \frac{s}{r_{in}} \qquad \theta_{out} = \frac{s}{r_{out}}$$

where r_{in} is the radius of the driving pulley and r_{out} is the radius of the driven pulley. Since

$$\frac{\theta_{in}}{\theta_{out}} = \frac{s/r_{in}}{s/r_{out}} = \frac{r_{out}}{r_{in}}$$

we have another formula for the ideal mechanical advantage of a belt drive system:

$$\text{IMA} = \frac{r_{out}}{r_{in}} = \frac{d_{out}}{d_{in}} \qquad\qquad \textit{Belt drive} \quad (9\text{–}14)$$

Here d_{in} and d_{out} are the pulley diameters. A driving pulley that is smaller than the driven one, as in Fig. 9–10, means a mechanical advantage greater than 1 and so an output torque greater than the input torque.

The smaller the driving pulley, the higher the ideal mechanical advantage

According to Eq. (8–2), angular speed ω is given by $\Delta\theta/\Delta t$. In a given time interval Δt the driving pulley turns through $\Delta\theta = \theta_{in}$ and the driven pulley turns through $\Delta\theta = \theta_{out}$. Hence

$$\omega_{in} = \frac{\theta_{in}}{\Delta t} \qquad \omega_{out} = \frac{\theta_{out}}{\Delta t}$$

and

$$\text{IMA} = \frac{\theta_{in}}{\theta_{out}} = \frac{\omega_{in}}{\omega_{out}} \qquad\qquad (9\text{–}15)$$

Ideal mechanical advantage and pulley speeds

Comparing this result with Eq. (9–14) leads to the conclusion that

$$\frac{\omega_{in}}{\omega_{out}} = \frac{d_{out}}{d_{in}} \qquad\qquad \textit{Belt drive} \quad (9\text{–}16)$$

The ratio of angular speeds is the inverse of the ratio of pulley diameters. When the motor (input) pulley is smaller than the machine (output) pulley, the machine turns at a smaller number of rpm than the motor but the torque provided to the machine shaft is higher than that of the motor; when the motor pulley is larger than the machine pulley, the machine speed is higher but the available torque is lower.

Step pulleys

A pair of step pulleys, as in Fig. 9–11, is widely used to couple machine tools with their motors. Shifting the belt from one set of grooves to another whose ratio of diameters is different provides a simple means of changing the machine speed.

Example A V-belt and a set of three-step pulleys are used to transmit power from a 1750-rpm electric motor to a band saw (Fig. 9–12). If the pulley diameters are 3 in., 4 in., and 5 in., and the diameter of the wheels over which the saw blade passes is 14 in., find the possible linear speeds of the blade in ft/min.

Solution When the belt joins the 3-in. motor pulley with the 5-in. saw pulley, the angular speed of the saw shaft is

$$\omega_{out} = \left(\frac{d_{in}}{d_{out}}\right)\omega_{in} = \left(\frac{3\,in.}{5\,in.}\right)1750\,rpm = 1050\,rpm$$

When the belt joins the two 4-in. pulleys, the machine speed is the same as the motor speed, and $\omega_{out} = 1750$ rpm. When the belt joins the 5-in. motor pulley with the 3-in. saw pulley,

$$\omega_{out} = \left(\frac{5\,in.}{3\,in.}\right)1750\,rpm = 2920\,rpm$$

The linear speed v of the saw blade is related to the angular speed ω of the saw shaft by the formula $v = \omega R$, where R is the radius of the saw wheel and ω is in radian measure. There are 2π radians in a revolution, so the three angular velocities above can also be expressed as

$$\omega_1 = \left(2\pi\,\frac{rad}{rev}\right)\left(1050\,\frac{rev}{min}\right) = 6600\,rad/min$$

FIG. 9–11 A pair of step pulleys makes it possible to change shaft speeds merely by shifting the belt.

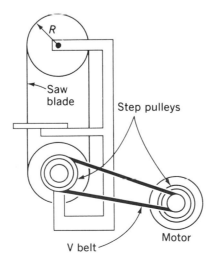

FIG. 9–12 A band saw. The blade speed is adjusted by shifting the V-belt on the step pulleys.

$$\omega_2 = \left(2\pi \frac{\text{rad}}{\text{rev}}\right)\left(1750 \frac{\text{rev}}{\text{min}}\right) = 11,000 \, \text{rad/min}$$

$$\omega_3 = \left(2\pi \frac{\text{rad}}{\text{rev}}\right)\left(2920 \frac{\text{rev}}{\text{min}}\right) = 18,300 \, \text{rad/min}$$

Since the wheel diameter is 14 in., its radius is $R = 7$ in. $= \frac{7}{12}$ ft $= 0.583$ ft. The possible linear speeds of the saw blade are therefore

$$v_1 = \omega_1 R = (6600 \, \text{rad/min})(0.583 \, \text{ft}) = 3850 \, \text{ft/min}$$
$$v_2 = \omega_2 R = (11,000 \, \text{rad/min})(0.583 \, \text{ft}) = 6410 \, \text{ft/min}$$
$$v_3 = \omega_3 R = (18,300 \, \text{rad/min})(0.583 \, \text{ft}) = 10,670 \, \text{ft/min} \quad \blacksquare$$

Gear drives

In its simplest form, a *gear* is just a notched wheel. When two gears whose teeth are the same size are placed so that the teeth mesh together, torque can be transferred from one gear to the other (Fig. 9–13). Gear drives have the advantage over belt drives of being able to transmit very high torques with no chance of slippage. They have the disadvantages of greater cost and weight and require more precise installation.

As in the case of a pulley and belt system, a small gear driving a larger one yields an increased torque and a lower speed, while a large gear driving a smaller one yields a reduced torque and a higher speed. A pair of meshing gears differs from a belt drive in that the gears rotate in opposite directions. Otherwise the same relationships among diameter, speed, torque, and mechanical advantage hold. Since the number of teeth N

(a) Spur gears

(b) Helical gears

(c) Bevel gears

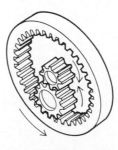

(d) Planetary gear

FIG. 9–13 Four common types of gears. (a) *Spur gears* are the most familiar, and have their teeth cut parallel to the axis of rotation. (b) *Helical gears* have curved teeth cut in a spiral pattern at an angle to the axis; several teeth of meshing helical gears are in contact at all times, which makes a set of these gears smoother in operation and able to bear heavier torques. (c) *Bevel gears* have straight teeth cut at 45° relative to the axis, and are used to change the direction of rotation by 90° with or without a change of speed. (d) A *planetary gear* is one whose teeth are cut on the inside of a wheel instead of the outside.

on the rim of a gear is proportional to its diameter, we can replace the ratio of diameters d_{out}/d_{in} by the ratio of gear teeth N_{out}/N_{in} in a pair of meshing gears. Accordingly

$$\text{IMA} = \frac{N_{out}}{N_{in}} = \frac{\omega_{in}}{\omega_{out}} \qquad\qquad \textit{Gear drive} \quad (9\text{--}17)$$

$$\text{IMA} = \frac{\text{output gear teeth}}{\text{input gear teeth}} = \frac{\text{input angular speed}}{\text{output angular speed}}$$

Example   A gasoline engine develops a power of 11.6 kW at an angular speed of 240 rad/s. What gear ratio is needed to provide an output torque of 200 N · m, assuming 92% efficiency in the gear system?

Solution   The engine produces the torque

$$\tau_{in} = \frac{P}{\omega} = \frac{11{,}600 \text{ W}}{240 \text{ rad/s}} = 48.3 \text{ N} \cdot \text{m}$$

and so the required actual mechanical advantage is, as in the case of belt drive,

$$\text{AMA} = \frac{\tau_{out}}{\tau_{in}} = \frac{200 \text{ N} \cdot \text{m}}{48.3 \text{ N} \cdot \text{m}} = 4.14$$

The corresponding ideal mechanical advantage is, from Eq. (9--17), equal to the gear ratio:

$$\text{IMA} = \frac{N_{out}}{N_{in}} = \frac{\text{AMA}}{\text{Eff}} = \frac{4.14}{0.92} = 4.5$$

The output gear should have 4.5 times as many teeth as the gear on the engine shaft.   ■

9--6   THE INCLINED PLANE

We are so accustomed to using the inclined plane in everyday life that it may be hard to think of it as a "machine." So instinctive an act as choosing a gradual slope of a hill to walk up instead of a steep slope is based upon the principle of the inclined plane. The most familiar adaptation of the inclined plane is a staircase, where the continuous surface of the plane is replaced by a series of steps for convenience in walking.

Figure 9--14 shows an inclined plane along which a crate of weight w is being pushed. The plane is L long and h high. From Eq. (9--3) we have

$$\text{IMA} = \frac{s_{in}}{s_{out}} = \frac{L}{h} \qquad\qquad \textit{Inclined plane} \quad (9\text{--}18)$$

The shallower the angle of the plane, the greater the mechanical advantage.

Example   A 150-kg safe on frictionless casters is to be raised 1.2 m off the ground to the bed of a truck. Planks 4.0 m long are available for the safe to be rolled along. How much force is needed to push the safe up to the truck?

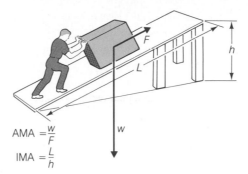

FIG. 9–14 The inclined plane.

$$AMA = \frac{w}{F}$$

$$IMA = \frac{L}{h}$$

Solution In general,

$$F_{in} = \frac{F_{out}}{IMA} = \frac{F_{out}}{(Eff)(IMA)}$$

Here the efficiency is 1, F_{out} is the weight mg of the safe, and $IMA = L/h$. Hence

$$F_{in} = \frac{mg}{(1)(L/h)} = \frac{mgh}{L} = \frac{(150\text{kg})(9.8\,\text{m/s}^2)(1.2\,\text{m})}{4.0\,\text{m}} = 441\,\text{N} \qquad \blacksquare$$

The simplest development of the inclined plane is the *wedge* (Fig. 9–15). The ideal mechanical advantage of a wedge L long and h thick is L/h just as for an inclined plane, but this mechanical advantage is never even approximately realized owing to friction. Aside from its use in splitting logs, leveling objects, and holding doors open, the wedge provides the operating principle of all cutting tools. A knife or chisel is obviously a wedge, but so are the teeth of a saw and the abrasive chips of a grindstone.

A wedge is an inclined plane

A *cam* is a kind of rotary wedge in which a turning rather than sliding motion produces the output force. From Fig. 9–16 we see that the mechanical advantage of a

A cam is a rotary wedge

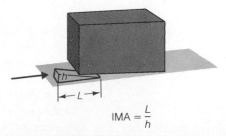

FIG. 9–15 *(Right)* the wedge. *(Left)* Hydraulic log splitter uses a wedge. (Photo © Roy King.)

$$IMA = \frac{L}{h}$$

FIG. 9–16 A cam is a rotary wedge.

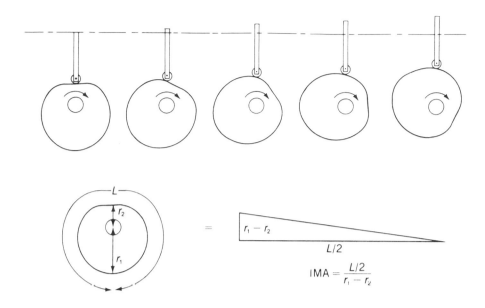

$$\text{IMA} = \frac{L/2}{r_1 - r_2}$$

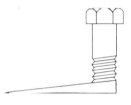

FIG. 9–17 A screw is an inclined plane wrapped around a cylinder.

cam is

$$\text{IMA} = \frac{L/2}{r_1 - r_2} \qquad\qquad Cam \quad (9\text{–}19)$$

where L is its circumference and r_1 and r_2 its maximum and minimum radii respectively. Cams are widely used in internal combustion engines to actuate cylinder valves normally held shut by strong springs.

9–7   THE SCREW

Right- and left-hand screws

Perhaps the most important application of the inclined plane is the *screw.* In essence a screw is an inclined plane wrapped around a cylinder to form a continuous helix (Fig. 9–17). The corrugations of a screw are called *threads,* and are usually (though not always) triangular in cross section. A *right-hand* screw is one that moves away from the viewer when turned clockwise, as in Fig. 9–18, while a *left-hand* screw moves away when turned counterclockwise. Standard screws are all right-handed, with left-hand ones employed only for special purposes.

Pitch of a screw

The *pitch p* of a screw is the distance between adjacent threads (Fig. 9–19), and the screw travels this far in each complete rotation. (In the British system, the number of threads per inch is specified, which is the reciprocal of p. Thus a screw that has 16 threads per inch has a pitch of $\frac{1}{16}$ in = 0.0625 in.) The theoretical mechanical advantage of a screw depends upon the lever arm L used to turn it. In a screwdriver L is the handle radius, while in a wrench it is the distance from where the wrench is grasped to the center of its jaws. When a screw of pitch p is turned through one rotation, the applied force moves through a circle of circumference $2\pi L$ and the screw advances the distance p. Hence

Right-hand screw

Left-hand screw

FIG. 9–18 A right-hand screw moves into the page when turned clockwise, whereas a left-hand screw moves into the page when turned counterclockwise.

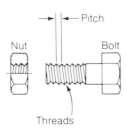

FIG. 9–19 A bolt is an example of the screw. The smaller its pitch, the greater the mechanical advantage.

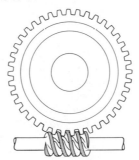

FIG. 9–20 A worm gear drive. Its ideal mechanical advantage equals the number of teeth in the output gear.

$$\text{IMA} = \frac{s_{\text{in}}}{s_{\text{out}}} = \frac{2\pi L}{p} \qquad \textit{The screw} \quad (9\text{–}20)$$

Owing to the considerable friction that occurs between threads in contact, a screw is a relatively inefficient machine, but this friction is normally an advantage inasmuch as it prevents the screw from backing out under load.

Example A wrench whose effective lever arm is 20 cm is used to tighten a nut whose pitch is 1.5 mm. The efficiency is 5%, and a force of 50 N (about 11 lb) is applied to the wrench. Find the force exerted by the nut as it is screwed down.

Solution Since 1 cm = 10 mm, $L = 200$ mm here, and so, from Eq. (9–5),

$$F_{\text{out}} = (\text{Eff})(\text{IMA})(F_{\text{in}}) = (\text{Eff})\left(\frac{2\pi L}{p}\right)(F_{\text{in}})$$

$$= (0.05)\left(\frac{2\pi \times 200\,\text{mm}}{1.5\,\text{mm}}\right)(50\,\text{N}) = 2094\,\text{N}$$

which is about 740 lb—42 times the force applied to the wrench. Evidently the screw is a very powerful device. ∎

Figure 9–20 shows a worm gear drive. Each turn of the input screw causes the output gear to rotate by one tooth, so that the mechanical advantage is equal to the number of teeth in the output gear. Worm gears are used when a large mechanical advantage is required, or when a nonreversible gear drive is required. Often an oil bath is provided to reduce friction, which otherwise is considerable.

Worm gear drive

IMPORTANT TERMS

A **machine** is a device that transmits force or torque. The three basic machines are the **lever,** the **inclined plane,** and the **hydraulic press.**

The **mechanical advantage** of a machine is the ratio between the output force (or torque) it exerts and the input force (or torque) that is furnished to it. The **ideal mechanical advantage** (IMA) is its value under ideal circumstances, while the **actual mechanical advantage** (AMA) is its value when friction is taken into account.

The **efficiency** of a machine is the ratio between its actual and ideal mechanical advantages; it is always less than 100%.

IMPORTANT FORMULAS

Actual mechanical advantage: $\text{AMA} = \dfrac{F_{out}}{F_{in}} = \dfrac{\tau_{out}}{\tau_{in}}$

Ideal mechanical advantage:

$$\text{IMA} = \frac{s_{in}}{s_{out}} = \frac{\theta_{in}}{\theta_{out}} = \frac{\omega_{in}}{\omega_{out}}$$

Efficiency: $\text{Eff} = \dfrac{\text{AMA}}{\text{IMA}}$

Output force: $F_{out} = (\text{Eff})(\text{IMA})(F_{in})$

Output torque: $\tau_{out} = (\text{Eff})(\text{IMA})(\tau_{in})$

MULTIPLE CHOICE

1. Which of the following cannot be increased by using a machine of some kind?
   a. force
   b. torque
   c. speed
   d. work

2. The three basic machines do not include the
   a. lever.
   b. wheel.
   c. inclined plane.
   d. hydraulic press.

3. The actual mechanical advantage of a machine
   a. is less than its ideal mechanical advantage.
   b. is equal to its ideal mechanical advantage.
   c. is more than its ideal mechanical advantage.
   d. may be any of the above, depending on its efficiency.

4. The output force produced by a lever does not depend upon
   a. the input force.
   b. the friction at the fulcrum.
   c. the ideal mechanical advantage of the arrangement.
   d. the class of the lever.

5. The ideal mechanical advantage of an inclined plane depends upon
   a. its length.
   b. its height.
   c. the product of its length and height.
   d. the ratio between its length and height.

6. The work needed to push something to the top of a frictionless inclined plane depends upon the plane's
   a. height.
   b. length.
   c. ideal mechanical advantage.
   d. actual mechanical advantage.

7. The ideal mechanical advantage of a screw depends upon
   a. its diameter.
   b. its length.
   c. its pitch.
   d. whether it is right-handed or left-handed.

8. A machine with an ideal mechanical advantage of 5 and an actual mechanical advantage of 4 is used to raise a 10-kg load by 4 m. The work input to the machine is
   a. 40 J.
   b. 160 J.
   c. 392 J.
   d. 490 J.

9. A person pries up one end of a 480-lb crate with a steel pipe 6 ft long. If the force exerted is 80 lb, the distance of the fulcrum from the crate is
   a. 6 in.
   b. 9 in.
   c. 18 in.
   d. 24 in.

10. A person pries up one end of a 200-kg crate with a steel pipe 2.0 m long. If the force exerted is 350 N, the distance of the fulcrum from the crate is
   a. 30 cm.
   b. 53 cm.
   c. 71 cm.
   d. 86 cm.

11. The minimum number of pulleys needed to achieve an ideal mechanical advantage of 6 is
   a. 3.
   b. 4.
   c. 5.
   d. 6.

12. The highest ideal mechanical advantage that can be obtained with a system of two pulleys is
   a. 1.
   b. 2.
   c. 3.
   d. 4.

13. A force of 50 N is needed to raise a 240-N load with a pulley system. The load goes up 1 m for every 5 m of rope pulled through the pulleys. The efficiency of the system is

a. 48%.
b. 50%.
c. 96%.
d. 104%.

14. A 600-rpm motor is coupled to an air compressor with a V-belt. The motor pulley is 3 in. in diameter and the compressor pulley is 9 in. in diameter. The compressor pulley rotates at

a. 200 rpm. b. 600 rpm.
c. 1800 rpm. d. 5400 rpm.

15. Two sets of spur gears, each set 96% efficient, drive a worm gear that is 80% efficient. The overall efficiency of the system is

a. 74%. b. 77%.
c. 80%. d. 96%.

16. One end of a plank 4 m long is supported 1 m above the ground. A 10-kg box on frictionless wheels is placed on the plank. The force parallel to the plane that must be exerted to keep the box stationary is

a. 24.5 N. b. 94.9 N.
c. 98 N. d. 392 N.

17. A screw jack has a pitch of 4 mm and a handle 50 cm long. Its efficiency is 20%. The force that must be applied to the handle in order to lift a load of 400 kg is

a. 1 N. b. 2.5 N.
c. 5 N. d. 25 N.

EXERCISES

9–1 Mechanical Advantage

1. What must be true of the mechanical advantage of a machine meant to increase force? Of a machine meant to increase speed?

2. What effect does proper lubrication have on the ideal mechanical advantage of a machine? On its actual mechanical advantage?

3. A diesel engine of 30% efficiency drives a 90%-efficient generator through a belt drive that is 95% efficient. Find the overall efficiency with which the chemical energy in the engine's fuel is converted to electrical energy.

9–2 The Lever

4. The human forearm is a Class III lever. The elbow is the fulcrum, the applied muscular force is exerted at a point perhaps 25 mm in front of it, and the load is at the hand, perhaps 33 cm from the elbow. How much force must the muscles of the arm develop to lift a 20-kg object held horizontally by the hand? How many pounds is this?

5. A pair of wire-cutting pliers is grasped with the hand 15 cm from the axis of its pivot. A wire is placed 2 cm from the axis. How much force is exerted on the wire when a 20-N force is applied by the hand?

6. How much force is needed to lift a 50-kg object using the arrangement shown here?

7. A steel rod 6.0 ft long is to be used to pry up one end of a crate weighing 800 lb. If a workman can exert a downward force of 100 lb on one end of the rod, where should he place a block of wood to act as the fulcrum?

8. A wheel and axle are used to raise water from a well. The wheel is 4 ft in diameter and the axle is 6 in. in diameter. If the efficiency is 60%, how much force must be applied to the wheel to lift a 4-lb bucket that contains 32 lb of water?

9. A vertical-axis capstan 60 cm in diameter was used to pull in the anchor chain of a whaler a century ago. Six arms 1.2 m long protruded from the capstan, and a man on each arm exerted a force of 300 N on it. The efficiency of the system was 85%. (a) Find the force applied to the anchor chain. (b) How far must each man have walked to raise 100 m of chain?

9–3 Block and Tackle

10. What is the IMA of the pulley system shown?

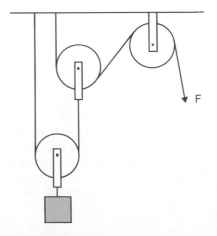

11. If the three pulleys of Exercise 10 were arranged as an ordinary block and tackle, find the maximum IMA possible.

12. In the frictionless pulley system shown, a force of 40 N is needed to support the 10-kg object. What is the mass of the lower pulley?

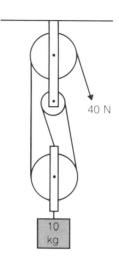

13. A block and tackle is used to pull a car out of the mud. A block with two pulleys is fastened to the car's bumper, and another block with three pulleys is tied to a tree. One end of a rope is made fast to the bumper and threaded through the various pulleys. Neglecting friction, how much force is applied to the car when two men exert a total force of 900 N on the free end of the rope?

9–4   Chain Hoist

14. A chain hoist with pulleys 20 cm and 17.5 cm in diameter is used to lift a 120-kg engine block by 1.5 m. In order to do this, a force of 80 N must be applied to the chain. Find the efficiency of the hoist.

15. The large pulley of a chain hoist is 15 in. in diameter and the small one is 13 in. in diameter. If its efficiency is 93%, find the force needed to lift one end of a 3200-lb car using the hoist. How much chain must be pulled through the system to raise the end of the car by 2 ft?

16. The hoist of the previous problem breaks down, and a block and tackle is rigged to substitute for it. What is the minimum number of pulleys the movable block must have if the ideal mechanical advantage of the block and tackle is to be the same as the ideal mechanical advantage of the chain hoist?

9–5   Torque Transmission

17. In a certain winch a spur gear with 10 teeth is used to drive another spur gear with 40 teeth. The efficiency of the winch is 80%. (a) If the input gear is turned at 8 revolutions/min, how fast does the output gear turn? (b) If a torque of 30 N · m is applied to the input gear, what is the output torque?

18. Find the force exerted on the free end of the rope by the geared windlass shown when a force of 150 N is applied to the crank handle. Neglect friction.

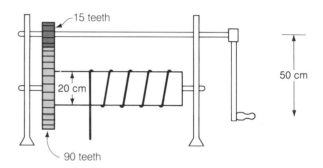

19. A diesel engine develops 61 hp at 3000 rpm. (a) What gear ratio is needed to provide an output torque of 350 lb · ft? (b) At how many rpm is this torque developed? Neglect friction.

20. A gasoline engine that develops 30 kW at 60 rps is coupled to a 4:1 reduction gearbox of 95% efficiency. Find its output torque.

21. A 0.5-hp 1750-rpm electric motor is coupled to a 5-in. diameter circular saw blade with a V-belt. The motor pulley is 6 in. in diameter and the saw pulley is 2 in. in diameter. With how much force do the teeth of the saw strike a piece of wood?

22. On a certain bicycle, the sprocket wheel to which the pedals are attached is 20 cm in diameter and the sprocket wheel on the rear axle is 10 cm in diameter. The wheels of the bicycle are 65 cm in diameter. How many times must the pedals be turned per minute in order that a rider travel at 6 m/s?

9–6   The Inclined Plane

23. A ramp 25 m long slopes down 1.2 m to the edge of a lake. What force is needed to pull a 300-kg boat on an 80-kg trailer up along the ramp if friction is negligible?

24. A force of 1.5 kN is needed to pull a wagon up an inclined plane 10 m long and 3 m high. If friction is negligible, find the mass of the wagon.

25. A block and tackle with one fixed and one movable pulley is used to drag a 40-kg dolly carrying a 200-kg load up a ramp 8 m long and 2 m high. If friction is negligible, find the force needed on the end of the rope.

26. A block and tackle with three fixed and two movable pulleys is used to pull a 1200-lb crate up a ramp 32 ft long and 8 ft high. If rollers are used to eliminate friction between the crate and the ramp, find the force needed on the end of the rope.

9-7 The Screw

27. A wood screw with a pitch of 2 mm is being turned by a screwdriver whose handle is 4 cm in diameter. A pilot hole has been drilled for the screw, and grease is used to increase the efficiency to 10%. If a force of 5 N is applied to the handle of the screwdriver, with how much force does the screw advance?

28. A machinist's vise has a screw of 3-mm pitch and a handle 13 cm long. In the absence of friction, how much force is developed between its jaws when a force of 40 N is applied to the end of the handle?

29. A screw jack whose pitch is 0.5 in. has a handle 3 ft long. A force of 17 lb must be applied to the handle in order to raise a load of 2500 lb. Find the efficiency of the jack.

30. A machinist's vise has a well-lubricated screw of 3-mm pitch and a handle 13 cm long. If a force of 15 N applied to the end of the handle produces a force of 3 kN between the jaws of the vise, find its efficiency.

31. A vertical screw jack has a pitch of 0.8 cm. Torque is transferred to it via a pair of bevel gears from a horizontal crank whose handle is 30 cm from the shaft; that is, the lever arm of the crank is 30 cm. The gear on the crankshaft is one-third the diameter of the gear on the screw jack. (a) If the efficiency of the system is 25%, find the load in kg that can be lifted when a 160-N force is applied to the crank handle. (b) How many turns of the crank are needed to raise this load 1 m?

32. The body of a turnbuckle consists of a hollow cylinder threaded internally with a right-hand thread at one end and a left-hand thread at the other, as shown below. Threaded rods fit into the ends; by turning the turnbuckle body appropriately, these rods can be pulled closer together with great force. A certain turnbuckle used on the backstay of a sailboat has threads of 2-mm pitch and is tightened with a 25-cm wrench. A force of 100 N is applied to the end of the wrench. How much tension is produced in the backstay if the efficiency of the turnbuckle is 40%?

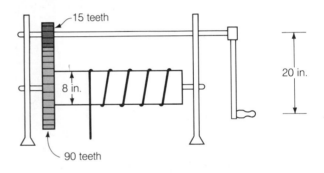

ANSWERS TO MULTIPLE CHOICE

1. d	6. a	11. c	16. a
2. b	7. c	12. c	17. d
3. a	8. d	13. c	
4. d	9. c	14. a	
5. d	10. b	15. a	

10

MECHANICAL PROPERTIES OF MATERIALS

The usual procedure of the physicist in approaching a complex situation is to first set up a simple model that represents the main features of the situation. Then, if predictions based on this model agree reasonably well with observation and experiment, the model is further refined until agreement is even better. Our work in physics thus far illustrates this procedure. We began by treating objects as though they are just particles, and later went on to consider them as rigid bodies. Actually, of course, there is no such thing as a rigid body: The strongest block of steel can be stretched, compressed, or twisted by applying suitable forces. In this chapter we shall come even closer to reality by examining some mechanical properties of materials that are important in technology.

10–1   DENSITY

Density is mass per unit volume

A characteristic property of every material is its *density*, which is its mass per unit volume. When we speak of lead as a "heavy" metal and of aluminum as a "light" one,

248

CHAPTER OBJECTIVES

Completing this chapter should enable you to:

1. Calculate the mass (or weight) of a body of matter given its volume and density, or calculate its volume given its mass (or weight) and density.

2. Distinguish between mass and weight densities.

3. Distinguish among tensile, compressive, and shear stresses.

4. Relate a stress of a particular kind to the strain it produces and to the appropriate modulus of elasticity of the material.

5. Predict whether or not an object will break or be permanently deformed when a certain stress is applied to it.

6. Determine the change in length of an object under tension or compression.

7. Determine the force needed to change the length of an object by a given amount.

8. Find the amount by which an object is distorted when it is under shear.

9. Determine the force needed to punch a hole in a sheet of metal.

10. Calculate the volume change in a body of matter subjected to a uniform compression.

what we really mean is that lead has a higher density than aluminum: A cubic meter of lead has a mass of 11,300 kg, whereas a cubic meter of aluminum has a mass of only 2700 kg.

The symbol for density is d, so that if a volume V of a certain substance has the mass m, its density is

$$d = \frac{m}{V} \qquad\qquad \text{Density} \quad (10\text{--}1)$$

$$\text{Density} = \frac{\text{mass}}{\text{volume}}$$

The densities of various common substances are given in Table 10–1. Although the correct SI unit of density is the kg/m^3, the g/cm^3 is often used as well.

Densities in the British system are properly expressed in $slugs/ft^3$. In practice, *weight density* $D = w/V$, the weight of a substance per unit volume, is more commonly

Density in the British system

TABLE 10–1
Densities of various substances at atmospheric pressure and room temperature

Substance	Mass Density, d			Weight Density, D
	kg/m^3	g/cm^3	$slugs/ft^3$	lb/ft^3
Air	1.3	1.3×10^{-3}	2.5×10^{-3}	8×10^{-2}
Alcohol (ethyl)	7.9×10^2	0.79	1.5	48
Aluminum	2.7×10^3	2.7	5.3	1.7×10^2
Balsa wood	1.3×10^2	0.13	0.25	8
Blood (37° C)	1.06×10^3	1.06	2.05	66
Bone	1.6×10^3	1.6	3.1	9.9×10^2
Brass	8.4×10^3	8.4	16	5.2×10^2
Carbon dioxide	2.0	2.0×10^{-3}	3.8×10^{-3}	0.12
Concrete	2.3×10^3	2.3	4.5	1.4×10^2
Gasoline	6.8×10^2	0.68	1.3	42
Gold	1.9×10^4	19	38	1.2×10^3
Helium	0.18	1.8×10^{-4}	3.5×10^{-4}	1.1×10^{-2}
Hydrogen	0.09	9×10^{-5}	1.8×10^{-4}	5.4×10^{-3}
Ice	9.2×10^2	0.92	1.8	58
Iron and steel	7.8×10^3	7.8	15	4.8×10^2
Lead	1.1×10^4	11	22	7×10^2
Mercury	1.4×10^4	14	26	8.3×10^2
Nitrogen	1.3	1.3×10^{-3}	2.4×10^{-3}	7.7×10^{-2}
Oak	7.2×10^2	0.72	1.4	45
Oxygen	1.4	1.4×10^{-3}	2.8×10^{-3}	9×10^{-2}
Pine	3.7×10^2	0.37	0.73	23
Silver	1.05×10^4	10.5	20	6.6×10^2
Water, pure	1.00×10^3	1.00	1.94	62
Water, sea	1.03×10^3	1.03	2.00	64

used than its density d, which is its mass per unit volume. The difference is a factor of $g = 32$ ft/s². Thus the density of pure water is $d = 1.94$ slug/ft³ and its weight density is $D = 62$ lb/ft³.

Specific gravity is density relative to water

The *specific gravity* of a substance is its density relative to that of water and so is a pure number. Since the density of water is almost exactly 1 g/cm³, the specific gravity of a substance is very nearly equal to the numerical value of its density when expressed in g/cm³. Thus, since the density of aluminum is 2.7 g/cm³, its specific gravity is 2.7. Specific gravity is also called *relative density*.

Example   A coil of sheet steel 0.80 mm thick and 500 mm wide has a mass of 156 kg. Find the length of the steel in the coil.

Solution   The volume of the steel is

$$V = \text{length} \times \text{width} \times \text{thickness} = LWT$$

Since $d = m/V$, $V = m/d$ and so, since $W = 0.50$ m and $T = 0.80$ mm $= 8.0 \times 10^{-4}$ m,

$$V = LWT = \frac{m}{d}$$

$$L = \frac{m}{dWT} = \frac{156\,\text{kg}}{(7.8 \times 10^3\,\text{kg/m}^3)(8.0 \times 10^{-4}\,\text{m})(0.50\,\text{m})} = 50\,\text{m}$$

The length of the steel in the coil is 50 m.   ■

Example A 50-g bracelet is suspected of being gold-plated lead instead of pure gold. When it is dropped into a full glass of water, 4.0 cm^3 of water overflows (Fig. 10–1). Is the bracelet pure gold? If not, what proportion of its mass is gold?

Solution The density of the bracelet is

$$d = \frac{m}{V} = \frac{50\,\text{g}}{4.0\,\text{cm}^3} = 12.5\,\text{g/cm}^3$$

This is less than the density of gold, which is 19 g/cm^3, so the bracelet is not pure gold. To find the proportion of gold it contains, we note that, since the bracelet's mass is 50 g,

$$m_{\text{gold}} + m_{\text{lead}} = 50\,\text{g}$$

and since its volume is 4.0 cm^3,

$$V_{\text{gold}} + V_{\text{lead}} = 4.0\,\text{cm}^3$$

From Eq. (10–1), $V = m/d$, which applies separately to the gold and lead contents of the bracelet. Therefore the second equation becomes

$$\frac{m_{\text{gold}}}{d_{\text{gold}}} + \frac{m_{\text{lead}}}{d_{\text{lead}}} = 4.0\,\text{cm}^3$$

From the first equation, $m_{\text{lead}} = 50\,\text{g} - m_{\text{gold}}$. Substituting this expression for m_{lead} and the values $d_{\text{gold}} = 19$ g/cm^3 and $d_{\text{lead}} = 11$ g/cm^3 in the third equation gives

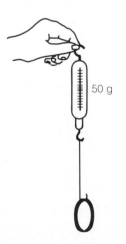

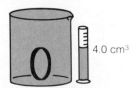

FIG. 10–1 The volume of this 50-g bracelet is 4.0 cm^3.

$$\frac{m_{\text{gold}}}{19\,\text{g/cm}^3} + \frac{50\,\text{g}}{11\,\text{g/cm}^3} - \frac{m_{\text{gold}}}{11\,\text{g/cm}^3} = 4.0\,\text{cm}^3$$

$$m_{\text{gold}} = 14\,\text{g}$$

Hence the proportion of gold in the bracelet is 14 g/50 g = 0.28 = 28%. ∎

Example   We want to pump water from a well 60 ft deep at the rate of 90 gal/min. Assuming 60% efficiency, how powerful should the pump motor be?

Solution   We begin with the definition of efficiency,

$$\text{Efficiency} = \frac{\text{power output}}{\text{power input}}$$

which means that

$$\text{Power input} = \frac{\text{power output}}{\text{efficiency}}$$

In the situation we have here,

$$\text{Power output} = \frac{\text{work}}{\text{time}} = \left(\frac{\text{work}}{\text{gal}}\right)\left(\frac{\text{gal}}{\text{s}}\right)$$

The work needed to raise a weight w through the height h is $wh = DVh$ since $w = DV$. Because 1 gal = 0.134 ft³ and the weight density D of water is 62 lb/ft³, the work needed to raise a gallon of water through 60 ft is

$$\frac{\text{Work}}{\text{gal}} = wh = DVh = \left(62\,\frac{\text{lb}}{\text{ft}^3}\right)\left(0.134\,\frac{\text{ft}^3}{\text{gal}}\right)(60\,\text{ft}) = 498\,\frac{\text{ft}\cdot\text{lb}}{\text{gal}}$$

The rate of flow in gal/s is

$$\frac{90\,\text{gal/min}}{60\,\text{s/min}} = 1.5\,\text{gal/s}$$

and so

$$\text{Power output} = \left(498\,\frac{\text{ft}\cdot\text{lb}}{\text{gal}}\right)\left(1.5\,\frac{\text{gal}}{\text{s}}\right) = 747\,\text{ft}\cdot\text{lb/s}$$

The efficiency of the system is 60% = 0.60, which means that the required power input is

$$\text{Power input} = \frac{\text{power output}}{\text{efficiency}} = \frac{747\,\text{ft}\cdot\text{lb/s}}{0.60} = 1245\,\text{ft}\cdot\text{lb/s}$$

The power input in horsepower is

$$\frac{1245\,\text{ft}\cdot\text{lb/s}}{(550\,\text{ft}\cdot\text{lb/s})/\text{hp}} = 2.26\,\text{hp}$$ ∎

10-2 ELASTICITY

While solid objects often seem perfectly rigid and unyielding, it is nevertheless possible to deform them either temporarily or permanently by applying stresses. Stress forces fall into three categories: *tensions, compressions,* and *shears.* These are illustrated in Fig. 10-2.

Categories of stress

A *tensile stress* is applied to an object when equal and opposite forces act away from each other on its ends, which tends to lengthen the object. A *compressive stress* is applied when such forces act toward each other, which tends to shorten the object. A *shearing stress* is applied when the equal and opposite forces have different lines of action, which tends to alter the shape of the object without changing its volume.

In simple terms, tension stretches an object, compression shrinks it, and shear twists it.

The response of an object to a given stress depends upon its composition, shape, temperature, and so on. Figure 10-3 shows the increases in length of identical steel, copper, and aluminum wires fixed in place at one end when a force of 100 N is applied to the other end.

The amount by which a particular object is deformed is directly proportional to the applied stress provided the stress does not exceed a certain limit. In the case of tension, for example, we might find that supporting a 20-kg mass with a certain thin wire stretches the wire by 1 mm (Fig. 10-4). Doubling the mass to 40 kg will stretch it by 2 mm, tripling the mass to 60 kg will stretch it by 3 mm, and so on. When the mass is removed, the wire returns to its original length.

Elastic deformation is proportional to applied stress

The above proportionality is called *Hooke's law,* and may be written

$$F = ks \qquad\qquad \textit{Hooke's law} \quad (10-2)$$

Hooke's law

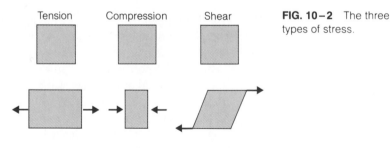

Tension   Compression   Shear

FIG. 10-2 The three types of stress.

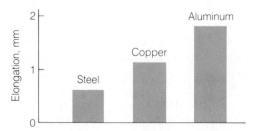

Elongation, mm

Steel   Copper   Aluminum

FIG. 10-3 The increases in length of three wires 1 m long and 1 mm in diameter when a tension force of 100 N is applied.

FIG. 10–4 The elongation of a wire is proportional to the stress applied to it, provided that the elastic limit is not exceeded. When the stress is removed, the wire returns to its original length.

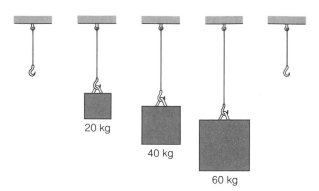

FIG. 10–5 Graph of the elongation of an iron rod as more and more tension is applied to it. When a tension below the elastic limit is applied and then removed, the rod returns to its original length, which is elastic behavior. When a tension exceeding the elastic limit is applied and removed, the rod does not contract fully but remains permanently longer, which is plastic behavior.

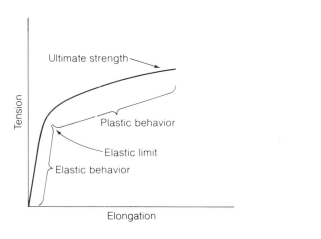

where F is the applied tensile force, s the resulting length increase, and k a constant whose value depends upon the nature and dimensions of the object under stress. The force constant k is higher for materials such as steel than it is for materials such as lead; it is directly proportional to the cross-sectional area of the object, so that a thick wire of a given material has a higher value of k than a thin wire of the same material. Relationships similar to Eq. (10–2) are found to apply to the behavior of solids under shear and to all states of matter under compression.

Elastic limit

The term *elastic limit* refers to the maximum stress that can be applied to an object without its being permanently deformed as a result. When its elastic limit is exceeded, the object may or may not be far from breaking. Brittle substances like glass or cast iron break at or near their elastic limits. Bone is another material that remains elastic until its breaking point is reached, and so it cannot be permanently altered in size or shape by applying a force. The elastic limit of many materials decreases with temperature, which is why metals are often heated before being formed into shape by bending or hammering.

Plastic deformation and ultimate strength

Figure 10–5 is a graph that shows how an iron rod stretches as more and more tension is applied to it. At first the graph is a straight line, which corresponds to Hooke's law. Past the elastic limit the graph flattens out, which means that each increase in tension by a given amount produces a greater increase in length than it did below the

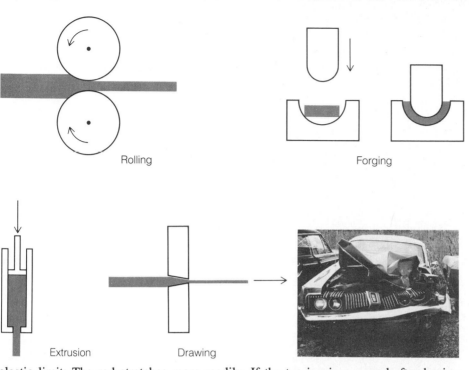

Rolling   Forging

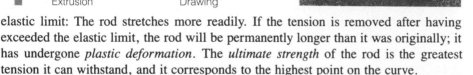

Extrusion   Drawing

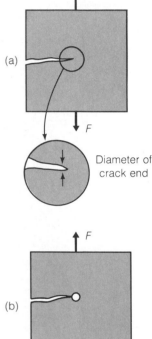

elastic limit: The rod stretches more readily. If the tension is removed after having exceeded the elastic limit, the rod will be permanently longer than it was originally; it has undergone *plastic deformation*. The *ultimate strength* of the rod is the greatest tension it can withstand, and it corresponds to the highest point on the curve.

Many metals (cast iron is a notable exception) can be deformed considerably beyond their elastic limits. A *malleable* metal is one that has a large plastic range in compression. A piece of such a metal can be formed into a desired shape by rolling, forging, or extrusion (Fig. 10-6). A *ductile* metal is one that has a large plastic range in tension and can be formed into rod or wire by being drawn through a die. The two terms are sometimes used interchangeably.

An object may fail through *fatigue* after repeated applications of stresses that are well under its original breaking strength. Tiny defects in the internal structure of the material grow a little each time a stress acts, and eventually cracks appear that lead to rupture (Fig. 10-7). Three factors are involved: the level of stress, the total number of stress cycles that occur, and corrosion. The number of cycles needed for failure decreases rapidly with increasing stress. A metal bar that can stand a million applications of a force equal to 20% of its ultimate strength may break after only 10,000 or 20,000 applications of a force twice as great. The problem is clearly most severe in machinery—a gasoline or diesel engine operating at 3000 rpm undergoes 180,000 stress cycles per hour. Stress concentrations occur wherever there is a sharp corner or notch, even a scratch, so parts designed for high-speed machines are always highly polished. Imperfections in a weld produce stress concentrations that promote fatigue failure, and welded parts therefore need generous margins of safety. Corrosion can accelerate fatigue by producing pits in a surface from which cracks can develop, and also by causing the cracks to grow faster.

FIG. 10-7 (a) A force applied to an object with a crack causes a concentration of stress at the end of the crack. The smaller the diameter of the end, the greater the stress concentration and the more likely the crack is to spread. (b) Drilling a hole at the end of the crack increases the diameter there and reduces the stress concentration.

10-3 STRUCTURE OF SOLIDS

Crystalline and amorphous solids

Most solid materials are *crystalline* in nature, with their atoms arranged in regular patterns. A few solids, such as pitch, wax, and glass, have irregular structures much like those of liquids; they are said to be *amorphous* ("without form") solids.

The atoms in a crystalline solid have a certain normal spacing and resist being pulled farther apart or being squeezed closer together. If we want a mental picture of the structure of a crystal, we can imagine tiny springs joining each atom to its neighbors (Fig. 10-8). Each spring represents a *bond* between a pair of atoms. In terms of this model, it is easy to see why even the smallest stress can deform a crystal, and why it returns to its original size and shape when a stress below the elastic limit is removed.

A dislocation is a missing line of atoms

Few crystals are perfect in structure, with completely regular arrangements of atoms. A common type of crystal defect is a missing line of atoms. Such a defect is called a *dislocation*. The presence of dislocations makes it possible to understand why a solid can be deformed permanently by bending, squeezing, or stretching without breaking. We might think such plastic deformations occur when a force is applied to a solid by the sliding of layers of atoms over one another, but calculations show that sliding of this kind—which involves the breaking of millions of bonds between atoms simultaneously—would require forces about a thousand times stronger than those actually found to produce the deformations.

How dislocations contribute to plastic deformation

Figure 10-9(a) shows a crystal with a missing row of atoms. When equal and opposite forces are applied to the crystal along different lines of action, as in Fig. 10-9(b), the dislocation shifts to the right as the atoms in the layer containing the missing row shift their bonds, one row at a time, with the atoms in the layer above. In Fig. 10-9(c) the dislocation has reached the end of the crystal, which is now permanently deformed. Because the bonds were shifted one row at a time, instead of all at once, much less force was needed than would have been required to deform a perfect solid.

Strengthening a metal

To increase the strength of a solid, it is necessary to impede the motion of dislocations in its structure. There are two chief ways to do this in a metal. One is to increase the number of dislocations by hammering the metal or squeezing it between rollers. The dislocations then become so numerous and tangled together that they interfere with each other's motion. This effect is called *work hardening*.

Another approach is to add foreign atoms to the metal that act as roadblocks to the progress of dislocations. Thus the addition of small amounts of carbon, chromium, manganese, and other elements to iron turns it into the much stronger steel by impeding the ability of dislocations to move through the material. The most suitable composition for steel depends upon how it is going to be used. For example, a carbon content of more than about 0.75% plus appropriate treatment gives a very hard steel (the steel used in files and razor blades typically contains 1.25% carbon), but such steels are difficult to weld. When welding is required, low-carbon steels are therefore preferred.

FIG. 10-8 A simple model of a crystalline solid. Each spring represents the bond between adjacent atoms, which resist being pulled farther apart or pressed closer together than their normal spacing. No such springs actually exist, but they help us to visualize the origins of elastic and plastic behavior in solids.

10-4 YOUNG'S MODULUS

Hooke's law for each kind of stress can be expressed in such a way that only a single constant need be known for a particular material in order to relate the force applied to *any* object of this material to the resulting elastic deformation, regardless of its size

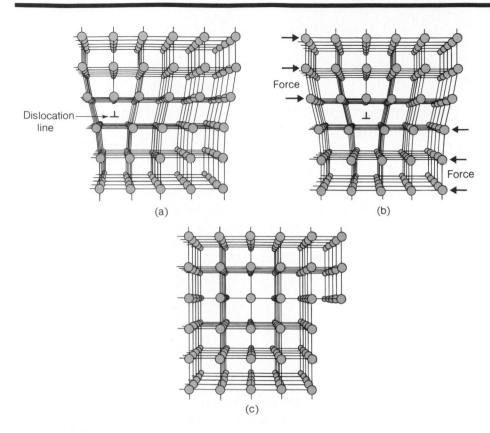

FIG. 10-9 The motion of a dislocation in crystal under stress results in a permanent deformation.

Dislocation line

(a)

(b)

(c)

and shape. Experimentally it is found that the relative change in size of an object is proportional to the ratio between the applied force and its cross-sectional area. This ratio is called the *stress* on the object: Stress is applied force per unit area. The resulting relative change in size is called *strain:* Strain is change in length per unit length, change in volume per unit volume, and so on.

Stress and strain

Thus we can summarize Hooke's law by saying that, below the elastic limit, *strain is proportional to stress*. Under a given force a thin rod will stretch more than a thick one, and a long rod will stretch more than a short one. The *modulus of elasticity* of a material subjected to a certain kind of stress is defined as the ratio between the stress and the strain that occurs because of it:

Stress is proportional to strain below the elastic limit

$$\text{Modulus of elasticity} = \frac{\text{stress}}{\text{strain}}$$

For tensile or compressive stresses, the modulus of elasticity is called *Young's modulus*. In the case of a rod of initial length L_0 and cross-sectional area A in which a tensile or compressive force F produces a change in ΔL in its length (Fig. 10-10),

Young's modulus applies to tensile and compressive stresses

$$\text{Stress} = \frac{\text{force}}{\text{area}} = \frac{F}{A}$$

$$\text{Strain} = \frac{\text{change in length}}{\text{original length}} = \frac{\Delta L}{L_0}$$

FIG. 10–10 An object under tension stretches by an amount that depends upon the value of Y for the material of which it is composed. Y is called *Young's modulus.*

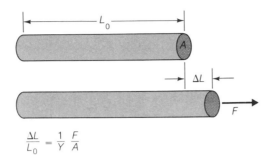

$$\frac{\Delta L}{L_0} = \frac{1}{Y}\frac{F}{A}$$

and so

$$\text{Young's modulus} = \frac{\text{stress}}{\text{strain}}$$

$$Y = \frac{F/A}{\Delta L/L_0} \tag{10–3}$$

Hence we have for the strain

$$\frac{\Delta L}{L_0} = \frac{1}{Y}\frac{F}{A} \qquad\qquad \textit{Tension or compression} \quad (10–4)$$

The rod increases in length by ΔL if it is in tension and decreases by that amount if it is in compression. The value of Y depends upon the composition of the rod. Young's moduli for a number of common substances are given in Table 10–2.

The pascal   The SI unit of stress is the N/m^2, which is called the *pascal* (Pa):

$$1 \text{ pascal} = 1 \text{ Pa} = 1 \text{ N/m}^2$$

In the British system, the customary unit of stress is the lb/in^2.

TABLE 10–2
Typical elastic moduli
$(1 \text{ Pa} = 1 \text{ N/m}^2)$

Material	Young's Modulus, Y $(\times 10^{10}\,Pa)$	$(\times 10^6\,lb/in.^2)$	Shear Modulus, S $(\times 10^{10}\,Pa)$	$(\times 10^6\,lb/in.^2)$	Bulk Modulus, B $(\times 10^{10}\,Pa)$	$(\times 10^6\,lb/in.^2)$
Aluminum	7.0	10	2.4	3.4	7.0	10
Brass	9.1	13	3.6	5.1	6.1	8.5
Concrete	2.0	2.9				
Copper	11	16	4.2	6.0	14	20
Glass	5.5	7.8	2.3	3.3	3.7	5.2
Granite	4.5	6.5			4.5	6.5
Iron	19	28	7.0	10	10	14
Lead	1.6	2.3	0.56	0.8	0.77	1.1
Pine wood (parallel to grain)	1.0	1.5				
Steel	20	29	8.4	12	16	23

Example A copper wire 1 mm in diameter and 2 m long is used to support a mass of 5 kg. By how much does the wire stretch under this load?

Solution From Eq. (10–4),

$$\Delta L = \frac{L_0 F}{Y A}$$

Here, since the radius of a wire 1 mm in diameter is 5×10^{-4}, we have

$$L_0 = 2\,\mathrm{m}$$
$$F = mg = (5\,\mathrm{kg})(9.8\,\mathrm{m/s^2}) = 49\,\mathrm{N}$$
$$Y = 1.1 \times 10^{11}\,\mathrm{Pa}$$
$$A = \pi r^2 = \pi(5 \times 10^{-4}\,\mathrm{m})^2 = 7.85 \times 10^{-7}\,\mathrm{m^2}$$

and so

$$\Delta L = \frac{(2\,\mathrm{m})(49\,\mathrm{N})}{(1.1 \times 10^{11}\,\mathrm{Pa})(7.85 \times 10^{-7}\,\mathrm{m^2})} = 1.1 \times 10^{-3}\,\mathrm{m} = 1.1\,\mathrm{mm} \quad\blacksquare$$

Example A sagging floor is jacked up and a steel girder 10 ft long whose cross-sectional area is 6 in.^{2} is put in place underneath. When the jack is removed, a sensitive strain gauge shows that the girder has been compressed by 0.008 in. Find the load the girder is supporting.

Solution In order to use Eq. (10–4), both L_0 and ΔL must be in the same units. Here it is convenient to express both in inches, with $L_0 = (10\,\mathrm{ft})(12\,\mathrm{in./ft}) = 120$ in. From Eq. (10–4) we find for the load on the girder

$$F = YA\frac{\Delta L}{L_0} = \left(29 \times 10^6 \frac{\mathrm{lb}}{\mathrm{in.^2}}\right)(6\,\mathrm{in.^2})\left(\frac{0.008\,\mathrm{in.}}{120\,\mathrm{in.}}\right) = 11{,}600\,\mathrm{lb} \quad\blacksquare$$

The proportionality between strain and stress is valid only when the elastic limit is not exceeded. Table 10–3 is a list of the elastic limits and ultimate strengths under tension of several materials; the elastic limit is the same in both tension and compression, but the ultimate strength may be much greater in compression. The compressive strength of cast iron, for instance, is more than three times its tensile strength, and the difference is a factor of five in the case of concrete. A material otherwise ideal for a certain purpose may have to be reinforced with something else to provide the missing tensile strength. Thus steel rods are commonly embedded in the concrete used for buildings and bridges, and glass fibers are embedded in the polyester resin used to make boat hulls.

Compressive strength usually exceeds tensile strength

In general, elastic limit and ultimate strength depend upon the history of an object as well as upon its composition: Hot-rolled and cold-rolled steel have different properties, as do annealed and tempered steel. Further, repeated cycles of stress tend to weaken an object through fatigue, as mentioned earlier. For these reasons the values in Table 10–3 are only approximate, although they are typical of each material.

Material	Elastic Limit ($\times 10^8$ Pa)	Elastic Limit ($\times 10^4$ lb/in.^{2})	Ultimate Strength Tension ($\times 10^8$ Pa)	Ultimate Strength Tension ($\times 10^4$ lb/in.^{2})	Ultimate Strength Compression ($\times 10^8$ Pa)	Ultimate Strength Compression ($\times 10^4$ lb/in.^{2})
Aluminum	1.8	2.6	2.0	2.9	2.0	2.9
Concrete	*		0.03	0.04	0.14	0.20
Copper	1.5	2.2	3.4	4.9	3.4	4.9
Granite	*		0.05	0.07	1.3	1.8
Iron (cast)	1.6	2.3	1.7	2.5	5.5	8.0
Pine wood (parallel to grain)	0.3	0.4	0.4	0.6	0.35	0.5
Steel	2.5	3.6	5.0	7.3	5.0	7.3

*Elastic limits in tension and compression are the same as the corresponding ultimate strengths.

TABLE 10–3
Typical elastic limits and
ultimate strengths (1 Pa =
1 N/m^2)

Example A nylon rope 0.75 in. in diameter has a breaking strength of 15,000 lb. Find the breaking strength of similar ropes (a) 0.50 in. and (b) 1.0 in. in diameter.

Solution (a) Since the breaking stress F/A is the same for all the ropes and $A = \pi d^2/4$,

$$\frac{F_2}{A_2} = \frac{F_1}{A_1}$$

$$F_2 = F_1 \left(\frac{A_2}{A_1}\right) = F_1 \left(\frac{\pi d_2^2/4}{\pi d_1^2/4}\right) = F_1 \left(\frac{d_2}{d_1}\right)^2$$

The ratio of breaking strengths is proportional to the square of the ratio of diameters. With the 0.75-in. rope as 1 and the 0.50-in. rope as 2, the breaking strength of the 0.50-in. rope is

$$F_2 = F_1 \left(\frac{d_2}{d_1}\right)^2 = 15,000\,\text{lb} \left(\frac{0.50\,\text{in.}}{0.75\,\text{in.}}\right)^2 = 6667\,\text{lb}$$

To two significant figures, this is 6700 lb.
   (b) Here $d_2 = 1.0$ in., so

$$F_2 = 15,000\,\text{lb} \left(\frac{1.0\,\text{in.}}{0.75\,\text{in.}}\right)^2 = 26,667\,\text{lb}$$

To two significant figures, this is 27,000 lb. ∎

Example What is the minimum radius of a copper wire able to support a 5-kg mass without exceeding its elastic limit?

Solution The applied force is

$$F = mg = (5\,\text{kg})(9.8\,\text{m/s}^2) = 49\,\text{N}$$

and the elastic limit of copper is, from Table 10–3, 1.5×10^8 Pa. Hence

$$\frac{F}{A} = 1.5 \times 10^8 \, \text{Pa}$$

$$A = \frac{49 \, \text{N}}{1.5 \times 10^8 \, \text{Pa}} = 3.27 \times 10^{-7} \, \text{m}^2$$

Since the cross-sectional area of a wire of radius r is $A = \pi r^2$, we have

$$r = \sqrt{\frac{A}{\pi}} = \sqrt{\frac{3.27 \times 10^{-7} \, \text{m}^2}{\pi}}$$

$$= \sqrt{1.04 \times 10^{-7} \, \text{m}^2} = \sqrt{10.4 \times 10^{-8} \, \text{m}^2} = 3.2 \times 10^{-4} \, \text{m} = 0.32 \, \text{mm} \quad ∎$$

10–5 SHEAR

Shear stresses change the shape of an object upon which they act. The volume of the object is not affected. The situation is much like that of a book whose covers are pushed out of line, as in Fig. 10–11(a): The layers of atoms, which are analogous to the pages of the book, are displaced sideways, but the spacing of the layers, which corresponds to the thickness of the pages, remains the same.

Let us consider a block of thickness d whose lower face is fixed in place and upon whose upper face the force F acts (Fig. 10–11(b)). A measure of the relative distortion of the block caused by the shear stress is the angle ϕ, called the *angle of shear*. Because this angle is always small, its value in radians is equal to the ratio s/d between the displacement s of the block's faces and the distance d between them. The greater the area A of these faces, the less they will be displaced by the shear force F. Therefore

Angle of shear is a measure of shear strain

$$\text{Shear stress} = \frac{\text{force}}{\text{area}} = \frac{F}{A}$$

$$\text{Shear strain} = \text{angle of shear} = \phi = \frac{s}{d}$$

and the stress-strain equation for shear is

$$\text{Shear modulus} = \frac{\text{shear stress}}{\text{shear strain}}$$

$$S = \frac{F/A}{\phi} = \frac{F/A}{s/d}$$

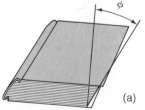

(a)

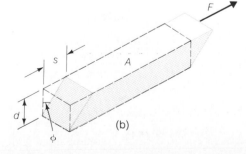

(b)

FIG. 10–11 (a) In shear there is a change in shape without a change in volume. The angle ϕ is the angle of shear. (b) The angle of shear ϕ (in radians) is equal to s/d. The greater the shear modulus S, the more rigid the material.

Hence we have

$$\phi = \frac{s}{d} = \frac{1}{S}\frac{F}{A} \qquad\qquad Shear \quad (10\text{--}5)$$

Here the applied forces are *parallel* to the faces upon which they act and not perpendicular as in the case of tension and compression.

Shear modulus

The quantity S is called the *shear modulus*, and the values of S for various substances are given in Table 10–2. The higher the value of S, the more rigid the material; S is sometimes referred to as the *modulus of rigidity* for this reason. It is interesting to note that the shear modulus of any material is usually a good deal less than its Young's modulus. This means that it is easier to slide the atoms of a solid past one another than it is to pull them apart or squeeze them together.

Shear strength

The shear strength of a material is the maximum shear stress an object of that material can withstand before breaking. The greater the shear strength of a material, the more force must be applied to cut a sheet of it with a pair of scissors (or their industrial equivalent) or to punch a hole in the sheet.

Example   Ordinary mild steel ruptures when a shear stress of about 3.5×10^8 Pa is applied. Find the force needed to punch a 1-cm diameter hole in a steel sheet 3 mm thick.

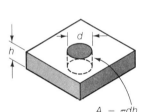

FIG. 10–12

Solution   The area across which the shear stress is exerted here is the cylindrical inner surface of the hole (Fig. 10–12), so that

$$A = \pi dh = \pi(10^{-2}\,\text{m})(3 \times 10^{-3}\,\text{m}) = 9.4 \times 10^{-5}\,\text{m}^2$$

Since we are given that $(F/A)_{\max} = 3.5 \times 10^8$ Pa, we have

$$F_{\max} = (3.5 \times 10^8\,\text{Pa})(9.4 \times 10^{-5}\,\text{m}^2) = 3.3 \times 10^4\,\text{N}$$

which is nearly 4 tons.   ■

10–6 BULK MODULUS

Uniform compression

When inward forces act over the entire surface of an object, its volume decreases by some amount ΔV from its original volume of V_0. Only those force components that are perpendicular to the object's surface where they act are effective in compressing it, since the parallel components lead only to shear stresses. If the compression force per unit area F/A is the same over the entire surface of the object, as in Fig. 10–13, then

$$\text{Volume stress} = \frac{\text{force}}{\text{area}} = \frac{F}{A}$$

$$\text{Volume strain} = \frac{\text{change in volume}}{\text{original volume}} = \frac{\Delta V}{V_0}$$

and

$$\text{Bulk modulus} = -\frac{\text{volume stress}}{\text{volume strain}}$$

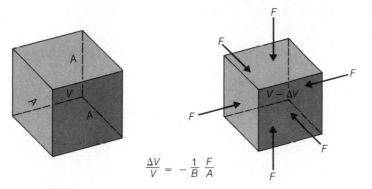

FIG. 10-13 Bulk compression.

$$B = -\frac{F/A}{\Delta V/V_0}$$

Thus we have for the relative change in volume

$$\frac{\Delta V}{V_0} = -\frac{1}{B}\frac{F}{A} \qquad\qquad \textit{Uniform compression} \quad (10\text{--}6)$$

The minus sign corresponds to the fact that an increase in force leads to a decrease in volume. The quantity B is called the *bulk modulus:* Typical values of B are given in Table 10–2.

The perpendicular stress F/A is usually called *pressure,* symbol p, as discussed in Chapter 11. Hence we can also write

Pressure is perpendicular force per unit area

$$\frac{\Delta V}{V_0} = -\frac{p}{B} \qquad\qquad\qquad\qquad (10\text{--}7)$$

Liquids can support neither tensions or shears, but they do tend to resist compression. Bulk moduli for several liquids are given in Table 10–4. Interatomic forces within a liquid are smaller than within a solid, which is reflected in the considerably smaller bulk moduli of liquids. Substantial forces are nevertheless needed to compress a liquid by more than a slight amount; to compress a volume of water by 1% requires an inward force per unit area of 3300 lb/in^2.

Example Verify the above statement.

Liquid	Bulk modulus, B	
	$(\times\ 10^9$ Pa$)$	$(\times\ 10^5$ lb/in.$^2)$
Alcohol, ethyl	0.90	1.3
Benzene	1.05	1.5
Kerosene	1.3	1.9
Mercury	26	38
Oil, lubricating	1.7	2.5
Water	2.3	3.3

TABLE 10-4
Bulk moduli of liquids at room temperature

Solution   The bulk modulus of water is 3.3×10^5 lb/in.^{2} and, by hypothesis, $\Delta V/V_0 = -0.01$. Hence

$$\frac{F}{A} = -B\frac{\Delta V}{V_0} = -\left(3.3 \times 10^5 \frac{\text{lb}}{\text{in.}^2}\right)(-0.01)) = 3300 \text{ lb/in.}^2$$

10−7   BUILDING OVER SPACE

FIG. 10−14   Three ways to build over space. (a) A horizontal beam experiences tensile as well as compressive stresses. (b) The principal stresses in an arch are compressive, so an arch made of stones or bricks, which are weak in tension but strong in compression, can support a considerable load. (c) Modern structural materials such as steel and reinforced concrete are strong in tension as well as compression and so can be used as cantilevers supported only at one end. (Photo © Roy King.)

Three successive inventions for building over space have shaped the course of architecture. The earliest, and still the most widely used, is the *post-and-beam* arrangement of Fig. 10−14(a), in which two vertical posts hold up a horizontal beam. Before steel came into general use in the last century, the width that could be spanned by a beam was severely limited by the nature of the materials available, which were mainly wood and stone. Wooden beams cannot support really large loads unless the span is narrow, and obtaining timber of adequate size and quality has always been a problem in much of the world. Stone can be quite strong in compression but is very much less so in tension, and the lower part of a beam is under tension. Narrow doorways and many interior supports were accordingly needed in buildings of even modest size that used post-and-beam construction until steel beams came into use in the nineteenth century.

The great advantage of the *arch* is that the principal stresses imposed on it by a load are compressive (Fig. 10−14(b)). Stone and brick are sufficiently strong in compression for quite large arches, and also for domes, which are the three-dimensional equivalent of arches. The arch first came into wide use in Roman times, although it had been known thousands of years earlier. A great many ancient structures based on the arch and the dome still stand, testimony to the sound engineering principle behind them.

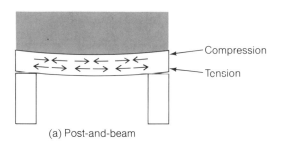

(a) Post-and-beam

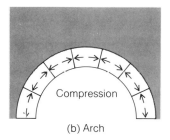

Compression

(b) Arch

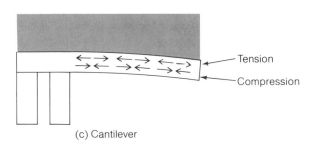

(c) Cantilever

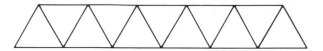

FIG. 10–15 The members of a truss are arranged in triangles. (Photo: Susan Lea)

The third, and most recent, development in building over space is the *cantilever,* a beam held in place at one end only (Fig. 10–14(c)). Reinforced concrete as well as steel can be used in cantilever construction. The tall buildings of today do not have the load-bearing walls of the past but have internal skeletons instead, with the outer parts of their floors and their thin "curtain" walls, often almost entirely glass, supported by cantilever beams that extend outward from central frames.

Solid beams are often replaced by *trusses*. A truss is a structure made up of straight members of wood or metal arranged in triangles (Fig. 10–15). A truss using the same amount of material as a solid beam is much more rigid; or, for the same rigidity, a truss is much lighter than the corresponding solid beam. The members of a truss are in tension or in compression, not in shear, and hence can be relatively thin. The longest truss in the world is a 1230-ft bridge span.

IMPORTANT TERMS

The **density** of a substance is its mass per unit volume. Its **weight density** is its weight per unit volume.

The **specific gravity** of a substance is its density relative to that of water. Specific gravity is also called **relative density.**

The three categories of stress forces are **tension,** in which equal and opposite forces that act away from each other are applied to a body; **compression,** in which equal and opposite forces that act toward each other are applied to a body; and **shear,** in which equal and opposite forces that do not act along the same line of action are applied to a body. A tensile stress tends to elongate a body, a compressive stress to shorten it, and a shearing stress to change its shape without changing its volume.

Hooke's law states that the amount of deformation experienced by a body under stress is proportional to the magnitude of the stress. Thus the elongation of a wire is proportional to the tension applied to it.

The **elastic limit** is the maximum stress a solid can be subjected to without being permanently altered. Hooke's law is valid only when the elastic limit is not exceeded.

The **stress** on an object is the applied force per unit area; the **strain** is the resulting change in a dimension of the object relative to its original value. A **modulus of elasticity** of a material is the ratio between a particular kind of applied stress and the resulting strain, provided that the elastic limit is not exceeded.

IMPORTANT FORMULAS

Density: $\quad d = \dfrac{m}{V}$

Weight density: $\quad D = \dfrac{w}{V}$

Hooke's law: $\quad F = ks$

Tension or linear compression: $\quad \dfrac{\Delta L}{L_0} = \dfrac{1}{Y} \dfrac{F}{A}$

Shear: $\quad \phi = \dfrac{s}{d} = \dfrac{1}{S} \dfrac{F}{A}$

Uniform compression: $\quad \dfrac{\Delta V}{V_0} = -\dfrac{1}{B} \dfrac{F}{A} = -\dfrac{p}{B}$

MULTIPLE CHOICE

1. Which of the following quantities is independent of the size and shape of an object composed of a given material?

   a. Volume   b. Mass
   c. Weight   d. Density

2. When equal and opposite forces are exerted on an object along different lines of action, the object is said to be under

   a. tension.   b. compression.
   c. shear.   d. elasticity.

3. Ductility and malleability refer to the ability of a metal to
  a. be deformed temporarily.
  b. be deformed permanently.
  c. shrink under compression.
  d. break under tension.

4. The stress on an object when a force acts on it is equal to
  a. the relative change in its dimensions.
  b. the applied force per unit area.
  c. Young's modulus.
  d. the elastic limit.

5. The stress on a wire supporting a load does not depend upon
  a. the wire's length.
  b. the wire's diameter.
  c. the mass of the load.
  d. the acceleration of gravity.

6. A shearing stress that acts on an object affects its
  a. length.   b. width.
  c. volume.   d. shape.

7. Another name for the shear modulus of a material is
  a. Young's modulus.
  b. modulus of rigidity.
  c. bulk modulus.
  d. ductility.

8. The only elastic modulus that applies to liquids is
  a. Young's modulus.
  b. shear modulus.
  c. modulus of rigidity.
  d. bulk modulus.

9. According to Hooke's law, the force needed to elongate an elastic object by an amount s is proportional to
  a. s.   b. $1/s$.
  c. s^2.   d. $1/s^2$.

10. Two wires are made of the same material, but wire A is half as long as and has twice the diameter of wire B. If they are to be stretched by the same amount, the required force on wire A must be
  a. one-eighth that on B.
  b. twice that on B.
  c. four times that on B.
  d. eight times that on B.

11. The force needed to punch a hole in a sheet of metal depends upon the metal's
  a. density.
  b. tensile strength.
  c. compressive strength.
  d. shear strength.

12. The density of air is $1.3 \ \text{kg/m}^3$. The air in a room 5 m long, 4 m wide, and 2.5 m high has a mass of
  a. 0.026 kg.   b. 6.5 kg.
  c. 38 kg.   d. 65 kg.

13. The density of brass is $8.4 \ \text{g/cm}^3$. The volume occupied by 200 g of brass is
  a. $0.042 \ \text{cm}^3$.   b. $0.41 \ \text{cm}^3$.
  c. $24 \ \text{cm}^3$.   d. $1680 \ \text{cm}^3$.

14. A quart of water weighs 2.1 lb. A quart of antifreeze whose specific gravity is 1.1 weighs
  a. 1.0 lb.   b. 1.1 lb.
  c. 1.9 lb.   d. 2.3 lb.

15. Gasoline has a specific gravity of 0.68. One L of gasoline weighs
  a. 0.68 N.   b. 1.4 N.
  c. 6.7 N.   d. 6.8 N.

16. A rubber strip 40 cm long has a cross-sectional area of $3 \ \text{mm}^2$ and a Young's modulus of $7 \ \text{MN/m}^2$. The force needed to stretch the strip to a length of 60 cm is
  a. 0.286 N.   b. 3.5 N.
  c. 10.5 N.   d. 3500 N.

17. A brass rod 5 ft long with a square cross section $\frac{1}{2}$ in. on a side is used to support a 520-lb load. Its elongation is
  a. 0.0008 in.   b. 0.0048 in.
  c. 0.0096 in.   d. 0.048 in.

18. An iron wire 1 m long with a square cross section 2 mm on a side is used to support a 100-kg load. Its elongation is
  a. 0.0027 mm.   b. 1.3 mm.
  c. 2.7 mm.   d. 3.7 mm.

19. A wire 10 m long with a cross-sectional area of $0.1 \ \text{cm}^2$ stretches by 13 mm when a load of 100 kg is suspended from it. The Young's modulus for this wire is
  a. 0.77×10^{10} Pa.
  b. 7.5×10^{10} Pa.
  c. 7.7×10^{10} Pa.
  d. 9.3×10^{10} Pa.

20. The ultimate strength of aluminum is $2 \times 10^8 \ \text{N/m}^2$. The maximum load that can be applied to an aluminum wire 3 mm in diameter and 4 m long is
  a. 0.144 kN.   b. 1.41 kN.
  c. 1.80 kN.   d. 13.9 kN.

21. The ultimate strength in compression of aluminum is 2×10^8 Pa. The maximum mass an aluminum cube 1 cm on each edge can support is approximately
  a. 10^2 kg.   b. 10^3 kg.
  c. 10^4 kg.   d. 10^5 kg.

22. Steel has a Young's modulus of $2 \times 10^{11} \ \text{N/m}^2$ and an elastic limit of 2.5×10^8 N/m^2. By how much can a steel rod 4 m long and 1 cm^2 in cross-sectional area be stretched before the elastic limit is exceeded?

 a. 1 mm   b. 2 mm
 c. 2.5 mm   d. 5 mm

23. The force needed to punch a hole 8 mm square in a steel sheet 3 mm thick whose shear strength is 2.5×10^8 N/m^2 is

 a. 2.67 kN.   b. 6 kN.
 c. 24 kN.   d. 48 kN.

24. When a pressure of 2×10^6 N/m^2 is applied to a sample of kerosene, it contracts by 0.15%. The bulk modulus of kerosene is

 a. 1.7×10^6 N/m^2.   b. 1.997×10^6 N/m^2.
 c. 2.003×10^6 N/m^2.   d. 1.3×10^9 N/m^2.

25. The deepest known point of the oceans, 11 km below the surface, is found in the Marianas Trench southwest of Guam in the Pacific. At that depth the water pressure exceeds sea-level pressure by 1.1×10^8 N/m^2. The density of sea water at sea level is 1.03×10^3 kg/m^3 and its bulk modulus is 2.3×10^9 N/m^3. The density of sea water at a depth of 11 km is

 a. 49 kg/m^3.   b. 215 kg/m^3.
 c. 936 kg/m^3.   d. 981 kg/m^3.

EXERCISES

10–1 Density

1. A 1200-kg concrete slab that measures 2 m × 1 m × 20 cm is delivered to a building under construction. Does it contain steel reinforcing rods or is it plain concrete?

2. A house has a volume of 20,000 ft^3. What is the weight of the air it contains?

3. Mammals have approximately the same density as fresh water. Find the volume in liters of a 55-kg woman and the volume in cubic meters of a 140,000-kg blue whale.

4. A mooring anchor is being made by pouring concrete into a cylindrical mold with an 18-in radius. How thick should the concrete be in order that the mooring weigh 3000 lb?

5. A helium-filled balloon used to carry scientific instruments to high altitudes has a radius of 2 m at sea level. What is the mass of the helium it contains?

6. If gold costs $300/oz, how many millimeters on a side does a $10,000 cube of gold measure?

7. The radius of the earth is 6.37×10^6 m and its mass is

5.98×10^{24} kg. (a) Find the average density of the earth. (b) The average density of rocks at the earth's surface is 2.7×10^3 kg/m^3. What must be true of the matter of which the earth's interior is composed? Is it likely that the earth is hollow and peopled by another species, as the ancients believed?

8. A 200-g bottle has a mass of 340 g when filled with water and 344 g when filled with blood plasma. What is the density of the plasma?

9. Water and oil are poured into the ends of a glass tube bent into a U-shape, as in the diagram. Find the density of the oil in grams per cubic centimeter.

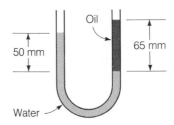

10. One gram of gold can be beaten out into a foil 1 m^2 in area. (Thus an ounce of gold can yield 300 ft^2 of foil.) How many atoms thick is such a foil? The mass of a gold atom is 3.27×10^{-25} kg.

11. A pump whose efficiency is 70% is to be used to raise water from a well 15 m deep at the rate of 60 L/s. At what rate must power be supplied to the pump?

12. A 1.0-kW motor drives a pump that raises gasoline from an underground tank through a height of 3.2 m. If the pump is 65% efficient, find the rate of flow of gasoline in L/min.

10–2 Elasticity

13. Two boys wish to break a string. Are they more likely to do this if each takes one end of the string and they pull against each other, or if they tie one end of the string to a tree and both pull on the free end? Why?

14. A rubber band is easy to stretch at first, but after it has been extended by a certain amount, additional stretching becomes more difficult at a greater rate. Make a rough stress-strain graph for rubber in tension.

15. What shape should the perforations in a sheet of postage stamps have in order to make the stamps separate most easily?

16. When a coil spring is used to support a 12-kg object, the spring stretches by 4 cm. What is the force constant of the spring?

17. A coil spring has a force constant of 1000 N/m. How

much will it stretch when it is used to support an object whose mass is 8 kg?

18. A coil spring has a force constant of 2 lb/ft. How much will it stretch when it is used to support an object whose mass is 0.5 slug?

10−4   Young's Modulus

19. A three-legged stool has one leg of aluminum, one of brass, and one of steel. The legs have the same dimensions. If the load on the stool is on its exact center, which leg is under the greatest stress and which under the least stress? Which leg experiences the greatest strain and which least strain?

20. A cable is replaced by another one of the same length and same material but of twice the diameter. How does this affect the maximum load that can be supported by the cable? What would be the result of using a cable of the same diameter and material but twice the length of the original one?

21. Two wires are made of the same material, but wire *A* is twice as long and has twice the diameter of wire *B*. Find the elongation of wire *B* relative to that of wire *A* when both are subjected to the same load.

22. The accompanying table shows the amounts by which a metal rod 120 cm long whose cross-sectional area is 16 mm^2 stretches when various loads are applied. (a) Draw a stress-strain graph for the metal. (b) Find its Young's modulus.

Load, kN	0	1	2	3	4	5
ΔL, mm	0	0.63	1.25	1.88	2.50	3.13

23. A typical sample of compact bone has a tensile strength of 1.2×10^8 Pa and a compressive strength of 1.7×10^8 Pa. (Bone is a heterogeneous substance in which protein fibers provide most of the tensile strength and inorganic salt crystals provide most of the compressive strength. The different properties of these materials lead to different values of Young's modulus for bone in tension and in compression as well as to different ultimate strengths.) Suppose you have cylinders of the same length and diameter of bone, aluminum, and steel. For each cylinder find the ratio between the maximum tensile and compressive forces it can withstand and its mass. How does bone compare with aluminum and steel as a structural material?

24. The actual cross-sectional dimensions of a "2 × 4" timber are 1.5×3.5 in. Find the maximum weight a pine 2 × 4 standing on end can support.

25. How safe is it for a 75-kg circus acrobat to balance on the index finger of his right hand, whose bones have a minimum cross-sectional area of 0.5 cm^2?

26. Find the maximum force a hammer can apply to a steel nail 2.5 mm in diameter if the nail's elastic limit is not to be exceeded.

27. A human hair 60 μm in diameter can just support a mass of 55 g (equivalent to about 2 oz of weight). How does the ultimate strength in tension of hair compare with that of aluminum?

28. In a test firing, the three main engines of the space shuttle *Columbia* developed a total thrust of 4.5×10^6 N. Eight steel bolts 90 mm in diameter were used to hold the *Columbia* to its launch pad during the firing. What was the ratio between the ultimate strength of the bolts and the applied stress?

29. A nylon rope 10 mm in diameter breaks when a load of 25 kN is applied to it. What would you estimate for the breaking strength of a nylon rope 6 mm in diameter? 14 mm in diameter?

30. A cube of pine wood 5 cm on an edge is held in the jaws of a vise by a force of 2000 N. By how much is the wood compressed?

31. A steel wire 1 m long and 1 mm square in cross section supports a mass of 6 kg. By how much does it stretch?

32. A steel post 4 m long and 3 cm in radius supports a load of 3000 kg. By how much is it shortened?

33. A steel column 12 ft long and of 4.3 in.^{2} cross-sectional area supports a load of 3 tons. By how much is it shortened?

34. A wire 10 ft long with a cross-sectional area of 0.12 in.^{2} stretches by 0.02 in. when a load of 600 lb is suspended from it. Find the value of Young's modulus for this wire.

35. An 80-kg man has femurs (the femur is the bone of the thigh) 42 cm long and 11 cm^2 in average cross-sectional area. Find the change in length of each femur as the man walks. Young's modulus for bone is 1.5×10^{10} Pa.

36. A sagging floor is jacked up and a steel girder 3 m long whose cross-sectional area is 40 cm^2 is put in place underneath. When the jack is removed, a sensitive strain gauge shows that the girder has been compressed by 0.2 mm. Find the load the girder is supporting.

37. A brass wire 2 m long whose cross-sectional area is 5 mm^2 stretches by 2.6 mm when a force of 600 N is applied. Find the value of Young's modulus for the wire.

38. By how much can a copper wire 2 m long be stretched before its elastic limit is exceeded?

39. An iron pipe 3 m long is used to support a sagging floor.

The inside diameter of the pipe is 10 cm and its outside diameter is 11 cm. When the force on it is 15 kN, by how much is it compressed?

40. A certain kind of brick has a density of 2400 kg/m^3 and an ultimate strength in compression of 9.6×10^7 N/m^2. Find the maximum height to which a wall of this brick can be built before the lowest brick disintegrates.

41. A wall of lead bricks 1 m high is used to shield a sample of radium. Each brick was originally a cube 10 cm on an edge. What is the height of the lowest brick when the wall has been erected?

42. A steel cable whose cross-sectional area is 2.5 cm^2 supports a 1000-kg elevator. The elastic limit of the cable is 3×10^8 N/m^2. What is the maximum upward acceleration that can be given the elevator if the tension in the cable is to be no more than 20% of the elastic limit?

43. A vise is used to hold a 2-in cube of pine wood while it is being worked on. The cube is in contact with the jaws of the vise, whose screw has a pitch of $\frac{3}{16}$ in. and whose handle is grasped 4 in. from the screw axis. If there is no friction and a force of 15 lb is applied to the handle, find the amount by which the cube is compressed.

44. The parachute of a 60-kg woman fails to open and she falls into a snowbank at a terminal speed of 60 m/s, coming to a stop 1.5 m below the surface of the snow. (a) Find the average force on her during the impact. (b) If the area of the woman's body that strikes the snow is 0.2 m^2 and the stress required for serious injury to body tissues is 5×10^5 N/m^2, is she likely to survive the impact?

45. A certain material has a density of d and a tensile strength of U. How long can a rod of this material be suspended from one end without breaking under its own weight? What is this length in the case of aluminum? Steel? (Assume the density of steel to be the same as that of iron.)

46. A copper wire and a steel wire are being used side by side to support a load. (a) If they have the same diameter, what proportion of the load does each one support? (b) What should the ratio of their diameters be if they are each to support half the load?

10–5   Shear

47. A 3-cm cube of raspberry gelatin on a table is subjected to a shearing force of 0.5 N. The upper surface is displaced by 5 mm. What is the shear modulus of the gelatin?

48. Two steel plates are riveted together with ten rivets each 5 mm in diameter. If the maximum shear stress the rivets can withstand is 3.5×10^8 N/m^2, how much force applied parallel to the plates is needed to shear off the rivets?

49. A punch press that exerts a force of 20 kN is employed to punch 1-cm-square holes in sheet aluminum. If the shear strength of aluminum is 70 MN/m^2, find the maximum thickness of aluminum sheet that can be used.

50. A slot 8 in. long and 0.5 in. wide is to be punched in a steel sheet 0.060 in. thick whose shear strength is 5×10^4 lb/in.^{2}. Find the force in tons that is needed.

51. When a torque τ is applied to a cylinder, the angle θ (in radians) through which it twists depends upon the cylinder's length L, radius r, and shear modulus S according to the formula

$$\theta = \frac{2\tau L}{\pi S r^4}$$

The geometry of the situation is shown in the figure below. The above formula makes it possible to determine S experimentally by applying a torque to a wire of a given material and measuring the twist that results, a much simpler procedure than one based directly on Eq. (10–5). Consider a rod 30 cm long that is suspended horizontally at its center by a wire 70 cm long and 1 mm in diameter. When a horizontal force of 0.02 N is applied to one end of the rod, it turns through 25°. Find the shear modulus of the wire.

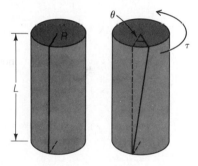

52. A solid steel drive shaft is 5 cm in diameter and 2.5 m long. If the angle of twist is not to exceed 2°, find (a) the maximum torque that can be applied to the shaft, and (b) the maximum power the shaft can transmit at 900 rpm.

10–6   Bulk Modulus

53. The elastic moduli of a particular material are related to one another, and each of them can be expressed in terms of the other two. For instance, the theory of elasticity shows that Young's modulus Y can be expressed in terms of the shear modulus S and the bulk modulus B by the formula

$$Y = \frac{9SB}{(3B + S)}$$

Check this formula for three of the materials listed in Table 10-2. Give several reasons why you would not expect perfect agreement.

54. If an aluminum object is placed in a vacuum, by what percentage will its volume increase? Atmospheric pressure is 1.013×10^5 N/m^2.

55. The pressure at a depth of 1 km in the ocean exceeds sea-level atmospheric pressure by about 10^7 N/m^2. If an iron anchor whose volume at the surface is 400 cm^3 is lowered to a depth of 1 km, by how much does its volume decrease?

10-7 Building over Space

56. The diagram shows the concrete floor of a building that is continued outside to form a cantilevered balcony. Because of their expense, steel rods are only used to reinforce con-

crete where tensile stresses occur. Show where reinforcing rods are most needed in the floor and balcony of the diagram.

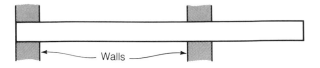

ANSWERS TO MULTIPLE CHOICE

1. d	**6.** d	**11.** d	**16.** b	**21.** c
2. c	**7.** b	**12.** d	**17.** c	**22.** d
3. b	**8.** d	**13.** c	**18.** b	**23.** c
4. b	**9.** a	**14.** d	**19.** b	**24.** d
5. a	**10.** d	**15.** c	**20.** b	**25.** d

11

FLUIDS

As the name implies, a fluid is a substance that flows readily. Gases and liquids are fluids, although the dividing line between solids and liquids is not always a sharp one. Because of its ability to flow, a fluid can exert a buoyant force on an immersed body, multiply an applied force, and provide "lift" to a properly shaped object moving through it—properties that have made possible such varied applications as the ship, the hydraulic press, and the airplane. In what follows we shall learn that there is nothing mysterious about these properties, which follow naturally from the laws of physics.

11–1 PRESSURE

Pressure is normal force per unit area

When a force **F** acts perpendicular to a surface whose area is A, the pressure p exerted on the surface is defined as the ratio between the magnitude F of the force and the area:

CHAPTER OBJECTIVES

Completing this chapter should enable you to:

1. Relate pressure to force and area.
2. Distinguish between gauge and absolute pressures.
3. Determine the mechanical advantage of a hydraulic press.
4. Find the pressure at a given depth in a fluid due to the weight of the overlying fluid.
5. Use Archimedes' principle to determine the buoyant force on an object immersed in a fluid and thereby establish whether or not it will float.
6. Distinguish between laminar and turbulent flow.
7. Calculate the rate of flow of a fluid through a pipe.
8. Use Bernoulli's equation to relate the pressure, height, and speed of a liquid at one place with their values at another place.

9. Calculate the speed at which a fluid emerges from a hole in a container.
10. Describe the origin of the lift produced by the flow of air past an airplane wing.
11. Explain the significance of the viscosity of a fluid and describe how it affects the flow of the fluid through a pipe.
12. Understand the significance of the Reynolds number and its importance in model experiments.
13. Explain what is meant by surface tension and how it originates.

$$p = \frac{F}{A} \qquad\qquad \textit{Pressure} \quad (11\text{–}1)$$

$$\text{Pressure} = \frac{\text{force}}{\text{area}}$$

Pressure is a scalar quantity. Often a perpendicular force is described as *normal,* which allows us to say that *pressure is the magnitude of normal force per unit area.*

Pressures may be measured in a number of ways, three of which are illustrated in Fig. 11–1. Usually what is directly determined is the difference between the unknown pressure and atmospheric pressure. This difference is *gauge pressure,* whereas the true pressure is called the *absolute pressure.* That is,

Gauge pressure and absolute pressure

$$p = p_{\text{gauge}} + p_{\text{atm}} \tag{11–2}$$

Absolute pressure = gauge pressure + atmospheric pressure

FIG. 11–1 Three types of pressure gauge. (a) An aneroid measures pressure in terms of the amount by which the thin, flexible ends of an evacuated metal chamber are pushed in or out by the external pressure. (b) A manometer measures pressure in terms of the difference in height *h* of two mercury columns, one open to the atmosphere and the other connected to the source of the unknown pressure. (c) A Bourdon tube straightens out when the internal pressure exceeds the external pressure.

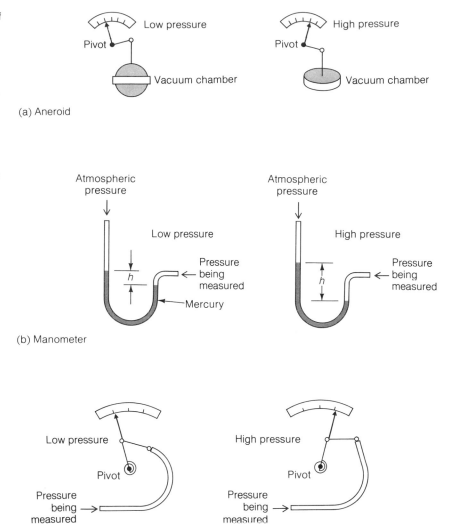

(a) Aneroid

(b) Manometer

(c) Bourdon tube

Thus a tire inflated to a gauge pressure of 28 lb/in.² contains air at an absolute pressure of 43 lb/in.², since sea-level atmospheric pressure is 14.7 lb/in.² (Fig. 11–2).

Pressure units

The SI unit of pressure is the *pascal* (Pa), as mentioned in Chapter 10, where 1 Pa = 1 N/m². The customary British unit of pressure is the lb/in.², or *psi* (for "pound per square inch"). The abbreviation *psig* stands for "pound per square inch gauge," and *psia* stands for "pound per square inch absolute." Unfortunately a number of other pressure units are also in common use. The chief ones are:

The *atmosphere* (atm) represents the average pressure exerted by the earth's atmosphere at sea level; 1 atm = 1.013 × 10⁵ Pa = 14.7 lb/in.²

The *bar* is equal to 10⁵ Pa and just slightly less than 1 atm. The *millibar* (mb), which is widely used in meteorology, is equal to 10⁻³ bar or 100 Pa. Average sea-

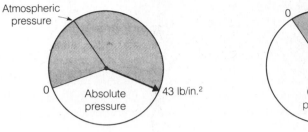

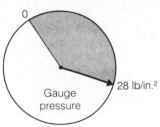

FIG. 11–2 On an absolute pressure scale, 0 corresponds to a perfect vacuum. On a gauge pressure scale, 0 corresponds to atmospheric pressure, which is 14.7 lb/in.² Shown is an absolute pressure of 43 lb/in.², which is equivalent to a gauge pressure of 28 lb/in.²

level atmospheric pressure is 1013 mb but actual pressures vary above and below this figure.

Example The weight of a 2400-lb car is supported equally by its four tires. The gauge pressure of the air in the tires is 28 lb/in.² Find the area of each tire that is in contact with the ground.

Solution The load on each tire consists of the portion of the car's weight it supports plus the weight of a column of the atmosphere whose cross-sectional area equals the area of the tire in contact with the ground, since this part of the tire has no air at atmospheric pressure under it to provide an equal upward force. Hence only the gauge pressure of the air in the tires, which is the excess over atmospheric pressure, is effective in supporting the car's weight. Each tire must support 600 lb. The area of each tire in contact with the ground is therefore

$$A = \frac{F}{P} = \frac{600\,\text{lb}}{28\,\text{lb/in.}^2} = 21.4\,\text{in.}^2 \qquad \blacksquare$$

Example The flat roof of a house is 10 m long and 8 m wide and has a mass of 7500 kg. Before a severe storm the people in the house closed its doors and windows so tightly that the air presure inside remained at 1013 mb even when the outside pressure dropped to 980 mb. (These are absolute pressures, of course.) Compare the upward force on the roof with its weight.

Solution The area of the roof is

$$A = \text{length} \times \text{width} = (10\,\text{m})(8\,\text{m}) = 80\,\text{m}^2$$

The difference Δp between the pressures on the lower and upper sides of the roof is

$$\Delta p = (1013\,\text{mb} - 980\,\text{mb})(100\,\text{Pa/mb}) = (33\,\text{mb})(100\,\text{Pa/mb})$$
$$= 3.3 \times 10^3\,\text{Pa}$$

Since the pressure on the lower side of the roof is greater, the net force on the roof is upward, and from Eq. (11–1) its magnitude is

$$F = A\Delta p = (80\,\text{m}^2)(3.3 \times 10^3\,\text{Pa}) = 2.64 \times 10^5\,\text{N}$$

This is 59,400 lb! The roof's weight is

$$mg = (7500\,\text{kg})(9.8\,\text{m/s}^2) = 7.35 \times 10^4\,\text{N}$$

Thus the upward force on the roof is nearly four times greater than its weight. If the roof is not well attached to the walls of the house, and if the windows do not break first, the roof will be lifted off during the storm. Clearly it is unwise to seal a building when a large drop in air pressure is coming. ■

The torr

The *torr* represents the pressure exerted by a column of mercury 1 mm high and is equal to 133 Pa. The torr was formerly referred to as the "millimeter of mercury," abbreviated mm Hg. In medicine, *blood pressures* are usually expressed in torr.

Blood pressure

A person's arterial blood pressures are usually measured with the help of an inflatable cuff wrapped around the upper arm at the level of the heart (Fig. 11–3). A stethoscope is used to monitor the sound of the blood flowing through an artery below the cuff. The cuff is first inflated until the flow of blood stops. Then the pressure of the cuff is gradually reduced until the blood just begins to flow, which is recognized by a gurgling sound in the stethoscope. This pressure, called *systolic,* represents the maximum pressure the heart produces in the artery. The pressure in the cuff is then further reduced until the gurgling stops, which corresponds to the restoration of normal blood flow. The pressure at this time, called *diastolic,* represents the pressure in the artery between the strokes of the heart. In a healthy person the systolic and diastolic pressures are respectively about 120 and 80 torr.

Three properties of pressure in a fluid

Pressure is a useful quantity because fluids flow under stress instead of being deformed elastically as solids are. This property of fluids has three important consequences:

(1) The forces a fluid at rest exerts on the walls of its container, and vice versa, always act perpendicular to the walls.

If this were not so, any sideways force by a fluid on a wall would be accompanied, according to the third law of motion, by a sideways force back on the fluid, which would cause the fluid to move parallel to the wall. But the fluid is at rest, so that any force it exerts must be perpendicular to the container walls. A *moving* fluid is another

FIG. 11–3 Arterial blood pressures are measured with the help of an inflatable cuff that is wrapped around the upper arm. A pump is used to inflate the cuff until the flow of blood stops, and air is then let out by the valve until the flow begins again. The stethoscope is used to monitor the blood flow.

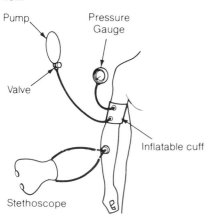

Pump

Pressure Gauge

Valve

Inflatable cuff

Stethoscope

FIG. 11–4 Variation of atmospheric pressure with height above sea level. At an altitude of 50 km, the pressure is down to about 1/1000 of its sea-level value. (1 mb = 100 Pa.)

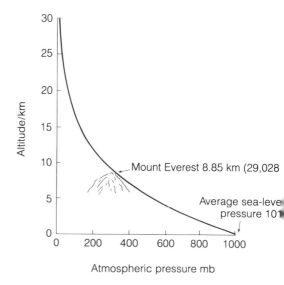

Mount Everest 8.85 km (29,028

Average sea-leve pressure 101

Altitude/km

Atmospheric pressure mb

matter; as we shall learn in Section 11–8, frictional forces act between a moving fluid and, for instance, the walls of a pipe or the bank of a river. The difference between a body of liquid in contact with a solid and two solids in contact is that in the first case there is no static friction between them.

(2) An external pressure exerted on a fluid is transmitted uniformly through-   Pascal's principle
out the volume of the fluid.

If this were not so, the fluid would flow from a region of high pressure to one of low pressure, which would equalize the pressure. We must keep in mind, however, that the above statement (known as *Pascal's principle*) refers to a pressure imposed from outside the fluid. The fluid at the bottom of a container is always under greater pressure than that at the top owing to the weight of the overlying fluid. A notable example is the earth's atmosphere (Fig. 11–4), although such pressure differences are ordinarily significant only for liquids.

(3) The pressure on a small surface in a fluid is the same regardless of the
orientation of the surface.

If this were not so, again, the fluid would flow in such a way as to equalize the pressure (Fig. 11–5).

A *pump* is a device for creating a pressure difference. The operation of the familiar   **Pumps**
piston pump is illustrated in Fig. 11–6. A pump of this kind can be used to provide high pressure (to inflate a tire or a football, for instance) or low pressure (to pump water from a well or the bilge of a boat, for instance). The heart circulates blood through the lungs and through the body by means of two pumps of this kind, with the contractions and relaxation of the muscular walls of the heart chambers taking the place of piston strokes. Rotary pumps are more suited to motor drive; two examples are shown in Fig. 11–6.

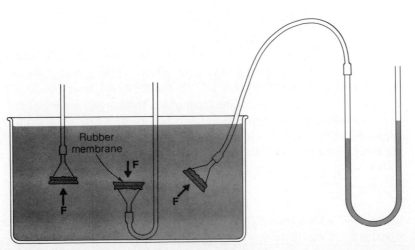

FIG. 11–5 The force exerted by the pressure in a fluid is the same in all directions at any depth.

Rubber
membrane

F

F

F

FIG. 11–6 (Below) A piston pump. In the suction stroke, the inlet valve opens and the fluid is drawn into the cylinder. When the piston is pushed down, the inlet valve closes and the outlet valve opens, and the fluid is forced out. (At right) Two types of rotary pump. In an impeller pump, the fluid is compressed when the rubber fins of the impeller are bent over at the flattened top of the chamber. In a centrifugal pump, the rotating blades set the fluid in motion.

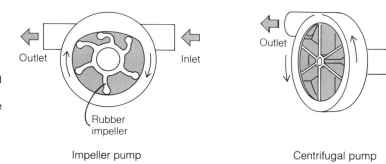

Impeller pump   Centrifugal pump

11–2 HYDRAULIC PRESS

In Chapter 9 we considered two of the three basic machines, the lever and the inclined plane. The third, the *hydraulic press,* is based upon property (2) of the previous section: An external pressure exerted on a fluid is transmitted uniformly throughout the volume of the fluid.

Figure 11–7 is a diagram of a hydraulic press being used in a garage to lift a car for servicing. When the input piston, whose area is A_{in}, is moved down through the distance L_{in} by an applied force F_{in}, the output piston, whose area is A_{out}, is raised through the distance L_{out}. Assuming that the fluid is incompressible, the volume of it transferred from one cylinder to the other as the pistons move is

$$\text{Volume} = \text{area} \times \text{distance} = A_{in}L_{in} = A_{out}L_{out}$$

From Eq. (9–3) the ideal mechanical advantage of the hydraulic press is L_{in}/L_{out}, so

$$\text{IMA} = \frac{L_{in}}{L_{out}} = \frac{A_{out}}{A_{in}} \qquad\qquad \textit{Hydraulic press} \quad (11\text{–}3)$$

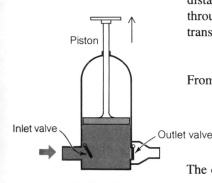

Piston

Inlet valve   Outlet valve

Suction stroke

The output force F_{out} is given by Eq. (9–5) as

$$F_{out} = (\text{Eff})\left(\frac{L_{in}}{L_{out}}\right)(F_{in}) = (\text{Eff})\left(\frac{A_{out}}{A_{in}}\right)(F_{in}) \qquad\qquad (11\text{–}4)$$

A small input force can be considerably increased simply by having the output piston much larger in area than the input piston. If the piston areas are in the ratio 100:1, and Eff = 100%, a force of 1 lb can lift a weight of 100 lb. Of course, for every inch the weight is raised the input piston must move down 100 in.

To understand how a hydraulic press achieves its multiplication of force, let us suppose that the press of Fig. 11–7 is 100% efficient. The input force F_{in} acting on the input piston produces the pressure

$$p_{in} = \frac{F_{in}}{A_{in}}$$

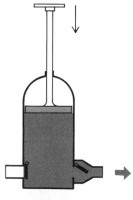

Pressure stroke

on the fluid in the system. This pressure is transmitted by the fluid to the output piston upon which it exerts the upward force F_{out}, where

$$\frac{F_{out}}{A_{out}} = p_{out}$$

Since the pressure is the same at both pistons, $p_{in} = p_{out}$ and

$$\frac{F_{out}}{A_{out}} = \frac{F_{in}}{A_{in}}$$

The pressure is the same on both pistons, but the force is not

$$F_{out} = \left(\frac{A_{out}}{A_{in}}\right)(F_{in}) \qquad \textit{Hydraulic press (100\% efficiency)} \quad (11\text{–}5)$$

This is the same as Eq. (11–4) with Eff = 1.

The purpose of the valves indicated in Fig. 11–7 is to permit the output piston to be raised by a series of short strokes of the input piston. When the latter moves downward, valve 1 closes and valve 2 opens, which allows fluid under pressure to move into the large cylinder. When the input piston is pulled, valve 1 opens and valve 2 closes, and additional fluid is drawn into the input cylinder to enable it to make another stroke.

The principle of the hydraulic press can be applied in a variety of ways. Common industrial uses include presses of various kinds, garage lifts, control systems in airplanes, and vehicle brakes (Fig. 11–8).

Example   The input and output pistons of a hydraulic jack are respectively 1 in. and 4 in. in diameter. A lever with a mechanical advantage of 6 is used to apply force to the input piston (Fig. 11–9). How much weight can the jack lift if a force of 40 lb is applied to the lever and friction is negligible in the system?

Solution   The mechanical advantage of the jack is

$$\text{MA} = \frac{A_{out}}{A_{in}} = \frac{\pi r_{out}^2}{\pi r_{in}^2} = \left(\frac{d_{out}}{d_{in}}\right)^2 = \left(\frac{4\,\text{in.}}{1\,\text{in.}}\right)^2 = 16$$

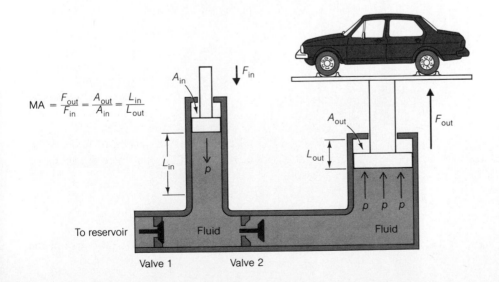

$$\text{MA} = \frac{F_{out}}{F_{in}} = \frac{A_{out}}{A_{in}} = \frac{L_{in}}{L_{out}}$$

FIG. 11–7 The hydraulic press makes use of the fact that pressure exerted on a fluid is transmitted equally throughout the fluid. Valve 1 is closed and valve 2 open on the downstroke of the input piston; valve 1 is open and valve 2 closed on the upstroke.

The total mechanical advantage of the lever + jack system is

$$MA_{total} = (MA_{lever})(MA_{jack}) = (6)(16) = 96$$

Hence the output force of the system when the input force is 40 lb is

$$F_{out} = (MA)(F_{in}) = (96)(40 \text{ lb}) = 3840 \text{ lb} \qquad \blacksquare$$

FIG. 11–8 A hydraulic vehicle brake. When the brake pedal is pushed, the pressure on the brake fluid in the master cylinder is transmitted to each brake cylinder where the force on the pistons pushes the brake shoes against the brake drums. The friction between the brake linings and the drums is what slows the car down. In a disk brake, the pistons of the brake cylinder press pads against the sides of a metal disk mounted on the inside of each wheel.

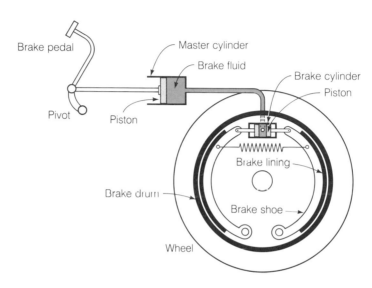

FIG. 11–9 Hydraulic jack. (Photo © Roy King.)

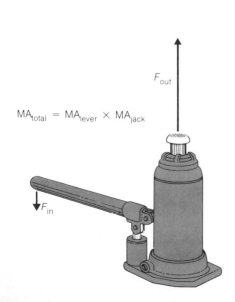

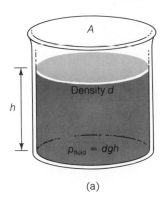

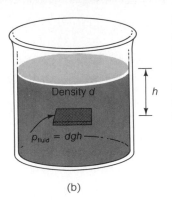

FIG. 11–10 (a) The gauge pressure a fluid in a tank exerts on the bottom is equal to dgh. The same formula holds for the fluid pressure at any depth h below the surface.

(a)   (b)

11–3   PRESSURE AND DEPTH

The pressure inside a volume of fluid depends upon the depth below the surface, since the deeper we descend, the greater the weight of the overlying fluid. Few submarines can descend more than about 100 m without the danger of collapsing. At a depth of 10 km in the ocean the pressure is about 1000 atm, enough to compress water by 3% of its volume.

Pressure in a fluid increases with depth

Suppose we have a tank of height h and cross-sectional area A that is filled with a fluid of density d (Fig. 11–10). The volume of the tank is

$$V = Ah$$

and the mass of fluid it contains is

$$m = dV = dAh$$

The weight of the fluid in the tank is therefore

$$w = mg = dgAh$$

The pressure p_{fluid} the fluid exerts on the bottom of the tank is its weight divided by the area of the bottom, with the result that

$$p_{\text{fluid}} = \frac{F}{A} = \frac{w}{A} = dgh \qquad (11\text{–}6)$$

The pressure difference between the top and the bottom of the tank is directly proportional to the height of the fluid column and to the fluid density.

The above result also applies to *any* depth h in a fluid, whether at the bottom or not, since the fluid beneath that depth does not contribute to the weight pressing down there.

The *total* pressure within a fluid, of course, also depends upon the pressure p_{external} exerted on its surface by the atmosphere or, perhaps, by a piston (Fig. 11–11). Thus, in general, the total pressure at a depth h in a fluid of density d is $p = p_{\text{external}} + p_{\text{fluid}}$:

$$p = p_{\text{external}} + dgh \qquad \textit{Pressure at depth h in a fluid} \quad (11\text{–}7)$$

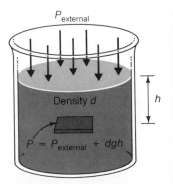

FIG. 11–11 The total pressure at a depth h in a fluid is the sum of the fluid pressure and the external pressure exerted on the fluid surface.

Example   Find the pressure on a scuba diver when she is 15 m below the surface of the sea.

Solution   The density of seawater is 1.03×10^3 kg/m³. Hence the pressure on the diver due to the sea alone is

$$dgh = \left(1.03 \times 10^3 \frac{\text{kg}}{\text{m}^3}\right)\left(9.8\frac{\text{m}}{\text{s}^2}\right)(15\,\text{m}) = 1.51 \times 10^5 \frac{\text{N}}{\text{m}^2} = 1.51 \times 10^5\,\text{Pa}$$

Since atmospheric pressure is 1.01×10^5 Pa, the total pressure on the diver is

$$p = p_{\text{external}} + dgh = (1.01 \times 10^5 + 1.51 \times 10^5)\,\text{Pa}$$

$$= 2.52 \times 10^5\,\text{Pa}$$

which is about $2\frac{1}{2}$ times atmospheric pressure. She is not crushed because the pressure within her body increases as she descends to match the pressure exerted on it. As the diver returns to the surface, it is essential that she breathe out continuously to allow the air pressure in her lungs to decrease at the same rate as the external pressure decreases. If this is not done, the pressure difference will cause her lungs to expand and rupture. Similarly a face mask that includes the nose as well as the eyes should be used instead of just goggles when scuba diving. Air exhaled through the nose can then keep the external pressure on the eyeballs the same as the internal pressure, which prevents damage to the blood vessels of the eyes. (See Sec. 26–6 for the reason why goggles or a face mask are needed for proper vision underwater.)

11–4   ARCHIMEDES' PRINCIPLE

Buoyancy

An object immersed in a fluid seems to weigh less than it does outside. This effect, known as *buoyancy,* permits people to swim, ships to float, and hot air or helium-filled balloons to rise through the atmosphere.

A very simple argument permits us to determine the buoyant force on an object. Suppose we have a solid object of volume V submerged in a fluid of density d. We begin by considering instead a body of fluid of the same size and shape as the object and located at the same depth, as in Fig. 11–12(a). This body of fluid is in equilibrium, which means that its weight of

$$w = Vdg$$

is supported by a buoyant force of this magnitude exerted by the rest of the fluid. The buoyant force is the vector sum of all the forces the rest of the fluid exerts on the body, and this force must be upward because the pressure (and hence the upward force) on the bottom of the body is greater than the pressure (and hence the downward force) on its top. (The forces on the sides of the body as a rule cancel out.) Therefore the buoyant force is Vdg, the weight of the body of fluid.

Origin of buoyant force

Now we replace the body of fluid by the solid object, as in Fig. 11–12(b). The various pressures remain the same, so the buoyant force of Vdg also remains the same. We conclude that

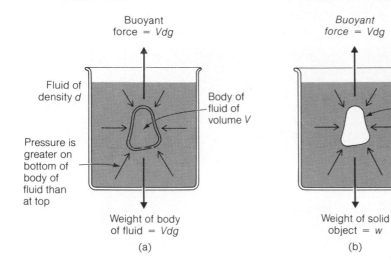

Buoyant force = Vdg

Fluid of density d

Pressure is greater on bottom of body of fluid than at top

Body of fluid of volume V

Weight of body of fluid = Vdg

(a)

Buoyant force = Vdg

Solid object of volume V

Weight of solid object = w

(b)

FIG. 11–12 The buoyant force on a submerged object is equal to the weight of the fluid it displaces.

$$F_{\text{buoyant}} = Vdg \qquad \textit{Archimedes' principle} \quad (11\text{–}8)$$

Buoyant force = weight of displaced fluid

The result is known as Archimedes' principle:

The buoyant force on a submerged object is equal to the weight of fluid displaced by the object.

Archimedes' principle allows us to determine only the buoyant force on a submerged object, not the net force on it. If the weight of the object is greater than the buoyant force on it, it will sink; if the weight is less than the buoyant force, it will rise; if the weight is equal to the buoyant force, it will float in equilibrium.

Net force on object is difference between buoyant force and its weight

Example An iceberg is a chunk of freshwater ice that has broken off from an ice cap, such as those that cover Greenland and Antarctica, or from a glacier at the edge of the sea. (More than three-quarters of the world's fresh water is in the form of ice.) Find the proportion of the volume of an iceberg that is submerged.

Solution If V_{ice} is the iceberg's total volume and V_{sub} is its submerged volume, then

Weight of iceberg = weight of displaced water,

$$V_{\text{ice}} d_{\text{ice}} g = V_{\text{sub}} d_{\text{seawater}} g$$

$$\frac{V_{\text{sub}}}{V_{\text{ice}}} = \frac{d_{\text{ice}}}{d_{\text{seawater}}} = \frac{9.2 \times 10^2 \, \text{kg/m}^3}{1.03 \times 10^3 \, \text{kg/m}^3} = 0.89$$

Thus 89% of the volume of an iceberg is below sea level (Fig. 11–13). ■

Example An iron anchor weighs 600 lb in air. How much force is needed to lift the anchor when it is submerged in seawater? The weight densities of iron and of seawater are respectively 480 and 64 lb/ft^3.

FIG. 11–13   Only 11% of the volume of an iceberg is above sea level.

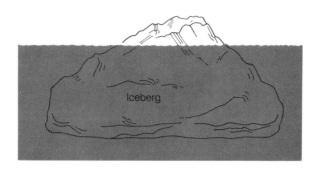

Solution   Since $D = w/V$, the anchor's volume is

$$V = \frac{w_i}{D_i} = \frac{600\,\text{lb}}{480\,\text{lb/ft}^3} = 1.25\,\text{ft}^3$$

The weight of seawater displaced by the anchor is

$$w_s = D_s V = 64\,\text{lb/ft}^3 \times 1.25\,\text{ft}^3 = 80\,\text{lb}$$

The buoyant force on the anchor is therefore 80 lb and the net force needed to lift the submerged anchor is

$$F = w_i - w_s = 600\,\text{lb} - 80\,\text{lb} = 520\,\text{lb}$$ ■

Hydrometer

The *hydrometer* is a simple instrument based on Archimedes' principle that can be used to find the density of a liquid. The principle of the hydrometer is illustrated in Fig. 11–14, where the same weighted rod is shown floating in gasoline and in water. The denser the liquid, the less the volume of the rod that must be immersed to provide enough buoyancy for it to float, and the greater the length that projects above the surface. Hence the rod floats higher in water ($d = 1.00$ g/cm^3) than in the less dense gasoline ($d = 0.68$ g/cm^3). The rod can be calibrated with a scale such that the liquid density can be read off directly from the mark at the liquid surface.

Since the density of the acid in a car's storage battery depends on its state of charge, a hydrometer can be used to establish the condition of the battery. A fully charged battery typically has a specific gravity of about 1.28 (that is, its density is 1.28 g/cm^3), which drops to about 1.15 when the battery is completely discharged. A

FIG. 11–14   The higher a hydrometer floats, the denser the liquid.

Gasoline

Water

hydrometer can also be used to determine the freezing point of the ethylene glycol antifreeze solution in the cooling system of a car, which depends upon the proportion of ethylene glycol it contains. Since ethylene glycol has a specific gravity of 1.11, the greater the specific gravity of the solution, the higher the proportion of antifreeze and the lower the freezing point. For example, a specific gravity of 1.04 corresponds to a 28% solution and to a freezing point of −14°C.

11–5 FLUID FLOW

The study of fluids in motion is one of the more difficult branches of mechanics because of the diversity of phenomena that may occur. However, the main features of fluid flow can be understood on the basis of a simple model that is reasonably realistic in many cases. This model concerns liquids that are incompressible and exhibit no viscosity. (*Viscosity* is the term used to describe internal friction in a fluid.) In the absence of viscosity, layers of fluid slide freely past one another and past other surfaces, so that it is reasonable to apply the model to such liquids as water but not to such liquids as molasses.

Another approximation we shall make is that the fluid undergoes *laminar* (or **Laminar flow** *streamline*) *flow* exclusively. In streamline flow, which is illustrated in Fig. 11–15, every particle of liquid passing a particular point follows the same path (called a *streamline*) as the particles that passed that point previously. Furthermore, the direction in which the individual fluid particles move is always the same as the direction in which the fluid as a whole moves.

At the other extreme is *turbulent flow*, which is characterized by the presence of **Turbulent flow** whirls and eddies, such as those in a cloud of cigarette smoke or at the foot of a waterfall. Turbulence generally occurs at high speeds and when there are obstructions or sharp bends in the path of the fluid.

The volume of liquid that flows through a pipe per unit time is easy to compute. **Rate of flow of liquid in** If the average speed of the liquid in the pipe of Fig. 11–16 is v, each part of the stream **a pipe** travels the distance vt in the time interval t. The volume V of liquid transported the distance vt in the time t is vt multiplied by the pipe's cross-sectional area A, or vtA. Therefore the rate of flow R of liquid through the pipe is

$$R = \frac{V}{t} = \frac{vtA}{t} = vA \qquad\qquad \textit{Rate of flow} \quad (11\text{--}9)$$

the product of the liquid speed and the cross-sectional area of the pipe. R is often

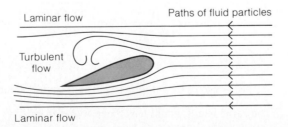

Laminar flow

Turbulent flow

Laminar flow

Paths of fluid particles

FIG. 11–15 Laminar and turbulent flows around an obstacle.

FIG. 11–16 The rate of flow of liquid through a pipe is equal to the product of the cross-sectional area of the pipe and the speed of the liquid.

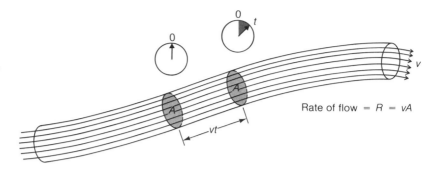

Rate of flow = $R = vA$

FIG. 11–17 In laminar flow, liquid speed is inversely proportional to the cross-sectional area of the pipe.

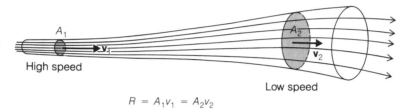

High speed

Low speed

$$R = A_1 v_1 = A_2 v_2$$

expressed in units such as L/min and gal/min instead of ft^3/s or m^3/s. Equation (11–9) holds only for streamline flow.

If the pipe size varies, the speed of the liquid also varies so as to keep R constant, so that

$$v_1 A_1 = v_2 A_2 \qquad \qquad \textit{Equation of continuity} \quad (11\text{--}10)$$

Hence a liquid flows faster through a constriction in a pipe and slower through a dilation. As in Fig. 11–17, streamlines drawn close together signify rapid motion while streamlines far apart signify slow motion.

Example   Water flows at 2.5 m/s through a garden hose whose inside radius is 6 mm. (a) What should the nozzle radius be for water to leave it at 10 m/s? (b) What is the rate of flow of water through the hose in liters per minute? (c) With how much force must the nozzle be held?

Solution   (a) Since $A = \pi r^2$, we have from Eq. (11–10)

$$v_1 (\pi r_1^2) = v_2 (\pi r_2^2)$$

$$r_2 = r_1 \sqrt{\frac{v_1}{v_2}} = 6\,\text{mm} \sqrt{\frac{2.5\,\text{m/s}}{10\,\text{m/s}}} = 3\,\text{mm}$$

(b) From Eq. (11–9) we have for the rate of flow

$$R = v_1 A_1 = v_1 (\pi r_1^2) = (2.5\,\text{m/s})\,(\pi)\,(0.006\,\text{m})^2 = 5.65 \times 10^{-4}\,\text{m}^3/\text{s}$$

The same result would have been found from $R = v_2 A_2$, of course. Since 1 m^3 = 10^3 L and 1 min = 60 s,

$$R = \left(5.65 \times 10^{-4} \frac{m^3}{s}\right)\left(10^3 \frac{L}{m^3}\right)\left(60 \frac{s}{min}\right) = 34 \, L/min$$

(c) From impulse = momentum change, if a mass of $m = dV$ of water flows through the nozzle in the time t, then

$$Ft = mv$$

$$F = \frac{mv}{t} = d\left(\frac{V}{t}\right)v = dRv$$

and so here

$$F = dRv_2 = \left(10^3 \frac{kg}{m^3}\right)\left(5.65 \times 10^{-4} \frac{m^3}{s}\right)\left(10 \frac{m}{s}\right)$$

$$= 5.65 \frac{kg \cdot m}{s} = 5.65 \, N$$

which is 1.27 lb. ∎

The rate at which a pump does work on a fluid passing through it is $P = Fv$, where the force applied is $F = pA$. Hence the power output of the pump is $P = pAv$, and, since the rate of flow is $R = vA$, we have the useful formula

$$P = pR \tag{11-11}$$
Power = (pressure)(rate of flow)

Example Oxygenated blood from the lungs is pumped by the left ventricle of the heart into a large artery called the aorta, whose diameter is typically 2 cm. When a person is resting, the rate of flow of blood might be 6 L/min. (a) Find the average speed of blood in the aorta under these circumstances. (b) Find the power output of the left ventricle of a person at rest assuming an average blood pressure of 100 torr at the aorta. (The right ventricle, which pumps blood at the same rate to the lungs where it gives up carbon dioxide and absorbs oxygen, has a smaller power output since there is less resistance to the flow of blood through the lungs than through the rest of the body.)

Solution (a) Since $1 \, L = 10^{-3} \, m^3$ and $1 \, min = 60 \, s$, $1 \, L/min = 1.667 \times 10^{-5}$ m^3/s, and $R = 1.00 \times 10^{-4} \, m^3/s$ here. The speed is therefore

$$v = \frac{R}{A} = \frac{1.00 \times 10^{-4} \, m^3/s}{\pi \times 10^{-4} \, m^2} = 0.318 \, m/s$$

(b) From Eq. (11-11) the power output of the left ventricle is

$$P = pR = (100 \, torr)(133 \, Pa/torr)(1.00 \times 10^{-4} m^3/s) = 1.33 \, W$$

Both the pressure and the rate of flow increase during physical activity. Because the power output of the heart increases faster than the output of the other muscles during exercise, a normally sedentary person should not attempt vigorous exercise without a gradual conditioning program first. ∎

11–6 BERNOULLI'S EQUATION

Changes in liquid speed arise from pressure differences

When a liquid flowing through a pipe enters a region where the pipe diameter is reduced, its speed increases. A change in speed involves an acceleration, which means that a net force must be acting upon the liquid. This force can only arise from a difference in pressure between the different parts of the pipe. Evidently the pressure in the part of the pipe having a large diameter is the greater, since the liquid increases in speed on its way to the constriction. Thus we expect a relationship between the pressure in a moving liquid and its speed, which turns out to be

$$p_1 + dgh_1 + \tfrac{1}{2}dv_1^2 = p_2 + dgh_2 + \tfrac{1}{2}dv_2^2 \qquad \textit{Bernoulli's equation} \quad (11\text{–}12)$$

Here p_1, h_1, and v_1 are respectively the pressure, height above some reference level, and speed of a liquid of density d at the point 1 in a body of the liquid, and p_2, h_2, and v_2 are the values of these quantities at another point 2.

Bernoulli's equation expresses conservation of energy in liquid flow

Equation (11–12) is known as *Bernoulli's equation,* after Daniel Bernoulli (1700–1782), who first derived it. According to Bernoulli's equation the quantity $p + dgh + \tfrac{1}{2}dv^2$ has the same value at all points in an incompressible liquid with negligible viscosity that undergoes laminar flow. The effect of viscosity is to dissipate mechanical energy into heat. If the viscosity of the liquid is not negligible, the quantity $p + dgh + \tfrac{1}{2}dv^2$ decreases in the direction of flow.

To derive Bernoulli's equation we consider a curved pipe of nonuniform cross section through which a liquid flows, as in Fig. 11–18. Let us apply the principle of conservation of energy to a parcel of the liquid of volume v_1tA_1 as it enters at the left in the time t and to the same parcel as it leaves at the right. The mass of the parcel is

FIG. 11–18 Bernoulli's equation.

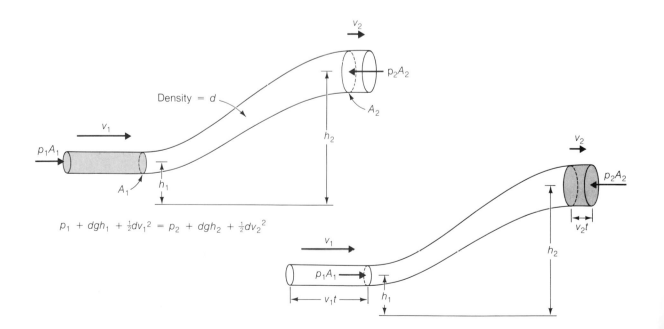

$$m = dV = dv_1tA_1 = dv_2tA_2$$ **Mass of liquid parcel**

since vA has the same value at 1 and 2. The net amount of work ΔW done on the liquid parcel as it passes from 1 to 2 must be equal to the net change in its potential energy ΔPE as its height goes from h_1 to h_2, plus the net change in its kinetic energy ΔKE as its speed goes from v_1 to v_2. That is,

$$\Delta W = \Delta PE + \Delta KE$$

The work done *on* the parcel at 1 is the force p_1A_1 on it multiplied by the distance v_1t through which the force acts. The work done *by* the parcel at 2 is the force p_2A_2 multiplied by the distance v_2t through which the force acts. The *net* work done on the parcel is therefore

$$\Delta W = p_1A_1v_1t - p_2A_2v_2t = \frac{p_1m}{d} - \frac{p_2m}{d}$$ **Work done on parcel**

The change in the potential energy of the parcel in going from 1 to 2 is

$$\Delta PE = mgh_2 - mgh_1$$ **Change in potential energy of parcel**

and the change in its kinetic energy is

$$\Delta KE = \tfrac{1}{2}mv_2^2 - \tfrac{1}{2}mv_1^2$$ **Change in kinetic energy of parcel**

Therefore

$$\Delta W = \Delta PE + \Delta KE$$ **Conservation of energy**

$$\frac{p_1m}{d} - \frac{p_2m}{d} = mgh_2 - mgh_1 + \tfrac{1}{2}mv_2^2 - \tfrac{1}{2}mv_1^2$$

When we divide through by the common factor m, multiply by d, and rearrange terms, the result is Bernoulli's equation,

$$p_1 + dgh_1 + \tfrac{1}{2}dv_1^2 = p_2 + dgh_2 + \tfrac{1}{2}dv_2^2$$

11-7   APPLICATIONS OF BERNOULLI'S EQUATION

In many situations the speed, pressure, or height of a liquid is constant, and simplified forms of Bernoulli's equation hold. Thus when a liquid column is stationary, we see that the pressure difference between two depths in it is **Liquid at rest**

$$p_2 - p_1 = dg(h_1 - h_2)$$ *Liquid at rest*

which is just what Eq. (11–6) states. Evidently the latter formula is included in Bernoulli's equation.

Another straightforward result occurs when $p_1 = p_2$. As an example, Fig. 11–19 **Torricelli's theorem** illustrates a liquid emerging from an orifice in a tank. The liquid pressure equals atmospheric pressure both at the top of the tank and at the orifice. If the orifice is small compared with the cross section of the tank, the liquid level in the tank will fall slowly enough for the liquid speed at the top of the tank to be assumed zero. If the speed of

FIG. 11–19 Torricelli's theorem.

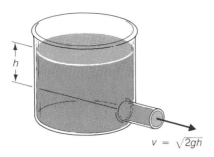

$v = \sqrt{2gh}$

the liquid as it leaves the orifice is v and the difference in height between the top of the liquid and the orifice is h, Bernoulli's equation reduces to

$$\tfrac{1}{2} dv^2 = dgh$$

Constant external pressure

$$v = \sqrt{2gh} \qquad\qquad\qquad \textit{Torricelli's theorem} \quad (11\text{--}13)$$

The speed with which the liquid is discharged is the same as the speed of a body falling from rest from the height h. This result is called *Torricelli's theorem*, and, like the relationship between pressure and depth, it is a special case of Bernoulli's equation. The rate at which liquid flows through the orifice may be found from Eq. (11–9) if the orifice area A is known. The volume of liquid being discharged per unit time is

$$R = vA = A\sqrt{2gh} \qquad\qquad\qquad \textit{Flow from orifice} \quad (11\text{--}14)$$

Example   How fast will water leak through a hole 1 cm^2 in area at the bottom of a tank in which the water level is 3 m high?

Solution   From the above formula

$$R = (1\,\text{cm}^2)\left(\frac{1}{10^4\text{cm}^2/\text{m}^2}\right)\sqrt{2 \times 9.8\,\text{m/s}^2 \times 3\,\text{m}} = 7.7 \times 10^{-4}\text{m}^3/\text{s}$$

which is almost a liter per second, an appreciable amount. ■

Constant height

The most interesting special case of Bernoulli's equation occurs when there is no change in height during the motion of the liquid (Fig. 11–20). Here

$$p_1 + \tfrac{1}{2} dv_1^2 = p_2 + \tfrac{1}{2} dv_2^2 \qquad\qquad\qquad \textit{Pressure and speed} \quad (11\text{--}15)$$

FIG. 11–20 In a horizontal pipe the pressure is greatest when the speed is least, and vice versa.

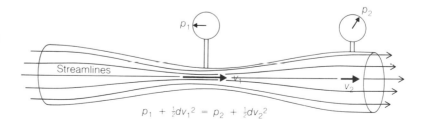

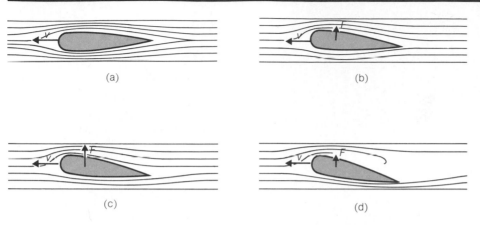

FIG. 11-21 Air flow past a wing. At (a) the flow is the same on both surfaces, so that no lift results. The lift at (c) is greater than that at (b) because of the greater pressure differences between upper and lower surfaces (the pressure is least where the streamlines are closest together). At (d) turbulence reduces the available lift.

which means that the pressure in the liquid is least where the speed is greatest, and vice versa.

A familiar application of Eq. (11–15) is the lifting force produced by the flow of air past the wing of an airplane, as in Fig. 11–21. (Air is, of course, a compressible fluid and so does not fit our model exactly, but the behavior predicted by Bernoulli's equation is not a bad approximation for gases at moderate speeds.) Air moving past the upper surface of the wing must travel faster than air moving past the lower surface; this is indicated by the closeness of the streamlines near the upper surface. The difference in speed leads to a decreased pressure over the top of the wing, a pressure that is equivalent to a suction force that lifts the wing. The greater the difference in air speeds around the upper and lower surfaces, the greater the lift that is produced, provided the wing angle is not so extreme that turbulence results (Fig. 11–21(d)). An additional source of lift is the deflection of air from the angled bottom of the wing, which leads to a reaction force on it that has an upward component.

How an airplane wing develops lift

The lift force F_{lift} on a wing of area A due to the difference in air speeds above and below it is, from Eq. (11–15),

Lift depends on wing area and square of speed

$$F_{lift} = (p_1 - p_2)A = \tfrac{1}{2}dA(v_2^2 - v_1^2)$$

Here v_1 is the air speed below the wing and v_2 is its speed above the wing. The relationships between these speeds and the speed v of the airplane depend on the shape of the wing and on the angle the wing makes with the direction of the airflow ahead of it, but in general $(v_2^2 - v_1^2)$ is proportional to v^2, so F_{lift} is proportional to Av^2. Thus the faster an airplane moves, the smaller the wing area needed to provide it with lift. Because a small wing also exhibits less resistance to motion through the air ("drag") than a large one, it is clear why high-speed airplanes have relatively small wings. The penalty such airplanes must pay is that their takeoff and landing speeds need to be correspondingly high.

At low speeds, most of the resistance a car offers to motion arises from the flexing of its tires. As the speed increases, aerodynamic drag becomes more and more important (Fig. 11–22), until finally the car's engine cannot provide enough power to overcome it and the car has reached its top speed. Since drag depends on the shape of a car as

Drag on a car

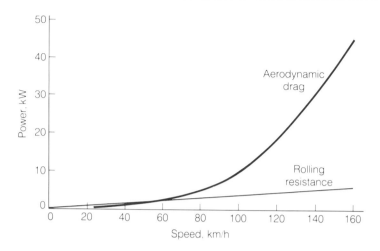

FIG. 11–22 The power needed to overcome the aerodynamic drag of a typical car increases with speed more rapidly than the rolling resistance of its tires does.

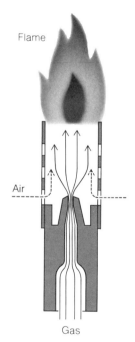

FIG. 11–23 Blowtorch nozzle.

well as upon its speed, much attention is paid nowadays to streamlining. However, a streamlined car is shaped much like a wing section, and in extreme cases (such as in racing cars) the resulting lift at high speeds significantly reduces the grip of the tires on the road. Because both steering and braking depend on this grip, such a situation is dangerous. Ridges ("spoilers") are often placed on the rear ends of fast cars to reduce lift by interfering with laminar flow, but an increase in drag is the inevitable result.

The nozzle of a blowtorch (Fig. 11–23) provides another application of the reduced pressure that accompanies an increase in fluid speed. The gas leaves the supply tube through a small orifice, which means that its speed is high as it escapes. The resulting low pressure sucks in air through holes in a sleeve around the end of the tube to mix the oxygen needed for burning with the gas. In this arrangement the flame is hotter than if the gas simply burns at the orifice without being mixed with air first. The laboratory Bunsen burner has a similar construction.

Still another application of Eq. (11–15) is the *Venturi meter,* which provides a convenient method for determining the rate of flow of a liquid through a pipe. As shown in Fig. 11–24, a Venturi meter consists of a constriction in a pipe with manometers connected to the wide and narrow sections. The liquid height in the manometer at the constriction is lower because the pressure there is lower, and the height difference H is proportional to the pressure difference, which in turn depends on the speed difference according to Eq. (11–15). The rate of flow R of the liquid is given by the formula

FIG. 11–24 The Venturi meter.

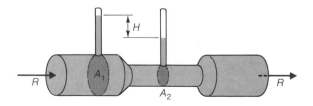

$$R = A_1 \sqrt{\frac{2gH}{(A_1/A_2)^2 - 1}} \tag{11–16}$$

Since A_1 and A_2 are fixed quantities, a measurement of H is all that is needed to find R.

11–8 VISCOSITY

The viscosity of a fluid is a kind of internal friction that prevents neighboring layers of the fluid from sliding freely past one another. Figure 11–25 shows how the speed of a fluid in a pipe varies with distance from the axis. The fluid in contact with the pipe's wall is stationary and, as we would expect, the speed increases to a maximum along the pipe's axis. The smaller the viscosity of the fluid, the greater the various speeds will be, but the characteristic way the speed varies will be maintained as long as the flow remains laminar.

Viscosity refers to internal friction in a fluid

The viscosity of a fluid is defined in terms of the experiment shown in Fig. 11–26, in which a plate of area A is being pulled across a layer of fluid s thick. (A hollow cylinder having a smaller rotating cylinder inside it, with a fluid layer between them, is a better arrangement for actual measurements.) For most fluids it is found that the force F required to pull the plate at the constant speed v is proportional to A and v and inversely proportional to s: The faster the motion and the thinner the layer of fluid, the more force is needed for a given plate area. Different fluids offer different degrees of resistance to the motion, but the force needed varies as Av/s for most of them provided that the speed is not so great that turbulence occurs. Thus we can write

How viscosity is defined

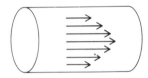

$$F = \frac{\eta A v}{s} \tag{11–17}$$

where η, the constant of proportionality, is called the *viscosity* of the fluid (η is the Greek letter *eta*). The SI unit of viscosity is evidently the N · s/m², which is known as the *poiseuille* (Pl). An older unit, the *poise,* remains in common use, where 10 poise = 1 N · s/m²; the *centipoise,* 0.01 poise, is equal to 10^{-3} N · s/m².

FIG. 11–25 Because of viscosity, the speed of fluid in a pipe varies from 0 at the pipe wall to a maximum along the axis.

The viscosities of some common fluids are listed in Table 11–1. The viscosity of a liquid decreases with temperature as its molecules become less and less tightly bound to one another (see Chapter 14), which is familiar to anyone who has heated honey or

Viscosity varies with temperature

FIG. 11–26 The viscosity of a fluid is defined in terms of the force needed to pull a flat plate at constant speed across a layer of the fluid.

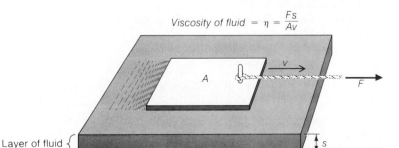

Viscosity of fluid $= \eta = \dfrac{Fs}{Av}$

Layer of fluid

Substance	Temperature (°C)	Viscosity (PI)
Gases		
Air	0	1.7×10^{-5}
	100	2.2×10^{-5}
Water vapor	100	1.3×10^{-5}
Liquids		
Alcohol (ethyl)	20	1.2×10^{-3}
Blood plasma	37	1.3×10^{-3}
Blood, whole (varies with speed)	37	$\sim 2 \times 10^{-3}$
Glycerin	20	0.83
Water	0	1.8×10^{-3}
	20	1.0×10^{-3}
	40	0.66×10^{-3}
	60	0.47×10^{-3}
	80	0.36×10^{-3}
	100	0.28×10^{-3}

molasses. One of the reasons a drop in body temperature is so dangerous is that the viscosity of the blood is thereby increased, impeding its flow. The viscosity of a gas, in contrast to that of a liquid, increases with temperature because the higher the temperature, the faster the gas molecules move and the more often they collide with one another (again, see Chapter 14).

The viscosity of motor oils is usually given in terms of SAE (Society of Automotive Engineers) index numbers. The higher the number, the greater the viscosity, so that SAE 30 oil is "thicker" than SAE 10 oil at a given temperature. Because viscosities of liquids decrease with temperature, SAE 30 oil has about the same viscosity at 20°C (68°F) as SAE 10 oil at 0°C (32°F). If SAE 30 oil were used in very cold weather, its resistance to flow would make an engine hard to start and it might not circulate adequately afterward; SAE 10 oil is better in such conditions. On the other hand, SAE 10 oil would be too thin on a hot summer day to provide proper lubrication, and SAE 30 oil would be appropriate then.

Let us return to a fluid moving through a pipe. If the pipe is cylindrical with the length L and inside radius r, and a fluid of viscosity η is flowing through it under the influence of a pressure difference $\Delta p = p_1 - p_2$ as in Fig. 11–27, the rate of flow is

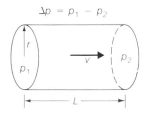

$\Delta p = p_1 - p_2$

FIG. 11–27

$$R = \frac{\pi r^4 \Delta p}{8\eta L} \qquad\qquad \textit{Poiseuille's law} \quad (11\text{–}18)$$

Equation (11–18) is known as *Poiseuille's law*. The dependence of the rate of flow on the viscosity η and on the pressure gradient $\Delta p/L$ are both about what we might expect, but the variation with r^4 is remarkable: Halving the radius of a pipe reduces R by a factor of 16 if Δp stays the same. The radius of the pipe plays a far more important role than its length does with respect to viscous resistance.

Example   Atherosclerosis is the medical term for a common condition in which arteries are narrowed by deposits of tissue called plaque. If the flow of blood is to continue at its usual rate, a higher pressure is required. A consequence of atherosclerosis is therefore an elevated blood presure, which has many undesirable effects on the body, one of them being that the heart must work harder to circulate the blood. (a) Find the increase in pressure needed to maintain R constant when the radius of an artery is decreased by 10%. (b) What is the corresponding increase in the power needed to maintain the flow of blood through that artery?

Atherosclerosis

Solution   (a) For a constant rate of flow R, the product $r^4 \Delta p$ must be constant. Here $r_2 = 0.9r_1$, so

$$r_1{}^4 \Delta p_1 = r_2{}^4 \Delta p_2$$

$$\frac{\Delta p_2}{\Delta p_1} = \left(\frac{r_1}{r_2}\right)^4 = \left(\frac{r_1}{0.9r_1}\right)^4 = 1.52$$

The blood pressure must increase by 52% if R is to be unchanged.

(b) From Eq. (11–11), $P = \Delta p R$. Here R is the same, so

$$\frac{P_2}{P_1} = \frac{\Delta p_2}{\Delta p_1} = 1.52$$

The power output of the heart must increase by 52% also. Narrowing of the arteries usually means a reduced flow of blood as well as an increase in blood pressure and the work done by the heart. ∎

11–9 REYNOLDS NUMBER

Whether fluid flow in a given situation is laminar or turbulent depends upon four factors: the density d and viscosity η of the fluid, its average speed v, and a characteristic dimension D of the system. These factors determine the type of flow according to the value of the *Reynolds number* N_R defined by

$$N_R = \frac{dvD}{\eta} \qquad\qquad \text{Reynolds number} \quad (11\text{–}19)$$

The Reynolds number is a dimensionless quantity; that is, N_R has no units associated with it and its value is a pure number that is the same regardless of which system of units is used for d, η, v, and D

For fluid flow in a pipe, D is the diameter of the pipe. Experiments show that in this case a Reynolds number of less than about 2000 corresponds to *laminar flow* and a Reynolds number of more than about 3000 corresponds to *turbulent flow*. Between 2000 and 3000 the flow may be of either type and may even shift back and forth between them.

Laminar and turbulent flow

Example In Section 11–5 an example considered a garden hose 6 mm in radius through which water pased at 2.5 m/s to provide a rate of flow of 34 L/min. (a) If the water temperature is 20°C, determine the nature of the flow in the hose from its Reynolds number. (b) What is the maximum speed for laminar flow here?

Solution (a) From Table 11–1 the viscosity of water 20°C is $\eta = 1.0 \times 10^{-3}$ PI $= 1.0 \times 10^{-3}$ N · s/m² $= 1.0 \times 10^{-3}$ kg/m · s. Since $D = 12$ mm $= 0.012$ m the Reynolds number is

$$N_R = \frac{dvD}{\eta} = \frac{(1.0 \times 10^3 \, \text{kg/m}^3)(2.5 \, \text{m/s})(0.012 \, \text{m})}{1.0 \times 10^{-3} \, \text{kg/m} \cdot \text{s}} = 30{,}000$$

We note that the units cancel one another out. Because N_R is greater than 3000, the motion of the water in the hose is turbulent.

(b) To find the maximum water speed for laminar flow, we simply solve Eq. (11–19) for v with $N_R = 2000$:

$$v = \frac{\eta N_R}{dD} = \frac{(1.0 \times 10^{-3} \text{PI})(2000)}{(1.0 \times 10^3 \, \text{kg/m}^3)(0.012 \, \text{m})} = 0.17 \, \text{m/s}$$

The corresponding rate of flow of water is, from Eq. (11–9),

$$R_2 = R_1 \left(\frac{v_2}{v_1} \right) = 34 \, \text{L/min} \left(\frac{0.17 \, \text{m/s}}{2.5 \, \text{m/s}} \right) = 2.3 \, \text{L/min} \qquad ■$$

Dynamic similarity

The Reynolds number is important because it provides a basis for experiments in which a small model system substitutes for a full-size system. Two systems are said to be *dynamically similar* when their Reynolds numbers are the same because the patterns of fluid flow in the systems will then be the same. A noteworthy example is the use of a wind tunnel to determine the areodynamic behavior of a model airplane. Here an appropriate dimension D in Eq. (11–19) would be the wingspan of the airplane. If the

Model airplane in wind tunnel. (Photo: NASA)

model is, say, one-third the size of the actual airplane, then the speed v of the air past the model must be three times the actual airplane's speed for N_R to be the same. Aerodynamic forces such as lift and drag can be measured on the model and converted to their equivalents for the full-size airplane. Another example is the towing of a model ship in a tank of water to establish the hydrodynamic behavior of the corresponding full-size ship before it is built.

11–10 SURFACE TENSION

FIG. 11–28 The surface tension of a water surface supplements buoyancy in supporting a steel needle. The elastic character of a liquid surface resembles that of a membrane under tension.

The surface of a liquid behaves much like a membrane under tension. Thus a steel sewing needle placed horizontally on the surface of some water in a dish does not sink even though its density is nearly eight times that of water (Fig. 11–28). The needle rests in a depression in the water surface just as if that surface were a sheet of rubber stretched across the dish. The term *surface tension* for this effect is quite appropriate. Another example of surface tension is the tendency of a liquid drop to assume a spherical shape, just as an inflated balloon does.

The origin of surface tension lies in the fact that molecules on the surface of a body of liquid are acted upon by a net inward force whereas those in the interior are acted upon by forces in all directions (Fig. 11–29). Since the surface molecules are all being pulled inward, the surface tends to contract to the minimum possible area. A sphere has the least area relative to its volume of any object, and so a liquid sample not acted upon by any external forces (such as gravitation or air resistance if it is moving) will take on a spherical form. A liquid surface can support a small object such as a needle because the weight of the object is insufficient to rupture the surface, which would involve first stretching it to a greater degree than that shown in Fig. 11–28.

Water has an exceptionally high surface tension, which is sometimes a handicap. A *surfactant* is a substance whose addition to a liquid reduces its surface tension. Detergents are efficient surfactants, which enables even an extremely dilute solution of detergent in water to penetrate small crevices and to be absorbed by tightly woven textiles. Surfactants are also called *wetting agents* because they allow a water solution to spread uniformly over a surface instead of forming into separate droplets.

Surfactants reduce surface tension

FIG. 11–29 Because a molecule on the surface of a body of liquid has no molecules above it to interact with, the net force on it is inwards.

A familiar phenomenon is the rise of most liquids in a capillary tube. Capillary action is responsible for many familiar effects, such as the ability of paper and cloth fibers to absorb water. Two factors are involved: the *cohesion* of the liquid, which refers to the attractive forces its molecules exert on one another, and the *adhesion* of the liquid to the surface of a solid, which refers to the attractive forces the solid exerts on the liquid molecules. If the adhesive forces are stronger than the cohesive ones, as they are in the case of water and glass, then the liquid tends to stick to the solid and will rise in a capillary tube of that material since the attraction of a liquid molecule to the wall of the tube is greater than the attraction of this molecule to its brethren. In such an event the liquid surface is concave, as in Fig. 11–30(a), with an angle of contact θ that is characteristic of the liquid-solid combination. For water and clean glass, $\theta \approx 0$, as it is for any liquid that "wets" a particular solid, because adhesion must be much greater than cohesion for this to occur. For kerosene and glass, $\theta = 26°$, which reflects the smaller amount by which adhesion exceeds cohesion for these substances. If cohesion is greater than adhesion, as it is for mercury and glass, the liquid level in a capillary tube is lower than the level of the surrounding liquid, and the liquid surface in the tube is convex, with an angle of contact greater than 90°; $\theta = 140°$ for mercury and glass.

FIG. 11–30

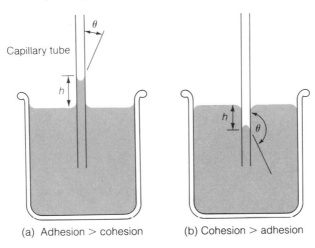

(a) Adhesion > cohesion   (b) Cohesion > adhesion

IMPORTANT TERMS

The **pressure** on a surface is the perpendicular force per unit area that acts upon it. **Gauge pressure** is the difference between true pressure and atmospheric pressure.

Pascal's principle states that an external pressure exerted on a fluid is transmitted uniformly throughout its volume.

The **hydraulic press** is a machine consisting of two fluid-filled cylinders of different diameters connected by a tube. The input force is applied to a piston in one of the cylinders, and the output force is exerted by a piston in the other cylinder.

Archimedes' principle states that the buoyant force on a submerged object is equal to the weight of fluid it displaces.

In **laminar** (or **streamline**) **flow** every particle of fluid passing a particular point follows the same path, whereas in **turbulent flow** irregular whirls and eddies occur. The greater the speed of a fluid in streamline flow, the lower its pressure.

The **viscosity** of a fluid is a measure of its internal friction.

The **Reynolds number** of a system involving fluid flow determines whether the flow will be laminar or turbulent. Systems with the same Reynolds number are **dynamically similar,** which means their patterns of fluid flow are the same.

The **surface tension** of a liquid refers to the tendency of its surface to contract to the minimum possible area in any situation.

IMPORTANT FORMULAS

Pressure: $p = \dfrac{F}{A} = p_{\text{gauge}} + p_{\text{atm}}$

Hydraulic press: $\text{IMA} = \dfrac{L_{\text{in}}}{L_{\text{out}}} = \dfrac{A_{\text{out}}}{A_{\text{in}}}$

Pressure at depth h in a fluid: $p = p_{\text{external}} + dgh$

Archimedes' principle: $F_{\text{buoyant}} = Vdg$

Rate of flow: $R = vA$

Equation of continuity: $R = v_1A_1 = v_2A_2$

Power: $P = pR$

Bernoulli's equation: $p + dgh + \tfrac{1}{2}dv^2 = constant$

Poiseuille's law: $R = \dfrac{\pi r^4 \, \Delta p}{8\eta L}$

Reynolds number: $N_R = \dfrac{dvD}{\eta}$

MULTIPLE CHOICE

1. Atmospheric pressure does not correspond to approximately
   a. 14.7 lb/in.^{2} b. 98 N/m^2.
   c. 1013 mb.   d. 1.013 × 10^5 Pa.

2. The hydraulic press is able to produce a mechanical advantage because
   a. the force a fluid exerts on a piston is always parallel to its surface.
   b. an external pressure exerted on a fluid is transmitted uniformly throughout its volume.
   c. at any depth in a fluid the pressure is the same in all directions.
   d. the pressure in a fluid varies with its speed.

3. In the operation of a hydraulic press, it is impossible for the output piston to exceed the input piston's
   a. displacement.
   b. speed.
   c. force.
   d. work.

4. The fluid at the bottom of a container is
   a. under less pressure than the fluid at the top.
   b. under the same pressure as the fluid at the top.
   c. under more pressure than the fluid at the top.
   d. any of the above, depending upon the circumstances.

5. The pressure at the bottom of a vessel filled with liquid does not depend on the
   a. acceleration of gravity.
   b. liquid density.
   c. height of the liquid.
   d. area of the liquid surface.

6. A man stands on a very sensitive scale and inhales deeply. The reading on the scale
   a. does not change.
   b. increases.
   c. decreases.
   d. depends on the expansion of his chest relative to the volume of air inhaled.

7. Buoyancy occurs because, as the depth in a fluid increases,
   a. its pressure increases.   b. its pressure decreases.
   c. its density increases.   d. its density decreases.

8. In order for an object to sink when placed in water, its specific gravity (relative density) must be
   a. less than 1.   b. equal to 1.
   c. more than 1.   d. any of the above, depending on its shape.

9. A cake of soap placed in a bathtub of water sinks. The buoyant force on the soap is
   a. zero.
   b. less than its weight.
   c. equal to its weight.
   d. more than its weight.

10. The density of fresh water is 1.00 g/cm^3 and that of seawater is 1.03 g/cm^3. A ship will float
   a. higher in fresh water than in seawater.
   b. at the same level in fresh water and in seawater.
   c. lower in fresh water than in seawater.
   d. any of the above, depending on the shape of its hull.

11. Bernoulli's equation is based upon
   a. the second law of motion.
   b. the third law of motion.
   c. conservation of momentum.
   d. conservation of energy.

12. An express train goes past a station platform at high speed. A person standing at the edge of the platform tends to be
   a. attracted to the train.
   b. repelled from the train.
   c. attracted or repelled, depending on the ratio between the speed of the train and the speed of sound.
   d. unaffected by the train's passage.

13. The volume of liquid flowing per second out of an orifice at the bottom of a tank does not depend on
   a. the area of the orifice.
   b. the height of liquid above the orifice.
   c. the density of the liquid.
   d. the value of the acceleration of gravity.

14. Of the following liquids, the one with the highest surface tension is
   a. cold water.   b. hot water.
   c. soapy water.   d. alcohol.

15. Water neither rises nor falls in a silver capillary. This suggests that the contact angle between water and silver is
   a. 0.   b. 45°.
   c. 90°.   d. 180°.

16. A 20-kg piston rests on a sample of a gas in a cylinder 10 cm in radius. The absolute pressure on the gas is
   a. 0.064 bar.   b. 0.62 bar.
   c. 1.064 bar.   d. 1.62 bar.

17. A 10-lb brick has the dimensions 4 in. × 6 in. × 12 in. The minimum pressure it can exert when one of its faces rests on a horizontal surface is
   a. 10 lb/ft^2.   b. 20 lb/ft^2.
   c. 30 lb/ft^2.   d. 60 lb/ft.^{2}

18. The input piston of a hydraulic press is 2 cm in diameter and the output piston is 1 cm in diameter. An input force of 1 N will produce an output force of

a. 0.25 N.   b. 0.50 N.
c. 2 N.   d. 4 N.

19. A manometer whose upper end is evacuated and sealed can be used to measure atmospheric pressure. Such an instrument is called a mercury barometer, and the average height of the mercury column in it is 760 mm. If water were used in a barometer instead of mercury, the height of the column of water would be about
a. 56 mm.   b. 760 mm.
c. 10 m.   d. 20 m.

20. A viewing window 30 cm in diameter is installed 3 m below the surface of an aquarium tank filled with seawater. The force the window must withstand is approximately
a. 22 N.   b. 218 N.
c. 2140 N.   d. 8562 N.

21. An aquarium tank 25 cm long and 15 cm wide contains water 10 cm deep but no fish. The gauge pressure on its bottom is
a. 0.98 Pa.   b. 98 Pa.
c. 980 Pa.   d. 9800 Pa.

22. If the tank in Exercise 21 contains fish as well as water, with the water depth the same 10 cm, the pressure at the bottom is
a. greater.
b. the same.
c. smaller.
d. any of the above, depending on the total mass of the fish.

23. An object suspended from a spring scale is lowered into a pail filled to the brim with a liquid and 4 N of the liquid overflows. The scale indicates that the object weighs 6 N in the liquid. The weight in air of the object is
a. 2 N.   b. 4 N.
c. 6 N.   d. 10 N.

24. A 22-g lead fishing sinker has an apparent mass of 20 g when immersed in water and 20.6 g when immersed in oil. The density of the oil is
a. 0.6 g/cm^3.   b. 0.7 g/cm^3.
c. 0.8 g/cm^3.   d. 1.4 g/cm^3.

25. A cube of oak (density 0.72 g/cm^3) 10 cm on each edge is floating in fresh water. The height of its upper surface above the water is
a. 1.4 cm.   b. 2.8 cm.
c. 5 cm.   d. 7.2 cm.

26. A wooden board 2 m long, 30 cm wide, and 4 cm thick floats in water with 1 cm of its thickness above the surface.

The mass of the board is
a. 1.8 kg.   b. 18 kg.
c. 24 kg.   d. 176 kg.

27. A force of 1000 lb is required to raise a concrete block to the surface of a freshwater lake. The force required to lift it out of the water is approximately
a. 700 lb.   b. 1062 lb.
c. 1140 lb.   d. 1800 lb.

28. The depth in fresh water at which the water density is 1% greater than its value at the surface is approximately
a. 2.3×10^2 m.   b. 2.3×10^3 m.
c. 2.3×10^4 m.   d. 2.3×10^5 m.

29. The total cross-sectional area of all the capillaries of a certain person's circulatory system is 0.25 m^2. If blood flows through the system at the rate of 100 cm^2/s, the average speed of blood in the capillaries is
a. 0.4 mm/s.   b. 4 mm/s.
c. 25 mm/s.   d. 400 mm/s.

30. A certain person's heart beats 1.2 times per second and pumps 10^{-4} m^3 of blood per beat against an average pressure of 14 kPa. The power output of the heart is
a. 1.17 W.   b. 1.4 W.
c. 1.68 W.   d. 11.7 W.

31. A water tank springs a leak from which water emerges at 5 m/s. The height of the water surface above the leak is
a. 0.51 cm.   b. 128 cm.
c. 196 cm.   d. 255 cm.

32. Water leaves the safety valve of a boiler at a speed of 30 m/s. The gauge pressure inside the boiler is
a. 4.5 millibars.   b. 1.5 bars.
c. 4.5 bars.   d. 450 bars.

33. A horizontal pipe of 10 cm^2 cross-sectional area is joined to another horizontal pipe of 50 cm^2 cross-sectional area. The speed of water in the small pipe is 6 m/s and the pressure there is 200 kPa. The speed of the water in the large pipe is
a. 0.24 m/s.   b. 1.2 m/s.
c. 3 m/s.   d. 13.4 m/s.

34. The pressure in the large pipe of the previous question is
a. 183 kPa.   b. 202 kPa.
c. 217 kPa.   d. 235 kPa.

35. The rate of flow of water through the above pipe is
a. 0.6 L/s.   b. 1.2 L/s.
c. 3 L/s.   d. 6 L/s.

EXERCISES

(Assume laminar flow and negligible viscosity unless otherwise noted.)

11–1 Pressure

1. A little water is boiled for a few minutes in a tin can, and the can is sealed while it is still hot. Why does the can collapse as it cools?

2. When a person drinks soda through a straw, where does the force come from that causes the soda to move upward?

3. What is the absolute pressure of the air in a flat tire?

4. A jar is filled to the top with water, and a piece of cardboard is slid over the opening so that there is only water in the jar. If the jar is now turned over, will the cardboard fall off? What will happen if there is any air in the jar?

5. A mercury barometer in Denver reads 720 torr. What is the atmospheric pressure there in millibars? In pascals?

6. The two upper front teeth of a person exert a total force of 30 N on a piece of steak. If the biting edge of each tooth measures 10 mm × 1 mm, find the pressure on the steak.

7. A 2-kg brick has the dimensions 7.5 cm × 15 cm × 30 cm. Find the pressures exerted by the brick on a table when it is resting on its various faces.

8. A tire inflated to a gauge pressure of 2 bars has a nail 3 mm in diameter embedded in it. What is the force tending to push the nail out of the tire?

9. A DC-9 airplane weighing 90,000 lb is in level flight. The area of its wings is 990 ft². What is the average difference in pressure between the upper and lower surfaces of its wings?

10. The weight of a car is equally supported by its four tires, each of which has an area of 150 cm² in contact with the ground. If the gauge pressure of the air in the tires is 1.8 bars, find the mass of the car.

11. The force on a phonograph needle whose point is 0.1 mm in radius is 0.2 N. What is the pressure in atmospheres that it exerts on the record?

12. A piston weighing 12 N rests on a sample of gas in a cylinder 5 cm in diameter. (a) What is the gauge pressure in the gas? (b) What is the absolute pressure in the gas?

13. A "plumber's helper" is a rubber cup with a wooden handle. If the diameter of the cup is 5 in., what is the maximum suction it can exert on a clogged pipe? How much force is needed on the handle to produce this suction?

14. A 70-kg swami lies on a bed of nails 1 cm apart whose points each have an area of 1 mm². If the area of the swami's body in contact with the bed is 0.5 m² and the threshold pressure for pain is 1 MPa (10^6 Pa), how disagreeable is the experience for the swami?

15. A hypodermic syringe whose cylinder is 1 cm in diameter and whose needle is 1 mm in diameter is used to inject a liquid into a vein in which the blood pressure is 15 torr. Find the minimum force needed on the plunger of the syringe.

16. A 7-g bullet leaves the 60 cm barrel of a 0.30-caliber rifle (whose bore diameter is 0.30 in. = 7.62 mm) at 600 m/s. Find the average pressure of the gases in the rifle barrel while the bullet is being accelerated.

11–2 Hydraulic Press

17. The input and output pistons of a hydraulic jack are respectively 5 mm and 30 mm in diameter. A lever with a mechanical advantage of 8 is used to apply force to the input piston. How much mass can the jack lift if a force of 150 N is applied to the lever and friction is negligible in the system?

18. A lever with a mechanical advantage of 5 is attached to the pump piston of a hydraulic press. The area of the pump piston is 1.6 in.² and that of the output piston is 20 in.² (a) If the press is perfectly efficient, find the force the output piston exerts when a force of 30 lb is applied to the pump lever. (b) If each stroke of the lever moves the pump piston 1 in., how many strokes are needed to move the output piston 1 ft?

11–3 Pressure and Depth

19. The height of water at two identical dams is the same, but dam *A* holds back a lake containing 2 km³ of water, whereas dam *B* holds back a lake containing 1 km³ of water. What is the ratio of the total force exerted on dam *A* to that exerted on dam *B?*

20. The three containers shown are filled with water to the same height. Compare the pressures at the bottoms of the containers.

A   *B*   *C*

21. In 1960 the U.S. Navy bathyscaphe *Trieste* descended to a depth of 10,920 m in the Pacific Ocean near Guam. Neglecting the increase in water density with depth, find the pressure on the *Trieste* at the bottom of its dive.

22. Find the pressure at a depth of 2.4 m in a swimming pool filled with fresh water.

23. A person sucking hard on a thin tube can reduce the pressure in it to 90% of atmospheric pressure. How high can the person suck water up the tube?

24. The average pressure on a dam is equal to the pressure at half the height H of the water. Show that the total force on the dam is proportional to H^2.

25. A submarine is at a depth of 100 ft in seawater. The interior of the submarine is maintained at normal atmospheric pressure. Find the force that must be withstood by a square hatch 2 ft on a side.

26. If the density of the atmosphere were uniform instead of decreasing with altitude, how thick a layer would it form?

27. A standing woman whose head is 35 cm above her heart bends over so that her head is 35 cm below her heart. What is the change (in torr) in the blood pressure in her head? (The arteries that lead to the head expand and contract as needed to keep the flow of blood to the brain constant despite changes in the elevation of the head relative to the heart. Such expansions and contractions require a few seconds to be completed, which explains why sitting up suddenly from a horizontal position may lead to a momentary dizzy sensation.)

28. The pressure of the blood in a patient's vein is 12 torr. Find the minimum height above the vein that a container of blood must be held for the blood to enter the vein. (The actual height is always much greater than this figure in order to achieve a reasonable flow.)

29. The average pressure of the blood in a person's arteries is 100 torr at the same elevation as the heart. Find the average pressure in the head of a standing person, say 40 cm above the heart, and in an artery in the foot, say 120 cm below the heart.

30. Water and olive oil are sucked up to the indicated heights in the device shown above. Find the density of olive oil.

31. Calculate the density of seawater at a depth of 5 km. Use the bulk modulus for water given in Table 9–4.

32. A person's head is 40 cm above his heart. The average arterial blood pressure at the level of the heart is 100 torr. Find the average arterial pressure in the person's head when standing and when lying down in an elevator accelerated upward at 3 m/s^2.

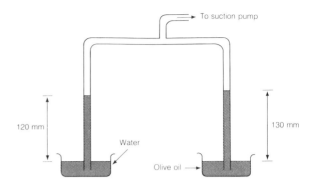

To suction pump

120 mm
130 mm
Water
Olive oil →

11–4 Archimedes' Principle

33. An ice cube floats in a glass of water filled to the brim. What will happen when the ice melts?

34. A wooden block is in such perfect contact with the bottom of a water tank that there is no water beneath it. Is there a buoyant force on the block?

35. Two spheres of the same diameter but of different mass are dropped from a tower. If air resistance is the same for both, which will reach the ground first? Why?

36. In central Sweden a bridge carries the Göta Canal over a highway. What, if anything, happens to the load on the bridge when a boat passes across it?

37. A ship catches fire and its steel hull expands as a result of the increase in temperature. What happens to the volume of water the ship displaces? What happens to the ship's freeboard, which is the height of its deck above sea level?

38. An aluminum canoe is floating in a swimming pool. After a while it begins to leak and sinks to the bottom of the pool. What happens to the water level in the pool?

39. A helium-filled balloon rises to a certain altitude in the atmosphere and floats there instead of rising indefinitely. Why?

40. A barge 35 m long and 6 m wide has a mass of 2×10^5 kg. What depth of seawater is needed to float it?

41. A woman whose mass is 60 kg stands on a rectangular swimming raft 3 m long and 2 m wide that is floating in fresh water. By how much does the raft rise when she dives off?

42. Six empty 55-gal drums that weigh 45 lb each are used as floats for a boat dock in a freshwater lake. The dock platform is 6 ft × 30 ft and is made of pine planks 1.5 in. thick. What is the maximum number of 170-lb people who can stand on the platform before it touches the water?

43. The densities of people are slightly less than that of water. Assuming that these densities are the same, compute the buoyant force of the atmosphere on a 160-lb man.

44. How much force is needed to raise a 200-kg iron anchor when it is submerged in seawater?

45. A 30-kg balloon is filled with 100 m^3 of hydrogen. How much force is needed to hold it down?

46. A design has been proposed for a modern helium-filled airship that is to have a useful lift of 5×10^5 kg at sea level. (a) How many m^3 of helium must the airship contain? (b) The length of the airship would be 400 m. If it were a cylinder, what would its diameter be?

47. The largest rigid airship ever built was the German *Hindenburg,* which was 245 m long and had a capacity of 2×10^5 m^3 of hydrogen. It was destroyed in a fire at Lakehurst, N.J., in 1937. What was the total payload of the *Hindenburg* including its structure but not including the hydrogen it contained?

48. A steel tank whose capacity is 200 L has a mass of 36 kg. (a) Will it float in seawater when empty? (b) When filled with fresh water? (c) When filled with gasoline?

49. A bracelet that appears to be gold is suspended from a spring scale that reads 50 g in air and 46 g when the bracelet is immersed in a glass of water. Is the bracelet pure gold?

50. A lead fishing sinker suspended from a spring scale is immersed in oil whose density is 820 kg/m^3. The reading on the scale is 0.2 N. Find the mass of the sinker.

51. When a 500-g statue of a falcon suspended from a spring scale is immersed in water, the scale reads 400 g, and when it is suspended in benzene, the scale reads 412 g. What is the density of benzene?

52. A weather balloon 2 m in radius whose total mass is 20 kg is tethered to the ground by a rope. A wind causes the rope to make an angle of 25° with the vertical. Find the tension in the rope.

53. A sailboat has 8000 kg of lead ballast attached to the bottom of its keel, the center of gravity of which is 1.5 m below the water surface when the boat is level. What is the torque exerted by this ballast when the boat is heeled by 20° from the vertical? Assume seawater. (*Hint:* Why is this problem here instead of in Chapter 4?)

11–5   Fluid Flow

54. A pipe whose inside diameter is 30 mm is connected to three smaller pipes whose inside diameters are 15 mm each.

If the liquid speed in the larger pipe is 1 m/s, find its speed in the smaller pipes.

55. Water flows through a hose whose internal diameter is 1 cm at a speed of 1 m/s. What should the diameter of the nozzle be if the water is to emerge at 5 m/s?

56. Twenty L of gasoline emerge per minute from a fuel hose at a speed of 0.4 m/s. What is the nozzle diameter?

57. The left ventricle of a certain running man pumps 20 L of blood per minute into his aorta at an average pressure of 140 torr. Find the total power output of his heart under the assumption that his right ventricle has an output 20% as great as that of his left ventricle.

58. The power output of an athlete's heart during strenuous activity might be 8 W. If the blood pressure increases by 150 torr when it passes through the athlete's heart, find the corresponding rate of flow of blood in L/min.

59. The nozzle of a garden hose has an inside diameter of 0.3 in. (a) If water flows through the hose at 10 gal/min, find the speed at which it leaves the nozzle. (b) With what force must the nozzle be held? (c) Find the power needed by the pump assuming 60% efficiency.

60. Water emerges from a fire hose at a speed of 20 m/s and a rate of flow of 50 L/s. (a) Find the force with which the nozzle must be held. (b) Find the required power of the pump motor assuming 50% overall efficiency.

11–7   Applications of Bernoulli's Equation

61. Will there be any difference between the rates of flow from pipes A and B in the figure? If so, from which pipe will the liquid flow faster?

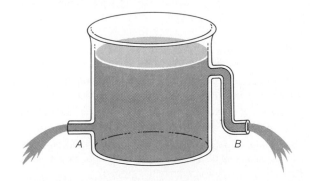

62. Why is it more dangerous when ice forms on an airplane's wings than when it forms on its fuselage?

63. During the pumping phase of the heart's action the pres-

sure of blood in the major arteries of a normal person at the level of the heart is about 120 torr. If one of these arteries is cut and blood spurts out vertically, how high will it go?

64. The figure shows a hose whose internal cross-sectional area is 2 cm² being used to siphon water out of a tank. Find the speed with which the water emerges and the rate of flow in L/s.

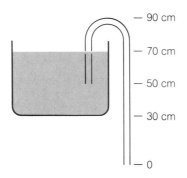

65. A boat strikes an underwater rock and punctures a hole 20 cm² in area in its hull 65 cm below the waterline. How many liters of water enter the hull per second?

66. Kerosene leaks out of a hole 0.5 in. in diameter at the bottom of a tank at a rate of 10 gal/min. How high is the kerosene in the tank?

67. (a) Milk flows through a Venturi meter whose cross-sectional areas are 6 cm² and 2 cm². Find the rate of flow in L/s when the difference between the manometer heights is 10 cm. (b) Find the rate of flow when the difference between the manometer heights is the same but the liquid is water.

68. A 1360-hp pump throws a jet of water 130 m into the air in Geneva, Switzerland. (a) With what speed does the water leave the mouth of the fountain? (b) If the overall efficiency is 60%, how many kilograms of water per minute are thrown into the air?

69. A horizontal stream of water leaves an orifice 1 m above the ground and strikes the ground 2 m away. (a) What is the speed of the water when it leaves the orifice? (b) What is the gauge pressure behind it?

70. A tank of height H is filled with water; it is open at the top. A hole is made a distance y from the top. How far from the tank does the water strike the ground?

71. A horizontal pipe 2 cm in radius at one end gradually increases in size so that it is 5 cm in radius at the other end. The pipe is 4 m long. Water is pumped into the small end of the pipe at a speed of 8 m/s and a pressure of 2 bars. Find the speed and pressure of the water at the pipe's large end.

72. The pipe of Exercise 71 is turned so as to be vertical with the small end underneath, so the water flows upward. Find the speed and pressure of the water at the large end of the pipe now.

73. Water flows upward through the pipe shown at 40 L/s. If the pressure at the lower end is 80 kPa, find the water speed at both ends and the pressure at the upper end.

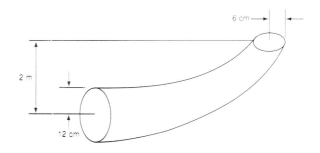

74. Gasoline enters the pipe shown from the left at 3 m/s and 150 kPa. Find the speed and pressure at which gasoline leaves the pipe and the rate of flow in kilograms per second.

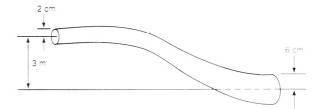

11–8 Viscosity

75. Does a fluid whose density is greater than that of another fluid necessarily have a greater viscosity as well?

76. A grease nipple on a car has a hole 0.5 mm in diameter and 5 mm long. If the viscosity of the grease used is 80 N · s/m², find the pressure needed to force 1 cm³ of grease into the nipple in 10 s. How many times atmospheric pressure is this?

77. Find the pressure gradient needed to pump 16 L of water at 20°C per minute through a pipe 6 mm in radius. Take the viscosity of the water into account.

78. A hypodermic syringe has a cylinder 8 mm in diameter and a needle 0.4 mm in diameter and 30 mm long. How much force is needed on the plunger to send water at 20°C through the needle at 0.1 cm³/s?

79. A typical capillary blood vessel is 1 mm long and has a radius of 2 μm. (a) If the pressure difference between its

ends is 20 torr, find the average speed of the blood that flows through such a capillary. (b) If the heart pumps 80 cm^3 of blood through a person's body per second, how many capillaries does the body contain?

11–9 Reynolds Number

80. Water at 40°C flows at 10 cm/s through a tube 1 cm in diameter. What is the Reynolds number of the system? Is the flow laminar or turbulent?

81. At what speed should air at 20°C flow through a tube for the Reynolds number to be the same as that of water at 20°C flowing through the tube at 20 cm/s? Assume that the viscosity of air at 20°C is 1.8×10^{-5} PI.

ANSWERS TO MULTIPLE CHOICE

1. b	**10.** c	**19.** c	**28.** b
2. b	**11.** d	**20.** c	**29.** a
3. d	**12.** a	**21.** c	**30.** c
4. c	**13.** c	**22.** b	**31.** b
5. d	**14.** a	**23.** d	**32.** c
6. d	**15.** c	**24.** b	**33.** b
7. a	**16.** d	**25.** b	**34.** c
8. c	**17.** b	**26.** b	**35.** d
9. b	**18.** a	**27.** d	

12

HARMONIC MOTION

Many events in both nature and technology are periodic, with a certain motion repeating itself over and over again. The most fundamental kind of oscillation is called *simple harmonic motion,* and all periodic events are either examples of simple harmonic motion or else the result of several such motions mixed together. In harmonic motion, the energy of the vibrations is continually transformed from kinetic to potential and back again. The potential energy may be elastic rather than gravitational; we shall find later that elasticity and gravitation are not the only phenomena in which potential energy is a useful concept. The electrical equivalent of potential energy, in particular, makes possible electrical oscillations that closely resemble harmonic motion.

12-1 ELASTIC POTENTIAL ENERGY

When we stretch a spring, it resists being lengthened, and if we then let it go, the spring returns to its original length. As we know, this is an example of *elastic* behavior. On

CHAPTER OBJECTIVES

Completing this chapter should enable you to:

1. Determine the elastic potential energy of a deformed object.

2. Recognize the circumstances needed for simple harmonic motion to take place.

3. Calculate the period and frequency of an oscillating system.

4. Calculate the speed and acceleration of an object in harmonic motion at any stage in a cycle.

5. Express the displacement, speed, and acceleration of an oscillating object using trigonometric notation.

6. Calculate the periods of simple torsion and physical pendulums.

7. Find the moment of inertia of an irregular object by the torsion-pendulum method.

8. Find the center of percussion of an object.

9. Describe the effects of damping on simple harmonic motion.

the other hand, when we stretch a piece of taffy, it also resists being lengthened, but if we then let it go, nothing happens: The deformation is permanent. This is an example of *plastic* behavior.

In the case of a stretched spring, a *restoring force* comes into being that tries to return the spring to its normal length. The farther we stretch the spring, the greater the restoring force we must overcome (Fig. 12–1). Exactly the same effect occurs when we compress the spring: It resists being shortened, and if we let it go, the spring returns to its normal length. Again a restoring force arises, and again the more the compression, the stronger the restoring force to be overcome.

A restoring force occurs when an elastic body is deformed

The amount s by which an elastic solid is stretched or compressed by a force F is directly proportional to F, provided the elastic limit is not exceeded. This proportionality is called Hooke's law, as mentioned in Chapter 10. Thus we can write

$$F = ks \qquad\qquad \text{\textit{Hooke's law}} \quad (12–1)$$

where k is a constant whose value depends upon the nature and dimensions of the object. A stiff spring has a higher value of k than a weak one.

FIG. 12–1 When a spring (or other elastic body) is stretched or compressed, a restoring force comes into being that tries to return the spring to its normal length.

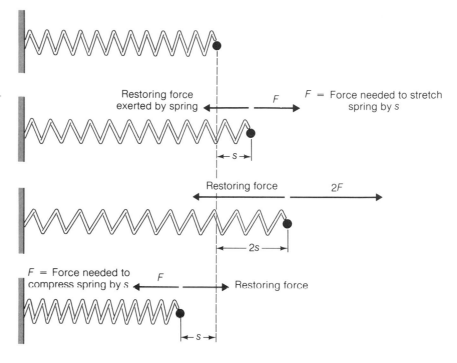

Restoring force exerted by spring

F

F = Force needed to stretch spring by s

Restoring force

$2F$

F = Force needed to compress spring by s

F

Restoring force

FIG. 12–2 To compute the work done in stretching a body that obeys Hooke's law, the varying force that actually acts during the expansion may be replaced by the average force.

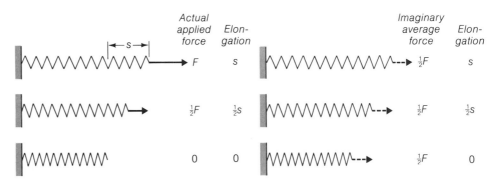

Actual applied force / Elongation

Imaginary average force / Elongation

F s $\frac{1}{2}F$ s

$\frac{1}{2}F$ $\frac{1}{2}s$ $\frac{1}{2}F$ $\frac{1}{2}s$

0 0 $\frac{1}{2}F$ 0

Work must be done to overcome the restoring force

The work needed to stretch (or compress) an object that obeys Hooke's law is easy to figure out. The work done by a force is the product of the force and the distance through which it acts. Here the force used in stretching the object is not constant but depends upon the elongation s at each point in the stretching process. Because F is proportional to s, the *average* force $\overline{F}$ applied while the body is stretched from its normal length by an amount s to its final length is

$$\text{Average force} = \overline{F} = \frac{F_{\text{initial}} + F_{\text{final}}}{2} = \frac{0 + ks}{2} = \tfrac{1}{2}\,ks$$

since the initial force is 0 and final force is ks (Fig. 12–2). The work done in stretching the spring is the product of the average force $\overline{F} = \tfrac{1}{2}\,ks$ and the total elongation s, so that

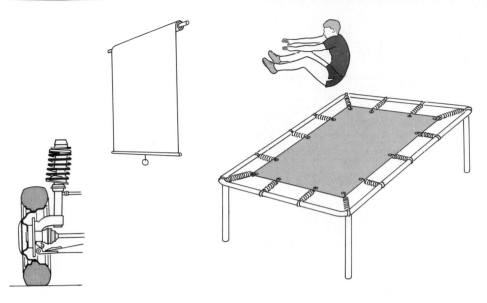

FIG. 12-3 Some devices that make use of elastic potential energy in their operation.

$$W = \mathrm{PE} = \tfrac{1}{2} ks^2 \qquad\qquad \textit{Elastic potential energy} \quad (12-2)$$

This formula is most often used in connection with springs: To stretch (or compress) a spring whose force constant is k by an amount s from its normal length requires $\tfrac{1}{2} ks^2$ of work to be done. This work goes into *elastic potential energy*. When the spring is released, its potential energy of $\tfrac{1}{2} ks^2$ is transformed into kinetic energy or into work done on something else (Fig. 12-3); work done against frictional forces within the spring itself always absorbs some of the available potential energy.

Work done to deform an elastic body equals potential energy of the deformed body

Example The horizontal spring shown in Fig. 12-4 has a force constant k of 90 N/m. Attached to the free end of the spring is a 1.4-kg block. If the spring is pulled out 50 cm from its equilibrium position and then released, what will the block's speed be when it returns to the equilibrium position?

Solution When the spring is released, its elastic potential energy starts to be converted into kinetic energy of the block. We shall assume that the spring's mass is small compared with that of the block and that its internal friction may be neglected. At the equilibrium position of the spring, $s = 0$, and all the initial potential energy of $\tfrac{1}{2} ks^2$ is now kinetic energy $\tfrac{1}{2} mv^2$. Hence

$$\mathrm{KE} = \mathrm{PE}$$

$$\tfrac{1}{2} mv^2 = \tfrac{1}{2} ks^2$$

$$v = \sqrt{\frac{k}{m}}\, s = \sqrt{\frac{90\,\mathrm{N/m}}{1.4\,\mathrm{kg}}} \times 0.50\,\mathrm{m} = 4.0\,\mathrm{m/s} \qquad\blacksquare$$

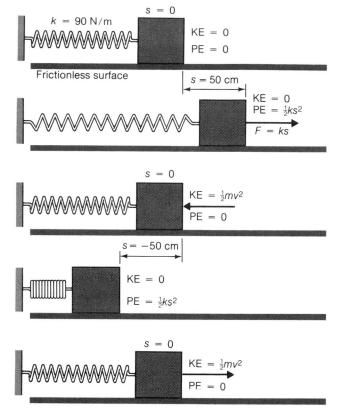

FIG. 12–4 A 1.4-kg block attached to a spring whose force constant is 90 N/m is pulled 50 cm from its equilibrium position. When the spring is released, its elastic potential energy of $\frac{1}{2}ks^2$ is converted into kinetic energy $\frac{1}{2}mv^2$, and as the block's momentum compresses the spring on the other side of the equilibrium position, the kinetic energy is converted back into elastic potential energy.

The energy of an oscillator shifts back and forth between potential energy and kinetic energy

12–2   SIMPLE HARMONIC MOTION

When a spring with an object attached to it is stretched and then released, it does not simply return to its equilibrium position and come to a stop there. What happens is that the elastic potential energy $\frac{1}{2}ks^2$ of the spring is changed into kinetic energy $\frac{1}{2}mv^2$ of the moving object, and as the object's momentum compresses the spring on the other side of the equilibrium position, this kinetic energy is changed back into elastic potential energy (Fig. 12–4). The amount of compression $-s$ will have the same magnitude as the original extension s, since

$$\tfrac{1}{2}ks^2 = \tfrac{1}{2}k(-s)^2$$

Left to itself in the absence of friction, the spring-object combination will continue oscillating back and forth indefinitely. The behavior of a system oscillating in this way is called *simple harmonic motion.*

Simple harmonic motion occurs whenever a force acts on a body in the opposite direction to its displacement from its normal position, with the force proportional to the displacement. The elastic restoring force of a stretched or compressed spring always tends to return the spring to its normal length, but the momentum associated with the moving mass compels it to overshoot and thus to oscillate.

Condition for simple harmonic motion

The *period T* of a body undergoing simple harmonic motion is the time required for it to make one complete oscillation. (A complete oscillation is often called a *cycle.*) In the case of a spring, the period is the time the spring spends in going from its maximum extension, say, through its maximum compression and back to its maximum extension once more, as in Fig. 12-5. To find out how the position of an oscillating object varies with time we can attach a pen to the object and pull a sheet of paper past it at a constant speed (Fig. 12-6). The graph has the same form as that of sin θ plotted against θ, and accordingly the curve is said to be *sinusoidal.* Sinusoidal variations are found elsewhere in physics also, for example in wave motion and in connection with alternating electric currents.

Period of harmonic motion

The maximum displacement A of an object undergoing harmonic motion on either side of its equilibrium position is called the *amplitude* of the motion:

Amplitude is maximum displacement

$$A = s_{max}$$

Amplitude = maximum displacement from equilibrium position

The period of all types of simple harmonic motion is given by the same formula,

$$T = 2\pi \sqrt{-\frac{s}{a}} \qquad \textit{Simple harmonic motion} \quad (12-3)$$

$$\text{Period} = 2\pi \sqrt{-\frac{\text{displacement}}{\text{acceleration}}}$$

where the acceleration a is that experienced by the body when it is at the displacement s from its equilibrium position. This formula is derived in Section 12-3.

To calculate the acceleration of a stretched spring we start with the second law of motion,

$$F = ma$$

and substitute the restoring force $F_r = -ks$ since it is the restoring force that causes the body to be accelerated. Thus

$$F_r = ma$$

$$a = \frac{F_r}{m} = \frac{-ks}{m} \qquad (12-4)$$

With this result we find that the period of a body of mass m attached to a spring of force constant k is

Period of system of spring + mass

FIG. 12–5 The period T of a body undergoing simple harmonic motion is the time required for it to make one complete oscillation. The maximum displacement on either side of the equilibrium position is called the *amplitude A* of the motion.

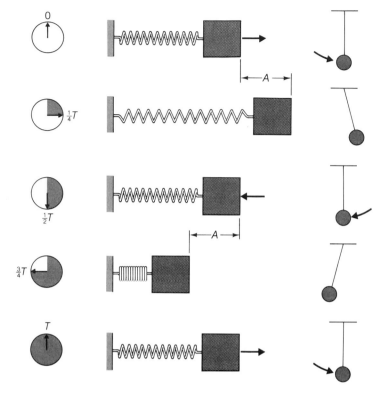

FIG. 12–6 A pen attached to an oscillating object can be used to trace out a graph of the position of the object versus time on a sheet of paper pulled past the pen at a constant speed. The resulting graph has the same form as that of sin θ plotted against θ, and accordingly the curve is said to be *sinusoidal*. Sinusoidal variations are found elsewhere in physics, for example, in wave motion and in alternating electric currents.

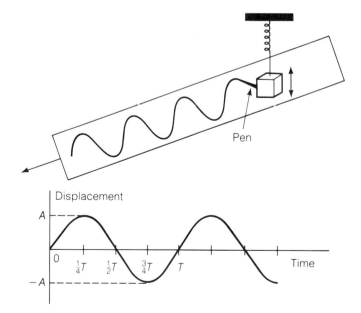

$$T = 2\pi \sqrt{-\frac{s}{a}} = 2\pi \sqrt{-\frac{s}{-ks/m}}$$

$$T = 2\pi \sqrt{\frac{m}{k}} \qquad\qquad \textit{Oscillating spring} \quad (12-5)$$

It is worth noting that the period T does not depend upon the amplitude A; no matter how much or how little the spring is initially pulled out, precisely the same amount of time is required for each cycle. If A is small, the maximum acceleration is also small and the body moves back and forth very slowly through its range, while if A is large, the acceleration is also large and the body moves correspondingly rapidly through the larger range. This peculiarity of simple harmonic motion is used in the design of mechanical clocks and watches, which use the rotational oscillations of a coil spring or the swings of a pendulum—both essentially simple harmonic motions—to maintain a constant rate independent of any changes in amplitude.

Period is independent of amplitude

A quantity often used in describing harmonic motion is *frequency*. The frequency is the number of cycles carried out per unit time. Hence frequency, whose symbol is f, is the reciprocal of period T,

The unit of frequency is the hertz

$$f = \frac{1}{T} \qquad\qquad \textit{Frequency} \quad (12-6)$$

The unit of frequency is the *hertz* (Hz), where 1 Hz = 1 cycle/s.

Example When a ball is suspended from a spring, the spring stretches by 7.0 cm (Fig. 12–7). If the ball oscillates up and down, what is its period? What is its frequency?

Solution The force exerted on the spring is the ball's weight of mg. Since $F = ks$, the force constant of the spring is

$$k = \frac{F}{s} = \frac{mg}{s}$$

The period of the oscillations is therefore

$$T = 2\pi \sqrt{\frac{m}{k}} = 2\pi \sqrt{\frac{ms}{mg}} = 2\pi \sqrt{\frac{s}{g}} = 2\pi \sqrt{\frac{0.070\,\text{m}}{9.8\,\text{m/s}^2}} = 0.53\,\text{s}$$

We did not have to know the ball's mass to find the period of its motion. The frequency that corresponds to a period of 0.53 s is

$$f = \frac{1}{T} = \frac{1}{0.53\,\text{s}} = 1.9\,\text{cycles/s} = 1.9\,\text{Hz} \qquad\qquad ■$$

FIG. 12–7

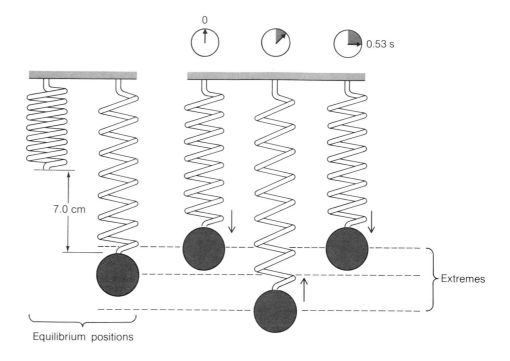

0

0.53 s

7.0 cm

Extremes

Equilibrium positions

12–3   A MODEL OF SIMPLE HARMONIC MOTION

This model makes it easy to find period of simple harmonic motion

Figure 12–8 shows a particle moving in a vertical circle at constant speed. The particle is illuminated from above, and it casts a shadow on a horizontal screen below its orbit. As the particle travels around the circle, its shadow oscillates back and forth. The shadow moves fastest at the center, slows down as it approaches each end of the path, comes to a stop, and then reverses its direction. We might suspect that the shadow is undergoing simple harmonic motion. To verify this suspicion, we must show that the acceleration a of the shadow at any time is proportional to its displacement s from the center of its path and opposite in direction.

Proof that the shadow undergoes simple harmonic motion

The acceleration of the shadow at any point is simply the horizontal component of the particle's acceleration a_c. As we know, a particle in uniform circular motion at the speed V in a circle of radius R experiences the centripetal acceleration

$$a_c = -\frac{V^2}{R}$$

where the minus sign indicates that the acceleration is inward toward the center of the circle. The horizontal component of this acceleration is the acceleration a of the shadow. Since corresponding sides of similar triangles are proportional, we see from Fig. 12–9 that

$$\frac{a}{a_c} = \frac{s}{R} \qquad \text{and so} \qquad a = \frac{s}{R} a_c$$

Light

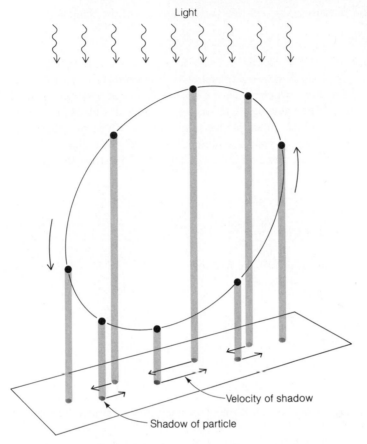

Velocity of shadow

Shadow of particle

FIG. 12-8 The shadow of a particle undergoing uniform circular motion executes simple harmonic motion.

Light

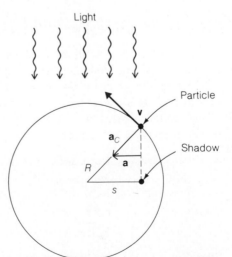

Particle

$\mathbf{v}$

$\mathbf{a}_C$

R $\mathbf{a}$

Shadow

s

FIG. 12-9 The acceleration of the shadow is the horizontal component of the particle's centripetal acceleration. The shadow's acceleration is proportional to its displacement s and is in the opposite direction.

Because $a_c = -V^2/R$, the acceleration of the shadow is

$$a = -\frac{s}{R}\frac{V^2}{R} = -\frac{V^2}{R^2}s \tag{12-7}$$

Thus the shadow's acceleration is proportional to its displacement s and in the opposite direction, which means the shadow is indeed undergoing simple harmonic motion.

The particle and its shadow have the same period

The value of the above analysis is that it gives us an easy way to find the period of an object in simple harmonic motion, since the particle and its shadow have the same period. The circumference of a circle of radius R is $2\pi R$, and a particle moving around the circle with the constant speed V covers this distance in the time

$$T = \frac{2\pi R}{V}$$

$$\text{Period} = \frac{\text{distance}}{\text{speed}}$$

From Eq. (12–7) we find that

$$\frac{R}{V} = \sqrt{-\frac{s}{a}}$$

and so

$$T = 2\pi\sqrt{-\frac{s}{a}} \tag{12-3}$$

This is the general formula for the period of simple harmonic motion given in Section 12–2.

12–4   POSITION, SPEED, AND ACCELERATION

How to find speed of oscillating object

The principle of conservation of energy permits us to express the speed of an object in simple harmonic motion in terms of its frequency f, amplitude A, and displacement s. The total energy of the oscillator (object plus spring) is the sum of its kinetic and potential energies at any time, which are respectively $\frac{1}{2}mv^2$ and $\frac{1}{2}ks^2$. At either extreme of the motion, when $s = +A$ or $s = -A$, the object is stationary and has only the potential energy $\frac{1}{2}kA^2$. Hence

Total energy of harmonic oscillator

$$\text{Total energy} = \text{KE} + \text{PE}$$

$$\tfrac{1}{2}kA^2 = \tfrac{1}{2}mv^2 + \tfrac{1}{2}ks^2$$

$$mv^2 = k(A^2 - s^2)$$

$$v = \sqrt{k/m}\,\sqrt{A^2 - s^2}$$

From Eqs. (12–5) and (12–6) we know that

$$f = \frac{1}{T} = \frac{1}{2\pi}\sqrt{\frac{k}{m}}$$

which can be rewritten as

$$\sqrt{k/m} = 2\pi f \qquad (12-8)$$

The speed of the object when it has the displacement s is accordingly

$$v = 2\pi f\sqrt{A^2 - s^2} \qquad \textit{Speed at a given displacement} \quad (12-9)$$

This formula gives only the absolute value of v; whether the sign of v is $+$ or $-$ depends upon whether the body is at $+s$ or $-s$ and upon whether it is on its way toward or away from the equilibrium position from there.

From Eq. (12–9) we see that the maximum speed v_{max} of the object, which occurs at the equilibrium position when $s = 0$, is

Maximum speed occurs at equilibrium position

$$v_{max} = 2\pi fA \qquad \textit{Maximum speed} \quad (12-10)$$

The maximum speed is proportional to both the frequency and the amplitude of the motion.

The energy of an oscillating object shifts back and forth between kinetic and potential forms. To find the total energy, we can calculate either KE_{max} or PE_{max}, with the help respectively of Eq. (12–10) or (12–8):

$$KE_{max} = \tfrac{1}{2}mv_{max}^2 = 2\pi^2 mf^2 A^2 \qquad \textit{Total energy} \quad (12-11)$$

$$PE_{max} = \tfrac{1}{2}kA^2 = 2\pi^2 mf^2 A^2 \qquad \textit{Total energy} \quad (12-12)$$

The total energy depends upon the square of the frequency and the square of the amplitude. Note that the total energy is *not* the sum of Eqs. (12–11) and (12–12) because when kinetic energy or potential energy reaches its maximum, the other is zero.

To find the acceleration of an object in simple harmonic motion, we refer back to Eq. (12–4), which states that

$$a = \frac{-ks}{m}$$

On the basis of Eq. (12–8) this formula becomes

$$a = -4\pi^2 f^2 s \qquad \textit{Acceleration at given displacement} \quad (12-13)$$

The acceleration is always opposite in direction to the displacement, which, of course, is one of the conditions for simple harmonic motion to occur. The maximum acceleration occurs at either extreme, when $s = \pm A$, and has the magnitude

Maximum acceleration occurs at endpoints

$$a_{max} = 4\pi^2 f^2 A \qquad \textit{Maximum acceleration} \quad (12-14)$$

The maximum acceleration is proportional to the square of the frequency and to the amplitude.

Although the above formulas were derived for the case of a vibrating spring, their validity is perfectly general and they apply to any type of harmonic oscillator, from the bob of a pendulum to an atom in a molecule.

Example   In an automobile engine, each piston moves up and down in an approximation of simple harmonic motion. If a certain piston has a mass of 0.5 kg, has a total travel ("stroke") of 12 cm, and has a frequency of oscillation of 60 Hz (corresponding to 3600 rpm), find the maximum force it experiences assuming simple harmonic motion.

Solution   The maximum force is given by $F_{max} = ma_{max}$. Here, since the amplitude is half the total travel, $A = 6$ cm $= 0.06$ m, and

$$F_{max} = ma_{max} = 4\pi^2 mf^2 A = 4\pi^2 (0.5 \text{ kg})(60 \text{ Hz})^2(0.06 \text{ m}) = 4264 \text{ N}$$

which is nearly 1000 lb.   ■

Example   A piston undergoes simple harmonic motion in a vertical direction with an amplitude of 2.5 in. A coin is placed on top of the piston (Fig. 12–10). What is the lowest frequency at which the coin will be left behind by the piston on its downstroke?

Solution   The coin will leave the piston when the downward acceleration of the latter exceeds the acceleration of gravity g. The maximum downward acceleration of piston occurs at the highest point of its motion, when $a_{max} = 4\pi^2 f^2 A$ according to Eq. (12–14). Hence we set a_{max} equal to g and solve for the frequency f with $A = 2.5$ in. $= 0.208$ ft:

$$a_{max} = g = 4\pi^2 f^2 A$$

$$f = \frac{1}{2\pi}\sqrt{\frac{g}{A}} = \frac{1}{2\pi}\sqrt{\frac{32 \text{ ft/s}^2}{0.208 \text{ ft}}} = 1.97 \text{ Hz}$$ ■

FIG. 12–10   When the downward acceleration of the piston exceeds g, the coin will be left behind by the piston on its downstroke.

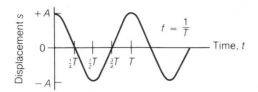

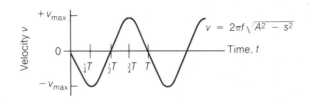

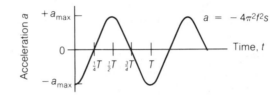

FIG. 12–11 Displacement, velocity, and acceleration in simple harmonic motion.

The variations of the displacement, velocity, and acceleration of a particle undergoing simple harmonic motion are plotted versus time in Fig. 12–11. The graphs are plotted on the assumption that the particle is at $s = +A$ when $t = 0$. This corresponds to pulling out the particle and letting it go at $t = 0$. At this instant the particle's acceleration is a maximum and is opposite in direction to s, while the velocity is zero since the particle has not yet started to move.

How s, v, and a vary with time in harmonic motion

When the particle is at the origin, $s = 0$ and the spring is at its normal length. Because the spring exerts no force on the particle at this time, its acceleration is zero. The speed of the particle is now a maximum, which follows from the fact that PE = 0 for a deformed spring.

When $s = -A$, the particle is at the other extreme of its range, and its acceleration, again a maximum, is positive, which means that it is once more in the direction of the origin, though now from the other side. All the energy of oscillation is potential, and the body is accordingly stationary at this instant.

12–5 TRIGONOMETRIC NOTATION

Sometimes it is useful to know the position, velocity, and acceleration of a harmonic oscillator at any time, not just at the middle or endpoints of its motion. We can obtain suitable formulas for these quantities with the help of Fig. 12–12. The model shown there is the same as the one in Fig. 12–8, where we saw that the shadow of a particle moving in a circle undergoes simple harmonic motion. The particle has an angular

Angular frequency of oscillation

FIG. 12–12 (a)–(c). The displacement, velocity, and acceleration of the shadow of the particle of Fig. 12–8 can be expressed in terms of the time t as shown here. The angular speed of the particle is ω, the magnitude of its linear velocity is ωA, and the magnitude of its centripetal acceleration is $\omega^2 A$.

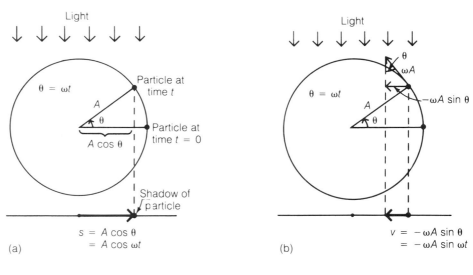

(a)

$$s = A \cos \theta$$
$$= A \cos \omega t$$

(b)

$$v = -\omega A \sin \theta$$
$$= -\omega A \sin \omega t$$

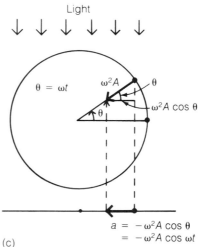

(c)

$$a = -\omega^2 A \cos \theta$$
$$= -\omega^2 A \cos \omega t$$

speed of $\omega = 2\pi f$, and ω is called the *angular frequency* of the harmonic motion of its shadow. The unit of ω is, as usual, the radian/s.

The angular displacement of the particle at any time t is $\theta = \omega t$, its linear speed in the circle is ωA, and its centripetal acceleration is $\omega^2 A$, where A is the radius of the circle. From Fig. 12–12 it is clear that

$$s = A \cos \theta = A \cos \omega t \qquad\qquad (12–15)$$

$$v = -\omega A \sin \theta = -\omega A \sin \omega t \qquad\qquad (12–16)$$

$$a = -\omega^2 A \cos \theta = -\omega^2 A \cos \omega t \qquad\qquad (12–17)$$

The sign convention used is that a quantity directed to the right is considered positive and one directed to the left is considered negative. Graphs of s, v, and a versus time were given in Fig. 12–11.

12–6   THE SIMPLE PENDULUM

A pendulum executes simple harmonic motion as it swings back and forth, provided that the arc through which the pendulum bob moves is a fairly small one. We shall see why this limitation arises if we use Eq. (12–3) to calculate the period of a pendulum.

Figure 12–13 shows a pendulum of length L whose bob has a mass m, together with a diagram of the forces acting on the bob. (It is assumed that the entire mass of the pendulum is concentrated in the bob.) The weight of the bob, $m\mathbf{g}$, which acts vertically downward, may be resolved into two forces, $\mathbf{T}$ and $\mathbf{F}$, which act respectively parallel to and perpendicular to the supporting string L. That is,

$$\mathbf{T} + \mathbf{F} = m\mathbf{g}$$

The force $\mathbf{F}$ is the restoring force that acts to return the bob to the midpoint of its motion. The space triangle hLx and the vector triangle $\mathbf{T}m\mathbf{g}\mathbf{F}$ are similar, since each contains a right angle and two sides of one are parallel to the two corresponding sides of the other, and so

$$\frac{F}{x} = \frac{mg}{L}$$

The restoring force acting on the bob is therefore

$$F = -\frac{mgx}{L}$$

where the minus sign indicates that $\mathbf{F}$ points in the direction of decreasing x.

If the bob is not far from the midpoint of its motion, the horizontal distance x is almost exactly equal to the actual path length s, and F then is given by

$$F = -\frac{mgs}{L}$$

Since F is proportional to $-s$, the motion is simple harmonic. The acceleration of the bob that results from this force is

A pendulum undergoes simple harmonic motion when it swings through a small arc

The restoring force is F

The condition for simple harmonic motion is that $x = s$

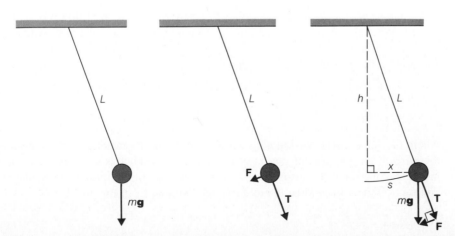

FIG. 12–13   A pendulum executes simple harmonic motion when its oscillations are so small in amplitude that the chord x is very nearly equal in length to the arc s.

$$a = \frac{F}{m} = -\frac{gs}{L}$$

Substituting in Eq. (12–3) we find that

$$T = 2\pi \sqrt{-\frac{s}{a}}$$

$$T = 2\pi \sqrt{\frac{L}{g}} \qquad\qquad\qquad \textit{Simple pendulum} \quad (12\text{–}18)$$

The period of a simple pendulum is independent of its mass

Provided that s is small enough so that it is close to x, the motion of a pendulum is simple harmonic in character with a period proportional to the square root of the pendulum's length and independent of the mass of the bob. If the arc through which the pendulum swings on either side of the vertical is 5°, a detailed calculation shows that the actual period will exceed that predicted by Eq. (12–18) by only 0.05%; if the arc is 10° the discrepancy will be 0.2%; and even if the arc is as much as 20° the discrepancy is only 0.8%. Only when the arc on either side of the vertical is about 50° does the discrepancy reach 5%.

Example   How long should a pendulum be for it to have a period of exactly 1.0 s?

Solution   We first solve Eq. (12–18) for L:

$$T = 2\pi \sqrt{\frac{L}{g}} \qquad T^2 = \frac{4\pi^2 L}{g} \qquad L = \frac{gT^2}{4\pi^2}$$

Inserting the values $g = 9.8 \text{ m/s}^2$ and $T = 1.0$ s, we find that

$$L = \frac{(9.8 \text{ m/s}^2)(1.0 \text{ s})^2}{4\pi^2} = 0.25 \text{ m} \qquad\qquad \blacksquare$$

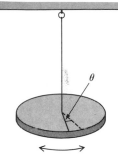

FIG. 12–14 A torsion pendulum. Twisting the wire leads to a restoring torque.

12–7   THE TORSION PENDULUM

The motion of a torsion pendulum is angular instead of linear

A *torsion pendulum* consists of an object suspended by a wire or thin rod, as in Fig. 12–14. When the object is turned through an angle and released, it will oscillate back and forth. The analogies between linear and angular quantities make it easy to find a formula for the period of these oscillations.

From Hooke's law the restoring torque τ that comes into being when the wire is twisted through an angle θ is

Hooke's law for angular motion

$$\tau = -K\theta$$

where the value of the torsion constant K depends on the material and dimensions of the wire. If the moment of inertia of the object about its axis of suspension is I (the moment of inertia of the suspending wire is usually negligible), then the angular acceleration α of the object when the torque τ acts on it is, from Eq. (8–21),

Acceleration of torsion pendulum

$$\alpha = \frac{\tau}{I} = -\frac{K}{I}\theta \qquad\qquad\qquad\qquad\qquad (12\text{–}19)$$

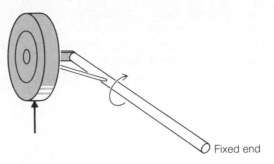

FIG. 12-15 A torsion bar spring absorbs a vertical impact on a car's wheel by twisting. The restoring torque of the twisted bar returns the wheel to its normal position afterward.

Fixed end

Comparing this formula with the equivalent result for a harmonic oscillator,

$$a = -\frac{ks}{m}$$

(12–4)   **Acceleration of harmonic oscillator**

suggests that the period of a torsion pendulum can be given by the general formula for the period of a harmonic oscillator,

$$T = 2\pi \sqrt{-\frac{s}{a}}$$

(12–3)

with θ/α replacing s/a. This idea turns out to be correct, and we have

$$T = 2\pi \sqrt{-\frac{\theta}{\alpha}} = 2\pi \sqrt{\frac{I}{K}}$$

Torsion pendulum   (12–20)

The simplest procedure for finding the moment of inertia of an irregular object is often to suspend it by a wire, measure the torsion constant K and the period of oscillation T, and then use Eq. (12–20) to find I.

How to find moment of inertia of irregular object

Example   A grindstone is suspended by a wire from its center. When a torque of 0.12 N · m is applied to the grindstone, it turns through 8°, and when it is released, it oscillates with a period of 1.0 s. Find the moment of inertia of the grindstone.

Solution   Since 360° $= 2\pi$ rad, $\theta = 0.14$ rad, and the torsion constant of the wire is

$$K = \frac{\tau}{\theta} = \frac{0.12\,\text{N} \cdot \text{m}}{0.14\,\text{rad}} = 0.86\,\text{N} \cdot \text{m/rad}$$

From Eq. (12–20),

$$I = \frac{KT^2}{4\pi^2} = \frac{(0.86\,\text{N} \cdot \text{m/rad})(1.0\text{s})^2}{4\pi^2} = 0.022\,\text{kg} \cdot \text{m}^2 \qquad \blacksquare$$

The ability of a twisted rod to provide a restoring torque is used in the torsion bar springs that replace coil or leaf springs in high-performance cars. As in Fig. 12–15,

Torsion bar springs

such a spring consists of a steel shaft attached at one end to the car's frame and at the other end to the control arm of a wheel. When the wheel goes over a bump in the road, the wheel is forced upward, and the impact is absorbed by the twisting of the torsion bar. A torsion bar needs less mass to store a given amount of potential energy than a coil or leaf spring, and occupies less space as well.

12–8   THE PHYSICAL PENDULUM

Any swinging object can be treated as a simple pendulum if L is properly interpreted

An object of any shape will oscillate back and forth when it is pivoted at some point other than its center of gravity and given a push to one side. Such an object is called a *physical pendulum*. The formula for the period of a simple pendulum also applies to a physical pendulum provided that the length L is properly interpreted.

Figure 12–16 shows a physical pendulum pivoted about a horizontal axis at O. The pendulum is displaced so that the line from O to its center of gravity is at the angle θ from the vertical. The pendulum's weight mg acts from the center of gravity and produces the restoring torque

$$\tau = -\, mgx$$

where x is the horizontal distance between O and the center of gravity. The minus sign reflects the fact that the restoring torque is always opposite in direction to the angular displacement θ of the pendulum.

When θ is small, the chord x is very nearly equal to the arc s. Assuming them to be equal is the same approximation made in analyzing the simple pendulum (see Fig. 12–13). Since h is the distance between the pivot point O and the center of gravity CG in Fig. 12–16, what we have for small θ is

$$x = s = h\theta$$

This means that

$$\tau = -mgh\theta$$

which is the condition for simple harmonic motion since the restoring torque is proportional to $-\theta$. If I is the moment of inertia of the pendulum about O and α is the angular acceleration produced by the restoring torque, then

$$\alpha = \frac{\tau}{I} = -\frac{mgh}{I}\,\theta$$

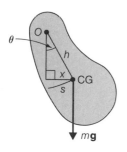

$$\tau = -(mg)(h\sin\theta)$$

FIG. 12–16   A physical pendulum pivoted at O. The center of gravity is marked CG.

Period of physical pendulum

By the same reasoning used in Section 12–7, the period of the physical pendulum is

$$T = 2\pi\sqrt{-\frac{\theta}{\alpha}} = 2\pi\sqrt{\frac{I}{mgh}} \qquad\qquad \textit{Physical pendulum}\quad (12\text{–}21)$$

The simple pendulum whose period is the same as that of a given physical pendulum may be found by setting equal the formulas for their respective periods of oscillation:

$$2\pi \sqrt{\frac{L}{g}} = 2\pi \sqrt{\frac{I}{mgh}}$$

$$L = \frac{I}{mh} \qquad\qquad (12\text{–}22)$$

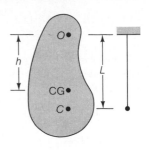

FIG. 12–17 A simple pendulum of length $L = I/mh$ has the same period as that of a physical pendulum. The point C is the center of oscillation.

Thus the mass of a physical pendulum can be regarded as concentrated at a point C the distance $L = I/mh$ from the point O. This point is called the *center of oscillation* (Fig. 12–17) and has two interesting properties:

1. If the pendulum is pivoted at C instead of at O, it will oscillate with the same period as before and O will be the new center of oscillation.

2. If the pendulum is struck along a line of action through C, there will be no reaction force on the pivot at O. A baseball that strikes a bat at the latter's center of oscillation does not produce a sting in the batter's hands, for example (Fig. 12–18). The center of oscillation is often called the *center of percussion* for this reason. The concept of center of percussion plays an important part in the design of many mechanical devices.

Example A uniform steel girder 15 ft long is suspended from one end by a crane. What is the period of its oscillations?

Solution From Fig. 8–9 the moment of inertia of a thin rod pivoted at one end is $I = \frac{1}{3}mL^2$, where L is the rod's length and m is its mass. The distance h from the pivot to the center of gravity of the girder is $L/2$. Hence the period of oscillation is

$$T = 2\pi \sqrt{\frac{I}{mgh}} = 2\pi \sqrt{\frac{mL^2/3}{mgL/2}} = 2\pi \sqrt{\frac{2L}{3g}}$$

$$= 2\pi \sqrt{\frac{(2)(15\,\text{ft})}{(3)(32\,\text{ft/s}^2)}} = 3.5\,\text{s}$$

The period does not depend upon the girder's mass, only upon its length. ∎

FIG. 12–18 When a physical pendulum is struck at its center of percussion, its pivot experiences no reaction force.

12–9 DAMPED HARMONIC OSCILLATOR

Once set in motion, an ideal harmonic oscillator should continue to oscillate indefinitely. Actual harmonic oscillators do not behave in this way; while in some cases they may oscillate for a long time, eventually the amplitude decreases and the motion comes to a stop. This effect is called *damping*. If the amount of damping is very large, as it is for a car whose springs are supplemented by good shock absorbers, no oscillation at all occurs; after being displaced, the object gradually returns to its equilibrium position and goes no further.

Damping is caused by frictional forces. The potential energy of a stretched spring is not completely converted into kinetic energy at $s = 0$, since some of it goes into work done against the frictional forces that are acting. At the end of each vibration the loss in energy leads to a shorter extension of the spring, until finally the amplitude

Actual harmonic oscillators are damped

The frequency of a damped oscillator is less than that of a comparable simple oscillator

FIG. 12–19 (a) Damped harmonic oscillator. (b) An overdamped oscillator returns to its equilibrium position very slowly. (c) A critically damped oscillator returns to its equilibrium position rapidly, but not so rapidly that it overshoots and begins to oscillate.

(a)

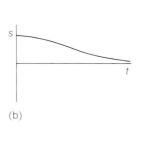

(b)

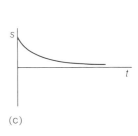

(c)

drops to zero. Damping also acts to reduce the frequency of the motion—the greater the damping, the slower the oscillations.

Damping may prevent oscillation

Figure 12–19 shows the effects of different amounts of damping on a harmonic oscillator. In (a) the damping is small, and the oscillations steadily decrease in amplitude. In (b) the damping is so great that the displaced object never oscillates but returns to its equilibrium position very slowly; such an oscillator is said to be *overdamped*. A *critically damped* oscillator, as in (c), falls right on the line between the other two situations; it returns to its equilibrium position as rapidly as possible without overshooting, which would mean oscillating. Critical damping is desirable in many situations, such as that of the suspension of a car.

IMPORTANT TERMS

A system under stress possesses **elastic potential energy,** which is equal to the work done in deforming it.

Simple harmonic motion is an oscillatory motion that occurs whenever a force acts on an object in the opposite direction to its displacement from its equilibrium position, with the magnitude of the force proportional to the magnitude of the displacement. Thus the force on an object in simple harmonic motion always tends to return it to its equilibrium position.

The **period** T of an object undergoing simple harmonic motion is the time required for it to make one complete oscillation. The **frequency** f of such an object is the number of complete oscillations it makes per unit time.

The **amplitude** of an object undergoing simple harmonic motion is its maximum displacement on either side of its equilibrium position. The period of the motion is independent of the amplitude.

The **center of percussion** of a pivoted object is that point at which it can be struck without producing a reaction force on its pivot.

In a **damped harmonic oscillator,** friction progressively reduces the amplitude of the vibrations.

IMPORTANT FORMULAS

Elastic potential energy: $\mathrm{PE} = \frac{1}{2} ks^2$

Harmonic oscillator: $T = \dfrac{1}{f} = 2\pi \sqrt{\dfrac{m}{k}}$

Simple pendulum: $T = 2\pi \sqrt{\dfrac{L}{g}}$

Torsion pendulum: $\quad T = 2\pi \sqrt{\dfrac{I}{K}}$

Physical pendulum: $\quad T = 2\pi \sqrt{\dfrac{I}{mgh}}$

MULTIPLE CHOICE

1. A spring whose force constant is k is cut in half. Each of the new springs has a force constant of

a. $\frac{1}{2}k$.
b. k.
c. $2k$.
d. $4k$.

2. The product of the period and the frequency of a harmonic oscillator is always equal to

a. 1.
b. π.
c. 2π.
d. the amplitude of the motion.

3. The period of a simple harmonic oscillator is independent of its

a. frequency.
b. amplitude.
c. force constant.
d. mass.

4. An object undergoes simple harmonic motion. Its maximum speed occurs when its displacement from its equilibrium position is

a. zero.
b. a maximum.
c. half its maximum value.
d. none of the above.

5. In simple harmonic motion, there is always a constant ratio between the displacement of the mass and its

a. speed.
b. acceleration.
c. period.
d. mass.

6. An object attached to a horizontal spring executes simple harmonic motion on a frictionless surface. The ratio between its kinetic energy when it passes through the equilibrium position and its potential energy when the spring is fully extended is

a. less than 1.
b. equal to 1.
c. more than 1.
d. equal to the ratio between its mass and the spring constant.

7. The amplitude of an object undergoing harmonic motion is

a. its total range of motion.
b. its maximum displacement on either side of the equilibrium position.

c. its minimum displacement on either side of the equilibrium position.
d. the number of cycles per second it describes.

8. The amplitude of a simple harmonic oscillator is doubled. Which of the following is also doubled?

a. its frequency.
b. its period.
c. its maximum speed.
d. its total energy.

9. The mass of a harmonic oscillator of period T is doubled. The new period is

a. $\sqrt{2}T$.
b. T.
c. $2T$.
d. $4T$.

10. An object suspended from a spring oscillates with a period of 1 s at the earth's surface. If the system were brought to the surface of Mars, where $g = 3.7 \text{ m/s}^2$, the period would be

a. less than 1 s.
b. 1 s.
c. more than 1 s.
d. any of the above, depending on the mass of the object.

11. The period of a simple pendulum depends on its

a. mass.
b. length.
c. total energy.
d. maximum speed.

12. A pendulum executes simple harmonic motion provided that

a. its bob is not too heavy.
b. the supporting string is not too long.
c. the arc through which it swings is not too small.
d. the arc through which it swings is not too large.

13. A pendulum is swinging in an elevator. Its period will be greatest when the elevator is

a. moving upward at constant speed.
b. accelerated upward.
c. moving downward at constant speed.
d. accelerated downward.

14. A lead sphere is suspended by a wire and set into rotational oscillation. If the sphere were flattened into a horizontal disk, the period of the oscillations would be

a. shorter.
b. the same.
c. longer.
d. any of the above, depending on the radius of the disk.

15. An object pivoted at an arbitrary point swings back and forth with the period T. The number of other points in the

object at which it can be pivoted and have the same period of oscillation T is
- a. 0.
- b. at least 1.
- c. at least 2.
- d. unlimited.

16. A force of 0.2 N is needed to compress a certain spring by 2 cm. Its potential energy when compressed is
- a. 2×10^{-3} J.
- b. 2×10^{-5} J.
- c. 4×10^{-5} J.
- d. 8×10^{-5} J.

17. A 250-g toy car is pressed against a horizontal spring of force constant 40 N/m. The spring is compressed 5 cm and then the car is released. If there is no friction, the car leaves the extended spring with a speed of
- a. 0.63 m/s.
- b. 1.6 m/s.
- c. 2.8 m/s.
- d. 8 m/s.

18. When a 1-kg mass is suspended from a spring, the spring stretches by 5 cm. The force constant of the spring is
- a. 0.2 N/m.
- b. 1.96 N/m.
- c. 49 N/m.
- d. 196 N/m.

19. If the suspended mass of Question 18 oscillates up and down, its period will be approximately
- a. 0.032 s.
- b. 0.071 s.
- c. 0.45 s.
- d. 4.5 s.

20. In order to oscillate at 10 Hz, a mass of 20 g should be suspended from a spring whose force constant is
- a. 2.5 N/m.
- b. 8.9 N/m.
- c. 12.6 N/m.
- d. 79 N/m.

21. A man walks to the end of a diving board, whose end moves down by 35 cm as a result. If the man bounces on the end of the board, the period of the oscillations will be
- a. 0.19 s.
- b. 1.2 s.
- c. 1.4 s.
- d. dependent on the man's mass.

22. The maximum speed of a particle that undergoes simple harmonic motion with a period of 0.5 s and an amplitude of 2 cm is
- a. π cm/s.
- b. 2π cm/s.
- c. 4π cm/s.
- d. 8π cm/s.

23. In a refrigeration compressor, a 2-lb piston undergoes 20 cycles/s in which its total travel is 6 in. The maximum force on the piston
- a. is 247 lb.
- b. is 494 lb.
- c. is 3948 lb.
- d. cannot be calculated from the given data.

24. A boy swings from a rope 4.9 m long. His approximate period of oscillation is
- a. 0.5 s.
- b. 3.1 s.
- c. 4.4 s.
- d. 12 s.

25. A pendulum whose period on the earth's surface is 4 s is installed in a satellite that circles the earth in an orbit of radius $1.25R$, where R is the earth's radius. The pendulum's period in the satellite is
- a. 0.
- b. 3.2 s.
- c. 4 s.
- d. 5 s.

EXERCISES

12–1   Elastic Potential Energy

1. The work needed to compress a spring by 4 cm is 0.8 J. What is the force constant of the spring?

2. A toy rifle employs a spring whose force constant is 200 N/m. In use, the spring is compressed 5 cm, and when released, it propels a 5-g rubber ball. What is the ball's speed when it leaves the rifle?

3. A force of 2 N is needed to push a 0.1-kg jack-in-the-box into its box, an operation in which the spring is compressed 10 cm. What will the maximum speed of the jack-in-the-box be when it pops out?

4. A 5-kg object is dropped on a vertical spring from a height of 2 m. If the maximum compression of the spring is 40 cm, what is its force constant?

5. A force of 5 lb compresses a spring by 3 in. Find the force constant of the spring and the elastic potential energy of the compressed spring.

6. A 0.5-lb ball is placed against the compressed spring of Exercise 5. When the spring is released, what is the initial speed of the ball?

7. A spring of force constant k_1 is connected end to end to a spring of force constant k_2. (a) What is the force constant k of the combination? (b) If $k_1 = 10$ N/m and $k_2 = 30$ N/m, find k.

8. A rubber band 30 cm in circumference has a cross section of 2 mm × 2 mm and a Young's modulus of 7 MN/m². Find the energy stored in the rubber band when it is stretched to twice its normal length, assuming that Y is constant over this extension.

12–2   Simple Harmonic Motion

9. Must a spring obey Hooke's law in order to oscillate?

10. At what point or points in its motion is the energy of a harmonic oscillator entirely potential? At what point or points is its energy entirely kinetic?

11. A wooden object is floating in a bathtub. It is pressed down and then released. Under what circumstances will its oscillations be simple harmonic in nature?

12. Two springs with the same force constant k are connected to an object of mass m as in the figures. Find the period of oscillation in each case in the absence of friction.

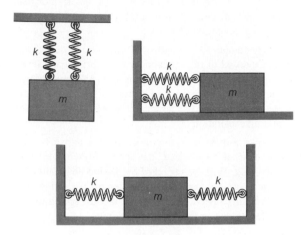

13. When a 1-kg mass is suspended from a spring, the spring stretches by 6 cm. If the mass oscillates up and down, what is its period? What is its frequency?

14. When a 2-lb weight is suspended from a spring, the spring stretches by 3 in. If the weight oscillates up and down, what is its period? What is its frequency?

15. A 1000-kg car has a period of vertical oscillation of 2 s. Find the effective force constant of its springs.

16. A 25-kg portable gasoline-powered generating set is mounted on four springs, which are depressed by 4 mm when the set is put down. Find the natural frequency of vibration of the system.

17. A spring has a 1-s period of oscillation when a 20-N weight is suspended from it. Find the elongation of the spring when a 50-N weight is suspended from it.

18. A body whose mass is 0.4 kg is suspended from a spring and oscillates with a period of 2 s. By how much will the spring contract when the body is removed?

19. A 70-kg gymnast drops on a trampoline from a height of 60 cm. The trampoline sags 40 cm when the gymnast strikes it, and he then bounces up and down. If the motion is simple harmonic, find its period.

20. A spring normally 10 cm long whose force constant is 150 N/m is connected to a spring normally 20 cm long whose force constant is 100 N/m. The combination is then pulled out so that its ends are 50 cm apart. (a) Find the amount by which each spring is stretched. (b) A 400-g block on a frictionless surface is attached between the springs. Find its period of oscillation.

21. A hole is bored through the earth along a diameter. Inside the hole the acceleration of gravity varies as rg/R, where R is the distance from the center of the earth, R is the earth's radius of 6.4×10^6 m, and g is the acceleration of gravity at the earth's surface. A stone is dropped into the hole and executes simple harmonic motion about the center of the earth. Why? Find the period of this motion.

22. A wooden cube of density d that is L long on each edge floats in a liquid density d' so that its upper and lower faces are horizontal. The cube is pushed down and released. Verify that the cube then oscillates up and down in simple harmonic motion with a period of $2\pi \sqrt{Ld/gd'}$.

12−4 Position, Speed, and Acceleration

23. Upon what, if anything, does the ratio between the maximum kinetic energy and maximum potential energy of a harmonic oscillator depend?

24. The total energy of a harmonic oscillator is doubled. How does this affect the frequency of the oscillation? Its amplitude? The maximum speed?

25. A harmonic oscillator has a period of 0.2 s and an amplitude of 10 cm. Find the speed of the moving object when it passes through the equilibrium position.

26. Atoms in a crystalline solid are in constant vibration at room temperature, where their motion has amplitudes in the neighborhood of 10^{-11} m. If the frequency of oscillation of one of the atoms in an iron bar is 2.5×10^{12} Hz, find its maximum speed and acceleration.

27. A rack that undergoes simple harmonic motion with an amplitude of 2 cm is used to test the ability of electronic equipment to survive accelerations. What frequency is needed to produce accelerations of $10g$?

28. The blade of a portable saber saw makes 3000 strokes per minute. Each stroke involves a total blade travel of $\frac{3}{4}$ in. Find the maximum speed of the blade.

29. The prongs of a tuning fork vibrate in simple harmonic motion at a frequency of 660 Hz and with an amplitude of 1 mm at their tips. Find the maximum speed and acceleration of the prong tips. Express the acceleration in terms of g.

30. A 3-kg object is suspended from a spring and oscillates at 5 Hz with an amplitude of 15 cm. (a) What is the total energy of the motion? (b) What is the object's maximum acceleration?

31. A 16-lb object is suspended from a spring and oscillates at 3 Hz with an amplitude of 6 in. (a) What is the total energy of the motion? (b) What is the object's maximum acceleration?

32. A 1.0-lb piston moves up and down in simple harmonic motion with a period of 0.02 s. If the amplitude of the motion is 2.5 in., find the maximum force the piston experiences.

33. An object whose mass is 1 kg hangs from a spring. When the object is pulled down 5 cm from its equilibrium position and released, it oscillates once per second. (a) What is the force constant of the spring? (b) What is the object's speed when it passes through its equilibrium position? (c) What is the maximum acceleration of the object?

34. An object whose mass is 0.005 kg is in simple harmonic motion with a period of 0.04 s and an amplitude of 0.01 m. (a) What is its maximum acceleration? (b) What is the maximum force on the object? (c) What is its acceleration when it is 0.005 m from its equilibrium position? (d) What is the force on it at that point?

35. An object is oscillating in simple harmonic motion with an amplitude of 20 mm and a period of 0.2 s. What is its speed when its displacement on either side of its equilibrium position is 5 mm?

36. What is the displacement on either side of its equilibrium position of the object of Exercise 35 when its speed is 0.5 m/s?

37. At what displacement relative to the amplitude is the kinetic energy of a harmonic oscillator three times the potential energy?

12—6 The Simple Pendulum

38. A body pivoted at some point is given an initial displacement and then released. Under what circumstances will it oscillate back and forth? Under what circumstances will the oscillations be simple harmonic in character? Under what circumstances will it behave like a simple pendulum?

39. What is the frequency of a pendulum whose normal period is T when it is in an elevator in free fall? What it is in an elevator descending at constant velocity? When it is in an elevator ascending at constant velocity?

40. A chandelier is suspended from a high ceiling with a cable 6 m long. What is its period of oscillation?

41. A chandelier is suspended from a high ceiling with a cable 20 ft long. What is its period of oscillation?

42. A pendulum whose length is 1.53 m oscillates 24 times per minute in a particular location. What is the acceleration of gravity there?

43. Find the length of a pendulum whose period is 1 s on the surface of Mars, where the acceleration of gravity is 38% of its value on the earth's surface.

44. A pendulum has a length of 1 ft. Find its period when it is suspended in (a) a stationary elevator; (b) an elevator falling at the constant speed of 20 ft/s; (c) an elevator falling at the constant acceleration of 7 ft/s²; (d) an elevator rising at the constant speed of 20 ft/s; (e) an elevator rising at the constant acceleration of 7 ft/s².

45. A pendulum has a length of 50 cm. Find its period when it is suspended in (a) a stationary elevator; (b) an elevator falling at the constant speed of 5 m/s; (c) an elevator falling at the constant acceleration of 2 m/s²; (d) an elevator rising at the constant speed of 5 m/s; (e) an elevator rising at the constant acceleration of 2 m/s².

46. When a mass m is suspended from a massless spring whose normal length is d, the spring stretches to a new length of $2d$. Will the period of oscillation be greater if the system is set into vertical oscillations or if it swings back and forth like a pendulum? Assume that the spring remains $2d$ in length during the pendulum swings.

12—7 The Torsion Pendulum

47. Find the period of the rotational oscillations of an aluminum sphere 8 cm in diameter whose mass is 725 g, which is suspended by a wire whose torsion constant is 0.1 N · m/rad.

48. A 300-g disk 8 cm in radius is suspended horizontally from its center by a wire. When a force of 5 N is applied to the edge of the disk, it turns through 15°. Find the period of oscillation of the disk.

49. A disk whose moment of inertia is 0.6 kg · m² is suspended horizontally from its center by a thin rod. When the disk is turned and then let go, it oscillates back and forth twice per second. Find the torsion constant of the rod.

50. A sphere suspended by a wire undergoes torsional oscillations with a period of 10 s. The sphere is then replaced by a cylinder of the same mass and radius that is suspended from the center of one face. What is the period of the new oscillator?

51. A wheel is suspended horizontally from its center by a wire. When a torque of 1.5 lb · ft is applied to the wheel, it turns through 10°, and when released, it oscillates with a period of 5 s. Find the wheel's moment of inertia.

12—8 The Physical Pendulum

52. Verify that T is expressed in units of time in (a) Eq. (12–5), (b) Eq. (12–18), (c) Eq. (12–20), and (d) Eq. (12–21).

53. Where should the center of percussion of an axe be located?

54. A 200-g brass hoop 40 cm in diameter is suspended on a knife edge on which it rocks back and forth. If the period of the oscillations is 1.27 s, find the moment of inertia of the hoop about an axis through its circumference perpendicular to its plane.

55. An iron bar 80 cm long is suspended from one end. What is the period of its oscillations? What would be the length of a simple pendulum with the same period?

56. A broomstick suspended from one end oscillates with the same period as a simple pendulum 0.8 m long. How long is the broomstick?

57. A leg may be approximated by a thin rod hinged at one end. (a) If its center of gravity is at its center, find the natural

period of oscillation of a leg 1.0 m long on this basis. (b) With a stride of 0.8 m, what walking speed does this correspond to?

ANSWERS TO MULTIPLE CHOICE

1. a	**6.** b	**11.** b	**16.** a	**21.** b
2. a	**7.** b	**12.** d	**17.** a	**22.** d
3. b	**8.** c	**13.** d	**18.** a	**23.** a
4. a	**9.** a	**14.** c	**19.** c	**24.** c
5. b	**10.** c	**15.** b	**20.** d	**25.** a

13

WAVES

The properties of a vibrating system and those of a wave traveling through a medium are similar in a number of respects. In both, certain characteristic motions recur at regular intervals; in both, the motions involve the continuous conversion of potential energy into kinetic energy and back; in both, the properties of matter play important roles in providing restoring forces. The chief difference is that the energy in a vibrating system remains localized in a single region, whereas waves carry energy from one place to another without any actual transport of matter.

13-1   WAVE MOTION

Waves transport energy

Energy can be transmitted from one place to another in a variety of ways. Suppose we wish to supply energy to a boat in the center of a lake from a position on the shore, and that the precise form in which the energy arrives does not matter. The most obvious

CHAPTER OBJECTIVES

Completing this chapter should enable you to:

1. Describe the reflection and transmission of waves at junctions between different media.

2. Use the principle of superposition to interpret what happens when two waves of the same kind meet.

3. Use the formula $v = f\lambda$ to relate the frequency and wavelength of a wave to its speed.

4. Relate the energy of a wave to its amplitude and frequency.

5. Distinguish between transverse and longitudinal waves.

6. Find the resonant frequencies of a vibrating string.

7. Understand why shock waves are produced by supersonic motion.

8. Distinguish between constructive and destructive interference.

9. Explain the origin of beats.

10. Describe the Doppler effect.

11. Use the decibel as a unit of sound intensity level.

thing to do is to throw a stone at the boat, thereby providing it with kinetic energy. Another method is to pour hot water into the lake, thereby providing the boat with thermal energy. Or we can simply drop a stone in the water near the shore; the waves that are produced transfer energy to the boat by causing it to move up and down (Fig. 13–1). When the stone strikes the water, a deformation of the water surface begins to spread. The energy that reaches the boat arrives as a periodic deformation that contains both kinetic and potential energy. Energy transport by means of the motion of a change in a medium is called *wave motion,* and it can occur in many forms.

13–2 PULSES IN A STRING

If we give one end of a stretched string a quick shake, a kink or *pulse* travels down the string at some speed v (Fig. 13–2). If the string is uniform and completely flexible, the pulse keeps the same shape as it moves. It is worth examining the behavior of pulses in a string both because this is the simplest kind of wave phenomenon and because it is easy to visualize what is going on.

Pulse speed in a string depends on properties of the string, not on properties of the pulse

The speed v of a pulse depends upon the properties of the string—how heavy it is and how tightly it is stretched—rather than upon the shape of the pulse or upon exactly how it is produced. Pulses move slowly down a slack, heavy rope; they move rapidly down a taut, light string. By "heavy" and "light" are meant the mass per unit length of a string, not its total weight; a pulse has the same speed in a long string under a given tension as in a short string of the same kind under the same tension. When the mass per unit length of a string is high, the pulse speed is low because the inertia of each segment of the string is large and it therefore responds slowly to the forces acting on it. When the string is tightly stretched, the pulse speed is high because the tendency of the string to straighten out is greater.

The above observations are reflected in the formula

$$v = \sqrt{\frac{T}{m/L}}$$ *Waves in a string*   (13–1)

for the speed of waves in a stretched string of mass m and length L that is under the tension T.

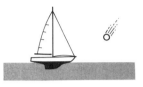

FIG. 13–1 Waves transmit energy from one place to another through the motion of a change in a medium.

Example   The stainless steel forestay of a racing sailboat is 20 m long and 1 cm in diameter, and its mass is 12 kg. In order to determine its tension, the stay is struck by a hammer at the lower end and the return of the pulse is timed. If the time interval is 0.20 s, what is the tension in the stay?

Solution   If L is the length of the stay, the pulse travels the distance $2L$ in the time t, so its speed is $v = 2L/t$. From Eq. (13–1),

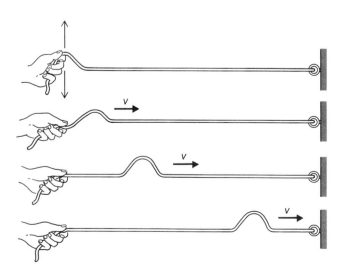

FIG. 13–2 A pulse moves along a stretched string with a constant speed v.

$$v = \sqrt{\frac{T}{m/L}} \qquad v^2 = \frac{T}{m/L} \qquad T = \frac{mv^2}{L}$$

Since $v = 2L/t$ and $L = 20$ m, $t = 0.20$ s, and $m = 12$ kg, the tension in the stay is

$$T = \frac{mv^2}{L} = \frac{m}{L}\left(\frac{2L}{t}\right)^2 = \frac{4mL}{t^2} = \frac{(4)(12\,\text{kg})(20\,\text{m})}{(0.20\,\text{s})^2} = 2.4 \times 10^4\,\text{N}$$

which is 5400 lb. ■

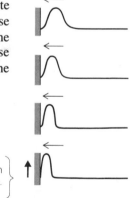

FIG. 13–3 The forward part of this traveling pulse is moving upward and the rear part is moving downward.

The energy content of a moving pulse is partly kinetic and partly potential. As the pulse travels, its forward part is moving upward and its rear part is moving downward; because the string has mass, there is a certain amount of kinetic energy associated with these up-and-down motions (Fig. 13–3). The potential energy is due to the tension in the string. Work had to be done in order to produce the pulse by pulling against the tension, and the deformed string accordingly has elastic potential energy.

A moving pulse has both kinetic energy and potential energy

When a pulse reaches the end of a string, it may be reflected and travel back toward its starting point. Depending upon how the end of the string is held in place, the reflected pulse may be inverted (upside down) or erect (right side up). Under just the right conditions, of course, the energy of the pulse may all be absorbed by the support and the pulse will then disappear.

Suppose the end of the string is held firmly in place. When the pulse arrives there, the string exerts an upward force on the support (assuming the pulse is upward, as in Fig. 13–4). By the third law of motion, the support then exerts an equal and opposite reaction force on the string. The effect of this reaction force is to produce a pulse whose displacement is opposite to that of the original pulse but otherwise with the same shape. The new inverted pulse proceeds back along the string in the reverse direction of that of the original pulse. Thus an upward pulse becomes a downward one upon reflection, and vice versa.

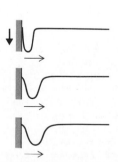

String exerts upward force on support when pulse arrives there. }

Support exerts downward reaction force on string to produce inverted pulse moving in opposite direction. }

FIG. 13–4 A pulse reaching a fixed end of the string is inverted upon reflection.

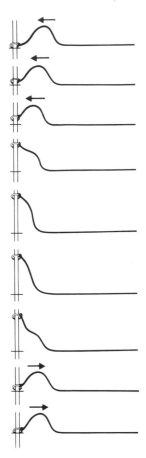

FIG. 13–5 A pulse reaching a free end of the string is not inverted on reflection.

If the end of the string is not held firmly in place, however, the reflected pulse is not inverted. Figure 13–5 shows the end of a string attached to a ring free to move up and down a frictionless rod. When the pulse arrives at this end, the string moves upward until its kinetic energy is completely converted into elastic potential energy. The end of the string then moves downward again to send out a pulse that is reversed in direction but otherwise the same as the original one. If the end of the string is held in a manner exactly in between complete rigidity and complete freedom, then the pulse will not be reflected at all but will disappear when it reaches the end.

A little experimentation with pulses in an actual string will show that, whereas it is very easy to hold the far end of the string so that the reflected pulse is smaller than the original one, it is not easy at all to keep some reflection from taking place. Similar difficulty is experienced in trying to construct surfaces that completely absorb sound, light, or water waves.

So far we have been considering pulses in a uniform string. Now let us connect two different strings together, one of them light (that is, with a low mass per unit length) and the other heavy (high mass per unit length). One end of the combination is fastened to something, and the other is given a shake to produce a pulse. Not surprisingly, the pulse passes from the first string to the second at the junction between them: The pulse is *transmitted*. But the transmission is not complete, since a reflected pulse also appears at the junction and proceeds in the opposite direction.

If the first string is the lighter one, the reflected pulse is inverted (Fig. 13–6). The greater inertia of the heavy string does not permit it to respond to the pulse as rapidly as the light string does, and an opposite reaction force occurs that causes the reflected pulse to be inverted, as though the junction were a rigid support. The energy of the original pulse is then split between the reflected and transmitted pulses. Since both strings have the same tension, the pulse travels more slowly in the heavy string. The length of the reflected pulse is the same as that of the original one, though its height is smaller since it has less energy. The transmitted pulse, however, is shorter, because its speed is less than that of the original pulse while the time interval in which it comes into being is the same as that of the original pulse.

On the other hand, if the first string is the heavy one, the reflected pulse is erect (Fig. 13–7). The smaller inertia of the light string permits it to follow the movements of the heavy one readily, and the situation is like that of a string whose end is able to move up and down freely. However, the light string does have some inertia, and so a reflected pulse again comes into being as well as a transmitted one. The pulse speed

FIG. 13–6 When a pulse passes from a light to a heavy string, reflection occurs with the reflected pulse being inverted. The transmitted pulse is right side up. Since both strings have the same tension, the pulse travels slower in the heavy string.

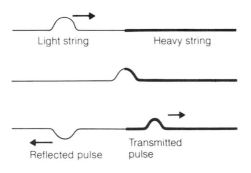

Light string   Heavy string

Reflected pulse   Transmitted pulse

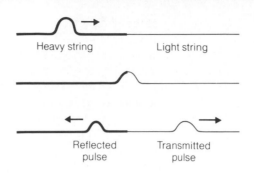

FIG. 13-7 When a pulse passes from a heavy to a light string, the reflected pulse stays right side up. The pulse speed is again less in the heavy string.

is higher in the light string, hence the pulse length is longer there than in the heavy one.

All types of waves, not just pulses in a stretched string, exhibit reflection and transmission at junctions between different media. For example, light waves are partially reflected and partially transmitted when they pass from air to glass, which is why we can see our images in a clear pane of glass (such as a shop window) even though the glass is transparent to light.

All waves are reflected at junctions

13-3 PRINCIPLE OF SUPERPOSITION

Each end of a stretched string is given an upward shake, and the pulses thus produced move along the string toward each other. What happens when they meet? The result is a larger pulse at the moment the pulses come together, and then the separate pulses reappear and continue unchanged in their original directions of motion. Each pulse proceeds as though the other does not exist (Fig. 13-8).

What happens when two pulses meet

The *principle of superposition* is a statement of the above behavior. This principle can be expressed as follows:

> **When two pulses travel past a point in a string at the same time, the displacement of the string at that point is the sum of the displacements each pulse would produce there by itself.**

Principle of superposition for pulses

What if one of the pulses is inverted relative to the other? According to the superposition principle, if the pulses have the same sizes and shapes, their displacements ought to cancel out when they meet, only to reappear later on after they have passed the crossing point. Such behavior is indeed observed in practice. At the instant

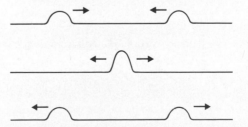

FIG. 13-8 Two pulses moving in opposite directions along a stretched string. The pulses are unaffected by their crossing.

FIG. 13–9 Complete cancellation occurs when two identical pulses with opposite displacements meet. At the instant of complete cancellation, the total energy of both pulses resides in the kinetic energy of the string segment where the cancellation occurs.

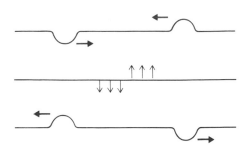

Where the wave energy goes during complete cancellation

of complete cancellation, the total energy of both pulses resides in the kinetic energy of the string segment where the cancellation occurs (Fig. 13–9).

13–4   PERIODIC WAVES

In a periodic wave, one pulse follows another in regular succession. Sound waves, water waves, and light waves are almost always periodic, although in each case a different quantity varies as the wave passes.

Sinusoidal waves

In periodic waves, a certain waveform—the shape of the individual waves—is repeated at regular intervals. Periodic waves of all kinds usually have sinusoidal waveforms; a stretched string down which such waves move presents exactly the same appearance as a graph of sin x (or cos x) versus x that is moved along the x-axis with the wave speed v (Fig. 13–10).

Why periodic waves are usually sinusoidal

Sinusoidal waves are common because the particles of matter in a medium that waves can travel through undergo simple harmonic motion when momentarily displaced from their equilibrium positions. The passage of a wave sets up coupled harmonic oscillations in the medium, with each particle behaving like a harmonic oscillator that has begun its cycle just a trifle later than the particle behind it (Fig. 13–11). The result in the case of waves in a stretched string is a waveform that is basically a graph of how the position of a harmonic oscillator varies with time, which is a sine curve. (Light waves are also sinusoidal in character even though their existence does not involve the motion of material particles and they can travel through empty space.)

Three related quantities are useful in describing periodic waves:

Speed, wavelength, and frequency

1. The *wave speed v,* which is the distance through which each wave moves per second.
2. The *wavelength* λ (Greek letter *lambda*), which is the distance between adjacent crests or troughs;
3. The *frequency f,* which is the number of waves that pass a given point per second.

FIG. 13–10 Most periodic waves have sinusoidal waveforms.

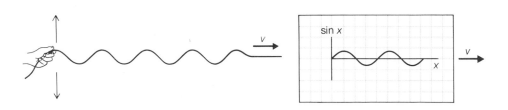

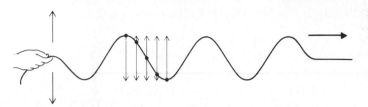

FIG. 13−11 Each particle in the path of a sinusoidal wave executes simple harmonic motion perpendicular to the wave direction.

As in the case of harmonic motion, the unit of wave frequency is the *cycle/s,* or the *hertz* (Hz) after Heinrich Hertz, one of the pioneers in the study of electromagnetic waves. Multiples of the cycle/s and of the hertz are used for high frequencies:

$$1 \text{ kilocycle/s (kc/s)} = 1 \text{ kilohertz (kHz)} = 10^3 \text{ cycles/s}$$

$$1 \text{ megacycle/s (Mc/s)} = 1 \text{ megahertz (MHz)} = 10^6 \text{ cycles/s}$$

The hertz is the unit of frequency

Thus a frequency of 50 MHz is equal to

$$(50 \text{ MHz})\left(10^6 \frac{\text{Hz}}{\text{MHz}}\right) = 5 \times 10^7 \text{ Hz} = 5 \times 10^7 \text{ cycles/s}$$

The wave speed, wavelength, and frequency of a train of waves are not independent of one another. In every second, f waves (by definition) go past a particular point, with each wave occupying a distance of λ (Fig. 13−12). Therefore a wave travels a total distance of $f\lambda$ per second, which is the wave speed v. Thus

How speed, wavelength, and frequency are related

$$v = f\lambda \qquad\qquad \text{\textit{Wave speed}} \quad (13-2)$$

Wave speed = frequency × wavelength

which is a basic formula that applies to all periodic waves, sinusoidal or not.

Example A marine radar operating at a frequency of 9400 MHz emits groups of radio waves 0.08 μs in duration. (The time needed for reflections of these groups to return indicates the distance of the target; see Sec. 25−2.) Radio waves, like light waves, are electromagnetic in nature and travel at 3.00×10^8 m/s. Find (a) the wavelength of these waves; (b) the length of each wave group, which is indicative of the precision with which the radar can measure distance; and (c) the number of waves in the group.

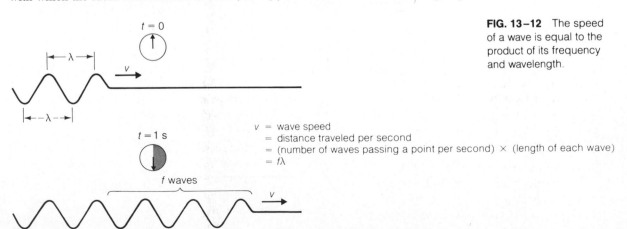

FIG. 13−12 The speed of a wave is equal to the product of its frequency and wavelength.

v = wave speed
  = distance traveled per second
  = (number of waves passing a point per second) × (length of each wave)
  = $f\lambda$

Solution (a) From Eq. (13–2) the wavelength is

$$\lambda = \frac{v}{f} = \frac{3.00 \times 10^8 \, \text{m/s}}{9.4 \times 10^9 \, \text{Hz}} = 3.19 \times 10^{-2} \, \text{m} = 3.19 \, \text{cm}$$

(b) We note that 1 μs = 1 microsecond = 10^{-6} s. Hence the duration of each group of waves is $t = 0.08 \, \mu s = 0.08 \times 10^{-6} \, s = 8 \times 10^{-8} \, s$, and the length of the group is

$$s = vt = (3.00 \times 10^8 \, \text{m/s})(8 \times 10^{-8} \, \text{s}) = 24 \, \text{m}$$

(c) We can find the number n of waves in each group in either of these ways:

$$n = ft = (9.4 \times 10^9 \, \text{Hz})(8 \times 10^{-8} \, \text{s}) = 752 \, \text{cycles} = 752 \, \text{waves}$$

$$n = \frac{s}{\lambda} = \frac{24 \, \text{m}}{3.19 \times 10^{-2} \, \text{m}} = 752 \, \text{wavelengths} = 752 \, \text{waves} \qquad \blacksquare$$

Wave period

Sometimes it is more useful to consider the *period T* of a wave, which is the time required for one complete wave to pass a given point (Fig. 13–13). Since f waves pass by per second, the period of each wave is

$$T = \frac{1}{f} \qquad\qquad\qquad \textit{Wave period} \quad (13–3)$$

If there are five waves per second passing by, for example, each wave has a period of $\frac{1}{5}$ s. In terms of period T, the formula for wave speed is

$$v = \frac{\lambda}{T} \qquad\qquad\qquad\qquad (13–4)$$

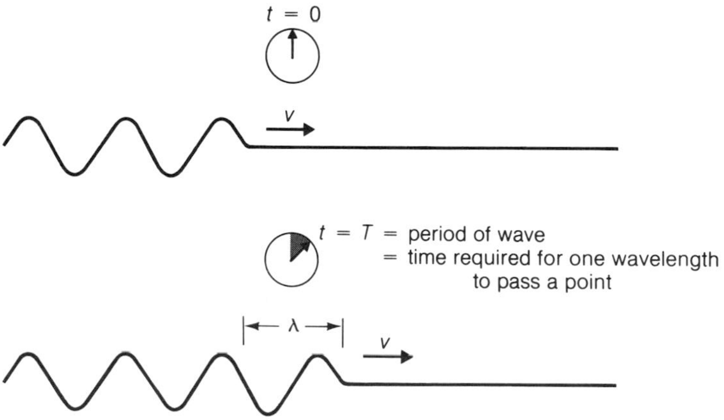

FIG. 13–13 The period of a wave is the time required for one complete wave to pass by a given point.

$t = 0$

v

$t = T$ = period of wave
= time required for one wavelength to pass a point

λ

v

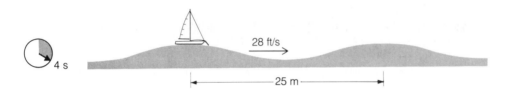

FIG. 13–14 Waves whose period is 4 s and whose wavelength is 25 m have a frequency of 0.25 Hz and a speed of 6.25 m/s.

Example An anchored boat is observed to rise and fall once every 4 s as waves whose crests are 25 m apart pass by it. Find the frequency and speed of the waves.

Solution The frequency of the waves is

$$F = \frac{1}{T} = \frac{1}{4\,\text{s}} = 0.25\,\text{Hz}$$

and their speed is

$$v = \frac{\lambda}{T} = \frac{25\,\text{m}}{4\,\text{s}} = 6.25\,\text{m/s}$$

See Fig. 13–14. ■

The *amplitude A* of a wave refers to the maximum displacement from their normal **Wave amplitude** positions of the particles that oscillate back and forth as the wave travels by (Fig. 13–15). The amplitude of a wave in a stretched string is the height of the crests above the original line of the string (or the depth of the troughs below the original line). The

FIG. 13–15 The quantity A is the amplitude of the wave.

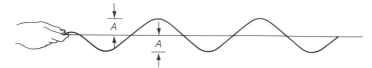

speed, frequency, and wavelength of a wave are independent of its amplitude, just as the period of a harmonic oscillator is independent of its amplitude.

Energy content of wave motion

The correspondence between harmonic and wave motion leads to an interesting result. As we found in Section 12–4, a particle of mass m that undergoes simple harmonic motion of frequency f and amplitude A has a total energy of

$$E = 2\pi^2 m f^2 A^2$$

This dependence of energy on f^2 and on A^2 is also true for mechanical waves of all kinds. (A mechanical wave is one that involves moving matter, in contrast to, say, an electromagnetic wave.) Waves in a string are an example: The energy per unit length in a string due to waves of frequency f and amplitude A is $2\pi^2 (m/L) f^2 A^2$, where m/L is the mass per unit length of the string.

Fourier's theorem

We have been considering sinusoidal waves thus far. However, in a medium whose properties do not vary with wave frequency, everything that is true for a sinusoidal wave is also true for all other periodic waves. An interesting and important theorem by Fourier shows why this should be so. What Fourier proved is that *any* periodic wave of frequency f, regardless of its waveform, can be thought of as a superposition of sinusoidal waves whose frequencies are f, $2f$, $3f$, and so on. The amplitudes of the various component waves depend on the precise character of the composite wave, and a mathematical procedure exists for finding these amplitudes. Fig. 13–16 shows how just three waves of appropriate frequency and amplitude add up to give an approximation of a square wave. The more the waves that are included, the better the approx-

FIG. 13–16 Fourier synthesis of a square wave.

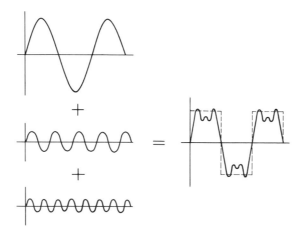

imation becomes. Even an isolated pulse can be represented by a superposition of sinusoidal waves, although here the frequencies must be very close to each other instead of being multiples of f, and a great many waves are needed.

13–5   TYPES OF WAVES

Waves in a stretched string are *transverse waves* since the individual segments of the string vibrate perpendicularly to the direction in which the waves travel, that is, from side to side. *Longitudinal waves* occur when the individual particles of a medium vibrate back and forth in the direction in which the waves travel (Fig. 13–17). Longitudinal waves are easy to produce in a long coil spring; each portion of the spring is alternately compressed and extended as the waves pass by. Longitudinal waves, then, are essentially density fluctuations.

Transverse and longitudinal waves

Waves on the surface of a body of water (or other liquid) are a combination of longitudinal and transverse waves. If we were somehow to tag individual water molecules and follow them when a train of waves passes by, we would find that their paths are like those shown in Fig. 13–18. Each molecule describes a circular orbit with a period equal to the period of the wave, and does not undergo a permanent displacement. Because successive molecules reach the tops of their orbits at slightly different times, the water surface takes the form of a series of crests and troughs. At the crest of a wave

Water waves combine transverse and longitudinal motions

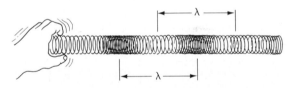

FIG. 13–17 Longitudinal waves in a coil spring.

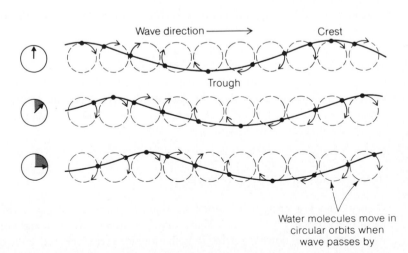

Water molecules move in circular orbits when wave passes by

FIG. 13–18 Water molecules move in circular orbits about their original positions when a typical deep-water wave passes by. At the crest of a wave the molecules are moving in the direction the wave is traveling, while in the trough the molecules are moving in the opposite direction. There is no net motion of water involved in the motion of such a wave.

FIG. 13–19

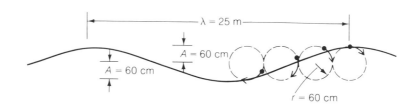

the molecules move in the direction the wave is traveling, while in a trough the molecules are moving in the opposite direction. The passage of a wave across the surface of a body of water, like the passage of a wave through any medium, involves the motion of a pattern: Energy is transported by virtue of the changing pattern, but there is no net transport of matter.

Example   The water waves in the previous example have an amplitude of 60 cm. Find the speed of an individual molecule of water on the surface.

Solution   The water molecules on the surface are moving in circles of radius 60 cm, so that as each wave passes by, the molecules travel a distance s equal to the circumference $2\pi r$ of the circle (Fig. 13–19). Hence

$$s = 2\pi r = 2\pi(0.6 \text{ m}) = 3.8 \text{ m}$$

Each wave takes $T = 4$ s to go past a given point, which means that the molecules must cover the 3.8-m circumference of their orbits in 4 s. The speed of each molecule is therefore

$$V = \frac{s}{T} = \frac{3.8 \text{ m}}{4 \text{ s}} = 0.95 \text{ m/s}$$

Wave speed can exceed particle speed in wave motion

The speed of the *wave*, however, is 6.25 m/s, more than six times greater. Thus the motion of the pattern that constitutes a wave in a medium can be much more rapid than the motions of the individual particles of the medium, and energy can be transported by wave motion faster than might be possible through the net transport of matter. ■

The speed of water waves in deep water depends upon their wavelength: The longer the waves, the faster they move. Thus waves 15 m long travel at 5 m/s but waves 150 m long travel at 15 m/s. In shallow water, the orbits of the water molecules touch the bottom, which slows down the progress of the waves: The shallower the water, the slower they move, regardless of wavelength. The wave speed in water 1 m deep is only half the wave speed in water 4 m deep.

13–6   STANDING WAVES

The vibrations of a string fixed at both ends constitute standing waves

When we pluck a string whose ends are fixed in place, the string starts to vibrate in one or more loops (Fig. 13–20). These *standing waves* may be thought of as the result of waves that travel down the string in both directions, are reflected at the ends, proceed across to the opposite ends and are again reflected, and so on.

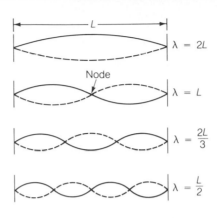

FIG. 13–20 Standing waves in a stretched string. A node is a point where no motion takes place. Strong winds set up standing waves in the Tacoma Narrows Bridge in Washington State soon after its completion in 1940. The bridge collapsed as a result. Today bridges are stiffened to prevent such disasters. (Photo: United Press International)

In order to understand how standing waves come into being, we must transfer to the case of waves our knowledge of how pulses in a string are reflected and of what happens when two pulses traveling in opposite directions meet. When a pulse in a string is reflected at a rigid support, the reflected pulse is inverted; a similar inversion occurs for periodic waves which means that, although their waveform and wavelength stay the same, the wave train is shifted by $\frac{1}{2}\lambda$. Thus a crest arriving at the end of the string is reflected as a trough and vice versa.

Now let us look into how two waves in the same string interact. Applied to waves, the principle of superposition states that

When two or more waves of the same nature travel past a point at the same time, the displacement at that point is the sum of the instantaneous displacements of the individual waves.

Principle of superposition for waves

The principle of superposition holds for all types of waves, including waves in a stretched string, sound waves, water waves, and light waves.

What the principle of superposition means is that every wave train proceeds independently of any others that may also be there. Should two waves with the same wavelength come together in such a way that crest meets crest and trough meets trough, the resulting composite wave will have an amplitude greater than that of either of the original waves. When this occurs the waves are said to *interfere constructively* with each other. Should the waves come together in such a way that crest meets trough and trough meets crest, the composite wave will have an amplitude less than that of the larger of the original waves. When this occurs the waves are said to *interfere destructively* with each other (Fig. 13–21).

Constructive and destructive interference

Let us apply these ideas to a stretched string whose ends are fixed in place. When the string is plucked, waves move back and forth between its ends, becoming inverted each time they are reflected, and they interfere with each other in such a way that destructive interference always occurs at the ends. If the string is L long, destructive interference will occur at the ends when the waves have wavelengths of $\lambda = 2L$, L, $2L/3$, $L/2$, and so on (Fig. 13–20). In the case of the shorter wavelengths, other points, called *nodes,* are present at intermediate positions where no motion of the string takes place.

A node is a place where no vibration occurs

FIG. 13–21 (a) Constructive interference. (b) Destructive interference.

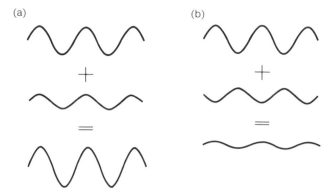

(a)   (b)

FIG. 13–22 The origin of a standing wave. Waves (broken lines) that move to the left and to the right add up to give a stationary wave pattern (solid line).

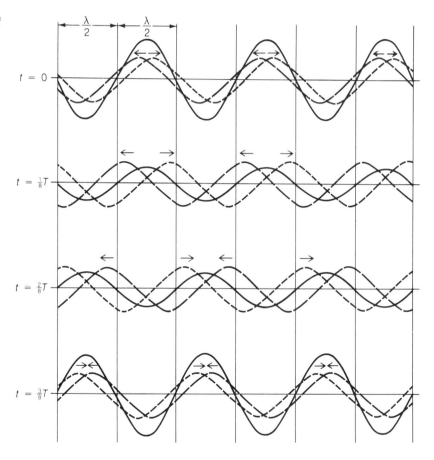

Origin of standing wave

Figure 13–22 shows how a standing wave pattern comes into being: The dotted curve represents a wave moving to the right, and the dashed curve a wave of the same wavelength and amplitude moving to the left. The sum of these waves is found by adding together their displacements at each point on the string, and is shown as a solid line. The addition is performed in Fig. 13–22 for four successive times one-eighth of a period apart. We note that the dotted and dashed curves do not always have displacements equal to zero at the ends of the string. There is no contradiction here since it is their resultant, the solid curve, that corresponds to reality, and this curve behaves properly.

Any type of wave can occur as a standing wave between suitable reflectors. The vibrating air columns in wind instruments and organ pipes—and in the throat, mouth, and nose of a person speaking—are standing sound waves, for instance. Standing light waves play an important role in the operation of lasers, as we shall find later.

Standing waves are a common phenomenon

13–7   RESONANCE

The condition that nodes occur at each end of the string of Fig. 13–20 restricts the possible wavelengths of standing waves to

The standing waves of a system can have only certain wavelengths

$$\lambda = \frac{2L}{n} \qquad n = 1, 2, 3, \ldots \qquad\qquad \textit{Standing waves} \quad (13-5)$$

The lowest possible frequency of oscillation f_1 of a stretched string corresponds to the longest wavelength, $\lambda = 2L$. Thus

$$f_1 = \frac{v}{\lambda} = \frac{v}{2L} \qquad\qquad\qquad (13-6)$$

Higher frequencies correspond to shorter wavelengths. From Eq. (13–5) we see that these higher frequencies are given in terms of f_1 by

$$f_n = nf_1 \qquad n = 2, 3, 4, \ldots \qquad\qquad (13-7)$$

The frequency f_1 is called the *fundamental frequency* of the string, and the higher frequencies f_2, f_3, and so on, are called *overtones*.

Since the wave speed v is $\sqrt{T/(m/L)}$ according to Eq. (13–1), we can express the fundamental frequency of a stretched string in terms of its length, linear density, and tension by the formula

Fundamental frequency of a stretched string

$$f_1 = \frac{1}{2L} \sqrt{\frac{T}{m/L}} \qquad\qquad\qquad (13-8)$$

This formula is the basis for the design of stringed musical instruments such as pianos and violins. A short, light, taut string means a high fundamental frequency of vibration, while a long, heavy, slack string means a low fundamental frequency. The tuning of a stringed instrument involves changing the tensions in the various strings until their fundamental frequencies are correct.

Example   The A-string of a violin has a linear density of 0.6 g/m and an effective length of 330 mm. (a) Find the tension required for its fundamental frequency to be 440 Hz. (b) If the string is under this tension, how far from one end should it be pressed against the fingerboard in order to have it vibrate at a fundamental frequency of 495 Hz, which corresponds to the note B?

Solution   (a) First we square both sides of Eq. (13−8) and then solve for T:

$$f_1^2 = \frac{T}{4L^2(m/L)}$$

$$T = 4L^2 f_1^2 (m/L) = 4(0.33\,\text{m})^2 (440\,\text{Hz})^2 (6 \times 10^{-4}\,\text{kg/m}) = 51\,\text{N}$$

This is about 11 lb.

(b) According to Eq. (13−8), $f_A/f_B = L_B/L_A$, so here

$$L_B = L_A \left(\frac{f_A}{f_B}\right) = (330\,\text{mm})\left(\frac{440\,\text{Hz}}{495\,\text{Hz}}\right) = 293\,\text{mm}$$

Hence the string should be pressed $(330 - 293)$ mm $= 37$ mm from one end.   ∎

The piano

The 88 "strings" of a piano are metal wires whose fundamental frequencies range from 27 Hz for the lowest bass note to 4186 Hz for the highest treble note. This span is so great that the density as well as the length and tension of the wires must be varied in order that the instrument be of reasonable size: The treble strings are thin steel wires, and the bass strings are wound with copper wire to increase their mass density m/L. The total tension of all the strings in a piano is about 20 tons and is supported by a heavy cast iron frame.

Resonance occurs when the frequency of an applied force equals one of the natural frequencies of a system

The fundamental frequency and overtones of a stretched string are its natural frequencies of oscillation: If the string is plucked, one or more of these frequencies will be excited. Eventually internal friction in the string will cause the various vibrations to die out. However, we can cause vibrations to persist by applying a periodic force to the string whose frequency is exactly the same as that of one of its natural frequencies. When this is done, the standing waves continue as long as the periodic force supplies energy to the string. If the energy provided exceeds that dissipated by internal friction, the amplitude of the standing wave will increase until the string may break. (A column of soldiers can destroy a flimsy bridge by marching across it in step with one of its natural frequencies, although the bridge may be capable of safely holding the static load of the soldiers.) This phenomenon is called *resonance*.

All rigid structures have resonant frequencies

When periodic impulses are given to a string at frequencies other than those of its fundamental frequency and overtones, hardly anything happens: The situation then is like pushing a child's swing at a frequency different from its natural one, which produces only minor oscillations. All rigid structures possess characteristic natural frequencies of oscillation, even though their vibrations may be more complex in character than those of a stretched string, and these vibrations can be excited by a periodic force of the right frequency. The rattling of a car at certain speeds is a familiar example of resonance.

13−8   SOUND

Sound waves are pressure fluctuations

Sound waves are longitudinal and consist of pressure fluctuations. The air (or other medium) in the path of a sound wave becomes alternately denser and rarer; the resulting changes in pressure cause our eardrums to vibrate with the same frequency, which produces the sensation of sound.

Most sounds are produced by vibrating objects. An example is the diaphragm of a loudspeaker (Fig. 13–23). When it moves outward, it pushes the air molecules directly in front of it closer together to form a region of high pressure that spreads out in front of the loudspeaker. The diaphragm then moves backward, thereby expanding the volume available to nearby air molecules. Air molecules now flow toward the diaphragm, and a region of low pressure spreads out directly behind the high-pressure region. The continued vibrations of the diaphragm thus send out successive layers of condensation and rarefaction.

The speed of sound in air at sea level and at 20°C is 343 m/s, which is 1125 ft/s. This speed increases at about 0.6 m/s per °C because the random speeds of air molecules increase with temperature and so make the passage of pressure fluctuations more rapid.

The musical note A, the frequency of which is 440 Hz, represents a wavelength λ in air of

$$\lambda = \frac{v}{f} = \frac{343 \, \text{m/s}}{440 \, \text{Hz}} = 0.78 \, \text{m}$$

which is about 31 in. A normal ear responds to sound waves with frequencies from about 20 Hz to about 20,000 Hz, which correspond to wavelengths in air from 17 m (54 ft) to 17 mm (0.65 in). The fundamental frequency of the speaking voice averages about 145 Hz in men and about 230 Hz in women. Even considering the overtones that are present, the frequencies contained in ordinary speech are, for the most part, below 1000 Hz. Sound waves whose frequencies are about 20,000 Hz are called *ultrasonic* and can be detected by appropriate electromechanical devices. Ultrasonic waves are audible to many animals whose hearing organs are smaller than those of human beings and so are better suited to cope with short wavelengths. Figure 13–24 shows how pulses of ultrasound are used to determine the depth of the water under a boat or ship.

Sound waves are transmitted by solids, liquids, and gases. In general, the stiffer

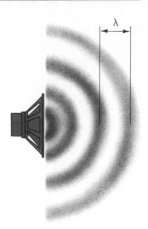

FIG. 13–23 Sound consists of longitudinal waves, representing condensations and rarefactions in the air in its path.

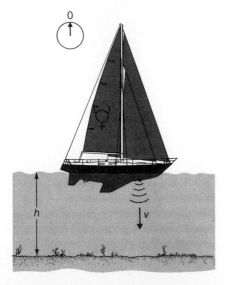

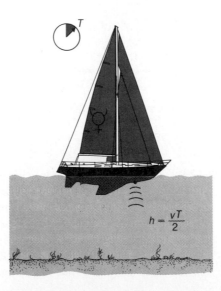

FIG. 13–24 Echo sounding permits water depth to be rapidly and continuously determined from a boat. (a) A pulse of high-frequency sound waves is sent out by a suitable device on the boat's hull. (b) The time *T* needed for the pulse to return after being reflected by the sea floor is a measure of the water depth *h*.

What the speed of sound depends upon

the material the faster the waves travel, which is reasonable when we reflect that stiffness implies particles that are tightly coupled together and therefore more immediately responsive to one another's motions. As in the case of waves in a stretched string, the less dense the material the less the inertia, and the higher the speed of the waves. A detailed analysis shows that the speed of sound in a fluid medium is given by

$$v = \sqrt{\frac{B}{d}}$$

where B is the bulk modulus of the fluid and d its density. In a solid

$$v = \sqrt{\frac{Y}{d}}$$

where Y is the Young's modulus of the material. Table 13–1 lists the speed of sound in various materials.

TABLE 13–1
Speed of sound

Medium	Speed, m/s	Speed, ft/s
Gases (0°C)		
Carbon dioxide	259	850
Air	331	1,086
Air, 20°C	343	1,125
Helium	965	3,165
Liquids (25°C)		
Ethyl alcohol	1,207	3,960
Water, pure	1,498	4,913
Water, sea	1,531	5,022
Solids		
Lead	1,200	4,000
Wood, mahogany	~4,300	~14,100
Iron and steel	~5,000	~16,400
Aluminum	5,100	16,730
Glass, pyrex	5,170	17,000
Granite	~6,000	~19,700

The frequency stays the same when a wave goes from one medium to another

The wavelength of the musical note A in seawater is

$$\lambda = \frac{v}{f} = \frac{1531 \text{ m/s}}{440 \text{ Hz}} = 3.5 \text{ m}$$

which is nearly five times the wavelength of the same note in air. When a sound wave (or any other type of wave) produced in one medium enters another in which its speed is different, the frequency of the wave remains the same while the wavelength changes. This is in accord with the behavior of pulses that pass from one stretched string to another, as discussed earlier.

The "sound barrier"

As an object moves through a fluid—for instance, an airplane through the atmosphere—it disturbs the fluid, and the resulting pressure waves spread out in spherical shells at the speed of sound. If the airplane itself is moving at the speed of sound, the waves pile up in front of it, as in Fig. 13–25(b), to create a wall of high pressure—the "sound barrier." For the airplane to go faster than sound, a considerable amount of force is needed to push it through this barrier, more than the force needed to give a comparable acceleration at speeds either less than (subsonic) or more than (supersonic) that of sound.

At supersonic speeds, an airplane outdistances the waves it produces, which gives rise to the pattern of crests shown in Fig. 13–25(c). Where successive crests overlap, constructive interference takes place, and the result is a conical shell of high pressure with the airplane at its apex. This shell is called a *shock wave* because the pressure increases sharply when it passes, and it is responsible for the sonic boom heard after a supersonic airplane has passed overhead. Because the rise in pressure is so sudden, such a shock wave can do physical damage to structures in its path even though the pressure change itself may not be very great. On a smaller scale, the crack of a whip is a sonic boom that occurs when the tip of the whip moves faster than sound.

A shock wave is produced by supersonic motion

Example The ratio between the speed of an airplane (or other moving object) and the speed of sound is called the *Mach number* after the Austrian physicist Ernst Mach.

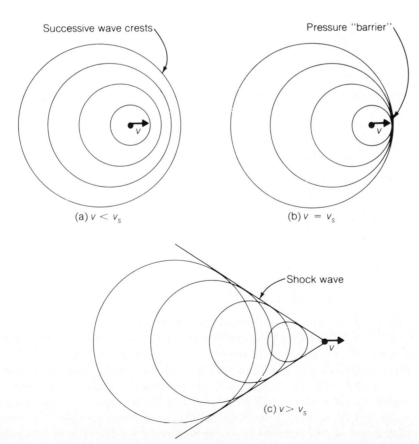

Successive wave crests

Pressure "barrier"

(a) $v < v_s$

(b) $v = v_s$

Shock wave

(c) $v > v_s$

FIG. 13–25 Waves produced by an object moving at (a) subsonic, (b) sonic, and (c) supersonic speeds.

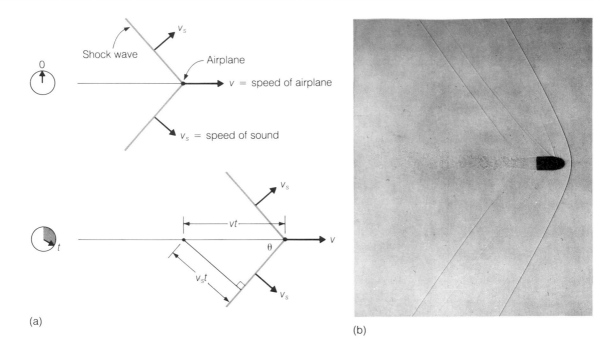

(a)

(b)

FIG. 13–26 (a) Geometry of shock wave produced by supersonic airplane. (b) Photograph of shock wave produced by supersonic bullet. (Photo by Dr. Harold Edgerton, M.I.T., Cambridge, Mass.)

Find the angle between the shock wave created by an airplane traveling at Mach 1.3 and the direction of the airplane.

Solution   As shown in Fig. 13–26, in the time t the airplane travels the distance vt and the shock wave travels the distance $v_s t$. Hence the angle θ between the directions of the airplane and of the shock wave is specified by

$$\sin \theta = \frac{v_s t}{vt} = \frac{v_s}{v}$$

Since a speed of Mach 1.3 means that $v = 1.3v_s$,

$$\sin \theta = \frac{v_s}{v} = \frac{v_s}{1.3v_s} = \frac{1}{1.3} = 0.769$$
$$\theta = \sin^{-1} 0.769 = 50°$$

13–9   INTERFERENCE OF SOUND WAVES

Pure sound waves are sinusoidal

Interference occurs in longitudinal as well as in transverse waves, as we just saw in the case of the shock waves produced in supersonic motion. Another example is the combination of sinusoidal sound waves to give the actual sounds we hear. A graph of pressure versus time at a certain place (or of pressure versus distance at a certain time) for a sound wave that contains a single frequency is sinusoidal in shape. A person singing the musical note "middle A" produces 440-Hz pressure waves whose amplitude might be 1 Pa, which is 0.001% of sea-level atmospheric pressure. The pressure in the path of such waves increases to 1 Pa above normal and decreases to 1 Pa below normal

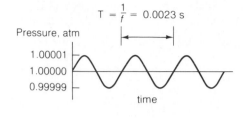

$$T = \frac{1}{f} = 0.0023 \text{ s}$$

Pressure, atm

1.00001
1.00000
0.99999

time

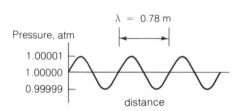

$\lambda = 0.78$ m

Pressure, atm

1.00001
1.00000
0.99999

distance

FIG. 13–27 The pressure variations that constitute the musical note A in air. The amplitude is approximately that which a singer would produce.

440 times per second. The graphs of Fig. 13–27 show the pressure variations of such waves under the assumption that they are purely sinusoidal.

Natural sounds are never as regular as that shown in Fig. 13–27 but consist of mixtures of different frequencies that combine to give complex waveforms. The wave-forms of sounds can be observed with the help of an oscilloscope, a device that displays electrical signals, such as those produced by a microphone in response to a sound, on the screen of a tube like a television picture tube. (The oscilloscope is described in Chapter 18.) Figure 13–28 shows several waveforms of musical notes, which are mixtures of a fundamental frequency and some of its overtones. The precise mixture determines the quality or timbre of the note, which is different for each instrument and for different human voices.

Certain mixtures of frequencies are pleasing to the ear, and sounds that incorporate them are considered "musical." For instance, a tone combined with its first overtone, whose frequency is twice as great, appears harmonious to a listener. In music, such an

Actual sounds have complex waveforms

Musical sounds

Flute

Violin

Clarinet

Human voice

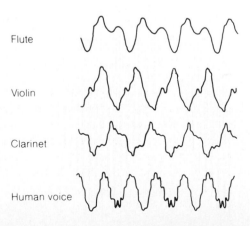

FIG. 13–28 Waveforms of some musical sounds.

FIG. 13–29 *Top*, the origin of beats in sound waves. *Bottom*, beats produced by waves of 20 and 21 Hz.

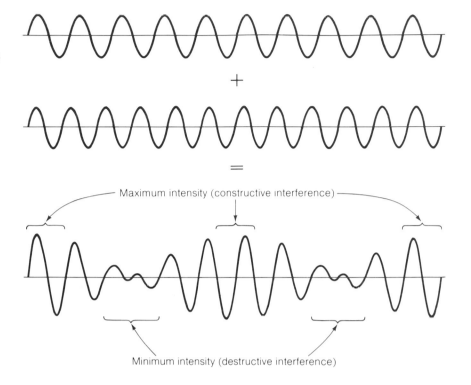

Maximum intensity (constructive interference)

Minimum intensity (destructive interference)

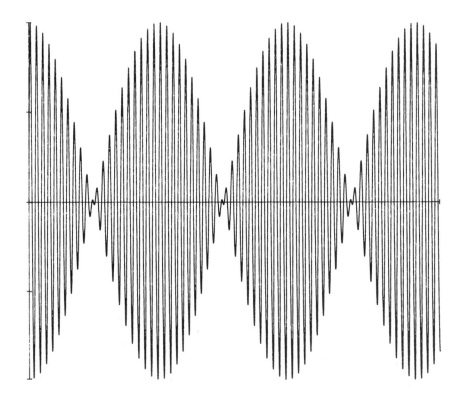

interval is called an *octave* because it includes eight notes. Another harmonious combination is a tone together with another whose frequency is 50% higher, so the frequencies are in the ratio 2:3. An example is C (264 Hz) plus G (396 Hz); such an interval includes five notes (here C, D, E, F, G) and so is called a *fifth*. Somewhat less agreeable are two notes whose frequencies are in the ratio 4:5, for instance C (264 Hz) and E (330 Hz), whose interval spans three notes and is called a *third*. The larger the numbers needed to express the ratio of frequencies, the less attractive the resulting sound is. Thus C and D are in the ratio 8:9 and seem discordant, and E and F, whose ratio is 15:16, are more discordant still. Ordinary sounds are mixtures of frequencies that have no special relationships with one another, and if the mixture seems particularly harsh, it qualifies as noise.

Let us look into what happens when two sounds of different frequency are combined. If two tuning forks (or other sources of single-frequency sound waves) whose frequencies are slightly different are struck at the same time, the sound that we hear fluctuates in intensity. At one instant we hear a loud tone, then virtual silence, then the loud tone again, then virtual silence, and so on. The origin of this behavior is shown schematically in Fig. 13–29; the loud periods occur when the waves from the two forks interfere constructively, thus reinforcing one another, and the quiet periods occur when the waves interfere destructively, thus partially or wholly canceling one another out. These regular loudness pulsations are called *beats*.

Beats result from the interference of sound waves

A piano tuner can tell whether a piano string is in tune by listening for beats between the sound of that string and the sound of a tuning fork; if beats occur, he adjusts the tension of the string until they disappear. The sound we hear when beats are produced has a frequency that is the average of the two original frequencies, and the number of beats per second equals the difference between the two original frequencies. Thus the simultaneous vibrations of a 440-Hz tuning fork and an out-of-tune A-string that vibrates at 444 Hz will produce a 442-Hz tone whose amplitude rises and falls four times per second.

The beat frequency is the difference between the two original frequencies

Beats are hard to distinguish as such when the two frequencies differ by more than perhaps 10 Hz. When the frequencies are far enough apart, however, they give rise to an audible "difference tone" responsible for the lack of harmony of certain combinations of musical notes that was mentioned above. For instance, the notes E and F have the respective frequencies 330 Hz and 352 Hz, and when they are emitted together, the result is an additional 22 Hz difference tone that makes the combination seem dissonant.

13–10 DOPPLER EFFECT

If we stand beside a road when a police car goes by with its siren blowing, we hear a sudden drop in the pitch of the siren as the car passes by. What happens is that the pitch of the siren as the car approaches is *higher* than the pitch when the car is stationary, and the pitch when the car recedes is *lower* than its normal one. We also hear a change of pitch if the siren is at rest and we go past it in a rapidly moving car: We find a higher pitch than usual as we approach the siren, and a lower one as we recede. The change in frequency of a sound brought about by relative motion between source and listener is called the *Doppler effect*.

Relative motion between source and observer changes the perceived frequency

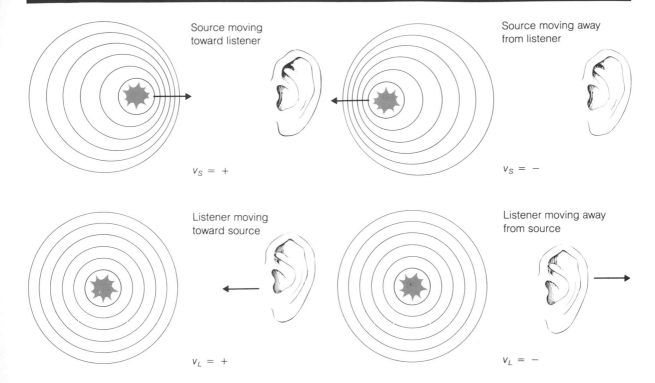

FIG. 13–30 The Doppler effect occurs when there is relative motion between a source of sound and a listener. The sign convention for Eq. (13–9) is shown.

The origin of the Doppler effect is straightforward. As a moving source emits sound waves, it is tending to overtake those traveling in the same direction (Fig. 13–30(a)). Hence the distance between successive waves is smaller than usual; since this distance is the wavelength of the sound, the corresponding frequency is higher than usual. At the same time the source is moving away from those of its waves that travel in the opposite direction, which increases the distance between successive waves behind it and thereby reduces their frequency.

If the source is stationary, a listener moving toward it intercepts more sound waves per unit time than if he were at rest, and accordingly he hears a higher frequency (Fig. 13–30(b)). When the observer moves away from the source, fewer of the waves catch up with him per unit time, and he hears a lower frequency.

The relationship between the frequency f_L the listener hears and the frequency f_S produced by the source is

$$f_L = f_S \left(\frac{v + v_L}{v - v_S} \right) \qquad \textit{Doppler effect in sound} \quad (13-9)$$

Here v is the speed of sound in air, v_L is the speed of the listener (reckoned as + if he moves toward the source, as − if he moves away from the source), and v_S is the speed of the source (reckoned as + if it moves toward the listener, as − if it moves away from the listener). If the listener is stationary, $v_L = 0$, while if the source is stationary, $v_S = 0$.

Example   The frequency of a train's whistle is 1000 Hz. (a) The train is approaching a stationary person at 40 m/s. What frequency does the person hear? (b) The train is stationary and the person is driving toward it in a car whose speed is 40 m/s. What frequency does the person hear now? (Assume that the speed of sound is $v = 343$ m/s, corresponding to room temperature at 20°C.)

Solution   (a) Here $f_S = 1000$ Hz, $v = 343$ m/s, $v_S = 40$ m/s, and $v_L = 0$. Hence the apparent frequency of the whistle is

$$f_L = f_S \left(\frac{v + v_L}{v - v_S} \right) = (1000 \, \text{Hz}) \frac{343 \, \text{m/s}}{(343 - 40) \text{m/s}} = 1132 \, \text{Hz}$$

(b) Again $f_S = 1000$ Hz and $v = 343$ m/s, but now $v_S = 0$ and $v_L = 40$ m/s. The apparent frequency of the whistle is therefore

$$f_L = f_S \left(\frac{v + v_L}{v - v_S} \right) = (1000 \, \text{Hz}) \frac{(343 + 40) \, \text{m/s}}{343 \, \text{m/s}} = 1117 \, \text{Hz} \qquad \blacksquare$$

Frequency measurements can be made so accurately that relatively small speeds can be detected by means of the Doppler effect. An example is the speed of blood in an artery, which does not exceed about 0.4 m/s even in the largest ones. When an ultrasound beam is directed at an artery, the waves reflected from the moving blood cells exhibit a Doppler shift because the cells then act as moving wave sources, and from this shift the speed of the blood can be found.

Doppler effect in light

The Doppler effect in light obeys the formula

$$f = f_s \sqrt{\frac{1 + v/c}{1 - v/c}} \qquad \textit{Doppler effect in light} \quad (13\text{–}10)$$

where f is the observed frequency, f_s is the frequency of the source, and v is the relative speed between source and observer. If source and observer are approaching each other, v is reckoned as $+$, and if they are receding from each other, v is reckoned as $-$. In a vacuum, the speed of light is $c = 3.00 \times 10^8$ m/s; its value in air is very close to this. Doppler shifts in the light from distant stars enables their motions to be detected, and such shifts in radar waves are widely used by police to determine vehicle speeds. The highly accurate Transit system of marine navigation is based on Doppler shifts in the radio waves emitted by a set of five earth satellites. (Light, radar, and radio waves are all electromagnetic in nature, but with very different frequencies.)

13–11   SOUND INTENSITY AND THE EAR

The rate at which a wave of any kind carries energy per unit cross-sectional area is called its *intensity I*. The unit of intensity is the watt/m^2. One joule of energy per second flows through a 1-m^2 surface perpendicular to the path of a wave whose intensity is 1 W/m^2 (Fig. 13–31). The minimum intensity a sound wave must have in order to be heard is about 10^{-12} W/m^2; at the other extreme, sound waves whose intensities exceed about 1 W/m^2 damage the ear.

The human ear does not respond linearly to sound intensity; doubling the intensity of a particular sound gives the sensation of a somewhat louder sound, but one that

The decibel is the unit of sound intensity level

FIG. 13–31 The intensity of a wave is a measure of the rate at which it transports energy.

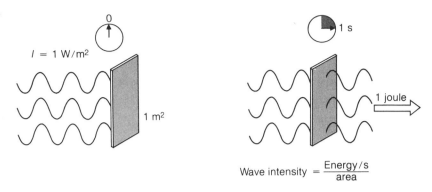

$I = 1 \text{ W/m}^2$

1 m^2

1 joule

$$\text{Wave intensity} = \frac{\text{Energy/s}}{\text{area}}$$

seems far less than twice as loud. For this reason the scale customarily used to measure the intensity level of a sound is logarithmic, which is a reasonable approximation of the actual response of the human ear. The unit is the *decibel* (dB), with a sound that can barely be heard ($I_0 = 10^{-12}$ W/m^2) being assigned a value of 0 dB. A 10-dB sound is, by definition, 10 times more intense than a 0-dB sound; a 20-dB sound is 10^2 (or 100) times more intense; a 30-dB sound is 10^3 (or 1000) times more intense; and so on. Formally, the *intensity level* β (Greek letter *beta*), in decibels, of a sound wave whose intensity in W/m^2 is I is given by

$$\beta = 10 \log \frac{I}{I_0} \qquad \text{Sound intensity level} \quad (13–11)$$

(If $a = 10^b$, then b is the logarithm of a. Thus 3 is the logarithm of 1000 since $1000 = 10^3$. Logarithms are discussed in Appendix B–5).

Ordinary conversation is usually about 40 dB, which is a sound intensity ten thousand times greater than the faintest sound that can be heard. City traffic noise is about 80 dB, a rock band using amplifiers produces an intensity of as much as 125 dB, and the noise of a jet airplane is about 140 dB at a distance of 100 ft (Fig. 13–32). An extended exposure to sound intensities of over 85 dB usually leads to permanent hearing damage.

Example The sound intensity level near a power lawn mower is 95 dB. What is the corresponding sound intensity?

Solution To solve Eq. (13–11) for I, we begin as follows:

$$\beta = 10 \log \frac{I}{I_0}$$

$$\frac{\beta}{10} = \log \frac{I}{I_0}$$

Now we take the antilogarithm of both sides,

$$\log^{-1}\left(\frac{\beta}{10}\right) = \log^{-1}\left(\log \frac{I}{I_0}\right) = \frac{I}{I_0}$$

which can be rewritten as

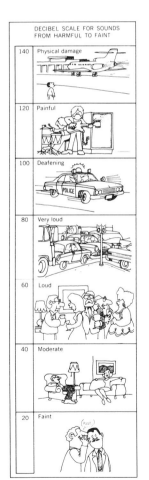

DECIBEL SCALE FOR SOUNDS
FROM HARMFUL TO FAINT

140	Physical damage
120	Painful
100	Deafening
80	Very loud
60	Loud
40	Moderate
20	Faint

FIG. 13–32 Decibel scale.

$$I = I_0 \log^{-1}\left(\frac{\beta}{10}\right)$$

Since $I_0 = 10^{-12}$ W/m^2 and here $\beta = 95$ dB, $\beta/10 = 9.5$ and

$$I = (10^{-12}\ \text{W/m}^2)(\log^{-1} 9.5)$$

To find $\log^{-1} 9.5$ with a calculator, enter 9.5 and press the 10^x key (INV LOG on some calculators). The result is $\log^{-1} 9.5 = 3.2 \times 10^9$, so

$$I = (10^{-12}\ \text{W/m}^2)(3.2 \times 10^9) = 3.2 \times 10^{-3}\ \text{W/m}^2 \qquad \blacksquare$$

Example A change in sound intensity level of 1 dB is about the minimum that can be detected by a person with good hearing; usually the change must be 2 or 3 dB to be readily apparent. What is the actual intensity ratio between two sounds that are 3 dB apart?

Solution If the respective intensities and intensity levels of the two sounds are I_1, β_1 and I_2, β_2, then

$$\beta_2 - \beta_1 = 10\left(\log\frac{I_2}{I_0} - \log\frac{I_1}{I_0}\right)$$

Since $\log x - \log y = \log x/y$ (see Appendix B–5),

$$\log\frac{I_2}{I_0} - \log\frac{I_1}{I_0} = \log\left(\frac{I_2/I_0}{I_1/I_0}\right) = \log\frac{I_2}{I_1}$$

and so

$$\beta_2 - \beta_1 = 10\log\frac{I_2}{I_1}$$

This relationship is plotted in Fig. 13–33. Here $\beta_2 - \beta_1 = 3$, hence

$$3 = 10\log\frac{I_2}{I_1} \qquad \text{or} \qquad 0.3 = \log\frac{I_2}{I_1}$$

Now we take the antilogarithm of both sides:

$$\log^{-1} 0.3 = \log^{-1}\left(\log\frac{I_2}{I_1}\right) = \frac{I_2}{I_1}$$

To find $\log^{-1} 0.3$ with a calculator, enter 0.3 and press the 10^x key (INV LOG on some calculators). The result is

$$\frac{I_2}{I_1} = 2$$

A 3 dB difference in intensity level corresponds to a factor of two in sound intensity. $\qquad \blacksquare$

Sensitivity of the ear

The ear is not equally sensitive to sounds of different frequencies. Maximum response occurs for sounds between 3000 and 4000 Hz, when the threshold for hearing is somewhat less than 0 dB. A 100-Hz sound, however, must have an intensity level

FIG. 13–33 Relationship between sound intensity ratio and difference in intensity level in decibels. The intensity ratio scale is not linear.

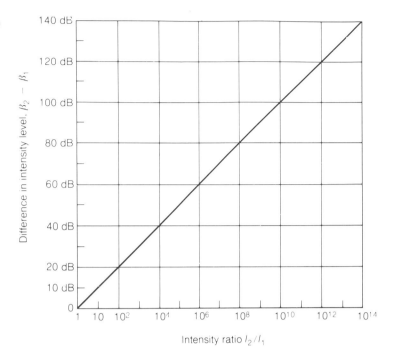

of at least 40 dB to be heard. Sounds whose frequencies are below about 20 Hz (*infrasound*) and above about 20,000 Hz (*ultrasound*) are inaudible to almost everybody regardless of intensity. The intensity level at which a sound produces a feeling of discomfort in the ear is relatively constant at approximately 120 dB for all frequencies. Figure 13–34 shows how the thresholds of hearing and of discomfort vary with frequency. Hearing deteriorates with age, most notably at the high-frequency end of the spectrum. A typical person 60 years old has a threshold of hearing about 10 dB higher than the lowest curve of Fig. 13–34 for 2000-Hz tones, nearly 30 dB higher for 8000-Hz tones, and nearly 70 dB higher for 12,000-Hz tones.

Decibel scale for power gain

The decibel scale is widely used in electronics as well as in acoustics. If the power input to an amplifier or other signal-processing device is P_{in} and the power output is P_{out}, the *power gain G* in decibels is defined as

$$G = 10 \log \frac{P_{out}}{P_{in}} \qquad \textit{Power gain} \quad (13\text{–}12)$$

Thus the power gain of an amplifier whose power input is 0.1 W and whose power output is 50 W is

$$G = 10 \log \frac{P_{out}}{P_{in}} = 10 \log \frac{50\,\text{W}}{0.1\,\text{W}} = 10 \log 500 = (10)(2.7) = 27\,\text{dB}$$

As discussed in Appendix B–5, the logarithm of a product is equal to the sum of the logarithms of the factors:

$$\log xyz = \log x + \log y + \log z$$

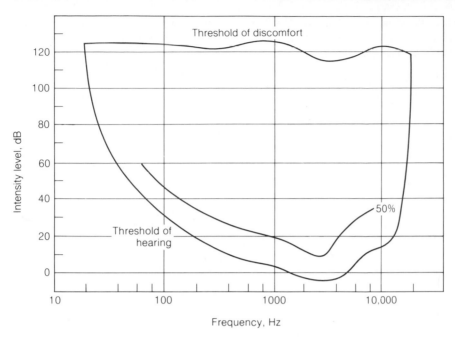

FIG. 13–34 The response of the ear to sound varies with frequency. Only 1% of the U.S. population can hear sounds with intensity levels that fall below the lower curve; 50% can hear sound with intensity levels that fall below the curve above it. Hearing acuity decreases with age.

Since power gains are logarithmic quantities, the overall gain in decibels of a system of several devices is just the sum of their separate gains expressed in decibels:

$$G \text{(overall)} = G_1 + G_2 + G_3 + \cdots \tag{13–13}$$

For example, an audio system might consist of a 35-dB preamplifier, a -10-dB attenuator (a negative dB rating means a negative power gain, so the output is less than the input), and a 70-dB amplifier. The overall gain of the system is

$$G \text{(overall)} = +35 \text{ dB} - 10 \text{ dB} + 70 \text{ dB} = +95 \text{ dB}$$

IMPORTANT TERMS

Wave motion is characterized by the propagation of a change in a medium. Waves transport energy from one place to another.

The **frequency** of a series of periodic waves is the number of waves that pass a particular point per unit time, while their **wavelength** is the distance between adjacent crests or troughs. The **period** is the time required for one complete wave to pass a particular point. The **amplitude** of a wave is the maximum displacement of a particle of the medium on either side of its normal position when the wave passes.

Longitudinal waves occur when the individual particles of a medium vibrate back and forth in the direction in which the waves travel. **Transverse waves** occur when the individual particles of a medium vibrate from side to side perpendicular to the direction in which the waves travel. The vibrations of a stretched string are transverse waves.

The **principle of superposition** states that when two or more waves of the same nature travel past a given point at the same time, the amplitude at the point is the sum of the amplitudes of the individual waves. The interaction of different wave trains is called **interference: Constructive interference** occurs when the resulting composite wave has an amplitude greater than that of either of the original waves, and **destructive interference** occurs when the resulting composite wave has an amplitude less than that of either of the original waves.

    Resonance occurs when periodic impulses are given to an object at a frequency equal to one of its natural frequencies of oscillation.

    Sound is a longitudinal wave phenomenon that results in periodic pressure variations. Sound intensity level is measured in **decibels.**

    A **shock wave** is a shell of high pressure produced by the motion of an object whose speed exceeds that of sound.

    The **Doppler effect** refers to the change in frequency of a wave when there is relative motion between its source and an observer.

IMPORTANT FORMULAS

Waves in a string: $\quad v = \sqrt{\dfrac{T}{m/L}}$

Wave motion: $\quad v = f\lambda = \dfrac{\lambda}{T}$

Standing waves: $\quad \lambda = \dfrac{2L}{n}$

$$n = 1, 2, 3, \ldots$$

Fundamental frequency and overtones: $\quad f_n = nf_1$

Doppler effect in sound: $\quad f_L = f_S\left(\dfrac{v + v_L}{v - v_S}\right)$

Doppler effect in light: $\quad f = f_S\sqrt{\dfrac{1 + v/c}{1 - v/c}}$

Sound intensity level: $\quad \beta\,(\text{dB}) = 10\log\dfrac{I}{I_0}$

MULTIPLE CHOICE

1. The speed of waves in a stretched string depends upon
    a. the tension in the string.
    b. the amplitude of the waves.
    c. the wavelength of the waves.
    d. the acceleration of gravity.

2. The higher the frequency of a wave,
    a. the lower its speed.
    b. the shorter its wavelength.
    c. the greater its amplitude.
    d. the longer its period.

3. Of the following properties of a wave, the one that is independent of the others is its
    a. amplitude.   b. speed.
    c. wavelength.   d. frequency.

4. The amplitude of a wave is related to its
    a. wavelength.   b. period.
    c. frequency.   d. intensity.

5. Waves transmit from one place to another
    a. mass.   b. amplitude.
    c. wavelength.   d. energy.

6. In a transverse wave, the individual particles of the medium
    a. move in circles.
    b. move in ellipses.
    c. move parallel to the direction of travel.
    d. move perpendicular to the direction of travel.

7. Water waves are
    a. longitudinal.
    b. transverse.
    c. partly longitudinal and partly transverse.
    d. sometimes longitudinal and sometimes transverse.

8. Sound waves are
    a. longitudinal.
    b. transverse.
    c. partly longitudinal and partly transverse.
    d. sometimes longitudinal and sometimes transverse.

9. Two waves meet at a time when one has the instantaneous amplitude A and the other has the instantaneous amplitude B. Their combined amplitude at this time is
    a. $A + B$.
    b. $A - B$.
    c. between $A + B$ and $A - B$.
    d. indeterminate.

10. A wave of amplitude A interferes with another wave of the same kind whose frequency is different but whose amplitude is also A. The resulting wave
    a. has an amplitude of $2A$.
    b. varies in amplitude between 0 and A.
    c. varies in amplitude between A and $2A$.
    d. varies in amplitude between 0 and $2A$.

11. Beats are the result of
    a. resonance.
    b. constructive interference.
    c. destructive interference.
    d. both constructive and destructive interference.

12. A frequency that is not an overtone of the fundamental frequency F is
    a. $F/2$.   b. $2F$.
    c. $3F$.   d. $4F$.

13. In order to double the fundamental frequency of a wire under the tension T, the tension must be changed to
  a. $0.5T$.   b. $1.4T$.
  c. $2T$.   d. $4T$.

14. Sound waves do not travel through
  a. solids.   b. liquids.
  c. gases.   d. a vacuum.

15. Sound travels fastest in
  a. air.   b. water.
  c. iron.   d. a vacuum.

16. The amplitude of a sound wave determines its
  a. pitch.   b. loudness.
  c. overtones.   d. resonance.

17. When a sound wave goes from air into water, the quantity that remains unchanged is its
  a. speed.   b. amplitude.
  c. frequency.   d. wavelength.

18. A pure musical tone causes a thin wooden panel to vibrate. This is an example of
  a. an overtone.   b. harmonics.
  c. resonance.   d. interference.

19. A sonic boom is heard after an airplane has passed overhead. This means that the airplane
  a. is accelerating.
  b. is climbing.
  c. was just then passing through the sound barrier.
  d. is traveling faster than sound.

20. Compared with the speed of sound in a given medium, the speed of a shock wave in that medium is
  a. smaller.
  b. the same.
  c. greater.
  d. any of the above, depending on the circumstances.

21. Sound waves whose frequency is 300 Hz have a speed relative to sound waves in the same medium whose frequency is 600 Hz that is
  a. half as great.
  b. the same.
  c. twice as great.
  d. four times as great.

22. Sound waves whose frequency is 300 Hz have a wavelength relative to sound waves in the same medium whose frequency is 600 Hz that is
  a. half as great.
  b. the same.
  c. twice as great.
  d. four times as great.

23. The Doppler effect occurs

  a. only in sound waves.
  b. only in longitudinal waves.
  c. only in transverse waves.
  d. in all types of waves.

24. Relative to the sound sent out by a siren on an ambulance headed away from an observer, the sound received by the observer has
  a. a lower speed.
  b. a lower frequency.
  c. a shorter wavelength.
  d. all of the above.

25. Radio amateurs are permitted to communicate on the "10-meter band." What frequency of radio waves corresponds to a wavelength of 10 meters?
  a. 3.3×10^{-8} Hz   b. 3.0×10^7 Hz
  c. 3.3×10^7 Hz   d. 3.0×10^9 Hz

26. A certain radio station broadcasts at a frequency of 660 kHz. The wavelength of the above waves is
  a. 2.2×10^{-3} m.   b. 4.55×10^2 m.
  c. 4.55×10^3 m.   d. 1.98×10^{14} m.

27. A boat at anchor is rocked by waves whose crests are 100 ft apart and whose speed is 25 ft/s. These waves reach the boat once every
  a. 2500 s.   b. 75 s.
  c. 4 s.   d. 0.25 s.

28. Waves in a lake are 5 m in length and pass an anchored boat 1.25 s apart. The speed of the waves is
  a. 0.25 m/s.
  b. 4 m/s.
  c. 6.25 m/s.
  d. impossible to find from the information given.

29. Two tuning forks of frequencies 310 and 316 Hz vibrate simultaneously. The number of times the resulting sound pulsates per second is
  a. 0.   b. 6.
  c. 313.   d. 626.

30. A string 180 cm long has a fundamental frequency of vibration of 300 Hz. The length of a similar string under the same tension whose fundamental frequency is 200 Hz is
  a. 120 cm.   b. 147 cm.
  c. 220 cm.   d. 270 cm.

31. How many times more intense is a 90-dB sound than a 40-dB sound?
  a. 5   b. 50
  c. 500   d. 10^5

32. Ten oboes produce a sound intensity level of 50 dB in a concert hall. The number of oboes needed to produce a sound intensity level of 60 dB there is

a. 15.
b. 20.
c. 100.
d. 200.

EXERCISES

13-2 Pulses in a String

1. A pulse sent down a long string eventually dies away and disappears. What happens to its energy?

2. A heavy rope hanging from the ceiling is given a transverse shake at its lower end. Show that the speed with which the resulting pulse travels up the rope increases as the square root of the height. (*Hint:* The tension at a given height equals the weight of the rope below that height.)

13-4 Periodic Waves

13-5 Types of Waves

3. The amplitude of a wave is doubled. If nothing else is changed, how is the flow of energy affected.

4. Which of the labeled quantities in the figure corresponds to the amplitude of the wave shown and which to its wavelength?

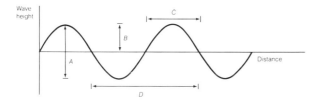

5. A wave of frequency f and wavelength λ passes from a medium in which its speed is v to another medium in which its speed is $2v$. What are the frequency and wavelength of the wave in the second medium? (*Hint:* Consider what happens at the interface between the two media when the wave passes through to decide whether the frequency or the wavelength or both change.)

6. A certain groove in a phonograph record moves past the needle at a speed of 0.4 m/s. The sound produced has a frequency of 3000 Hz. What is the wavelength of the wiggles in the groove?

7. Water waves are observed approaching a lighthouse at a speed of 6 m/s. There is a distance of 7 m between adjacent crests. (a) What is the frequency of the waves? (b) What is their period?

8. A certain groove in a phonograph record moves past the needle at a speed of 12 in./s. If the wiggles in the groove are 0.003 in. apart, what is the frequency of the sound that is produced?

9. Radio waves of very long wavelength can penetrate farther into seawater than those of shorter wavelength. The U.S. Navy communicates with submerged submarines using radio waves whose frequency is 76 Hz. What is the wavelength in air of these waves in kilometers? In miles?

10. In a ripple tank 90 cm across, a 6-Hz oscillator produces waves whose wavelength is 5 cm. Find the time needed for the waves to cross the tank.

11. An anchored boat rises by 4 ft from its normal level and sinks by the same amount every 5.7 s as waves whose crests are 160 ft apart pass it. Find the speed of the waves and the speed of the water molecules on the surface.

13-6 Standing Waves

13 7 Resonance

12. The vibrating part of a violin string whose linear density is 4.7 g/m is 30 cm long. What tension should the string be under if its fundamental frequency is to be 440 Hz, the musical note A?

13. A stretched wire 1 m long has a fundamental frequency of 300 Hz. (a) What is the speed of the waves in the wire? (b) What are the frequencies of the first three overtones?

14. A steel wire 1 m long whose mass is 10 g is under a tension of 400 N. (a) What is the wavelength of its fundamental mode of vibration? (b) What is the frequency of this mode? (c) What is the wavelength of the sound waves produced when the string vibrates in this mode?

15. The G and A strings of a violin have the respective fundamental frequencies of 196 Hz and 440 Hz. If the G string has a linear density of 3 g/m, what should the linear density of the A string be if both are to be under the same tension?

16. One of the strings of a violin has an effective length of 1 ft and weighs 0.05 oz. What tension should the string be under if its fundamental frequency is to be 440 Hz, the musical note A?

17. The vibrating part of the E string of a violin is 330 mm long and has a fundamental frequency of 659 Hz. What is its fundamental frequency when the string is pressed against the fingerboard at a point 60 mm from its end? What are the first and second overtones of the string under these circumstances?

18. The vibrating part of the G string of a certain violin is 330 mm long and has a fundamental frequency of 196 Hz when under a tension of 50 N. (a) Find the linear density of the string. (b) Where should the string be pressed in order for it to vibrate at 220 Hz?

13–8 Sound

(Assume that the speed of sound in sea-level air is 343 m/s, which corresponds to a temperature of 20°C.)

19. In general, in what state of matter does sound travel fastest? Why?

20. As a clarinet is played, the temperature of the air column inside it rises. What effect does this have on the frequencies of the notes it produces?

21. What is the wavelength in feet of sound waves of frequency 8000 Hz in (a) air, (b) seawater, (c) steel?

22. Verify that v has the dimensions of length/time in the formulas $v = \sqrt{T/(m/L)}$, $v = \sqrt{B/d}$, and $v = \sqrt{Y/d}$, which respectively give the speed of waves in a stretched string, in a fluid, and in a solid.

23. A worker strikes a steel rail with a hammer, and the sound reaches an observer 0.5 km away both through the air and through the rail. How much time separates the two sounds?

24. An ultrasonic beam used in scanning body tissue has a frequency of 1.2 MHz. If the speed of sound in a particular tissue is 1540 m/s and the limit of resolution is equal to one wavelength, what is the size of the smallest detail that can be resolved?

25. There are two mechanisms by which a person determines the direction from which a sound comes. One compares the loudness of the sound at the left ear with that at the right ear, which is most effective at low frequencies. The other compares the phases of the waves that arrive at the two ears, which is most effective at high frequencies. (The *phase* of a wave is the part of its cycle it is in at a particular time and place.) The crossover point of equal effectiveness occurs at about 1200 Hz, and as a result sounds with frequencies in the neighborhood of 1200 Hz are difficult to locate. How does the wavelength of a 1200-Hz sound compare with the distance between your ears?

26. A mine explodes at sea, and there is an interval of 5 s between the arrival of the sound through the water and its arrival through the air at a nearby ship. How far away is the ship?

27. An airplane is flying at 800 km/h at an altitude of 2.5 km. When the sound of the airplane's engines appears to a person on the ground to come from directly overhead, how far away is a point directly under the airplane?

28. A violin string is set in vibration at a frequency of 440 Hz. How many vibrations does it make while its sound travels 200 m in air?

29. A tuning fork vibrating at 440 Hz is placed in distilled water. (a) What are the frequency and wavelength of the waves produced within the water? (b) What are the frequency and wavelength of the waves produced in the surrounding air when the water waves reach the surface?

30. Find the angle between the shock wave created by an airplane traveling at 1500 ft/s just above sea level and the direction of the airplane.

31. The angle between the shock wave created by an airplane and the direction of the airplane is 60°. Find the Mach number corresponding to the speed of the airplane.

32. The speed of surface waves in shallow water h deep is $\sqrt{gh}$. How fast is a boat moving in water 1.5 m deep if the total angular width of its wake is 50°?

13–9 Interference of Sound Waves

33. How can constructive and destructive interference be reconciled with the principle of energy conservation?

34. A man has two tuning forks, one marked "256 Hz" and the other of unknown frequency. He strikes them simultaneously, and hears 10 beats per second. "Aha," he says, "the other tuning fork has a frequency of 266 Hz." What is wrong with his conclusion?

35. Determine whether the difference tone produced by the musical notes C (264 Hz) and G (396 Hz) will be consonant (pleasing to the ear) or dissonant.

36. Two identical steel wires have fundamental frequencies of vibration of 400 Hz. The tension in one of the wires is increased by 2%, and both wires are plucked. How many beats per second occur?

13–10 Doppler Effect

37. A fire engine has a siren whose frequency is 500 Hz. What frequency is heard by a stationary observer when the engine moves toward him at 12 m/s? When it moves away from him at 12 m/s?

38. A police car moving at 80 mi/h is chasing a truck moving in the same direction at 60 mi/h. The police car has its siren on. If the truck driver finds the siren's frequency to be

600 Hz, what frequency does the driver of the police car hear?

39. A person in a car is driving at 60 km/h toward a ferry whose whistle is blowing at 400 Hz. (a) What frequency does she hear? (b) The ferry leaves the dock and heads directly away from the motorist at 15 km/h, still blowing its whistle. What frequency does she hear now?

40. A spacecraft moving away from the earth at 97% of the speed of light transmits data at the rate of 10^4 pulses/s. At what rate are they received?

41. A motorist goes through a red light and, when he is arrested, claims that the color he actually saw was green ($\lambda = 5.4 \times 10^{-7}$ m) and not red ($\lambda = 6.2 \times 10^{-7}$ m) because of the Doppler effect. The judge accepts this explanation and instead fines him for speeding at the rate of $1.00 for each km/h he exceeded the speed limit of 80 km/h. What was his fine?

42. The smallest frequency change a person with normal hearing can detect is about 1% for frequencies over 500 Hz. What is the speed of a car with a siren if a person at rest can just detect a change in the siren's pitch as it passes by? (*Hint: v_s is much smaller than v, so $v^2 - v_s^2 \approx v^2$.*)

43. A latecomer to a concert hurries down the aisle toward his seat so fast that the note middle C (256 Hz) appears 1 Hz higher in frequency. How fast is he going?

44. Train A is heading east at the speed v_A and train B is heading west at the speed v_B on an adjacent track. The locomotive on train A is blowing its whistle continuously. A passenger on train B observes the frequency of the sound from the whistle to be 307 Hz when the trains approach each other, 256 Hz when they are abreast, and 213 Hz when they recede from each other. From these figures find the speed of each train relative to the ground.

45. A moving reflector approaches a stationary source of sound of frequency f with the speed u. A listener nearby hears both the original sound waves and the reflected sound waves. Obtain a formula for the number of beats per second the listener hears.

46. The speed with which blood flows through an artery can be determined from the Doppler shift in high-frequency sound waves sent into the artery and detected after reflection from the moving blood cells. What is done is to aim the source and receiver of the sound so that the reflections occur from cells moving away from them. Thus the sound waves striking a cell are Doppler shifted because of its motion away from the source, and the waves that reach the receiver are

further shifted because they come from a reflector moving away from it. (a) Derive a formula for the frequency f_r of the waves that reach the receiver in terms of the source frequency f_s, the speed v of sound in blood (which is 1570 m/s), and the speed v_b of the blood. (b) Solve this formula for v_b. (c) Find the blood speed when the source frequency is 10^6 Hz and the Doppler shift is 40 Hz.

13–11   Sound Intensity

47. The faintest audible sound has an amplitude of about 2×10^{-5} N/m². What percentage of sea-level atmospheric pressure is this?

48. How many times more intense is a 60-dB sound than a 50-dB sound? Than a 40-dB sound? Than a 20-dB sound?

49. What is the equivalent in decibels of a sound whose intensity is 5×10^{-6} W/m²?

50. The sound intensity level near a power lawn mower is 95 dB. What is the corresponding sound intensity?

51. A siren produces a 120-dB sound. What is its intensity in watts per square meter?

52. The sound intensity level in a busy street is 70 dB. At what rate does sound energy enter a room through an open window 0.8 m square?

53. An amplifier has a power gain of 40 dB. What is the actual ratio between its output and input powers?

54. A microphone with an output of 10^{-9} W is connected to a 20-dB preamplifier and a 90-dB amplifier. What is the power output of the system?

55. According to government regulations, the maximum permitted daily exposure time in a workplace to noise of 90 dB is 8 h; to noise of 95 dB, 4 h; to noise of 100 dB, 2 h; and so on. Thus each increase of 5 dB means halving the permitted exposure time. What intensity ratio corresponds to a 5-dB change in intensity level?

56. A riveting gun in a shipyard is producing 95 dB of noise. What is the intensity level when a second riveting gun begins to operate?

57. A chorus of 20 voices is singing at an intensity level of 70 dB. If the voices all have the same intensity level, what is it?

58. At a party, the intensity level of the conversation is 65 dB when a record player is switched on and set to an intensity level of 70 dB. What is the intensity level in the room now?

ANSWERS TO MULTIPLE CHOICE

1. a	**9.** a	**17.** c	**25.** b
2. b	**10.** d	**18.** c	**26.** b
3. a	**11.** d	**19.** d	**27.** c
4. d	**12.** a	**20.** b	**28.** b
5. d	**13.** d	**21.** b	**29.** b
6. d	**14.** d	**22.** c	**30.** d
7. c	**15.** c	**23.** d	**31.** d
8. a	**16.** b	**24.** b	**32.** c

14

TEMPERATURE AND MATTER

Nearly everything expands when heated. Why does this happen? Because something is familiar does not mean that its explanation is obvious, and, as we shall see in this chapter, it is necessary to look deeply into the structure of matter in order to understand its thermal behavior. What we shall find is a close relationship between molecular energy and temperature. This relationship is the key to understanding a great many physical phenomena, for instance why gases become hot when compressed, why evaporation cools liquids, and the significance of relative humidity.

14–1   TEMPERATURE

Temperature, like force, is a key concept in physics that needs an elaborate definition in order to be specified precisely, even though we have a clear idea of its meaning in terms of our sense impressions. We do not need such precision yet, and so we shall

CHAPTER OBJECTIVES

Completing this chapter should enable you to:

1. Convert temperatures from the Celsius to the Fahrenheit scale and vice versa.
2. Calculate the change in length of an object when it is heated or cooled.
3. Calculate the change in volume of a solid or liquid when it is heated or cooled.
4. Use Boyle's law to relate pressure and volume changes in a gas at constant temperature.
5. Use Charles' law to relate temperature and volume changes in a gas at constant pressure.
6. Explain the significance of the absolute temperature scale.
7. Convert Celsius temperatures to absolute temperatures and absolute temperatures to Celsius temperatures.

8. Use the ideal gas law to relate changes in the temperature, pressure, and volume of a gas to each other.
9. Distinguish among elements, compounds, and solutions.
10. Distinguish between atoms and molecules.
11. Calculate the mass of a molecule of given composition.
12. Relate the properties of gases and liquids to the behaviors of their constituent particles.
13. Describe how the relative humidity of a volume of air is related to its degree of saturation with water vapor and how relative humidity varies with temperature.

dodge the issue for the moment and consider temperature merely as that which is responsible for sensations of hot and cold.

There are a number of properties of matter that vary with temperature, and these can be used to construct *thermometers,* devices for measuring temperature. For example, when an object is heated sufficiently, it glows—at first a dull red, then bright red, orange, yellow, and finally, at a high enough temperature, it becomes "white hot." By measuring the color of the light it gives off, we can accurately determine the temperature of an object. This method can only be used at rather high temperatures, however.

The color of a hot object varies with its temperature

Of wider application is the fact that matter usually expands when its temperature is increased and contracts when its temperature is decreased. Concrete roads must be laid with regular gaps to allow for expansion in the summer; heated air above a radiator rises as it expands and becomes lighter than the surrounding air; a column of mercury in a glass tube changes length with a change in temperature. All three of these observations have resulted in practical thermometers (Fig. 14–1).

Thermal expansion

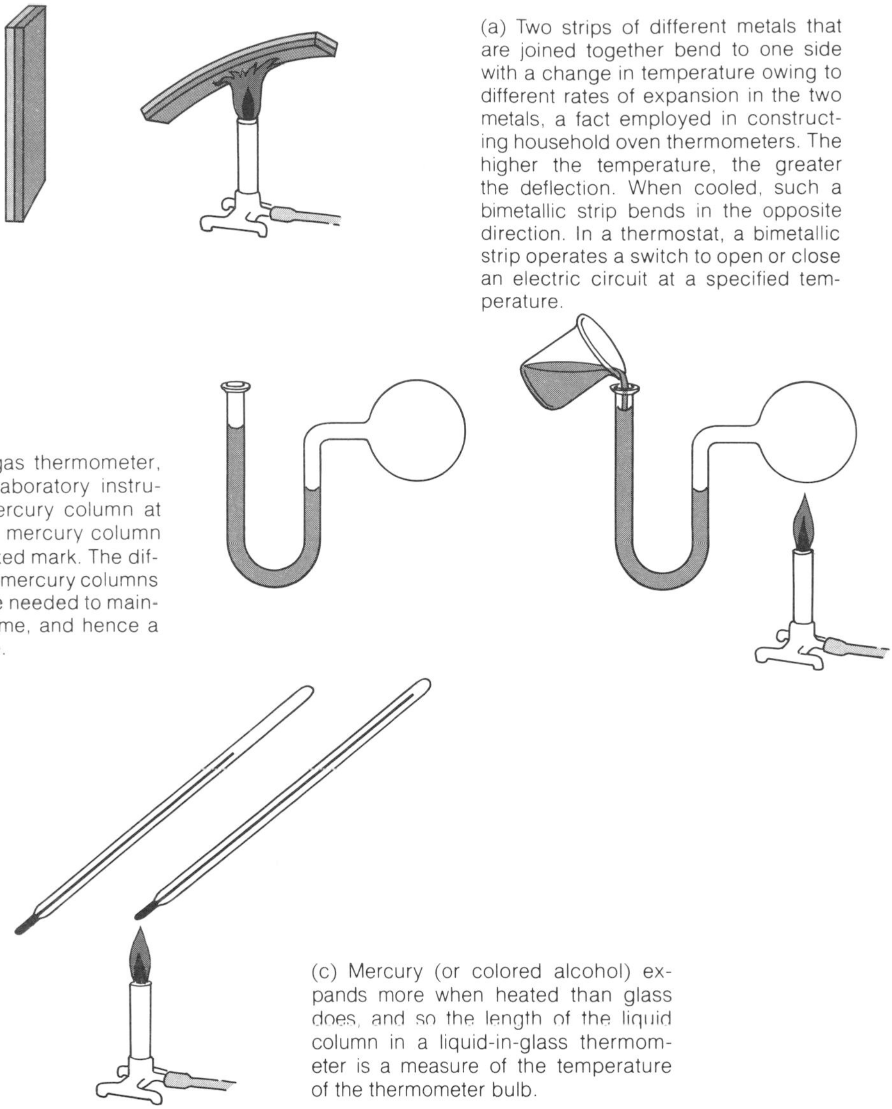

FIG. 14–1   Three types of thermometer.

(a) Two strips of different metals that are joined together bend to one side with a change in temperature owing to different rates of expansion in the two metals, a fact employed in constructing household oven thermometers. The higher the temperature, the greater the deflection. When cooled, such a bimetallic strip bends in the opposite direction. In a thermostat, a bimetallic strip operates a switch to open or close an electric circuit at a specified temperature.

(b) In a constant-volume gas thermometer, which is a very sensitive laboratory instrument, the height of the mercury column at the left is adjusted until the mercury column at the right just touches a fixed mark. The difference in heights of the two mercury columns is a measure of the pressure needed to maintain the gas in a fixed volume, and hence a measure of the temperature.

(c) Mercury (or colored alcohol) expands more when heated than glass does, and so the length of the liquid column in a liquid-in-glass thermometer is a measure of the temperature of the thermometer bulb.

The Celsius (centigrade) temperature scale

Before we can use any of these or other thermal properties of matter to construct a practical thermometer, we must begin by specifying a temperature scale and the method by which we shall calibrate the thermometer in terms of this scale. Water is a readily available liquid that freezes into a solid, ice, and vaporizes into a gas, steam, at definite temperatures at sea-level atmospheric pressure. We can establish a temperature scale by defining the freezing point of water at 1 atm pressure (or, more exactly, the point at which a mixture of ice and water is in equilibrium, with exactly as much ice melting as water freezing) as 0°, and defining the boiling point of water at 1 atm pressure (or, more exactly, the point at which a mixture of steam and water is in equilibrium) as 100°. This scale is called the *Celsius* scale, and temperatures measured

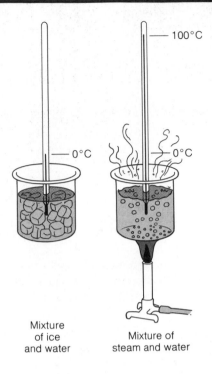

FIG. 14–2 Calibrating a thermometer on the Celsius scale. A mixture of ice and water at atmospheric pressure is, by definition, at 0°C, and a mixture of steam and water at atmospheric pressure is, again by definition, at 100°C.

Mixture of ice and water

Mixture of steam and water

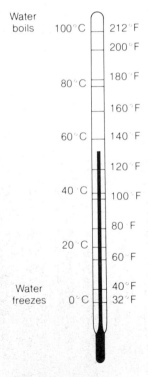

in it are written, for example, "40°C." In the United States, the Celsius scale is sometimes called the *centigrade* scale.

To calibrate a thermometer, say an ordinary mercury thermometer, we first put it into a mixture of ice and water. When the mercury has come to rest we mark the position of its top 0°C on the glass (Fig. 14–2). Then we put it into a mixture of steam and water, and when the mercury column has again come to rest, we mark the new position of the top of the mercury column 100°C. Finally we divide the interval between the 0°C and 100°C markings into 100 equal parts, each representing a change in temperature of 1°C, and extend the scale with divisions of the same length beyond 0°C and 100°C as far as is convenient. In doing this we have, of course, assumed that changes in the length of the mercury column are always directly proportional to the changes in temperature that brought them about.

Although the Celsius scale is used in most of the world, a different temperature scale called the *Fahrenheit* scale is commonly used for nonscientific purposes in the United States. In the Fahrenheit scale the freezing point of water is 32°F and the boiling point of water is 212°F (Fig. 14–3). This means that 180°F separates the freezing and boiling points of water, whereas 100°C separates them in the Celsius scale. Therefore Fahrenheit degrees are 100/180 or 5/9 as large as Celsius degrees:

5 Celsius degrees = 9 Fahrenheit degrees

We can convert temperatures from one scale to the other with the help of the formulas

$$T_F = \tfrac{9}{5} T_C + 32° \qquad \text{\textit{Celsius to Fahrenheit}} \quad (14\text{–}1)$$

$$T_C = \tfrac{5}{9}(T_F - 32°) \qquad \text{\textit{Fahrenheit to Celsius}} \quad (14\text{–}2)$$

FIG. 14–3 The Fahrenheit and Celsius temperature scales. Normal body temperature is 37°C = 98.6°F.

Example   The air temperature on a hot summer day is 95°F. What is the Celsius equivalent of 95°F?

Solution   Since $T_F = 95°$

$$T_C = \tfrac{5}{9}(T_F - 32°) = \tfrac{5}{9}(95° - 32°) = \tfrac{5}{9}(63°) = 35°C$$ ■

Example   Nitrogen freezes at $-210°C$. What is the Fahrenheit equivalent of $-210°C$?

Solution   Since $T_C = -210°$,

$$T_F = \tfrac{9}{5}T_C + 32° = \tfrac{9}{5}(-210°) + 32° = -378° + 32° = -346°F$$ ■

14−2   THERMAL EXPANSION

Nearly all substances expand when they are heated and contract when they are cooled. A long steel bridge may vary in length by over a meter between summer and winter, and allowance must be made for this in its design.

Thermal expansion is proportional to temperature change

Experiments show that, to a good approximation, a change in temperature of ΔT causes most solids to change in length by an amount proportional to their original lengths and to ΔT. If the original length of a rod of a certain material is L_0, its change in length ΔL after its temperature changes by ΔT is

$$\Delta L = aL_0\,\Delta T \qquad\qquad \text{\textit{Thermal expansion}} \quad (14-3)$$

Change in length = a × original length × temperature change

Coefficient of linear expansion

The quantity a, called the *coefficient of linear expansion*, is a constant whose value depends upon the nature of the material. Different substances expand (and contract) to different extents; a lead rod, for example, changes in length by 60 times as much as a quartz rod of the same initial length when both are heated or cooled through the same temperature interval. Table 14−1 lists coefficients of linear expansion for various materials.

Allowance must be made in the design of a bridge for its thermal expansion and contraction. The Golden Gate Bridge in San Francisco varies by more than a meter in length between summer and winter. (Photo: Tom Stack & Assoc.)

TABLE 14−1
Coefficients of linear expansion

Material	Coefficient, $\times 10^{-5}/°C$	Coefficient, $\times 10^{-5}/°F$
Aluminum	2.4	1.3
Brass	1.8	1.0
Concrete	0.7–1.2	0.4–0.7
Copper	1.7	0.94
Iron	1.2	0.67
Lead	3.0	1.7
Quartz	0.05	0.028
Silver	2.0	1.1
Steel	1.2	0.67

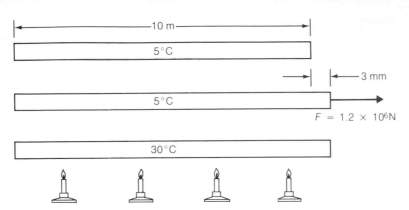

Example What is the increase in length of a steel girder that is 10 m long at 5°C when its temperature rises to 30°C?

Solution The coefficient of linear expansion of steel is $1.2 \times 10^{-5}/°C$, and so, from Eq. (14–2),

$$\Delta L = aL_0\,\Delta T = (1.2 \times 10^{-5}/°C)(10\text{ m})(25°C) = 3 \times 10^{-3}\text{ m} = 3\text{ mm}$$

which is about $\frac{1}{8}$ in. (Fig. 14–4). ∎

Example How much force is associated with the expansion of the girder if its cross-sectional area is 200 cm²?

Solution Since the girder increases in length by 3 mm, the force is the same as that required to stretch it by 3 mm. Equation (10–4) gives the change in length ΔL of a rod that is subjected to a tension or compression force F as

$$\Delta L = \frac{L_0 F}{Y A}$$

where A is the cross-sectional area of the rod and Y is Young's modulus for the material of the rod. Since the girder has $A = 200$ cm² $= 0.02$ m² and, from Table 10–2, Young's modulus for steel is 2×10^{11} N/m²,

$$F = \frac{YA\,\Delta L}{L_0} = \frac{(2 \times 10^{11}\text{ N/m}^2)(0.02\text{ m}^2)(3 \times 10^{-3}\text{ m})}{10\text{ m}} = 1.2 \times 10^{6}\text{ N}$$

A force of 1.2 million newtons, which is equivalent to 135 tons, is associated with the expansion of the girder. Clearly, thermal expansion can involve quite considerable forces. ∎

A formula similar to Eq. (14–3) holds for the changes in volume, ΔV, of a solid or liquid whose temperature changes by an amount ΔT. Here we have **Volume expansion**

$$\Delta V = bV_0\,\Delta T \qquad\qquad \textit{Volume expansion} \quad (14\text{–}4)$$

where V_0 is the original volume and b is the *coefficient of volume expansion*. Table 14–2 is a list of coefficients of volume expansion for various substances. In general, the coefficients of linear and volume expansion are related by

$$b = 3a$$

so that we can readily determine the values of b for the materials of Table 14–1.

TABLE 14–2
Coefficients of volume expansion

Substance	Coefficient, $\times\ 10^{-4}/°C$	Coefficient, $\times\ 10^{-4}/°F$
Ethyl alcohol	7.5	4.2
Glass (average)	0.2	0.1
Glycerin	5.1	2.8
Ice	0.5	0.3
Mercury	1.8	1.0
Pyrex glass	0.09	0.05
Water	2.1	1.2

Example Calculate the volume of water that overflows when a Pyrex beaker filled to the brim with 250 cm³ of water at 20°C is heated to 60°C.

Solution First, we note that a hole in a body expands or contracts by just as much as a solid object whose composition is that of the body and whose volume is that of the hole (Fig. 14–5). The reason is that the material around the hole is not affected by whether the hole is there or not. The boundary of the hole, and therefore the size and shape of the hole, thus behaves as it would if the hole were filled. This means that we can write for the increase in capacity of the beaker

$$\Delta V_b = b_P V_b\, \Delta T = (0.09 \times 10^{-4}/°C)(250\,\text{cm}^3)(40°C) = 0.09\,\text{cm}^3$$

The increase in the volume of the water is

$$\Delta V_w = b_w V_w\, \Delta T = (2.1 \times 10^{-4}/°C)(250\,\text{cm}^3)(40°C) = 2.1\,\text{cm}^3$$

and so the volume of water that overflows is

$$\Delta V_w - \Delta V_b = 2.0\,\text{cm}^3 \qquad \blacksquare$$

FIG. 14–5 A cavity in a body expands or contracts with a change in temperature precisely as a solid object of the same size, shape, and composition would.

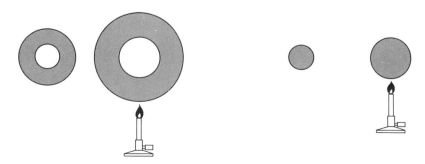

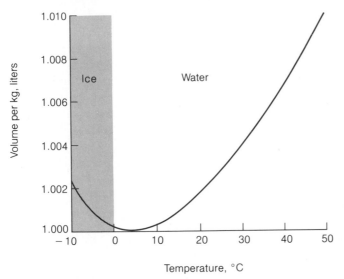

The thermal expansion of water is unusual. Above 4°C, water expands when heated, just as most other substances do. From 0°C to 4°C, however, the volume of a water sample *decreases* with increasing temperature (Fig. 14–6); thus water has its maximum density at 4°C. Equally unusual is the expansion of water when it freezes, so that ice floats and exposed water pipes may burst in a severe winter. (The reason for this behavior is discussed in Section 29–3.) Because ice floats, a body of water freezes in winter from the top down, not from the bottom up as it would if ice were denser than water and cold water always denser than warm water. Ice is a much poorer conductor of heat than water (the difference in heat conductivity is nearly a factor of 4); hence the layer of ice that forms initially on the surface of a body of water helps prevent further freezing. Ice forming at the bottom would not have this effect. Many lakes, rivers, and arms of the sea are therefore able to escape total freezing in winter, which enables their plant and animal life to survive.

FIG. 14–6 Water has its greatest density at 4°C. Because water expands when it freezes, ice floats. (Photo: Owen Franken/Stock, Boston, Inc.)

14–3 BOYLE'S LAW

A peculiar difficulty arises when we attempt to measure the coefficient of volume expansion of a gas. Unlike solids and liquids, gases do not have specific volumes at a given temperature, but expand to fill their containers. The only way to change the volume of a gas is to change the capacity of its container. However, even though its volume may remain the same, another property of a confined gas varies with its temperature, namely the pressure it exerts on the container walls. The air pressure in an automobile tire drops in cold weather and increases in warm, an illustration of this property.

At constant volume, the pressure of a gas increases with temperature

When the temperature of a sample of gas is held constant, the absolute pressure it exerts on its container is very nearly inversely proportional to the volume of the container. Expanding the container lowers the pressure; shrinking the container raises

Boyle's law holds at constant temperature

FIG. 14–7 Boyle's law states that the volume of a gas sample is inversely proportional to its pressure at constant temperature. Thus $p_1V_1 = p_2V_2 = p_3V_3$ as shown.

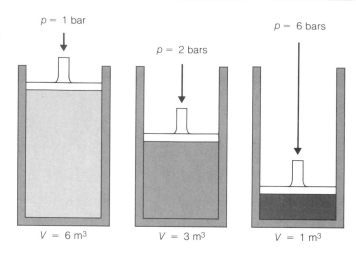

$p = 1$ bar

$p = 2$ bars

$p = 6$ bars

$V = 6$ m³ $V = 3$ m³ $V = 1$ m³

the pressure. Conversely, increasing the pressure on a gas sample reduces its volume; decreasing the pressure increases its volume (Fig. 14–7). This relationship is called *Boyle's law* after its discoverer, Robert Boyle (1627–1691). Though not exact, Boyle's law is an excellent approximation over a wide range of temperatures and pressures.

Boyle's law can be expressed in the form

$$p_1V_1 = p_2V_2 \qquad (T = \text{constant}) \qquad\qquad \textit{Boyle's law} \quad (14\text{–}5)$$

where p_1 is the absolute gas pressure when the volume of the gas is V_1, and p_2 is the gas pressure when the volume is V_2.

Example How much air at the sea-level atmospheric pressure of 15 lb/in.² can be stored in the 12-ft³ tank of an air compressor which can withstand an absolute pressure of 100 lb/in.²?

Solution Letting state 1 represent the air in the compressor tank and state 2 the air at atmospheric pressure, we have from Eq. (14–5)

$$V_2 = \frac{p_1V_1}{p_2} = \frac{(100\,\text{lb/in.}^2)(12\,\text{ft}^3)}{15\,\text{lb/in.}^2} = 80\,\text{ft}^3$$

A total of 80 ft³ of air at atmospheric pressure can be stored in the 12-ft³ compressor tank at the high pressure.   ■

Example A scuba diver's 12-L tank is filled with air at a gauge pressure of 150 bars. If the diver uses 30 L of air per minute at the same pressure as the water pressure at her depth below the surface, how long can she remain at a depth of 15 m in seawater?

Solution Since atmospheric pressure is ~1.0 bar, the absolute pressure of the air in the tank is

$$p_1 = p_{\text{atm}} + p_{\text{gauge}} = 1\,\text{bar} + 150\,\text{bars} = 151\,\text{bars}$$

The absolute pressure at a depth of 15 m is, according to the example in Section 11–3,

$$p_2 = p_{atm} + dgh = 1.0 \text{ bar} + 1.5 \text{ bars} = 2.5 \text{ bars}$$

From Eq. (14–5) we have for the volume of air available at a pressure of 2.5 bars

$$V_2 = \frac{p_1 V_1}{p_2} = \frac{(151 \text{ bars})(12 \text{ L})}{2.5 \text{ bars}} = 725 \text{ L}$$

However, 12 L of air remain in the tank, so 713 L arc usable and

$$t = \frac{713 \text{ L}}{30 \text{ L/min}} = 24 \text{ min} \qquad \blacksquare$$

14–4 CHARLES'S LAW

Now let us see what happens to a gas when its temperature is changed. As mentioned earlier, if the volume of a gas is held constant, the pressure it exerts on its container depends upon its temperature. According to Boyle's law, then, if we hold the gas pressure constant, its volume should vary with temperature. When this prediction is experimentally tested, which was first done over 150 years ago by Charles and Gay-Lussac, it is found that the change in volume ΔV of a gas sample is in fact related to a change ΔT in its temperature by the same formula, Eq. (14–4),

At constant pressure, the volume of a gas increases with temperature

$$\Delta V = bV_0 \, \Delta T$$

that holds for solids and liquids.

The significant thing about gases at constant pressure is that they *all* have very nearly the same coefficient of volume expansion b; by contrast, as Tables 14–1 and 14–2 indicate, the thermal coefficients for solids and liquids may have markedly different values for different substances. At 0°C the coefficient of volume expansion b_0 for all gases is very close to

All gases have the same coefficient of volume expansion at constant pressure

$$b_0 = \frac{0.0037}{°C} = \frac{1/273}{°C}$$

If we vary the temperature of a gas sample while holding its pressure constant, its volume changes by 1/273 of its volume at 0°C for each 1°C temperature change. A balloon filled with air whose volume at 0°C is 1.000 m^3 has a volume of 1.037 m^3 at 10°C and 0.963 m^3 at −10°C (Fig. 14–8).

What happens when the balloon is cooled to −273°C? At that temperature the air in the balloon should have lost 273/273 of its volume at 0°C, and therefore should have vanished entirely! Actually, all gases condense into liquids at temperatures above −273°C, so the question has no physical meaning. But −273°C is still an important temperature. Let us set up a new temperature scale, the *absolute temperature scale,* and designate −273°C as the zero point (Fig. 14–9). Temperatures in the absolute scale are expressed in *kelvins,* denoted K, after Lord Kelvin (1824–1907), a noted British physicist. To convert temperatures from one scale to the other we note that

Absolute temperature scale

FIG. 14–8 Charles's law states that the volume of a gas sample is directly proportional to its absolute temperature at constant pressure. Thus $V_1/T_1 = V_2/T_2 = V_3/T_3$ as shown.

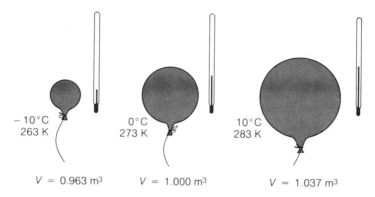

FIG. 14–9 The absolute temperature scale.

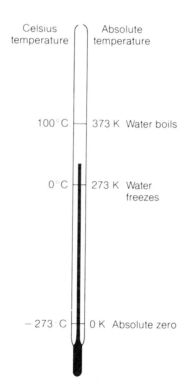

$$T_K = T_C + 273 \qquad\qquad \textit{Celsius to absolute} \quad (14-6)$$

$$T_C = T_K - 273 \qquad\qquad \textit{Absolute to Celsius} \quad (14-7)$$

Thus the equivalent on the absolute scale of room temperature, 20°C, is

$$T_K = T_C + 273 = (20 + 273)K = 293\ K$$

Charles's law holds at constant pressure

The reason for setting up the absolute temperature scale is that, provided the pressure is constant, *the volume of a gas sample is directly proportional to its absolute temperature* (Fig. 14–10). This relationship is called *Charles's law*. Like Boyle's law,

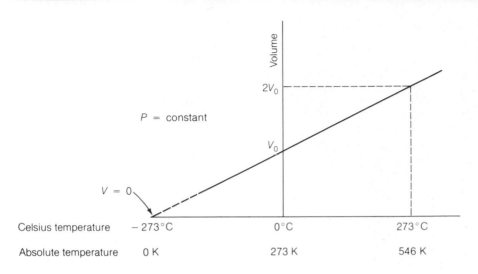

FIG. 14-10 The volume of a gas at constant pressure is directly proportional to its absolute temperature.

Charles's law is not a basic physical principle but deviations from it are usually quite small.

We can express Charles's law in the form

$$\frac{V_1}{T_1} = \frac{V_2}{T_2} \qquad (p = \text{constant}) \qquad\qquad \textit{Charles's law} \quad (14-8)$$

where V_1 is the volume of a gas sample at the absolute temperature T_1 and V_2 is its volume at the absolute temperature T_2.

If there were a gas that did not liquify before reaching 0 K, then at 0 K its volume would shrink to zero. Since a negative volume has no meaning, it is natural to think of 0 K as *absolute zero*. Actually, 0 K is indeed the lower limit to temperatures capable of being attained, but on the basis of a stronger argument than one based on imaginary gases. This argument is discussed later in this chapter. To five significant figures the Celsius equivalent of absolute zero is −273.16°C.

Absolute zero

An absolute temperature scale, called the *Rankine scale,* may be constructed based upon the Fahrenheit scale. Temperatures in this scale are designated °R. Absolute zero on the Rankine scale is

The Rankine scale

$$0°R = \tfrac{9}{5}(-273°C) + 32° = -460°F$$

Hence the conversion formulas for °F to °R and for °R to °F are

$$T_R = T_F + 460° \qquad\qquad \textit{Fahrenheit to Rankine} \quad (14-9)$$

$$T_F = T_R - 460° \qquad\qquad \textit{Rankine to Fahrenheit} \quad (14-10)$$

Charles's law holds when temperatures are expressed in the Rankine scale as well as when they are expressed in the Kelvin scale.

Example   To what temperature must a gas sample originally at 40°F and atmospheric pressure be heated if its volume is to double while its pressure remains the same?

Solution   The Rankine equivalent of 40°F is

$$T_1 = T_F + 460° = 40° + 460° = 500°R$$

Since $V_2 = 2V_1$, we have from Eq. (14–8)

$$T_2 = \frac{V_2 T_1}{V_1} = \frac{(2V_1)T_1}{V_1} = 2T_1 = 2(500°R) = 1000°R$$

The Fahrenheit equivalent of 1000°R is

$$T_2 = T_R - 460° = 1000° - 460° = 540°F$$ ■

14–5   IDEAL GAS LAW

The ideal gas law combines Boyle's law and Charles's law

Boyle's law and Charles's law for the same gas sample can be combined into a single formula called the *ideal gas law:*

$$\frac{p_1 V_1}{T_1} = \frac{p_2 V_2}{T_2} \qquad \qquad \textit{Ideal gas law} \quad (14-11)$$

When $T_1 = T_2$, the ideal gas law becomes Boyle's law,

$$p_1 V_1 = p_2 V_2 \qquad (T = \text{constant})$$

and when $p_1 = p_2$ it becomes Charles's law,

$$\frac{V_1}{T_1} - \frac{V_2}{T_2} \qquad (p = \text{constant})$$

Ideal gas

The ideal gas law is obeyed approximately by all gases. The significant thing is not that the agreement with experiment is never perfect, but that *all* gases, no matter what kind, behave in almost exactly the same way. An *ideal gas* is defined as one that obeys Eq. (14–11) exactly. While no ideal gases actually exist, they do provide a target for theories of the gaseous state to aim at. It is reasonable to suppose that the ideal gas law is a consequence of the basic nature of gases. Hence the next step is to account for this law and only afterward to seek reasons for its failure to be completely correct.

Example   (a) A tank with a capacity of 1 m³ contains helium gas at 27°C under a pressure of 20 atm. The helium is used to fill a balloon. When the balloon is filled, the gas pressure inside it is 1 atm, and its temperature has dropped to −33°C. What is the volume of the balloon at this time? (b) After a while the helium in the balloon is warmed by the atmosphere and returns to its original temperature of 27°C, and it expands further to maintain its pressure at 1 atm. What is the final volume of the balloon? (The gas pressure in the balloon is actually slightly greater than 1 atm to balance the tendency of the rubber to contract, but this is ignored here for convenience.)

Solution   (a) The equivalents of 27°C and −33°C on the absolute scale are 300 K and 240 K respectively. Applying the ideal gas law to the initial expansion, we obtain

$$V_2 = \frac{T_2}{T_1}\frac{p_1}{p_2}V_1 = \left(\frac{240\,\text{K}}{300\,\text{K}}\right)\left(\frac{20\,\text{atm}}{1\,\text{atm}}\right)(1\,\text{m}^3) = 16\,\text{m}^3$$

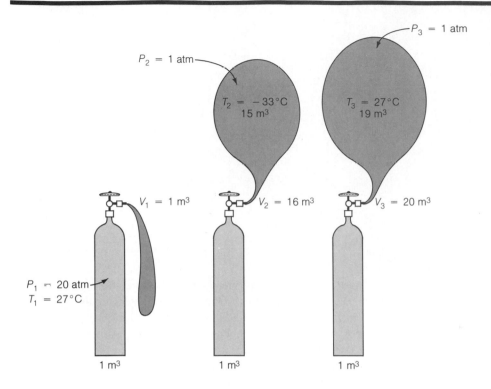

Because the tank's capacity is 1 m³, the balloon's volume after the initial expansion is 15 m³.

(b) When the helium has reached the outside air temperature of 27°C, which we shall call state 3, then $T_1 = T_3$. Hence we need only apply Boyle's law to states 1 and 3 to obtain the eventual volume of the helium:

$$V_3 = \frac{p_1}{p_3}V_1 = \left(\frac{20\,\text{atm}}{1\,\text{atm}}\right)(1\,\text{m}^3) = 20\,\text{m}^3$$

Again we subtract the 1 m³ volume of the tank to find the volume of the balloon itself, which is 19 m³ (Fig. 14–11).   ∎

14–6   STRUCTURE OF MATTER

Before we go on to see how the ideal gas law is explained, let us review the notions of element, compound, and solution, which apply to bulk matter, and those of atom and molecule, which apply to matter on a very small scale of size.

    Liquids and gases are almost always *homogeneous,* which means that every portion of a particular sample is exactly like every other portion (Fig. 14–12). Solids may be either homogeneous or *heterogeneous;* if the latter, some portions of a particular sample may be different from others. A bar of gold, for example, is a homogeneous solid, while a piece of wood is a heterogeneous one. A heterogeneous solid is not always easy to recognize as such, and instruments such as the microscope may be required for definite identification.

Homogeneous and heterogeneous substances

FIG. 14–12 Classification of bulk matter. A heterogeneous substance is a mixture of separate portions of two or more homogeneous substances. Elements are the simplest substances and cannot be further broken down by ordinary means. Two or more elements can combine to form a compound, which has a definite structure and properties of its own. A solution is a uniform mixture of elements or compounds that retain their separate identities and can easily be separated.

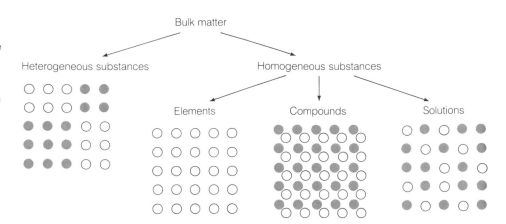

Homogeneous substances may be further classified into *elements, compounds,* and *solutions. Elements* are the simplest substances we encounter in bulk; they cannot be decomposed or changed into one another by ordinary chemical or physical means. There are over 100 known elements (see Appendix C), of which 92 have been found in nature and the rest artificially prepared. At room temperature and sea-level atmospheric pressure, 10 elements are gases, namely argon, chlorine, fluorine, helium, hydrogen, krypton, nitrogen, oxygen, radon, and xenon, and two are liquids, namely bromine and mercury. The rest are in the solid state, the majority being metals.

Two or more elements may combine chemically to form a *compound,* a new substance whose properties are different from those of the elements that compose it. Each ingredient of a *solution,* in contrast, retains its characteristic properties (except, of course, for the mechanical properties of solids and gases dissolved in liquids), and may be separated from the other ingredients by relatively simple methods. Boiling and freezing are examples of such methods, since these changes of state occur at specific temperatures for each element or compound. Air, for instance, is a solution of several gases, chiefly nitrogen and oxygen. Oxygen boils at −183°C whereas nitrogen boils at −196°C, which is 13° lower. If we heat a sample of liquid air to a temperature over −196°C but under −183°C, the nitrogen will vaporize and we will be left ideally with oxygen alone. Under certain circumstances, nitrogen and oxygen unite to form the compound nitric oxide; the boiling point of nitric oxide is −152°C, and heating a sample of liquid nitric oxide to this temperature will vaporize the entire sample. The constituents of a solution may be elements or compounds or both.

Another distinction between compounds and solutions is that the elements in a compound are present in certain definite proportions, always the same for a particular compound, whereas the constituents of a solution may be present in a wide range of proportions. At sea level the mass ratio of the nitrogen and oxygen in the atmosphere varies slightly about an average of 3.2:1, and is several percent greater at high elevations; the mass ratio of the nitrogen and oxygen in nitric oxide is always exactly 0.88:1. If there is too much of either nitrogen or oxygen when nitric oxide is being made, the extra amount will not combine but will be left over and can be separated out at an appropriate temperature (Fig. 14–13).

Compounds and solutions

Law of definite proportions

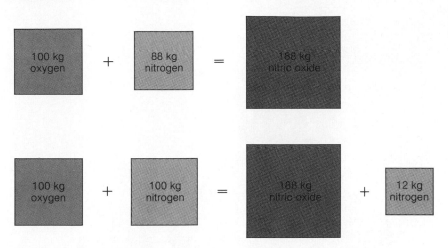

FIG. 14-13 The law of definite proportions.

The idea that matter cannot be divided indefinitely, that all substances are com- **Atoms and molecules** posed of characteristic individual particles, is an ancient one. The basic particles of many compounds are called *molecules*. (In Chapter 29 the structure of compounds is discussed more completely.) Although molecules may be further broken down, when this happens they no longer are representative of the original substance. The molecules of a compound consist of the *atoms* of its constituent elements joined together in a definite ratio. Thus each molecule of water contains two hydrogen atoms and one oxygen atom. While the ultimate particles of elements are atoms, many elemental gases consist of molecules rather than atoms. Oxygen molecules, for instance, contain two oxygen atoms each. The molecules of other gases, such as helium and argon, are single atoms. Fig. 14-14 shows the compositions of some common molecules.

Atoms and most molecules are very small and are present in huge numbers in even a tiny bit of matter. If each atom in a penny were worth 1¢, all the money in the world would not be enough to pay for it.

The masses of atoms and molecules are usually expressed in *atomic mass units* **Atomic mass unit** (u), where

$$1 \text{ atomic mass unit} = 1 \text{ u} = 1.66 \times 10^{-27} \text{ kg}$$

A list of the atomic masses of the elements is given in Appendix C; if we know the composition of a compound, we can calculate the corresponding molecular mass.

Example How many H_2O molecules are present in 1 g of water? The atomic mass of hydrogen is 1.008 u and that of oxygen is 16.00 u.

Solution We begin by finding the mass of the H_2O molecule in u:

$$2H = 2 \times 1.008 \text{ u} = 2.02 \text{ u}$$
$$O = 1 \times 16.00 \text{ u} = \underline{16.00 \text{ u}}$$
$$18.02 \text{ u}$$

The mass of the H_2O molecule in kg is therefore

$$m = (18.02 \text{ u})(1.66 \times 10^{-27} \text{ kg/u}) = 2.99 \times 10^{-26} \text{ kg}$$

FIG. 14–14 Molecular structures of several common substances.

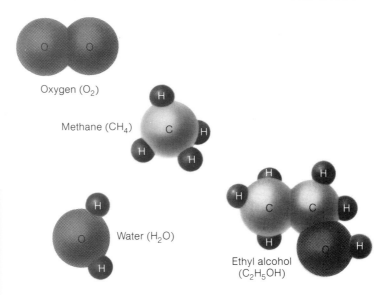

Oxygen (O$_2$)

Methane (CH$_4$)

Water (H$_2$O)

Ethyl alcohol (C$_2$H$_5$OH)

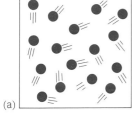

(a)

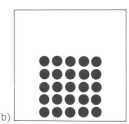

(b)

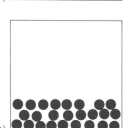

(c)

FIG. 14–15 (a) The molecules of a gas are in constant, random motion. (b) The constituent particles of a solid are also in motion, but oscillate about definite equilibrium positions. (c) The molecules of a liquid keep a more or less constant distance apart, but move about freely.

and so the number of H$_2$O molecules in 1 g $= 10^{-3}$ kg of water is

$$\text{Molecules of H}_2\text{O} = \frac{\text{mass of H}_2\text{O}}{\text{mass of H}_2\text{O molecule}} = \frac{10^{-3}\,\text{kg}}{2.99 \times 10^{-26}\,\text{kg}}$$

$$= 3.34 \times 10^{22}\,\text{molecules} \qquad \blacksquare$$

A considerable amount of experimentation and analysis was needed to establish the reality of atoms and molecules. Although the full story of the kinetic-molecular theory of matter, a large part of which involves chemistry, will not be gone into here, it is easy to show that it can account for the ideal gas law.

14–7   KINETIC THEORY OF GASES

According to the assumptions of the *kinetic theory of gases,* a gas consists of a great many tiny individual molecules that do not interact with one another except when collisions occur. The molecules are supposed to be far apart compared with their dimensions and to be in constant motion, as in Fig. 14–15. They are kept from escaping into space only by the solid walls of a container (or, in the case of the earth's atmosphere, by gravity). A natural result of the random motion and large molecular separation is the tendency of a gas to completely fill its container and to be easily compressed or expanded.

The particles of a solid, on the other hand, are close together, and are held in place by attractive and repulsive forces between them that provide the solid with its characteristic rigidity. In a liquid the intermolecular forces are able to keep the volume of a sample constant but are not strong enough to prevent nearby molecules from sliding past one another, which results in the ability of liquids to flow.

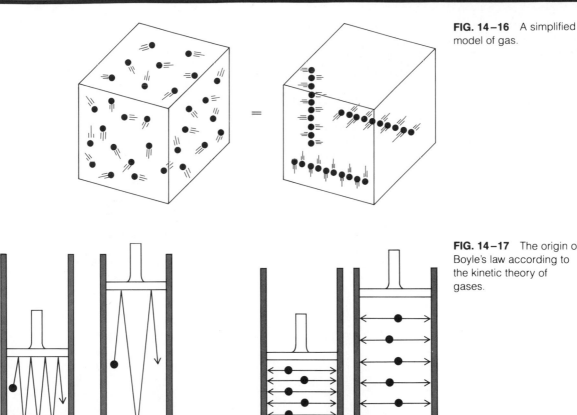

FIG. 14–16 A simplified model of gas.

FIG. 14–17 The origin of Boyle's law according to the kinetic theory of gases.

Pressure falls on top and bottom of expanded cylinder because molecules spend more time in transit between collisions

Pressure falls on sides of expanded cylinder because molecules spread their impacts over a larger area

Origin of Boyle's law

Boyle's law follows directly from the picture of a gas as a group of randomly moving molecules. The pressure the gas exerts comes from the impacts of its molecules; the great number of molecules in even a tiny gas sample means that their separate blows appear as a continuous force to our senses and measuring instruments. Fig. 14–16 shows a simplified model of a gas confined to a box. Although the molecules are actually traveling about in all directions, the effects of their collisions with the walls of the box are the same as if one-third of them were moving back and forth between each pair of opposite walls.

When a cylinder containing a gas is doubled in volume, as in Fig. 14–17, those molecules moving up and down have twice as far to go between impacts. Since their speed is unchanged, the time between impacts is also doubled, and the pressure they exert on the top and bottom of the cylinder falls to half its original value. The expansion of the cylinder also means that the molecules moving horizontally are now spread over twice their former area, and the pressure on the sides of the cylinder therefore falls to half its original value as well. Thus doubling the volume means halving the pressure,

FIG. 14–18 The temperature of a gas increases when it is compressed because the average energy of its molecules increases; the temperature of a gas decreases when it is expanded because the average energy of its molecules decreases.

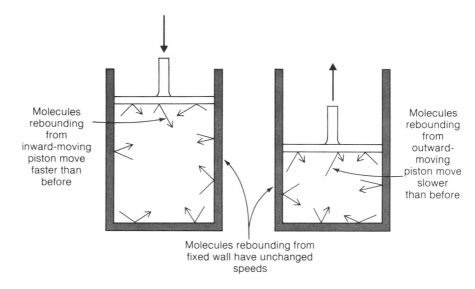

Molecules rebounding from inward-moving piston move faster than before

Molecules rebounding from outward-moving piston move slower than before

Molecules rebounding from fixed wall have unchanged speeds

FIG. 14–19 At absolute zero, the kinetic theory of gases predicts that molecular translational motion in a gas will cease. In reality, at absolute zero the molecules would retain a small minimum amount of kinetic energy.

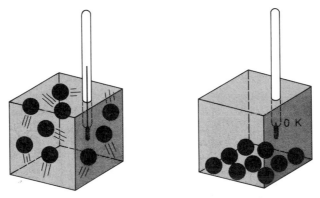

0 K

which is Boyle's law. Similar reasoning accounts for a rise in pressure when the volume is reduced.

Charles's law follows from the kinetic theory of gases when a further assumption is made:

Temperature is a measure of molecular kinetic energy

> **The average kinetic energy of the random translational motions of the molecules of a gas is proportional to the absolute temperature of the gas.**

This assumption is reasonable, since we observe that compressing a gas quickly (so no heat can enter or leave the container) raises its temperature, and such a compression must increase the average energy of the molecules because they bounce off the inward-moving piston more rapidly than they approach it (Fig. 14–18). A familiar example of this effect is a baseball rebounding with greater speed when struck by a bat. On the other hand, expanding a gas lowers its temperature, and such an expansion reduces molecular energies since molecules lose speed in bouncing off an outward-moving piston. The association between molecular energy and temperature agrees with experience.

The interpretation of absolute zero in terms of the elementary kinetic theory of gases is simple: It is that temperature at which all molecular translational movement in a gas stops (Fig. 14−19). A more advanced analysis shows that the molecules cannot be completely at rest, but the difference is small and not important for the discussion here.

Absolute zero

14−8 MOLECULAR MOTION IN GASES

The precise relationship between the average molecular kinetic energy KE_{av} and absolute temperature T is found to be

Molecular kinetic energy

$$KE_{av} = \tfrac{3}{2} kT \qquad\qquad\qquad \textit{Molecular energy} \quad (14-12)$$

where k, known as Boltzmann's constant, has the value

$$k = 1.38 \times 10^{-23} \text{ J/K} \qquad\qquad \textit{Boltzmann's constant}$$

Equation (14−12) holds for the molecules of all gases regardless of the masses of their molecules and has been verified by direct measurements of molecular speeds.

Thus we have an interpretation of temperature in terms of molecular motion that is much more precise and definite than simply describing temperature as that which is responsible for sensations of hot and cold.

The molecules in a gas at a particular temperature do not all have the same speed. The graph in Fig. 14−20 shows the distribution of molecular speeds in oxygen at 273 K and in hydrogen at 273 K. The mass of an O_2 molecule is 16 times that of an H_2 molecule. Average molecular speed decreases with molecular mass, hence at the same temperature molecular speeds in hydrogen are on the average greater than in oxygen. At the same temperature the average molecular *energy* is the same for all gases, however.

Variation of molecular speeds

In Fig. 14−21 we see the distributions of molecular speeds in oxygen at 73 K and at 273 K. The average molecular speed increases with temperature, as predicted. The curves of Figs. 14−20 and 14−21 are not symmetrical because the lower limit to v is fixed at $v = 0$ whereas there is, in principle, no upper limit; actually, as the curves show, the likelihood of speeds many times greater than the average is small.

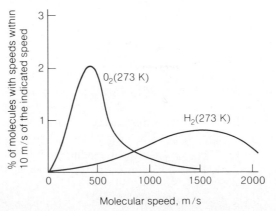

FIG. 14−20 Molecular speeds in oxygen and hydrogen at 273 K (0°C). The smaller masses of H_2 molecules means that they have higher average speeds than O_2 molecules at the same temperature, since the average kinetic energy depends only on temperature.

FIG. 14–21 Molecular speeds in oxygen at 73 K (−200°C) and 273 K (0°C). The higher the temperature, the greater the average kinetic energy.

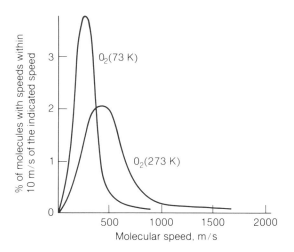

14–9   MOLECULAR MOTION IN LIQUIDS

Liquids as well as gases consist of molecules in constant motion. At its boiling point, the molecules of a liquid have enough energy on the average to break loose from the forces that hold them together, and the liquid becomes a gas. What about the evaporation of a liquid at temperatures under its boiling point? A bowl of water at less than 100°C gradually turns into vapor, and the remaining water grows colder as a result. The faster the evaporation, the stronger the cooling effect. Alcohol is so effective at chilling the skin on contact because it evaporates rapidly.

Evaporation on the basis of kinetic theory

Evaporation can be understood on the basis of the distribution of molecular speeds in a liquid. While not exactly the same as that found in a gas, this distribution resembles those shown in Figs. 14–20 and 14–21 in that a certain fraction of the molecules in any sample have much greater and much smaller speeds than the average. The fastest molecules have enough energy to escape through the liquid surface despite the attractive forces of the other molecules. The molecules left behind redistribute the available energy in collisions among themselves, but, because the most energetic ones escape, the average energy that remains is less than before and the liquid is now at a lower temperature (Fig. 14–22).

FIG. 14–22 After evaporation, the remaining liquid is cooler than before.

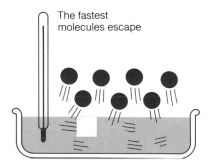

The fastest molecules escape

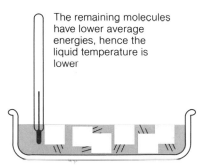

The remaining molecules have lower average energies, hence the liquid temperature is lower

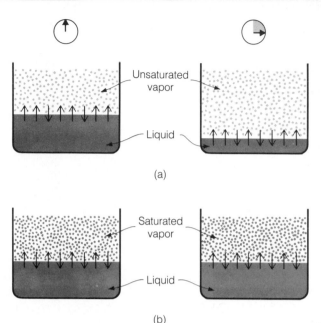

FIG. 14–23 (a) When the region over a liquid is not saturated with vapor, more molecules leave the liquid than return to it, and the liquid gradually evaporates. (b) When the region is saturated with vapor, as many molecules return to the liquid as leave it, and the liquid level remains the same.

When molecules from the vapor above a liquid surface happen to strike the surface, they may be trapped there, so that a constant two-way traffic of molecules to and from the liquid occurs. If the density of the vapor above the liquid is sufficiently great, just as many molecules return as leave it at any time. This situation is described by saying that the region is *saturated* with the vapor (Fig. 14–23). The higher the temperature, the greater the maximum vapor density: At 0°C the density of water vapor at saturation is 5 g/m³, at 20°C it is 17 g/m³, at 100°C it is 598 g/m³, and at 300°C it is all the way up to 45.6 kg/m³ (Fig. 14–24). If for any reason (such as a sudden drop in temperature) the vapor density is more than the saturation value, condensation will be more rapid than evaporation until equilibrium is established again. It is for this reason that on a hot day moisture condenses on the outside of a glass that contains a cold drink. It is also why cloud caps form over mountains (Fig. 14–25).

The *relative humidity* of a volume of air describes its degree of saturation with water vapor. Relative humidities of 0, 50%, and 100% mean respectively that no water vapor is present; that the air contains half as much moisture as the maximum possible; and that the air is saturated. On a hot day the evaporation of sweat from the skin is the chief way in which the human body dissipates heat, and a high relative humidity is uncomfortable because it slows down the process. On the other hand, a low relative humidity is not desirable because it leads to the drying of the skin and mucous membranes. The regulation of relative humidity is as important a function of a heating or of an air-conditioning system as the regulation of temperature.

From Fig. 14–24 we can see to what extent heating decreases the relative humidity of an air sample and cooling increases its relative humidity. For instance, between 10°C and 20°C the saturated vapor density (which corresponds to 100% relative humidity) just about doubles. This means that if outside air at 10°C whose relative humidity is, say, 70% is taken inside a house and heated to 20°C, the relative humidity indoors will only be 35% since the actual vapor density stays the same. A way to humidify heated air in winter is clearly desirable. If the outside air is at 30°C with 70% relative humidity,

Saturation vapor density increases with temperature

Relative humidity is water vapor density relative to saturation density

Changing the temperature of a volume of air also changes its relative humidity

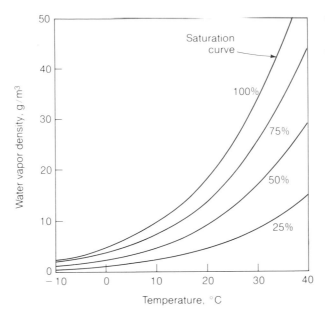

Temperature, °C	Water vapor density, g/m³
−10	2.36
−5	3.41
0	4.85
5	6.80
10	9.41
15	12.8
20	17.3
25	23.0
30	30.4
35	39.6
40	51.1

then cooling it down to 24°C is enough to bring it to saturation, which is 100% relative humidity. Further cooling will cause water to condense out. An air-conditioning system therefore should include a way to remove water vapor from the air being cooled.

The *dew point* of air with a given water vapor content is the temperature at which the air would be saturated. For instance, suppose we have a sample of air at 20°C that contains 9.4 g/m³ of water vapor. Saturated air at 20°C contains 17.3 g/m³, so the relative humidity of the sample is

$$\text{Relative humidity} = \frac{\text{actual vapor density}}{\text{saturated vapor density}}$$
$$= \frac{9.4 \text{ g/m}^3}{17.3 \text{ g/m}^3} = 0.54 = 54\%$$

The dew point of the air sample is 10°C since, from Fig. 14-24, saturated air at 10°C contains 9.4 g/m³ of water vapor. Dew point is used in meteorology as a guide to the possibility of fog. When air temperature and dew point are far apart, there is no danger of fog, whereas when air temperature is near the dew point and is decreasing, fog is likely.

Clouds form when moist air cools enough for its moisture to condense.

Air expands and cools as it rises.

IMPORTANT TERMS

The **temperature** of a body of matter is a measure of the average kinetic energy of random translational motion of its constituent particles.

A **thermometer** is a device for measuring temperature. The two temperature scales in common use are the **Celsius** (centigrade) scale, in which the freezing point of water is assigned the value 0°C and its boiling point the value 100°C, and the **Fahrenheit** scale, in which these points are assigned the values 32°F and 212°F respectively.

The **coefficient of linear expansion** is the ratio between the change in length of a solid rod of a particular material and its original length per 1° change in temperature. The **coefficient of volume expansion** is the ratio between the change in volume of a sample of a particular solid or liquid and its original volume per 1° change in temperature.

Boyle's law states that, at constant temperature, the absolute pressure of a sample of a gas is inversely proportional to its volume, so that $pV = $ constant at that temperature regardless of changes in either p or V individually.

Charles's law states that, at constant pressure, the volume of a sample of a gas is directly proportional to its absolute temperature, so that $V/T = $ constant at that pressure regardless of changes in either V or T individually.

The **Kelvin absolute temperature scale** has its zero point at $-273°$C; temperatures in this scale are designated K. The **Rankine absolute temperature scale** has its zero point at $-460°$F; temperatures in this scale are designated °R. **Absolute zero** is $0\,\text{K} = 0°\text{R} = -273°\text{C} = -460°\text{F}$.

The equation $pV/T = $ constant, a combination of Boyle's and Charles's laws, is called the **ideal gas law** and is obeyed approximately by all gases.

According to the **kinetic theory of gases,** a gas consists of a great many tiny individual molecules that do not interact with one another except when collisions occur. The molecules are far apart compared with their dimensions and are in constant random motion. The ideal gas law may be derived from the kinetic theory of gases. The average kinetic energy of gas molecules is proportional to the absolute temperature of the gas. At absolute zero, gas molecules would have virtually no kinetic energy of translational motion.

The **relative humidity** of a volume of air is the ratio between the amount of water vapor it contains and the amount that would be present at saturation.

IMPORTANT FORMULAS

Celsius and Fahrenheit scales:

$$T_F = \tfrac{9}{5} T_C + 32°$$

$$T_C = \tfrac{5}{9}(T_F - 32°)$$

Thermal expansion: $\Delta L = aL_0 \Delta T$
$$\Delta V = bV_0 \Delta T$$

Boyle's law:

$$pV = \text{constant} \qquad (T = \text{constant})$$

Absolute temperature scale:

$$T_K = T_C + 273$$

Rankine scale:

$$T_R = T_F + 460°$$

Charles's law:

$$\frac{V}{T} = \text{constant} \qquad (p = \text{constant})$$

Ideal gas law:

$$\frac{pV}{T} = \text{constant}$$

Molecular kinetic energy: $\text{KE}_{av} = \tfrac{3}{2} kT$

MULTIPLE CHOICE

1. Two thermometers, one calibrated in the Celsius scale and the other in the Fahrenheit scale, are used to measure the same temperature. The numerical reading on the Fahrenheit thermometer
   a. is proportional to that on the Celsius thermometer.
   b. is greater than that on the Celsius thermometer.
   c. is less than that on the Celsius thermometer.
   d. may be greater or less than that on the Celsius thermometer.

2. An iron bar is heated from 20°C to 100°C.
   a. Its volume increases and its density decreases.
   b. Its volume increases and its density remains the same.
   c. Its volume increases and its density increases.
   d. Its volume remains the same and its density increases.

3. The volume of a gas sample is proportional to its
   a. Fahrenheit temperature.
   b. Celsius temperature.
   c. absolute temperature.
   d. pressure.

4. Absolute zero may be regarded as that temperature at which
   a. water freezes.
   b. all gases become liquids.

   c. all substances are solid.

   d. molecular motion in a gas would be the minimum possible.

5. A Celsius thermometer and an absolute thermometer are used to measure the temperature of the same gas sample. The readings on the thermometers are respectively T_C and T_K.

   a. T_C is smaller than T_K.

   b. T_C is larger than T_K.

   c. T_C is equal to T_K.

   d. Any of the above is true, depending on the gas temperature.

6. Which of the following formulas expresses the relationship between the pressure and absolute temperature of a gas sample whose volume is fixed?

   a. $\dfrac{p_1}{T_2} = \dfrac{p_2}{T_1}$ b. $\dfrac{p_1}{T_1} = \dfrac{p_2}{T_2}$

   c. $\dfrac{p_1}{p_2} = \dfrac{T_2}{T_1}$ d. $\dfrac{p_1}{T_1} = \dfrac{T_2}{p_2}$

7. Two elements *cannot* be combined chemically to make

   a. a compound.   b. another element.

   c. a gas.   d. a liquid.

8. The relative proportions of the elements in a compound

   a. may vary considerably.

   b. may vary only slightly.

   c. do not vary.

   d. may or may not vary, depending on the compound.

9. Which of the following statements is not correct?

   a. Matter is composed of tiny particles called molecules.

   b. These molecules are in constant motion.

   c. All molecules have the same size and mass.

   d. The differences between the solid, liquid, and gaseous states can be attributed to the relative freedom of motion of their respective molecules.

10. The number of known elements is approximately

   a. 50.

   b. 100.

   c. 200.

   d. 500.

11. Rust is an example of

   a. an element.

   b. a compound.

   c. a mixture.

   d. a solution.

12. Of the following, a molecule is best described as

   a. any very tiny particle.

   b. the smallest particle found in nature.

   c. the smallest particle of a substance that is representative of the substance.

   d. the ultimate particle of which all matter is composed.

13. The molecules of a compound

   a. are all the same.

   b. may be different.

   c. always consist of exactly two atoms.

   d. never contain more than one atom of a given kind.

14. Which of the following is not true of molecular motion in a gas?

   a. There is no order in the motion.

   b. There is no uniformity of speed or direction.

   c. There is a definite average speed at a given temperature.

   d. There is a definite average direction of motion at a given temperature.

15. The kinetic-molecular theory of gases predicts that, at a given temperature,

   a. all of the molecules in a gas have the same average speed.

   b. all of the molecules in a gas have the same average energy.

   c. light gas molecules have lower average energies than heavy gas molecules.

   d. light gas molecules have higher average energies than heavy gas molecules.

16. The volume of a gas is held constant while its temperature is raised. The pressure the gas exerts on the walls of its container increases because

   a. the masses of the molecules increase.

   b. each molecule loses more kinetic energy when it strikes the wall.

   c. the molecules are in contact with the wall for a shorter time.

   d. the molecules have higher speeds and strike the wall more often.

17. The temperature of a gas is held constant while its volume is reduced. The pressure the gas exerts on the walls of its container increases because its molecules

   a. strike the container walls more often.

   b. strike the container walls with higher speeds.

   c. strike the container walls with greater force.

   d. have more energy.

18. Dry air has a relative humidity of

   a. 0.

   b. 1%.

   c. 50%.

   d. 100%.

19. When saturated air is cooled,
   a. the relative humidity goes down.
   b. the relative humidity goes up.
   c. it becomes able to take up more water vapor.
   d. some of its water content condenses out.

20. The higher the temperature of a volume of air, the
   a. more water vapor it can hold.
   b. less waver vapor it can hold.
   c. greater its possible relative humidity.
   d. lower its possible relative humidity.

21. Oxygen boils at $-183°C$. This temperature is
   a. $-215°F$.
   b. $-297°F$.
   c. $-329°F$.
   d. $-361°F$.

22. A temperature of $100°F$ is almost exactly
   a. $38°C$.
   b. $56°C$.
   c. $122°C$.
   d. $212°C$.

23. A copper bar is 1 m long at $20°C$. At what temperature will it be shorter by 1 mm?
   a. $-17°C$.
   b. $-39°C$.
   c. $-59°C$.
   d. $-79°C$.

24. A silver bar increases in length by 0.1% when its temperature increases by $50°C$. The coefficient of linear expansion of silver is
   a. $2 \times 10^{-5}/°C$.
   b. $2.5 \times 10^{-5}/°C$.
   c. $5 \times 10^{-5}/°C$.
   d. $2 \times 10^{-4}/°C$.

25. A certain container holds 1 kg of air at atmospheric pressure. When an additional kg of air is pumped into the container, the new pressure is
   a. $\frac{1}{2}$ atm.
   b. 1 atm.
   c. 2 atm.
   d. 4 atm.

26. If the absolute pressure on 10 ft^3 of air is increased from 30 lb/in.^{2}. to 120 lb/in.^{2}, the new volume of the air will be
   a. 2.5 ft^3.
   b. 5 ft^3.
   c. 40 ft^3.
   d. 900 ft^3.

27. An absolute temperature of 100 K is the same as a Celsius temperature of
   a. $-173°C$.
   b. $32°C$.
   c. $212°C$.
   d. $373°C$.

28. The temperature of an object is increased by $50°C$. This is equivalent to an increase in its absolute temperature by
   a. 27.8 K.
   b. 50 K.
   c. 90 K.
   d. 323 K.

29. The boiling point of water on the Rankine scale is
   a. $-248°R$.
   b. $-61°R$.
   c. $485°R$.
   d. $672°R$.

30. A 1-L sample of neon at $0°C$ and 100 kPa pressure is compressed to 0.25 L and its temperature is increased to $273°C$. The new pressure of the gas is

   a. 50 kPa.
   b. 200 kPa.
   c. 400 kPa.
   d. 800 kPa.

31. The average speed of the molecules in a bottle of gas at the pressure p and absolute temperature T is doubled. The new pressure and temperature are, respectively,
   a. $2p$, $2T$.
   b. $2p$, $4T$.
   c. $4p$, $2T$.
   d. $4p$, $4T$.

32. A sample of hydrogen gas is compressed to half its original volume while its temperature is held constant. If the average speed of the hydrogen molecules was originally v, their new average speed is
   a. $4v$.
   b. $2v$.
   c. v.
   d. $\frac{1}{2}v$.

33. At which of the following temperatures would the molecules of a gas have twice the average kinetic energy they have at room temperature, $20°C$?
   a. $40°C$
   b. $80°C$
   c. $313°C$
   d. $586°C$

34. The mass of a nitrogen molecule is 14 times greater than that of a hydrogen molecule. The temperature of a sample of hydrogen whose average molecular energy is equal to that in a sample of nitrogen at 300 K is
   a. 6.5 K.
   b. 21 K.
   c. 300 K.
   d. 4200 K.

EXERCISES

14−1 Temperature

1. The melting point of lead is $330°C$ and its boiling point is $1170°C$. Express these temperatures on the Fahrenheit scale.

2. The normal temperature of the human body is $98.6°F$. What is this temperature on the Celsius scale?

3. At what temperature would Celsius and Fahrenheit thermometers give the same reading?

4. Mercury freezes at $-40°C$. What is this temperature on the Fahrenheit scale?

5. Dry ice (solid carbon dioxide) vaporizes at $-112°F$. What is this temperature on the Celsius scale?

14−2 Thermal Expansion

6. An aluminum nut is screwed tightly on a brass bolt. Should the assembly be heated or cooled to make it easier to remove the nut? What if a brass nut were screwed tightly on an aluminum bolt?

7. In the construction of a light bulb, wires are led through the glass at the base by means of airtight seals. If the wires were made of copper, what would happen when the light is turned on and the bulb heats up? What must be true for a wire to be successfully used for this purpose?

8. Verify that the force associated with the thermal expansion or contraction of a solid object depends on its cross-sectional area but not on its length.

9. A steel bridge is 100 m long at 30°C. What is its change in length when the temperature falls to −10°C?

10. How large a gap should be left between steel rails that are 10 m long when laid at 20°C if they are to just barely touch at 30°C?

11. A rod 2 m long expands by 1 mm when heated from 8°C to 70°C. What is the coefficient of linear expansion of the material from which the rod is made?

12. The outside diameter of a wheel is 1.000 m. An iron tire for this wheel has an inside diameter of 0.992 m at 20°C. To what temperature must the tire be heated in order for it to fit over the wheel?

13. A steel tape measure is calibrated at 70°F. A reading of 120 ft 0 in. is obtained when it is used to determine the width of a building at 0°F. What is the true width of the building at 0°F?

14. A 1-pt aluminum bottle is filled with ethyl alcohol at 40°F. To what temperature must the bottle be heated in order that 0.25 oz of alcohol overflow?

15. A Pyrex beaker is filled to the brim with 250 cm³ of glycerin at 15°C. How much glycerin overflows at 25°C?

16. A load of 4000 kg is placed on a vertical steel column 3 m long and cross-sectional area 50 cm² when the temperature is 20°C. To what should the temperature be increased if the length of the column is to be the same after the load is applied as it was originally?

17. An aluminum mast whose cross-sectional area is 0.05 ft² is 60 ft long at 60°F. (a) By how much does it increase in length when its temperature rises to 100°F? (b) How much force is associated with the expansion?

18. The density of lead is 11.0 g/cm³ at 20°C. Find its density at 200°C.

19. Vodka that is "100 proof" is a mixture of half ethyl alcohol and half water. How much profit per liter will a merchant make if he buys vodka at $10.00 per liter at 0°C and sells it at $10.00 per liter at 25°C?

20. An aluminum piston 80 mm in diameter moves inside a steel cylinder with a clearance of 0.1 mm at 20°C. At what temperature will the two have exactly the same diameter, so the piston can no longer move?

21. A concrete swimming pool 12 m × 6 m × 2.5 m is filled with water to within 6 mm of the top when the temperature is 10°C. The coefficient of linear expansion of the concrete used is 0.9×10^{-5}/°C. What will happen to the water level as the temperature increases? If it rises, at what temperature will the water begin to overflow?

22. A thermometer is being made from a glass tube whose inside diameter is 0.3 mm. The distance between the 0°C and 100°C marks is to be 300 mm. What volume (in cubic millimeters) should the mercury bulb at the end of the tube have?

23. The rails of a railway are sometimes welded together instead of being laid with gaps to allow for thermal contraction and expansion. What percentage of the ultimate strength of steel is the thermal stress associated with a temperature change of 40°C?

24. A sign is suspended from the middle of a steel cable attached to poles 20.00 m apart. When the temperature is 30°C, the cable sags so that the sign is 2.00 m below a horizontal line between the ends of the cable. Find the amount of sag on a winter day when the temperature is −10°C.

14−3 Boyle's Law

25. If the pressure on 10 m³ of nitrogen is increased from 200 kPa to 800 kPa, what will the new volume of the nitrogen be?

26. How much air at sea-level atmospheric pressure can be stored in the 350-L tank of an air compressor that can withstand an absolute pressure of 750 kPa?

27. An oxygen cylinder used for welding contains 1.3 ft³ of oxygen at 70°F and a gauge pressure of 2500 lb/in². Find the weight of the oxygen given that its density is 0.09 lb/ft³ at 70°F and 1 atm pressure.

28. The containers shown are filled with air at the same temperature and at the pressures indicated. What will the pressure in the system become when the valve is opened?

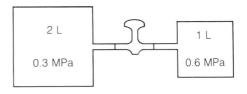

29. A flat tire contains 5 L of air at atmospheric pressure, which means an absolute pressure of very nearly 1 bar. How

many strokes of a pump whose capacity is 0.5 L are needed to raise the pressure in the tire to a gauge pressure of 2 bars? Assume the temperature and volume of the tire do not change in the process.

30. A glass tube closed at its lower end contains a 100-mm column of mercury above 100 mm of trapped air, as shown below. When the tube is inverted, the mercury column shifts so that the trapped air occupies 131 mm of the tube. Find the atmospheric pressure at the time these measurements were made.

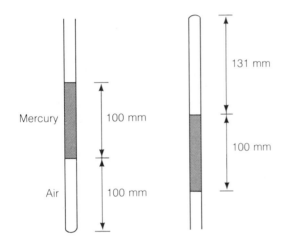

31. A bubble of methane ("marsh gas") doubles in volume when it rises from the bottom of a pond to the surface. If the water temperature is the same throughout the pond, how deep is it?

14−4   Charles's Law

14−5   Ideal Gas Law

32. Actual molecules attract one another slightly. Does this tend to increase or decrease gas pressures from values computed from the ideal gas law? Why?

33. At absolute zero, an ideal gas sample would occupy zero volume. Why would an actual gas not occupy zero volume at absolute zero?

34. (a) Starting from the ideal gas law, obtain an equation relating the pressure and temperature of a gas at constant volume. (b) A gas sample initially at 27°C is heated at constant volume until its pressure increases by 20%. Find its new temperature.

35. To what Celsius temperature must a gas sample initially at 20°C be heated if its volume is to double while its pressure remains the same?

36. An air tank used for scuba diving has a safety valve set to open at an absolute pressure of 280 bars. The normal absolute pressure of the full tank at 20°C is 200 bars. If the tank is heated after being filled to the latter pressure, at what temperature will the safety valve open?

37. The tires of a stationary car contain air at a gauge pressure of 24 lb/in.² and a temperature of 40°F. When the car is driven for a while, the temperature of the tires increases to 80°F. If the tire volumes remain unchanged, find their new gauge pressure.

38. A sample of gas occupies 100 cm³ at 0°C and 1 atm pressure. What is its volume (a) at 50°C and 1 atm pressure; (b) at 0°C and 2.2 atm pressure; (c) at 50°C and 2.2 atm pressure?

39. A sample of gas occupies 2 m³ at 300K and an absolute pressure of 2×10^5 Pa. (a) What is its pressure at the same temperature when it has been compressed to a volume of 1 m³? (b) What is its volume at the same temperature when its pressure has been decreased to 1.5×10^5 Pa? (c) What is its volume at a temperature of 400 K and a pressure of 2×10^5 Pa?

40. A sample of gas occupies 10 ft³ at 30°F and 1 atm pressure. What is its volume (a) at 120°F and 1 atm pressure; (b) at 30°F and 2.2 atm pressure; (c) at 120°F and 2.2 atm pressure?

41. In a certain diesel engine, air at 20°C and 1 atm pressure is drawn into each cylinder in turn and compressed by a piston to $\frac{1}{20}$ of its original volume and a pressure of 55 atm. Diesel oil is then injected into the cylinder and is ignited by the high temperature of the air. What is this temperature?

42. The density of air is 1.293 kg/m³ at 0°C and 1 atm pressure. Find its density at 100°C and 2 atm pressure.

43. A diver blows an air bubble 1 cm in diameter at a depth of 10 m in a freshwater lake where the temperature is 5°C. What is the diameter of the bubble when it reaches the surface of the lake where the temperature is 20°C?

14−6   Structure of Matter

44. Which of the following homogeneous liquids are elements, which are compounds, and which are solutions? Alcohol, mercury, liquid hydrogen, pure water, sea water, rum.

45. Which of the following substances are homogeneous and which are heterogeneous? Blood, carbon dioxide gas, solid carbon dioxide, rock, steak, iron, rust, concrete, air, oxygen, salt, milk.

46. Find the mass of a molecule of ethyl alcohol, C_2H_6O.

47. Find the mass of a molecule of glucose, $C_6H_{12}O_6$.

48. How many hydrogen atoms are present in 100 g of boric acid, H_3BO_3?

49. How many lead atoms are present in 50 g of lead?

14−7   Kinetic Theory of Gases

14−8   Molecular Motion in Gases

50. When they are close together, molecules attract one another slightly (the reason for this is discussed in Chap. 29). As a result of this attraction, are gas pressures higher or lower than expected from the ideal gas law?

51. Molecular speeds are comparable with those of rifle bullets, yet a gas with a strong odor, such as ammonia, takes a few minutes to diffuse through a room. Why?

52. According to the kinetic theory of gases, molecular motion virtually ceases only at absolute zero. How can this be reconciled with the definite shape and volume of a solid at temperatures well above absolute zero?

53. Is it meaningful to say that an object at a temperature of 200°C is twice as hot as one at 100°C?

54. What is the average kinetic energy of the molecules of a gas (a) at 0°C? (b) at 100°C?

55. To what temperature must a gas sample initially at 27°C be raised in order for the average energy of its molecules to double?

56. To what temperature must a gas sample initially at 27°C be raised in order for the average speed of its molecules to double?

57. The average speed of a hydrogen molecule at room temperature is about 1.6 km/s. What is the average speed of an oxygen molecule, whose mass is 16 times greater, at this temperature?

58. Consider the following gases: CO_2, UF_6, H_2, He, Xe, NH_3. (a) Which has the highest average molecular speed at a given temperature? (b) Which has the lowest average molecular speed?

59. (a) Find the average speed of carbon dioxide (CO_2) molecules at 0°C. (b) At what temperature would this speed be doubled?

60. Silver is a vapor at 1500 K. What is the average speed of silver atoms in a vapor at this temperature?

61. Two vessels of the same size are at the same temperature. One of them holds 1 kg of H_2 gas and the other holds 1 kg of N_2 gas. (a) Which vessel contains more molecules? How many times more? (b) Which vessel is under the greater pressure? How many times greater? (c) In which vessel are the average molecular speeds greater? How many times greater?

13−9   Molecular Motion in Liquids

62. What does it mean to say that a certain volume of air at 22°C has a relative humidity of 60%? What would happen to the relative humidity if the temperature of the air were increased? Decreased?

63. The air in a closed container is saturated with water vapor at 20°C. What is the relative humidity?

64. Why does the air in a heated room tend to be dry?

65. What is the approximate dew point of air at 25°C whose relative humidity is 75%?

66. Outdoor air at 0°C and relative humidity 50% is brought into a house and heated to 20°C. What is the relative humidity of the air in the house?

67. A basement pipe through which water at 5°C is flowing just begins to sweat. If the air in the basement is at 15°C, what is its relative humidity?

68. Moisture condenses on the outside of a glass that contains water at 5°C when it is brought into a room at 20°C. What is the minimum relative humidity of the air in the room?

ANSWERS TO MULTIPLE CHOICE

1. d	**8.** c	**15.** b	**22.** a	**29.** d
2. a	**9.** c	**16.** d	**23.** b	**30.** d
3. c	**10.** b	**17.** a	**24.** a	**31.** d
4. d	**11.** b	**18.** a	**25.** c	**32.** c
5. a	**12.** c	**19.** d	**26.** a	**33.** c
6. b	**13.** a	**20.** a	**27.** a	**34.** c
7. b	**14.** d	**21.** b	**28.** b	

15

HEAT

The higher the temperature of a body of matter, the faster its particles move in a random manner, and the more internal energy it contains. Heat may be thought of as internal energy in transit: When heat is added to a body, its temperature generally increases, and when heat is removed, its temperature generally decreases. The temperature change that goes with addition or removal of a given amount of heat varies with the mass and nature of the body. Changes of state—from solid to liquid, liquid to gas, solid to gas, and their reverses—involve changes in internal energy without changes in temperature as the atoms or molecules of the substance alter their relationships with one another.

15–1 INTERNAL ENERGY AND HEAT

Every body of matter contains a certain amount of internal energy in addition to any kinetic or potential energy it may have because of its motion or position. This internal

CHAPTER OBJECTIVES

Completing this chapter should enable you to:

1. Distinguish between temperature and heat.

2. Distinguish between internal energy and heat.

3. Use the kilocalorie, joule, kilowatt-hour, Btu, and ft · lb interchangeably as energy units.

4. Calculate the heat gain or loss in a given mass of water that corresponds to a certain change in its temperature.

5. Use the notion of specific heat capacity to relate the change in temperature of a given body of matter to the heat added to or removed from it.

6. Describe what is meant by heat of fusion and heat of vaporization.

7. Calculate the heat needed to melt a given mass of a certain solid or to vaporize a given mass of a certain liquid.

8. Describe the effects of pressure on changes of state.

9. Predict on the basis of its triple-point diagram whether a certain substance will be a solid, a liquid, or a vapor at a given temperature and pressure.

---

energy belongs to the individual atoms or molecules of which the body is composed and which are in constant random motion.

The total amount of internal energy a body contains depends upon its temperature, upon its composition, upon its mass, and upon its physical state (solid, liquid, or gas). However, the temperature of the body alone is what determines whether internal energy will be transferred from it to another body with which it is in contact, or vice versa. A large block of ice at 0°C has far more internal energy than a cup of hot water. However, when the water is poured on the ice some of the ice melts and the water becomes cooler, which means that energy has passed from the water to the ice (Fig. 15–1).

When the temperature of a body increases, it is customary to say that *heat* has been added to it; when the temperature of a body decreases, it is customary to say that heat has been removed from it. We can therefore define heat as follows:

Temperature difference determines direction of internal energy flow

Heat is internal energy in transit between two bodies of matter

FIG. 15–1 The internal energy of a body depends upon its temperature and mass. However, the direction of internal energy flow depends only on the temperatures of the bodies involved.

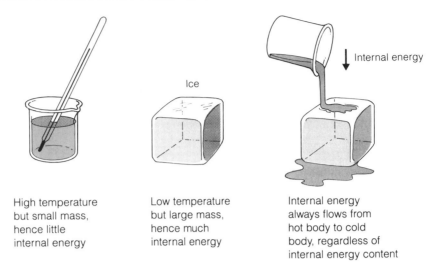

Ice

Internal energy

High temperature but small mass, hence little internal energy

Low temperature but large mass, hence much internal energy

Internal energy always flows from hot body to cold body, regardless of internal energy content

Heat is internal energy in transit from one body of matter to another by virtue of a temperature difference between them.

Changes of state (for instance, from ice to water or from water to steam) involve the transfer of heat to or from a body of matter without any change in its temperature. We should keep in mind that heat transfer is not the only way to change the temperature of a body of matter. A body that has work done on it may become hotter as a result, and a body that does work on something else may become cooler.

The term *heat* remains in the vocabulary of physics partly because of convenience and partly because of tradition. Temperature, on the other hand, is a unique concept both as an indicator of the direction of internal energy flow and as a measure of average molecular kinetic energy.

15–2   UNITS OF HEAT

The kilocalorie

Since heat is a form of energy, the correct SI unit of heat is the joule. However, two older units, the *kilocalorie* (kcal) and the *British thermal unit* (Btu) are still widely used, and for the time being it is necessary to be familiar with them. The kilocalorie is the amount of heat required to raise the temperature of 1 kg of water through 1°C. Similarly, 1 kcal of heat must be removed from 1 kg of water to reduce its temperature by 1°C. Because this amount of heat actually varies slightly with temperature, the kilocalorie is formally defined as the amount of heat involved in changing the temperature of 1 kg of water from 14.5°C to 15.5°C; the difference can be ignored for most purposes, however. The relationship between the joule and the kilocalorie is as follows:

$$1 \text{ J} = 2.39 \times 10^{-4} \text{ kcal} \qquad 1 \text{ kcal} = 4185 \text{ J}$$

The kilojoule and the kilowatt-hour

Because the joule is so small a unit, the kilojoule (kJ) is often used instead in heat calculations, where $1 \text{ kJ} = 10^3$ J. Another energy unit in common use is the kilowatt-hour (kWh), which is equivalent to the work done in 1 h by a 1 kW source of power. Hence

$$1 \text{ kWh} = (1000 \text{ J/s})(60 \text{ s/min})(60 \text{ min/h}) = 3,600,000 \text{ J}$$

We therefore have the following additional conversion factors:

$$1 \text{ kJ} = 0.239 \text{ kcal} = 2.78 \times 10^{-4} \text{ kWh}$$
$$1 \text{ kWh} = 3.60 \text{ MJ} = 860 \text{ kcal}$$
$$1 \text{ kcal} = 4.185 \text{ kJ} = 1.16 \times 10^{-3} \text{ kWh}$$

The *calorie* that dieticians use is the same as the kilocalorie. Thus the energy content of a 150-calorie cupcake is really 150 kcal. The carbohydrate content of foods averages 4.1 kcal/g, the protein content 4.2 kcal/g, and the fat content 9.3 kcal/g; pure ethanol (ethyl alcohol) has an energy content of 7.1 kcal/g. (A heat unit once widely used is also called the calorie. This is equal to the heat needed to raise the temperature of 1 g of water by 1°C, so that 1000 cal = 1 kcal. The dietician's calorie is sometimes written "Calorie" to distinguish it from the ordinary, smaller calorie.) **The dietician's calorie is the kcal**

Example   A woman eats a cupcake and proposes to work off its 150 kcal energy content by exercising with a 10-kg barbell. If each lift of the barbell from the floor to over her head is through 2.0 m and the efficiency of her body is 10% under these circumstances, how many times must she lift the barbell?

Solution   The work done in each lift of the barbell is

$$W_{\text{out}} = mgh = (10 \text{ kg})(9.8 \text{ m/s}^2)(2.0 \text{ m}) = 196 \text{ J}$$

At 10% efficiency the energy used is

$$W_{\text{in}} = \frac{W_{\text{out}}}{\text{Eff}} = \frac{196 \text{ J}}{0.10} = 1960 \text{ J} = 1.96 \text{ kJ}$$

which is equivalent to

$$W_{\text{in}} = (1.96 \text{ kJ})(0.239 \text{ kcal/kJ}) = 0.468 \text{ kcal}$$

The barbell must therefore be lifted

$$\frac{W_{\text{total}}}{W_{\text{in}}} = \frac{150 \text{ kcal}}{0.468 \text{ kcal}} = 320 \text{ times}$$ ■

The British thermal unit is the amount of heat needed to raise the temperature of 1 lb of water by 1°F; when 1 Btu of heat is removed from 1 lb of water, its temperature falls by 1°F. The Btu is related to the other units of energy and heat as follows: **The Btu**

$$1 \text{ Btu} = 1054 \text{ J} = 0.252 \text{ kcal} = 778 \text{ ft} \cdot \text{lb} = 2.93 \times 10^{-4} \text{ kWh}$$

We note that weight rather than mass is specified in defining the Btu. In practice this is not an important difference because what is being considered is the quantity of matter on the earth's surface without regard to its dynamical properties. All the formulas and procedures that follow can be used with British units by expressing temperature in degrees Fahrenheit and heat in Btu, and letting *m* represent weight in pounds.

15-3   SPECIFIC HEAT CAPACITY

Water changes least in temperature when heat is added or removed

Samples of other substances respond to the addition or removal of a given amount of heat with temperature changes greater than that of an equal mass of water. One kg of water increases in temperature by 1°C when 1 kcal of heat is added to it, but 1 kcal of heat increases the temperature of 1 kg of helium (its volume held constant) by 1.3°C, of 1 kg of ice by 2°C, and of 1 kg of gold by 33°C (Fig. 15–2).

The *specific heat capacity* (symbol *c*) of a substance refers to what we might think of as its thermal inertia:

Specific heat capacity

The specific heat capacity of a substance is the amount of heat that must be added or removed from a unit mass of it to change its temperature by 1°.

A high specific heat capacity means a relatively small change in temperature for a given change in internal energy content, just as a large mass means a relatively small acceleration when a given force is applied.

Units of specific heat capacity

When the kcal is being used as the heat unit, the corresponding unit of *c* is the kcal/kg · °C. Thus the specific heat capacity of water is $c = 1.00$ kcal/kg · °C since 1 kcal of heat is involved when 1 kg of water changes in temperature by 1°C. In the British system, the unit of *c* is the Btu/lb · °F. Because of the way the kcal and the Btu are defined, the numerical value of *c* for a substance is the same whether the kcal/kg · °C or the Btu/lb · °F is the unit; thus the specific heat capacity of water is 1.00 Btu/lb · °F.

In the SI system, the unit of *c* is, of course, the J/kg · °C, and in this system the specific heat capacity of water is 4185 J/kg · °C. Usually it is more convenient to use the kilojoule in connection with heat, in which case $c = 4.19$ kJ/kg · °C. Table 15–1 is a list of specific heats for some common substances. The actual values vary somewhat with temperature, and the ones given in the table represent averages.

FIG. 15–2 When 1 kcal of heat is added to 1 kg of each of the substances shown, their respective rises in temperature differ considerably.

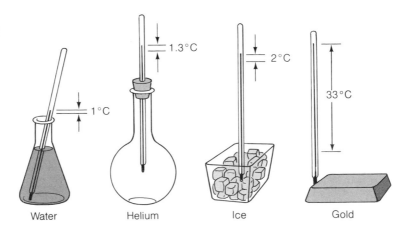

Water   Helium   Ice   Gold

Substance	Specific Heat Capacity (kcal/kg · °C) (Btu/lb · °F)	(kJ/kg · °C)
Air (1 atm)	0.17	0.70
Alcohol (ethyl)	0.58	2.43
Aluminum	0.22	0.92
Concrete	0.7	2.9
Copper	0.093	0.39
Glass	0.20	0.84
Gold	0.030	0.13
Granite	0.19	0.80
Human body	0.83	3.47
Ice	0.50	2.09
Iron and steel	0.11	0.46
Lead	0.030	0.13
Mercury	0.033	0.14
Silver	0.056	0.23
Steam	0.48	2.01
Water	1.00	4.19
Wood, pine	0.47	1.95

TABLE 15–1
Average specific heat capacities of various substances

With the help of specific heat capacity we can write a formula for the quantity of heat Q involved when a quantity m of a substance undergoes a change in temperature of ΔT. This formula is simply

Quantity of heat

$$Q = mc\,\Delta T \qquad \text{Quantity of heat} \quad (15-1)$$

Heat transferred = (mass)(specific heat capacity)(temperature change)

Example How much heat must be removed from 14 lb of aluminum in order to cool it from 80°F to 15°F?

Solution From Table 15–1 the specific heat capacity of aluminum is 0.22 Btu/lb · °F. Since $\Delta T = -65°F$ here,

$$Q = mc\,\Delta T = (14\,\text{lb})\left(0.22\frac{\text{Btu}}{\text{lb}\cdot°F}\right)(-65°F) = -200\,\text{Btu}$$

The minus sign means that this quantity of heat is to be removed to achieve the temperature change of $-65°F$. ∎

Example Water flows through a 3.5-kW instantaneous-type water heater at a rate of 1.5 L/min. If the initial water temperature is 20°C and all the heat produced is added to the water, find the temperature of the hot water.

Solution Since 3.5 kW = 3.5 kJ/s, the heat provided to the water in 1 min is

$$Q = (3.5\,\text{kJ/s})(60\,\text{s}) = 210\,\text{kJ}$$

The mass of 1 L of water is 1 kg, so the mass of water that flows through the heater per minute is $m = 1.5$ kg. From Eq. (15–1) we have for the rise in temperature

20°C

1.5 L/min.

3.5 kW

53°C

FIG. 15–3

$$\Delta T = \frac{Q}{mc} = \frac{210\,kJ}{(1.5\,kg)(4.19\,kJ/kg \cdot °C)} = 33°C$$

The final water temperature is 20°C + 33°C = 53°C (Fig. 15–3).   ∎

Example   If 0.20 kg of coffee at 90°C is poured into a 0.30-kg cup at 20°C, and we assume that no heat is transferred to or from the outside, what is the final temperature of the coffee?

Solution   We shall take the specific heat of coffee to be that of water and the specific heat of the cup to be that of glass. To solve the problem, we begin by noting that

$$Q_{gained} = Q_{lost}$$

Heat gained by cup = heat lost by coffee

In this formula, both Q_{gained} and Q_{lost} are considered positive quantities. This means that the ΔT values on both sides must also be positive quantities, and we accordingly proceed as follows. If the final temperature of both coffee and cup is T, then the heat gained by the cup is

$$Q_{gained} = m_{cup}\, c_{cup}\, (T - 20°C) = (0.30\,kg)\left(0.20\,\frac{kcal}{kg \cdot °C}\right)(T - 20°C)$$

$$= (0.06\,T - 1.2)\,kcal$$

The heat lost by the coffee is

$$Q_{lost} = m_{coffee}\, c_{coffee}\, (90°C - T)$$

$$= (0.20\,kg)\left(1.0\,\frac{kcal}{kg \cdot °C}\right)(90°C - T) = (18 - 0.20\,T)\,kcal$$

Now we set the heat gained by the cup to the heat lost by the coffee and solve for T:

$$Q_{gained} = Q_{lost}$$
$$0.06\,T - 1.2 = 18 - 0.20\,T$$
$$0.26\,T = 19.2$$
$$T = 74°C$$

The temperature of the coffee drops by 16°C as it warms the cup (Fig. 15–4). Evidently it is necessary to preheat the cup if one wants really hot coffee.   ∎

FIG. 15–4

90°C   +   20°C   =   74°C

15-4 CHANGE OF STATE

Not always does the addition or removal of heat from a sample of matter lead to a change in its temperature. Instead the sample may change its state from solid to liquid or from liquid to gas when heat is added, or it may change from gas to liquid or from liquid to solid when heat is taken away. Such changes of state take place at definite temperatures for most substances at any given pressure, but for a few (glass or wax, for instance) there is only a gradual softening or hardening over a range of temperatures. Substances of the latter kind are not true solids, however; their structures are really those of liquids, and their hardness at room temperature is a kind of exaggerated viscosity.

Changes of state involve changes in internal energy

Figure 15-5 shows what happens when we add heat at a constant rate to 1 kg of ice that is initially at −50°C. The specific heat capacity of ice is 0.5 kcal/kg · °C, and so 25 kcal is needed to bring the ice to 0°C.

At 0°C the ice begins to melt. The temperature remains constant until all the ice has melted, which requires a total of 80 kcal. Thus 80 kcal/kg is the *heat of fusion* of water: The amount of heat needed to convert 1 kg of ice into 1 kg of water at its melting point of 0°C.

The melting of ice

When all the ice has turned to water, the temperature goes up once more as further heat is supplied. Since the specific heat of water is 1 kcal/kg · °C, there is now a rise of 1°C per kcal of heat. This rate of change is less than that of ice, since the specific heat of water is greater than that of ice, and so the slope of the graph is less steep.

When 100°C is reached, the water begins to turn into steam. The temperature stays constant until a total of 540 kcal is added, at which time all the water has become steam. Thus 540 kcal/kg is the *heat of vaporization* of water: the amount of heat needed to convert 1 kg of water into 1 kg of steam at its boiling point of 100°C (at atmospheric pressure).

The boiling of water

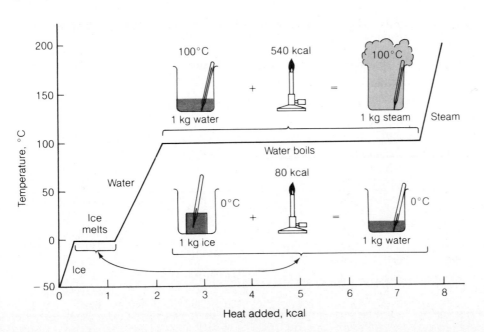

FIG. 15-5 A graph of the temperature of 1 kg of water, initially ice at −50°C, as heat is added to it.

After the water has become steam, its temperature rises again. The specific heat of steam is 0.48 kcal/kg · °C, so the temperature increase is 2.1°C per kcal of heat, and the slope of the graph is therefore steeper than it was for ice or water.

Heat of fusion

In general, the *heat of fusion* of a substance is the amount of heat that must be supplied to change 1 kg (or 1 lb) of the substance at its melting point from the solid to the liquid state; the same amount of heat must be removed from 1 kg (or 1 lb) of the substance in the liquid state at its melting point to change it to a solid. The usual symbol for heat of fusion is L_f.

Heat of vaporization

The *heat of vaporization* of a substance is the amount of heat that must be supplied to change 1 kg (or 1 lb) of the substance at its boiling point from the liquid to the gaseous (or vapor) state; the same amount of heat must be removed from 1 kg (or 1 lb) of the substance in the gaseous state at its boiling point to change it into a liquid. The usual symbol for heat of vaporization is L_v. The heats of fusion and vaporization for a number of substances are listed in Table 15–2 together with their melting and boiling points. To convert a heat of fusion or heat of vaporization expressed in kcal/kg to its equivalent in Btu/lb, we note that 1 kcal/kg = 1.80 Btu/lb.

Why steam is dangerous

The high heat of vaporization of water is what makes steam so dangerous. If a gram of water (about a third of a teaspoonful) at its boiling point of 100°C falls on a person's skin, the heat given to the skin will be

$$Q = mc\Delta T = (0.001 \text{ kg})(1.00 \text{ kcal/kg} \cdot °C)(100°C - 37°C) = 0.063 \text{ kcal}$$

since body temperature is 37°C. However, if a gram of steam strikes the person's skin, the heat of vaporization given up as the steam condenses into water will be

$$Q = mL_v = (0.001 \text{ kg})(540 \text{ kcal/kg}) = 0.540 \text{ kcal}$$

for a total of 0.603 kcal—nearly ten times more heat. Great care must always be taken when working with steam.

TABLE 15–2
Heats of fusion and vaporization and melting and boiling points of various substances at atmospheric pressure

Example   How much steam is produced when 1000 Btu of heat is added to 2.00 lb of water at 120°F?

Solution   The heat needed to raise the temperature of the water from 120°F to its boiling point of 212°F is, since $\Delta T = 212°F - 120°F = 92°F$,

$$Q_1 = mc\Delta T = (2.00 \text{ lb})(1 \text{ Btu/lb} \cdot °F)(92°F) = 184 \text{ Btu}$$

Substance	*Melting Point* °C	°F	*Heat of Fusion, L_f* kcal/kg	kJ/kg	Btu/lb	*Boiling Point* °C	°F	*Heat of Vaporization, L_v* kcal/kg	kJ/kg	Btu/lb
Alcohol (ethyl)	−114	−173	25	105	45	78	172	204	854	367
Lead	330	626	5.9	25	10.6	1170	2138	175	732	315
Mercury	−39	−38	2.8	12	5.0	358	676	71	297	128
Nitrogen	−210	−346	6.1	26	11	−196	−320	48	201	86
Oxygen	−219	−362	3.3	14	5.9	−183	−306	51	213	92
Silver	961	1762	21	88	38	2193	3979	558	2335	1004
Water	0	32	80	335	144	100	212	540	2260	972

FIG. 15-6

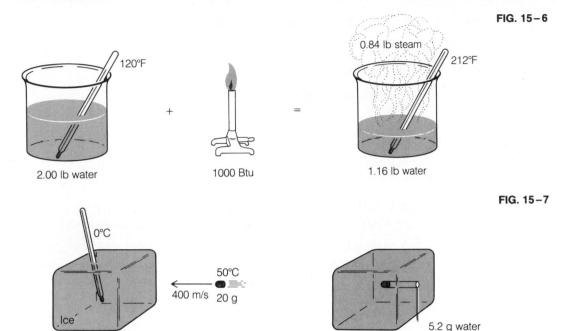

FIG. 15-7

Thus $Q_2 = 1000$ Btu $- Q_1 = 816$ Btu of heat is available to convert water at 212°F to steam at the same temperature. From

$$Q_2 = m_{\text{steam}} L_v$$

we have

$$m_{\text{steam}} = \frac{Q_2}{L_v} = \frac{816\,\text{Btu}}{972\,\text{Btu/lb}} = 0.84\,\text{lb}$$

Of the 2.00 lb of water we started with, 0.84 lb becomes steam at 212°F and 1.16 lb remains as water at 212°F (Fig. 15–6). (As before, for the sake of convenience we use m in the formulas for heat with British units even though weights are specified rather than masses when using this system of units.) ∎

Example A 20-g lead bullet whose temperature is 50°C and whose speed is 400 m/s strikes a large block of ice at 0°C and stops inside it (Fig. 15–7). Assuming that all the energy lost by the bullet goes into melting ice, how much ice melts?

Solution This problem is most easily worked out with the kJ as the heat unit, since mechanical and heat quantities are both involved. As it cools to 0°C the bullet loses the heat

$$Q_{\text{lost}} = mc\Delta T = (0.02\,\text{kg})(0.13\,\text{kJ/kg}\cdot\text{°C})(50\text{°C}) = 0.13\,\text{kJ}$$

The kinetic energy lost by the bullet in coming to a stop is

$$KE_{\text{lost}} = \tfrac{1}{2}mv^2 = \tfrac{1}{2}(0.02\,\text{kg})(400\,\text{m/s})^2 = 1600\,\text{J} = 1.60\,\text{kJ}$$

If M is the mass of ice that melts, the heat gained by the ice is

$$Q_{\text{gained}} = ML_f = (M)(335 \text{ kJ/kg}) = 335M \text{ kJ/kg}$$

From conservation of energy,

$$\text{Heat lost by bullet} + \text{KE lost by bullet} = \text{heat gained by ice}$$

$$Q_{\text{lost}} + \text{KE}_{\text{lost}} = Q_{\text{gained}}$$

$$0.13 \text{ kJ} + 1.60 \text{ kJ} = 335M \text{ kJ/kg}$$

$$M = \frac{(0.13 + 1.60) \text{ kJ}}{335 \text{ kJ/kg}} = \frac{1.73 \text{ kJ}}{335 \text{ kJ/kg}} = 5.2 \times 10^{-3} \text{ kg} = 5.2 \text{ g} \quad \blacksquare$$

Example   What is the minimum amount of ice at $-10°C$ that must be added to 0.50 kg of water at 20°C in order to bring the temperature of the water down to 0°C?

Solution   In a problem like this, it does not matter whether we use the kcal or the kJ as the unit of heat. If we use the kcal, then the heat Q_1 absorbed by the unknown mass of ice in going from $-10°C$ to its melting point of 0°C is

$$Q_1 = m_{\text{ice}} c_{\text{ice}} \Delta T_{\text{ice}} = m_{\text{ice}} \left(0.50 \frac{\text{kcal}}{\text{kg} \cdot °\text{C}} \right)(10°C) = m_{\text{ice}} (5 \text{ kcal/kg})$$

and the heat Q_2 absorbed by the ice in melting at 0°C is

$$Q_2 = m_{\text{ice}} L_{f\,\text{ice}} = (m_{\text{ice}})(80 \text{ kcal/kg})$$

Hence the total heat absorbed by the ice is

$$Q_{\text{gained}} = Q_1 + Q_2 = (5 + 80) m_{\text{ice}} \text{ kcal/kg}$$

The heat lost by the water in cooling from 20°C to 0°C is

$$Q_{\text{lost}} = m_{\text{water}} c_{\text{water}} \Delta T_{\text{water}} = (0.50 \text{ kg}) \left(1.0 \frac{\text{kcal}}{\text{kg} \cdot °\text{C}} \right)(20°C) = 10 \text{ kcal}$$

FIG. 15–8

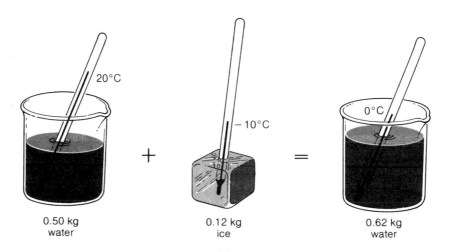

0.50 kg
water

+

0.12 kg
ice

=

0.62 kg
water

Setting the heat gained equal to the heat lost and then solving for m_{ice} yields (Fig. 15–8)

$$Q_{gained} = Q_{lost}$$

$$85m_{ice} \, kcal/kg = 10 \, kcal$$

$$m_{ice} = 0.12 \, kg$$

Under certain circumstances most substances can change directly from the solid to the vapor state, or vice versa. Both processes are called *sublimation*. For example, "dry ice" (solid carbon dioxide) evaporates directly to gaseous carbon dioxide at temperatures above $-78.5°C$ and does not pass through the liquid state. With the exception of carbon dioxide and a few other substances, however, sublimation does not occur except at pressures well below that of the atmosphere.

Sublimation refers to direct change from solid to vapor or vapor to solid

15–5 THE TRIPLE POINT

Changes of state are affected by pressure. Figure 15–9 shows how the boiling point of water varies with pressure: the higher the pressure, the higher the boiling point. This is the principle that underlies the pressure cooker. By heating water in a sealed container, the pressure can be raised well above 1 atm, and the temperature at which the water inside boils will be correspondingly higher than 100°C. In this way food can be cooked more rapidly than in an open pan. At a pressure of 2 atm, for instance, water boils at 120°C. The converse is also true: Lowering the pressure below 1 atm reduces the boiling point below 100°C. Atmospheric pressure decreases with altitude, so the boiling point of water is lower than 100°C in regions located above sea level. Water boils at 96°C in Denver, for instance.

The boiling point of water increases with pressure

The upper limit of the *vaporization curve* of Fig. 15–9, which occurs at a temperature of 374°C and a pressure of 218 atm, is known as the *critical point*. A substance cannot exist in the liquid state at a temperature above that of its critical point, regardless of how great the pressure may be. The gas becomes more and more dense with increas-

Critical point

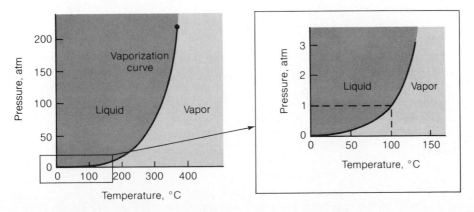

FIG. 15–9 The vaporization curve of water, showing how the boiling point of water varies with pressure. Above the temperature of its critical point a substance cannot exist in the liquid state.

FIG. 15–10   The fusion curve of water, showing how the melting point of ice varies with pressure

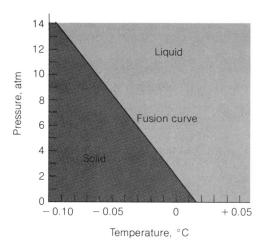

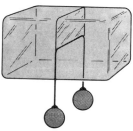

FIG. 15–11   Because the melting point of ice decreases with pressure, a thin wire with weights at both ends can melt its way through a block of ice.

ing pressure but does not condense into a liquid with a definite interface between liquid and gas. Helium has the lowest critical temperature, $-268°C$, and is therefore a gas at all temperatures above that. The critical temperature for water is 374°C.

The melting points of solids also depend upon pressure, although to a smaller extent than the boiling points. The variation of the melting point of ice with pressure is shown in the *fusion curve* of Fig. 15–10. Ice, together with gallium and bismuth, is unusual in that its melting point *decreases* with increasing pressure; the melting points of all other substances increase with increasing pressure. Hence it is possible to melt ice by applying pressure to it as well as by heating it (Fig. 15–11). An ice skater makes

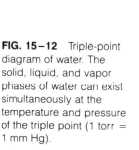

FIG. 15–12   Triple-point diagram of water. The solid, liquid, and vapor phases of water can exist simultaneously at the temperature and pressure of the triple point (1 torr = 1 mm Hg).

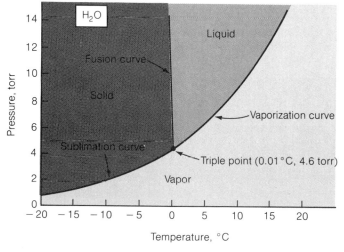

use of this fact in an interesting way. His entire weight is supported by skate blades of very small area, and the resulting pressure on the ice may exceed 1000 atm. The ice under the blades melts because of the great pressure, which creates a thin film of water that acts as an efficient lubricant. On unusually cold days even such pressures may not be sufficient to melt the ice, and skating then becomes impossible.

The fusion and vaporization curves of water intersect at a temperature of 0.01°C and a pressure of 4.6 torr, as shown on the combined plot of Fig. 15–12. Along the fusion curve both ice and water can simultaneously exist, and along the vaporization curve both water and water vapor can simultaneously exist. Under conditions corresponding to those of the intersection of the two curves, the solid, liquid, and vapor states of water can all exist together. This intersection is accordingly called the *triple point* of water.

At its triple point, the solid, liquid, and vapor states of a substance can exist together

At pressures below that of its triple point, no substance can exist as a liquid. The dividing line on a pressure-temperature graph between the solid and vapor states is called the *sublimation curve,* since it represents the conditions required for a solid to vaporize directly or a vapor to solidify directly. At atmospheric pressure the addition of heat causes ordinary ice to melt, since the triple point of water lies well below 1 atm. However, the addition of heat at that pressure causes solid carbon dioxide to sublime, since its triple point lies above 1 atm (Fig. 15–13).

Sublimation curve

At a temperature below that of its triple point, a substance passes directly from the solid state to the vapor state if the pressure is sufficiently low. This phenomenon is the basis of the *freeze-drying* process widely used for the preservation of foods, blood plasma, and biological samples. The usual procedure is to cool the material below the triple point of water in a gas-tight chamber, and then to evacuate the chamber with a vacuum pump. Freeze drying affects the structure of a material of biological origin less than other methods of dehydration do.

Freeze drying

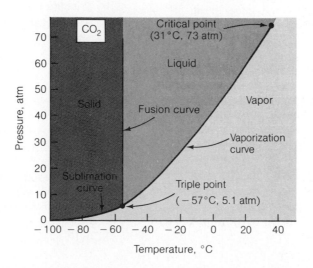

FIG. 15–13 Triple-point diagram of carbon dioxide.

IMPORTANT TERMS

Heat is internal energy in transit from one body of matter to another by virtue of a temperature difference between them. If a body of matter does not change state during the addition or removal of heat, and neither does work nor has work done on it, the change in its internal energy results in a corresponding change in temperature. The SI unit of heat is the joule. Another unit of heat in common use is the **kilocalorie** (kcal), which is the amount of heat required to change the temperature of 1 kg of water by 1°C. The unit of heat in the British system is the **British thermal unit** (Btu), which is that amount of heat required to change the temperature of 1 lb of water by 1°F.

The **specific heat capacity** of a substance is the amount of heat required to change the temperature of 1 kg (or 1 lb) of it by 1°.

The **heat of fusion** of a substance is the amount of heat that must be supplied to change 1 kg (or 1 lb) of it at its melting point from the solid to the liquid state; the same amount of heat must be removed from the substance in the liquid state at its melting point to change it to a solid.

The **heat of vaporization** of a substance is the amount of heat that must be supplied to change 1 kg (or 1 lb) of it at its boiling point from the liquid to the gaseous (or vapor) state; the same amount of heat must be removed from the substance at its boiling point to change it into a liquid.

Sublimation is the direct conversion of a substance from the solid to the vapor state, or vice versa, without its first becoming a liquid.

The **critical point** is the upper limit of the vaporization curve of a substance; a substance cannot exist in the liquid state at a temperature above that of its critical point.

The **triple point** is the intersection of the vaporization, fusion, and sublimation curves of a substance. All three states of a substance may exist in equilibrium at the temperature and pressure of its triple point.

IMPORTANT FORMULAS

Heat and temperature change: $Q = mc \, \Delta T$

Heat and change of state: $Q = mL$

MULTIPLE CHOICE

1. Two blocks of lead, one twice as heavy as the other, are both at 50°C. The ratio of the internal energy of the heavier block to that of the lighter block is

a. $\frac{1}{2}$.

b. 1.

c. 2.

d. 4.

2. Of the following substances, the one that requires the greatest amount of heat per kilogram for a given increase in temperature is

a. ice.

b. water.

c. steam.

d. copper.

3. A cup of hot coffee can be cooled by placing a cold spoon in it. A spoon of which of the following materials would be most effective for this purpose, assuming the spoons all have the same mass?

a. aluminum

b. copper

c. iron

d. silver

4. When a vapor condenses into a liquid,

a. it absorbs heat.

b. it evolves heat.

c. its temperature rises.

d. its temperature drops.

5. The freezing point of a substance is always lower than its

a. melting point.

b. boiling point.

c. heat of fusion.

d. heat of vaporization.

6. The heat of vaporization of a substance is

a. less than its heat of fusion.

b. equal to its heat of fusion.

c. greater than its heat of fusion.

d. any of the above, depending on the nature of the substance.

7. Water does not have an unusually high

a. boiling point.

b. heat of fusion.

c. heat of vaporization.

d. specific heat capacity.

8. The freezing point of water is 0°C. Its melting point is

a. slightly less than 0°C.

b. 0°C.

c. slightly more than 0°C.

d. 32°C.

9. One gram of steam at 100°C causes a more serious burn than 1 g of water at 100°C because the steam

a. is less dense.

b. strikes the skin with greater force.

c. has a higher specific heat capacity.

d. contains more energy.

10. Sublimation refers to

a. the vaporization of a solid without first becoming a liquid.

b. the melting of a solid.

c. the vaporization of a liquid.

d. the condensation of a gas into a liquid.

11. A pressure cooker cooks food more rapidly than an ordinary pot with a loose lid because

a. the pressure forces heat into the food.

b. the higher pressure lowers the boiling point of water.

c. the higher pressure raises the boiling point of water.

d. the higher pressure increases the specific heat capacity of water.

12. Ice skates are sharpened in order to

a. increase the pressure on the ice and thereby lower its melting point.

b. increase the pressure on the ice and thereby raise its melting point.

c. decrease the pressure on the ice and thereby lower its melting point.

d. decrease the pressure on the ice and thereby raise its melting point.

13. Under conditions corresponding to its triple point, a substance

a. is in the solid state.

b. is in the liquid state.

c. is in the gaseous state.

d. may be in any or all of the above states.

14. In an hour the heat produced by a 1-kW heating element equals

a. 3.4 Btu.   b. 1054 Btu.

c. 3416 Btu.   d. 3600 Btu.

15. A 61-kg woman eats a banana whose energy content is 100 kcal. If this energy were used to raise her from the ground, her approximate height would be

a. 0.7 m.   b. 7 m.

c. 70 m.   d. 700 m.

16. Three hundred kcal of heat is added to 20 kg of water at 10°C. The final temperature of the water is

a. 11.5°C.   b. 15°C.

c. 25°C.   d. 150°C.

17. Thirty kilojoules of heat is removed from 1.4 kg of ice initially at −10°C. Its new temperature is

a. −10.2°C.   b. −20.2°C.

c. −30°C.   d. −98°C.

18. Hot water at 100°C is added to 300 g of water initially at 0°C until the mixture is at 40°C. The minimum mass of added hot water is

a. 60 g.   b. 75 g.

c. 120 g.   d. 200 g.

19. Fifty kJ of heat is added to a 10-kg piece of lead at its

melting point of 330°C and 2 kg of lead melts. The heat of fusion of lead is

a. 2.5 kJ/kg.   b. 3.3 kJ/kg.

c. 5 kJ/kg.   d. 25 kJ/kg.

20. Fifty Btu of heat is added to a 50-lb block of ice at 32°F. The amount of ice that melts is

a. 0.347 lb.   b. 0.625 lb.

c. 1.0 lb.   d. 25 lb.

21. If 1 kg of punch of specific heat capacity 3.4 kJ/kg · °C at a temperature of 5°C is poured into a 1-kg glass punch bowl that is at 20°C, the final temperature of the punch is

a. 6°C.   b. 8°C.

c. 10°C.   d. 12°C.

22. If 10 kg of ice at 0°C is added to 2 kg of steam at 100°C, the temperature of the resulting mixture is

a. 0°C.   b. 23°C.

c. 28°C.   d. 40°C.

23. A 1-kg lead bar at 80°C is placed in 2 kg of water at 20°C. The final temperature of the lead bar is

a. 22°C.   b. 28°C.

c. 40°C.   d. 50°C.

24. Ten lb of ice at 0°F is added to 100 lb of water at 50°F. The temperature of the resulting mixture is

a. 19°F.   b. 31°F.

c. 32°F.   d. 34°F.

25. Water at 50°C can be obtained by mixing together which one or more of the following combinations?

a. 1 kg of ice at 0°C and 1 kg of steam at 100°C

b. 1 kg of ice at 0°C and 1 kg of water at 100°C

c. 1 kg of water at 0°C and 1 kg of steam at 100°C

d. 1 kg of water at 0°C and 1 kg of water at 100°C

26. Fifteen megajoules of heat is removed from 5 kg of steam initially at 200°C. The steam afterward is in the form of

a. a mixture of steam and water.

b. water.

c. a mixture of water and ice.

d. ice.

EXERCISES

15–1   Internal Energy and Heat

15–2   Units of Heat

15–3   Specific Heat Capacity

1. A jar of water is shaken vigorously. What becomes of the work that is done?

2. Why will the engine of a car whose cooling system is filled with an alcohol antifreeze be more likely to overheat in summer than one whose cooling system is filled with water?

3. If 200 g of water is to be heated from 15°C to 100°C to make a cup of tea, how much heat is needed?

4. Two lb of water is to be heated from 70°F to 212°F to make a pot of coffee. How much heat is needed?

5. Twenty kilojoules of heat is added to a 400-g piece of wood at 20°C. The final temperature of the wood is 50°C. Find its specific heat capacity.

6. How much heat must be added to 1 g of silver to raise its temperature from −5°C to 65°C?

7. If 10^4 kcal of heat is removed from a metric ton (10^3 kg) of iron at 300°C, what is its final temperature?

8. Eight Btu of heat is added to a 10-lb copper bar at 60°F. Find its final temperature.

9. How much heat must be removed from 60 lb of ice at 20°F to lower its temperature to −20°F?

10. The specific heat capacity of air at constant pressure is approximately 1.0 kJ/kg·°C. Find the heat needed to raise the temperature of the air in a room 4 m × 5 m × 3 m from 5°C to 20°C.

11. A 25-kg storage battery has an average specific heat of 0.84 kJ/kg·°C. When fully charged, the battery contains 1.4 MJ of electrical energy. If all of this energy were dissipated within the battery, find the increase in its temperature.

12. In an effort to lose weight, a person runs 5 km per day at a speed of 4 m/s. While running, the person's body processes consume energy at a rate of 1.4 kW. Fat has an energy content of about 40 kJ/g. How much fat is metabolized during each run?

13. Another person decides to lose weight by eating only cold food. A 100-g piece of apple pie yields about 350 kcal of energy when eaten. If its specific heat capacity is 0.4 kcal/kg·°C, how much greater is its energy content at 50°C than at 20°C? What percentage difference is this?

14. A 150-L water heater is rated at 8 kW. If 20% of the heat escapes, how long does the heater take to raise the temperature of 150 L of water from 10°C to 60°C?

15. A 50-gal water heater is rated at 55,000 Btu/h. If 20% of the heat escapes, how long does the heater take to raise the temperature of 50 gal of water from 50° to 140°F?

16. A 60-kg woman does 1.2 MJ of work on a certain day at an average efficiency of 15%. (a) How many kcal of energy must be provided by her food if she is to neither store energy nor use stored energy on that day? (b) If she were so

well insulated that only 90% of the waste heat is lost to the outside world, what would her temperature be at the end of the day? Normal body temperature is 37°C.

17. Radiant energy from the sun arrives at the earth at the rate of about 1.4 kW per square meter of surface perpendicular to the sun's rays. On a clear day a reflector 1 m^2 in area is used to concentrate sunlight on a 1-kg lead bar initially at 20°C. How long will it take the lead to reach its melting point of 330°C under the assumption that it absorbs energy at the rate of 0.5 kW?

18. A space vehicle returning to earth is first slowed by its motors to a speed of 450 m/s and then is slowed further to 100 m/s by air resistance in the atmosphere. If its average specific heat capacity is 0.6 kJ/kg·°C, find the increase in its temperature on the assumption that half the heat produced is absorbed by the vehicle and that this heat is distributed uniformly throughout its volume.

19. An essential part of a residential solar heating system is a way to store heat for use at night and on cloudy days. In a particular system, the water used for storage is initially at 160°F and it is required to provide an average of 25,000 Btu/h to keep the house at an average of 65°F for three days. (a) What is the minimum amount of water required? (b) What volume will it occupy? (c) How many 55-gal drums would be needed to hold the water?

20. Air is blown past the 2-kW element of an electric heater at a rate of 0.05 m^3/s. If the density of air is 1.3 kg/m^3 and its specific heat capacity is 1 kJ/kg·°C, what is the increase in temperature of the air that passes through the heater?

21. A 300-g aluminum pot containing 0.5 L of water is placed on the 1-kg iron hot plate of an electric stove. The heating element of the hot plate is rated at 1.5 kW. If the hot plate, the pot, and the water are all at 20°C when the heating element is switched on, how long will it take until the water begins to boil? Neglect heat losses to the room.

22. The 30-kg copper tank of a 5-kW water heater has a capacity of 100 kg of water. The water temperature in the tank is 15°C when it is switched on and is 50°C an hour later. What percentage of the energy input was lost?

23. A bathtub contains 70 kg of water at 26°C. If 10 kg of water at 90°C is poured in, what is the final temperature of the mixture?

24. A 0.1-kg piece of silver is taken from a bath of hot oil and placed in a 0.08-kg glass jar containing 0.2 kg of water at 15°C. The temperature of the water increases by 8°C. What was the temperature of the oil?

25. A 0.6-kg copper container holds 1.5 kg of water at 20°C. A 0.1-kg iron ball at 120°C is dropped into the water. What is the final temperature of the water?

26. (a) Show that the increase in length of a metal bar when it absorbs an amount of heat Q does not depend on the original length of the bar. (b) How much heat in kJ must be added to an iron bar 4 mm in diameter if its length is to increase by 1 mm?

27. A 1200-kg car whose transmission is in neutral is coasting down a 10° hill with its brakes being used to maintain a constant speed. If the total mass of the car's iron brake drums is 20 kg, find the increase in their temperature per 100 m of travel. Assume that the only friction acting occurs in the brakes and that all the heat produced is absorbed by the drums, so the resulting figure is a maximum.

15–4 Change of State

28. Answer the following questions about the substance whose heating curve is shown. (a) What are its melting and boiling points? (b) Along what portions of the curve is it a solid, a liquid, and a gas? (c) Along what portion of the curve is it a mixture of solid and liquid? (d) Along what portion of the curve is it a mixture of liquid and gas?

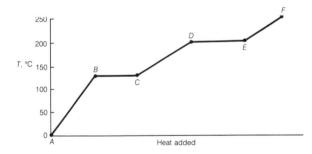

29. Which is more effective in cooling a drink, 10 g of water at 0°C or 10 g of ice at 0°C?

30. When a certain quantity of a vapor condenses into a liquid, what happens to its internal energy content and to its temperature?

31. How much heat (in kilocalories) is needed to change 4 kg of water at 10°C into ice at −10°C?

32. How much heat (in British thermal units) is needed to change 3 lb of water at 80°F to steam at 250°F?

33. How much steam is produced when 1 MJ of heat is added to 1 kg of water at 50°C?

34. One hundred kilojoules of heat is added to 600 g of zinc at 20°C and 75 g of it melts. If the specific heat capacity of zinc is 0.385 kJ/kg·°C and it melts at 420°C, find its heat of fusion.

35. One way to produce fresh water from seawater is to boil seawater and condense the steam, which leaves the salt behind. (a) Find the minimum amount of heat (in joules) needed to produce 1 L of fresh water in this way from seawater at 18°C. Assume that the specific heat capacity and heat of vaporization of seawater are the same as those of fresh water. (b) At what rate (in megawatts) must energy be supplied to produce 100,000 L of fresh water per day by this method?

36. Banana ice cream freezes at −2°C and has a heat of fusion of 290 kJ/kg. Its specific heat capacity is 3.3 kJ/kg·°C when melted and 1.9 kJ/kg·°C when frozen. How much refrigeration capacity, in kilowatts, is needed to make 500 kg of banana ice cream at −12°C in an 8-h working day from a mix initially at 18°C?

37. If all the heat lost by 1 kg of water at 0°C when it turns into ice at 0°C could be turned into kinetic energy, what would the speed of the ice be?

38. A 1-kg block of ice at 0°C falls into a lake whose water is also at 0°C, and 0.01 kg of ice melts. What was the minimum altitude from which the ice fell?

39. A lead bullet at 100°C strikes a steel plate and melts. What was its minimum speed?

40. A man is sitting in the shade in an ambient air temperature of 37°C, which is the same as his body temperature. Under these circumstances the chief way his body gets rid of the 120 W his metabolic processes liberate is through the evaporation of sweat. How much sweat per hour is required? At 37°C the heat of vaporization of water is 2430 kJ/kg. (In Section 14–9 we saw how it is possible for a liquid to evaporate at a temperature below its boiling point.)

41. A 50-kg block of ice at 0°C is pushed across a wooden floor also at 0°C for a distance of 20 m. A total of 25 g of ice melts as a result of the friction of the block on the floor. What is the minimum coefficient of friction in this case?

42. How much steam at 150°C is needed to melt 200 g of ice at 0°C?

43. How much ice at −10°C is required to cool a mixture of 0.1 kg ethyl alcohol and 0.1 kg water from 20°C to 5°C?

44. Water at 10°C from a river is used to condense spent steam at 120°C from an electric generating plant to water at 50°C. If the cooling water leaves the condenser at 30°C, how many kilograms of river water are needed per kilogram of steam?

45. By mistake, 0.2 kg of water at 0°C is poured into a vessel containing liquid nitrogen at −196°C. How much nitrogen vaporizes?

46. If 6 kg of ice at −10°C is added to 6 kg of water at +10°C, find the temperature of the resulting mixture.

47. Ten pounds of steam at 220°F is passed through 10 lb of water at 200°F. What is the temperature and physical state of the mixture afterward?

48. A 5-kg iron bar is taken from a forge at a temperature of 1000°C and plunged into a pail containing 10 kg of water at 60°C. How much steam is produced?

15–5   The Triple Point

49. If you are unhappy at the rate at which eggs cook in a pan of boiling water, would it make sense to turn up the gas flame? To use a pressure cooker?

50. The pressure and temperature of the atmosphere at 35,000 m are 4 torr and −23°C respectively. What is the state of water under those conditions? Of carbon dioxide?

51. Carbon dioxide is usually shipped in tanks under a pressure of approximately 70 atm. At 20°C, is the carbon dioxide a solid, a liquid, or a gas?

ANSWERS TO MULTIPLE CHOICE

1. c	**7.** a	**12.** a	**17.** b	**22.** d
2. b	**8.** b	**13.** d	**18.** d	**23.** a
3. a	**9.** d	**14.** c	**19.** d	**24.** d
4. b	**10.** a	**15.** d	**20.** a	**25.** d
5. b	**11.** c	**16.** c	**21.** b	**26.** c
6. c				

16

THERMODYNAMICS

Thermodynamics has as its basic concern the transformation of heat into mechanical energy. Thermodynamics thus plays a central role in technology, since almost all the "raw" energy available for our use is liberated in the form of heat. A device or system that converts heat into mechanical energy is called a heat engine, and the principles that govern its operation are the same whether it is an automobile engine whose heat source is the burning of gasoline, a steam turbine whose heat source is a nuclear reactor, the earth's atmosphere whose heat source is the sun, or a human being whose heat source is the food he or she eats.

16-1   FIRST LAW OF THERMODYNAMICS

Heat engines

Three characteristic processes take place in all heat engines:

1. Heat is absorbed from a source at a high temperature.
2. Mechanical work is done.
3. Heat is given off at a lower temperature.

CHAPTER OBJECTIVES

Completing this chapter should enable you to:

1. Describe the significance of the first and second laws of thermodynamics.

2. Calculate the work done by an expanding gas.

3. Use the indicator (pressure-volume) diagram of an engine to find the work done per cycle.

4. Distinguish between isothermal and adiabatic processes.

5. Calculate the maximum efficiency of a heat engine.

6. Find the power output of a piston engine.

7. Describe the operation of gasoline, diesel, and jet engines.

8. Explain why the second law of thermodynamics is relevant to the problem of energy supply in the future.

Different heat engines carry out these processes in different ways, but the general pattern of operation is always the same (Fig. 16–1). Both hot and cold reservoirs are needed in order to produce a flow of heat from which work can be extracted. Some of the heat input to the engine must pass through it unused; otherwise there will be no heat flow and hence no work output.

Two general principles have been found to apply to all heat engines. The *first law of thermodynamics* expresses the conservation of energy: Energy cannot be created or destroyed, but may be converted from one form to another. In general terms, this law states that, in any process a system of some kind undergoes,

First law of thermodynamics

Net heat input = change in internal energy + net work output

$$Q = \Delta U + W \qquad \textit{First law of thermodynamics} \quad (16\text{--}1)$$

In Eq. (16–1), Q is the net heat added to the system during the process; if the system gives off heat, Q is negative. When the internal energy U of the system increases, ΔU is positive; when U decreases, ΔU is negative. The net work done *by* the system on something else during the process is W, so W is negative if work is done *on* the system.

FIG. 16−1 The work
output of a heat engine is
the difference between the
amount of heat it takes in
from a high-temperature
reservoir and the heat it
exhausts to a low-
temperature reservoir. The
low-temperature reservoir
is needed for a flow of
heat to occur, part of
which the engine converts
into work.

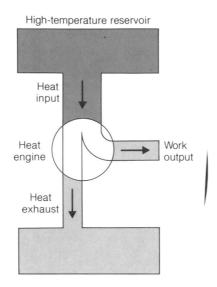

High-temperature reservoir

Heat
input

Heat
engine

Work
output

Heat
exhaust

If the system is an engine that operates in a cycle, energy may be stored and
released from storage, but the engine does not undergo a net change in its internal
energy. In this case,

Net heat input = work output

The net heat input equals the amount of heat the engine takes in from a reservoir at
high temperature minus the amount of heat it exhausts to a reservoir at low temperature.
In a steam engine the high-temperature reservoir is the boiler, and the low-temperature
reservoir is the escaping steam; in a gasoline engine the high-temperature reservoir is
the exploding mixture of air and gasoline vapor in each cylinder, and the low-temperature reservoir is the exhaust gas; in the earth's atmosphere the ultimate high-temperature reservoir is the sun, and the ultimate low-temperature reservoir is the rest of the
universe.

16−2 WORK DONE BY AND ON A GAS

The work output of most heat engines is produced by an expanding gas: Steam in the
case of a steam engine or turbine, a mixture of burning gases in the case of gasoline,
diesel, and jet engines. It is therefore important to know what the factors are that
govern the work done by a gas when it expands.

**Work is done by an
expanding gas**

Figure 16−2 shows a gas-filled cylinder with a movable piston. The cross-sectional area of the cylinder is A and the gas pressure is p. From the definition of pressure
as $p = F/A$, the force on the piston is

$$F = pA$$

If the distance Δs through which the piston moves is small, the pressure p remains very
nearly constant during the expansion, and the work ΔW done by the expanding gas is

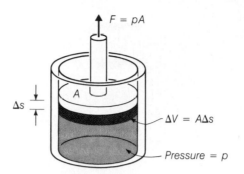

FIG. 16–2 The work ΔW done by a gas under the pressure p in expanding by the amount ΔV is $\Delta W = p\Delta V$.

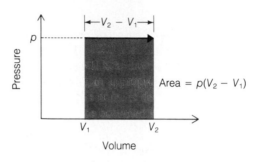

FIG. 16–3 The work done by a gas at the constant pressure p when it expands from V_1 to V_2 is $W = p(V_2 - V_1)$, which equals the shaded area under the curve that represents the process.

Work = force × distance

$$\Delta W = F\Delta s = pA\,\Delta s$$

Since $A\Delta s$ is the change in volume ΔV of the gas,

$$\Delta W = p\Delta V \qquad\qquad \textit{Work done by expanding gas} \quad (16-2)$$

If the gas is compressed rather than expanded, ΔV is negative, and ΔW is negative also: Work is done *on* a gas during a compression.

Work is done on a gas being compressed

    An *isobaric* process is one during which a system remains under constant pressure. Hence an isobaric expansion at the pressure p from an initial volume V_1 to a final volume V_2 involves the work

$$W = p(V_2 - V_1) \qquad\qquad \textit{Isobaric expansion} \quad (16-3)$$

This expansion is plotted on a graph of pressure versus volume in Fig. 16–3. The expansion is the horizontal line at the pressure p from V_1 to V_2. The shaded region under this line has an area of (height)(width) = $(p)(V_2 - V_1)$. But this is the same as the work done in the expansion! We therefore have the very useful result that

The work done by an expanding gas equals the area under the corresponding curve on a p-V diagram.

How to find work done by expanding gas

    What about an expansion in which the pressure changes? In such a case, we can imagine the expansion to consist of a series of small expansions. The first expansion involves a volume change of ΔV_1 at the pressure p_1, the second a volume change of ΔV_2 at the pressure p_2, and so on (Fig. 16–4). The total work done is

$$W = p_1\Delta V_1 + p_2\Delta V_2 + p_3\Delta V_3 + \cdots$$

FIG. 16-4 Even when the pressure of an expanding gas varies, the work it does equals the area under the curve of the process on a p–V graph. To verify this, the region under the curve can be imagined as divided into thin strips each of which represents a small expansion at constant pressure.

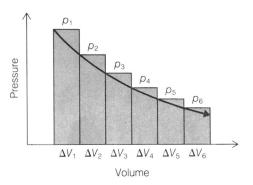

The actual smooth curve of pressure versus volume has been replaced by a jagged curve, but by making the volume changes ΔV small enough, we can get as close as we want to the actual curve. Since the work done in each little expansion ΔV equals the area under the horizontal line segment at its top, the total work done in the expansion equals the area under the p-V curve that represents it, in accord with the above statement.

Example The heat of vaporization of water at atmospheric pressure is $L_v = 2260$ kJ/kg. (a) How much of this heat represents work done to expand the water into steam against the pressure of the atmosphere? (b) What becomes of the rest of the heat? At $T = 100°C$ and $p = 1$ atm the density of water is 10^3 kg/m^3 and the density of steam is 0.60 kg/m^3.

Solution (a) The volumes of 1 kg of water and 1 kg of steam under the specified conditions are respectively

$$V_1 = \frac{m}{d_{water}} = \frac{1\,\text{kg}}{10^3\,\text{kg/m}^3} = 10^{-3}\,\text{m}^3$$

$$V_2 = \frac{m}{d_{steam}} = \frac{1\,\text{kg}}{0.60\,\text{kg/m}^3} = 1.67\,\text{m}^3$$

Atmospheric pressure is $p = 1.013 \times 10^5$ Pa. Hence the work done during the expansion of 1 kg of water to steam is, from Eq. (16–3),

$$W = p(V_2 - V_1) = (1.013 \times 10^5\,\text{Pa})(1.67\,\text{m}^3 - 0.001\,\text{m}^3) = 1.69 \times 10^5\,\text{J}$$
$$= 169\,\text{kJ}$$

This is

$$\frac{W}{L_v} = \frac{169\,\text{kJ}}{2260\,\text{kJ}} = 0.075 = 7.5\%$$

of the heat of vaporization.

(b) The rest of the heat of vaporization becomes internal energy of the steam. From the first law of thermodynamics the internal energy ΔU added to the water as it becomes steam is

$$\Delta U = Q - W = 2260\,\text{kJ} - 169\,\text{kJ} = 2091\,\text{kJ}$$ ∎

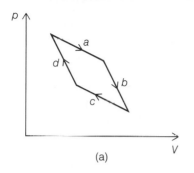

(a)

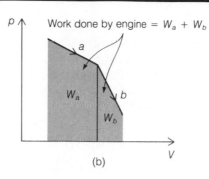

Work done by engine = $W_a + W_b$

(b)

FIG. 16–5 Indicator diagram of the operating cycle of an imaginary heat engine. The work done in an entire cycle is equal to the area enclosed by the curve *abcd* of the cycle.

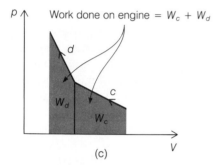

Work done on engine = $W_c + W_d$

(c)

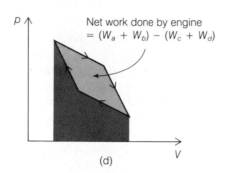

Net work done by engine
= $(W_a + W_b) - (W_c + W_d)$

(d)

The difference between a heat engine that runs continuously and a one-time-only event such as a dynamite blast is that the engine is not permanently changed as it turns heat into work. A sequence of processes brings the engine back to its starting point after each cycle. The *p-V* graph of a heat engine is therefore a closed curve that the engine traces over and over again in its operation, as in Fig. 16–5(a).

The cycle shown in Fig. 16–5(a) has four parts, each representing a different process. In *a* and *b* the gas in the engine expands, thereby doing the work W_a and W_b on whatever the engine is connected to. This work is equal to the area under the curves *a* and *b* on the graph (Fig. 16–5(b)). The compressions *c* and *d* then bring the engine back to its original state. In these compressions the work W_c and W_d is done on the engine itself (Fig. 16–5(c)). Thus the net work done in each cycle is

Work done in an engine cycle

Net work = work done by engine − work done on engine

$$W = (W_a + W_b) - (W_c + W_d)$$

As we see in Fig. 16–5(d), W is equal to the area enclosed by the cycle *abcd*. The larger this area, the more work is done. A graph of this kind for a heat engine is called its *indicator diagram*. Indicator diagrams for various engines are given later in this chapter.

16–3   SECOND LAW OF THERMODYNAMICS

Heat is the easiest and cheapest form of energy to obtain, since all we need do to liberate it is to burn a fuel such as wood, coal, or oil. The real problem is to turn heat

Heat is disordered energy

into mechanical energy so it can power cars, ships, airplanes, electric generators, and machines of all kinds. To appreciate the problem, we recall that heat consists of the kinetic energies of moving atoms and molecules. In order to change heat into a more usable form, we must extract some of the energy of the random motions of atoms and molecules and convert it into regular motions of a piston or a wheel. Such conversions cannot take place efficiently, for the same reason that it is easier to break a wineglass than to put the fragments together again: The natural tendency of all physical systems is toward increasing disorder. The *second law of thermodynamics* is an expression of this tendency.

Many processes permitted by conservation laws nevertheless do not occur

The first law of thermodynamics forbids an engine from operating without a source of energy, but it does not tell us anything about the character of possible sources of energy. For instance, there is an immense amount of internal energy in the atmosphere, yet it is impossible to run a car by just taking in air, extracting some of its internal energy, and then exhausting liquid air. Or, to give an even more extreme case, it is energetically possible for a puddle of water to rise by itself into the air, cooling and freezing into ice as its internal energy changes into potential energy (Fig. 16–6). After all, a block of ice dropped from a sufficient height melts when it strikes the ground, with its initial potential energy first being converted to kinetic energy and then into heat. Needless to say, water does not rise upward of its own accord, and we must find a way to express this conclusion.

Second law of thermodynamics

The second law of thermodynamics is the physical principle, independent of the first law and not derivable from it, that supplements the first law in limiting our choice of heat sources for our engines. It can be stated in a number of equivalent ways, a common one being as follows:

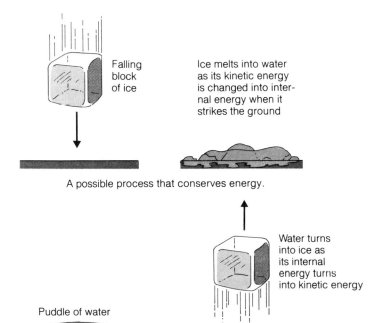

FIG. 16–6 The second law of thermodynamics provides a way to identify impossible processes that are in accord with all other physical principles.

Falling block of ice

Ice melts into water as its kinetic energy is changed into internal energy when it strikes the ground

A possible process that conserves energy.

Water turns into ice as its internal energy turns into kinetic energy

Puddle of water

An impossible process that conserves energy.

It is impossible to construct an engine, operating in a cycle (that is, continuously), that does nothing other than take heat from a source and perform an equivalent amount of work.

According to the second law of thermodynamics, then, no engine can be completely efficient—some of its heat input *must* be wasted. As we shall see, the greatest efficiency any heat engine is capable of depends upon the temperatures of its heat source and of the reservoir to which it exhausts heat. The greater the difference between these temperatures, the more efficient the engine. The second law is a consequence of the fact that

A heat engine extracts energy from the flow of heat through it

The natural direction of heat flow is from a reservoir of internal energy at a high temperature to a reservoir of internal energy at a low temperature, regardless of the total energy content of each reservoir.

This statement may be regarded as an alternative expression of the second law (Fig. 16–7).

If we are to make use of the internal energy content of the atmosphere or the oceans, we must first provide a reservoir at a lower temperature than theirs in order to extract heat from them. There is no reservoir in nature suitable for this purpose; if there were, heat would flow into it until its temperature reached that of its surroundings. To set up a low-temperature reservoir, we must use a refrigerator (which is a heat engine running in reverse by using up energy to extract heat), and in this way we will perform more work than we can obtain from the heat of the atmosphere or oceans.

A low-temperature reservoir is essential for a heat engine

To sum up the discussion, the first law of thermodynamics tells us that we can't get something for nothing. The second law singles out heat from other kinds of energy and recognizes that all conversions of heat into any other form must be inefficient: We can't break even, either.

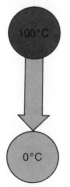

Heat flows naturally from hot reservoir to a cold one.

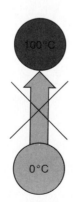

Heat cannot flow by itself from a cold reservoir to a hot one.

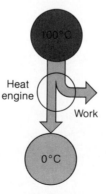

Some of the heat flow can be converted to work by an engine.

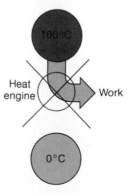

All of the heat that leaves a reservoir cannot be converted into work; some of the heat must flow into a cold reservoir.

FIG. 16–7 The second law of thermodynamics.

16−4 CARNOT ENGINE

Every heat engine behaves in the same general way: It absorbs heat at a certain temperature, converts some of the heat into work, and exhausts the rest at a lower temperature. Because of the second law of thermodynamics, we cannot expect a heat engine to be 100% efficient in turning heat into work. But suppose we have an engine that is not subject to friction, to the loss of stored heat by conduction or radiation, or to other kinds of energy loss that can, in principle, be reduced as much as we like by careful construction. What is the maximum efficiency of such an ideal engine?

Reversible engines are the most efficient

It is clear that maximum efficiency can be reached only if all the processes that occur in the engine's operation do not involve changes in the engine that need extra work to reverse, because any such changes obviously mean the waste of energy. That is, *every process in the engine must be reversible without losing any work.*

An isothermal process takes place at constant temperature

The flow of heat from a hot reservoir to a cooler one is not reversible in this way because work would have to be done to "push" the heat back the other way. However, heat flow at constant temperature *is* reversible. An *isothermal* process is one in which the temperature of the substance that undergoes a change of some kind stays constant. If the substance is a gas confined to a container, we can imagine that the container is a good conductor of heat and is surrounded by a constant-temperature reservoir of heat. No process that occurs in the real world is ever wholly isothermal, just as no actual gas exactly resembles an ideal gas, but many processes are quite close to being isothermal.

In an isothermal process in a system of some kind, all the heat absorbed by the system is turned into work. If the system has work done on it, all the work is given off as heat.

No heat enters or leaves a system during an adiabatic process

At the other extreme from being in such close contact with a heat reservoir that its temperature never changes, a system might be so completely isolated from its surroundings that heat can neither enter nor leave it. Any process undergone by a system in this situation is called *adiabatic*. An adiabatic process can involve a temperature change and still be reversible since no heat transfer is involved that would be

FIG. 16−8 Isobaric (p = constant), isothermal (T = constant), and adiabatic (Q = 0) processes.

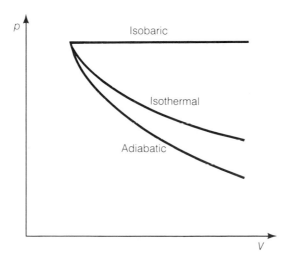

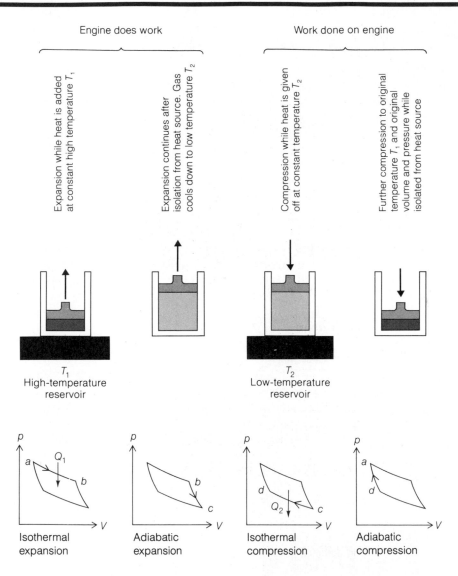

Engine does work | Work done on engine

Expansion while heat is added at constant high temperature T_1

Expansion continues after isolation from heat source. Gas cools down to low temperature T_2

Compression while heat is given off at constant temperature T_2

Further compression to original temperature T_1 and original volume and pressure while isolated from heat source

T_1
High-temperature reservoir

T_2
Low-temperature reservoir

Isothermal expansion | Adiabatic expansion | Isothermal compression | Adiabatic compression

FIG. 16–9 The Carnot cycle. This imaginary engine, which uses an ideal gas as its working substance, is the most efficient possible operating between reservoirs at the temperatures T_1 and T_2.

subject to the second law of thermodynamics. Most rapid thermodynamic processes are approximately adiabatic, since heat transfer takes time and the process may be completed before an appreciable amount of heat has passed through the walls of the system.

In an adiabatic process, any work done by the system comes from the internal energy of the system. Usually this means that the temperature of the system decreases. If the system has work done on it, its internal energy increases by the same amount. Usually this means that the temperature of the system increases. Figure 16–8 compares isobaric, isothermal, and adiabatic processes on a p-V diagram.

A heat engine that employs only isothermal and adiabatic processes was devised in 1824 by the French engineer Sadi Carnot. A *Carnot engine* consists of a cylinder that is filled with an ideal gas and has a movable piston at one end. The four stages in

Operating cycle of a Carnot engine

its operating cycle are shown in Fig. 16–9, together with a graph of each stage on a pressure-volume diagram. These stages are as follows:

1. An amount of heat Q_1 is added to the gas, which expands isothermally at its initial temperature T_1. The heat added equals exactly the work done by the gas, which is why its temperature does not change.

2. The heat source is removed, and the expansion is allowed to continue. The second expansion is adiabatic and takes place at the expense of the energy stored in the gas, and so the gas temperature falls from T_1 to T_2. During expansions 1 and 2 the piston exerts a force on whatever it is attached to, and thereby performs work.

3. Having done work in pushing the piston outward, the engine must now be returned to its initial state in order for it to be able to do further work. The third stage involves an isothermal compression of the gas at the constant temperature T_2 during which an amount of heat Q_2 is given off. The heat given off exactly equals the work done on the gas by the piston, which is why its temperature does not change.

4. The gas is returned to its initial temperature, pressure, and volume by an adiabatic compression in which heat is neither added to it nor removed from it. Work is done on the gas in this compression, which is why its temperature rises.

16–5 ENGINE EFFICIENCY

In each cycle a Carnot engine performs some net amount of work W, which is the difference between the work it does during the two expansions and the work done on it during the two compressions. It has taken in the heat Q_1 and ejected the heat Q_2; we note that the heat Q_2 *must* be ejected in order that the engine return to its initial state from which it can begin another cycle (Fig. 16–10). According to the first law of thermodynamics,

$$W = Q_1 - Q_2 \qquad\qquad (16-4)$$

FIG. 16–10 The efficiency of a Carnot engine is equal to $1 - T_2/T_1$.

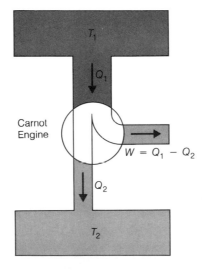

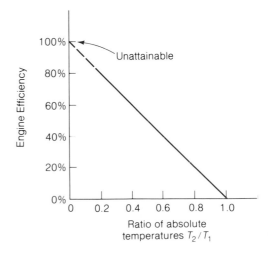

The efficiency of the engine is the ratio between its work output W and its heat input Q_1, so that

$$\text{Eff} = \frac{W}{Q_1} = \frac{Q_1 - Q_2}{Q_1} = 1 - \frac{Q_2}{Q_1} \qquad (16-5)$$

The smaller the ratio of the ejected heat Q_2 to the absorbed heat Q_1, the more efficient the engine.

As it happens, the heat Q transferred to or from a Carnot engine is directly proportional to the absolute temperature T of the reservoir with which it is in contact. That is, for a given Carnot engine,

Heat transfer to and from a Carnot engine is proportional to temperature

$$\frac{Q}{T} = \text{constant} \qquad\qquad\qquad \textit{Carnot engine} \quad (16-6)$$

(The argument that leads to this conclusion is rather long, although it does not involve any physical principles we have not yet encountered, and so is omitted here.) According to Eq. (16-6), the ratio Q_2/Q_1 between the amounts of heat ejected and absorbed per cycle by a Carnot engine is equal to the ratio T_2/T_1 between the temperatures of the respective reservoirs. The efficiency of such an engine is therefore

$$\text{Eff} = 1 - \frac{T_2}{T_1} \qquad\qquad\qquad \textit{Carnot efficiency} \quad (16-7)$$

The smaller the ratio between the absolute temperatures T_2 and T_1, the more efficient the engine (Fig. 16-10). No engine can be 100% efficient because no reservoir can have an absolute temperature of 0 K. (Even if such a reservoir could somehow be created, the exhaust of heat to it by the engine would raise its temperature above 0 K immediately.)

Ideal engine efficiency depends upon temperatures at which heat is absorbed and exhausted

As mentioned earlier, the significance of the Carnot engine is that, being reversible, it has the highest efficiency permitted by the laws of thermodynamics. A real engine is never exactly reversible because of such irreversible transformations as those involved in friction and in heat losses through the engine walls, and its efficiency is less than that of a Carnot engine. We therefore have a third way to express the second law of thermodynamics:

No engine operating between the absolute temperatures T_1 and T_2 can be more efficient than a Carnot engine operating between the same temperatures, whose efficiency is $1 - T_2/T_1$.

A Carnot engine is the most efficient possible

Example Steam enters a certain steam turbine (Fig. 16-11) at a temperature of 570°C and emerges into a partial vacuum at a temperature of 95°C. What is the upper limit to the efficiency of this engine?

Solution The absolute temperatures equivalent to 570°C and 95°C are respectively 843 K and 368 K. The efficiency of a Carnot engine operating between these two absolute temperatures is

$$\text{Eff} = 1 - \frac{T_2}{T_1} = 1 - \frac{368\,\text{K}}{843\,\text{K}} = 0.56$$

which is 56%. The efficiency of a Carnot engine is the maximum possible for an engine

FIG. 16–11 (a) A primitive steam turbine. (b) In a modern turbine, steam flows past a dozen or more sets of blades on the same shaft to extract as much power as possible. Stationary blades are interleaved between the moving blades to direct the flow of steam in the most advantageous way. (Photo: Electric Power Research Institute (EPRI))

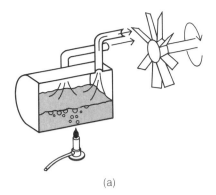

(a)

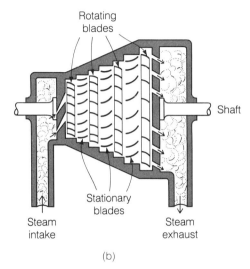

(b)

operating between a given pair of temperatures. An actual steam turbine operating between 570°C and 95°C would have an efficiency of no more than about 40% because of the inevitable presence of friction and heat losses to the atmosphere. Practical problems rule out steam temperatures higher than about 570°C, which would otherwise improve the efficiency. One of these problems is that, at such temperatures, steam can decompose into hydrogen and oxygen. Steel absorbs hydrogen and becomes brittle as a result, which can reduce the lifetime of the steel components of a turbine to an uneconomic extent. ∎

16–6 HEAT OF COMBUSTION

The *heat of combustion* of a substance is the amount of heat given off when a unit quantity (1 kg or 1 lb) of it is completely burned. Table 16–1 lists heats of combustion for a number of common fuels.

Substance		Heat of Combustion		TABLE 16–1
Solids	kcal/kg	MJ/kg	Btu/lb	Typical heats of combustion for common fuels
Charcoal	8,100	33.9	14,600	
Coal	7,800	32.6	14,000	
Wood	4,500	18.8	8,000	
Liquids	kcal/kg	MJ/kg	Btu/lb	
Diesel oil	10,700	44.8	19,400	
Domestic fuel oil	10,800	45.2	19,500	
Ethyl alcohol	7,800	32.6	14,000	
Gasoline	11,300	47.3	20,400	
Kerosene	11,000	46.0	19,800	
Gases*	kcal/m^3	MJ/m^3	Btu/ft^3	
Acetylene	12,900	54.0	1,450	
Coal gas	4,300	18.0	490	
Hydrogen	2,445	10.2	275	
Natural gas	8,000–17,000	33–71	900–2000	
Propane	20,600	86.2	2,330	

*Volumes at 0°C and atmospheric pressure.

Example A coal-fired power station whose electrical output is 200 MW has an overall efficiency of 35%. How much coal does it consume per day?

Solution The power input to the station is

$$P_{input} = \frac{P_{output}}{Eff} = \frac{200\,MW}{0.35} = 571\,MW = 571\,MJ/s$$

Since the heat of combustion of coal is $Q/m = 32.6$ MJ/kg, the mass of coal consumed per second is

$$\frac{m}{t} = \frac{Q/t}{Q/m} = \frac{P_{input}}{Q/m} = \frac{571\,MJ/s}{32.6\,MJ/kg} = 17.5\,kg/s$$

A day contains 8.64×10^4 s, so the mass of coal consumed per day is

$$m = (17.5\,kg/s)(8.64 \times 10^4\,s) = 1.5 \times 10^6\,kg \qquad \blacksquare$$

Example The diesel engine of a fishing boat consumes 90 lb/h of fuel when its power output is 200 hp. Find the efficiency of the engine.

Solution Since the heat of combustion of diesel oil is 19,400 Btu/lb, the power input of the engine is

$$P_{input} = (90\,lb/h)(19,400\,Btu/lb) = 1.746 \times 10^6\,Btu/h$$

To convert this figure into hp we note that 1 Btu = 778 ft · lb, 1 h = 3600 s, and 1 hp = 550 ft · lb/s. Hence

$$P_{input} = \frac{(1.746 \times 10^6 \, Btu/h)(778 \, ft \cdot lb/Btu)}{(3600 \, s/h)(550 \, ft \cdot lb/s)/hp} = 686 \, hp$$

The efficiency of the engine is therefore

$$Eff = \frac{P_{output}}{P_{input}} = \frac{200 \, hp}{686 \, hp} = 0.29 = 29\%$$ ■

Hydrogen as a fuel

From Table 16–1 it would seem that hydrogen has a very low heat of combustion, well under those of the other gaseous fuels. However, hydrogen is also much less dense than the other gases, and on a mass basis its heat of combustion of 143 MJ/kg (61,400 Btu/lb) is actually by far the highest of any of the fuels listed. The exceptional heat of combustion of hydrogen is the reason for the use of liquid hydrogen in spacecraft engines, where fuel weight is a critical factor. Liquid hydrogen would be an ideal fuel for many other purposes as well, except for two factors. First, its safe storage and transport in large quantities is not easy. Second, although there is almost limitless hydrogen in the waters of the earth, a great deal of expensive electric energy is needed to extract it, so the overall efficiency of hydrogen as a fuel is today quite low. Research is under way on methods to use sunlight as the energy source for decomposing water into hydrogen and oxygen, and if it is successful, and the safety problems can be overcome, hydrogen may well be widely used in the future.

The amount of air needed for the complete combustion of a fuel depends upon its heat of combustion and is about 0.33 kg per MJ (0.77 lb per 1000 Btu). Thus nearly 15 kg of air, about 11 m^3 at atmospheric pressure, is needed for each kg of diesel oil used by an engine. A conventional submarine cannot use its diesel engines when submerged because enough air cannot be stored for this purpose. Such a submarine uses electric motors powered by storage batteries when it is underwater, and must return to the surface periodically to recharge the batteries using generators connected to its diesel engines. Submarines powered by nuclear reactors need air only for their crews and can travel submerged for long distances.

Pollution by gasoline engines

A kilogram of gasoline requires 15.6 kg of air for complete combustion, whose products are carbon dioxide (CO_2) and water vapor. These gases are odorless and harmless. A "rich" mixture of gasoline and air contains a greater proportion of gasoline than this, and a "lean" mixture contains a smaller proportion; both mixtures produce polluting exhaust gases. If the mixture that enters the cylinders of an engine from its carburetor is rich, the amount of air is not enough for complete combustion. As a result the exhaust will contain unburned hydrocarbons and carbon monoxide (CO), which are harmful substances. The hazard of carbon monoxide arises from its tendency to combine with the hemoglobin of the blood in place of oxygen, thereby depriving the body of the oxygen it needs and leading to death if too much CO is inhaled. If the mixture of gasoline and air is lean, there will be no unburned hydrocarbons or carbon monoxide because combustion will be complete, but in this case the resulting temperature will be high enough to oxidize some of the nitrogen in the air. Nitrogen oxides are poisonous and combine with water to form corrosive nitric acid. It is very difficult to build an engine in which just the right mixture of gasoline and air is burned at just the right temperature throughout all the variations in speed and power output involved in driving a motor vehicle. For this reason special devices are needed to treat most vehicle exhausts to prevent excessive pollution of the atmosphere.

16–7   INTERNAL COMBUSTION ENGINES

An internal combustion engine is able to achieve a relatively high operating efficiency by generating the input heat within the engine itself. In a gasoline engine a mixture of air and gasoline vapor is ignited in each cylinder by a spark plug, and the evolved heat is converted into mechanical energy by the pressure of the hot gases on a piston. The greater the ratio between the initial and final volumes of the expanding gases, the greater the engine efficiency. In a gasoline engine this ratio is limited to about 8 to 1, since the gasoline-air mixture in the cylinder will otherwise ignite by itself during its compression before the end of the stroke is reached. The result is a theoretical efficiency of about 55% and an actual efficiency of about 30%.

The more efficient diesel engine avoids the difficulty of preignition by compressing only air and injecting fuel oil into the hot, compressed air at the instant the piston has reached the top of its travel. No spark plug is required. The compression ratio in a diesel engine is typically 20 to 1, for a theoretical efficiency of as much as 70% and an actual efficiency of about 35%.

Diesel engines are more efficient than gasoline engines

Figure 16–12 shows the operating cycle of a typical four-stroke gasoline engine. In the intake stroke a mixture of gasoline vapor and air from the carburetor is drawn into the cylinder through the intake valve by the suction of the downward-moving piston. In the compression stroke both valves are closed and the upward-moving piston compresses the fuel-air mixture. At the top of the stroke the spark plug is fired, which ignites the fuel-air mixture. The burning fuel expands and forces the piston down in the power stroke. At the end of the power stroke the exhaust valve opens and the upward-moving piston expels the waste gases.

Gasoline engine cycle

The operating cycle of a gasoline engine (often called the *Otto cycle* after the inventor of the first practical such engine) is shown on a *p-V* diagram in Fig. 16–13. The heat Q_1 enters the engine from c to d when the gasoline-air mixture is ignited by the spark plug. The power stroke de is an adiabatic expansion, and the heat Q_2 is ejected from e to b when the exhaust valve is open at the end of the power stroke. The heat

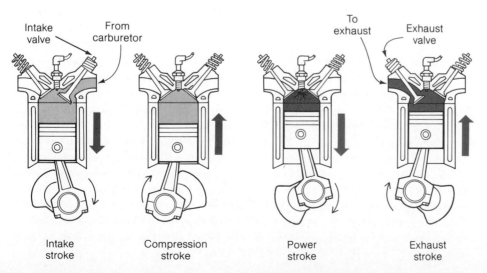

Intake valve   From carburetor   To exhaust   Exhaust valve

Intake stroke   Compression stroke   Power stroke   Exhaust stroke

FIG. 16–12   The operating cycle of a four-stroke gasoline engine

FIG. 16–13 Idealized indicator diagram of the Otto cycle of a four-stroke gasoline engine.

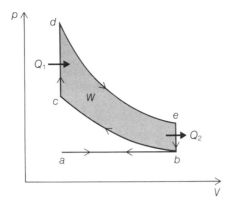

ab: intake of fuel-air mixture
bc: compression stroke
cd: ignition of fuel-air mixture
de: power stroke
e: exhaust valve opens
ba: exhaust stroke

FIG. 16–14 Idealized indicator diagram of a four-stroke diesel engine.

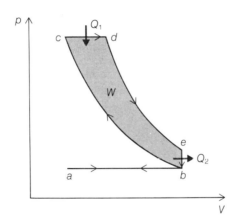

ab: intake of air
bc: compression stroke
cd: fuel injection and burning
ce: power stroke
e: exhaust valve opens
ba: exhaust stroke

that leaves during the exhaust stroke itself from *b* to *a* is smaller and is ignored in the ideal cycle of the diagram. The work *W* done in each cycle is equal to the area enclosed by *bcdeb*. The greater the compression ratio, the larger the area, and the more work is done per cycle.

Diesel engine cycle

A diesel engine has a somewhat different operating cycle whose indicator diagram is shown in Fig. 16–14. Air is drawn into the cylinder in the intake stroke *ab* and is compressed adiabatically, which raises its temperature, in the compression stroke *bc*. The heat Q_1 enters the engine in the constant-pressure expansion *cd* during which fuel is injected into the hot air and burned. Then comes an adiabatic expansion to *e* when the exhaust valve opens and the heat Q_2 is ejected.

Two-stroke and four-stroke engines

Each cylinder in a four-stroke engine has one power stroke in every two shaft revolutions. When a lightweight engine is necessary a two-stroke cycle can be used, which increases the power output by permitting a power stroke in each cylinder in every shaft revolution. Usually a two-stroke cycle means reduced efficiency in the case of a gasoline engine, which is unimportant in such applications as outboard motors for boats. It means greater complexity in the case of a diesel engine, which is a fair price to pay for decreased weight in an engine that must be heavily built to withstand 20-to-1 compressions.

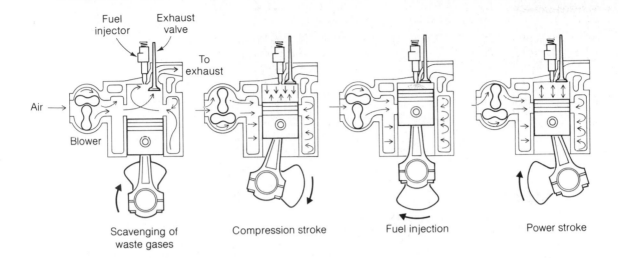

Scavenging of waste gases   Compression stroke   Fuel injection   Power stroke

Figure 16–15 illustrates the operation of a two-stroke diesel engine. When the piston is at the bottom of its path, air from a blower enters to flush waste gases from the previous power stroke out through the exhaust valve. As the piston moves upward it compresses the fresh air to a fraction of its initial volume. At the top of the compression stroke the air temperature is perhaps 550°C, and fuel oil sprayed in by the injector is ignited at once. The burning fuel presses down on the piston during the power stroke that follows.

FIG. 16–15 The operating cycle of a two-stroke diesel engine.

The power output of each cylinder of a gasoline or diesel engine depends upon four factors: the average pressure p on the piston during a power stroke; the length L of piston travel; the area A of the piston; and the number N of power strokes per second. Since the force on the piston during a power stroke is $F = pA$, the work done per power stroke is $W = FL = pLA$. Hence

Power output of piston engine

$$\text{Power output} = (\text{work done per stroke})\left(\frac{\text{power strokes}}{\text{second}}\right)$$

$$P = pLAN \qquad\qquad \textit{Power output} \quad (16\text{–}8)$$

In SI units, with p in Pa, L in m, and A in m^2, P is in watts. In British units, with p in lb/in.^2, L in ft, and A in in.^2, p is in ft · lb/s. If we want P in horsepower, it is necessary to divide $pLAN$ by 746 W/hp or by 550 (ft · lb/s)/hp, respectively.

Example The Mercedes-Benz OM314 is a four-cylinder, four-stroke diesel engine that develops 80 hp at 2600 rpm. The pistons of this engine are 97 mm in diameter and their length of travel is 128 mm. Find the average pressure on the pistons during each power stroke.

Solution The area of each piston is

$$A = \frac{\pi d^2}{4} = \frac{\pi (0.097 \text{ m})^2}{4} = 0.0074 \text{ m}^2$$

In a four-stroke engine, a power stroke occurs in each cylinder once every two revolutions, so here there are 1300 power strokes per minute or $1300/60 = 21.7$ per second. Since the engine has four cylinders, $P = 4pLAN$ and

$$p = \frac{P}{4LAN} = \frac{(80\,\text{hp})(746\,\text{W/hp})}{(4)(0.128\,\text{m})(0.0074\,\text{m}^2)(21.7/\text{s})} = 7.26 \times 10^5\,\text{Pa} = 7.26\,\text{bars}$$

This is equivalent to $105\,\text{lb/in}^2$. Of course, the pressure calculated in this way is an average: The actual pressure is greater at the beginning of the stroke just after the fuel is ignited, and less at the end of the stroke. ■

Gas turbines

The outstanding efficiency of conventional gasoline and diesel engines (nearly as high as those of steam turbines) has to some extent retarded the development of the still more efficient gas turbine. A gas turbine is similar to the steam turbine of Fig. 16–11(b) except that hot gases from the burning fuel pass through its sets of blades instead of steam. A gas turbine is lighter in weight and has fewer moving parts than a

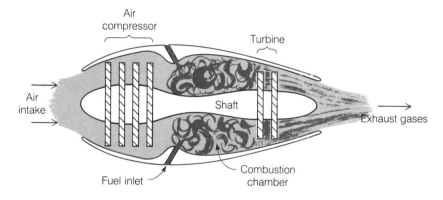

FIG. 16–16   Turbojet engine. The reaction force of the exhaust gases pushes the aircraft forward. (Bottom drawing courtesy of United Technologies, Pratt & Whitney.)

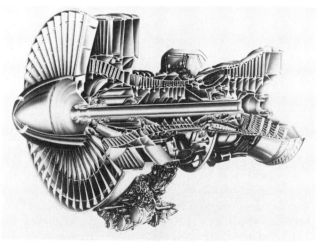

piston engine, but the high temperature and high rotational speeds at which it operates present difficulties in manufacture. Gas turbines are nevertheless coming into wider and wider use: "Turboprop" aircraft engines are gas turbines, for instance, and a number of ships are already powered by gas turbines.

The rapidly rotating shaft of a turboprop engine is coupled to a propeller through a reduction gear. In a turbojet engine the propeller is eliminated and the hot gases from the burning fuel are ejected at high speed from the rear of the engine to furnish a reaction force that pushes the aircraft forward (Fig. 16–16). The energy liberated by the burning fuel is thus converted directly into propulsion with no moving parts except for a turbine that powers the necessary air compressor. Rocket motors are jet engines in which the required oxygen or other oxidizing agent for fuel combustion comes from an internal reservoir instead of from the atmosphere (Fig. 6–7). In a solid-fuel rocket, the ultimate in simplicity, both the components required for combustion are combined in a stable mixture whose reaction rate when ignited is relatively slow and steady rather than explosive.

Jet engines

16–8   ENERGY AND CIVILIZATION

The development of modern civilization has been paralleled by a steady increase in the world's use of energy. This is no accident: All our activities require energy, and the more energy that is readily available in convenient form, the more effectively we can satisfy our desires for food, clothing, shelter, warmth, light, transport, communication, and manufactured goods. Our earliest ancestors had only food as an energy source, and utilized a total of perhaps 8 MJ/day; the mastery of fire and the harnessing of domestic animals yielded an approximately sixfold increase in this figure; the flowering of the industrial revolution a century ago meant a rise in energy consumption to about 300 MJ/day in the more advanced countries; and today in the United States each person uses an average of nearly 1000 MJ/day. Energy production involves one-third of the country's industrial plant today.

The more advanced a society, the more energy it uses

A single source has provided almost all the energy available to us today—the sun. Light and heat arrive directly from the sun; food and wood owe their energy contents to photosynthesis in plants; water power exists because solar heat evaporates water from the oceans to fall later as rain on high ground; wind power is a consequence of motions in the atmosphere whose energy source is solar radiation. The fossil fuels coal, oil, and natural gas were formed from plants and animals that lived and stored energy derived from sunlight millions of years ago.

The sun is the source of most of the energy available on the earth's surface

Figure 16–17 shows how rapidly the world's use of energy has increased in recent times and also indicates the relative importance of the chief energy sources. Existing trends have been projected to the year 2000, but it is not possible to go beyond that with any confidence for a number of reasons. An obvious one is that fossil fuels, which today furnish nearly 98% of our energy, cannot last much longer at current rates of consumption. Natural gas, the least polluting of them, will be the first to run out, perhaps soon after the turn of the century. Oil will be next, some decades later: Our descendants will find hard to believe the exhaustion by mere burning of these magnif-

Fossil fuels furnish most energy today but are being rapidly depleted

Geothermal power station at Roosevelt Hot Springs, Utah uses the heat of the earth's interior as its energy source. (Photo: EPRI)

For two decades the ebb and flow of tides in the Rance River estuary in northern France have provided energy for generators whose output is 240 MW. The lock at right permits ships to enter and leave the river. (Photo: Courtesy French Embassy)

(Photo: Las Vegas News Bureau)

Windmill "farm" at Altamont Pass, California. More than 13,000 windmills have been installed in California in the past few years to generate electricity. (Photo: EPRI)

Water is used to absorb and store solar energy for heating this house in Albuquerque, New Mexico. (Photo: EPRI)

Coal, the most abundant fossil fuel, produces about half the electric energy in the United States. Shown is a coal-fired power plant at Pleasant Prairie, Wisconsin. (Photo: EPRI)

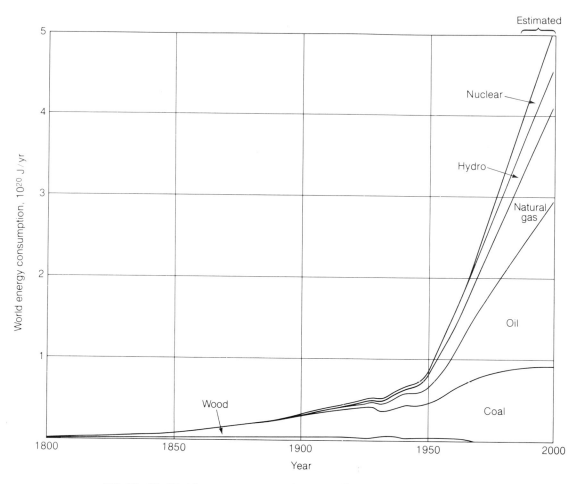

FIG. 16–17   World energy sources and consumption.

icent feedstocks for synthetic materials of all kinds. Even though the coal currently consumed each year took about two million years to accumulate, enough remains to last another two or three centuries. But coal is far from being a desirable fuel. Not only does its mining leave large tracts of land unfit for further use, but the air pollution due to coal burning is responsible for harming the health of millions of people through cancer and respiratory diseases. Coal-burning power plants actually expose their surrounding populations to more radioactivity than do normally operating nuclear plants because of the radon gas they liberate.

Nuclear fuel reserves are large but nuclear power also has large drawbacks

Nuclear fuel reserves are greater than those of fossil fuels, especially since "breeder" reactors already have been built that convert the normally unusable ^{238}U form of uranium into plutonium, which can serve as a reactor fuel (see Chapter 30). The energy content of the U.S. stockpile of refined ^{238}U already about equals that in all the country's coal reserves. To be sure, nuclear energy has serious drawbacks. Although the safety record of nuclear installations has been good on all overall basis, even a single malfunction has the potential of doing enormous harm. A power station reactor produces several tons of wastes each year whose radioactivity will remain high for thousands of years, and their safe disposal remains a major question. And the widespread production and use of plutonium as a fuel cannot ignore its suitability for nuclear weapons, in many people's eyes a sufficient reason to avoid this technology.

In the long run, practical ways to utilize the energy of nuclear fusion seem certain to be developed. As described in Chapter 30, a fusion reactor will obtain its fuel from the sea, will be safe and nonpolluting, and cannot be adapted for military purposes. But nobody can predict when this ultimate source of energy will become an everyday reality or how expensive it will be. **Nuclear fusion is a long-term prospect**

The big question today is how to bridge the shift to fusion energy, a shift whose time scale can only be guessed at. Expanded use of coal can help, but only at great cost—either in terms of environmental damage and human suffering, or in terms of the expense of minimizing them. And any acceleration in the rate of burning fossil fuels will bring with it an increase in the amount of carbon dioxide in the atmosphere, which may well have serious consequences for the world's weather because of the role of this gas in absorbing heat reradiated from the earth's surface. Since 1880 the carbon dioxide content of the atmosphere has gone up by 12%, and currently about 12 billion tons of carbon dioxide pour from our chimneys and exhaust pipes per year. Like coal, nuclear energy can fill the coming energy gap, but also not without hazards that would best be avoided. **Coal is plentiful but is not an ideal fuel**

What about the energy of trees and plants, of winds and tides, of the sun's radiation, and of the heat of the earth's interior? The technology already exists, and no resources are used up. A brief look shows that such energy sources are unlikely to provide more than a small fraction of future energy needs. In the case of burning wood and vegetable matter in general ("biomass"), the pollution produced (especially of carcinogens) would exceed that due to burning coal of equal energy content, and taking agricultural land away from food production would be catastrophic in a time of rising population. The land needed to provide enough alcohol by conversion of plant sugar or starch to operate one car could provide enough food for 8 to 16 people. Windmills to generate electricity certainly represent a better approach, but locations with sufficiently strong and reliable winds are few and seldom near the locations where the electricity is needed most. The ebb and flow of tides in the Rance River estuary in northern France drive generators of 240-MW capacity, and a few other places in the world (such as the Bay of Fundy between Maine and Nova Scotia) promise the capture of as much or more tidal energy, but again the potential contribution on a global basis is not impressive. Geothermal plants have operated for some time in California, in Italy, and in New Zealand, but the geological conditions needed for the successful extraction of the earth's heat are not very common. **Other energy sources**

The direct use of solar radiation is more promising, but problems of cost and scale exist here, too. Even if a substantial fraction of American homes were to be fitted for solar space heating, the energy saving relative to the total energy used in the country would be only a percent or two—most desirable, but hardly a solution to the overall problem. The economic production of electricity from sunlight needs more efficient photovoltaic cells and better methods of storing electric energy than now exist, but even if they are developed, the size of the required investment in technology and land area means that only a gradual shift to their use is to be expected. Figure 16–18 shows how the amount of solar energy received per square meter per year varies in the United States. **Solar energy**

FIG. 16–18 Solar energy received per year in the United States expressed in kilowatt-hours per square meter (1 kW·h = 3.6 MJ).

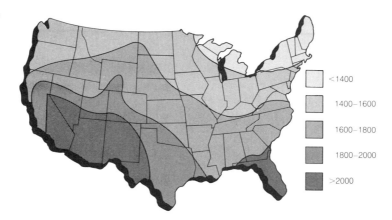

<1400

1400–1600

1600–1800

1800–2000

>2000

As energy consumption increases, the disposal of waste heat becomes more difficult

Clearly there is no magic solution possible in the near future to the problem of safe, cheap, and abundant energy. The only sensible course is to try to get the best from each available technology while proceeding as rapidly as possible to perfect fusion energy. But will the advent of fusion energy solve the energy problem forever? The ultimate limit to world energy production is set by the inefficiency of the ways by which thermal energy is converted into mechanical energy and thence into electrical energy—in other words, by the second law of thermodynamics. The best of today's power stations have overall efficiencies of only about 35%, and this figure is not likely to improve by much in the foreseeable future. The waste heat must go somewhere, and even today heavily industrialized countries find its disposal difficult if environmental damage is to be avoided. In the United States about 10% of the flow of all rivers and streams is already being used to provide cooling water for generating plants. The biological consequences of large-scale heating of inland waters are considerable. The oceans can absorb vast amounts of waste heat with minimal side effects, but if most power plants were located on their shores the transmission of electricity inland would then be a major problem. More and more power plants are discharging waste heat into the atmosphere through cooling towers, but here too there are snags in the long run, since local heating of the atmosphere alters the weather and climate of a region, not necessarily for the better.

The real problem is population growth

The situation is far from hopeless with respect to both resources and environmental damage, provided the world's total energy requirements do not continue to increase much longer at the present rate. It is precisely here that social rather than technical considerations enter the picture, because to slow down the growth of energy consumption without at the same time limiting population growth (which is by far the most critical problem of the modern world) means a decrease in average living standards, which are too low already in most of the world. No comfortable resolution is in sight to the fundamental conflict between a rising demand for energy from an exploding population and the inability of our planet to provide an unlimited supply.

IMPORTANT TERMS

A **heat engine** is any device that converts heat into mechanical energy or work.

The **first law of thermodynamics** states that the heat input to a heat engine is equal to its work output plus any change in its internal energy.

An **isobaric process** takes place at constant pressure; an **isothermal process** takes place at constant temperature. No heat enters or leaves a system during an **adiabatic process.**

The **second law of thermodynamics** states that it is impossible to construct an engine, operating in a repeatable cycle, that does nothing besides taking energy from a source and performing an equivalent amount of work.

A **Carnot engine** is an idealized engine that is not subject to such practical difficulties as friction but that obeys all physical laws. No engine operating between two given temperatures can be more efficient than a Carnot engine operating between them.

IMPORTANT FORMULAS

First law of thermodynamics: $Q - \Delta U + W$

Work done by expanding gas:

$$\Delta W = p\Delta V$$

Engine efficiency: $\quad \text{Eff} = \dfrac{W}{Q_1} = 1 - \dfrac{Q_2}{Q_1}$

Carnot efficiency: $\quad \text{Eff} = 1 - \dfrac{T_2}{T_1}$

MULTIPLE CHOICE

1. A heat engine operates by taking in heat at a particular temperature and
  a. converting it all into work.
  b. converting some of it into work and exhausting the rest at a lower temperature.
  c. converting some of it into work and exhausting the rest at the same temperature.
  d. converting some of it into work and exhausting the rest at a higher temperature.

2. In order to operate, a heat engine must have
  a. either a hot or a cold reservoir.
  b. both a hot and a cold reservoir.
  c. a boiler.
  d. a supply of fuel that can be burned, such as coal or oil.

3. The natural direction of heat flow is from a high-temperature reservoir to a low-temperature reservoir, regardless of their respective heat contents. This fact is incorporated in the
  a. first law of thermodynamics.
  b. second law of thermodynamics.
  c. law of conservation of energy.
  d. principle of superposition.

4. A type of process that does not need outside energy to reverse is one that occurs at constant
  a. temperature.
  b. pressure.
  c. volume.
  d. speed.

5. An adiabatic process in a system is one in which
  a. no heat enters or leaves the system.
  b. the system does no work nor is work done on it.
  c. the temperature of the system remains constant.
  d. the pressure of the system remains constant.

6. The work output of every heat engine
  a. equals the difference between its heat intake and heat exhaust.
  b. equals that of a Carnot engine with the same intake and exhaust temperatures.
  c. depends only upon its intake temperature.
  d. depends only upon its exhaust temperature.

7. In any process, the maximum amount of heat that can be converted to mechanical energy
  a. depends on the amount of friction present.
  b. depends on the intake and exhaust temperatures.
  c. depends on whether kinetic or potential energy is involved.
  d. is 100%.

8. In any process, the maximum amount of mechanical energy that can be converted to heat
  a. depends on the amount of friction present.
  b. depends on the intake and exhaust temperatures.
  c. depends on whether kinetic or potential energy is involved.
  d. is 100%.

9. A frictionless heat engine can be 100% efficient only if its exhaust temperature is
  a. equal to its input temperature.
  b. less than its input temperature.
  c. 0°C.
  d. 0 K.

10. A Carnot engine turns heat into work

a. with 100% efficiency.

b. with 0% efficiency.

c. without itself undergoing a permanent change.

d. with the help of expanding steam.

11. A Carnot engine operating between the absolute temperatures T_1 and T_2

a. is 100% efficient.

b. has the maximum efficiency possible under these circumstances.

c. has an efficiency of T_2/T_1.

d. has the same efficiency as an actual engine operating between T_1 and T_2.

12. Which of the following engines is normally the least efficient?

a. gasoline engine   b. diesel engine

c. gas turbine   d. Carnot engine

13. Which of the following engines is the most efficient?

a. gasoline engine   b. diesel engine

c. gas turbine   d. Carnot engine

14. The fuel in a diesel engine is ignited by

a. a spark plug.

b. being compressed until its temperature is high enough.

c. the hot compressed air into which it is injected.

d. exhaust gases remaining from the previous cycle.

15. The chief source of energy in the world today is

a. coal.   b. oil.

c. natural gas.   d. uranium.

16. Assuming the most efficient technologies for the extraction of energy, the fuel with the smallest known reserves is

a. coal.   b. oil.

c. natural gas.   d. nuclear.

17. Assuming the most efficient technologies for the extraction of energy, the fuel with the largest known reserves is

a. coal.   b. oil.

c. natural gas.   d. nuclear.

18. A gas is compressed from 5 L to 3 L under an average pressure of 2×10^5 Pa. During the compression 300 J of heat is given off. The internal energy of the gas

a. decreases by 100 J.   b. is unchanged.

c. increases by 100 J.   d. increases by 400 J.

19. A Carnot engine absorbs heat at a temperature of 127°C and exhausts heat at a temperature of 77°C. Its efficiency is

a. 13%.   b. 39%.

c. 61%.   d. 88%.

20. If a Carnot engine exhausting heat at 140°F is to have an efficiency of 33%, it must take in heat at

a. 200°F.   b. 440°F.

c. 660°F.   d. 900°F.

21. A Carnot engine operates between 800 K and 200 K. If it absorbs 8 kJ of heat in each cycle, the work it does per cycle is

a. 1 kJ.   b. 2 kJ.

c. 2.67 kJ.   d. 6 kJ.

EXERCISES

16–2 Work Done by and on a Gas

1. A gas sample expands from V_1 to V_2. Does it perform the most work when the expansion takes place at constant pressure, at constant temperature, or adiabatically? In which process does the gas perform the least work? Why?

2. The operation of a steam engine proceeds approximately as follows: (1) water under pressure is heated to the boiling point; (2) the water turns into steam and expands at constant pressure at its boiling point; (3) the steam enters the cylinder of the engine and expands adiabatically against the piston; (4) the spent steam condenses into water at constant pressure; (5) the water is pumped back into the boiler. Plot the entire cycle on a p-V-diagram and indicate in what parts of the cycle heat is absorbed, heat is rejected, and work is done on the outside world.

3. A gas expands by 1 L at a constant pressure of 2 bars. During the expansion 500 J of heat is added. Find the change in the internal energy of the gas.

4. A gas expands isothermally from 10 ft³ to 15 ft³ while absorbing 50 Btu of heat. Find the average pressure (in pounds per square inch) on the gas during the expansion.

5. A steam engine built in 1712 had a piston 1.8 ft in diameter that moved back and forth through a distance of 8 ft. If the pressure on the piston was a constant 15 lb/in.² during each expansion, find the work done.

6. The indicator diagram of a certain engine is shown below (1 kPa $= 10^3$ Pa). (a) Find the work done in each part of the cycle, namely $a \rightarrow b$, $b \rightarrow c$, $c \rightarrow d$, and $d \rightarrow a$, and identify it as work done by or on the engine. (b) How much net work is done per cycle by the engine?

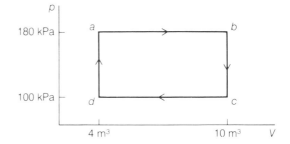

7. The engine of Exercise 6 is operated in reverse, with each cycle proceeding as *adcba*. Answer the same questions for the engine in this case.

16–4   Carnot Engine

16–5   Engine Efficiency

8. The sun's corona is a very dilute gas at a temperature of about 10^6 K that is believed to extend into interplanetary space at least as far as the earth's orbit. Why can we not use the corona as the high-temperature reservoir of a heat engine in an earth satellite?

9. An engine operating between 300°C and 50°C is 15% efficient. What would its efficiency be if it were a Carnot engine?

10. A typical temperature for surface water in a tropical ocean is 25°C, whereas at a depth of a kilometer or so it is only 7°C. It has been proposed to operate heat engines using surface water as the hot reservoir and deep water as the cold reservoir. What would the maximum efficiency of such an engine be? Why might such an engine be a practical proposition even with so low an efficiency? (An experimental plant of this kind built on the Japanese island of Nauru in 1981 takes in 1400 tons each of hot and cold water per hour to produce 100 kW of electricity. Operating the plant itself uses 90 kW, so the net output is only 10 kW. A full-size 400-MW plant would have to pump 10 million tons of water per hour, the flow of a major river. Despite this, the economics of ocean thermal energy do not seem hopeless.)

11. One of the most efficient engines ever developed operates between about 2000 K and 700 K. Its actual efficiency is 40%. What percentage of its maximum possible efficiency is this?

12. An engine is proposed that is to operate between 200°C and 50°C with an efficiency of 35%. Will the engine perform as predicted? If not, what would its maximum efficiency be?

13. A Carnot engine absorbs 200 kcal of heat at 500 K and exhausts 150 kcal. What is the exhaust temperature?

14. A steam engine takes in steam at 200°C. What is the maximum temperature at which spent steam can leave the engine if its efficiency is to be 20%.

15. A Carnot engine takes in 1 MJ of heat from a reservoir at 327°C and exhausts heat to a reservoir at 127°C. How much work does it do?

16. A Carnot engine operating between 580°F and 120°F has a power output of 120 hp. How much heat per minute does it absorb? How much does it exhaust?

17. A Carnot engine absorbs 500 Btu of heat at 500°F and performs 100,000 ft·lb of work. What is the exhaust temperature?

18. Three designs for a heat engine to operate between 450 K and 300 K are proposed. Design *A* is claimed to require a heat input of 0.2 kcal for each 1000 J of work output, design *B* a heat input of 0.6 kcal, and design *C* a heat input of 0.8 kcal. Which design would you choose and why?

19. A Carnot engine whose efficiency is 35% takes in heat at 500°C. What must the intake temperature be if the efficiency is to be 50% with the same exhaust temperature?

20. A coal-fired electric power plant of 400 MW output operates at an efficiency of 39%, which is about three-quarters of its Carnot efficiency. Water from a river flowing at 50 m^3/s is used to absorb the waste heat. By how much does the river's temperature rise as a result?

21. The total drop of the Wollomombi Falls in Australia is 482 m. What would be the Carnot efficiency of an engine operating between the top and bottom of the falls if the water temperature at the top were 10°C and all the potential energy of the water at the top were converted to heat at the bottom?

16–6   Heat of Combustion

22. In a certain power station coal is consumed at the rate of 0.4 kg/h for each kilowatt of electrical output. Find the overall efficiency of the power station.

23. A turboprop aircraft engine consumes 250 g/h of kerosene for each horsepower it develops. Find the efficiency of the engine.

24. How many cubic meters of propane must be burned to heat 50 L of water from 5°C to 90°C? Assume that 25% of the heat is wasted.

25. A 4250-hp aircraft engine is used in a power station in England to run a 3-MW generator to supply electricity during peak-load periods. The engine consumes 0.82 lb of kerosene per kilowatt-hour of energy produced. What is the overall efficiency of the installation?

26. A certain jet airplane uses 20,000 lb of kerosene in traveling 900 mi at an average speed of 500 mi/h. If the engines of the airplane develop an average of 15,000 hp during the flight, what is their percentage efficiency?

27. At 800 rpm a diesel engine develops 60 kW and consumes 18 kg of fuel per hour. At 1800 rpm the same engine develops 120 kW and consumes 32 kg of fuel per hour. Find its efficiency at each speed.

28. A certain power station consumes 2000 metric tons (1 metric ton = 10^3 kg) of coal per day. The overall effi-

ciency of the station is 40%. Find the number of megawatts of electricity the station produces.

16–7   Internal Combustion Engines

29. A four-cylinder, two-stroke diesel engine has pistons 108 mm in diameter whose travel is 127 mm. The engine develops 100 kW at 33 rev/s. Find the average pressure on the pistons during the power stroke.

30. The six-cylinder, four-stroke gasoline engine of a car has pistons 3.5 in. in diameter whose travel is 4 in. If the average pressure on the pistons during the power stroke is 70 lb/in.^{2}, find the number of horsepower developed by the engine when it operates at 3000 rpm.

ANSWERS TO MULTIPLE CHOICE

1. b	**6.** a	**10.** c	**14.** c	**18.** c
2. b	**7.** b	**11.** b	**15.** b	**19.** a
3. b	**8.** d	**12.** a	**16.** c	**20.** b
4. a	**9.** d	**13.** d	**17.** d	**21.** d
5. a				

17

HEAT TRANSFER

Heat can be transferred from one place to another in three different ways: conduction, convection, and radiation. When we put one end of an iron poker in a fire, the other end becomes warm as a result of the conduction of heat through the iron. Conduction is a very slow process in air; a stove warms a room chiefly through the actual movement of heated air, a process called convection. Neither conduction nor convection can take place to any real extent in space, which is virtually empty of matter. Instead, the heat the earth receives from the sun arrives in the form of radiation. These mechanisms all embody a fundamental fact: The natural direction of heat flow is from hot bodies to cold ones. A refrigerator is a device that transfers heat in the opposite direction, from cold to hot, and it needs a source of work in order to do so. The principles of refrigeration are examined in the latter part of this chapter.

17–1 CONDUCTION

Mechanisms of heat conduction

The three basic mechanisms of heat transfer are illustrated in Fig. 17–1. In most materials, conduction is a simple consequence of the kinetic behavior of matter. Mol-

CHAPTER OBJECTIVES

Completing this chapter should enable you to:

1. Calculate the rate at which heat is conducted through a layer of a given material.

2. Find the thermal resistance of a sandwich of different materials and use this to calculate its rate of heat conduction.

3. Compare the rates at which an object loses heat by convection for different values of ΔT.

4. Calculate the rate at which an object emits radiation.

5. Describe the operating principles of a refrigerator.

6. Find the coefficient of performance of actual and ideal refrigerators.

7. Describe how a heat pump operates.

8. Predict the performance of a refrigeration system in a given situation on the basis of its capacity rating.

ecules (or atoms, depending upon the nature of the rod) at the hot end of a rod vibrate faster and faster as the temperature there increases. When these molecules collide with their less energetic neighbors, some kinetic energy is transferred to them (Fig. 17–2). Through such molecular collisions energy travels down the rod, and, since the energy of random molecular motion is heat, we equally well describe the situation by saying that heat travels down the rod. The average positions of the molecules themselves do not change in conduction.

For heat to be conducted through a body, its ends must be at different temperatures. If the entire body is at the same temperature, all its molecules have the same average energy, and any molecule has as much chance of losing energy in a collision with a nearby molecule as it has of gaining energy. In this case there is no flow of energy from a region of rapidly moving molecules to an adjacent one of slowly moving molecules.

There are wide differences in the ability of various substances to conduct heat. Gases are poor conductors, because their molecules are relatively far apart and so do not collide often. The molecules of liquids and nonmetallic solids are closer together,

Metals are the best conductors of heat

FIG. 17–1 Mechanisms of heat transfer.

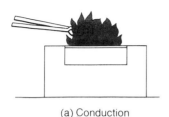

(a) Conduction

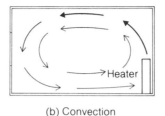

→Heater

(b) Convection

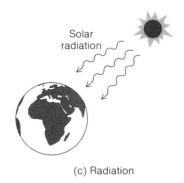

Solar radiation

(c) Radiation

FIG. 17–2 Molecules at the hot end of a rod vibrate faster than those at the cold end.

leading to somewhat higher thermal conductivities. Metals exhibit by far the greatest ability to conduct heat; this is why saucepans are made of metal but have handles of wood or plastic.

Free electrons in a metal enable it to conduct heat and electricity well

The reason for the exceptional thermal conductivity of metals is the same as for their exceptional electrical conductivity: Many of their electrons are able to move about freely instead of being bound permanently to particular atoms. Picking up kinetic energy at the hot end of a metal object, the free electrons can travel past many atoms before giving up their energy in collisions, and thereby can speed up the rate of energy transport toward the cold end of the object. Heat conduction by free electrons in a metal compares with heat conduction by molecular interactions as travel by express train compares with travel by local train. We shall further explore the important subject of free electrons in metals in Chapter 29.

Factors that determine rate of heat conduction

The rate at which heat flows through a slab of some material depends upon four quantities (Fig. 17–3):

1. The temperature difference ΔT between the faces of the slab; the greater the temperature difference, the faster the heat transfer.
2. The thickness d of the slab; the thicker the slab, the slower the heat transfer.
3. The area A of the slab; the greater the area, the faster the heat transfer.
4. The *thermal conductivity k* of the material, which is a measure of its ability to conduct heat; the larger the value of k, the faster the heat transfer.

If Q is the amount of heat that passes through the slab in the time t, then

$$\frac{Q}{t} = \frac{kA\,\Delta T}{d} \qquad\qquad \textit{Heat conduction} \quad (17\text{–}1)$$

In SI units, Q/t is expressed in watts, ΔT is in °C, d in m, A in m^2, and k in W/m · °C.

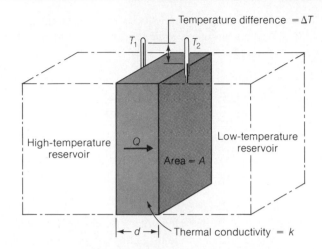

FIG. 17–3 The rate Q/t at which heat flows through the slab is equal to $kA\,\Delta T/d$.

When Q/t is expressed in kcal/s, the corresponding unit for k is the kcal/m · s · °C. In the British system, it is customary in problems that involve heat transfer to have Q/t in Btu/h, ΔT in °F, d in in., and A in ft². The British unit for k is therefore the Btu/(ft² · h · °F/in.). Table 17–1 is a list of the thermal conductivities of a number of common materials.

TABLE 17–1
Thermal conductivities

Material	k, $\dfrac{kcal}{m \cdot s \cdot °C}$	k, $\dfrac{W}{m \cdot °C}$	k, $\dfrac{Btu}{ft^2 \cdot h \cdot °F/in.}$
Metals			
Aluminum	4.9×10^{-2}	205	1.4×10^3
Brass	2.6×10^{-2}	109	0.76×10^3
Copper	9.2×10^{-2}	385	2.7×10^3
Iron and steel	1.2×10^{-2}	50	0.35×10^3
Silver	9.7×10^{-2}	406	2.8×10^3
Other materials			
Body fat	0.4×10^{-4}	0.17	1.2
Brick	1.5×10^{-4}	0.6	4.4
Concrete	2×10^{-4}	0.8	5.8
Glass	1.9×10^{-4}	0.80	5.5
Ice	4×10^{-4}	1.6	12
Water	1.4×10^{-4}	0.60	4.2
Wood (pine)	0.3×10^{-4}	0.13	0.87
Snow	2×10^{-4}	0.8	5.8
Insulating materials			
Cork	0.1×10^{-4}	0.04	0.3
Glass wool	0.1×10^{-4}	0.04	0.3
Down	0.06×10^{-4}	0.02	0.17
Kapok	0.08×10^{-4}	0.03	0.23
Gases			
Hydrogen	0.3×10^{-4}	0.13	0.9
Air	0.06×10^{-4}	0.024	0.17

Example   A door of pine wood is to be installed in a brick wall 10 in. thick. If the rate of heat conduction is to be no greater than before, what minimum thickness should the door have?

Solution   Since A and ΔT are equal in both cases, the ratio k/d must be the same for Q/t to be the same. Hence

$$\frac{k_{\text{brick}}}{d_{\text{brick}}} = \frac{k_{\text{pine}}}{d_{\text{pine}}}$$

$$d_{\text{pine}} = \left(\frac{k_{\text{pine}}}{k_{\text{brick}}}\right)(d_{\text{brick}}) = \left(\frac{0.87}{4.4}\right)(10\text{ in.}) = 2.0\text{ in.}$$ ■

Example   An electric water heater has a cylindrical tank 60 cm in diameter and 50 cm long that is insulated with 2.0 cm of glass wool. If the water in the tank is at 80°C and the outside of the insulating jacket is at 25°C, at what rate does the tank lose heat to the air around it? How many kWh of electric energy must be provided each day to keep the water at 80°C even if no water is drawn from the tank?

Solution   (a) Each end of the tank is a circle 30 cm in radius (see Fig. 17–4) whose area is

$$A_{\text{end}} = \pi r^2 = \pi(0.30\text{ m})^2 = 0.283\text{ m}^2$$

The body of the tank is a cylinder 30 cm in radius and 50 cm long whose area is

$$A_{\text{cylinder}} = (\text{circumference})(\text{length}) = (2\pi r)(L) = (2\pi)(0.30\text{ m})(0.50\text{ m})$$
$$= 0.942\text{ m}^2$$

The total area of the tank is

$$A = 2A_{\text{end}} + A_{\text{cylinder}} = 2(0.283\text{ m}^2) + 0.942\text{ m}^2 = 1.51\text{ m}^2$$

Since $k = 0.04$ W/m · °C for glass wool, $\Delta T = 55$°C, and $d = 0.020$ m, the rate at which heat is lost from the tank by conduction is

$$\frac{Q}{t} = \frac{kA\Delta T}{d} = \frac{(0.04\text{ W/m} \cdot \text{°C})(1.51\text{ m}^2)(55\text{°C})}{0.020\text{ m}} = 166\text{ W}$$

(b) In the course of 24 h the total energy needed to make up the heat loss is

$$Q = \left(\frac{Q}{t}\right)(t) = (0.166\text{ kW})(24\text{ h}) = 4.0\text{ kWh}$$

FIG. 17–4

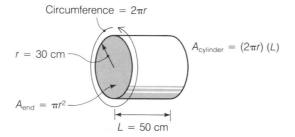

Circumference = $2\pi r$

$r = 30$ cm

$A_{\text{cylinder}} = (2\pi r)\ (L)$

$A_{\text{end}} = \pi r^2$

$L = 50$ cm

It is worth noting that the statement of the problem gives the temperature of the outside of the tank. If instead we were only given the room temperature, we would not know by just how much the insulating effect of the air layers outside the tank affects the actual temperature at the tank's surface. This insulating effect can be large enough to reduce the heat flow considerably. Only if there is a very vigorous circulation of air past the tank will the temperature of the outside of the tank equal room temperature. ■

Because the rate of heat flow Q/t through an insulating layer depends upon $1/d$, in any particular application it usually does not make sense to increase the insulation thickness d beyond a certain value. Consider a freezer that absorbs 2000 kcal/day through 5 cm of insulation. Doubling the thickness to 10 cm will decrease the heat flow to 1000 kcal/day. A further doubling to 20 cm of insulation will reduce the heat flow to 500 kcal/day. As the thickness increases, the rate of change of heat flow goes down: Going from 5 cm to 10 cm cuts the heat flow by 1000 kcal/day, but going from 10 cm to 20 cm only brings it down by an additional 500 kcal/day. Clearly there is a point at which the cost of more insulation averaged over an appropriate period of time will exceed the savings in operating costs. Just what this point is depends, of course, on the relative costs involved, including that of the space occupied by the insulation.

Choosing insulation thickness

17–2   THERMAL RESISTANCE

The insulating ability of a layer of some material is often given in terms of its *thermal resistance R*. If the thermal conductivity of the material is k and it is d thick,

The greater the thermal resistance, the greater the insulating ability

$$\text{Thermal resistance} = \frac{\text{thickness}}{\text{thermal conductivity}}$$

$$R = \frac{d}{k} \qquad\qquad \textit{Thermal resistance} \quad (17\text{–}2)$$

The greater the value of R, the greater the resistance to the flow of heat. In British units, thermal resistance is expressed in $\text{ft}^2 \cdot {}^\circ\text{F}/(\text{Btu/h})$. Thus the thermal resistance of 2.0 in. of pine wood is

$$R = \frac{d}{k} = \frac{2.0\,\text{in.}}{0.87\,\text{Btu/(ft}^2 \cdot \text{h} \cdot {}^\circ\text{F/in.)}} = 2.3\frac{\text{ft}^2 \cdot {}^\circ\text{F}}{\text{Btu/h}}$$

We can think of this result in two ways (Fig. 17–5). (1) If the temperature difference across the wood is 2.3°F, then 1 Btu/h of heat flows through each ft^2 of it. (2) If the temperature difference is 1°F, then 1 Btu/h of heat flows through each 2.3 ft^2. Usually R values are given without units, so that the above result would be expressed as "R-2.3." In terms of R rather than k/d, Eq. (17–1) becomes

$$\frac{Q}{t} = \frac{A\Delta T}{R} \qquad\qquad \textit{Heat conduction} \quad (17\text{–}3)$$

Example   Find the rate at which the heat flows through a 2.0-in. pine door 7.0 ft high and 34 in. wide when the inside temperature is 70°F and the outside temperature is 80°F.

FIG. 17–5 Two ways to interpret a thermal resistance of $R = 2.3$ ft²·°F/(Btu/h).

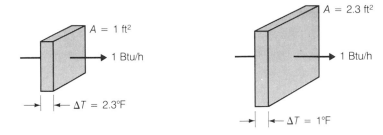

Solution   The width of the door is $(34 \text{ in.})/(12 \text{ in./ft}) = 2.83$ ft, so its area is

$$A = (7.0 \text{ ft})(2.83 \text{ ft}) = 20 \text{ ft}^2$$

From Eq. (17–3) the heat flow is therefore

$$\frac{Q}{t} = \frac{A\Delta T}{R} = \frac{(20 \text{ ft}^2)(10°F)}{2.3 \text{ ft}^2 \cdot °F/(Btu/h)} = 87 \text{ Btu/h} \qquad \blacksquare$$

Thermal resistance of a composite material

The advantage of using R values is that, when we have a sandwich of two or more layers of different materials, the R value for the sandwich is the sum of the R values of the individual layers:

$$R = R_1 + R_2 + R_3 + \cdots \qquad\qquad \textit{Composite material} \quad (17–4)$$

Example   Compare the thermal resistance of a pane of glass $\frac{1}{4}$ in. thick with that of a sandwich consisting of two panes of glass $\frac{1}{8}$ in. thick separated by $\frac{1}{4}$ in. of air (Fig. 17–6).

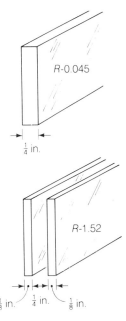

FIG. 17–6 A double-glazed window has a much higher thermal resistance than a single-glazed one.

Solution   (a) The R value of the single pane of glass is

$$R = \frac{d}{k} = \frac{0.25 \text{ in.}}{5.5 \text{ Btu}/(\text{ft}^2 \cdot \text{h} \cdot °F/\text{in.})} = 0.045 \frac{\text{ft}^2 \cdot °F}{\text{Btu/h}} = R - 0.045$$

(b) The R value of glass $\frac{1}{8}$ in. thick is half that of glass $\frac{1}{4}$ in. thick, so here $R_1 = R_3 = 0.023 \text{ ft}^2 \cdot °F/(Btu/h)$. The R value of the air gap is

$$R_2 = \frac{d}{k} = \frac{0.25 \text{ in.}}{0.17 \text{ Btu}/(\text{ft}^2 \cdot \text{h} \cdot °F/\text{in.})} = 1.47 \frac{\text{ft}^2 \cdot °F}{\text{Btu/h}} = R - 1.47$$

Thus the thermal resistance of the glass-air-glass sandwich is

$$R = R_1 + R_2 + R_3 = (0.023 + 1.47 + 0.023) \frac{\text{ft}^2 \cdot °F}{\text{Btu/h}} = 1.52 \frac{\text{ft}^2 \cdot °F}{\text{Btu/h}} = R - 1.52$$

This is nearly 34 times greater than the R value of the same total thickness of glass found in (a). The advantage of double-glazed windows in regions with cold winters is obvious. For even better insulation, triple-glazed windows can be used. Such a window consists of three panes of glass separated by two air spaces. Why not just have a thicker air space between two panes of glass? The reason is that heat flow through a thick layer of air occurs by means of convection as well as conduction, and the insulating effect is thereby reduced (see Section 17–3). $\blacksquare$

17–3   CONVECTION

Convection is a much simpler physical process than conduction, since it consists of the actual motion of a volume of hot fluid from one place to another. The hot fluid pushes aside cold fluid in its path, thereby setting up a *convection current*. Convection is the chief mechanism of heat transfer in fluids under most circumstances.

Convection involves the motion of a heated fluid

Convection may be either *natural* or *forced*. In natural convection, the buoyancy of a heated fluid leads to its motion; when a portion of a fluid (either gas or liquid) is heated, it expands to become of lower density than the surrounding, cooler fluid and hence rises upward (Fig. 17–7). A steam or hot-water heating system uses radiators in each room which heat the rooms with the help of the convection currents they set up.

Natural convection

In forced convection, a blower or pump sends the heated fluid to its destination. In the cooling system of most cars, water is circulated between the engine block and the radiator by a pump; in the radiator, heat is transferred to the atmosphere by conduction through the thin-walled metal tubes of which it is made. A hot-air household heating system uses a fan to blow the air from the furnace through ducts to outlets in each room.

Forced convection

The rate Q/t at which a hot object transfers heat to a surrounding fluid by convection is approximately proportional to the area A of the object in contact with the fluid and to the temperature difference ΔT between them:

Rate of heat transfer by convection

$$\frac{Q}{t} = hA \; \Delta T \qquad\qquad\qquad\qquad Convection \quad (17\text{–}5)$$

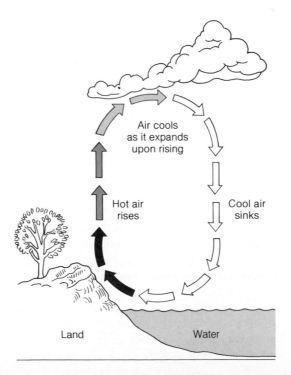

FIG. 17–7 The temperature of a land surface rises more rapidly than that of a water surface in sunlight. The resulting convection is responsible for the sea breeze experienced on sunny days near the shore of a body of water.

Air cools as it expands upon rising

Hot air rises

Cool air sinks

Land

Water

The convection coefficient h depends on the form and orientation of the object (it is nearly twice as great for the upper surface of a horizontal plate as it is for the lower surface) and on the properties and state of motion of the fluid (in the case of convective loss from the skin to air, h is twice as great when the air is moving at 1 m/s than it is in quiet air).

Example A photographic enlarger has as its light source a 25-W lamp in a metal hood that keeps light from escaping except through the enlarger lens. When the darkroom temperature is 20°C, the metal hood is at a temperature of 50°C. If the 25-W lamp is replaced by a 50-W lamp, what is the new temperature of the hood?

Solution Since h and A are each the same in both cases,

$$\frac{(Q/t)_1}{(Q/t)_2} = \frac{\Delta T_1}{\Delta T_2}$$

When $(Q/t)_1 = 25$ W, the temperature difference ΔT_1 between the hood and the surrounding air is 50°C − 20°C = 30°C. When $(Q/t)_2 = 50$ W,

$$\Delta T_2 = \Delta T_1 \frac{(Q/t)_2}{(Q/t)_1} = (30°\text{C})\left(\frac{50\text{ W}}{25\text{ W}}\right) = 60°\text{C}$$

Hence the new hood temperature is 20°C + 60°C = 80°C. ∎

17–4   RADIATION

Electromagnetic waves

In the process of radiation, energy is transported by means of *electromagnetic waves*. These waves travel at the speed of light (3×10^8 m/s = 186,000 mi/s), and require no material medium for their passage. Radio and radar waves, light waves, and X and gamma rays are all electromagnetic waves; they differ only in their wavelength. Figure 17–8 shows the classification of electromagnetic waves according to wavelength. Visible light ranges in wavelength from about 400 nm for violet light through blue, green, yellow, and orange light to about 700 nm for red light (1 nm = 1 nanometer = 10^{-9} m).

All objects radiate electromagnetic waves

Every object radiates electromagnetic waves of all wavelengths, though the intensities of the different wavelengths vary considerably. We are all familiar with the glow of a hot piece of metal, where enough radiation is emitted as visible light for our eyes to respond, but other wavelengths are also given off. An object need not be so hot that it gives off visible light for it to be radiating electromagnetic energy—the radiation from an object at room temperature, for instance, is mainly in the infrared part of the spectrum to which the eye is not sensitive.

A good absorber is a good radiator

The ability of an object to emit radiation is proportional to its ability to absorb radiation: A good absorber is a good emitter, and vice versa. This conclusion follows from the fact that something at the same temperature as its environment must be absorbing and emitting radiation at exactly the same rates. When an object is warmer than its environment, it emits more radiation than it absorbs, and it is this difference we detect. A perfect absorber is called a *blackbody*, and it is accordingly the best possible radiator as well.

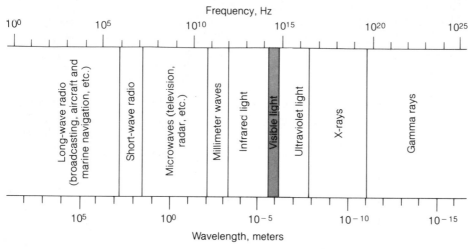

Frequency, Hz

Wavelength, meters

FIG. 17–8 The electromagnetic wave spectrum. The boundaries of the various categories are not sharp except in the case of visible light.

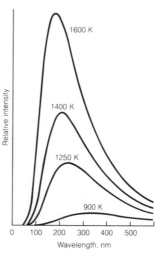

FIG. 17–9 The intensity of the electromagnetic radiation emitted by an object at various temperatures, as a function of wavelength. Visible light ranges in wavelength from 400 to 700 nm (1 nm = 10^{-9} m).

The rate R (in W/m²) at which an object of surface area A and absolute temperature T emits radiation is given by the *Stefan-Boltzmann law* as

$$R = \frac{P}{A} = e\sigma T^4 \qquad \textit{Stefan-Boltzmann law} \quad (17–6)$$

Radiation rate varies as T^4

The value of the constant σ (Greek letter *sigma*) is

$$\sigma = 5.67 \times 10^{-8} \text{ W/m}^2 \cdot \text{K}^4$$

The *emissivity e* depends on the nature of the radiating surface and ranges from 0, for a perfect reflector that does not radiate at all, to 1, for a blackbody. Some typical values of e are 0.07 for polished steel, 0.6 for oxidized copper and brass, and 0.97 for matte black paint.

Figure 17–9 shows how the intensity of the radiation emitted by an object varies with wavelength at various temperatures. The total emitted radiation (which is proportional to the area under each curve) increases with increasing temperature, while the

The higher an object's temperature, the shorter the predominant wavelength of its radiation

FIG. 17–10 (a) The color of an object hot enough to glow varies with its temperature approximately as shown here. (b) This thermogram of a healthy woman shows skin temperatures as shades of gray. Lighter regions at the center of a breast would correspond to the higher temperatures associated with cancer. (Photo: R. Clark/M. Goff)

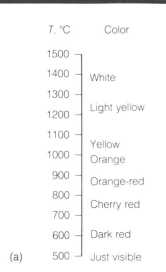

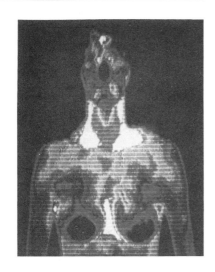

(a) (b)

wavelength corresponding to the peak of each curve decreases. An object that glows red is not as hot as one that glows white, since red light has the longer wavelength (see Chapter 27). Figure 17–10(a) shows how the color of a hot object varies with its temperature.

Thermograph

Example A *thermograph* is a device that measures the amount of infrared radiation each small portion of a person's skin emits and presents this information in pictorial form by different shades of gray or different colors in a *thermogram* (Fig. 17–10b). The skin over a tumor is warmer than elsewhere (perhaps because of increased blood flow or a higher rate of metabolism), and thus a thermogram is a valuable diagnostic aid for detecting such diseases as breast and thyroid cancer. To verify that a small difference in skin temperature leads to a significant difference in radiation rate, calculate the percentage difference between the radiation from skin at 34°C and at 35°C.

Solution The emissivity of the skin is the same at both temperatures, whose values on the absolute scale are $T_1 = 34°C + 273 = 307$ K and $T_2 = 35°C + 273 = 308$ K. Since R_1 is proportional to T_1^4 and R_2 is proportional to T_2^4,

$$\frac{R_2 - R_1}{R_1} = \frac{T_2^4 - T_1^4}{T_1^4} = \frac{(308\,\text{K})^4 - (307\,\text{K})^4}{(307\,\text{K})^4} = 0.013$$

which is 1.3%. ■

FIG. 17–11 A small hole in the wall of a cavity acts as a blackbody since any radiation that falls on it is trapped inside until absorbed. The emissivity of such a hole is $e = 1.00$.

Example A small hole in the wall of a cavity in an object of any kind behaves like a blackbody because any radiation that falls on it is trapped inside by reflections from the cavity wall until it is absorbed (Fig. 17–11). At what rate does radiation escape from a hole 10 cm² in area in the wall of a furnace whose interior is at a temperature of 700°C?

Solution The absolute temperature of the interior of the furnace is $T = 700°C + 273 = 973$ K. Since the hole acts as a blackbody, its emissivity is $e = 1.00$. The hole's area is $A = (10\,\text{cm}^2)(10^{-4}\,\text{m}^2/\text{cm}^2) = 10^{-3}\,\text{m}^2$. From Eq. (17–6) the hole radiates energy at the rate

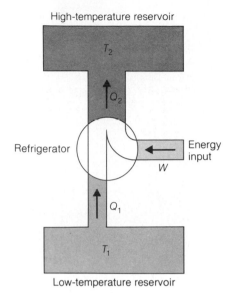

High-temperature reservoir

Low-temperature reservoir

FIG. 17-12 (Left) A refrigerator takes heat from a low-temperature reservoir and transfers it to a high-temperature one. Energy must be supplied to permit this reverse flow of heat to occur.

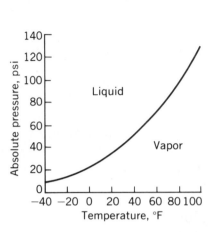

FIG. 17-13 (Right) Vaporization curve of Freon 12, a common refrigerant. Freon 12 is a trade name for dichlorodifluoromethane, CCl_2F_2.

$$P = RA = e\sigma AT^4 = (1.00)(5.67 \times 10^{-8}\,\text{W/m}^2 \cdot \text{K}^4)(10^{-3}\,\text{m}^2)(973\,\text{K})^4$$
$$= 51\,\text{W} \qquad \blacksquare$$

17-5 THE REFRIGERATOR

A *refrigerator* is a heat engine operating in reverse. Ordinarily a heat engine absorbs heat from a high-temperature reservoir and gives it off to a low-temperature one, with an output of mechanical work being provided in the process. In a refrigerator, on the other hand, heat is transferred from a low-temperature reservoir (typically a storage chamber) to a high-temperature one (the outside world), and mechanical energy must be supplied in order to do this. In effect, heat must be "pushed uphill" if it is to go from a cold region to a warm one (Fig. 17–12). In a typical refrigerator, two to three times as much heat is extracted from the storage chamber as the amount of external energy provided.

It is important to keep in mind that a refrigerator does not "produce cold," since cold is a relative deficiency of internal energy and not something in its own right, as heat is. What a refrigerator does is to remove internal energy from a certain place and transport it elsewhere.

In nearly all refrigerators the working substance (or *refrigerant*) is a gas that is easily liquified. Common refrigerants are ammonia, Freon 12, methyl chloride, and sulfur dioxide. Figure 17–13 shows the vaporization curve of Freon 12. Under conditions of temperature and pressure corresponding to points above the curve only liquid Freon is present, while under conditions corresponding to points below the curve only Freon vapor is present. Other refrigerants have different vaporization curves. At atmospheric pressure Freon 12 boils at $-22°F$, while at this pressure ammonia boils at

Energy is needed to reverse the natural flow of heat

Refrigerant

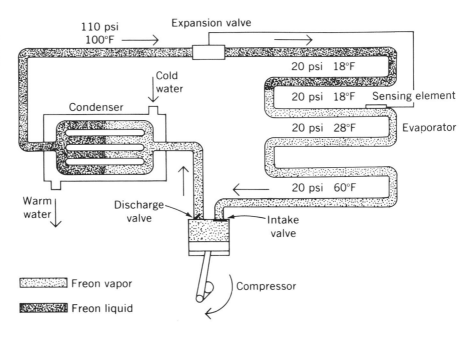

FIG. 17–14 A refrigeration system using Freon 12. 1 psi = 1 lb/in². Heat from the refrigerated space enters the system through the evaporator and leaves it through the condenser. Gauge pressures are shown.

28°F, methyl chloride at − 11°F, and sulfur dioxide at 14°F. The choice of a refrigerant depends upon the precise kind of refrigerator involved and the temperatures between which it is to operate.

Typical refrigeration system

Let us examine a typical refrigeration system that uses Freon 12. As in Fig. 17–14, this system consists of a *compressor,* a *condenser,* an *expansion valve,* and an *evaporator.* When the piston of the compressor moves downward, Freon vapor at 20 psi (gauge pressure) and approximately room temperature is sucked into the cylinder. As the piston reaches the bottom of its stroke and begins to move upward, the intake valve closes and the discharge valve opens. The compressed Freon, which emerges at a pressure of 110 psi and a high temperature, passes into the condenser in which it is cooled until it liquifies. The condenser may be water cooled, as shown in Fig. 17–14, or air cooled. It is in this stage that the heat extracted from the refrigerated space passes to the outside world.

The liquid Freon then goes into the expansion valve from which it emerges at a lower pressure (20 psi) and temperature (18°F). The amount of Freon supplied by the valve is regulated by a sensing element placed in the refrigerated space; this may be simply a gas-filled bulb that responds to temperature changes by pressure changes that operate a bellows in the valve. As the cold liquid Freon flows through the evaporator tubes it absorbs heat from the region being cooled. From Fig. 17–13 we see that, at an absolute pressure of 35 psi, the boiling point of Freon 12 is 18°F, so that the heat absorbed by the liquid Freon in the evaporator causes it to vaporize. Farther along in the evaporator the Freon vapor itself absorbs heat, rising in temperature to perhaps 28°F. The amount of this temperature rise (called *super heat* by refrigeration engineers) is critical for the efficiency of the system, and is usually 10°F or less. Finally the Freon vapor leaves the evaporator and enters the compressor to begin another cycle.

17–6 THERMODYNAMICS OF REFRIGERATION

A refrigerator absorbs heat at a low temperature and gives it off at a high temperature. According to the second law of thermodynamics, such a cycle cannot take place without an external source of energy to force heat to flow in the reverse of its usual direction. In the most common type of refrigerator, the compressor is the means through which work is done on the refrigerant as it circulates between the evaporator, where it picks up heat, and the condenser, where the heat is given up. If Q_1 is the heat absorbed at the low temperature T_1, and Q_2 is the heat ejected at the high temperature T_2 (Fig. 17–12), then the work done on the refrigerator is

$$W = Q_2 - Q_1$$

The *coefficient of performance* is the ratio between the absorbed heat and the work required:

Coefficient of performance

$$\text{Coefficient of performance} = \frac{\text{heat absorbed}}{\text{work done}}$$

$$CP = \frac{Q_1}{W}$$

$$= \frac{Q_1}{Q_2 - Q_1} \qquad \textit{Coefficient of performance} \quad (17\text{–}7)$$

The higher the coefficient of performance, the less work is needed to remove a given amount of heat from the cold chamber. Typical coefficient of performance values range from 2 to 3: Two to three times as much heat is removed as the amount of work done.

A quantity called *energy efficiency ratio* (EER) is sometimes used instead of coefficient of performance. The energy efficiency ratio of a refrigerator or air conditioner is the ratio between the heat in Btu extracted per hour and the power in watts the machine consumes:

Energy efficiency ratio

$$\text{EER} = \frac{\text{heat absorbed per hour (Btu/h)}}{\text{power consumption (watts)}} = 3.42 \, CP$$

Thus a 3.5-kW air conditioner that removes 25,000 Btu per hour has an energy efficiency ratio of 7.1 Btu/Wh, which is equivalent to a coefficient of performance of 2.1.

An *ideal refrigerator* is just a Carnot engine operating backwards. As in a Carnot engine, the heat transferred to or from a Carnot refrigerator is proportional to the absolute temperature T at which the transfer takes place, and so

Ideal refrigerator is Carnot engine running backwards

$$\frac{Q_2}{Q_1} = \frac{T_2}{T_1}$$

The coefficient of performance of a Carnot refrigerator is accordingly

$$CP = \frac{Q_1}{Q_2 - Q_1} = \frac{T_1}{T_2 - T_1} \qquad \textit{Ideal refrigerator} \quad (17\text{–}8)$$

The more nearly equal the two temperatures are, the less work is needed: The temperature "hill" is less steep. Here is where practical difficulties arise, because a low output temperature T_2 is not always easy to arrange. For one thing, the higher T_2 is, the faster is the exchange of heat between refrigerant and cooling water or air in the condenser. A high ideal coefficient of performance in a refrigerator may therefore mean a low actual coefficient of performance when the power consumed by the water pump or fan in the condenser is taken into account. Equation (17–8) applies to an ideal refrigerator, and actual refrigerators, in which friction and unwanted heat transfer from the environment are present, are less efficient owing to these factors as well.

Example   A 500-W refrigerator with a coefficient of performance of 2.0 extracts heat from a storage chamber at $-20°C$ and exhausts it at $40°C$. What would the coefficient of performance of an ideal refrigerator operating between the same two temperatures be? How many joules per hour can the refrigerator extract from the storage chamber?

Solution   (a) The absolute temperatures that correspond to $-20°C$ and $40°C$ are respectively

$$T_1 = -20°C + 273 - 253 \text{ K} \qquad T_2 = 40°C + 273 = 313 \text{ K}$$

Hence the coefficient of performance of the ideal refrigerator would be

$$\text{CP} = \frac{T_1}{T_2 - T_1} = \frac{253 \text{ K}}{313 \text{ K} - 253 \text{ K}} = 4.2$$

The actual refrigerator is a little less than half as efficient as an ideal one would be.
    (b) From Eq. (17–7), $Q_1 = (\text{CP})(W)$. Since $P = W/t$,

$$\frac{Q_1}{t} = (\text{CP})\left(\frac{W}{t}\right) = (\text{CP})(P) = (2.0)(500 \text{ W}) = 1000 \text{ W}$$

To express this result in J/h we note that 1000 W = 1000 J/s and 1 h = 3600 s. Hence

$$\frac{Q_1}{t} = \left(1000 \frac{\text{J}}{\text{s}}\right)\left(3600 \frac{\text{s}}{\text{h}}\right) = 3.6 \times 10^6 \text{J/h} = 3.6 \text{ MJ/h} \qquad \blacksquare$$

Heat pump

   A *heat pump* is a refrigeration system that can extract heat from the cold outdoors in winter and deliver it to the interior of a house. In summer, the same system can serve as an air conditioner to take heat from the house and exhaust it to the warmer outdoors (Fig. 17–15). The concept of the heat pump is interesting because more heat is transferred than the work done. From Eq. (17–7), $\text{CP} = Q_1/W$, and so, since $Q_1 = Q_2 - W$, we have $\text{CP} = (Q_2 - W)/W$ and

$$Q_2 = (\text{CP} + 1)W \qquad\qquad\qquad \textit{Heat pump} \quad (17–9)$$

The heat Q_2 delivered by a heat pump in a certain period of time is $(\text{CP} + 1)$ times greater than the energy W supplied to it. For example, a 5-kW heat pump with a coefficient of performance of 3 can provide a house with heat at a rate of $(3 + 1)$ (5 kW) = 20 kW. An ordinary furnace would have to burn fuel at the same rate of 20 kW and so would require four times as much energy input as the heat pump. However, heat pumps have a cost problem to overcome: Energy in the form of fossil fuel (coal, oil, gas) is still quite a bit cheaper than the electrical energy needed to operate a heat pump, and in addition a heat pump is a more expensive installation than an ordi-

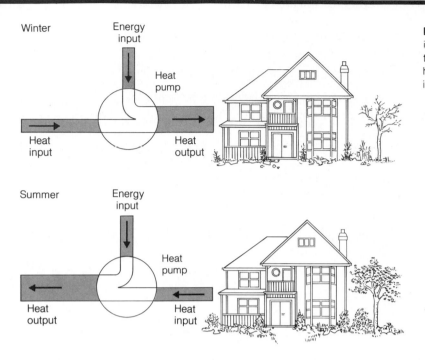

nary heating plant. Only when a heat pump is needed for air conditioning as well as for space heating can it at present be a practical proposition.

17–7   REFRIGERATION CAPACITY

The *capacity* of a refrigerator is the rate at which it is able to remove heat. The conventional unit of refrigeration capacity is the *ton,* which is defined as that rate of heat removal that can freeze 1 ton of water at 32°F to ice at 32°F per day. Because the heat of fusion of water is 144 Btu/lb,

The refrigeration ton is a unit of power

$$1 \text{ refrigeration ton} = (144 \text{ Btu/lb})(2000 \text{ lb/day}) = 2.88 \times 10^5 \text{ Btu/day}$$

In terms of heat removed per hour,

$$1 \text{ refrigeration ton} = \frac{2.88 \times 10^5 \text{ Btu/day}}{24 \text{ h/day}} = 12{,}000 \text{ Btu/h}$$

This is equivalent to 12.65 MJ/h or 3.51 kW.

Example   The freezing point of beef is 28°F and its heat of fusion is 100 Btu/lb. The specific heat of beef is normally 0.75 Btu/lb · °F and drops to 0.40 Btu/lb · °F when it is frozen. Find the refrigeration capacity required to freeze 800 lb of beef per hour that is initially at 50°F and to cool it further to 5°F.

Solution   The heat Q_a that must be removed to bring 800 lb of beef from 50°F to 28°F is

$$Q_1 = mc\,\Delta T = (800\,\text{lb})\left(0.75\frac{\text{Btu}}{\text{lb}\cdot{}^\circ\text{F}}\right)(22^\circ\text{F}) = 13{,}200\,\text{Btu}$$

To freeze the beef at 28°F requires the removal of

$$Q_2 = mL_f = (800\,\text{lb})(100\,\text{Btu/lb}) = 80{,}000\,\text{Btu}$$

while to cool it further to 5°F after it has been frozen requires the removal of

$$Q_3 = mc\,\Delta T = (800\,\text{lb})\left(0.40\frac{\text{Btu}}{\text{lb}\cdot{}^\circ\text{F}}\right)(23^\circ\text{F}) = 7360\,\text{Btu}$$

The total heat to be extracted is

$$Q = Q_1 + Q_2 + Q_3 = 1.01 \times 10^5\,\text{Btu}$$

The refrigeration capacity required to cool, freeze, and further cool 800 lb of beef per hour from 50°F to 5°F is therefore

$$\frac{Q}{t} = \frac{1.01 \times 10^5\,\text{Btu/h}}{1.2 \times 10^4\,(\text{Btu/h})/\text{ton}} = 8.4\,\text{tons}$$ ∎

IMPORTANT TERMS

Heat can be transferred from one place to another by means of **conduction, convection,** or **radiation.** In conduction heat is transported by successive molecular collisions, and in convection by the motion of a volume of hot fluid from one place to another. The **thermal conductivity** of a material is a measure of its ability to conduct heat. The **thermal resistance** of a layer of a material is a measure of its insulating ability.

Heat transfer by radiation takes place by means of **electromagnetic waves** that require no material medium for their passage.

A **refrigerator** is a device that transfers heat from a cold reservoir to a hot one, and it must perform work to do this. In essence, it is a heat engine operating in reverse.

The **coefficient of performance** of a refrigerator is the amount of heat removed per unit of work done.

The **capacity** of a refrigerator is the rate at which it can remove heat. The conventional unit of refrigeration capacity is the **ton,** which is that rate of heat removal that can freeze 1 ton of water at its freezing point to ice at the same temperature per day; one refrigeration ton = 12,000 Btu/h.

IMPORTANT FORMULAS

Heat conduction: $\dfrac{Q}{t} = \dfrac{kA\,\Delta T}{d} = \dfrac{A\,\Delta T}{R}$

Thermal resistance: $R = \dfrac{d}{k}$

$$= R_1 + R_2 + R_3 + \cdots$$

Convection: $\dfrac{Q}{t} = hA\,\Delta T$

Stefan-Boltzmann law: $\dfrac{P}{A} = e\sigma T^4$

Refrigeration coefficient of performance:

$$\text{CP} = \frac{Q_1}{W} = \frac{Q_1}{Q_2 - Q_1}$$

$$= \frac{T_1}{T_2 - T_1}\,(\text{ideal refrigerator})$$

MULTIPLE CHOICE

1. The natural direction of the heat flow between two reservoirs depends on
 a. their temperatures.
 b. their internal energy contents.
 c. their pressures.
 d. whether they are in the solid, liquid, or gaseous state.

2. Body *A* is at a higher temperature than body *B*. When they are placed in contact, heat will flow from *A* to *B*

a. only if A has the greater internal energy content.

b. only if both are fluids.

c. only if A is on top of B.

d. until both have the same temperature.

3. The rate at which heat flows through a slab of some material does not depend on

a. the temperature difference between the faces of the slab.

b. the thickness of the slab.

c. the area of the slab.

d. the specific heat of the material.

4. Metals are good conductors of heat because

a. they contain free electrons.

b. their atoms are relatively far apart.

c. their atoms collide infrequently.

d. they have reflecting surfaces.

5. The materials with the highest heat conductivities are the

a. gases.   b. liquids.

c. woods.   d. metals.

6. Heat transfer by conduction occurs

a. only in liquids.

b. only in solids.

c. only in liquids and solids.

d. in gases, liquids, and solids.

7. Heat transfer by convection occurs

a. only in gases.

b. only in liquids.

c. only in gases and liquids.

d. in gases, liquids, and solids.

8. In natural convection, a heated portion of a fluid moves because

a. its molecular motions become aligned.

b. of molecular collisions within it.

c. its density is less than that of the surrounding fluid.

d. of currents in the surrounding fluid.

9. Electromagnetic radiation is emitted

a. only by radio and television antennas.

b. only by objects hotter than their surroundings.

c. only by objects hot enough to glow.

d. by all objects.

10. The rate at which an object radiates electromagnetic energy does not depend upon its

a. surface area.

b. mass.

c. temperature.

d. ability to absorb radiation.

11. Four pieces of iron are heated in a furnace to different temperatures. The one at the highest temperature appears

a. white.   b. yellow.

c. orange.   d. red.

12. The physics underlying the operation of a refrigerator more closely resembles the physics underlying

a. a heat engine.

b. the melting of ice.

c. the freezing of water.

d. the evaporation of water.

13. A refrigerator

a. produces cold.

b. causes heat to vanish.

c. removes heat from a region and transports it elsewhere.

d. changes heat to cold.

14. A refrigerator exhausts

a. less heat than it absorbs from its contents.

b. the same amount of heat it absorbs from its contents.

c. more heat than it absorbs from its contents.

d. any of the above, depending on the circumstances.

15. Heat is absorbed by the refrigerant in a refrigeration system when it

a. melts.   b. vaporizes

c. condenses.   d. is compressed.

16. A heat pump is basically

a. a pump for hot air.

b. a pump for hot water.

c. a heat engine.

d. a refrigerator.

17. A refrigeration ton is the rate of heat removal that can

a. cool 1 ton of water by 1°F per hour.

b. cool 1 ton of water by 1°F per day.

c. freeze 1 ton of water at 32°F per hour.

d. freeze 1 ton of water at 32°F per day.

18. The 3-mm iron door of an oven is replaced by a brass door 9 mm thick. With the oven and room temperatures the same as before, the rate of heat flow through the door relative to its original rate is

a. 46%.   b. 72%.

c. 138%.   d. 300%.

19. A brick wall is 5 m long, 3 m high, and 30 cm thick. The conductivity of brick is 0.6 W/m · °C. When the outside temperature of the wall is 0°C and the inside temperature is 20°C, the rate at which heat flows through the wall is

a. 54 W.   b. 60 W.

c. 0.6 kW.   d. 1.67 kW.

20. When the difference in temperature between the sides of a wooden board 80 cm long, 50 cm wide, and 10 mm

thick is 40°C, heat flows through the board at the rate of 240 W. The thermal conductivity of the wood is
   a. 0.0015 W/m·°C.
   b. 0.15 W/m·°C.
   c. 6.67 W/m·°C.
   d. 6.67 J/m²·°C.

21. The thermal conductivity of cork is 0.3 Btu (ft² · h · °F/in.). An R-2.5 layer of cork has a thickness of
   a. 0.12 in.
   b. 0.75 in.
   c. 2.5 in.
   d. 8.3 in.

22. A brass rod at a temperature of 150°C radiates energy at a rate of 20 W. If its temperature is increased to 300°C, at what rate does it radiate now?
   a. 27 W
   b. 40 W
   c. 67 W
   d. 320 W

23. The average temperature of the earth's surface is 15°C. Assuming that the earth radiates as a blackbody, the rate at which it loses energy by radiation is
   a. 0.0029 W/m².
   b. 251 W/m².
   c. 288 W/m².
   d. 390 W/m².

24. A refrigerator gives off 3 J of heat for every joule of mechanical energy input. Its coefficient of performance is
   a. 1.
   b. 2.
   c. 3.
   d. 4.

EXERCISES

17–1   Conduction

1. Solid wood has a thermal conductivity about twice that of sawdust of the same kind. Can you account for the difference?

2. In the winter, why does the steel blade of a shovel seem colder than its wooden handle?

3. A pine wood door 60 mm thick is to be removed and the opening bricked up. How thick should the brick wall be if the rate of heat transfer through this part of the building is to be unchanged?

4. What thicknesses of (a) concrete, (b) brick, and (c) pine wood have the same insulating ability as 10 cm of glass wool?

5. The walls of a certain refrigerator are equivalent to a 60-ft² layer of glass wool 3 in. thick. If the interior of the refrigerator is at an average of 40°F and the kitchen temperature is 70°F, what is the rate of heat flow through the refrigerator walls?

6. A house window consists of a pane of glass 1 m wide, 1.2 m high, and 5 mm thick. If the inner and outer faces of the glass are at temperatures of 15°C and 5°C, respectively, at what rate is heat lost by conduction through the window?

7. The steel hull of a boat is 6 mm thick and its underwater area is 120 m². If the water temperature is 26°C and the temperature of the boat's interior is 20°C, how much heat per day enters the boat through the hull?

8. The neoprene foam material of a scuba diver's wet suit has a thermal conductivity of 0.075 W/m·°C. (a) Find the rate of heat loss of a diver whose skin area is 1.7 m² wearing a suit of such material 6 mm thick who is diving in 10°C water. Assume a skin temperature of 32°C. (b) How much heat does the diver lose in a dive lasting 1 h? (The suit material when submerged is squeezed by water pressure to a smaller thickness than at the surface. At a depth of 20 m the thickness is about half that at the surface, for instance. The figure quoted above refers to the actual thickness at the diver's depth.)

9. A double boiler has an aluminum upper pan whose bottom is 20 cm in diameter and 0.9 mm thick. If the lower pan contains boiling water and the upper one milk at 5°C, compute the rate of heat transfer to the milk. What happens to this rate as the milk warms up?

10. The hinges of a freezer door 15 cm thick are attached by four bolts 4 mm in diameter that pass through the entire door and are secured on the inside by nuts. If the interior of the freezer is at −18°C and the room it is in is at 18°C, find the rate in watts at which heat is lost through the bolts.

11. An oven has a brick wall 5 in. thick through which heat is found to pass at a rate of 5 Btu/ft² per minute. If the outside of the oven is at 200°F, find the temperature of the inside of the oven.

12. An insulated copper rod 5 mm on a side and 1 m long has one end in a steam bath and the other in contact with a block of ice at 0°C. How much ice melts per hour?

13. An igloo of compacted snow has a surface area of 24 m² and a wall thickness of 40 cm. The thermal conductivity of the snow is 2×10^{-4} kcal/m·s·°C. If the temperature inside the igloo is 20°C and the outside temperature is −10°C, how much blubber (heat of combustion = 9500 kcal/kg) must be burned per hour to keep the interior temperature constant? Neglect the melting of the igloo wall and the heat evolved by the igloo's inhabitants.

14. Water in an outdoor swimming pool 5 m × 8 m is covered by a 32-mm layer of ice. If the water is at 0°C and the surrounding air is at −10°C, how many millimeters of ice form per hour?

15. The layer of ice on a frozen lake is 2 in. thick at 5 P.M. During the night the air temperature is 28°F and the temperature of the water beneath the ice is 32°F. How thick is the ice at 8 A.M. the next morning?

17-2   Thermal Resistance

16. A quantity called the *heat transmission coefficient, U,* is sometimes used to describe the insulating ability of a building structure such as a wall, roof, or floor. It is equal to $1/R$. Find the amount of heat that passes through a wall 10 ft high and 18 ft long for which $U = 0.15$ Btu/ft$^2 \cdot$ h $\cdot$ °F when the inside temperature is 70°F and the outside temperature is 20°F.

17. A sheet of R-0.08 plasterboard is 0.5 in. thick. Find the thermal conductivity of the plasterboard.

18. A double-glazed window 6 ft × 6 ft consists of two panes of glass 0.25 in. thick separated by 0.25 in. of air. The temperature of the outside of the window is 40°F and that of the inside is 65°F. (a) Find the heat that passes through the window per hour. (b) The window leaks and the space between the panes has filled up with water. If the outside and inside temperatures are the same as before, find the heat that passes through the window per hour now. The thermal conductivity of water is 4.4 Btu/(ft$^2 \cdot$ h $\cdot$ °F/in.).

19. A 12-cm brick wall between two rooms in a house has a 1 cm layer of plaster ($k = 0.5$ W/m $\cdot$ °C) on each side. If the temperature difference between the sides of the wall is 3°C, find the rate at which heat is conducted through each square meter of the wall.

17-3   Convection

20. Why is it advantageous to install the heating element of an electric hot water heater near the bottom of the tank?

21. A roast turkey cools from 150° to 140°F in 5 min when it is in a room whose temperature is 70°F. How long will it take to cool from 110°F to 100°F in the same room?

22. A glass of beer warms from 7°C to 10°C in 3 min when the air temperature is 35°C. How long will it take to warm from 12°C to 15°C?

17-4   Radiation

23. Under what circumstances does an object radiate electromagnetic waves? How is the predominant wavelength in the radiation related to the temperature of the object?

24. By what mechanism or mechanisms does a person seated in front of a fire receive heat from it? A person seated in front of a radiator through which hot water is circulated?

25. A bare copper pipe carries hot water to a faucet. What effect, if any, will polishing the pipe have on the rate of heat flow from the pipe to the room?

26. If a red star and a white star radiate energy at the same rate, can they be the same size? If not, which must be the larger? Why?

27. A Thermos bottle consists of two glass vessels, one inside the other, with the space between them evacuated. The vessels are both coated with thin films of silver. Why is this device so effective in keeping the contents of the bottle at a constant temperature?

28. A sufficiently hot piece of metal glows brightly but a transparent piece of quartz at the same temperature hardly glows at all. What is the reason for the difference?

29. Sunspots appear dark, although their temperatures are typically 5000 K, because the rest of the sun's surface is even hotter, about 5800 K. Compare the radiation rates of surfaces of the same emissivity whose temperatures are respectively 5000 K and 5800 K.

30. The sun radiates approximately as a blackbody at a temperature of 5800 K, and solar energy arrives at the earth at a rate of about 1.4 kW $\cdot$ m^2. At what rate would solar energy arrive if the sun were 10% cooler?

31. At what rate does radiation escape from a hole 2 cm^2 in area in the wall of a furnace whose interior is at a temperature of 600°C?

32. A copper sphere 5 cm in diameter whose emissivity is 0.3 is heated in a furnace to 400°C. At what rate does it radiate energy?

33. An object is at a temperature of 400°C. At what temperature would it radiate energy twice as fast?

34. The color of an object hot enough to glow varies with its temperature approximately as follows:

Temperature, °C	Appearance
500	Just visible
750	Cherry red
1000	Orange
1250	Light yellow

If a certain blackbody radiates at 1 kW when its temperature is 500°C, what will it radiate at the other temperatures listed above?

35. A woman is taking a sunbath with 0.5 m^2 of her skin exposed to solar radiation, which reaches her at the rate of 1.2 kW/m^2. Her skin temperature is the same as the air temperature. She has light skin whose emissivity is 0.7, so that 0.7 of the radiation reaching her is absorbed. Her metabolic processes result in the liberation of heat at a rate of 80 W. How much sweat must she evaporate per hour in order to get rid of the metabolic heat and absorbed solar energy? The heat of vaporization of sweat is 2430 kJ/kg.

36. A naked person whose skin area is 1.5 m² and whose skin temperature is 33°C is sitting in a room at 25°C. Find the rates at which the person loses heat by convection and by radiation. The convection coefficient *h* for such a person is about 7 W/m² · °C, and at such temperatures the emissivity of human skin is very nearly 1 (the radiation is chiefly in the infrared part of the spectrum).

17–5   The Refrigerator

37. An attempt is made to cool a kitchen in the summer by switching on an electric fan and closing the kitchen door and windows. What will happen?

38. In another attempt to cool the kitchen, the refrigerator door is left open, again with the kitchen door and windows closed. Now what will happen?

17–6   Thermodynamics of Refrigeration

39. A refrigerator with a coefficient of performance of 2.4 extracts heat from a freezer at −10°C and exhausts it at 25°C. How much work is required per J of heat extracted?

40. Find the coefficient of performance of a refrigerator whose energy supply equals half the amount of heat it extracts from a cold chamber.

41. A 5-ton air conditioner has a coefficient of performance of 2.5. How many hp are needed to operate its compressor?

42. How much power in kilowatts must be supplied to a central air conditioning system with a coefficient of performance of 2.3 that removes heat from a house at the rate of 50,000 Btu/h? What is its energy efficiency ratio?

43. A Carnot refrigerator extracts heat from a freezer at −5°C and exhausts it at 25°C. How much work per joule of heat extracted is required?

44. A Carnot refrigerator is used to make 1 kg of ice at −10°C from 1 kg of water at 20°C, which is also the temperature of the kitchen. How many joules of work must be done?

45. A certain refrigerator that absorbs heat at 5°F and exhausts it at 85°F has a coefficient of performance of 3.0. What would the coefficient of performance of a Carnot refrigerator be that operated between the same two temperatures?

46. In its operation, a refrigerator absorbs heat from a storage chamber at −23°C and exhausts heat at 27°C. The rate

at which heat is exhausted is found to be 60 kcal/min, and the coefficient of performance is half that of a Carnot refrigerator. (a) How much heat is absorbed from the storage chamber per minute? (b) How much power does the refrigerator require?

47. Three designs for a refrigerator to operate between −20°C and 40°C are proposed. Design *A* is claimed to require 300 J of work for each kcal of heat extracted, design *B* to require 950 J, and design *C* to require 2000 J. Which design would you choose and why?

17–7   Refrigeration Capacity

48. A ¾-ton window air conditioner is found able to maintain a constant temperature of 70°F in an empty room when the outside temperature is 90°F by running half the time. A sitting person liberates about 400 Btu/h. How many people can sit reading in the room, each with his own 75-W lamp, without exceeding the capacity of the air conditioner?

49. An opening 8 ft long and 5 ft high is cut in the west wall of a house and a picture window is installed. The window consists of two panes of glass separated by an air space, and its thermal conductivity is the same as that of the wall material it replaces. In the afternoon an average of 150 Btu/h of solar radiation passes through each ft² of window area. What is the additional load (in refrigeration tons) on the house's air conditioning system?

50. Ice cream has a specific heat of 0.78 Btu/lb · °F before it freezes and 0.45 Btu/lb · °F afterward. Its heat of fusion is 126 Btu/lb, and it freezes at 28°F. What refrigeration capacity is required to produce 60 lb of ice cream at 8°F from a mix at 60°F in 1 h?

51. An ice-making plant produces 50 tons of ice per day at 15°F from water at 60°F. Assuming no losses, what must be the refrigeration capacity of the plant?

ANSWERS TO MULTIPLE CHOICE

1. a	**7.** c	**13.** c	**19.** c
2. d	**8.** c	**14.** c	**20.** b
3. d	**9.** d	**15.** b	**21.** b
4. a	**10.** b	**16.** d	**22.** c
5. d	**11.** a	**17.** d	**23.** d
6. d	**12.** a	**18.** b	**24.** b

18

ELECTRICITY

The success of the laws of motion, of the law of gravitation, and of the kinetic-molecular theory of matter might lead us to think that we now have, at least in outline, a complete picture of the workings of the physical universe. To see how false this notion is all we need do is perform a simple experiment: On a dry day, we run a hard rubber comb through our hair, and find that the comb is now able to pick up small bits of paper and lint. The attraction cannot be due to gravity, because the gravitational force between comb and paper is far too small and should not, in any case, depend upon whether the comb is run through our hair or not. What has been revealed by this experiment is an electrical phenomenon, so called after *elektron,* the Greek word for amber, a substance used in the earliest studies of electricity.

18–1   ELECTRIC CHARGE

Electric forces are responsible for most aspects of the behavior of matter

Electricity is familiar to all of us as the name for that which causes our light bulbs to glow, many of our motors to turn, our telephones and radios to transmit sounds, our television screens to transmit pictures. But there is more to electricity than its tech-

CHAPTER OBJECTIVES

Completing this chapter should enable you to:

1. Distinguish between positive and negative electric charges.

2. Calculate the magnitude of the force one charge exerts on another and determine the direction of this force.

3. Describe the concept of a field and what is meant by an electric field.

4. Calculate the force on a charge located in an electric field.

5. Determine the magnitude and direction of the electric field produced by a given distribution of charges.

6. Use lines of force to visualize an electric field.

7. Determine the potential energy of a charge in an electric field.

8. Determine the potential energy of a system of two charges.

9. Describe what is meant by potential difference.

10. Find the potential difference between two points in an electric field.

11. Relate the work done by or on a charge to the potential difference through which it moves.

12. Use the electron volt as a unit of energy.

nological uses. Electric forces hold electrons to nuclei to form atoms, and they hold atoms together to form molecules, solids, and liquids. All of the chief properties of matter in bulk—except mass—can be traced to the electrical nature of its basic particles.

Let us begin our study of electricity by examining three experiments. The first experiment is shown in Fig. 18–1. By convention, we call whatever it is that a rubber rod possesses by virtue of having been stroked with a piece of fur *negative electric charge*. Part of the negative charge on the rubber rod of Fig. 18–1 flowed to the pith ball when it was touched, and the fact that the ball then flew away from the rod suggests that negative electric charges repel each other.

Negative charges repel each other

The next experiment, shown in Fig. 18–2, is very similar. By convention, we call whatever it is that a glass rod possesses by virtue of having been stroked with a silk cloth *positive electric charge*. Part of the positive charge on the glass rod of Fig. 18–2 flowed to the pith ball when it was touched, and the fact that the ball then flew away from the rod suggests that positive electric charges repel each other.

Positive charges repel each other

Why is it assumed that the electric charge on the glass rod is different from that on the rubber rod? The reason lies in the result of the third experiment, which is shown in Fig. 18–3. The attraction of the two pith balls means (1) that the charges they carry

Unlike charges attract each other

FIG. 18–1 A rubber rod stroked with fur becomes negatively charged; two negatively charged objects repel each other.

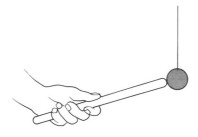

A pith ball suspended by a fine string is touched by a hard rubber rod. Nothing happens.

The rubber rod is stroked with a piece of fur.

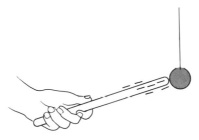

The pith ball is again touched by the rubber rod.

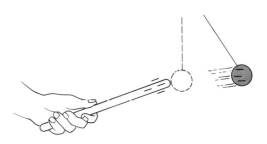

After the touch, the ball flies away from the rod.

FIG. 18–2 A glass rod stroked with silk becomes positively charged; two positively charged objects repel each other.

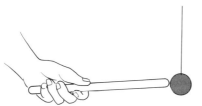

A second pith ball is touched by a glass rod. Nothing happens.

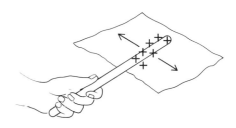

The glass rod is stroked with a silk cloth.

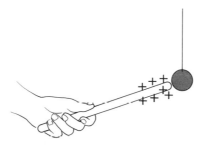

The pith ball is again touched by the glass rod.

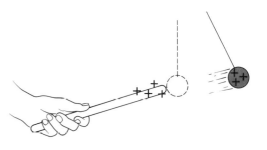

After the touch the ball flies away from the rod.

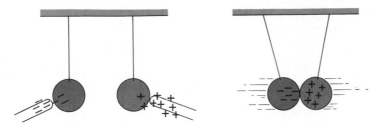

FIG. 18–3 Objects with unlike charges attract each other.

are different, since like charges have already been observed to repel, and (2) that unlike charges attract each other.

The preceding results can be summarized very simply:

Like charges repel; unlike charges attract.

Behavior of charges

Where do the charges come from when one substance is stroked with another? When we charge one pith ball with a rubber rod and another with the fur the rod was stroked with, we find (Fig. 18–4) that the two balls attract; since the rubber rod is negatively charged, this experiment indicates that the fur is positively charged. A similar experiment with a glass rod and a silk cloth indicates that the cloth picks up a negative charge during the stroking. What the process of stroking does is to *separate* charges. We conclude that rubber has an affinity of some kind for negative charges and fur an affinity for positive charges, so that, when rubbed together, each tends to pick up a different kind of charge.

A great many experiments with a variety of substances have shown that there are only the two kinds of electric charge, positive and negative, that we have spoken of. All electrical phenomena involve either or both kinds of charge. An "uncharged" body of matter actually possesses equal amounts of positive and negative charge, so that appropriate treatment—mere rubbing will do for some substances—can leave an excess of either kind on the body and thereby cause it to show electrical effects.

Only two kinds of charge exist

What is electric charge? All that can be said is that charge, like mass, is a fundamental property of certain of the elementary particles of which all matter is composed. Three types of particles are found in atoms, the positively charged *proton,* the negatively charged *electron,* and the neutral (that is, uncharged) *neutron.* The proton and electron have exactly equal amounts of charge, though of opposite sign. Neutrons

Ordinary matter consists of neutrons, protons, and electrons

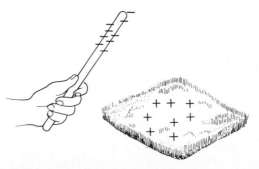

FIG. 18–4 The process of stroking a rubber rod with a piece of fur serves to separate charges so that the rod becomes negatively charged and the fur positively charged.

TABLE 18–1
Neutrons, protons, and electrons are the constituents of atoms. An atom contains equal numbers of protons and electrons, so it is electrically neutral. The neutrons and protons are responsible for most of an atom's mass.

	Neutron	Proton	Electron
Charge	0	$+e = 1.60 \times 10^{-19}$ C	$-e = 1.60 \times 10^{-19}$ C
	Uncharged	Equal in magnitude, opposite in sign	
Mass	1.675×10^{-27} kg	1.673×10^{-27} kg	9.11×10^{-31} kg
	Nearly equal		Electron mass is 1/1836 of proton mass

FIG. 18–5 Structure of the helium atom. The diameter of the atom is much greater than the diameter of its nucleus. Nearly all the mass of an atom is concentrated in its nucleus.

Helium atom

Helium nucleus

• Electron   (−)
● Proton   (+)
○ Neutron   (0)

and protons have nearly the same mass, which is almost 2000 times greater than the electron mass (Table 18–1). An atom normally contains equal numbers of protons and electrons, so it is electrically neutral unless disrupted in some way. Fig. 18–5 shows the structure of a helium atom, whose nucleus consists of two protons and two neutrons. The two electrons that circle the nucleus balance the charges of the protons. Atoms and nuclei are discussed in detail in later chapters.

Conservation of charge   The *principle of conservation of charge* states that

The net electric charge in an isolated system remains constant.

By "net charge" is meant the algebraic sum of the charges present—the total positive charge minus the total negative charge. Net charge can be positive, negative, or zero. Every known physical process in the universe conserves electric charge.

18–2   COULOMB'S LAW

Law of electric force between charges   The law of force between charges was first published by the eighteenth-century French scientist Charles Coulomb, and is called Coulomb's law in his honor. If we use the symbol Q for electric charge, Coulomb's law for the magnitude F of the force $\mathbf{F}$ between two charges Q_A and Q_B the distance r apart states that

$$F = k\frac{Q_A Q_B}{r^2} \qquad\qquad \textit{Coulomb's law} \quad (18–1)$$

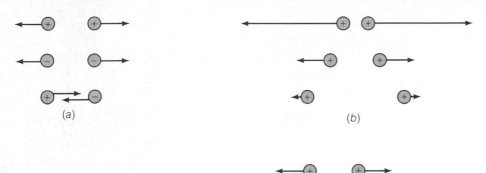

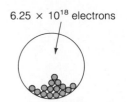

FIG. 18-6 (a) Like charges repel, unlike charges attract. (b) The force between two charges varies inversely as the square of their separation. (c) The force is proportional to the product of the charges.

The force between two charges is proportional to both of the charges and is inversely proportional to the square of the distance between them. The quantity k is a constant whose value depends upon the units employed and upon the medium (air, vacuum, oil, and so forth) in which the charges are located.

Electric force is, of course, a vector quantity, and the formula above gives only its magnitude. The direction of **F** is always along the line joining Q_A and Q_B, and the force is attractive if Q_A and Q_B have opposite signs and repulsive if they have the same signs (Fig. 18-6).

The unit of electric charge is the *coulomb* (abbreviated C). The formal definition of the coulomb, given in Chapter 23, is expressed in terms of magnetic forces. A more realistic way to think of the coulomb is in terms of the number of individual elementary charges that add up to this amount of charge. All charges, both positive and negative, occur only in multiples of 1.60×10^{-19} C. No elementary particle has ever been found with a charge different from $\pm 1.60 \times 10^{-19}$ C or 0. The electron has a charge of -1.60×10^{-19} C and the proton has a charge of $+1.60 \times 10^{-19}$ C. If 6.25×10^{18} electrons were assembled, the total charge would be -1 C; if 6.25×10^{18} protons were assembled, the total charge would be $+1$ C (Fig. 18-7).

Because electric charge always occurs in multiples of $\pm 1.60 \times 10^{-19}$ C, this amount of charge has been given a special name, the electron (or electronic) charge, and a special symbol, e:

$$e = 1.60 \times 10^{-19} \text{ C} \qquad \qquad \textit{Electron charge}$$

Thus a charge of $+1.60 \times 10^{-19}$ C is abbreviated $+e$, and one of -3.20×10^{-19} C is abbreviated $-2e$.

In most processes that lead to a net charge on some object, electrons are either added to it or removed from it. Hence we can think of an object whose charge is -1 C as having 6.25×10^{18} electrons more than its normal number, and of an object whose charge is $+1$ C as having 6.25×10^{18} electrons less than its normal num-

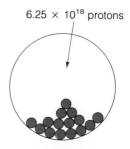

6.25×10^{18} protons

$Q = +1$ coulomb

6.25×10^{18} electrons

$Q = -1$ coulomb

FIG. 18-7

ber. (By "normal number" is meant a number of electrons equal to the number of protons present, so that the object has no net charge.)

Coulomb constant

When Q_A and Q_B in Coulomb's law are expressed in coulombs and r in meters, the constant k has the value in vacuum of

$$k = 9.0 \times 10^9 \text{ N} \cdot \text{m}^2/\text{C}^2 \qquad\qquad \textit{Coulomb constant}$$

The value of k in air is very slightly greater.

The constant k is often written

$$k = \frac{1}{4\pi\epsilon_0}$$

where ϵ_0, called the permittivity of free space, is equal to

$$\epsilon_0 = 8.85 \times 10^{-12} \text{ C}^2/\text{N} \cdot \text{m}^2$$

Example   Find the force between two charges of 1 C each that are 1 m apart (Fig. 18–8).

Solution   From Coulomb's law,

$$F = k\frac{Q_A Q_B}{r^2} = \left(9 \times 10^9 \frac{\text{N} \cdot \text{m}^2}{\text{C}^2}\right)\left(\frac{1 \text{ C} \times 1 \text{ C}}{1 \text{ m}^2}\right) = 9 \times 10^9 \text{N}$$

The coulomb is a very large unit

This force is equal to about 2 billion lb. Even the most highly charged objects that can be produced seldom contain more than a tiny fraction of a coulomb of net charge of either sign.   ∎

At the beginning of this chapter a familiar observation was noted: A hard rubber comb that has been charged by being passed through someone's hair on a dry day is able to attract small bits of paper. Since the paper bits were originally uncharged, how could the comb exert a force on them?

How a charge attracts an uncharged object

The explanation depends upon Coulomb's law (Fig. 18–9). When the negatively charged comb is brought near the paper, some of the negative charges in the paper that are not tightly bound in place move as far away as they can from the comb, while some of the positive charges that are not tightly bound move toward the comb. Because electric forces vary inversely with distance, the attraction between the comb and the

FIG. 18–8

FIG. 18–9

closer positive charges is greater than the repulsion between the comb and the farther negative charges, and so the paper moves toward the comb. Only a small amount of charge separation actually occurs, and so, with little force available, only very light objects can be attracted in this way.

18–3   ELECTRICITY AND MATTER

Coulomb's law for the electric force between charges is very similar to Newton's law for the gravitational force between masses. The most striking difference is that electric forces may be either attractive or repulsive, whereas gravitational forces are always attractive. The latter fact means that matter in the universe tends to come together to form large bodies, such as stars and planets, and these bodies are always found in groups, such as galaxies of stars and families of planets.

Gravitational forces dominate on a large scale

There is no comparable tendency for electric charges of either sign to come together; quite the contrary. Unlike charges attract strongly, which makes it hard to separate neutral matter into portions of opposite signs. Furthermore, like charges repel, so it becomes harder and harder to add further charge to an already charged object. Hence the large-scale structure of the universe is chiefly governed by gravitational forces.

On an atomic scale, though, the relative importance of gravity and electricity is reversed. Elementary particles are so tiny that the gravitational forces between them are insignificant, whereas their electric charges are sufficiently great for electric forces to govern the structures of atoms, molecules, liquids, and solids.

Electric forces dominate on a small scale

18–4   ELECTRIC FIELD

Electric forces, like gravitational forces, act between objects that may be widely separated. A *force field* is a concept that provides a framework for understanding how forces are transmitted from one object to another across empty space.

When an electric charge is present somewhere, the properties of space around it can be considered to be altered in such a way that another charge brought to this region will experience a force there. The "alteration in space" caused by a stationary charge is called its *electric field,* and any other charge is thought of as interacting with the field and not directly with the charge that gives rise to it. All forces, not just electric ones, can be interpreted as arising through the action of a force field of one kind or another. Thus the sun is regarded as being surrounded by a gravitational field, and it is the forces exerted by this field on the planets that are the centripetal forces that hold them in their orbits. In the case of the "direct contact" forces between solid objects in everyday life, the force fields involved are electric fields.

An electric force on a charge may be regarded as due to the action of an electric field on the charge

An electric field is a force field that exists wherever an electric force acts on a charge. The source of the field may be a single charge or many charges. We would like to specify electric field in such a way that we can then determine the force that would act on any charge placed anywhere in the field. The electric field $\mathbf{E}$ at a point in space is therefore defined as the ratio between the force $\mathbf{F}$ on a charge Q (assumed positive) at that point and the magnitude of Q:

Definition of electric field

$$E = \frac{F}{Q} \qquad \qquad \textit{Electric field} \quad (18-2)$$

The units of E are newtons per coulomb. Electric field is a vector quantity that possesses both magnitude and direction.

Once we know what the electric field E is at some point, from the definition we see that the force that the field exerts on a charge Q at that point is

$$\mathbf{F} = Q\mathbf{E} \qquad \qquad \textit{Electric force} \quad (18-3)$$

Force = charge × electric field

Example   A small copper ball of mass 5 g is given a charge of $+3 \times 10^{-8}$ C. What is its acceleration when placed in an electric field whose magnitude is 10^4 N/C?

Solution   The force on the ball has the magnitude

$$F = QE = (3 \times 10^{-8}\ \text{C})(10^4\ \text{N/C}) = 3 \times 10^{-4}\ \text{N}$$

Hence the ball's acceleration is

$$a = \frac{F}{m} = \frac{3 \times 10^{-4}\,\text{N}}{5 \times 10^{-3}\,\text{kg}}$$
$$= 0.06\ \text{m/s}^2 \qquad \blacksquare$$

Electric field of a single charge

We can use Coulomb's law to determine the magnitude of the electric field around a single charge Q. First we find the force F that Q exerts upon a test charge q at the distance r away, which is

$$F = k\frac{Qq}{r^2}$$

Since $E = F/q$ by definition, we have

$$E = \frac{F}{q} = k\frac{Q}{r^2} \qquad \qquad \textit{Electric field of a charge} \quad (18-4)$$

This formula tells us that the electric field magnitude the distance r from a point charge is directly proportional to the magnitude Q of the charge and is inversely proportional to the square of r.

Electric field of many charges

When more than one charge contributes to the electric field at a point P, the net field E is the *vector sum* of the fields of the individual charges. That is,

$$\mathbf{E} = \mathbf{E}_1 + \mathbf{E}_2 + \mathbf{E}_3 + \cdots = \Sigma\, \mathbf{E_n} \qquad \qquad \textit{Total field} \quad (18-5)$$

Example   Two charges, $Q_1 = +2 \times 10^{-8}$ C and $Q_2 = +3 \times 10^{-8}$ C, are 50 mm apart, as in Fig. 18–10. What is the electric field halfway between them?

Solution   At a point halfway between the charges, $r_1 = r_2 = 25$ mm. Hence the electric fields of the two charges have the respective magnitudes

$$E_1 = k\frac{Q_1}{r_1^2} = \frac{(9 \times 10^9\,\text{N}\cdot\text{m}^2/\text{C}^2)(2 \times 10^{-8}\,\text{C})}{(2.5 \times 10^{-2}\,\text{m})^2} = 2.9 \times 10^5\,\text{N/C}$$

$$E_2 = k\frac{Q_2}{r_2^2} = \frac{(9 \times 10^9 \text{N} \cdot \text{m}^2/\text{C}^2)(3 \times 10^{-8}\text{C})}{(2.5 \times 10^{-2}\text{m})^2} = 4.3 \times 10^5 \text{N/C}$$

The direction of an electric field at a point is the direction in which a positive charge placed there would move. Hence the direction of E_1 is to the right in Fig. 18–10 and that of E_2 is to the left. Since E_1 and E_2 are in opposite directions, the magnitude of their sum E is

$$E = E_2 - E_1 = 1.4 \times 10^5 \text{ N/C}$$

The direction of E is toward Q_1.

If the charges had both been negative rather than positive, the field halfway between them would have the same magnitude but would be in the opposite direction. If one charge had been negative and the other positive, E_1 and E_2 would point in the same direction along the line between the charges. The total field E there would then have the magnitude $E_1 + E_2$ and its direction would be toward the negative charge. ■

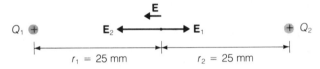

FIG. 18–10

Example   Where is the electric field equal to zero in the neighborhood of the charges of the previous example?

Solution   Since the fields E_1 and E_2 are in opposite directions along a line between Q_1 and Q_2, the point where they cancel each other out must lie on this line. The cancellation occurs where the magnitudes E_1 and E_2 are equal:

$$E_1 = E_2 \qquad k\frac{Q_1}{r_1^2} = k\frac{Q_2}{r_2^2} \qquad \frac{Q_1}{r_1^2} = \frac{Q_2}{r_2^2}$$

This last is an equation with two unknown quantities, r_1 and r_2. We need another equation that relates r_1 and r_2 in order to find their values. Such an equation is provided by the fact that the distance between Q_1 and Q_2 is $s = 50$ mm, so that

$$r_1 + r_2 = s \qquad r_2 = (s - r_1)$$

Hence

$$\frac{Q_1}{r_1^2} = \frac{Q_2}{r_2^2} = \frac{Q_2}{(s - r_1)^2}$$

To solve for r_1, the first step is to rewrite the equation as follows:

$$\frac{(s - r_1)^2}{r_1^2} = \frac{Q_2}{Q_1}$$

Now we take the square root of both sides:

$$\frac{s - r_1}{r_1} = \sqrt{\frac{Q_2}{Q_1}}$$

Finally we multiply both sides by r_1 and solve for r_1:

$$s - r_1 = \sqrt{\frac{Q_2}{Q_1}}\, r_1$$

$$s = r_1 + \sqrt{\frac{Q_2}{Q_1}}\, r_1 = r_1\left(1 + \sqrt{\frac{Q_2}{Q_1}}\right)$$

$$r_1 = \frac{s}{1 + \sqrt{Q_2/Q_1}}$$

Here $s = 50$ mm and

$$\sqrt{\frac{Q_2}{Q_1}} = \sqrt{\frac{3 \times 10^{-8}\,\text{C}}{2 \times 10^{-8}\,\text{C}}} = 1.22$$

Therefore

$$r_1 = \frac{s}{1 + \sqrt{Q_2/Q_1}} = \frac{50\,\text{mm}}{1 + 1.22} = \frac{50\,\text{mm}}{2.22} = 22.5\,\text{mm}$$

The point where $\mathbf{E} = 0$ lies between the charges the distance 22.5 mm from the charge Q_1.

If the charges had both been negative rather than positive, the point where $\mathbf{E} = 0$ would be the same. If one charge had been negative and the other positive, the point where $\mathbf{E} = 0$ would be on an extension of the line between Q_1 and Q_2 beyond the smaller charge. ∎

18–5   ELECTRIC LINES OF FORCE

An electric field is a region of space in which an object by virtue of its electric charge is acted upon by a force. To visualize an electric field, we may use *lines of force*. These are constructed as follows. At several points in space we draw arrows in the direction a positive charge there would experience a force (Fig. 18–11(a)). Then these arrows are connected to form smooth curves called lines of force (Fig. 18–11(b)). Two rules should be kept in mind:

Rules for lines of force of electric field

1. Lines of force leave positive charges and enter negative ones.
2. The spacing of lines of force is such that they are close together where the field is strong and far apart where the field is weak.

If we place a positive charge in an electric field, it will experience a force in the direction of the line of force it is on, and the magnitude of the force will be proportional to the concentration of lines of force in its vicinity. Figure 18–12 shows the patterns of lines of force around a positive and a negative charge and around pairs of charges.

Lines of force are imaginary

It is important to keep in mind that lines of force do not actually exist as threads in space; they are simply a device to help our thinking about force fields. The use of

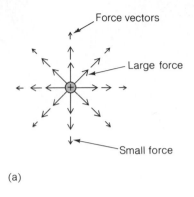

Force vectors

Large force

Small force

(a)

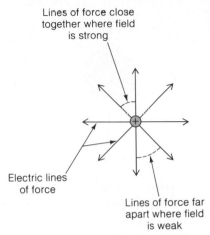

Lines of force close together where field is strong

Electric lines of force

Lines of force far apart where field is weak

(b)

FIG. 18–11 How electric lines of force around a positive charge are drawn. In (a) the forces whose vectors are shown are those exerted by the charge on a positive test charge.

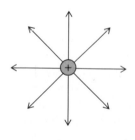

(a)

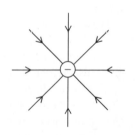

(b)

FIG. 18–12 Lines of force near separate charges and near pairs of charges. At large distances relative to the separation of the charges, the field in (c) approaches that of a single charge.

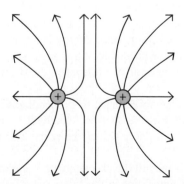

(c)

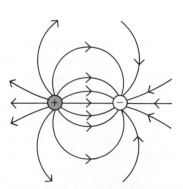

(d)

FIG. 18–13 (a) Electric lines of force around a uniformly charged sphere. (b) Electric lines of force around a point charge.

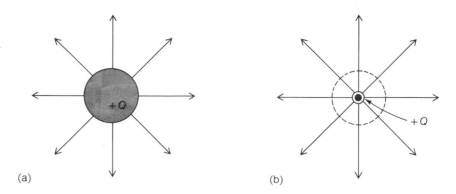

(a)   (b)

lines of force is not limited to electric fields. Gravitational and magnetic fields, for instance, are often described in this manner.

The notion of lines of force is often useful as a guide to our thinking when electric fields of charge distributions are concerned. For example, let us examine the electric field around a spherical distribution of charge whose total amount is Q. Such a distribution is quite easy to arrange: All we need do is to add the charge Q to a metal sphere. Since metals are good conductors of electricity, the mutual repulsion of the individual charges that make up Q causes them to spread out uniformly over the surface of the sphere so that the charges are as far apart as possible. Evidently it doesn't matter whether the sphere is hollow or solid.

Electric field around a charged sphere

Whatever the number of lines of force we choose to represent the electric field around a point charge Q, the same number must be associated with a body that carries the same net charge Q. In the case of a charged sphere, the lines of force are symmetrically arranged, and therefore the pattern of lines of force outside the sphere is identical with the pattern around a point charge of the same magnitude (Fig. 18–13). The same is true for the electric field $\mathbf{E}$ itself, of course.

E = 0 inside a charged conductor

Three important properties of a charged metal object are illustrated in Fig. 18–14. First, inside the object the electric field is 0 everywhere; the electric fields due to the individual charges on the surface all cancel out in the interior, although outside the object they do not. We can see why this must be true even without a formal calculation. Suppose there *were* an electric field $\mathbf{E}$ in the interior of the object. Then the charges inside the object that are free to move (electrons in the case of a metal) would do so under the influence of the field $\mathbf{E}$. But no currents are observed in a charged conducting object except for a moment after the charge is placed on it; since energy is needed to maintain an electric current, a supply of energy would be needed for currents to continue in such an object. The only conclusion is that the interior of an isolated conducting object is always free of electric field.

E is perpendicular to the surface of a charged conductor

A similar argument shows that $\mathbf{E}$ just outside the object must be perpendicular to the object's surface. If $\mathbf{E}$ had a component parallel to the surface, electrons on the surface would be in constant motion. Because they are not, $\mathbf{E}$ must be perpendicular to the surface.

Any excess charge (+ or −) on a conductor resides on its surface

The third property of a charged metal object is that all the excess charge (of either sign) on the object must reside on its surface. If any point in the interior had a net charge, an electric field would be present around it. But there are no electric fields inside a metal object, hence there can be no excess charge there.

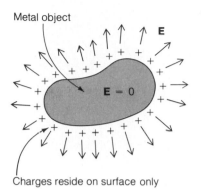

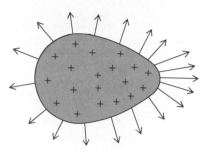

Metal object

E

E = 0

Charges reside on surface only

FIG. 18–14 (Left) There are neither charges nor electric field inside a charged metal object. Just outside the object the field is perpendicular to its surface.

FIG. 18–15 (Right) The electric field near a charged nonspherical metal object is strongest near the parts of the object that have the smallest radii of curvature.

The greater the curvature, the stronger the field

When a nonspherical metal body is charged, the individual charges do not distribute themselves uniformly on its surface. As a general rule, the more highly curved parts have greater concentrations of charge than the gently curved parts, and the electric field is accordingly most intense near the highly curved parts (Fig. 18–15). We can see why this must be true from the above discussion of the electric field around a charged metal object. There is no field inside the object, and the field just outside it is perpendicular to its surface. In order to bring this about, the charges must be arranged on the object's surface so that the inward-directed parts of the fields of charges on one side of the object exactly balance the inward-directed parts of the fields on the opposite side. Hence the charges at the small end of the egg-shaped object in Fig. 18–15 must be closer together than those at the large end for their respective fields to cancel out in the interior.

At a metal point, the electric field may become so great that it causes a separation of charge in nearby air molecules. The resulting electrical "discharge" is visible as a luminous glow. If two oppositely charged pointed rods are brought close together, charge flows between them via disrupted air molecules and a "spark" occurs. If more gently curved charged bodies, such as large spheres, are similarly brought near each other, their separation must be much smaller before a spark can occur.

18–6 ELECTRIC POTENTIAL ENERGY

In our study of mechanics we found the related concepts of work and potential energy to be useful in analyzing a wide variety of situations. These concepts are equally useful in the study of electrical phenomena, in particular electric current. Instead of potential energy itself, a related quantity called *potential difference* turns out to be especially appropriate in electrical problems.

Let us examine the potential energy of a charge in a uniform electric field together with a gravitational analogy. At the left in Fig. 18–16(a) is a uniform electric field **E** between two parallel, uniformly charged plates *A* and *B,* and at the right is a region near the earth's surface in which the gravitational field is also uniform. Now we place a particle of charge *Q* in the electric field and a particle of mass *m* in the gravitational field. As shown in Fig. 18–16(b) the charge is acted upon by the electric force

FIG. 18–16   Analogy between electric and gravitational potential energy.

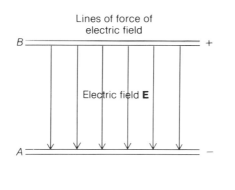
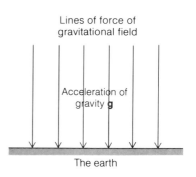

Lines of force of electric field

B ————————————— +

Electric field **E**

A ————————————— −

Lines of force of gravitational field

Acceleration of gravity **g**

The earth

(a)

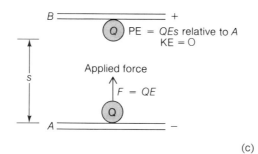
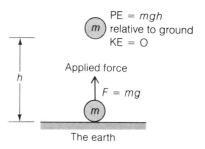

B ═══════════════ +

Q

F = Q**E**   Force exerted on Q by electric field

A ═══════════════ −

m

Force exerted on m by gravitational field   **F** = m**g**

The earth

(b)

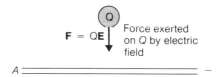
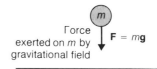

B ———————————— +

Q PE = QEs relative to A
KE = O

Applied force

$F = QE$

Q

s

A ———————————— −

PE = mgh
m relative to ground
KE = O

Applied force

$F = mg$

m

h

The earth

(c)

B ═══════════════ +

Q

m

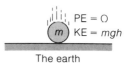

Q PE = O
KE = QEs

A ═══════════════ −

m PE = O
KE = mgh

The earth

(d)

$$\mathbf{F}_{elec} = Q\mathbf{E}$$

and the mass is acted upon by the gravitational force

$$\mathbf{F}_{grav} = m\mathbf{g}$$

If the charge is on plate A and we want to move it to plate B, we must apply a force of magnitude QE to it because we have to push against a force of this magnitude (and opposite direction) exerted by the electric field. When the charge is at B, we will have performed on it the amount of work

Work = force $\times$ distance

$$W = QEs$$

where s is the distance the charge has moved (Fig. 18–16(c)). Similarly, to raise the mass from the ground to a height h, we must apply a force of magnitude mg to it, and the work we do is

$$W = mgh$$

At plate B the charge has the potential energy

$$PE = QEs \qquad\qquad \textit{PE of charge in uniform electric field} \quad (18-6)$$

with respect to A. If we let it go, the potential energy will become kinetic energy as the electric field $\mathbf{E}$ accelerates the charge, and when the charge is back at A it will have a kinetic energy equal to QEs (Fig. 18–16(d)). In the same way, the mass has potential energy *with respect to the ground* in its new location. This potential energy is equal to the work done in raising it through the height h, and is

$$PE = mgh$$

If we let the mass go, it will fall to the ground with a final kinetic energy of *mgh*.

To summarize: The amount of work that must be performed to move a charge Q from A to B, the distance s apart, in a uniform electric field $\mathbf{E}$ is QEs. At B the charge accordingly has the potential energy QEs. If the charge is released at B, the force $Q\mathbf{E}$ acting on it produces an acceleration such that the charge has the kinetic energy QEs when it is back at A. Thus the work done in moving the charge in the field becomes potential energy, which in turn becomes kinetic energy when it is released.

There is no change in the energy of a charge moved perpendicular to an electric field, just as there is no change in the energy of a mass moved perpendicular to a gravitational field (for instance, along the earth's surface).

18–7   POTENTIAL ENERGY OF TWO CHARGES

A particle of charge Q_A that is located the distance r from another particle of charge Q_B has electrical potential energy because there is a force exerted on it by the electric field of Q_B. When Q_A and Q_B have the same sign, the force is repulsive; when the charges have opposite signs, the force is attractive. In either case, when Q_A is released, it will begin to move and acquire kinetic energy at the expense of its original potential energy (Fig. 18–17).

We have been assuming that Q_B is fixed in place. If instead Q_A is fixed in place, then we can speak of the potential energy of Q_B in the electric field of Q_A. This potential

FIG. 18–17 A charge has potential energy when it is in the electric field of another charge. Actually, the potential energy belongs to the system of the two charges. Here Q_B is assumed to be fixed in place, so the potential energy of the system becomes kinetic energy of Q_A when Q_A is released.

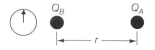

At rest, Q_A has PE only

When released, Q_A gains KE at the expense of its original PE

energy is exactly the same as before, since by Newton's third law of motion the force one object exerts on another is equal in magnitude to the force the second object exerts on the first.

The potential energy really belongs to the *system* of the two particles. If the particles are both released, *both* of them begin to move, and they share the original potential energy between them. The relative velocities of the two particles will be such as to conserve linear momentum, so the lighter particle will move faster than the heavier one. Thus total amount of energy available depends upon the charges of the particles, and the division of the energy depends upon their masses.

Potential energy of two charges is a property of the system of both of them

Potential energy = 0 for two charges infinitely far apart

A potential energy of any kind must be specified relative to a reference location. Near the earth, for example, the gravitational potential energy of an object is usually given with the earth's surface as the reference location for PE = 0. In the case of individual charges interacting with one another, the reference location is chosen to be infinity, since the electric field of a charge falls to zero an infinite distance away.

Because the electric field **E** of the charge Q_B (which we assume to be fixed in place) is not uniform, it is not easy to calculate the potential energy of Q_A when it is the distance r away. The result turns out to be

$$\text{PE} = k\frac{Q_A Q_B}{r} \qquad \textit{Potential energy of system of two charges} \quad (18\text{–}7)$$

Potential energy > 0 for a repulsive force; potential energy < 0 for an attractive force

If the charges have the same sign, their potential energy is positive; thus a positive potential energy corresponds to a repulsive force. If the charges have opposite signs, their potential energy is negative; thus a negative potential energy corresponds to an attractive force. The potential energy of a charge decreases as it moves away from another charge of the same sign and increases as it moves away from another charge of opposite sign.

Example   An electron of initial speed 10^3 m/s is aimed at another electron, whose position is fixed, from a distance of 1 mm. How close to the stationary electron will the other one approach before it comes to a stop and reverses its direction?

Solution   Figure 18–18 illustrates the situation. Since $Q_A = Q_B = -e$ here, the potential energies of the moving electron at the initial distance r_1 ($= 10^{-3}$ m) and the final distance r_2 are respectively

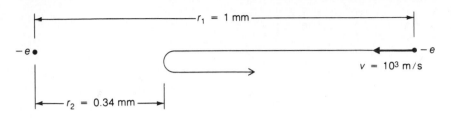

FIG. 18-18

$$PE_1 = \frac{ke^2}{r_1} \qquad PE_2 = \frac{ke^2}{r_2}$$

The potential energies are both positive, corresponding to a repulsive force between the electrons. The difference between the two potential energy values is equal to the initial kinetic energy $\frac{1}{2}mv^2$ of the moving electron, and therefore

$$KE = PE_2 - PE_1$$

$$\frac{1}{2}mv^2 = ke^2 \left(\frac{1}{r_2} - \frac{1}{r_1} \right)$$

$$\frac{1}{r_2} = \frac{mv^2}{2ke^2} + \frac{1}{r_1}$$

$$r_2 = \frac{1}{(mv^2/2ke^2) + (1/r_1)} = 3.4 \times 10^{-4}\,m = 0.34\,mm \qquad \blacksquare$$

18-8   POTENTIAL DIFFERENCE

The quantity *potential difference* is introduced to describe the situation of a charge in an electric field in an especially convenient way. The potential difference V_{AB} between two points A and B is defined as the ratio between the work that must be done to take charge Q from A to B and the value of Q:

$$V_{AB} = \frac{W_{AB}}{Q} \qquad\qquad \textit{Potential difference} \quad (18-8)$$

Potential difference = work per unit charge

The unit of potential difference is the joule per coulomb. Because this quantity is so frequently used, its unit has been given a name of its own, the *volt* (V). Thus

The volt is the unit of potential difference

1 volt = 1 joule/coulomb

In a uniform electric field, $W_{AB} = QEs$, with the result that the potential difference between A and B is QEs/Q or

$$V_{AB} = Es \qquad \textit{Potential difference in a uniform electric field} \quad (18-9)$$

In a uniform electric field, the potential difference between two points is the product of the field magnitude E and the separation s of the two points in a direction parallel to that of $\mathbf{E}$ (Fig. 18-19).

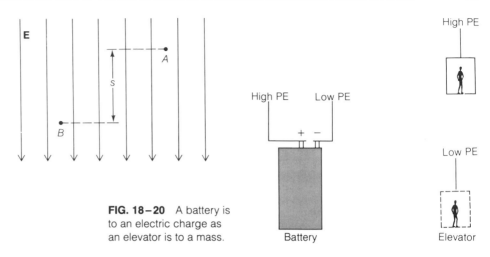

FIG. 18–19 The potential difference between two points in a uniform electric field **E** is equal to Es, where s is the component parallel to **E** of the distance between the points.

FIG. 18–20 A battery is to an electric charge as an elevator is to a mass.

Positive and negative potential differences

A positive potential difference means that the energy of the charge is *greater* at B than at A; a negative potential difference means that its energy is *less* at B than at A. If V_{AB} is positive, then a charge at B tends to return to A, whereas if V_{AB} is negative, the charge tends to move further away from A.

Batteries and generators are sources of potential difference

One advantage of specifying the potential difference V between two points in an electric field, rather than the magnitude E of the field between them, is that an electric field is normally created by imposing a difference of potential between two points in space. A battery is a device that uses a chemical process to produce a potential difference between two terminals (Fig. 18–20). A "6-volt" battery is one that has a potential difference of 6 V between its terminals. A generator is another device for producing a potential difference. Batteries and generators are to electric charges what elevators are to masses: All of them increase the potential energy of what they act upon.

Work done on a charge

When a charge Q goes from one terminal of a battery whose potential difference is V to the other, Eq. (18–8) tells us that the work

$$W = QV \qquad\qquad \textit{Work done on a charge} \quad (18\text{–}10)$$

is done on it regardless of the path taken by the charge and regardless of whether the actual electric field on the charge is strong or weak. Given V we can find W at once, no matter what the details of the process are. Hence the notion of potential difference makes our study of electrical phenomena easier, just as the notion of potential energy made our study of mechanical phenomena easier.

Example   Figure 18–21 shows a tube that has a source of electrons at one end and a metal plate at the other. A 100-V battery is connected between the electron source and the plate, so that there is a potential difference of 100 V between them. The negative terminal of the battery is connected to the electron source. What is the speed of the electrons when they arrive at the metal plate? (The tube is evacuated to prevent collisions between the electrons and air molecules.)

Solution   To find the kinetic energy, we note that the work done by the electric field within the tube on an electron is $W = QV = eV$. Since the kinetic energy of the electron

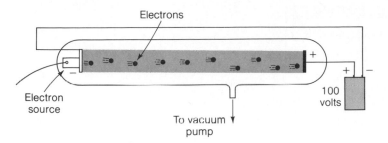

FIG. 18–21

is equal to the work done on it, its kinetic energy after passing through the entire field is eV. Hence

$$\text{KE} = W = eV = \tfrac{1}{2}mv^2$$

$$v = \sqrt{\frac{2eV}{m}} = \sqrt{\frac{2(1.6 \times 10^{-19}\,\text{C})(100\,\text{V})}{9.1 \times 10^{-31}\,\text{kg}}} = 5.9 \times 10^{6}\,\text{m/s} \qquad \blacksquare$$

Example The electron source and the positive electrode are 40 cm apart in the tube of the preceding problem. What is the magnitude E of the electric field (assumed uniform) between them? If this distance is reduced to 20 cm, what is the new value of E and what effect does this change have on the final speed of the electrons?

Solution (a) The electric field magnitude in a uniform field is given by $E = V/s$ from Eq. (18–9). Hence

$$E = \frac{100\,\text{V}}{0.4\,\text{m}} = 250\,\text{V/m}$$

If s is reduced to 20 cm, the field magnitude increases to

$$E = \frac{100\,\text{V}}{0.2\,\text{m}} = 500\,\text{V/m}$$

(b) The energy given to an electron by this field when it travels through the entire 20 cm is the same as the energy given to an electron that travels through the 40-cm length of the previous, weaker field, since in both cases QEs is the same. We reach the same conclusion by noting that $\text{KE} = QV$ and the potential difference V in this problem is independent of the spacing of the electrodes. The final electron speed is therefore unchanged. $\qquad \blacksquare$

> The final electron speed depends on the potential difference, not on the details of the field

18–9 CATHODE-RAY TUBE

Before their nature was understood, beams of electrons were known as "cathode rays." Even today, a tube in which an electron beam can be controlled in such a way as to trace a desired pattern on a fluorescent screen is called a *cathode-ray tube*. Figure 18–22 shows how such a tube is constructed. Electrons emitted from the heated cathode (negative electrode) are accelerated through a potential difference of 5 kV to 50 kV between the cathode and the anode (positive electrode) and emerge from a hole in the

> In a cathode-ray tube, an electron beam traces out an image on a fluorescent screen

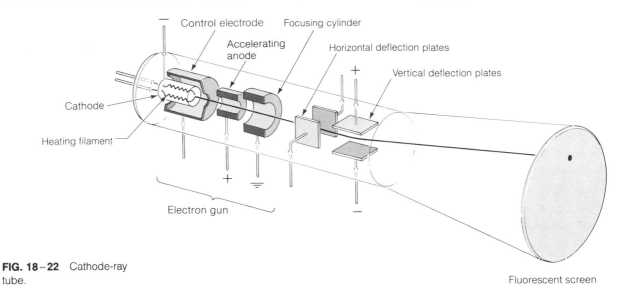

FIG. 18-22   Cathode-ray tube.

Fluorescent screen

latter. Near the cathode is an electrode whose potential difference relative to the cathode controls the number of electrons in the beam and therefore the brightness of the spot on the screen. The entire assembly, which also incorporates another anode to focus the beam, is called an *electron gun*.

The electron beam then passes through a horizontal electric field set up between a pair of parallel metal plates. The direction of the field determines whether the beam is deflected to the right or to the left, and its magnitude (which is proportional to the voltage applied to the plates) governs the amount of the deflection. The vertical electric field between the next set of plates similarly governs the vertical position of the electron beam when it reaches the screen, which is coated with a substance called a *phosphor* that emits light when struck by the electrons. The screen continues to glow for a short time after the electron beam has moved away from each point, which permits a picture to be built up that lasts long enough to be viewed as a whole.

An oscilloscope displays electrical signals for visual inspection

An *oscilloscope* uses a cathode-ray tube to display electrical signals that vary rapidly, since the small mass of the electron permits a beam of them to respond at once to a changing voltage applied to a pair of deflecting plates. The horizontal deflection plates are connected to a source of sawtooth pulses (Fig. 18-23(a)), which rise steadily from a certain voltage $-V$ through 0 to $+V$ and then return rapidly to $-V$ to start a new cycle. In the figure the pulses take 0.1 s to go from $-V$ to $+V$, where V is chosen to correspond to a full horizontal deflection in a particular tube. Thus each sawtooth pulse sweeps the electron beam from left to right across the screen in 0.1 s and then returns the beam to the left to start another sweep. If we connect the vertical deflection plates to a voltage source that alternates in polarity 50 times per second (in other words, whose frequency is 50 Hz), what we will see are five complete cycles on the oscilloscope screen, as in Fig. 18-23(b), because each cycle involves 1/50 s = 0.02 s. A wide range of sweep times is provided on an oscilloscope, and the input signal can be amplified as necessary to give a trace large enough to be inspected and measured.

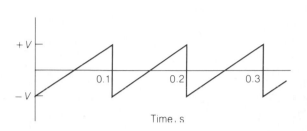

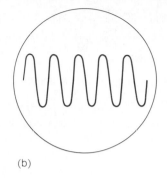

(a)

(b)

FIG. 18–23 (a) A sawtooth voltage with a period of 0.1 s. (b) Oscilloscope trace of a 50-Hz alternating signal applied to the vertical deflection plates when the sawtooth voltage of (a) is applied to the horizontal deflection plates. Each cycle takes 0.02 s, so five appear on the screen. (Photo courtesy of Heath Company/Heath-Zenith.)

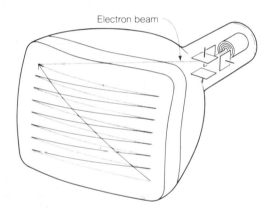

Electron beam

The picture tube of a television set is a cathode-ray tube, but in this application it is more convenient to use magnetic rather than electric fields to deflect the electron beam; magnetic fields are considered in later chapters. The electron beam is moved horizontally across the screen just as in an oscilloscope, but after each sweep the beam is shifted downward in order to cover a new line on the screen (Fig. 18–24). In the United States, a complete image is divided into 525 lines, but in many other countries more lines are used to give better picture quality. A complete image is built up by two scans of the screen, each covering alternate lines, a process that takes place 30 times per second. During the scans the beam intensity changes to produce the variations in brightness that make up the image. In color television separate signals are transmitted that correspond to the red, green, and blue contents of the image. The receiver screen consists of red, green, and blue phosphor dots that flash accordingly when the electron beam strikes them.

FIG. 18–24 The electron beam of a television picture tube covers the screen in a pattern of horizontal lines starting at the upper left. The returns from right to left take about one-tenth as long as the sweeps from left to right that produce the image.

18–10 THE ELECTRON VOLT

The electron volt (abbreviated eV) is a widely used energy unit in atomic and nuclear physics. By definition, 1 eV is the energy acquired by an electron that has been accelerated through a potential difference of 1 volt. Hence

The electron volt is an energy unit

$$W = QV$$
$$1 \text{ eV} = (1.60 \times 10^{-19} \text{C})(1.00 \text{ V})$$

and so

$$1 \text{ eV} = 1.60 \times 10^{-19} \text{ J}$$ *Electron volt*

Typical quantities expressed in electron volts are the ionization energy of an atom (which is the work needed to remove one of its electrons) and the binding energy of a molecule (which is the work needed to break it apart into separate atoms). Thus the ionization energy of nitrogen is usually given as 14.5 eV and the binding energy of the hydrogen molecule, which consists of two hydrogen atoms, is usually given as 4.5 eV.

MeV and GeV
The eV is too small a unit for nuclear physics, where its multiples the MeV (10^6 eV) and the GeV (10^9 eV) are commonly used. The M and G respectively signify *mega* ($= 10^6$) and *giga* ($= 10^9$) and are used in connection with other units as well, for instance the megabuck ($\$10^6$) and the gigawatt ($10^9$ watts). A typical quantity expressed in MeV is the energy liberated when the nucleus of a uranium atom splits into two parts. Such *fission* of a uranium nucleus releases an average of 200 MeV; this is the process that powers nuclear reactors and atomic bombs.

Example   What is the speed of a neutron whose kinetic energy is 50 eV?

Solution   The kinetic energy of the neutron is

$$\text{KE} = (50 \text{ eV})(1.6 \times 10^{-19} \text{ J/eV}) = 8 \times 10^{-18} \text{ J}$$

Since $\text{KE} = \frac{1}{2} mv^2$ and, from Table 18–1, the neutron mass is 1.67×10^{-27} kg, the speed of the neutron is

$$v = \sqrt{\frac{2\,\text{KE}}{m}} = \sqrt{\frac{(2)(8 \times 10^{-18} \text{ J})}{1.67 \times 10^{-27} \text{ kg}}} = 9.8 \times 10^4 \text{ m/s} \qquad \blacksquare$$

IMPORTANT TERMS

Electric charge, like mass, is a basic property of certain of the elementary particles of which all matter is composed. There are two kinds of electric charge, **positive charge** and **negative charge;** charges of like sign repel, unlike charges attract. The unit of charge is the **coulomb.** All charges, of either sign, occur in multiples of the fundamental **electron charge** of 1.6×10^{-19} coulomb.

The principle of **conservation of charge** states that the net electric charge in an isolated system remains constant.

Coulomb's law states that the force one charge exerts upon another is directly proportional to the magnitudes of the charges and inversely proportional to the square of the distance between them.

An **atom** consists of a tiny, positively charged nucleus surrounded at some distance by electrons. The nucleus con-

sists of protons and neutrons, and the number of electrons equals the number of protons so the atom as a whole is electrically neutral.

A **force field** is a region of space at every point of which an appropriate test object would experience a force.

An **electric field** exists wherever an electric force acts on a charged particle. The magnitude of an electric field at a point is equal to the force that would act on a charge of $+1$ C placed there; the direction of the field is the direction of the force on the charge. The unit of electric field is the V/m, which is equal to 1 N/C.

The electric **potential difference** between two points is the work that must be done to take a charge of 1 C from one of the points to the other. The unit of potential difference is the **volt,** which is equal to 1 J/C.

In a **cathode-ray tube,** an electron beam controlled by electric (or magnetic) fields traces out an image on a fluorescent screen.

The **electron volt** is the energy acquired by an electron that has been accelerated by a potential difference of 1 V. It is equal to 1.6×10^{-19} J.

IMPORTANT FORMULAS

Coulomb's law: $\quad F = k\dfrac{Q_A Q_B}{r^2}$

Electric field: $\quad \mathbf{E} = \dfrac{\mathbf{F}}{Q}$

Electric field of a charge: $\quad E = k\dfrac{Q}{r^2}$

Potential energy of a charge in uniform field: $\quad PE = QEs$

Potential energy of two charges: $\quad PE = k\dfrac{Q_A Q_B}{r}$

Potential difference: $\quad V_{AB} = \dfrac{W_{AB}}{Q}$

$\qquad\qquad\qquad\quad = Es \quad$ *(in uniform electric field)*

MULTIPLE CHOICE

1. Electric charge
   a. is a continuous quantity that can be subdivided indefinitely.
   b. is a continuous quantity but it cannot be subdivided into smaller parcels than $\pm 1.6 \times 10^{-19}$ C.
   c. occurs only in separate parcels, each of $\pm 1.6 \times 10^{-19}$ C.
   d. occurs only in separate parcels, each of ± 1 C.

2. An object has a positive electric charge whenever
   a. it has an excess of electrons.
   b. it has a deficiency of electrons.
   c. the nuclei of its atoms are positively charged.
   d. the electrons of its atoms are positively charged.

3. A negative electric charge
   a. interacts only with positive charges.
   b. interacts only with negative charges.
   c. interacts with both positive and negative charges.
   d. may interact with either positive or negative charges, depending on circumstances.

4. Which of the following statements is not true?
   a. The positive charge in an atomic nucleus is due to the protons it contains.

b. All protons have the same charge.
   c. Protons and electrons have charges equal in magnitude although opposite in sign.
   d. Protons and electrons have equal masses.

5. An atom consists of
   a. a uniform distribution of positive charge in which electrons are embedded.
   b. a uniform distribution of negative charge in which protons are embedded.
   c. a small negative nucleus surrounded at a distance by protons.
   d. a small positive nucleus surrounded at a distance by electrons.

6. Which of the following statements is not true?
   a. The positive charge in an atomic nucleus is due to the protons it contains.
   b. All protons have the same charge.
   c. Protons and electrons have charges equal in magnitude although opposite in sign.
   d. Protons and electrons have equal masses.

7. Coulomb's law belongs in the same general category as
   a. the law of gravitation.
   b. the laws of motion.
   c. the laws of thermodynamics.
   d. the conservation principles of mechanics.

8. In the formula $F = kQ_A Q_B/r^2$, the value of the constant k
   a. is the same under all circumstances.
   b. depends upon the medium the charges are located in.
   c. is different for positive and negative charges.
   d. has the numerical value 1.6×10^{-19}.

9. Relative to the electric force between two protons, the gravitational force between them is
   a. weaker.
   b. equal in magnitude.
   c. stronger.
   d. any of the above, depending on how far apart the protons are.

10. Of the following quantities, the one that is vector in character is electric
    a. charge.
    b. field.
    c. energy.
    d. potential difference.

11. The electric field at a point in space is equal in magnitude to
    a. the potential difference there.
    b. the electric charge there.

c. the force a charge of one coulomb would experience there.

d. the force an electron would experience there.

12. The magnitude of the electric field in the region between two parallel oppositely charged metal plates is

a. zero.

b. uniform throughout the region.

c. greatest near the positive plate.

d. greatest near the negative plate.

13. From its definition, the unit of electric field E is the N/C. An equivalent unit of E is the

a. $V \cdot m$.   b. $V \cdot m^2$.

c. V/m.   d. V/m^2.

14. It is not true that electric lines of force

a. leave positive charges and enter negative ones.

b. are close together where an electric field is strong and far apart where the field is weak.

c. have the same pattern around a charged sphere as they do around a point charge.

d. actually exist.

15. Ten million electrons are placed on a solid copper sphere. The electrons become

a. uniformly distributed on the sphere's surface.

b. uniformly distributed in the sphere's interior.

c. concentrated at the center of the sphere.

d. concentrated at the bottom of the sphere.

16. A system of two charges has a positive potential energy. This signifies that

a. both charges are positive.

b. both charges are negative.

c. both charges are positive or both are negative.

d. one charge is positive and the other is negative.

17. The electron volt is a unit of

a. charge.

b. potential difference.

c. energy.

d. momentum.

18. If 10,000 electrons are removed from a neutral pith ball, its charge is now

a. $+1.6 \times 10^{-15}$ C.   b. $+1.6 \times 10^{-23}$ C.

c. -1.6×10^{-15} C.   d. -1.6×10^{-23} C.

19. Two charges of $+Q$ are 1 cm apart. If one of the charges is replaced by a charge of $-Q$, the magnitude of the force between them is

a. zero.   b. smaller.

c. the same.   d. larger.

20. A charge of $+q$ is placed 2 cm from a charge of $-Q$. A second charge of $+q$ is then placed next to the first. The force on the charge of $-Q$

a. decreases to half its former magnitude.

b. remains the same.

c. increases to twice its former magnitude.

d. increases to four times its former magnitude.

21. Two charges repel each other with a force of 10^{-6} N when they are 10 cm apart. When they are brought closer together until they are 2 cm apart, the force between them becomes

a. 4×10^{-8} N.   b. 5×10^{-6} N.

c. 8×10^{-6} N.   d. 2.5×10^{-5} N.

22. Two charges, one positive and the other negative, are initially 2 cm apart and are then pulled away from each other until they are 6 cm apart. The force between them is now smaller by a factor of

a. $\sqrt{3}$.   b. 3.

c. 9.   d. 27.

23. The force between two charges of -3×10^{-9} C that are 5 cm apart is

a. 1.8×10^{-16} N.   b. 3.6×10^{-15} N.

c. 1.6×10^{-6} N.   d. 3.2×10^{-5} N.

24. Two equal charges attract each other with a force of 10^{-5} N. When they are moved 4 mm farther apart, the force between them becomes 2.5×10^{-6} N. The original separation of the charges was

a. 1 mm.   b. 2 mm.

c. 4 mm.   d. 8 mm.

25. The magnitude of the above charges is

a. 6.67×10^{-11} C.   b. 1.33×10^{-10} C.

c. 1.05×10^{-9} C.   d. 2.11×10^{-9} C.

26. The force on an electron in an electric field of 200 V/m is

a. 8×10^{-22} N.   b. 3.2×10^{-21} N.

c. 3.2×10^{-17} N.   d. 6.4×10^{-15} N.

27. The electric field 2 cm from a certain charge has a magnitude of 10^5 V/m. The value of E 1 cm from the charge is

a. 2.5×10^4 V/m.   b. 5×10^4 V/m.

c. 2×10^5 V/m.   d. 4×10^5 V/m.

28. An electric field of magnitude 200 V/m can be produced by applying a potential difference of 10 V to a pair of parallel metal plates separated by

a. 2 cm.   b. 5 cm.

c. 20 m.   d. 2000 m.

29. A charge of 10^{-10} C between two parallel metal plates 1 cm apart experiences a force of 10^{-5} N. The potential difference between the plates is

a. 10^{-5} V.   b. 10 V.

c. 10^3 V.   d. 10^5 V.

30. In charging a certain storage battery, a total of 2×10^5 C is transferred from one set of electrodes to another. The potential difference between the electrodes is 12 V. The energy stored in the battery is

    a. 1.7×10^4 J.   b. 2.4×10^6 J.

    c. 2.4×10^7 J.   d. 2.9×10^7 J.

31. The potential difference between two metal plates 2 cm apart needed to accelerate an electron to a speed of 5×10^6 m/s is

    a. 2.84×10^{-5} V.   b. 0.355 V.

    c. 1.42 V.   d. 71.1 V.

32. The KE of the electron of the previous question is

    a. 1.82×10^{-36} eV.   b. 1.14×10^{-17} eV.

    c. 1.42 eV.   d. 71.1 eV.

33. The rest energy of the proton is 938 MeV, which is equal to

    a. 1.67×10^{-27} J.   b. 1.50×10^{-10} J.

    c. 1.04×10^{-8} J.   d. 1.50×10^{-8} J.

34. The speed of an electron whose speed is 10^7 m/s has an energy of

    a. 4.6×10^{-17} eV.   b. 160 eV.

    c. 284 eV.   d. 568 eV.

35. The speed of an electron whose KE is 100 eV is

    a. 5.9×10^6 m/s.   b. 3.5×10^7 m/s.

    c. 1.5×10^{16} m/s.   d. 3.7×10^{25} m/s.

EXERCISES

18–1 Electric Charge

1. (a) When two objects attract each other electrically, must both of them be charged? (b) When two objects repel each other electrically, must both of them be charged?

2. What reasons might there be for the universal belief among scientists that there are only two kinds of electric charge?

3. How can the principle of charge conservation be reconciled with the fact that a rubber rod can be charged by stroking it with a piece of fur?

4. An insulating rod has a charge of $+Q$ at one end and a charge of $-Q$ at the other. How will the rod behave when it is placed near a fixed positive charge that is initially equidistant from the ends of the rod?

5. Nearly all the mass of an atom is concentrated in its nucleus. Where is its charge located?

6. Electricity was once regarded as a weightless fluid, an excess of which was "positive" and a deficiency of which was "negative." What phenomena can this hypothesis still explain? What phenomena can it not explain?

18–2 Coulomb's Law

7. How do we know that the inverse square force holding the earth in its orbit around the sun is not an electrical force?

8. A certain raindrop picks up 30 electrons in falling through the air. What is its charge? By how much has its mass increased?

9. Find the charge of 1 μg of protons.

10. Two charges attract each other with a force of 4×10^{-6} N when they are 4 mm apart. Find the force between them when their separation is increased to 5 mm.

11. Two electric charges originally 8 cm apart are brought closer together until the force between them is greater by a factor of 16. How far apart are they now?

12. Two charges of unknown magnitude and sign are observed to repel one another with a force of 0.1 N when they are 5 cm apart. What will the force be when they are (a) 10 cm apart? (b) 50 cm apart? (c) 1 cm apart?

13. The nucleus of a hydrogen atom is a single proton. Find the force between the two protons in a hydrogen molecule, H_2, that are 7.42×10^{-11} m apart. (The two electrons in the molecule spend more time between the protons than outside them, which leads to attractive forces that balance the repulsion of the protons and permit a stable H_2 molecule; see Chapter 29.)

14. A charge of -5×10^{-7} C is 10 cm from a charge of $+6 \times 10^{-6}$. Find the magnitude and direction of the force on each charge.

15. A charge of $+5 \times 10^{-9}$ C is attracted by a charge of -3×10^{-7} C with a force of 0.135 N. How far apart are they?

16. Two metal spheres, one with a charge of $+2 \times 10^{-5}$ C and the other with a charge of -1×10^{-5} C, are 10 cm apart. (a) What is the force between them? (b) The two spheres are brought into contact, and then separated again by 10 cm. What is the force between them now?

17. How far apart should two electrons be if the force each exerts on the other is to equal the weight of an electron?

18. At what distance apart (if any) are the electric and gravitational forces between two electrons equal in magnitude? Between two protons? Between an electron and a proton?

19. According to one model of the hydrogen atom, it consists of a proton circled by an electron whose orbit has a radius of 5.3×10^{-11} m. How fast must the electron be moving if the required centripetal force is provided by the electric force exerted by the proton?

20. As mentioned in the text, the permittivity of free space is $\epsilon_0 = 8.85 \times 10^{-12}$ $C^2/N \cdot m^2$. The permittivity of air is

only a trifle greater than this, but it is considerably greater for some other materials. For example, the permittivity of water is $\epsilon = 80\epsilon_0$. Find the force, in air and in water, between two charges of $+2 \times 10^{-10}$ C that are 1 mm apart.

21. A particle carrying a charge of $+6 \times 10^{-9}$ C is located halfway between two other charges, one of $+1 \times 10^{-8}$ C and the other of -1×10^{-8} C, that are 40 cm apart. All three charges lie on the same straight line. What is the magnitude and direction of the force on the $+6 \times 10^{-9}$ C charge?

22. A test charge of -5×10^{-8} C is placed between two other charges so that it is 5 cm from a charge of -3×10^{-8} C and 10 cm from a charge of -6×10^{-8} C. The three charges lie along a straight line. What is the magnitude and direction of the force on the test charge?

23. Two charges, one of -1×10^{-8} C and the other of -3×10^{-8} C are 0.4 m apart. (a) Where should a charge of -1×10^{-9} C be placed on the line between the other charges in order that there be no resultant force on it? (b) Where should a charge of $+1 \times 10^{-9}$ C be placed in order that there be no resultant force on it?

24. Two charges, one of $+2 \times 10^{-8}$ C and the other of $+1 \times 10^{-8}$ C are 0.2 m apart. Where should an electron be placed in order that there be no resultant force on it?

25. Two 3-g balloons are suspended from a nail by strings 50 cm long. Each balloon has a charge of $+Q$, and there is an angle of 40° between the strings. Find Q.

26. Four charges of $+1 \times 10^{-8}$ C are at the corners of a square 0.2 m on each side. Find the magnitude and direction of the force on one of them.

27. Three charges, $+Q$, $+Q$, and $-Q$, are at the vertexes of an equilateral triangle a long on each side. Find the magnitude and direction of the force on one of the positive charges.

18–4   Electric Field

28. An insulating rod has a charge of $+Q$ at one end and a charge of $-Q$ at the other. How will the rod behave when it is placed in a uniform electric field whose direction is (a) parallel to the rod; (b) perpendicular to the rod?

29. Find the electric field 40 cm from a charge of $+7 \times 10^{-5}$ C.

30. The electron in a hydrogen atom averages 5.3×10^{-11} m away from the proton that is the nucleus of this atom. How strong is the electric field the electron experiences?

31. Four charges of $+1$ μC are at the corners of a square that measures 1 m on each side. Find the electric field at the center of the square.

32. A particle carrying a charge of 10^{-5} C starts moving from rest in a uniform electric field whose intensity is 50 N/C. (a) What is the force on the particle? (b) How much kinetic energy will the particle have after it has moved 1 m?

33. How strong an electric field is needed to support a proton against gravity at sea level?

34. Two charges of $+4$ μC and $+8$ μC are 2 m apart. What is the electric field halfway between them?

35. Two charges, one of $+1.5$ μC and the other of $+3$ μC, are 0.2 m apart. Where is the electric field along the line joining them equal to zero?

36. What is the electric field at one vertex of an equilateral triangle whose sides are 1 m long if there are charges of $+20$ μC at the other vertexes?

18–5   Electric Lines of Force

37. Can lines of force ever intersect in space? Explain.

38. What can you tell about the force a charged object would experience at a given point in an electric field by looking at a sketch of the lines of force of the field?

39. Sketch the pattern of electric lines of force in the neighborhood of two charges, $+Q$ and $+2Q$, that are a short distance apart.

40. Sketch the pattern of electric lines of force in the neighborhood of two charges, $-Q$ and $+2Q$, that are a short distance apart.

41. A charge of $+Q$ is placed on a cubical copper box 20 cm on an edge. What is the magnitude and direction of the electric field at the center of the box?

18–7   Potential Energy of Two Charges

42. The potential energy of a certain system of two charges increases as the charges are moved farther apart. What does this tell us about the signs of the charges?

43. Find the potential energy of the hydrogen atom described in Exercise 19.

44. Two electrons are 10^{-9} m apart when both are released. What is the speed of the electrons when they are 10^{-8} m apart?

18–8   Potential Difference

45. Twelve joules of work are needed to transfer 2 C of charge from one terminal of a storage battery to the other. What is the potential difference between the terminals?

46. An electric field stronger than 3×10^6 V/m will cause sparks to occur in air. What is the maximum potential difference that can be applied across two metal plates 1 mm apart before sparking begins?

47. A potential difference of 50 V is applied across two parallel metal plates and an electric field of 10^4 V/m is produced. How far apart are the plates?

48. The potential difference between two parallel metal plates that are 0.5 cm apart is 10^4 V. Find the force on an electron located between the plates.

49. A cloud is at a potential of 8×10^6 V relative to the ground. A charge of 40 C is transferred in a lightning stroke between the cloud and the ground. Find the energy dissipated.

50. The electrodes in a neon sign are 1.2 m apart and the potential difference across them is 8000 V. (a) Find the acceleration of a neon ion of mass 3.3×10^{-26} kg and charge $+e$ in the field. (b) If the ion starts at the positive electrode of the sign and moves unimpeded to its negative electrode, how much energy would it gain? (c) Why is it extremely unlikely that the ion would actually acquire this much energy?

51. Two parallel metal plates are 4 cm apart. If the force on an electron between the plates is to be 10^{-4} N, what should the potential difference between them be?

52. The storage battery of a car is being charged by an alternator at the rate of 10 C/s. If the potential difference across the alternator's terminals is 14 V and the alternator is 90% efficient, find the power supplied by the car's engine to charge the battery.

53. In charging a certain 20-kg storage battery, a total of 2×10^5 C is transferred from one set of electrodes to another. The potential difference between the electrodes is 14 V. (a) How much energy is stored in the battery? (b) If this energy were used to raise the battery above the ground, how high would it go? (c) If this energy were used to provide the battery with kinetic energy, what would its speed be?

54. A potential difference of 1000 V is applied across two parallel metal plates 100 mm apart. An electron leaves the negative plate at the same time as a proton leaves the positive plate. (a) Find the speeds and kinetic energies of the particles when they reach the opposite plates. (b) At what distance from the positive plate do the electron and proton pass each other?

55. A potential difference of 10 V is applied across two parallel metal plates 2 cm apart. An electron is projected at a speed of 10^7 m/s halfway between the plates and parallel to them. How far will the electron travel before striking the positive plate?

18–9   Cathode-Ray Tube

56. A potential difference of 20 kV is used to accelerate electrons in the electron gun of a cathode-ray tube. How much kinetic energy do the electrons have when they leave the gun? What is their speed?

57. What effect does doubling the accelerating voltage of the electron gun in a cathode-ray tube have on the speed of the electrons that reach the screen?

58. The electron gun of a television picture tube has an accelerating potential difference of 15 kV and a power rating of 25 W. How many electrons reach the screen per second? At what speed?

18–10   The Electron Volt

59. What is the kinetic energy in electron volts of a potassium atom of mass 6.5×10^{-26} kg whose speed is 10^6 m/s?

60. What is the kinetic energy in electron volts of an electron whose speed is 10^6 m/s?

61. What is the speed of an electron whose kinetic energy is 50 eV?

62. What is the speed of a neutron whose kinetic energy is 50 eV?

63. Typical chemical reactions absorb or release energy at the rate of several eV per molecular change. What change in mass is associated with the absorption or release of 1 eV?

64. Find the rest energy of the electron in MeV.

ANSWERS TO MULTIPLE CHOICE

1. c	**8.** b	**15.** a	**22.** c	**29.** c
2. b	**9.** a	**16.** c	**23.** d	**30.** b
3. c	**10.** b	**17.** c	**24.** c	**31.** d
4. d	**11.** c	**18.** a	**25.** b	**32.** d
5. d	**12.** b	**19.** c	**26.** c	**33.** b
6. d	**13.** c	**20.** c	**27.** d	**34.** c
7. a	**14.** d	**21.** d	**28.** b	**35.** a

19

ELECTRIC CURRENT

An electric current consists of a flow of charge from one place to another. Currents and not stationary charges are involved in nearly all practical applications of electricity. In this chapter we begin by considering the mechanisms by which various materials conduct current and the factors that govern direct currents in metal conductors. Then we go on to the important topic of the transport of energy by electric current. Finally we examine the related topics of electrolysis, which is widely used to refine such elements as aluminum and to plate such metals as chromium on objects made of other metals, and batteries and fuel cells, which use chemical reactions to produce electric current.

19–1   ELECTRICAL CONDUCTION

Most substances conduct electricity either very well or very badly

An electric current is a flow of charge. Nearly all substances fall into two categories: *conductors,* through which charge can flow easily, and *insulators,* through which charge can flow only with great difficulty. Metals, many liquids, and plasmas (gases

CHAPTER OBJECTIVES

Completing this chapter should enable you to:

1. Describe the mechanisms by which electric current is conducted through various substances.
2. Describe electric current and potential difference by analogy with the flow of water in a pipe.
3. Use Ohm's law, $I = V/R$, to solve problems that involve the current in a circuit, its resistance, and the potential difference across it.
4. Calculate the resistance of a given conductor.
5. Take into account the variation of resistance with temperature for a given conductor.
6. Determine the correct wire size for a certain application.

7. Relate the power consumed by an electrical appliance to its resistance, the current in it, and the voltage across it.
8. Distinguish between polar and nonpolar molecules.
9. Describe the behavior of electrolytes when dissolved in water.
10. Understand how elements are liberated by the passage of an electric current in electrolysis.
11. Relate the capacity rating of a battery to the energy stored in it and to the combinations of current and discharge time it can provide.

whose molecules are charged) are conductors. Nonmetallic solids, certain liquids, and gases whose molecules are electrically neutral are insulators. Several substances, called *semiconductors,* are intermediate in their ability to conduct charge.

Conduction in a metal

In a solid metal, each atom gives up one or more electrons to a common "gas" of freely moving electrons. These electrons can move easily through the crystal structure of the metal, so if one end of a metal wire is given a positive charge and the other end a negative charge, electrons will flow through the wire from the negative to the positive end. This flow, of course, constitutes an electric current. By supplying new electrons to the negative end of the wire and removing electrons from the positive end as they arrive there—which can be done by connecting the wire to a battery or to a generator—a constant current can be maintained in the wire.

Insulators

In nonmetallic solids, such as salt, glass, rubber, minerals, wood, and plastics, all the atomic electrons are fixed in particular atoms or groups of atoms and cannot move from place to place. Such solids are therefore insulators. Actually, nonmetallic solids do conduct very small amounts of current, but their abilities to do this are very poor compared with those of metals. For instance, when identical bars of copper and sulfur are connected to the same battery, about 10^{23} times more current flows in the copper bar.

Semiconductors

As mentioned earlier, there are a few substances called *semiconductors* through which current flows more readily than through insulators but still with more difficulty than through conductors. Thus about 10^7 times more current flows in a germanium bar connected to a battery than in a sulfur bar of the same size, but this is still about 10^{16} times less current than in a copper bar. The electric conductivity of solids is discussed in more detail in Chapter 29.

Superconductivity is a low-temperature effect

At temperatures near absolute zero (0 K, which we recall is $-273°C$) certain metals, alloys, and chemical compounds lose all of their resistance to the flow of electric current. This effect, called *superconductivity,* was discovered by Kamerlingh Onnes in Holland in 1911. For example, aluminum is superconducting at temperatures under 1.20 K, lead at temperatures under 7.22 K, and CuS (copper sulfide) at temperatures under 1.6 K. If a current is set up in a closed wire loop at room temperature, it will die out in less than a second even if the wire is made of a good conductor such as copper or silver. However, if the wire is made of a superconducting material and is kept cold enough, the current will continue indefinitely. Currents have persisted in superconducting loops for years at a time.

Superconductivity is of immense potential importance for the transmission of electric energy and in applications where strong magnetic fields are required. Already laboratory electromagnets with superconducting coils are in use, and experimental electric motors whose windings are superconducting have been built. There is no basic reason why superconducting magnets cannot be used to support trains and thereby both increase their speeds and reduce their power requirements. The immediate problem is that the best materials for the purpose thus far discovered exhibit superconductivity only at temperatures under about 20 K, although it is possible that eventually superconductors will be developed that function at more practical temperatures.

Conduction in liquids and gases

The mechanism of electrical conduction in liquids and gases is different from that in metals. The current in a metal consists of a flow of electrons past the stationary atoms in its structure. The current in a fluid medium other than a liquid metal, however, consists of a flow of entire atoms or molecules that are electrically charged. An atom or molecule that carries a net charge is called an *ion,* and both positive and negative ions participate in the conduction process in liquids and gases.

A positive ion has a deficiency of electrons; a negative ion has a surplus of electrons

An atom or molecule becomes a positive ion when it loses one or more of its electrons; if it gains one or more electrons in addition to its usual complement, it becomes a negative ion. The fundamental positive charges in matter are protons, which are very tightly bound in the nucleus of every atom. Atomic electrons, however, are held more loosely, and one or two of them can be detached from an atom with relative ease. Thus the oxygen and nitrogen gases in ordinary air become ionized when a spark occurs, in the presence of a flame, and by the passage of X rays or even ultraviolet light. These processes so disturb the air molecules that some electrons are dislodged, leaving behind positive ions. The liberated electrons almost at once become attached to other nearby molecules to create negative ions (Fig. 19–1).

Recombination

The electrical attraction between positive and negative charges sooner or later brings the ions together, and the extra electrons on the negative ions become reattached to the positive ions. The gas molecules are then neutral, as they were originally. This *recombination* is rapid at normal atmospheric pressure and temperature.

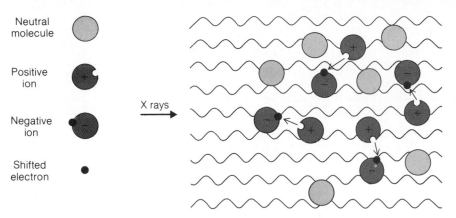

Neutral molecule

Positive ion

Negative ion

Shifted electron

X rays

FIG. 19–1 Schematic representation of the ionization of air by X rays. A molecule losing an electron becomes a positive ion; a molecule gaining an electron becomes a negative ion.

In the upper atmosphere, where air molecules are so far apart that the recombination of ions is a slow process, the continual bombardment of X rays and ultraviolet light from the sun maintains a certain proportion of ions at all times. The layers of ions in the upper atmosphere constitute the *ionosphere,* and they make possible long-range radio communication by their ability to reflect radio waves (see Fig. 25–8). The ionosphere is an example of a *plasma,* which is a gas whose constituent particles are electrically charged. The behavior of a plasma, unlike that of an ordinary gas, is strongly influenced by electric and magnetic forces. Most of the universe is in the plasma state.

Origin of the ionosphere

The earth as a whole is a fairly good conductor. Hence if a charged object is connected to the earth by a piece of metal, the charge is conducted away from the object to the earth: This convenient way to remove charge from an object is called *grounding* the object. As a safety measure, the metal shells of electrical appliances are grounded through special wires that give electric charges in the shells paths to the earth. The round post in the familiar three-prong electric plug is the ground connection.

Grounding

19–2 ELECTRIC CURRENT

The magnitude of an electric current, denoted I, is the rate at which charge passes a given point. If the net charge Q goes past in the time interval t, then the average current is

$$I = \frac{Q}{t}$$

Electric current (19–1)

$$\text{Current} = \frac{\text{charge}}{\text{time interval}}$$

The unit of electric current is the *ampere* (A), where

The ampere

1 ampere = 1 coulomb/second

The direction of a current is, by convention, taken as that in which *positive* charges would have to move in order to produce the same effects as the observed current. Thus

FIG. 19–2 The ampere is the unit of electric current. The flow of charge in a circuit is like the flow of water in a pipe except that a return wire is necessary in order to have a complete conducting path. By convention, an electric current is assumed to flow from the positive terminal of a battery or generator to its negative terminal in an external circuit. Actual currents in metals consist of electrons that move in the opposite direction.

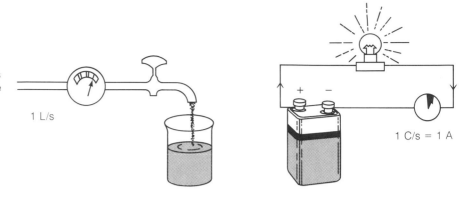

a current is always assumed to proceed from the positive terminal of a battery or generator to its negative terminal in an external circuit (Fig. 19–2).

Despite the above convention, actual electric currents in metals consist of flows of electrons, which carry negative charges. However, a current that consists of negative particles moving in one direction is electrically the same as a current that consists of positive particles moving the other way. Since there is no overwhelming reason to prefer one way of designating current to the other, we shall follow the usual practice of considering current as a flow of positive electric charge.

Two conditions must be met in order for an electric current to exist between two points. These are:

A conducting path is needed for a current to occur

1. There must be a path between the two points along which charge can flow. As was discussed earlier, metals, many liquids, and plasmas allow charge to pass through them readily and are classed as conductors. Nonmetallic solids, certain liquids, and gases whose molecules are electrically neutral allow charge to pass through them only with great difficulty and are classed as insulators. A few substances have an intermediate ability to permit the flow of charge and are classed as semiconductors.

A potential difference is also needed

2. There must be a difference of potential between the two points. (A superconductor is an exception to this requirement.) Just as the rate of flow of water between the ends of a pipe depends upon the difference of pressure between them, so the rate of flow of charge between two points depends upon the difference of potential between them. A large potential difference means a large "push" given to each charge.

Potential difference is analogous to water pressure

The analogy between electric current and water flow is a close one. The rate of flow of water in a pipe may be increased by having the water fall through a greater height, which increases the pressure in the pipe and thereby leads to a greater force on each parcel of water. Similarly the current in a wire may be increased by increasing the potential difference across it, which means a stronger electric field in the wire and thus more force on the moving charges that constitute the current (Fig. 19–3).

Definition of resistance

A particular conducting path—for instance, a copper wire, a light bulb, an electric heater, a transistor—is usually called a conductor, even though this is also the name of the class of substances through which current flows readily. The *resistance* of a conductor is the ratio between the potential difference *V* across it and the resulting current *I* that flows:

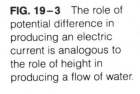

FIG. 19-3 The role of potential difference in producing an electric current is analogous to the role of height in producing a flow of water.

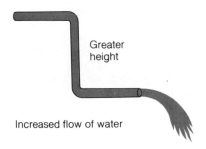

Greater height

Increased flow of water

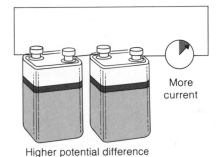

More current

Higher potential difference

$$R = \frac{V}{I} \qquad\qquad\qquad Resistance \quad (19-2)$$

$$Resistance = \frac{\text{potential difference}}{\text{current}}$$

The unit of resistance is the *ohm* (Ω), where   **The ohm**

1 ohm = 1 volt/ampere

$1\,\Omega = 1\,V/A$

A conductor in which there is a current of 1 A when a potential difference of 1 V is present across it has a resistance of 1 Ω. (The symbol Ω is the Greek capital letter *omega*.)

Example   A 120-V electric heater draws a current of 15 A. Find its resistance.

Solution   From the definition of resistance,

$$R = \frac{V}{I} = \frac{120\,V}{15\,A} = 8\,\Omega \qquad\qquad\qquad \blacksquare$$

19-3 OHM'S LAW

The resistance of a conductor depends in general both upon its properties—its nature and its dimensions—and upon the potential difference applied across it. In some con-

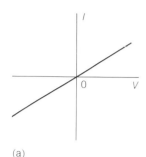

(a)

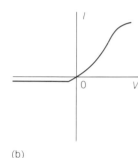

(b)

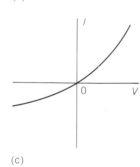

(c)

FIG. 19-4 The relationship between current and voltage for (a) a metal, (b) a vacuum tube, and (c) a semiconductor diode. Only in (a) is *I* proportional to *V*, a relationship known as Ohm's law.

ductors *R* increases when *V* increases, in others *R* decreases when *V* increases, and in still others *R* depends upon the direction of the current (Fig. 19–4).

Metallic conductors usually have constant resistances (at constant temperature), so that *I* is directly proportional to *V* in them. This relationship is called *Ohm's law,* since it was first verified experimentally by the German physicist Georg Ohm (1787–1854). Ohm's law states that

$$I = \frac{V}{R} \qquad (R = \text{constant}) \qquad\qquad Ohm's\ law \quad (19-3)$$

Despite its name, Ohm's law is not a true physical principle, but it is obeyed by most metals under a wide range of circumstances.

Ohm's law must be distinguished from the definition of resistance,

$$R = \frac{V}{I}$$

Ohm's law only holds for conductors in which the ratio *V/I* is always constant. The resistance of a certain conductor is given by $R = V/I$ whether or not it is constant as the voltage is changed.

Example A light bulb has a resistance of 240 Ω. Find the current in it when it is placed in a 120-V circuit (Fig. 19–5).

Solution From Ohm's law,

$$I = \frac{V}{R} = \frac{120\ \text{V}}{240\ \Omega} = 0.5\ \text{A} \qquad\qquad\blacksquare$$

Example The current in the coil of an 8-Ω loudspeaker is 0.5 A. Find the voltage across its terminals.

Solution We rewrite Ohm's law in the form $V = IR$ and obtain

$$V = IR = (0.5\ \text{A})(8\ \Omega) = 4\ \text{V} \qquad\qquad\blacksquare$$

An electric current in body tissue affects it both by stimulating nerves and muscles and through the heat produced. Tissue is a fairly good conductor because of the ions in solution it contains. Dry skin has a higher resistance and so is able to protect to some extent the rest of the body in the event of accidental exposure to a high potential difference, protection that disappears when the skin is wet. A current of as little as 0.5 mA (that is, 0.0005 ampere) is perceptible to most people, one of 5 mA is painful, and one of 10 mA or more causes muscle contractions. Such contractions may prevent the person involved from letting go of the source of the current, and breathing becomes impossible when the current exceeds about 18 mA.

Since a closed conducting path is necessary for a current to occur, touching a single "live" conductor has no effect if the body is isolated. However, if a person at the same time is in contact with a water pipe, or is standing on wet soil, or otherwise is grounded, a current will pass through his or her body. The significance of being grounded is that the earth is able to neutralize any amount of charge. If a positive electrode is grounded, electrons flow from the earth until it is neutral; if a negative electrode is grounded, electrons flow from it to the earth until it is neutral. The resistance of the human body itself is of the order of magnitude of 1000 Ω, so contact via wet skin with a 120-V line will lead to a current in the neighborhood of $I = $ 120 V/1000 Ω = 0.12 A = 120 mA. Such a current is extremely dangerous because it is likely to cause the heart muscles to contract rapidly and irregularly, and, if allowed to persist, will cause death.

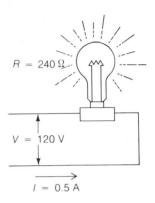

FIG. 19-5

Because the water in a bathtub is grounded via the tub's drainpipe, a person in the tub is at risk if he or she touches any electrical device, even a switch; the moisture on a wet finger may be sufficient to provide a conducting path to the interior of the device. Another situation of great potential danger occurs in hospitals where electrical appliances are often attached to patients to monitor various functions, or to control them as in the case of cardiac pacemakers. Even a minor malfunction of such an appliance, or the failure to properly ground it, may have fatal consequences because the conducting path in the body is then short compared with what it is when the contact is made via a finger.

It is sometimes thought that a fuse or circuit breaker in a circuit eliminates any danger. Such a device "opens" a circuit when the current is greater than a certain value and thus prevents damage to the equipment being used or to the wiring that carries the current. For instance, a 15-A fuse contains a length of special wire that melts when the current is 15 A or more. However, since as little as about 50 mA (1/300 of 15 A) is enough to kill, fuses and circuit breakers evidently provide no protection to people.

19-4 RESISTIVITY

The resistance of a conductor that obeys Ohm's law depends upon three factors:

1. The material of which it is composed; the ability to carry an electric current varies more than almost any other physical property of matter.
2. Its length L; the longer the conductor, the greater its resistance.
3. Its cross-sectional area A; the thicker the conductor, the less its resistance.

Factors that govern resistance

Once again we note the correspondence to water flowing through a pipe: The longer the pipe, the more chance friction against the pipe wall has to slow down the water, and the wider the pipe, the larger the volume of water that can pass through per second when everything else is the same (Fig. 19-6).

The simple formula

$$R = \rho \frac{L}{A}$$

Resistance of ohmic conductor (19-4)

has been found to hold for the resistance of a conductor that obeys Ohm's law. The quantity ρ (the Greek letter *rho*) is called the *resistivity* of the material from which the

Resistivity

FIG. 19-6 The way in which the dimensions of a conductor affect the flow of charge in it is analogous to the way in which the dimensions of a pipe affect the flow of water in it.

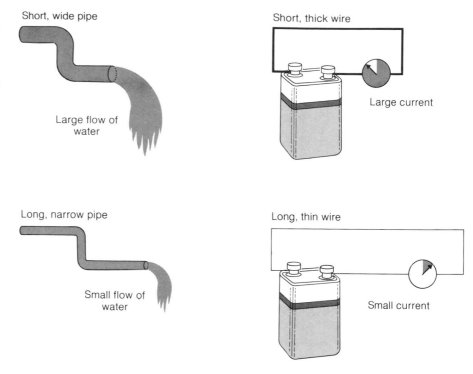

Short, wide pipe

Large flow of water

Short, thick wire

Large current

Long, narrow pipe

Small flow of water

Long, thin wire

Small current

conductor is made. Table 19–1 lists the resistivities of various substances at room temperature (20°C). Given the nature of a conductor and its dimensions, the value of R can be calculated at once. In the SI system, lengths are measured in meters and areas in square meters, and the unit of resistivity is accordingly the ohm-meter ($\Omega \cdot$ m). Metric wire sizes are usually specified by their cross-sectional areas in square millimeters, so a particular wire might be referred to as 2.5 mm^2, for instance, instead of having its diameter of 1.8 mm quoted.

Example What length of copper wire whose area is 0.1 mm^2 is needed to provide a resistance of 3 Ω?

Solution Since 1 mm $= 10^{-3}$ m, 1 mm$^2 = 10^{-6}$ m^2 and

$$L = \frac{RA}{\rho} = \frac{(3\,\Omega)(0.1\,\text{mm}^2)(10^{-6}\,\text{m}^2/\text{mm}^2)}{1.7 \times 10^{-8}\,\Omega \cdot \text{m}}$$
$$= 18\,\text{m} \quad\blacksquare$$

The circular mil is a unit of area

In engineering practice using the British system it is customary to express the cross-sectional area of a round conductor in terms of a unit called the *circular mil* (cmil). A circular mil is the area of a circle whose diameter is 1 mil, where 1 mil $=$ 0.001 in. (Fig. 19–7). The area of a circle in cmil is equal to the square of its diameter in mils:

$$A_{\text{cmil}} = (d_{\text{mil}})^2 \qquad\qquad\qquad \textit{Circular mil}$$

Substance	$\rho(\Omega \cdot m)$	$\rho(\Omega \cdot cmil/ft)$	$\alpha(/^\circ C)$
Conductors			
Aluminium	2.6×10^{-8}	16	0.0039
Constantan (60% Cu, 40% Ni)	49×10^{-8}	295	0.000002
Copper	1.7×10^{-8}	10.4	0.0039
Iron	12×10^{-8}	72	0.0050
Lead	21×10^{-8}	126	0.0043
Manganin (84% Cu, 12% Mn, 4% Ni)	44×10^{-8}	265	0.000000
Mercury	98×10^{-8}	590	0.00088
Nichrome	112×10^{-8}	676	0.0002
Platinum	11×10^{-8}	66	0.0036
Silver	1.6×10^{-8}	9.6	0.0038
Semiconductors			
Carbon	3.5×10^{-5}	2.1×10^4	−0.0005
Germanium	0.5	3×10^8	
Copper oxide (CuO)	1×10^3	6×10^{11}	
Insulators			
Glass	$10^{10} - 10^{14}$	$10^{19} - 10^{23}$	
Quartz	7.5×10^{17}	4.5×10^{26}	
Sulfur	10^{15}	10^{24}	

TABLE 19–1
Approximate resistivities (at 20°C) and their temperature coefficients

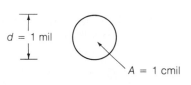
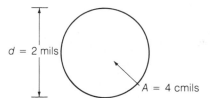

FIG. 19–7 The circular mil (cmil) is a unit of area equal to the area of a circle whose diameter is 1 mil, where 1 mil = 0.001 in. The area of a circle whose diameter is d in mils is given by d^2 in cmils.

Since nearly all wires have circular cross sections, using the cmil as the unit of area simplifies calculations by eliminating the factor $\pi/4$. (We recall that the area in ordinary units of a circle of diameter s is $\pi d^2/4$.) When the length of a wire is in ft and its area in cmils, the unit of resistivity is the $\Omega \cdot$ cmil/ft, as in Table 19–1.

Example Find the resistance of 20 ft of constantan wire 0.005 in. in diameter.

Solution Since 1 mil = 0.001 in., the wire is 5 mils in diameter and has a cross-sectional area of

$$A_{cmil} = (d_{mil})^2 = (5)^2 \text{ cmils} = 25 \text{ cmils}$$

Hence the resistance of the wire is

$$R = \rho \frac{L}{A} = \frac{(295 \, \Omega \cdot cmil/ft)(20\,ft)}{25\,cmils} = 236 \, \Omega \quad \blacksquare$$

Example   The maximum resistance of a copper wire 800 ft long is to be 2.0 Ω. What should the minimum diameter of the wire be?

Solution   The minimum area of the wire in cmils is

$$A = \frac{\rho L}{R} = \frac{(10.4\,\Omega \cdot \text{cmil/ft})(800\,\text{ft})}{2.0} = 4160\,\text{cmils}$$

Since $A_{\text{cmil}} = (d_{\text{mil}})^2$,

$$d_{\text{mil}} = \sqrt{A_{\text{cmil}}} = \sqrt{4160}\,\text{mils} = 64\,\text{mils}$$

This corresponds to No. 14 wire in the American Wire Gage system.   ■

Temperature variation of resistivity

The resistivities of nearly all substances vary with temperature. In general, metals increase in resistivity with an increase in temperature while nonmetals decrease in resistivity. If R is the resistance of a conductor at a particular temperature, then the change ΔR in its resistance when the temperature changes by ΔT is approximately proportional to both R and ΔT, and therefore

$$\Delta R = \alpha R\,\Delta T \qquad\qquad \textit{Temperature and resistance} \quad (19\text{--}5)$$

The quantity α is the temperature coefficient of resistivity of the material. In Table 19–1 the temperature coefficient of carbon is labeled negative because its resistivity decreases with increasing temperature.

Example   A *resistance thermometer* makes use of the temperature variation of resistivity. When a coil of platinum wire whose resistance at 20°C is 11 Ω is placed in a furnace, its resistance doubles to 22 Ω. What is the temperature of the furnace, assuming that α remains constant?

Solution   Since $R = 11\,\Omega$ and $\Delta R = 22\,\Omega - 11\,\Omega = 11\,\Omega$,

$$\Delta T = \frac{\Delta R}{\alpha R} = \frac{11\,\Omega}{(0.0036/°\text{C})(11\,\Omega)} = 278°\text{C}$$

The temperature of the furnace is $T + \Delta T = 20°\text{C} + 278°\text{C} = 298°\text{C}$.   ■

19–5   DETERMINING WIRE SIZE

The choice of the wire size for a particular application is determined by either or both of two factors:

1. The maximum current the wire may have to carry.
2. The maximum permissible voltage drop in the wire.

The procedure is to find the smallest wire that each factor permits, and then to use the larger of the two.

   The maximum current a wire may safely carry depends both upon how its temperature varies with current and upon the nature of its insulation: A wire that gets too hot may be a fire hazard, and its insulation may melt to produce an electrical hazard

| | Ampacity, A | | | **TABLE 19-2** |
AWG No.	Rubber-Covered, Type RH	Varnished Cambric, Type V	Asbestos, Type A	
14	15	25	30	
12	20	30	40	
10	30	40	55	
8	45	50	70	
6	65	70	95	
4	85	90	120	
3	100	105	145	
2	115	120	165	
1	130	140	190	
0	150	155	225	
00	175	185	250	
000	200	210	285	
0000	230	235	340	

TABLE 19-2
Allowable current-carrying capacities (ampacities) of copper wires whose sizes are given in the American Wire Gage (AWG) system

as well. The larger the diameter of a wire, the greater its current-carrying capacity, partly because its resistance is lower and partly because it has more surface area to dissipate heat. Table 19–2 shows the allowable currents (called *ampacities*) for copper wires of standard sizes. For example, asbestos-covered No. 14 wire (diameter 64.08 mils, which is 1.63 mm) can safely carry 30 A, whereas rubber-covered No. 14 wire is limited to 15 A.

Current-carrying capacity is sometimes called ampacity

For the second factor, it is necessary to begin by calculating the highest resistance the wire can have in order that the voltage drop not exceed the specified limit. Given the resistance and the length of the wire, the wire size can be found in the usual way.

Example A water heater draws 30 A from a 120-V power source 10 m away. What is the minimum cross-section of the wire in mm^2 that can be used if the voltage is not to be lower than 115 V at the heater?

Solution The permissible voltage drop is 5 V, and the resistance that corresponds to this drop when the current is 30 A is

$$R = \frac{V}{I} = \frac{5\,V}{30\,A} = 0.167\,\Omega$$

The total length of wire involved is twice the distance between the source and heater, so $L = 2 \times 10$ m $= 20$ m. From Eq. (19–4) we have

$$A = \frac{\rho L}{R} = \frac{(1.7 \times 10^{-8}\,\Omega \cdot m)(20\,m)}{0.167\,\Omega} = 2.04 \times 10^{-6}\,m^2 = 2.04\,mm^2 \qquad \blacksquare$$

In the United States, wire is manufactured in the standard sizes specified by the American Wire Gage (AWG) system (Table 19–3). The largest wire in this system is AWG No. 0000, which is nearly $\frac{1}{2}$ in. in diameter, and the smallest is the hair-thin No. 52, which is 0.00079 in. in diameter. The sequence of diameters is such that every third

American Wire Gage system

TABLE 19–3
American Wire Gage
table. The resistances
given are for copper wire
at 20°C

AWG No.	Diameter, mils	Resistance, $\Omega/1000$ ft
0000	460.0	0.0490
000	409.6	0.0618
00	364.8	0.0779
0	324.9	0.0983
1	289.3	0.1239
2	257.6	0.1563
3	229.4	0.1970
4	204.3	0.2485
5	181.9	0.3133
6	162.0	0.3951
7	144.3	0.4982
8	128.5	0.6282
9	114.4	0.7921
10	101.9	0.9989
11	90.74	1.260
12	80.81	1.588
13	71.96	2.003
14	64.08	2.525
15	57.07	3.184
16	50.82	4.016
17	45.26	5.064
18	40.30	6.385
19	35.89	8.051
20	31.96	10.15
21	28.46	12.80
22	25.35	16.14
23	22.57	20.36
24	20.10	25.67
25	17.90	32.37
26	15.94	40.81
27	14.20	51.47
28	12.64	64.90
29	11.26	81.83
30	10.03	103.2
31	8.928	130.1
32	7.950	164.1
33	7.080	206.9
34	6.305	260.9
35	5.615	329.0
36	5.000	414.8
37	4.453	523.1
38	3.965	659.6
39	3.531	831.8
40	3.145	1049

gage number means a cross-sectional area half as great and so a resistance twice as great for the same wire length. Thus No. 17 copper wire has a resistance of 5.06 Ω/ 1000 ft, double the 2.53-Ω/1000 ft resistance of No. 14 wire. Even sizes are normally employed for wiring purposes, with the odd sizes finding use in coils of various kinds, such as the windings of motors and transformers. No. 14 wire is the smallest permitted by the National Electrical Code for residential, farm, and industrial wiring.

19–6 ELECTRIC POWER

Electric energy in the form of electric current is converted into heat in an electric stove, into radiant energy in a light bulb, into chemical energy when a storage battery is charged, and into mechanical energy in an electric motor. The widespread use of electric energy is due as much to the ease with which it can be transformed into other kinds of energy as to the ease with which it can be carried through wires.

Electric energy is readily transformed into other forms of energy

The work that must be done to take a charge Q through the potential difference V is, by definition,

Work done by a current

$$W = QV$$

Since a current I carries the amount of charge $Q = It$ in the time t, the work done is

$$W = IVt \qquad\qquad \textit{Work done by a current} \quad (19\text{–}6)$$

The energy input to a device of any kind through which the current I flows when the potential difference V is placed across it is equal to the product of the current, the potential difference, and the time span.

We recall from Chapter 5 that *power* is the term given to the rate at which work is being done, so that

$$P = \frac{W}{t}$$

$$\text{Power} = \frac{\text{work done}}{\text{time interval}}$$

When the work is done by an electric current, $W = IVt$, and so

$$P = IV \qquad\qquad \textit{Electric power} \quad (19\text{–}7)$$

Electric power = (current)(potential difference)

The unit of power is the watt, and when I and V are in amperes and volts, respectively, P will be in watts.

Watts = amperes × volts

Example A solar cell 10 cm in diameter produces a current of 2.15 A at 0.45 V in bright sunlight whose intensity is 0.1 W/cm^2. Find the efficiency of the cell.

Solution The area of the cell is $A = \pi d^2/4 = 78.5$ cm^2, so it receives solar energy at the rate of

$$P_{\text{input}} = (0.1 \text{ W/cm}^2)(78.5 \text{ cm}^2) = 7.85 \text{ W}$$

The power output of the cell is

$$P_{\text{output}} = IV = (2.15\ \text{A})(0.45\ \text{V}) = 0.97\ \text{W}$$

The efficiency of the cell is therefore

$$\text{Eff} = \frac{P_{\text{output}}}{P_{\text{input}}} = \frac{0.97\ \text{W}}{7.85\ \text{W}} = 0.12 = 12\%$$ ■

Example   How much current is drawn by a $\frac{1}{2}$-hp electric motor operated from a 120-V source of electricity? Assume that 80% of the electric energy absorbed by the motor is turned into mechanical work.

Solution   The power input to the motor is $P_{\text{input}} = IV$ and its power output is $P_{\text{output}} = (0.5\ \text{hp})(746\ \text{W/hp}) = 373\ \text{W}$. Hence

$$P_{\text{input}} = IV = \frac{P_{\text{output}}}{\text{Eff}}$$

$$I = \frac{P_{\text{output}}}{(\text{Eff})(V)} = \frac{373\ \text{W}}{(0.8)(120\ \text{V})} = 3.9\ \text{A}$$ ■

Formulas for electric power   The power consumed by a resistance that obeys Ohm's law ($I = V/R$) through which current passes may be expressed in the alternative forms

$$P = IV \qquad\qquad (19-7)$$

$$P = I^2R \qquad\qquad (19-8)$$

$$P = \frac{V^2}{R} \qquad\qquad (19-9)$$

Equation (19–7) holds regardless of the nature of the current-carrying device. Depending upon which quantities are known in a specific case, any of the above expressions for P may be used.

Example   Find the power consumed by a 240-Ω light bulb when the current through it is 0.5 A.

Solution   The formula $P = I^2R$ is easiest to use here. We have

$$P = I^2R = (0.5\ \text{A})^2(240\ \Omega) = 60\ \text{W}$$ ■

Fuses and circuit breakers protect circuits from overloads   Owing to the resistance that all conductors offer to the flow of charge through them, electric power is dissipated whenever a current exists regardless of whether the current also supplies energy that is converted to some other form. Electrical resistance is much like friction: The power consumed in causing a current to flow is dissipated as heat. If too much current flows in a particular wire, it becomes so hot that it may start a fire or even melt. To prevent this from happening, as mentioned earlier, nearly all electric circuits are protected by fuses or circuit breakers, which interrupt the current when I exceeds a safe value. For example, a 15-A fuse in a 120-V power line means that the maximum power that can be carried is

$$P = IV = (15\ \text{A})(120\ \text{V}) = 1800\ \text{W}$$

TABLE 19-4

Unknown Quantity	Known Quantities					
	V and I	I and R	V and R	P and I	P and V	P and R
$V =$		IR		$\dfrac{P}{I}$		$\sqrt{PR}$
$I =$			$\dfrac{V}{R}$		$\dfrac{P}{V}$	$\sqrt{\dfrac{P}{R}}$
$R =$	$\dfrac{V}{I}$			$\dfrac{P}{I^2}$	$\dfrac{V^2}{P}$	
$P =$	IV	I^2R	$\dfrac{V^2}{R}$			

Table 19–4 summarizes the various formulas for potential difference V, current I, resistance R, and power P that follow from Ohm's law, $I = V/R$, and from the power formula, $P = IV$.

19–7 IONS IN SOLUTION

Many liquids contain positive and negative ions at all times and hence are able to conduct electricity. Let us look into how the ions in a liquid come into being and how they are able to resist the recombination that occurs so readily in a gas.

When atoms join together to form a molecule, their electrons are shifted in such a manner that electric forces hold the atoms together. We shall consider the details of the binding process in Chapter 29, but for the moment it is sufficient for us to note that certain molecules have asymmetrical (nonsymmetrical) distributions of charge and behave as though negatively charged at one end and positively charged at the other. A molecule of this kind is called a *polar molecule;* the water molecule is an example (Fig. 19–8). A *nonpolar molecule,* on the other hand, has a uniform distribution of charge. All molecules are normally electrically neutral, and the distinction between the polar and nonpolar varieties lies in the way their electrons are arranged.

Polar and nonpolar molecules

The fact that polar molecules exist helps to explain a number of familiar phenomena. The behavior of compounds in solution is a good example. Water readily dissolves such compounds as salt and sugar, but cannot dissolve fats or oils. Gasoline readily dissolves fats and oils, but cannot dissolve salt or sugar. The key to these differences lies in the strongly polar nature of water molecules and the nonpolar nature of gasoline

Water is a polar liquid; gasoline is a nonpolar liquid

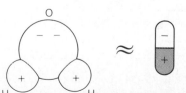

FIG. 19–8 The end of a water molecule where the hydrogen atoms are attached behaves as if positively charged, and the opposite end behaves as if negatively charged. The water molecule is therefore polar.

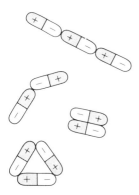

FIG. 19–9 Because water molecules are polar, they tend to clump together under the influence of electric forces.

molecules. Water molecules tend to form clumps under the influence of the electric forces between the ends of adjacent molecules, as shown in Fig. 19–9.

Polar molecules of other substances, such as sugar, can join in the clumps of water molecules, and are therefore easily dissolved by water (Fig. 19–10). The nonpolar molecules of fats and oils, however, do not interact with water molecules. If samples of oil and water are mixed together, the attraction of water molecules for one another acts to squeeze out the oil molecules, and the mixture soon separates into layers of each substance. Fat and oil molecules dissolve only in liquids whose molecules are similar to theirs, which is why gasoline is a solvent for these compounds (Fig. 19–11). In general, then, "like dissolves like."

Soaps and detergents are effective cleansing agents because their molecules are polar at one end but nonpolar at the other. With their help dirt particles of all kinds can be loosened and then washed away from a surface on which they are present.

Many solid compounds have structures that consist of ions rather than of neutral atoms. Thus the sodium chloride (NaCl) of ordinary salt consists of Na^+ and Cl^- ions in the regular geometrical array shown in Fig. 19–12. (The symbol Na^+ refers to a sodium atom that has lost an electron to leave it with a net charge of $+e$, and the symbol Cl^- refers to a chlorine atom that has gained an electron to give it a net charge of $-e$.)

When a crystal of an ionic compound such as NaCl is placed in water, the water molecules cluster around the crystal's ions with their positive ends toward negative ions

FIG. 19–10 Polar compounds such as sugar dissolve in water because their molecules can link up with water molecules.

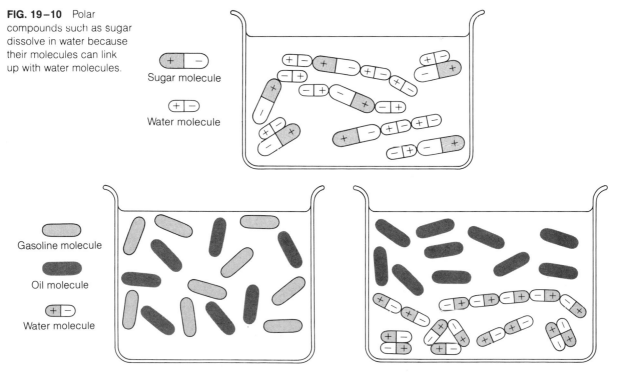

FIG. 19–11 Nonpolar compounds dissolve only in nonpolar liquids. Thus oil dissolves in gasoline but not in water.

FIG. 9–12 A sodium chloride crystal consists of Na$^+$ and Cl$^-$ ions in a regular geometrical arrangement.

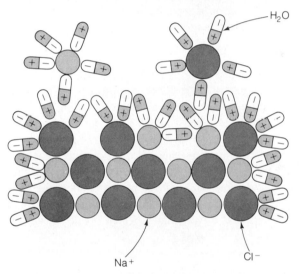

FIG. 19–13 The solution of solid NaCl.

and their negative ends toward positive ions. The attraction of several water molecules is usually sufficient to pull an ion from the rest of the crystal, and it moves away surrounded by water molecules (Fig. 19–13). The resulting solution contains ions rather than molecules of the dissolved compound.

Substances that separate into free ions when dissolved in water are called *electrolytes* since they are able to conduct electric current by the migration of positive and negative ions. All ionic compounds soluble in water and certain other soluble compounds, such as HCl, are electrolytes. Still other compounds, such as sugar, are nonelectrolytes even though they are soluble in water.

Since the outer electron structure of an ion may be very different from that of the corresponding neutral atom, it is not surprising that the ions of an element may behave very differently from its atoms or molecules. Thus gaseous chlorine is greenish in color, has a strong, irritating taste, and is very active chemically, whereas a solution of chlorine ions is colorless, has a mild, pleasant taste, and is only feebly active.

Electrolytes separate into free ions when dissolved in water

Ions in solution have their own characteristic properties

19-8   ELECTROLYSIS

In electrolysis, an electric current in a liquid liberates free elements

Let us look into what happens when an electric current is passed through a liquid containing ions. Because water itself may participate in the events that occur in a solution, for simplicity we shall consider molten NaCl rather than a NaCl solution. When a current flows through a bath of molten NaCl, as in Fig. 19–14, metallic sodium is observed to deposit out at the cathode (negative electrode) and gaseous chlorine to bubble up from the anode (positive electrode). (At the temperature of molten NaCl, metallic sodium is also in the liquid state.)

These results are not hard to understand in view of the presence of free Na^+ and Cl^- ions in the bath. The negative electrode attracts Na^+ ions and, when they arrive, neutralizes them by transferring an electron to each one:

$$Na^+ + e^- \rightarrow Na$$

The resulting sodium atoms, unlike sodium ions, are not soluble and appear at this electrode as ordinary metallic sodium. The positive electrode at the same time attracts Cl^- ions and neutralizes them by absorbing an electron from each one:

$$Cl^- \rightarrow Cl + e^-$$

The insoluble chlorine atoms are evolved as chlorine gas. The entire phenomenon is an example of *electrolysis,* the process by which free elements are liberated from a liquid by the passage of an electric current.

Electroplating is an example of electrolysis

A widely used application of electrolysis is the depositing, or *plating,* of a thin layer of one metal on an object made of another metal. Sometimes this is done because the plating metal is expensive, for instance, gold or silver; in other cases the reason is to protect the base metal from corrosion, as in the chromium plating of steel. Nonmetallic items can be plated by first coating them with a conducting substance such as graphite. Figure 19–15 shows an arrangement for silver plating a spoon. The bath is a solution of silver nitrate, which breaks down into Ag^+ and NO_3^- (nitrate) ions.

FIG. 19-14 The electrolysis of molten sodium chloride.

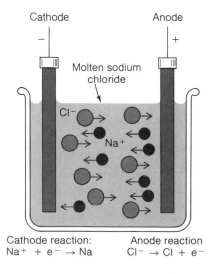

Cathode   Anode

Molten sodium chloride

Cathode reaction: Anode reaction
$Na^+ + e^- \rightarrow Na$ $Cl^- \rightarrow Cl + e^-$

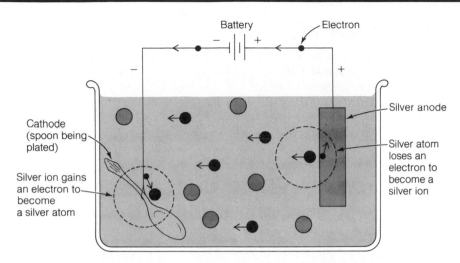

FIG. 19–15 Silver plating.

Silver atoms enter the solution as Ag^+ ions at the anode, and these ions are attracted to the spoon where they gain electrons to become silver atoms once more. Because an Ag atom loses an electron to become Ag^+ more readily than an NO_3^- ion loses its extra electron, the NO_3^- ions stay in solution and do not participate in the plating process.

19–9   THE BATTERY

A chemical reaction that involves the transfer of electrons from one substance to another can be used to produce an electric current if matters are so arranged that the electrons do not move directly between the reacting substances but instead pass through an external circuit. The dry cell of a flashlight, the storage battery of a car, and the fuel cell of a spacecraft are all based on reactions of this kind and are collectively called *electrochemical cells*.

In a battery, chemical energy becomes electric energy

A storage battery is "charged" by the electrolytic reactions that occur when a current is passed through it from an outside source; when it acts as a battery, the same reactions occur backwards to produce a current in the reverse direction as the ingredients of the cell restore themselves to their initial state. In principle, all batteries can be reversed to charge them, but in practice the construction of certain types of cells prevents this, and such cells must be discarded when they are exhausted.

Each cell of the lead-acid storage battery of a car in its charged state has positive plates of lead dioxide, PbO_2, interleaved with negative plates of pure lead. The electrolyte is sulfuric acid, H_2SO_4, which in solution dissociates into H^+ and SO_4^{2-} (sulfate) ions. When the cell provides current, electrons leave the negative plates and travel through the external circuit to the positive plates. The reactions that occur at each set of plates are shown in Fig. 19–16. The effect of drawing current from the cell is evidently to deposit lead sulfate, which is nearly insoluble, on both sets of plates. When no more Pb and PbO_2 are accessible to the electrolyte, the cell is "dead" and can produce no further current. To recharge the cell, a current must be passed through

Lead-acid storage battery

FIG. 19–16 (a) The lead-acid storage battery. The reactions that occur at each electrode when the battery provides current are shown. To charge the battery, a current is passed through it in the opposite direction, which reverses these reactions. (b) Each cell of an actual storage battery contains alternate plates of lead and of lead dioxide. The plates of each kind are connected together and to an outside terminal.

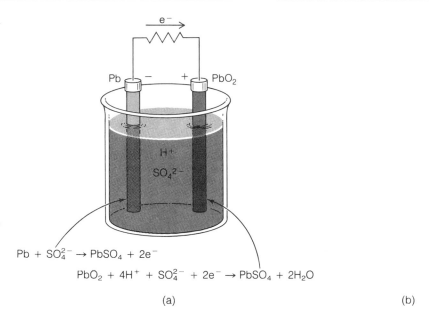

$$Pb + SO_4^{2-} \rightarrow PbSO_4 + 2e^-$$

$$PbO_2 + 4H^+ + SO_4^{2-} + 2e^- \rightarrow PbSO_4 + 2H_2O$$

(a)　　　　　　　　　　　　　　　　　　　(b)

it in the opposite direction, which reverses the original reactions and restores the plates to their original compositions.

Specific gravity of a battery's electrolyte

Since the hydrogen and sulfate ions of the electrolyte become part of the lead sulfate deposits on plates as the cell is discharged, the sulfuric acid content of the electrolyte falls and its density decreases. We recall that the specific gravity of a substance is its density relative to that of water. A fully charged cell has a specific gravity of about 1.26, which falls to about 1.19 when the cell is half discharged and to 1.11 when it is completely discharged.

Nickel-cadmium storage battery

Other types of storage cells have been developed for special purposes (Table 19–5). An example is the nickel-cadmium cell, which consists of nickel and cadmium elec-

TABLE 19–5
Battery cells.

Type	Emf	Properties
Lead-acid	2.1 V	Rechargeable; can provide the high currents needed to start gasoline and diesel engines
Edison	1.4 V	Nickel and iron electrodes in hydroxide electrolyte; rechargeable; lighter and more durable than lead-acid cell; widely used in industry
Nickel-cadmium	1.25 V	Rechargeable; rugged; used for portable power tools and electrical appliances
Carbon-zinc	1.5 V	Cheap but low capacity and short shelf life; used for flashlights, portable radios, etc.
Mangancse-alkaline	1.5 V	Higher capacity and longer shelf life than carbon-zinc cell
Mercury	1.35 V	Long shelf life; constant emf; rugged; used in hearing aids, cameras, watches

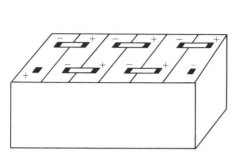

FIG. 19–17 A 12-V storage battery consists of six 2-V cells connected together.

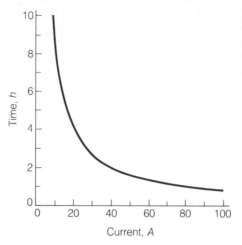

Fig. 19–18 Combinations of current and discharge time available from a battery rated at 80 A·h.

trodes in an electrolyte of potassium hydroxide. Such cells have good low-temperature performance and keep their charge for long periods, but are expensive. A nickel-cadmium battery may retain 50% of its initial charge after a year of inactivity, whereas a lead-acid battery may lose as much as 1% of its charge per day. Several new types of storage cells with high energy densities have been proposed for powering electric cars and as adjuncts of electric power stations for peak loads, but they are still in the experimental stage.

The potential difference that exists across a battery when it is not connected to any external circuit is called its *electromotive force* (emf), which is the electrical equivalent of pressure in a liquid. The main factor that determines the emf of a cell is the chemical nature of the reaction that takes place in it. For instance, a fully charged nickel-cadmium cell has an emf of 1.25 V, whereas it is 2.1 V for a lead-acid cell; the "12-volt" batteries used in cars consist of six lead-acid cells connected together (Fig. 19–17). If we continue to apply a charging current to a fully charged cell, its emf will not continue to increase. What will happen in the case of a lead-acid cell is that water in the electrolyte will undergo electrolysis and decompose into hydrogen and oxygen gases. The result is that the electrolyte level will go down and the acid will become more concentrated, which can damage the plates. In addition, if the battery is not in a well-ventilated location, the mixture of hydrogen and oxygen that is given off can be ignited by a spark and explode; this is a matter of importance in boats.

The emf of a particular cell is affected by its state of charge. The 2.1-V emf of a fully charged lead-acid cell falls to about 1.8 V when it is completely discharged. Of course, a cell that is supplying current has a lower potential difference across its terminals than its emf because of the *IR* potential drop inside the cell itself. This subject is further considered in Chapter 20.

A battery is rated according to the amount of charge it can deliver, expressed in ampere-hours (A · h). Thus a battery whose capacity is 80 A · h can supply a current of 1 A for 80 h, a current of 2 A for 40 h, a current of 80 A for 1 h, and so forth. The lesser the current, the longer the time (Fig. 19–18).

Electromotive force

Ampere-hour rating

Example   (a) How many coulombs of charge can a 12-V, 80-A · h battery transfer? (b) How much energy is stored in the battery?

Solution   (a) From Eq. (19–1), $I = Q/t$, so $Q = It$. The charge transferred by a current of 80 A that flows for 1 h = 3600 s is

$$Q = It = (80 \text{ A})(3600 \text{ s}) = 2.88 \times 10^5 \text{ C}$$

(b) From Eq. (19–6), the energy stored in the battery is

$$W = IVt = (80 \text{ A})(12 \text{ V})(3600 \text{ s}) = 3.46 \times 10^6 \text{ J} = 3.46 \text{ MJ} \qquad \blacksquare$$

19–10   THE FUEL CELL

FIG. 19–19 A hydrogen-oxygen fuel cell. The reactions in the electrolyte are promoted by catalysts. The reaction at the negative electrode must occur twice for each time the reaction at the positive electrode occurs. The net result is the formation of water and the liberation of energy. Photo shows a 4.8-MW fuel cell stack (courtesy Department of Energy)

A *fuel cell* is a type of battery in which a continuous flow of the initial reactant chemicals is possible, which means that the cell need never be exhausted (like a dry battery) or need recharging (like a storage battery). The efficiency with which chemical energy is converted to electrical energy in a fuel cell is not limited by the second law of thermodynamics, as would be the case if the same fuel were burned to power an internal combustion engine or a steam turbine that drives an electric generator. The absence of moving parts is another advantage. The only mechanism of energy loss in a fuel cell

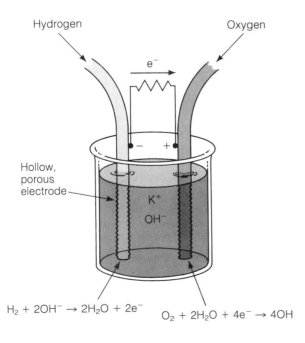

Hydrogen   Oxygen

e^-

Hollow, porous electrode

K^+

OH^-

$H_2 + 2OH^- \rightarrow 2H_2O + 2e^-$ $O_2 + 2H_2O + 4e^- \rightarrow 4OH$

is heat production by electric currents within the cell, which is minor compared with the energy losses in conventional generating plants. Fuel cells are already used to supply electricity in space vehicles, where their high power/weight ratio is important, and are being developed as power sources for electric cars and as self-contained units to furnish electricity to homes.

The hydrogen-oxygen cell (Fig. 19–19) is convenient to illustrate how a fuel cell works. The hollow electrodes are made of inert conducting materials with small pores that permit the gases to come in contact with the electrolyte at a gradual rate. The electrolyte of the cell is a potassium hydroxide (KOH) solution that contains K^+ and OH^- ions. The net effect of the reactions at the electrodes is the combination of hydrogen and oxygen to form water, together with the flow of electrons between the electrodes:

$$2H_2 + O_2 \rightarrow 2H_2O + \text{flow of 4 electrons}$$

If two volumes of hydrogen gas are added to one of oxygen and the mixture ignited, a violent explosion occurs, with water as the product. In the hydrogen-oxygen fuel cell the same chemical process occurs, but the liberated energy appears in the form of electric current.

IMPORTANT TERMS

An **ion** is an atom or group of atoms that carries a net electric charge. An atom or group of atoms becomes a negative ion when it picks up one or more electrons in addition to its normal number, and becomes a positive ion when it loses one or more of its usual number.

A flow of electric charge from one place to another is called an **electric current.** The unit of electric current is the **ampere,** which is equal to a flow of 1 coulomb/second.

Ohm's law states that the current in a metallic conductor is proportional to the potential difference between its ends. The **resistance** of a conductor is the ratio between the potential difference across its ends and the current that flows. The unit of resistance is the **ohm,** which is equal to 1 volt/ampere. The resistance of a conductor is proportional to its length and to the **resistivity** ρ of the material of which it is made, and inversely proportional to its cross-sectional area.

A **polar molecule** is one whose charge distribution is not uniform, so that one end is positive and the other negative even though the molecule as a whole is electrically neutral.

A substance that separates into free ions when dissolved in water is called an **electrolyte** since the resulting solution is able to conduct electric current.

Electrolysis is the process by which free elements are liberated from a liquid by the passage of an electric current.

A **battery** is a device in which chemical reactions produce an electric current. A **fuel cell** is a type of battery in which a continuous supply of the initial reactant chemicals is possible.

IMPORTANT FORMULAS

Electric current: $I = \dfrac{Q}{t}$

Resistance: $R = \dfrac{V}{I}$

Ohm's law: $I = \dfrac{V}{R}$ *(holds for most metals)*

Resistance of ohmic conductor: $R = \rho \dfrac{L}{A}$

Power: $P = IV$

$\qquad = I^2R = \dfrac{V^2}{R}$ *(ohmic conductor)*

MULTIPLE CHOICE

1. The particle easiest to remove from an atom is the
  a. electron.   b. proton.
  c. neutron.   d. nucleus.

2. A solid conductor is one
   a. whose electrons are firmly bound in place.
   b. in which some electrons are free to move about.
   c. which has no electrons.
   d. which has too many electrons.

3. Superconductivity occurs in certain substances
   a. only at very low temperatures.
   b. only at very high temperatures.
   c. at all temperatures.
   d. only when they are in the liquid state.

4. The term ionosphere refers to
   a. a layer of ions in the upper atmosphere.
   b. the polar molecules that surround an ion in solution.
   c. the ions of one sign that surround an ion of the other sign in solution.
   d. a charged metal sphere.

5. The resistance of a conductor does not depend on its
   a. mass.   b. length.
   c. cross-sectional area.   d. resistivity.

6. A certain wire has a resistance R. The resistance of another wire, identical with the first except for having twice its diameter, is
   a. $\frac{1}{4}R$.   b. $\frac{1}{2}R$.
   c. $2R$.   d. $4R$.

7. A certain piece of copper is to be shaped into a conductor of minimum resistance. Its length and cross-sectional area
   a. should be, respectively, L and A.
   b. should be, respectively, $2L$ and $\frac{1}{2}A$.
   c. should be, respectively, $\frac{1}{2}L$ and $2A$.
   d. do not matter, since the volume of copper remains the same.

8. Which of the following materials has the smallest resistivity?
   a. carbon   b. iron
   c. germanium   d. silver

9. The temperature of a copper wire is raised. Its resistance
   a. decreases.
   b. remains the same.
   c. increases.
   d. any of the above, depending upon the temperatures involved.

10. If the diameter d of a circle is expressed in mils, its area in units of circular mils is equal to
    a. d.   b. d^2.
    c. πd.   d. πd^2.

11. Of the following combinations of units, the one that is not equal to the watt is the
    a. J/s.   b. AV.
    c. $A^2 \Omega$.   d. Ω^2/V.

12. The unit of emf is the
    a. ohm.   b. ampere.
    c. volt.   d. watt.

13. All molecules are normally
    a. neutral.   b. charged.
    c. polar.   d. nonpolar.

14. A molecule to which an electron is added becomes
    a. a negative ion.   b. a positive ion.
    c. a polar molecule. d. an electrolyte.

15. A molecule whose charge distribution is not perfectly symmetrical is called
    a. a polar molecule.   c. an electrolyte.
    b. a nonpolar molecule. d. an organic molecule.

16. Water is an excellent solvent because its molecules are
    a. neutral.   b. highly polar.
    c. nonpolar.   d. anodes.

17. Oil does not mix with water because
    a. their respective molecules are different in mass.
    b. their respective molecules are different in size.
    c. water molecules are polar whereas oil molecules are nonpolar.
    d. oil molecules are polar whereas water molecules are nonpolar.

18. Substances that separate into free ions when dissolved in water are called
    a. polar.   b. nonpolar.
    c. anodes.   d. electrolytes.

19. Crystalline solids such as NaCl that consist of ions dissolve only in liquids that are
    a. polar.   b. nonpolar.
    c. ionized.   d. oily.

20. A fuel cell does not require
    a. an electrolyte.   b. an anode.
    c. a cathode.   d. recharging.

21. A flow of 1000 electrons/s constitutes a current of
    a. 1.6×10^{-16} A.   b. 2.7×10^{-18} A.
    c. 2.7×10^{-19} A.   d. 4.5×10^{-20} A.

22. An ordinary size "D" cell can provide a current of 50 mA for about 100 h provided it is used no more than 2 h per day. Under these conditions such a cell has a rating of
    a. 0.05 A · h.   b. 0.5 A · h.
    c. 5 A · h.   d. 50 A · h.

23. The current in a 12-Ω toaster operated at 120 V is
    a. 0.1 A.   b. 10 A.
    c. 12 A.   d. 1440 A.

24. The 8-Ω coil of a loudspeaker carries a current of 0.8 A. The potential difference across its terminals is

a. 0.1 V. b. 5.12 V.
c. 6.4 V. d. 10 V.

25. An electric iron draws a current of 15 A when connected to a 120-V power source. Its resistance is
   a. 0.125 Ω. b. 8 Ω.
   c. 16 Ω. d. 1800 Ω.

26. A wire has a diameter of 0.05 in. Its cross-sectional area in circular mils is
   a. 0.0196 cmil. b. 0.0025 cmil.
   c. 50 cmil. d. 2500 cmil.

27. A 200-m length of copper wire has a resistance of 2 Ω. Its cross-sectional area is
   a. 0.0017 mm^2. b. 1.7 mm^2.
   c. 3.4 mm^2. d. 5.3 mm^2.

28. The power rating of an electric motor that draws a current of 3 A when operated at 120 V is
   a. 40 W. b. 360 W.
   c. 540 W. d. 1080 W.

29. When a 100-W, 240-V light bulb is operated at 200 V, the current that flows in it is
   a. 0.35 A. b. 0.42 A.
   c. 0.50 A. d. 0.58 A.

30. A storage battery is being charged at a rate of 60 W at a potential difference of 14 V. The number of electrons transferred per hour from one set of the battery's plates to the other set is
   a. 1.54×10^4. b. 7.44×10^{29}.
   c. 2.68×10^{31}. d. 9.64×10^{34}.

31. The energy content of a fully charged 12-V storage battery of capacity 50 A · h is
   a. 50 J. b. 600 J.
   c. 36 kJ. d. 2.16 MJ.

EXERCISES

19–1 Electrical Conduction

1. How does electrical conduction in a metal differ from that in an ionized gas?

2. What property of the electrons in a metal enables it to conduct electric current readily? What property of the electrons in an insulator prevents it from conducting electric current readily?

3. List several good conductors of electricity and several good insulators. How well do these substances conduct heat? What general relationship between an ability to conduct electricity and an ability to conduct heat can you infer?

4. What aspect of superconductivity has prevented its large-scale application thus far?

19–2 Electric Current

5. Why are two wires used to carry electric current instead of a single one?

6. Sensitive instruments can detect the passage of as few as 60 electrons/s. To what current does this correspond?

7. Approximately 10^{20} electrons/cm participate in conducting electric current in a certain wire. (That is, 10^{20} electrons in each centimeter of the wire are in motion when a current is being carried by the wire.) What is the average speed of the electrons when there is a current of 1 A in the wire?

19–3 Ohm's Law

8. Which of the *I-V* graphs shown corresponds to a substance that obeys Ohm's law?

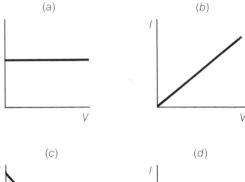

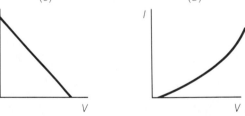

9. In the process of resistance welding, a pair of electrodes presses two sheets of metal together. Heat from the passage of current melts the metal sheets where they are in contact between the electrodes to form a weld. What potential difference is needed to produce a current of 10 kA when the resistance of the metal between the electrodes is 0.0002 Ω?

10. What potential difference must be applied across a 1500-Ω resistor in order that the resulting current be 50 mA?

11. Find the resistance of a 120-V electric toaster that draws a current of 8 A.

12. A 240-V water heater has a resistance of 24 Ω. What must be the minimum rating of the fuse in the circuit to which the heater is connected?

13. A Wheatstone bridge (shown below) provides a convenient means for measuring an unknown resistance R in terms of the known resistances A and B and the calibrated variable resistance C. The resistance of C is varied until no current flows through the meter G, in which case the bridge is said to be *balanced*. Show that $R = AC/B$ when the bridge is balanced.

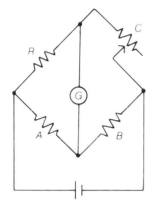

19−4   Resistivity

14. When a metal object is heated, both its dimensions and its resistivity increase. Is the increase in resistivity likely to be a consequence of the increase in length?

15. How does the resistance per meter of 5-mm² copper wire compare with that of 2-mm² copper wire?

16. The resistance of No. 8 copper wire is 0.21 Ω/100 m. What is the resistance per 100 m of No. 8 aluminum wire?

17. AWG 000 wire has a cross-sectional area of 85 mm². Find the resistance of 10 km of copper wire of this size at 20°C.

18. Find the resistance of a carbon cylinder 2.0 mm in diameter and 20 mm long.

19. A silver wire 2 m long is to have a resistance of 0.5 Ω. What should its diameter be?

20. A 40-Ω resistor is to be wound from platinum wire 0.1 mm in diameter. How much wire is needed?

21. A silicon wafer 1 cm square and 2 mm thick has a resistance of 1.2 kΩ. What is the resistivity of silicon?

22. Find the diameter of a wire whose cross-sectional area is 6400 cmil.

23. No. 18 wire has a diameter of 0.0403 in. What is its cross-sectional area in circular mills?

24. The maximum resistance of an aluminum wire 2200 ft long is to be 1.8 Ω. Find the minimum diameter of the wire in mils.

25. How long should a copper wire 20 mils in diameter be for it to have a resistance of 5 Ω?

26. How long should a platinum wire 6 mils in diameter be for it to have a resistance of 10 Ω?

27. The resistance of a copper wire is 100 Ω at 20°C. Find its resistance at 0°C and at 80°C.

28. Motors, generators, transformers, and other electrical devices should not be operated above certain temperatures, whose values depend on their construction. The temperature in the interior of such a device can be found by measuring the resistance of one of its windings before it is run and after it has been run for some time. If the copper field windings of an electric motor have resistances of 100 Ω at 20°C and of 115 Ω when the motor is operating, what is their temperature under the latter conditions?

29. An iron wire has a resistance of 2.00 Ω at 0°C and a resistance of 2.46 Ω at 45°C. Find the temperature coefficient of resistivity of the wire.

30. A copper wire 1 mm in diameter carries a current of 12 A. Find the potential difference between two points in the wire that are 100 m apart.

31. A metal rod 1 m long and 1 cm in diameter is drawn out into a wire 1 mm in diameter. (a) What is the length of the wire? (b) Compare the resistance of the rod with the resistance of the wire.

32. Aluminum wires are sometimes used to transmit electric power instead of copper wires. What is the ratio between the masses of an aluminum and a copper wire of the same length whose cross-sectional areas are such that they have the same resistance? (The density of aluminum is 2.70 × 10³ kg/m³ and that of copper is 8.89 × 10³ kg/m³.)

19−5   Determining Wire Size

33. The maximum allowable voltage drop for a certain 70-m length of cable that carries a current of 20 A to a shed is 3 V. What is the minimum cross-sectional area of the copper wires of the cable in square millimeters?

34. An electric motor draws 10 A from a 240-V power source 120 ft away. What is the minimum resistance per 1000 ft for the wire that can be used if the voltage drop is not to exceed 2%? To what diameter of copper wire does this correspond?

35. Find the maximum length of AWG No. 12 copper wire that can be used to carry a current of 12 A if the voltage drop is not to exceed 3 V.

36. An electric pump draws 32 A from a 120-V power source 240 ft away. (a) Find the minimum AWG size of rubber-insulated wire that can be safely used. (b) What is the voltage drop when this wire is used?

19–6 Electric Power

37. It is sometimes said that an electrical appliance "uses up" electricity. What does such an appliance actually use in its operation?

38. Alice and Fred are discussing whether an electric heater with a large or with a small resistance will yield the greater heat output. Alice favors a small value of R because $P = V^2/R$; Fred favors a large value of R because $P = I^2R$. What is your conclusion?

39. Currents of 3 A flow through two wires, one that has a potential difference of 60 V across its ends and another that has a potential difference of 120 V across its ends. Compare the rates at which charge and energy pass through each wire.

40. Find the maximum current in a 50-Ω, 10-W resistor if its power rating is not to be exceeded.

41. An electric drill rated at 400 W is connected to a 240-V power line. How much current does it draw?

42. How many electrons flow through the filament of a 120-V, 60-W electric light bulb per second?

43. An electric motor whose power output is 0.5 hp draws 4 A at 120 V. What is its efficiency?

44. The alternator on a car's engine is 83% efficient. Find the mechanical power the alternator takes when it is delivering 20 A at 14 V.

45. A 240-V electric motor whose efficiency is 80% draws 10 A when it is used to operate a milling machine. Find the power output of the motor.

46. A 240-V clothes dryer draws a current of 15 A. How much energy, in kilowatt-hours and in joules, does it use in 45 min of operation?

47. A trolley car of mass 10^4 kg takes 10 s to reach a speed of 8 m/s starting from rest. If it operates from a 5-kV power line and is 100% efficient, find the average current it draws during the acceleration.

48. When a certain 1.5-V battery is used to power a 3-W flashlight bulb, it is exhausted after an hour's use. (a) How much charge has passed through the bulb in this period of time? (b) If the battery costs $0.50, find the cost of a kilowatt-hour of electric energy obtained in this way. How does this compare with the cost of the electric energy supplied to your home?

49. A 12-V storage battery with an internal resistance of 0.012 Ω delivers 80 A when used to crank a gasoline engine. If the battery mass is 20 kg and it has an average specific heat of 0.2 kcal/kg · °C, what is its rise in temperature during 1 min of cranking the engine?

50. An electric water heater has a resistance of 12 Ω and is operated from a 120-V power line. If no heat escapes from it, how much time is required for it to raise the temperature of 40 kg of water from 15°C to 80°C?

51. In a Van de Graaff generator an insulating belt is used to carry charges to a large metal sphere. In a typical generator of this kind, the potential difference between the sphere and the source of the charges is 5×10^6 V. (a) If the belt carries charge to the sphere at a rate of 10^{-3} A, how much power is required? (b) How much energy in eV will an electron have if it is accelerated by such a potential difference? (c) Express the answer to (b) in joules.

52. Large electromagnets are sometimes wound with copper tubing through which water is circulated to carry away the heat produced. If such an electromagnet has a coil whose resistance is 0.8 Ω and through which 30 L/min of water flows, find the rise in temperature of the water when a current of 300 A is present in the coil. Assume that all the heat produced is absorbed by the water.

19–7 Ions in Solution

53. When two charges are separated by a medium, the force between them is reduced by a factor K, called the *dielectric constant* of the medium. Hence Coulomb's law in general is $F = kQ_1Q_2/Kr^2$. Why do you think $K = 80$ for water but only 2.3 for benzene?

54. How could you experimentally distinguish between a solution of an electrolyte and one of a nonelectrolyte?

55. Distinguish between a molecular ion and a polar molecule.

56. A calcium atom has 20 protons in its nucleus. (a) How many electrons does this atom contain? (b) How many protons are in the nucleus of a Ca^{++} ion? (c) How many electrons does this ion contain?

57. An oxygen atom has eight protons in its nucleus. (a) How many electrons does this atom contain? (b) How many protons are in the nucleus of an O^{--} ion? (c) How many electrons does this ion contain?

19–9 The Battery
19–10 The Fuel Cell

58. In what fundamental way or ways is a fuel cell different from a battery?

59. A mercury cell with a capacity of 1.5 A·h and an emf of 1.35 V is to be used to power a cardiac pacemaker. (a) If the power required is 0.1 mW, how long will the cell last? (b) What will be the average current?

60. The 12-V battery of a car is required to be able to operate its 1.5-kW starting motor for a total of at least 10 min. (a) What should the minimum capacity of the battery be in A·h? How much energy is stored in such a battery?

ANSWERS TO MULTIPLE CHOICE

1. a	**8.** d	**14.** a	**20.** d	**26.** d
2. b	**9.** c	**15.** a	**21.** a	**27.** b
3. a	**10.** b	**16.** b	**22.** c	**28.** b
4. a	**11.** d	**17.** c	**23.** b	**29.** a
5. a	**12.** c	**18.** d	**24.** c	**30.** d
6. a	**13.** a	**19.** a	**25.** b	**31.** d
7. c				

20

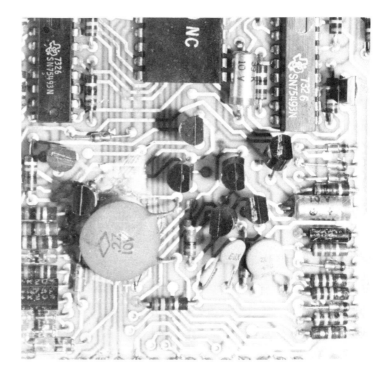

DIRECT CURRENT CIRCUITS

An electric circuit in its simplest form consists of a source of electric energy connected to a *load*, which may be a lamp, a motor, or any other device that absorbs electric energy and converts it into some other form of energy or into work. Many circuits are more complicated than this, with several loads that may be connected in a variety of ways and perhaps with more than one source of electric energy. The physical principles needed to analyze a circuit are straightforward, and practice makes their application to a complex circuit a routine matter.

20–1 RESISTORS IN SERIES

Equivalent resistance

In analyzing a direct-current circuit, it is convenient to group its various components into individual *resistors* that we imagine to be joined together by resistanceless wires. The *equivalent resistance* of a set of interconnected resistors is the value of the single resistor that can replace the entire set without affecting the current in the rest of the circuit.

CHAPTER OBJECTIVES

Completing this chapter should enable you to:

1. Calculate the equivalent resistance of two or more resistors connected in series and find the total current and the current in each one when a voltage is applied across the combination.

2. Calculate the equivalent resistance of two or more resistors connected in parallel and find the total current and the current in each one when a voltage is applied across the combination.

3. Analyze series-parallel combinations of resistors.

4. Find the series resistor needed to reduce the current in a given circuit to a certain value.

5. Find the parallel resistor needed to reduce the resistance of a given circuit to a certain value.

6. Distinguish between electromotive force and potential difference.

7. Calculate the voltage needed to charge a storage battery of given internal resistance at a certain rate.

8. Express the condition for maximum power transfer from a source of electrical energy to an external circuit.

9. Use Kirchhoff's rules to analyze a complex circuit.

10. Describe the design and limitations of ammeters, voltmeters, and ohmmeters.

Resistors in series and parallel

The symbol for a resistor, ⏤\/\/\/⏤ , represents any circuit component that has electrical resistance and obeys Ohm's law. An individual resistor used in eletrical and electronic circuits usually consists either of a coil of high-resistivity wire or of a carbon rod. Wire-wound resistors range in value up to 100 kΩ and can be constructed to dissipate hundreds of watts but are fairly expensive. Carbon resistors are cheap and go up to 20 MΩ but can dissipate only a few watts. A variable resistor (called a *rheostat* or a *potentiometer* depending on how it is connected) can be of either kind and has a movable contact whose position on the wire coil or carbon rod governs its resistance.

Many ways in which resistors can be put together are merely combinations of the basic *series* and *parallel* arrangements (Fig. 20–1). Resistors in series are connected consecutively so the same current is present in all of them. On the other hand, resistors in parallel have their ends connected together, so the total current is split among them.

The potential difference V across the ends of a series set of resistors is the sum of the potential differences V_1, V_2, V_3, . . . , across each one. This statement follows from the principle of conservation of energy: If V is the work done per coulomb in "pushing" a charge through the set of resistors, then it must equal the sum $V_1 +$

FIG. 20-1

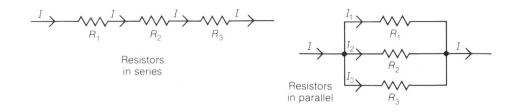

Resistors
in series

Resistors
in parallel

FIG. 20-2 Resistors in series. The same current I passes through each resistor.

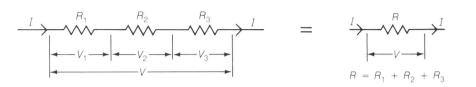

$R = R_1 + R_2 + R_3$

$V_2 + V_3 + \cdots$ of the amounts of work done per coulomb in pushing the charge through each resistor in turn. In the case of the three resistors in series of Fig. 20-2,

$$V = V_1 + V_2 + V_3$$

Resistors in series have the same current

Since the same current I passes through all the resistors, the individual potential drops are

$$V_1 = IR_1 \qquad V_2 = IR_2 \qquad V_3 = IR_3$$

If we denote the equivalent resistance of the set by R, the potential difference across it is

$$V = IR$$

We therefore have

$$V = V_1 + V_2 + V_3$$
$$IR = IR_1 + IR_2 + IR_3$$

Dividing through by the current I gives

$$R = R_1 + R_2 + R_3$$

Equivalent resistance of series resistors

In general, *the equivalent resistance of a set of resistors connected in series is equal to the sum of the individual resistances:*

$$R = R_1 + R_2 + R_3 + \cdots \qquad\qquad \textit{Resistors in series} \quad (20\text{--}1)$$

Example A 5-Ω resistor and a 20-Ω resistor are connected in series and a potential difference of 100 volts is applied across them by means of a generator. Find (a) the equivalent resistance of the circuit, (b) the current that flows in it, (c) the potential difference across each resistor, (d) the power dissipated by each resistor, and (e) the power dissipated by the entire circuit.

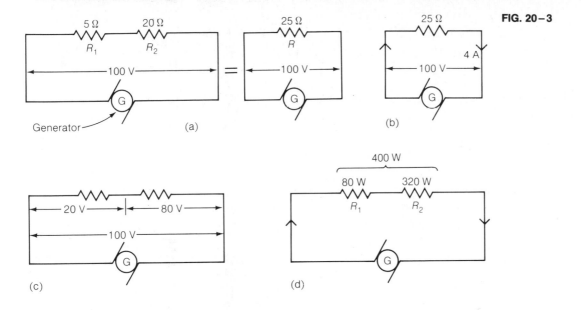

FIG. 20-3

Solution Successive stages in the solution of this problem are shown in Fig. 20–3.

(a) The equivalent resistance of the two resistors is

$$R = R_1 + R_2 = 5\ \Omega + 20\ \Omega = 25\ \Omega$$

This resistance is more than that of either of the resistors.

(b) The current in the circuit is, from Ohm's law,

$$I = \frac{V}{R} = \frac{100\ \text{V}}{25\ \Omega} = 4\ \text{A}$$

The current in both resistors is 4 A.

(c) The potential difference across R_1 is

$$V_1 = IR_1 = (4\ \text{A})(5\ \Omega) = 20\ \text{V}$$

and the potential difference across R_2 is

$$V_2 = IR_2 = (4\text{A})(20\ \Omega) = 80\ \text{V}$$

We note that the sum of V_1 and V_2 is 100 V, the same as the impressed potential difference, as of course it must be.

(d) The power dissipated by R_1 is

$$P_1 = IV_1 = (4\ \text{A})(20\ \text{V}) = 80\ \text{W}$$

We can also find P_1 from the alternative formulas $P = I^2R$ and $P = V^2/R$. In each case we get the same answer:

$$P_1 = I^2R_1 = (4\ \text{A})^2(5\ \Omega) = 80\ \text{W} \qquad P_1 = \frac{V^2}{R} = \frac{(20\ \text{V})^2}{5\ \Omega} = 80\ \text{W}$$

FIG. 20–4

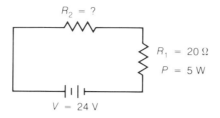

$R_2 = ?$

$R_1 = 20\ \Omega$

$P = 5\ W$

$V = 24\ V$

The power dissipated by R_2 is

$$P_2 = IV_2 = (4\ A)(80\ V) = 320\ W$$

(e) The power dissipated by the entire circuit is

$$P = IV = (4\ A)(100\ V) = 400\ W$$

and is equal to $P_1 + P_2$. The dissipated power appears as heat. ■

Example A 20-Ω load whose maximum power rating is 5 W is to be connected to a 24-V battery. What is the minimum resistance of the series resistor (Fig. 20–4) that is required?

Solution The maximum allowable current in the resistor R_1 may be found from the formula $P = I^2 R_1$ to be

$$I = \sqrt{\frac{P}{R_1}} = \sqrt{\frac{5\ W}{20\ \Omega}} = 0.5\ A$$

For a current of 0.5 A, the equivalent resistance R of the circuit should be

$$R = \frac{V}{I} = \frac{24\ V}{0.5\ A} = 48\ \Omega$$

Since $R = R_1 + R_2$, the minimum resistance R_2 of the required series resistor is

$$R_2 = R - R_1 = 48\ \Omega - 20\ \Omega = 28\ \Omega$$ ■

20–2 RESISTORS IN PARALLEL

Resistors in parallel have the same voltage across them

When two or more resistors are connected in parallel, the same potential difference is applied across all of them. As we shall find, the current that passes through each resistor is inversely proportional to its resistance: the less the resistance, the more the current.

Let us consider three resistors, R_1, R_2, and R_3, that are connected in parallel, as in Fig. 20–5. The total current I through the set is equal to the sum of the currents through the separate resistors, so that

$$I - I_1 + I_2 + I_3$$

The potential difference V is the same across all the resistors, and by applying Ohm's law to each of them in turn we find that

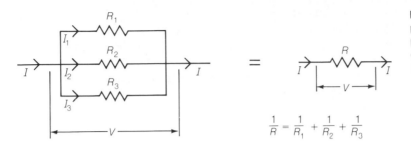

$$I_1 = \frac{V}{R_1} \qquad I_2 = \frac{V}{R_2} \qquad I_3 = \frac{V}{R_3}$$

The smaller the resistance, the greater the proportion of the total current that flows through it.

The total current flowing through the set of three resistors is given in terms of their equivalent resistance R by

$$I = \frac{V}{R}$$

Hence we have

$$I = I_1 + I_2 + I_3$$

$$\frac{V}{R} = \frac{V}{R_1} + \frac{V}{R_2} + \frac{V}{R_3}$$

The final step is to divide through by V, which gives

$$\frac{1}{R} = \frac{1}{R_1} + \frac{1}{R_2} + \frac{1}{R_3}$$

In general, *the reciprocal of the equivalent resistance of a set of resistors connected in parallel is equal to the sum of the reciprocals of the individual resistances:*

Equivalent resistance of parallel resistors

$$\frac{1}{R} = \frac{1}{R_1} + \frac{1}{R_2} + \frac{1}{R_3} + \cdots \qquad \textit{Resistors in parallel} \quad (20\text{–}2)$$

The formulas for the equivalent resistances of series and parallel arrangements of resistors are in accord with the basic formula

$$R = \rho \frac{L}{A}$$

for the resistance of a conductor. When several resistors are connected in series, the effect is the same as increasing the length L. When they are connected in parallel, the effect is the same as increasing the cross-sectional area A. Thus a series set of resistors lets through *less* current than any of the individual resistors, and a parallel set of resistors lets through *more* current than any of the individual resistors, in each case assuming the same potential difference.

As a check after making a calculation, it is worth recalling that in a series circuit, the equivalent resistance is always *greater* than any of the individual resistances. In a parallel circuit, on the other hand, the equivalent resistance is always *smaller* than any of the individual resistances.

Two resistors in parallel   In the case of two resistors in parallel, the equivalent resistance is

$$\frac{1}{R} = \frac{1}{R_1} + \frac{1}{R_2}$$

The lowest common denominator of the right-hand side of this formula is $R_1 R_2$, which permits us to write

$$\frac{1}{R} = \frac{R_1 + R_2}{R_1 R_2}$$

Taking the reciprocal of both sides of this equation yields the convenient result

$$R = \frac{R_1 R_2}{R_1 + R_2} \qquad \qquad \textit{Two resistors in parallel} \quad (20-3)$$

When using a calculator that has a reciprocal [1/X] key, it is easier to use Eq. (20-2) directly, as shown in part (a) of the example below.

Example   A 5-Ω resistor and a 20-Ω resistor are connected in parallel and a potential difference of 100 V is applied across them by means of a generator. Find (a) the equivalent resistance of the circuit, (b) the current that flows in each resistor and in the circuit as a whole, (c) the power dissipated by each resistor, and (d) the power dissipated by the entire circuit.

Solution   Successive stages in the solution of this problem are shown in Fig. 20-6.
(a) The equivalent resistance of the resistors is

$$R = \frac{R_1 R_2}{R_1 + R_2} = \frac{(5\,\Omega)(20\,\Omega)}{5\,\Omega + 20\,\Omega} = 4\,\Omega$$

Alternatively, we can find R from

$$\frac{1}{R} = \frac{1}{5\,\Omega} + \frac{1}{20\,\Omega}$$

How to use a calculator to find R for parallel resistors

with the following key sequence on a calculator: [5] [1/X] [+] [20] [1/X] [=] [1/X]. The generalization to three or more resistors is obvious. For three resistors, for instance, the sequence would be: [R_1] [1/X] [+] [R_2] [1/X] [+] [R_3] [1/X] [=] [1/X].
(b) The current that flows in R_1 is

$$I_1 = \frac{V}{R_1} = \frac{100\,\text{V}}{5\,\Omega} = 20\,\text{A}$$

and the current that flows in R_2 is

$$I_2 = \frac{V}{R_2} = \frac{100\,\text{V}}{20\,\Omega} = 5\,\text{A}$$

FIG. 20–6

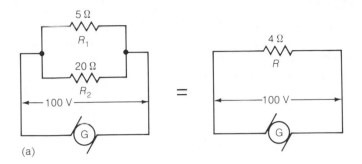

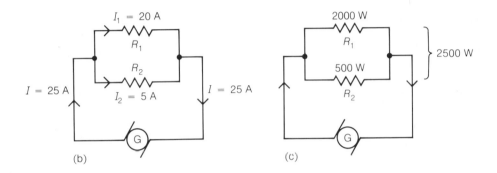

(a)

(b)   (c)

The total current in the circuit is

$$I = \frac{V}{R} = \frac{100\ \text{V}}{4\ \Omega} = 25\ \text{A}$$

and is equal to $I_1 + I_2$.

(c) The power dissipated by R_1 is

$$P_1 = I_1 V = (20\ \text{A})(100\ \text{V}) = 2000\ \text{W}$$

and that dissipated by R_2 is

$$P_2 = I_2 V = (5\ \text{A})(100\ \text{V}) = 500\ \text{W}$$

(d) The power dissipated by the entire circuit is

$$P = IV = (25\ \text{A})(100\ \text{V}) = 2500\ \text{W}$$

and is equal to $P_1 + P_2$.

We recall from the previous section that, when a 100-V potential difference is applied across the same two resistors connected in series, the power dissipated is only 400 W. The difference is due to the lower equivalent resistance of the parallel combination (4 Ω instead of 25 Ω), which permits more current to flow. Since $P = IV$ and V is the same in both cases, the greater current in the parallel circuit leads to a greater power dissipation. ■

FIG. 20–7

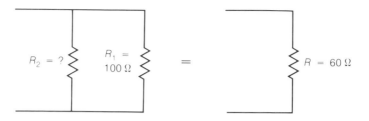

Example   A circuit has a resistance of 100 Ω. How can it be reduced to 60 Ω?

Solution   In order to obtain an equivalent resistance of $R = 60 \ \Omega$, a resistor R_2 must be connected in parallel with the circuit of $R_1 = 100 \ \Omega$ (Fig. 20–7). To find the value of R_2 we proceed as follows:

$$\frac{1}{R} = \frac{1}{R_1} + \frac{1}{R_2}$$

$$\frac{1}{R_2} = \frac{1}{R} - \frac{1}{R_1} = \frac{R_1 - R}{R_1 R}$$

$$R_2 = \frac{R_1 R}{R_1 - R} = \frac{(100 \ \Omega)(60 \ \Omega)}{100 \ \Omega - 60 \ \Omega} = 150 \ \Omega$$

With a calculator, it is easier to work from

$$\frac{1}{R_2} = \frac{1}{60 \ \Omega} - \frac{1}{100 \ \Omega}$$

and use this key sequence:

[60] [1/X] [−] [100] [1/X] [=] [1/X]   ■

20–3   SERIES-PARALLEL COMBINATIONS

Complicated combinations of resistors may often be decomposed into series and parallel groupings that can be analyzed in the usual way. An example will make clear the procedure.

Example   Find the equivalent resistance of the set of resistors shown in Fig. 20–8.

Solution   Figure 20–9 shows how the original circuit is broken up into series and parallel groupings. The first step is to find the equivalent resistance R' of the parallel resistors R_2 and R_3:

$$\frac{1}{R'} = \frac{1}{R_2} + \frac{1}{R_3} = \frac{1}{12 \ \Omega} + \frac{1}{8 \ \Omega}$$

$$R' = 4.8 \ \Omega$$

This pair of resistors is in series with R_1, and the equivalent resistance R'' of the upper branch of the circuit is therefore

FIG. 20-8

$R_1 = 3\,\Omega$
$R_2 = 12\,\Omega$
$R_3 = 8\,\Omega$
$R_4 = 6\,\Omega$

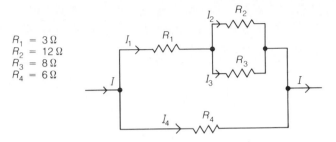

FIG. 20-9 Successive steps in determining the equivalent resistance R of the resistor network shown in Fig. 20-8.

(a) $\quad = \quad$ $R' = \dfrac{R_2 R_3}{R_2 + R_3}$

(b) $\quad = \quad$ $R'' = R_1 + R'$

(c) $\quad = \quad$ $R = \dfrac{R'' R_4}{R'' + R_4}$

$R'' = R' + R_1 = 4.8\,\Omega + 3\,\Omega = 7.8\,\Omega$

The upper and lower branches of the circuit are in parallel, which means that the equivalent resistance R of the entire circuit is

$$\frac{1}{R} = \frac{1}{R''} + \frac{1}{R_4} = \frac{1}{7.8\,\Omega} + \frac{1}{6\,\Omega}$$

$$R = 3.4\,\Omega$$

■

Example   A potential difference of 12 V is applied across the set of resistors shown in Fig. 20-8. Find the current that flows through each resistor.

Solution   (a) The full potential difference is applied across R_4, and so, by Ohm's law, the current I_4 in it is

$$I_4 = \frac{V}{R_4} = \frac{12\,\text{V}}{6\,\Omega} = 2.0\,\text{A}$$

(b) The current I_1 that flows through R_1 also flows through the entire upper branch of the circuit. Since the equivalent resistance of this branch is R'',

$$I_1 = \frac{V}{R''} = \frac{12\ \text{V}}{7.8\ \Omega} = 1.54\ \text{A}$$

(c) The potential difference V' across the parallel resistors R_2 and R_3 is equal to the current I_1 through the equivalent resistance R' multiplied by R':

$$V' = I_1 R' = (1.54\ \text{A})(4.8\ \Omega) = 7.4\ \text{V}$$

Another way to find V' is to subtract the potential difference

$$V_1 = I_1 R_1 = (1.54\ \text{A})(3\ \Omega) = 4.6\ \text{V}$$

from the total of 12 volts to obtain

$$V' = V - V_1 = 12\ \text{V} - 4.6\ \text{V} = 7.4\ \text{V}$$

Hence the currents I_2 and I_3 through resistors R_2 and R_3 are respectively

$$I_2 = \frac{V'}{R_2} = \frac{7.4\ \text{V}}{12\ \Omega} = 0.62\ \text{A} \qquad I_3 = \frac{V'}{R_3} = \frac{7.4\ \text{V}}{8\ \Omega} = 0.92\ \text{A} \qquad \blacksquare$$

20−4   ELECTROMOTIVE FORCE

Emf is no-load voltage of a source

As mentioned in Chapter 19, the electromotive force of a source of electric energy such as a battery or generator is equal to its potential difference when it is not connected to an external circuit. Electromotive force is usually referred to simply as emf, and its symbol is $\mathscr{E}$.

As charges pass through a source of electric energy, work is done on them, and the emf of the source is the work done per coulomb on the charges. The emf of an automobile storage battery is 12 V, which means that 12 J of work are done on each coulomb of charge that passes through the battery. In the case of a battery, chemical energy is converted into electric energy by means of the work done on the charges that move through it; in a generator, mechanical energy is converted into electric energy; in a thermocouple, heat energy is converted into electric energy; and so on. The emf of an electrical source bears a relationship to its power output similar to that of applied force to mechanical power in a machine, which is the reason for its name.

Internal resistance of an emf source

When a source of electric energy is part of a complete circuit, a current I flows, and the potential difference across the terminals of the source is always *less* than its emf owing to its *internal resistance*. Every electrical source has a certain amount of internal resistance r, which means that a potential drop Ir occurs *within* the source. Hence the actual terminal voltage V across a source of emf $\mathscr{E}$ and internal resistance r is

$$V = \mathscr{E} - Ir \qquad\qquad \textit{Terminal voltage} \quad (20-4)$$

Terminal voltage = emf − potential drop within source

If the source is disconnected, no current flows, and $V = \mathscr{E}$; the existence of a current lowers the value of V by an amount proportional to I.

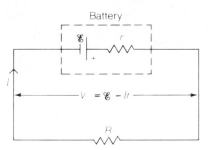

FIG. 20–10 The terminal voltage *V* of a battery in a circuit is always less than its emf $\mathscr{E}$ because of the potential drop *Ir* within the battery itself.

Figure 20–10 shows a battery of emf $\mathscr{E}$ connected to an external circuit whose equivalent resistance is R. The total resistance in the entire circuit is R plus the internal resistance r of the battery, so that the current I that flows is

$$I = \frac{\mathscr{E}}{R + r} \qquad\qquad (20\text{–}5)$$

The actual potential difference across the battery is given by Eq. (20–4).

The internal resistance of a battery governs the maximum current it can supply. One of the major advantages of lead-acid batteries is their low internal resistances, which permit large currents to be drawn for short periods to operate the starting motors of gasoline and diesel engines. The lower the temperature of the electrolyte in a battery, the more slowly its ions move and the higher the internal resistance. As a result the current available for the starting motor in freezing weather may be less than half that available on a warm day.

How cold weather affects a storage battery

Example A "D" cell of emf 1.5 V and internal resistance 0.3 Ω is connected to a flashlight bulb whose resistance is 3.0 Ω. Find the current in the circuit and the terminal voltage of the cell.

Solution The current in the circuit is

$$I = \frac{\mathscr{E}}{R + r} = \frac{1.5\,\text{V}}{3.0\,\Omega + 0.3\,\Omega} = 0.45\,\text{A}$$

The terminal voltage of the cell is

$$V = \mathscr{E} - Ir = 1.5\,\text{V} - (0.45\,\text{A})(0.3\,\Omega) = 1.37\,\text{V} \qquad \blacksquare$$

Series

Example A storage battery whose emf is 12 V and whose internal resistance is 0.2 Ω is to be charged at a rate of 20 A. What applied voltage is required?

Solution The applied voltage V must exceed the battery's emf $\mathscr{E}$ by the amount Ir to provide the charging current I. Hence

$$V = \mathscr{E} + Ir = 12\,\text{V} + (20\,\text{A})(0.2\,\Omega) = 16\,\text{V} \qquad \blacksquare$$

When the emf of a single cell is too small for a particular application, two or more can be connected in series (Fig. 20–11). The emf of the set is the sum of the emf's of the individual cells, and the internal resistance of the set is the sum of the individual internal resistances:

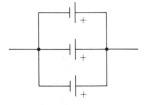

Parallel

FIG. 20–11 Cells in series and parallel. In series, terminals of opposite polarity are connected together. In parallel, terminals of the same polarity are connected together.

$$\mathscr{E}_{\text{series}} = \mathscr{E}_1 + \mathscr{E}_2 + \mathscr{E}_3 + \cdots$$

$$r_{\text{series}} = r_1 + r_2 + r_3 + \cdots$$

A familiar example of such an arrangement is the use of six lead-acid cells in series to make the 12-V battery of a car.

Cells in parallel

When the emf of a cell or battery is sufficient but its capacity is too small, two or more with the same emf can be connected in parallel. (If the cells have different emf's, currents will circulate among them that waste energy.) The emf and internal resistance of the set are now as follows:

$$\mathscr{E}_{\text{parallel}} = \mathscr{E}_1 = \mathscr{E}_2 = \mathscr{E}_3 = \cdots$$

$$\frac{1}{r_{\text{parallel}}} = \frac{1}{r_1} + \frac{1}{r_2} + \frac{1}{r_3} + \cdots$$

If all the cells have the same internal resistance r, r_{parallel} is just r divided by the number of cells. The reduced internal resistance of a parallel set of cells means that more current can be drawn by a given load. When a car's battery is too weak to start the car on a cold day, the remedy is another battery put in parallel with it using jumper cables, which will increase the available current. The ampere-hour capacity of a series set of cells is the same as that of each cell; the capacity of a parallel set is the sum of the capacities of the individual cells.

20–5   IMPEDANCE MATCHING

Energy transfer is a maximum when $R = r$

When a source of electric energy is connected to a load, as in Fig. 20–12, the rate at which energy is transferred to the load is a maximum when both source and load have the same resistance. If R is the load resistance and I is the current, then according to Eq. (20–5), $I = \mathscr{E}/(R + r)$ and the power in the load is

$$P = I^2 R = \frac{\mathscr{E}^2 R}{(R + r)^2} \qquad \textit{Power transfer} \quad (20\text{–}6)$$

The maximum value of P corresponds to $R = r$, when

$$P_{\text{max}} = \frac{\mathscr{E}^2 R}{(R + R)^2} = \frac{\mathscr{E}^2 R}{(2R)^2} = \frac{\mathscr{E}^2}{4R} \qquad (R = r) \tag{20–7}$$

Figure 20–13 shows how P/P_{max} varies with the resistance ratio R/r.

Example   Verify that the power transfer from a 100-V source of internal resistance $r = 10\ \Omega$ to a load resistance R is a maximum when $R = 10\ \Omega$.

Solution   Table 20–1 shows the current I and power P calculated from Eqs. (20–5) and (20–6) for values of R from 2 Ω to 20 Ω. The maximum power of 250 W evidently occurs for $R = r = 10\ \Omega$. Although the current is greater when R is less than r, the smaller R means that P is less than P_{max}. When R is greater than r, the current is then sufficiently low that P is less than P_{max} in this case as well. ■

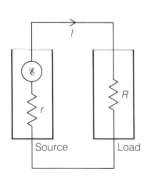

FIG. 20–12

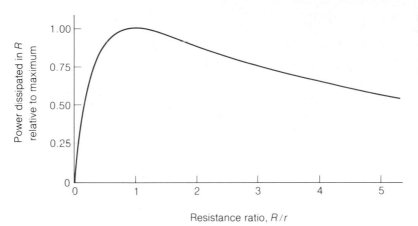

FIG. 20-13 The power transferred to an external load of resistance R is a maximum when R equals the internal resistance r of the source.

R, Ω	I, A	P, W
2	8.3	139
4	7.1	204
6	6.3	234
8	5.6	247
10	6.0	250
12	4.5	248
14	4.2	243
16	3.8	237
18	3.6	230
20	3.3	222

TABLE 20-1
Current I and load power P for various values of R for the circuit of Fig. 20-12 when $\mathscr{E} = 100$ V, $r = 10$ Ω

The conclusion reached in the above discussion is an example of the phenomenon of *impedance matching:* When energy is being transferred from one system to another (here from the battery to the external resistance), the efficiency is greatest when both systems have the same impedance, which is a general term for resistance to the flow of energy in whatever form it may take in a particular case. We encountered impedance matching earlier, the first time in Section 6-7. There we saw that when a moving object strikes a stationary one, the maximum transfer of energy takes place when both have the same mass. In this situation the inertia of each object, as measured by its mass, represents its impedance. The curve of relative energy transfer versus mass ratio in Fig. 6-17 is identical in form to that of Fig. 20-13.

Impedance matching is a general concept

Another example of impedance matching occurs when a pulse or wave moves down a stretched string, as discussed in Section 13-2. When two strings having the same mass per unit length are joined together, a pulse in one passes to the other with no reflection. However, if the second string has either a greater or a smaller mass per unit length, reflection occurs at the junction and not all the energy of the pulse is transmitted (Figs. 13-6 and 13-7). Though best known for its application to electric circuits, impedance matching is an important concept in many other branches of physics and engineering as well.

Impedance matching in wave motion

20–6 KIRCHHOFF'S RULES

It is often difficult or impossible to determine the currents that flow in the various branches of a complex network merely by computing equivalent resistances. Two rules formulated by Gustav Kirchhoff (1824–1887) make it possible to find the current in each part of a direct-current circuit, no matter how complicated, if we are given the emf's of the sources of potential difference and the resistances of the various circuit elements. These rules apply to *junctions,* which are points where three or more wires come together, and to *loops,* which are closed conducting paths that are part of the circuit (Fig. 20–14).

Junctions and loops

Kirchhoff's first rule follows from the conservation of electric charge. Charge is never found to accumulate at any point in a circuit, nor can it be created there, and so there cannot be any net current into or out of a junction. Hence the first rule:

Kirchhoff's first rule

1. The sum of the currents flowing into a junction is equal to the sum of the currents flowing out of the junction.

The second rule is a consequence of the conservation of energy. The sum of the emf's in a loop equals the amount of work done per coulomb by the sources of emf *on* a charge that moves once around the loop. The work done per coulomb *by* the charge as it moves around the loop equals the sum of the *IR* potential drops along the way. To conserve energy, the work done on the charge must be the same as the work done by the charge, and so we have Kirchhoff's second rule:

Kirchhoff's second rule

2. The sum of the emf's around a loop is equal to the sum of the *IR* potential drops around the loop.

(1) Assume the branch currents

A definite procedure must be followed when Kirchhoff's rules are applied to a network. Let us call a section of a loop between two junctions a *branch.* The first step is to give the current in each branch a symbol of its own (I_1, I_2, I_3, and so on) and to assume a direction for each current. These symbols and directions should be marked on the circuit diagram, as in Fig. 20–14. The current is the same in all the resistors and sources of emf in a given branch, but of course the currents are different in the

FIG. 20–14 A *junction* is a point where three or more wires come together; a *loop* is any closed conducting path in the network. Here the internal resistances of the batteries are included in R_1 and R_3. The directions of the currents I_1, I_2, and I_3 are chosen arbitrarily; the value of I_3 turns out to be negative, meaning that its actual direction is opposite to the one shown.

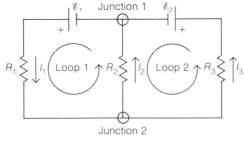

$\mathcal{E}_1 = 6V$ $R_1 = 8\,\Omega$ $I_1 = ?$
$\mathcal{E}_2 = 10V$ $R_2 = 4\,\Omega$ $I_2 = ?$
$R_3 = 7\,\Omega$ $I_3 = ?$

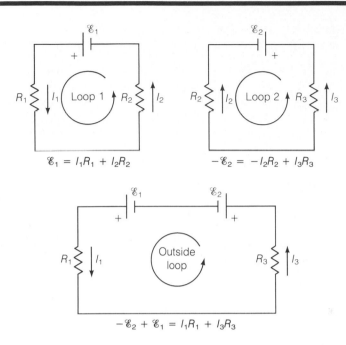

FIG. 20-15 The three loops of the circuit shown in Fig. 20-14. Loops 1 and 2 are used in solving for the unknown currents. The outside loop can be used to check the results.

$$\mathscr{E}_1 = I_1 R_1 + I_2 R_2$$

$$-\mathscr{E}_2 = -I_2 R_2 + I_3 R_3$$

$$-\mathscr{E}_2 + \mathscr{E}_1 = I_1 R_1 + I_3 R_3$$

different branches. It does not matter which direction is chosen for each current. If we have guessed correctly about the direction of the current in a branch, the solution of the problem will give a positive value for the current. If we have guessed wrong, a negative value will indicate that the actual current in the branch is in the opposite direction.

When the currents have been assigned to each branch, we then apply Kirchhoff's first rule to the junctions in the network. This will give us as many equations as the number of junctions. However, we will always find that one of these equations is a combination of the others and so contains no new information. Thus the number of junction equations we can use is always one less than the number of junctions.

(2) Apply the first rule to the junctions

The third step is to apply Kirchhoff's second rule to the loops in the network. In going around a loop, we must follow a consistent path, either clockwise or counterclockwise. These paths should be marked on the diagram, as in Fig. 20-14. An emf is reckoned positive if we meet the negative terminal of its source first; if instead we meet the positive terminal first, the emf is reckoned negative. An *IR* drop is considered positive if the assumed current in the resistor is in the same direction as the path we are following; the drop is considered negative if the current is opposite in direction to that of our path. The internal resistance of an emf source is treated as if it is a separate resistor in series with the source. As in the case of the junction equations, it is important to avoid loop equations that merely duplicate information in the other loop equations. As a guide, the number of useful loop equations will equal the number of separate areas enclosed by branches of the network. (We can think of these areas as adjacent plots of land.)

(3) Apply the second rule to the loops

Example Find the current in each of the resistors of the network shown in Fig. 20-14.

Solution   The assumed directions of the unknown currents I_1, I_2, and I_3 are shown in the figure. When we apply Kirchhoff's first rule to junction 1, we obtain

$$I_1 = I_2 + I_3 \qquad\qquad\qquad\qquad \textit{Junction 1}$$

This rule applied to junction 2 gives the same result, so we have just one junction equation, as we should.

In applying Kirchhoff's second rule to the two loops we shall follow counterclockwise routes. For loop 1,

$$\mathscr{E}_1 = I_1R_1 + I_2R_2 \qquad\qquad\qquad\qquad \textit{Loop 1}$$

and for loop 2,

$$-\mathscr{E}_2 = -I_2R_2 + I_3R_3 \qquad\qquad\qquad\qquad \textit{Loop 2}$$

In loop 2 we consider $\mathscr{E}_2$ as negative because we encounter its positive terminal first, and I_2 as negative because its direction is opposite to our counterclockwise path.

There is also a third loop, namely the outside one in Fig. 20–14, which similarly must obey Kirchhoff's second rule. Proceeding counterclockwise yields

$$-\mathscr{E}_2 + \mathscr{E}_1 = I_1R_1 + I_3R_3 \qquad\qquad\qquad\qquad \textit{Outside loop}$$

We note that this last equation is just the sum of the two preceding loop equations. (With two enclosed areas in the network, we expect only two independent loop equations.) Thus we may use the junction equation and any two of the loop equations to solve for the unknown currents; nothing will be gained by using all three loop equations.

There are three unknown currents, and we have three separate equations relating them. Hence the problem can be solved by routine algebra. As an example of how this might be done, we shall start by substituting $I_2 + I_3$ for I_1 in the first loop equation to obtain

$$\mathscr{E}_1 = I_2R_1 + I_3R_1 + I_2R_2$$

We now solve both this equation and the second loop equation for I_3:

$$I_3 = \frac{\mathscr{E}_1 - I_2R_1 - I_2R_2}{R_1} \qquad\qquad I_3 = \frac{-\mathscr{E}_2 + I_2R_2}{R_3}$$

Setting the two expressions for I_3 equal enables us to solve for I_2 as follows:

$$\frac{\mathscr{E}_1 - I_2R_1 - I_2R_2}{R_1} = \frac{-\mathscr{E}_2 + I_2R_2}{R_3}$$

$$\mathscr{E}_1R_3 - I_2R_1R_3 - I_2R_2R_3 = -\mathscr{E}_2R_1 + I_2R_1R_2$$

$$I_2(R_1R_3 + R_2R_3 + R_1R_2) = \mathscr{E}_2R_1 + \mathscr{E}_1R_3$$

$$I_2 = \frac{\mathscr{E}_2R_1 + \mathscr{E}_1R_3}{R_1R_3 + R_2R_3 + R_1R_2}$$

Finally we insert the values of $\mathscr{E}_1$, $\mathscr{E}_2$, R_1, R_2, and R_3 given in Fig. 20–14 to obtain

$$I_2 = 1.05 \text{ A}$$

With the value of I_2 known we can find the value of I_3 from the second loop equation:

$$I_3 = \frac{-\mathscr{E}_2 + I_2R_2}{R_3} = -0.83 \, \text{A}$$

The minus sign means that the direction of I_3 is opposite to that shown in Fig. 20–14. The junction equation finally provides us with the value of I_1.

$$I_1 = I_2 + I_3 = 0.22 \, \text{A}$$

The same results would have been obtained had we chosen other directions for the currents or taken other routes around the loops in applying Kirchhoff's second rule. The important thing is not what choice of directions is made but to follow that choice consistently in working out the problem.

The outside loop equation, which was not used in the calculation, provides us with a handy way to check the results. We have

$$-\mathscr{E}_2 + \mathscr{E}_1 = I_1R_1 + I_3R_3$$
$$-10 \text{ V} + 6 \text{ V} = (0.22 \, \text{A})(8 \, \Omega) + (-0.83 \, \text{A})(7 \, \Omega) \qquad \textit{Outside loop}$$
$$-4 \text{ V} = -4 \text{ V}$$

In the above calculation symbols for the various quantities were used until the very end to make clear the operations involved. However, in working out problems of this kind it is usually easier and faster to substitute numerical values from the start. ∎

20–7 AMMETERS, VOLTMETERS, AND OHMMETERS

The traditional instrument for measuring electrical quantities is the *galvanometer,* whose operation is described in Chapter 21. The deflection of a galvanometer's pointer is proportional to the current that passes through it. A galvanometer can be used to measure potential difference as well as current. It has a certain resistance, and according to Ohm's law the current in it is directly proportional to the potential difference across its terminals. Hence the meter's scale can be calibrated equally well in volts as in amperes. A meter that has a full-scale reading of, say, 1 mA (10^{-3} A) and a resistance of 50 Ω can also be used to measure potential differences of up to

$$V_{\text{max}} = I_{\text{max}}R_{\text{coil}} = (10^{-3} \, \text{A})(50 \, \Omega) = 0.05 \, \text{V}$$

A galvanometer employed to measure the current in a circuit is called an *ammeter* and is inserted in series in the circuit. The basic range of a galvanometer (0 to 1 mA in the case of the above meter) can be extended by connecting a low resistance shunt in parallel with it to carry the bulk of the current, leaving a known fraction of the current to be registered by the meter itself.

An ammeter is used in series

As an example, let us convert the above galvanometer to measure currents of up to 1 A. What we must do is place a resistor in parallel with the meter that will divert

$$1.000 \, \text{A} - 0.001 \, \text{A} = 0.999 \, \text{A}$$

of the maximum current, leaving 0.001 A to give a full-scale reading on the meter (Fig. 20–16). To obtain the value of R_{shunt} we note that, since the potential difference V is the same across each branch of a parallel circuit,

FIG. 20–16 The range of current a galvanometer can measure can be extended by placing a shunt resistor in parallel with it. The equivalent resistance of such an ammeter should be very low.

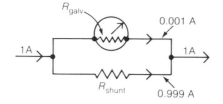

$$V = I_{shunt}R_{shunt} = I_{galv}R_{galv}$$

and so

$$R_{shunt} = R_{galv}\frac{I_{galv}}{I_{shunt}} = (50\ \Omega)\left(\frac{0.001\ A}{0.999\ A}\right) = 0.050\ \Omega$$

An ammeter should have a low resistance

An ideal ammeter has a very low equivalent resistance so that its presence in any circuit alters the properties of the circuit as little as possible. The above ammeter has an equivalent resistance of very nearly 0.050 Ω; unless the equivalent resistance of a circuit in which it is inserted is large compared with 0.050 Ω, the meter reading will not be an accurate measure of the current in the absence of the ammeter.

A voltmeter is used in parallel

A galvanometer employed to measure potential difference is called a *voltmeter*. The basic voltage range of a certain galvanometer can be extended by connecting a resistor in series with it in order to hold the maximum current down to whatever figure is appropriate for the meter's movement.

We might wish to convert the above 0–1-mA galvanometer to a voltmeter with a full-scale reading of 1 V. This means that when a potential difference of 1 V is across the voltmeter's terminals, a current of 1 mA flows in the meter's coil. The equivalent resistance of the voltmeter must therefore be

$$R = R_{series} + R_{galv} = \frac{V}{I}$$

and so

$$R_{series} = \frac{V}{I} - R_{galv} = \frac{1\ V}{0.001\ A} - 50\ \Omega = 950\ \Omega$$

A 950-Ω resistor connected in series with the galvanometer converts it to a voltmeter whose range is 0 to 1 V (Fig. 20–17).

A voltmeter should have a high resistance

Placing a voltmeter across a circuit element is the same thing as connecting a resistor in parallel with it, which changes the properties of the circuit and so leads to an incorrect potential difference measurement. For this reason an ideal voltmeter has a very high resistance. A voltmeter rated at 10,000 Ω/V, which is typical of inexpensive combination volt-ohm-milliammeters, has a resistance of 10,000 Ω on the 0–1-V scale, 100,000 Ω on the 0–10-V scale, and so forth. Better voltmeters of the conventional type have ratings of 50,000 Ω/V, and electronic voltmeters may go as high as 10^7 Ω/V, which means that they have essentially no effect on a circuit they are connected to.

Ohmmeter

A galvanometer can be used to measure an unknown resistance R with the help of a battery and a series resistor, as in Fig. 20–18. Such a device is called an *ohmmeter*.

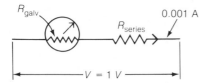

FIG. 20–17 The range of voltage a galvanometer can measure can be extended by placing a resistor in series with it. The equivalent resistance of such a voltmeter should be very high.

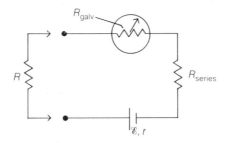

FIG. 20–18 How a galvanometer can be connected to measure resistance.

The meter's scale is calibrated backwards, so that a full-scale deflection corresponds to $R = 0$, when the current is a maximum; no deflection corresponds to $R = \infty$, when the current is 0. The purpose of the series resistor is to have $I = I_{max}$ for the galvanometer when the only resistance in the circuit is that of the resistor, that of the meter, and that of the battery. In the case of a 0–1-mA, 50-Ω galvanometer used with a 1.5-V, 0.05-Ω battery,

$$R_{series} + R_{galv} + r_{bat} = \frac{\mathcal{E}}{I}$$

or

$$R_{series} = \frac{\mathcal{E}}{I} - R_{galv} - r_{bat} = \frac{1.5\,V}{0.001\,A} - 50\,\Omega - 0.05\,\Omega = 1450\,\Omega$$

An ohmmeter is not a precision instrument, but is handy and often sufficient. A Wheatstone bridge (Exercise 13 of Chap. 19) offers greater accuracy.

IMPORTANT TERMS

The **equivalent resistance** of a set of interconnected resistors is the value of the single resistor that can be substituted for the entire set without affecting the current that flows in the rest of any circuit of which it is a part. Resistors in **series** are connected consecutively so that the same current flows through all of them, whereas resistors in **parallel** have their terminals connected together so that the total current is split up among them.

The **electromotive force** (emf) of a battery, generator, or other source of electric energy is the potential difference across its terminals when no current flows. When a current is flowing, the terminal voltage is less than the emf owing to the potential drop in the **internal resistance** of the source.

Impedance matching refers to the fact that a source of electric energy transfers maximum power to a load when the resistance of the load equals the internal resistance of the source.

Kirchhoff's rules for network analysis are: (1) The sum of the currents flowing into a junction of three or more wires is equal to the sum of the currents flowing out of the junction; (2) the sum of the emf's around a closed conducting loop is equal to the sum of the IR potential drops around the loop.

IMPORTANT FORMULAS

Resistors in series: $\quad R = R_1 + R_2 + R_3 + \cdots$

Resistors in parallel:

$$\frac{1}{R} = \frac{1}{R_1} + \frac{1}{R_2} + \frac{1}{R_3} + \cdots$$

Emf and terminal voltage: $V = \mathscr{E} - IR$

MULTIPLE CHOICE

1. A resistor R_1 dissipates the power P when connected to a certain generator. If a resistor R_2 is inserted in series with R_1, the power dissipated by R_1
   a. is less than P.
   b. is more than P.
   c. is P.
   d. may be any of the above, depending on the values of R_1 and R_2.

2. If the resistor R_2 of Question 1 is placed in parallel with R_1, the power dissipated by R_1
   a. is less than P.
   b. is more than P.
   c. is P.
   d. may be any of the above, depending on the values of R_1 and R_2.

3. The unit of emf is the
   a. ohm.   b. ampere.
   c. volt.   d. watt.

4. A battery is connected to an external circuit. The potential drop within the battery is proportional to
   a. the emf of the battery.
   b. the equivalent resistance of the circuit.
   c. the current in the circuit.
   d. the power dissipated in the circuit.

5. A battery of emf $\mathscr{E}$ and internal resistance r is connected to an external circuit of equivalent resistance R. If $R = r$,
   a. the current in the circuit will be a minimum.
   b. the current in the circuit will be a maximum.
   c. the power dissipated in the circuit will be a minimum.
   d. the power dissipated in the circuit will be a maximum.

6. Which of the following is neither a basic physical law nor derivable from one?
   a. Coulomb's law
   b. Ohm's law
   c. Kirchhoff's first law
   d. Kirchhoff's second law

7. If the wrong direction is assumed for a current I in a network being analyzed by Kirchhoff's rules, the value of the current that is obtained will be
   a. I.   b. $-I$.
   c. incorrect.   d. 0.

8. An ammeter has a
   a. shunt resistor of small R.

   b. shunt resistor of large R.
   c. series resistor of small R.
   d. series resistor of large R.

9. A voltmeter has a
   a. shunt resistor of small R.
   b. shunt resistor of large R.
   c. series resistor of small R.
   d. series resistor of large R.

10. The potential drop across a resistor can be measured by connecting
   a. a voltmeter in series with it.
   b. a voltmeter in parallel with it.
   c. an ammeter in series with it.
   d. an ammeter in parallel with it.

11. A 20-V potential difference is applied across a series combination of a 10-Ω resistor and a 30-Ω resistor. The current in the 10-Ω resistor is
   a. 0.5 A.   b. 0.67 A.
   c. 1 A.   d. 2 A.

12. The potential difference across the 10-Ω resistor of Question 11 is
   a. 5 V.   b. 10 V.
   c. 15 V.   d. 20 V.

13. The equivalent resistance of a 10-Ω resistor and a 30-Ω resistor connected in parallel is
   a. 0.13 Ω.   b. 7.5 Ω.
   c. 20 Ω.   d. 40 Ω.

14. A 20-V potential difference is applied across the resistors of Question 13. The current in the 10-Ω resistor is
   a. 0.5 A.   b. 1 A.
   c. 2 A.   d. 2.67 A.

15. The equivalent resistance of a network of three 2-Ω resistors cannot be
   a. 0.67 Ω.   b. 1.5 Ω.
   c. 3 Ω.   d. 6 Ω.

16. A 12-V potential difference is applied across a series combination of four 6-Ω resistors. The current in each resistor is
   a. 0.5 A.   b. 2 A.
   c. 8 A.   d. 18 A.

17. A 12-V potential difference is applied across a parallel combination of four 6-Ω resistors. The current in each resistor is
   a. 0.5 A.   b. 2 A.
   c. 8 A.   d. 18 A.

18. Two identical resistors connected in parallel have an equivalent resistance of 2 Ω. If the resistors were connected in series, their equivalent resistance would be

a. 2 Ω. b. 4 Ω.
c. 8 Ω. d. 16 Ω.

19. A resistor of unknown resistance is connected in parallel with a 12-Ω resistor and a battery of emf 24 V and negligible internal resistance is connected across the combination. The battery provides a current of 3 A. The unknown resistance is

a. 8 Ω. b. 12 Ω.
c. 24 Ω. d. 36 Ω.

20. A 10-Ω resistor and a 20-Ω resistor are connected in parallel to a battery. If heat is produced in the 10-Ω resistor at the rate P, the rate at which heat is produced in the 20-Ω resistor is

a. $P/4$. b. $P/2$.
c. P. d. $2P$.

21. A 10-Ω resistor and a 20-Ω resistor are connected in series to a battery. If heat is produced in the 10-Ω resistor at the rate P, the rate at which heat is produced in the 20-Ω resistor is

a. $P/4$. b. $P/2$.
c. P. d. $2P$.

22. A 50-V battery is connected across a 10-Ω resistor and a current of 4.5 A flows. The internal resistance of the battery is

a. 0. b. 0.5 Ω.
c. 1.1 Ω. d. 5 Ω.

23. A 4-Ω resistor is connected to a battery of emf 20 V and internal resistance 1 Ω. The potential difference across the resistor is

a. 4 V. b. 16 V.
c. 19 V. d. 20 V.

24. Two batteries of emf 6 V and internal resistance 1 Ω are connected in parallel. The load that will absorb the greatest power from the combination has a resistance of

a. 0.5 Ω. b. 1 Ω.
c. 2 Ω. d. 6 Ω.

25. A 0–10-mA galvanometer with a coil resistance of 20 Ω is converted to a 0–10-A ammeter by using a

a. 0.02-Ω shunt resistor.
b. 0.2-Ω shunt resistor.
c. 9.99-Ω shunt resistor.
d. 9.99-Ω series resistor.

26. A 0–10-mA galvanometer with a coil resistance of 20 Ω is converted to a 0–50-V voltmeter by using a

a. 50-Ω shunt resistor.
b. 4980-Ω series resistor.
c. 5000-Ω series resistor.
d. 5020-Ω series resistor.

EXERCISES

20–1 Resistors in Series

1. A 200-Ω and a 500-Ω resistor are in series as part of a larger circuit. If the voltage across the 200-Ω resistor is 2 V, find the voltage across the 500-Ω resistor.

2. It is desired to limit the current in an 80-Ω resistor to 0.5 A when it is connected to a 50-V power source. What is the value of the series resistor that is needed?

3. How can the resistance of a 20-Ω circuit be reduced to 5 Ω?

4. A stingy American moves to Europe, where 240 V is the normal household voltage, and proposes to use his old 120-V light bulbs by connecting them two at a time in series. How good is this idea when both bulbs of each pair have the same power rating? When the bulbs have different ratings—for instance, 10 W and 100 W at 120 V?

5. A set of Christmas tree lights consists of 12 bulbs connected in series to a 120-V power source. Each bulb has a resistance of 5 Ω. (a) What is the current in the circuit? (b) How much power is dissipated in the circuit?

6. A 5-Ω light bulb and a 10-Ω light bulb are connected in scrics with a 12-V battery. (a) What is the current in each bulb? (b) What is the voltage across each bulb? (c) What is the power dissipated by each bulb and the total power dissipated by the circuit?

20–2 Resistors in Parallel

20–3 Series-Parallel Combinations

7. The light bulbs in the circuit of the figure below are identical. Which bulb gives off the most light? The least light?

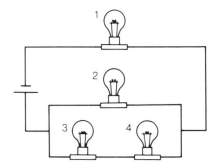

8. List the resistances that can be obtained by combining three 100-Ω resistors in all possible ways.

9. A 20-Ω resistor, a 40-Ω resistor, and a 50-Ω resistor are

connected in parallel across a 60-V power source. Find the equivalent resistance of the set and the current in each resistor.

10. In a certain bus, forty 15-W, 30-V light bulbs are connected in parallel to a 30-V power source. (a) What is the current provided by the source? (b) What is the current in each bulb? (c) How much power is provided by the source?

11. The light bulbs of Exercise 6 are connected in parallel across the same battery. Answer the same questions for this arrangement.

12. Each of the resistors in the circuit of the figure below can safely dissipate 10 W. What is the maximum power the entire circuit can dissipate?

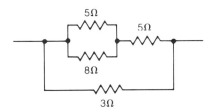

13. (a) Find the equivalent resistance of the circuit below. (b) What is the current in the 8-Ω resistor when a potential difference of 12 V is applied to the circuit?

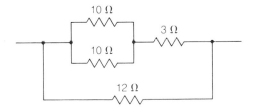

14. (a) Find the equivalent resistance of the circuit shown below. (b) What is the total current and the current in each resistor when a potential difference of 20 V is applied to the circuit?

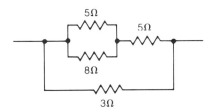

15. A 5-Ω and a 10-Ω resistor are connected in parallel. This combination is connected in series with another pair of parallel resistors whose resistances are both 8 Ω. (a) What is the equivalent resistance of the network? (b) The network

is connected to a 24-V battery whose internal resistance is 1.5 Ω. Find the current in each of the resistors.

16. (a) Find the equivalent resistance of the circuit below. (b) What is the current in the 12-V resistor when a potential difference of 100 V is applied to the circuit?

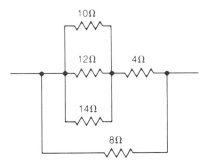

17. A 60-V potential difference is applied to the circuit below. Find the current in the 10-Ω resistor. (*Hint:* Redraw the circuit to bring out the series and parallel combinations of resistors more clearly.)

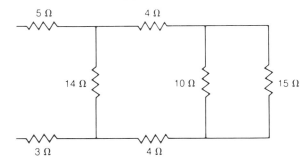

20−4   Electromotive Force

18. Why is it undesirable to connect cells of different emf in parallel?

19. A source of what potential difference is needed to charge a battery of emf 24 V and internal resistance 0.1 Ω at a rate of 70 A?

20. A 12-V battery of internal resistance 1.5 Ω is connected to an 8-Ω resistor. Find the total power produced by the battery and the percentage of this power dissipated as heat within it.

21. A battery having an emf of 24 V is connected to a 10-Ω load, and a current of 2.2 A flows. Find the internal resistance of the battery and its terminal voltage.

22. A generator has an emf of 240 V and an internal resistance of 0.3 Ω. When the generator is supplying a current

of 20 A, find (a) its terminal voltage, (b) the power supplied to the load, and (c) the power dissipated in the generator itself.

23. The brightness of a light bulb depends on the power dissipated by its filament. As a dry cell ages, its internal resistance increases while its emf remains approximately unchanged at 1.5 V. A fresh No. 6 dry cell might have an internal resistance of 0.05 Ω and an old one an internal resistance of 0.20 Ω. Find the ratio between the powers dissipated in a 0.25-Ω bulb when it is connected to a fresh and to an old dry cell.

24. Four batteries, each of emf 6.0 V and internal resistance 0.30 Ω, are connected in series with a load of 2.0 Ω. Find the current in the load.

25. If the batteries of Exercise 24 are connected in parallel with the same load, find the current in the load.

26. Twelve cells, each of emf 2.1 V and internal resistance 0.2 Ω, are connected as shown below. Find the emf and internal resistance of the combination.

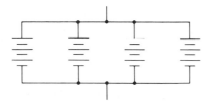

27. A 12-volt storage battery with an internal resistance of 0.012 Ω delivers 80 A when used to crank a gasoline engine. If the battery mass is 20 kg and it has an average specific heat of 0.2 kcal/kg·°C, what is its rise in temperature during 1 min of cranking the engine?

28. A certain 16-cell 32-V storage battery has an emf of 33.5 V when charged to 75% of its 250 A-h capacity. The internal resistance of the battery is 0.1 Ω. It is desired to charge the battery to its full capacity, when its emf will be 34.3 V. (a) What potential difference must be applied to the battery if it is to be charged at the initial rate of 40 A? (b) If this potential difference is held constant, what will be the rate of charge at the end of the process? (c) The emf of the battery arises from the conversion of chemical to electrical energy; the higher potential difference is required for charging in order to pass a current through the battery and thereby produce chemical changes that store energy. Find the proportion of the power supplied during the charging process that is stored as chemical energy and the proportion that is dissipated as heat. (Assume an average emf during charging of 33.9 V.)

20-6 Kirchhoff's Rules

29. Find the values of R_1 and R_2 in the circuit below.

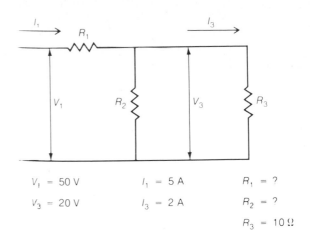

$V_1 = 50$ V $\qquad I_1 = 5$ A $\qquad R_1 = ?$

$V_3 = 20$ V $\qquad I_3 = 2$ A $\qquad R_2 = ?$

$\qquad \qquad \qquad \qquad \qquad \qquad \qquad R_3 = 10\,\Omega$

30. Three identical 1.5-V dry cells with internal resistances of 0.15 Ω are connected in parallel with an external 0.5-Ω resistor. How much current flows through the resistor?

31. Find the current in the 20-Ω resistor in the circuit below.

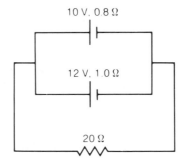

10 V, 0.8 Ω

12 V, 1.0 Ω

20 Ω

32. Find the current in the 10-Ω resistor in the circuit below.

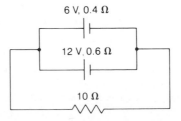

6 V, 0.4 Ω

12 V, 0.6 Ω

10 Ω

33. Find the currents in each of the resistors of the circuit below. The internal resistances of the batteries are included in the values of the resistors.

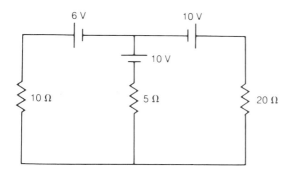

34. Find the currents in the three resistors in the circuit below. The internal resistances of the batteries are included in the values of the resistors.

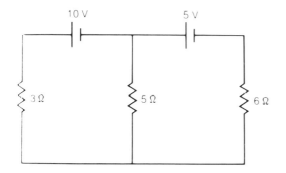

35. Find the potential differences between *a* and *b* and between *a* and *c* in the circuit below.

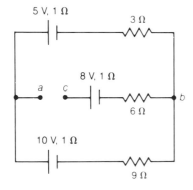

36. If *a* and *c* in the circuit shown above are connected, find the potential difference between *a* and *b*.

37. (a) Find the current in the 5-Ω resistor in the circuit below. (b) Find the potential difference between the points *a* and *b*.

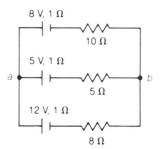

38. A galvanometer whose coil has a resistance of 12 Ω requires a current of 1.7 mA for full-scale deflection. (a) What shunt resistance is needed to convert the meter to a 0–10-mA ammeter? (b) What series resistance is needed to convert the meter to a 0–10-V voltmeter?

39. A galvanometer whose coil has a resistance of 60 Ω requires a current of 0.02 mA for full-scale deflection. (a) What shunt resistance is needed to convert the meter to a 0–1-A ammeter? (b) What series resistance is needed to convert the meter to a 0–5-V voltmeter?

40. An ammeter has a full-scale reading of 100 mA. The potential difference across the meter when it reads 50 mA is 0.02 V. What must be done to convert the meter to have a full-scale reading of 1.0 A?

41. A voltmeter has a full-scale reading of 10 V. The current through the meter when it reads 10 V is 0.06 mA. What must be done to convert the meter to have a full-scale reading of 100 V?

42. The coil of an ammeter has a resistance of 4 Ω and its shunt has a resistance of 0.02 Ω. What is the current in the coil when the meter reads 15 A?

43. A voltmeter whose resistance is 2000 Ω is placed across a resistor of unknown resistance, and the combination is connected in series with an ammeter. The ammeter reads 0.040 A when the voltmeter reads 12 V. Find the unknown resistance.

ANSWERS TO MULTIPLE CHOICE

1. a	**7.** b	**12.** a	**17.** b	**22.** c
2. c	**8.** a	**13.** b	**18.** c	**23.** b
3. c	**9.** d	**14.** c	**19.** c	**24.** a
4. c	**10.** b	**15.** b	**20.** b	**25.** a
5. d	**11.** a	**16.** a	**21.** d	**26.** b
6. b				

21

MAGNETISM

Magnetism and electricity both arise from the same basic interaction between electric charges. Charges at rest relative to an observer appear to him or her to exert only electric forces upon one another. When the charges are in motion relative to the observer, however, the forces acting between them seem different from before, and these differences are traditionally attributed to "magnetic" forces. In reality, magnetic forces represent modifications to electric forces that come about because of the motions of the charges involved. It is convenient to consider magnetic and electric forces separately and to think in terms of separate magnetic and electric fields, but we should keep in mind that these distinctions are artificial.

21–1   NATURE OF MAGNETISM

The forces that moving charges exert upon one another are different from those the same charges exert when at rest. For instance, if we place a current-carrying wire

CHAPTER OBJECTIVES

Completing this chapter should enable you to:

1. Describe the nature of a magnetic field and explain how its magnitude and direction are defined.

2. Calculate the magnetic field near a long, straight current.

3. Calculate the magnetic field at the center of a flat coil of one or more turns.

4. Calculate the magnetic field inside a solenoid.

5. Describe the properties of ferromagnetic materials.

6. Predict the behavior of a ferromagnetic material on the basis of its hysteresis loop.

7. Determine the direction and magnitude of the force on a charge moving in a magnetic field.

8. Calculate the orbit radius of a charge moving in a magnetic field.

9. Describe how a mass spectrometer measures atomic masses.

10. Determine the direction and magnitude of the force on a current in a magnetic field.

11. Calculate the force between two parallel currents and establish whether the force is attractive or repulsive.

12. Calculate the torque on a current loop in a magnetic field.

13. Describe how a galvanometer and an electric motor operate.

14. Interpret the behavior of permanent magnets in terms of the behavior of current loops.

parallel to another current-carrying wire, with the currents in the same direction, we find that the wires attract each other (Fig. 21–1). If the currents are in opposite directions, the forces on the wires are repulsive.

Gravitational forces cannot be responsible for these observations since they are never repulsive, and electric forces cannot be responsible since there is no net charge on a wire when a current is present in it. The forces that come into being when electric currents interact are called *magnetic forces*. All magnetic effects can ultimately be traced to currents or, more exactly, to moving electric charges. Of course, the word *magnetic* suggests ordinary magnets and their familiar attraction for iron objects, but, as we shall see, this is but one aspect of the whole subject of magnetism.

Magnetic forces arise from the interactions of moving charges

The gravitational force between two masses and the electric force between two charges at rest are both *fundamental forces* in the sense that they cannot be accounted for in terms of anything else. On the other hand, the force a bat exerts on a ball is not fundamental because it can be traced to the electric forces between the atomic electrons of the bat and the atomic electrons of the ball.

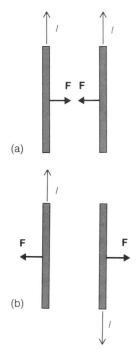

FIG. 21-1 (a) Parallel electric currents in the same direction attract each other. (b) When the currents are in opposite directions, they repel each other.

(a)

(b)

Magnetic field is defined in terms of the force on a moving charge

No force acts on a charge moving in the direction of B

What about magnetic forces? It is an important fact that whatever it is in nature that manifests itself as an electric force between stationary charges also manifests itself as a magnetic force between moving charges. One effect is not possible without the other. There is only a single fundamental interaction between charges, the *electromagnetic interaction,* which has two aspects, electric and magnetic.

Thus the proper way to interpret what we perceive as separate electric and magnetic fields is that they are both manifestations of a single electromagnetic field that surrounds every electric charge. The electric field is always there, but the magnetic field only appears when relative motion is present. In the case of a wire that carries an electric current, there is only a magnetic field because the wire itself is electrically neutral. The electric field of the electrons is canceled out by the opposite electric field of the positive ions in the wire, but the ions are stationary and have no magnetic field to cancel the magnetic field of the moving electrons. If we simply move a wire that has no current flowing in it, the electric and magnetic fields of the electrons are canceled by the electric and magnetic fields of the positive ions.

21-2   MAGNETIC FIELD

We recall from Chapter 18 that the electric field **E** at a given place is defined in terms of the force **F** the field exerts on a stationary positive charge Q placed there. Because the electric force on a charge is always found to be proportional to Q, the magnitude of **E** is appropriately specified by the ratio

$$E = \frac{F}{Q} \qquad\qquad\qquad \textit{Electric field magnitude}$$

The direction of **E** is taken as the same as the direction of **F**. Once we know **E**, we can readily find the magnitude and direction of the electric force on *any* charge at that place.

The symbol for *magnetic field* is **B**. Because magnetic forces only act on moving charges, **B** is defined in terms of the magnetic force **F** exerted on a positive charge Q whose velocity is **v.** Experiment and theory both show that the magnetic force on a moving charge is proportional to two factors. One is the product Qv: The larger the charge and the faster it moves, the greater the magnetic force.

The other factor concerns direction. When a charge is in a magnetic field, there is always a certain line along which it can move with no magnetic force acting on it. The direction of **B** is taken to lie along this line. When **v** is at the angle θ with respect to **B,** the magnetic force on the charge is found to be proportional to $\sin \theta$, so that F is a maximum at $\theta = 90°$; that is, the maximum force occurs for motion perpendicular to **B.** Because F depends on $Qv \sin \theta$, the magnitude B of the magnetic field responsible for **F** is defined, by analogy with the definition of electric field $E = F/Q$, as

$$B = \frac{F}{Qv \sin \theta} \qquad\qquad \textit{Magnetic field magnitude} \quad (21\text{--}1)$$

When **v** is perpendicular to **B,** $\sin \theta = \sin 90° = 1$ and

$$B = \frac{F}{Qv} \qquad (\mathbf{v} \perp \mathbf{B})$$

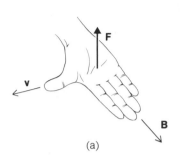

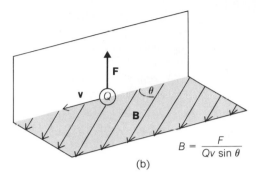

$$B = \frac{F}{Qv \sin \theta}$$

(a)　　　　　　　　　　　　　　　(b)

FIG. 21–2 (a) Right-hand rule for the direction of the magnetic field **B** that exerts the force **F** on a positively charged particle of velocity **v**. If the particle has a negative charge, the force is in the opposite direction. (b) The magnitude of **B** is defined in terms of F, Q, and v.

Although **B** lies along the line where **F** = 0, two opposite directions are possible along that line, so the direction of **B** needs further specification. Let us consider a charge moving perpendicular to that line. The direction of **B** is, by convention, given by a right-hand rule (Fig. 21–2):

> **Open your right hand so that the fingers are together and the thumb sticks out. When your thumb is in the direction of v and your palm faces in the direction of F, your fingers are in the direction of B.**

Right-hand rule for direction of B given F and v

(An easy way to remember this rule is to associate the outstretched thumb with hitch-hiking and so with velocity, the palm with pushing on something and so with force, and the parallel fingers with magnetic lines of force.)

This completes an operational definition of **B,** since we now have an unambiguous way to establish **B** in any region of space by performing suitable experiments. Such an experiment might use a cathode-ray tube in which the electron beam is deflected by the magnetic field as well as by an electric field, as described in the illustrative problem at the end of this section.

The unit of magnetic field is, from the above definition, the newton/ampere-meter (N/A · m), since the units of QV are coulomb-meter/second = ampere-meter. The name *tesla,* abbreviated T, has been given to this unit:

The tesla is the unit of magnetic field

1 tesla = 1 T = 1 N/A · m　　　　　　　　　　　*The tesla*

Thus a force of 1 N will be exerted on a charge of 1 C when it is moving at 1 m/s perpendicular to a magnetic field whose magnitude is 1 T. (When **v** is perpendicular to **B**, $\theta = 90°$ and $\sin \theta = 1$.)

The tesla is also referred to as the *weber*/m^2. Another unit of **B** in common use is the *gauss,* where

1 gauss = 10^{-4} T　　　or　　　1 T = 10^4 gauss

Figure 21–3 contains some representative values of magnetic field that may help in acquiring a feeling for the magnitude of the tesla.

Example The electrons in the beam of the cathode-ray tube shown in Fig. 21–4 are accelerated through a potential difference V of 1000 V. A magnetic field **B** applied to

FIG. 21–3 Some representative values of magnetic field.

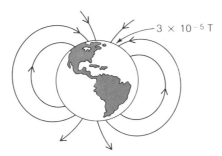

The magnitude of the earth's magnetic field at sea level is about 3×10^{-5} T.

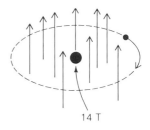

The magnetic field produced at the nucleus of a hydrogen atom by the electron circling around it is about 14 T.

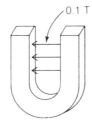

The magnetic field near a strong permanent magnet is about 0.1 T.

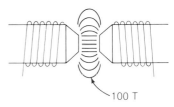

The most powerful magnetic fields achieved in the laboratory have magnitudes in the neighborhood of 100 T.

FIG. 21–4 Experimental arrangement to determine **B**. The electric field between the plates is adjusted until the electron beam is undeflected. The same procedure can be used with a known magnetic field to determine the charge-to-mass ratio e/m of the electron.

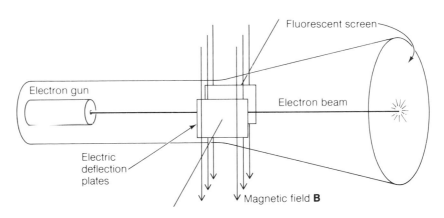

the region between the deflection plates bends the beam out of the paper, but this effect is canceled out when the electric field E between the plates is 10^4 V/m. Find the direction and magnitude of **B**.

Solution   Since electrons have negative charges, the direction of **B** must be opposite to that given by the right-hand rule, so it is downward. To find B, we start with the speed v of the electrons. Since the electron kinetic energy $\frac{1}{2} mv^2$ is equal to the energy eV they gain in the electron gun,

$$\tfrac{1}{2}mv^2 = eV$$

$$v = \sqrt{\frac{2eV}{m}}$$

Electron speed

When the electric force eE on the electron balances the magnetic force $evB \sin\theta = evB$ (since $\theta = 90°$ here),

$$eE = evB$$

Force balance

$$B = \frac{E}{v}$$

Substituting for v gives

$$B = \frac{E}{v} = E\sqrt{\frac{m}{2eV}} = (10^4\,\text{V/m})\sqrt{\frac{9.1 \times 10^{-31}\,\text{kg}}{2(1.6 \times 10^{-19}\,\text{C})(10^3\,\text{V})}} = 5.3 \times 10^{-4}\,\text{T}$$

(We must be careful not to confuse eV, the product of the electron charge e and the potential difference V, with eV, the abbreviation for the electron volt, which is a unit of energy.)

The same experimental arrangement can be used with a known B to find the ratio e/m between the charge and the mass of the electron. This was first done in 1897 by the English physicist J. J. Thomson, whose finding that e/m is always the same provided the first definite evidence that "cathode rays" are actually streams of particles. Later work by R. A. Millikan in the United States showed that electrons all have the same charge $-e = -1.60 \times 10^{-19}$ C, which permitted the electron mass to be established. ■

21–3 MAGNETIC FIELD OF A CURRENT

Now that both the direction and magnitude of magnetic field **B** have been specified in terms of procedures for finding them, we can go on to the magnetic fields produced by various electric currents. Unfortunately the calculations needed to determine **B** in a given situation are usually fairly difficult, so only the results will be given here.

Figure 21–5 shows the configuration of the magnetic field around a long, straight wire that carries the current I. The lines of force take the form of a series of concentric

Field around long, straight current

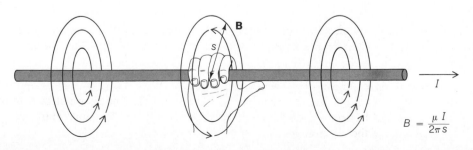

FIG. 21–5 The lines of force of the magnetic field around a long, straight current consist of concentric circles. The sense of the field is given by the right-hand rule.

$$B = \frac{\mu I}{2\pi s}$$

circles with the current at the center. The magnitude of the field a distance s from the wire is given by

$$B = \frac{\mu I}{2\pi s} \qquad \qquad \textit{Long, straight current} \quad (21\text{--}2)$$

The greater the current and the closer one is to it, the stronger the magnetic field.

Magnetic permeability

The constant μ is called the *permeability* of the medium in which the magnetic field exists. In free space,

$$\mu_0 = 4\pi \times 10^{-7} \text{ T} \cdot \text{m/A} = 1.257 \times 10^{-6} \text{ T} \cdot \text{m/A}$$

so that

$$\frac{\mu_0}{4\pi} = 10^{-7} \text{ T} \cdot \text{m/A}$$

The value of μ in air is very close to μ_0; they will be assumed to be the same here. The unit of permeability is sometimes expressed in other ways, for instance as N/A^2.

As shown in Fig. 21–5, the sense of the field around a straight current is given by another right-hand rule:

Direction of B around straight current

Grasp the wire with the right hand so that the thumb points in the direction of the current; the curled fingers of that hand point in the direction of the magnetic field.

Example Find the magnetic field in air 1 cm from a wire that carries a current of 1 A.

Solution Since 1 cm $= 10^{-2}$m, we have (Fig. 21–6)

$$B = \frac{\mu_0 I}{2\pi s} = \frac{(4\pi \times 10^{-7}\text{T} \cdot \text{m/A})(1\,\text{A})}{(2\pi)(10^{-2}\,\text{m})} = 2 \times 10^{-5}\,\text{T}$$

This is only a little smaller than the magnitude of the earth's magnetic field. For this reason great care is taken aboard ships to keep current-carrying wires away from magnetic compasses. ∎

FIG. 21–6

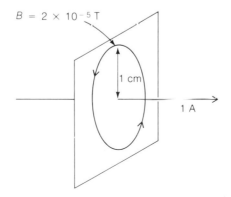

$B = 2 \times 10^{-5}\,\text{T}$

1 cm

1 A

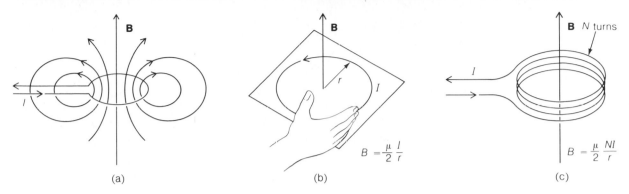

(a)   (b)   (c)

FIG. 21–7 (a) The magnetic field around a circular current loop. (b) At the center of the loop, **B** is perpendicular to the plane of the loop and its direction is given by the right-hand rule shown. (c) If there are N loops, the magnetic fields of the individual loops add up to give a field N times as strong as each one produces by itself.

The magnetic field around a circular current loop has the configuration shown in Fig. 21–7. At the center of the loop **B** is perpendicular to the plane of the loop and has the magnitude

$$B = \frac{\mu I}{2r} \qquad\qquad \textit{Center of current loop} \quad (21\text{–}3)$$

Field inside current loop

where I is the current in the loop and r is its radius. The direction of **B** is given by still another right-hand rule:

> **Grasp the loop so that the curled fingers of the right hand point in the direction of the current; the thumb of that hand then points in the direction of B.**

Direction of B inside loop

In the case of a flat coil of more than one loop, as in Fig. 21–7(c), the magnetic fields of each individual loop add up to give a proportionately stronger field. If there are N turns, then,

$$B = \frac{\mu NI}{2r} \qquad\qquad \textit{Center of flat coil} \quad (21\text{–}4)$$

A *solenoid* is a coil of wire in the form of a helix (Fig. 21–8). If the turns are close together and the solenoid is long relative to its diameter, then the magnetic field within it is uniform and parallel to its axis except near the ends. The direction of the field inside a solenoid is given by the same right-hand rule that gives the direction of **B** inside a current loop. The magnetic field in the interior of a solenoid l long that has N turns of wire and carries the current I has the magnitude

The field inside a solenoid is uniform

$$B = \mu \frac{N}{l} I \qquad\qquad \textit{Interior of solenoid} \quad (21\text{–}5)$$

The diameter of the solenoid does not matter, provided it is small compared with the length l.

(a)

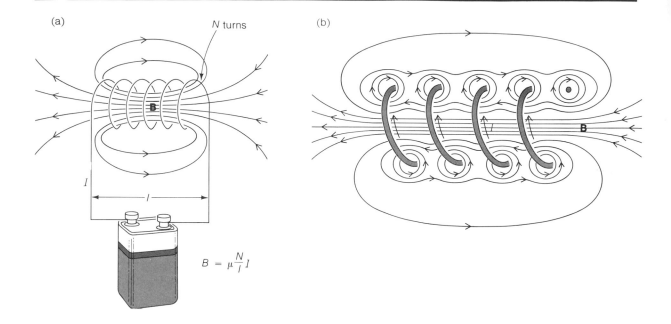

(b)

$$B = \mu \frac{N}{l} I$$

FIG. 21–8 (a) The magnetic field inside a solenoid is uniform except near its end if the solenoid is long relative to its diameter and if its turns are close together. (b) An expanded view of a solenoid showing how the magnetic fields of the individual turns add together to yield a uniform field inside it.

Example A solenoid 20 cm long and 4 cm in diameter with an air core is wound with a total of 200 turns of wire. The solenoid is aligned with its axis parallel to the earth's magnetic field at a place where the latter is 3×10^{-5} T in magnitude. What should the current in the solenoid be in order for its field to exactly cancel the earth's field inside the solenoid?

Solution In air $\mu = \mu_0$, and so, since $l = 0.2$ m here, the required current is

$$I = \frac{Bl}{\mu_0 N} = \frac{(3 \times 10^{-5}\,\text{T})(0.2\,\text{m})}{(4\pi \times 10^{-7}\,\text{T} \cdot \text{m/A})(2 \times 10^2)} = 0.024\,\text{A} = 24\,\text{mA}$$

The solenoid diameter here has no significance except as a check that the solenoid is long relative to its diameter. ■

Field of a bar magnet

The magnetic field of a bar magnet is identical with that of a solenoid, which is not surprising since all permanent magnets owe their character to an alignment of atomic current loops no different in principle from the alignment of the current loops in a solenoid (Fig. 21–9). The behavior of permanent magnets is discussed later in this chapter.

FIG. 21–9 The magnetic fields of a bar magnet and of a solenoid are the same.

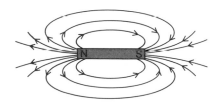
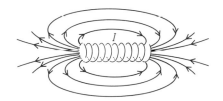

21–4   FERROMAGNETISM

The magnetic field produced by a current-carrying solenoid is changed in strength when a rod of almost any material is inserted in it. Some materials increase B (for instance, aluminum), others decrease B (for instance, bismuth), but in almost all cases the difference is very small. However, a few substances yield a dramatic increase in B when placed in a solenoid—the new field may be hundreds or thousands of times greater in magnitude than before (Fig. 21–10). Such substances are called *ferromagnetic*. Iron is the most familiar example, but nickel, cobalt, and certain alloys are also ferromagnetic, as are ceramic materials called *ferrites*.

In a ferromagnetic material, each atom acts as a tiny bar magnet, with adjacent atoms tending to be locked together with their fields aligned. Atoms in such a material are grouped together in assemblies called *domains*, each about 5×10^{-5} m across and just visible in a microscope. In an unmagnetized sample, the directions of magnetization of the domains are randomly oriented, though within each domain the atomic magnets are parallel. When such a sample is placed in an external magnetic field, either the atomic magnets within the domains turn to line up with the field or, in pure and homogeneous materials, the domain walls change so that those domains already lined up with the field grow at the expense of the others (Fig. 21–11). The first process requires stronger fields in order to occur than the second. Hence good "permanent" magnets are irregular in structure—for instance steel rather than pure iron—and once magnetized, cannot change their magnetization by the easy process of domain wall motion. When all the atomic magnets in a ferromagnetic sample are lined up, no further increase in B is possible, and the sample is said to be *saturated*.

Above a certain temperature (770°C in the case of iron), the atoms in a domain pick up enough kinetic energy to overcome the interatomic forces that hold their magnetic fields in alignment. The fields then become randomly oriented, and the ferromagnetic material loses its special magnetic properties. Thus heating a "permanent

Ferromagnetism increases B greatly

Magnetic domains

FIG. 21–10 (a) Solenoid with no core. (b) Solenoid with ferromagnetic core.

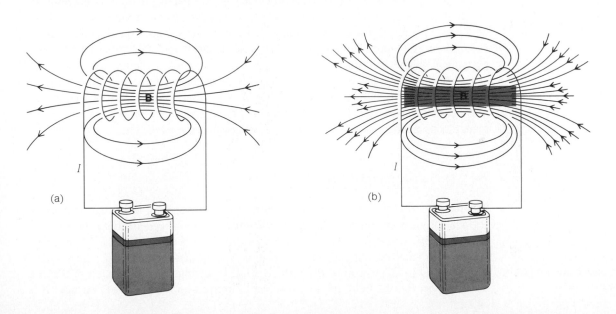

(a)   (b)

FIG. 21–11　(a)
Magnetization of a
ferromagnetic material by
domain alignment (b)
Magnetization by domain
growth

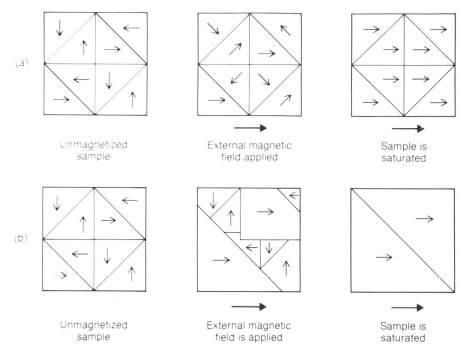

FIG. 21–11 (a) Magnetization of a ferromagnetic material by domain alignment (b) Magnetization by domain growth

(a)

Unmagnetized
sample　　　　External magnetic
field applied　　　　Sample is
saturated

(b)

Unmagnetized
sample　　　　External magnetic
field is applied　　　　Sample is
saturated

Ferromagnetism disappears at high temperatures

Why electromagnets have iron cores

magnet" sufficiently will cause it to become demagnetized. Hammering a permanent magnet also tends to disturb the alignment of atomic fields, though some ferromagnetic materials are able to retain their magnetization despite almost any mechanical disturbance.

Iron, unlike steel (which is an alloy, or mixture, of iron with carbon and other elements), tends to lose its magnetization when an external magnetic field is removed. Hence if an iron rod is placed inside a solenoid, we have a very strong magnet that can be turned on and off just by switching the current in the solenoid on and off. Such an *electromagnet* is much stronger than the solenoid itself and can be stronger than a permanent magnet as well; also, unlike a permanent magnet, its field can be controlled at will by adjusting the current in the solenoid. Electromagnets are among the most widely used electrical devices. They range in size from the tiny one in a telephone receiver that causes a steel plate to vibrate and thus produce the sounds we hear to the giant electromagnets used to pick up automobiles in scrap yards.

21–5　HYSTERESIS

The magnetic field inside a solenoid with no core is, according to Eq. (21–5), $B_0 = \mu_0(N/l)I$. When an iron core is inside the solenoid, the magnetic field increases to $B = \mu(N/l)I$, where μ is the permeability of the iron. Let us examine the magnetic behavior of annealed (soft) iron by comparing B and B_0 as the current I is increased.

Magnetic behavior of annealed iron

Figure 21–12 is a plot of B (the total field, including the contribution of the iron core) against B_0 (the field due to the solenoid itself, without the core). Because B is

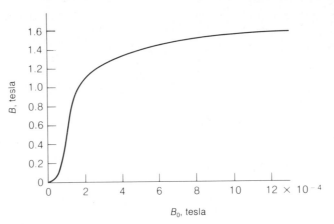

FIG. 21–12 The magnetization curve of annealed iron. B_0 is the magnetic field without the iron present, and B is the total field including the effect of the iron.

so much greater than B_0, the respective scales on the graph are different. At first, B_0 is too weak to align the domains in the iron to any great extent, so B increases slowly with increasing B_0. When B_0 is made stronger, it is more effective in aligning the domains, and B rises rapidly until it is over 5500 times more than B_0. Eventually, when the domains are virtually all aligned with B_0, B levels off to a nearly constant saturated value. Clearly μ, the permeability of the iron, is not a fixed quantity but varies with the magnetizing field B_0.

A ferromagnetic material tends to retain a degree of magnetization even when the magnetic field that originally aligned its domains is removed. Hence μ does not even have a fixed value at a given B_0 but may take on different values depending on its past history. This phenomenon is called *hysteresis*.

Hysteresis refers to the magnetic "memory" of a ferromagnetic material

Suppose that we place a sample of unmagnetized iron in a coil and vary the current from zero to a maximum in one direction, down through zero to a maximum in the other direction, back through zero to the first maximum, and so on. Figure 21–13 is a plot of B versus B_0 for this cycle. When we first turn on the current, the B-B_0 curve is the same as that of Fig. 21–12. At the point b we reduce the current, so that B_0 drops, but now B does *not* retrace its original path. From b to c the values of B are higher than they were from a to b at corresponding values of B_0 owing to the magnetic "memory" of ferromagnetic materials. At c there is no current in the coil and $B_0 = 0$, yet B nevertheless has the magnitude B_c; the iron sample is now permanently magnetized.

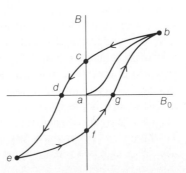

FIG. 21–13 Hysteresis loop.

FIG. 21–14 Successive hysteresis loops during the demagnetization of a ferromagnetic sample.

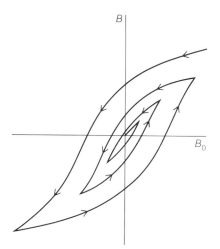

Reversing the direction of B_0 does not at first reverse B but merely reduces it, until finally, at d, B_0 is sufficiently negative to bring B to $B = 0$. A further increase in $-B_0$ takes the B-B_0 curve to e, where B and B_0 are equal in magnitude and opposite in sign to what they were at b. When B_0 is again brought to $B_0 = 0$, B is at f, where $B_f = -B_c$. Increasing B_0 in the positive sense returns B_0 to point b, but along a curve on which B is always less than it was from a to b. Further cycles simply retrace the curve $bcdefgb$.

Demagnetization

From Fig. 21–14 it would seem that, once a ferromagnetic material is magnetized, it cannot be demagnetized (without heating) to make $B = 0$ when $B_0 = 0$. There is a procedure that can do this, however. What is done is to carry a magnetized sample through a succession of hysteresis curves, each with a smaller $B_{0\text{max}}$. Figure 21–15 shows that, as $B_{0\text{max}}$ is brought closer and closer to zero, the curve approaches the origin where $B_0 = B = 0$. The method by which jewelers demagnetize wristwatches is based upon this procedure: The watch is placed in a coil connected to a source of alternating current, and the current is gradually decreased to zero.

Retentivity and coercive force

The shape of the hysteresis loop of a particular material provides important information about its magnetic behavior. For instance, what properties should we look for in choosing a material for a permanent magnet? Obviously, we want the residual magnetization to be a maximum, which means that the point c in Fig. 21–14 should represent as high a value of B as possible. The value of B corresponding to this point is called the *retentivity* of the material involved. Thus we might regard an alloy of 98% Fe, 0.86% C, and 0.9% Mn as ideal, since its retentivity of 0.95 T is quite high. However, a permanent magnet should also be able to keep its magnetization despite stray magnetic fields from nearby currents, a factor as significant as its retentivity. The quantity in a hysteresis curve that is a measure of the ability of a material to resist changes in its magnetization is the value of B_0 at the point d, which is known as the *coercive force*. The steel alloy mentioned above happens to have a coercive force of only about 0.0045 T, which is not particularly large. The alloy Alnico 2 (55% Fe, 10% Al, 17% Ni, 12% Co, 6% Cu) has a retentivity of 0.76 T, somewhat lower than that of the steel alloy above, but its coercive force is about 0.053 T, more than ten times greater. Which alloy is best in a particular permanent magnet application must be decided on the basis of a comparision of both retentivity and coercive force.

Under other circumstances the area enclosed by the hysteresis loop is of interest, since this area is proportional to the energy dissipated as heat when a sample is carried through an entire magnetization cycle. The heat may be thought of as arising from the work done by the minute magnetic elements within the material as they shift their directions. In a transformer, for instance, a current whose direction is periodically reversed passes through a coil with a ferromagnetic core. A core with a large hysteresis loop becomes very hot in a transformer, evidence of inefficiency since it is electric energy that is being wasted. Hence a narrow hysteresis loop is desirable in selecting alloys for transformer cores.

The area of a hysteresis loop represents dissipated energy

21–6 FORCE ON A MOVING CHARGE

The defining property of a magnetic field is its ability to exert a force on an electric current, whether it is a current in a wire, a moving charged particle, or an atomic current as in an iron bar. This property has been exploited both technologically, as in the electric motor, and scientifically, as in such research tools as the mass spectrometer and various kinds of particle accelerators. The law that governs the magnetic force on a current element in a magnetic field is a straightforward one, and in the remainder of this chapter we shall see how it is applied in a number of situations.

According to Eq. (21–1), the force on a particle of charge Q and velocity $\mathbf{v}$ in the magnetic field $\mathbf{B}$ has the magnitude

$$F = QvB \sin \theta \qquad \textit{Force on moving charge} \quad (21\text{–}6)$$

where θ is the angle between $\mathbf{v}$ and $\mathbf{B}$ (Fig. 21–2(b)). The direction of the force $\mathbf{F}$ is given by the right-hand rule shown in Fig. 21–2(a).

The work done by a force on a body upon which it acts depends upon the component of the force in the direction the body moves. Because the force on a charged particle in a magnetic field is perpendicular to its direction of motion, the force does no work on it. Hence the particle keeps the same speed v and energy it had when it entered the field, even though it is deflected. On the other hand, the speed and energy of a charged particle in an *electric* field are always affected by the interaction between the field and the particle, unless $\mathbf{v}$ is perpendicular to $\mathbf{E}$.

A constant magnetic field does no work on a charged particle

A particle of charge Q and velocity $\mathbf{v}$ that is moving in a uniform magnetic field so that $\mathbf{v}$ is perpendicular to $\mathbf{B}$ experiences a force of magnitude

$$F = QvB \qquad (\mathbf{v} \perp \mathbf{B}) \qquad (21\text{–}7)$$

since $\sin 90° = 1$. This force is directly perpendicular to both $\mathbf{v}$ and $\mathbf{B}$, so the particle travels in a circular path (Fig. 21–15).

A charged particle moving perpendicular to a magnetic field follows a circular path

To find the radius R of the circular path of the charged particle, we note that the magnetic force QvB provides the particle with the centripetal force mv^2/R that keeps it moving in a circle. Equating the magnetic and centripetal forces yields

FIG. 21–15 The path of
a charged particle moving
perpendicular to a uniform
magnetic field is a circle.

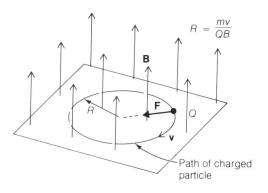

FIG. 21–15 The path of
a charged particle moving
perpendicular to a uniform
magnetic field is a circle.

$$R = \frac{mv}{QB}$$

Path of charged
particle

$$F_{\text{magnetic}} = F_{\text{centripetal}}$$

$$QvB = \frac{mv^2}{R}$$

and so, solving for R, we obtain

$$R = \frac{mv}{QB}$$ *Orbit radius in magnetic field*   (21–8)

The radius of a charged particle's orbit in a uniform magnetic field is directly proportional to its momentum mv and inversely proportional to its charge and to the magnitude of the field. The greater the momentum, the larger the circle, and the stronger the field, the smaller the circle.

   A charged particle moving parallel to a magnetic field experiences no force and is not deflected; the same particle moving perpendicular to the field follows a circular path. Hence a charged particle whose direction of motion is oblique with respect to **B** follows a helical (corkscrew) path. If we call $v_\parallel$ the component of the particle's velocity **v** that is parallel to **B** and $v_\perp$ the component of **v** perpendicular to **B,** then the motion of the particle is the resultant of a forward motion at the velocity $v_\parallel$ and a circular motion perpendicular to this whose radius is $mv_\perp/QB$ (Fig. 21–16).

   An extremely interesting effect occurs when a charged particle moving in a magnetic field approaches a region where the field becomes stronger. The magnetic lines of force that describe such a field converge (come together), since their spacing is

FIG. 21–16 A charged
particle that has velocity
components both parallel
and perpendicular to a
magnetic field follows a
helical path in the field.

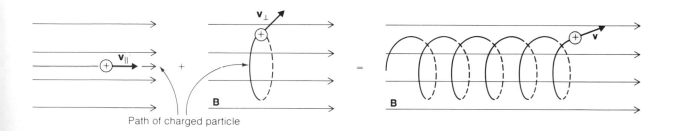

Path of charged particle

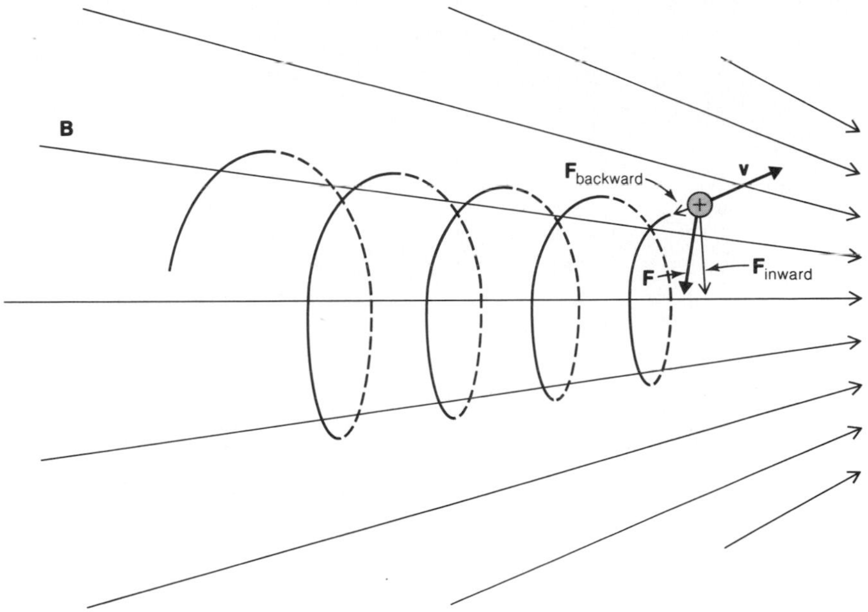

FIG. 21–17 The principle of the magnetic mirror.

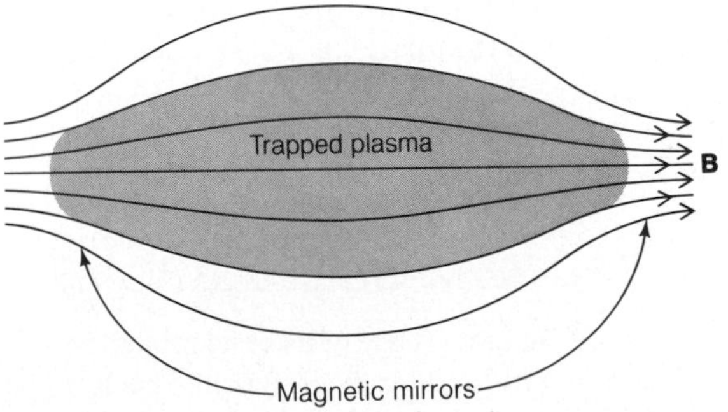

FIG. 21–18 A "magnetic bottle."

always proportional to the magnitude of the field they describe. The force the particle experiences now has a backward component as well as the inward component that leads to its helical path, as shown in Fig. 21–17. The backward force may be strong enough and extend over a long enough distance to reverse the particle's direction of motion. A converging magnetic field can thus act as a *magnetic mirror.*

Magnetic mirrors are found both in the laboratory and in nature. In the laboratory a pair of them can be used as a "magnetic bottle," as in Fig. 21–18, to contain a hot plasma (highly ionized gas) in research on thermonuclear fusion. If a solid container were used, contact with its walls would contaminate the plasma and also cool it so that the ions would not have enough energy to interact. Magnetic bottles of this kind are somewhat leaky, because ions moving along the axis of a magnetic mirror experience no backward force and hence are able to escape.

Magnetic bottle

FIG. 21–19 Protons and electrons are trapped by the earth's magnetic field.

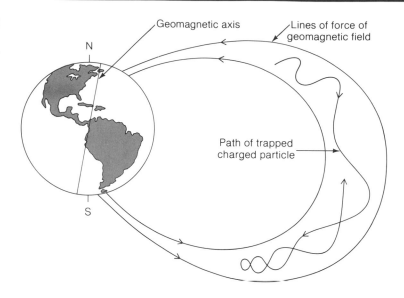

Geomagnetic axis

Lines of force of geomagnetic field

N

Path of trapped charged particle

S

Magnetosphere

The earth's magnetic field traps electrons and protons from space in the *magnetosphere*, a giant doughnut-shaped magnetic bottle that surrounds the earth and extends from about 1000 km above the equator out to perhaps 65,000 km. The magnetosphere contains large numbers of particles with relatively high energies (100 MeV, for instance). Figure 21–19 shows a typical particle path in the magnetosphere.

21–7   THE MASS SPECTROMETER

A mass spectrometer measures atomic masses

The mass of an atom is one of its basic properties and an accurate knowledge of atomic masses provides considerable insight into nuclear phenomena. A variety of instruments called *mass spectrometers* have been devised to measure atomic masses, and we shall consider the operating principles of the particularly simple one shown in Fig. 21–20.

The first step in the operation of this spectrometer is to produce ions of the substance under study. If the substance is a gas, ions can be formed by electron bombardment; if it is a solid, it is often incorporated in an electrode that is used as one terminal of an electric arc discharge. The ions emerge from their source through a slit with the charge $+e$ and are then accelerated by an electric field. (Ions with other charges are sometimes present but are easily taken into account.)

Velocity selector

When the ions enter the spectrometer, as a rule they are traveling in slightly different directions with slightly different speeds. A pair of slits serves to collimate the beam, that is, to eliminate those ions not moving in the desired direction. Then the beam passes through a *velocity selector.* The velocity selector consists of uniform electric and magnetic fields that are perpendicular to each other and to the beam of ions. The electric field **E** exerts the force $F_{\text{electric}} = eE$ on the ions to the right, whereas the magnetic field **B** exerts the force $F_{\text{magnetic}} = evB$ on those to the left. In order for

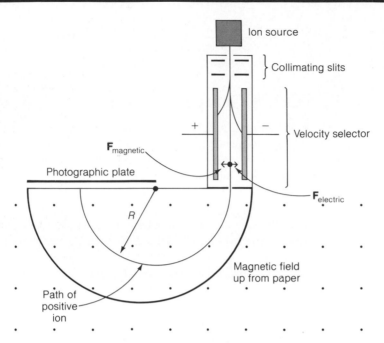

FIG. 21–20 A simple mass spectrometer. Modern instruments use electrical ion detectors and more complicated fields.

Ion source

Collimating slits

Velocity selector

$\mathbf{F}_{magnetic}$

Photographic plate

$\mathbf{F}_{electric}$

R

Magnetic field up from paper

Path of positive ion

an ion to reach the slit at the far end of the velocity selector it must suffer no deflection inside the selector, which means that the condition for escape is

$$F_{electric} = F_{magnetic}$$

$$eE = evB$$

Hence the ions that escape all have the speed

$$v = \frac{E}{B}$$

Once past the velocity selector the ions enter a uniform magnetic field and follow circular paths whose radius is given by Eq. (21–8). Since v, e, and B are known, a measurement of R yields a value for m, the ion mass.

Example The velocity selector of a mass spectrometer consists of an electric field of $E = 40{,}000$ V/m perpendicular to a magnetic field of $B = 0.0800$ T. The same magnetic field is used to deflect the ions that have passed through the velocity selector. Ions of a certain isotope of lithium are found to have radii of curvature in the magnetic field of 390 mm. What is their mass?

Solution The speed of the ions is

$$v = \frac{E}{B} = \frac{4.00 \times 10^4 \text{V/m}}{8.00 \times 10^{-2}\text{T}} = 5.00 \times 10^5 \text{m/s}$$

From Eq. (21–8) we obtain

$$m = \frac{QBR}{v} = \frac{(1.60 \times 10^{-19}\,\mathrm{C})(8.00 \times 10^{-2}\,\mathrm{T})(0.390\,\mathrm{m})}{5.00 \times 10^5\,\mathrm{m/s}} = 9.98 \times 10^{-27}\,\mathrm{kg}\quad\blacksquare$$

21–8   FORCE ON A CURRENT

Since an electric current is a flow of charge we would expect a current-carrying wire to be affected by a magnetic field in a manner similar to that of a moving charged particle. According to Eq. (21–6), the force on a charge Q whose velocity is $\mathbf{v}$ when it is in the magnetic field $\mathbf{B}$ has the magnitude

$$F = QvB \sin \theta \qquad\qquad (21\text{--}6)$$

where θ is the angle between $\mathbf{v}$ and $\mathbf{B}$. What we must do to find an expression for the force on a current is to replace the Qv of the above formula with the quantity appropriate for a current.

Equivalence of current element and moving charge

Figure 21–21(a) shows a particle of charge Q and speed v. In the time t the particle travels the distance

$$\Delta L = vt$$

and while it does so it is equivalent to a current of

FIG. 21–21 The force on a charge Q moving with the speed v in a magnetic field is the same as that on a wire ΔL long carrying the current I, where $I\,\Delta L = Qv$.

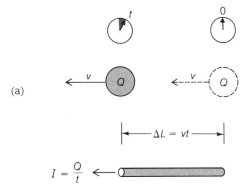

(a)

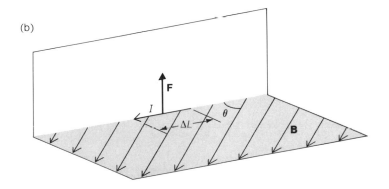

(b)

$$I = \frac{Q}{t}$$

Hence

$$v = \frac{\Delta L}{t} \quad \text{and} \quad Q = It$$

so that

$$Qv = I \, \Delta L$$

We conclude that the force on an element ΔL long of current I when it is in a magnetic field **B** has the magnitude

$$F = I \, \Delta LB \sin \theta \qquad \textit{Force on current element} \quad (21–9)$$

where θ is the angle between the direction of I and that of **B** (Fig. 21–21(b)).

    There are two simple ways to determine the direction of the force on a current element in a magnetic field. Both give the same result, of course, and deciding which one to use in a particular case is largely a matter of personal preference. The first is essentially the same as the right-hand rule used for a moving charge in Section 21–2, except that now the thumb points in the direction of the current. This version of the rule is illustrated in Fig. 21–22.

    Another method for finding the direction of **F** is based upon the pattern of lines of force around a current in a magnetic field. Figure 21–23(a) shows the lines of force of a uniform field **B** in the absence of a current, and (b) shows the lines of force around a wire carrying a current I in the absence of a magnetic field of external origin. Since the current is into the paper, the lines of force are concentric circles in the clockwise sense. When field (a) is added vectorially to field (b), the resulting pattern of lines of force is like that shown in (c): The lines are closer together in the region above the wire where the field of the current is in the same direction as **B** and farther apart under the wire where the field of the current is opposite to **B**. *The direction of the force on the current element is from the region of strong field to the region of weak field,* as though the lines of force were rubber bands that try to straighten out when distorted by the presence of the current.

Right-hand rule for force on a current

Lines-of-force rule for force on a current

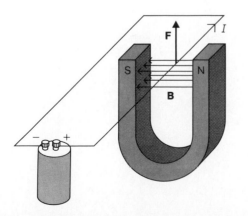

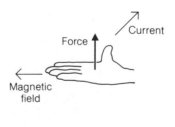

FIG. 21–22 The right-hand rule for the direction of the force on a current-carrying wire in a magnetic field.

FIG. 21–23 The
direction of the force on a
current element in a
magnetic field is from the
region of strong field to
the region of weak field.

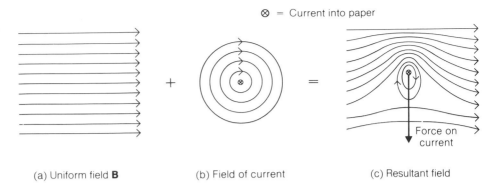

⊗ = Current into paper

(a) Uniform field **B**   (b) Field of current   (c) Resultant field

While the pictorial method for establishing the direction of **F** is easy to use and appeals to the intuition, it must be kept in mind that lines of force do not in fact exist but are only a device for visualizing the magnitude and direction of a force field. It is the field itself that exists in space as a continuous property of the region it occupies, not a series of strings. However, despite the artificial nature of lines of force, they can be very helpful in representing various aspects of the interaction between magnetic fields and electric currents, and we can freely make use of them for this purpose.

Example   A wire carrying a current of 100 A due west is suspended between two towers 50 m apart. The lines of force of the earth's magnetic field enter the ground there in a northerly direction at a 45° angle; the magnitude of the field at that location is 5×10^{-5} T. Find the force on the wire exerted by the earth's field.

Solution   The wire is perpendicular to **B,** and so the magnitude of the force is

$$F = I \, \Delta L B \sin \theta = (100 \text{ A})(50 \text{ m})(5 \times 10^{-5} \text{ T})(\sin 90°) = 0.25 \text{ N}$$

Since 1 N = 0.225 lb, the force is 0.056 lb, about an ounce. By either of the methods described above, the force acts downward at a 45° angle with the ground toward the south (Fig. 21–24).   ■

FIG. 21–24

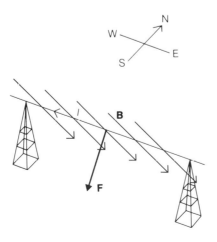

21–9 FORCE BETWEEN TWO CURRENTS

Every current is surrounded by a magnetic field, and because of this nearby currents exert forces upon one another. The forces are magnetic in origin; a current-carrying wire has no net electric charge and hence cannot interact electrically with another such wire.

Figure 21–25 shows two parallel wires a distance s apart that carry the currents I_1 and I_2 respectively. The magnetic fields a distance s from each of the wires are, from Eq. (21–2),

$$B_1 = \frac{\mu I_1}{2\pi s} \qquad B_2 = \frac{\mu I_2}{2\pi s}$$

Magnetic fields of the currents

The fields are perpendicular to the wires, which means that $\theta = 90°$ and $\sin \theta = 1$, and therefore the force F_{12} on a length L of current 1 exerted by the magnetic field of current 2 is

$$F_{12} = I\,\Delta LB \sin\theta = \frac{\mu I_1 I_2}{2\pi s}L \qquad\qquad (21\text{–}10)$$

Force on current 1 due to field of current 2

The force F_{21} on current 2 that is exerted by the magnetic field of current 1 has exactly the same magnitude, and we may express both of them in the alternative form

$$\frac{F}{L} = \frac{\mu I_1 I_2}{2\pi s} \qquad \textit{Force between long parallel currents} \quad (21\text{–}11)$$

where F/L is the *force per unit length* each wire exerts on the other by virtue of its magnetic field. From the pattern of lines of force around each wire, it is clear that the forces are always opposite in direction and hence obey Newton's third law of motion, as they must. The forces are attractive when the currents are in the same direction and repulsive when they are in opposite directions.

Parallel currents attract, antiparallel currents repel

Example The cables that connect the starting motor of a car with its battery are 1 cm apart for a distance of 40 cm. Find the forces between the cables when the current in them is 300 A.

Solution The currents in the cables are opposite in direction, and hence the forces are repulsive. Their magnitudes, taking $\mu = \mu_0$, are

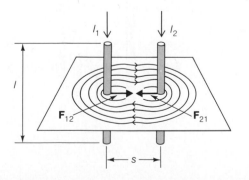

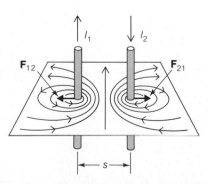

FIG. 21–25 Equal and opposite forces are exerted by parallel currents on each other. The forces are attractive when the currents are in the same direction, repulsive when they are in opposite directions.

$$F = \frac{\mu_0 I^2}{2\pi s} L = \frac{(4\pi \times 10^{-7}\,\mathrm{T \cdot m/A})(300\,\mathrm{A})^2(0.4\,\mathrm{m})}{2\pi \times 10^{-2}\,\mathrm{m}} = 0.72\,\mathrm{N}$$

which is 0.16 lb, a perceptible amount. ∎

Definition of ampere

Equation (21–11) is used to define the ampere: An ampere is that current in each of two parallel wires 1 m apart in free space that produces a force on each wire of exactly 2×10^{-7} N per meter of length. (Thus $\mu_0 = 4\pi \times 10^{-7}\,T \cdot m/A$.) In turn, the coulomb is defined in terms of the ampere as that amount of charge transferred per second by a current of 1 A. The ampere is chosen as the primary electrical unit instead of the coulomb because it can be defined in terms of a more direct experiment than would be possible with the coulomb.

21–10 TORQUE ON A CURRENT LOOP

A straight current-carrying wire is acted upon by a force when it is in a magnetic field, provided that it is not parallel to the direction of **B**. A loop of current in a uniform magnetic field experiences no net force, but instead a torque occurs that tends to rotate the loop to bring its plane perpendicular to **B**. This is the principle that underlies the operation of all electric motors, from the tiniest one in a clock to the many-thousand-horsepower giant in a locomotive.

FIG. 21–26 (a) A current-carrying wire loop whose plane is parallel to a magnetic field experiences a torque. (b) If the plane of the loop is perpendicular to the magnetic field, there is no torque on the loop. In both cases there is no net force on the loop.

Let us examine the forces on each side of a rectangular current-carrying wire loop whose plane is parallel to a uniform magnetic field **B,** as in Fig. 21–26(a). The sides A and C of the loop are parallel to **B** and so there is no magnetic force on them. Sides B and D are perpendicular to **B**, however, and each therefore experiences a force. To find the directions of the forces on B and D we can use the right-hand rule: With the

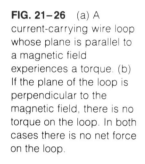

⊙ Current out of paper
⊗ Current into paper

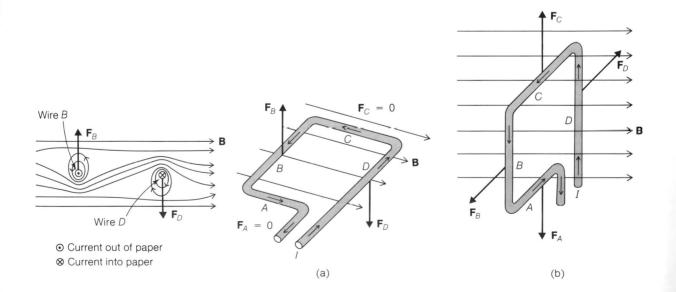

(a)   (b)

fingers of the right hand in line with B and the outstretched thumb in line with $\mathbf{I}$, the palm faces the same way as $\mathbf{F}$. What we find is that $\mathbf{F}_B$ is opposite in direction to $\mathbf{F}_D$. The same conclusion can be obtained by examining the pattern of lines of force.

The forces $\mathbf{F}_B$ and $\mathbf{F}_D$ are the same in magnitude, so there is no net force on the current loop. But $\mathbf{F}_B$ and $\mathbf{F}_D$ do not act along the same line, and hence they exert a torque on the loop that tends to turn it. This is a perfectly general conclusion that holds for a current loop of any shape in a magnetic field. **Origin of torque on a current loop**

If the plane of the loop is perpendicular to the magnetic field instead of parallel to it, there is neither a net force nor a net torque on it. This is easy to verify from Fig. 21–26(b), bearing in mind the right-hand rule for the direction of the force on each side of the loop. Evidently $\mathbf{F}_A$ and $\mathbf{F}_C$ cancel each other out, and $\mathbf{F}_B$ and $\mathbf{F}_D$ also cancel each other out. There is no torque now because $\mathbf{F}_A$ and $\mathbf{F}_C$ have the same line of action, and $\mathbf{F}_B$ and $\mathbf{F}_D$ have the same line of action.

The above results can be summarized by saying that

A current loop in a magnetic field always tends to turn so that its plane becomes perpendicular to the field. **Current loop in magnetic field**

A detailed calculation shows that the torque τ on a wire loop in a magnetic field is given by

$$\tau = IAB \cos \theta \qquad \text{Torque on current loop} \quad (21–12)$$

where

   I = current in loop

   A = area enclosed by loop

   B = magnetic field

   θ = angle between plane of loop and direction of $\mathbf{B}$

Although a rectangular loop is shown in Fig. 21–26, the loop's shape does not matter, only its area A.

If a coil of N loops is in the magnetic field instead of a single loop, the torque on it is

$$\tau = INAB \cos \theta \qquad \text{Torque on coil} \quad (21–13)$$

The value of $\cos \theta$ is 1 when $\theta = 0$, and the torque has its maximum value of $INAB$ when the plane of the coil is parallel to $\mathbf{B}$, as in Fig. 21–26(a). When $\theta = 90°$, the plane of the coil is perpendicular to $\mathbf{B}$, as in Fig. 21–26(b), and $\cos \theta = \cos 90° = 0$, $\tau = 0$. The quantity INA is sometimes called the *magnetic moment* of the coil.

21–11   GALVANOMETERS AND MOTORS

We have seen that a current-carrying wire loop tends to rotate in a magnetic field. The *galvanometer* uses this behavior to measure current. Figure 21–27 shows the basic construction of a galvanometer. A U-shaped permanent magnet is used to provide a magnetic field, and between its poles is a small coil wound on an iron core to increase **The galvanometer**

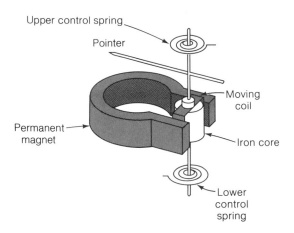

FIG. 21-27 The construction of a common type of galvanometer.

the torque developed when the unknown current is passed through it. The coil assembly is held in place by two bearings that permit it to rotate, and a pair of hairsprings keeps the pointer at 0 when there is no current in the coil.

When a current flows, there is a torque on the coil because of the interaction between the current and the magnetic field, and the coil rotates as far as it can against the opposing torque of the springs. The more the current, the stronger the torque, and the farther the coil turns. The restoring torque of the hairsprings is proportional to the angle through which they are twisted, and as a result the deflection of the pointer is directly proportional to the current I in the coil.

Galvanometers of the above type can be constructed that are able to respond to currents of as little as 0.1 microampere (10^{-7} A), though ordinary commercial meters are less sensitive. Even greater sensitivity can be attained if the moving coil is suspended by a thin wire to which a small mirror is attached: Bearing friction is avoided in this way, and the mirror deflects a light beam so that the "pointer" may be a meter or more long instead of a few centimeters. Laboratory galvanometers like this can be used to measure currents of 10^{-10} A.

How an electric motor achieves continuous rotation

The torque that a magnetic field exerts on a current loop disappears when the loop turns so that its plane is perpendicular to the field direction. If the loop swings past this position, the torque on it will be in the opposite sense and will return the loop to the perpendicular orientation. In order to construct a motor capable of continuous rotation, then, the current in the loop must be automatically reversed each time it turns through 180°. The method by which this reversal is accomplished is shown in Fig. 21-28. The current is led to the loop by means of graphite rods called *brushes* which press against a split ring called a *commutator*. As the loop rotates, the current is reversed twice per turn as the commutator segments make contact alternately with the brushes. The torque is always in the same direction, except at the moments of switching when it is zero because the loop is perpendicular to the field. However, the angular momentum of the loop carries it past this point, and it can continue to turn indefinitely.

While actual direct-current electric motors, such as the starter motor of a car, are the same in principle as the simple device of Fig. 21-28, they employ a number of methods to increase the available torque. Electromagnets rather than permanent magnets provide the field, and there are six or more different coils with many turns each on a slotted iron core called an *armature,* instead of a single loop (Fig. 21-29). A

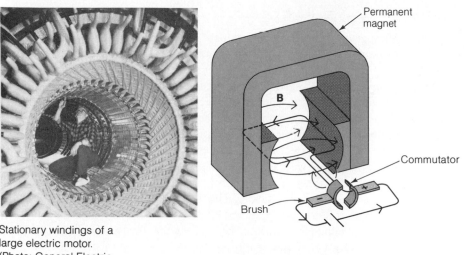

FIG. 21—28 A simple direct-current electric motor. The commutator automatically reverses the current in the rotating loop twice per rotation so that the torque will stay in the same direction.

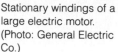

Stationary windings of a large electric motor. (Photo: General Electric Co.)

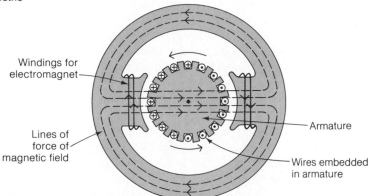

FIG. 21—29 Actual direct-current electric motors employ various means to increase the available torque.

commutator with a pair of segments for each coil is provided so that only those coils approximately parallel to the magnetic field receive current at any time, which means that maximum torque is developed continuously.

Electric energy for industrial and domestic purposes is usually transmitted by *alternating current* (ac) whose direction periodically reverses itself. In the United States the frequency of ordinary alternating current is 60 Hz, which means that the current changes direction 120 times per second; this frequency is 50 Hz in much of the rest of the world. In alternating-current electric motors, commutators and brushes are not needed because the current itself does the required reversing, which makes such motors easier to build and more reliable than direct-current motors. In order to start the armature of an alternating-current electric motor turning, and to ensure that it begins to turn the desired way, an auxiliary stationary winding is used. Together with the operating winding, this creates a magnetic field whose direction rotates about the motor's axis. The rotating field pulls the armature around when the motor is switched on. As the armature approaches its normal running speed, the alternating magnetic fields of the operating windings are then able to keep it going, and (depending on the motor design) the starting winding may be cut out of the circuit.

Alternating-current motors

21–12   MAGNETIC POLES

It may seem strange that there has been no mention until now of the "magnetic poles" that figure in elementary discussions of magnetism. The reason is that all magnetic fields, including those of permanent magnets, originate in electric currents (or, more precisely, in moving charges); and all magnetic forces arise from interactions between currents and magnetic fields. To understand electromagnetic phenomena of any kind, it is necessary to start directly from these fundamental concepts.

The field of a bar magnet is like that of a solenoid

The magnetic field of a bar magnet is identical with that of a solenoid, as we saw in Fig. 21–9, because in a permanent magnet atomic current loops are aligned by their mutual interactions. Hence we can use the ideas of this chapter to understand the behavior of permanent magnets.

Because the external magnetic field of a bar magnet seems to originate in its ends, these are by custom called its *poles*. At one time it was believed that the poles were "magnetic charges" analogous to electric charges, and that the field of the magnet was due to these poles. This belief was reinforced by the repulsion of like poles and the attraction of unlike ones, phenomena that have their true explanation in the forces between parallel and antiparallel currents (Fig. 21–30).

Magnetic poles are basically different from electric charges

An important difference between magnetic poles and electric charges is that poles always occur in pairs of equal strength and opposite polarity. If a magnet is sawed in half, the poles are not separated but instead two new magnets are created, as in Fig. 21–31. Magnetic poles are therefore not really like electric charges, and all effects they are supposed to cause can be explained in terms of the behavior of current-carrying solenoids.

FIG. 21–30 Interactions between magnets can be traced to interactions between current loops.

Measurements of the earth's magnetic field show that it is very much like the field that would be produced by a powerful current loop whose center is a few hundred kilometers from the earth's center and whose plane is tilted by 11° from the plane of the earth's equator (Fig. 21–32). On the basis of geological evidence the earth is thought to have a core of molten iron 3470 km (2160 mi) in radius, a little over half the earth's radius, and there is no doubt today that electric currents in this core are

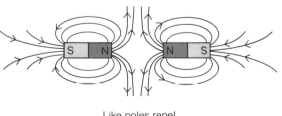

Like poles repel

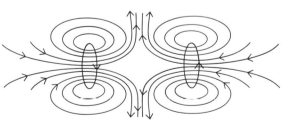

Parallel loops with opposite currents repel

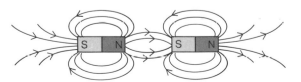

Unlike poles attract

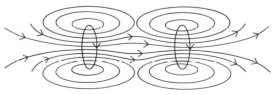

Parallel loops with similar currents attract

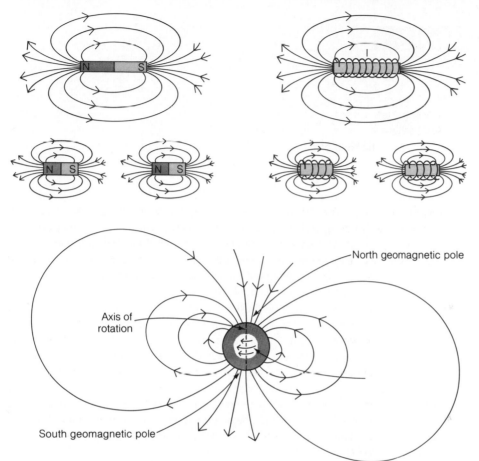

FIG. 21–31 Cutting a magnet in half produces two new magnets.

FIG. 21–32 The earth's magnetic field originates in currents in its core of molten iron. The magnetic axis is tilted by 11° from the axis of rotation.

North geomagnetic pole

Axis of rotation

South geomagnetic pole

responsible for the observed geomagnetic field; the details of how these currents came into being and how they are maintained are still uncertain, however.

As we saw, a current loop tends to rotate in a magnetic field until its axis is parallel to the field. A bar magnet, too, tends to rotate in a magnetic field until it is aligned with the field direction. Because of the earth's magnetic field, a magnet suspended by a string turns so as to line up in an approximately north-south direction. A compass consists of a pivoted magnetized iron needle together with a card that permits directions relative to magnetic north to be determined (Fig. 21–33). The end of a freely swinging magnet that points toward the north is called its north-seeking pole, usually shortened to just *north pole,* and the other end is its south-seeking pole, or *south pole.* Magnetic lines of force leave the north pole of a magnet and enter its south pole. (The north geomagnetic pole is thus in reality a south pole, and the south geomagnetic pole is a north pole; this has been a source of confusion for several hundred years.)

The mechanism by which a magnet or a solenoid attracts an iron object (or an object of any other ferromagnetic material such as cobalt or nickel) is very similar to the way in which an electric charge attracts an uncharged object. First the presence of the magnet causes the atomic magnets in the iron object to line up with its field by one

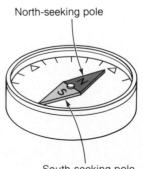

North-seeking pole

South seeking pole

A magnetic

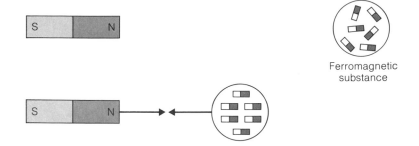

Ferromagnetic substance

of the mechanisms shown in Fig. 21–11, and then the attraction of opposite poles leads to a force on the object that draws it toward the magnet, as in Fig. 21–34.

The relay

A *relay* is a switch operated by an electromagnet. Figure 21–35 shows the construction of a typical relay. A small current in the solenoid is enough to pull the armature down against the spring and close the contact points. When the current in the solenoid is cut off, the spring pushes the armature up and breaks the secondary circuit.

The advantage of a relay is that it enables one circuit to control one or more others without an electrical connection between them. Thus a very small current (a few mA perhaps) can control a much larger one (many amperes). An example is the relay used to switch a heating system on and off. The relay's coil is actuated by a thermostat whose contacts close when the room temperature drops below a preset value. Because little current is needed to operate the relay, the thermostat can be quite small and light wiring is sufficient between it and the relay. Large contact points can be provided on the relay to carry the current needed by the oil burner and the pumps that circulate the hot air or hot water it produces. Some relays are made with multiple contacts so that several different secondary circuits can be switched on or off at the same time.

FIG. 21-35 A switch operated by an electromagnet is called a *relay*.

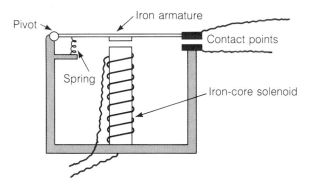

IMPORTANT TERMS

Charged particles in motion relative to an observer exert forces upon one another that are different from the electric forces they exert when at rest. These differences are by custom said to arise from **magnetic forces.** In reality, magnetic forces represent modifications of electric forces due to the motion of the charges involved.

A **magnetic field** exists wherever a magnetic force would act on a moving charged particle. The direction of a magnetic field **B** at a point is such that a charged particle would experience no force if it moves in that direction at the point. The magnitude of **B** is numerically equal to the force that would act on a charge of 1 C moving at 1 m/s perpendicular to **B.** The unit of magnetic field is the **tesla** (T), equal to 1 N/A · m.

When different substances are inserted in a current-carrying wire coil, the magnetic field in its vicinity changes. Those substances that lead to a great increase in B are called **ferromagnetic.** A **permanent magnet** is an object composed of ferromagnetic material whose atomic current loops have been aligned by an external current.

The **permeability** of a medium is a measure of its magnetic properties. Ferromagnetic substances have higher permeabilities than free space. The permeability of a ferromagnetic material in a given magnetizing field B_0 depends upon its past history as well as upon B_0, a phenomenon called **hysteresis.**

A **galvanometer** is a sensitive current-measuring device that is based upon the tendency of a coil carrying a current to rotate in a magnetic field until its axis is parallel to the field.

The ends of a permanent magnet are called its **poles.** Magnetic lines of force leave the **north pole** of a magnet and enter its **south pole.**

IMPORTANT FORMULAS

Magnetic field: $\quad B = \dfrac{F}{Qv \sin \theta}$

Field around long, straight current: $\quad B = \dfrac{\mu I}{2\pi s}$

Field at center of flat coil: $\quad B = \dfrac{\mu NI}{2r}$

Field in interior of solenoid: $\quad B = \mu \dfrac{N}{l} I$

Force on moving charge: $\quad F = QvB \sin \theta$

Orbit radius: $\quad R = \dfrac{mv}{QB}$

Force on current element: $\quad F = I\, \Delta L\, B \sin \theta$

Force between parallel currents: $\quad \dfrac{F}{L} = \dfrac{\mu I_1 I_2}{2\pi s}$

MULTIPLE CHOICE

1. All magnetic fields originate in
   a. iron atoms.
   b. permanent magnets.
   c. magnetic domains.
   d. moving electric charges.

2. An observer moves past a stationary electron. His instruments measure
   a. an electric field only.
   b. a magnetic field only.
   c. both electric and magnetic fields.
   d. any of the above, depending upon his speed.

3. Magnetic fields do not interact with
   a. stationary electric charges.
   b. moving electric charges.
   c. stationary permanent magnets.
   d. moving permanent magnets.

4. Magnetic lines of force provide a convenient way to visualize a magnetic field. Which of the following statements is not true?
   a. The path followed by an iron particle released in a magnetic field corresponds to a line of force.
   b. The path followed by an electric charge released in a magnetic field corresponds to a line of force.
   c. A compass needle in a magnetic field lines up parallel to the lines of force around it.
   d. Lines of force do not actually exist.

5. A drawing of the lines of force of a magnetic field provides information on
   a. the direction of the field only.
   b. the magnitude of the field only.
   c. both the direction and magnitude of the field.
   d. the source of the field.

6. In a drawing of magnetic lines of force, the stronger the field is,
   a. the closer together the lines of force are.
   b. the farther apart the lines of force are.
   c. the more nearly parallel the lines of force are.
   d. the more divergent the lines of force are.

7. The magnetic field near a strong permanent magnet might
be
    a. 10^{-9} T.   b. 10^{-5} T.
    c. 0.1 T.   d. 100 T.

8. A typical value for the earth's magnetic field at sea level
is
    a. 3×10^{-9} T.   b. 3×10^{-5} T.
    c. 3×10^{5} T.   d. 3×10^{9} T.

9. The magnetic field a distance d from a long, straight wire
is proportional to
    a. d.   b. d^2.
    c. $1/d$.   d. $1/d^2$.

10. A current is flowing east along a power line. If we
neglect the earth's field, the direction of the magnetic field
below it is
    a. north.   b. east.
    c. south.   d. west.

11. Inside a solenoid the magnetic field
    a. is zero.
    b. is uniform.
    c. increases with distance from the axis.
    d. decreases with distance from the axis.

12. The magnitude of the magnetic field inside a solenoid
of N turns does not depend upon
    a. the nature of the medium inside the solenoid.
    b. the solenoid's length.
    c. the solenoid's diameter.
    d. the current in the solenoid.

13. Which of the following statements about the magnetic
field of a solenoid with an iron core is (are) not *always* true?
    a. Increasing I increases B.
    b. Decreasing I decreases B.
    c. $B = 0$ when $I = 0$.
    d. Changing the direction of I changes the direction of
       B.

14. The magnetic field inside a current-carrying spring has
the magnitude B. If the spring is pulled out to double its
original length, the field inside is now
    a. $0.5\,B$.   b. B.
    c. $2B$.   d. $4B$.

15. When a magnetized iron bar is strongly heated, its magnetic field
    a. becomes weaker.   b. becomes stronger.
    c. reverses its direction.   d. is unchanged.

16. An electron enters a magnetic field parallel to **B**. The
electron's
    a. motion is unaffected.
    b. direction is changed.

    c. speed is changed.
    d. energy is changed.

17. An electron enters a magnetic field perpendicular to **B**.
The electron's
    a. motion is unaffected.
    b. direction is changed.
    c. speed is changed.
    d. energy is changed.

18. The right-hand rule for the direction of the force on a
charged particle in a magnetic field applies
    a. only to positive charges.
    b. only to negative charges.
    c. to both positive and negative charges.
    d. only when the particle is moving parallel to the field.

19. An ion moves in a circular orbit of radius R in a magnetic field. If the particle's speed is doubled, the orbit radius
will become
    a. $R/2$.   b. R.
    c. $2R$.   d. $4R$.

20. An ion that moves through crossed electric and mag-
netic fields (that is, so that $\mathbf{E} \perp \mathbf{B}$) perpendicular to both **E**
and **B** is not deflected when its speed is equal to
    a. EB.   b. E/B.
    c. B/E.   d. B/E^2.

21. A "magnetic mirror" that can reflect approaching charged
particles is a
    a. sheet of ferromagnetic material.
    b. uniform magnetic field.
    c. magnetic field whose lines of force converge.
    d. magnetic field whose lines of force diverge.

22. A current-carrying wire is in a uniform magnetic field
with the direction of the current the same as that of the field.
    a. There is a force on the wire that tends to move it
       parallel to the field.
    b. There is a force on the wire that tends to move it
       perpendicular to the field.
    c. There is a torque on the wire that tends to rotate it
       until it is perpendicular to the field.
    d. There is neither a force nor a torque on the wire.

23. A current-carrying loop in a magnetic field always tends
to rotate until the plane of the loop is
    a. parallel to the field.
    b. perpendicular to the field.
    c. either parallel or perpendicular to the field, depend-
       ing on the direction of the current.
    d. at a 45° angle with the field.

24. The nature of the force responsible for the operation of
an electric motor is

a. electric.

b. magnetic.

c. a combination of electric and magnetic.

d. either electric or magnetic depending on the design of the motor.

25. The magnetic field of a bar magnet most closely resembles the magnetic field of

a. a straight current-carrying wire.

b. a stream of electrons moving parallel to one another.

c. a current-carrying wire loop.

d. a horseshoe magnet.

26. A permanent magnet does not exert a force on

a. an unmagnetized iron bar.

b. a magnetized iron bar.

c. a stationary electric charge.

d. a moving electric charge.

27. The needle of a magnetic compass

a. is affected only by permanent magnets.

b. rotates continuously in the magnetic field of an electric current.

c. aligns itself parallel to a magnetic field.

d. aligns itself perpendicular to a magnetic field.

28. At different places on the earth's surface, the earth's magnetic field

a. is the same in direction and magnitude.

b. may be different in direction but not in magnitude.

c. may be different in magnitude but not in direction.

d. may be different in both magnitude and direction.

29. The earth's magnetic field is believed to arise from

a. a magnetized solid iron core.

b. magnetized iron deposits.

c. electric currents in a molten iron core.

d. electric currents in the ionosphere.

30. In a certain electric motor, wires that carry a current of 6 A are perpendicular to a magnetic field of 0.5 T. The force per cm on these wires is

a. 0.03 N.   b. 0.18 N.

c. 3 N.   d. 300 N.

31. The magnetic field 2 cm from a long, straight wire is 10^{-6} T. The current in the wire is

a. 0.01 A.   b. 0.1 A.

c. 1 A.   d. 10 A.

32. The magnetic field inside a 100-turn solenoid 2 cm long that carries a 10-A current is

a. 0.00063 T.   b. 0.00126 T.

c. 0.063 T.   d. 0.126 T.

33. Two parallel wires in free space are 10 cm apart and

carry currents of 10 A each in the same direction. The force each wire exerts on the other per meter of length is

a. 2×10^{-7} N, attractive.

b. 2×10^{-7} N, repulsive.

c. 2×10^{-4} N, attractive.

d. 2×10^{-4} N, repulsive.

EXERCISES

21–2 Magnetic Field

21–3 Magnetic Field of a Current

1. An experimenter is able to measure electric, magnetic, and gravitational fields. Which of these will be detected when (a) a proton moves past her, and (b) when she moves past a proton?

2. When you face the screen of a TV tube, in what direction is the magnetic field of the electron beam?

3. A current is flowing north along a power line. What is the direction of the magnetic field above it? Below it?

4. Approximately how strong would you expect the magnetic field near a powerful permanent magnet to be: 0.001 T? 0.1 T? 10 T? 1000 T?

5. At what distance from a long, straight wire carrying a current of 12 A does the magnetic field equal that of the earth, approximately 3×10^{-5} T?

6. A power line 10 m above the ground carries a current of 5000 A. Find the magnetic field of the current on the ground directly under the cable.

7. What should the current be in a wire loop 1 cm in diameter if the magnetic field at the center of the loop is to be 0.001 T?

8. A 10-turn circular coil of radius 2 cm carries a current of 0.5 A. Find the magnetic field at its center.

9. A long solenoid is wound with 30 turns/cm. What is the magnetic field inside the solenoid when the current is 3.8 A?

10. A solenoid 10 cm long is meant to have a magnetic field of 0.002 T inside it when the current is 3 A. How many turns are needed?

11. Two parallel wires 10 cm apart carry currents in the same direction of 8 A. Find the magnetic field midway between the wires.

12. Two parallel wires 10 cm apart carry currents in opposite directions of 8 A. Find the magnetic field midway between the wires.

13. Two parallel wires 20 cm apart carry currents in the same direction of 5 A. Find the magnetic field between the wires 5 cm from one of them and 15 cm from the other.

14. Two parallel wires 20 cm apart carry currents in opposite directions of 5 A. Find the magnetic field between the wires 5 cm from one of them and 15 cm from the other.

15. A tube 20 cm long is wound with 400 turns of wire in one layer and with 200 turns in a second layer in the opposite sense to the first. A current of 0.1 A is passed through the coils. What is the magnetic field in the interior of the tube?

16. A solenoid is wound with 18 turns/cm and carries a current of 1.2 A. Another layer of turns is wound over the solenoid with 10 turns/cm, and a current of 5.0 A is passed through the new coil, opposite to the direction of the current in the solenoid. What is the magnetic field in the interior of the solenoid?

17. A current of 0.5 A is passed through a solenoid wound with 20 turns/cm. A single loop of wire 1 cm in radius is bent around the middle of the solenoid. What should the current in this loop be in order for the magnetic field at its center to cancel out the field of the solenoid there?

18. A toroidal coil is a solenoid bent around into a doughnut shape with its axis in a circle of radius R. Find the magnetic field inside a toroidal coil of $R = 10$ cm that has 800 turns when the current is 0.5 A.

19. Two long, parallel wires a distance s apart each carry the current I in opposite directions. Verify that the magnetic field at a point equidistant from the wires and x away from the plane in which they lie is given by $2\mu_0 Is/\pi(4x^2 + s^2)$.

21–5   Hysteresis

20. Hysteresis loops for three ferromagnetic materials are shown below. (a) Which material would be most suitable for the core of an electromagnet used in a telephone receiver? (b) Which would be most suitable for use as a permanent magnet?

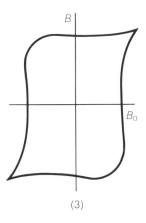

(3)

21. An alternating current whose variation with time follows a sine curve is sent through an iron-core solenoid. Does the resulting magnetic field also vary sinusoidally with time?

22. The alloy Alnico 2 has a retentivity of 0.76 T and a coercive force of 0.053 T. The corresponding figures for a certain cobalt steel alloy are 0.95 T and 0.023 T. (a) Which material would make the initially stronger permanent magnet? (b) Which would be more likely to retain its magnetization?

23. The table below lists corresponding values of B_0 and B for a type of carbon steel, where B_0 is the magnetic field of a current-carrying coil with no core and B is the magnetic field of the same coil with a carbon steel core. (a) Plot a graph of B versus B_0 for this material. (b) Plot a graph of the relative permeability μ/μ_0 of this material versus B_0. Approximately what is the maximum value of μ/μ_0, and at approximately what value of B_0 is this maximum reached?

B_0	B
0.0×10^{-5} T	0.0 T
4.2	0.2
6.3	0.4
7.7	0.6
9.0	0.8
12	1.0
20	1.2
36	1.4
75	1.6

21–6   Force on a Moving Charge

24. What aspect of the magnetic field of a "magnetic mirror" enables it to reflect approaching charged particles?

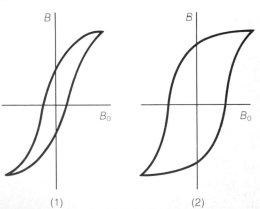

(1)   (2)

25. An electron is moving vertically upward when it enters a magnetic field directed to the east. In what direction is the force on the electron?

26. What path is followed by an electron sent into a current-carrying solenoid along its axis? What is its path when the electron enters the solenoid at an angle relative to its axis?

27. What is the radius of the path of an electron whose speed is 10^7 m/s in a magnetic field of 0.02 T when the electron's path is perpendicular to the field?

28. Compute the minimum radius of curvature in the earth's magnetic field at sea level, assuming $B = 3 \times 10^{-5}$ T, of (a) a proton whose speed is 2×10^7 m/s, and (b) an electron of the same speed.

29. An electron in a television picture tube travels at 3×10^7 m/s and is acted on both by gravity and by the earth's magnetic field. Which exerts the greater force on the electron?

30. A charge of $+ 10^{-6}$ C is moving at 500 m/s long a path parallel to a long, straight wire and 0.1 m from it. The wire carries a current of 2 A in the same direction as that of the charge. What is the magnitude and direction of the force on the charge?

31. A long, straight wire carries a current of 100 A. (a) What is the force on an electron traveling parallel to the wire, in the opposite direction to the current, at a speed of 10^7 m/s when it is 10 cm from the wire? (b) Find the force on the electron under the above circumstances when it is traveling perpendicularly toward the wire.

32. A charge of $+ 2 \times 10^{-6}$ C is moving at 10^3 m/s at a distance of 12 cm away from a straight wire carrying a current of 4 A. Find the magnitude and direction of the force on the charge when it is moving parallel to the wire (a) in the same direction as the current, and (b) in the opposite direction.

33. An electron is moving at 6×10^7 m/s at a distance of 5 cm from a long, straight wire carrying a current of 40 A. Find the magnitude and direction of the force on the electron when it is moving parallel to the wire (a) in the same direction as the current, and (b) in the opposite direction to the current.

34. The charge of Exercise 32 is moving in a direction perpendicular to the same wire. Find the magnitude and direction of the force on the charge when it is moving (a) toward the wire, and (b) away from the wire.

35. The electron of Exercise 33 is moving in a direction perpendicular to the same wire. Find the magnitude and direction of the force on the electron when it is moving (a) toward the wire, and (b) away from the wire.

36. According to the Bohr model, a hydrogen atom consists of an electron that circles a proton at a speed of 2.2×10^6 m/s in an orbit of radius 5.3×10^{-11} m. (a) Find the magnetic field the proton experiences as a result of the motion of the electron around it. (b) Find the magnetic field that would be needed by an electron with this speed to move in an orbit of this radius.

37. Prove that the time required for a charged particle in a magnetic field to make a complete revolution is independent of its speed and the radius of its orbit.

21–7 The Mass Spectrometer

38. An electron moving through an electric field of 500 V/m and a magnetic field of 0.1 T experiences no force. The two fields and the electron's direction of motion are all mutually perpendicular. What is the speed of the electron?

39. A velocity selector uses a magnet to produce a 0.05-T magnetic field and a pair of parallel metal plates 1 cm apart to produce a perpendicular electric field. What potential difference should be applied to the plates to permit singly charged ions of speed 5×10^6 m/s to pass through the selector?

40. A mass spectrometer employs a velocity selector consisting of a magnetic field of 0.0400 T perpendicular to an electric field of 50,000 V/m. The same magnetic field is then used to deflect the ions. Find the radius of curvature of singly charged lithium ions of mass 1.16×10^{-26} kg in this spectrometer.

21–8 Force on a Current

41. A current-carrying wire is in a magnetic field. (a) What angle should the wire make with **B** for the force on it to be zero? (b) What should the angle be for the force to be a maximum?

42. A current is flowing vertically upward when it enters a magnetic field directed to the east. In what direction is the force on the current?

43. A horizontal north-south wire 5 m long is in a 0.02 T magnetic field whose direction is northeast. (a) What is the magnitude and direction of the force on the wire when a 4-A current flows north in it? (b) When the same current flows south in it?

44. A vertical wire 2 m long is in a 10^{-2} T magnetic field whose direction is northeast. (a) What is the magnitude and direction of the force on the wire when a 5-A current flows upward in it? (b) When the same current flows downward in it?

45. In a typical loudspeaker, a permanent magnet creates a radial magnetic field in which a wire coil attached to the

apex of a paper cone can move perpendicular to the field. An alternating current in the coil causes the cone to oscillate and thereby produce sound waves. If the 40-turn coil of a certain loudspeaker is 8 mm in radius and is located in a 0.4-T magnetic field, find the force on the coil when the current in it is 0.05 A.

46. A copper wire whose linear density is 10 g/m is stretched horizontally perpendicular to the direction of the horizontal component of the earth's magnetic field at a place where the magnitude of that component is 2×10^{-5} T. What must the current in the wire be for its weight to be supported by the magnetic force on it? What do you think would happen to such a wire if this current were passed through it?

21–9   Force Between Two Currents

47. A stream of protons is moving parallel to a stream of electrons. Is the force between the two streams necessarily attractive? Explain.

48. A beam of protons, initially moving slowly, is accelerated to higher and higher speeds. What happens to the diameter of the beam during this process?

49. The electron beam in a cathode-ray tube is aimed parallel to a nearby wire carrying a current in the same direction. Is the electron beam deflected? If so, in what direction?

50. What, if anything, happens to the length of a helical spring when a current is passed through it?

51. A current is passed through a loop of highly flexible wire. What shape does the loop assume? Why?

52. A certain electric transmission line consists of two wires 4 m apart that carry currents of 10^4 A. If the towers supporting the wires are 200 m apart, how much force does each current exert on the other between the towers?

53. The parallel wires in a lamp cord are 2.0 mm apart. What is the force per meter between them when the cord is used to supply power to a 120-V, 200-W light bulb?

54. Two parallel wires 1 m long each of mass 20 g are suspended by strings 10 cm long from an overhead rod. When a current I is passed through the wires in opposite directions, the wires swing out so that the strings are 5° apart. Find I.

21–10   Torque on a Current Loop

55. A current-carrying wire loop is in a uniform magnetic field. Under what circumstances, if any, will there be no

torque on the loop? No net force? Neither torque nor net force?

56. What is the direction of the magnetic field that causes the wire loop below to experience a clockwise torque about the z-axis?

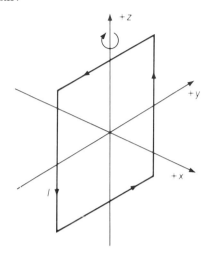

21–11   Magnetic Poles

57. Would you expect a compass to be more accurate near the equator or in the polar regions. Why?

58. Why is a piece of iron attracted by *either* pole of a magnet?

59. The magnetic field shown below is produced by which of the following: a north pole and a south pole; two north poles; two south poles; a south pole and an unmagnetized iron bar?

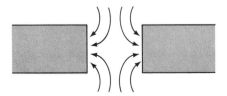

60. A bar magnet is placed in a uniform magnetic field so that it is parallel to the field with its north pole in the field direction. Sketch the lines of force of the resulting magnetic field **B** around the magnet. Indicate where **B** = 0.

61. The bar magnet of Exercise 60 is turned so that its north pole is now opposite to the direction of the uniform field. Sketch the lines of force of the resulting magnetic field **B** around the magnet for this situation. Indicate where **B** = 0.

ANSWERS TO MULTIPLE CHOICE

1. d	**10.** a	**18.** a	**26.** c
2. c	**11.** b	**19.** c	**27.** c
3. a	**12.** c	**20.** b	**28.** d
4. b	**13.** a,b,c,d	**21.** c	**29.** c
5. c	**14.** a	**22.** d	**30.** a
6. a	**15.** a	**23.** b	**31.** b
7. c	**16.** a	**24.** b	**32.** c
8. b	**17.** b	**25.** c	**33.** c
9. c			

22

ELECTROMAGNETIC INDUCTION

We have seen that, by causing a current to flow, an electric field is able to produce a magnetic field. Is there any way in which a magnetic field can produce an electric field? This problem was unsuccessfully tackled by many of the early workers in electricity and magnetism. Finally, in 1831, Michael Faraday in England and Joseph Henry in the United States independently discovered the phenomenon of electromagnetic induction. Two familiar applications of electromagnetic induction are the electric generator, which is the source of most electric power, and the transformer, which enables the emf of an alternating current to be increased or decreased easily.

22–1 ELECTROMAGNETIC INDUCTION

There is no current in a stationary wire in an unchanging magnetic field when the wire is not connected to a source of emf. What Faraday and Henry found is that *moving* the wire produces a current (Fig. 22–1). It does not matter whether the wire or the source of the magnetic field is being moved, provided that a component of the motion is

CHAPTER OBJECTIVES

Completing this chapter should enable you to:

1. Describe the origin of electromagnetic induction.

2. Find the emf induced in a wire moved through a magnetic field.

3. Determine the emf induced in a wire loop when the magnetic flux threading it changes.

4. Use Lenz's law to predict the direction of an induced current.

5. Describe how a generator converts mechanical energy into electric energy.

6. Understand how direct-current generators differ from

alternating-current generators and how an alternating current can be rectified into a direct current.

7. Describe the role of back emf in the operation of an electric motor.

8. Find the secondary emf and current of a transformer given its primary emf and current and the ratio of turns.

9. Explain how induction heating occurs.

10. Describe how sounds can be recorded and reproduced.

---

perpendicular to the field; the origin of the current lies in the *relative motion* between a conductor and a magnetic field. This effect is known as *electromagnetic induction*.

Faraday generalized his observations with the help of the notion of lines of force:

> **An electromotive force is produced in a conductor whenever it cuts across magnetic lines of force.**

Electromagnetic induction

It is this electromotive force that leads to the current that flows whenever there is relative motion between a conductor and a magnetic field. In fact, it is not even necessary for there to be actual motion of either a wire or of a source of magnetic field, because a magnetic field that changes in strength has moving lines of force associated with it.

Moving a wire *parallel* to a magnetic field does not give rise to a current: Electromagnetic induction occurs only when there is a component of the wire's velocity perpendicular to the lines of force.

No emf arises from motion parallel to a magnetic field

FIG. 22–1 (a) There is no current in a stationary wire in a magnetic field. (b) When the wire is moved across the field, a current is produced. Reversing the direction of motion also reverses the direction of the current.

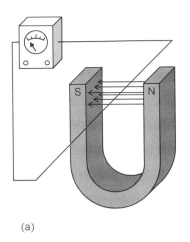

(a)

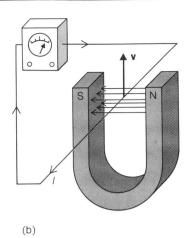

(b)

FIG. 22–2 Right-hand rule for the direction of an induced current.

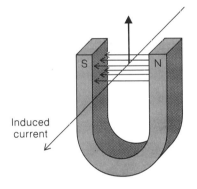

Induced current

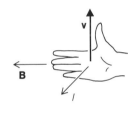

Origin of electromagnetic induction

It is not hard to show that electromagnetic induction is in accord with the force exerted by a magnetic field on a moving charge. As we saw in the previous chapter, a wire carrying a current is pushed sideways in a magnetic field because of the forces exerted on the moving electrons. In Faraday's and Henry's experiments electrons are also moved through a magnetic field, but now by shifting the entire wire. As before, the electrons are pushed to the side, and as a result move along the wire. The motion of these electrons along the wire is the electric current we measure.

Direction of induced current

The right-hand rule for the force on a moving positive charge in a magnetic field can also be used to give the direction of the induced current in a wire moving across a magnetic field. Hold your right hand so that the fingers point in the direction of **B** and the outstretched thumb is in the direction of **v.** The palm then faces in the same direction as the force on the positive charges in the wire, and hence in the direction of the conventional current (Fig. 22–2).

22–2 MOVING WIRE IN A MAGNETIC FIELD

When a wire moves through a magnetic field **B** with a velocity **v** that is not parallel to **B,** an electromotive force $\mathscr{E}$ comes into being between the ends of the wire. It is not

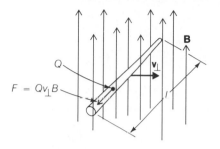

FIG. 22-3 The force on a charge Q within a conductor when the conductor is moved through a magnetic field is $Qv_\perp B$, where $v_\perp$ is the component of the conductor's velocity perpendicular to **B**.

hard on the basis of what we already know to determine the magnitude of $\mathscr{E}$ in terms of **B**, **v** and the length L of the wire. We shall assume that the wire, its direction of motion, and the magnetic field are all perpendicular to one another as in the previous diagrams.

According to Eq. (21–7) the force on a charge Q in a moving wire when the wire has the velocity component $\mathbf{v}_\perp$ perpendicular to a magnetic field **B** has the magnitude

$$F = Qv_\perp B$$

The direction of this force is along the wire, as in Fig. 22–3.

Let us consider a charge Q that moves from one end of the wire to the other under the influence of the magnetic force $Qv_\perp B$. The work done on the charge by the agent that moves the wire is

$$\text{Work} = \text{force} \times \text{distance}$$
$$W = FL = Qv_\perp BL$$

since the wire is L long. By definition the potential difference between two points is the work done in moving a unit charge between these points, so the potential difference between the ends of the wire is W/Q or

$$\mathscr{E} = BLv_\perp \qquad\qquad\qquad\qquad\qquad \textit{Motional emf} \quad (22\text{–}1)$$

This potential difference is the emf induced in the wire by its motion through the magnetic field (Fig. 22–4).

How large will the current that flows in the wire be? The answer is given by Ohm's law, since an induced emf is like any other emf in its ability to cause a current to flow. If the total resistance in the circuit is R, the resulting current will be equal to $I = \mathscr{E}/R$ as long as the emf exists.

Voltage across ends of wire moving in a magnetic field

Example Find the potential difference between the wing tips of a jet airplane induced by its motion through the earth's magnetic field. The total wing span of the airplane is 40 m and its speed is 300 m/s in a region where the vertical component of the earth's field has the magnitude 3×10^{-5} T (Fig. 22–5).

Solution Substituting the given values of $v_\perp$, L, and B in Eq. (22–1) for the induced emf yields

$$\mathscr{E} = BLv_\perp = (3 \times 10^{-5}\,\text{T})(40\,\text{m})\left(300\,\frac{\text{m}}{\text{s}}\right) = 0.36\,\text{V} \qquad\qquad \blacksquare$$

FIG. 22−4 The magnitude of the motional emf induced in a conductor moving through a magnetic field depends upon the magnetic field strength, upon the length and speed of the conductor, and upon the direction of **v** relative to **B**. The conductors shown here are oriented perpendicular to **B**; that is, perpendicular to the page.

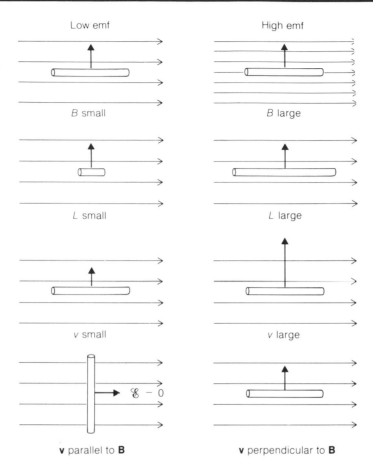

FIG. 22−5

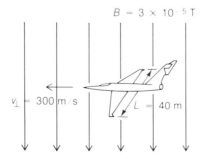

$B = 3 \times 10^{-5}$ T

$v_\perp = 300$ m/s $L = 40$ m

22−3   FARADAY'S LAW

Magnetic flux

A convenient approach to electromagnetic induction makes use of the notion of *magnetic flux*. If the area of a wire loop is A and it is perpendicular to a magnetic field **B,** the total magnetic flux Φ (Greek letter *phi*) through the loop is defined as

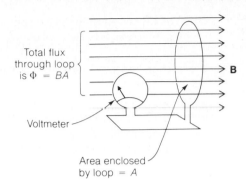

FIG. 22–6 The emf induced in the wire loop is equal to the rate at which the flux Φ it encloses is changing.

Total flux through loop is $\Phi = BA$

Voltmeter

Area enclosed by loop = A

$$\Phi = BA_\perp \qquad\qquad \textit{Magnetic flux} \quad (22\text{–}2)$$

The loop can have any shape, but its area must be taken from its projection on a plane perpendicular to **B**; see Fig. 20–6. The unit of flux is the *weber* (Wb), where 1 Wb = $1 \text{ T} \cdot \text{m}^2$.

Faraday found that the electromotive force $\mathscr{E}$ in such a wire loop is *equal to the rate of change of the flux through it.* That is,

$$\mathscr{E} = -\frac{\Delta\Phi}{\Delta t} \qquad\qquad \textit{Induced emf} \quad (22\text{–}3)$$

Faraday's law of electromagnetic induction

where $\Delta\Phi$ is the change in the flux Φ that takes place during a period of time Δt. The values of the magnetic field B and the loop area A are, in themselves, irrelevant; only the rate at which either or both of them changes is important. Equation (22–3) is called Faraday's law of electromagnetic induction. The reason for the minus sign is given below.

The induced emf in a wire loop is equal to the potential difference that would be found between the ends of an identical *open* wire loop. The notion of emf is useful here because it is meaningless to speak of the potential difference around a closed circuit; the emf in a closed circuit is the potential difference that would be found between the ends of the circuit if it were cut anywhere.

If a coil of N turns replaces the single loop of Fig. 22–6, the induced emf's in the turns are added together, and the total emf in the entire coil is

$$\mathscr{E} = -N\frac{\Delta\Phi}{\Delta t} \qquad\qquad (22\text{–}4)$$

The minus sign in Eqs. (22–3) and (22–4) is a consequence of the law of conservation of energy. If the sign of $\mathscr{E}$ were the same as that of $\Delta\Phi/\Delta t$, the induced electric current would be in such a direction that its own magnetic field would *add* to that of the external field **B**; this additional changing field would then augment the existing rate of change of the flux Φ, and more and more current would flow even if the external contribution to Φ were to stay constant. But energy is associated with every current, and no current can increase without external energy being supplied to it. The only possibility, then, is that $\mathscr{E}$ be opposite in sign to $\Delta\Phi/\Delta t$, which means that

Lenz's law follows from conservation of energy

The direction of an induced current is always such that its own magnetic field opposes the change in flux responsible for producing it.

This observation is known as *Lenz's law*. An example of Lenz's law is illustrated in Fig. 22–7.

Example   A 12-turn coil 10 cm in diameter has its axis parallel to a magnetic field of 0.5 T which is produced by a nearby electromagnet. The current in the electromagnet is cut off, and as the field collapses an average emf of 8 V is induced in the coil. What was the length of time required for the field to disappear?

Solution   The change in the flux through the coil is

$$\Delta\Phi = BA = B\pi r^2 = (-0.5 \text{ T})(\pi)(0.05 \text{ m})^2 = -0.0039 \text{ Wb}$$

If we assume that the flux drops to zero at a uniform rate,

$$\Delta t = -\frac{N\Delta\Phi}{\mathcal{E}} = -\frac{(12)(-0.0039 \text{ Wb})}{8\text{V}} = 0.0059 \text{ s} \qquad\blacksquare$$

Derivation of Faraday's law

It is not hard to derive Faraday's law of electromagnetic induction from the formula for the emf induced in a wire moved across a magnetic field. Figure 22–8 shows a

FIG. 22–7 The current induced in the wire loop is such that its own magnetic field opposes the field change due to the motion of the permanent magnet, in accordance with Lenz's law.

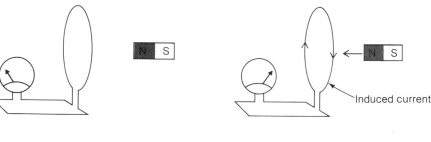

Induced current

FIG. 22–8 The flux enclosed by the wire frame and movable wire increases as the wire moves to the right. An emf is induced in the wire by its motion through the magnetic field, which is equal to the rate of change of the enclosed flux, in agreement with Faraday's law of electromagnetic induction.

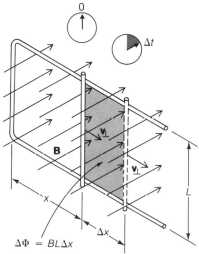

$\Delta\Phi = BL\Delta x$

moving wire that slides across the legs of a U-shaped metal frame, so that the frame completes the loop. At the moment (say $t = 0$) when the moving wire is the distance x from the closed end of the frame, the flux enclosed by the loop is

$$\Phi = BA = BLx$$

At the later time $t = \Delta t$ the moving wire is $x + \Delta x$ from the closed end of the frame, so the flux now enclosed by the loop is

$$\Phi + \Delta\Phi = BLx + BL\,\Delta x$$

Hence the increase in the enclosed flux during the time Δt is

$$\Delta\Phi = BL\,\Delta x$$

The speed of the moving wire is $v_\perp$, and so

$$\Delta x = v_\perp\,\Delta t$$

and

$$\Delta\Phi = BLv_\perp\,\Delta t$$

$$\frac{\Delta\Phi}{\Delta t} = BLv_\perp$$

But $BLv_\perp$ is the emf induced in the moving wire. Hence we have, inserting a minus sign to remind ourselves of Lenz's law,

$$\mathscr{E} = -\frac{\Delta\Phi}{\Delta t}$$

which is Faraday's law.

22-4 THE GENERATOR

Electromagnetic induction is of immense practical importance since it is the means whereby nearly all the world's electric power is produced. In a generator a coil of wire is rotated in a magnetic field so that the flux through the coil changes constantly. The resulting potential difference across the ends of the coil causes a current to flow in an external circuit, and this current can be transmitted by a suitable system of wires for long distances from its origin.

Electromagnetic induction underlies the operation of the generator

Despite its name, a generator does not *create* electric energy; what it does is *convert* mechanical energy into electric energy, just as a battery converts chemical energy into electric energy.

The construction of an idealized simple generator is shown in Fig. 22–9. A wire loop is rotated in a magnetic field, and the ends of the loop are connected to two *slip rings* on the shaft. Brushes pressing against the slip rings permit the loop to be connected to an external circuit. Only sides *B* and *D* of the loop contribute to the induced emf. Because the sides of the loop reverse their direction of motion through the magnetic field twice per rotation, the induced current is also reversed twice per rotation and for

Alternating-current generator

FIG. 22–9 A simple alternating-current generator.

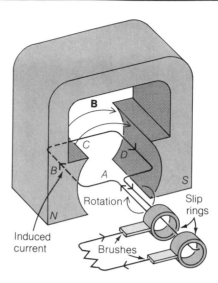

FIG. 22–10 The variation with time of the emf of a simple alternating-current generator.

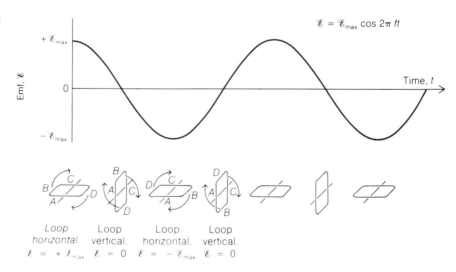

this reason is called an *alternating current*. The variation with time of the emf of this generator is shown in Fig. 22–10. The shape of the curve is that of a sine (or cosine) curve. In a complete cycle, which corresponds to one turn of the loop, the emf increases to a maximum, falls to zero, continues to a negative maximum, and then returns to zero. If the loop rotates f times per second, the generator output will go through f cycles per second. An alternating-current generator is often called an *alternator*.

In nearly all actual generators, electromagnets produce the magnetic field. (If a permanent magnet is used, the generator is called a *magneto*.) In an alternator, the armature and field windings are often exchanged so that, in effect, the field rotates inside a stationary armature. This makes no difference to the physics involved since

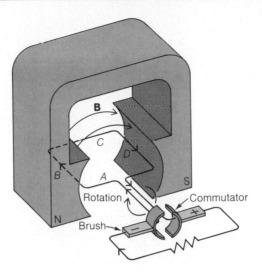

FIG. 22-11 A simple direct-current generator. The current in the brushes is always in the same direction.

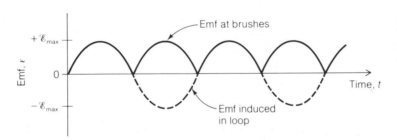

FIG. 22-12 The variation with time of the emf of a simple direct-current generator.

there is still relative movement of a conductor and a magnetic field. The advantage of a stationary armature is that a high-voltage output can be taken directly from it without any moving electrical contacts, with the field current being supplied at a modest voltage.

The alternating-current output of a simple generator can be changed to direct current by substituting a commutator for the slip rings that connect the rotating coil with the external circuit (Fig. 22-11). The emf of such a direct-current generator varies with time as shown in Fig. 22-12; the emf is always in the same direction, but it rises to a maximum and drops to zero twice per complete rotation. The resemblance between the direct-current generator of Fig. 22-11 and the direct-current motor of Fig. 21-28 is not accidental: One is the inverse of the other. If the output terminals of a direct-current generator are connected to a battery, it will run as a motor.

The emf of Fig. 22-12, though unidirectional, pulsates too much for most applications. A steadier emf can be obtained by using a pair of coils on the armature set perpendicular to each other. When one of the coils has its maximum emf, the other has an emf of zero. Figure 22-13 shows the emf's of each coil and also what happens when the outputs of the two are connected in series: The resulting emf has only a moderate ripple. Commercial generators have many coils in their armatures to produce very nearly constant emf's.

Direct-current generator

FIG. 22-13 The output of a direct-current generator whose armature contains two perpendicular coils has only a slight ripple.

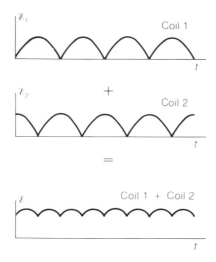

22-5 RECTIFIERS

A rectifier converts alternating current to direct current

Another approach to obtaining direct current from an alternator is to use one or more *rectifiers* in its output circuit. A rectifier is a device that permits current to pass through it in only one direction; the symbol of a rectifier is ——▶|—— . (More precisely, the resistance of a rectifier is very low for a current in one direction, very high for a current in the opposite direction.) Nowadays most rectifiers are semiconductor diodes whose operation is described in Chapter 29. Because alternators are simpler to construct than direct-current generators and are more reliable, they are often used in combination with rectifiers to produce the direct current needed in cars and other vehicles.

Half- and full-wave rectifiers

The simplest way to use a rectifier is shown in Fig. 22-14. Because half the output of the alternator is wasted in such a *half-wave* rectifier circuit, a *full-wave* circuit is more common. A full-wave circuit usually employs four rectifiers in the "bridge" arrangement of Fig. 22-15. Current flows through one pair of rectifiers when it is in

FIG. 22-14 Half-wave rectifier circuit.

FIG. 22-15 Full-wave rectifier circuit.

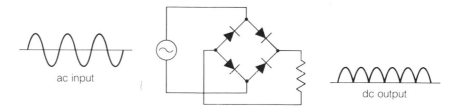

one direction and through the other pair when it is in the opposite direction to give the pulsating output shown. Ripple-free direct current can be obtained by using a combination of capacitors and inductors (see Section 24–10) to absorb energy at the top of each half cycle and then release it at the bottom.

22–6 BACK EMF

In an electric motor a torque is produced through the interaction of the current in its armature with a magnetic field. As we learned in Section 22–4, a coil rotated in a magnetic field is a source of emf, which means that an emf is generated in the rotating armature of an electric motor. This induced emf exists simply because wires are being moved through a magnetic field: The presence of a current of external origin in the armature at the same time, which is what causes it to turn in the first place, does not alter the situation. Every electric motor is also a generator.

Induced emf's occur in all electric motors

By Lenz's law all induced emf's are such as to oppose the changes that bring them into being. The induced emf in a motor armature is therefore opposite in direction to the external voltage applied to the armature. The effect of this *back emf* (or *counter emf*) is to reduce the current in the armature to

Back emf reduces the armature current

$$I = \frac{V - \mathscr{E}_b}{R} \qquad (22-5)$$

where V is the external voltage, $\mathscr{E}_b$ the back emf, and R the resistance of the armature windings. (If the armature and field windings are connected in series, as they are in some motors, R is the sum of their resistances.)

Because the back emf is proportional to the speed of the armature, it is zero when the motor is turned on and increases as the speed increases. The current is therefore a maximum when the motor starts and drops afterward. The existence of back emf in a motor helps to keep its speed constant. At the normal speed,

Back emf and motor speed

$$V = \mathscr{E}_b + IR \qquad (22-6)$$

Impressed voltage = back emf + potential drop in armature

If the speed falls, $\mathscr{E}_b$ decreases and I goes up, and the greater current means more torque on the armature to restore the speed to its original value. If the speed rises above normal, $\mathscr{E}_b$ increases and I goes down, with the result that the torque is reduced and the motor loses speed until it assumes its normal value.

Example A 120-V direct-current electric motor has its armature and field windings connected in parallel. (Such a motor is said to be "shunt wound." When armature and field are in series, the motor is said to be "series wound." See Fig. 22–16.) The armature resistance is 2.0 Ω and the field resistance is 200 Ω. When the motor is at its operating speed of 1800 rpm, the total current is 3.0 A. Find (a) the back emf of the motor at its operating speed, (b) the power output and efficiency of the motor, and (c) the current in the motor at the moment its switch is turned on.

Shunt- and series-wound motors

FIG. 22–16 (a) Shunt-wound direct-current motor. (b) Series-wound direct-current motor.

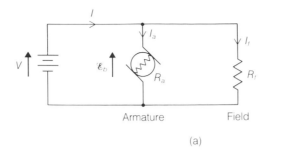

$$V - \mathcal{E}_b = I_a R_a$$
$$V = I_f R_f$$

(a)

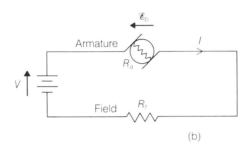

$$V - \mathcal{E}_b = IR_a + IR_f$$

(b)

Solution   (a) The current in the field winding is

$$I_f = \frac{V}{R_f} = \frac{120\,\text{V}}{200\,\Omega} = 0.6\,\text{A}$$

Since the total current is 3.0 A, the armature current is

$$I_a = I - I_f = 3.0\,\text{A} - 0.6\,\text{A} = 2.4\,\text{A}$$

The magnitude of the back emf is therefore

$$\mathcal{E}_b = V - I_a R_a = 120\,\text{V} - (2.4\,\text{A})(2.0\,\Omega) = 115\,\text{V}$$

The direction of the back emf is, of course, opposite to that of the impressed potential difference V.

Power output of motor is product of back emf and armature current   (b) The mechanical power output of the motor (ignoring friction) is equal to the product $\mathcal{E}_b I_a$ of the back emf and the armature current. Hence

$$P_{\text{output}} = \mathcal{E}_b I_a = (115\,\text{V})(2.4\,\text{A}) = 276\,\text{W}$$

The electrical power input of the motor is equal to IV, the total current multiplied by the applied potential difference:

$$P_{\text{input}} = IV = (3.0\,\text{A})(120\,\text{V}) = 360\,\text{W}$$

Hence the motor's efficiency is

$$\text{Eff} = \frac{\text{power output}}{\text{power input}} = \frac{276\,\text{W}}{360\,\text{W}} = 0.77 = 77\%$$

Thus 23% of the power input is dissipated as heat in the armature and field windings. As a check, we can calculate the power loss directly:

$$P_{\text{loss}} = I_f^2 R_f + I_a^2 R_a = (0.6\,\text{A})^2 (200\,\Omega) + (2.4\,\text{A})^2 (2.0\,\Omega) = 83.5\,\text{W}$$

This is

$$\frac{P_{\text{loss}}}{P_{\text{input}}} = \frac{83.5\,\text{W}}{360\,\text{W}} = 0.23 = 23\%$$

of the power input, in agreement with the above figure.

(c) At the moment the motor is started, there is no back emf because the armature is still stationary. The total resistance of the armature and field windings is

$$R = \frac{R_a R_f}{R_a + R_f} = \frac{(2.0\,\Omega)(200\,\Omega)}{2.0\,\Omega + 200\,\Omega} = 1.98\,\Omega$$

The current is therefore

$$I = \frac{V}{R} = \frac{120\,\text{V}}{1.98\,\Omega} = 61\,\text{A}$$

The starting current is more than 20 times greater than the operating current! ■

22-7 THE TRANSFORMER

Earlier it was mentioned that a current can be induced in a wire or other conductor by a changing magnetic field as well as by relative motion between a wire and a constant magnetic field. The essential condition for an induced emf is that magnetic lines of force cut across the wire (or vice versa), and it does not matter exactly how this comes about.

A change in the current in a wire loop is accompanied by a change in its magnetic field. If there is another wire loop nearby, an emf will therefore be induced in it. If an alternating current flows in the first loop, the magnetic field around it will vary periodically and an alternating emf will be induced in the second loop. A *transformer* is a device based upon this effect which is used to produce an emf in a secondary alternating-current circuit larger or smaller than the emf in the primary circuit. A transformer thus permits electric energy to be transferred from one alternating-current circuit to another without a direct connection between them.

The output emf of a transformer may be greater or smaller than its input emf

With the help of Fig. 22-17 we can see how a changing current in a wire loop induces a changing current in another wire loop not connected to it.

The transformer is based upon electromagnetic induction

(a) The switch connecting loop 1 with the battery is open, and no current is present in either loop.

(b) The switch has just been closed, and the expanding lines of force resulting from the increasing current in loop 1 cut across loop 2, inducing a current in it. The current in loop 2 is in the *opposite* direction to that in loop 1 because of Lenz's law: The direction of an induced current is always such that its own magnetic field opposes the change in flux that is inducing it.

FIG. 22–17 When the current in a wire loop is turned on or off, the magnetic field around it changes. The changing field can induce a current in another nearby loop, even though both loops are stationary.

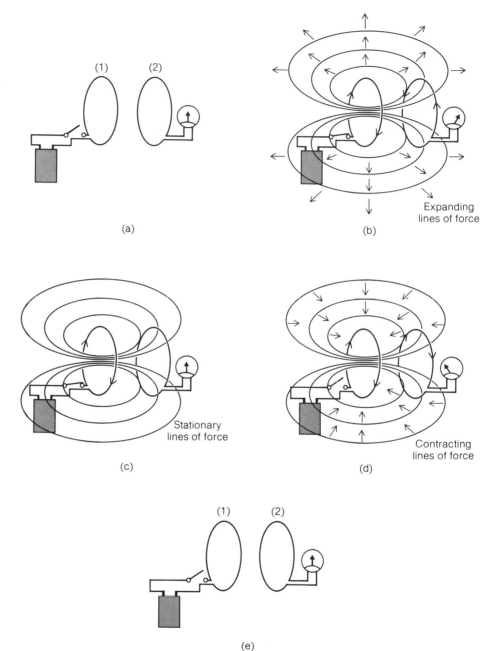

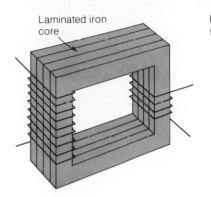

FIG. 22-18 A simple transformer.

Laminated iron core

(c) Now there is a constant current in loop 1, and since the flux through it does not change, no current is induced in loop 2.

(d) The switch has just been opened, and the contracting lines of force resulting from the decreasing current in loop 1 cut across loop 2, inducing a current in it. The current in loop 2 is now in the *same* direction as that in loop 1, since it is the decreasing current in loop 1 that leads to the current in loop 2 and Lenz's law requires that an induced current oppose the change in flux that brings it about.

(e) Finally the current in loop 1 disappears, and no current is present in either loop.

Instead of being switched on and off, the current input to a transformer is alternating current that reverses its direction regularly. The construction of a simple transformer is shown in Fig. 22-18. Coils are used instead of single loops, and they are wound on a common iron core so that the alternating magnetic flux set up by an alternating current in one coil links with the other coil. This alternating flux induces an alternating emf in the latter coil. The *primary winding* of a transformer is the coil that is fed with an alternating current, and the *secondary winding* is the coil to which power is transferred via the changing magnetic flux. The designations of primary and secondary depend only upon how the transformer is connected; either winding may be the primary, and the other is then the secondary. A transformer may have more than one secondary winding so that it can provide several different secondary emf's.

A transformer operates on alternating current

When a piece of metal is placed in a coil that carries an alternating current, the changing magnetic flux causes *eddy currents* to flow in the metal. The energy of these currents is dissipated as heat, and the metal becomes hot as a result. This effect is called *induction heating*. Induction furnaces are rapid and are often used in foundries to melt casting metal before it is poured into molds.

Eddy currents and induction heating

22-8 STEP-UP AND STEP-DOWN TRANSFORMERS

Let us consider an ideal transformer in which there is no "leakage" of flux outside the iron core and no losses within it, so that $\Delta\Phi/\Delta t$ is the same for each of the turns of both windings. The emf *per turn* is therefore the same in both windings, and the total emf in each winding is proportional to the number of turns it contains. Hence

$$\frac{\mathcal{E}_1}{\mathcal{E}_2} = \frac{N_1}{N_2} \tag{22-7}$$

$$\frac{\text{Primary emf}}{\text{Secondary emf}} = \frac{\text{total primary turns}}{\text{total secondary turns}}$$

When the secondary winding has more turns than the primary winding, $\mathcal{E}_2$ is greater than $\mathcal{E}_1$ and the result is a *step-up transformer;* when the reverse is the case and $\mathcal{E}_2$ is less than $\mathcal{E}_1$, the result is a *step-down transformer.*

In an ideal transformer the power output is equal to the power input. If the primary current is I_1 and the second current is I_2, then

$$\mathcal{E}_1 I_1 = \mathcal{E}_2 I_2$$

Power input = power output

and

$$\frac{I_1}{I_2} = \frac{\mathcal{E}_2}{\mathcal{E}_1} = \frac{N_2}{N_1} \qquad\qquad \textit{Transformer} \quad (22\text{-}8)$$

Increasing the voltage with a step-up transformer means a proportionate decrease in the current, while dropping the voltage leads to a greater current.

Example   A transformer connected to a 120-V alternating-current power line has 200 turns in its primary winding and 50 turns in its secondary winding. The secondary is connected to a 100-Ω light bulb. How much current is drawn from the 120-V power line?

Solution   The voltage across the secondary of the transformer is

$$V_2 = \frac{N_2}{N_1} V_1 = \left(\frac{50\,\text{turns}}{200\,\text{turns}}\right)(120\,\text{V}) = 30\,\text{V}$$

and so the current in the secondary circuit is

$$I_2 = \frac{V_2}{R} = \frac{30\,\text{V}}{100\,\Omega} = 0.3\,\text{A}$$

Hence the current in the primary circuit is

$$I_1 = \frac{N_2}{N_1} I_2 = \left(\frac{50\,\text{turns}}{200\,\text{turns}}\right)(0.3\,\text{A}) = 0.075\,\text{A} \qquad\blacksquare$$

Why transformer cores are laminated

A transformer core is itself an electrical conductor, and the changing magnetic flux within it leads to the eddy currents mentioned above. Eddy currents are wasteful partly because of the power lost as heat and partly because the flux they establish interferes with the proper operation of the transformer. To minimize eddy currents, actual transformer cores are usually laminated, with many thin sheets arranged parallel to the flux. The laminations are insulated from each other by natural oxide layers on their surfaces or by varnish coatings; only very weak eddy currents are induced in a core of this kind. In small transformers an alternative technique is to use a core of

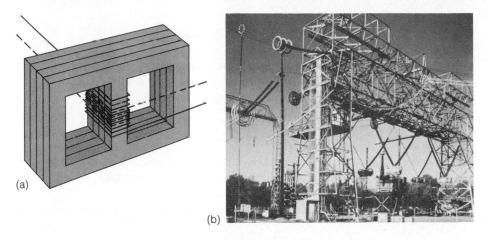

(a)

(b)

FIG. 22–19 (a) Most transformers have concentric windings to minimize flux leakage. (b) Transformers at a power station step up the voltage of the electric power generated there for transmission over long distances. The higher the voltage V, the lower the current I for the same power P, since P = IV. The advantage of a low current is that less energy is lost as heat in the transmission lines. Other transformers step down the voltage for the consumer to the usual 240 V or 120 V. (Photo: EPRI)

compressed powdered iron, so that each grain of iron is fairly well insulated from the others, although magnetically the core behaves as if it were solid.

There are still other sources of power loss in a transformer. The windings themselves have a certain amount of resistance, and part of the energy that flows through them is dissipated as heat. Also, in all magnetic materials some energy is lost as heat per hysteresis cycle in the atomic rearrangements that accompany flux changes (see Section 21–5). Leakage of flux from the core of Fig. 22–18 between the separated primary and secondary windings is more easily remedied. Instead, transformers normally have concentric windings as in Fig. 22–19, and in this arrangement the flux through each winding is virtually the same. Efficiencies as high as 99% are achieved in large transformers.

Alternating current owes its wide use largely to the ability of transformers to change the voltage at which it is transmitted from one value to another. Electricity generated at perhaps 11,000 volts is stepped up by transformers at the power station to as much as 750,000 volts (or even more) for transmission, and subsequently is stepped down by transformers at substations near the point of consumption to a few thousand volts. Local transformers then step this down further to 240 and 120 volts. High voltages are desirable for transmission because, since $P = IV$, the higher the voltage, the smaller the current for a given amount of power. Power is dissipated as heat at the rate I^2R, and so the smaller the current, the less the power lost due to the resistance of the transmission line.

High-voltage transmission minimizes power losses

Example Find the power lost as heat when a 10-Ω cable is used to transmit 1 kW of electricity at 240 V and at 240,000 V.

Solution The current in the cable in each case is

$$I_a = \frac{P}{V_a} = \frac{1000\,\text{W}}{240\,\text{V}} = 4.17\,\text{A}$$

$$I_b = \frac{P}{V_b} = \frac{1000\,\text{W}}{240,000\,\text{V}} = 4.17 \times 10^{-3}\,\text{A}$$

The respective rates of heat production per kilowatt are therefore

$$I_a^2 R = (4.17\,\text{A})^2 (10\,\Omega) = 174\,\text{W}$$
$$I_b^2 R = (4.17 \times 10^{-3}\,\text{A})^2 (10\,\Omega) = 1.74 \times 10^{-4}\,\text{W}$$

which is a difference of a factor of 10^6. Transmission at 240 V therefore means a million times more power lost as heat than does transmission at 240,000 V. ■

22−9 SOUND REPRODUCTION

What we have learned thus far about electricity and magnetism enables us to understand how sounds can be converted into electrical signals by a microphone, recorded on a disk or magnetic tape, and then reproduced by a loudspeaker.

Moving coil microphone

A microphone is a device that changes the pressure variations of sound waves into alternating current. Figure 22−20(a) shows a moving coil microphone. The diaphragm is a cone of corrugated paper that vibrates in response to a sound wave. As the diaphragm vibrates, a coil of fine wire at its neck moves back and forth in the field of a permanent magnet, which induces a current in the coil whose frequency and amplitude correspond to those of the sound wave.

FIG. 22−20 Three types of microphone. (a) Moving coil microphone, (b) crystal microphone, (c) carbon microphone. A fourth is shown in Fig. 23−4. A loudspeaker is similar in construction to (a) except that an alternating current in the coil causes the diaphragm to vibrate and thus produce sound waves.

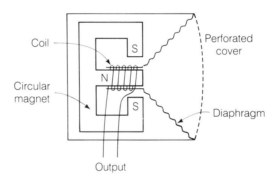

(a) *Moving coil microphone*

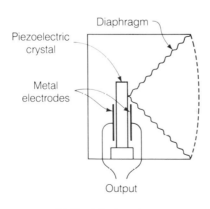

(b) *Crystal microphone*

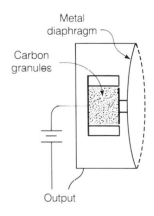

(c) *Carbon microphone*

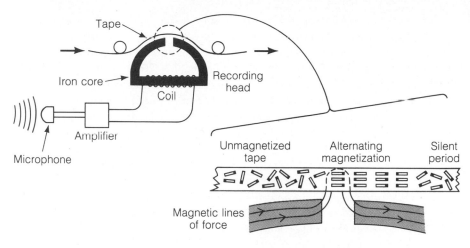

Tape

Iron core

Coil

Recording head

Microphone

Amplifier

Unmagnetized tape

Alternating magnetization

Silent period

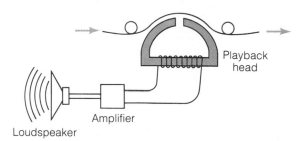

Magnetic lines of force

(a) *Recording*

FIG. 22–21 A tape recorder. (a) Recording, (b) playback. The polarity and degree of magnetization of the ferromagnetic coating on the tape correspond to the pressure fluctuations of the original sound wave. Photo shows a multitrack tape recorder with separate recording and playback heads. (Photo courtesy of Ampex Corporation.)

Playback head

Loudspeaker

Amplifier

(b) *Playback*

Crystal and carbon microphones

Another common microphone is based on *piezoelectricity,* which is the property certain crystals have of becoming electrically polarized when a mechanical stress is applied. In a crystal microphone, one end of a crystal of this kind is fixed in place and the other end is attached to a diaphragm, as in Fig. 22–20(b). When the diaphragm vibrates in response to a sound wave, a varying potential difference appears across the crystal and is picked up by metal electrodes. The microphone used in telephones has a small cup filled with carbon granules that a metal diaphragm presses against (Fig. 22–20 (c)). The greater the pressure on the diaphragm, the better the electrical contact between the granules, and the lower the resistance of the unit. By placing a battery in the circuit, a pulsating current is produced when a sound wave reaches the diaphragm. The capacitor microphone, still another type, is described in Section 23–2.

Disk recording

To record sound waves on a disk, the electrical output of a microphone is amplified and then connected to an electromagnet that vibrates an iron armature in accord with the alternating current from the amplifier. A sharp stylus attached to the armature cuts a wavy groove in a rotating wax disk; the waves in the groove are a mechanical replica of the original sound waves. A metal die is then made from the wax disk and is used to produce the plastic disks sold commercially. A record player reverses the process, with a crystal pickup that operates in the same way as a crystal microphone but is actuated by a stylus that vibrates as the wavy groove in the record passes it. The output of the pickup is amplified and used to operate a loudspeaker. A loudspeaker is very similar to the moving coil microphone of Fig. 22–20 (a) except that an alternating current in the coil causes the diaphragm to vibrate and thus produce a sound wave in the air around it.

In a recent development, the output of the microphone is converted to digital form so that a coded series of pulses replaces the original continuously varying current. These pulses are then recorded on a plastic disk in the form of tiny pits. To play such a "compact disk" a narrow light beam from a laser is directed at the rotating disk and the presence or absence of a pit determines whether or not the light is reflected. A light-sensitive detector picks up the pulses of reflected light which are converted back into an alternating current, amplified, and finally fed into a loudspeaker.

Tape recording

In a tape recorder, the electrical signal from the amplifier is converted into a magnetic pattern on the ferromagnetic coating of a plastic tape. As in Fig. 22–21, the tape is magnetized by being passed over an electromagnet connected to the amplifier. The air gap in the electromagnet of the recording head allows its magnetic field to reach the tape. The playing head is similar in design but operates on the basis of electromagnetic induction. The changing magnetic fields of the moving tape induce a corresponding alternating current in the coil of the playing head, which is amplified and fed into a loudspeaker to reconstruct the original sound.

A videotape recorder uses recording and playback heads like those in Fig. 22–21, but the electrical signals it records come from a television camera (or receiver) and are played back through a television set.

IMPORTANT TERMS

Electromagnetic induction refers to the production of an electric field in a conductor whenever magnetic lines of force move across it.

Lenz's law states that the direction of an induced current must be such that its own magnetic field opposes the changes in flux that are inducing it.

A **generator** is a device that converts mechanical energy into electric energy.

The direction of an **alternating current** reverses itself periodically.

A **back** (or **counter**) **emf** is induced in the rotating coils of an electric motor and is opposite in direction to the external voltage applied to them.

An alternating current flowing in the primary coil of a **transformer** induces another alternating current in the secondary coil. The ratio of the emf's is proportional to the ratio of turns in the coils.

IMPORTANT FORMULAS

Motional emf: $\mathscr{E} = BLv_\perp$

Magnetic flux: $\Phi = BA_\perp$

Induced emf: $\mathscr{E} = -N\dfrac{\Delta\Phi}{\Delta t}$

Transformer: $\dfrac{I_1}{I_2} = \dfrac{\mathscr{E}_2}{\mathscr{E}_1} = \dfrac{N_2}{N_1}$

MULTIPLE CHOICE

1. Electromotive force is most closely related to
   a. electric field.
   b. magnetic field.
   c. potential difference.
   d. mechanical force.

2. The emf produced in a wire by its motion across a magnetic field does not depend on
   a. the length of the wire.
   b. the diameter of the wire.
   c. the orientation of the wire.
   d. the magnitude B of the field.

3. A bar magnet is passed through a coil of wire. The induced current is greatest when
   a. the magnet moves slowly, so that it is inside the coil for a long time.
   b. the magnet moves fast, so that it is inside the coil for a short time.

   c. the north pole of the magnet enters the coil first.
   d. the south pole of the magnet enters the coil first.

4. The magnetic flux through a wire loop in a magnetic field **B** does not depend upon
   a. the area of the loop.
   b. the shape of the loop.
   c. the angle between the plane of the loop and the direction of **B**.
   d. the magnitude of B of the field.

5. The unit of magnetic flux is the weber, where 1 Wb =
   a. $1\ T \cdot m^2$.
   b. $1\ T/m^2$.
   c. $1\ A \cdot m^2$.
   d. $1\ A/m^2$.

6. A wire loop is moved parallel to a uniform magnetic field. The induced emf in the loop
   a. depends upon the area of the loop.
   b. depends upon the shape of the loop.
   c. depends upon the magnitude of the field.
   d. is 0.

7. The direction of an induced current is always such that its own magnetic field
   a. reinforces the change in flux producing it.
   b. opposes the change in flux producing it.
   c. has no effect on the change in flux producing it.
   d. any of the above, depending on the orientation of the induced current relative to the flux whose change is producing it.

8. When a wire loop is rotated in a magnetic field, the direction of the induced emf changes once in each
   a. $\frac{1}{4}$ revolution.
   b. $\frac{1}{2}$ revolution.
   c. 1 revolution.
   d. 2 revolutions.

9. A generator most closely resembles
   a. a transformer.
   b. an electric motor.
   c. a betatron.
   d. a battery.

10. The emf produced by an alternating-current generator varies between $\mathscr{E}_{max}$ and
   a. 0.
   b. $\frac{1}{2}\mathscr{E}_{max}$.
   c. $-\mathscr{E}_{max}$.
   d. the back emf.

11. The back emf in an electric motor is a maximum when the motor
   a. is switched on.
   b. is increasing in speed.
   c. has reached its maximum speed.
   d. is decreasing in speed.

12. When the speed of a direct-current motor drops because of an increased load, there is also a drop in the
   a. impressed voltage.
   b. back emf.
   c. current.
   d. armature resistance.

13. The alternating-current output of a simple generator can be changed to direct current by connecting the output of the rotating coil to the external circuit with
a. slip rings.
b. a commutator.
c. a resistor.
d. a transformer.

14. The alternating current in the secondary coil of a transformer is induced by
a. a varying electric field.
b. a varying magnetic field.
c. the iron core of the transformer.
d. motion of the primary coil.

15. The ratio between primary and secondary currents in a transformer does not depend on the
a. ratio of turns in the two windings.
b. resistance of the windings.
c. nature of the core.
d. primary voltage.

16. Eddy currents occur only in
a. ferromagnetic materials.
b. insulators.
c. metals.
d. coils.

17. High voltages are used in electric transmission lines
a. to compensate for the IR drops over long distances.
b. because the resulting low currents mean less energy loss.
c. because energy travels faster at high voltages.
d. to keep birds from perching on them.

18. A loudspeaker is closest in construction to
a. a moving coil microphone.
b. a crystal microphone.
c. a carbon granule microphone.
d. an electric motor.

19. The aluminum mainboom of a sailboat sailing downwind at 5 m/s is 7 m long. The boom is at an angle of 75° relative to the boat's direction of motion. If the vertical component of the earth's magnetic field there is 4×10^{-5} T, the potential difference between the ends of the boom is
a. 0.03 mV.
b. 0.36 mV.
c. 1.35 mV.
d. 1.40 mV.

20. A square wire loop 5 cm on a side is perpendicular to a magnetic field of 0.08 T. If the field drops to 0 in 0.2 s, the average emf induced in the loop during that time is
a. 0.04 mV.
b. 0.5 mV.
c. 1 mV.
d. 8 V.

21. A 400-turn coil of resistance 5 Ω encloses an area of 30 cm². How rapidly should a magnetic field parallel to the coil axis change in order to induce a current of 0.3 A in the coil?
a. 0.05 T/s
b. 0.25 T/s
c. 0.8 T/s
d. 1.25 T/s

22. The primary winding of a transformer has 200 turns and its secondary winding has 50 turns. If the current in the secondary winding is 40 A, the current in the primary is
a. 10 A.
b. 80 A.
c. 160 A.
d. 8000 A.

23. The primary winding of a transformer has 200 turns and its secondary winding has 50 turns. If an alternating-current emf of 12 V is applied to the primary, the secondary emf will be
a. 3 V.
b. 6 V.
c. 24 V.
d. 48 V.

24. A 100% efficient transformer has 100 turns in its primary winding and 300 turns in its secondary. If the power input to the transformer is 60 W, the power output is
a. 20 W.
b. 60 W.
c. 180 W.
d. 540 W.

25. A 60-W, 24-V light bulb is to be powered from a 120-V ac line. Which of the following combinations of primary and secondary transformer windings is needed?
a. 20 turns primary, 100 turns secondary
b. 50 turns primary, 100 turns secondary
c. 100 turns primary, 20 turns secondary
d. 100 turns primary, 50 turns secondary

EXERCISES

22-1   Electromagnetic Induction

22-2   Moving Wire in a Magnetic Field

1. A car is traveling from New York to Florida. Which of its wheels have a positive charge and which a negative charge?

2. A train is traveling at 130 km/h in a region where the vertical component of the earth's magnetic field is 3×10^{-5} T. The railway is standard gauge with its rails 1.435 m apart. Find the potential difference between the wheels on each axle of the train's cars.

3. A car is traveling at 30 m/s on a road where the vertical component of the earth's magnetic field is 3×10^{-5} T. What is the potential difference between the ends of its axles, which are 2 m long?

4. A potential difference of 1.8 V is found between the ends of a 2-m wire moving in a direction perpendicular to a magnetic field at a speed of 12 m/s. What is the magnitude of the field?

22–3 Faraday's Law

5. Lenz's law is a consequence of which conservation principle?

6. Explain what is meant by the emf in a wire loop.

7. A bar magnet held vertically with its north pole downward is dropped through a wire loop whose plane is horizontal. Is the current induced in the loop just before the magnet enters it clockwise or counterclockwise as seen by an observer above? What is the direction of the current when the magnet passes through the center of the loop? Just after it leaves the loop?

8. One end of a bar magnet is thrust into a coil, and the induced current is clockwise as seen from the side of the coil into which it was thrust. (a) Was the end of the magnet its north or south pole? (b) What will be the direction of the induced current when the magnet is withdrawn?

9. A loop of copper wire is rotated in a magnetic field about an axis along a diameter. (a) Why does the loop resist this rotation? (b) If the loop were made of aluminum wire, would the resistance to rotation be different?

10. A wire loop is 5 cm in diameter and is oriented with its plane perpendicular to a magnetic field. How rapidly should the field change if a potential difference of 1 V is to appear across the ends of the loop?

11. A rectangular wire loop 5 cm × 10 cm in size is perpendicular to a magnetic field of 10^{-3} T. (a) What is the flux through the loop? (b) If the magnetic field drops to zero in 3 s, what is the potential difference induced between the ends of the loop during that period?

12. A flat 100-turn coil that encloses an area of 50 cm^2 is turned so that its plane goes from being parallel to a magnetic field of 10^{-3} T to being perpendicular to the field in 0.04s. Find the average emf induced in the coil.

13. The magnetic field inside a solenoid 15 mm in radius is increasing at a rate of 0.06 T/s. Find the emf induced in a 400-turn coil wound around the solenoid.

14. A circular wire loop 10 cm in radius with a resistance of 0.4 Ω is in a magnetic field of 0.6 T whose direction is perpendicular to the plane of the loop. The loop is grasped at opposite ends of a diameter and pulled out into a rectangle 30 cm long and 1.4 cm wide in a time of 0.2 s. Find the average current in the wire while it is being deformed.

15. A square wire loop 10 cm on a side is oriented with its plane perpendicular to a magnetic field. The resistance of the loop is 5 Ω. How rapidly should the magnetic field change if a current of 2 A is to flow in the loop?

16. An air-core solenoid 20 cm long and 2 cm in radius is wound with 1000 turns and has a secondary coil of 300 turns wound over it. (a) If the current in the solenoid is changing at a rate of 0.5 A/s, find the emf in the secondary coil. (b) An iron core with a constant permeability of $\mu = 400\mu_0$ is placed in the solenoid. What is the induced emf now?

22–4 The Generator
22–6 Back Emf

17. Why is it easy to turn the shaft of a generator when it is not connected to an outside circuit, but much harder when such a connection is made?

18. Why does an electric motor require more current when it is turned on than when it is running continuously?

19. A coil is rotated 100 times per second in a magnetic field. What time interval separates successive instants when the induced emf is zero?

20. How is the back emf in the armature of a dc motor at a given speed of rotation related to the emf developed in the same armature when the motor is used as a generator at the same speed of rotation and with the same current in its field coils?

21. For which of the following can alternating current be used without first being rectified into direct current: an electromagnet; a light bulb; an electric heater; electroplating; charging a storage battery?

22. In a series-wound generator the magnetic field is produced by an electromagnet whose windings are connected in series with the armature windings. What happens to the emf in such a generator when the current drawn by the external circuit increases? When the external circuit current decreases? Why?

23. In a shunt-wound generator the electromagnet windings are connected in parallel with the armature windings. What happens to the emf in such a generator when the current drawn by the external circuit increases? When the current decreases? Why?

24. A 1.5-kW dc generator has a full-load potential difference of 35 V and a no-load potential difference of 38 V. How much power is dissipated as heat in the armature at full load?

25. The armature of a dc generator has a resistance of 0.20 Ω. The terminal potential difference of the generator is 120 V with no load and 115 V on full load. How much power is delivered at full load?

26. A square wire loop 10 cm on a side rotates at a constant angular velocity in a uniform magnetic field **B** so that it turns through 90° in 0.05 s, as shown below. Initially the loop lies in the xy plane and afterwards in the yz plane. The field **B** has a magnitude of 0.4 T and points in the $+z$ direction. (a) Find the average emf $\mathscr{E}_{av}$ induced in the loop during the turn. (b) In which part of the turn is $\mathscr{E}$ a minimum? A maximum? What are the values of $\mathscr{E}_{min}$ and $\mathscr{E}_{max}$?

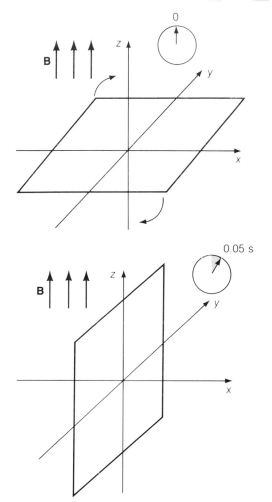

27. A shunt-wound 120-V, 3-kW generator has an armature resistance of 0.1 Ω and a field resistance of 120 Ω. Find its efficiency.

28. A series-wound 24-V dc electric motor has an armature resistance of 0.1 Ω and a field resistance of 0.4 Ω. The motor draws 20 A at its normal operating speed. Find the power output, power input, and efficiency of the motor.

29. A shunt-wound 120-V dc electric motor has an armature resistance of 1.0 Ω and a field resistance of 200 Ω. The motor has a power input of 420 W at 2000 rpm. (a) Find the power output and efficiency of the motor. (b) What is the starting current? (c) What series resistance is required if the starting current is to be limited to 25 A?

30. Find the power output, power input, and efficiency of the motor of Exercise 29 when it is operating at 1800 rev/min.

22-7 The Transformer

22-8 Step-up and Step-down Transformers

31. Why must alternating current be used in a transformer?

32. What would happen if the primary winding of a transformer were connected to a battery?

33. What is it whose action on the secondary winding of a transformer causes an alternating potential difference to occur across its ends even though the primary and secondary windings are not connected?

34. What is the difference between a step-up transformer and a step-down transformer? Is there any way to use a step-up transformer as a step-down transformer and vice versa? If so, how can this be done?

35. What is the relationship between the frequency of the input current to a transformer and the frequency of the output current?

36. Why are transformer cores made from thin sheets of steel? Would there be any advantage in making permanent magnets in this way?

37. A transformer has 100 turns in its primary winding and 500 turns in its secondary winding. If the primary voltage and current are respectively 120 V and 3 A, find the secondary voltage and current. (Assume 100% efficiency in Exercises 37–40.)

38. An electric welding machine employs a current of 400 A. The device uses a transformer whose primary winding has 400 turns and which draws 4 A from a 220-V power line. (a) How many turns are there in the secondary winding of the transformer? (b) What is the potential difference across the secondary?

39. A transformer rated at a maximum power of 10 kW is used to couple a 5000-V transmission line to a 240 V circuit. (a) What is the ratio of turns in the windings of the transformer? (b) What is the maximum current in the 240-V circuit?

40. A transformer connected to a 120-V power line has 100

turns in its primary winding and 40 turns in its secondary winding. An 80-Ω light bulb is connected to the secondary. What is the current in the primary winding?

22–9 Sound Reproduction

41. What is the purpose of the air gaps in the recording and playback heads of a tape recorder?

ANSWERS TO MULTIPLE CHOICE

1. c	**6.** d	**11.** c	**16.** c	**21.** d
2. b	**7.** b	**12.** b	**17.** b	**22.** a
3. b	**8.** b	**13.** b	**18.** a	**23.** a
4. b	**9.** b	**14.** b	**19.** c	**24.** b
5. a	**10.** c	**15.** d	**20.** c	**25.** c

23

CAPACITANCE AND INDUCTANCE

Energy is needed to bring an electric field into existence. This energy does not disappear but is available for doing work under the right conditions. A capacitor is a device that stores energy in the form of electric field. Energy is also associated with a magnetic field and is thus present around every electric current. Devices called *inductors* are designed to make use of this property of currents. When a charged capacitor is connected to an inductor, energy flows back and forth between them in the same way that kinetic energy and potential energy are interchanged in a harmonic oscillator. As in the case of a harmonic oscillator, the alternations in energy flow in a given capacitor-inductor combination occur at a certain natural frequency, a property that is made use of in practically every branch of electronics.

23-1 ELECTRIC FIELD ENERGY

A capacitor stores energy in the form of electric field

Work must be done to create an electric field, since work must be done to separate positive and negative charges against the Coulomb forces attracting them together. Where does the work go? The answer is that it is stored as potential energy, which we

616

CHAPTER OBJECTIVES

Completing this chapter should enable you to:

1. Calculate the energy density of an electric field.

2. Determine the capacitance of a system of conductors.

3. Find the charge on a capacitor that has a given potential difference across it.

4. Find the potential difference across a capacitor that has a given charge on its plates.

5. Calculate the potential energy of a charged capacitor.

6. Describe what happens when a slab of insulating material is placed between the plates of a capacitor.

7. Calculate the equivalent capacitance of two or more capacitors connected in series and find the charge on each one and the voltage across it when a potential difference is applied across the combination.

8. Calculate the equivalent capacitance of two or more capacitors connected in parallel and find the charge on each one and the voltage across it when a potential difference is applied across the combination.

9. Describe the significance of the time constant of a circuit that contains capacitance and resistance.

10. Calculate the energy density of a magnetic field.

11. Determine the inductance of a solenoid.

12. Find the potential energy of a current-carrying inductor.

13. Find the self-induced emf in an inductor when the current through it changes at a given rate.

14. Describe the operation of an electrical oscillator and determine its frequency.

15. Describe the significance of the time constant of a circuit that contains inductance and resistance.

---

can think of as residing in the electric field between the charges. A system of conductors that stores energy in the form of an electric field is called a *capacitor.*

An example of a capacitor is a pair of parallel metal plates of area A that are a distance d apart, as in Fig. 23–1. We imagine that we create an electric field between them by bringing electrons from one plate to the other until a total charge of Q has been transferred. When we are finished, there will be a potential difference of V between the plates.

At the beginning, the potential difference between the plates is small, and little **Charging a capacitor** work is needed to move each electron. As the charge builds up, the potential difference becomes greater, and more work is needed per electron. The average potential difference $\overline{V}$ during the charge transfer is

$$\overline{V} = \frac{V_{\text{final}} + V_{\text{initial}}}{2} = \frac{V + 0}{2} = \tfrac{1}{2} V$$

Since the total charge transferred is Q, the work W that has been done is the product of Q and the average potential difference $\overline{V}$, namely,

FIG. 23–1   As a capacitor is charged, the potential difference between its plates builds up and more work must be done to transfer each successive charge. (In an actual capacitor, the electrons are transferred by means of an electric circuit connecting the plates, not directly.)

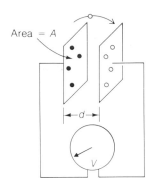
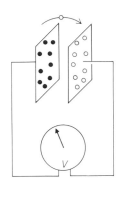
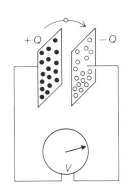

• $+e$ ○ $-e$

$$W = Q\overline{V} = \tfrac{1}{2} QV \qquad\qquad \textit{Energy of charged capacitor} \quad (23-1)$$

Because the metal plates are good conductors, the $+$ and $-$ charges on them are spread out evenly, and the field between the plates (except near the edges) is uniform with the magnitude

$$E = 4\pi k \frac{Q}{A} \qquad\qquad \textit{Electric field inside capacitor} \quad (23-2)$$

The derivation of this formula, which follows from the considerations discussed in Chapter 18, is not given here because it involves calculus. We can rewrite Eq. (23–2) in the form

$$Q = \frac{AE}{4\pi k} \qquad\qquad\qquad\qquad (23-3)$$

The potential difference between the plates is equal to

$$V = Ed \qquad\qquad\qquad\qquad (23-4)$$

Hence the electric potential energy of the system of two charged plates is

$$W = \tfrac{1}{2} QV = \frac{Ad}{8\pi k} E^2 \qquad\qquad\qquad (23-5)$$

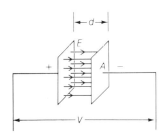

FIG. 23–2   The energy density of the electric field in a region is the total energy in the region divided by its volume.

Energy density of electric field

We note that Ad, the product of the area of each plate and the distance between them, is the volume occupied by the electric field $\mathbf{E}$ (Fig. 23–2). If we define the *energy density w* of an electric field as the electric potential energy per unit volume associated with it, we have here

$$w = \frac{W}{Ad} = \frac{E^2}{8\pi k} \qquad\qquad \textit{Electric energy density} \quad (23-6)$$

This important formula states that the energy density of an electric field is directly proportional to the square of its magnitude E. Even though this formula was derived for a special situation, it is a completely general result.

Example   Dry air is an insulator provided the electric field in it does not exceed about 3×10^6 V/m. What energy density does this correspond to?

Solution   Here $E = 3 \times 10^6$ V/m, and so

$$w = \frac{E^2}{8\pi k} = \frac{(3 \times 10^6 \,\text{V/m})^2}{8\pi(9 \times 10^9 \,\text{N} \cdot \text{m}^2/\text{C}^2)} = 40 \,\text{J/m}^3$$

At sea level 1 m³ of air has a mass of 1.3 kg, so if this amount of energy were gravitational potential energy, it would correspond to an elevation of about 3 m.   ∎

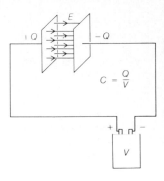

FIG. 23–3 The ratio between the charge on a particular capacitor and the potential difference across it is a constant called its *capacitance*.

23–2   CAPACITANCE

The potential difference V across a capacitor is always directly proportional to the charge Q on either of its plates (Fig. 23–3): The more the charge, the stronger the electric field between the plates, and the greater the potential difference. The ratio between Q and V is therefore a constant for any capacitor and is known as its *capacitance* (symbol C):

$$C = \frac{Q}{V} \qquad\qquad \textit{Capacitance} \quad (23\text{–}7)$$

$$\text{Capacitance} = \frac{\text{charge on either conductor}}{\text{potential difference between conductors}}$$

The unit of capacitance is the *farad,* abbreviated F, where

1 farad = 1 coulomb/volt

The farad, microfarad, and picofarad

The farad is so large a unit that, for practical purposes, it is usually replaced by the *microfarad* (μF) or *picofarad* (pF), whose values are

$1\mu\text{F} = 10^{-6} \,\text{F} \qquad 1 \,\text{pF} = 10^{-12} \,\text{F}$

The capacitance of a pair of separated conductors depends solely upon their geometry and upon the material between them. In the case of a parallel-plate capacitor in vacuum, from Eqs. (23–3) and (23–4) we have

Parallel-plate capacitor

$$C = \frac{Q}{V} = \frac{AE/4\pi k}{Ed}$$

$$= \frac{1}{4\pi k}\frac{A}{d} \qquad\qquad \textit{Parallel-plate capacitor} \quad (23\text{–}8)$$

To a good degree of approximation the same formula can be used for such a capacitor in air as well.

Example   The plates of a parallel-plate capacitor are 10 cm square and 1 mm apart. Find its capacitance in air and the charge each plate will acquire when a potential difference of 100 V is applied.

Solution   Here $A = 10^{-2}$ m² and $d = 10^{-3}$ m. In air the capacitance is

$$C = \frac{1}{(4\pi)(9 \times 10^9\,\text{N} \cdot \text{m}^2/\text{C}^2)} \times \frac{10^{-2}\,\text{m}^2}{10^{-3}\,\text{m}} = 8.85 \times 10^{-11}\,\text{F} = 88.5\,\text{pF}$$

If a potential difference of 100 V is placed across the plates of this capacitor, they will acquire charges of

$$Q = CV = (8.85 \times 10^{-11}\,\text{F})(100\,\text{V}) = 8.85 \times 10^{-9}\,\text{C} \qquad \blacksquare$$

Capacitor microphone

A parallel-plate capacitor can be used as a microphone if one of its plates is light and flexible enough to respond to the changing air pressure of a sound wave (Fig. 23–4). When the pressure increases, the diaphragm moves closer to the fixed plate, and C increases; when the pressure drops, the diaphragm moves outward, and C decreases. Because the voltage across the capacitor is fixed, the changes in C vary the charge on it, and the result is an electrical output whose variations match those in the incoming sound wave.

Energy of charged capacitor

The formula $\frac{1}{2} QV$ for the potential energy of a charged capacitor derived in the previous section can be written in three equivalent ways:

$$W = \tfrac{1}{2} QV \qquad\qquad \textit{Potential energy of charged capacitor} \quad (23\text{--}9)$$

$$W = \tfrac{1}{2} CV^2 \qquad\qquad\qquad\qquad\qquad\qquad\qquad\qquad\qquad\quad (23\text{--}10)$$

$$W = \frac{1}{2} \frac{Q^2}{C} \qquad\qquad\qquad\qquad\qquad\qquad\qquad\qquad\qquad\quad (23\text{--}11)$$

These equations hold for all capacitors, regardless of their construction.

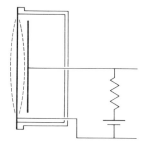

FIG. 23–4 Sound waves cause the diaphragm of a capacitor microphone to vibrate, and the resulting changes in its capacitance are transformed into electrical signals.

Example   A heart attack often leads to a condition called *fibrillation* in which the heart's actions lose their synchronization and it is unable to pump blood effectively. This condition can often be corrected by an electric shock to the heart that completely stops it for a moment; the heart may then start again spontaneously in its normal rhythm. An appropriate such shock can be provided by the discharge of a 10-μF capacitor that has been charged to a potential difference of 6000 V. (a) How much energy is released in the current pulse? (b) How much charge passes through the patient's body? (c) If the pulse lasts 5 ms, what is the average current?

Solution

$$(a)\ W = \tfrac{1}{2} CV^2 = \tfrac{1}{2}(10^{-5}\,\text{F})(6000\,\text{V})^2 = 180\,\text{J}$$

$$(b)\ Q = CV = (10^{-5}\,\text{F})(6000\,\text{V}) = 0.06\,\text{C}$$

$$(c)\ \bar{I} = \frac{Q}{t} = \frac{0.06\,\text{C}}{5 \times 10^{-3}\,\text{s}} = 12\,\text{A} \qquad \blacksquare$$

Table 23–1 is a summary of the various formulas for charge Q, potential difference V, capacitance C, and energy W that follow from the definition of capacitance $C = Q/V$ and from the energy formula $W = \frac{1}{2} QV$.

Arcing in a switch can be minimized by a capacitor

Although capacitors have their most important application in alternating-current circuits, as we shall find in the next chapter, they have direct-current uses as well, such as the ones described above. Another is to minimize the effects of arcing across switch contacts. Immediately after a switch is opened in a current-carrying circuit, the

Unknown Quantity	Known Quantities					
	C and V	C and Q	Q and V	W and C	W and V	W and Q
$Q =$	CV			$\sqrt{2WC}$	$\dfrac{2W}{V}$	
$V =$		$\dfrac{Q}{C}$		$\sqrt{\dfrac{2W}{C}}$		$\dfrac{2W}{Q}$
$C =$			$\dfrac{Q}{V}$		$\dfrac{2W}{V^2}$	$\dfrac{Q^2}{2W}$
$W =$	$\dfrac{CV^2}{2}$	$\dfrac{Q^2}{2C}$	$\dfrac{QV}{2}$			

TABLE 23-1

air gap between the switch contacts is so small that the voltage across them exceeds the insulating ability of the gap and an arc discharge occurs briefly. If the original current in the circuit was large, the contacts can be damaged. The problem is especially severe in direct-current motors and generators, whose brushes and commutators are really switches that may open and close hundreds of times per second. A capacitor connected across a switch prevents such arcing (Fig. 23–5). When the switch is closed, the capacitor is uncharged since its plates are at the same potential. When the switch is opened, the uncharged capacitor provides a path for current whose resistance is lower than that of the air gap, so the momentary current after the switch is opened goes to charge the capacitor rather than to produce an arc in the air gap.

FIG. 23–5 A capacitor connected across a switch prevents arcing when the switch is opened by providing an alternate path of lower resistance than the initial small air gap.

23–3 DIELECTRIC CONSTANT

Let us now examine what happens when a slab of an insulating material is placed between the plates of a capacitor.

Although an insulator cannot conduct electric current, it can respond to an electric field in another way. The molecules of all substances either normally have a nonuniform distribution of electric charge within them or assume such a distribution under the influence of an electric field. As we know, a molecule of the former kind is called a *polar molecule,* and behaves as though one end is positively charged and the other negatively charged. In an assembly of polar molecules when there is no external electric field, the molecules are randomly oriented as in Fig. 23–6(a). When an electric field is present, it acts to align the molecules opposite to the field, as in Fig. 23–6(b).

How polar molecules respond to electric fields

While nonpolar molecules ordinarily have symmetric charge distributions, an electric field is able to distort their arrangements of electrons so that an effective separation of charge takes place (Fig. 23–7). Again the molecules have their charged ends aligned opposite to the external field. In either case, then, the net electric field between the plates of the capacitor is *less* than it would be with nothing between them.

How nonpolar molecules respond to electric fields

The *dielectric constant,* symbol K, of a substance is a measure of how effective it is in reducing an electric field set up across a sample of it. For a capacitor with a given charge Q, reducing the electric field means reducing V as well, and since $C = Q/V$, this means an increase in its capacitance. If the capacitance of a capacitor is C_0 when there is a vacuum between its plates, its capacitance will be

Dielectric constant

$$C = KC_0 \qquad\qquad (23–12)$$

FIG. 23–6 An electric field tends to align polar molecules opposite to the field.

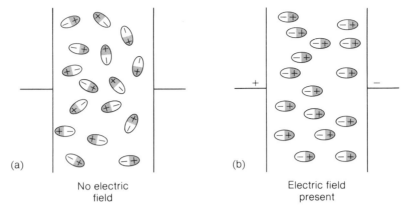

(a)   No electric field

(b)   Electric field present

FIG. 23–7 An electric field tends to distort the charge distributions in molecules that are ordinarily nonpolar.

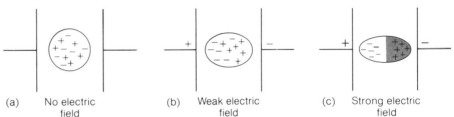

(a)   No electric field

(b)   Weak electric field

(c)   Strong electric field

when a substance of dielectric constant K is between the plates (Fig. 23–8). Table 23–2 is a list of dielectric constants for various substances. Water and alcohol molecules are highly polar, and the values of K for water, ice, and ethyl alcohol are accordingly high.

Example   A capacitor with air between its plates is connected to a 50-V source and then disconnected. The space between the plates of the charged capacitor is filled with Teflon ($K = 2.1$). What is the potential difference across the capacitor now?

Solution   The initial charge on the capacitor is $Q = C_1 V_1$. When the Teflon dielectric is inserted, the charge remains the same but the capacitance increases to $C_2 = KC_1$. The new voltage across the capacitor is therefore

$$V_2 = \frac{Q}{C_2} = \frac{C_1 V_1}{KC_1} = \frac{V_1}{K} = \frac{50\,\text{V}}{2.1} = 23.8\,\text{V}$$

The ratio V_2/V_1 is independent of the original capacitance of the capacitor.   ■

Dielectric strength

Besides its dielectric constant, an important property of a material used between the plates of a capacitor is its *dielectric strength,* which is the maximum electric field that can safely be applied to it before it breaks down and loses its insulating ability. Air has a dielectric strength of about 3 MV/m; thus a voltage of 300 V will produce a spark in an air gap of 0.1 mm. Most materials used as separators in capacitors and to insulate wires have dielectric strengths that exceed 10 MV/m. Rubber and Teflon, for instance, have dielectric strengths of 28 MV/m and 59 MV/m, respectively.

Capacitor construction

Commercial capacitors consist of many interleaved plates to make possible a high

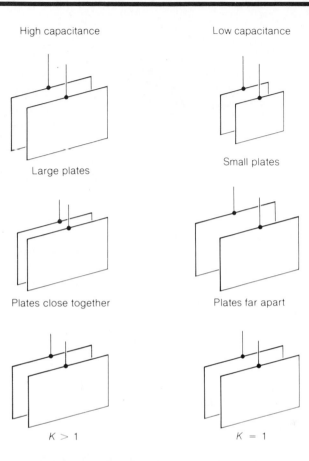

High capacitance

Large plates

Low capacitance

Small plates

Plates close together

Plates far apart

$K > 1$

$K = 1$

FIG. 23–8 The capacitance of a parallel-plate capacitor depends upon the area and spacing of its plates and upon the dielectric constant K of the medium between them.

Substance	K	Substance	K
Air	1.0006	Mica	2.5–7
Air, liquid ($-191°C$)	1.4	Neoprene	6.7
Alcohol, ethyl	26	Sulfur	3.9
Benzene	2.3	Teflon	2.1
Glass	5–8	Water	80
Ice ($-2°C$)	94	Waxed paper	2.2

TABLE 23–2

capacitance in a small unit (Fig. 23–9). Solid dielectrics are usually used, both to increase C and to maintain a fixed distance between the plates. With sheets of a solid dielectric such as waxed paper or mica between them, the plates can be of inexpensive metal foil, without the rigidity required to prevent accidental contact if only air separated them. In an *electrolytic capacitor* the plates have extremely thin dielectric layers formed on their surfaces by chemical action. A conducting paste between these plates constitutes the other electrode of the capacitor. The very small thickness of the dielectrics in electrolytic capacitors permits them to have capacitances of over 100 μF without excessive bulk.

Variable capacitors normally have two sets of rigid aluminum plates that are inter-

Variable capacitors

FIG. 23–9 Most fixed capacitors are made of interleaved sheets of metal foil separated by layers of dielectric material.

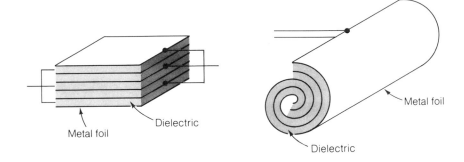

leaved with air as the dielectric, as in Fig. 23–10. One of the sets is mounted on a shaft, and by rotating the shaft the amount of overlap between the plates can be adjusted. Since the overlapped area of the plates is the chief contributor to their capacitance, turning the shaft varies C.

As discussed in Chapter 22, the process of induction heating uses an alternating magnetic field to induce eddy currents in a metal object. The resulting I^2R losses heat the object. A corresponding effect occurs in nonconducting materials when an alternating electric field is applied. Here the heat is produced as the polarizations of the molecules change back and forth. *Dielectric* (or *capacitive*) *heating* has the same basic advantage as induction heating: The heat is produced uniformly in the interior of the object, rather than being transferred from the outside. Dielectric heating makes it possible to bring the object to a high temperature rapidly with no danger of burning its exterior.

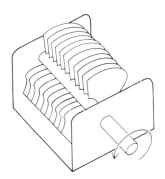

FIG. 23–10 Variable capacitors of this type are widely used to tune radio receivers.

23–4 CAPACITORS IN COMBINATION

The equivalent capacitance of two or more capacitors connected together can be determined in a manner analogous to that used in the case of resistors in combination, though the results are different.

Capacitors in series

Figure 23–11 shows three capacitors in series. Each has charges of the same magnitude Q on its plates, in agreement with the principle of conservation of charge. Hence the potential differences across the capacitors are respectively

$$V_1 = \frac{Q}{C_1} \qquad V_2 = \frac{Q}{C_2} \qquad V_3 = \frac{Q}{C_3}$$

and so, if C is the equivalent capacitance of the set, we have

$$V = V_1 + V_2 + V_3$$

$$\frac{Q}{C} = \frac{Q}{C_1} + \frac{Q}{C_2} + \frac{Q}{C_3}$$

$$\frac{1}{C} = \frac{1}{C_1} + \frac{1}{C_2} + \frac{1}{C_3}$$

For any number of capacitors in series,

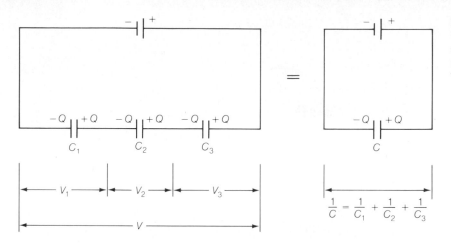

FIG. 23–11 Capacitors in series. A charge of the same magnitude Q is present on all the plates of the capacitors.

$$\frac{1}{C} = \frac{1}{C_1} + \frac{1}{C_2} + \frac{1}{C_3} + \cdots \qquad \textit{Capacitors in series} \quad (23\text{–}13)$$

The reciprocal of the equivalent capacitance of a series arrangement of capacitors is equal to the sum of the reciprocals of the individual capacitors. Evidently C is smaller than the capacitance of any of the individual capacitors.

If there are only two capacitors in series, Eq. (23–13) becomes

Two capacitors in series

$$C = \frac{C_1 C_2}{C_1 + C_2} \qquad\qquad (23\text{–}14)$$

As in the case of the similar Eq. (19–11) for resistors in parallel, a calculator that has a reciprocal [1/X] key makes it easy to solve Eq. (23–13) for C in an actual problem. If three capacitors are in series, the key sequence would be: $[C_1][1/X][+][C_2]$ $[1/X][+][C_3][1/X][=][1/X]$.

Example Two capacitors, one of 10 μF and the other of 20 μF, are connected in series across a 12-V battery, as in Fig. 23–12. Find the equivalent capacitance of the combination, the charge on each capacitor, and the potential difference across it.

Solution The equivalent capacitance of the two capacitors is

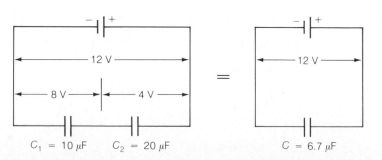

FIG. 23–12

$$C = \frac{C_1 C_2}{C_1 + C_2} = \frac{(10\,\mu F)(20\,\mu F)}{10\,\mu F + 20\mu F} = 6.7\,\mu F$$

and so the charges on them are

$$Q_1 = Q_2 = Q = CV = (6.7 \times 10^{-6}\,F)(12\,V) = 8.0 \times 10^{-5}\,C$$

The potential differences across the capacitors are respectively

$$V_1 = \frac{Q}{C_1} = \frac{8.0 \times 10^{-5}\,C}{1.0 \times 10^{-5}\,F} = 8.0\,V$$

$$V_2 = \frac{Q}{C_2} = \frac{8.0 \times 10^{-5}\,C}{2.0 \times 10^{-5}\,F} = 4.0\,V$$

The potential difference is greatest across the capacitor of smaller C. The sum of V_1 and V_2 is 12 V, as it should be. ∎

Capacitors in parallel   Figure 23–13 shows three capacitors connected in parallel. The same potential difference is across all of them, so that the charges on their plates have the respective magnitudes

$$Q_1 = C_1 V \qquad Q_2 = C_2 V \qquad Q_3 = C_3 V$$

The total charge $Q_1 + Q_2 + Q_3$ on either the positive or negative plates of the capacitors is equal to the charge Q on the corresponding plate of the equivalent capacitor, and hence

$$Q = Q_1 + Q_2 + Q_3$$
$$CV = C_1 V + C_2 V + C_3 V$$
$$C = C_1 + C_2 + C_3$$

FIG. 23–13 Capacitors in parallel. The same potential difference V is across all of them, but each has a charge on its plates whose magnitude is proportional to its capacitance.

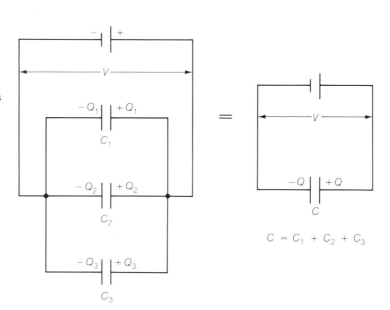

This result can be generalized to

$$C = C_1 + C_2 + C_3 + \cdots \qquad \textit{Capacitors in parallel} \quad (23\ 15)$$

More energy is stored in a set of capacitors when they are connected in parallel across a given potential difference than when they are connected in series, because the electric fields in them are stronger in the parallel case.

Example   The capacitors of Fig. 23–12 are reconnected in parallel across the same battery. Compare the energies of the capacitors now with what they were when connected in series.

Solution   In the series connection, the capacitors had the energies

$$W_1\,(\text{series}) = \tfrac{1}{2}\,QV_1 = \tfrac{1}{2}\,(8.0 \times 10^{-5}\text{C})(8.0\,\text{V}) = 3.2 \times 10^{-4}\text{J}$$

$$W_2\,(\text{series}) = \tfrac{1}{2}\,QV_2 = \tfrac{1}{2}\,(8.0 \times 10^{-5}\text{C})(4.0\,\text{V}) = 1.6 \times 10^{-4}\text{J}$$

The capacitor of smaller C had the greater energy. In the parallel connection, the capacitors have the same potential difference $V = 12$ V, and their energies are

$$W_1(\text{parallel}) = \tfrac{1}{2}\,C_1 V^2 = \tfrac{1}{2}\,(1.0 \times 10^{-5}\text{F})(12\,\text{V})^2 = 7.2 \times 10^{-4}\text{J}$$

$$W_2(\text{parallel}) = \tfrac{1}{2}\,C_2 V^2 = \tfrac{1}{2}\,(2.0 \times 10^{-5}\text{F})(12\,\text{V})^2 = 14 \times 10^{-4}\text{J}$$

In this parallel connection, the stored energy is 4.4 times as great as it was in the series connection. ∎

23–5 MAGNETIC FIELD ENERGY

Energy is associated with magnetic as well as with electric fields. To appreciate why, let us look at what happens when a wire loop is connected to a battery. As the current starts to flow, a magnetic field begins to build up. But this changing magnetic field induces an emf in the *same* wire loop whose current causes the field in the first place. By Lenz's law the self-induced emf is such as to oppose the change in flux that is responsible for it. The battery must therefore push electrons through the wire loop against the self-induced emf due to the increasing magnetic field, which means that work has to be done in order to produce the magnetic field. (The work needed to overcome the effect of the self-induced emf has nothing to do with the resistance of the wire; it is a consequence of electromagnetic induction only.)

A self-induced emf occurs whenever the current in a wire loop changes

When the current in the wire loop reaches its final value, there is no more self-induced emf. The work performed to establish the magnetic field has become magnetic energy.

The amount of energy contained in the magnetic field of a solenoid when a steady current I flows through it is equal to the work that had to be done against the self-induced emf $\mathscr{E}$ in order to establish the current starting from $I = 0$ (Fig. 23–14). This amount of energy turns out to be

Magnetic energy of current-carrying solenoid

$$W = \frac{lA}{2\mu}\,B^2 \qquad \textit{Magnetic field energy of solenoid} \quad (23\text{–}16)$$

FIG. 23–14 A self-induced emf opposes the establishment of a current in a solenoid. The work done against this emf becomes magnetic energy.

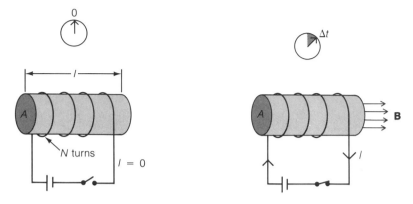

Energy density of magnetic field

The quantity lA is the volume occupied by the magnetic field **B,** since it is the product of the length of the solenoid and its cross-sectional area. (Since we are considering an ideal solenoid whose length is large relative to its diameter, we can ignore the magnetic field escaping from the ends.) The *magnetic energy density* of the field is therefore

$$w = \frac{W}{lA} = \frac{B^2}{2\mu} \qquad\qquad\qquad \textit{Magnetic energy density} \quad (23\text{–}17)$$

The energy density of a magnetic field is directly proportional to the square of its magnitude B. This formula holds for all magnetic fields, not just for those inside solenoids. We recall from Eq. (23–6) that the energy density of an electric field **E** is similarly proportional to E^2.

Example   The earth's magnetic field in a certain region has the magnitude 6×10^{-5} T. Find the magnetic energy per cubic kilometer in this region.

Solution   The magnetic energy density corresponding to $B = 6 \times 10^{-5}$ T is, since here $\mu = \mu_0$,

$$w = \frac{B^2}{2\mu_0} = \frac{(6 \times 10^{-5}\,\text{T})^2}{2(1.26 \times 10^{-6}\,\text{T} \cdot \text{m/A})} = 1.4 \times 10^{-3}\,\text{J/m}^3$$

A cubic kilometer is equal to $(10^3\,\text{m})^3 = 10^9\,\text{m}^3$, so the total magnetic energy in a cubic kilometer is

$$W = wV = (1.4 \times 10^{-3}\,\text{J/m}^3)(10^9\,\text{m}^3) = 1.4 \times 10^6\,\text{J}$$

This amount of energy is enough to raise the 365,000-ton Empire State Building by 0.42 mm. ■

23–6   INDUCTANCE

A circuit element in which a self-induced emf accompanies a changing current is called an *inductor.* A solenoid is an example of an inductor.

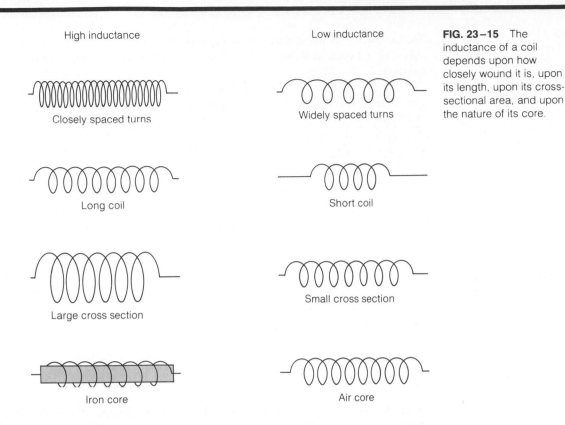

High inductance

Closely spaced turns

Long coil

Large cross section

Iron core

Low inductance

Widely spaced turns

Short coil

Small cross section

Air core

FIG. 23–15 The inductance of a coil depends upon how closely wound it is, upon its length, upon its cross-sectional area, and upon the nature of its core.

Normally the energy content W of an inductor is most conveniently expressed in terms of the current I present in it. Since the magnetic field B in an inductor is proportional to I in the absence of ferromagnetic materials, W is proportional to I^2. The constant of proportionality is written as $L/2$, where L is called the *inductance* of the inductor. Thus

$$W = \tfrac{1}{2} LI^2 \qquad\qquad \text{Potential energy of inductor} \quad (23\text{–}18)$$

An inductor stores energy in the form of magnetic field

The unit of inductance is the *henry* (H), where

$$1\,\text{H} = 1\,\frac{\text{J}}{\text{A}^2} = 1\,\frac{\text{V} \cdot \text{s}}{\text{A}}$$

The henry, millihenry, and microhenry

The henry, like the farad, is usually too large a unit for convenience. Inductances are accordingly often expressed in *millihenries* (mH) or *microhenries* (μH), where

$$1\,\text{mH} = 10^{-3}\,\text{H} \qquad 1\,\mu\text{H} = 10^{-6}\,\text{H}$$

The inductance of an inductor depends upon its geometry and upon the presence of a core with magnetic properties. A coil consisting of many turns has a greater inductance than one consisting of a few turns because its magnetic field is stronger for a given current (Fig. 23–15). The longer the coil and the greater its cross-sectional area, the greater its inductance because it contains a larger volume of magnetic field. A core in a coil changes its inductance by changing the flux through the coil (see

Factors that influence inductance

Inductance of a solenoid

Section 21–4). In particular, a ferromagnetic core increases the inductance considerably—by a factor of as much as 10^4 in some cases—but by an amount that is not constant.

Let us calculate the inductance of a solenoid. From Eqs. (23–16) and (23–18), which are different ways to express the energy of a solenoid when it carries the current I, we have

$$\frac{lA}{2\mu} B^2 = \tfrac{1}{2} LI^2$$

$$L = \frac{lA}{\mu I^2} B^2$$

Since the magnetic field within the solenoid is, from Eq. (21–5),

$$B = \mu \frac{N}{l} I$$

we see that

$$L = \left(\frac{lA}{\mu I^2} \right) \left(\mu \frac{N}{l} I \right)^2$$

$$= \mu N^2 \frac{A}{l} \qquad\qquad\qquad \textit{Inductance of solenoid} \quad (23\text{–}19)$$

Example   Find the inductance in air of a 1000-turn solenoid 10 cm long that has a cross-sectional area of 20 cm^2. How much energy is stored in the magnetic field of the solenoid when it carries a 0.01-A current?

Solution   Since 10 cm $= 10^{-1}$ m, 20 cm$^2 = 2 \times 10^{-3}$ m^2, and $\mu = \mu_0$ in air (very nearly),

$$L = \mu_0 N^2 \frac{A}{l} = \left(1.26 \times 10^{-6} \frac{\text{T} \cdot \text{m}}{\text{A}} \right) (10^3)^2 \left(\frac{2 \times 10^{-3} \text{m}^2}{10^{-1} \text{m}} \right)$$

$$= 2.5 \times 10^{-2} \text{H} = 25 \,\text{mH}$$

The energy stored in the solenoid's magnetic field when $I = 0.01$ A $= 10^{-2}$ A is

$$W = \tfrac{1}{2} LI^2 = \tfrac{1}{2} (2.5 \times 10^{-2} \text{H})(10^{-2} \text{A})^2 = 1.3 \times 10^{-6} \text{J}$$

■

23–7   SELF-INDUCED EMF

It is easy to show that the inductance L of an inductor determines the magnitude of the self-induced emf $\mathscr{E}$ that accompanies a changing current in the inductor. Let us consider a current that increases uniformly from $I = 0$ to $I = \Delta I$ in a time interval Δt. The average current $\overline{I}$ during Δt is $\tfrac{1}{2} \Delta I$ and the total charge Q that passes through the coil while the current is building up to its final value of ΔI is

$$Q = \bar{I}\,\Delta t = \tfrac{1}{2}\,\Delta I\,\Delta t$$

The work that was done during the buildup of the current is the product of the total charge Q and the emf $\mathscr{E}$ and is negative since it is done *against* the emf:

Work done against self-induced emf

$$W = -Q\mathscr{E} = -\tfrac{1}{2}\,\Delta I\mathscr{E}\,\Delta t$$

From Eq. (23–18) we have another expression for W, namely $W = \tfrac{1}{2}L(\Delta I)^2$, so that

$$\tfrac{1}{2}L(\Delta I)^2 = -\tfrac{1}{2}\Delta I\mathscr{E}\,\Delta t$$

Solving for the self-induced emf $\mathscr{E}$ gives the important formula

$$\mathscr{E} = -L\frac{\Delta I}{\Delta t} \qquad\qquad \textit{Self-induced emf} \quad (23\text{–}20)$$

A high value of L means a large $\mathscr{E}$ for a given rate of change $\Delta I/\Delta t$, and a low value of L means a small $\mathscr{E}$. Because $\mathscr{E}$ is opposite in sign to $\Delta I/\Delta t$, Eq. (23–20) is in accord with Lenz's law.

Example   An average self-induced emf of -0.75 V is produced in a 25-mH coil when the current in it falls to 0 in 0.01 s. What was the original current in the coil?

Solution   From Eq. (23–20),

$$\Delta I = \frac{-\mathscr{E}\,\Delta t}{L} = \frac{-(-0.75\ \text{V})(0.01\ \text{s})}{2.5 \times 10^{-2}\,\text{H}} = 0.3\ \text{A}$$

Since $\Delta I = I_0 - 0$, the original current was $I_0 = 0.3$ A.   ■

23–8   ELECTRICAL OSCILLATIONS

Let us connect a charged capacitor to an inductor, as in Fig. 23–16. The following sequence of events occurs:

Oscillations of an LC circuit

(a) Initially the capacitor's plates each possess the charge Q and its electric energy is $\tfrac{1}{2}Q^2/C$, while the inductor has no energy since $I = 0$.
(b) The capacitor is partially discharged, and the current in the inductor leads to a magnetic field whose energy is $\tfrac{1}{2}Li^2$.
(c) No charge is left on the capacitor's plates, and all the energy of the circuit is magnetic energy $\tfrac{1}{2}LI^2$.
(d) Now charge begins to build up on the capacitor's plates with polarities the reverse of those originally present, and the electric energy of the capacitor grows at the expense of the magnetic energy of the inductor as the current drops.
(e) At the instant the capacitor is fully charged again, the current is 0 and the energy of the circuit is once more entirely electric.
(f) Next the capacitor begins to discharge, and eventually the entire circuit returns to its original state at (a). The cycle will continue to repeat itself indefinitely if no resistance is present.

FIG. 23–16 The various stages in the oscillation of a circuit containing the inductance L and the capacitance C. The maximum charge on the capacitor is Q, and the maximum current in the circuit is I.

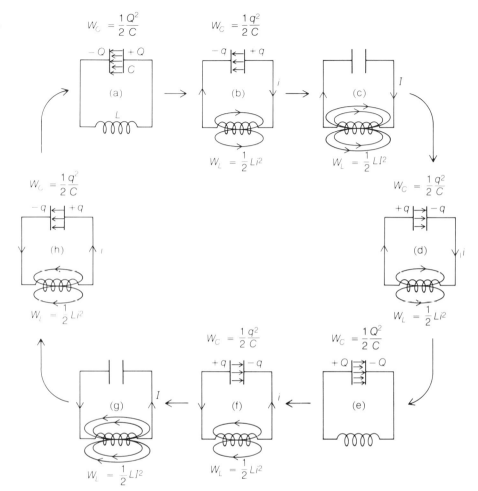

Electrical oscillator is analog of harmonic oscillator

The above sequence has an exact counterpart in the behavior of a harmonic oscillator. As we learned in Chapter 12, the basic process involved in the operation of a harmonic oscillator is the continual interchange of energy between kinetic energy and potential energy. Here the same sort of interchange occurs, and we can regard the electric energy of a charged capacitor as corresponding to potential energy and the magnetic energy of an inductor through which a current flows as corresponding to kinetic energy.

A harmonic oscillator whose mass is m and whose spring constant is k has the potential and kinetic energies

Harmonic oscillator

$$\text{PE} = \tfrac{1}{2}ks^2 \qquad \text{KE} = \tfrac{1}{2}mv^2$$

where s is the displacement of the object from its equilibrium position and v is its speed. The corresponding formulas for the electric and magnetic energies in an electrical oscillator are

Electrical oscillator

$$W_e = \frac{1}{2}\frac{1}{C}Q^2 \qquad W_m = \frac{1}{2}LI^2$$

We note that $1/C$ corresponds to k, since both determine the amount of energy present at the moment when nothing is moving in the respective oscillators. Similarly L corresponds to m. Both are measures of inertia—a large inductance tends to slow down changes in current, and a large m tends to slow down changes in speed. The frequency of a harmonic oscillator is

$$f = \frac{1}{2\pi}\sqrt{\frac{k}{m}}$$

Harmonic oscillator frequency

and so it is not surprising that the frequency of an electrical oscillator is

$$f = \frac{1}{2\pi\sqrt{LC}}$$

Oscillator frequency   (23–21)

Electrical oscillator frequency

The larger the inductance L and the capacitance C, the lower the frequency.

The parallel between electrical and mechanical oscillations is a very close one: Just as every mechanical system has certain specific natural frequencies of vibration that depend upon its properties, so every electric circuit has a natural frequency of oscillation that depends upon L and C. Just as the energy of every actual mechanical vibration is eventually dissipated as heat owing to the inevitable presence of friction, so the energy of every electrical oscillation is eventually dissipated as heat owing to the inevitable presence of resistance in the wires of the circuit. Electrical resonance is much like mechanical resonance: If a source of potential that alternates at the natural frequency of a circuit is connected to it, energy is fed into the oscillations that can maintain or increase their amplitude. At any other frequency the energy absorbed is small.

Radio communication is made possible by the ability of an LC circuit to oscillate only at the frequency given by Eq. (23–21). Such a circuit is used to "tune" a transmitter so that only a single frequency of radio waves is produced, and another similar circuit is used to "tune" a receiver to that same frequency. Electrical oscillations are discussed further in the next chapter.

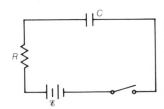

FIG. 23–17 A circuit that contains a capacitor, a resistor, and a source of emf in series. When the switch is closed, the charge on the capacitor increases gradually to its ultimate value of $C\mathscr{E}$.

Example   In the antenna circuit of a certain radio receiver, the inductance is fixed at 4 mH but the capacitance can be varied. When the capacitance is 10 pF, what is the frequency of the radio waves the receiver responds to?

Solution   From Eq. (23–21),

$$f = \frac{1}{2\pi\sqrt{LC}} = \frac{1}{2\pi\sqrt{(4\times10^{-3}\,\text{H})(10\times10^{-12}\,\text{F})}}$$
$$= 7.96\times10^5\,\text{Hz} = 796\,\text{kHz}$$

23–9   TIME CONSTANTS

Circuits that contain resistance as well as capacitance or inductance do not respond instantaneously to changes in the applied emf. For instance, when a capacitor is connected to a battery, as in Fig. 23–17, it does not immediately become fully charged. At first the only limit to the current that flows to the capacitor is the resistance R in

Charging or discharging a capacitor takes time

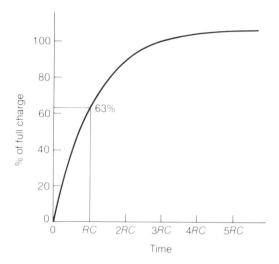

FIG. 23–18 The growth of charge in a capacitor.

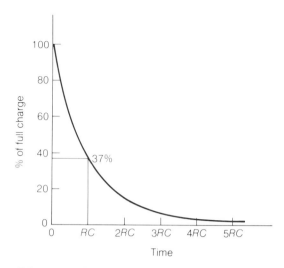

FIG. 23–19 The decay of charge in a circuit containing capacitance and resistance when the battery is short-circuited.

the circuit, so that the initial current is $I = \mathcal{E}/R$, where $\mathcal{E}$ is the emf of the battery. As the capacitor becomes charged, however, a potential difference appears across it, whose polarity is such as to tend to oppose the further flow of current. When the charge on the capacitor has built up to some value Q, this opposing potential difference is $V = Q/C$. Hence the net potential difference is $\mathcal{E} - Q/C$, and the current is

$$I = \frac{\mathcal{E} - Q/C}{R}$$
(23–22)
$$\text{Current} = \frac{\text{impressed emf} - \text{voltage across capacitor}}{\text{resistance}}$$

As Q increases, then, its *rate* of increase drops. This gives a steadily increasing slope in the curve of Fig. 23–18, which is a graph showing how Q varies with time when a capacitor is being charged; the capacitor is connected to the battery at $t = 0$.

Time constant of capacitive circuit

A mathematical analysis of Eq. (23–22) shows that after a time interval of RC (the product of the resistance R in the circuit and the capacitance C of the capacitor), the charge on the capacitor reaches 63% of its ultimate value of $Q_0 = C\mathcal{E}$. The time RC is therefore a convenient measure of how rapidly the capacitor becomes charged and is accordingly called the *time constant* of the circuit:

$$T_C = RC \qquad\qquad \textit{Time constant of RC circuit} \quad (23–23)$$

In principle the capacitor acquires its ultimate charge Q_0 only after an infinite time has elapsed, but, as we can see from Fig. 23–18, this value is very nearly reached after only a few time constants; at $t = 3RC$, the charge is 95% of Q_0, and at $t = 4RC$ it is 98% of Q_0.

If a capacitor with an initial charge is discharged through a resistance, its charge decreases with time as shown in Fig. 23–19. After the time RC the charge on the capacitor is reduced to 37% of its original value, a drop of 63%.

Example A 5-μF capacitor is charged by being connected to a 1.5-V dry cell. The total resistance of the circuit is 2 Ω. (a) What is the final charge on the capacitor and how long does it take to reach 63% of this charge? (b) The battery is then disconnected from the capacitor. If the resistance of the dielectric between the capacitor plates is 10^{10} Ω, after what period of time will the charge on the capacitor drop to 37% of its initial value?

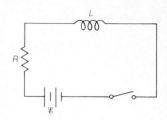

FIG. 23–20 A circuit with an inductor, a resistor, and a source of emf in series. When the switch is closed, the current increases gradually to its ultimate value of $\mathscr{E}/R$ because of the opposing self-induced emf in the inductor.

Solution (a) The ultimate charge on the capacitor is

$$Q = CV = (5 \times 10^{-6}\text{ F})(1.5\text{ V}) = 7.5 \times 10^{-6}\text{ C}$$

The time constant of the circuit is

$$T_C = RC = (2\Omega)(5 \times 10^{-6}\text{ F}) = 10^{-5}\text{ s}$$

and so this is the time needed for the capacitor to acquire 63% of its ultimate charge.

(b) When the battery is disconnected from the capacitor, charge gradually leaks through the dielectric of the latter. The time constant for discharge is

$$T_C = RC = (10^{10}\Omega)(5 \times 10^{-6}\text{ F}) = 5 \times 10^{4}\text{s} = 14\text{ h}$$

and in this period the charge on the capacitor will drop to 37% of its original value.   ∎

When a circuit containing inductance is connected to a battery, the current in the circuit does not rise instantly to its ultimate value $I = \mathscr{E}/R$, where $\mathscr{E}$ is the emf of the battery and R is the total resistance in the circuit. As the switch in Fig. 23–20 is closed, the current I starts to grow, and as a result the induced emf $- L(\Delta I/\Delta t)$ comes into being in the opposite direction to the battery emf $\mathscr{E}$. The net emf acting to establish current in the circuit is therefore $\mathscr{E} - L(\Delta I/\Delta t)$, and the current reaches its final value of I_0 in a gradual manner.

When the current in the circuit of Fig. 23–20 is I and is changing at the rate $\Delta I/\Delta t$,

A current in an inductor takes time to build up or to disappear

$$\mathscr{E} - \text{L}\frac{\Delta I}{\Delta t} = IR \tag{23–24}$$

Impressed emf − induced emf = net voltage

The rate at which the current is increasing is therefore given by

$$\frac{\Delta I}{\Delta t} = \frac{\mathscr{E} - IR}{L}$$

The larger the inductance L, the more gradually the current increases. At the moment the switch is closed, $I = 0$ and $\Delta I/\Delta t$ has its maximum value of

$$\left(\frac{\Delta I}{\Delta t}\right)_{\text{max}} = \frac{\mathscr{E}}{L}$$

FIG. 23–21   The growth of current in a circuit containing inductance and resistance.

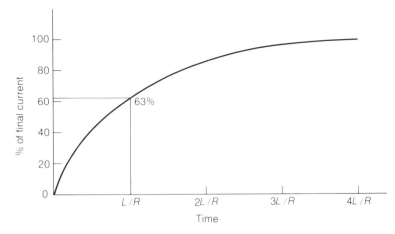

Eventually, the current reaches its final value of I_0 and $\Delta I/\Delta t = 0$. From then on

$$\frac{\Delta I}{\Delta t} = 0 = \frac{\mathscr{E} - I_0 R}{L}$$

$$I_0 = \frac{\mathscr{E}}{R}$$

Thus the effect of having inductance in the circuit is to delay the establishment of the final current.

Time constant of inductive circuit

The graph in Fig. 23–21 shows how I varies with time when a current is being established in a circuit containing inductance. A mathematical analysis shows that, after a time interval of L/R, the current reaches 63% of I_0. The time L/R is therefore a convenient measure of how rapidly a current rises in a circuit containing inductance, and, like RC in a circuit containing capacitance, is called the *time constant* of the circuit.

$$T_L = \frac{L}{R} \qquad\qquad \textit{Time constant of RL circuit}\quad (23\text{--}25)$$

When the battery in Fig. 23–20 is short-circuited by a wire, the current I drops slowly in the manner shown in Fig. 23–22, since the induced emf now tends to maintain the existing current. The current falls to 37% of its original value in the time L/R after the battery is short-circuited.

A coil of inductance L has the magnetic energy of $\frac{1}{2} LI^2$ stored in it when a current I is present in it. When the potential difference across the coil that is responsible for the current is shorted out, the energy $\frac{1}{2} LI^2$ is what powers the self-induced emf that retards the drop in current. The gradual rise of current in a circuit containing inductance may be thought of as the result of the initial absorption of $\frac{1}{2} LI^2$ of potential energy by the circuit, and its gradual drop may be thought of as the result of the transfer of the inductor's potential energy to the circuit.

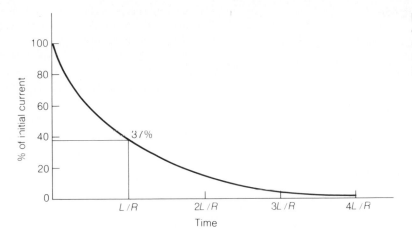

Example A 2-H inductor whose resistance is 100 Ω is connected to a 24-V battery of negligible internal resistance. Find the initial current in the circuit and the initial rate at which the current is increasing, the time required for the current to reach 63% of its ultimate value, and the magnitude of the final current.

Solution The initial current is 0, and the initial rate of current increase is

$$\frac{\Delta I}{\Delta t} = \frac{\mathscr{E}}{L} = \frac{24\,\text{V}}{2\,\text{H}} = 12\,\text{A/s}$$

The time needed for the current to rise to 63% of the final value is

$$T_L = \frac{L}{R} = \frac{2\,\text{H}}{100\,\Omega} = 0.02\,\text{s}$$

and the final current will be

$$I_0 = \frac{\mathscr{E}}{R} = \frac{24\,\text{V}}{100\,\Omega} = 0.24\,\text{A}$$ ∎

IMPORTANT TERMS

The **energy density** of an electric field is the electric potential energy per unit volume associated with it. The **energy density** of a magnetic field is the magnetic energy per unit volume associated with it.

A **capacitor** is a device that stores electrical energy in the form of an electric field. The ratio between the charge on either plate of a capacitor and the potential difference between the plates is called its **capacitance**. The unit of capacitance is the **farad**, which is equal to 1 coulomb per volt.

The **dielectric constant** K of a particular material is a measure of how effective it is in reducing an electric field set up across a sample of it.

The **inductance** L of a circuit is the ratio between the magnitude of the self-induced emf $\mathscr{E}$ due to a changing current in it and the rate of change $\Delta I / \Delta t$ of the current. The unit of inductance is the **henry**.

IMPORTANT FORMULAS

Electric energy density: $w = \dfrac{E^2}{8\pi k}$

Capacitance: $C = \dfrac{Q}{V}$

Parallel-plate capacitor:

$C = \dfrac{KA}{4\pi kd}$ $(K = \text{dielectric constant})$

Potential energy of charged capacitor:

$W = \frac{1}{2} QV = \frac{1}{2} CV^2 = \dfrac{Q^2}{2C}$

Capacitors in series: $\dfrac{1}{C} = \dfrac{1}{C_1} + \dfrac{1}{C_2} + \dfrac{1}{C_3} + \cdots$

Capacitors in parallel: $C = C_1 + C_2 + C_3 + \cdots$

Magnetic energy density: $w = \dfrac{B^2}{2\mu}$

Inductance of solenoid: $L = \mu N^2 \dfrac{A}{l}$

Potential energy of inductor: $W = \frac{1}{2} LI^2$

Self-induced emf: $\mathcal{E} = -L \dfrac{\Delta I}{\Delta t}$

Oscillator frequency: $f = \dfrac{1}{2\pi \sqrt{LC}}$

Time constants: $T_C = RC$

$T_L = \dfrac{L}{R}$

MULTIPLE CHOICE

1. The energy content of a charged capacitor resides in its
  a. plates.
  b. potential difference.
  c. charge.
  d. electric field.

2. One picofarad (pF) is equal to
  a. 10^{-6} F.
  b. 10^{-9} F.
  c. 10^{-12} F.
  d. 10^{-3} μF.

3. The potential energy stored in a charged capacitor is not equal to
  a. $V/2Q^2$.
  b. $QV/2$.
  c. $CV^2/2$.
  d. $Q^2/2C$.

4. A material with a high dielectric constant K is placed between the plates of a charged capacitor. As a result

  a. the capacitance of the capacitor decreases.
  b. the electric field between the plates decreases.
  c. the charge on the plates increases.
  d. the energy stored in the capacitor increases.

5. Doubling the distance between the plates of a parallel-plate capacitor of capacitance C changes its capacitance to
  a. $C/4$.
  b. $C/2$.
  c. $2C$.
  d. $4C$.

6. Magnetic fields invariably contain
  a. a ferromagnetic material.
  b. inductance.
  c. electric current.
  d. energy.

7. The unit of inductance is the henry, where 1 H =
  a. $1\ \text{J} \cdot \text{A}^2$.
  b. $1\ \text{J/A}^2$.
  c. $1\ \text{V} \cdot \text{A}$.
  d. $1\ \text{V/A}$.

8. A wire coil carries the current I. The potential energy of the coil does not depend upon
  a. the value of I.
  b. the number of turns in the coil.
  c. whether the coil has an iron core or not.
  d. the resistance of the coil.

9. A large increase in the inductance of a coil can be achieved by using a core that is
  a. conducting.
  b. insulating.
  c. ferromagnetic.
  d. dielectric.

10. An LC circuit has the initial capacitance C. In order to double its natural frequency of oscillation, the capacitance must be changed to
  a. $\frac{1}{4} C$.
  b. $\frac{1}{2} C$.
  c. $2C$.
  d. $4C$.

11. The time constant of an RL circuit is the time needed for the current to reach which percentage of its final value?
  a. 50%
  b. 63%
  c. 90%
  d. 100%

12. A capacitor acquires a charge of 0.002 C when connected across a 50-V battery. Its capacitance is
  a. 1 μF.
  b. 2 μF.
  c. 4 μF.
  d. 40 μF.

13. A 50-μF capacitor has a potential difference of 8 V across it. Its charge is
  a. 4×10^{-3} C.
  b. 4×10^{-4} C.
  c. 6.25×10^{-5} C.
  d. 6.25×10^{-6} C.

14. The plates of a parallel-plate capacitor of capacitance C are brought together to one-third their original separation. The capacitance is now
  a. $\frac{1}{9} C$.
  b. $\frac{1}{3} C$.
  c. $3C$.
  d. $9C$.

15. If a 20-μF capacitor is to have an energy content of 2.5 J, it must be placed across a potential difference of
- a. 150 V.
- b. 350 V.
- c. 500 V.
- d. 250,000 V.

16. A parallel-plate capacitor has a capacitance of 50 pF in air and 110 pF when immersed in turpentine. The dielectric constant of turpentine is
- a. 0.45.
- b. 0.55.
- c. 1.1.
- d. 2.2.

17. Two 50-μF capacitors are connected in series. The equivalent capacitance of the combination is
- a. 25 μF.
- b. 50 μF.
- c. 100 μF.
- d. 200 μF.

18. The capacitor combination of Question 17 is connected across a 100-V battery. The potential difference across each capacitor is
- a. 25 V.
- b. 50 V.
- c. 100 V.
- d. 200 V.

19. Two 50-μF capacitors are connected in parallel. The equivalent capacitance of the combination is
- a. 25 μF.
- b. 50 μF.
- c. 100 μF.
- d. 200 μF.

20. The capacitor combination of Question 19 is connected across a 100-V battery. The potential difference across each capacitor is
- a. 25 V.
- b. 50 V.
- c. 100 V.
- d. 200 V.

21. One hundred 8-μF capacitors are connected in parallel to a 120-V source. The energy stored in the capacitors could keep a 5-W lamp lit for
- a. 9.6 ms.
- b. 1.15 s.
- c. 2.3 s.
- d. 1.8×10^{10} s.

22. A 1-μF capacitor and a 3-μF capacitor are both charged to 100 V. They are then connected with like charges together. The energy stored in the system is
- a. 0.1 mJ.
- b. 0.01 J.
- c. 0.02 J.
- d. 0.04 J.

23. If the above capacitors were instead connected with unlike charges together, the energy stored in the system would be
- a. 0.005 J.
- b. 0.01 J.
- c. 0.013 J.
- d. 0.02 J.

24. The energy contained in a cubic meter of space in which the magnetic induction is 1 T is
- a. 6.3×10^{-7} J.
- b. 3.97×10^{-5} J.
- c. 3.97×10^{5} J.
- d. 7.94×10^{5} J.

25. The self-induced emf in a 0.1-H coil when the current in it is changing at the rate of 200 A/s is
- a. 125 V.
- b. 20 V.
- c. 8×10^{-4} V.
- d. 8×10^{-5} V.

26. The current in a circuit falls to 0 from 16 A in 0.01 s. The average emf induced in the circuit during the drop is 64 V. The inductance of the circuit is
- a. 0.032 H.
- b. 0.04 H.
- c. 0.25 H.
- d. 4 H.

27. A 2-mH coil carries a current of 10 A. The energy stored in its magnetic field is
- a. 0.05 J.
- b. 0.1 J.
- c. 1.0 J.
- d. 100 J.

28. What current should flow in a 50-mH coil in order that its energy be 1 J?
- a. 3.2 A
- b. 4.5 A
- c. 6.3 A
- d. 40 A

29. An *LC* circuit contains a 10-pF capacitor and a 4-mH inductor. Its natural frequency of oscillation is
- a. 8×10^{5} Hz.
- b. 5×10^{6} Hz.
- c. 4×10^{12} Hz.
- d. 2.5×10^{13} Hz.

30. A capacitor has a potential difference of 50 V when its charge is 1 mC. If the capacitor is discharged through a 400-Ω resistor, the time needed for its charge to fall to 37% of its initial value is
- a. 0.008 s.
- b. 0.05 s.
- c. 20 s.
- d. 125 s.

31. A 0.05-H inductor of resistance 20 Ω is connected to a 24-V battery of negligible internal resistance. The initial rate at which the current in the inductor rises when the connection is made is
- a. 0.06 A/s.
- b. 1.2 A/s.
- c. 480 A/s.
- d. 9600 A./s.

32. The time required for the current in the above inductor to reach 63% of its final value is
- a. 1.6 ms.
- b. 2.5 ms.
- c. 1 s.
- d. 400 s.

EXERCISES

23–1 Electric Field Energy

1. A potential difference of 300 V is applied across a pair of parallel metal plates 1 cm apart. What is the energy density of the electric field between the plates?

2. The electric field near the earth's surface is about 100 V/m. How much electrical energy is stored in the lowest kilometer of the atmosphere?

23–2 Capacitance

3. A 25-μF capacitor is connected to a source of potential difference of 1000 V. What is the resulting charge on the capacitor? How much energy does it contain?

4. What is the potential difference between the plates of a 20-μF capacitor whose charge is 0.01 C? How much energy does it contain?

5. What potential difference must be applied across a 10-μF capacitor if it is to have an energy content of 1 J?

6. A 0.5-μF capacitor has plates 0.5 mm apart in air. What is the maximum charge the capacitor can have before sparks occur between the plates?

7. The plates of a parallel-plate capacitor are 50 cm^2 in area and 1 mm apart. (a) What is its capacitance? (b) When the capacitor is connected to a 45-V battery, what is the charge on either plate? (c) What is the energy of the charged capacitor?

8. A parallel-plate capacitor with plates 0.2 mm apart in air has a charge on each plate of 4×10^{-8} C when the potential difference is 250 V. Find its capacitance, the area of each plate, and the stored energy.

9. A parallel-plate capacitor of capacitance C is given the charge Q and then disconnected from the circuit. How much work is required to pull the plates of this capacitor to twice their original separation?

10. A variable capacitor set at 200 pF is connected to a 100-V battery. The battery is then disconnected, and the capacitor is adjusted to a capacitance of 10 pF. (a) What is the potential difference across the capacitor now? (b) In the absence of friction, how much work had to be done to turn the shaft of the capacitor to the new setting?

11. A potential difference of 100 V is applied across a pair of parallel metal plates 5 cm square and 1 mm apart. (a) What is the force between the plates? (b) Is the force attractive or repulsive? (c) What is the energy density in the region between the plates?

12. A capacitor is charged by connecting it through a resistance to a battery. Verify that half the work done by the battery is dissipated as heat.

23–3   Dielectric Constant

13. Is there any kind of material that, when inserted between the plates of a capacitor, reduces its capacitance?

14. A sheet of mica whose dielectric constant is 5 is placed between the plates of a charged, isolated parallel-plate capacitor. How is the potential difference across the capacitor affected? How is the charge on the capacitor affected?

15. A parallel-plate capacitor with air between its plates is charged until a potential difference of V appears across it. Another capacitor, having hard rubber (dielectric constant

= 3) between its plates but otherwise identical, is also charged to the same potential difference. If the energy of the first capacitor is W, what is that of the second?

16. What effect does placing a slab of a material of dielectric constant K between the plates of a charged capacitor have on the energy content of the capacitor? (The capacitor is disconnected from the charging circuit before the dielectric is inserted.) If the energy is greater that before, where does the additional energy come from? If the energy is less than before, where does the lost energy go?

17. The capacitance of a parallel-plate capacitor is increased from 8 μF to 50 μF when a sheet of glass is inserted between its plates. What is the dielectric constant of the glass?

18. A capacitor with air between its plates is connected to a battery and each of its plates receives a charge of 10^{-4} C. While still connected to the battery, the capacitor is immersed in oil, and a further charge of 10^{-4} C is added to each plate. What is the dielectric constant of the oil?

19. The space between the plates of the capacitor of Exercise 7 is filled with sulfur. Answer the same questions for this case.

23–4   Capacitors in Combination

20. Find the equivalent capacitance of a 20-μF capacitor and a 50-μF capacitor that are connected in series.

21. Find the equivalent capacitance of a 20-μF capacitor and a 50-μF capacitor that are connected in parallel.

22. List the capacitances that can be obtained by combining three 10-μF capacitors in all possible ways.

23. Three capacitors whose capacitances are 5, 10, and 50 μF are connected in series across a 12-V battery. Find the charge on each capacitor and the potential difference across it.

24. Three capacitors whose capacitances are 2, 4, and 5 μF are connected in series across a 100-V battery. Find the charge on each capacitor and the potential difference across it.

25. The three capacitors of Exercise 23 are connected in parallel across the same battery. Find the charge on each capacitor and the potential difference across it.

26. The three capacitors of Exercise 24 are connected in parallel across the same battery. Find the charge on each capacitor and the potential difference across it.

27. Find the equivalent capacitance of the system shown on next page.

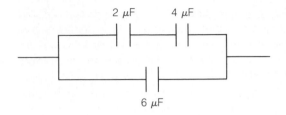

2 µF 4 µF

6 µF

28. The dielectric of the 2-µF capacitor in the system of Exercise 27 breaks down and becomes conducting. Find the equivalent capacitance of the system now.

29. If the dielectric of the 6-µF capacitor of the system of Exercise 27 had broken down instead, what would the equivalent capacitance of the system be?

30. A 1-µF capacitor and a 2-µF capacitor are each charged across a potential difference of 1200 V. The capacitors are then connected with terminals of the same sign together. What is the final charge of each capacitor?

31. The charged capacitors of Exercise 30 are connected together with terminals of opposite sign together. Now what is the final charge of each capacitor?

23–5 Magnetic Field Energy

32. The strongest magnetic fields that have been produced in the laboratory have been about 10^2 T. (a) How much energy is contained in 1 L of such a field? (b) What electric field would have the same energy density?

23–6 Inductance

33. The greater its capacitance, the less energy is stored in a capacitor when it is given a certain charge. On the other hand, the greater its inductance, the more energy is stored in an inductor when a certain current is present in it. Explain the difference.

34. Find the inductance of a coil 40 cm long and 4 cm in diameter that has 1000 turns of wire.

35. A solenoid 20 cm long and 2 cm in diameter has an inductance of 0.178 mH. How many turns of wire does it contain?

36. A 20-mH coil carries a current of 0.2 A. (a) How much energy is stored in its magnetic field? (b) What should the current be in order that it contain 1 J of energy?

37. A solenoid 20 cm long and 2.4 cm in diameter is wound with 1200 turns of wire whose total resistance is 40 Ω. The solenoid is connected to a 12-V battery whose internal resis-

tance is 2 Ω. Find the inductance of the solenoid, the final current in it, and its energy content when the final current flows.

38. The solenoid of Exercise 39 is immersed in liquid oxygen whose permeability is 1.0049 times greater than that of free space. Find the inductance of the solenoid, the final current in it, and its energy content when the final current flows.

39. An inductor consists of an iron ring 5 cm in diameter and 1 cm² in cross-sectional area that is wound with 1000 turns of wire. (Such an inductor is essentially a solenoid bent into a circle.) If the permeability of the iron is constant at 400 times that of free space at the magnetic fields at which the inductor will be used, find its inductance.

40. Show that the total inductance L of three inductors in parallel is given by $1/L = 1/L_1 + 1/L_2 + 1/L_3$. (*Hint:* The self-induced emf is the same for all the inductors since they are connected in parallel.)

41. Show that the total inductance L of three inductors in series is given by $L = L_1 + L_2 + L_3$. (*Hint:* The total self-induced emf of the inductors is equal to the sum of the self-induced emfs of the individual inductors.)

23–7 Self-Induced Emf

42. What becomes of the work done against the back emf in an inductive circuit when a current is being established in it?

43. What is the direction of the self-induced emf in a coil when the current in it increases? When the current decreases? What is the reason in each case?

44. The current in a circuit drops from 5 A to 1 A in 0.1 s. If an average emf of 2 V is induced in the circuit while this is happening, find the inductance of the circuit.

45. What is the self-induced emf in a 0.4-H coil when the current in it is changing at a rate of 500 A/s?

46. A 2-H coil carries a current of 0.5 A. (a) How much energy is stored in it? (b) In how much time should the current drop to 0 if an emf of 100 V is to be induced in it?

23–8 Electrical Oscillations

47. Find the natural frequency of an LC circuit in which $L = 12$ mH and $C = 5$ µF.

48. What inductance is needed in a circuit in which $C = 60$ µF if its natural frequency is to be 30 Hz?

49. What capacitance is needed in a circuit in which $L = 2$ H if its natural frequency is to be 200 Hz?

50. The frequencies used in commercial radio broadcasting range from 550 to 1600 kHz. What range of capacitance should a variable capacitor have if it is connected to a coil of inductance 1 mH in a circuit designed to respond to frequencies in this band?

51. A 1-μF capacitor is charged by being connected to a 10-V battery. The battery is then removed and the capacitor connected to a 10-mH coil. (a) Find the frequency of the resulting oscillations. (b) Find the maximum value of the charge on the capacitor. (c) Find the maximum current that flows through the inductor.

52. A 2-μF capacitor is charged by being connected to a 24-V battery. The battery is then removed and the capacitor connected to a 50-mH coil. After $\frac{1}{8}$ cycle, the initial energy will be equally divided between the electric field in the capacitor and the magnetic field in the coil, as in Fig. 21–16(b). (a) Find the time interval required for this to occur. (b) Find the charge on the capacitor and the current in the inductor at this moment.

23–9   Time Constants

53. A reusable flash bulb requires an energy of 100 J for its discharge. A 450-V battery is used to charge a capacitor for this purpose. The resistance of the charging circuit is 15 Ω. (a) What is the required capacitance? (b) What is the time constant of the circuit?

54. A 100-μF electrolytic capacitor has a leakage current of 5 μA when the potential difference across its terminals is 12 V. If the capacitor is connected to a 12-V battery and then removed, how long will it take for the charge to fall to 37% of its original value?

55. A 5-μF capacitor is connected across a 1000-V battery with wires whose resistance is a total of 5000 Ω. (a) What is the time constant of this circuit? (b) What is the initial current that flows when the battery is connected? (c) How long would it take to charge the capacitor if this current remained constant?

56. A 50-mH coil with a resistance of 20 Ω is connected to a 90-V battery of negligible internal resistance. (a) What is the time constant of the circuit? (b) How much energy is stored in the magnetic field of the coil when the current has reached its final value?

57. What is the inductance of a coil whose resistance is 14 Ω and whose time constant is 0.1 s?

58. A 2-μF capacitor is connected in series with a 6-μF capacitor and a 100-Ω resistor. (a) What is the time constant of the combination? (b) The combination is connected to a 24-V source for a time equal to the time constant, and then disconnected. What is the voltage across each capacitor?

59. A potential difference of 50 V is suddenly applied across a 12-mH, 8-Ω coil. Find (a) the initial current and the initial rate of change of current, (b) the current when the rate of change of current is 2000 A/s, and (c) the final current and final rate of change of current.

60. A potential difference of 100 V is suddenly applied across a 0.5-H, 20-Ω inductor. Find (a) the initial rate of increase of current, (b) the rate of increase of current when the current is 3 A, (c) the final current, and (d) the energy content of the inductor when the final current flows in it.

61. A coil 20 cm long and 3 cm in diameter is tightly wound with one layer of copper wire 1 mm in diameter. Find its time constant.

ANSWERS TO MULTIPLE CHOICE

1. d	**8.** d	**15.** c	**21.** b	**27.** b
2. c	**9.** c	**16.** d	**22.** c	**28.** c
3. a	**10.** a	**17.** a	**23.** a	**29.** a
4. b	**11.** b	**18.** b	**24.** c	**30.** a
5. b	**12.** d	**19.** c	**25.** b	**31.** c
6. d	**13.** b	**20.** c	**26.** b	**32.** b
7. b	**14.** c			

24

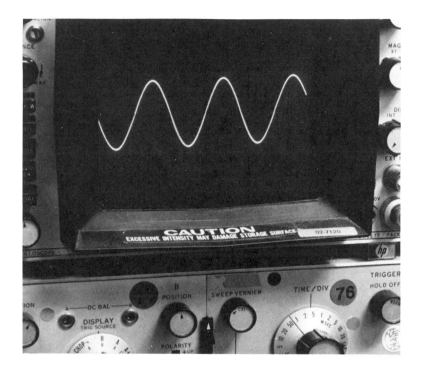

ALTERNATING CURRENT

Nearly all the world's electrical energy is carried by alternating current. The preference for alternating current is chiefly due to the economy of high-voltage transmission, which minimizes I^2R heat losses; with step-up transformers at the production end and step-down transformers at the consumption end, the transmission voltage is limited only by insulation and atmospheric discharge problems. Alternating current is also preferred in industry because alternating-current electric motors are, as a class, cheaper, more durable, and less in need of maintenance than direct-current motors; direct-current motors, however, have certain characteristics, such as better speed regulation, that make them more suitable for specialized duties. Alternating currents are involved in all aspects of modern communication: The electrical equivalent of a sound wave of a certain frequency is an alternating current of that frequency, and radio waves are produced by antennas fed with high-frequency alternating current. Alternating current behaves in a circuit in a very different way from direct current, and some knowledge of alternating-current circuit behavior is necessary to understand much of modern technology.

CHAPTER OBJECTIVES

Completing this chapter should enable you to:

1. Calculate the effective values of current and voltage for an alternating current.

2. Describe the phase relationships between current and voltage in a resistor, in an inductor, and in a capacitor.

3. Calculate the reactance of an inductor.

4. Calculate the reactance of a capacitor.

5. Calculate the impedance of a series *RLC* circuit and determine the phase relationship between its current and voltage.

6. Distinguish among resistance, reactance, and impedance.

7. Distinguish between actual power and apparent power in an alternating-current circuit and explain why the volt-ampere is a useful unit.

8. Determine the ratio of turns in a transformer to be used for impedance matching.

9. Understand the differences in behavior of series and parallel alternating-current circuits at resonance.

10. Describe the operation of high-pass, low-pass, band-pass, and band-reject filters.

24–1   EFFECTIVE CURRENT AND VOLTAGE

A direct current is described in terms of its direction and magnitude. In a direct-current circuit there might be a current of 6 A that flows from the positive terminal of a battery through a resistance network to its negative terminal. An alternating current has neither a constant direction nor a constant magnitude. How shall we describe it?

Since alternating current flows back and forth in a circuit, it has no "direction" in the same sense as a direct current has. However, the oscillations have a certain frequency in each case, and the value of this frequency—that is, how many times per second the current goes through a complete cycle—forms part of the description of the current.

An alternating current varies with time in the manner shown in Fig. 24–1, and it would seem natural to specify its maximum value, I_{max}, as well as its frequency. The trouble with doing this is that I_{max} is not a measure of the ability of a current to do work or produce heat. A 6-A direct current is not equivalent to an alternating current

The effective value of an alternating-current current is a measure of its ability to do work or produce heat

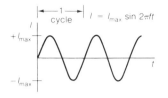

FIG. 24–1 The variation of an alternating current with time. The frequency of the current is the number of cycles that occur per second.

in which $I_{max} = 6$ A. A better procedure is to define an *effective current* I_{eff} such that a direct current of this magnitude produces heat in a resistor at the same rate as the alternating current.

The variation of an alternating current with time obeys the formula

$$I = I_{max} \sin 2\pi ft \qquad (24-1)$$

where f is the frequency of the current. In this formula, the current is assumed to be $I = 0$ and is increasing when $t = 0$. Figure 24–1 is a graph of this formula. The rate at which heat is dissipated in a resistance R by an alternating current is, at any time t, given by

$$I^2 R = I^2_{max} R \sin^2 2\pi ft$$

The average value of $I^2 R$ over a complete cycle is

$$[I^2 R]_{av} = I^2_{eff} R = I^2_{max} R [\sin^2 2\pi ft]_{av} \qquad (24-2)$$

What we must find is the average value of $\sin^2 2\pi ft$ over a complete cycle. (The average value of I over a cycle is 0, since I is positive for half the cycle and negative for the other half. However, I^2 is always positive, and its average is a positive quantity. See Fig. 24–2.)

How effective current is calculated

We begin with the trigonometric identity

$$\sin^2 \theta = \tfrac{1}{2} (1 - \cos 2\theta)$$

The average value of $\sin^2 \theta$ over a complete cycle is therefore

$$[\sin^2 \theta]_{av} = \tfrac{1}{2}[1 - \cos 2\theta]_{av} = \tfrac{1}{2} - \tfrac{1}{2}[\cos 2\theta]_{av} = \tfrac{1}{2}$$

since, over a complete cycle, the average value of $\cos 2\theta$ is 0 by the same reasoning as in the case of $\sin \theta$. Hence we see that

$$I^2_{eff} R = I^2_{max} R [\sin^2 2\pi ft]_{av} = \tfrac{1}{2} I^2_{max} R$$

and

Effective current and voltage are 70.7% of maximum values

$$I_{eff} = \frac{I_{max}}{\sqrt{2}} = 0.707 I_{max} \qquad \qquad \textit{Effective current} \quad (24-3)$$

The effective magnitude of an alternating current is 70.7% of its maximum value.

FIG. 24–2 (a) The average value of I in an alternating-current circuit is zero. (b) The average value of I^2 in an alternating-current circuit is $\tfrac{1}{2} I^2_{max}$.

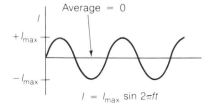

(a)

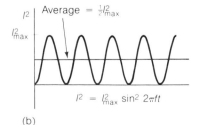

(b)

In a similar way the effective voltage in an alternating-current circuit turns out to be

$$V_{eff} = \frac{V_{max}}{\sqrt{2}} = 0.707\,V_{max} \qquad\qquad \textit{Effective voltage} \quad (24-4)$$

It is customary to express currents and voltages in alternating-current circuits in terms of their effective values. Thus the potential difference across a "120-volt, 60-Hz" power line actually varies from

$$+ V_{max} = +\frac{V_{eff}}{0.707} = +\frac{120\,\text{volts}}{0.707} = +170\,\text{volts}$$

through 0 to -170 volts and back to $+170$ volts a total of 60 times per second.

In what follows, when current, potential difference, and emf values are given for an alternating-current circuit without other qualification, they will refer to the effective magnitudes of these quantities.

By analogy with circular motion, angular frequency ω (in radians per second) is often used instead of frequency f (in hertz) in discussing alternating currents, where

$$\omega = 2\pi f$$

Angular frequency of alternating current

In this notation, which will not be used here, the instantaneous current in an alternating-current circuit is written

$$I = I_{max}\sin\omega t$$

What are called here "effective" values of current and voltage are elsewhere sometimes called "root-mean-square" or "rms" values from their definitions as the square roots of the average values of I^2 and V^2.

24–2 PHASORS

A convenient way to represent alternating-current currents and voltages is in terms of *phasors*. This scheme is based on the fact that the component in any direction of a uniformly rotating vector varies sinusoidally—that is, in the manner that sin θ varies with θ—with time. In the case of an alternating current, the length of the phasor $\mathbf{I}_{max}$ corresponds to I_{max} and we imagine it to rotate f times per second in a counterclockwise sense (Fig. 24–3). Since the angle θ is equal to $2\pi ft$, the vertical component of the phasor at any time t corresponds to the instantaneous current I of Eq. (24–1):

A phasor is a rotating vector that can represent an alternating current or voltage

$$I = I_{max}\sin\theta = I_{max}\,2\pi ft$$

In a similar way, a phasor $\mathbf{V}_{max}$ can be used to represent an alternating-current voltage V.

Phasors are helpful because the current and voltage in an alternating-current circuit or circuit element always have the same frequency but may differ in phase, so that the maxima in each quantity do not occur at the same times. Suppose, for instance, that the voltage in a certain circuit leads the current by $\frac{1}{6}$ cycle, as in Fig. 24–4(a). In a phasor diagram, such as Fig. 24–4(b), this situation is shown by having the phasor

Phasors are useful when I and V are out of phase in a circuit

FIG. 24–3 Phasor representation of I_{max} and I in an alternating current of frequency f. The vertical component of the phasor I_{max} at any time t is equal to the instantaneous value I of the current at that time.

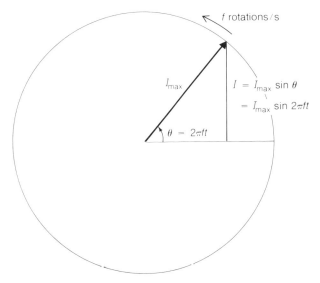

FIG. 24–4 (a) In a certain alternating-current circuit, the oscillations in V occur earlier than those in I by $\frac{1}{6}$ cycle. (b) In a phasor diagram, the phasor V_{max} leads the phasor I_{max} by $360°/6 = 60°$. The orientations of the phasors correspond to the time T in the graphs of part (a).

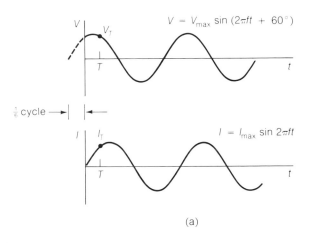

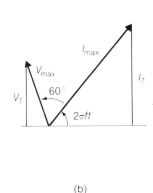

(a)   (b)

$\mathbf{V}_{max}$ at an angle of 60° counterclockwise from the phasor $\mathbf{I}_{max}$, since $360°/6 = 60°$. We must imagine the two phasors rotating together, always with the same angle of 60° between them, and generating the changing values of V and I shown in Fig. 24–4(a) as time goes on.

24–3   PHASE RELATIONSHIPS

All actual electric circuits exhibit resistance, capacitance, and inductance to some degree. When direct current flows through a circuit, only its resistance is significant, but all three properties of the circuit affect the flow of alternating current.

In a resistor, V is in phase with I

We shall first consider a pure-resistance alternating-current circuit, an idealized circuit whose capacitance and inductance are negligible. The instantaneous values of

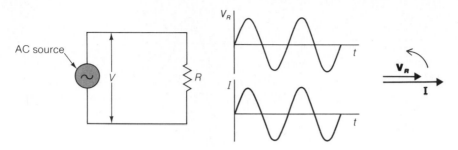

the voltage and current are *in phase* at all times in such a circuit: Both V and I are 0 at the same time, both V and I pass through their maximum values at the same time, and so on (Fig. 24-5). The phasors for V and I therefore remain together.

In an inductor, V leads I

Now let us look at a pure-inductance alternating-current circuit, an idealized circuit whose resistance and capacitance are negligible. There is no IR potential drop across the inductance, and the potential difference across it is therefore proportional to $\Delta I/\Delta t$, the rate of change of the current (see Section 23-7). In this situation V and I cannot be in phase with each other: I changes most rapidly when $I = 0$, so $V = \pm V_{max}$ when $I = 0$, while $\Delta I/\Delta t = 0$ at $I = I_{max}$, so $V = 0$ when $I = \pm I_{max}$. As shown in Fig. 24-6,

The voltage across a pure inductor leads the current in the inductor by $\frac{1}{4}$ cycle.

That is, the variations in the voltage occur $\frac{1}{4}$ cycle *earlier than* the corresponding variations in the current.

In a phasor diagram, a difference in phase of $\frac{1}{4}$ cycle means that the angle between the phasors **V** and **I** is $360°/4 = 90°$. Since **V** leads **I** in a pure inductor, **V** is $90°$ counterclockwise from **I**, as in Fig. 24-6.

In a capacitor, V lags behind I

The potential difference across a capacitor depends upon the amount of charge stored on its plates. The charge is a maximum at each moment that $I = 0$, which is when the current is about to reverse direction and carry away the stored charge. The potential difference across a capacitor is

$$V = \frac{Q}{C}$$

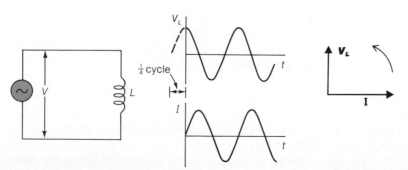

FIG. 24–7 The voltage across a pure capacitor lags behind the current into and out of the capacitor by $\frac{1}{4}$ cycle.

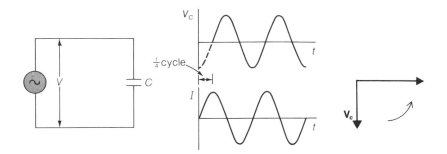

and so $V = \pm V_{max}$ when $I = 0$. The stored charge is 0 at each moment that $I = \pm I_{max}$, because at these times the former stored charge is all gone and charge of the opposite sign is about to build up. Hence $V = 0$ when $I = \pm I_{max}$. As shown in Fig. 24–7,

> **The voltage across a pure capacitor lags behind the current into and out of the capacitor by $\frac{1}{4}$ cycle.**

That is, the variations in the voltage occur $\frac{1}{4}$ cycle *later than* the corresponding variation in the current. In a phasor diagram, **V** is 90° clockwise from **I,** as in Fig. 24–7.

How alternating current passes through a capacitor

In a pure-capacitance circuit the phase relationship between current and voltage is different from what it is in pure-resistance and pure-inductance circuits. Although alternating current does not flow *through* a capacitor, it does flow *into one plate and out of the other* since changes in the amount of charge stored on one of the plates are mirrored by changes in the charge stored on the other plate. A flow of $+Q$ into one plate means that $+Q$ flows out of the other plate to leave the latter with a net charge of $-Q$. If direct current were involved, eventually enough charge would accumulate to stop the arrival of any more, and the current would cease. In the case of an alternating current, however, the current always stops and reverses itself periodically anyway, so the presence of a series capacitor does not prevent alternating current from flowing in the circuit.

24–4   INDUCTIVE REACTANCE

Resistors, inductors, and capacitors all impede the flow of alternating current in a circuit. The effect of resistance is to dissipate part of the electric energy that passes through it into heat. In those conductors in which Ohm's law is valid for direct current, it is valid for alternating current as well, and

$$I = \frac{V_R}{R} \tag{24–5}$$

Here V_R represents the effective potential difference across the resistance R, and I is the effective current through it.

Origin of inductive reactance

The opposition an inductor offers to the flow of alternating current arises from the self-induced back emf produced in it by the changing current. The back emf rep-

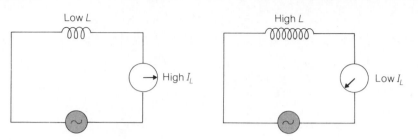

FIG. 24–8 At a given frequency, the lower the inductance L, the lower the inductive reactance X_L and the higher the current I_L.

resents a potential drop across the inductor, and the current in the circuit is correspondingly reduced.

The *inductive reactance X_L* of an inductor is a measure of its effect on an alternating current passing through it. The effective current I in an inductor is related to the effective potential difference V_L across it and the inductive reactance X_L by

$$I = \frac{V_L}{X_L} \tag{24–6}$$

The unit of inductive reactance is the ohm. Equation (24–6) is analogous to Ohm's law, but there is a basic distinction between reactance and resistance in that there is no power loss in an inductor whereas power is dissipated as heat in a resistor.

No power is lost in a pure inductor

The inductive reactance of an inductor is given by the formula

$$X_L = 2\pi fL \qquad\qquad\qquad \textit{Inductive reactance} \quad (24–7)$$

where f is the frequency of the current in hertz (cycles/second) and L is the inductance in henries. The direct dependence of X_L on f and L is reasonable: The self-induced back emf, which is what opposes the current, is proportional to both $\Delta I/\Delta t$ and L, and so the more rapidly the current changes and the larger the value of L, the greater the back emf (Fig. 24–8).

Example What is the current in a coil of negligible resistance and inductance 0.40 H when it is connected to a 120-V, 60-Hz power line?

Solution The reactance of the coil is

$$X_L = 2\pi fL = (2\pi)(60 \text{ Hz})(0.4 \text{ H}) = 151 \ \Omega$$

and the current in it is accordingly

$$I = \frac{V_L}{X_L} = \frac{120 \text{ V}}{151 \ \Omega} = 0.80 \text{ A}$$

Both the 120-V and 0.80-A figures represent effective values. ∎

24–5 CAPACITIVE REACTANCE

The extent to which a capacitor opposes the flow of alternating current depends upon its *capacitive reactance X_C*. If V_C is the effective potential difference across a capacitor whose reactance is X_C, the effective current into and out of the capacitor is

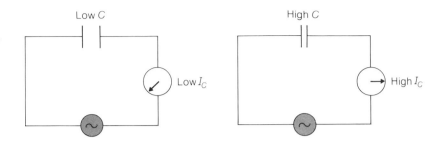

FIG. 24–9 At a given frequency, the higher the capacitance C, the lower the capacitive reactance X_c and the higher the current I_c.

$$I = \frac{V_C}{X_c} \tag{24–8}$$

The capacitive reactance of a capacitor is given by the formula

$$X_C = \frac{1}{2\pi fC} \qquad \qquad \textit{Capacitive reactance} \quad (24–9)$$

If f is in hertz and the capacitance C in farads, the unit of X_C is the ohm.

Origin of capacitive reactance

A capacitor impedes the flow of alternating current by virtue of the reverse potential difference that appears across it as charge builds up on its plates. Thus there is a potential drop across a capacitor in an alternating-current circuit that affects the current just as the potential drop in a resistor does.

The inverse dependence of X_C upon f and C can be understood from the following argument. If the charge on the plates of a capacitor is changed by ΔQ in the time Δt, then the instantaneous current is $I_{\text{inst}} = \Delta Q/\Delta t$. If V_C is the change in the potential difference across the capacitor that leads to ΔQ, then $\Delta Q = C(\Delta V_C)$, and $I_{inst} = C(\Delta V_C/\Delta t)$. The higher the frequency, the greater the rate of change $\Delta V_C/\Delta t$ of the potential difference, and the greater the current. The larger C is, also, the greater the current (Fig. 24–9). Since a high current means a low reactance, increasing f and C decreases X_C.

Example   A capacitor whose reactance is 80 Ω at 50 Hz is used in a 60-Hz circuit. What is its reactance in the latter circuit?

Solution   The capacitance of the capacitor, from Eq. (24–9), is

$$C = \frac{1}{2\pi fX_C} = \frac{1}{(2\pi)(50\,\text{Hz})(80\,\Omega)} = 4 \times 10^{-5}\,\text{F}$$

which is 40μF. Its reactance at 60 Hz is

$$X_C = \frac{1}{2\pi fC} = \frac{1}{(2\pi)(60\,\text{Hz})(4 \times 10^{-5}\,\text{F})} = 66\,\Omega \qquad ■$$

Capacitive and inductive reactances vary differently with frequency, as Fig. 24–10 shows: X_C decreases with increasing f, whereas X_L increases with increasing f.

Only alternating current can pass through a capacitor

In the limit of $f = 0$, which means direct current, $X_L = 0$ and $X_C = \infty$. When the current does not vary, there is no self-induced back emf in an inductor, and no inductive reactance to impede current. On the other hand, a capacitor completely

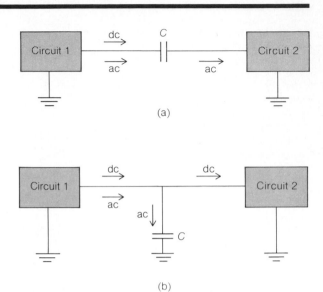

FIG. 24–11 A capacitor can be used to favor the passage of either alternating current or direct current between two circuits, depending upon how it is connected.

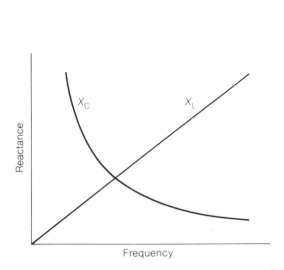

FIG. 24–10 The reactance X_L of an inductor increases with frequency, whereas the reactance X_c of a capacitor decreases with frequency. In the direct-current limit of $f = 0$, $X_L = 0$ and $X_c = \infty$.

obstructs direct current because the charge that builds up on its plates remains there instead of surging back and forth as it does when an alternating-current potential is applied.

Because capacitors pass alternating current but not direct current, they can be used to favor the transmission of signals of one kind or the other between two circuits. In Fig. 24–11(a), the capacitor permits an alternating-current signal to go from circuit 1 to circuit 2 while stopping any direct current. In Fig. 24–11(b), a direct current can pass between the circuits, but the capacitor provides a path to ground for alternating current, and most of the alternating current will take this path if the reactance is low.

Capacitors as filters for alternating current or direct current

24–6 IMPEDANCE

A series circuit that contains resistance, inductance, and capacitance can be represented as in Fig. 24–12, where each of these circuit properties is considered as lumped into a single resistor, inductor, and capacitor. If an alternating-current source of emf is connected to the circuit, at any instant the applied voltage V is equal to the sum of the voltage drops across the various circuit elements:

How to add instantaneous voltages

$$V = V_R + V_L + V_C \qquad \textit{Instantaneous voltage} \quad (24\text{–}10)$$

However, V_R, V_L, and V_C are *out of phase with one another.* The voltage across the resistor, V_R, is always in phase with the current I, but V_L is $\frac{1}{4}$ cycle ahead of I, and V_C is $\frac{1}{4}$ cycle behind I. This situation is shown in Fig. 24–12. While Eq. (24–10) is always correct when V, V_R, V_L, and V_C refer to the instantaneous values of the various voltages,

FIG. 24–12 The instantaneous value of the voltage V applied to a series RLC circuit is equal to the sum of the instantaneous values of V_R, V_L, and V_C. This relationship does not hold for the effective values of the various voltages because of the phase differences among them.

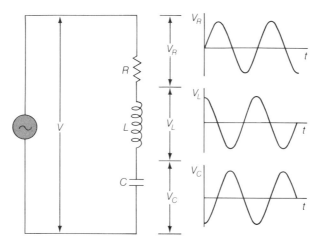

FIG. 24–13 Phasor diagram of the effective voltages across the resistor, inductor, and capacitor of Fig. 24–12. The magnitude of the vector sum of the voltage phasors equals the effective voltage across the entire circuit. The angle ϕ is the phase angle. The direction of the effective current phasor **I** is the same as that of $\mathbf{V}_R$.

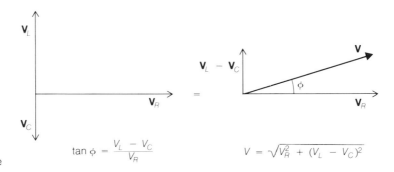

$$\tan \phi = \frac{V_L - V_C}{V_R} \qquad V = \sqrt{V_R^2 + (V_L - V_C)^2}$$

it is *not* correct when effective (or maximum) values are involved, and we must use a vectorial approach to take the phase differences into account.

How to add effective voltages

Figure 24–13 is a phasor diagram that shows the phase differences between $\mathbf{V}_R$, $\mathbf{V}_L$, and $\mathbf{V}_C$: $\mathbf{V}_L$ is 90° ahead of $\mathbf{V}_R$ (hence 90° counterclockwise from $\mathbf{V}_R$, since phasors rotate counterclockwise) and $\mathbf{V}_C$ is 90° behind $\mathbf{V}_R$. The vector sum $\mathbf{V}$ of $\mathbf{V}_R$, $\mathbf{V}_L$, and $\mathbf{V}_C$ represents the effective voltage across the terminals of the circuit, and its magnitude is

$$V = \sqrt{V_R^2 + (V_L - V_C)^2} \qquad\qquad \textit{Effective voltage} \quad (24\text{–}11)$$

Phase angle

The angle ϕ between $\mathbf{V}$ and $\mathbf{V}_R$ is called the *phase angle* because it is a measure of how much the voltage in the circuit leads or lags behind the current. The direction of the effective current phasor **I** is always the same as the direction of $\mathbf{V}_R$.

In Fig. 24–13, V_L is greater than V_C, and $(V_L - V_C)$ is a positive quantity. If instead V_L is the smaller quantity, $(V_L - V_C)$ is negative and the vector $\mathbf{V}$ is below the x-axis (Fig. 24–14). However, Eq. (24–11) still applies, since $(V_L - V_C)^2 = (V_C - V_L)^2$. The phase angle in either case is specified by

$$\tan \phi = \frac{V_L - V_C}{V_R} \qquad\qquad \textit{Phase angle} \quad (24\text{–}12)$$

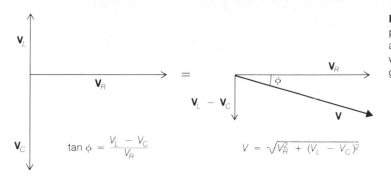

If $V_C > V_L$, the result will be a negative value for ϕ, which signifies that **V** lags behind **V**$_R$ (and behind the current **I**).

Example A resistor, a capacitor, and an inductor are connected in series across an alternating-current power source. The effective voltages across the circuit components are $V_R = 5$ V, $V_C = 10$ V, and $V_L = 12$ V. Find the effective voltage of the source and the phase angle in the circuit.

Solution   (a) From Eq. (24–11),

$$V = \sqrt{V_R^2 + (V_L - V_C)^2} = \sqrt{(5\,\text{V})^2 + (12\,\text{V} - 10\,\text{V})^2} = 5.4\,\text{V}$$

(b) Since

$$\tan \phi = \frac{V_L - V_C}{V_R} = \frac{12\,\text{V} - 10\,\text{V}}{5\,\text{V}} = 0.4$$

we have for the phase angle $\phi = 22°$.   ∎

Because

$$V_R = IR \qquad V_L = IX_L \qquad \text{and} \qquad V_C = IX_C$$

we can rewrite Eq. (24–11) in the form

$$V = I\sqrt{R^2 + (X_L - X_C)^2}$$

The quantity

$$Z = \sqrt{R^2 + (X_L - X_C)^2} \qquad\qquad \textit{Impedance} \quad (24\text{–}13)$$

is known as the *impedance* of a series circuit containing resistance, inductance, and capacitance. The unit of impedance is evidently the ohm. Impedance in an alternating-current circuit plays the same role that resistance does in a direct-current circuit, and in an alternating-current circuit

$$I = \frac{V}{Z} \qquad\qquad \textit{Current in ac circuit} \quad (24\text{–}14)$$

is the effective current that flows when the effective voltage V is applied. It is important

Impedance is alternating-current equivalent of resistance; its unit is the ohm

FIG. 24–15 The phasor impedance diagram that corresponds to the phasor voltage diagram shown in Fig. 24–13.

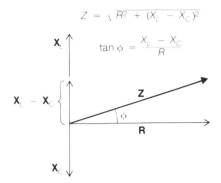

$$Z = \sqrt{R^2 + (X_L - X_C)^2}$$

$$\tan \phi = \frac{X_L - X_C}{R}$$

to keep in mind that Z not only depends upon the circuit parameters R, L, and C but also varies with the frequency f.

Because the current I is the same in all parts of the circuit at all times and

$$Z = \frac{V}{I} \qquad\qquad R = \frac{V_R}{I}$$

$$X_L = \frac{V_L}{I} \qquad\qquad X_C = \frac{V_C}{I}$$

the phasor voltage diagram of Fig. 24–13 can be replaced by the phasor impedance diagram of Fig. 24–15. The phase angle ϕ is the same in both cases, of course, and can be calculated from the relation

$$\tan \phi = \frac{X_L - X_C}{R} \qquad\qquad\qquad \textit{Phase angle} \quad (24\text{–}15)$$

Example   Analyze in detail a series circuit that consists of a 10-mH inductor, a 10-μF capacitor, and a 30-Ω resistor connected to a source of 100-V, 400-Hz alternating current.

Solution   The reactances of the inductor and capacitor at 400 Hz are

$$X_L = 2\pi f L = (2\pi)(400\,\text{Hz})(0.01\,\text{H}) = 25\,\Omega$$

$$X_C = \frac{1}{2\pi f C} = \frac{1}{(2\pi)(400\,\text{Hz})(10^{-5}\,\text{F})} = 40\,\Omega$$

The vector impedance diagram for this circuit is shown in Fig. 24–16. The impedance Z is

$$Z = \sqrt{R^2 + (X_L - X_C)^2} = \sqrt{(30\,\Omega)^2 + (25\,\Omega - 40\,\Omega)^2} = 34\,\Omega$$

The phase angle ϕ is found as follows:

$$\tan \phi = \frac{X_L - X_C}{R} = -\frac{15\,\Omega}{30\,\Omega} = -0.50 \qquad \phi = -27°$$

A negative phase angle signifies that the voltage lags behind the current. Here the lag is 27°, which is $\frac{27}{360}$ or 0.075 of a complete cycle.

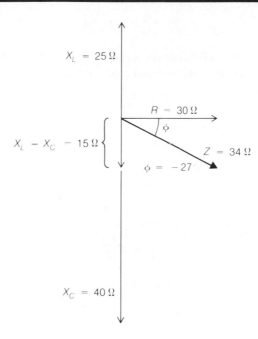

FIG. 24–16 A negative phase angle occurs when X_C is greater than X_L, and signifies that the voltage in the circuit lags behind the current.

The current in the circuit is

$$I = \frac{V}{Z} = \frac{100\,\text{V}}{34\,\Omega} = 2.94\,\text{A}$$

The effective potential difference across each of the circuit elements is

$$V_L = IX_L = (2.94\,\text{A})(25\,\Omega) = 74\,\text{V}$$

$$V_C = IX_C = (2.94\,\text{A})(40\,\Omega) = 118\,\text{V}$$

$$V_R = IR = (2.94\,\text{A})(30\,\Omega) = 88\,\text{V}$$

Evidently the effective voltage across an inductor or capacitor in an alternating-current circuit can exceed the effective voltage applied to the entire circuit.

The arithmetic sum of the above voltages is 280 V, but this sum means nothing because the voltages are not in phase with one another. The vector sum of the voltage, which takes into account the phase differences among them, is

$$V = \sqrt{V_R^2 + (V_L - V_C)^2} = \sqrt{(88\,\text{V})^2 + (74\,\text{V} - 118\,\text{V})^2} = 98\,\text{V}$$

The difference between this figure and the applied voltage of 100 V is entirely due to the rounding off of V_L, V_C, and V_R when they were calculated. ∎

24–7 RESONANCE

When an alternating-current voltage is applied to a series circuit, the current that flows depends upon the frequency. The greatest current flows when the impedance Z is a minimum. Since

Impedance is a minimum when $X_L = X_C$

$$Z = \sqrt{R^2 + (X_L - X_C)^2}$$

the condition for minimum impedance in a given circuit is that the frequency be such that the inductive and capacitive reactances are equal. When this is true,

$$X_L = X_C$$

$$2\pi f_0 L = \frac{1}{2\pi f_0 C}$$

and thus

$$f_0 = \frac{1}{2\pi \sqrt{LC}} \qquad \qquad \textit{Resonance frequency} \quad (24-16)$$

When the impressed voltage has this frequency, the current is a maximum and is limited only by the resistance R. The frequency f_0 is called the *resonance frequency* of the circuit, and *resonance* occurs in a series circuit when the impressed voltage oscillates with the resonance frequency.

Resonance frequency is natural frequency of oscillation

We note that Eq. (24–16) is the same as Eq. (23–21) for the "natural" frequency at which an LC circuit will oscillate if the capacitor is initially charged. When a circuit is in resonance with an applied voltage, the energy alternately stored and discharged from the capacitor is precisely equal to the energy stored and discharged from the inductor. At other frequencies the energy contents of the capacitor and inductor are different, which interferes with the back-and-forth flow of power and gives rise to an impedance that exceeds the resistance R. The current that flows in an RLC circuit at resonance is

$$I = \frac{V}{R}$$

just as in a direct-current circuit.

Sharpness of tune depends upon the X_L/R ratio

Figure 24–17 shows how the current in a series circuit varies with frequency. The less the resistance R relative to the inductive reactance X_L, the sharper the peak in the curve. The antenna circuit of a radio receiver is tuned to respond to a particular frequency of radio waves by adjusting a variable capacitor or inductor until the resonance frequency of the circuit is equal to the signal frequency. A circuit in which X_L/R is

FIG. 24–17 The variation with frequency of the current in a series alternating-current circuit. The maximum current occurs at the resonance frequency f_0.

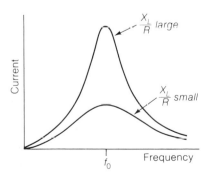

large has a narrow response curve and can separate signals from two stations very close together in frequency.

At resonance, $X_L = X_C$, and there is no phase difference between current and voltage. Another way to specify the condition for resonance, then, is to require that current and voltage be in phase at all times, just as they are in a pure-resistance circuit.

Current and voltage are in phase at resonance

Example The antenna circuit of a radio receiver consists of a 10-mH coil and a variable capacitor; the resistance in the circuit is 50 Ω. An 880-kHz (kilohertz) radio wave produces a potential difference of 10^{-4} V across the circuit. Find the capacitance required for resonance and the current at resonance.

Solution The capacitance required for resonance at 880 kHz is

$$C = \frac{1}{(2\pi f)^2 L} = \frac{1}{(2\pi \times 8.8 \times 10^5 \, \text{Hz})^2 (10^{-2} \, \text{H})} = 3.3 \times 10^{-12} \, \text{F} = 3.3 \, \text{pF}$$

At resonance, inductive and capacitive reactances cancel each other out, and the current that flows is

$$I = \frac{V}{R} = \frac{10^{-4} \, \text{V}}{50 \, \Omega} = 2 \times 10^{-6} \, \text{A} \qquad \blacksquare$$

24–8 POWER IN AC CIRCUITS

No power is consumed in a pure inductor or capacitor in an alternating-current circuit, since these elements act merely as temporary reservoirs of energy and return whatever energy they absorb in one quarter of a cycle to the circuit in the next quarter of that cycle. No power is therefore needed to maintain an alternating-current current in the inductive and capacitive parts of a circuit. The resistance in the circuit, however, dissipates power as heat at the rate $P = IV_R$, where, as usual, I and V_R are effective values. From Fig. 24–13 we see that

Inductors and capacitors do not absorb power

$$V_R = V \cos \phi \qquad\qquad\qquad\qquad (24\text{--}17)$$

since $V \cos \phi$ is the component of $\mathbf{V}$ that is in phase with the current phasor $\mathbf{I}$. Hence the effective power absorbed in an alternating-current circuit is

$$P = IV \cos \phi \qquad\qquad\qquad \textit{Power in ac circuit} \quad (24\text{--}18)$$

The quantity $\cos \phi$ is called the *power factor* of the circuit. The power factor is equal to 1 only at resonance, when current and voltage are in phase; at resonance, $\phi = 0$, and $\cos \phi = 1$. Under other circumstances the voltage is not in phase with the current, and the actual power in the circuit is less than IV.

Power factor

There is an interesting analogy between Eq. (24–18) and the mechanical power $P = Fv \cos \theta$ developed by a force $\mathbf{F}$ that acts on an object moving at the velocity $\mathbf{v}$ when the angle between $\mathbf{F}$ and $\mathbf{v}$ is θ. When $\mathbf{F}$ is parallel to $\mathbf{v}$, $\cos \theta = \cos 0 = 1$ and the power is a maximum. When $\mathbf{F}$ is perpendicular to $\mathbf{v}$, $\cos \theta = \cos 90° = 0$ and

$P = 0$. In the electrical case, we can think of V as corresponding to F and of I as corresponding to v.

From Fig. 24–15 we see that

$$\cos \phi = \frac{R}{Z} = \frac{R}{\sqrt{R^2 + (X_L - X_C)^2}} \qquad \textit{Power factor} \quad (24\text{–}19)$$

The power factor of a circuit is the ratio between its resistance and its impedance. Often power factors are expressed as percentages rather than as decimals or fractions. Thus a phase angle of 60° means a power factor of

$$\cos 60° = 0.50 = 50\%$$

Instruments called wattmeters have been devised which respond directly to the effective product of V and I. An alternating-current wattmeter connected in a circuit gives a lower value for the power than the product of the effective values of V and I obtained from a separate voltmeter and ammeter in the same circuit (except at resonance), because the separate meters are not affected by the phase difference between current and voltage. Alternating-current generators, transformers, and power lines are usually rated in *volt-amperes,* the product of effective voltage and current without regard to actual power. This is because higher values of V and/or I than are reflected in its power consumption must be supplied to a circuit whose power factor is less than 1. It is convenient to think of the power factor as having the unit of watts per volt-ampere.

The volt-ampere is the unit of apparent power; apparent power may exceed consumed power in an alternating-current circuit

Alternating-current devices of various kinds—for example, alternating-current electric motors and fluorescent lamps—may have net inductive or capacitive reactances and consequently have power factors of less than 100%. A power factor of 70%, for instance, means that 1 kVA (kilovolt-ampere) of apparent power must be supplied for every 700 W of power actually consumed. This is an uneconomical situation because of the additional generator capacity required as well as because of the additional heat losses in the transmission lines as the unused power circulates between the generator and the device. The remedy is to introduce capacitors or inductors into the power line to increase the power factor to an acceptable figure.

Example   An inductive load connected to a 24-V, 400-Hz power source draws a current of 2 A and dissipates 40 W. (a) Find the power factor of the load and the phase angle ϕ. (b) What series capacitance is needed to make the power factor 100%? (c) What would the current then be? (d) How much power would the load then dissipate?

Solution   (a) Since $P = IV \cos \phi$, the power factor is

$$\cos \phi = \frac{P}{IV} = \frac{40\,\text{W}}{(2\,\text{A})(24\,\text{V})} = 0.83 = 83\%$$

The phase angle is therefore $\phi = 34°$.

(b) The power factor will be 100% when $X_C = X_L$. In the original circuit, $X_C = 0$ and so, from Eq. (24–15), $X_L = R \tan \phi$. The resistance in the circuit is

$$R = \frac{P}{I^2} = \frac{40\,\text{W}}{(2\,\text{A})^2} = 10\,\Omega$$

Hence the inductive reactance is

$$X_L = R \tan \phi = (10\ \Omega)(\tan 34°) = 6.75\ \Omega$$

Since the required capacitive reactance X_C must equal X_L when $f = 400$ Hz,

$$C = \frac{1}{2\pi f X_C} = \frac{1}{(2\pi)(400\ \text{Hz})(6.75\ \Omega)} = 5.9 \times 10^{-5}\ \text{F} = 59\ \mu\text{F}$$

(c) When $X_C = X_L$, $Z = R$ and

$$I = \frac{V}{Z} = \frac{V}{R} = \frac{24\ \text{V}}{10\ \Omega} = 2.4\ \text{A}$$

(d) $P = I^2 R = (2.4\ \text{A})^2(10\ \Omega) = 57.6\ \text{W}.$ ∎

24–9   IMPEDANCE MATCHING

In Chapter 19 we saw that the maximum power transfer between two direct-current circuits occurs when their resistances are equal. The same conclusion applies to alternating-current circuits with the impedance of the circuits rather than their resistances as the quantities to be matched. An additional consideration with alternating-current circuits is that an impedance mismatch may result in a distorted signal by altering the properties of the circuits, for instance by changing their resonance frequencies.

Power transfer is a maximum between circuits with the same impedance

Alternating-current circuits have the advantage that a transformer can be used to correct impedance mismatches. A common use of a transformer for this purpose occurs in an audio system where the amplifier circuit might have an impedance of several thousand ohms whereas the voice coil of the loudspeaker has an impedance of only a few ohms. Connecting the voice coil directly to the amplifier would be grossly inefficient.

A transformer can be used to match impedances

Let us calculate the ratio of turns N_1/N_2 between the primary and secondary windings of a transformer needed to couple a circuit of impedance Z_1 to a circuit of impedance Z_2. If the voltages and currents in the circuits are respectively V_1, I_1 and V_2, I_2, then from Eq. (22–8)

$$\frac{V_1}{V_2} = \frac{N_1}{N_2} \quad \text{and} \quad \frac{I_2}{I_1} = \frac{N_1}{N_2}$$

Since $Z_1 = V_1/I_1$ and $Z_2 = V_2/I_2$,

$$\frac{Z_1}{Z_2} = \frac{V_1 I_2}{V_2 I_1} = \left(\frac{N_1}{N_2}\right)^2$$

and the ratio of the turns is

$$\frac{N_1}{N_2} = \sqrt{\frac{Z_1}{Z_2}} \qquad \qquad \textit{Impedance matching} \quad (24–20)$$

Thus if a loudspeaker whose voice has an impedance of 10 Ω is to be used with an amplifier whose load impedance is 9000 Ω, a transformer should be used whose ratio of turns is

$$\frac{N_1}{N_2} = \sqrt{\frac{Z_1}{Z_2}} = \sqrt{\frac{9000\ \Omega}{10\ \Omega}} = 30$$

24–10 PARALLEL AC CIRCUITS AND FILTERS

The branch currents in a parallel circuit are not in phase

When a resistor, an inductor, and a capacitor are connected in parallel, as in Fig. 24–18, the potential difference is the same across each circuit element:

$$V = V_R = V_L = V_C$$

The total instantaneous current is the sum of the instantaneous currents in each branch, as in the case of a direct-current parallel circuit, but this is not true of the total effective current because the branch currents are not in phase. The current I_R is always in phase with the voltage V, but I_C leads V by 90° and I_L lags behind V by 90°. A phasor diagram of the situation is shown in Fig. 24–19. The magnitudes of the currents in the branches are

$$I_R = \frac{V}{R} \qquad I_C = \frac{V}{X_C} \qquad I_L = \frac{V}{X_L}$$

and their vector sum is

$$I = \sqrt{I_R^2 + (I_C - I_L)^2} \qquad\qquad \textit{Effective current}\quad (24\text{–}21)$$

The phase angle ϕ specified by

$$\tan \phi = \frac{I_C - I_L}{I_R} \qquad\qquad \textit{Phase angle}\quad (24\text{–}22)$$

is the angle between the current and the voltage. A positive phase angle means that the current leads the voltage, a negative one means that the current lags behind the voltage. The power dissipated in a parallel alternating-current circuit is given by the same formula as in the case of a series circuit, namely $P = IV \cos \phi$.

In a series *RLC* circuit, the impedance is a minimum at resonance and increases at higher and lower frequencies than f_0. At resonance, $X_L = X_C$, $Z = R$, and $I = V/R$. In a parallel *RLC* circuit, resonance again corresponds to $X_L = X_C$, $Z = R$, and

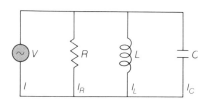

FIG. 24–18 A parallel *RLC* circuit. The potential differences across the circuit elements are the same.

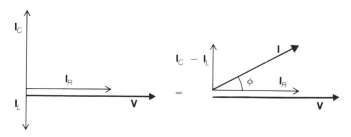

FIG. 24–19 A phasor diagram of the effective currents in the resistor, inductor, and capacitor of Fig. 24–18. The magnitude *I* of the vector sum of the current vectors is equal to the effective current through the entire circuit.

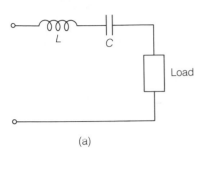

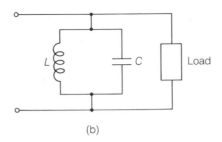

(a)

(b)

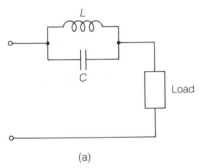

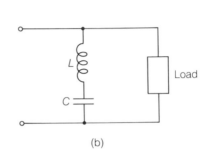

(a)

(b)

FIG. 24–21 Band-reject filters.

$I = V/R$, but now the impedance is a *maximum* at f_0 since at higher and lower frequencies some current can pass through the inductor and capacitor as well as through the resistor. Thus a series circuit can be used as a selector to favor a particular frequency, and a parallel circuit with the same L and C can be used as a selector to discriminate against the same frequency. Such circuits form the basis of "band-pass" and "band-reject" filters.

In a parallel circuit, impedance is a maximum at resonance

Figure 24–20 shows two kinds of band-pass filters. At frequencies near f_0 the impedance of the series LC circuit of Fig. 24–20(a) is low, so the load current is high. (From Fig. 24–17 we can see why a band of frequencies rather than a single one is favored.) At frequencies much above or below f_0 the impedance is high, so the load current is low. In Fig. 24–20(b) the impedance of the parallel LC circuit is high for frequencies near f_0, and most of the current flows through the load. At other frequencies, current is diverted through the LC circuit and thus reduces the load current. The band-reject filters of Fig. 24–21 behave in just the opposite ways to minimize the load current for frequencies near f_0.

Band-pass and band-reject filters

Band-pass filters make it possible to send a number of different signals at the same time through a single pair of wires. What is done is to have each signal modulate a "carrier" alternating current of constant frequency, so that the resulting variations in the amplitude of the carrier correspond to the signal. This procedure is illustrated in Fig. 25–6 for radio transmission. Modulated carriers of different frequencies can be transmitted at the same time from one place to another along the same pair of wires. At the receiving end band-pass filters separate the carriers and special circuits then demodulate them to reconstruct the original signals. Telephone systems have used this technique for over 60 years. Up to 32 simultaneous telephone conversations can be carried on an ordinary pair of wires, and a coaxial cable, in which one of the conductors is in the form of a tube insulated from a central wire, can carry thousands of them.

This is a 22-conductor coaxial cable that can carry as many as 90,000 telephone conversations at the same time. (Photo: Bell Labs)

FIG. 24–22 (a) High-pass filter. At low frequencies, X_C is large and X_L is small. (b) Low-pass filter. At high frequencies, X_C is small and X_L is large.

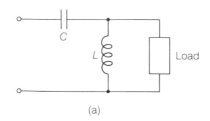

(a)

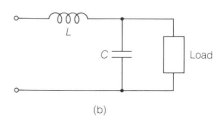

(b)

High- and low-pass filters

Other arrangements of inductors and capacitors can act as "high-pass" filters to discriminate against low-frequency currents and as "low-pass" filters to favor them. In the high-pass filter of Fig. 24–22(a), since $X_C = 1/(2\pi fC)$, the lower the frequency, the greater the reactance of the capacitor. In addition, since $X_L = 2\pi fl$, the lower the frequency, the less the reactance of the inductor and the more the current diverted through it instead of passing through the load. These effects are used in the opposite ways in the low-pass filter of Fig. 24–22(b), which is widely used to smooth out ripples in the output of alternating-current to direct-current rectifiers (see Section 22–5). More elaborate circuits than those of Figs. 24–20 to 24–22 can be designed that provide sharper cutoffs between the passed and rejected frequency bands.

IMPORTANT TERMS

The **effective value** of an alternating current is such that a direct current of this magnitude produces heat in a resistor at the same rate as the alternating current.

A **phasor** is a rotating vector whose projection can represent either current or voltage in an alternating-current circuit.

The **phase relationships** between the instantaneous voltage and instantaneous current in alternating-current circuit components are as follows: The voltage across a pure resistor is in phase with the current; the voltage across a pure inductor leads the current by $\frac{1}{4}$ cycle; the voltage across a pure capacitor lags behind the current by $\frac{1}{4}$ cycle.

The **inductive reactance** X_L of an inductor is a measure of its effect on an alternating current. The **capacitive reactance** X_C of a capacitor is a measure of its effect on an alternating current. Both X_L and X_C vary with the frequency of the current.

The **impedance** Z of an alternating-current circuit is analogous to the resistance of a direct-current circuit. The **resonance frequency** of an alternating-current circuit is that frequency for which the impedance is a minimum.

The **power factor** of an alternating-current circuit is the ratio between the power consumed in the circuit and the product of the effective current and voltage there; this ratio

is equal to that between the resistance and the impedance of the circuit, and is less than 1, except at resonance. The unit of the apparent power $V_{eff}I_{eff}$ is the **volt-ampere,** as distinct from the watt, which is the unit of consumed power.

A **filter** is a circuit that discriminates against the passage of currents in a certain frequency range.

IMPORTANT FORMULAS

Effective current: $I_{eff} = \dfrac{I_{max}}{\sqrt{2}} = 0.707\, I_{max}$

Effective voltage: $V_{eff} = \dfrac{V_{max}}{\sqrt{2}} = 0.707\, V_{max}$

Inductive reactance: $X_L = 2\pi fL$

Capacitive reactance: $X_C = \dfrac{1}{2\pi fC}$

Impedance: $Z = \sqrt{R^2 + (X_L - X_C)^2}$

Current in ac circuit: $I = \dfrac{V}{Z}$

Phase angle: $\tan \phi = \dfrac{X_L - X_C}{R}$ $\cos \phi = \dfrac{R}{Z}$

Resonance frequency: $f_0 = \dfrac{1}{2\pi \sqrt{LC}}$

Power in ac circuit: $P = IV \cos \phi$

Impedance-matching transformer: $\dfrac{N_1}{N_2} = \sqrt{\dfrac{Z_1}{Z_2}}$

MULTIPLE CHOICE

1. The effective voltage in an alternating-current circuit is equal to
   a. $0.5V_{max}$.
   b. $0.707V_{max}$.
   c. $V_{max}/0.707$.
   d. $I_{max}/0.707$.

2. The current in an alternating-current circuit varies between
   a. 0 and I_{eff}.
   b. 0 and I_{max}.
   c. $-I_{eff}$ and $+I_{eff}$.
   d. $-I_{max}$ and $+I_{max}$.

3. In an alternating-current circuit, the voltage
   a. leads the current.
   b. lags the current.
   c. is in phase with the current.
   d. is any of the above, depending on the circumstances.

4. The voltage cannot be exactly in phase with the current in a circuit that contains
   a. only resistance.
   b. only inductance.
   c. inductance and capacitance.
   d. inductance, capacitance, and resistance.

5. The voltage lags behind the current by $\frac{1}{4}$ cycle in
   a. a pure capacitor.
   b. a pure inductor.
   c. a pure resistor.
   d. a circuit with capacitance and inductance.

6. The reactance of a capacitor is X when an ac voltage of frequency f is applied to it. If the frequency is changed to $f/2$, the reactance becomes
   a. $X/2$.
   b. X.
   c. $2X$.
   d. $4X$.

7. The reactance of an inductor is X when an ac voltage of frequency f is applied to it. If the frequency is changed to $f/2$, the reactance becomes
   a. $X/2$.
   b. X.
   c. $2X$.
   d. $4X$.

8. The unit of inductive reactance is the
   a. henry.
   b. tesla.
   c. weber.
   d. ohm.

9. When voltage and current are in phase in an ac circuit, the

a. impedance is 0.
b. reactance is 0.
c. resistance is 0.
d. phase angle is 90°.

10. The impedance of a circuit does not depend on
   a. I.
   b. f.
   c. R.
   d. C.

11. A resistor, a capacitor, and an inductor are connected in series to a source of ac power. If the inductance is decreased, the impedance of the circuit
   a. decreases.
   b. increases.
   c. decreases or increases.
   d. decreases, increases, or remains the same.

12. The power dissipated in an alternating-current circuit depends on its
   a. resistance.
   b. inductive reactance.
   c. capacitive reactance.
   d. impedance.

13. A coil of inductance L has an inductive reactance of X_L in an alternating-current circuit in which the effective current is I. The coil is made from a superconducting material and has no resistance. The rate at which power is dissipated in the coil is
   a. 0.
   b. IX_L.
   c. I^2X_L.
   d. IX_L^2.

14. The power factor of a circuit is equal to
   a. RZ.
   b. R/Z.
   c. X_L/Z.
   d. X_C/Z.

15. At resonance, it is *not* true that
   a. $R = Z$.
   b. $X_L = 1/X_C$.
   c. $P = IV$.
   d. $I = V/R$.

16. The impedance of a parallel RLC circuit at resonance is
   a. less than R.
   b. equal to R.
   c. more than R.
   d. any of the above, depending on the circumstances.

17. Impedance is a maximum at resonance in
   a. a series RLC circuit.
   b. a parallel RLC circuit.
   c. all RLC circuits.
   d. no RLC circuits.

18. The power factor of a circuit in which $X_L = X_C$
   a. is 0.
   b. is 1.
   c. depends on the ratio X_L/X_C.
   d. depends on the value of R.

19. A voltmeter across an alternating-current circuit reads 50 V and an ammeter in series with the circuit reads 5 A.

The power consumption of the circuit
a. is less than or equal to 250 W.
b. is exactly equal to 250 W.
c. is equal to or more than 250 W.
d. may be less than, equal to, or more than 250 W.

20. The power factor of a certain circuit in which the voltage lags behind the current is 80%. To increase the power factor to 100%, it is necessary to add to the circuit additional
a. resistance.   b. capacitance.
c. inductance.   d. impedance.

21. A simple low-pass filter can be made by using
a. an inductor in series with the load.
b. a capacitor in series with the load.
c. a resistor in series with the load.
d. an inductor in parallel with the load.

22. The inductive reactance of a 1-mH coil in a 5-Hz circuit is
a. 3.1 Ω.   b. 6.3 Ω.
c. 10 Ω.   d. 31 Ω.

23. A current of 0.5 A flows in an electromagnet of negligible resistance when it is connected to a source of 120-V, 60-Hz alternating current. The inductance of the electromagnet is
a. 0.011 mH.   b. 0.16 H.
c. 0.64 H.   d. 240 H.

24. The capacitive reactance of a 5-μF capacitor in a 20-kHz circuit is
a. 0.63 Ω.   b. 1.6 Ω.
c. 5 Ω.   d. 16 Ω.

25. A 2-μF capacitor is connected to a 50-V, 400-Hz power source. The current that flows is
a. 0.2 mA.   b. 0.25 A.
c. 2.5 A.   d. 3.5 A.

26. A resistor, a capacitor, and an inductor are connected to an ac power source of frequency f. The effective voltages across the circuit components are $V_R = 10$ V, $V_C = 20$ V, and $V_L = 14$ V. The effective voltage of the source is
a. 8 V.   b. 11.7 V.
c. 16 V.   d. 35.4 V.

27. The phase angle in the circuit of Question 26 is
a. $-31°$.   b. $+31°$.
c. $-53°$.   d. $+53°$.

28. The resonance frequency of the circuit of Question 26 is
a. less than f.
b. equal to f.
c. more than f.

d. any of the above, depending on the values of the components.

29. In a series alternating-current circuit $R = 10$ Ω, $X_L = 8$ Ω, and $X_C = 6$ Ω when the frequency is f. The impedance at this frequency is
a. 10.2 Ω.   b. 12 Ω.
c. 24 Ω.   d. 104 Ω.

30. The phase angle in the circuit of Question 29 is
a. 0.2°.   b. 2°.
c. 11°.   d. 45°.

31. The resonance frequency of the circuit of Question 29 is
a. less than f.
b. equal to f.
c. more than f.
d. any of the above, depending on the applied voltage.

32. The resonance frequency of a circuit that contains a 50-mH inductor and a 0.2-μF capacitor is
a. 16 kHz.   b. 10 kHz.
c. 63 kHz.   d. 16 MHz.

33. A coil connected to a 120-V ac source draws a current of 0.5 A and dissipates 50 W. If a capacitor were connected in series to bring the power factor to 1, the circuit would dissipate
a. 50 W.   b. 60 W.
c. 72 W.   d. 100 W.

34. A 2-μF capacitor, a 5-mH inductor, and a 30-Ω resistor are connected in series across a 100-V, 1-kHz power source. The current in the circuit is
a. 1.76 A.   b. 2.65 A.
c. 3.33 A.   d. 5.5 A.

35. The power dissipated in the circuit of Question 34 is
a. 93 W.   b. 176 W.
c. 202 W.   d. 211 W.

36. The minimum rating in volt-amperes needed by the power source of the circuit of Question 34 is
a. 93 V $\cdot$ A.   b. 176 V $\cdot$ A.
c. 265 V $\cdot$ A.   d. 333 V $\cdot$ A.

37. What should the capacitance of the circuit of Question 34 be in order that it have a resonance frequency of 1 kHz?
a. 5.07 μF.   b. 31.8 μF.
c. 50.7 μF.   d. 71 μF

38. With the capacitance of Question 37, the power dissipated in the circuit of Question 34 would be
a. 144 W.   b. 176 W.
c. 230 W.   d. 333 W.

EXERCISES

24–1 Effective Current and Voltage

1. The graph below shows how the current in a certain ac circuit varies with time. What is the frequency of the current?

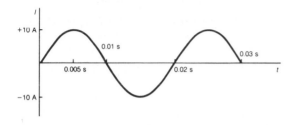

2. What is the effective current in the circuit shown above?

3. An ammeter in series with an ac circuit reads 10 A and a voltmeter across the circuit reads 60 V. (a) What is the maximum current in the circuit? (b) What is the maximum potential difference across the circuit? (c) Does the maximum current necessarily occur at the same moments as the maximum voltage?

4. The dielectric used in a certain capacitor breaks down at a voltage of 300 V. Find the highest effective sinusoidal ac voltage that can be applied to it.

5. The potential difference across a source of alternating current varies sinusoidally with time with $V_{max} = 100$ V. Find the value of the instantaneous potential difference (a) $\frac{1}{8}$ cycle, (b) $\frac{1}{4}$ cycle, (c) $\frac{3}{8}$ cycle, and (d) $\frac{1}{2}$ cycle after $V = 0$.

6. A certain 250-Hz ac power source has a maximum potential difference of 24 V. Find the value of the instantaneous potential difference (a) 0.0005 s, (b) 0.001 s, (c) 0.002 s, (d) 0.0025 s, (e) 0.004 s, and (f) 0.005 s after $V = 0$.

24–4 Inductive Reactance

7. What happens to the impedance of a coil when an iron core is inserted in it?

8. A coil has the reactance X_L when an ac voltage of frequency f is appled to it. What is the reactance when the frequency is changed to $2f$?

9. What is the reactance of a 5-mH inductor at 10 Hz? At 10 kHz?

10. The reactance of an inductor is 80 Ω at 500 Hz. Find its inductance.

11. Find the current that flows when a 3.0-mH inductor is connected to a 15-V, 5-kHz power source.

12. A current of 0.8 A flows through a 50-mH inductor of negligible resistance that is connected to a 120-V power source. What is the frequency of the source?

13. An inductor of negligible resistance whose reactance is 120 Ω at 200 Hz is connected to a 240-V, 60-Hz power line. What is the current in the inductor?

14. A 30-mH, 60-Ω inductor is connected to a 20-V, 400-Hz power source. Find the current in the inductor and the power dissipated in it.

15. The current in a resistor is 2 A when it is connected across a 240-V, 50-Hz line. How much inductance should be connected in series with the resistor to reduce the current to 1 A?

16. A coil of unknown inductance and resistance is observed to draw 50 mA when a dc potential difference of 5 V is applied. When the coil is connected to a source of 400-Hz ac, however, a potential difference of 8 V is required to yield the same current. (a) Find the inductance and resistance of the coil. (b) Find the power dissipated in the coil in each situation.

17. A coil of unknown inductance and resistance is observed to draw 8 A when a 40-V dc potential difference is applied, and 5 A when a 40-V, 60-Hz ac potential difference is applied. (a) Find the inductance and resistance of the coil. (b) Find the power dissipated in the coil in each situation.

18. A circuit that contains inductance and resistance has an impedance of 50 Ω at 100 Hz and an impedance of 100 Ω at 500 Hz. What are the values of the inductance and the resistance of the circuit?

24–5 Capacitive Reactance

19. A parallel-plate capacitor with air as the dielectric has a reactance of X_C when an ac voltage of frequency f is applied to it. What happens to the reactance when the frequency is changed to $2f$?

20. What happens to the reactance of the capacitor of Exercise 19 when the space between its plates is filled with a substance of dielectric constant $K = 4$ and the applied voltage has the frequency f?

21. What is the reactance of an 80-pF capacitor at 10 kHz? At 10 MHz?

22. Find the current that flows when a 10-μF capacitor is connected to a 15-V, 5-kHz power source.

23. The reactance of a capacitor is 50 Ω at 200 Hz. What is its capacitance?

24. A capacitor of unknown capacitance is found to have a reactance of 120 Ω at 200 Hz. The capacitor is connected to a 240-V, 60-Hz power line. (a) What is the current in the circuit? (b) Why is this current not the same as that of Exercise 13?

25. A 5-μF capacitor is connected in series with a 300-Ω resistor, and a 120-V, 50-Hz potential difference is applied. Find the current in the circuit and the power dissipated.

26. The current in a resistor is 2 A when it is connected across a 240-V, 50-Hz line. How much capacitance should be connected in series with the resistor to reduce the current to 1 A?

27. A capacitor of unknown capacitance is connected in series with an 80-Ω resistor. The combination is found to draw 0.5 A when connected to a 120-V, 60-Hz power source. Find (a) the capacitance of the capacitor, (b) the power dissipated in the capacitor, and (c) the power dissipated in the resistor.

28. A circuit that contains capacitance and resistance has an impedance of 30 Ω at 80 Hz and an impedance of 14 Ω at 240 Hz. What are the values of the capacitance and resistance of the circuit?

24–6   Impedance

29. What properties of an ac circuit are described by its resistance, capacitive reactance, inductive reactance, and impedance? What are the similarities and differences among these quantities?

30. You have dc and ac sources of the same emf and dc and ac ammeters. How would you use them to decide whether a circuit element in a black box with two terminals is a resistor, a capacitor, an inductor, or a rectifier?

31. The frequency of the alternating potential difference applied to a series RLC circuit is halved. What happens to the resistance, the inductive reactance, and the capacitive reactance of the circuit? What further information is needed to establish what happens to the impedance of the circuit?

32. A pure capacitor, a pure inductor, and a pure resistor are connected in series across an ac power source. A voltmeter placed in turn across each circuit element reads 15 V, 20 V, and 20 V, respectively. What is the potential difference of the source?

33. A pure capacitor, a pure inductor, and a pure resistor are connected in series across a 60-V ac power source. The potential difference across the capacitor is 60 V, and that across the inductor is also 60 V. What is the potential difference across the resistor?

34. A series circuit has a resistance of 40 Ω, an inductive reactance of 30 Ω, and a capacitive reactance of 50 Ω when connected to a certain ac source. Find (a) the impedance of the circuit, (b) the phase angle, and (c) the potential difference required for a current of 1.5 A to flow.

35. A series circuit has a resistance of 20 Ω, an inductive reactance of 20 Ω, and a capacitive reactance of 20 Ω when connected to an ac source. Find (a) the impedance of the circuit, (b) the phase angle, and (c) the potential difference required for a current of 20 A to flow.

36. The voltage leads the current in a certain ac circuit by 30°. The effective current in the circuit is 5 A. (a) Is the capacitive reactance greater than or less than the inductive reactance? (b) What is the value of I when $V = 0$?

37. A 10-μF capacitor, a 0.10-H inductor, and a 60-Ω resistor are connected in series across a 120-V, 60-Hz power line. Find (a) the current in the circuit, (b) the power dissipated in it, and (c) the potential difference across each of the circuit elements.

38. The circuit of Exercise 37 is connected across a 120-V, 30-Hz power line. Answer the same questions for this situation.

24–7   Resonance

39. An alternating potential difference whose frequency is higher than the resonance frequency of a series RLC circuit is applied to it. Does the voltage in the circuit lead or lag the current?

40. In the antenna circuit of a radio receiver that is tuned to a certain station, $R = 5$ Ω, $L = 5$ mH, and $C = 5$ pF. (a) Find the frequency of the station. (b) If the voltage applied to the circuit is 0.5 mV, find the resulting current. (c) What should the capacitance be in order to receive an 800-kHz radio signal?

41. In a series ac circuit, $R = 20$ Ω, $X_L = 10$ Ω, and $X_C = 25$ Ω when the frequency is 400 Hz. Find the resonance frequency of the circuit.

24–8   Power in AC Circuits

42. What is the significance of the power factor of a circuit? Is it independent of frequency? Under what circumstances (if any) can it be zero? Under what circumstances (if any) can it be 100%?

43. A 10-kW electric motor has an inductive power factor of 70%. What minimum rating in kVA must the power line

have? A capacitor is connected in series with the motor to increase the power factor to 100%. How does this affect the required rating of the power line?

44. A 10-μF capacitor, a 30-mH inductor, and a 15-Ω resistor are connected in series with a 10-V, 250-Hz power source. Find (a) the impedance of the circuit, (b) the current in it, (c) the power factor, (d) the power dissipated, and (e) the minimum apparent-power rating of the source. (f) What current will flow if the circuit is connected to a 10-V power line whose frequency is equal to its resonance frequency?

45. A 60-μF capacitor, a 0.3-H inductor, and a 50-Ω resistor are connected in series with a 120-V, 60-Hz power source. Find (a) the impedance of the circuit, (b) the current in it, (c) the power factor, (d) the power dissipated, and (e) the minimum apparent-power rating of the source. (f) What current will flow if the circuit is connected to a 120-V power lines whose frequency is equal to its resonance frequency?

46. A circuit consisting of a capacitor in series with a resistor draws 3.6 A from a 50-V, 100-Hz power line. The circuit dissipates 120 W. (a) What is the precise phase relationship between voltage and current in the circuit? (b) What series inductance should be inserted in the circuit if the current and voltage are to be in phase? (c) How much current would the circuit then draw? (d) How much power would it then dissipate?

47. An inductor dissipates 75 W of power when it draws 1.0 A from a 120-V, 60-Hz power line. (a) What is its power factor? (b) What capacitance should be connected in series with it to increase its power factor to 100%? (c) How much current would the circuit then draw? (d) How much power would it then dissipate? (e) What should the minimum volt-ampere rating of the power line be then?

24–9 Impedance Matching

48. A microphone whose impedance is 20 Ω is to be used with an amplifier whose input impedance is 50,000 Ω. What should the ratio of turns be in the required transformer?

24–10 Parallel AC Circuits and Filters

49. Loudspeakers are limited in their frequency responses, with large speakers ("woofers") providing more accurate sound reproduction at low frequencies and small ones ("tweeters") at high frequencies. A crossover circuit that divides the output of an audio amplifier into a low-frequency and a high-frequency component is needed when a pair of such speakers is used: Which output of the circuit shown below should be connected to the woofer and which to the tweeter?

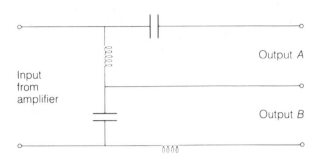

50. What kind of filter is shown below?

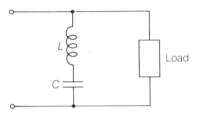

51. What kind of filter is shown below?

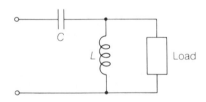

52. A 10-Ω resistor and an 8-μF capacitor are connected in parallel across a 10-V, 1-kHz power source. Find the current in each component, the total current, the impedance of the circuit, the phase angle, and the power dissipated by the circuit.

53. A 10-Ω resistor and a 2-mH inductor are connected in parallel across a 10-V, 1-kHz power source. Answer the same questions for this circuit.

54. A 10-Ω resistor, an 8-μF capacitor, and a 2-mH inductor are connected in parallel across a 10-V, 1-kHz power source. Answer the same questions for this circuit.

55. The figure below shows a simple high-pass filter. (a) What inductance is needed to discriminate against frequencies under 20 kHz when the load is 50 Ω? (b) What proportion of the total current passes through the load when the frequency of the applied signal is 2 kHz? (c) When it is 200 kHz?

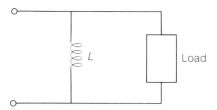

56. The figure below shows a simple low-pass filter. (a) What capacitance is needed to discriminate against frequencies above 20 kHz when the load is 50 Ω? (b) What proportion of the total current passes through the load when the frequency of the applied signal is 2 kHz? (c) When it is 200 kHz?

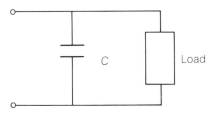

ANSWERS TO MULTIPLE CHOICE

1. b	**9.** b	**17.** b	**25.** b	**32.** a
2. d	**10.** a	**18.** b	**26.** b	**33.** c
3. d	**11.** c	**19.** a	**27.** a	**34.** a
4. b	**12.** a	**20.** c	**28.** c	**35.** a
5. a	**13.** a	**21.** a	**29.** a	**36.** b
6. c	**14.** b	**22.** d	**30.** c	**37.** a
7. a	**15.** b	**23.** c	**31.** a	**38.** d
8. d	**16.** b	**24.** b		

25

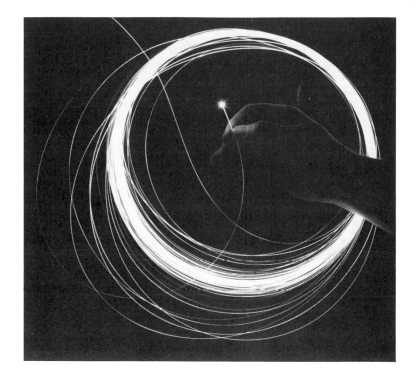

LIGHT

The discovery that light consists of electromagnetic waves is among the most note-worthy achievements of science. That electromagnetic waves should exist was proposed in 1864 by James Clerk Maxwell, who went on to conclude that light waves were of this nature. Experimental confirmation came some years later. Although the wave behavior of light can be analyzed directly from Maxwell's theory, such calculations are quite difficult. A less comprehensive but simpler approach to optics was suggested in 1678 by Christian Huygens. In this chapter the reflection and refraction of light at plane surfaces will be considered with the help of Huygens' method. We should note that in certain important respects light has the character of a stream of particles rather than that of a series of waves, a duality examined in Chapter 28.

25−1   ELECTROMAGNETIC WAVES

A changing electric field is equivalent to a magnetic field

In electromagnetic induction, a changing magnetic field induces an emf in a nearby wire loop or other conducting path. Thus a changing magnetic field is equivalent in its effects to an electric field. The converse is also true: A changing electric field is equiv-

CHAPTER OBJECTIVES

Completing this chapter should enable you to:

1. Describe the nature of electromagnetic waves.

2. Use the formula $c = f\lambda$ to relate the frequency and wavelength of an electromagnetic wave to the speed of light.

3. Use Huygens' principle to investigate the motion of wavefronts.

4. Distinguish between light waves and light rays.

5. Find the direction of a reflected ray of light.

6. Distinguish between real and virtual images.

7. Describe the origin of refraction and the significance of the index of refraction of a medium.

8. Find the direction of a refracted ray of light by using Snell's law.

9. Relate the actual depth of a submerged object to its apparent depth.

10. Explain how total internal reflection occurs.

11. Understand what is meant by a solid angle and how to calculate one.

12. Relate the luminous flux emitted by a light source and the illumination it produces to the luminous intensity of the source.

---

alent in its effects to a magnetic field. This is true even in empty space, where electric currents cannot flow. No simple experiment can directly demonstrate the latter equivalence (unlike electromagnetic induction, which is very easy to exhibit), and it was first proposed by Maxwell on the basis of an indirect argument. Electromagnetic waves occur as a consequence of these two effects—a changing magnetic field produces an electric field, and a changing electric field produces a magnetic field. The two constantly varying fields are coupled together as they travel through space.

What came from Maxwell's analysis was that electromagnetic waves spread out in space from an initial disturbance in the same manner that waves spread out from a disturbance in a body of water. If we throw a stone into a pond, oscillations occur in which energy is continually interchanged between the kinetic energy of moving water and the potential energy of water higher than its normal level. These oscillations begin where the stone lands, and spread out as waves across the surface of the pond. The wave speed depends upon the properties of the pond water, varying with temperature, impurity content, and so on, but it is independent of the wave amplitude. As we saw in Chapter 13, this is typical wave behavior. When electromagnetic waves spread out from an electric or magnetic disturbance, their energy is constantly being interchanged

Electromagnetic waves consist of coupled electric and magnetic fields

between the fluctuating electric field and the fluctuating magnetic field of the waves.

Formation of an electromagnetic wave

Let us connect a pair of metal rods to a source of alternating emf, as in Fig. 25–1. (Such a source is called an oscillator, as we know.) For clarity we will imagine that there is only a single charge in each rod at any time.

(a) When the oscillator is switched on, a positive charge in the upper rod begins to move upward and a negative charge in the lower rod begins to move downward. The electric lines of force around the charges are indicated by the heavy lines, and the magnetic lines of force due to the motion of the charges (which are concentric circles perpendicular to the paper) are indicated by crosses when their direction is into the paper and by dots when their direction is out of the paper. (The dots represent arrowheads and the crosses represent the tail feathers of arrows.)

(b) The charges have reached the limit of their motion and have stopped, so that they stop producing a magnetic field. The outer magnetic lines of force do not disappear but continue to travel outward.

(c) The emf of the oscillator now begins to decrease, and the charges move toward each other. The result is a magnetic field in the opposite direction to the earlier field. The electric field is in the same direction as before.

FIG. 25–1 A pair of metal rods connected to an electrical oscillator emit electromagnetic waves.

(d) The emf has passed through 0 and begun to increase in the opposite sense. As a result there is a negative charge in the upper rod and a positive one in the lower rod which begin to move apart. The electric field is therefore opposite in direction to the

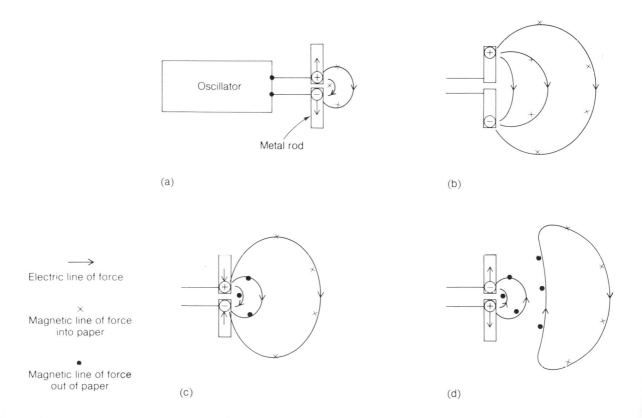

→
Electric line of force

×
Magnetic line of force into paper

●
Magnetic line of force out of paper

(a)   (b)

(c)   (d)

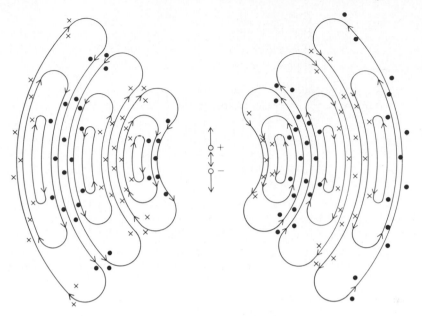

FIG. 25–2 The configuration of electric and magnetic fields that spread outward from a pair of oscillating charges.

earlier field, but the magnetic field is in the same direction since magnetically a positive charge moving downward is equivalent to a negative charge moving upward.

Owing to this sequence of changes in the fields, the outermost electric and magnetic lines of force respectively form into closed loops. These loops of force, which lie in perpendicular planes, are divorced from the oscillating charges that gave rise to them and continue moving outward, constituting an electromagnetic wave. As the charges continue oscillating back and forth, further associated loops of electric and magnetic lines of force are emitted, forming an expanding pattern of loops.

The electric and magnetic fields of an electromagnetic wave form closed loops

Figure 25–2 shows the configuration of the electric and magnetic fields that spread outward from a pair of oscillating charges. The actual fields are three dimensional, so that the magnetic lines of force form loops in planes perpendicular to the line joining the charges.

Three properties of electromagnetic waves are worth noting:

1. The variations occur simultaneously in both fields (except close to the oscillating charges), so that the electric and magnetic fields have maxima and minima at the same times and in the same places.

Properties of electromagnetic waves

2. The directions of the electric and magnetic fields are perpendicular to each other and to the direction in which the waves are moving. Light waves are therefore transverse.

3. The speed of the waves depends only upon the electric and magnetic properties of the medium they travel in, and not upon the amplitudes of the field variations.

Figure 25–3 is an attempt at portraying properties 1 and 2 above in terms of lines of force of **E** and **B** a long distance from a source of electromagnetic waves. Closer to the source the lines of force are curved, as in Fig. 25–2. It is worth keeping in mind

FIG. 25–3 The electric and magnetic fields of an electromagnetic wave far from its source vary as shown here. The field directions are perpendicular to each other and to the direction of propagation.

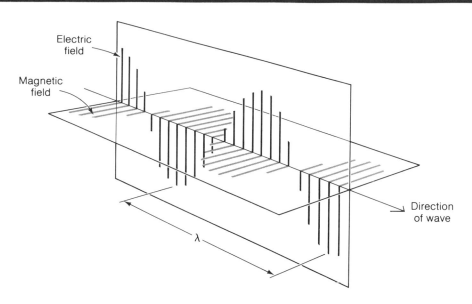

that, unlike the other types of waves considered in Chapter 13—waves in a stretched string, water waves, sound waves—nothing material moves in the path of an electromagnetic wave. The only changes are in electric and magnetic fields.

All accelerated charges produce electromagnetic waves

We have been considering a special kind of electromagnetic wave source. Actually, *all* accelerated charges radiate electromagnetic waves, regardless of the manner in which the acceleration occurs.

Maxwell's theory of electromagnetic waves showed that their speed c in free space depends solely upon ϵ_0 and μ_0, the permittivity (Sec. 18–2) and permeability (Sec. 21–3) of free space. Maxwell found that the speed c is given by

Speed of light

$$c = \frac{1}{\sqrt{\epsilon_0 \mu_0}}$$

$$= \frac{1}{\sqrt{(8.85 \times 10^{-12}\,\text{C}^2/\text{N} \cdot \text{m}^2)(1.26 \times 10^{-6}\,\text{T} \cdot \text{m/A})}}$$

$$= 3.00 \times 10^8\,\text{m/s}$$

which is the same speed that had been experimentally measured for light waves in free space! The correspondence was too great to be accidental, and, as further evidence became known, the electromagnetic nature of light found universal acceptance. To five significant figures the value of c is 2.9979×10^8 m/s.

25–2   TYPES OF ELECTROMAGNETIC WAVES

All electromagnetic waves have the same basic nature

Light is not the only example of an electromagnetic wave. Although all electromagnetic waves share certain basic properties, other features of their behavior depend upon their frequencies. Light waves themselves span a brief frequency interval, from about

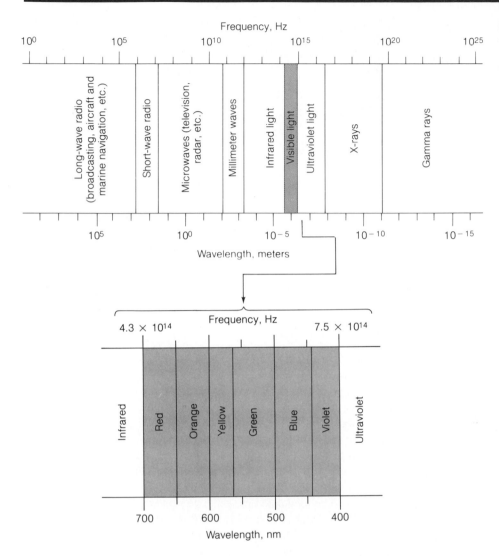

FIG. 25–4 The electromagnetic wave spectrum. The boundaries of the various categories are not sharp. (1 nm = 1 nanometer = 10^{-9} m.)

4.3×10^{14} Hz for red light to about 7.5×10^{14} Hz for violet light. Electromagnetic waves with frequencies between these limits are the only ones that the eye responds to, and specialized instruments of various kinds are required to detect waves with higher and lower frequencies. Figure 25–4 shows the electromagnetic wave *spectrum* from the low frequencies used in radio communication to the high frequencies found in X rays and gamma rays (which are considered in later chapters). The wavelengths corresponding to the various frequencies are also shown.

The colors of things we see by reflected light depend on the kind of light that falls on them and on the nature of their surfaces. If a surface reflects all light that falls on it, the color of the surface will be white when white light illuminates it, red when red light illuminates it, and so on. A surface that reflects only, say, green light will appear green only when the light illuminating it contains green; otherwise it will appear black. A surface that absorbs all light that falls on it appears black.

FIG. 25–5 In radiotelegraphy, a constant-frequency electromagnetic wave is switched on and off in a coded sequence to transmit information. The letter "v," which is ··· — in Morse code, is shown.

As we recall, the product of the frequency f of a wave and its wavelength λ is just the wave speed, here c. Hence, given the wavelength or frequency of a particular electromagnetic wave, we can immediately find the other quantity.

Example   Find the wavelengths of yellow light whose frequency is 5×10^{14} Hz and of radio waves whose frequency is 1 MHz. (1 MHz = 1 megahertz = 10^6 Hz.)

Solution   The wavelength of the yellow light is

$$\lambda = \frac{c}{f} = \frac{3 \times 10^8 \, \text{m/s}}{5 \times 10^{14} \, \text{Hz}} = 6 \times 10^{-7} \, \text{m} = 600 \, \text{nm}$$

which is less than 1/1000 of a millimeter. (1 nm = 1 nanometer = 10^{-9} m. The nanometer is often used for expressing the wavelengths found in light.) By the same procedure we find that a 1-MHz radio wave has a wavelength of 300 m.   ■

Amplitude modulation

Electromagnetic waves provide a means for transmitting information from one place to another without wires or other material links between them. There are two principal methods of incorporating information in an electromagnetic wave, *amplitude modulation* and *frequency modulation*.

In amplitude modulation, a "carrier wave" of constant frequency is varied in amplitude, with the variations constituting the signal. The simplest example is a flashlight switched on and off to give a series of dots and dashes. Another example is the radiotelegraph. Sequences of dots and dashes are used in the Morse code to represent letters of the alphabet and numbers (Fig. 25–5).

A carrier wave can also be modulated in such a way that the amplitude variations correspond to sound waves, as shown schematically in Fig. 25–6. A microphone converts sound waves to an equivalent electric signal, which is then amplified and combined with the carrier. The final wave is broadcast from an antenna, and at the receiving station the carrier is removed to leave an audio signal. The latter is then amplified and used to generate sound waves in a loudspeaker.

Frequency modulation

One difficulty with amplitude modulation is that such sources of random electromagnetic waves as electric storms and electric machinery can interfere with the broadcast waves to cause "static." In frequency modulation the frequency, not the amplitude, of the carrier is varied in accordance with the audio signal (Fig. 25–7). The information content of a frequency-modulated electromagnetic wave is virtually immune to disturbance.

In television the desired scene is focused on the light-sensitive screen of a special tube which is then scanned in a zigzag fashion by an electron beam. The signal extracted from the screen varies with the image brightness at each successive point that is scanned, and this signal is then used to modulate a carrier wave for broadcasting. At the receiver, suitable circuits demodulate the wave, and in the picture tube the image is reproduced

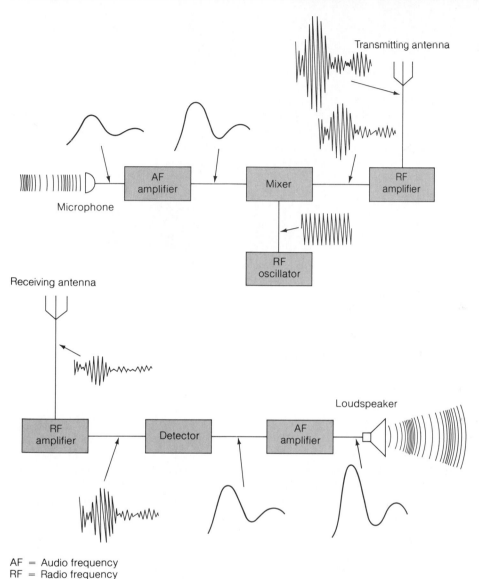

FIG. 25–6 In amplitude modulation, the amplitude of a constant-frequency carrier wave is varied in accordance with an audio signal. The transmitting and receiving systems shown are highly simplified.

AF = Audio frequency
RF = Radio frequency

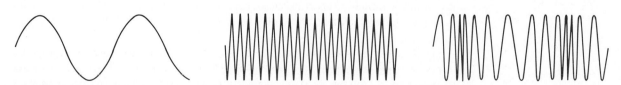

FIG. 25–7 The information content of a frequency-modulated wave resides in its frequency variations rather than in its amplitude variations.

FIG. 25–8 (a) The ionosphere is a region in the upper atmosphere whose ionized layers make possible long-range radio communication by their ability to reflect radio waves. (b) Only radio waves in the medium- and high-frequency (MF and HF) bands are reflected by ionosphere. Low-frequency (LF) waves are absorbed there. Very high- and ultra high-frequency (VHF and UHF) waves pass through the ionosphere and so are used to communicate with satellites and spacecraft.

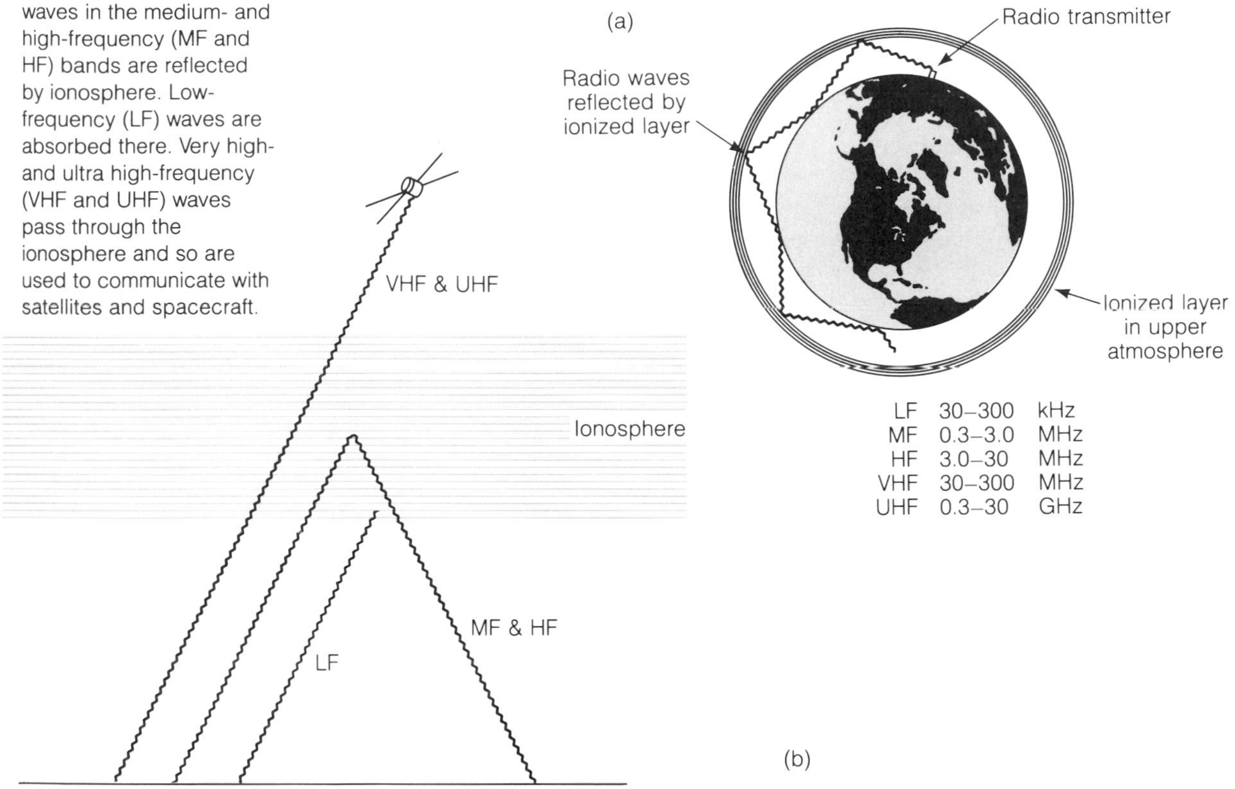

(a)

Radio transmitter

Radio waves reflected by ionized layer

Ionized layer in upper atmosphere

VHF & UHF

Ionosphere

LF	30–300	kHz
MF	0.3–3.0	MHz
HF	3.0–30	MHz
VHF	30–300	MHz
UHF	0.3–30	GHz

MF & HF

LF

(b)

with the help of an electron beam that moves in synchronism with the electron beam in the camera tube (See Fig. 18–23.)

Reflection from the ionosphere permits long-range radio communication

Radio waves in the short-wave band of Fig. 25–4 have the useful property of being reflected from the earth's ionosphere (Section 19–1), a region high in the atmosphere where ions are relatively abundant. Without the ionosphere, radio communication would be limited to short distances since electromagnetic waves travel in straight lines and would be shielded from remote receivers by the curvature of the earth. However, since radio waves of appropriate frequencies can bounce one or more times between the ionosphere and the earth's surface, long-range transmission is possible, even to the opposite side of the earth (Fig. 25–8). Because the ion content of the ionosphere and the heights of the individual ion layers vary with time of day and are influenced by streams of fast protons and electrons emitted by the sun during periods of sunspot activity, there are no absolute rules for which frequencies are best for which distances. Generally speaking, though, the higher the frequency, the greater the range. Thus a frequency of 10 MHz would ordinarily give good results between 300 and 2400 km whereas 20 MHz would be better for 2400 to 11,000 km.

The higher-frequency waves used for FM and television broadcasting are not reflected by the ionosphere, so their reception is limited to the line of sight between transmitting and receiving antennas. When the frequencies are up around 10^{10} Hz, corresponding to wavelengths in the centimeter range, the waves can readily be focused into narrow beams. Such beams are reflected by objects such as ships and airplanes, which is the basis of *radar* (from *ra*dio *d*etection *a*nd *r*anging). A rotating antenna is used to send out a pulsed beam, and the distance of a particular target is established by the time needed for the echo to return to the antenna. A cathode-ray tube is used to display the echoes, with the center of the picture representing the position of the antenna.

Radar uses a beam of short-wavelength radio waves to detect targets

The rotating scanner of a radar set sends out pulses of high-frequency radio waves in a narrow beam and then detects their reflections from objects in their paths. These reflections are then displayed on a screen. Shown here is the image of a harbor entrance. The center of the screen corresponds to the position of the ship carrying the radar. (Photo: Raytheon)

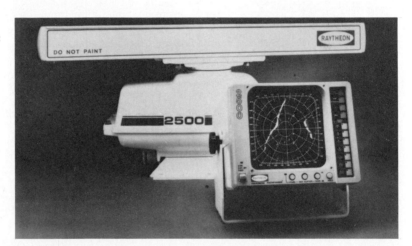

25–3 HUYGENS' PRINCIPLE

Most optical phenomena can be understood on the basis of the wave nature of light alone, without any direct reference to its electromagnetic character. As mentioned earlier, three centuries ago Huygens devised an approach to the behavior of waves of all kinds that is particularly convenient to use. His method concerns *wavefronts*. A wavefront is an imaginary surface that joins points where all of the waves involved are in the same phase of oscillation. As in Fig. 25–9, waves from a point source spread out in a series of spherical wavefronts. (In the case of water waves, the wavefronts that result when a stone is dropped in a lake are circular.) At a long distance from a point source, the curvature of the wavefronts is so small that they can be considered as a series of planes.

Wavefronts

Huygens' principle states that

FIG. 25–9 The spherical wavefronts of a point source become plane wavefronts at a long distance from the source.

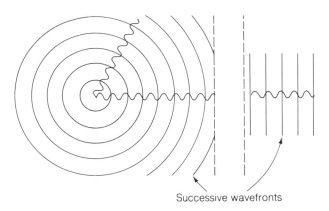

Successive wavefronts

FIG. 25–10 Huygens' principle applied to the motion of (a) plane and (b) spherical wavefronts.

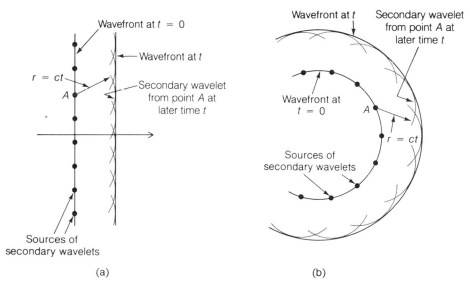

(a)   (b)

Huygens' principle

Every point on a wavefront can be considered as a source of secondary wavelets that spread out in all directions with the wave speed of the medium. The wavefront at any time is the envelope of these wavelets.

Figure 25–10 shows how Huygens' principle is applied to the motion of plane and spherical wavefronts in a uniform medium. It does not matter just where on the initial wavefront we imagine the sources of the secondary wavelets to be.

Despite its name, Huygens' principle is not in the same category as such fundamental principles as those of conservation of mass, energy, momentum, and electric charge, but is rather a convenient means for studying wave motion. We shall find Huygens' principle useful in understanding a variety of optical phenomena.

Rays are a convenient way to represent paths taken by light waves

Although we shall use the notion of wavefronts for the actual analysis of wave propagation, the results are commonly represented in terms of *rays*. A light ray is simply an imaginary line in the direction in which the wavefronts advance, and so it is perpendicular to the wavefronts. The picture most of us have of a light ray is a narrow pencil of light, which is perfectly legitimate. However, an approach based only on rays does not reveal such characteristic wave behavior as diffraction (the bending of waves around an obstacle into the "shadow" region). We must remember that the motion of wavefronts is what is really significant, although it is both proper and convenient to use rays to summarize our conclusions.

25−4 REFLECTION

Every object reflects a certain proportion of the light falling upon it, and this reflected light is what enables us to see it. In most cases the surface of the object has irregularities that spread out an initially parallel beam of light in all directions to produce *diffuse reflection* (Fig. 25−11(a)). A surface so smooth that any irregularities in it are small relative to the wavelength of the light falling upon it behaves differently. When a parallel beam of light is directed at such a surface, it is *specularly reflected* in only one direction (Fig. 25−11(b)). Reflection from a brick wall is diffuse whereas reflection from a mirror is specular. Often a mixture of both kinds of reflection occurs, as in the case of a surface coated with varnish or glossy enamel. Our concern here is with specular reflection only.

Diffuse and specular reflection

When we see an object, what enters our eyes are light waves reflected from its surface. What we perceive as the object's color therefore depends upon two things: the kind of light falling on it and the nature of its surface. If white light is used to illuminate an object that absorbs all colors other than red, the object will appear red. If green light is used instead, the object will appear black because it absorbs green light. A white object reflects light of all wavelengths equally well, and its apparent color is the same as the color of the light reaching it. A black object, on the other hand, absorbs light of all wavelengths, and it appears black no matter what color light reaches it.

Appearance of a colored object

Let us examine the specular reflection of a series of plane wavefronts, which corresponds to a parallel beam of light. At $t = 0$ in Fig. 25−12 the wavefront AB just touches the mirror, and secondary wavelets start to spread out from A. After a time t the end B of the wavefront reaches the mirror at C. The secondary wavelets from A are now at the point D. The wavefront which was AB at $t = 0$ is therefore CD at the later time t.

The angle i between an approaching wavefront and the reflecting surface is called the *angle of incidence,* and the angle r between a receding wavefront and the reflecting surface is called the *angle of reflection.* Here

Angles of incidence and reflection

FIG. 25−11

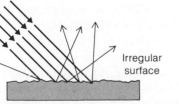

(a)   Diffuse reflection

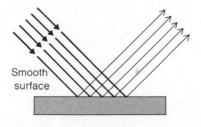

(b)   Specular reflection

FIG. 25–12 The behavior of successive wavefronts during reflection from a plane mirror. The angle of incidence *i* and the angle of reflection *r* are equal.

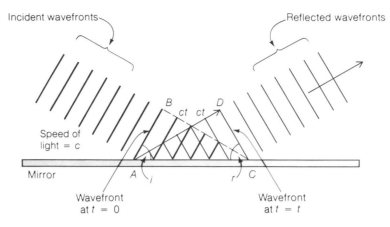

FIG. 25–13 Ray representation of reflection. The angles of incidence and reflection are measured with respect to the normal to the reflecting surface.

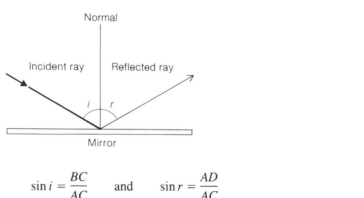

$$\sin i = \frac{BC}{AC} \qquad \text{and} \qquad \sin r = \frac{AD}{AC}$$

The wavelet that originates at *A* takes the time *t* to reach *D*. Hence $AD = ct$, where *c* is the speed of light. The point *B* of the wavefront *AB* also takes the time *t* to reach *C*, and so $BC = ct$ as well. Therefore $AD = BC$ and sin *i* = sin *r*, from which we conclude that the angles *i* and *r* are equal. Thus we have the basic law of reflection:

Law of reflection

The angle of reflection of a plane wavefront with a plane mirror is equal to the angle of incidence.

The normal to a surface is a line perpendicular to the surface

In the ray model of light propagation, the angles of incidence and reflection are measured with respect to the *normal* to the reflecting surface at the point where the light strikes it (Fig. 25–13). The normal is a line drawn perpendicular to the surface at that point. In this representation, too, the angles of incidence and reflection are equal. The incident ray, the reflected ray, and the normal all lie in the same plane.

The image of an object we see in a plane mirror appears to be the same size as the object and as far behind the mirror as the object is in front of it. Let us investigate how this situation arises.

Real and virtual images

In Fig. 25–14 three typical light rays from a point object at *A* impinge on a mirror and are reflected, in each case with an angle of reflection equal to the angle of incidence. To the eye, the three diverging rays apparently come from the point *A'* behind the mirror. The rays that seem to come from the image do not actually pass through it, and

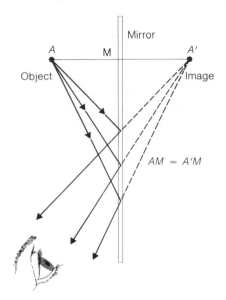

FIG. 25–14 Rays from a point object at A appear to come from A' after reflection from a plane mirror. The image is as far behind the mirror as the object is in front of it.

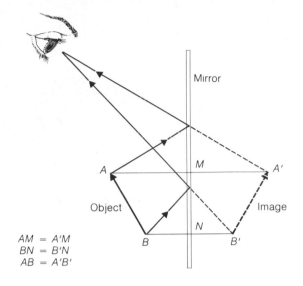

FIG. 25–15 Image formation by a plane mirror. The image is erect and is the same size and shape as the object.

for this reason the image is said to be *virtual*. (A *real image* is formed by light rays that pass through it.) From the geometry of the figure it is clear that $AM = A'M$, so that the object and the image are the same distance on either side of the mirror.

In Fig. 25–15, an actual object is being reflected in a plane mirror. Again simple geometry shows that the object and its virtual image have the same dimensions. Every point in the image is directly behind the corresponding point on the object, and so the orientation of the image is the same as that of the object; the image is therefore *erect*. However, left and right are interchanged in a mirror image, as in Fig. 25–16, because front and back have been reversed by the reflection. Thus a printed page appears backward in a mirror, and what seems to be one's left hand in a mirror is actually one's right hand.

25–5 REFRACTION

It is a matter of experience that a beam of light passing at a slanting angle from one medium to another, say from air to water, is deflected at the surface between the two media. The bending of a light beam when it passes from one medium to another is called *refraction,* and it is responsible for such familiar effects as the apparent distortion of objects partially submerged in water.

Several important aspects of refraction are illustrated in Fig. 25–17. When a light beam goes from air into water along the normal to the surface between them, it simply continues along the same path. However, when it enters the water at any other angle, it is bent toward the normal. The paths are reversible; thus a light beam emerging from the water is bent *away* from the normal as it enters the air.

FIG. 25–16 Left and right are interchanged in reflection.

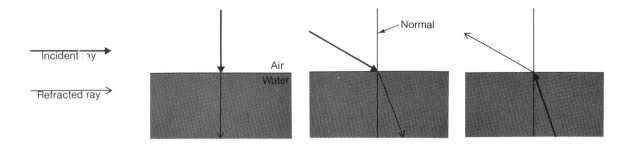

Incident ray →

Refracted ray →

FIG. 25-17 The bending of a light beam when it passes obliquely from one medium to another is called *refraction*.

Refraction occurs because light travels at different speeds in the two media. Let us see what happens to the plane wavefront AB in Fig. 25–18 when it passes at a slanting angle from a medium in which its speed is v_1 to a medium in which its speed is v_2, where v_2 is less than v_1. At $t = 0$ the wavefront AB just comes in contact with the interface between the two media. After a time t, the end B of the wavefront reaches the interface at C, and the secondary wavelets from A are now at D. Since v_2 is less than v_1, the secondary wavelets generated at the interface travel a shorter distance in the same time interval than do wavelets in the first medium, and the distance AD is shorter than BC. The refracted wavefronts accordingly move in a different direction from that of the incident wavefronts.

Angles of incidence and refraction

The angle i between an approaching wavefront and the interface between two media is called the *angle of incidence* of the wavefront, and the angle r between a receding wavefront and the interface it has passed through is called the *angle of refraction*. From Fig. 25–18 we see that

$$\sin i = \frac{BC}{AC} \qquad \text{and} \qquad \sin r = \frac{AD}{AC}$$

Because $BC = v_1 t$ and $AD = v_2 t$,

$$\frac{\sin i}{\sin r} = \frac{BC}{AD} = \frac{v_1 t}{v_2 t}$$

and so we have

FIG. 25-18 The behavior of successive wavefronts during refraction. The speed of light v_1 in the first medium is greater than that in the second medium.

$$\frac{\sin i}{\sin r} = \frac{v_1}{v_2}$$

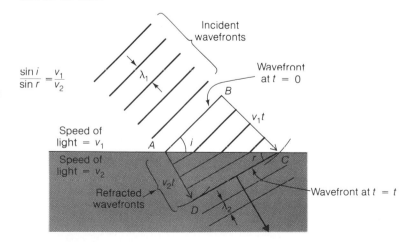

$$\frac{\sin i}{\sin r} = \frac{v_1}{v_2} \qquad\qquad \textit{Snell's law} \quad (25–1)$$

This useful result is known as *Snell's law* after its discoverer, the seventeenth-century Dutch astronomer Willebrord Snell. It states that

Snell's law relates angles of incidence and refraction

The ratio of the sines of the angles of incidence and refraction is equal to the ratio of the speeds of light in the two media.

Refraction, like reflection, can be described in terms of the ray model of light. As in Fig. 25–19, the angles *i* and *r* are taken with respect to the normal to the interface at the point where the rays meet it.

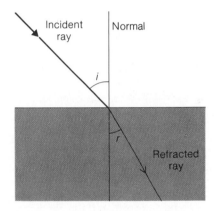

FIG. 25–19 Ray representation of refraction.

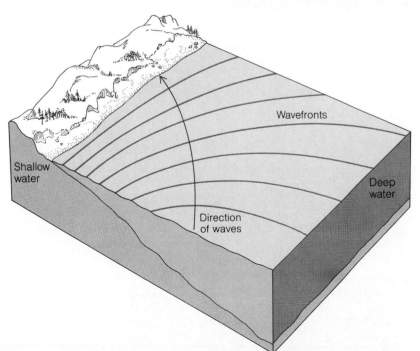

FIG. 25–20 Water waves that approach the shore obliquely are refracted because they move more slowly in shallow water.

Why does light travel more slowly in a material medium than in a vacuum? The medium consists of atoms that contain electrons. When a light wave arrives at an atom, its electrons absorb energy from the wave as they begin to oscillate at the same frequency. These oscillations then reradiate light waves with this frequency. Because of the inertia of the electrons and because they are bound to atoms, their oscillations lag slightly behind those of the incoming wave. The result is a wave speed in the medium that is less than its speed in free space.

Refraction of water waves

Refraction is not confined to light waves. A conspicuous example of refraction can be observed in water waves that approach a sloping beach. Regardless of the direction of the waves in open water, their direction becomes more and more perpendicular to the shoreline as they come nearer (Fig. 25–20). The reason is that the speed of a water wave decreases in shallow water because of friction with the bottom, so a wavefront moving shoreward is progressively affected by the change in speed and swings around until it is parallel to the shore.

25–6 INDEX OF REFRACTION

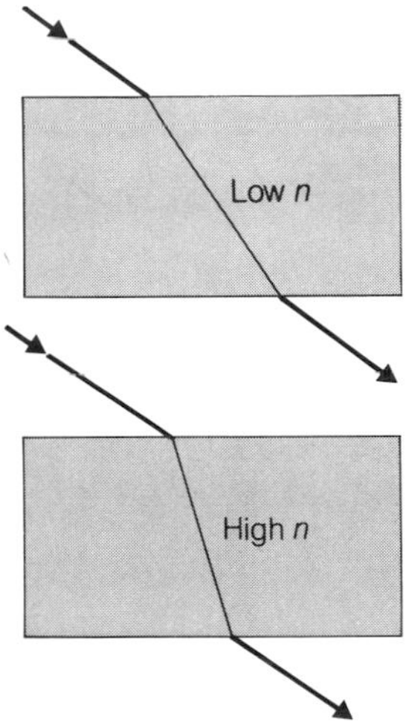

The ratio between the speed of light c in free space and its speed v in a particular medium is called the *index of refraction* of the medium. The greater the index of refraction, the greater the extent to which a light beam is deflected upon entering or leaving the medium (Fig. 25–21). The symbol for index of refraction is n, so that

$$n = \frac{c}{v}$$

Index of refraction (25–2)

$$\text{Index of refraction} = \frac{\text{speed of light in free space}}{\text{speed of light in medium}}$$

Table 25–1 is a list of the values of n for a number of substances.

It is easy to rewrite Snell's law in terms of the indexes of refraction n_1 and n_2 of two successive media. In these media light has the respective speeds

$$v_1 = \frac{c}{n_1} \quad \text{and} \quad v_2 = \frac{c}{n_2}$$

and so Snell's law becomes

$$\frac{\sin i}{\sin r} = \frac{v_1}{v_2} = \frac{c/n_1}{c/n_2} = \frac{n_2}{n_1}$$

FIG. 25–21 The greater the index of refraction n of a medium, the greater the deflection of a light beam on entering or leaving it.

This is usually written in the form

$$n_1 \sin i = n_2 \sin r$$

Snell's law (25–3)

TABLE 25–1
Indexes of refraction. The values of n vary slightly with the frequency of the light

Substance	n	Substance	n
Air	1.0003	Glass, flint	1.63
Benzene	1.50	Ice	1.31
Carbon disulfide	1.63	Lucite™ and Plexiglas™	1.51
Diamond	2.42	Quartz	1.46
Ethyl alcohol	1.36	Water	1.33
Glass, crown	1.52	Zircon	1.92

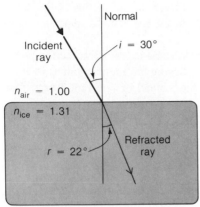

FIG. 25-22

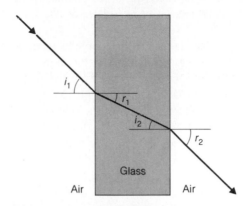

FIG. 25-23 Light entering a glass plate with parallel sides emerges parallel to its original direction but displaced to one side.

Example   A beam of parallel light enters a block of ice at an angle of incidence of 30° (Fig. 25–22). What is the angle of refraction in the ice?

Solution   The indexes of refraction of air and ice are respectively 1.00 and 1.31. From Snell's law,

$$\sin r = \frac{n_1}{n_2} \sin i = \frac{1.00}{1.31} \sin 30° = 0.382$$

$$r = 22°$$

Example   A ray of light is incident at an angle of 45° on one side of a glass plate of index of refraction 1.6. Find the angle at which the ray emerges from the other side of the plate.

Solution   The geometry of this problem is shown in Fig. 25–23. The angle of refraction r_1 at the first side of the plate is found as follows:

$$\sin r_1 = \frac{n_{air}}{n_{glass}} \sin i_1 = \frac{1.0}{1.6} \sin 45° = 0.442$$

$$r_1 = 26°$$

Because the sides of the plate are parallel, the angle of incidence i_2 at the second side is equal to r_1. If we let r_2 be the angle of refraction at which the ray emerges from the plate,

$$\sin r_2 = \frac{n_{glass}}{n_{air}} \sin i_2 = \frac{1.6}{1.0} \sin 26° = 0.701$$

$$r_2 = 45°$$

The ray leaves the glass plate parallel to its original direction but displaced to one side.

This is a general result for parallel-sided plates that holds for all angles of incidence and all indexes of refraction; of course, when $i_1 = 0$, the ray is not displaced.   ■

Dispersion by a prism separates white light into its component colors

The index of refraction of a medium depends to some extent upon the frequency of the light involved, with the highest frequencies having the highest values of n. In ordinary glass the index of refraction for violet light is about one percent greater than that for red light, for example. Since a different index of refraction means a different degree of deflection when a light beam enters or leaves a medium, a beam containing more than one frequency is split into a corresponding number of different beams when it is refracted. This effect, called *dispersion,* is illustrated in Fig. 25–24, which shows

FIG. 25–24 Dispersion of white light by a prism. The various colors blend into one another smoothly.

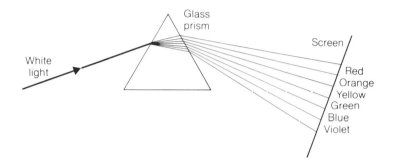

FIG. 25–25 Rainbows are created by the dispersion of sunlight by raindrops. Red light arrives at the eye of the observer from the upper drop shown here; violet light arrives from the lower drop. Other raindrops yield the other colors and produce a continuous arc in the sky.

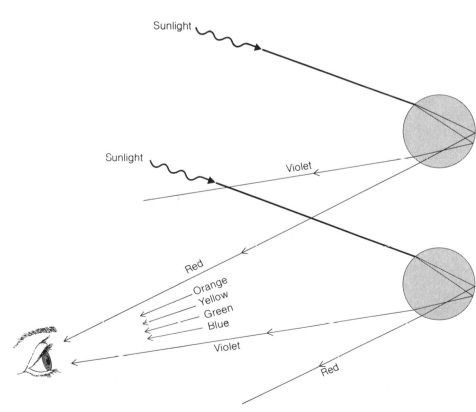

the result of directing a narrow pencil of white light at one face of a glass prism. The initial beam separates into beams of various colors, from which we conclude that white light is actually a mixture of light of these different colors. The band of colors that emerges from the prism is known as a *spectrum*.

Dispersion in water droplets is responsible for rainbows, which are seen when the sun is behind an observer who is facing falling rain. Figure 25–25 shows what happens when a ray of sunlight enters a raindrop, where it is first refracted, then reflected at the far surface, and finally refracted again when it emerges. Dispersion occurs at each refraction, with the result that the angles between the incoming sunlight and the violet and red ends of the spectrum are respectively 40° and 42°. Thus the light that reaches the observer comes from a ring in the sky between 40° and 42° from the direction of the sunlight. In this ring red light is on the outside because, being deviated most, it originates in droplets farther away. Between the red outside of the rainbow and the violet inside appear all the other colors of the spectrum. A person in an airplane can see the entire ring, but from the ground only the upper part appears to give the familiar arc-shaped spectrum.

Light arriving at raindrops at a larger angle than that shown in Fig. 25–25 can undergo two reflections inside each drop before emerging, which leads to an outer secondary rainbow whose colors are in the reverse order. Because some light is refracted out of a water drop at each reflection, the double reflection makes the secondary rainbow much fainter than the primary one.

Origin of the rainbow

25–7 APPARENT DEPTH

A familiar example of refraction is the apparent reduction in the depth of an object submerged in water or other transparent liquid. Thus an oar dipped in a lake seems bent upward where it enters the water because its submerged portion appears to be

Submerged objects seem closer to the surface than they actually are

FIG. 25–26 An oar appears bent when partly immersed in water because of refraction.

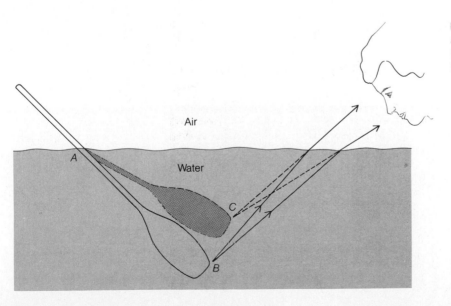

closer to the surface than it actually is. Figure 25–26 shows how this effect comes about. Light leaving the tip of the oar at B is bent away from the normal upon entering the air. To an observer above, who instinctively interprets what he sees in terms of the straightline propagation of light, the tip of the oar is at C, and the submerged part of the oar seems to be AC and not AB.

It is not difficult to relate the apparent and actual depths of a submerged object. Figure 25–27 shows a fish at the point F, an actual depth h below the surface of a body of water, whereas to an observer in the air the fish is at F', only h' below the surface. If we restrict ourselves to rays that are nearly vertical,

$$\frac{\sin i}{\sin r} \approx \frac{\tan i}{\tan r} \tag{25–4}$$

since $\sin \theta \approx \tan \theta$ when θ is small. From Fig. 25–27 we see that

$$\tan i = \frac{x}{h} \qquad \text{and} \qquad \tan r = \frac{x}{h'}$$

where x is the horizontal distance between the position of the fish and the point O where the ray under consideration leaves the water. Hence

$$\frac{\tan i}{\tan r} = \frac{x/h}{x/h'} = \frac{h'}{h}$$

and so, from Eq. (25–4) and Snell's law,

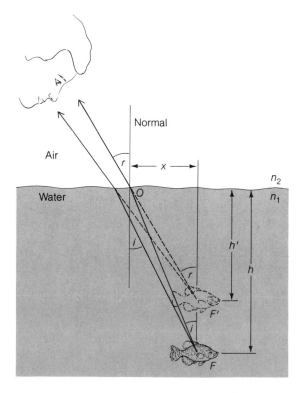

FIG. 25–27 Submerged bodies seem closer to the surface than they actually are.

$$\frac{h'}{h} = \frac{n_2}{n_1}$$
Apparent depth (25−5)

Example To a sailor standing on the deck of a yacht, the water depth appears to be 3.0 m. If this estimate is correct in terms of what he sees, what is the true depth?

Solution Here, with $n_1 = 1.33$ and $n_2 = 1.00$ (the indexes of refraction of water and air respectively),

$$\frac{h'}{h} = \frac{1.00}{1.33} = 0.752$$

The apparent depth h' is only about three-quarters of the true depth h. Hence

$$h = \frac{h'}{0.752} = \frac{3.0\,\text{m}}{0.752} = 4.0\,\text{m}$$ ∎

25−8 TOTAL INTERNAL REFLECTION

An interesting phenomenon known as *total internal reflection* can occur when light passes from one medium to another which has a *lower* index of refraction, for instance from water or glass to air. In this case the angle of refraction is greater than the angle of incidence, and a light ray is bent *away* from the normal (ray 2 in Fig. 25−28). As the angle of incidence is increased, a certain *critical angle* i_c is reached for which the angle of refraction is 90°. The "refracted" ray now travels along the interface between the two media and cannot escape (ray 3). A ray approaching the boundary at an angle of incidence greater than the critical angle is reflected back into the medium it comes from, with the angle of reflection being equal to the angle of incidence as in any other instance of reflection (ray 4).

Past the critical angle, light is reflected rather than refracted

To find the value of the critical angle, we set $i = i_c$ and $r = 90°$ in Snell's law, and we obtain, since $\sin 90° = 1$,

$$n_1 \sin i_c = n_2 \sin 90°$$

$$\sin i_c = \frac{n_2}{n_1}$$
Critical angle (25−6)

For a ray going from water to air,

$$\sin i_c = \frac{n_{\text{air}}}{n_{\text{water}}} = \frac{1.00}{1.33} = 0.752$$

$$i_c = 49°$$

What does a person (or a fish) who is underwater see when looking upward? By reversing all the rays of light in Fig. 25−28 (the paths taken by light rays are always reversible), it is clear that light from everywhere above the water's surface reaches the eyes through a circle on the surface. The rays are all concentrated in a cone whose angular width is twice i_c, which is 98°. Outside this cone is darkness (Fig. 25−29).

What a fish sees

The sharpness and brightness of a light beam are better preserved by total internal reflection than by reflection from an ordinary mirror, and optical instruments accord-

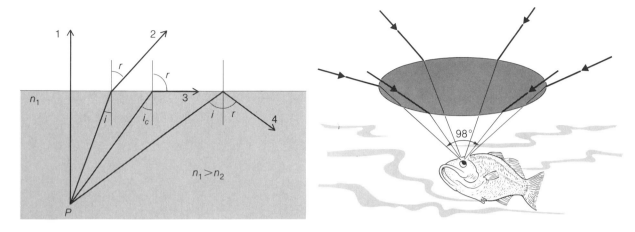

FIG. 25–28 Total internal reflection occurs when the angle of refraction of a light ray going from one medium to another of lower index of refraction equals or exceeds 90°.

FIG. 25–29 An underwater observer sees a circle of light at the surface. All the light reaching him or her is concentrated in a cone 98° wide.

FIG. 25–30 Three applications of total internal reflection.

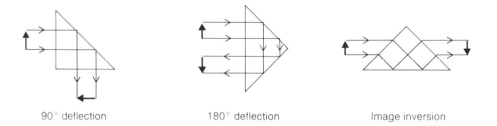

90° deflection   180° deflection   Image inversion

ingly use prisms in preference to mirrors whenever light is to be changed in direction Figure 25–30 shows three types of totally reflecting prisms in common use.

How to pipe light

Total internal reflection makes it possible to "pipe" light from one place to another with a rod of glass or transparent plastic. As in Fig. 25–31, successive internal reflections occur at the surface of the rod, and nearly all the light entering at one end emerges at the other. If a cluster of narrow glass fibers is used instead of a single thick rod, an image can be transferred from one end to the other since each fiber carries intact a part of the image. Because a tube of glass fibers is flexible, it can be used for such purposes as examining a person's stomach by passing the tube in through the mouth. Some of the fibers are used to provide light for illumination, and the others carry the reflected light back outside for viewing.

Fiber optics in telephone systems

Glass fibers are coming into increasing use in telephone systems, where they have a number of advantages over electric wires for carrying information. In a fiber-optic system, the electric signals that would normally be sent directly through copper wires are first converted into a series of pulses according to a suitable code. The pulses are then transmitted as flashes of light down a thin glass fiber (0.005–0.1 mm in diameter), and at the other end the flashes are reconverted into electrical signals. Modern electronic methods permit a maximum of 32 telephone conversations to be carried at the same

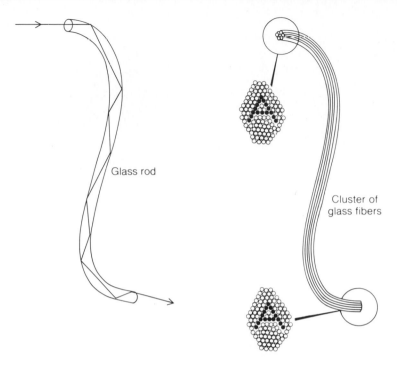

FIG. 25-31 Light can be "piped" from one place to another by means of successive internal reflections in a glass rod.

Glass rod

Cluster of glass fibers

time by a pair of wires, but the theoretical maximum for a glass fiber is in the millions and, as a practical matter, more than 10 thousand conversations can today be carried on the same fiber. In addition, optical signals fade less rapidly in a glass fiber than electrical ones do in a copper wire, so fewer amplifiers are needed over long distances, and electrical interference from nearby circuits is impossible. Thousands of kilometers of optical cable already link telephone exchanges in the United States and in other countries, and optical cables will soon span the Atlantic and Pacific oceans.

25-9 LUMINOUS FLUX

Visible light represents only part of the electromagnetic radiation emitted by most light sources, perhaps 10% in the case of an ordinary light bulb. Also, the eye does not respond with equal sensitivity to light of different colors; the sensitivity is a maximum for green light and it decreases toward both ends of the visible spectrum (see Section 26-6). Simply stating the power output of a light source in watts therefore does not convey much information about the illumination it actually provides, and a separate system of units has been devised for this purpose that is more closely related to the visual repose of the human eye.

The brightness of a light source is referred to as its *luminous intensity I,* for which the *candela* (cd) is the SI unit. The candela is defined in terms of the light emitted by a small pool of platinum at its melting point. A candle has a luminous intensity of

Eye sensitivity varies with color of light

Luminous intensity

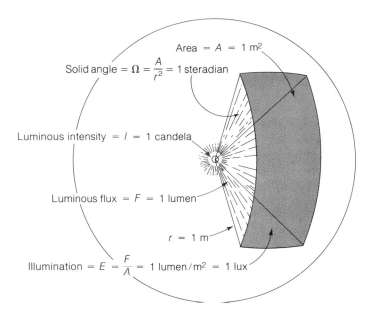

FIG. 25-32 How the luminous flux emitted by an isotropic light source and the illumination it produces are related to the intensity of the source.

about 1 cd, and in fact the former standard of this quantity was an actual candle of specified composition and dimensions.

Luminous flux

The *luminous flux F* emitted by a light source describes the total amount of visible light it gives off; a small bright source may give off less light than a large dim source, just as a small hot object may give off less heat than a large warm one. The unit of luminous flux is the *lumen* (lm), which equals the luminous flux that falls on each square meter of a sphere 1 m in radius at whose center is a 1-cd light source that radiates equally well in all directions (Fig. 25–32). Such a source is said to be *isotropic*. A lumen of flux is equivalent to about 0.0015 W of light of wavelength 555 nm, which corresponds to the yellow-green light the eye is most sensitive to. Since the area of a sphere of radius r is $4\pi r^2$, the above sphere has a total area of 4π m^2, and the total luminous flux radiated by a 1-cd source is thus 4π lm. The luminous flux F radiated by an isotropic source whose intensity is I is accordingly given by

$$F = 4\pi I \qquad\qquad\qquad \textit{Isotropic source} \quad (25\text{–}7)$$

Luminous flux = $4\pi \times$ luminous intensity

Solid angle

The above definition of luminous flux can be generalized to nonisotropic light sources, which are sources that do not radiate equally in all directions. To do this we need the concept of the *solid angle,* which is the three-dimensional counterpart of an ordinary angle in two dimensions. As in Fig. 25–32, the solid angle Ω (Greek capital letter *omega*) subtended by an area A on the surface of a sphere of radius r is defined as

$$\Omega = \frac{A}{r^2} \qquad\qquad\qquad \textit{Solid angle} \quad (25\text{–}8)$$

Solid angle = $\dfrac{\text{area on surface of sphere}}{(\text{radius of sphere})^2}$

The unit of solid angle is the *steradian* (sr). Since the total surface area of a sphere is $4\pi r^2$, the total solid angle subtended by a sphere is $4\pi r^2/r^2$ sr $= 4\pi$ sr. Like the degree and the radian, the steradian is a ratio that has no dimensions and so disappears in calculations.

The steradian is the unit of solid angle

In terms of solid angle, the definition of luminous flux becomes

$$F = I\Omega \qquad\qquad\qquad \textit{Luminous flux} \quad (25-9)$$

Luminous flux = (luminous intensity)(solid angle)

Thus a lumen equals the luminous flux that a 1-cd source radiates per steradian of solid angle, and the candela is equivalent to the lumen per steradian.

Example A spotlight concentrates the light from a 200-cd isotropic lamp into a circle 1.2 m in radius at a distance of 20 m. Find the luminous intensity of the source looking into the beam. (This is the luminous intensity an isotropic source would need to provide the same luminous flux on the illuminated circle.)

Solution The total luminous flux emitted by the lamp is

$$F = 4\pi I = (4\pi)(200 \text{ cd}) = 2513 \text{ lm}$$

Since the radius $R = 1.2$ m of the illuminated circle is small compared with the 20-m radius of a sphere centered on the spotlight, we can ignore the difference between the circle's plane area and its area as measured on the surface of the sphere. Hence

$$A = \pi R^2 = \pi (1.2 \text{ m})^2 = 4.52 \text{ m}^2$$

and the solid angle of the spotlight beam is

$$\Omega = \frac{A}{r^2} = \frac{4.52 \text{ m}^2}{(10 \text{ m})^2} = 0.0452 \text{ sr}$$

The luminous intensity of the light source looking into the beam is, from Eq. (25-9),

$$I' = \frac{F}{\Omega} = \frac{2513 \text{ lm}}{0.0452 \text{ sr}} = 5.56 \times 10^4 \text{ cd}$$

The optical system of the spotlight has therefore produced an increase in source brightness of $I'/I = 278$. ∎

The luminous flux emitted by a light source per watt of power input is its *luminous efficiency*. Ordinary light sources fall into three categories:

Luminous efficiency

1. An *incandescent lamp* has a filament of tungsten (a high melting point metal) that is heated to a high temperature by the passage of electric current. Most of the power supplied to an incandescent lamp is wasted as heat. Such a lamp increases in efficiency with power, because the higher the power, the higher the filament temperature and the greater the proportion of visible light in the total radiation. A 10-W lamp, for instance, has a luminous efficiency of about 8 lm/W, whereas it is 16 lm/W for a 100-W lamp.

2. A *vapor lamp* contains a gas or vapor that is caused to radiate by the passage of an electric current. The color of the light depends on the nature of the gas or vapor, as discussed in Chapter 28. Vapor lamps are more efficient than incandescent lamps; a typical 100-W mercury vapor lamp emits 3600 lm for an efficiency of 36 lm/W.

3. A *fluorescent lamp* consists of a glass tube filled with a mixture of mercury vapor and an inert gas such as argon. When a current is passed through the mixture, ultraviolet radiation is produced. The inside of the tube is coated with a material called a *phosphor* that emits visible light when it is excited by the ultraviolet radiation. Fluorescent lamps produce little heat and so are very efficient; a 100-W fluorescent lamp might give off 5500 lm for an efficiency of 55 lm/W, figures that are 3.4 times greater than those for a 100-W tungsten-filament lamp.

25–10   ILLUMINATION

Illumination

The more the luminous flux that reaches a given surface, such as a book, the more brightly it is lit. To put this idea on a definite basis, the *illumination E* of a surface is defined as the luminous flux per unit area that reaches it:

$$E = \frac{F}{A} \qquad\qquad\qquad \textit{Illumination} \quad (25-8)$$

$$\text{Illumination} = \frac{\text{luminous flux}}{\text{area}}$$

Table 25–2 summarizes the various quantities introduced in this section and the previous one. In the SI system the unit of illumination is the lumen per square meter, or *lux* (lx). In the British system it is the lumen per square foot, or *footcandle,* where 1 footcandle = 10.8 lux. About 500 lx is a reasonable figure for general interior lighting, with at least 800 lx recommended for reading and 1100 for electrical and machine shop work.

Isotropic source

The illumination on a surface a distance r away from an isotropic light source of intensity I is given by

$$E = \frac{I \cos \theta}{r^2} \qquad\qquad \textit{Illumination due to isotropic source} \quad (25-9)$$

In this formula θ is the angle between the direction of the light and a normal to the surface where the light strikes it (Fig. 25–33). The cos θ factor arises because E depends on the component $E \cos \theta$ of the flux F that is perpendicular to the surface. Thus the illumination is a maximum of I/r^2 for light striking a surface at an angle of $\theta = 0$

TABLE 25–2
For an isotropic source,
$F = 4\pi I$.

Quantity	Symbol	Meaning	Formula	Unit
Luminous intensity	I	Brightness of light source		candela (cd)
Solid angle	Ω	Three-dimensional equivalent of an angle in a plane	$\Omega = \dfrac{A}{r^2}$	steradian (sr)
Luminous flux	F	Quantity of visible light	$F = I\Omega$	lumen (lm)
Luminous efficiency		Luminous flux emitted relative to power input to light source	F/P	lumen/watt (lm/W)
Illumination	E	Luminous flux per unit area	$E = F/A$	lux ($= $ lm/m^2) footcandle ($= $ lm/ft^2)

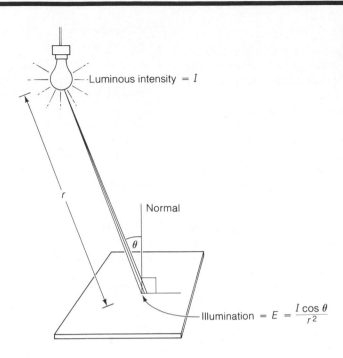

FIG. 25–33 The illumination of a surface is equal to the luminous flux per unit area that strikes it.

since $\cos 0 = 1$, and it drops to $E = 0$ for light parallel to the surface when $\theta = 90°$ and $\cos 90° = 0$.

The illumination from an isotropic source varies as $1/r^2$, just as in the case of sound waves. Doubling the distance from such a source reduces the illumination to $1/(2)^2 = \frac{1}{4}$ of its original value; cutting the distance in half increases the illumination to $1/(\frac{1}{2})^2 = (2)^2 = 4$ times its original value.

Example   (a) A certain 60-W light bulb has a luminous intensity of 70 cd. A reflector concentrates all the light from the bulb into a circle whose area is 0.8 m² when the bulb is 1 m from a book that is being held perpendicular to the light rays. Find the illumination of the book. (b) If the lamp had no reflector and radiated isotropically, how far from the book would it have to be located to produce the same illumination?

Solution   (a) The bulb emits a total luminous flux of

$$F = 4\pi I = (4\pi)(70 \text{ cd}) = 880 \text{ lm}$$

The illumination of the book is therefore

$$E = \frac{F}{A} = \frac{880 \text{ lm}}{0.8 \text{ m}^2} = 1100 \text{ lx}$$

(b) Here $\theta = 0$, $\cos \theta = 1$, and $E = I/r^2$. Hence

$$r = \sqrt{\frac{I}{E}} = \sqrt{\frac{70 \text{ cd}}{1100 \text{ lx}}} = 0.25 \text{ m}$$

The reflection of light from the walls of the room is neglected here, but obviously it would contribute to the illumination of the book. ■

IMPORTANT TERMS

Electromagnetic waves consist of coupled electric and magnetic field oscillations. Radio waves, microwaves, light waves, X rays, and gamma rays are all electromagnetic waves differing only in their frequency.

In **amplitude modulation,** information is contained in variations in the amplitude of a constant-frequency carrier wave. In **frequency modulation,** information is contained in variations in the frequency of the carrier wave.

A **wavefront** is an imaginary surface that joins points where all the waves from a source are in the same phase of oscillation. According to **Huygens' principle,** every point on a wavefront can be considered as a point source of secondary wavelets that spread out in all directions with the wave speed of the medium. The wavefront at any time is the envelope of these wavelets.

In **diffuse reflection** an incident beam of parallel light is spread out in many directions, while in **specular reflection** the angle of reflection is equal to the angle of incidence. A **mirror** is a specular reflecting surface of regular form that can produce an image of an object placed before it.

A **real image** of an object is formed by light rays that pass through the image; the image would therefore appear on a properly placed screen. A **virtual image** can only be seen by the eye because the light rays that seem to come from the image actually do not pass through it.

The bending of a light beam when passing from one medium to another is called **refraction.** The quantity that governs the degree to which a light beam will be deflected in entering a medium is its **index of refraction,** defined as the ratio between the speed of light in free space and its speed in the medium. **Dispersion** refers to the splitting up of a beam of light containing different frequencies by passage through a substance whose index of refraction varies with frequency.

Snell's law states that the ratio between the sine of the angle of incidence of a light ray upon an interface between two media and the sine of the angle of refraction is equal to the ratio of the speeds of light in the two media.

In **total internal reflection,** light arriving at a medium of lower index of refraction at an angle greater than the critical angle of incidence is reflected back into the medium it came from.

A **solid angle** is the counterpart in three dimensions of an angle in two dimensions.

The **luminous intensity** of a light source refers to its brightness with respect to the visual response of the eye, and the **luminous flux** it emits is a measure of its light output.

The **illumination** of a surface is equal to the flux reaching it per unit area.

IMPORTANT FORMULAS

Index of refraction: $n = \dfrac{c}{v}$

Snell's law: $n_1 \sin i = n_2 \sin r$

Apparent depth: $\dfrac{h'}{h} = \dfrac{n_2}{n_1}$

Critical angle: $\sin i_c = \dfrac{n_2}{n_1}$

Solid angle: $\Omega = \dfrac{A}{r^2}$

Luminous flux: $F = I\Omega$

$\qquad\qquad\quad F = 4\pi I$ (isotropic source)

Illumination: $E = \dfrac{F}{A}$

$\qquad\quad E = \dfrac{I \cos \theta}{r^2}$ (isotropic source)

MULTIPLE CHOICE

1. A changing electric field gives rise to
a. a magnetic field.   b. electromagnetic waves.
c. sound waves.   d. nothing in particular.

2. Electromagnetic waves transport
a. wavelength.   b. frequency.
c. charge.   d. energy.

3. The direction of the magnetic field in an electromagnetic wave is
a. parallel to the electric field.
b. perpendicular to the electric field.
c. parallel to the direction of propagation.
d. random.

4. Which of the following are not electromagnetic in nature?
a. infrared rays   b. ultraviolet rays
c. radar waves   d. sound waves

5. In a vacuum, the speed of an electromagnetic wave
a. depends upon its frequency.
b. depends upon its wavelength.
c. depends upon its electric and magnetic fields.
d. is a universal constant.

6. The highest frequencies are found in
    a. X rays.   b. ultraviolet light.
    c. radio waves.   d. radar waves.

7. The ionosphere is a region of ionized gas in the upper atmosphere. The ionosphere is responsible for
    a. the blue color of the sky.
    b. rainbows.
    c. long-distance radio communication.
    d. the ability of satellites to orbit the earth.

8. Light of which of the following colors has the lowest frequency?
    a. red   b. blue
    c. yellow   d. green

9. Light of which of the following colors has the shortest wavelength?
    a. red   b. blue
    c. yellow   d. green

10. The Spanish flag is yellow and red. When viewed with yellow light it appears
    a. all yellow.   b. yellow and red.
    c. yellow and black.   d. white and red.

11. The Danish flag is red and white. When viewed with red light it appears
    a. all red.   b. all white.
    c. red and white.   d. red and black.

12. Reflection from a mirror is said to be
    a. specular.   b. diffuse.
    c. real.   d. virtual.

13. All real images
    a. are erect.
    b. are inverted.
    c. can appear on a screen.
    d. cannot appear on a screen.

14. When you look at yourself in a plane mirror, what you see is a
    a. real image behind the mirror.
    b. real image in front of the mirror.
    c. virtual image behind the mirror.
    d. virtual image in front of the mirror.

15. When a beam of light enters one medium from another, a quantity that never changes is its
    a. direction.   b. speed.
    c. frequency.   d. wavelength.

16. The bending of a beam of light when it passes from one medium to another is known as
    a. refraction.   b. reflection.
    c. diffraction.   d. dispersion.

17. The index of refraction of a material medium
    a. is always less than 1.
    b. is always equal to 1.
    c. is always greater than 1.
    d. may be less than, equal to, or greater than 1.

18. Relative to the angle of incidence, the angle of refraction
    a. is smaller.
    b. is the same.
    c. is larger.
    d. may be any of the above.

19. A light ray enters one medium from another along the normal. The angle of refraction
    a. is 0.
    b. is 90°.
    c. equals the critical angle.
    d. depends upon the indexes of refraction of the two media.

20. Dispersion is the term used to describe
    a. the splitting of white light into its component colors in refraction.
    b. the propagation of light in straight lines.
    c. the bending of a beam of light when it goes from one medium to another.
    d. the bending of a beam of light when it strikes a mirror.

21. According to Snell's law,
    a. $\sin i = \sin r$.   b. $v_1 \sin i = v_2 \sin r$.
    c. $v_2 \sin i = v_1 \sin r$.   d. $\sin i / \sin r = v_2 v_2$.

22. Light goes from medium A to medium B at an angle of incidence of 40°. The angle of refraction is 30°. The speed of light in B
    a. is less than that in A.
    b. is the same as that in A.
    c. is greater than that in A.
    d. could be any of the above, depending on the specific mediums.

23. The depth of an object submerged in a transparent liquid
    a. always seems less than its actual depth.
    b. always seems more than its actual depth.
    c. may seem less or more than its actual depth, depending on the index of refraction of the liquid.
    d. may seem less or more than its actual depth, depending on the angle of view.

24. Total internal reflection can occur when light passes from one medium to another
    a. that has a lower index of refraction.
    b. that has a higher index of refraction.
    c. that has the same index of refraction.
    d. at less than the critical angle.

25. When a light ray approaches a glass-air interface from the glass side at the critical angle, the angle of refraction is
   a. 0.
   b. 45°.
   c. 90°.
   d. equal to the angle of incidence.

26. The brightness of a light source is called its luminous intensity, whose unit is the
   a. candela.　　　　　b. lumen.
   c. lux.　　　　　　　d. footcandle.

27. Luminous efficiency is least for a
   a. low-wattage light bulb.
   b. high-wattage light bulb.
   c. mercury vapor lamp.
   d. fluorescent tube.

28. The minimum illumination recommended for reading is
   a. 800 cd.　　　　　b. 800 lm.
   c. 800 lx.　　　　　d. 800 W.

29. Radio amateurs are permitted to communicate on the "10-m band." What frequency of radio waves corresponds to a wavelength of 10 m?
   a. 3.3×10^{-8} Hz　　　b. 3.0×10^{7} Hz
   c. 3.3×10^{7} Hz　　　d. 3.0×10^{9} Hz

30. The index of refraction of benzene is 1.5. The speed of light in benzene is
   a. 1.5×10^{8} m/s.　　　b. 2×10^{8} m/s.
   c. 3×10^{8} m/s.　　　d. 4.5×10^{8} m/s.

31. Light enters a glass plate at an angle of incidence of 40° and is refracted at an angle of refraction of 25°. The index of refraction of the glass is
   a. 0.625.　　　　　b. 0.66.
   c. 1.52.　　　　　　d. 1.6.

32. A clear lake is 2 m deep. Assuming that the index of refraction of water is $n = \frac{4}{3}$, the apparent depth of the lake to a person looking down into it is
   a. $\frac{2}{3}$ m.　　　　　b. $1\frac{1}{2}$ m.
   c. 2 m.　　　　　　d. $2\frac{2}{3}$ m.

33. The solid angle subtended by a hemisphere about its center
   a. is $\pi/2$ sr.
   b. is π sr.
   c. is 2π sr.
   d. depends on the radius of the hemisphere.

34. The luminous flux emitted by a 60-cd isotropic light source is concentrated on an area of 0.5 m². The illumination of the area is
   a. 9.6 lx.　　　　　b. 120 lx.
   c. 377 lx.　　　　　d. 1508 lx.

EXERCISES

25–1　Electromagnetic Waves

1. Why was electromagnetic induction discovered much earlier than its converse, the production of a magnetic field by a varying electric field?

2. Under what circumstances does a charge radiate electromagnetic waves?

3. Why are light waves able to travel through a vacuum whereas sound waves cannot?

4. Light is said to be a transverse wave phenomenon. What is it that varies at right angles to the direction in which a light wave travels?

5. In an electromagnetic wave, what is the relationship, if any, between the variations in the electric and magnetic fields?

6. A radio transmitter has a vertical antenna. Does it matter whether the receiving antenna is vertical or horizontal? Explain.

25–2　Types of Electromagnetic Waves

7. The marine radiotelephone station on Tahiti transmits at the frequency 4390.2 kHz. What wavelength does this correspond to?

8. A radar sends out 0.05-μs pulses of microwaves whose wavelength is 2.5 cm. What is the frequency of these microwaves? How many waves does each pulse contain?

9. A nanosecond is 10^{-9} s. (a) What is the frequency of an electromagnetic wave whose period is 1 ns? (b) What is its wavelength? (c) To what class of electromagnetic waves does it belong?

25–3　Huygens' Principle

10. What is a wavefront? What is the relationship between a light ray and the wavefronts whose motion it is used to describe?

25–4　Reflection

11. What is the difference between a real image and a virtual image?

12. Under what circumstances do the incident ray, the reflected ray, and the normal all lie in the same plane?

13. A plane mirror is mounted on the back of a truck that is traveling at 30 km/h. How fast does the image of a man standing in the road behind the truck seem to be moving away from him?

14. What is the height of the smallest mirror in which a woman 160 cm tall can see herself at full length? Does it matter how far from the mirror she stands?

25–5   Refraction

25–6   Index of Refraction

15. Do electromagnetic waves have the same speed in all transparent media? If not, how do their speeds in a transparent medium compare with their speed in free space?

16. Flint glass and carbon disulfide have almost the same index of refraction. How does this explain the fact that a flint-glass rod immersed in carbon disulfide is nearly invisible?

17. Explain why a cut diamond held in white light shows flashes of color. What would happen if it were held in red light?

18. Why is a beam of white light not dispersed into its component colors when it passes perpendicularly through a pane of glass?

19. What is the angle of incidence of a ray of light that goes from air into water without being deflected?

20. The figure below shows a ray of light going from water to air. Which is the refracted ray?

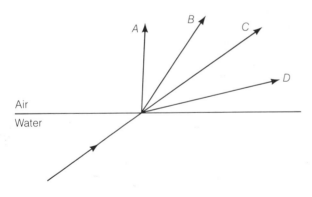

21. Using the indexes of refraction given in Table 25–1 and taking the speed of light in free space as 3×10^8 m/s, find the speed of light in (a) air, (b) diamond, (c) crown glass, and (d) water.

22. The frequency of the light in a particular beam depends solely on its source, whereas the wavelength depends on the speed of light in the medium it travels through. Find the ratio between the wavelengths of a light beam that passes from one medium to another in terms of their indexes of refraction.

23. A flashlight is frozen into a block of ice. If its beam strikes the surface of the ice at an angle of incidence of 37°, what is the angle of refraction?

24. What is the angle of refraction of a beam of light that enters the surface of a lake at an angle of incidence of 50°?

25. A beam of light strikes a pane of glass at an angle of incidence of 60°. If the angle of refraction is 35°, find the index of refraction of the glass.

26. A beam of light enters a liquid of unknown composition at an angle of incidence of 30° and is deflected by 5° from its original path. Find the index of refraction of the liquid.

27. The index of refraction of ordinary crown glass for red light is 1.51 and for violet light is 1.53. A beam of white light falls on a cube of such glass at an incident angle of 40°. What is the difference between the angles of refraction of the red and the violet light?

28. A ray of light passes through a plane boundary separating two media whose indexes of refraction are $n_1 = 1.5$ and $n_2 = 1.3$. (a) If the ray goes from medium 1 to medium 2 at an angle of incidence of 45°, what is the angle of refraction? (b) If the ray goes from medium 2 to medium 1 at the same angle of incidence, what is the angle of refraction?

29. A double-glazed window consists of two panes of glass separated by an air space for better insulation. (a) If the glass has an index of refraction of 1.5, find the angle at which a ray of sunlight enters a room if it arrives at the window at an angle of incidence of 70°. (b) The window leaks, and the space between the panes becomes filled with water. What is the angle at which the above ray enters the room now?

30. A ray of light strikes a glass plate at an angle of incidence of 55°. If the reflected and refracted rays are perpendicular to each other, find the index of refraction of the glass.

25–7   Apparent Depth

31. The olive in a martini cocktail ($n = 1.35$) appears to be 4 cm below the surface. What is the actual depth of the olive?

32. A glass paperweight in the form of a 6-cm cube is placed on a letter. How far underneath the top of the cube does the letter appear if the index of refraction of the glass is 1.5?

33. A barrel of ethyl alcohol is 0.80 m high. How high does it appear to somebody looking into it from above?

25–8   Total Internal Reflection

34. The critical angle for total internal reflection in Lucite is 41°. Find its index of refraction.

35. Find the critical angle for total internal reflection in a diamond when the diamond is (a) in air, and (b) immersed in water.

36. A light bulb is on the bottom of a swimming pool 1.8 m below the water surface. A person on a diving board above the pool sees a circle of light. What is its diameter?

37. Prisms are used in optical instruments such as binoculars instead of mirrors because total internal reflection better preserves the sharpness and brightness of a light beam. Find the minimum index of refraction of the glass to be used in a prism that is meant to change the direction of a light beam by 90°.

25–9   Luminous Flux

25–10   Illumination

38. Find the solid angle subtended at the center of a sphere of radius 8 m by an area of 200 cm^2 on its surface.

39. About 29% of the earth's surface is above sea level. If all the land areas were concentrated in a single region, what solid angle would it subtend at the center of the earth?

40. What area on the surface of a sphere of radius 4 m is cut by a solid angle of 0.2 sr?

41. A projector produces an illumination of 10,000 lx on a wall perpendicular to its beam that is 6 m away. What is the luminous intensity looking into the beam?

42. A reflector concentrates the light emitted by a 60-W lamp whose luminous efficiency is 14 lm/W into a beam of solid angle 0.1 sr. What is the luminous intensity of the source looking into its beam?

43. A reflector concentrates the light emitted by an isotropic 20-W lamp on an area perpendicular to the beam of 5 ft^2 at a distance of 10 ft. If the luminous intensity looking into the beam is 2500 cd, find the luminous efficiency of the lamp.

44. A certain 100-W light bulb has a luminous intensity of 120 cd. (a) What is its luminous efficiency? (b) If the bulb radiates isotropically and is suspended 1.2 m above a table, what is the illumination directly below it?

45. A spotlight concentrates all the light from a 200-cd bulb in a beam 1 m in diameter. Find the illumination it produces on a surface perpendicular to the beam.

46. On a certain night the full moon produced an illumination of 0.6 lx. Find the luminous intensity of the moon that night. The distance between the earth and the moon is 3.84×10^8 m.

47. A skating rink 30 m by 60 m is to have an average illumination at night of 300 lx. How many 1-kW, 2500-cd lamps are needed if they are equipped with reflectors that direct $\frac{2}{3}$ of their luminous flux on the rink?

48. A 250-cd unshaded light bulb is 1.4 m above a table. Find the illumination of the table directly under the bulb and at a point 1.0 m to one side.

49. An unshaded light bulb 0.8 m above a workbench produces an illumination of 1200 lx directly below it. What is the illumination at a point on the bench 1.0 m to one side?

ANSWERS TO MULTIPLE CHOICE

1. a	**8.** a	**15.** c	**22.** a	**29.** b
2. d	**9.** b	**16.** a	**23.** a	**30.** b
3. b	**10.** c	**17.** c	**24.** a	**31.** c
4. d	**11.** a	**18.** d	**25.** c	**32.** b
5. d	**12.** a	**19.** a	**26.** a	**33.** c
6. a	**13.** c	**20.** a	**27.** a	**34.** d
7. c	**14.** c	**21.** c	**28.** c	

26

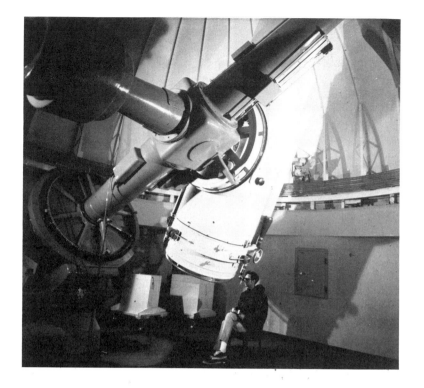

LENSES AND MIRRORS

A lens refracts light from an object that passes through it in such a way as to be able to form an image of the object. The image may be real or virtual, erect or inverted, and larger, smaller, or the same size as the object. In this chapter we shall investigate not only how individual lenses produce images but also how this is done by such systems of lenses as the microscope and the telescope. The discussion will be limited to lenses whose thicknesses are small relative to their diameters. Mirrors with spherically curved reflecting surfaces are also able to form images of an object. As we shall find, there are many similarities between the optics of lenses and mirrors.

26-1   LENSES

A converging lens brings parallel rays to a real focal point

A *lens* is a transparent object of regular form that alters the shape of wavefronts of light that pass through it. A simple way to appreciate how a lens produces its effects is to begin with a pair of prisms, as in Fig. 26-1(a). When the prisms have their bases together, parallel light that approaches is deviated so that the various rays intersect. However, they do not intersect at a single focal point because each prism merely changes

CHAPTER OBJECTIVES

Completing this chapter should enable you to:

1. Distinguish between converging and diverging lenses and mirrors.

2. Calculate the focal length of a lens in air and in a liquid of a given index of refraction.

3. Calculate the focal length of a spherical mirror.

4. Use diagrams to show how converging and diverging lenses and mirrors form images.

5. Determine from a diagram whether an image produced by a lens or mirror will be real or virtual, erect or inverted, and larger, smaller, or the same size as the object.

6. Use the lens equation to relate object and image distances to the focal length of a lens.

7. Use the mirror equation to relate object and image distances to the focal length of a mirror.

8. Calculate the magnification produced by a lens or mirror.

9. Describe the significance of the *f*-numbers of a camera lens.

10. Describe the origins of common defects of vision and how they can be corrected.

11. Analyze image formation in a microscope and in a telescope.

12. Describe the origins of the chromatic and spherical aberrations of a lens and how they can be corrected.

13. Describe the origin of the spherical aberration of a mirror and how it can be eliminated.

the direction of the rays without affecting the shape of the wavefronts. If properly curved rather than flat surfaces are employed, as in Fig. 26–1(b), the object is called a *converging lens,* and it acts to bring an incoming parallel beam of light to a single focal point *F.*

A *diverging lens* may similarly be thought of as a development of a pair of prisms with their apexes together (Fig. 26–2). The prisms deviate incoming parallel light

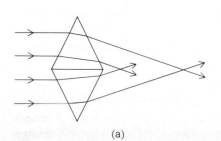

(a)

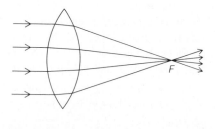

(b)

FIG. 26–1 A converging lens may be thought of as a development of two prisms placed base to base.

FIG. 26–2 A diverging lens may be thought of as a development of two prisms placed apex to apex.

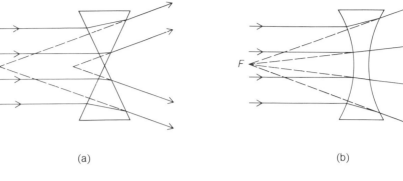

(a)   (b)

FIG. 26–3 How converging and diverging lenses affect plane wavefronts.

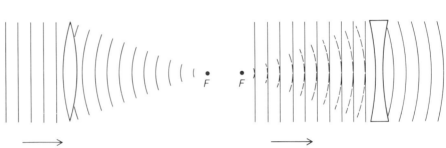

A diverging lens spreads out parallel rays as though they came from a virtual focal point

outward, but the diverging rays cannot be projected back to a single point. The corresponding lens has curved surfaces so shaped that the diverging rays have a single virtual focal point F; they *seem* to come from F, but actually do not.

Figure 26–3 shows how lenses affect wavefronts. The speed of light in glass (or other suitable transparent material) is less than in air, and light takes more time to pass through a certain thickness of glass than through the same thickness of air. Light that enters the center of a converging lens is accordingly slowed down more than light that enters toward the edge, with the result that the wavefronts converge after passing through the lens. Exactly the opposite effect occurs with a diverging lens.

Thin lenses with spherical surfaces produce undistorted images

A thin lens with spherical surfaces, or one plane and one spherical surface, produces an undistorted image of something placed in front of it. Such lenses are also the easiest to manufacture. *Aspherical* (nonspherical) lens surfaces are used in special applications to avoid certain kinds of distortion that occur with spherical lenses. The various forms of simple lenses are shown in Fig. 26–4; converging lenses are always thickest in the center, whereas diverging lenses are always thinnest in the center. A meniscus lens has one concave and one convex surface.

Focal length

The distance from a lens to its focal point is called its *focal length f*. The focal length of a particular lens depends both upon the index of refraction n of its material relative to that of the medium it is in, and upon the radii of curvature R_1 and R_2 of its surfaces. In the case of a lens whose thickness is small compared with R_1 and R_2, these quantities are related by the *lensmaker's equation:*

$$\frac{1}{f} = (n-1)\left(\frac{1}{R_1} + \frac{1}{R_2}\right) \qquad \textit{Lensmaker's equation} \quad (26\text{–}1)$$

Sign conventions for lensmaker's equation

In using this equation, it does not matter which surface of the lens is considered as 1 and which as 2. However, the sign given to each radius of curvature *is* important. We

Converging lenses ($f = +$)   Diverging lenses ($f = -$)

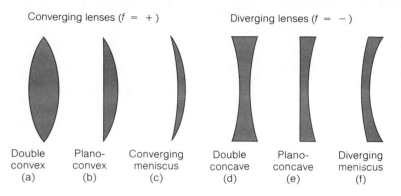

| Double convex (a) | Plano-convex (b) | Converging meniscus (c) | Double concave (d) | Plano-concave (e) | Diverging meniscus (f) |

FIG. 26–4 Some simple lenses. The focal length of a converging lens is reckoned positive, that of a diverging lens is reckoned negative.

shall reckon a radius as $+$ if the surface is convex (curved outward), and as $-$ if the surface is concave (curved inward). A plane surface has, in effect, an infinite radius of curvature, and $1/R$ for a plane surface is therefore 0. In Fig. 26–4(a), both radii are $+$; in (b), the first radius is infinite and the second is $+$; in (c), the first radius is $-$ and the second is $+$; in (d), both radii are $-$; and so on.

Depending upon its shape, a lens may have a positive or a negative focal length. A positive focal length signifies a converging lens, and a negative one signifies a diverging lens.

Focal length is $+$ for a converging lens, $-$ for a diverging lens

Example   A meniscus has a convex surface whose radius of curvature is 25 cm and a concave surface whose radius of curvature is 15 cm. The index of refraction is 1.52. Find the focal length of the lens and whether it is converging or diverging.

Solution   Here $R_1 = +25$ cm and $R_2 = -15$ cm in accord with the sign convention. From the lensmaker's equation,

$$\frac{1}{f} = (n - 1)\left(\frac{1}{R_1} + \frac{1}{R_2}\right) = (1.52 - 1.00)\left(\frac{1}{25\,\text{cm}} - \frac{1}{15\,\text{cm}}\right)$$

$$= (0.52)(0.040 - 0.067)\,\text{cm}^{-1} = -0.014\,\text{cm}^{-1}$$

and so

$$f = -71 \text{ cm}$$

The negative focal length indicates a diverging lens.   ∎

Example   The above lens is placed in water, whose index of refraction is 1.33. Find its focal length there.

Solution   The index of refraction of the glass relative to water is

$$n' = \frac{\text{index of refraction of glass}}{\text{index of refraction of water}} = \frac{1.52}{1.33} = 1.14$$

From the lensmaker's equation, since R_1 and R_2 are the same in both air and water, the ratio between the focal length f' of the lens in water and its focal length f in air is

$$\frac{f'}{f} = \frac{n - 1}{n' - 1} = \frac{1.52 - 1}{1.14 - 1} = 3.7$$

Hence

$$f' = 3.7f = (3.7)(-71 \text{ cm}) = -263 \text{ cm}$$

The focal length of *any* lens made of this glass is 3.7 times longer in water than in air. ■

26–2   IMAGE FORMATION

Ray tracing

A scale drawing provides a handy way to find the size and position of an image formed by a lens. The procedure is to consider two different light rays that originate at a certain point on an object and to trace their paths until they (or their extensions) come together again after being refracted by the lens. In this connection it is worth noting that a lens has *two* focal points, one on each side of the lens the distance *f* from its center. We shall call the focal point on the side of the lens from which the light comes the *near focal point*, and the one on the other side of the lens the *far focal point*.

Three rays that are particularly easy to trace are shown in Fig. 26–5. They are:

Rays for finding an image produced by a lens

1. A ray that leaves the object parallel to the lens axis. When this ray is refracted by the lens, it passes through the far focal point of a converging lens or seems to come from the near focal point of a diverging lens.

2. A ray that leaves the object and passes through the near focal point of a converging lens or is directed toward the far focal point of a diverging lens. When this ray is refracted, it proceeds parallel to the axis.

3. A ray that leaves the object and proceeds through the center of the lens. If the lens is thin relative to the radii of curvature of its surfaces, this ray is not deviated.

FIG. 26–5 The position and size of an image produced by a thin lens can be determined by tracing any two of the rays shown. In tracing the rays, any deviations produced by the lens are assumed to occur at its central plane.

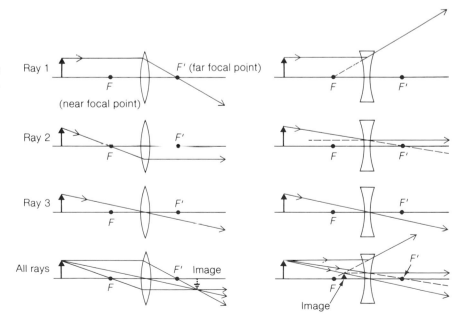

Ordinarily only two of these rays are required to establish the image of an object. Figure 26–6 shows how the properties of the image produced by a converging lens depend upon the position of the object. As usual, solid lines represent the actual paths taken by light rays, and dashed lines represent virtual paths.

Images produced by a converging lens

When the object is closer to the lens than F, the near focal point, the image is erect, enlarged, and virtual. The image seems to be behind the lens because the refracted rays diverge as though coming from a point behind it. (A virtual image can be seen by the eye, but it cannot appear on a screen because no light rays actually pass through such an image.)

No image is formed of an object precisely at the focal point because the refracted rays are all parallel and hence never intersect. The image in this case may be said to be at infinity.

FIG. 26–6 Image formation by a converging lens.

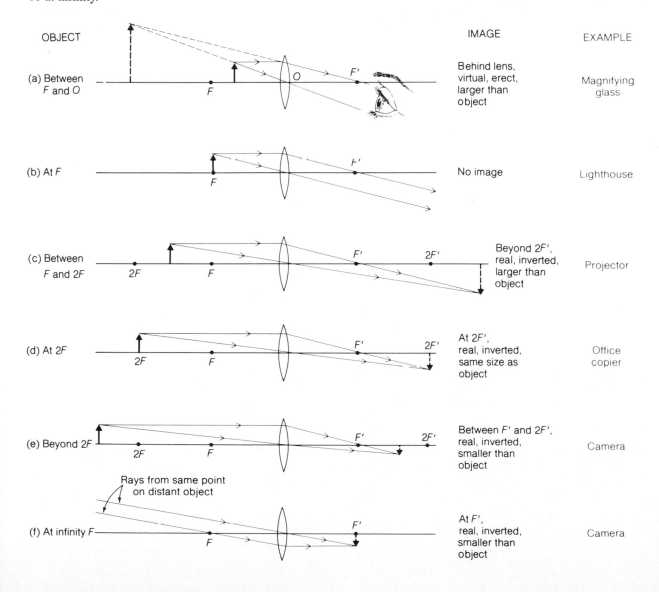

OBJECT	IMAGE	EXAMPLE
(a) Between F and O	Behind lens, virtual, erect, larger than object	Magnifying glass
(b) At F	No image	Lighthouse
(c) Between F and 2F	Beyond 2F', real, inverted, larger than object	Projector
(d) At 2F	At 2F', real, inverted, same size as object	Office copier
(e) Beyond 2F	Between F' and 2F', real, inverted, smaller than object	Camera
(f) At infinity	At F', real, inverted, smaller than object	Camera

FIG. 26–7 The image of a real object formed by a diverging lens is always virtual, erect, and smaller than the object.

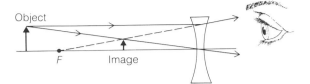

An object farther from the lens than F always has an inverted real image that may be smaller than, the same size as, or larger than the object, depending upon whether the object distance is between f and $2f$, equal to $2f$, or greater than $2f$.

In contrast to the variety of image sizes, natures, and locations produced by a converging lens, the image of a real object formed by a diverging lens is always virtual, erect, and smaller than the object (Fig. 26–7).

Images produced by a diverging lens

26–3   THE LENS EQUATION

A simple formula relates the positions of the image and the object of a thin lens to the lens's focal length f. In terms of the symbols

p = distance of object from lens

q = distance of image from lens

f = focal length of lens

this formula is

$$\frac{1}{p} + \frac{1}{q} = \frac{1}{f} \qquad\qquad \text{\textit{Lens equation}} \quad (26–2)$$

When any two of the three quantities are known, the third can be calculated. The sign conventions to be observed when using the lens equation are:

Sign conventions for lens equation

1. The focal length f is considered $+$ for a converging lens, $-$ for a diverging lens.
2. Object distance p and image distance q are considered $+$ for real objects and images, $-$ for virtual objects and images.

Derivation of lens equation

The lens equation can be derived geometrically with the help of Fig. 26–8. We observe that the triangles ABO and $A'B'O$ are similar, which means that corresponding sides of these triangles are proportional. Hence

FIG. 26–8 A ray diagram for deriving the lens equation.

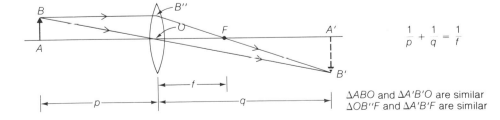

$$\frac{1}{p} + \frac{1}{q} = \frac{1}{f}$$

ΔABO and $\Delta A'B'O$ are similar
$\Delta OB''F$ and $\Delta A'B'F$ are similar

$$\frac{A'B'}{AB} = \frac{A'O}{AO} = \frac{q}{p} \qquad\qquad (26\text{–}3)$$

The triangles $OB''F$ and $A'B'F$ are also similar, and

$$\frac{A'B'}{OB''} = \frac{A'F}{OF} = \frac{q-f}{f} = \frac{q}{f} - 1 \qquad\qquad (26\text{–}4)$$

Now, it is clear that

$$OB'' = AB$$

since the light ray BB'' is parallel to the axis. Therefore

$$\frac{A'B'}{OB''} = \frac{A'B'}{AB}$$

which permits us to set the right-hand sides of Eqs. (26–3) and (26–4) equal:

$$\frac{q}{f} - 1 = \frac{q}{p}$$

Dividing each term of this equation by q yields

$$\frac{1}{f} - \frac{1}{q} = \frac{1}{p}$$

and, rearranging terms, we obtain

$$\frac{1}{p} + \frac{1}{q} = \frac{1}{f}$$

The above derivation involved a converging lens with an object distance greater than the focal length, but the resulting formula is valid for diverging lenses as well and for any object distance.

It is often convenient to solve the lens equation for p or q before using it in a calculation. We find that

Alternate forms of lens equation

$$\frac{1}{p} = \frac{1}{f} - \frac{1}{q} = \frac{q-f}{qf}$$

$$p = \frac{qf}{q-f} \qquad\qquad (26\text{–}5)$$

and, in a similar way, that

$$q = \frac{pf}{p-f} \qquad\qquad (26\text{–}6)$$

$$f = \frac{pq}{p+q} \qquad\qquad (26\text{–}7)$$

Equation (26–2) can be used directly for problems with the help of a calculator that has a reciprocal [1/X] key. The formulas for p, q, and f and the corresponding sequences of key strokes are as follows:

Using a calculator

$$\frac{1}{p} = \frac{1}{f} - \frac{1}{q} \qquad (f)\,[1/X][-]\,(q)\,[1/X][\,=\,][1/X]$$

$$\frac{1}{q} = \frac{1}{f} - \frac{1}{p} \qquad (f)\,[1/X][-]\,(p)[1/X][\,=\,][1/X]$$

$$\frac{1}{f} = \frac{1}{p} + \frac{1}{q} \qquad (p)\,[1/X][+]\,(q)[1/X][\,=\,][1/X]$$

Converging lens

Example   A plano-convex lens of focal length 5 cm is used in a reading lamp to focus light from a bulb on a book. If the lens is 60 cm from the book, how far from the bulb's filament should it be?

Solution   Here the focal length of the lens is $+5$ cm and the image distance is $+60$ cm. From Eq. (26–5) the object distance is

$$p = \frac{qf}{q-f} = \frac{(60\,\text{cm})(5\,\text{cm})}{60\,\text{cm} - 5\,\text{cm}} = 5.45\,\text{cm}$$

With the help of a calculator we would start with

$$\frac{1}{p} = \frac{1}{f} - \frac{1}{q} = \frac{1}{5\,\text{cm}} - \frac{1}{60\,\text{cm}}$$

and find $p = 5.45$ cm with fewer key strokes.   ∎

Diverging lens

Example   Motion picture directors often use diverging lenses for preliminary inspection of scenes to see how they would appear on the screen without having to move the camera into place first. If a double-concave lens of focal length -20 in. is used for this purpose, where does the image of a scene 30 ft away seem to be located?

Solution   Since $f = -20$ in. and $p = 30$ ft $= 360$ in.,

$$\frac{1}{q} = \frac{1}{f} - \frac{1}{p} = \frac{1}{-20\,\text{in.}} - \frac{1}{360\,\text{in.}}$$
$$q = -19\,\text{in.}$$

The image is located 19 in. behind the lens. The negative image distance signifies a virtual image, which is always on the same side of a lens as its object. This situation is pictured in Fig. 26–7.   ∎

26–4   THE CAMERA

A camera contains a converging lens that forms images on a light-sensitive photographic film in an arrangement like that of Fig. 26–9. The distance between the lens and the film is adjustable to permit bringing objects at varying distances to a sharp focus on the film. The film is exposed by opening the shutter for a fraction of a second. The shorter the exposure time, the better the ability to "freeze" a moving object; the longer the exposure time, the more light reaches the film and hence the better the ability to photograph a poorly lit scene.

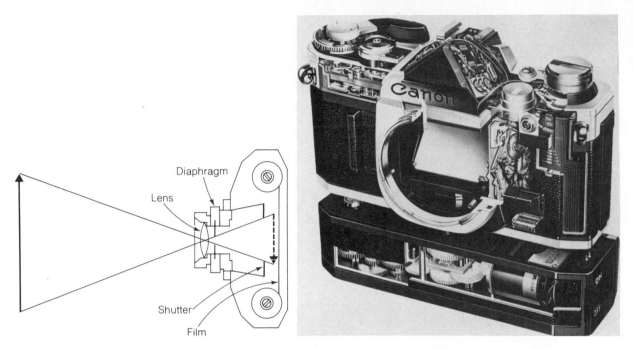

FIG. 26–9 *(Left)* The light-sensitive film is exposed by opening the shutter for a fraction of a second. The adjustable diaphragm permits the intensity of the light entering the camera to be varied to suit the shutter speed and film used. *(Right)* A modern 35-mm reflex camera with built-in exposure meter and motor drive. The mirror reflects the image via the prism at the top of the camera to the eye; when the shutter lever is pressed, the mirror swings up out of the way. (Courtesy of Canon USA, Inc.)

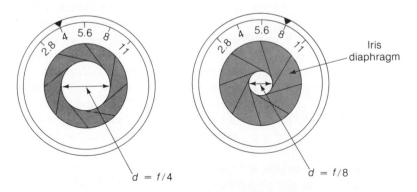

FIG. 26–10 The larger the *f*-number of a camera diaphragm, the smaller the aperture. The sequence of values is such that changing the aperture by one *f*-number changes its area and the amount of light reaching the film by a factor of 2. An aperture of *f*/4 admits four times as much light as one of *f*/8.

For a given shutter speed, the amount of light reaching the film from a particular scene depends upon the area of the lens aperture, which is regulated by an iris diaphragm (Fig. 26–10). Apertures are conventionally specified in terms of the focal length f of the lens: $f/8$ means that the diameter d of the opening is $\frac{1}{8}$ of f. The "8" in this case is referred to as the "*f*-number" of the aperture, so that, in general,

The larger the *f*-number, the smaller the aperture and the fainter the light reaching the film

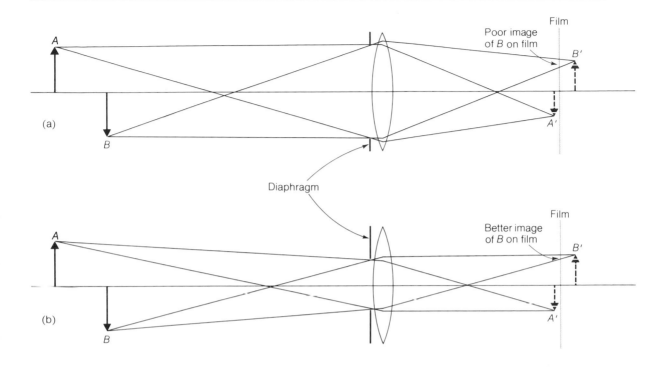

FIG. 26–11 Reducing the aperture of a lens increases the depth of field. Here arrowhead A is in focus on the film, whereas arrowhead B is in focus behind the film because it is closer to the lens. (a) With a large aperture, the image of B on the film is very indistinct. (b) With a small aperture, the image of B on the film is sharper.

$$f\text{-number} = \frac{\text{focal length}}{\text{aperture diameter}} = \frac{f}{d}$$

The larger the f-number, the smaller the aperture and the less the light reaching the film. Because the area of the aperture is $\pi d^2/4$, the illumination of the film is proportional to the square of d, and it is also inversely proportional to the square of the image distance q. Since q is usually quite close to f in a camera, an aperture of a given f-number therefore gives essentially the same illumination whatever the focal length of the lens. This is the reason for the f-number scheme. The diaphragm of a camera lens is calibrated using a set of f-number values (for instance, 2.8, 4, 5.6, 8, 11, 16) such that their squares are approximately in the ratios 1:2:4:8 and so on, which means that changing the setting by one f-number changes the illumination of the film by the factor 2. The same amount of light reaches the film of a camera during exposures of $\frac{1}{50}$ s at $f/8$, $\frac{1}{100}$ s at $f/5.6$, and $\frac{1}{200}$ s at $f/4$ with all lenses.

A small aperture provides greater depth of field

In addition to helping determine the amount of light reaching the film, the size of a camera's lens aperture affects the *depth of field,* which is the range of object distances at which reasonably sharp images are produced. As Fig. 26–11 shows, reducing the aperture (increasing the f-number) leads to an increased depth of field. Thus more of a scene will appear sharp if a slow shutter speed and small aperture are used (provided motion in the scene is not a problem, of course).

Wide-angle and telephoto lenses

The film size and the desired angle of view are what determine the appropriate focal length of a camera lens. A "normal" lens usually provides an angle of view of about 45°, which means a focal length of 50 mm for a camera using 35-mm film on which the image size is 24 × 36 mm. A lens of shorter focal length is a *wide-angle lens* because it captures more of a given scene, though at the expense of reducing the sizes of details in the scene. Typical wide-angle lenses for a 35-mm camera have focal lengths of 35 and 28 mm, which give angles of view respectively of 63° and 75°. A *telephoto lens* has a long focal length to provide larger images of distant objects, but this means a reduced angle of view: A 135-mm telephoto lens for a 35-mm camera has an angle of view of 18°, and it is only 8° for a 300-mm telephoto lens.

Example What range of motion should the 50-mm lens of a 35-mm camera have if the camera is to be capable of photographing objects as close as 50 cm from the lens? What is the aperture diameter when the diaphragm is set at $f/2.8$? At $f/16$?

Solution   (a) To find the image distance q for an object distance of $p = 50$ cm $= 500$ mm we proceed in the usual way:

$$\frac{1}{q} = \frac{1}{f} - \frac{1}{p} = \frac{1}{50\,\text{mm}} - \frac{1}{500\,\text{mm}}$$

$$q = 55.6\,\text{mm}$$

An object at infinity is brought to a sharp focus on the film when the image distance is equal to the focal length (see Fig. 26–6(f)), which in this case is 50 mm. Hence a range of adjustment of 5.6 mm will permit the camera to photograph objects from 50 cm away to infinity.

(b) At $f/2.8$ the aperture diameter of the lens is

$$d = \frac{f}{2.8} = \frac{50\,\text{mm}}{2.8} = 17.9\,\text{mm}$$

and at $f/16$ it is

$$d = \frac{50\,\text{mm}}{16} = 3.13\,\text{mm}$$ ∎

26–5 MAGNIFICATION

The *linear magnification m* of an optical system of any kind is the ratio between the size of the image and that of the object. A magnification of exactly 1 means that the image and the object are the same size; a magnification of more than 1 means that the image is larger than the object; and a magnification of less than 1 means that the image is smaller than the object. By "size" is meant any transverse linear dimension, for instance height or width. Thus we can write

Definition of linear magnification

$$m = \frac{h'}{h}$$

$$\text{Linear magnification} = \frac{\text{height of image}}{\text{height of object}}$$

where $h = $ height of object and $h' = $ height of image.

FIG. 26–12   The magnification of a lens is equal to minus the ratio of image and object distances. A positive magnification means an erect image, while a negative magnification means an inverted image.

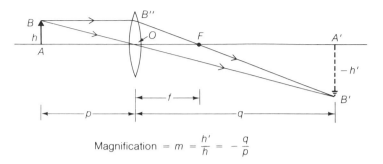

$$\text{Magnification} = m = \frac{h'}{h} = -\frac{q}{p}$$

Figure 26–12 shows a converging lens that forms the image $A'B'$ of the object AB. The object height is $AB = h$, and that of the image is $A'B' = h'$. The image is inverted and therefore its height is considered negative. The triangles ABO and $A'B'O$ are similar, and their corresponding sides are proportional. Hence

$$\frac{A'B'}{AB} = \frac{A'O}{AO}$$

Because $A'B' = -h'$, $AB = h$, $A'O = q$, and $AO = p$, we find that

$$\frac{h'}{h} = -\frac{q}{p}$$

and the magnification produced by the lens is

$$m = \frac{h'}{h} = -\frac{q}{p} \qquad\qquad \textit{Linear magnification of lens} \quad (26–8)$$

The magnification formula holds for all lenses and object distances

This formula is a general one that holds for diverging as well as converging lenses and for any object distance.

A useful feature of Eq. (26–8) is that it automatically indicates whether an image is erect (m positive) or inverted (m negative) with respect to the object. Table 26–1 summarizes the various sign conventions we have been using.

Magnifying glass

Example   A "magnifying glass" is a converging lens held less than its focal length from an object being examined (see Fig. 26–6(a)). How far should a double-convex lens whose focal length is 6 in. be held from an object to produce an erect image three times larger?

Solution   The required magnification is $+3$ since an erect image is required. From Eq. (26–8),

TABLE 26–1
Sign conventions for lenses

Quantity	Positive	Negative
Focal length f	Converging lens	Diverging lens
Object distance p	Real object	Virtual object
Image distance q	Real image	Virtual image
Magnification m	Erect image	Inverted image

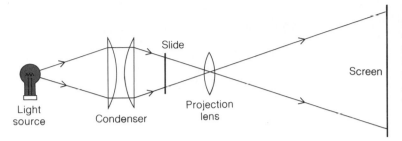

FIG. 26–13 An optical projection system. The condenser causes the slide to be evenly illuminated, thereby making possible an image of uniform brightness on the screen.

$$m = -\frac{q}{p}$$

$$q = -mp = -3p$$

Since $q = -3p$ and $f = +6$ in., we have from the thin lens equation

$$\frac{1}{f} = \frac{1}{p} + \frac{1}{q} = \frac{1}{p} - \frac{1}{3p} = \frac{3}{3p} - \frac{1}{3p} = \frac{2}{3p}$$

$$3p = 2f$$

$$p = \tfrac{2}{3}f = \tfrac{2}{3}(6\,\text{in.}) = 4\,\text{in.}$$

When the lens is held 4 in. from the object, the image will be virtual, erect, and enlarged three times. ∎

Example A slide projector is to be used with its lens 6 m from a screen. If a projected image 1.5 m square of a slide 5 cm square is desired, what should the focal length of the lens be? **Projector**

Solution Projectors use converging lenses to produce enlarged, inverted, real images of transparencies that are illuminated from behind (Fig. 26–13). In the present case, the image distance is given as $q = 6$ m, and the required magnification is

$$m = \frac{h'}{h} = \frac{-1.5\,\text{m}}{0.05\,\text{m}} = -30$$

The image height is reckoned negative as it is inverted. From Eq. (26–8),

$$p = -\frac{q}{m} = -\frac{6\,\text{m}}{-30} = 0.2\,\text{m}$$

Equation (26–2) now gives us the focal length of the lens:

$$\frac{1}{f} = \frac{1}{p} + \frac{1}{q} = \frac{1}{0.2\,\text{m}} + \frac{1}{6\,\text{m}}$$

$$f = 19.4\,\text{cm} \qquad \blacksquare$$

26–6 THE EYE

The structure of the human eye is shown in Fig. 26–14. Its diameter is typically a little less than 3 cm, and its approximately spherical shape is maintained by internal pressure. **Structure of the eye**

FIG. 26–14 The human eye slightly larger than life size.

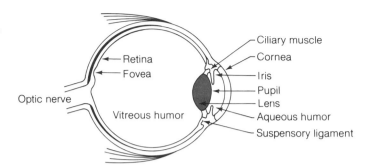

Incoming light first passes through the *cornea*, the transparent outer membrane, and then enters a liquid called the *aqueous humor.* The *lens* of the eye is a jellylike assembly of tiny transparent fibers that slide over one another when the shape of the lens is altered by the *ciliary muscle* to which it is attached by ligaments. In front of the lens is the colored *iris* whose aperture is the *pupil*. Behind the lens is a cavity filled with another liquid, the *vitreous humor,* and the lining of this cavity is the *retina*.

The retina

The retina contains millions of tiny structures called *rods* and *cones* that are sensitive to light; the cones are responsible for vision in bright conditions, the rods for vision in dim conditions. Because the sensation of color is produced by the cones, colors are difficult to distinguish in a faint light. The rods and cones respond to the image formed by the lens on the retina and transmit the information to the brain through the *optic nerve*. There are no photoreceptors near the optic nerve and that region is accordingly called the "blind spot." The blind spot covers a field of view about 8° high and 6° wide. Its existence is not conspicuous for two reasons: The blind spots of the two eyes obscure different fields of vision, and the eyes are never completely at rest but are in constant scanning motion. Vision is most acute for the image focused on the *fovea,* a small region near the center of the retina, whereas the rest of the field of view appears less distinct. Since the fovea contains only cones, it is easier to see something in a dim light by looking a bit to one side instead of directly at it.

The pupil changes size according to light intensity

In bright light the pupil contracts to reduce the amount of light reaching the retina, and it opens wider in faint light. A fully opened pupil admits about 16 times as much light as a fully contracted one, since the respective pupil diameters are about 2 and 8 mm for a ratio of 4:1. The retina itself is also able to cope with a considerable range of brightnesses. The sensitivity of the eye varies with wavelength, so light of the same intensity but of different wavelengths will give different impressions of brightness. Maximum sensitivity in bright light occurs in the middle of the visible spectrum at about $\lambda = 555$ nm, which corresponds to green (Fig. 26–15). Maximum sensitivity in dim light, when the rods rather than the cones are chiefly involved, occurs at about 510 nm, which is closer to the blue end of the spectrum and accounts for the reduced ability of the eye to respond to red light against a dark background. For this reason red illumination is provided when accommodation to darkness must be preserved, for instance on the bridge of a ship at night.

The focal length of the eye's lens changes to suit object distance

Refraction at the cornea is responsible for most of the focusing power of the eye; the lens is less effective because its index of refraction is not very different from those of the aqueous and vitreous humors on both sides of it. The shape of the lens is

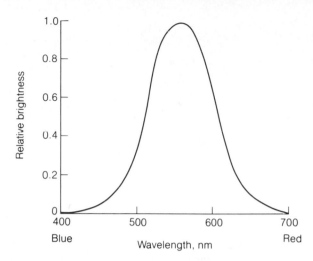

FIG. 26-15 How the sensitivity of a normal human eye to bright light varies with wavelength. In dim light, the wavelength of maximum sensitivity shifts from about 555 nm (green light) to about 510 nm.

controlled by the ciliary muscle. When this relaxes, the lens brings objects at infinity to a sharp focus on the retina. To permit a closer object to be viewed, the ciliary muscle forces the lens into a more convex shape whose focal length is appropriate for the object distance involved. The image distance in this situation, which is the lens-retina distance, is fixed, and focusing is thus carried out by changing the focal length of the lens. The latter process is called *accommodation*. In the case of a camera, on the other hand, the focal length of the lens is fixed, and focusing is done by changing the image distance.

The limited range of accommodation of the lens is not sufficient to make up for the loss of refractive power that occurs when the eye is immersed in water, which is considerable since the index of refraction of the cornea (1.38) is close to that of water (1.33). Hence underwater objects cannot be brought to a sharp focus unless goggles or a face mask are used to keep water away from the cornea. Fish are able to see clearly when submerged because the lenses of their eyes are spherical and have high indexes of refraction. Focusing in most fish eyes is carried out by shifting the lens closer to or farther from the retina, in the same way a camera lens is focused.

Underwater vision

26-7 DEFECTS OF VISION

Two common defects of vision are *myopia* (nearsightedness) and *hyperopia* (farsightedness). In myopia, the eyeball is too long, and light from an object at infinity comes to a focus in front of the retina, as in Fig. 26-16(b). Accommodation permits nearby objects to be seen clearly, but not more distant ones. A diverging lens of the proper focal length can correct this condition. In hyperopia, the eyeball is too short, and light from an object at infinity does not come to a focus within the eyeball at all. Its power of accommodation permits a hyperopic eye to focus on distant objects, but the range of accommodation is not enough for nearby objects to be seen clearly. The correction for hyperopia is a converging lens, as in Fig. 26-16(c).

Nearsightedness and farsightedness

FIG. 26–16 Myopia and hyperopia are common defects of vision that can be remedied with corrective lenses.

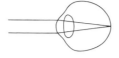

(a) Normal eye

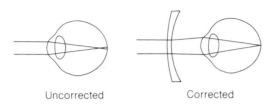

Uncorrected   Corrected

(b) Myopia (nearsightedness)

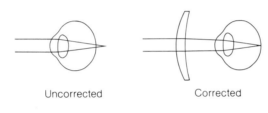

Uncorrected   Corrected

(c) Hyperopia (farsightedness)

Dioptric power

Opticians often use the *power D* of a lens, expressed in *diopters,* in place of its focal length. If the focal length f is given in meters, then

$$D \text{ (diopters)} = \frac{1}{f} \tag{26–9}$$

Thus a converging lens of $f = +50$ cm $= +0.50$ m could also be described as having a power of $+2$ diopters.

Correcting for nearsightedness

Example   A certain nearsighted eye cannot see objects distinctly when they are more than 25 cm away. Find the power in diopters of a correcting lens that will enable this eye to see distant objects clearly.

Solution   The purpose of the lens is to form an image 25 cm in front of the eye of an object that is infinitely far away. Since the image is to be on the same side of the lens as the object, the image must be virtual, so $q = -25$ cm. The object distance is $p = \infty$; hence

$$\frac{1}{f} = \frac{1}{p} + \frac{1}{q} = \frac{1}{\infty} - \frac{1}{25 \text{ cm}} = 0 - \frac{1}{25 \text{ cm}}$$
$$f = -25 \text{ cm} = -0.25 \text{ m}$$

The minus sign means that the lens is diverging. Its power in diopters is

$$D = \frac{1}{f} = \frac{-1}{0.25\,\text{m}} = -4\,\text{diopters}$$ ∎

Example   A certain farsighted eye cannot see objects distinctly when they are closer than 1 m away. Find the power in diopters of a correcting lens that will enable this eye to read a letter 25 cm away.

Correcting for
farsightedness

Solution   The purpose of the lens is to form an image 1 m in front of the eye of an object that is 25 cm away. Since the image is to be on the same side of the lens as the object, the image must be virtual, and so $q = -1$ m. The object distance is $p = 25$ cm $= 0.25$ m, hence

$$\frac{1}{f} = \frac{1}{p} + \frac{1}{q} = \frac{1}{0.25\,\text{m}} - \frac{1}{1\,\text{m}}$$
$$f = +0.33\,\text{m}$$

The plus sign means that the lens is converging. Its power in diopters is

$$D = \frac{1}{f} = \frac{1}{0.33\,\text{m}} = +3\,\text{diopters}$$ ∎

The range of accommodation decreases with age as the lens hardens, a condition known as *presbyopia*. This does not affect distant vision, but the "near point" of the eye (the closest position at which objects can be seen distinctly) gets increasingly far away. From perhaps 7 cm at age 10, the near point recedes until it is often 2 m or more at age 60. The correction for presbyopia is a converging lens; if the range of accommodation is very limited, more than one set of corrective lenses may be required.

Presbyopia refers to the gradual loss of accommodation

The closer an object is to the eye, the larger it seems and the more the detail that can be made out. Since the eye cannot focus on objects closer than the near point, this sets a limit to the magnification of which the eye is capable. By convention, the distance of most distinct vision is taken as 25 cm (about 10 in.), which is a comfortable object distance for most people. For purposes of calculation, an optical instrument such as a microscope or a telescope is therefore assumed to form a virtual image 25 cm behind the lens (or lens system) nearest the eye.

Astigmatism is a defect of vision caused by the cornea (or sometimes the lens) having different curvatures in different planes. When light rays that lie in one plane are in focus on the retina of an astigmatic eye, those in other planes will be in focus either in front or in back of the retina. As a result only one of the bars of a cross can be in focus at the same time (Fig. 26–17). Astigmatism is a source of eyestrain because the mechanism of accommodation continually varies the focus of the lens in an effort

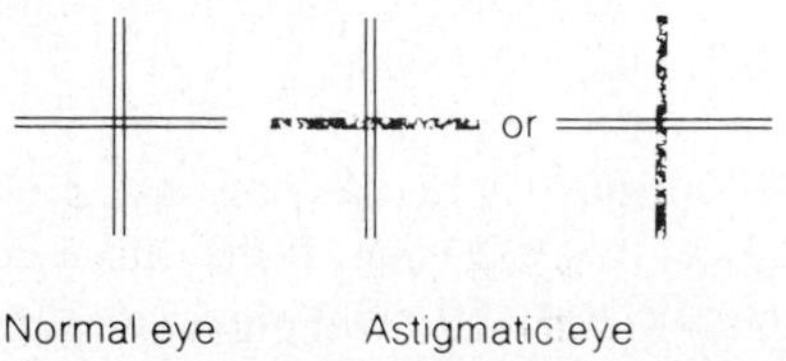

Astigmatism can be corrected by a cylindrical lens

Normal eye   Astigmatic eye

FIG. 26–17 How a cross is seen by a normal and by an astigmatic eye.

FIG. 26–18 How a cylindrical lens can improve the image formed by an astigmatic eye.

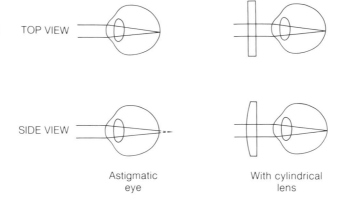

TOP VIEW

SIDE VIEW

Astigmatic eye

With cylindrical lens

to produce a completely sharp image. The remedy is a corrective lens that has a cylindrical curvature, as in Fig. 26–18.

26–8   THE MICROSCOPE

A microscope produces enlarged images of nearby objects

In many applications a system of two or more lenses is superior to a simple lens. An ordinary magnifying glass, for instance, cannot produce images enlarged beyond 3 × or so without severe distortion. Greater magnifications of nearby objects can be satisfactorily obtained by using instead two converging lenses arranged as a *microscope*. The optical system of a microscope is shown in Fig. 26–19. The *objective* is a lens of short focal length that forms an enlarged real image of the object. This image is further enlarged by the *eyepiece,* which acts as a simple magnifier to form a virtual final image. The image produced by the objective is thus the object of the eyepiece, and the final image has been magnified twice.

The total magnification produced by a two-lens system such as a microscope is the product of the magnification m_1 of the first lens (the objective) and the magnification m_2 of the second lens (the eyepiece):

$$m = m_1 m_2 \qquad\qquad\qquad \textit{Total magnification} \quad (26\text{–}10)$$

The limit of useful magnification is about 500 × with a microscope

Typical objectives yield magnifications of 10 × to 100 ×, and standard eyepiece magnifications are 5 ×, 10 ×, and 15 ×. Total magnifications of up to 1500 × are therefore possible with a good laboratory microscope. However, as discussed in the next chapter, the wave nature of light limits the useful magnification of a microscope to a maximum of perhaps 500 ×; higher magnifications give larger images (which can be helpful) but do not show finer details. Actual microscopes use compound lenses that consist of two to six elements in place of the single lenses shown in Fig. 26–19, but their optical behavior is the same.

Example   A microscope has an objective of focal length 4 mm and an eyepiece of focal length 20 mm. If the image distance of the objective is 160 mm and that of the eyepiece is 250 mm (which are the usual figures for these quantities), find the mag-

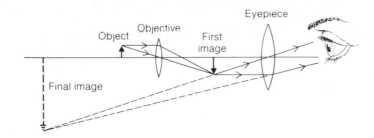

FIG. 26-19 In a microscope, the image formed by the objective is further magnified by the eyepiece. In this diagram the rays used to locate the final image are not the same as those used to locate the first image.

nification produced by each lens and by the entire microscope. What is the distance between the objective and the eyepiece?

Solution (a) The object distance of the objective is

$$\frac{1}{p_1} = \frac{1}{f_1} - \frac{1}{q_1} = \frac{1}{4\,\text{mm}} - \frac{1}{160\,\text{mm}}$$

$$p_1 = 4.10\,\text{mm}$$

and so its magnification is

$$m_1 = -\frac{q_1}{p_1} = -\frac{160\,\text{mm}}{4.10\,\text{mm}} = -39$$

The minus sign means the image is inverted.

(b) With $q_2 = -250$ mm and $f_2 = 20$ mm the same procedure yields for the eyepiece

$$\frac{1}{p_2} = \frac{1}{f_2} - \frac{1}{q_2} = \frac{1}{20\,\text{mm}} - \frac{1}{-250\,\text{mm}} = \frac{1}{20\,\text{mm}} + \frac{1}{250\,\text{mm}}$$

$$p_2 = 18.5\,\text{mm}$$

$$m_2 = -\frac{q_2}{p_2} = -\frac{-250\,\text{mm}}{18.5\,\text{mm}} = 13.5$$

(c) The magnification of the microscope is

$$m = m_1 m_2 = (-39)(13.5) = -527$$

and the distance between the objective and the eyepiece (see Fig. 26-19) is

$$L = q_1 + p_2 = 160\,\text{mm} + 18.5\,\text{mm} = 178.5\,\text{mm}$$ ∎

26-9 THE TELESCOPE

A telescope is a lens system used to examine distant objects. As in a microscope, two lenses are involved, with an eyepiece to enlarge the image produced by the objective. A telescope objective, however, has a long focal length whereas that of a microscope is very short.

The objective of a telescope has a long focal length

Figure 26-20 shows a simple telescope. The image produced by the objective is

FIG. 26–20   Ray
diagram of a simple
telescope. The rays used
to locate the final image
are not the same as those
used to locate the first
image.

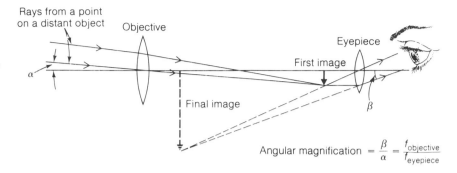

Angular magnification $= \dfrac{\beta}{\alpha} = \dfrac{f_{\text{objective}}}{f_{\text{eyepiece}}}$

real, inverted, and smaller than the object. The eyepiece then acts as a simple magnifier
to form an enlarged virtual image whose object is the initial image. In practice, the
object distance is usually very long relative to the focal length of the objective, and
the final image is smaller than the object itself. However, the image seen by the eye is
larger than it would be without the telescope, so the effect is the same as if the object
were closer to the eye than it actually is.

**Angular magnification
equals ratio of focal
lengths of objective and
eyepiece**

Telescopes are described in terms of the *angular magnification* they produce. This
is the ratio between the angle β subtended at the eye by the image and the angle α
subtended at the eye by the object seen directly. For distant objects ($p > > f$) the
angular magnification of a telescope is simply the ratio between the focal lengths of
its objective and eyepiece:

$$m_{\text{ang}} = \frac{f_{\text{objective}}}{f_{\text{eyepiece}}} = \frac{\beta}{\alpha} \qquad\qquad \textit{Angular magnification} \quad (26\text{–}11)$$

Angular magnification is sometimes called *magnifying power.*

The higher the magnification, the greater in diameter the objective must be in
order to gather in enough light for the image to be visible. This sets a limit to the size
of a refracting telescope, since a large glass lens tends to distort under its own weight.
The largest refracting telescope in the world is the 1.02-m-diameter instrument at Yerkes
Observatory in Wisconsin, whose objective has a focal length of nearly 20 m.

Example   A telescope with an objective of focal length 60 cm and an eyepiece of
focal length 1.5 cm is used to examine a hummingbird 5 cm long that is 20 m away.
If the image distance of the eyepiece is 25 cm (the distance of most distinct vision),
what is the apparent length of the hummingbird?

Solution   The angular magnification of the telescope is

$$m_{\text{ang}} = \frac{f_{\text{objective}}}{f_{\text{eyepiece}}} = \frac{60\,\text{cm}}{1.5\,\text{cm}} = 40$$

The angle α subtended by the bird from the location of the telescope is

$$\alpha = \frac{\text{object length}}{\text{object distance}} = \frac{0.05\,\text{m}}{20\,\text{m}} = 0.0025\,\text{radian}$$

If L is the length of the bird's image as seen through the telescope, the angle β this
image subtends is

$$\beta = \frac{\text{image length}}{\text{image distance}} = \frac{L}{25 \text{ cm}}$$

Since the angular magnification of the telescope is 40, from Eq. (26–11) we have

$$\beta = m_{\text{ang}}\alpha$$

$$\frac{L}{25 \text{ cm}} = (40)(0.0025 \text{ radian})$$

$$L = 2.5 \text{ cm}$$

The hummingbird seems to be 2.5 cm long, half its actual length, and to be located 25 cm from the viewer's eye. ■

Modern astronomical telescopes always use concave parabolic mirrors as their objectives, since such a mirror produces a real image of a distant object and can be adequately supported from behind. A further advantage is that a parabolic mirror is not subject to the aberrations mentioned in the next section. A small secondary mirror reflects the image outside the telescope tube for viewing or, more often, for photographing (Fig. 26–21). When used as a camera, a telescope needs no eyepiece: Its objective lens or mirror simply acts as a giant telephoto lens. One of the largest reflecting telescopes in the world is at Mount Palomar in California and has a mirror 5.08 m in diameter.

Example The diameter of the planet Mars is 6.8×10^6 m. What focal length must a telescope objective have in order to produce a photographic image of Mars 1 mm in diameter at a time when Mars is 8×10^{10} m from the earth?

Solution Here $h = 6.8 \times 10^6$ m and $h' = 1$ mm $= 10^{-3}$ m, so the required magnification is

$$m = \frac{h'}{h} = -\frac{10^{-3} \text{ m}}{6.8 \times 10^6 \text{ m}} = -1.5 \times 10^{-10}$$

A minus sign is used because the image will be inverted. The object distance p is the Mars-earth distance of 8×10^{10} m. Since the focal length f is going to be much smaller than p, $p - f \approx p$ and the image distance q is

$$q = \frac{pf}{p - f} \approx f$$

The formula $m = -q/p$ therefore becomes $m = -f/p$ here, and

$$f = -mp = -(-1.5 \times 10^{-10})(8 \times 10^{10} \text{ m}) = 12 \text{ m}$$

The telescope objective should have a focal length of 12 m. Of course, further enlargement of the image can be made from the negative. ■

A telescope can be constructed that produces an erect image if a third lens is introduced between the objective and the eyepiece (Fig. 26–22). The additional lens merely inverts the initial image, but it has the disadvantage of lengthening the telescope tube. A better scheme employs a pair of prisms that serve both to invert the image so

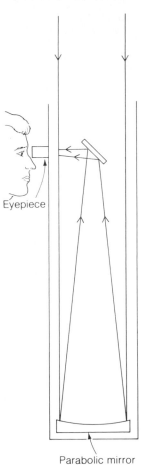

Eyepiece

FIG. 26–21 One type of reflecting telescope.

Parabolic mirror

How erect images can be produced by a telescope

FIG. 26–22   In a terrestrial telescope, an intermediate lens is used to produce an erect image.

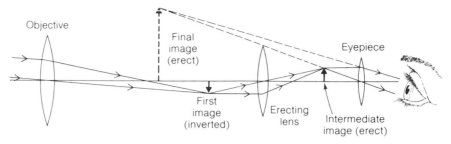

FIG. 26–23   *(Left)* A pair of prisms is used to erect the image in each half of a binocular. *(Right)* Cutaway of a binocular that employs a different prism arrangement. (Courtesy of Bushnell Division of Bausch & Lomb Inc.)

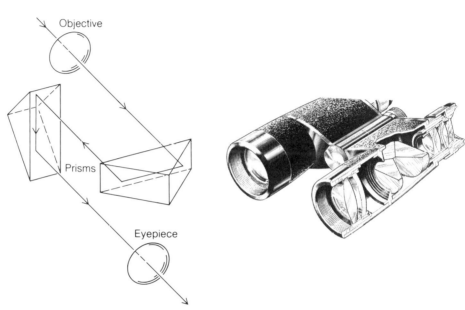

that it is erect and to shorten the instrument length. The latter method is employed in prism binoculars (Fig. 26–23).

26–10   LENS ABERRATIONS

Chromatic aberration

The image formed by a single lens is never a perfect replica of its object. Of the variety of aberrations such an image is subject to, perhaps the most familiar is the presence of fringes of color around whatever is being viewed. This *chromatic aberration* is a consequence of the variation with wavelength of the index of refraction of glass (Fig. 26–24). Because of this variation, the focal length of a lens is slightly different for light of different colors, and the fringes are the result. The remedy for chromatic aberration is to combine a converging and a diverging lens made of different glass so that the dispersion produced by one is canceled by the other while leaving a net converging or diverging power. Such an *achromatic lens* is illustrated in Fig. 26–24.

Spherical aberration

The lens equation was derived on the basis of light rays that made only small angles with the axis. When a simple lens is used to form images of objects some

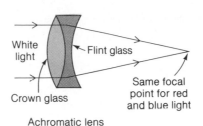

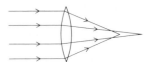

FIG. 26–24 A compound lens made of different types of glass can correct for chromatic aberration.

distance from the axis, a variety of aberrations arise even if the light used is mono-chromatic. One of them is *spherical aberration,* in which rays passing near the lens rim come to a focus closer to the lens than rays near the axis (Fig. 26–25). Another is distortion of the image because the magnification of a simple lens varies with the distance of an object from the axis. By using several lenses of different types of glass and different curvatures to replace a single lens, the chief aberrations can be minimized. Compound lenses that consist of two or more elements are always used in high-per-formance optical systems such as those in microscopes, prism binoculars, and quality cameras.

FIG. 26–25 Spherical aberration.

26–11 SPHERICAL MIRRORS

Mirrors with spherical reflecting surfaces form images in much the same way that lenses do. Figure 26–26(a) shows how a concave mirror converges a parallel beam of light to a real focal point, and Fig. 26–26(b) shows how a convex mirror diverges such a beam so that the reflected rays seem to come from a virtual focal point behind the mirror. If R is the radius of the reflecting surface in each case, the focal lengths of these mirrors are

Concave mirrors are converging, convex mirrors are diverging

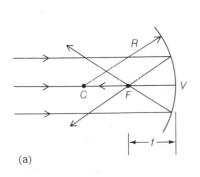

(a)

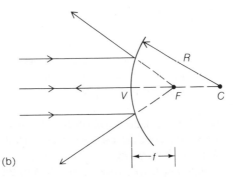

(b)

FIG. 26–26 (a) A concave spherical mirror converges a parallel beam of incident light. (b) A convex spherical mirror diverges a parallel beam of incident light.

R = radius of curvature
V = vertex
C = center of curvature
F = focal point
f = focal length

FIG. 26–27 The position and size of an image produced by a spherical mirror can be determined by tracing any two of the rays shown.

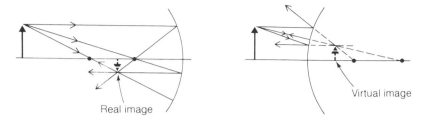

Real image   Virtual image

$$f = \frac{R}{2} \qquad\qquad\qquad\qquad\qquad \textit{Concave mirror} \quad (26\text{–}12)$$

$$f = -\frac{R}{2} \qquad\qquad\qquad\qquad\qquad \textit{Convex mirror} \quad (26\text{–}13)$$

The position, size, and nature of the image produced by a spherical mirror of an object in front of it can be determined with the help of a scale drawing. As in the case of a lens, the procedure is to consider two different light rays that come from a certain point on the object and to trace their paths until they (or their backward extensions) come together again after reflection. There are three rays that are especially useful because they are so easily traced (Fig. 26–27):

Rays for finding the image produced by a spherical mirror

1. A ray that leaves the object parallel to the mirror axis. When this ray is reflected, it passes through the focal point of a concave mirror or seems to come from the focal point of a convex mirror.

2. A ray that leaves the object and passes through the focal point of a concave mirror, or is directed toward the focal point of a convex mirror. When this ray is reflected, it proceeds parallel to the axis.

3. A ray that leaves the object along a radius of the mirror. When this ray is reflected, it returns along its original path.

Images produced by a concave mirror

In any given situation only two of these rays are needed to locate the image of a reflected object. Figure 26–28 shows how the properties of the image produced by a concave mirror vary with the position of the object. Solid lines represent the actual paths taken by light rays, and dashed lines represent virtual paths.

When the object is closer to the mirror than the focal point *F*, the image is erect (right side up), enlarged, and virtual. The image *seems* to be behind the mirror because the reflected light rays from the mirror diverge as though coming from a point behind the mirror.

When the object is precisely at the focal point, no image is formed because the reflected rays are all parallel and so do not intersect. The image in this case is sometimes said to be at infinity.

An object between the focal point *F* and the center of curvature *C* has an inverted, enlarged image that is real. The image would appear on a screen placed at its position.

When the object is at *C* its image is at the same place and is the same size but is inverted. An object past *C* has a real image that is reduced in size and is inverted.

Images produced by a convex mirror

The image formed by a convex mirror of a real object is always erect, smaller than the object, and virtual (Fig. 26–29). The field of view of a convex mirror is wider

OBJECT

IMAGE

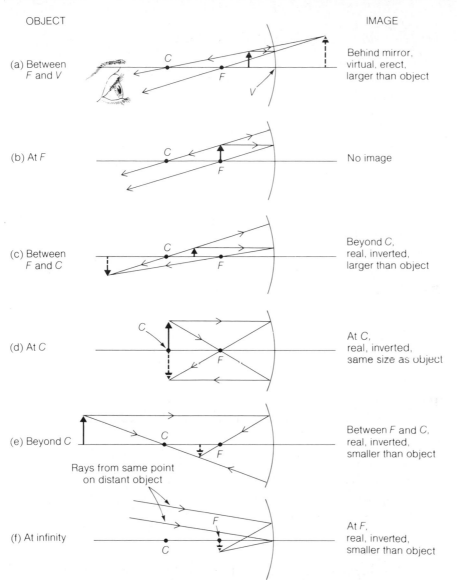

FIG. 26–28 Image formation by a concave mirror.

(a) Between F and V — Behind mirror, virtual, erect, larger than object

(b) At F — No image

(c) Between F and C — Beyond C, real, inverted, larger than object

(d) At C — At C, real, inverted, same size as object

(e) Beyond C — Between F and C, real, inverted, smaller than object

Rays from same point on distant object

(f) At infinity — At F, real, inverted, smaller than object

than that of a plane mirror, which accounts for a number of its applications, such as at blind corners in roads.

The object and image distances p and q of a spherical mirror are related to its focal length by the same formula that holds for a thin lens:

Formulas for spherical mirrors

$$\frac{1}{p} + \frac{1}{q} = \frac{1}{f} \qquad \qquad \textit{Mirror equation} \quad (26\text{–}14)$$

Equations (26–5), (26–6), and (26–7) hold here as well as for lenses. The linear magnification of a mirror also follows the same formula as for a lens:

FIG. 26–29 The image of a real object formed by a convex mirror is always virtual, erect, and smaller than the object.

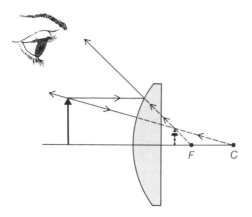

$$m = \frac{h'}{h} = -\frac{q}{p} \qquad \qquad \textit{Magnification of a mirror} \quad (26\text{–}15)$$

The sign conventions for a mirror are given in Table 26–2.

TABLE 26–2
Sign conventions for spherical mirrors

Quantity	Positive	Negative
Focal length f	Concave mirror	Convex mirror
Object distance p	Real object	Virtual object
Image distance q	Real image	Virtual image
Magnification m	Erect image	Inverted image

Example A candle 5 cm high is placed 40 cm from a concave mirror whose radius of curvature is 60 cm. Find the position, size, and nature of the image.

Solution The focal length of the mirror is

$$f = \frac{R}{2} = \frac{60\,\text{cm}}{2} = 30\,\text{cm}$$

The situation therefore corresponds to that shown in Fig. 26–28(c). The image distance q is

$$\frac{1}{q} = \frac{1}{f} - \frac{1}{p} = \frac{1}{30\,\text{cm}} - \frac{1}{40\,\text{cm}}$$

$$q = 120\,\text{cm}$$

The image distance is positive, so the image is a real one. The size of the image is, from Eq. (26–15),

$$h' = -h\frac{q}{p} = (-5\,\text{cm})\left(\frac{120\,\text{cm}}{40\,\text{cm}}\right) = -15\,\text{cm}$$

The image of the candle is three times as large as the candle itself and is inverted. ∎

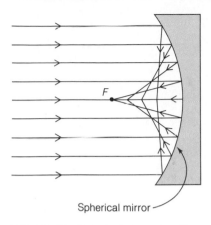

FIG. 26–30 Spherical aberration. The more highly curved the mirror, the greater the aberration.

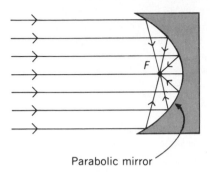

FIG. 26–31 A parabolic mirror does not exhibit spherical aberration.

Example A concave shaving mirror has a focal length of 18 in. How far away from it should one's face be for the reflected image to be erect and twice its actual size?

Solution Here the magnification is $+2$ since the image is erect. Hence

$$m = -\frac{q}{p} = 2 \qquad q = -2p$$

The negative image distance signifies that the image is virtual; see Fig. 26–28(a). Now we substitute $f = 18$ in. and $q = -2p$ into the mirror equation and solve for the object distance p:

$$\frac{1}{p} + \frac{1}{q} = \frac{1}{f}$$

$$\frac{1}{p} - \frac{1}{2p} = \frac{1}{18 \text{ in.}}$$

$$\frac{1}{2p} = \frac{1}{18 \text{ in.}}$$

$$p = 9 \text{ in.}$$

When one's face is 9 in. from the mirror, the image one sees is magnified twice and is erect. ■

Mirrors do not suffer from chromatic aberration but do exhibit spherical aberration. In the case of a concave mirror, rays reflected from the outer part of the mirror cross the axis closer to the vertex than rays reflected from the central part (Fig. 26–30). When a high-quality image is required, for example in the case of an astronomical telescope (Fig. 26–21), a concave mirror whose cross-sectional shape is a parabola is the answer, as in Fig. 26–31. However, despite the presence of spherical aberration, spherical mirrors are common because they are the easiest to manufacture.

Spherical aberration in a mirror

IMPORTANT TERMS

A **lens** is a transparent object of regular form that can produce an image of an object placed before it. A **converging lens** brings parallel light to a single **real focal point,** while a **diverging lens** deviates parallel light outward as though it originated at a single **virtual focal point.** The distance from a lens to its focal point is its **focal length.**

The **magnification** of an optical system is the ratio between the size of the image and that of the object.

A **microscope** is a lens system used to produce enlarged images of nearby objects. A **telescope** is a lens system used to produce larger images of distant objects than would be seen by the unaided eye, although the image itself may be smaller than the actual object.

The **distance of most distinct vision** is 25 cm (10 in.) for most people; this distance is used as the image distance of the eyepieces of microscopes and telescopes.

A **concave mirror** curves inward toward its center and converges parallel light to a single real focal point.

A **convex mirror** curves outward toward its center and diverges parallel light as though the reflected light came from a single virtual focal point behind the mirror. The distance from a mirror to its focal point is the focal length of the mirror.

IMPORTANT FORMULAS

Lensmaker's equation:

$$\frac{1}{f} = (n - 1)\left(\frac{1}{R_1} + \frac{1}{R_2}\right) \quad \begin{array}{l}(R = +\text{ for convex,}\\ -\text{ for concave surface,}\\ \infty \text{ for flat surface)}\end{array}$$

Focal length of concave mirror: $f = \dfrac{R}{2}$

Focal length of convex mirror: $f = -\dfrac{R}{2}$

Lens and mirror equation: $\dfrac{1}{p} + \dfrac{1}{q} = \dfrac{1}{f}$

Alternate forms of lens and mirror equation: $p = \dfrac{qf}{q - f}$

$$q = \frac{pf}{p - f}$$

$$f = \frac{pq}{p + q}$$

Linear magnification: $m = -\dfrac{q}{p}$

$$m = m_1 m_2$$

Angular magnification of telescope: $m_{ang} = \dfrac{f_{objective}}{f_{eyepiece}}$

MULTIPLE CHOICE

1. In order to calculate the focal length of a glass lens, it is not necessary to know
   a. the index of refraction of the glass.
   b. the index of refraction of the medium in which the lens is located.
   c. the radii of curvature of the lens surfaces.
   d. the diameter of the lens.

2. A converging lens may not have
   a. a positive focal length.
   b. a negative focal length.
   c. one plane surface.
   d. one concave surface.

3. A negative focal length corresponds to which one or more of the following?
   a. double-convex lens   b. plano-concave lens
   c. convex mirror   d. concave mirror

4. Relative to its object, a real image formed by a lens is always
   a. erect.   b. inverted.
   c. smaller.   d. larger.

5. Relative to its object, a real image formed by a spherical mirror is always
   a. erect.   b. inverted.
   c. smaller.   d. larger.

6. An object infinitely far from a converging lens has an image that is
   a. real.
   b. virtual.
   c. erect.
   d. larger than the object.

7. An object farther from a converging lens than its focal point always has an image that is
   a. inverted.   b. virtual.
   c. the same in size.   d. smaller in size.

8. An object closer to a converging lens than its focal point always has an image that is
   a. inverted.   b. virtual.
   c. the same in size.   d. smaller in size.

9. The image of a real object formed by a diverging lens is always
   a. real.
   b. virtual.

c. inverted.

d. larger than the object.

10. A positive magnification signifies an image that is

a. erect.

b. inverted.

c. smaller than the object.

d. larger than the object.

11. A negative image distance signifies an image that is

a. real.   b. virtual.

c. erect.   d. inverted.

12. A converging lens of focal length F produces an image of a real object that is smaller than the object. The object distance must be

a. less than F.   b. between F and $2F$.

c. more than F.   d. any of the above.

13. The image in the above situation is

a. real and erect.   b. real and inverted.

c. virtual and erect.   d. virtual and inverted.

14. When a converging lens of focal length F is used as a magnifying glass, the object distance must be

a. less than F.   b. equal to F.

c. between F and $2F$.   d. more than $2F$.

15. The image a camera forms on the film is

a. always real.

b. always virtual.

c. sometimes real and sometimes virtual.

d. neither real nor virtual.

16. The pupil of the eye controls

a. the focal length of the eye.

b. the range of accommodation of the eye.

c. the distance of most distinct vision.

d. the amount of light reaching the eye.

17. The lens of the eye forms an image on the retina that is

a. real and erect.   b. real and inverted.

c. virtual and erect.   d. virtual and inverted.

18. Which of the following combinations of shutter speed and lens opening will admit the most light to the film of a camera?

a. $\frac{1}{125}$ s at $f/8$ b. $\frac{1}{125}$ s at $f/16$

c. $\frac{1}{250}$ s at $f/4$ d. $\frac{1}{250}$ s at $f/5.6$

19. An exposure meter indicates that a camera should be set for a shutter speed of $\frac{1}{125}$ s and an aperture of $f/8$ to photograph a certain scene. If the shutter speed is changed to $\frac{1}{500}$ s, the aperture should be changed to

a. $f/2$.   b. $f/4$.

c. $f/5.6$.   d. $f/16$.

20. An object is located 10 in. from a converging lens of focal length 12 in. The image distance is

a. $+5.45$ in.   b. -5.45 in.

c. $+60$ in.   d. -60 in.

21. An object is located 12 in. from a converging lens of focal length 10 in. The image distance is

a. $+5.45$ in.   b. -5.45 in.

c. $+60$ in.   d. -60 in.

22. The image of an object 10 cm from a lens is located 10 cm behind the object. The focal length of the lens is

a. $+6.7$ cm.   b. -6.7 cm.

c. $+20$ cm.   d. -20 cm.

23. A pencil 10 cm long is placed 70 cm in front of a lens of focal length $+50$ cm. The image is

a. 4 cm long and erect.

b. 4 cm long and inverted.

c. 25 cm long and erect.

d. 25 cm long and inverted.

24. A pencil 10 cm long is placed 100 cm in front of a lens of focal length $+50$ cm. The image is

a. 5 cm long and erect.

b. 5 cm long and inverted.

c. 10 cm long and erect.

d. 10 cm long and inverted.

25. A pencil 10 cm long is placed 175 cm in front of a lens of focal length $+50$ cm. The image is

a. 4 cm long and erect.

b. 4 cm long and inverted.

c. 25 cm long and erect.

d. 25 cm long and inverted.

26. A magnifying glass is to be used at the fixed object distance of 1 in. If it is to produce an erect image magnified 5 times, its focal length should be

a. $+0.2$ in.   b. $+0.8$ in.

c. $+1.25$ in.   d. $+5$ in.

27. Four lenses with the listed focal lengths are being considered for use as a microscope objective. The one that will produce the greatest magnification with a given eyepiece has the focal length

a. -5 mm.   b. $+5$ mm.

c. -5 cm.   d. $+5$ cm.

28. Four lenses with the listed focal lengths are being considered for use as a telescope objective. The one that will produce the greatest magnification with a given eyepiece has the focal length

a. -1 m.   b. $+1$ m.

c. -2 m.   d. $+2$ m.

29. A concave mirror produces an erect image when the object distance is

a. less than f.   b. equal to f.

c. between f and $2f$.   d. greater than $2f$.

30. The image forced by a concave mirror is larger than the object
   a. when p is less than $2f$.
   b. when p is more than $2f$.
   c. for no values of p.
   d. for all values of p.

31. The image formed by a convex mirror is larger than the object
   a. when p is less than $2f$.
   b. when p is more than $2f$.
   c. for no values of p.
   d. for all values of p.

32. A convex mirror is ground with a radius of curvature of 12 in. Its focal length is
   a. 6 in.   b. 24 in.
   c. -6 in.   d. -24 in.

33. A pencil 10 cm long is placed 30 cm in front of a mirror of focal length $+50$ cm. The image is
   a. 2.5 cm long and erect.
   b. 25 cm long and erect.
   c. 250 cm long and erect.
   d. 25 cm long and inverted.

34. A pencil 10 cm long is placed 1 m in front of a mirror of focal length $+50$ cm. The image is
   a. 3 cm long and erect.
   b. 10 cm long and erect.
   c. 3 cm long and inverted.
   d. 10 cm long and inverted.

35. A pencil 10 cm long is placed 30 cm in front of a mirror of focal length -50 cm. The image is
   a. 25 cm long and erect.
   b. 6.25 cm long and erect.
   c. 25 cm long and inverted.
   d. 6.25 cm long and inverted.

EXERCISES

26–1   Lenses

1. A fortune-teller's crystal ball is 15 cm in diameter. (a) Would you expect the lensmaker's equation to hold for the focal length of this ball? (b) Would you expect the ball to form undistorted images?

2. Does a spherical bubble of air in a volume of water act to converge or diverge light passing through it?

3. A double-convex lens made of crown glass is placed in a tank of benzene. Will it act as a converging or as a diverging lens there?

4. A double-concave lens has surfaces whose radii of curvature are both 40 cm. The lens is made from flint glass whose index of refraction is 1.55. Find the focal length.

5. A converging meniscus lens has surfaces whose radii of curvature are 20 cm and 30 cm. The lens is made from crown glass whose index of refraction is 1.50. Find its focal length.

6. The index of refraction of crown glass is 1.523 for blue light and 1.517 for red light. How far apart are the focal points for blue and red light of a plano-convex lens of crown glass whose radius of curvature is 200 mm?

7. A diverging meniscus lens has surfaces whose radii of curvature are 20 cm and 30 cm. The lens is made from crown glass whose index of refraction is 1.50. Find its focal length.

8. A double-convex lens whose focal length is 35 cm has surfaces whose radii of curvature are 25 cm and 50 cm, respectively. Find the index of refraction of the glass.

9. A plano-concave lens of focal length 9 in. is to be ground from quartz of index of refraction 1.55. Find the required radius of curvature.

10. A diverging lens made from glass of $n = 1.55$ has a focal length of -8 cm in air. (a) Is it converging or diverging when immersed in water? (b) What is its focal length in water?

11. A converging lens made from glass of $n = 1.60$ has a focal length in air of 15 cm. (a) Is it converging or diverging when immersed in water? (b) What is its focal length in water?

26–2   Image Formation

12. Under what circumstances, if any, is a light ray that passes through a converging lens not deflected? Under what circumstances, if any, is a light ray that passes through a diverging lens not deflected?

13. Under what circumstances, if any, will a diverging lens form an inverted image of a real object? Under what circumstances, if any, will a converging lens form an erect image?

14. Is the mercury column in a thermometer wider or narrower than it appears?

15. Is there any way in which a diverging lens, used by itself, can form a real image of a real object? Is there any way in which a converging lens, used by itself, can form a virtual image of a real object?

26–3   The Lens Equation

16. A slide is in sharp focus on a screen. The screen is then moved farther away from the projector. Should the projector's lens be moved closer to or farther from the slide to bring it back into focus on the screen?

17. A lens is to be used to focus sunlight on a piece of paper to ignite it. What kind of lens should be used? If the focal length of the lens is 8 cm, how far should it be held from the paper?

18. In order to photograph herself for a passport picture, a woman stands beside her camera 0.6 m from a mirror. If the focal length of the camera lens is 40 mm, how far from the film should it be?

19. A lens held 20 cm from a sardine produces a real, inverted image of it 30 cm on the other side. What is the focal length of the lens? Is it converging or diverging?

20. A lens held 20 cm from a sardine produces a virtual, erect image of it that appears to originate 10 cm in front of the sardine. What is the focal length of the lens? Is it converging or diverging?

21. Verify that the effective focal length f of two thin lenses of focal lengths f_1 and f_2 that are in contact is given by

$$\frac{1}{f} = \frac{1}{f_1} + \frac{1}{f_2}$$

(To do this, let the image produced by the first lens be the object of the second.)

26–4 The Camera

22. What is the diameter of an $f/8$ telephoto lens whose focal length is 300 mm?

23. An exposure meter indicates that a camera should be set for a shutter speed of $\frac{1}{125}$ s and an aperture of $f/8$ to photograph a certain scene. If the shutter speed is changed to $\frac{1}{500}$ s, what should the aperture be changed to?

24. The settings of a camera are changed from $\frac{1}{125}$ s at $f/16$ to $\frac{1}{250}$ s at $f/5.6$. What is the difference in the amount of light reaching the film?

26–5 Magnification

25. What can you say about the properties of an image when the magnification is (a) less than -1? (b) between -1 and 0? (c) between 0 and $+1$? (d) greater than $+1$?

26. A candle is placed with its flame 10 cm from a lens whose focal length is $+15$ cm. Find the location of the image of the flame. Is the image larger or smaller than the actual flame? What is the character of the image; that is, is it erect or inverted, real or virtual?

27. The flame of the candle of Exercise 26 is 15 cm from the lens. Answer the same questions for this case.

28. The flame of the candle of Exercise 26 is 25 cm from the lens. Answer the same questions for this case.

29. The flame of the candle of Exercise 26 is 30 cm from the lens. Answer the same questions for this case.

30. The flame of the candle of Exercise 26 is 50 cm from the lens. Answer the same questions for this case.

31. A motion picture director holds a diverging lens 15 ft from an actress and sees her one-tenth of her normal size. What is the focal length of the lens?

32. A camera whose lens has a focal length of 90 mm is used to photograph a person 4 m away. (a) How far in front of the film should the lens be placed? (b) If the film is 40 mm square, what is the area of the subject that it covers?

33. An aerial camera whose lens has a focal length of 1 m is used to photograph a military base from an altitude of 7 km. How long is the image on the film of a tank 9 m long?

34. The moon's diameter is 3476 km and its average distance from the earth is 3.8×10^5 km. (a) What is the diameter of its image when a 35-mm camera with a 300-mm telephoto lens is used to photograph it? (b) What would the image diameter be if a camera using 9×12-cm film were used with the same lens?

35. A magnifying glass of 10-cm focal length is held 8 cm from a stamp. What is the actual length of a feature of the stamp that appears to be 1 cm long?

36. A magnifying glass of 50-mm focal length is held 35 mm from a spider egg 0.5 mm long. What is the apparent size of the egg?

37. A photographic enlarger is being designed to produce prints 16 in. $\times$ 20 in. from negatives 4 in. $\times$ 5 in. (a) If the maximum distance from negative to print paper is to be 2 ft, what should the focal length of the lens be? (b) How far would a 4 in. $\times$ 5 in. negative be from the print paper using the same lens if the enlargement size were 8 in. $\times$ 10 in.?

38. The actual size of each frame of 16-mm motion picture film is 7.5 mm $\times$ 10.5 mm. A projector whose lens has a focal length of 25 mm is to be used with a screen 1.5 m wide. How far from the projector should the screen be located?

39. An object should be about 25 cm from a normal eye for maximum distinctness of vision. (a) Find a formula for the magnification of a converging lens when it is used as a magnifying glass with an image distance of -25 cm; see Fig. 26–6(a). (b) Find the magnification of a lens whose focal length is 5 cm.

26–6 The Eye

26–7 Defects of Vision

40. What are the characteristics of the image formed on the retina by the lens of the eye?

41. The image distance in a certain normal eye is 20 mm. What range of focal lengths does the eye have if its near point is 20 cm?

42. A myopic eye cannot bring to a focus objects farther than 15 cm away. What type of lens is needed to permit this eye to see clearly objects at infinity, and what focal length should such a lens have?

43. A presbyopic eye has a near point 1.0 m away. What type of lens is needed to permit the eye to see clearly objects 25 cm away, and what focal length should such a lens have?

44. A hyperopic eye has a near point of 60 cm. What is its near point when a correcting lens of $+3.33$ diopters is used?

45. A myopic person whose eyes have far points of 60 cm is lent a pair of glasses whose power is -1.5 diopters. How far can she see clearly with these glasses?

46. A myopic eye with near and far points of 12 and 20 cm, respectively, is given a corrective lens that permits distant vision. What is the new near point?

47. A hyperopic person whose eyeglass lenses are converging with a power of $+2.5$ diopters can see objects distinctly as close as 25 cm away. What is the distance of the near point without the eyeglasses?

48. A nearsighted person whose eyeglass lenses are diverging with a power of -2.5 diopters can see distant objects distinctly. What is the distance of the far point without the eyeglasses?

26–8 The Microscope

49. In a certain microscope an objective of 4-mm focal length is used with a $10\times$ eyepiece. (a) What is the magnification of the microscope? (b) How far should the objective be from the specimen being examined?

50. In a certain microscope an objective of 10-mm focal length is used with a $5\times$ eyepiece. (a) What is the magnification of the microscope? (b) How far should the objective be from the specimen being examined?

26–9 The Telescope

51. A telescope has an objective whose focal length is 60 cm. (a) Find the eyepiece focal length needed for an angular magnification of 20. (b) The telescope is used to watch a young zebra 500 m away. If the zebra's image is 20 mm long and is located 25 cm in front of the eyepiece, find the zebra's actual length.

52. A telescope has an objective of 30 in. focal length and an eyepiece of 1.0-in. focal length. The telescope is focused on a distant albatross. (a) What is the angular magnification? (b) How far apart are the lenses? Assume that the final image is located 10 in. in front of the eyepiece.

53. A telescope with an objective of focal length 1.0 m and an eyepiece of focal length 5.0 cm is used to examine a penguin 40 cm high. Find the apparent height of the penguin when it is (a) 50 m, and (b) 5 m away from the telescope. Assume the image to be 25 cm in front of the eyepiece.

54. A Galilean telescope has a converging lens as its objective and a diverging lens as its eyepiece. Its advantages are that it produces an erect image and is short in length; its main disadvantage is a narrow field of view that limits it to low magnifications. Inexpensive opera glasses consist of a pair of Galilean telescopes. Show that the distance between the objective and eyepiece in a Galilean telescope when it is focused on a distant object with the final image at infinity is the difference between the absolute values $|f_1|$ and $|f_2|$ of their focal lengths (absolute value is the value without regard to sign).

26–11 Spherical Mirrors

55. Is there any way in which a convex mirror, used by itself, can form a real image of a real object? Is there any way in which a concave mirror, used by itself, can form a virtual image of a real object?

56. A convex mirror has a radius of curvature of 40 cm. What is its focal length? Where is its focal point?

57. A butterfly is 20 cm in front of a concave mirror whose focal length is 40 cm. Find the location of the image. Is the image larger or smaller than the butterfly? What is the character of the image—that is, is it erect or inverted, real or virtual?

58. A dime is 40 cm in front of the mirror of Exercise 57. Answer the same questions for the image of the dime.

59. A peanut is 50 cm in front of the mirror of Exercise 57. Answer the same questions for the image of the peanut.

60. A caterpillar is 80 cm in front of the mirror of Exercise 57. Answer the same questions for the image of the caterpillar.

61. A button is 100 cm in front of the mirror of Exercise 57. Answer the same questions for the image of the button.

62. An object 4 in. high is 9 in. in front of a convex mirror

whose focal length is 12 in. What is the height and character of the image?

63. A dentist's concave mirror has a diameter of 1 cm and a focal length of 2.5 cm. What magnification does it produce when held 1.8 cm from a tooth?

64. A man stands 20 ft from a concave mirror, and an inverted image of himself of the same height is formed on a screen beside him. What is the radius of curvature of the mirror?

65. A worm crawls toward a polished metal ball 60 cm in diameter lying on a lawn. How far from the surface of the ball is the worm when its image appears to be 10 cm behind the surface?

66. A mirror in an amusement park produces an erect image four times enlarged of anyone standing 3 m away. (a) Is the mirror concave or convex? (b) What is its radius of curvature?

67. What should the radius of curvature of a convex mirror be if it is to produce an image one-fifth the size of an object 150 cm away?

68. The moon is 3476 km in diameter. What radius of curvature should a concave mirror have if it is to produce a lunar image 1 cm in diameter when the moon is 3.84×10^5 km away?

69. A virtual image 6 cm long is formed of a paper clip 2 cm long placed 10 cm in front of a concave mirror of unknown curvature. Where else can the paper clip be placed for an image 6 cm long to be formed? What is the nature of the image in the latter case?

70. A pawn 5 cm high is placed in front of a concave mirror whose radius of curvature is 1 m. What are the two object distances that will lead to images 20 cm high? What is the character of the image in each case?

ANSWERS TO MULTIPLE CHOICE

1. d	**8.** b	**15.** a	**22.** c	**29.** a
2. b	**9.** b	**16.** d	**23.** d	**30.** a
3. b, c	**10.** a	**17.** b	**24.** d	**31.** c
4. b	**11.** b	**18.** c	**25.** b	**32.** c
5. b	**12.** c	**19.** b	**26.** c	**33.** b
6. a	**13.** b	**20.** d	**27.** b	**34.** d
7. a	**14.** a	**21.** c	**28.** d	**35.** b

27

PHYSICAL OPTICS

In the previous chapter we had no need to invoke the wave character of light. But, although the properties of lenses and mirrors are more readily analyzed in terms of rays than in terms of waves, there are other optical phenomena in which the wave nature of light is directly involved. The study of such phenomena is called physical optics, whereas the study of those aspects of light behavior that can be understood using a ray treatment is geometrical optics. Since the basic laws of refraction and reflection follow from Huygens' principle, geometrical optics is evidently an approximation of physical optics whose usefulness comes from its simplicity. In this chapter we shall find the wave approach of physical optics necessary for explaining interference, diffraction, polarization, and scattering.

27–1 INTERFERENCE OF LIGHT

Constructive and destructive interference

When light waves from one source are mixed with those from another source, the two wave trains are said to *interfere*. We recall from Chapter 13 the principle of superposition, which governs interference: When two or more waves of the same nature travel

CHAPTER OBJECTIVES

Completing this chapter should enable you to:

1. Use the principle of superposition to interpret constructive and destructive interference.

2. Distinguish between coherent and incoherent sources of light.

3. Describe the origins of the interference patterns produced by a double slit and by a diffraction grating.

4. Describe the origin of the interference pattern produced by a thin film.

5. Explain why a thin coating of a suitable substance minimizes the reflection of light from a glass surface.

6. Use Huygens' principle to interpret the diffraction of waves around the edge of an obstacle.

7. Calculate the resolving power of an optical system.

8. Distinguish between polarized and unpolarized light and account for basic polarization effects.

9. Explain why the sky is blue.

past a point at the same time, the amplitude at that point is the sum of the instantaneous amplitudes of the individual waves. Constructive interference refers to the reinforcement of waves in phase (in step) with one another, and destructive interference refers to the partial or complete cancellation of waves out of phase with one another (Fig. 27–1).

Anyone with a pan of water can see how interference between water waves can lead to a water surface disturbed in a variety of characteristic patterns. Two people who

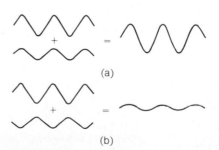

FIG. 27–1 (a) Constructive interference. (b) Destructive interference.

hum fairly pure tones slightly different in frequency will hear beats as the result of interference in the sound waves. But if we shine light from two flashlights at the same place on a screen, there is no evidence of interference: The region of overlap is merely uniformly bright.

Why interference is hard to demonstrate in light

There are two reasons why it is hard to observe interference in light. First, light waves have very short wavelengths—the visible part of the spectrum extends only from 400 nm for violet light to 700 nm for red light. Second, every natural source of light emits light waves only as short trains of random phase, so that any interference that occurs is averaged out during even the briefest period of observation by the eye or photographic film unless special procedures are used. Interference in light is nevertheless just as real an effect as interference in water or sound waves, and there is one example of it familiar to everybody—the bright colors of a thin film of oil spread out on a water surface.

Coherent and incoherent light sources

Two sources of waves are said to be *coherent* if there is a fixed phase relationship between the waves they emit during the time the waves are being observed. It does not matter whether the waves are exactly in step when they leave the sources, or exactly out of step, or anything in between; the important thing is that the phase relationship stays the same. If the sources shift back and forth in relative phase while the observation is made, the phase differences average out, and there will be no interference pattern. The latter sources are *incoherent*.

Ordinary light sources do not emit continuous wave trains

The question of coherence is especially significant for light waves because an excited atom usually radiates for no more than 10^{-8} s. Therefore a monochromatic light source such as a gas discharge tube (a neon sign is an example) does not emit a continuous wave train, as a radio antenna does, but instead a series of individual wave trains whose phases are random. The light from such a tube actually comes from a great many individual, uncoordinated sources, namely the gas atoms, and these individual sources are in effect being switched on and off rapidly and irregularly (Fig. 27–2).

Coherence depends on the time scale

Suppose we have two point sources of monochromatic light, for instance a discharge tube with a cover that has two pinholes close together. Different atoms are

FIG. 27–2   (a) Radio waves from an antenna are coherent. (b) Light waves from a gas discharge tube are incoherent.

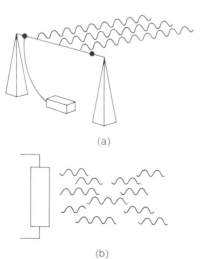

(a)

(b)

behind each pinhole, so they are independent sources. Therefore we have at most 10^{-8} s to observe the interference of waves from the two sources. If our detecting instruments are fast enough, as some modern electronic devices are, interference can be demonstrated and the sources can be considered coherent. If we are limited to the eye and to photographic film, which average arriving light signals over times far greater than 10^{-8} s, no interference can be observed in the light from the two sources, and they must then be considered incoherent. Like beauty, coherence lies in the eye of the beholder.

Does a brief lifetime of an excited atom mean that interference patterns can never be literally seen but can only be recorded by instruments? Not at all. There are three ways to construct separate sources of light coherent for long enough periods of time to produce visible interference patterns. These are:

1. Illuminate two (or more) slits with light from one slit behind them. Then the light waves from the secondary slits are automatically coordinated.   **Coherent light sources**
2. Obtain coherent virtual sources from a single source by reflection or refraction. This is how interference is produced by thin films of oil.
3. Coordinate the radiating atoms in each separate source so that they always have the same phase even though different atoms are radiating at successive instants. This is done in the *laser.*

We shall examine the first two of these methods here; the laser is discussed in Chapter 28.

27–2 DOUBLE SLIT

The interference of light waves was demonstrated in 1801 by Thomas Young, who used an arrangement similar to that shown in Fig. 27–3. A source of monochromatic light (that is, light consisting of only a single wavelength) is placed behind a narrow slit S in an opaque screen, and another screen with two similar slits A and B is placed on the other side. Light from S passes through both A and B and then to the viewing screen.   **If light were not a wave, the screen in Young's experiment would be totally dark**

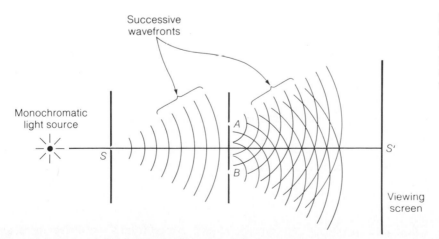

FIG. 27–3 Young's double-slit experiment. In accord with Huygens' principle, each slit acts as a source of secondary wavelets.

FIG. 27–4 The appearance of the screen in Young's experiment.

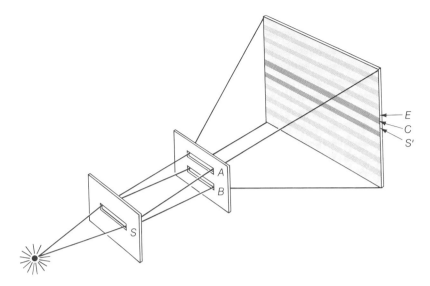

If light were not a wave phenomenon, we would expect to find the viewing screen completely dark, since no light ray can reach it from the source along a straight path. What actually happens is that each slit acts as a source of secondary wavelets—we recall Huygens' principle from Section 25–3—so that the entire screen is illuminated. Even the point S', separated from S by the opaque barrier between the slits A and B, turns out to be bright rather than dark (Fig. 27–4).

An interference pattern consists of bright and dark lines

Owing to interference the screen is not evenly illuminated but shows a pattern of alternate bright and dark lines. Light waves from slits A and B are exactly in phase, since A and B are the same distance from S. The centerline S' of the screen is equally distant from A and B, so light waves from these slits interfere constructively there to produce a bright line.

Let us next see what happens at the position C on the screen located to one side of S'. The distance BC is longer than the distance AC by the amount BD, which is equal to exactly half a wavelength of the light being used. That is,

Condition for the first dark line

$$BD = \tfrac{1}{2}\lambda$$

When a crest from A reaches C, this difference in path length means that a trough from B arrives there at the same time, since $\tfrac{1}{2}\lambda$ separates a crest and a trough in the same wave. The two cancel each other out, the light intensity at C is zero, and a dark line results on the screen there. At S' the equality of path length gives rise to constructive interference; at C the difference of $\tfrac{1}{2}\lambda$ in path length gives rise to destructive interference (Fig. 27–5).

If we go past C on the screen we will come to a point E such that the distance BE is greater by exactly one wavelength than the distance AE. That is, the difference BF between BE and AE is

Condition for the next bright line

$$BF = \lambda$$

Consequently, when a crest from A reaches E, a crest from B also arrives there, although

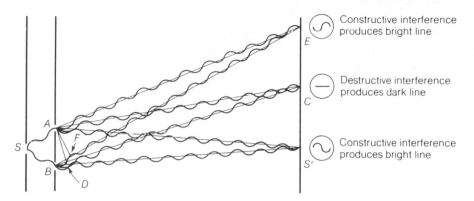

Constructive interference produces bright line

Destructive interference produces dark line

Constructive interference produces bright line

FIG. 27–5 Origin of the double-slit interference pattern.

the latter crest left B earlier than that from A owing to the longer path it had to cover. Because $BF = \lambda$, waves arriving at E from both slits are always in the same part of their cycles, and they constructively interfere to produce a bright line at E.

By continuing the same analysis, we find that the alternate bright and dark lines actually observed on the screen correspond respectively to locations where constructive and destructive interference occurs. Waves reaching the screen from A and B along paths that are equal or differ by a whole number of wavelengths (λ, 2λ, 3λ, and so on) reinforce, while those whose paths differ by an odd number of half wavelengths ($\frac{1}{2}\lambda$, $\frac{3}{2}\lambda$, $\frac{5}{2}\lambda$, and so on) cancel. At intermediate locations on the screen the interference is only partial, so that the line intensity on the screen varies gradually between the bright and dark lines.

Origin of pattern of bright and dark lines

27–3   DIFFRACTION GRATING

There are two difficulties in using a double slit for measuring wavelengths. First, the "bright" lines on the screen are actually extremely faint and an intense light source is therefore required; second, the lines are relatively broad and it is hard to locate their centers accurately. A *diffraction grating* that consists, in essence, of a large number of parallel slits overcomes both of these difficulties. Gratings are made by ruling grooves on a glass or metal plate with a diamond; the clear bands between the grooves are the "slits" (Fig. 27–6). Replica gratings, made by allowing a transparent liquid plastic to harden in contact with an original grating, are ordinarily used in practice. Replica gratings are often given a thin coating of silver or aluminum and produce their characteristic diffraction patterns by the interference of reflected rather than transmitted light. A phonograph record held at a glancing angle acts as a reflecting grating by virtue of its closely spaced grooves. Opals are natural diffraction gratings whose constituent particles (tiny spheres of silicon dioxide) are spaced apart about 250 nm in regular arrays.

Gratings are ruled with from 200 to 1000 lines/mm, and a lens is used to focus the light from the slits between them on a screen. The effect of phase differences among the rays from the various slits is accentuated by their great number, and as a result the intensity of light on the screen falls rapidly on either side of the center of each bright

A diffraction grating uses interference to disperse light

A grating produces sharper and brighter interference maxima than a double slit

line. The bright lines are therefore sharp, and, because there are so many slits, they are also bright in a literal sense. Very accurate wavelength determinations can be made with the help of a grating, and wavelengths that are close together can be resolved. The analysis of a light beam in terms of the particular wavelengths it contains is today almost invariably carried out by using a grating.

Figure 27–7 shows a diffraction grating that forms a bright line on a screen at a deviation angle of θ from the original beam of light. The condition for a bright line is that $s = n\lambda$, where $n = 1, 2, 3,$ and so on. Since $s = d \sin \theta$, where d is the spacing of the slits, bright lines occur at those angles for which

$$\sin \theta = n\frac{\lambda}{d} \qquad n = 1, 2, 3, \cdots \qquad \textit{Bright lines} \quad (27\text{--}1)$$

Spectral orders

When light with different wavelengths is directed on a grating, a series of spectra is formed on each side of the original beam corresponding to $n = 1$, $n = 2$, and so on. The *first-order spectrum* contains bright lines for which $n = 1$, the *second-order spectrum* contains bright lines for which $n = 2$, and so on (Fig. 27–8). In some gratings the higher-order spectra overlap, so that, for example, the blue end of the third-order spectrum may be deviated by less than the red end of the second-order spectrum. According to Eq. (27–1), the angle θ increases with wavelength, so that

FIG. 27–6 (a) A diffraction grating. Each space between the ruled grooves acts as a slit. (b) The greater the number of slits, the sharper the lines a single wavelength produces.

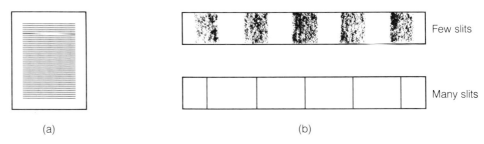

(a)   (b)

FIG. 27–7 The plane diffraction grating.

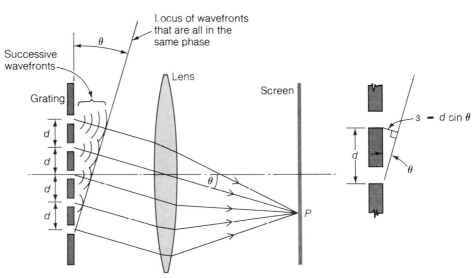

blue light is deviated least and red light most, which is the reverse of what occurs when a prism is used to form a spectrum.

Example   Visible light includes wavelengths from approximately 4×10^{-7} m (violet light) to 7×10^{-7} m (red light). Find the angular width of the first-order spectrum produced by a grating ruled with 800 lines/mm.   **Width of grating spectrum**

Solution   The slit spacing d corresponding to 800 lines/mm is

$$d = \frac{10^{-3} \text{ m/mm}}{800 \text{ lines/mm}} = 1.25 \times 10^{-6} \text{ m}$$

Since $n = 1$ for a first-order spectrum, the angular deviations of violet and red light respectively are given by

$$\sin \theta_b = \frac{\lambda_b}{d} = \frac{4 \times 10^{-7} \text{ m}}{1.25 \times 10^{-6} \text{ m}} = 0.32 \qquad \theta_b = 19°$$

and

$$\sin \theta_r = \frac{\lambda_r}{d} = \frac{7 \times 10^{-7} \text{ m}}{1.25 \times 10^{-6} \text{ m}} = 0.56 \qquad \theta_r = 34°$$

The total width of the spectrum is therefore $34° - 19° = 15°$ (Fig. 27–9).   ■

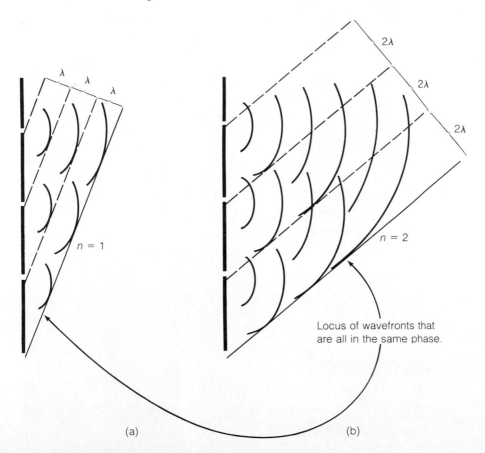

(a)

(b)

Locus of wavefronts that are all in the same phase.

FIG. 27–8   (a) In a first-order spectrum, the paths of diffracted rays from successive slits differ by λ in length. (b) In a second-order spectrum, the path differences are 2λ.

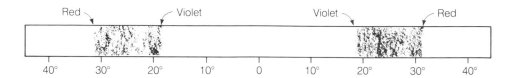

FIG. 27–9 First-order spectra of white light produced by a grating ruled with 800 lines/mm.

27–4   THIN FILMS

Thin films produce interference patterns

We have all seen the marvelous rainbow colors that appear in soap bubbles and thin oil films. Some of us may also have observed the patterns of light and dark bands that occur when two glass plates are almost (but not quite) in perfect contact. Both phenomena owe their origins to a combination of reflection and interference.

Let us consider a beam of monochromatic light that strikes a thin film of soapy water. Figure 27–10 shows a ray picture of what happens. We notice that some reflection takes place at both the air-soap and soap-air interfaces. This is a general result: Waves are always partially reflected when they go from one medium to another in which their speed is different. (See the discussion in Section 13–2 of the reflection of pulses in a string under similar circumstances.)

A light ray actually consists of a succession of wavefronts. Figure 27–11 is the same diagram with the wavefronts drawn in. In (a) the two reflected wave trains are out of phase and they interfere destructively to partially or completely cancel out. Most or all of the light reaching this part of the soap bubble therefore passes right through.

Another part of the soap film may have a different thickness. When the film is a little thinner than in (a), the waves in the two reflected trains are exactly in phase, and they interfere constructively to reinforce one another, as in Fig. 27–11(b). Light reaching this part of the soap bubble is strongly reflected. Shining monochromatic light on a soap bubble therefore yields a pattern of light and dark that results from the varying thickness of the bubble.

Why thin films appear colored in white light

When white light is directed at a soap bubble, light waves of each wavelength present pass through the soap film without reflection at those places where the film is exactly the right thickness for the two reflected rays to destructively interfere. Light waves of the other wavelengths are reflected to at least some extent, and give rise to

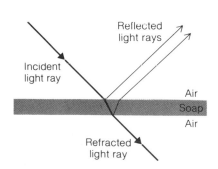

FIG. 27–10 Reflection occurs at both surfaces of a soap film. Photo shows interference of light in soap bubbles. (Fundamental Photographs, New York.)

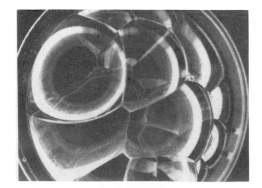

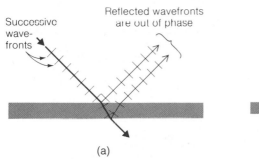

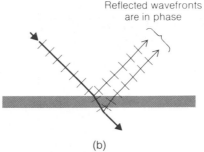

FIG. 27–11
(a) Destructive and
(b) constructive
interference in a thin film.

the vivid colors seen. The varying thickness of the bubble means that the color of the light reflected from the bubble changes from place to place. Exactly the same effect is responsible for the coloration of thin oil films. Generally speaking, soap or oil films whose thickness is comparable with the wavelengths in visible light give rise to the most striking color effects.

A thin film of air between two sheets of glass or transparent plastic also yields a pattern of colored bands when illuminated with white light. A notable example is *Newton's rings,* which occur when a slightly curved lens is placed on a flat glass plate (the curvature is exaggerated in Fig. 27–12). Again reflection takes place at both the top and bottom of the film, and again the result is constructive or destructive interference, depending upon the film thickness. Because the thickness of the air film increases with distance from the central point of contact, the pattern of light and dark bands consists of concentric circles.

Newton's rings

The dark spot at the center of a set of Newton's rings is not what we might expect to find (Fig. 27–13). At the center, where the air film between the pieces of glass is extremely small, the path difference between the waves reflected from the upper and lower surfaces of the film is negligible. Hence there ought to be constructive interference and reinforcement of the light to yield a bright spot. What this analysis overlooks is the fact that a wave reflected at the surface of a new medium in which its speed is less (in optical terms, a medium of higher index of refraction) is shifted by half a wavelength. That is, a positive displacement of the wave variable is reflected as a negative one, and vice versa. The same effect was noted in the discussion of pulses in

Why the center of Newton's rings is a dark spot

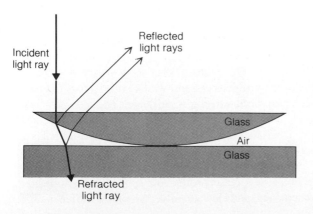

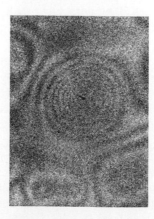

FIG. 27–12 Newton's rings. (Photo by Richard Megna/Fundamental Photographs, New York.)

FIG. 27–13 Origin of dark spot at center of Newton's rings.

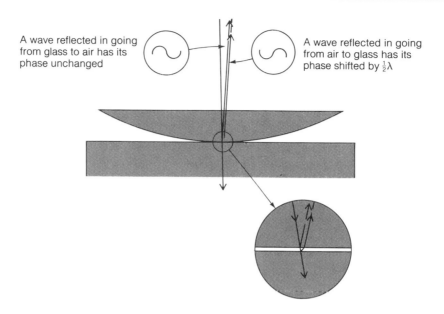

A wave reflected in going from glass to air has its phase unchanged

A wave reflected in going from air to glass has its phase shifted by $\frac{1}{2}\lambda$

a stretched string and is shown in Fig. 13–6. Thus a $\frac{1}{2}\lambda$ shift occurs when light waves are reflected in going from air to glass, but not in going from glass to air. In consequence the two wave trains reflected at the center of a Newton's ring pattern exactly cancel each other out to yield the dark spot actually observed.

Lens reflection is a problem in optical instruments

About 4% of the light striking a glass-air interface is reflected. This is not a lot, but there may be many glass-air interfaces in an optical instrument and the total amount of light lost through reflection may be considerable. There are ten such interfaces in each of the optical systems in a pair of binoculars, for instance, so only about two-thirds of the incoming light actually gets through to the observer's eyes. Even in a camera, where there are fewer glass-air interfaces, reflections are a nuisance because they may lead to secondary images that blur the picture.

Coated lenses reduce reflection

To reduce reflections at a glass-air interface, the glass can be coated with a very thin layer of a transparent substance (usually magnesium fluoride) whose index of refraction is intermediate between those of glass and of air. If the layer is exactly $\frac{1}{4}\lambda$ thick, light reflected at its bottom will have traveled $\frac{1}{2}\lambda$ farther when it rejoins light reflected at the top of the layer, and the two will cancel out exactly, as in Fig. 27–11(a).

Why coated lenses appear colored

But the cancellation described above is exact only for a particular wavelength λ, whereas white light contains a range of wavelengths. What is therefore done is to choose a wavelength in the middle of the visible spectrum, which corresponds to green light, so that at least partial cancellation occurs over a wide range. The red and violet ends of the spectrum are accordingly least affected and the light reflected from a coated lens is a mixture of these colors, a purplish hue. The average reflectivity of a glass surface coated in this way is only about 1%. Multiple coatings are sometimes used to reduce reflection even further; a triple coating brings the average reflectivity below 0.5%. Despite their lack of perfection at suppressing reflections, coated lenses transmit appreciably more light than uncoated ones, and they are universally used in fine optical instruments.

27-5 DIFFRACTION

Waves are able to bend around the edge of an obstacle in their path, a property called *diffraction*. We all have heard sound that originated around the corner of a building from where we were standing, for example. The sound waves cannot have traveled in a straight line from their source to our ears, and refraction cannot account for their behavior. Water waves, too, diffract, as the simple experiment illustrated in Fig. 27–14 shows. The waves on the far side of the gap spread out into the geometrical "shadow" of the gap's edges, though with reduced amplitude. The diffracted waves spread out as though they originated at the gap, in accord with Huygens' principle.

Waves diffract around the edge of an obstacle

If we look closely at the edge of the shadow cast by an obstruction in the path of the light spreading out from a pinhole or other point source, we will see that it is not sharp but smeared out (Fig. 27–15). The fuzzy edges of shadows are not easy to observe because the wavelengths in visible light are so short, less than 10^{-6} m, and the extent of diffraction into the shadow zone is correspondingly small. (In contrast, a typical audible sound wave might have a wavelength of 1 m and a typical wave in a pan of water might have a wavelength of 10 cm, and it is easy to observe diffraction effects with such waves.) In fact, because he was not able to see any diffraction with his relatively crude apparatus, Newton felt sure that light could not consist of waves.

Shadows do not have sharp edges

A broad light source such as a light bulb or the sun does not produce sharp shadows for another reason. In this case light from different parts of the source passes the edge of the obstacle at different angles, which is not true of light from a point source.

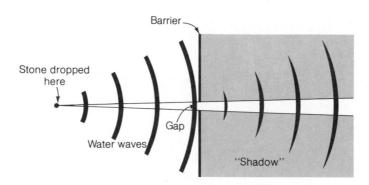

FIG. 27–14 Diffraction in water waves. The waves on the far side of the gap spread out as though they had originated at the gap.

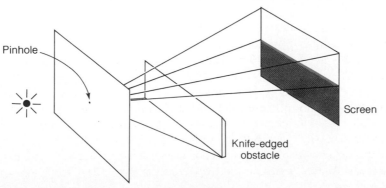

FIG. 27–15 Even under ideal conditions, the edge of a shadow is never completely sharp.

FIG. 27–16 The shadow of a razor blade.

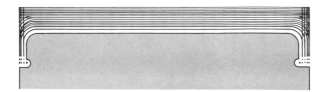

Origin of diffraction fringes

We can refine our observation of diffraction further by using a monochromatic light source and a sheet of photographic film instead of a screen. When we enlarge the developed image on the film, we find a pattern of light and dark fringes at the edge of the shadow. Figure 27–16 shows what the shadow caused by a razor blade looks like. Patterns like this are the result of interference between secondary wavelets from different parts of the same wavefront, not from different sources as in Young's double-slit experiment. The wavefronts in a beam of unobstructed light produce secondary wavelets that interfere in such a way as to produce new wavefronts exactly like the old ones. By obstructing part of the wavefronts, points in the shadow region are not reached by secondary wavelets from the entire initial wavefronts but only from part of them, and the result is an interference pattern.

Diffraction limits the useful magnification of an optical system

Diffraction sets a limit to the useful magnification of an optical system such as that of a telescope or microscope. Diffraction occurs whenever wavefronts of light are obstructed, and the light that enters a lens (or mirror) is affected by the limited opening that admits only part of each incident wavefront. No matter how perfect a lens is, the image of a point source of light it produces is always a tiny disk of light with bright and dark fringes around it (Fig. 27–17). Only if the lens has an infinite diameter can a point source give rise to a point image. The smaller the lens, the larger the image of a point source. The angular width of the radius of this disk of light is about

$$\theta_0 = 1.22 \frac{\lambda}{D} \qquad \qquad \textit{Diffraction disk} \quad (27\text{–}2)$$

in radians, where λ is the wavelength of the light used and D is the lens diameter.

Criterion for resolving nearby objects

Two objects separated by less than θ_0 cannot be *resolved,* that is, distinguished apart, no matter how high the magnification employed, because their images will overlap (Fig. 27–18). Hence there is no advantage in using a higher magnification than will just reveal features that subtend the angle θ_0 at the position of the lens. Although Eq. (27–2) was derived for the image of a point source of light an infinite distance

FIG. 27–17 The image a lens produces of a point source of light is always a tiny disk with bright and dark fringes around it. The smaller the lens, the larger the image.

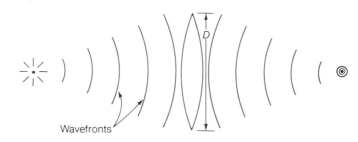

Wavefronts

from a lens, it is a reasonable approximation of the resolving power of a telescope or microscope when D is taken as the diameter of the objective lens.

If two objects d apart are the distance L from an observer, the angle between them, in radians, is

$$\theta = \frac{d}{L}$$

Hence Eq. (27–2) can be rewritten

$$d_0 = 1.22 \frac{\lambda L}{D}$$
Resolving power   (27–3)

In this formula

d_0 = minimum separation of objects that can be resolved

λ = wavelength of the light used

L = object distance

D = diameter of objective lens or mirror

Example   The pupils of a person's eyes under ordinary conditions of illumination are about 3 mm in diameter, and the distance of most distinct vision is 25 cm for most people. What is the resolving power of the eye at this distance under the assumption that it is limited only by diffraction?

Solution   Using $\lambda = 550$ nm, which is in the middle of the visible spectrum,

$$d_0 = 1.22 \frac{\lambda L}{D} = \frac{(1.22)(5.5 \times 10^{-7}\,\text{m})(0.25\,\text{m})}{0.003\,\text{m}} = 5.6 \times 10^{-5}\,\text{m} = 0.056\,\text{mm}$$

The photoreceptors in the retina are not quite close enough together to permit this degree of resolution, and 0.1 mm is a more realistic figure under ideal conditions. In terms of angular resolving power, $\theta_0 \approx 5 \times 10^{-4}$ radians for the human eye. ∎

Large lens

Small lens

Very small lens

FIG. 27–18   A large lens or mirror is better able to resolve nearby objects than a small one.

Resolving power of the eye

27–6   POLARIZATION

A *polarized* beam of transverse waves is one whose vibrations occur in only a single direction perpendicular to the direction in which the beam travels, so that the entire wave motion is confined to a plane called the *plane of polarization* (Fig. 27–19). When many different directions of polarization are present in a beam of transverse waves, vibrations occur equally often in all directions perpendicular to the direction of motion, and the beam is then said to be *unpolarized*. Since the vibrations that constitute longitudinal waves can take place in only one direction, namely that in which the waves travel, longitudinal waves cannot be polarized.

Light waves are transverse, and it is possible to produce and detect polarized light. To clarify the ideas involved, let us first consider the behavior of transverse waves in a stretched string. If the string passes through a tiny hole in a fence, as in Fig. 27–20(a), waves traveling down the string are stopped since the string cannot vibrate

Polarized transverse waves lie in a single plane of polarization

FIG. 27–19 An
unpolarized and a
polarized beam of
transverse waves.

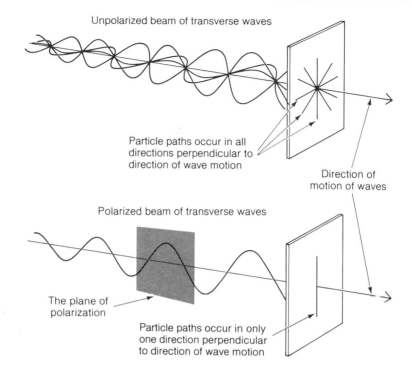

Unpolarized beam of transverse waves

Particle paths occur in all
directions perpendicular to
direction of wave motion

Direction of
motion of waves

Polarized beam of transverse waves

The plane of
polarization

Particle paths occur in only
one direction perpendicular
to direction of wave motion

there. When the hole is replaced by a vertical slot, waves whose vibrations are vertical can get through the fence, but waves with vibrations in other directions cannot; see Fig. 27–20(b) and (c). In a situation in which several waves vibrating in different directions move down the string, the slot stops all but vertical vibrations (d): An initially unpolarized series of waves has become polarized.

The above approach can be used to determine whether a particular kind of wave can be polarized or not. In the case of a stretched string, what we do is erect another fence a short distance from the first, as in Fig. 27–20(e). If the slot in the new fence is also vertical, those waves that can get through the first fence can also get through the second. If the slot in the new fence is horizontal, however, it will stop all waves that reach it from the first fence (f).

**Longitudinal waves
cannot be polarized**

Should longitudinal waves (say in a spring) go through the fence, it is possible that their amplitudes might decrease in passing through the slots, but the relative alignments of the slots would not matter; see Fig. 27–20(g) and (h). On the other hand, the alignment of the slots is the critical factor in the case of transverse waves.

The preceding chain of reasoning made it possible for the polarization of light waves to be demonstrated in the last century. A number of substances—for instance, quartz, calcite, and tourmaline—have different indexes of refraction for light with different planes of polarization relative to their crystal structures, and prisms can be made from them that transmit light in only a single plane of polarization. When a beam of unpolarized light is incident upon such a prism, only those of its waves whose planes of polarization are parallel to a particular plane in the prism emerge from the other side. The remainder of the waves are absorbed or deflected.

Polaroid is an artificially made polarizing material in wide use that only transmits

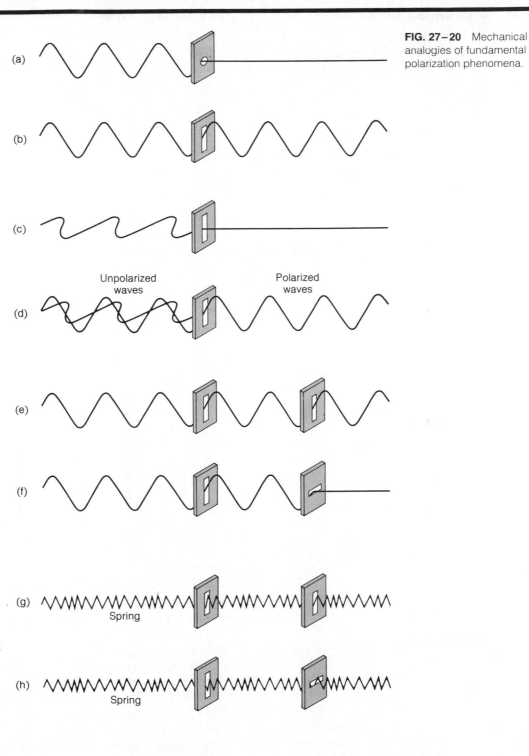

FIG. 27–20 Mechanical analogies of fundamental polarization phenomena.

FIG. 27–21 Experiment showing the transverse nature of light waves.

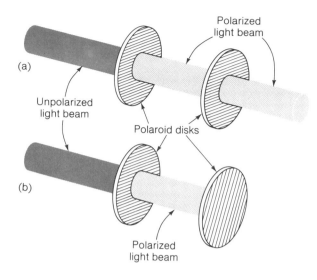

FIG. 27–22 The electric field of a light wave defines its plane of polarization.

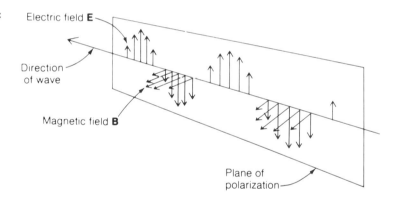

How polarization in light can be demonstrated

Plane of polarization of a light wave

light with a single plane of polarization. To exhibit the transverse nature of light waves, we first place two Polaroid disks in line so that their axes of polarization are parallel (Fig. 27–21(a)), and note that all light passing through one disk also passes through the other. Then we turn one disk until its axis of polarization is perpendicular to that of the other (Fig. 27–21(b)), and note that all light passing through one disk is now *stopped* by the other.

Just what is it whose vibrations are aligned in a beam of polarized light? As discussed in Chapter 25, light waves consist of oscillating electric and magnetic fields perpendicular to each other. Because it is the electric fields of light waves whose interactions with matter produce nearly all common optical effects, the plane of polarization of a light wave is considered to be that in which both the direction of its electric field and the direction of the wave lie (Fig. 27–22). Even though nothing material moves during the passage of a light wave, it is possible to establish its transverse nature and identify its plane of polarization.

27–7   SCATTERING

When light waves encounter an obstacle of some sort, they are diffracted around its edges. If the size of the obstacle is small relative to the wavelength of the light, the diffracted wavefronts are more or less spherical, and they spread out as though they originated in the obstacle. The incoming light is said to be *scattered* by the obstacle.

In general, the intensity of light of wavelength λ scattered by an object small compared with λ is proportional to λ^{-4}. The shorter the wavelength, the greater the proportion of the incoming light that is scattered. This is the reason that the sky is blue. When we look at the sky, what we see is light from the sun that has been scattered by molecules in the upper atmosphere. Blue light, which consists of the shortest wavelengths, is scattered about ten times more readily than red light, so the scattered light is chiefly blue in color as in Fig. 27–23(a). At sunrise or sunset, when sunlight must make a very long passage through the atmosphere to reach an observer, much of its short-wavelength content is scattered out along the way, and the sun accordingly appears red in color as in Fig. 27–23(b). The water droplets and ice crystals in clouds are larger than λ and the scattering they produce is independent of λ; hence clouds do not appear colored. Above the atmosphere the sky appears black, and the moon, stars, and planets are visible to astronauts in the daytime.

Scattering of light of the same kind as that producing the blue sky can be shown by directing a narrow beam of white light at a glass of unhomogenized milk diluted

Why the sky is blue

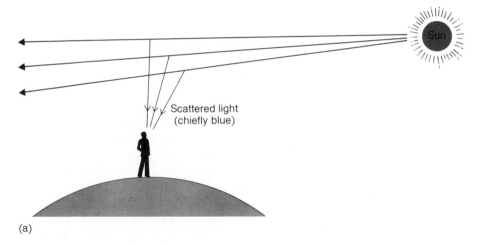

Scattered light
(chiefly blue)

(a)

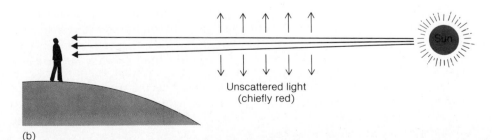

Unscattered light
(chiefly red)

(b)

FIG. 27–23 (a) Blue light is scattered the most by the earth's atmosphere, red light the least. Hence the sky appears blue. (b) Why the sun appears red at sunrise and sunset.

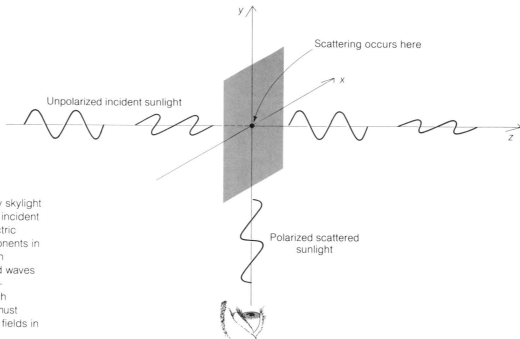

FIG. 27–24 Why skylight is polarized. Only incident waves whose electric fields have components in the x-direction can produce scattered waves that move in the y-direction, and such scattered waves must have their electric fields in the x-direction.

with water and shaken up. The scattering centers here are fat globules. The milk appears bluish from the side and the beam that emerges appears reddish, as in the case of sunlight.

Skylight is polarized

Skylight is not only blue but is partly polarized as well. To verify this statement, all one has to do is hold up a piece of Polaroid against the sky and rotate it. Sunglasses are often made with Polaroid lenses so oriented as to discriminate against the polarized sunlight, which reduces glare while affecting other light to a lesser extent. The eye itself responds equally to all states of polarization.

Origin of skylight polarization

Figure 27–24 shows why scattered light is polarized. A beam of unpolarized light heading in the $+z$-direction strikes some air molecules and is scattered. In this process electrons in the molecules are set in vibration by the electric fields of the light waves, and the vibrating electrons then reradiate. Because the electric field of an electromagnetic wave is perpendicular to its direction of motion (Fig. 27–22), the initial beam contains electric fields that lie in the xy-plane only. A scattered wave that proceeds downward, which is the $-y$-direction in the figure, can have its electric field in the x-direction only, so it is polarized. Only two planes of polarization for the incident light are shown for clarity; in the case of intermediate planes of polarization, the components of **E** in the x-direction can also lead to scattered waves traveling downward that are similarly polarized. Because skylight arrives at our eyes from a variety of directions, the polarization is not complete, but enough occurs to be easily demonstrated.

IMPORTANT TERMS

Two sources of waves are **coherent** if there is a fixed phase relationship between the waves they emit during the time the waves are being observed. Interference can be observed only in waves from coherent sources.

The ability of waves to bend around the edges of obstacles in their paths is called **diffraction**. A **diffraction grating** is a series of parallel slits that produces a spectrum through the interference of light that is diffracted by them.

The **resolving power** of an optical system refers to its ability to produce separate images of nearby objects; resolving power is limited by diffraction, and the larger the objective lens of an optical system, the greater its resolving power.

A **polarized** beam of transverse waves is one whose vibrations occur in only a single direction perpendicular to the direction in which the beam travels, so that the entire wave motion is confined to a plane called the **plane of polarization**. An unpolarized beam of transverse waves is one whose vibrations occur equally often in all directions perpendicular to the direction of motion.

IMPORTANT FORMULAS

Diffraction grating: $\sin \theta = n \dfrac{\lambda}{d}$ (bright lines)

Resolving power: $d_0 = 1.22 \dfrac{\lambda L}{D}$

MULTIPLE CHOICE

1. In order to produce an interference pattern, the waves used must
   a. all have the same wavelength.
   b. be in a narrow beam.
   c. be coherent.
   d. be electromagnetic.

2. Coherent electromagnetic waves are not emitted by
   a. two antennas connected to the same radio transmitter.
   b. two pinholes in an opaque shield over a sodium-vapor lamp.
   c. a pinhole in an opaque shield over a sodium-vapor lamp and its reflection in a mirror.
   d. two lasers.

3. In a double-slit experiment, the maximum intensity of the first bright line on either side of the central one occurs on the screen at locations where the arriving waves differ in path length by
   a. $\lambda/4$.   b. $\lambda/2$.
   c. λ.   d. 2λ.

4. Wavelength determinations cannot be made with the help of
   a. a glass prism.
   b. a pair of narrow parallel slits.
   c. a diffraction grating.
   d. a pair of Polaroid disks.

5. An interference pattern is produced whenever
   a. reflection occurs.   b. refraction occurs.
   c. diffraction occurs.   d. polarization occurs.

6. A characteristic property of the spectra produced by a diffraction grating is the
   a. sharpness of the bright lines.
   b. diffuseness of the bright lines.
   c. absence of bright lines.
   d. absence of dark lines.

7. The greater the number of lines that are ruled on a grating of given width,
   a. the shorter the wavelengths that can be diffracted.
   b. the longer the wavelengths that can be diffracted.
   c. the narrower the spectrum that is produced.
   d. the broader the spectrum that is produced.

8. When a spectrum is formed by a double slit or a grating, the light deviated most
   a. is red.   b. is green.
   c. is blue.   d. depends on the slit spacing.

9. Thin films of oil and soapy water owe their brilliant colors to a combination of reflection and
   a. refraction.   b. interference.
   c. diffraction.   d. polarization.

10. The minimum separation of two features of a distant object that can be discerned by a telescope does *not* depend on
   a. the diameter of the objective lens.
   b. the focal length of the objective lens.
   c. the wavelength of the light being used.
   d. the distance to the object.

11. Lenses are coated in order to reduce
   a. reflection.   b. refraction.
   c. diffraction.   d. polarization.

12. Diffraction refers to
   a. the splitting of a beam of white light into its component colors.
   b. the interference of light that produces bright colors in thin oil films.
   c. the bending of waves around the edge of an obstacle in their path.
   d. the increase in frequency due to motion of a wave source toward an observer.

13. The wavelength of light plays no role in
    a. interference.
    b. diffraction.
    c. resolving power.
    d. polarization.

14. An unpolarized beam of transverse waves is one whose vibrations
    a. are confined to a single plane.
    b. occur in all directions.
    c. occur in all directions perpendicular to their direction of motion.
    d. have not passed through a Polaroid disk.

15. Longitudinal waves do not exhibit
    a. refraction.
    b. reflection.
    c. diffraction.
    d. polarization.

16. It is impossible to polarize
    a. white light.
    b. radio waves.
    c. X rays.
    d. sound waves.

17. Light scattered perpendicularly to an incident beam is partly
    a. diffracted.
    b. polarized.
    c. monochromatic.
    d. coherent.

18. The sky is blue because
    a. air molecules are blue.
    b. the lens of the eye is blue.
    c. the scattering of light is more efficient the shorter its wavelength.
    d. the scattering of light is more efficient the longer its wavelength.

19. The wavelength of light that is deviated in the first order by 15° by a 5000-line/cm grating is
    a. 52 nm.
    b. 259 nm.
    c. 518 nm.
    d. 773 nm.

20. A pair of binoculars designated "7 × 50" has a magnification of 7 and an objective lens diameter of 50 mm. The smallest detail that in principle can be perceived with such an instrument when viewing an object 1 km away in light of wavelength $\times\ 5^{-7}$ has a linear dimension of approximately
    a. 1 mm.
    b. 1 cm.
    c. 10 cm.
    d. 1 m.

EXERCISES

27–1   Interference of Light

27–2   Double Slit

1. Can light from incoherent sources interfere? If so, then why is a distinction made between coherent and incoherent sources?

2. What becomes of the energy of the light waves whose destructive interference leads to dark lines in an interference pattern?

3. The interference of water waves is very conspicuous on the surfaces of all bodies of water, from puddles to oceans. Why is the interference of light waves usually so much harder to detect?

27–3   Diffraction Grating

4. What advantages has a diffraction grating over a double slit for determining wavelengths of light?

5. What is the difference between the first-order and the second-order spectra produced by a grating? Which is wider? Does a prism produce spectra of different orders?

6. What governs the angular width of the first-order spectrum of white light produced by a grating?

7. Light of wavelength 750 nm is directed on a grating ruled with 4000 lines/cm. What is the angular deviation of this light in (a) the first order, and (b) the third order?

8. A 5500-line/cm diffraction grating produces an image deviated by 27° in the second order. Find the wavelength of the light.

9. How many diffracted images are formed on either side of the central image when radiation of wavelength 600 nm falls on a 4000-line/cm grating?

10. Light containing wavelengths of 500 and 550 nm is directed at a 2000-line/cm grating. How far apart are the lines formed by these wavelengths on a screen 4 m away in the second order?

11. White light that contains wavelengths from 400 to 700 nm is directed at a 3000-line/cm grating. How wide is the first-order spectrum on a screen 2 m away?

12. White light that contains wavelengths from 400 to 700 nm is directed at a 5000-line/cm grating. Do the first- and second-order spectra overlap? The second- and third-order spectra? If there is an overlap, will the use of a grating with a different number of lines/cm change the situation?

27–4   Thin Films

13. As a soap bubble is blown up, its wall becomes thinner and thinner. Just before the bubble breaks, the thinnest part of its wall turns black. Why?

14. The index of refraction of a certain soap bubble illuminated by white light is $n = 1.35$. If the bubble appears orange ($\lambda = 630$ nm) at a point nearest the viewer, what is the minimum thickness of the bubble at that point? Be sure to take into account any phase shifts on reflection.

15. The index of refraction of magnesium fluoride is 1.38. How thick should an antireflection coating of this material be for maximum cancellation at 550 nm, which is the middle of the visible spectrum?

27–5   Diffraction

16. The waves used to carry television signals cannot reach receivers beyond the visual horizon of their transmitting antennas, whereas ordinary radio waves readily travel beyond the visual horizon of their transmitting antennas. Can you think of a reason for this contradictory behavior?

17. Radio waves diffract pronouncedly around buildings, whereas light waves, which are also electromagnetic waves, do not. Why?

18. What do diffraction and interference have in common? How do they differ? In Young's double-slit experiment, which effects are due to diffraction and which to interference?

19. Explain the peculiar appearance of a distant light source when seen through a piece of finely woven cloth.

20. A camera can be made by using a pinhole instead of a lens. What happens to the sharpness of the picture if the hole is too large? If it is too small?

21. Give two advantages a large-diameter telescope objective has over a small-diameter one.

22. A radar has a resolving power of 30 m at a range of 1 km. What is the minimum width of its antenna if its operating frequency is 9500 MHz?

23. The Hale telescope at Mount Palomar in California has a concave mirror 5 m in diameter. How many meters apart must two features of the moon's surface be in order to be resolved by this telescope? Take the distance from the earth to the moon as 386,000 km and use 500 nm for the wavelength of the light.

24. An astronaut circles the earth in a satellite at an altitude of 150 km. If the diameter of his pupils is 2 mm and the average wavelength of the light reaching his eyes is 550 nm, is it conceivable that he can distinguish sports stadiums on the earth's surface? Private houses? Cars?

25. The Jodrell Bank radiotelescope has a parabolic reflecting "dish" 76 m in diameter. (a) What is the angular diameter in degrees of a point source of radio waves of wavelength 21 cm as seen by this telescope? (b) How does the above figure compare with the angular width of the moon, whose diameter is 3476 km and whose average distance from the earth is 3.8×10^5 km?

26. According to a famous battle command, "Don't fire until you see the whites of their eyes." (a) If the diameter of

the white of an eye is 20 mm, the diameter of the pupil in bright sunlight is 2 mm, and the light has a wavelength of 500 nm, find this distance on the basis of the theoretical resolution formula. (b) The actual resolving power of the eye, which is limited by the structure of the retina, is about 5×10^{-4} rad. Find the maximum distance at which the white of an eye can be distinguished on the basis of this figure.

27. At night, the pupils of a person's eyes are 8 mm in diameter. (a) How many km away from a car facing a woman will she be able to distinguish its headlights from each other? (b) If her pupils were 4 mm in diameter (say at twilight), how far away from the car could she distinguish its headlights from each other? Assume the headlights are 1.5 m apart, that the average wavelength of their light is 600 nm, and that her eyes are capable of attaining their theoretical resolving power.

28. A certain laser produces a beam of monochromatic light whose wavelength is 550 nm and whose initial diameter is 1 mm. (a) In what distance will diffraction have caused the beam to double its diameter? (b) If the initial diameter of the beam were 1 cm, would the doubling distance be different? If so, what would it be?

29. The smaller the aperture of a camera lens, the greater the depth of field. However, a small aperture means reduced resolution. The criterion for an enlarged print to show sharp detail from a small negative is that the image of a point object on it be not more than about 0.01 mm across. Find the maximum f-number of a camera lens in order that this criterion be met for a distant object in 550-nm light.

27–6   Polarization

30. Which of the following can occur in (a) transverse waves and (b) longitudinal waves: refraction, dispersion, interference, diffraction, polarization?

31. Since light consists of transverse waves, why is not every light beam polarized?

32. What is the relationship between the plane of polarization of a transverse wave and its direction of propagation?

ANSWERS TO MULTIPLE CHOICE

1. c	**5.** c	**9.** b	**13.** d	**17.** b
2. b	**6.** a	**10.** b	**14.** c	**18.** c
3. c	**7.** d	**11.** a	**15.** d	**19.** c
4. d	**8.** a	**12.** c	**16.** d	**20.** b

28

ATOMIC PHYSICS

Particles and waves are separate concepts in everyday life. It is impossible to confuse a stone, for instance, with the waves it produces when it is thrown into a lake. But in the world of the atom, the situation is very different. Electromagnetic waves—which, as we know, show such typical wave behavior as interference—have important properties in common with particles. And electrons—whose particle nature is clearly shown in the operation of a television picture tube—nevertheless have important properties in common with waves. On a very small scale of size, then, a wave-particle duality replaces the distinction between waves and particles so obvious on a large scale. This duality turns out to be the key to understanding the structures of atoms and why they behave as they do.

28-1   PHOTOELECTRIC EFFECT

Photoelectric effect

When light (particularly ultraviolet light) falls on a metal surface, electrons are observed to be given off (Fig. 28-1). This phenomenon is known as the *photoelectric effect*. It is not, at first glance, anything to surprise us, for light waves carry energy, and some of the energy absorbed by the metal may somehow concentrate on individual electrons

CHAPTER OBJECTIVES

Completing this chapter should enable you to:

1. Describe the origin of the photoelectric effect.

2. Calculate the energy of a photon of light of given frequency.

3. Relate the voltage across an X-ray tube to the frequency of the X rays produced.

4. Describe the wave nature of a moving particle in terms of waves of probability.

5. Find the wavelength of the matter waves of a moving particle.

6. Relate the mass of a moving object to its rest mass and its speed with respect to an observer.

7. Describe the relationship between the wave nature of atomic electrons and atomic structure.

8. Calculate the energy levels of a hydrogen atom.

9. Distinguish between the ground state and the excited states of an atom.

10. Determine the ionization energy of an atom.

11. Understand the significance of the probability cloud of an atomic electron.

12. Describe the origins of absorption and emission spectra.

13. Describe the operation of a laser and the properties of laser beams.

and reappear as kinetic energy. Upon closer inspection of the data, however, we find that the photoelectric effect can hardly be explained in so straightforward a manner.

The first peculiarity of the photoelectric effect is that, even when the metal surface is only faintly illuminated, the emitted electrons (which are called *photoelectrons*) leave the surface immediately. But according to the electromagnetic theory of light, the energy content of light waves is spread out across the width of the wavefronts of the light beam involved. Calculations show that a definite period of time—several months in the case of a beam of very low intensity—should go by before any individual electrons gather enough energy to leave the metal. Instead, the electrons are found to be emitted as soon as the light is turned on.

Photoelectrons are emitted immediately

Another unexpected discovery is that the energy of the photoelectrons does not depend upon the intensity of the light. A bright light yields more electrons than a dim one, but their average energy remains the same. This behavior contradicts the electromagnetic theory of light, which predicts that the energy of photoelectrons should depend upon the intensity of the light beam responsible for them.

The energies of the photoelectrons emitted from a given metal surface turn out, most surprisingly of all, to depend upon the *frequency* of the light employed. At frequencies below a certain critical one (which is characteristic of the particular metal),

Photoelectron energy increases with frequency of incident light

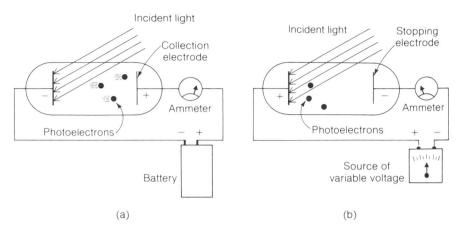

FIG. 28–1   (a) A method of detecting the photoelectric effect. The photoelectrons ejected from the irradiated metal plate are attracted to the positive collection electrode at the other end of the tube, and the current that results is measured with an ammeter. (b) A method of detecting the maximum energy of the photoelectrons. Note polarity opposite to that in (a). As the stopping electrode is made more negative, the slower photoelectrons are repelled before they can reach it. Finally a voltage will be reached at which no photoelectrons whatever are received at the stopping electrode, as indicated by the current dropping to zero, and this voltage corresponds to the maximum photoelectron energy.

All light-sensitive detectors (including the eye) are based upon the absorption by electrons in the target atoms of energy from photons of light. In a camera, the absorbed energy alters silver halide crystals in the film so that chemical treatment can convert them into grains of pure silver. (Photo: Nikon, Inc.)

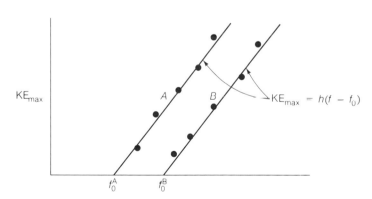

Frequency of incident light, f

FIG. 28–2   The variation of maximum photoelectron energy with the frequency of the incident light for two target metals. No photoelectrons are emitted for frequencies below f_0^A in the case of metal A and below f_0^B in the case of metal B. In both cases, however, the angle between the experimental line and either axis is the same. Hence we may write the equation of the lines as $KE_{max} = h(f - f_0)$, where h has the same value in all cases but where f_0, the minimum frequency required for photoelectric emission to occur, depends upon the nature of the target metal.

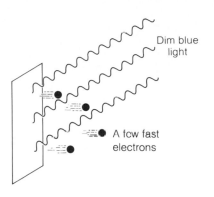 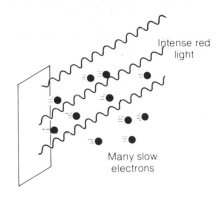

Dim blue light

A few fast electrons

Intense red light

Many slow electrons

FIG. 28-3 The higher the frequency of the light, the more KE the photoelectrons have. The brighter the light, the more photoelectrons are emitted. Blue light has a higher frequency than red light.

no electrons whatever are given off. Above this threshold frequency the photoelectrons have a range of energies from zero to a certain maximum value, and *this maximum energy increases with increasing frequency* (Fig. 28-2). High-frequency light yields high maximum photoelectron energies; low-frequency light yields low maximum photoelectron energies. Thus dim blue light produces electrons with more energy than those produced by intense red light, although the latter results in a greater number of them (Fig. 28-3).

28-2 QUANTUM THEORY OF LIGHT

The photoelectric effect is not the only optical phenomenon that is at odds with the idea that light consists of electromagnetic waves, or, indeed, with the idea that light consists of waves of any kind. All of these other phenomena, however, can be understood if light is regarded as consisting instead of little bursts of energy. Each burst is called a *quantum* (plural *quanta*) and its energy content E is related to the light frequency f by the equation

The energy of light waves travels in bursts called quanta

$$E = hf \qquad \qquad \text{Quantum energy} \quad (28-1)$$

The quantity h is a universal constant, called *Planck's constant,* whose value is

Planck's constant

$$h = 6.63 \times 10^{-34} \text{ J} \cdot \text{s} \qquad \qquad \textit{Planck's constant}$$

Example The eye can detect as little as 10^{-18} J of electromagnetic energy. How many photons of orange light whose wavelength is 6×10^{-7} m does this energy represent?

Solution The frequency of this light is

$$f = \frac{c}{\lambda} = \frac{3 \times 10^8 \text{ m/s}}{6 \times 10^{-7} \text{ m}} = 5 \times 10^{14} \text{ Hz}$$

Each photon has the energy hf. Hence

$$\text{Number of photons} = \frac{\text{detectable energy}}{\text{energy/photon}}$$

$$= \frac{10^{-18}\,\text{J}}{(6.63 \times 10^{-34}\,\text{J} \cdot \text{s})(5 \times 10^{14}\,\text{Hz})} = 3\ \text{photons} \qquad \blacksquare$$

Because the electron volt (eV) is a common unit of energy in atomic physics, it is useful to have h expressed in eV · s as well as in J · s:

$$h = 4.14 \times 10^{-15}\ \text{eV} \cdot \text{s} \qquad\qquad\qquad \textit{Planck's constant}$$

The h of the photoelectric effect formula $\text{KE}_{\text{max}} = h(f - f_0)$ given in Fig. 28–2 is Planck's constant. The significance of this formula becomes clear when it is rewritten

$$hf = \text{KE}_{\text{max}} + hf_0 \qquad\qquad\qquad \textit{Photoelectric effect} \quad (28\text{–}2)$$

What this equation states is that

Quantum energy = maximum electron energy + energy required to

eject an electron

The work function is the minimum energy needed to remove an electron from a surface

The reason for a threshold energy hf_0 (often called *work function*) is clear: It is the minimum energy required to dislodge an electron from the metal surface, as in Fig. 28–4. (There must be such a minimum energy, or electrons would leave metals all the time.) There are several reasons why not all photoelectrons should have the same energy even though a single frequency of light is used. For instance, not all the quantum energy hf may be transferred to a single electron, and an electron may lose some of its initial energy in collisions with other electrons within the metal before it actually emerges from the surface.

Example   The photoelectric work function for copper is 4.5 eV. Find the maximum energy of the photoelectrons when ultraviolet light of 1.5×10^{15} Hz falls on a copper surface.

Solution   The quantum energy of the incident photons in eV is

FIG. 28–4   If the energy hf_0 is required to remove an electron from a metal surface, the maximum electron kinetic energy will be $hf - hf_0$ when light of frequency f is directed at the surface.

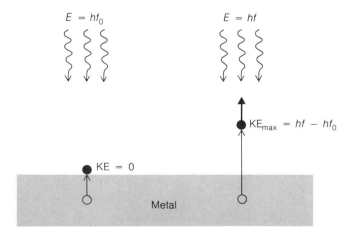

$$hf = (4.14 \times 10^{-15} \text{ eV} \cdot \text{s})(1.5 \times 10^{15} \text{ Hz}) = 6.2 \text{ eV}$$

Hence the maximum photoelectron energy is

$$\text{KE}_{\text{max}} = hf - hf_0 = (6.2 - 4.5)\text{eV} = 1.7 \text{ eV} \qquad \blacksquare$$

The notion that light travels as a series of little packets of energy (sometimes referred to as *quanta,* sometimes as *photons*) is in complete contradiction with the wave theory of light (Fig. 28–5). And the wave theory, as we know, has some powerful observational evidence on its side. There is no other way to explain interference effects, for example. According to the wave theory, light spreads out from a source just as ripples spread out on the surface of a lake when a stone is dropped into it, with the energy of the light distributed continuously throughout the wave pattern. According to the quantum theory, light spreads out from a source as a succession of tiny packets of energy, each sufficiently small to permit its being absorbed by a single electron. Yet, despite the particle picture of light that it presents, the quantum theory requires a knowledge of the light frequency f, a wave quantity, in order to determine the energy of each quantum.

Photons are quanta of light

Light as a wave phenomenon

On the other hand, the quantum theory of light is able to explain the photoelectric effect. It predicts that the maximum photoelectron energy should depend upon the frequency of the incident light and not upon its intensity, precisely the opposite of what the wave theory suggests, and it is able to explain why even the weakest light can lead to the immediate emission of photoelectrons. The wave theory can give no reason why there should be a threshold frequency below which no photoelectrons are observed, no matter how strong the light beam, something that follows naturally from the quantum theory.

Light as a quantum phenomenon

Which theory is correct? The history of physics is filled with examples of physical ideas that required revision or even replacement when new data conflicted with them, but this is the first occasion in which two completely different theories are both required to explain a single phenomenon. In thinking about this, it is important for us to note that, in a particular situation, light behaves *either* as though it has a wave nature *or* a particle nature. While light sometimes assumes one form and sometimes the other, there is no physical process in which both are simultaneously exhibited. The same light beam can diffract around an obstacle and then impinge on a metal surface to eject photoelectrons, but these two processes occur separately.

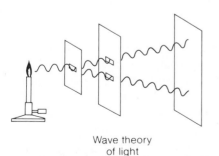

Wave theory
of light

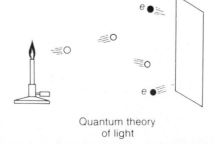

Quantum theory
of light

FIG. 28–5 The wave theory of light is needed to explain diffraction and interference phenomena, which the quantum theory cannot explain. The quantum theory of light is needed to explain the photoelectric effect, which the wave theory cannot explain.

The wave and quantum theories of light are complementary

The electromagnetic theory of light and the quantum theory of light complement each other; by itself, each theory is "correct" in certain experiments, and there are no relevant experiments that neither can account for. Light must be thought of as a phenomenon that incorporates both particle and wave characters. Although we cannot visualize its "true nature," these complementary theories of light are able to account for its behavior, and we have no choice but to accept them both.

28–3   X RAYS

In 1895 Wilhelm Roentgen found that a mysterious, highly penetrating radiation is emitted when high-speed electrons impinge on matter. The X rays (so called because their nature was then unknown) caused phosphorescent substances to glow, exposed photographic plates, traveled in straight lines and were not affected by electric or magnetic fields. The more energetic the electrons, the more penetrating the X rays, and the greater the number of electrons, the greater the density of the resulting X-ray beam.

X rays are short-wavelength electromagnetic waves

After over ten years of study, it was finally established that X rays exhibit, under certain circumstances, both interference and polarization effects, leading to the conclusion that they are electromagnetic waves. From the interference experiments their wavelengths were found to be very short, shorter than those in ultraviolet light. Electromagnetic radiation in the approximate wavelength interval from 0.01 to 10 nm is today classed as X-radiation.

How X rays are produced

Figure 28–6 is a diagram of an X-ray tube. Battery A sends a current through the filament, heating it until it emits electrons. These electrons are then accelerated toward a metallic target by the potential difference V provided by battery B. The tube is evacuated to permit the electrons to reach the target freely. The impact of the electrons causes the target to give off X rays.

What is the physical process involved in the production of X rays? It is known that charged particles emit electromagnetic waves whenever they are accelerated, and so we may reasonably identify X rays as the radiation accompanying the slowing down of fast electrons when they strike matter. The great majority of the incident electrons, to be sure, lose their kinetic energy too gradually for X rays to be evolved, and merely act to heat the target. (Consequently the targets in X-ray tubes are made of metals with high melting points, and a means for cooling the target is often provided.) A few electrons, however, lose much or all their energy in single collisions with target atoms, and this is the energy that appears as X rays. In other words, we may regard X-ray production as an inverse photoelectric effect.

Since the threshold energy hf_0 needed to remove an electron from a metal is only a few electron volts, whereas the accelerating potential V in an X-ray tube usually exceeds 10,000 volts, we can neglect the work function hf_0 here. The highest frequency f_{max} found in the X rays emitted from a particular tube should therefore correspond to a quantum energy of hf_{max}, where hf_{max} equals the kinetic energy

$$KE = Ve$$

of an electron that has been accelerated through a potential difference of V (see Chapter 18). We conclude that

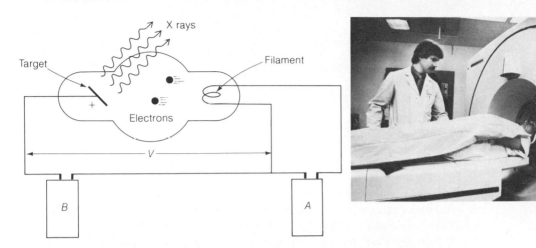

FIG. 28–6 *(Left)* An X-ray tube. *(Right)* In a "CAT scanner," a series of X-ray exposures of a patient taken from different directions are combined by a computer to give cross-sectional images of the part of the body being examined. The tissue is, in effect, sliced up by the computer on the basis of the X-ray exposures, and any desired slice can be displayed. This technique permits an abnormality to be detected and its exact location found, which might be impossible to do from an ordinary X-ray picture. (Photo: Univ. of AZ News/Comm. Health Science Center, Tucson, Jan Rooney)

$$hf_{\max} = Ve \qquad\qquad X\ ray\ energy \quad (28-3)$$

in the operation of an X-ray tube.

Example Find the highest frequency present in the radiation from an X-ray machine whose operating potential is 50,000 volts.

Solution From the formula $hf_{\max} = Ve$ we find that

$$f_{\max} = \frac{Ve}{h} = \frac{(5.0 \times 10^4\,\text{V})(1.6 \times 10^{-19}\,\text{C})}{6.6 \times 10^{-34}\,\text{J} \cdot \text{s}} = 1.2 \times 10^{19}\,\text{Hz}$$

The corresponding wavelength is $\lambda = c/f = 2.5 \times 10^{-11}$ m $= 0.025$ nm. ■

28–4 MATTER WAVES

As we have seen, electromagnetic waves under certain circumstances have properties characteristic of particles. In 1924 Louis de Broglie wondered whether what we normally think of as particles might not have wave properties, too. Soon afterward de Broglie's idea was taken up and developed by a number of other physicists into the elaborate, mathematically difficult—but very beautiful—theory called *quantum mechanics*.

The coming of quantum mechanics did more than provide an accurate and complete description of atomic phenomena. What it also did was to change the way in which physicists regard nature, so that they now think in terms of probabilities instead of in terms of certainties. The universe is in many respects closer to a roulette wheel than to a clock, but it is a roulette wheel that obeys certain rules, and the laws of physics are these rules.

Reasoning by analogy with the properties of photons, de Broglie suggested that any object whose momentum is *mv* behaves like a wave whose wavelength is

Quantum mechanics is a probabilistic description of nature

De Broglie proposed that moving objects have wave properties

$$\lambda = \frac{h}{mv}$$ *De Broglie wavelength*   (28−4)

The more momentum something has, the shorter its wavelength.

Matter waves exhibit diffraction and interference

How can de Broglie's idea be verified? Perhaps the most striking example of wave behavior is a diffraction pattern, which depends upon the ability of waves both to bend around obstacles and to interfere constructively and destructively with one another. Several years after de Broglie's work, a number of experiments confirmed that a stream of electrons is diffracted by the regular arrays of atoms in a crystal. The diffraction patterns that were observed were in complete accord with the electron wavelengths predicted by Eq. (28−4).

Criterion for type of behavior

In certain aspects of its behavior, a moving object resembles a wave, and in other aspects it resembles a particle. Which type of behavior is most conspicuous depends upon how the object's de Broglie wavelength compares with its dimensions and with the dimensions of whatever it interacts with. Two calculations will help us appreciate this statement.

Example   Find the de Broglie wavelengths of a 1500-kg car whose speed is 30 m/s and of an electron whose speed is 10^7 m/s.

Solution   The car's de Broglie wavelength is

$$\lambda = \frac{h}{mv} = \frac{6.63 \times 10^{-34}\,\text{J} \cdot \text{s}}{(1.5 \times 10^3\,\text{kg})(30\,\text{m/s})} = 1.5 \times 10^{-38}\,\text{m}$$

This wavelength is so small relative to the dimensions of the car that no wave behavior can possibly be detected. For the electron, however,

$$\lambda = \frac{h}{mv} = \frac{6.63 \times 10^{-34}\,\text{J} \cdot \text{s}}{(9.1 \times 10^{-31}\,\text{kg})(10^7\,\text{m/s})} = 7.3 \times 10^{-11}\,\text{m}$$

which is comparable in size to atomic dimensions. It is not surprising, then, that electrons with speeds of about 10^7 m/s exhibit diffraction effects when directed at crystals.   ■

The electron microscope achieves high magnifications because of the short wavelength of electron waves

Not long after their discovery, matter waves were used for a very practical purpose, namely as replacements for light waves in microscopes. The resolving power of a microscope is proportional to the wavelength of whatever is used to illuminate the specimen, which limits optical microscopes to useful magnifications of less than 500×. Moving electrons, however, may have wavelengths much shorter than those of light waves. X rays also have short wavelengths, but it has proved difficult to focus them, whereas electrons, by virtue of their charge, are easily controlled by electric and magnetic fields. In an electron microscope, a beam of electrons passes through a very thin specimen in a vacuum chamber, to prevent scattering of the beam and hence blurring of the image, and is brought to a focus on a fluorescent screen or a photographic plate by a system of magnetic fields that act as lenses (Fig. 28−7). For technical reasons the full theoretical resolution of electron waves cannot be realized in practice. As an example, a 100-keV electron has a wavelength of 3.7×10^{-12} m, which is 0.0037 nm, but the actual resolution available ordinarily would not exceed perhaps 0.1 nm.

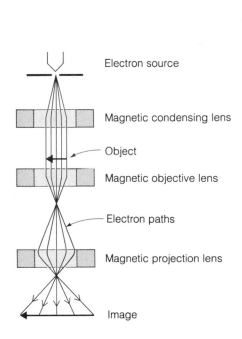

Electron source

Magnetic condensing lens

Object

Magnetic objective lens

Electron paths

Magnetic projection lens

Image

FIG. 28-7 An electron microscope uses electrons instead of light waves to produce an enlarged image. Because electron de Broglie wavelengths are shorter than optical wavelengths, an electron microscope has a higher resolving power and hence can be used to produce higher magnifications. The magnetic "lenses" are current-carrying coils whose magnetic fields focus an electron beam just as the glass lenses in an ordinary microscope focus a light beam. (Photo courtesy of The Applied Research Laboratories Division of Bausch & Lomb Inc.)

However, this is still a great improvement on the 200 nm or more resolution of an optical microscope, and magnifications of $100,000 \times$ or more are commonly achieved with electron microscopes.

In water waves, the quantity that varies periodically is the height of the water surface. In sound waves, the variable quantity is pressure in the medium the waves travel through. In light waves, the variable quantities are the electric and magnetic fields. What is it that varies in the case of matter waves?

The quantity whose variations constitute the matter waves of a moving object is known as its *wave function*. The symbol for wave function is ψ, the Greek letter *psi*.

Wave function and probability density

The value of ψ^2 for a particular object at a certain place and time is proportional to the probability of finding the object at that place at that time.

For this reason the quantity ψ^2 is called the *probability density* of the object. A large value of ψ^2 signifies that the body is likely to be found at the specified place and time; a small value of ψ^2 signifies that the body is unlikely to be found at that place and time. Thus matter waves may be regarded as waves of probability.

A group or packet of matter waves is associated with every moving object. The packet travels with the same speed as the object does. The waves in the packet have the average wavelength $\lambda = h/mv$ given by de Broglie's formula (Fig. 28-8). Even though we cannot visualize what is meant by ψ and so cannot form a mental image of matter waves, the agreement between theory and experiment signifies that the notion of matter waves is a meaningful way to describe moving objects.

FIG. 28–8 (a) Particle description of moving object. (b) Wave description of moving object.

(a)

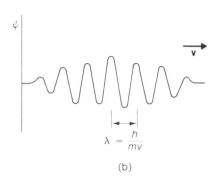

$$\lambda = \frac{h}{mv}$$

(b)

28–5   RELATIVITY OF MASS

In 1905 a young physicist of 26 named Albert Einstein published three short papers that were to change decisively the course not only of physics, but of modern civilization as well. The first paper concerned the nature of light and proposed that light has a dual character with particle as well as wave properties. This work was described earlier in this chapter, and from it flowed the quantum theory of the atom that is the subject of the remainder of the chapter. The second paper dealt with *Brownian motion,* the irregular zigzag movement of tiny bits of suspended matter such as smoke particles in air **Brownian motion** or pollen grains in water. Einstein arrived at a formula that related Brownian motion to the bombardment of the particles by randomly moving molecules of the fluid in which they were suspended. Although the kinetic-molecular theory of matter had been proposed many years before (see Chapter 14), this was the long-awaited definite link with experiment that convinced the remaining doubters.

Einstein's third paper introduced the theory of relativity, an analysis of how measurements of time and space are affected by relative motion between an observer and what he or she is studying. From this theory have come some remarkable predictions, all of them confirmed by experiment. The equivalence of matter and energy according to the formula $E_0 = m_0 c^2$ (Eq. 5–13) is one of these predictions. Another is that the mass of an object is not constant but depends on the object's speed with respect to the observer. The greater the speed, the greater the mass. The variation of mass with speed is not noticeable for ordinary moving things, but in the subatomic world the variation is conspicuous because such tiny particles as electrons, protons, and neutrons may move very fast.

The variation of mass with speed obeys the formula

$$m = \frac{m_0}{\sqrt{1 - v^2/c^2}} \qquad\qquad \textit{Mass increase} \quad (28\text{–}5)$$

whose symbols have these meanings:

m_0 = mass measured when object is at rest ("rest mass")

m = mass measured when object is in relative motion

v = speed of relative motion

c = speed of light

Since the denominator of Eq. (28–5) is always less than one, an object will always appear more massive when in relative motion than when at rest. This mass increase is reciprocal; to a measuring device on the rocket ship in flight, its twin ship on the ground also appears to have a mass m greater than its own mass m_0.

An object is more massive when moving than when at rest

Relativistic mass increases are significant only at speeds approaching that of light (Fig. 28–9). The rest mass of the Apollo 11 spacecraft (apart from its launch vehicle, which dropped away after accelerating the spacecraft) was about 63,070 kg. On its way to the moon, the spacecraft's speed was about 10,840 m/s and so its mass in flight, as measured by an observer on the earth, increased to

Speeds near that of light are needed for appreciable mass increases

$$m = \frac{m_0}{\sqrt{1 - \dfrac{v^2}{c^2}}} = \frac{63{,}070\,\text{kg}}{\sqrt{1 - \dfrac{(1.084 \times 10^4 \text{m/s})^2}{(3 \times 10^8 \text{m/s})^2}}} = 63{,}070.000041\,\text{kg}$$

This is not much of a change.

Smaller objects can be given much higher speeds. It is not very difficult to accelerate electrons (rest mass 9.109×10^{-31} kg) to speeds of, say, $0.9999c$. The mass of such an electron is

$$m = \frac{m_0}{\sqrt{1 - \dfrac{v^2}{c^2}}} = \frac{9.109 \times 10^{-31}\,\text{kg}}{\sqrt{1 - \dfrac{(0.9999c)^2}{c^2}}} = 644 \times 10^{-31}\,\text{kg}$$

The electron's mass is 71 times its rest mass! The mass increases predicted by the relativistic mass formula, even such remarkable ones as this, have been experimentally verified without exception.

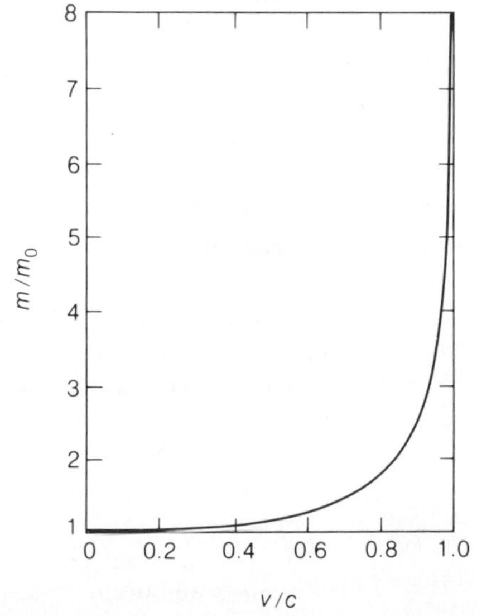

FIG. 28–9 The relativity of mass.

No object can move faster than light

Equation (28–5) has something interesting to say about the greatest speed an object can have. The closer v approaches c, the closer v^2/c^2 approaches one, and the closer $\sqrt{1 - (v^2/c^2)}$ approaches zero. As the denominator of Eq. (28–5) becomes smaller, the mass m becomes larger, so that if the relative speed v actually were equal to the speed of light, the object's mass would be infinite. The concept of an infinite mass anywhere in the universe is, of course, nonsense on many counts; it would have required an infinite force to have accelerated it to the speed of light, for instance, and it would exert an infinite gravitational force on all other bodies in the universe. Hence we interpret Eq. (28–5) to mean that no material body can ever equal or exceed the speed of light.

The mass given by Eq. (28–5) must be used to calculate the de Broglie wavelength of a moving particle when its speed is great enough for an appreciable mass increase. For a speed of $0.1c$ the mass increase is only 0.5%, and so for speeds less than this the mass increase can often be ignored.

28–6   THE HYDROGEN ATOM

Every atom consists of a tiny, massive, positively charged nucleus surrounded by enough negatively charged electrons to leave the atom as a whole electrically neutral. These electrons cannot be stationary, since the electric attraction of the nucleus would pull them in at once. If the electrons are in motion around the nucleus, however, stable orbits like those of the planets about the sun would seem to be possible.

Mechanical model of hydrogen atom

A hydrogen atom, the simplest of all, has a single electron and a nucleus that consists of a single proton. Experiments indicate that 13.6 eV of work must be performed to break apart a hydrogen atom into a proton and electron that go their separate ways (Fig. 28–10). With the help of calculations based upon what we already know about mechanics and electricity this figure leads to an orbital radius of $r = 5.3 \times 10^{-11}$ m and an orbital speed of $v = 2.2 \times 10^6$ m/s for the electron in the hydrogen atom. The proton mass is 1836 times the electron mass, so we can consider the proton as stationary with the electron revolving around it in such a way that the required centripetal force is provided by the electric force exerted by the proton (Fig. 28–11(a)).

Mechanical model is in conflict with electromagnetic theory

What such an analysis overlooks is that, according to electromagnetic theory, all accelerated electric charges radiate electromagnetic waves—and an electron moving in a circular path is certainly accelerated. Thus an atomic electron circling its nucleus to keep from falling into it by electrical attraction cannot help radiating away energy, which means that it must spiral inward until it is swallowed up by the nucleus (Fig. (28–11(b)). Clearly the ordinary laws of physics cannot account for the stability of the hydrogen atom, the simplest atom of all, whose electron must be whirling around the nucleus to keep from being pulled into it and yet must be radiating electromagnetic energy continuously. However, since other phenomena similarly impossible to under-

FIG. 28–10 To break a hydrogen atom apart requires 13.6 eV.

Hydrogen atom   +   13.6 eV   =   proton   +   electron

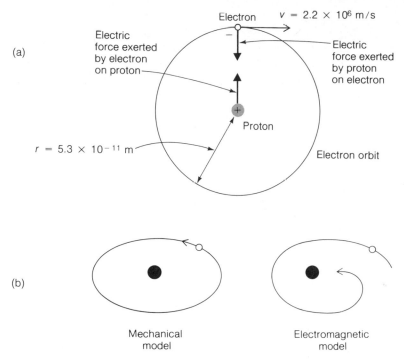

(a)

Electron $v = 2.2 \times 10^6$ m/s

Electric force exerted by electron on proton

Electric force exerted by proton on electron

Proton

$r = 5.3 \times 10^{-11}$ m

Electron orbit

(b)

Mechanical model

Electromagnetic model

FIG. 28–11 (a) The hydrogen atom consists of an electron circling a proton. The electric force exerted by the proton on the electron provides the centripetal force required to hold it in a circular path. The proton is nearly 2000 times as heavy as the electron, so its motion under the influence of the electric force that the electron exerts is insignificant. (b) The mechanical and electromagnetic models of the hydrogen atoms are in conflict. Only quantum theory can account for the stability of atoms.

stand—the photoelectric effect, for instance—find explanation in terms of quantum concepts, it is appropriate for us to inquire whether this might also be true for the atom.

In the discussion that follows of the Bohr model of the atom, the initial argument is somewhat different from that of Niels Bohr, who in 1913 did not have the notion of matter waves to guide his thinking. The results are exactly the same as Bohr obtained, however.

Let us begin by looking into the wave properties of the electron in the hydrogen atom. The de Broglie wavelength λ of an object of mass m and speed v is $\lambda = h/mv$, where h is Planck's constant. The speed of the electron in a hydrogen atom is $v = 2.2 \times 10^6$ m/s, and so the wavelength of its matter waves is

Wavelength of electron in hydrogen atom

$$\lambda = \frac{h}{mv} = \frac{6.63 \times 10^{-34}\,\text{J} \cdot \text{s}}{(9.1 \times 10^{-31}\,\text{kg})(2.2 \times 10^6\,\text{m/s})} = 3.3 \times 10^{-10}\,\text{m}$$

This is a most exciting result, because the electron's orbit has a circumference of exactly

$$2\pi r = 2\pi \times 5.3 \times 10^{-11}\,\text{m} = 3.3 \times 10^{-10}\,\text{m}$$

We therefore conclude that *the orbit of the electron in a hydrogen atom corresponds to one complete electron wave joined on itself* (Fig. 28–12).

The fact that the electron orbit in a hydrogen atom is one electron wavelength in circumference is just the clue we need to construct a theory of the atom. If we examine the vibrations of a wire loop, as in Fig. 28–13, we see that their wavelengths always fit a whole number of times into the loop's circumference, each wave joining smoothly with the next. In the absence of dissipative effects, such vibrations would persist

The ground-state electron orbit is one wavelength in circumference

FIG. 28-12 The electron orbit in a ground-state hydrogen atom is exactly one de Broglie wavelength in circumference.

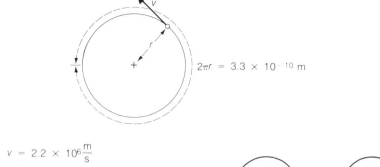

$2\pi r = 3.3 \times 10^{-10}$ m

$v = 2.2 \times 10^{6} \dfrac{m}{s}$

$m = 9.1 \times 10^{-31}$ kg

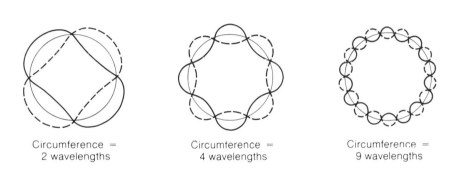

$\lambda = 3.3 \times 10^{-10}$ m

FIG. 28-13 Three possible modes of vibration of a wire loop.

Circumference = 2 wavelengths

Circumference = 4 wavelengths

Circumference = 9 wavelengths

FIG. 28-14 Unless a whole number of wavelengths fits into the wire loop, destructive interference causes the vibrations to die out rapidly.

Each orbit is characterized by a quantum number n

indefinitely. Why are these the only vibrations possible in a wire loop? A fractional number of wavelengths cannot be fitted into the loop and still allow each wave to join smoothly with the next (Fig. 28–14); the result would be destructive interference as the waves travel around the loop, and the vibrations would die out rapidly.

By considering the behavior of electron waves in the hydrogen atom as analogous to the vibrations of a wire loop, then, we may postulate that

An electron can circle an atomic nucleus only if its orbit is a whole number of electron wavelengths in circumference.

The above postulate is the decisive one in our understanding of the atom. We note that it combines both the particle and wave characters of the electron into a single statement. Although we can never observe these different characters at the same time in an experiment, they are inseparable in nature.

It is easy to express in a formula the condition that a whole number of electron wavelengths fit into the electron's "orbit." The circumference of a circular orbit of radius r is $2\pi r$, and so the condition for orbit stability is

$$n\lambda = 2\pi r_n \qquad n = 1, 2, 3, \ldots \qquad \textit{Condition for orbit stability} \quad (28-6)$$

where r_n is the radius of the orbit that contains n wavelengths. The quantity n is called the *quantum number* of the orbit.

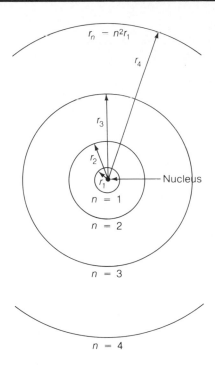

FIG. 28-15 Electron orbits in the hydrogen atom according to the Bohr model. The orbit radii are proportional to n^2.

$r_n = n^2 r_1$

r_4

r_3

r_2

r_1

Nucleus

$n = 1$

$n = 2$

$n = 3$

$n = 4$

A straightforward calculation shows that the stable electron orbits are those whose radii are given by the formula

$$r_n = n^2 r_1 \qquad n = 1, 2, 3, \ldots \qquad \textit{Orbital radii in Bohr atom} \quad (28-7)$$

where $r_1 = 5.3 \times 10^{-11}$ m is the radius of the innermost orbit (Fig. 28–15).

28-7 ENERGY LEVELS

The total energy of a hydrogen atom is not the same in the various permitted orbits. The energy E_n of a hydrogen atom whose electron is in the nth orbit is given by

Each orbit has a different energy

$$E_n = \frac{E_1}{n^2} \qquad n = 1, 2, 3, \ldots \qquad \textit{Energy levels of hydrogen atom} \quad (28-8)$$

where $E_1 = -13.6$ eV $= -2.18 \times 10^{-18}$ J is the energy corresponding to the innermost orbit. The energies specified by the above formula are called the *energy levels* of the hydrogen atom and are the only ones possible. An analogy might be with a person on a ladder, who can stand only on its steps and not in between them.

The energy levels of the hydrogen atom are all less than zero, which signifies that the electron does not have enough energy to escape from the atom. The lowest energy level E_1, corresponding to the quantum number $n = 1$, is called the *ground state* of the atom; the highest levels E_2, E_3, E_4, and so on are called *excited states* (Fig. 28–16).

Ground and excited states

As the quantum number n increases, the energy E_n approaches closer and closer to zero. In the limit of $n = \infty$, $E_\infty = 0$ and the electron is no longer bound to the proton to form an atom. An energy greater than zero signifies an unbound electron

The higher its quantum number, the less tightly bound an atomic electron is

FIG. 28–16 Energy levels of the hydrogen atom.

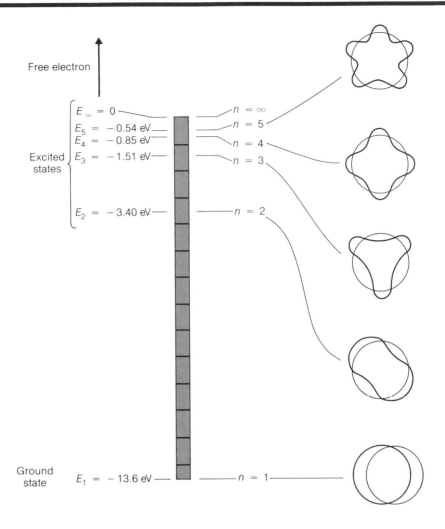

which, since it has no closed orbit that must satisfy quantum conditions, may have any positive energy whatever.

Ionization energy

The work needed to remove an electron from an atom in its ground state is called its *ionization energy*. The ionization energy is therefore equal to $-E_1$, the amount of energy that must be provided to raise an electron from its ground state to an energy of $E = 0$, when it is free. In the case of hydrogen, the ionization energy is 13.6 eV, since the ground-state energy of the hydrogen atom is -13.6 eV.

Quantization is characteristic of many quantities in nature

The presence of definite energy levels in an atom—which is true for all atoms, not just the hydrogen atom—is another example of the fundamental graininess of physical quantities on an atomic scale. In the everyday world, matter, electric charge, energy, and so on seem continuous and capable of being cut up, so to speak, into parcels of any size we like. In the world of the atom, however, matter consists of elementary particles of fixed rest masses that join together to form atoms of fixed rest masses; electric charge always comes in multiples of $+e$ and $-e$; energy in the form of electromagnetic waves of frequency f always comes in separate photons of energy hf; and stable systems of particles, such as atoms, can have only certain energies and no others.

Other quantities in nature are also grainy, or *quantized,* and it has turned out that this graininess is the key to understanding how the properties of matter we are familiar with in everyday life originate in the interactions of elementary particles. In the case of the atom, the quantization of energy is a consequence of the wave nature of moving bodies: In an atom the electron wave functions can only occur in the form of standing waves, much as a violin string can vibrate only at those frequencies that give rise to standing waves.

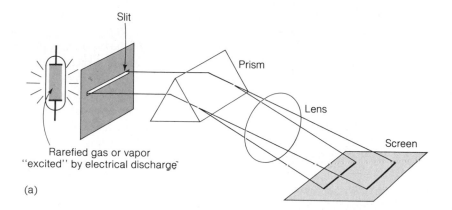

(a)

FIG. 28–17 (a) The spectrum of a substance may be obtained by "exciting" a sample of its vapor in a tube by means of an electrical discharge and then directing the light it gives off through a prism or grating. (b) The most prominent lines in the spectra of hydrogen, helium, and mercury in the visible region.

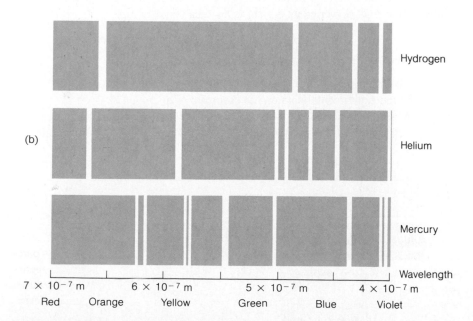

28–8   ATOMIC SPECTRA

The Bohr theory is strikingly confirmed in its successful explanation of atomic spectra. When a gas or vapor at somewhat less than atmospheric pressure is "excited" by the passage of an electric current through it, light whose spectrum consists of a limited number of individual wavelengths is emitted (Fig. 28–17). The characteristic red-orange color of a neon sign is an example of this phenomenon.

Every element has a unique line spectrum

Atomic spectra are called *line spectra* from their appearance. Every element exhibits a unique line spectrum when a sample of it is suitably excited, and the presence of any element in a substance of unknown composition can be established by the appearance of its characteristic wavelengths in the spectrum of the substance. Spectral lines are found in the infrared and ultraviolet as well as in the visible region.

Absorption and emission spectra

It is worth noting that whereas unexcited gases and vapors do not radiate their characteristic spectral lines, they do *absorb* light of certain of those wavelengths when white light is passed through samples of them. In other words, the *absorption spectrum* of an element is closely related to its *emission spectrum*. Emission spectra consist of bright lines on a dark background; absorption spectra consist of dark lines on a bright background. Figure 28–18 shows the absorption and emission spectra of sodium vapor.

A photon is emitted when an atom in an excited state falls to a lower state

The wavelengths present in atomic spectra fall into definite series. The spectral series of hydrogen are shown in Fig. 28–19. The presence of a sequence of separate energy levels in the hydrogen atom suggests a connection with line spectra. Let us suppose that when an electron in an excited state drops to a lower state, the difference in energy between the states is emitted as a single photon of light. Because electrons cannot, according to our model, exist in an atom except in certain specific energy levels, a rapid "jump" from one level to the other, with the energy difference being given off all at once in a photon rather than in some gradual manner, fits in well with this model.

If the quantum number of the initial (higher energy) state is n_i and the quantum number of the final (lower energy) state is n_f, then

$$E_i - E_f = hf \qquad \textit{Origin of spectral lines} \quad (28-9)$$

Initial energy − final energy = quantum energy

where f is the frequency of the emitted photon and h is Planck's constant. This formula agrees with the experimental data on the spectrum of hydrogen.

Figure 28–20 is an energy-level diagram of the hydrogen atom showing the possible transitions from initial quantum states to final ones. Each transition—or "jump"—

(a)

(b)

FIG. 28–18   (a) Absorption spectrum of sodium vapor. (b) Emission spectrum of sodium vapor. Each dark line in the absorption spectrum corresponds to a bright line in the emission spectrum.

Auroras like this one are caused by streams of fast protons and electrons from the sun that excite gases in the upper atmosphere to emit light whose wavelengths are characteristic of the atoms present. (Photo: Robert A. Bumpas, National Center for Atmospheric Research/National Science Foundation)

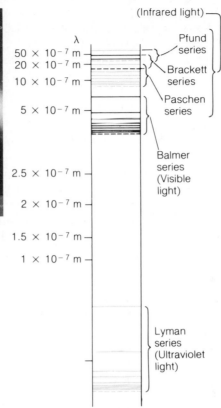

FIG. 28–19 The line spectrum of hydrogen with the various series of spectral lines indicated. The wavelength scale is not linear in order to cover the entire spectrum.

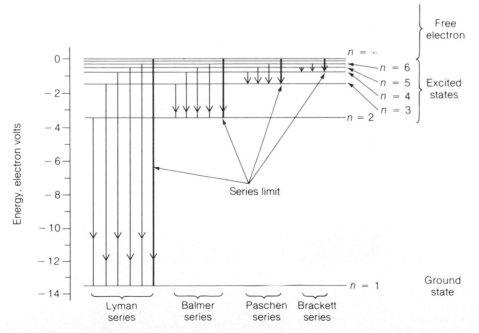

FIG. 28–20 Energy-level diagram of the hydrogen atom with some of the transitions that give rise to spectral lines indicated.

involves a characteristic amount of energy, and hence a photon of a certain characteristic frequency. The larger the energy difference between initial and final energy levels, as indicated by the lengths of the arrows, the higher the frequency of the emitted photon. The origins of the various series of spectral lines are indicated on the diagram.

28–9 COMPLEX ATOMS

Quantum theory of the atom

The Bohr theory of the atom is indeed impressive in its agreement with experiment, but it has certain serious limitations. These limitations are absent from the more general *quantum theory of the atom,* which was developed a decade after Bohr's work, and whose refinement and application to new problems continues to the present day. Not only has the quantum theory of the atom provided the theoretical framework for understanding the structure of the atom itself, but it has also furnished key insights that explain how and why atoms join together to form molecules, solids, and liquids.

Electron probability cloud in an atom

In the Bohr model of the hydrogen atom, the electron is regarded as being confined to a definite circular orbit, and the only quantity that changes as it revolves around its nucleus is its position on this fixed circle. A single quantum number is all that is required to specify the physical state of such an electron.

In the quantum theory of the atom, an electron is not limited to a specific orbit but moves about in three dimensions. We can think of the electron as circulating in a probability cloud: Where the cloud is most dense (ψ^2 has the highest value), the electron is most likely to be found, and where the cloud is least dense (ψ^2 has the lowest value), the electron is least likely to be found. Figure 28–21 shows the probability cloud for the ground state of the hydrogen atom, which is the state in which the electron has the lowest energy and is the one usually occupied.

Three quantum numbers, not the single one of the Bohr theory, are needed to specify the size and shape of the probability cloud of an atomic electron. One of them, designated n as in the Bohr theory, is the chief factor that governs the electron's energy and its average distance from the nucleus. The other two, designated l and m_l, together with n determine the details of the probability cloud.

FIG. 28–21 Probability cloud for the ground state of the hydrogen atom.

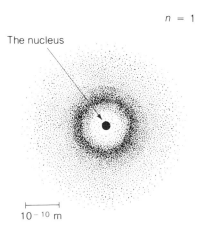

$n = 1$

The nucleus

10^{-10} m

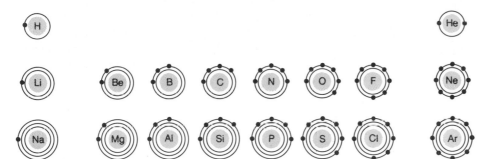

The electrons in an atom that share the same quantum number n are said to occupy the same *shell*. These electrons average about the same distance from the nucleus and usually have comparable, though not identical, energies. The number of electrons in a shell of given n is limited to $2n^2$, which means that there can be at most 2 electrons in the $n = 1$ shell, 8 electrons in the $n = 2$ shell, 18 electrons in the $n = 3$ shell, and so on. The electrons in an atom in its ground state occupy the innermost shells to the maximum extent possible. For example, the chlorine atom, which contains a total of 17 electrons, has two of them in its $n = 1$ shell, eight in its $n = 2$ shell, and seven in its $n = 3$ shell. Figure 28–22 is a highly schematic representation of the electron structures of the 18 lightest elements.

Atomic electrons in a shell have the same quantum number n

A fourth quantum number, m_s, also turns out to be needed to completely specify the state of an atomic electron. An electron behaves in certain respects as though it is a spinning charged sphere, so that it can be thought of as a tiny bar magnet. The quantum number m_s describes how the electron aligns itself in a magnetic field, either parallel ("spin up," corresponding to $m_s = +\frac{1}{2}$) or antiparallel ("spin down," corresponding to $m_s = -\frac{1}{2}$). According to the *exclusion principle,* no two electrons in an atom can have the same sets of quantum numbers, which means that at most two electrons, one with each direction of spin, can occupy a particular probability cloud.

Electron spin

In Chapter 29 we shall see how these ideas make it possible to understand how atoms can stick together to form molecules, solids, and liquids.

28–10 ATOMIC EXCITATION

There are two principal ways in which an atom may be excited to an energy level above that of its ground state and thereby become able to radiate. One of these ways is a collision with another atom during which part of their kinetic energy is transformed into electron energy within either or both of the participating atoms. An atom excited in this way will then lose its excitation energy by emitting one or more photons in the course of returning to its ground state (Fig. 28–23). In an electric discharge in a gas, an electric field accelerates electrons and charged atoms and molecules (whose charge arises from either too many or too few electrons to neutralize the positive charge of their nuclei) until their kinetic energies are enough to excite atoms with which they

Excitation by collision

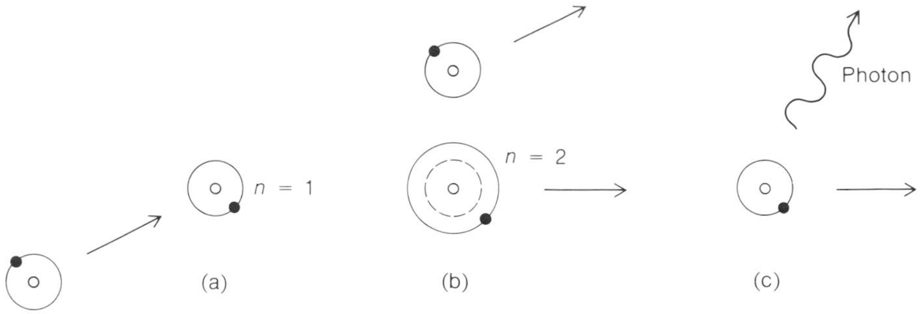

FIG. 28–23 In (a) both atoms are in their ground states. During the collision some kinetic energy is transformed into excitation energy, and in (b), the target atom is in an excited state. In (c) the target atom has returned to its ground state by emitting a photon.

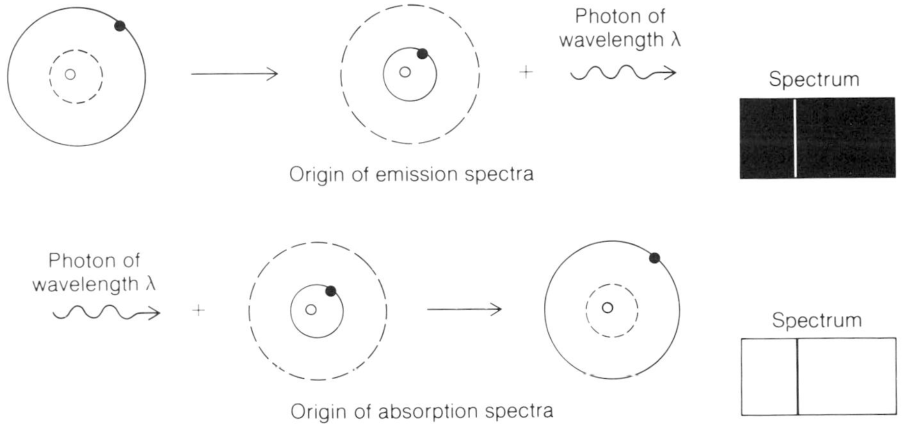

FIG. 28–24 The origins of emission and absorption spectra.

Photon of wavelength λ

Spectrum

Origin of emission spectra

Photon of wavelength λ

Spectrum

Origin of absorption spectra

happen to collide. A neon sign is a familiar example of how applying a strong electric field between electrodes in a gas-filled tube leads to the emission of the characteristic spectral radiation of that gas, which happens to be orange light in the case of neon.

Excitation by absorption of radiation

Another excitation mechanism is the absorption by an atom of a photon of light whose energy is just the right amount to raise it to a higher energy level. A photon of wavelength 121.7 nm is emitted when a hydrogen atom in the $n = 2$ state drops to the $n = 1$ state; hence the absorption of a photon of wavelength 121.7 nm by a hydrogen atom initially in the $n = 1$ state will bring it up to the $n = 2$ state. This process explains the origin of absorption spectra (Fig. 28–24).

When white light (in which all wavelengths are present) is passed through hydrogen gas, photons of those wavelengths that correspond to jumps between hydrogen energy levels are absorbed. The resulting excited hydrogen atoms reradiate their excitation energy almost at once, but these photons come off in random directions, not all in the same direction as in the original beam of white light (Fig. 28–25). The dark lines in an absorption spectrum are therefore never totally dark, but only appear so by contrast with the bright background of transmitted light. We would expect the lines in the absorption spectrum of a particular substance to be the same as lines in its emission spectrum, which is what is found. As Fig. 28–18 shows, only a few of the emission lines appear in absorption. The reason is that reradiation is so rapid that practically all the atoms in an absorbing gas or vapor are initially in their lowest ($n = 1$) energy

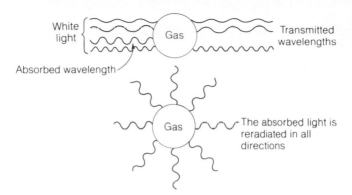

Gas — Transmitted wavelengths

Absorbed wavelength

Gas — The absorbed light is reradiated in all directions

FIG. 28–25 The dark lines in an absorption spectrum are never totally dark.

states and therefore take up and reemit only photons that represent transitions to and from the $n = 1$ state.

28–11 THE LASER

The *laser* is a device for producing a light beam with a number of remarkable properties:

1. The beam is extremely intense, more intense by far than the light from any other source. To achieve an energy density equal to that in some laser beams, a hot object would have to be at a temperature of 10^{30} K.

2. There is very little spreading out, so that a laser beam from the earth that was reflected by a mirror left on the moon during the Apollo 11 expedition remained narrow enough to be detected upon its return to the earth—a total journey of more than 750,000 km.

3. The light is almost perfectly monochromatic.

4. The light is coherent, with the waves all exactly in step with one another (Fig. 28–26). It is possible to obtain interference patterns not only by merely placing two slits in a laser beam but also by using beams from two separate lasers.

The term *laser* stands for *l*ight *a*mplification by *s*timulated *e*mission of *r*adiation. Let us look into how a laser functions.

Left to itself, an atom always remains in its ground state, the quantum state in which it has the least possible energy. The atom can be raised to an excited state by picking up energy in a collision with an electron or another atom or by absorbing a photon of exactly the right frequency. An excited atom normally falls to its ground state almost at once, sometimes first dropping to an intermediate excited state, with the emission of a photon during each transition. The lifetime of most excited states is only about 10^{-8} s, but certain states are *metastable* (temporarily stable). An atom may remain in a metastable state for 10^{-3} second or more before radiating, provided it does not undergo a collision in the meantime in which its excitation energy would be lost (Fig. 28–27). The operation of lasers depends upon the existence of metastable states in atoms.

Three kinds of transition involving electromagnetic radiation are possible between two energy levels in an atom whose energy difference is $E_2 - E_1 = hf$. In *induced*

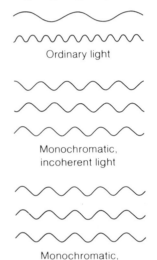

Ordinary light

Monochromatic, incoherent light

Monochromatic, coherent light

FIG. 28–26 A laser produces a beam of light whose waves all have the same frequency (monochromatic) and are in phase with one another (coherent). The beam is also well collimated and spreads out very little even over a long distance.

Processes of absorption and emission

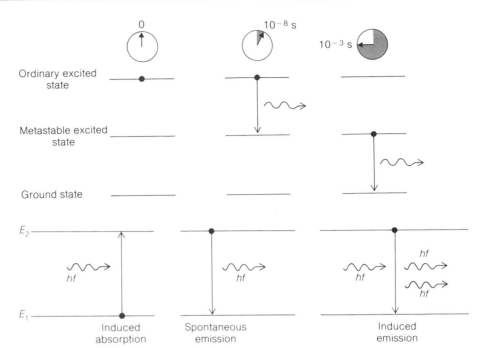

FIG. 28-27 An atom can exist in a metastable energy level for a longer time before radiating than it can in an ordinary energy level.

FIG. 28-28 The three types of transition between energy levels in an atom.

absorption the atom when in the lower level absorbs a photon of energy hf and is elevated to the upper level (Fig. 28–28). In *spontaneous emission* the atom when in the upper level emits a photon of energy hf and falls to the lower one. There is also a third possibility, *induced emission,* in which the presence of radiation of frequency f causes a transition from the upper level to the lower one. In induced emission, the emitted light waves are exactly in step with the incident ones, so that the result is a stronger beam of coherent light.

Now let us suppose we have a group of atoms of some kind which have metastable states of excitation energy hf. If we can somehow raise a majority of the atoms to the metastable level and then shine light of frequency f on the group, there will be more induced emission than induced absorption and the result will be an amplification of the original light. This is the concept that underlies the operation of the laser. The term *population inversion* is given to a group of atoms in which a majority are in excited states, because under normal circumstances the ground states are the more highly populated.

Optical pumping

There are a number of ways in which population inversions can be produced. One of them, called *optical pumping,* employs an external light source some of whose photons have the right frequency to raise the atoms to excited states that decay into the desired metastable ones. This is the method used in the ruby laser, in which the chromium ions Cr^{+++} in a ruby crystal are excited by light from a xenon-filled flash lamp (Fig. 28–29). (A ruby is a crystal of Al_2O_3, a small number of whose Al^{+++} ions are replaced by Cr^{+++} ions, which are responsible for the reddish color. Such ions are Cr atoms that have lost three electrons each.)

Operation of ruby laser

The Cr^{+++} ions are raised in the pumping process to a level E_3 from which they decay to the metastable level E_2 by losing energy to other atoms in the crystal. This metastable level has a lifetime of about 0.003 s. Because there are comparatively few

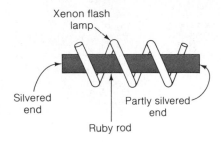

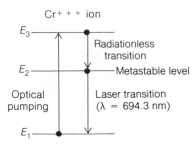

FIG. 28–29 The ruby laser produces pulses of light.

Cr^{+++} ions in the ruby rod and a xenon flash lamp emits a great deal of light, most of the Cr^{+++} ions can be pumped to the E_2 level in this way. A few Cr^{+++} ions in the E_2 level then spontaneously fall to the E_1 level, and some of the resulting photons bounce back and forth between the reflecting ends of the ruby rod. The presence of light of exactly the right frequency now stimulates the other Cr^{+++} ions in the E_2 level to radiate, and the result is an avalanche process that produces a large pulse of red light in a few microseconds that emerges from the partly silvered end of the rod. The length of the rod is made to be exactly a whole number of half-wavelengths long, so the radiation trapped in it forms an optical standing wave. Since the induced emissions are stimulated by the standing wave, their waves are all in step with it, which is why the final pulse of light is both coherent and collimated.

The helium-neon gas laser achieves a population inversion in a different way. A mixture of about 7 parts of helium and 1 part of neon at a low pressure (1 mm of mercury) is placed in a glass tube that has parallel mirrors at both ends. An electric discharge is then produced in the gas by electrodes connected to a source of high-frequency alternating current, and collisions with electrons from the discharge excite He and Ne atoms to metastable states respectively 20.61 and 20.66 eV above their ground states (Fig. 28–30). Some of the excited He atoms transfer their energy to ground-state Ne atoms in collisions, with the 0.05 eV of additional energy being provided by the kinetic energy of the atoms. The purpose of the He atoms is thus to help in producing a population inversion in the Ne atoms.

An excited Ne atom emits a photon of wavelength $\lambda = 632.8$ nm in the transition to another excited state 18.70 eV above the ground state; this is one of the transitions that leads to laser action. Then another photon is spontaneously emitted in a transition to a lower metastable state (this transition produces only incoherent radiation because it is spontaneous rather than induced), and the remaining excitation energy is lost in collisions with the tube walls. Because the electron impacts that produce a population inversion in the Ne atoms are not intermittent but occur all the time, a helium-neon laser operates continuously.

Operation of helium-neon laser

The helium-neon laser operates continuously

FIG. 28–30 Energy levels involved in the helium-neon laser, which operates continuously.

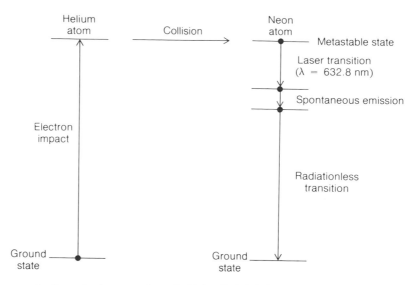

In the ruby laser, at least half the Cr^{+++} ions must be in the metastable state or absorption rather than induced emission will predominate and laser amplification will not occur. In the helium-neon laser, the final state of the laser transition is 18.70 eV above the ground state and hence almost entirely unpopulated at ordinary temperatures, which means that very few Ne atoms will be able to absorb photons with the laser wavelength of 632.8 nm. For this reason the fraction of Ne atoms that must be excited in order for laser amplification to occur can be much less than half the total, which is a definite advantage.

Many other kinds of lasers of have been built and still more are under development. Some are very small and produce the coded flashes of light that carry telephone signals along glass fiber transmission lines. More powerful lasers are used in surgery where a laser beam has the advantage of sealing small blood vessels it cuts through. Lasers find numerous applications in industry, from cutting fabric for clothing and making holes in nipples for babies' bottles to welding pipelines and heat-treating the surfaces of engine crankshafts and cylinder walls to harden them.

IMPORTANT TERMS

The **photoelectric effect** is the emission of electrons from a metal surface when light shines on it.

The **quantum theory of light** states that light travels in tiny bursts of energy called **quanta** or **photons**. The quantum theory of light is required to account for the photoelectric effect.

X rays are high-frequency electromagnetic waves emitted when fast electrons impinge on matter.

A moving body behaves as though it has a wave character. The waves representing such a body are **matter waves,** also called **de Broglie waves.** The wave variable in a matter wave is its **wave function,** whose square is the **probability density** of the body. The value of the probability density of a particular body at a certain place and time is proportional to the probability of finding the body at that place at that time. Matter waves may thus be regarded as waves of probability.

The **relativity of mass** refers to the increase in the measured mass of an object when it is moving relative to an observer. The object's **rest mass** is its mass measured when it is at rest relative to the observer.

An **emission spectrum** consists of the various wavelengths of light emitted by an excited substance. An **absorption spectrum** consists of the various wavelengths of light absorbed by a substance when white light is passed through it.

According to the **Bohr theory of the atom,** an electron can circle an atomic nucleus indefinitely without radiating energy if its orbit is a whole number of electron wavelengths in circumference. The number of wavelengths that fit into a particular permitted orbit is called the **quantum number** of that orbit. The electron energies corresponding to the various quantum numbers constitute the energy levels of the atom, of which the lowest is the **ground state** and the rest are **excited states.** When an electron in an excited state drops to a lower state, the difference in energy between the states is emitted as a single photon of light; when a photon of the same wavelength is absorbed, the electron goes from the lower to the higher state. The above two processes account for the properties of emission and absorption spectra respectively.

In the **quantum theory of the atom,** three quantum numbers are needed to specify the probability cloud that an electron occupies in an atom. The greater the probability density in a certain part of the cloud, the more likely the electron is to be found there. The electron's energy and average distance from the nucleus are chiefly determined by the quantum number n. The electrons in an atom that have the same value of n are said to occupy the same *shell.*

Every electron has a certain amount of angular momentum called its **spin.** The quantum number m_1 that governs electron spin has two possible values, $+\frac{1}{2}$ and $-\frac{1}{2}$. Owing to its spin, every electron acts like a tiny bar magnet.

According to the **exclusion principle,** each electron in an atom must have a different set of quantum numbers. Because of this principle, atoms with different numbers of electrons have different properties.

A **laser** is a device for producing a narrow, monochromatic, coherent beam of light. The term stands for light amplification by stimulated emission of radiation. Laser operation depends upon the existence of **metastable states,** which are excited atomic states that can persist for unusually long periods of time.

IMPORTANT FORMULAS

Quantum energy: $E = hf$

Photoelectric effect: $hf = \mathrm{KE}_{max} + hf_0$

X-ray energy: $hf_{max} = Ve$

De Broglie wavelength: $\lambda = \dfrac{h}{mv}$

Mass increase $m = \dfrac{m_0}{\sqrt{1 - v^2/c^2}}$

Condition for orbit stability: $n\lambda = 2\pi r_n$

Energy levels of hydrogen atom:

$$E_n = \frac{E_1}{n^2} \qquad n = 1,2,3, \ldots$$

Origin of spectral lines: $E_i - E_f = hf$

MULTIPLE CHOICE

1. Photoelectrons are emitted by a metal surface only when the light directed at it exceeds a certain minimum
  a. wavelength.
  b. frequency.
  c. speed.
  d. charge.

2. When light is directed on a metal surface, the maximum energies of the emitted electrons
  a. vary with the intensity of the light.
  b. vary with the frequency of the light.
  c. vary with the speed of the light.
  d. are random.

3. The photoelectric effect can be understood on the basis of
  a. the electromagnetic theory of light.
  b. Huygens' principle.
  c. the principle of superposition.
  d. none of the above.

4. According to the theories of modern physics, light
  a. is exclusively a wave phenomenon.
  b. is exclusively a particle phenomenon.
  c. combines wave and particle properties.
  d. has neither wave nor particle properties.

5. The mass of a photon
  a. is 0.
  b. is the same as that of an electron.
  c. depends on its frequency.
  d. depends on its energy.

6. When the speed of the electrons striking a metal surface is increased, the result is an increase in
  a. the number of X rays emitted.
  b. the frequency of the X rays emitted.
  c. the speed of the X rays emitted.
  d. the size of the X rays emitted.

7. According to the theories of modern physics,
  a. all particles exhibit wave behavior.
  b. only moving particles exhibit wave behavior.
  c. only charged particles exhibit wave behavior.
  d. only uncharged particles exhibit wave behavior.

8. The description of a moving body in terms of matter waves is legitimate because
   a. it is based upon common sense.
   b. matter waves have been actually seen.
   c. the analogy with electromagnetic waves is plausible.
   d. theory and experiment agree.

9. The wave packet that corresponds to a moving particle
   a. has the same size as the particle.
   b. has the same speed as the particle.
   c. has the speed of light.
   d. consists of X rays.

10. A moving body is described by the wave function ψ at a certain time and place. The value of ψ^2 is proportional to the body's
   a. electric field.
   b. speed.
   c. energy.
   d. probability of being found.

11. The classical model of the hydrogen atom fails because
   a. a moving electron has more mass than an electron at rest.
   b. a moving electron has more charge than an electron at rest.
   c. the attractive force of the nucleus is not enough to keep an electron in orbit around it.
   d. an accelerated electron radiates electromagnetic waves.

12. In the Bohr model of the hydrogen atom, the electron revolves around the nucleus in order to
   a. emit spectral lines.
   b. produce X rays.
   c. form energy levels that depend on its speed.
   d. keep from falling into the nucleus.

13. According to the Bohr model, an electron can revolve around the nucleus of a hydrogen atom indefinitely if its orbit is
   a. a perfect circle.
   b. sufficiently far from the nucleus to avoid capture.
   c. less than one de Broglie wavelength in circumference.
   d. exactly one de Broglie wavelength in circumference.

14. A hydrogen atom is in its ground state when its orbital electron
   a. is within the nucleus.
   b. has escaped from the atom.
   c. is in its lowest energy level.
   d. is stationary.

15. With increasing quantum number, the energy difference between adjacent energy levels
   a. decreases.

   b. remains the same.
   c. increases.
   d. sometimes decreases and sometimes increases.

16. An atom emits a photon when one of its electrons
   a. collides with another of its electrons.
   b. is removed from the atom.
   c. undergoes a transition to a quantum state of lower energy.
   d. undergoes a transition to a quantum state of higher energy.

17. The bright-line spectrum produced by the excited atoms of an element contains wavelengths that
   a. are the same for all elements.
   b. are characteristic of the particular element.
   c. are evenly distributed throughout the entire visible spectrum.
   d. are different from the wavelengths in its dark-line spectrum.

18. A neon sign does not produce
   a. a line spectrum.
   b. an emission spectrum.
   c. an absorption spectrum.
   d. photons.

19. When an atom absorbs a photon of light, which one or more of the following can happen?
   a. An electron shifts to a state of smaller quantum number.
   b. An electron shifts to a state of higher quantum number.
   c. An electron leaves the atom.
   d. An X-ray photon is emitted.

20. In an emission spectral series,
   a. the lines are evenly spaced.
   b. the excited electrons all started from the same energy level.
   c. the excited electrons all return to the same energy level.
   d. the excited electrons leave the atoms.

21. The sun's spectrum consists of a light background crossed by dark lines. This suggests that the sun
   a. is a hot object surrounded by a hot atmosphere.
   b. is a hot object surrounded by a cool atmosphere.
   c. is a cool object surrounded by a hot atmosphere.
   d. is a cool object surrounded by a cool atmosphere.

22. The quantum number n of the lowest energy state of a hydrogen atom
   a. is 0.
   b. is 1.
   c. depends on the orbit size.
   d. depends on the electron speed.

23. Most excited states of an atom have lifetimes of about
   a. 10^{-8} s.   b. 10^{-3} s.
   c. 1 s.   d. 10 s.

24. The exclusion principle states that no two electrons in an atom can
   a. be present in the same probability cloud.
   b. have the same spin.
   c. have the same quantum numbers.
   d. interact with each other.

25. The operation of the laser is based upon
   a. The exclusion principle.
   b. the interference of de Broglie waves.
   c. induced emission of radiation.
   d. spontaneous emission of radiation.

26. The light produced by a laser is not
   a. incoherent.
   b. monochromatic.
   c. in the form of a narrow beam.
   d. electromagnetic.

27. Light of wavelength 5×10^{-7} m consists of photons whose energy is
   a. 1.1×10^{-48} J.   b. 1.3×10^{-27} J.
   c. 4×10^{-9} J.   d. 1.7×10^{-5} J.

28. A lamp emits light of frequency 5×10^{15} Hz at a power of 25 W. The number of photons given off per second is
   a. 1.3×10^{-19}.   b. 8.3×10^{-17}.
   c. 7.6×10^{18}.   d. 1.9×10^{50}.

29. An X-ray photon has an energy of 6.6×10^{-15} J. The frequency that corresponds to this energy is
   a. 4.4×10^{-48} Hz.   b. 10^{-19} Hz.
   c. 10^{15} Hz.   d. 10^{19} Hz.

30. An electron has a de Broglie wavelength of 0.1 nm. Neglecting its relativistic mass change, the electron's speed is
   a. 6.6×10^{-24} m/s.   b. 3.8×10^{3} m/s.
   c. 7.3×10^{6} m/s.   d. 10^{10} m/s.

31. A particle of rest mass 1 g is moving relative to an observer at $0.6c$. The observer measures its mass to be
   a. 0.63 g.   b. 1.29 g.
   c. 1.58 g.   d. 2.5 g.

32. When an object whose rest mass is 1 kg approaches the speed of light, its mass approaches
   a. 0.   b. 0.5 kg.
   c. 2 kg.   d. ∞.

33. The speed of an electron whose mass is ten times its rest mass is
   a. 2×10^{8} m/s.   b. 2.98×10^{8} m/s.
   c. 4×10^{8} m/s.   d. 3×10^{9} m/s.

34. Which of the following transitions in a hydrogen atom emits the photon of highest frequency?
   a. $n = 1$ to $n = 2$ b. $n = 2$ to $n = 1$
   c. $n = 2$ to $n = 6$ d. $n = 6$ to $n = 2$

35. Which of the following transitions in a hydrogen atom absorbs the photon of highest frequency?
   a. $n = 1$ to $n = 2$ b. $n = 2$ to $n = 1$
   c. $n = 2$ to $n = 6$ d. $n = 6$ to $n = 2$

36. Which of the following transitions in a hydrogen atom emits the photon of lowest frequency?
   a. $n = 1$ to $n = 2$ b. $n = 2$ to $n = 1$
   c. $n = 2$ to $n = 6$ d. $n = 6$ to $n = 2$

EXERCISES

28–1   Photoelectric Effect

28–2   Quantum Theory of Light

1. Why do you think the wave aspect of light was discovered earlier than its particle aspect?

2. If Planck's constant were smaller than it is, would quantum phenomena be more or less conspicuous than they are now in everyday life?

3. The eye can detect as little as 10^{-18} J of electromagnetic energy. How many photons of wavelength 600 nm does this energy represent?

4. A detached retina is being "welded" back using 20-ms pulses from an 0.5-W laser operating at a wavelength of 643 nm. How many photons are in each pulse?

5. Find the frequency and wavelength corresponding to a 5-MeV photon.

6. The threshold frequency for photoelectric emission in calcium is 7.7×10^{14} Hz. Find the maximum energy in electron volts of the photoelectrons when light of frequency 1.20×10^{15} Hz is directed on a calcium surface.

7. What is the maximum wavelength of light that will lead to photoelectric emission from platinum, whose work function is 5.6 eV? In what part of the spectrum is such light?

8. A silver ball is suspended by a string in a vacuum chamber and ultraviolet light of wavelength 200 nm is directed at it. What electric potential will the ball acquire as a result? The work function of silver is 4.7 eV.

9. Light from the sun reaches the earth at a rate of about 1.4×10^{3} W/m² of area perpendicular to the direction of the light. Assume sunlight is monochromatic with a frequency of 5×10^{14} Hz. (a) How many photons fall per second on each square meter of the earth's surface directly facing the sun? (b) How many photons are present in each m^3 near the earth on the sunlit side?

10. Light of wavelength 420 nm falls on the cesium surface of a photoelectric cell at the rate of 5 mW. If one photoelectron is emitted for every 10^4 incident photons, find the current produced by the cell. The work function of cesium is 1.9 eV.

28-3 X Rays

11. Electrons are accelerated in television tubes through potential differences of about 10,000 V. Find the highest frequency of the electromagnetic waves that are emitted when these electrons strike the screen of the tube. What type of waves are these?

12. What voltage must be applied to an X-ray tube for it to emit X rays with a minimum wavelength of 3×10^{-11} m?

28-4 Matter Waves

13. Must a particle have an electric charge in order for matter waves to be associated with its motion?

14. A photon and a proton have the same wavelength. What can be said about how their linear momenta compare? About how the photon's energy compares with the particle's kinetic energy?

15. What is the simplest experimental procedure that can distinguish between a gamma ray whose wavelength is 10^{-11} m and an electron whose de Broglie wavelength is also 10^{-11} m?

16. A proton and an electron have the same de Broglie wavelength. How do their speeds compare?

17. What is the de Broglie wavelength of a 1-mg grain of sand blown by the wind at a speed of 20 m/s?

18. Find the de Broglie wavelength of a 1-MeV proton.

19. Green light has a wavelength of about 550 nm. Through what potential difference must an electron be accelerated to have this wavelength?

20. Show that the de Broglie wavelength of an oxygen molecule in thermal equilibrium in the atmosphere at 20°C is smaller than its diameter of about 4×10^{-10} m.

28-5 Relativity of Mass

21. Can the rest mass of a moving particle be determined by measuring its de Broglie wavelength?

22. Find the mass of an object whose rest mass is 1000 g when it is traveling at 10%, 90%, and 99% of the speed of light.

23. A man has a mass of 100 kg on the ground. When he is in a spacecraft in flight, an observer on the earth measures his mass to be 101 kg. How fast is the spacecraft moving?

24. What speed must a particle have if its mass is to be triple its rest mass?

25. Calculate the de Broglie wavelength of (a) an electron whose speed is 1×10^8 m/s, and (b) an electron whose speed is 2×10^8 m/s. Use relativistic formulas.

28-6 The Hydrogen Atom
28-7 Energy Levels

26. In the Bohr model of the hydrogen atom, why is the electron pictured as revolving around the nucleus?

27. In the Bohr theory of the hydrogen atom, the electron is in constant motion. How can such an electron have a negative amount of energy?

28. How does the energy difference between adjacent energy levels in the hydrogen atom vary with quantum number?

29. How much energy is needed to ionize a hydrogen atom when it is in the $n = 4$ state?

30. Calculate the average kinetic energy per molecule in a gas at room temperature (20°C), and show that this is much less than the energy required to raise a hydrogen atom from its ground state ($n = 1$) to its first excited state ($n = 2$).

31. To what temperature must a hydrogen gas be heated if the average molecular kinetic energy is to equal the ionization energy of the hydrogen atom?

32. Find the quantum number of the Bohr orbit in a hydrogen atom that is 1 mm in radius. What is the energy (in electron volts) of an atom in this state? Why is such an orbit unlikely to be occupied?

33. (a) Calculate the de Broglie wavelength of the earth. (b) What is the quantum number that characterizes the earth's orbit about the sun? (The earth's mass is 6.0×10^{24} kg, its orbital radius is 1.5×10^{11} m, and its orbital speed is 3×10^4 m/s.)

34. (a) Derive a formula for the frequency of revolution of an electron in the nth orbit of the Bohr atom. (b) Find the frequencies of revolution of the electron when it is in the $n = 1$ and $n = 2$ orbits. (c) An electron spends about 10^{-8} s in an excited state before it drops to a lower state by giving up energy in the form of a photon. How many revolutions does an electron in the $n = 2$ state of the Bohr atom make before dropping to the $n = 1$ state? How does this compare with the number of revolutions the earth has made around the sun in the 4.5×10^9 years of its existence?

35. Show that the angular momentum of an electron in the nth orbit of a Bohr atom is equal to $nh/2\pi$. (In fact, this was the starting point of Bohr's original formulation of his theory of the hydrogen atom, which was carried out before de Broglie waves had been discovered.)

28–8   Atomic Spectra

28–10   Atomic Excitation

36. When radiation with a continuous spectrum is passed through a volume of hydrogen gas whose atoms are all in the ground state, which spectral series will be present in the resulting absorption spectrum?

37. Explain why the spectrum of hydrogen has many lines, although a hydrogen atom contains only one electron.

38. Would you expect the fact that the atoms of an excited gas are in rapid random motion to have any effect on the sharpness of the spectral lines they produce?

39. What kind of spectrum is observed in (a) light from the hot filament of a light bulb; (b) light from a sodium-vapor highway lamp; (c) light from an electric light bulb that has passed through cool sodium vapor?

40. Find the wavelength of the spectral line that corresponds to a transition in hydrogen from the $n = 6$ state to the $n = 3$ state. In what part of the spectrum is this?

41. Find the wavelength of the spectral line that corresponds to a transition in hydrogen from the $n = 10$ state to the ground state. In what part of the spectrum is this?

42. A proton and an electron, both at rest initially, combine to form a hydrogen atom in the ground state. A single photon is emitted in this process. What is its wavelength?

43. What is the shortest wavelength present in the Brackett series of spectral lines?

44. What is the shortest wavelength present in the Paschen series of spectral lines?

45. A beam of electrons is used to bombard gaseous hydrogen. What is the minimum energy in electron volts the electrons must have if the second line of the Paschen series, corresponding to a transition from the $n = 5$ state to the $n = 3$ state, is to be emitted?

46. A beam of electrons whose energy is 13 eV is used to bombard gaseous hydrogen. What series of wavelengths will be emitted?

28–11   The Laser

47. Why is the length of the optical cavity of a laser so important?

48. Steam at 100°C can be thought of as an excited state of water at 100°C. Suppose that a laser could be built based on the transition from steam to water, with the energy lost per molecule of steam appearing as a photon. What would the frequency of such a photon be? To what region of the spectrum does this correspond? The heat of vaporization of water is 6.77×10^{-20} J/molecule.

ANSWERS TO MULTIPLE CHOICE

1. b	**9.** b	**16.** c	**23.** a	**30.** c
2. b	**10.** d	**17.** b	**24.** c	**31.** c
3. d	**11.** d	**18.** c	**25.** c	**32.** d
4. c	**12.** d	**19.** b, c	**26.** a	**33.** b
5. a	**13.** d	**20.** c	**27.** c	**34.** b
6. b	**14.** c	**21.** b	**28.** c	**35.** a
7. b	**15.** a	**22.** b	**29.** d	**36.** d
8. d				

29

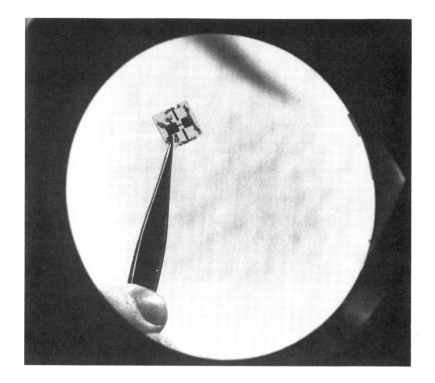

ATOMS IN COMBINATION

Individual atoms are rare on the earth and in its lower atmosphere; only the inert gas atoms (such as those of helium and argon) occur by themselves. All other atoms are found linked in small groups called *molecules* or in larger ones as liquids or solids. Some of these groups consist only of atoms of the same element, others consist of atoms of different elements, but in every case the arrangement is favored over separate atoms because of interactions between the atoms that reduce the total energy of the system. The quantum theory of the atom is able to account for these interactions in a natural way, with no special assumptions, which is further testimony to the power of this approach.

29-1   THE HYDROGEN MOLECULE

What is a molecule?

A molecule is a group of atoms that stick together strongly enough to act as a single particle. A molecule always has a certain definite structure; hydrogen molecules always consist of two hydrogen atoms each, for instance, and water molecules always consist of one oxygen atom and two hydrogen atoms each (see Fig. 14-14). A piece of iron

CHAPTER OBJECTIVES

Completing this chapter should enable you to:

1. Explain why the electron sharing of a covalent bond leads to a lower energy for a system of atoms and thus to a stable molecule.

2. Distinguish between amorphous and crystalline solids.

3. Distinguish between covalent and ionic solids.

4. Describe the nature of the van der Waals bonds between molecules.

5. Explain how the properties of a metal are related to the electron gas in its structure.

6. Explain how the electrical behavior of a solid is determined by the character of its energy bands.

7. Understand why metals are opaque to visible light whereas insulators are basically transparent.

8. Distinguish between *n*- and *p*-type semiconductors.

9. Explain why a semiconductor diode permits current to pass through it in one direction only.

10. Understand how a transistor amplifies weak signals into strong ones.

is not a molecule because, even though its atoms stay together, any number of them do so to form an object of any size or shape.

A molecule of a given kind is complete in itself with little tendency to gain or lose atoms. If one of its atoms is somehow removed or another atom is somehow attached, the result is a molecule of a different kind with different properties. A liquid or a solid, on the other hand, can gain or lose additional atoms of the kinds already present without changing its character.

The chief mechanism that bonds atoms together to form molecules involves the sharing of electrons by the atoms involved. As the shared electrons circulate around the atoms, they are more often between the atoms than they are on the outside, which produces an attractive force. To see why this force should occur, we may think of the atoms that are sharing the electrons as positive ions, so the presence of the shared electrons between them means a negative charge that holds the positive ions together. This is especially easy to see in the case of the hydrogen molecule, H_2, which consists of two hydrogen atoms.

The atoms in a molecule are held together by electron sharing

In the hydrogen molecule, the two protons are 7.42×10^{-11} m apart and the two electrons, one contributed by each atom, belong to the entire molecule rather than

FIG. 29–1 (a) Orbit model of the hydrogen molecule. (b) Probability cloud model of the hydrogen molecule.

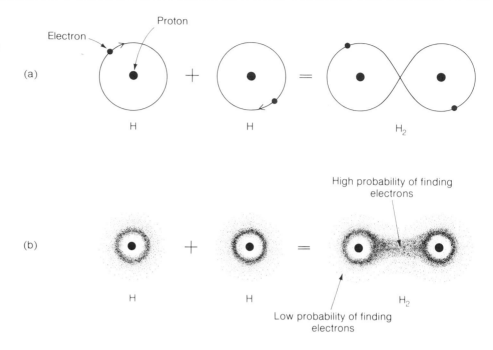

to their parent nuclei. Figure 29–1(a) is a picture of the H_2 molecule in terms of electron orbits. A more realistic picture of this molecule is provided by the quantum theory, in which the notion of electrons with definite, predictable positions and speeds at every moment is replaced by the notion of probability clouds. Figure 29–1(b) shows how the electron probability clouds of two hydrogen atoms join to form the probability cloud of a hydrogen molecule.

Origin of bonding force in H_2

Because the electrons spend more time on the average between the protons than they do on the outside, there is effectively a net negative charge between the protons. The attractive force this charge exerts on the protons is more than enough to counterbalance the direct repulsion between them. If the protons are too close together, however, their repulsion becomes dominant and the molecule is not stable. The balance between attractive and repulsive forces occurs at a separation of 7.42×10^{-11} m, where the total energy of the H_2 molecule is -4.5 eV. Hence 4.5 eV of work must be done to break an H_2 molecule into two H atoms:

$$H_2 + 4.5 \text{ eV} \rightarrow H + H$$

By comparison, the binding energy of the hydrogen atom is 13.6 eV:

$$H + 13.6 \text{ eV} \rightarrow p^+ + e^-$$

Atoms are more stable than molecules

This is an example of the general rule that it is easier to break up a molecule than to break up an atom.

Why do only two hydrogen atoms join together to form a molecule? Why not three, or four, or a hundred?

Molecular size and the exclusion principle

The basic reason for the limited size of molecules is the exclusion principle (Section 28–8), which forbids more than one electron in an atomic system from having the same set of quantum numbers. If a third H atom were to be brought up to an H_2 molecule, its electron would have to leave the $n = 1$ shell and go to the $n = 2$ shell in order for an H_3 molecule to be formed, since only two electrons (one with its spin up, the other with its spin down) can occupy the $n = 1$ shell. But an $n = 2$ electron in the hydrogen atom has over 10 eV more energy than an $n = 1$ electron, and the binding energy in H_2 is only 4.5 eV. Hence an H_3 molecule, if it could somehow be put together, would immediately break apart into $H_2 + H$ with the release of energy. Similar considerations hold for other molecules.

The exclusion principle is also responsible for the fact that the inert gases—helium, neon, argon, and so on—do not occur as molecules. The electron shells of the inert gas atoms are all filled to capacity, so in order for two of them to share an electron pair, one of the electrons would have to go into an empty shell of higher energy. The increase in energy involved would be more than the energy decrease produced by sharing the electrons, and therefore no such molecules as He_2 or Ne_2 occur.

Why the inert gases do not form molecules

29-2 COVALENT BOND

The mechanism by which electron sharing holds atoms together to form molecules is known as *covalent bonding*. Often it is convenient to think of the atoms as being held together by *covalent bonds*, with each shared pair of electrons making up a bond.

A shared pair of electrons constitutes a covalent bond

More complex atoms than hydrogen also join together to form molecules by sharing electrons. Depending upon the electron structures of the atoms involved, there may be one, two, or three covalent bonds—shared electron pairs—between the atoms. For example, in the oxygen molecule O_2 there are two bonds between the O atoms, and in the nitrogen molecule N_2 there are three bonds between the N atoms. Covalent bonds are represented either by a pair of dots or a single dash for each shared pair of electrons. Thus the H_2, O_2, and N_2 molecules can be represented as follows:

More than one pair of electrons may be shared

H:H   or   H—H

O::O   or   O=O

N:::N   or   N≡N

Covalent bonds are not limited to atoms of the same element nor to only two atoms per molecule. Here are two examples of more complicated molecules, with a line representing each covalent bond:

```
                            H
                            |
                  H         N—H
Water, H₂O        |    Ammonia, NH₃        |
                  O—H                      H
```

We notice that oxygen participates in two bonds in H_2O and nitrogen in three bonds in NH_3, whereas H participates in only one bond in both cases. This behavior is in accord with the fact that the hydrogen, oxygen, and nitrogen molecules respectively have one, two, and three bonds between their atoms.

Carbon atoms tend to form four covalent bonds at the same time, since they have

A carbon atom participates in four covalent bonds

four electrons in their outer shells and these shells lack four electrons to be complete. Various distributions of these bonds are possible, including bonds between adjacent carbon atoms in a complex molecule. The structures of the common covalent molecules methane, carbon dioxide, and acetylene illustrate the different bonds in which carbon atoms can participate to form molecules.

$$\begin{array}{ccc} & \text{H} & \\ & | & \\ \text{H}\!-\!\text{C}\!-\!\text{H} & \text{O}=\text{C}=\text{O} & \text{H}-\text{C}\equiv\text{C}-\text{H} \\ & | & \\ & \text{H} & \end{array}$$

Methane   Carbon dioxide   Acetylene

Organic chemistry is the study of carbon compounds

Carbon atoms are so versatile in forming covalent bonds with each other as well as with other atoms that literally millions of carbon compounds are known, some whose molecules contain tens of thousands of atoms. Such compounds were once thought to originate only in living things, and their study is accordingly known even today as organic chemistry.

When atoms join together to form a molecule, their inner, complete electron shells undergo little change. In molecules whose atoms are the same, such as H_2, O_2, and N_2, the atoms share their outer electrons evenly, and on the average have uniform distributions of electric charge.

Polar and nonpolar molecules

In a molecule composed of different atoms, the shared electrons favor one or another of the atoms, depending upon which elements are involved. The resulting molecule has a nonuniform distribution of electric charge, with some parts of the molecule being positively charged and other parts being negatively charged. A molecule of this kind, as we learned earlier, is called a *polar molecule,* whereas a molecule whose charge distribution is symmetric is called a *nonpolar molecule.* Thus the covalent molecule HCl is polar because a chlorine atom has greater attraction for an electron than a hydrogen atom, a situation we can represent by placing the pair of dots that indicate a covalent bond closer to the Cl atom:

Hydrochloric acid molecule   H :Cl

Water molecules are highly polar

The water molecule is highly polar because the two O—H bonds are 104.5° apart (not 180° apart), and the two shared electron pairs spend more time near the O atom than near the H atoms:

Water molecule   H :O̤

H

29–3   SOLIDS AND LIQUIDS

The same covalent bonds that can tie several atoms together into a molecule can in certain circumstances also tie an unlimited number of them together into a solid or a liquid. Other bonding mechanisms are found in solids and liquids as well, one of which

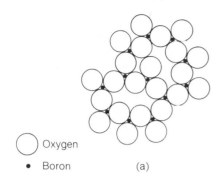

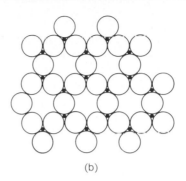

◯ Oxygen

• Boron   (a)   (b)

FIG. 29-2 Structure of B_2O_3. (a) Amorphous B_2O_3, a glass, has short-range order only in its structure. (b) Crystalline B_2O_3 has long-range order as well.

is responsible for the ability of metals to conduct electric current. Although only a very small proportion of the universe is in the solid state and even less is in the liquid state, matter in these forms constitutes much of the physical world of our experience, and modern technology is to a large extent based upon the unique characteristics of various solid materials.

Most solids are crystalline in nature, with their constituent atoms or molecules arranged in regular, repeated patterns. A crystal is thus characterized by the presence of *long-range order* in its structure.

Crystals have long-range order

Other solids lack the definite arrangements of atoms and molecules so conspicuous in crystals. They can be thought of as liquids whose stiffness is due to an exaggerated viscosity. Examples of such *amorphous* ("without form") solids are pitch, glass, and many plastics. The structures of amorphous solids exhibit *short-range order* only.

Amorphous solids have short-range order only

Some substances, for instance B_2O_3, can exist in either crystalline or amorphous forms. In both cases each boron atom is surrounded by three larger oxygen atoms, which represents a short-range order. In a B_2O_3 crystal a long-range order is also present, as shown in a two-dimensional representation in Fig. 29-2, whereas amorphous B_2O_3, a glassy material, lacks this additional regularity.

The lack of long-range order in amorphous solids means that the various bonds vary in strength. When an amorphous solid is heated, the weakest bonds break at lower temperatures than the others and the solid softens gradually, whereas in a crystalline solid the bonds break simultaneously and melting is a sudden process.

Liquids have more in common with solids than with gases, even though liquids share with gases the ability to flow from place to place. Because the density of a given liquid is usually approximately the same as that of the same substance in solid form, we conclude that the bonding mechanism is similar in both cases. When a solid is heated to its melting point, its atoms or molecules pick up enough energy to shift the bonds holding them together so that they form into separate clusters, but not until the liquid is heated to its vaporization point are the atoms or molecules able to break loose completely and form a gas. This interpretation of the liquid state is confirmed by X-ray studies that reveal definite clusters of atoms or molecules in a liquid (that is, short-range order like that in amorphous solids), but the clusters are constantly shifting their arrangements unlike the permanent arrangements in a solid.

Nature of the liquid state

The unusual behavior of water near the freezing point that was mentioned in Chapter 14 can be traced to the above effect. Ice crystals have very open structures because each H_2O molecule can participate in only 4 bonds with other H_2O molecules;

Why ice floats

in other solids, each atom or molecule may have as many as 12 nearest neighbors, which allows the assemblies to be more compact. Because clusters of molecules are smaller and less stable in the liquid state, water molecules are on the average packed more closely together than are ice molecules, and water has the higher density: Hence ice floats. The density of water increases from 0°C to a maximum at 4°C as large clusters of H_2O molecules break up into smaller ones that occupy less space in total, and only above 4°C does the normal thermal expansion of a liquid show up as a decreasing density with increasing temperature.

29-4   COVALENT AND IONIC SOLIDS

Four bonding mechanisms in solids

Molecular bonds are all basically covalent in character, even though in a few highly polar molecules the shared electrons stay so close to one of the partner atoms that there is effectively a transfer rather than a sharing of them. In solids there are four bonding mechanisms that occur, depending upon the nature of the solid: covalent, ionic, van der Waals, and metallic.

Covalent bonding in solids

The covalent bonds between atoms that are responsible for molecules also act to hold certain crystalline solids together. Figure 29-3 shows the array of carbon atoms in a diamond crystal, with each carbon atom sharing electron pairs with the four other carbon atoms adjacent to it. All the electrons in the outer shells of the carbon atoms participate in the bonding, and it is therefore not surprising that diamond is extremely hard and must be heated to over 3500°C before its crystal structure is disrupted and it melts. Purely covalent crystals are relatively few in number. In addition to diamond, some examples are silicon, germanium, and silicon carbide ("Carborundum"). Like diamond, all are hard and have high melting points.

Ionic bonding

In covalent bonding, two atoms share one or more pairs of electrons. In *ionic bonding,* one or more electrons from one atom transfer to another atom, producing a positive ion and a negative ion that attract each other. Let us see how ionic bonding works in the case of NaCl, whose crystals constitute ordinary table salt. We shall consider a hypothetical NaCl "molecule" for convenience; salt crystals are assemblies of Na and Cl atoms which, although they do not pair off into individual molecules, do interact through ionic bonds whose nature can be most easily examined in terms of a single NaCl unit.

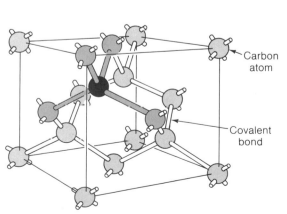

FIG. 29-3 Covalent structure of diamond. Each carbon atom shares electron pairs with four other carbon atoms. (Photo: Smithsonian Institution)

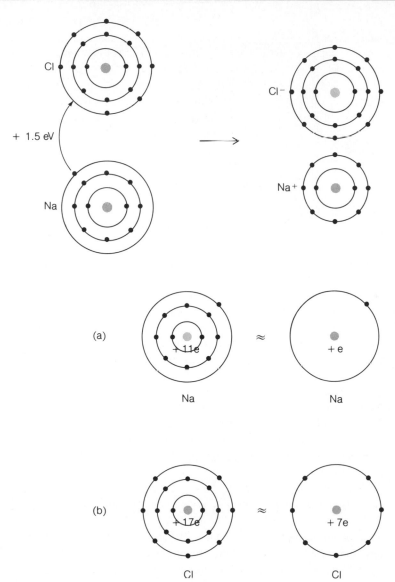

FIG. 29–4 Electron transfer in NaCl. The resulting ions attract each other electrically.

FIG. 29–5 (a) A sodium atom. The presence of 10 electrons in the $n = 1$ and $n = 2$ shells effectively shields the outer $n = 3$ electron from all but $+e$ of the nuclear charge. In consequence an Na atom tends to lose its outer electron to become an Na^+ ion. (b) A chlorine atom. The inner electrons leave unshielded $+7e$ of the nuclear charge. In consequence a Cl atom tends to pick up another electron to become a Cl^- ion.

Figure 29–4 shows schematically an electron shifted from a sodium atom, which becomes an Na^+ ion, to a chlorine atom, which becomes a Cl^- ion. The ions then attract each other electrically. The combination is stable because more energy is needed to pull the Na^+ and Cl^- ions apart than will be supplied by the return of the shifted electron from the Cl^- ion to the Na^+ ion. Not all pairs of atoms can interact by electron transfer to form stable combinations; let us see why Na and Cl have this ability.

The single outermost electron of the Na atom is relatively easy to detach because of the presence of the 10 inner electrons, which shield the outer electron from all but $+e$ of the actual nuclear charge of $+11e$ (Fig. 29–5(a)). In the Cl atom, on the other hand, the 10 inner electrons leave $+7e$ of the nuclear charge unshielded, so the attractive force on the outer electrons is much greater (Fig. 29–5(b)). Accordingly, a Cl

Why an Na atom tends to lose an electron and a Cl atom tends to gain an electron

FIG. 29–6   An NaCl crystal is composed of Na$^+$ and Cl$^-$ ions in an array such that each ion is surrounded by six ions of the other kind. Na$^+$ ions are small because they have lost their outer electrons.

atom tends to pick up an additional electron to become a Cl$^-$ ion. Thus Na and Cl are an ideal match: One readily loses an electron, the other readily gains an electron. The same is true for the partners in other ionic solids, which owe these tendencies to their electron structures.

The structure of an NaCl crystal is shown in Fig. 29–6. Each ion behaves like a point charge and thus tends to attract to itself as many ions of opposite sign as can fit around it. In an NaCl crystal each Na$^+$ ion is surrounded by six Cl$^-$ ions and vice versa. In crystals having different structures the number of "nearest neighbors" around each ion may be 3, 4, 6, 8, or 12. Ionic bonds are usually fairly strong, and consequently ionic crystals are strong, hard, and have high melting points.

Many crystalline bonds are partly ionic and partly covalent in origin. An example of such mixed bonding is quartz, SiO_2.

29–5   VAN DER WAALS BONDS

A number of molecules and nonmetallic atoms exist whose electronic structures do not lend themselves to either of the above kinds of bonding. The inert gas atoms, which have filled outer electron shells, and molecules such as methane,

$$
\begin{array}{c}
H \\
| \\
H - C - H \\
| \\
H
\end{array}
$$

whose outer electrons are fully involved in the molecular bond itself, fall into this category. However, even these virtually noninteracting substances condense into solids and liquids at low enough temperatures through the action of what are known collectively as *van der Waals forces*.

Polar-polar attraction   The electric attraction between polar molecules that was discussed in Section 19–7 is an example of a van der Waals force. These molecules have asymmetric distributions

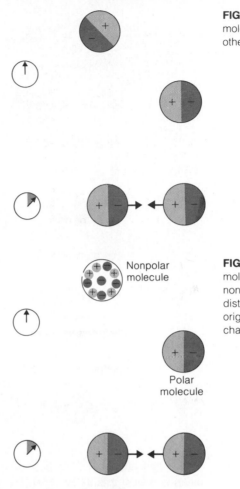

FIG. 29−7 Polar molecules attract each other electrically.

FIG. 29−8 A polar molecule attracts a nonpolar one by first distorting the latter's originally symmetric charge distribution.

Nonpolar molecule

Polar molecule

of charge, with one end positive and the other negative. When one such molecule is near another, the ends of opposite polarity attract each other to hold the molecules together (Fig. 29−7).

A somewhat similar effect occurs when a polar molecule is near a nonpolar one. The electric field of the polar molecule distorts the initially symmetric charge distribution of the nonpolar one, as in Fig. 29−8, and the two then attract each other in the same way as any other pair of polar molecules. This phenomenon is similar to that involved in the attraction of bits of paper by a charged rubber comb shown in Fig. 18−9.

Polar-nonpolar attraction

Nonpolar molecules attract one another in much the same way that polar molecules attract nonpolar ones. In a nonpolar molecule, the electrons are distributed symmetrically *on the average,* but *at any moment* one part of the molecule contains more electrons than usual and the rest of the molecule contains fewer. Thus *every* molecule (and atom) behaves as though it is polar, though the direction and magnitude of the polarization vary constantly. The fluctuations in the charge distributions of nearby nonpolar molecules keep in step through the action of electric forces, and these forces also hold the molecules together (Fig. 29−9).

Nonpolar-nonpolar attraction

FIG. 29−9 Nonpolar molecules have, on the average, symmetric charge distributions, but at any instant the distributions are not necessarily symmetric. The fluctuations in the charge distributions of adjacent nonpolar molecules keep in step, which leads to an attractive force between them.

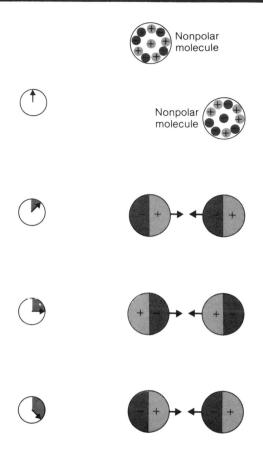

Van der Waals bonds are weak

In general, van der Waals bonds are considerably weaker than ionic, covalent, and metallic bonds; usually less than 1% as much energy is needed to remove an atom or molecule from a van der Waals solid as is required in the case of ionic or covalent crystals. As a result the inert gases and compounds with symmetric molecules liquify and vaporize at rather low temperatures. Thus the boiling point of argon is −186°C and the boiling point of methane is −161°C. Molecular crystals, whose lattices consist of individual molecules held together by van der Waals forces, generally lack the mechanical strength of other kinds of crystals.

29−6 METALLIC BOND

A fourth important type of cohesive force in crystalline solids is the *metallic bond*, which has no molecular counterpart.

An electron "gas" bonds metals

A characteristic property of all metal atoms is the presence of only a few electrons in their outer shells, and these electrons can be detached relatively easily to leave behind positive ions. According to the theory of the metallic bond, a metal in the solid state consists of an assembly of atoms that have given up their outermost electrons to a common "gas" of freely moving electrons that circulates through the entire metal.

The electric interaction between the positive ions and the negative electron gas holds the metal together.

This theory has much in its favor. The high electrical and thermal conductivity of metals follows from the ability of the free electrons to migrate through their crystal structures, in contrast to the electrons in ionic and covalent crystals which are bound to particular atoms or pairs of atoms. Also, since the atoms in a metal interact through the medium of a common electron gas, the properties of mixtures of different metal atoms should not depend very much on the relative proportions of each kind of atom, provided that their sizes are similar. This prediction is fulfilled in the observed behavior of alloys, as opposed to the specific atomic proportions characteristic of ionic and covalent solids.

Table 29–1 summarizes the properties of the four types of crystalline solids that have been discussed here.

29–7 ENERGY BANDS

The notion of energy bands provides a useful framework for understanding the electrical behavior of solids. The nature and properties of semiconductors in particular are clarified with the help of an energy-band analysis.

When atoms are brought as close together as those in a crystal, they interact with one another to such an extent that their outer electron shells constitute a single system of electrons common to the entire array of atoms. The exclusion principle prohibits more than two electrons (one with each spin) in any energy level of a system. This

TABLE 29–1
Types of crystalline solids

Type	Covalent	Ionic	Molecular	Metallic
Lattice	Shared electrons	Negative ion / Positive ion	Instantaneous charge separation in molecule	Metal ion / Electron gas
Bond	Shared electrons	Electric attraction	Van der Waals forces	Electron gas
Properties	Very hard; high melting point; soluble in very few liquids	Hard; high melting point; may be soluble in polar liquid such as water	Soft; low melting and boiling points; soluble in covalent liquids	Ductile; metallic luster; ability to conduct heat and electric current readily
Example	Diamond, C	Sodium chloride, NaCl	Methane, CH_4	Iron, Fe

FIG. 29–10 Energy bands replace energy levels of outer electrons in an assembly of atoms that are close together.

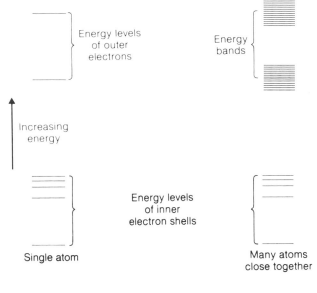

FIG. 29–11 (a) Overlapping energy bands. (b) The gap between energy bands that do not overlap is called a *forbidden band*.

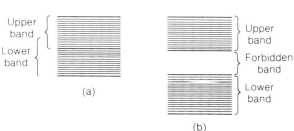

Origin of energy bands in a crystal

principle is obeyed in a crystal because the energy levels of the outer electron shells of the various atoms are all slightly altered by their mutual interaction. (The inner shells do not interact and therefore do not undergo a change.) As a result of the shifts in the energy levels, an *energy band* exists in a crystal in place of each sharply defined energy level of its component atoms (Fig. 29–10). While these bands are actually composed of a great many individual energy levels, as many as there are atoms in the crystal, the levels are so near one another as to form a continuous distribution.

A forbidden band may separate energy bands

The energy bands in a crystal correspond to energy levels in an atom, and an electron in a crystal can only have an energy that falls within one of these bands. The various energy bands in a crystal may or may not overlap, depending upon the composition of the crystal (Fig. 29–11). If they do not overlap, the gaps between them represent energy values that electrons in the crystal cannot have. The gaps are accordingly known as *forbidden bands*.

The energy bands we have been speaking of contain all the possible energies that can be possessed by electrons. The electrical properties of a crystalline solid depend upon both its energy-band structure and the way in which the bands are normally occupied by electrons. We shall examine a few specific cases to see how the energy-band approach accounts for the observed electrical behavior of such solids.

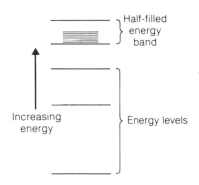

FIG. 29–12 Energy levels and bands in solid sodium, which is a metal. (Not to scale.)

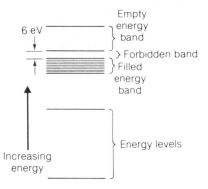

FIG. 29–13 Energy levels and bands in diamond, which is an electrical insulator. (Not to scale.)

Figure 29–12 shows the energy bands of solid sodium. A sodium atom has only one electron in its outer shell. This means that the upper energy band in a sodium crystal is only half filled with electrons, since each level within the band, like each level in the atom, is capable of containing *two* electrons. When an electric field is established in a sodium crystal, electrons readily acquire the small additional energy they need to move up in their energy band. The additional energy is in the form of kinetic energy, and the moving electrons constitute an electric current. Sodium is therefore a good conductor, as are other crystalline solids with energy bands that are only partially filled. Such solids are metals.

Energy bands of a metal

Figure 29–13 shows the energy-band structure of diamond. The two lower energy bands are completely filled, and there is a gap of 6 eV between the top of the higher of these bands and the empty band above it. Hence a minimum of 6 eV of additional energy must be given to an electron in a diamond crystal if it is to be capable of free motion, since it cannot have an energy lying in the forbidden band. Such an energy boost cannot readily be imparted to an electron in a crystal by an electric field. Diamond, like other solids with similar energy-band structures, is therefore an electrical insulator.

Energy bands of an insulator

Silicon has a crystal structure similar to that of diamond, and, as in diamond, a gap separates the top of a filled energy band from an empty higher band. However, whereas the gap is 6 eV wide in diamond, it is only 1.1 eV wide in silicon. At very low temperatures silicon is little better than diamond as a conductor, but at room temperature a small proportion of its electrons can possess enough kinetic energy of thermal origin to exist in the higher band. These few electrons are sufficient to permit a limited amount of current to flow when an electric field is applied. Thus silicon has a resistivity intermediate between those of conductors (such as sodium) and those of insulators (such as diamond), and is classified as a *semiconductor*.

Energy bands of a semiconductor

The optical properties of solids and their energy-level structures are closely related. Photons of visible light have energies from about 1 to 3 eV. A free electron in a metal can readily absorb such a photon since its allowed energy band is only partly filled, and metals are accordingly opaque. The characteristic luster of a metal is due to the

Optical properties of solids

reradiation of light absorbed by its free electrons. If the metal surface is smooth, the reradiated light appears as a reflection of the original incident light.

For an electron in an insulator to absorb a photon, on the other hand, the photon energy must be more than 3 eV in order for the electron to jump across the forbidden band to the next allowed band. Insulators are therefore unable to absorb photons of visible light and are transparent. To be sure, most samples of insulating materials do not appear transparent, but this is usually due to the scattering of light by irregularities in their structures. Insulators are opaque to ultraviolet light, whose higher frequencies mean high enough photon energies to enable electrons to cross the forbidden band. Because the forbidden bands in semiconductors are comparable in width to the photon energies of visible light, they are usually opaque to visible light but are transparent to infrared light whose lower frequencies mean photon energies too low to be absorbed. Thus infrared lenses can be made from the semiconductor germanium, whose appearance is that of a solid metal.

29–8   IMPURITY SEMICONDUCTORS

Electrons carry current in an *n*-type semiconductor

The conductivity of semiconductors can be markedly affected by slight amounts of impurity. Suppose we add several arsenic atoms to a silicon crystal. Arsenic atoms have five electrons in their outermost shells, while silicon atoms have four. When an arsenic atom replaces a silicon atom in a silicon crystal, four of its electrons join in covalent bonds with its nearest neighbors. The fifth outer electron needs very little energy to be detached and move about freely in the crystal. Such a solid is called an *n-type* semiconductor because electric current in it is carried by the motion of negative charges toward the positive end of a sample of it (Fig. 29–14). In an energy-band diagram, as in Fig. 29–15, the effect of arsenic as an impurity in silicon is to supply occupied energy levels just below an empty energy band. These levels are called *donor levels*.

A "hole" is a missing electron

If we add gallium atoms to a silicon crystal, a different effect occurs. Gallium atoms have only three electrons in their outer shells, and their presence leaves vacancies

FIG. 29–14 Current in an *n*-type semiconductor is carried by excess electrons that do not fit into the electron-bond structure of the crystal.

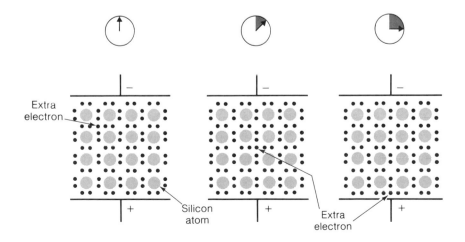

Extra electron

Silicon atom

Extra electron

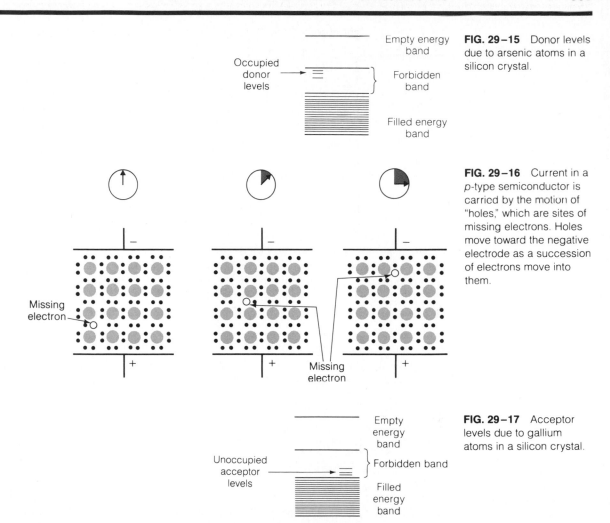

FIG. 29-15 Donor levels due to arsenic atoms in a silicon crystal.

FIG. 29-16 Current in a *p*-type semiconductor is carried by the motion of "holes," which are sites of missing electrons. Holes move toward the negative electrode as a succession of electrons move into them.

FIG. 29-17 Acceptor levels due to gallium atoms in a silicon crystal.

called *holes* in the electron structure of the crystal. An electron requires little energy to move into a hole, but as it does so it leaves a new hole in its previous location. When an electric field is applied to a silicon crystal in which gallium is present as an impurity, electrons move toward the positive electrode by successively filling holes. The flow of current here is best described in terms of the motion of the holes, which behave as though they are positive charges since they move toward the negative electrode (Fig. 29-16). A material of this kind is therefore called a *p-type* semiconductor. In the energy-band diagram of Fig. 29-17 we see that the effect of gallium as an impurity is to provide energy levels, called *acceptor levels,* just above the highest filled band. Electrons that enter these levels leave behind unoccupied levels in the formerly filled band which make possible the conduction of current.

Holes carry current in a *p*-type semiconductor

Adding an impurity to a semiconductor is called *doping*. Phosphorus, antimony, and bismuth as well as arsenic have atoms with five outer electrons and so can be used as donor impurities in doping silicon and germanium to yield an *n*-type semiconductor. Similarly indium and tellurium as well as gallium have atoms with three outer electrons and so can be used as acceptor impurities. A very small amount of impurity can lead to a dramatic change in the conductivity of a semiconductor; for instance, 1 part of a donor impurity per 10^8 parts of germanium increases its conductivity by a factor of 12. Silicon and germanium are not the only semiconducting materials with practical applications: Another important class of semiconductors consists of such compounds as GaAs, GaP, InSb, InAs, and InP.

29−9   SEMICONDUCTOR DIODES

Semiconductor junctions

The significance of semiconductors in technology arises from the degree of control of electric current that can be accomplished by combining n- and p-type semiconductors in various ways. Most semiconductor devices depend for their action on the properties of junctions between n- and p-type materials. Such junctions can be produced in several ways. One of them is to gradually pull out a crystal that is forming in molten silicon that contains, say, a donor impurity, and quickly add an acceptor impurity to the melt during the process. The first part of the resulting crystal to solidify will be n type and the rest will be p type if the proportion of acceptor impurity exceeds that of donor impurity. Another technique, which is especially adapted to the manufacture of integrated circuits, involves diffusing impurities in gas form into a semiconductor wafer in regions defined by masks. A series of diffusion steps using donor and acceptor impurities and other materials can produce circuits that contain as many as 50,000 resistors, capacitors, diodes, and transistors on a silicon chip 5 mm square.

Reverse bias

An important property of a p-n semiconductor junction is that electric current can pass through it much more readily in one direction than in the other. In the crystal shown in Fig. 29–18, the left-hand end is a p-type region (current carried by holes)

FIG. 29−18   (a) Current is carried in a p-type semiconductor by the motion of holes and in an n-type semiconductor by the motion of electrons. (b) Reverse bias, little current flows. (c) Forward bias, much current flows.

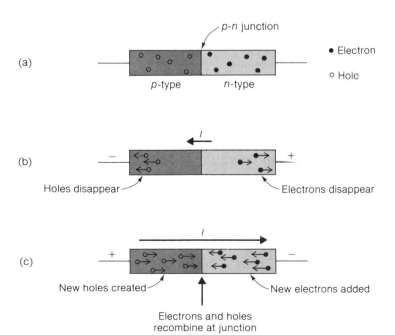

and the right-hand end is an *n*-type region (current carried by electrons). When a voltage is applied across the crystal so that the *p* end is negative and the *n* end positive, the holes in the *p* region migrate to the left and the electrons in the *n* region migrate to the right. Only a limited number of holes and electrons are in the respective regions and new ones appear spontaneously at only a very slow rate; the current through the entire crystal is therefore negligible. This situation is called *reverse bias*.

Figure 29–18(c) shows the same crystal with the connections changed so that the *p* end is positive and the *n* end negative. Now new holes are created continuously by the removal of electrons at the *p* end while new electrons are fed into the *n* end of the crystal. The holes migrate to the right and the electrons to the left to produce a net positive current flowing from + to −. The electrons and holes meet at the junction between the *p* and *n* regions and recombine there: A hole is a missing electron, and when electrons and holes come together they disappear into the regular structure of the crystal and can no longer act as current carriers. Thus current can flow readily through a semiconductor junction from the *p* to the *n* region, but hardly at all in the opposite direction. Figure 29–19 (a) shows how *I* varies with *V* for a *p-n* rectifier. Such devices, called *diodes*, are widely used today. The symbol of a diode is ⭢⊢ .

Although the reverse current in many semiconductor diodes remains very small, in certain diodes the reverse current rises sharply when a particular voltage is reached, as in Fig. 29–19 (b). Such diodes are called *Zener diodes* and are particularly useful in voltage regulation circuits. The symbol of a Zener diode is ⭢⊦ .

One effect that contributes to the sudden increase in reverse current in a Zener diode is the pulling away of electrons on the *p* side of the junction to the *n* side by the reverse voltage. Another is the acceleration of electrons near the junction by the electric field there, which gives them enough energy to knock out other electrons from atoms in their paths and thus create new electron-hole pairs. The voltage at which Zener breakdown occurs depends on the construction of the diode and can be as little as 6 V. Figure 29–20 shows a simple voltage regulation circuit. No current passes through the diode until the breakdown voltage is reached. After this, the diode continues to maintain the breakdown voltage across its terminals, so the output voltage has this value even if the input voltage varies.

Energy is needed to create an electron-hole pair, and this energy is released when an electron and a hole recombine. In silicon and germanium the recombination energy is absorbed by the crystal as heat, but in certain other semiconductors, notably gallium

Light-emitting diodes

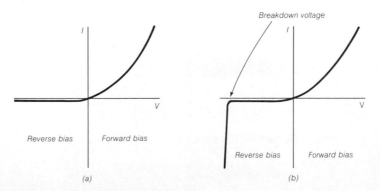

FIG. 29–19 (a) The relationship between voltage and current in an ordinary semiconductor diode. The reverse current is extremely small and is exaggerated here for clarity. (b) Voltage-current relationship in a Zener diode. The reverse current increases sharply at a particular voltage.

FIG. 29–20 A simple voltage-regulator circuit using a Zener diode. The resistor is needed to prevent too great a current in the diode in the breakdown region.

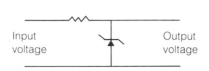

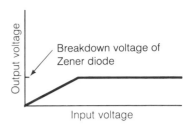

Input voltage

Output voltage

Output voltage

Input voltage

Breakdown voltage of Zener diode

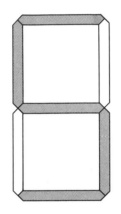

FIG. 29–21 A typical digital display uses seven light-emitting diodes to form two squares. Shown is how the number five is formed.

Junction transistors

arsenide, a photon is emitted when recombination occurs. This is the basis of the *light-emitting diode* (Fig. 29–21). Solid-state lasers have also been made that make use of this phenomenon.

29–10 TRANSISTORS

Transistors are semiconductor devices that are able to amplify weak signals into strong ones. Figure 29–22 shows an *n-p-n* junction transistor, one of a variety of such devices, which consists of a *p*-type material sandwiched between *n*-type materials. The *p*-type region is called the *base*, and the two *n*-type regions are the *emitter* and the *collector.* In the figure the transistor is connected as a simple amplifier. There is a forward bias across the emitter-base junction in this circuit, so electrons pass readily from the emitter to the base. Depending upon the rate at which holes are produced in the base, which in turn depends upon its potential relative to the emitter and hence upon the signal input, a certain proportion of the electrons from the emitter will recombine there. The rest of the electrons migrate across the base to the collector to complete the emitter-collector circuit. If the base is at a high positive potential relative to the emitter, many holes are produced there to recombine with electrons from the emitter, and little current can flow through the transistor. If the relative positive potential of the base is low, the number of electrons arriving from the emitter will exceed the number of holes being formed in the base, and the surplus electrons will continue across the base to the collector. Changes in the input-circuit current are thus mirrored by changes in the output-circuit current, which is only a few percent smaller.

The ability of the transistor of Fig. 29–22 to produce amplification comes about

FIG. 29–22 A simple transistor amplifier.

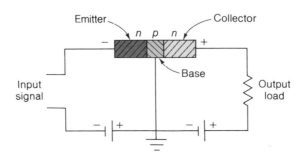

Emitter

Collector

n p n

Base

Input signal

Output load

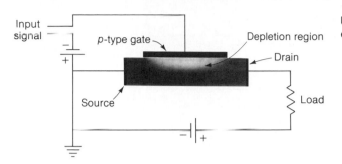

Input signal

p-type gate

Source

Depletion region

Drain

Load

FIG. 29–23 Λ field-effect transistor.

because the reverse bias across the base-collector junction permits a much higher voltage in the output circuit than that in the input circuit. Since electric power = current × voltage, the power of the output signal can greatly exceed the power of the input signal.

One problem with the junction transistor is that it is difficult to incorporate large numbers of them in an integrated circuit. The *field-effect transistor* (FET) lacks this disadvantage and is widely used today. As in Fig. 29–23, an *n*-channel field-effect transistor consists of a strip of *n*-type material with contacts at each end together with a strip of *p*-type material, called the *gate,* on one side. When connected as shown, electrons move from the *source* terminal to the *drain* terminal through the *n*-type channel. The *p-n* junction is given a reverse bias, and as a result both the *n* and *p* materials near the junction are depleted of charge carriers (see Fig. 29–17(b)). The higher the reverse potential on the gate, the larger the depleted region in the channel, and the fewer the electrons available to carry the current. Thus the gate voltage controls the channel current.

In a metal oxide semiconductor field-effect transistor (MOSFET), the semiconductor gate is replaced by a metal film separated from the channel by an insulating layer of silicon dioxide. The metal film is thus capacitively coupled to the channel, and its potential controls the drain current through the number of induced charges in the channel. A MOSFET is easier to manufacture than a FET and occupies only a few percent of the area needed for a junction transistor.

Field-effect transistors

A silicon "chip" smaller than a fingernail that contains thousands of resistors, capacitors, diodes, and transistors. The basic circuits of modern electronic devices, notably computers, are incorporated in such chips.

IMPORTANT TERMS

A **molecule** is a group of atoms that stick together strongly enough to act as a single particle. A molecule of a given compound always has a certain definite structure and is complete in itself with little tendency to gain or lose atoms.

In a **covalent bond** between atoms in a molecule or a solid, the atoms share one or more electron pairs.

Solids whose constituent atoms or molecules are arranged in regular, repeated patterns are called **crystalline.** When only short-range order is present, the solid is **amorphous.**

In an **ionic bond,** electrons are transferred from one atom to another, and the two then attract each other electrically.

Van der Waals forces arise from the electric attraction between asymmetrical charge distributions in atoms and molecules.

The **metallic bond** that holds metal atoms together in the solid state arises from a "gas" of freely moving electrons that pervades the entire metal.

Because the atoms in a crystal are so close together, the energy levels of their outer electron shells are altered slightly to produce **energy bands** characteristic of the entire crystal, in place of the individual sharply defined energy levels of the separate atoms. Gaps between energy bands represent energies forbidden to electrons in the crystal and are called **forbidden bands.**

Semiconductors are intermediate in their ability to carry electric current between conductors and insulators. An ***n*-type semiconductor** is one in which electric current is carried by the motion of electrons. A ***p*-type semiconductor** is one in which electric current is carried by the motion of holes, which are vacancies in the electron structure that behave like positive charges.

A semiconductor **diode** permits current to flow through it only in one direction. Signals can be amplified by means of a **transistor,** another semiconductor device.

MULTIPLE CHOICE

1. Relative to the energy needed to separate the electrons of an atom from its nucleus, the energy needed to separate the atoms of a molecule is
     a. smaller.   b. about the same.
     c. larger.   d. much larger.

2. The reason only two H atoms join to form a hydrogen molecule is a consequence of
     a. the exclusion principle.
     b. the size of the H atom.
     c. the shape of the H atom.
     d. the mass of the H atom.

3. The hydrogen atoms in a hydrogen molecule stick together because
     a. their electrons spend all their time between the nuclei.
     b. their electrons spend more time between the nuclei than outside them.
     c. their electrons spend more time outside the nuclei than between them.
     d. their nuclei attract each other.

4. In a covalent bond,
     a. electrons are shifted from one atom to another.
     b. only atoms of the same element are present.
     c. there must be at least one carbon atom.
     d. adjacent atoms share electron pairs.

5. In a molecule that consists of two atoms,
     a. only a single pair of electrons can be shared.
     b. more than one pair of electrons can be shared.
     c. one of the atoms must be a hydrogen atom.
     d. one of the atoms must be a carbon atom.

6. The number of covalent bonds a hydrogen atom forms when it combines chemically is
     a. 1.   b. 2.
     c. 3.   d. 4.

7. The number of covalent bonds a carbon atom usually forms when it combines chemically is
     a. 1.   b. 2.
     c. 3.   d. 4.

8. Compared with the total number of electrons in the atoms that come together to form a molecule, the number of electrons in the molecule is
     a. less.
     b. the same.
     c. more.
     d. any of the above, depending on the type of molecule.

9. When two or more atoms join to form a stable molecule,
     a. energy is absorbed.
     b. energy is given off.
     c. there is no energy change.
     d. any of the above, depending upon the circumstances.

10. Short-range order is never found in
     a. crystalline solids.
     b. amorphous solids.
     c. liquids.
     d. gases.

11. A crystalline solid always
     a. is transparent.
     b. is held together by covalent bonds.
     c. is held together by ionic bonds.
     d. has long-range order in its structure.

12. An amorphous solid is closest in structure to
    a. a covalent crystal.
    b. an ionic crystal.
    c. a semiconductor.
    d. a liquid.

13. The lowest melting points are usually found in solids held together by
    a. covalent bonds.
    b. ionic bonds.
    c. metallic bonds.
    d. van der Waals bonds.

14. Van der Waals forces arise from
    a. electron transfer.
    b. electron sharing.
    c. symmetrical charge distributions.
    d. asymmetrical charge distributions.

15. The particles that make up the lattice of the van der Waals crystal of a compound are
    a. electrons.   b. atoms.
    c. ions.   d. molecules.

16. The particles that make up the lattice of an ionic crystal are
    a. electrons.   b. atoms.
    c. ions.   d. molecules.

17. The particles that make up the lattice of a covalent crystal are
    a. electrons.   b. atoms.
    c. ions.   d. molecules.

18. A property of metals that is not due to the electron "gas" that pervades them is their unusual ability to
    a. conduct electricity.
    b. conduct heat.
    c. reflect light.
    d. form oxides.

19. The weakest bonds are found in
    a. covalent solids.   b. ionic solids.
    c. metallic solids.   d. molecular solids.

20. Ice is an example of a
    a. covalent solid.
    b. ionic solid.
    c. metallic solid.
    d. molecular solid.

21. Diamond is an example of a
    a. covalent solid.
    b. ionic solid.
    c. metallic solid.
    d. molecular solid.

22. A crystal whose upper energy band is partly occupied by electrons is a
    a. conductor.
    b. insulator.
    c. *n*-type semiconductor.
    d. *p*-type semiconductor.

23. A crystal with a wide forbidden band between a filled lower band and an empty upper band is
    a. a conductor.
    b. an insulator.
    c. an *n*-type semiconductor.
    d. a *p*-type semiconductor.

24. Diamond is transparent because
    a. its atoms are far apart.
    b. it is held together by covalent bonds.
    c. the forbidden band above its highest filled energy band is narrow.
    d. the forbidden band above its highest filled energy band is wide.

25. A hole in a *p*-type semiconductor is
    a. an excess electron.
    b. a missing electron.
    c. a missing atom.
    d. a donor level.

26. Current in a *p*-type semiconductor is carried by
    a. electrons.   b. holes.
    c. positive ions.   d. negative ions.

27. Current in an *n*-type semiconductor is carried by
    a. electrons.   b. holes.
    c. positive ions.   d. negative ions.

28. A junction between *n*- and *p*-type semiconductors
    a. acts as an insulator for currents in both directions.
    b. conducts current in both directions equally well.
    c. conducts current readily when the *p* end is positive and the *n* end is negative.
    d. conducts current readily when the *p* end is negative and the *n* end is positive.

29. A light-emitting diode gives off a photon of light when
    a. two holes collide.
    b. two electrons collide.
    c. a hole is created by the freeing of an electron.
    d. a hole is filled by recombination with an electron.

EXERCISES

29–1   The Hydrogen Molecule

29–2   Covalent Bond

1. What is wrong with the model of a hydrogen molecule in which the two electrons are supposed to follow figure-eight orbits that encircle the two protons?

2. What must be true of the spins of the two electrons in the H_2 molecule?

3. The atoms in a molecule are said to share electrons, yet some molecules are polar. Explain.

4. The energy needed to detach the electron from a hydrogen atom is 13.6 eV, but the energy needed to detach an electron from a hydrogen molecule is 15.7 eV. Why do you think the latter energy is greater?

5. What property of carbon atoms enables them to form so many varied and complex molecules?

6. At what temperature would the average kinetic energy of the molecules in a hydrogen gas be equal to their binding energy?

29–4   Covalent and Ionic Solids

7. The separation between Na^+ and Cl^- ions in an NaCl crystal is 2.8×10^{-10} m, whereas it is 2.4×10^{-10} m in an NaCl molecule such as might exist in the gaseous state. Why is it reasonable that these separations be different?

29–5   Van der Waals Bonds

8. The temperature of a gas falls when it passes slowly from a full container to an empty one through a porous plug. Since the expansion is into a rigid container, no mechanical work is done. What is the origin of the fall in temperature?

9. Van der Waals forces can hold inert gas atoms together to form solids, but they cannot hold such atoms together to form molecules in the gaseous state. Why not?

29–6   Metallic Bond

10. Does the "gas" of freely moving electrons in a metal include all the electrons present? If not, which electrons are members of the "gas"?

11. What is the connection between the ability of a metal to conduct electricity and its ability to conduct heat?

29–7   Energy Bands
29–8   Impurity Semiconductors

12. What kind of solid is one whose upper energy band is only partly filled with electrons?

13. What is the basic reason that energy bands rather than specific energy levels exist in a solid?

14. The energy gap in silicon is 1.1 eV, and in diamond it is 6 eV. How transparent would you expect these substances to be to visible light?

15. How does the energy-band structure of a solid determine whether it is a conductor, a semiconductor, or an insulator of electricity?

16. The forbidden energy band in germanium that lies between the highest filled band and the empty band above it has a width of 0.7 eV. Compare the conductivity of germanium with that of silicon at (a) very low temperatures and (b) room temperature.

17. Does the addition of a small amount of indium to germanium result in the formation of an *n*- or a *p*-type semiconductor? An indium atom has three electrons in its outermost shell; a germanium atom has four.

ANSWERS TO MULTIPLE CHOICE

1. a	**9.** b	**16.** c	**23.** b
2. a	**10.** d	**17.** b	**24.** d
3. b	**11.** d	**18.** d	**25.** b
4. d	**12.** d	**19.** d	**26.** b
5. b	**13.** d	**20.** d	**27.** a
6. a	**14.** d	**21.** a	**28.** c
7. d	**15.** d	**22.** a	**29.** d
8. b			

30

THE NUCLEUS

Until now we have not had to consider the nucleus of an atom as anything but a tiny positively charged object whose only functions are to provide the atom with most of its mass and to hold its electrons in place. Since the behavior of atomic electrons is responsible for the behavior of matter in bulk, the properties of matter we have been exploring, except for mass, have nothing directly to do with atomic nuclei. Nevertheless, the nucleus turns out to be of supreme importance in the universe: The elements exist by virtue of the ability of nuclei to hold multiple electric charges, and the energy involved in nearly all natural processes has its ultimate origin in nuclear reactions and transformations.

30–1 NUCLEAR STRUCTURE

The forces holding a nucleus together are much stronger than electric or gravitational forces

The nature and behavior of the electron structure of the atom was understood before even the composition of its nucleus was known. The reason is that the nucleus is held together as a unit by forces vastly stronger than the electric forces that hold the electrons to the nucleus, and it is correspondingly harder to break apart a nucleus to find out what is inside. Changes in the electron structure of an atom, such as those that occur

CHAPTER OBJECTIVES

Completing this chapter should enable you to:

1. Describe the composition of a nucleus and what is meant by the isotopes of an element.

2. Find the binding energy of a nucleus.

3. Calculate the energy given off in nuclear fission and fusion reactions.

4. State the various kinds of radioactive decay and explain why they occur.

5. Describe how the various kinds of radiation can be detected.

6. Interpret the changing rate of decay of a radioisotope in terms of half-life.

7. Understand the hazards of ionizing radiation.

8. Discuss nuclear fission and the conditions needed for a chain reaction to occur.

9. Describe the construction and operation of a nuclear reactor.

10. Explain why there is so much interest in breeder reactors.

11. Discuss the potential of thermonuclear energy.

---

in the emission of photons or in the formation of chemical bonds, involve energies of only several eV, whereas changes in nuclear structure involve energies of several MeV, a million times more.

Let us first inquire into the composition of the nucleus: What is it that gives a nucleus its characteristic mass and charge?

The nucleus of the hydrogen atom consists of a single proton, whose charge is $+e$ and whose mass is

$$m_{\text{proton}} = 1.673 \times 10^{-27} \text{ kg}$$

The proton mass is 1836 times that of the electron, so by far the major part of the hydrogen atom's mass resides in its nucleus. This is true of all other atoms as well.

Elements more complex than hydrogen have nuclei that contain *neutrons* as well as protons. The neutron, as its name suggests, is uncharged. The neutron mass is slightly more than that of the proton:

Atomic nuclei consist of protons and neutrons, jointly called *nucleons*

$$m_{\text{neutron}} = 1.675 \times 10^{-27} \text{ kg}$$

Neutrons and protons are jointly called *nucleons*.

Every neutral atom contains the same number of protons and electrons; this number

Significance of atomic number

Atomic mass unit

is the *atomic number Z* of the element involved. Except in the case of ordinary hydrogen atoms, the number of neutrons in a nucleus equals or, more often, exceeds the number of protons. The compositions of atoms of the four lightest elements are illustrated in Fig. 30–1.

The mass of an atom can be determined with the help of a mass spectrometer, such as the one described in Section 21–7. As mentioned earlier, atomic masses are conventionally expressed in terms of the atomic mass unit, abbreviated u, whose value is

$$1 \text{ u} = 1.66 \times 10^{-27} \text{ kg} \qquad \textit{Atomic mass unit}$$

The electron, proton, neutron, and hydrogen atom masses in u are respectively

$$m_e = 0.000549 \text{ u} \qquad m_p = 1.007277 \text{ u}$$

$$m_n = 1.008665 \text{ u} \qquad m_{\text{H}} = 1.007825 \text{ u}$$

Because m_{H} and m_n are so close to 1 u, we would expect atomic masses expressed in u to be very nearly whole numbers. This is often true. For example, the atomic mass of helium is 4.003 u, that of lithium is 6.939 u, and that of beryllium is 9.012 u. (Mass values such as these always include the masses of the surrounding electrons in the neutral atom.) However, the chlorine found in nature has an atomic mass of 35.46 u, which does not fit in with this picture.

The isotopes of an element differ in number of nuclear neutrons

Chlorine is an example of an element whose nuclei do not all have the same composition. The several varieties of an element are called its *isotopes*. The number of protons is always equal to the atomic number Z of the element, of course, but the number of neutrons may be different. Thus chlorine consists of two isotopes, one whose nuclei contain 17 protons and 18 neutrons and another whose nuclei contain 17 protons and 20 neutrons. There are about three times as many nuclei of the former type as there are of the latter, which yields an average atomic mass of 35.46 u (Table 30–1).

Deuterium and tritium are hydrogen isotopes

All elements have isotopes, even hydrogen. The most common hydrogen isotope has nuclei that each consist of a single proton. Less abundant is the isotope *deuterium,* whose nuclei each consist of a proton and a neutron, and the isotope *tritium,* whose nuclei each consist of a proton and two neutrons (Fig. 30–2). Deuterium is stable, but tritium is radioactive and a sample of it gradually changes to an isotope of helium. The stream of cosmic rays from space continually replenishes the earth's tritium by nuclear reactions in the atmosphere; only about 2 kg of tritium of natural origin is present on the earth's surface, nearly all of it in the oceans.

FIG. 30–1 The electronic and nuclear compositions of hydrogen, helium, lithium, and beryllium atoms.

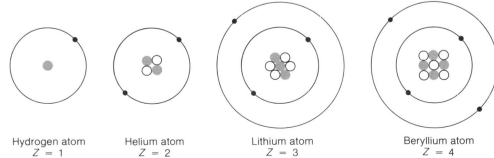

● Proton
○ Neutron
• Electron

Hydrogen atom
Z = 1

Helium atom
Z = 2

Lithium atom
Z = 3

Beryllium atom
Z = 4

Element	Properties of Element		Properties of Isotope			
	Atomic Number	Average Atomic Mass, u	Protons in Nucleus	Neutrons in Nucleus	Atomic Mass, u	Relative Abundance
Hydrogen	1	1.008	1	0	1.008	99.985%
			1	1	2.014	0.015%
			1	2	3.016	Very small
Chlorine	17	35.46	17	18	34.97	75.53%
			17	20	36.97	24.47%

TABLE 30–1
The isotopes of hydrogen and chlorine found in nature

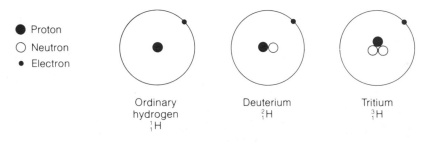

Proton
Neutron
Electron

Ordinary hydrogen
^1_1H

Deuterium
^2_1H

Tritium
^3_1H

FIG. 30–2 The three isotopes of hydrogen.

Because the chemical properties of an element depend upon the distribution of the electrons in its atoms, which in turn depends upon the nuclear charge, nuclear structure beyond the number of protons present has little significance for the chemist. The physical properties of an element, however, depend strongly on the nuclear structures of its isotopes, whose behavior may be very different from one another although chemically they are indistinguishable.

Isotopes of an element have almost identical chemical behavior

The conventional symbols for nuclear species, or *nuclides,* all follow the pattern $^A_Z X$, where

Symbols of nuclides

X = chemical symbol of the element

Z = atomic number of the element

= number of protons in the nucleus

A = mass number of the nuclide

= number of protons and neutrons in the nucleus

Hence ordinary hydrogen is designated ^1_1H, since its atomic number and mass number are both 1, while tritium is designated ^3_1H. The two isotopes of chlorine mentioned above are designated $^{35}_{17}Cl$ and $^{37}_{17}Cl$ respectively.

30–2 BINDING ENERGY

As mentioned above, the nucleus of a deuterium atom consists of a proton and a neutron. Thus we would expect that the mass of a deuterium atom, ^2_1H, should be equal to the mass of an ordinary hydrogen atom, ^1_1H, plus the mass of a neutron. However, it turns

FIG. 30–3 The mass of every atom is less than the total of the masses of its constituent neutrons, protons, and electrons. This phenomenon is illustrated here for the deuterium atom.

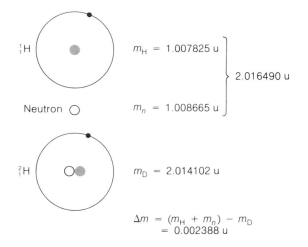

out that the mass of ^2_1H is 0.002388 u *less* than the combined masses of a ^1_1H atom and a neutron (Fig. 30–3).

The case of deuterium is an example of a general observation: *Stable atoms always have less mass than the combined masses of their constituent particles.* The energy equivalent of the "missing" mass of a nucleus is called its *binding energy.*

The binding energy of a nucleus is the energy needed to break it up into separate neutrons and protons

In order to break a nucleus apart into its constituent nucleons, an amount of energy equal to its binding energy must be supplied either in a collision with another particle or by the absorption of a sufficiently energetic photon. Binding energics are due to the action of the nuclear forces that hold nuclei together, just as ionization energies of atoms, which must be supplied to remove electrons from them, are due to the action of the electric forces that hold them together.

The difference between the ^2_1H atomic mass and the combined masses of ^1_1H and neutron is 0.002388 u. The usual energy unit in nuclear physics is the MeV, where

$$1 \text{ MeV} = 10^6 \text{ eV} = 1.60 \times 10^{-13} \text{ J}$$

In terms of MeV, the energy equivalent of 1 u of mass is

FIG. 30–4 The binding energy of the deuteron is 2.22 MeV, which means that this much energy is required to make up the difference between the deuteron's mass and the combined mass of its constituent neutron and proton. Absorbing 2.22 MeV, for instance by being struck by a 2.22-MeV gamma-ray photon, gives a deuteron enough additional mass to split into a neutron and a proton.

$$E_u = 931 \text{ MeV} \qquad\qquad \textit{Energy per atomic mass unit}$$

The binding energy of the *deuteron* (as the deuterium nucleus is called) is therefore

$$(0.002388 \text{ u})(931 \text{ MeV/u}) = 2.22 \text{ MeV}$$

This figure is confirmed by experiments that show that the minimum energy a photon must have in order to disrupt a deuteron is 2.22 MeV (Fig. 30–4).

Nuclear binding energies are strikingly high. The range for stable nuclei is from 2.2 MeV for ^2_1H (deuterium) to 1640 MeV for $^{209}_{83}\text{Bi}$ (an isotope of the metal bismuth).

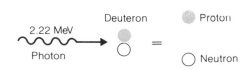

Larger nuclei are all unstable and decay radioactively. To appreciate how high binding energies are, we can compare them with more familiar energies in terms of kilojoules of energy per kilogram of mass. In these units, a typical binding energy is 8×10^{11} kJ/kg—800 billion kJ/kg. By contrast, to boil water involves a heat of vaporization of a mere 2260 kJ/kg, and even the heat given off by burning gasoline is only 4.7×10^4 kJ/kg, 17 million times smaller.

Nuclei of intermediate size are the most stable

The binding energy per nucleon in a nucleus is equal to the total binding energy (calculated from the mass deficiency of the nucleus) divided by the number of neutrons and protons it contains. A very interesting and important curve results when we plot binding energy per nucleon versus mass number, as in Fig. 30–5. Except for the high peak for $^4_2$He, the curve is a quite regular one. Nuclei of intermediate size have the highest binding energies per nucleon, which means that their nucleons are held together more securely than the nucleons in both heavier and lighter nuclei. The maximum in the curve is 8.8 MeV/nucleon at $A = 56$, which corresponds to the iron nucleus $^{56}_{26}$Fe.

Nuclear fission

A remarkable feature of nuclear structure is illustrated by this curve. Suppose that we split the nucleus $^{235}_{92}$U, whose binding energy is 7.6 MeV/nucleon, into two fragments. Each fragment will be the nucleus of a much lighter element, and therefore will have a higher binding energy per nucleon than the uranium nucleus. The difference is about 0.8 MeV/nucleon, and so, if such *nuclear fission* were to take place, an energy of

$$\left(0.8 \, \frac{\text{MeV}}{\text{nucleon}}\right)(235 \text{ nucleons}) = 188 \, \text{MeV}$$

would be given off per splitting. This is a truly immense amount of energy to be

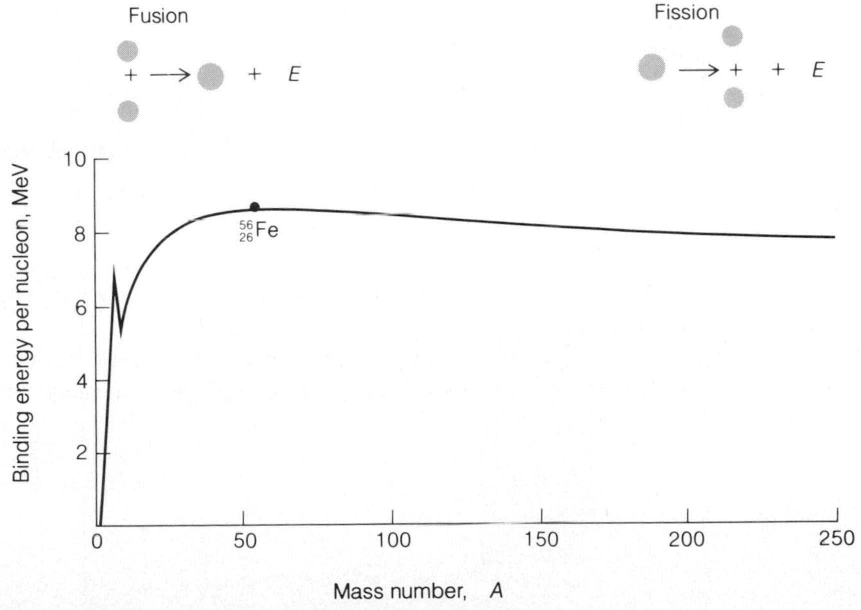

FIG. 30–5 The binding energy per nucleon versus mass number. The higher the binding energy per nucleon, the more stable the nucleus. When a heavy nucleus is split into two lighter ones, a process called *fission*, the greater binding energy of the product nuclei causes the liberation of energy. When two very light nuclei join to form a heavier one, a process called *fusion*, the greater binding energy of the product nucleus again causes the liberation of energy.

produced in a single atomic event. As a comparison, chemical processes involve energies of the order of magnitude of 1 electron volt per reacting atom, 10^{-8} the energy involved in fission.

Nuclear fusion

Figure 30–5 also shows that if two of the extremely light nuclei are combined to form a heavier one, the higher binding energy of the latter will also result in the evolution of energy. For instance, if two deuterons were to join to make a ^4_2He nucleus, more than 23 MeV would be released. This process is known as *fusion,* and, together with fission, it promises to be the source of more and more of the world's energy as reserves of fossil fuels are depleted. Nuclear fusion is the means by which the sun and stars obtain their energy.

The binding energy curve is important because it is the key to energy production in the universe

The graph of Fig. 30–5 has a good claim to being the most significant in all of science. The fact that binding energy exists at all means that nuclei more complex than the single proton of hydrogen can be stable. Such stability in turn accounts for the existence of the various elements and so for the existence of the many and diverse forms of matter we see around us. Because the curve peaks in the middle, we have the explanation for the energy that powers, directly or indirectly, the evolution of the entire universe: It comes from the fusion of protons and neutrons to form heavier nuclei. And the harnessing of nuclear fission in reactors and weapons has irreversibly changed modern civilization.

30–3   RADIOACTIVITY

Not all atomic nuclei are stable. At the beginning of the twentieth century it became known, as the result of research by Becquerel, the Curies, and others, that some nuclei exist that spontaneously transform themselves into other nuclear species with the emission of radiation. Such nuclei are said to be *radioactive*.

Unstable nuclei emit three kinds of radiation:

FIG. 30–6 (a) The radiations from a radium sample may be analyzed with the help of a magnetic field. In the figure the direction of the field is into the paper; hence the positively charged alpha particles (which are helium nuclei) are deflected to the left and the negatively charged beta particles (which are electrons), to the right. Gamma rays (which are energetic photons) carry no charge and are not affected by the magnetic field. (b) Alpha particles are stopped by a piece of cardboard. Beta particles penetrate the cardboard but are stopped by a sheet of aluminum. Even a thick slab of lead may not stop all the gamma rays.

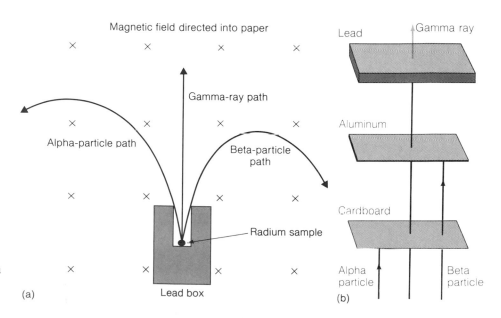

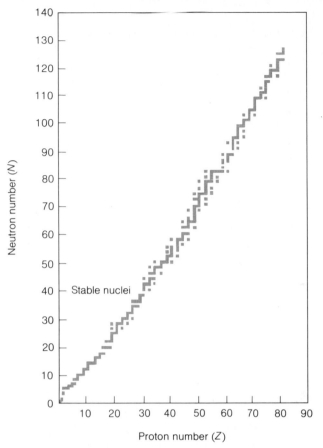

FIG. 30–7 The number of neutrons versus the number of protons in stable nuclei. The larger the nucleus, the greater the proportion of neutrons.

1. *Alpha particles,* which are the nuclei of $^4_2$He atoms.
2. *Beta particles,* which are electrons or positrons (positively charged electrons).
3. *Gamma rays,* which are photons of high energy.

Radiations from unstable nuclei

The early experimenters identified these radiations with the help of a magnetic field. Figure 30–6 shows a radium sample in a magnetic field directed into the paper: The positively charged alpha particles are deflected to the left and the negatively charged beta particles are deflected to the right. Gamma rays carry no charge and are not affected by the magnetic field.

To understand why alpha, beta, and gamma decays take place, let us look at Fig. 30–7 which is a plot of the number of neutrons versus the number of protons in stable nuclei. For light, stable nuclei the number of neutrons and protons are about the same, while for heavier nuclei somewhat more neutrons than protons are required for stability. It is evident that an element of a given atomic number has only a very narrow range of possible numbers of neutrons if it is to be stable.

Suppose now that a nucleus exists that has too many neutrons for stability relative to the number of protons present. If one of the excess neutrons transforms itself into a proton, this will simultaneously reduce the number of neutrons while increasing the

Negative beta decay occurs in nuclei with too many neutrons

FIG. 30-8 How alpha and beta decays tend to bring an unstable nucleus to a stable configuration.

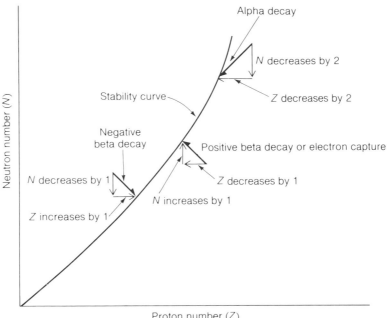

number of protons (Fig. 30-8). To conserve electric charge, such a transformation requires the emission of a negative electron, and we may write it in equation form as

$$n^0 \rightarrow p^+ + e^- \hspace{4cm} \textit{Electron emission} \quad (30\text{--}1)$$

The electron leaves the nucleus, and is detectable as a "beta particle." The nucleus left behind may have some extra energy as a consequence of its shifted binding energy, and this energy is given off in the form of gamma rays. Sometimes more than one such *beta decay* is required for a particular unstable nucleus to reach a stable form.

Should the nucleus have too few neutrons, the inverse reaction

Positron emission occurs in nuclei with too few neutrons

$$p^+ \rightarrow n^0 + e^+ \hspace{4cm} \textit{Positron emission} \quad (30\text{--}2)$$

in which a proton becomes a neutron with the emission of a *positron* (positive electron), may take place. This is also called beta decay, since it resembles the emission of negative electrons from an unstable nucleus in every way save for the difference in charge.

A process that competes with positron emission is the capture by a nucleus with too small a neutron/proton ratio of one of the electrons in its innermost atomic shell. The electron is absorbed by a nuclear proton that becomes a neutron in so doing. This process can be expressed as

Electron capture is an alternative to positron emission

$$p^+ + e^- \rightarrow n^0 \hspace{4cm} \textit{Electron capture} \quad (30\text{--}3)$$

Electron capture does not lead to the emission of a particle, but can be detected by the X-ray photon that is produced when one of the atom's outer electrons falls into the vacancy left by the absorbed electron.

Alpha decay

Another way of altering its structure to achieve stability may involve a nucleus in *alpha decay,* in which an alpha particle consisting of two neutrons and two protons is emitted. Thus negative beta decay increases the number of protons by one and decreases the number of neutrons by one; positive beta decay and electron capture decrease the number of protons by one and increase the number of neutrons by one; and alpha decay

Decay	Nuclear Transformation	Example
Alpha	$^A_Z X \rightarrow ^{A-4}_{Z-2} Y + ^4_2 He$	$^{238}_{92} U \rightarrow ^{234}_{90} Th + ^4_2 He$
Electron emission	$^A_Z X \rightarrow _{z+1}^A Y + e^-$	$^{14}_6 C \rightarrow ^{14}_7 N + e^-$
Positron emission	$^A_Z X \rightarrow _{z-1}^A Y + e^+$	$^{64}_{29} Cu \rightarrow ^{64}_{28} Ni + e^+$
Electron capture	$^A_Z X + e^- \rightarrow _{z-1}^A Y$	$^{64}_{29} Cu + e^- \rightarrow ^{64}_{28} Ni$
Gamma	$^A_Z X^* \rightarrow ^A_Z X + \gamma$	$^{87}_{38} Sr^* \rightarrow ^{87}_{38} Sr + \gamma$

TABLE 30-2
Radioactive decay. The asterisk (*) denotes an excited nuclear state and γ denotes a gamma-ray photon

decreases both the number of protons and the number of neutrons by two. These processes are summarized in Table 30-2 and shown schematically in Fig. 30-9. Very often a succession of alpha and beta decays, with accompanying gamma decays to carry off excess energy, is required before a nucleus reaches stability.

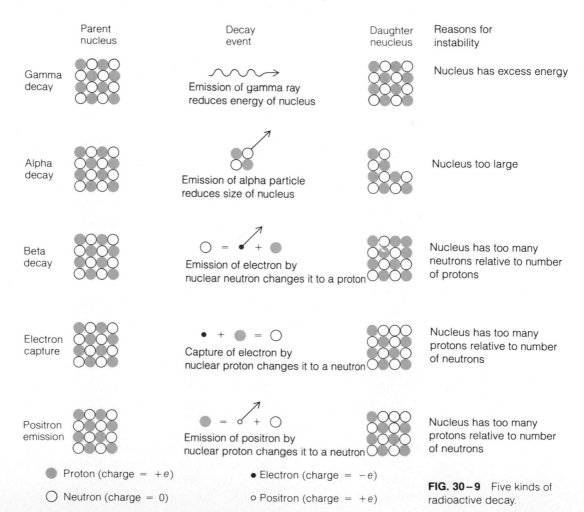

FIG. 30-9 Five kinds of radioactive decay.

Why alpha particles are emitted

It is appropriate to ask why it is that only alpha particles are given off by nuclei that are too heavy and not, for example, individual protons or $^3_2$He nuclei. The reason is the high binding energy of the alpha particle (see Fig. 30–5), which means that it has much less mass than four individual nucleons. Because of this small mass, an alpha particle can be ejected by a heavy nucleus with energy to spare. Thus the alpha particle released in the decay of $^{232}_{92}$U has a kinetic energy of 5.4 MeV, while 6.1 MeV would have to be supplied from the outside to this nucleus if it is to release a proton, and 9.6 MeV supplied if it is to release a $^3_2$He nucleus.

30–4 RADIATION DETECTION

A variety of methods have been developed for detecting photons and subatomic particles such as those emitted in radioactive decay. Nearly all of them are based on the ionization that results, directly or indirectly, from the passage of a photon or particle through matter.

Heavy charged particles

Heavy charged particles, such as protons and alpha particles, lose energy chiefly by electric interactions with atomic electrons. These electrons are either raised to excited states or, more often, are pulled away from their parent atoms entirely. The ejected electrons may have energies of several hundred electron volts and so are able to ionize atoms themselves. The incident particle slows down gradually until it comes to a stop or reacts with a nucleus in its path.

Electrons

Electrons, too, lose energy by ionization, but for them another mechanism may also be important. As mentioned in Section 25–2, electromagnetic radiation is given off whenever an electric charge is accelerated. Energy loss by radiation is more significant for electrons than for heavier particles because they are more violently accelerated when passing near nuclei in their paths. The more energetic the electron and the greater the nuclear charge of the atoms it encounters, the more it radiates. In lead the rate of energy loss by radiation becomes equal to that by ionization for an electron energy of about 10 MeV, whereas in air ionization remains dominant below an electron energy of about 100 MeV.

Photons

The three principal ways in which the energetic photons of X and gamma rays lose energy when they pass through matter are the *photoelectric effect, Compton scattering,* and *pair production* (Fig. 30–10). In each case photon energy is conveyed to electrons, which in turn lose energy mainly by exciting or ionizing atoms in the absorbing material.

At low photon energies the photoelectric effect, which was discussed in Section

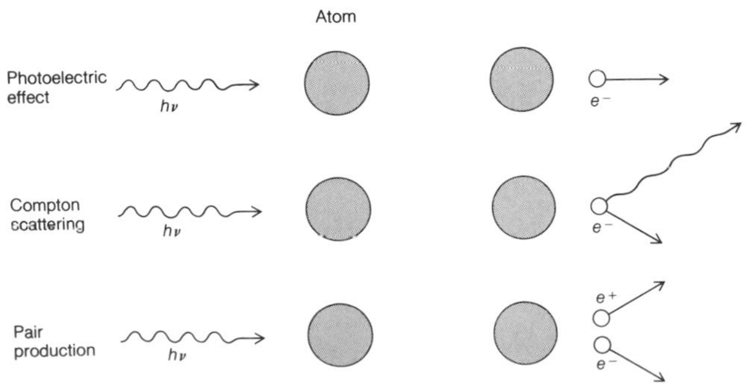

FIG. 30–10 X and gamma rays interact with matter chiefly through the photoelectric effect, Compton scattering, and pair production. Pair production requires a photon energy of at least 1.02 MeV.

28–1, is the principal mechanism of energy loss. At higher energies, a photon may transfer only part of its energy hv to an electron it collides with. The scattered photon has the lower energy hv', and the electron flies off with the kinetic energy KE = $hv - hv'$. This process is called *Compton scattering*.

At still higher energies, it is possible for a photon to materialize into a positron and an electron when it passes near a nucleus. Since the energy equivalent of the mass of either particle is 0.51 MeV, the minimum photon energy needed for such *pair production* is 1.02 MeV. In pair production, the electromagnetic energy of a photon is turned into matter, with any photon energy in excess of 1.02 MeV appearing as kinetic energy of the newly created positron and electron. Electric charge as well as energy is conserved in pair production since the positron $(Q = +e)$ and electron $(Q = -e)$ have opposite charges. Gamma rays from radioactive decay and X rays interact with matter largely through Compton scattering.

Pair production

A number of devices can be used to detect the ionization produced by an energetic photon or charged particle. Some, such as the *ionization chamber* and the *Geiger counter,* are gas-filled tubes with electrodes that attract the electrons and ions to produce voltage pulses in external circuits (Fig. 30–11). Another device, the *scintillation counter,* is based on the emission of a flash of light by certain substances (such as those used to make the screen of a television picture tube) when struck by ionizing radiation. The light is then picked up by a sensitive photomultiplier tube (Fig. 30–12). *Semiconductor detectors* make use of the fact that a thin layer on both sides of a *p-n* junction is normally depleted in charge carriers when the crystal has an external voltage across it. Electrons and holes produced there by an ionizing particle lead to a measurable pulse.

Ionization detectors

Although neutrons do not themselves produce ionization and so cannot be detected directly, indirect techniques have been devised. For example, a counter whose sensitive volume is rich in hydrogen will respond to fast neutrons through the ionization produced by protons recoiling after collisions with the neutrons. In the case of slow neutrons,

Neutron detectors

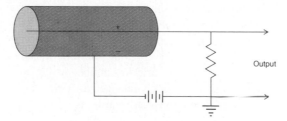

Output

FIG. 30–11 A gas-filled detector of this kind can be used either as an ionization chamber or as a Geiger counter. When it is to be an ionization chamber, the applied voltage is low, and the output pulse that occurs when a photon or charged particle passes through the tube is a measure of the amount of ionization produced in the gas. When it is to be a Geiger counter, the applied voltage is high enough to give the initial ions enough energy of their own to cause further ionization, and the result is a large output pulse that is the same for all incident photons or particles.

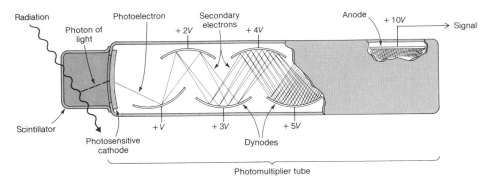

FIG. 30–12 Essential features of a scintillation counter. The scintillator may be a crystalline or plastic solid or a liquid. The number of dynodes in the photomultiplier tube may be ten or more, with each dynode several hundred volts higher in potential than the one before it. Each dynode produces 2 to 5 secondary electrons per incident electron for a total amplification of as much as 10^6 or 10^7. A scintillation counter is more rapid in its operation than an ionization chamber or Geiger counter.

one method is to employ the nuclear reaction

$$\,_0^1 n + \,_5^{10}\text{B} \rightarrow \,_2^4\text{He} + \,_3^7\text{Li}$$

in which a neutron reacts with a nucleus of the boron isotope $\,_5^{10}\text{B}$ to produce an alpha particle and a nucleus of the lithium isotope $\,_3^7\text{Li}$. If $\,_5^{10}\text{B}$ is incorporated in the sensitive volume of a suitable counter (the gas BF_3 is often used), the appearance of ionization that corresponds to an alpha particle signifies the capture of a slow neutron.

30–5   HALF-LIFE

The becquerel and the curie are units of activity

The rate at which a sample of radioactive material decays is called its *activity*. The SI unit of activity is the *becquerel*, where 1 Bq = 1 event per second. The activities encountered in practice are usually so high that the MBq (10^6 Bq) and GBq (10^9 Bq) are more suitable. The traditional unit of activity, which is still in common use, is the *curie*. Originally the curie was defined as the activity of 1 g of radium ($\,_{88}^{226}\text{Ra}$), and its precise value accordingly changed as measuring techniques improved. For this reason the curie is now defined arbitrarily as

$$1 \text{ curie} = 3.70 \times 10^{10} \text{ events/s} = 37 \text{ GBq}$$

The activity of 1 g of radium is a few percent smaller. The *millicurie* (10^{-3} curie) and *microcurie* (10^{-6} curie) are frequently employed to supplement the curie. A luminous watch dial contains several microcuries of $\,_{88}^{226}\text{Ra}$; ordinary potassium has an activity of about 1 millicurie/kg owing to the presence of the radioactive isotope $\,_{19}^{40}\text{K}$; "cobalt-60" sources of 1 or more curies are widely used in medicine for radiation therapy and industrially for the inspection of metal castings and welded joints.

One of the characteristics of all types of radioactivity is that the rate at which the nuclei in a given sample decay always follows a curve whose shape is like that shown in Fig. 30–13. If we start with a sample whose rate of decay is, say, 100 events/s, it

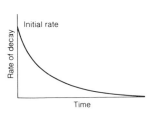

FIG. 30–13 The rate at which a sample of radioactive substance decays is not constant but varies with time in the manner shown in the curve.

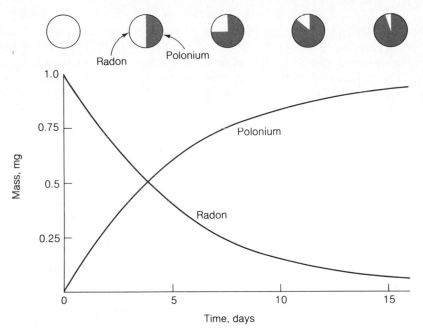

FIG. 30–14 The alpha decay of radon ($^{222}_{86}$Rn) to polonium ($^{218}_{84}$Po) has a half-life of 3.8 days. The sample of radon whose decay is graphed here had an initial mass of 1.0 mg.

will not continue to decay at that rate but instead fewer and fewer disintegrations will occur in each successive second.

Half-life

Some radioactive nuclides decay faster than others, but in each case a certain definite time is required for half of an original sample to decay. This time is called the *half-life* of the nuclide. For instance, the radon isotope $^{222}_{86}$Rn undergoes alpha decay to the polonium isotope $^{218}_{84}$Po with a half-life of 3.8 days. Should we start with 1 mg of radon in a closed container (since it is a gas), $\frac{1}{2}$ mg will remain undecayed after 3.8 days, $\frac{1}{4}$ mg will remain undecayed after 7.6 days; $\frac{1}{8}$ mg will remain undecayed after 11.4 days; and so on (Fig. 30–14).

Example The hydrogen isotope tritium, $^3_1$H, is radioactive and emits an electron with a half-life of 12.5 years. (a) What does $^3_1$H become after beta decay? (b) What percentage of an original sample of tritium will remain 25 years after its preparation?

Solution (a) When a nucleus emits an electron, its atomic number increases by 1 unit (corresponding to an increase in nuclear charge of $+e$) and its mass number is unchanged. Helium has the atomic number 2, and so

$$^3_1\text{H} \rightarrow {}^3_2\text{He} + e^-$$

(b) Twenty-five years represents two half-lives of tritium; hence $\frac{1}{2} \times \frac{1}{2} = \frac{1}{4} = 25\%$ of the original sample remains undecayed. ∎

Half-lives vary widely

Half-lives range from billionths of a second to billions of years. Samples of radioactive isotopes decay in the manner illustrated because a great many individual nuclei are involved, each having a certain probability of decaying. The fact that radon has a half-life of 3.8 days signifies that every radon nucleus has a 50% chance of decaying in any 3.8-day period. Because a nucleus does not have a memory, this does *not* mean that a radon nucleus has a 100% chance of decaying in 7.6 days: The likelihood of decay of a given nucleus stays the same until it actually does decay. Thus a half-life of 3.8 days means a 75% probability of decay in 7.6 days, an 87.5% probability of decay in 11.4 days, and so on, because in each interval of 3.8 days the probability is 50%.

Half-lives are unaffected by external conditions

The above discussion suggests that radioactive decay involves individual events that take place within individual nuclei, rather than collective processes that involve more than one nucleus in interaction. This idea is confirmed by experiments that show that the half-life of a particular isotope is unaffected by changes of pressure, temperature, electric and magnetic fields, and so on, which might, if strong enough, influence internuclear phenomena.

30 – 6   RADIATION HAZARDS

Ionizing radiation is harmful to living things

Like X rays, the various radiations from radioactive nuclei are able to ionize matter through which they pass. All ionizing radiation is harmful to living tissue, although if the damage is slight, the tissue can often repair itself with no permanent effect. It is easy to underestimate radiation hazards because there is usually a delay, sometimes of many years, between an exposure and some of its possible consequences, which include cancer, leukemia, and genetic changes that may lead to children handicapped in various ways.

Many useful processes involve ionizing radiation. Some employ such radiation directly, as in the X rays and gamma rays used in medicine and industry. In other cases the radiation is unwanted but inescapable, notably in the operation of nuclear reactors and in the disposal of their wastes. It is not always easy to find an appropriate balance between risk and benefit where radiation is concerned. This seems particularly true for medical X rays, which constitute the largest single source of radiation dosage for an average person in the United States (Table 30 – 3).

All X-ray exposures are dangerous; their benefits do not always outweigh the risks

It seems to be an unfortunate fact that many X-ray exposures are made for no strong reason and do more harm than good. In this category are "routine" chest X rays upon hospital admission, "routine" X rays as part of regular physical examinations, and "routine" dental X rays. The once-routine mass screening by X ray of symptomless young women for breast cancer is now generally thought to have increased, not decreased, the overall death rate due to cancer. Especially dangerous is the X raying of pregnant women, until recently another "routine" procedure, which dramatically increases the chances of cancer in their children.

Since the carcinogenic properties of X rays have been known since 1902, all this is hard to excuse. X rays unquestionably are valuable in medicine. The point is that every exposure should have a definite justification that outweighs the risk.

Radiation dosage

The SI unit of radiation dosage is the *sievert,* which is the amount of any radiation that has the same biological effect as that produced by the absorption of 1 J of X or gamma rays by 1 kg of body tissue. (An older unit, the *rem,* is equal to 0.01 sievert.) Although there is no general agreement, many radiobiologists think that a cancer occurs once for every 10 to 70 sieverts of radiation exposure of a population. Such unavoidable natural sources of radiation as cosmic rays and radioactive materials in the earth and in the body itself lead to a dosage rate per person of about 0.8 millisievert/year. Averaged over the U.S. population, other sources of radiation add 1.06 millisieverts/

Percentage	Source
42	Medical X rays
15	Occupational exposures, consumer products (such as TV sets and smoke detectors), fallout from past tests of nuclear weapons
0.2	Nuclear reactors
15	Cosmic rays (see Sec. 30−9) ⎫
14	Radionuclides in rocks and soil ⎬ Natural background
14	Radionuclides in the body ($^{40}_{19}K$, $^{14}_{6}C$, and so on) ⎭

TABLE 30−3 Sources of radiation dosage for an average person in the U.S. The total is 1.86 millisieverts/year, which is about equivalent to two dental X rays. Actual dosages vary considerably. For instance, some people receive more medical X rays than others; cosmic rays are more intense at high altitudes; concentrations of radioactive minerals exist in some regions; and so on.

year, with diagnostic X rays contributing the largest amount. Nuclear power stations are responsible for only 0.003 millisievert/year. The total is 1.86 millisieverts/year, about the dose received from two dental X rays.

30−7 RADIOMETRIC DATING

Radioactive decay is the basis of a valuable method for establishing the ages of a variety of geological and biological specimens. Because the decay of a given radioactive nuclide proceeds at a constant rate regardless of its environment, the ratio between the amounts of that nuclide and of its stable daughter in a specimen depends upon the specimen's age. The greater the proportion of the daughter nuclide, the older the specimen. Let us see how this procedure is used to date objects of biological origin using *radiocarbon,* the beta-active carbon isotope $^{14}_{6}C$.

Cosmic rays are high-energy atomic nuclei, mainly protons, that circulate through the Milky Way galaxy of which the sun is a member. About 10^{18} of them arrive at the earth each second and disrupt the nuclei of atoms they encounter in the atmosphere to produce showers of secondary particles. Among these secondaries are neutrons that can react with nitrogen nuclei in the atmosphere to form radiocarbon with the emission of a proton:

Radiocarbon is a by-product of cosmic-ray bombardment

$$^{14}_{7}N + ^{1}_{0}n \rightarrow ^{14}_{6}C + ^{1}_{1}H$$

The proton picks up an electron and becomes a hydrogen atom. The ^{14}C isotope has too many neutrons for stability and ultimately undergoes beta decay to become ^{14}N with a half-life of 5760 years.

$$^{14}_{6}C \rightarrow ^{14}_{7}N + e^{-}$$

At the present time a total of about 90 tons of radiocarbon is distributed around the world, which is replaced by the cosmic-ray bombardment as it decays so that the amount remains constant.

Shortly after their formation, radiocarbon atoms combine with oxygen molecules to form carbon dioxide (CO_2) molecules. Green plants convert water and carbon dioxide into carbohydrates in the process of photosynthesis, which means that every plant contains a certain amount of radiocarbon. Animals eat plants, and so they too contain radiocarbon. Thus every living thing on the earth has a very small proportion of radiocarbon in its tissues. Because the mixing of radiocarbon is relatively efficient,

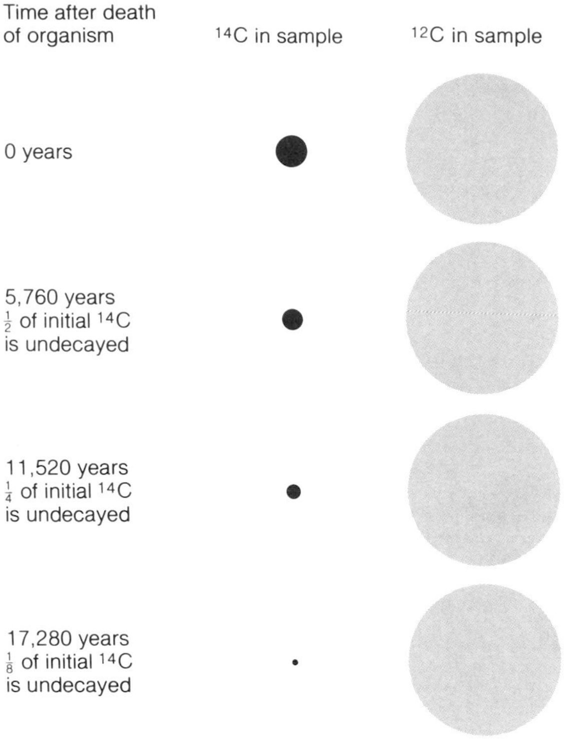

FIG. 30–15 The radioactive ^{14}C content of a specimen of plant or animal tissue decreases steadily, whereas its ^{12}C content does not change. Hence the $^{14}C/^{12}C$ ratio indicates the time that has passed since the organism died. The half-life of ^{14}C is 5760 years.

Time after death of organism	^{14}C in sample	^{12}C in sample
0 years		
5,760 years $\frac{1}{2}$ of initial ^{14}C is undecayed		
11,520 years $\frac{1}{4}$ of initial ^{14}C is undecayed		
17,280 years $\frac{1}{8}$ of initial ^{14}C is undecayed		

living plants and animals all have the same ratio of radiocarbon to ordinary carbon, which is ^{12}C.

Principle of radiocarbon dating

When a plant or animal dies, however, it no longer takes in radiocarbon atoms, but the radiocarbon atoms it already contains continually decay to nitrogen. After 5760 years a dead organism has left only half as much radiocarbon as it had while alive, after 11,520 years only a quarter as much, and so on. By determining the ratio of radiocarbon to ordinary carbon it is therefore possible to evaluate the ages of ancient objects and remains of organic origin (Fig. 30–15). This technique permits the dating of mummies, wooden implements, cloth, leather, charcoal from campfires, and similar artifacts from ancient civilizations as much as 50,000 years old, about nine half-lives of ^{14}C.

Geological dating methods

Because the earth's history goes back 4.5 or so billion years, geologists use radioactive nuclides of much longer half-lives than that of radiocarbon to date rocks. The most common geological dating methods use the nuclides ^{40}K, ^{87}Rb, and ^{238}U (Table 30–3). In each case the assumption is made that all of the stable daughter nuclide found in a particular rock sample originated from the decay of the parent nuclide. Although ^{40}K and ^{87}Rb decay into stable nuclides in a single step, ^{238}U requires no less than 14 successive alpha and beta decays to become ^{206}Pb. However, the half-lives of the intermediate products are so short compared with the 4.5-billion-year half-life of ^{238}U itself that only the ^{238}U and ^{206}Pb contents of a particular sample need be considered. What is meant by the age of a rock depends upon the nature of the rock; it may refer to the time at which the minerals of the rock crystallized, for example.

Method	Parent Nuclide	Daughter Nuclide	Half-Life
Potassium-argon	^{40}K	^{40}Ar	1.3×10^9 years
Rubidium-strontium	^{87}Rb	^{87}Sr	4.7×10^{10} years
Uranium-lead	^{238}U	^{206}Pb	4.5×10^9 years

TABLE 30-4
Geological dating methods

Meteorites and lunar rocks as well as terrestrial rocks have been dated by the methods of Table 30-4.

30-8 FISSION

When two nuclei approach close enough together, it is possible for a rearrangement of their constituent nucleons to occur with one or more new nuclei formed. Such a process is called a *nuclear reaction,* by analogy with chemical reactions in which two or more compounds may combine to form new ones.

Atoms and molecules are neutral, so it is easy for them to come together and react. Nuclei all have positive charges, as much as $+92e$ in nuclei found in nature, and the electric repulsion between them is sufficient to keep them beyond the range where they can interact unless they are moving very fast to begin with. In the sun and other stars, whose interior are at temperatures of many millions of K, the nuclei present are moving fast enough on the average for nuclear reactions to be frequent, and indeed nuclear reactions provide the energy that maintains these temperatures.

Mutual repulsion hinders nuclear reactions

In the laboratory it is easy enough to produce nuclear reactions on a very small scale, either with alpha particles from radioactive substances or with protons, deuterons, or even heavier nuclei accelerated in cyclotrons and similar devices. But only one type of nuclear reaction has as yet proved to be a practical source of energy on the earth, namely the fission that occurs when neutrons strike the nuclei of certain very heavy nuclei.

In nuclear fission, which can take place only in certain very heavy nuclei such as $^{235}_{92}$U, the absorption of an incoming neutron causes the target nucleus to split into two smaller nuclei called *fission fragments* (Fig. 30-16). Because stable light nuclei have relatively fewer neutrons than do heavy nuclei, the fragments are unbalanced when they are formed and at once release one or two neutrons each. Usually the fragments are still unstable and undergo radioactive decay to achieve appropriate neutron-proton ratios. The products of fission, such as the wastes from a nuclear reactor and the fallout from a nuclear bomb burst, are therefore highly radioactive. Since some of these products have long half-lives, their radioactivity will persist for many generations and their disposal remains a problem without a really satisfactory solution as yet.

Fission products are highly radioactive and may remain so for long periods

A variety of nuclear species may appear as fission fragments. An example of a typical fission reaction is

A typical fission reaction

$$^{235}_{92}\text{U} + ^1_0n \rightarrow ^{236}_{92}\text{U} \rightarrow ^{140}_{54}\text{Xe} + ^{94}_{38}\text{Sr} + ^1_0n + ^1_0n + 200\,\text{MeV} \qquad (30-4)$$

The $^{236}_{92}$U that is first formed lasts for only a small fraction of a second before it splits into two parts. About 84% of the total energy liberated during fission appears as kinetic energy of the fission fragments, about 2.5% as kinetic energy of the neutrons, and

FIG. 30–16   In nuclear fission an absorbed neutron causes a heavy nucleus to split in two parts, with the emission of several neutrons and gamma rays.

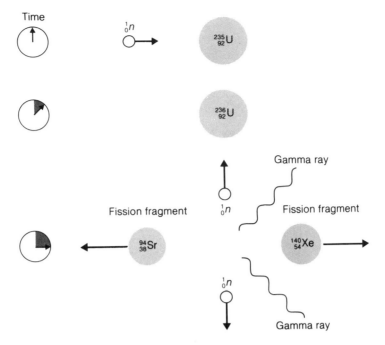

FIG. 30–17   Simplified sketch of a chain reaction in $^{235}_{92}$U.

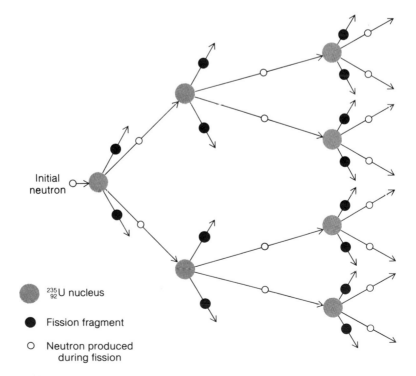

about 2.5% in the form of instantaneous emitted gamma rays, with the remaining 11% being given off in the decay of the fission fragments.

Because each fission event liberates two or three neutrons while only one neutron is required to initiate it, a rapidly multiplying sequence of fissions can occur in a lump of suitable material (Fig. 30–17). When uncontrolled, such a *chain reaction* gives off an immense amount of energy in a short time. If we assume that two neutrons emitted in each fission are able to induce further fissions and that the time between the emission of a neutron and its subsequent absorption is 10^{-8} s, a chain reaction starting with a single fission will release 2×10^{13} J of energy in less than 10^{-6} s! An uncontrolled chain reaction evidently can cause an explosion of exceptional magnitude and in fact is the basis of the "atomic" bomb.

Chain reaction

The destructive power of nuclear weapons does not stop with their detonation but continues long afterward through the radioactive debris that is produced and widely dispersed. Tens of thousands of nuclear weapons of various kinds now exist. The explosion of even a few of them would, by causing dust clouds that block sunlight, probably lead to months of darkness and bitter cold that might well doom those living things that were able to survive the deadly effects of the blasts themselves and the subsequent radioactive fallout.

In an atomic bomb, uncontrolled fission produces an explosion

30–9 NUCLEAR REACTORS

When controlled so that exactly one neutron per fission causes another fission, a chain reaction occurs at a constant power output. Such a reaction makes a very efficient source of energy: An output of about 1 MW (1000 kW) is produced by the fission of 1 g of a suitable nuclide per day, as compared with the consumption of 2.6 tons of coal per day in a conventional power plant. A device in which a chain reaction can be started and controlled is called a *nuclear reactor.*

Fission occurs at a steady rate in a nuclear reactor

The energy liberated in a nuclear reactor appears as heat, and it is extracted by circulating a coolant liquid or gas. The hot coolant can then be used as the heat source of a conventional steam turbine, which in turn may power an electric generator, a ship, or a submarine. In 1978, 236 reactors in 22 countries (72 of them in the United States) rated at a total of 125,000 MW were responsible for producing about 8% of the world's electrical energy. A somewhat larger number of reactors were in use in various naval vessels in the same year.

Each fission in ^{235}U liberates an average of 2.5 neutrons, hence no more than 1.5 neutrons per fission can be lost if a continuing chain reaction is to take place. Neutrons can be lost in two ways: through the reactor surface and by absorption within the reactor without inducing fission. Since a large object has less surface area relative to its volume than a small object of the same shape, neutron escape can be minimized simply by increasing the reactor size. It is also helpful to surround the reactor with a material that scatters neutrons without much tendency to absorb them so that some of the emerging neutrons will be reflected back.

Neutron absorption inside the reactor is a more difficult problem because natural uranium contains only 0.7% of the fissionable isotope ^{235}U. The more abundant ^{238}U readily captures the fast neutrons liberated in fission, but usually does not undergo fission itself. The neutrons absorbed by ^{238}U are therefore wasted. However, ^{238}U has little ability to capture *slow* neutrons, whereas slow neutrons are much more apt to

Why neutrons must be slowed down in a reactor

induce fission in ^{235}U than fast ones. Slowing down the fast neutrons produced by fission will thus prevent them from being taken up unproductively by ^{238}U and at the same time promote further fissions in ^{235}U.

The moderator

In order to slow down fission neutrons, the uranium fuel in a reactor is surrounded by a suitable substance called a *moderator.* When a fast neutron collides with a nucleus, some of its initial kinetic energy is given to the target nucleus. Just how much energy is lost by the neutron depends on the details of the interaction, but in general the more nearly equal the masses of the particles involved are, the more the energy transferred (see Section 6–7). Hence hydrogen would seem the best moderator, since hydrogen nuclei are simply protons whose mass is nearly identical to the neutron mass, and indeed ordinary ("light") water is the most common moderator in nuclear reactors since each water molecule contains two hydrogen atoms. The water also serves as the coolant in such reactors. The slowed-down neutrons are in thermal equilibrium with the moderator, which means energies of less than 1 eV at typical reactor operating temperatures.

Enriched uranium contains more ^{235}U than does natural uranium

Unfortunately a neutron colliding with a proton may well become attached to it to form a deuteron, $^2_1$H. The neutron losses in a light-water reactor make it necessary to use *enriched* uranium whose ^{235}U content has been increased fourfold to about 3%. Most enriched uranium is produced by the gaseous diffusion process in which uranium hexafluoride (UF_6) gas is exposed to a succession of semipermeable barriers. Because of their smaller mass, molecules of $^{235}UF_6$ are slightly more likely to diffuse through each barrier than molecules of $^{238}UF_6$, and any desired degree of enrichment can be achieved in this way. Gaseous diffusion requires a large plant and much energy, so it is an expensive procedure. Other approaches, such as the use of gas centrifuges, are under development.

The actual operation of a reactor begins when enough fissionable material is brought together with a moderator. A stray neutron—from a spontaneous fission, perhaps, or a cosmic-ray secondary—causes a ^{235}U nucleus to split, which releases two or three other neutrons. These neutrons are slowed down from energies of several MeV to thermal energies in 0.001 s or so by collisions with moderator nuclei and then are absorbed by other ^{235}U nuclei to induce further fissions. To control the rate of the resulting chain reaction, the reactor has movable rods of cadmium or boron, which are good absorbers of slow neutrons. As these rods are inserted further and further into the reactor, the reaction rate is progressively reduced.

Pressurized-water reactor

The fuel for a light-water reactor consists of uranium oxide (UO_2) pellets sealed in long, thin zirconium alloy tubes that are assembled along with control rods into a core that is enclosed in a steel pressure vessel. A typical pressure vessel is 10 m high, has an inside diameter of 3 m, and has walls 20 cm thick. In a *pressurized-water reactor* (PWR), the most common type, the water that circulates through the core is kept at a sufficiently high pressure, about 155 atm, to prevent boiling. The water enters the pressure vessel at perhaps 280°C and leaves at 320°C, passing through a heat exchanger to produce steam that drives a turbine (Fig. 30–18). Such a pressurized-water reactor used to produce electricity might contain 70 tons of UO_2 and operate at 2700 MW to yield 900 MW of electric power.

Boiling-water reactor

In a *boiling-water reactor* (BWR), a lower pressure of about 70 atm inside the pressure vessel allows steam to form, and this steam is sent directly to a turbine without using a separate steam generator. The boiling-water reactor has the advantage over the pressurized-water reactor of greater simplicity, but it has the disadvantage that, if the

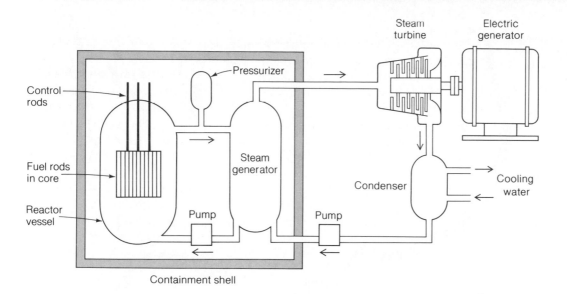

Control rods

Fuel rods in core

Reactor vessel

Pressurizer

Steam generator

Pump

Steam turbine

Electric generator

Condenser

Cooling water

Pump

Containment shell

FIG. 30-18 The basic design of a pressurized-water nuclear reactor. Water serves both as moderator and as coolant for the core. In a boiling-water reactor, steam is allowed to form inside the reactor vessel and then is used directly to power the turbine.

coolant becomes contaminated by fission products from a leaking fuel rod or by corrosion products that have become radioactive due to neutron bombardment, there is one less barrier to the possible contamination of the outside world. Several hundred pressurized-water reactors and boiling-water reactors are today in use for generating electricity and for warship propulsion. Although their capital costs exceed those of fossil-fuel plants, their fuel costs are lower.

"Heavy" water, whose molecules contain deuterium ($_1^2$H) atoms instead of ordinary hydrogen ($_1^1$H) atoms, is a better moderator than light water because deuterons are less likely to capture neutrons. Although heavy water is more expensive than light water, natural uranium can be used as the fuel instead of enriched uranium. Reactors of this kind are the most efficient of nonbreeder reactors.

Early reactors, including the very first one (which was built in 1942 at the University of Chicago by Enrico Fermi, Walter Zinn, and their collaborators) used natural uranium with graphite as the moderator. Graphite, a form of pure carbon, has little tendency to pick up neutrons, has good mechanical properties, and is readily available, all of which help compensate for its relative inefficiency at slowing down neutrons. (A neutron loses 28% of its kinetic energy in a head-on collision with a ^{12}C nucleus, compared with 89% when the collision is with a deuteron and nearly 100% when it is with a proton.) More recently graphite has been used as the moderator in high-temperature gas-cooled reactors (HTGRs). A high-temperature gas-cooled reactor uses a gas such as helium or carbon dioxide for heat transfer and operates at a much higher temperature than a water-cooled reactor, which means a higher thermal efficiency (see Chapter 16). In a typical high-temperature gas-cooled reactor, helium enters at 319°C and emerges at 756°C for an overall efficiency of 38%, as compared with 32% to 33% for a pressurized-water reactor or boiling-water reactor. High-temperature gas-cooled reactors are more difficult to build than light-water reactors, but a number of them are operating in various countries.

Over 300 nuclear power stations in 32 countries provide about one-eighth of the

High-temperature gas-cooled reactor

FIG. 30-19 Nuclear power plants in operation or under construction in the United States. The total is 142.

world's electricity. The distribution of such plants in the United States is shown in Fig. 30–19. No new plants have been ordered since 1979, when failures in its cooling system disabled one of the reactors at Three Mile Island in Pennsylvania with the escape of a certain amount of radioactive material. Public unease with the safety record of the nuclear industry together with a leveling off of demand for electricity led to a halt in plans to expand nuclear energy further in the United States. In other countries nuclear reactors continue to be built; about half of France's electricity, for example, already comes from nuclear power stations, and more are planned there.

30-10 BREEDER REACTORS

Neptunium and plutonium are transuranic elements not found in nature

Certain nonfissionable nuclides can be transformed into fissionable ones by absorbing neutrons. A notable example is ^{238}U, which becomes ^{239}U when it captures a fast neutron. This uranium isotope beta decays with a half-life of 24 min into $^{239}_{93}Np$, an isotope of the element *neptunium*, which is also beta-radioactive. The decay of $^{239}_{93}Np$ has a half-life of 2.3 days and yields $^{239}_{94}Pu$, a fissionable isotope of *plutonium* whose half-life against alpha decay is 24,000 years. The entire sequence is thus

$$^{238}_{92}U + {}^{1}_{0}n \rightarrow {}^{239}_{92}U \xrightarrow[\text{24 min}]{} {}^{239}_{93}Np + e^{-}$$

$$\xrightarrow[\text{2.3 d}]{} {}^{239}_{94}Pu + e^{-}$$

Because plutonium is chemically different from uranium, the separation of ^{239}Pu from the ^{238}U remaining after neutron bombardment is easier to carry out than the separation of ^{235}U from the more abundant ^{238}U in natural uranium. Both neptunium and plutonium are *transuranic elements* ($Z > 92$), none of which are found on the earth because their half-lives are too short for them to have survived even if they had been present when the earth came into being 4.5 billion years ago. Transuranic elements up to $Z = 105$ have been produced by the neutron bombardment of lighter nuclides.

Another "fertile" nuclide besides ^{238}U is the thorium isotope ^{232}Th, whose transmutation into the fissionable ^{233}U proceeds via the protactinium isotope ^{233}Pa:

$$^{232}_{90}\text{Th} + {}^1_0n \rightarrow {}^{233}_{90}\text{Th} \xrightarrow{\hspace{2cm}} {}^{233}_{91}\text{Pa} + e^-$$

$$22 \text{ min} \qquad\qquad \xrightarrow{\hspace{2cm}} {}^{233}_{92}\text{U} + e^-$$

$$27 \text{ d}$$

A *breeder reactor* is one designed to produce as much or more fissionable material, in the form of ^{239}Pu or ^{233}U, than the ^{235}U it consumes. Because the otherwise useless ^{238}U makes up 99.3% of natural uranium, and ^{232}Th is plentiful, the significance of breeder reactors in a world of dwindling energy resources is obvious. The widespread use of breeder reactors would mean that known reserves of nuclear fuel would be adequate for centuries to come, rather than decades, which would give enough time for alternative energy technologies to be perfected and implemented. On the other hand, the problems of reactor safety, nuclear waste disposal, and the proliferation of nuclear weapons would be more severe.

Breeder reactors

The thorium breeding cycle is most effective with the slow neutrons of the reactors described above. To convert ^{238}U to ^{239}Pu efficiently, on the other hand, requires fast neutrons, and a *fast breeder reactor* makes use of fissions induced by fast rather than slow neutrons. Present breeder reactors are all of the fast type and use liquid sodium as the coolant; because they operate at high temperatures, their overall efficiencies are 40% or more. Such reactors yield as much as 60 to 70 times more energy from a given amount of natural uranium than conventional reactors do.

30–11 FUSION ENERGY

The basic energy-producing process in stars—and hence the source, direct or indirect, of nearly all the energy in the universe—is the fusion of hydrogen nuclei into helium nuclei. This can take place under stellar conditions in two different series of nuclear reactions. In stars whose internal temperatures are below 1.6×10^7 K, such as the sun, the *proton-proton cycle* predominates:

The sun obtains its energy from the proton-proton cycle of fusion reactions

$$^1_1\text{H} + {}^1_1\text{H} \rightarrow {}^2_1\text{H} + e^+ + 0.4 \text{ MeV}$$

Proton-proton cycle

$$^1_1\text{H} + {}^2_1\text{H} \rightarrow {}^3_2\text{He} + 5.5 \text{ MeV}$$

$$^3_2\text{He} + {}^3_2\text{He} \rightarrow {}^4_2\text{He} + 2\,{}^1_1\text{H} + 12.9 \text{ MeV}$$

The first two of these reactions must each occur twice for every synthesis of $^4_2$He, so the total energy liberated is 24.7 MeV.

The energy liberated by nuclear fusion is often called *thermonuclear energy*. High temperatures and densities are necessary for fusion reactions to occur often enough for a substantial amount of thermonuclear energy to be produced. The high temperature assures that the initial light nuclei have enough thermal energy to overcome their mutual electric repulsion and come close enough together to react, and the high density assures that such collisions are frequent. A further condition for the proton-proton and similar cycles is a large reacting mass, such as that of a star, since a number of separate steps is involved in each cycle and much time may go by between the initial fusion of a particular proton and its eventual incorporation in an alpha particle.

Practical fusion reactions for the earth

On the earth, where any reacting mass must be very limited in size, an efficient fusion process cannot involve more than a single step. Two reactions that appear promising as sources of commercial power involve the combination of two deuterons to form a triton and a proton,

$$_1^2\text{H} + {_1^2\text{H}} \rightarrow {_1^3\text{H}} + {_1^1\text{H}} + 4.0\,\text{MeV}$$

or their combination to form a $_2^3\text{H}$ nucleus and a neutron,

$$_1^2\text{H} + {_1^2\text{H}} \rightarrow {_2^3\text{He}} + {_0^1n} + 3.3\,\text{MeV}$$

Both reactions have about equal probabilities. A major advantage of these reactions is that deuterium is present in seawater. Although its concentration is only 0.015%, this adds up to a total of about 10^{15} tons in the world's oceans.

The first fusion reactors are more likely to employ a deuterium-tritium mixture because the reaction

$$_1^3\text{H} + {_1^2\text{H}} \rightarrow {_2^4\text{He}} + {_0^1n} + 17.6\,\text{MeV}$$

has a higher yield than the others at relatively low temperatures. Seawater contains too little tritium for its economical extraction, but it can be produced by the neutron bombardment of the two isotopes of natural lithium:

$$_3^6\text{Li} + {_0^1n} \rightarrow {_1^3\text{H}} + {_2^4\text{He}}$$
$$_3^7\text{Li} + {_0^1n} \rightarrow {_1^3\text{H}} + {_2^4\text{He}} + {_0^1n}$$

In fact, tritium can be made by the fusion reactor itself by providing a lithium shield to absorb the neutrons liberated during the reactor's operation. Liquid lithium, heated by energetic neutrons, could also be used to extract the energy produced by being circulated between the reactor and a steam generator. The steam then would power a turbine connected to an electric generator, as in the case of fossil-fuel and fission electric plants.

Possible methods for producing fusion energy

The big problem in making fusion energy practical is to achieve the required combination of temperature, density, and containment time for a deuterium or deuterium-tritium mixture to react sufficiently to produce a net energy yield. Such a combination occurs in the explosion of a fission ("atomic") bomb, and incorporating the ingredients for fusion in such a bomb leads to a much more destructive weapon, the "hydrogen" bomb. One approach to the more controlled release of fusion energy involves using energetic beams to both heat and compress tiny deuterium-tritium pellets to produce what are, in effect, miniature hydrogen bomb explosions. A succession of

(a)

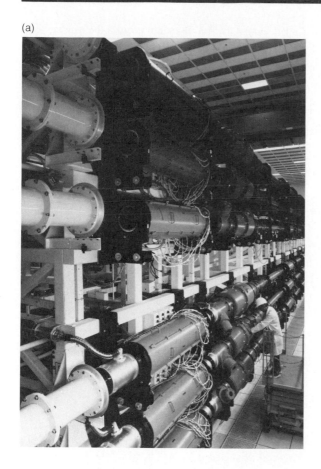

(b)

these bursts would furnish a steady supply of energy. Laser beams (Fig. 30–20(a)) have received the most attention for this purpose, but electron and proton beams also show promise. However, the greatest amount of development has been concentrated on containing an extremely hot, dense plasma (fully ionized gas) of the reactants by means of a strong magnetic field, and considerable progress has been made in this direction (Fig. 30–20(b)).

Although practical fusion reactors are still in the future (how far, nobody knows), the advantages of thermonuclear energy—cheap, essentially limitless fuel; much less hazard than for fission energy; relatively minor production of radioactive wastes; no possibility of military use—are such that eventually it will almost certainly supply a large part of the world's energy needs.

FIG. 30–20 (a) Part of the Shiva laser system that can deliver thousands of joules of energy in less than 10^{-9} s to a fusion target the size of a grain of sand. (b) Magnetic fields, which have superconducting windings, will confine plasmas in fusion research; see Fig. 21–18. (Photos courtesy of Lawrence Livermore National Laboratory.)

IMPORTANT TERMS

The **atomic number** of an element is the number of electrons in each of its atoms or, equivalently, the number of protons in each of its atomic nuclei.

The **neutron** is an electrically neutral particle, slightly heavier than the positively charged **proton,** that is present in nuclei together with protons. Neutrons and protons are jointly called **nucleons.** The **mass number** of a nucleus is the number of nucleons it contains.

Isotopes of an element have the same atomic number but different mass numbers. Symbols of **nuclides** (nuclear species) follow the pattern

$$^A_Z X$$

where X is the chemical symbol of the element, Z its atomic number, and A the mass number of the particular nuclide.

The **binding energy** of a nucleus is the energy equivalent of the difference between its mass and the sum of the masses of its individual constituent nucleons. This amount of energy must be supplied to the nucleus if it is to be completely disintegrated.

Radioactive nuclei spontaneously transform themselves into other nuclear species with the emission of radiation. The radiation may consist of **alpha particles,** which are the nuclei of helium atoms, or **beta particles,** which are positive or negative electrons. Positive electrons are known as **positrons.** Electron capture by a nucleus is an alternative to positron emission. The emission of **gamma rays,** which are energetic photons, enables an excited nucleus to lose its excess energy.

The time required for half of a given sample of a radioactive nuclide to decay is called its **half-life.**

In **nuclear fission,** the absorption of neutrons by certain heavy nuclei causes them to split into smaller **fission fragments.** Because each fission also liberates several neutrons, a rapidly multiplying sequence of fissions called a **chain reaction** can occur if a sufficient amount of the proper material is assembled. A **nuclear reactor** is a device in which a chain reaction can be initiated and controlled.

In **nuclear fusion,** two light nuclei combine to form a heavier one with the emission of energy. The energy liberated in the process when it takes place on a large scale is called **thermonuclear energy.**

MULTIPLE CHOICE

1. The chemical behavior of an atom is determined by its
   a. atomic number.
   b. mass number.
   c. binding energy.
   d. number of isotopes.

2. The atomic number of an atom equals the number of its
   a. protons.
   b. neutrons.
   c. protons and neutrons.
   d. protons and electrons.

3. The mass number of an atom equals the number of its
   a. protons.
   b. neutrons.
   c. protons and neutrons.
   d. protons and electrons.

4. The number of protons in the nucleus of the aluminum isotope $^{27}_{13}$Al is
   a. 13.
   b. 14.
   c. 27.
   d. 40.

5. The number of neutrons in the nucleus of the aluminum isotope $^{27}_{13}$Al is
   a. 13.
   b. 14.
   c. 27.
   d. 40.

6. The atoms of the isotopes of an element have different
   a. half-lives.
   b. numbers of electrons.
   c. numbers of protons.
   d. numbers of neutrons.

7. Which of the following is not an isotope of hydrogen?
   a. $^1_0$H
   b. $^1_1$H
   c. $^2_1$H
   d. $^3_1$H

8. Relative to the sum of the masses of its constituent nucleons, the mass of a nucleus is
   a. greater.
   b. the same.
   c. smaller.
   d. sometimes greater and sometimes smaller.

9. The ionization energy of an atom relative to the binding energy of its nucleus is
   a. greater.
   b. the same.
   c. smaller.
   d. sometimes greater and sometimes smaller.

10. The binding energy per nucleon is
    a. the same for all nuclei.
    b. greatest for very small nuclei.
    c. greatest for nuclei of intermediate size.
    d. greatest for very large nuclei.

11. The element whose nuclei contain the most tightly bound nucleons is
    a. helium.
    b. carbon.
    c. iron.
    d. uranium.

12. Nuclear fission and fusion reactions give off energy because
    a. the binding energy per nucleon is least for nuclei of intermediate size.
    b. the binding energy per nucleon is greatest for nuclei of intermediate size.

c. they result in the production of neutrons.

d. they result in the production of protons.

13. In a stable nucleus the number of neutrons is always

a. less than the number of protons.

b. less than or equal to the number of protons.

c. more than the number of protons.

d. more than or equal to the number of protons.

14. If radium and chlorine combine to form radium chloride, the compound is

a. no longer radioactive.

b. half as radioactive as its radium content.

c. as radioactive as its radium content.

d. twice as radioactive as its radium content.

15. Which of these types of radiation has the greatest ability to penetrate matter?

a. alpha particles   b. beta particles

c. gamma rays   d. X rays

16. Which of these types of radiation has the least ability to penetrate matter?

a. alpha particles   b. beta particles

c. gamma rays   d. X rays

17. A nucleus increases in atomic number when it emits

a. a proton.   b. a neutron.

c. an electron.   d. a positron.

18. When a nucleus undergoes radioactive decay, its new mass number is

a. always less than its original mass number.

b. always more than its original mass number.

c. never less than its original mass number.

d. never more than its original mass number.

19. A nucleus with an excess of neutrons may decay radioactively with the emission of

a. a neutron.   b. a proton.

c. an electron.   d. a positron.

20. Gamma rays do not lose energy in passing through matter by

a. materializing in electron-positron pairs.

b. giving up all their energy to atomic electrons in the photoelectric effect.

c. giving up part of their energy to atomic electrons in the Compton effect.

d. undergoing radioactive decay.

21. The half-life of a radioactive nuclide is

a. half the time needed for a sample to decay entirely.

b. half the time a sample can be kept before it begins to decay.

c. the time needed for half a sample to decay.

d. the time needed for the remainder of a sample to decay after half of it has already decayed.

22. As a sample of a radioactive nuclide decays, its half-life

a. decreases.

b. remains the same.

c. increases.

d. any of the above, depending upon the isotope.

23. The largest amount of radiation received by an average person in the United States comes from

a. medical X rays.

b. nuclear reactors.

c. fallout from past weapons tests.

d. natural sources.

24. The radioisotope used for dating that has the shortest half-life is

a. carbon-14.

b. potassium-40.

c. rubidium-87.

d. uranium-235.

25. A suitable object for radiocarbon dating would be one whose age is

a. 10,000 years.

b. 100,000 years.

c. 1 million years.

d. 10 million years.

26. The process by which a heavy nucleus splits into two lighter nuclei is known as

a. fission.

b. fusion.

c. alpha decay.

d. a chain reaction.

27. By "chain reaction" is meant

a. the joining together of protons and neutrons to form atomic nuclei.

b. the joining together of light nuclei to form heavy ones.

c. the successive fissions of heavy nuclei induced by neutrons emitted in the fissions of other heavy nuclei.

d. the burning of uranium in a special type of furnace called a nuclear reactor.

28. Most of the energy liberated during fission appears

a. as KE of the neutrons that are emitted.

b. as KE of the fission fragments.

c. in the decay of the fission fragments.

d. as gamma rays.

29. Enriched uranium is a better fuel for nuclear reactors than natural uranium because it has a greater proportion of

a. slow neutrons.   b. deuterium.
c. $^{235}_{92}$U.   d. $^{238}_{92}$U.

30. From the point of view of a power plant engineer, a nuclear reactor is a source of
   a. heat.   b. electric current.
   c. slow neutrons.   d. gamma rays.

31. The sun's energy comes from
   a. nuclear fission.
   b. radioactivity.
   c. the conversion of hydrogen to helium.
   d. the conversion of helium to hydrogen.

32. Fusion reactions on the earth are likely to use as fuel
   a. ordinary hydrogen.   b. deuterium.
   c. plutonium.   d. uranium.

33. The mass of a $^7_3$Li nucleus is 0.042 u less than the sum of the masses of 3 protons and 4 neutrons. The binding energy per nucleon in $^7_3$Li is
   a. 5.6 MeV.   b. 10 MeV.
   c. 13 MeV.   d. 39 MeV.

34. The product of the alpha decay of the polonium isotope $^{210}_{84}$Po is
   a. $^{208}_{80}$Hg.   b. $^{206}_{82}$Pb.
   c. $^{206}_{84}$Po.   d. $^{214}_{86}$Rn.

35. The helium isotope $^6_2$He undergoes beta decay with the emission of an electron. The product of the decay is
   a. $^6_1$H.   b. $^5_2$He.
   c. $^7_2$He.   d. $^6_3$He.

36. When a nucleus of the copper isotope $^{64}_{29}$Cu captures an electron, it becomes
   a. $^{63}_{28}$Ni.   b. $^{64}_{28}$Ni.
   c. $^{63}_{29}$Cu.   d. $^{64}_{30}$Zn.

37. When a nucleus of the copper isotope $^{64}_{29}$Cu emits a positron, it becomes
   a. $^{63}_{28}$Ni.   b. $^{64}_{28}$Ni.
   c. $^{63}_{29}$Cu.   d. $^{64}_{30}$Zn.

38. The product of the gamma decay of the strontium isotope $^{87}_{38}$Sr is
   a. $^{86}_{37}$Rb.   b. $^{87}_{37}$Rb.
   c. $^{86}_{38}$Sr.   d. $^{87}_{38}$Sr.

39. The lead isotope $^{214}_{82}$Pb decays into the bismuth isotope $^{214}_{83}$Bi by emitting
   a. an alpha particle.   b. a proton.
   c. an electron.   d. a positron.

40. The half-life of a certain radioactive nuclide is 6 h. If we start out with 10 g of the nuclide, after 1 day there will be
   a. none left.   b. 0.625 g left.
   c. 1.6 g left.   d. 2.5 g left.

41. After 2 hours, $\frac{1}{16}$ of the initial amount of a certain radioactive nuclide remains undecayed. The half-life of the nuclide is
   a. 15 min.   b. 30 min.
   c. 45 min.   d. 60 min.

EXERCISES

30–1   Nuclear Structure

1. In what ways are the isotopes of an element similar to one another? In what ways are they different?

2. State the number of neutrons and protons in each of the following nuclei: $^6_3$Li; $^{13}_6$C; $^{31}_{15}$P; $^{94}_{40}$Zr; $^{137}_{56}$Ba.

3. State the number of neutrons and protons in each of the following nuclei: $^{10}_5$Be; $^{22}_{10}$Ne; $^{36}_{16}$S; $^{88}_{38}$Sr; $^{180}_{72}$Hf.

4. The first nuclear reaction ever observed occurred when $^{14}_7$N was bombarded with alpha particles and protons were ejected. Give the atomic number, mass number, and chemical symbol of the resulting nucleus.

5. A reaction often used to detect neutrons occurs when a neutron strikes a $^{10}_5$B nucleus, with the subsequent emission of an alpha particle. What is the atomic number, mass number, and chemical name of the remaining nucleus?

6. A nucleus of $^{15}_7$N is struck by a proton. A nuclear reaction can take place with the emission of either a neutron or an alpha particle. Give the atomic number, mass number, and chemical name of the remaining nucleus in each of the above cases.

7. Complete the following nuclear reactions:
$^6_3$Li + ? → $^7_4$Be + ^1_0n
$^{10}_5$B + ? → $^7_3$Li + $^4_2$Hc
$^{35}_{17}$Cl + ? → $^{32}_{16}$S + $^4_2$He

8. Using the figures in Table 30–1, verify that the average atomic mass of chlorine is 35.46 u.

9. Ordinary boron is a mixture of the $^{10}_5$B and $^{11}_5$B isotopes and has a composite atomic mass of 10.82 u. Find the percentage of each isotope present in ordinary boron. The atomic masses of $^{10}_5$B and $^{11}_5$B are, respectively, 10.01 u and 11.01 u.

30–2   Binding Energy

10. The binding energy of $^{20}_{10}$Ne is 160.64 MeV. Find its atomic mass.

11. The mass of $^4_2$He is 4.002603 u. Find its binding energy and binding energy per nucleon.

12. The neutron decays in free space into a proton and an

electron. What must be the minimum binding energy contributed by a neutron to a nucleus in order that the neutron not decay inside the nucleus? How does this figure compare with the observed binding energies per nucleon in stable nuclei?

13. The nuclear reaction

$$^6_3\text{Li} + {}^2_1\text{H} \rightarrow 2{}^4_2\text{He}$$

evolves 22.4 MeV. Calculate the mass of ^6_3Li in u.

14. The atomic masses of $^{15}_7\text{N}$, $^{15}_8\text{O}$, and $^{16}_8\text{O}$ are, respectively, 15.0001, 15.0030, and 15.9949 u. (a) Find the average binding energy per nucleon in $^{16}_8\text{O}$. (b) How much energy is needed to remove one proton from $^{16}_8\text{O}$? (c) How much energy is needed to remove one neutron from $^{16}_8\text{O}$? (d) Why are these figures different from one another?

30–3 Radioactivity

15. Radium undergoes spontaneous decay into helium and radon. Why is radium regarded as an element rather than as a chemical compound of helium and radon?

16. What happens to the atomic number and mass number of a nucleus that emits a gamma ray? What happens to the actual mass of the nucleus?

17. The nuclide $^{238}_{92}\text{U}$ decays into a lead isotope through the successive emissions of eight alpha particles and six electrons. What is the symbol of the lead isotope?

18. A $^{80}_{35}\text{Br}$ nucleus can decay by beta decay, positron emission, or electron capture. What is the daughter nucleus in each case?

19. The nucleus $^{233}_{90}\text{Th}$ undergoes two negative beta decays in becoming an isotope of uranium. What is the symbol of the isotope?

20. (a) Under what circumstances does a nucleus emit an electron? A positron? (b) The nuclei $^{14}_8\text{O}$ and $^{19}_8\text{O}$ both undergo beta decay in order to become stable nuclei. Which would you expect to emit a positron and which an electron?

21. The atomic mass of $^{226}_{88}\text{Ra}$ is 226.0254 u, and the energy liberated in its alpha decay is 4.87 MeV. (a) Identify the daughter nucleus and find its atomic mass. (b) The alpha particle emitted in the decay is observed to have a KE of 4.78 MeV. Where do you think the other 0.09 MeV goes?

22. The nuclide $^{232}_{92}\text{U}$ (mass 232.0372 u) alpha decays into $^{228}_{90}\text{Th}$ (mass 228.0287 u). (a) Find the amount of energy released in the decay. (b) Is it possible for $^{232}_{92}\text{U}$ to decay into $^{231}_{92}\text{U}$ (mass 231.0264 u) by emitting a neutron? Why? (c) Is it possible for $^{232}_{92}\text{U}$ to decay into $^{231}_{91}\text{Pa}$ (mass 231.0359 u) by emitting a proton? Why?

30–4 Radiation Detection

23. What is the chief way in which heavy charged particles, such as protons and alpha particles, lose energy in passing through matter?

24. In what ways do electrons lose energy in passing through matter?

25. In what ways do gamma rays lose energy in passing through matter?

30–5 Half-Life

26. How many disintegrations per second occur in a 25 millicurie sample of thorium?

27. The half-life of radium is 1600 years. How long will it take for $\frac{15}{16}$ of a given sample of radium to decay?

28. The activity of a sample of $^{227}_{90}\text{Th}$ is observed to be 12.5% of its original amount 54 days after the sample was prepared. What is the half-life of $^{227}_{90}\text{Th}$?

29. The half-life of the alpha-emitter $^{210}_{84}\text{Po}$ is 138 days. This nuclide is used to power a thermoelectric cell whose initial output is 1 W. What will the output be after 3 years?

30–7 Radiometric Dating

30. What are the two basic conditions that must be met by a radionuclide in order that it be useful in dating a particular kind of rock?

31. (a) What is the basis of the radiocarbon dating procedure? (b) What steps would you take to find the age of an ancient piece of wood by radiocarbon dating?

30–8 Fission
30–9 Nuclear Reactors

32. What is the limitation on the fuel that can be used in a reactor whose moderator is ordinary water? Why is the situation different if the moderator is heavy water?

33. Can you think of any reason why a gas-cooled reactor with a graphite moderator might be safer than a reactor that uses water as coolant and moderator?

34. (a) How much mass is lost per day by a nuclear reactor operated at a 1 GW (10^9 W) power level? (b) If each fission in $^{235}_{92}\text{U}$ releases 200 MeV, how many fissions must occur per second to yield a power level of 1 GW?

35. $^{235}_{92}\text{U}$ loses about 0.1% of its mass when it undergoes fission. (a) How much energy is released when 1 kg of $^{235}_{92}\text{U}$ undergoes fission? (b) One ton of TNT releases about

4×10^9 J when it is detonated. How many tons of TNT are equivalent in destructive power to a bomb that contains 1 kg of $^{235}_{92}$U?

30–11   Fusion Energy

36. What are the differences and similarities between nuclear fission and nuclear fusion? Where does the evolved energy come from in each case?

37. Why are the conditions in the interior of a star favorable for nuclear fusion reactions?

38. Fusion reactions involving deuterons can take place when the deuterons have energies of about 10 keV or more. Find the temperature of a deuterium plasma in which the average KE of the deuterons is 10 keV.

39. Old stars obtain part of their energy by the fusion of three alpha particles to form a $^{12}_{6}$C nucleus, whose mass is 12.0000 u. How much energy is evolved in each such reaction?

ANSWERS TO MULTIPLE CHOICE

1. a	**10.** c	**18.** d	**26.** a	**34.** b
2. a	**11.** c	**19.** c	**27.** c	**35.** d
3. c	**12.** b	**20.** d	**28.** b	**36.** b
4. a	**13.** d	**21.** c	**29.** c	**37.** b
5. b	**14.** c	**22.** b	**30.** a	**38.** d
6. d	**15.** c	**23.** d	**31.** c	**39.** c
7. a	**16.** a	**24.** a	**32.** b	**40.** b
8. c	**17.** c	**25.** a	**33.** a	**41.** b
9. c				

APPENDIX A
USEFUL MATHEMATICS

A–1 ALGEBRA

Algebra is a generalized arithmetic in which symbols are used in place of numbers. Algebra thus provides a language in which general relationships can be expressed among quantities whose numerical values need not be known in advance.

The arithmetical operations of addition, subtraction, multiplication, and division have the same meanings in algebra. The symbols of algebra are normally letters of the alphabet. If we have two quantities a and b and add them to give the sum c, we would write

$$a + b = c$$

If we subtract b from a to give the difference d, we would write

$$a - b = d$$

Multiplying a and b together to give e may be written in any of these ways:

$$a \times b = e \qquad ab = e \qquad (a)(b) = e$$

Whenever two algebraic quantities are written together with nothing between them, it is understood that they are to be multiplied.

Dividing a by b to give the quotient f is usually written

$$\frac{a}{b} = f$$

but it may sometimes be more convenient to write

$$a/b = f$$

which has the same meaning.

Parentheses and brackets are used to show the order in which various operations are to be performed. Thus

$$\frac{(a + b)c}{d} - e = f$$

means that, in order to find f, we are first to add a and b together, then multiply their sum by c and divide by d, and finally subtract e.

A–2 EQUATIONS

An *equation* is simply a statement that a certain quantity is equal to another one. Thus

$$7 + 2 = 9$$

which contains only numbers, is an arithmetical equation, and

$$3x + 12 = 27$$

which contains a symbol as well, is an algebraic equation. The symbols in an algebraic equation usually cannot have any arbitrary values if the equality is to hold. Finding the possible values of these symbols is called *solving* the equation. The *solution* of the latter equation above is

$$x = 5$$

since only when x is 5 is it true that $3x + 12 = 27$.

In order to solve an equation, a basic principle must be kept in mind.

Any operation performed on one side of an equation must be performed on the other.

An equation therfore remains valid when the same quantity, numerical or otherwise, is added to or subtracted from both sides, or when the same quantity is used to multiply or divide both sides. Other operations, for instance squaring or taking the square root, also do not alter the equality if the same thing is done to both sides. As a simple example, to solve $3x + 12 = 27$, we first subtract 12 from both sides:

$$3x + 12 - 12 = 27 - 12$$
$$3x = 15$$

To complete the solution we divide both sides by 3:

$$\frac{3x}{3} = \frac{15}{3}$$
$$x = 5$$

To check a solution, we substitute it back in the original equation and see whether the equality is still true. Thus we can check that $x = 5$ by reducing the original algebraic equation to an arithmetical one:

$$3x + 12 = 27$$
$$(3)(5) + 12 = 27$$
$$15 + 12 = 27$$
$$27 = 27$$

Two helpful rules follow directly from the principle stated above. The first is,

Any term on one side of an equation may be transposed to the other side by changing its sign.

To verify this rule, we subtract b from each side of the equation

$$a + b = c$$

to obtain

$$a + b - b = c - b$$
$$a = c - b$$

We see that b has disappeared from the left-hand side and $-b$ is now on the right-hand side.

The second rule is,

A quantity which multiplies one side of an equation may be transposed in order to divide the other side, and vice versa.

To verify this rule, we divide both sides of the equation

$$ab = c$$

by b. The result is

$$\frac{ab}{b} = \frac{c}{b}$$
$$a = \frac{c}{b}$$

We see that b, a multiplier on the left-hand side, is now a divisor on the right-hand side.

EXAMPLE Solve the following equation for x:

$$4(x - 3) = 7$$

SOLUTION The above rules are easy to apply here:

$$4(x - 3) = 7$$
$$x - 3 = \frac{7}{4}$$
$$x = \frac{7}{4} + 3 = 1.75 + 3 = 4.75$$

When each side of an equation consists of a fraction, all we need to do to remove the fractions is to *cross multiply*:

$$\frac{a}{b} = \frac{c}{d}$$
$$ad = bc$$

What was originally the denominator (lower part) of each fraction now multiplies the numerator (upper part) of the other side of the equation.

EXAMPLE Solve the following equation for y:

$$\frac{5}{y + 2} = \frac{3}{y - 2}$$

SOLUTION First we cross multiply to get rid of the fractions, and then solve in the usual way:

$$5(y - 2) = 3(y + 2)$$
$$5y - 10 = 3y + 6$$
$$5y - 3y = 6 + 10$$
$$2y = 16$$
$$y = 8$$

EXERCISES

1. Solve each of the following equations for x. The answers are given at the end of Appendix A.

(a) $\dfrac{x + 7}{6} = x + 2$

(b) $\dfrac{3x - 42}{9} = 2(7 - x)$

(c) $\dfrac{1}{x + 1} = \dfrac{1}{2x - 1}$

(d) $\dfrac{8}{x} = \dfrac{1}{4 - x}$

(e) $\dfrac{x}{2x - 1} = \dfrac{5}{7}$

A–3   SIGNED NUMBERS

The rules for multiplying and dividing positive and negative quantities are straightforward. First, perform the indicated operation on the absolute value of the quantity (the absolute value of -7 is 7, for instance). If the quantities are both positive or both negative, the result is positive:

$$(+a)(+b) = (-a)(-b) = +ab$$

$$\frac{+a}{+b} = \frac{-a}{-b} = +\frac{a}{b}$$

If one quantity is positive and the other negative the result is negative:

$$(-a)(+b) = (+a)(-b) = -ab$$

$$\frac{-a}{+b} = \frac{+a}{-b} = -\frac{a}{b}$$

A few examples might be helpful:

$$(-6)(-3) = 18 \qquad \frac{-20}{-5} = 4$$

$$(4)(-5) = -20 \qquad \frac{6}{-2} = -3$$

$$(-10)(7) = -70 \qquad \frac{-30}{15} = -2$$

EXAMPLE   Find the value of

$$z = \frac{xy}{x - y}$$

when $x = -12$ and $y = 4$.

SOLUTION   We begin by evaluating xy and $x - y$, which are

$$xy = (-12) \times 4 = -48$$
$$x - y = (-12) - 4 = -16$$

Hence

$$z = \frac{xy}{x - y} = \frac{-48}{-16} = 3$$

EXERCISES

2. Evaluate the following:

(a) $\dfrac{3(x + y)}{2}$ when $x = 5, y = -2$

(b) $\dfrac{1}{x - y} - \dfrac{1}{x + y}$ when $x = 3, y = -5$

(c) $\dfrac{3(x + 7)}{y + 2}$ when $x = 3, y = -6$

(d) $\dfrac{x + y}{2z} + \dfrac{z}{x - y}$ when $x = -2, y = 2, z = 4$

(e) $\dfrac{x + z}{y} - \dfrac{xy}{2}$ when $x = 2, y = -8, z = 10$

A–4   EXPONENTS

It is often necessary to multiply a quantity by itself a number of times. This process is indicated by a superscript number called the exponent, according to the following scheme:

$$A = A^1$$
$$A \times A = A^2$$
$$A \times A \times A = A^3$$
$$A \times A \times A \times A = A^4$$
$$A \times A \times A \times A \times A = A^5$$

We read A^2 as "A squared" because it is the area of a square of length A on a side; similarly A^3 is called "A cubed" because it is the volume of a cube each of whose sides is A long. More generally we speak of A^n as "A to the nth power." Thus A^5 is read as "A to the fifth power."

When we multiply a quantity raised to some particular power (say A^n) by the same quantity raised to another power (say A^m), the result is that quantity raised to a power equal to the sum of the original exponents. That is,

$$A^n A^m = A^{(n + m)}$$

For example,

$$A^2 A^5 = A^7$$

which we can verify directly by writing out the terms:

$$(A \times A)(A \times A \times A \times A \times A)$$

$$= A \times A \times A \times A \times A \times A \times A = A^7$$

From the above result we see that when a quantity raised to a particular power (say A^n) is to be multiplied by itself a total of m times, we have

$$(A^n)^m = A^{nm}$$

For example,

$$(A^2)^3 = A^6$$

since

$$(A^2)^3 = A^2 \times A^2 \times A^2 = A^{(2 + 2 + 2)} = A^6$$

Reciprocal quantities are expressed in a similar way with the addition of a minus sign in the exponent, as follows:

$$\frac{1}{A} = A^{-1} \quad \frac{1}{A^2} = A^{-2} \quad \frac{1}{A^3} = A^{-3} \quad \frac{1}{A^4} = A^{-4}$$

Exactly the same rules as before are used in combining quantities raised to negative powers with one another and with some quantity raised to a positive power. Thus

$$A^5 A^{-2} = A^{(5-2)} = A^3$$
$$(A^{-1})^{-2} = A^{-1(-2)} = A^2$$
$$(A^3)^{-4} = A^{-4 \times 3} = A^{-12}$$
$$AA^{-7} = A^{(1-7)} = A^{-6}$$

It is important to remember that any quantity raised to the zeroth power, say A^0, is equal to 1. Hence

$$A^2 A^{-2} = A^{(2-2)} = A^0 = 1$$

This is more easily seen if we write A^{-2} as $1/A^2$:

$$A^2 A^{-2} = A^2 \times \frac{1}{A^2} = \frac{A^2}{A^2} = 1$$

A–5   ROOTS

The *square root* of A, $\sqrt{A}$, is that quantity which, when multiplied by itself, is equal to A:

$$\sqrt{A} \times \sqrt{A} = A$$

The square root has that name because the length of each side of a square of area A is given by $\sqrt{A}$. Some examples of square roots are as follows:

$$\sqrt{1} = 1 \qquad \text{because } 1 \times 1 = 1$$
$$\sqrt{16} = 4 \qquad \text{because } 4 \times 4 = 16$$
$$\sqrt{100} = 10 \qquad \text{because } 10 \times 10 = 100$$
$$\sqrt{42.25} = 6.5 \qquad \text{because } 6.5 \times 6.5 = 42.25$$
$$\sqrt{9A^2} = 3A \qquad \text{because } 3A \times 3A = 9A^2$$

The square root of a number less than 1 is larger than the number itself:

$$\sqrt{0.01} = 0.1 \qquad \text{because } 0.1 \times 0.1 = 0.01$$
$$\sqrt{0.49} = 0.7 \qquad \text{because } 0.7 \times 0.7 = 0.49$$

The square root of a quantity may be either positive or negative because $(+A)(+A) = (-A)(-A) = A^2$. Whether the positive or negative root (or even both) is correct in a given problem depends upon the details of the problem and is usually obvious. Using exponents we see that, because

$$(A^{1/2})^2 = A^{2 \times (1/2)} = A^1 = A$$

we can express square roots by the exponent $\frac{1}{2}$:

$$\sqrt{A} = A^{1/2}$$

Other roots may be expressed similarly. The *cube root* of a quantity A, written $\sqrt[3]{A}$, when multiplied by itself twice equals A. That is,

$$\sqrt[3]{A} \times \sqrt[3]{A} \times \sqrt[3]{A} = A$$

which may be more conveniently written

$$(A^{1/3})^3 = A$$

where $\sqrt[3]{A} = A^{1/3}$.

In general the nth root of a quantity, $\sqrt[n]{A}$, may be written $A^{1/n}$, which is a more convenient form for most purposes. Some examples may be helpful:

$$\sqrt{A^4} = (A^4)^{1/2} = A^{1/2 \times 4} = A^2$$
$$(\sqrt{A})^4 = (A^{1/2})^4 = A^{4 \times 1/2} = A^2$$
$$\sqrt{\sqrt{A}} = (A^{1/2})^{1/2} = A^{1/2 \times 1/2} = A^{1/4}$$
$$(A^{1/4})^{-7} = A^{-7 \times (1/4)} = A^{-7/4}$$
$$(A^3)^{-1/3} = A^{-(1/3) \times 3} = A^{-1}$$
$$(A^{1/4})^{1/4} = A^{(1/4) \times (1/4)} = A^{1/16}$$

Although procedures exist for finding roots arithmetically with pencil and paper, in practice it is much easier to use an electronic calculator. Calculators for science and engineering usually have a square root [$\sqrt{}$] key. For other roots, such as $A^{1/3}$, and for nonintegral exponents in general, such as $A^{3.27}$, logarithms must be used, as described in Appendix B-5. A calculator with [LOG] and [10^x] or [INV LOG] keys makes such quantities easy to evaluate.

A–6   EQUATIONS WITH POWERS AND ROOTS

An equation that involves powers or roots or both is subject to the same basic principle that governs the manipulation of simpler equations: Whatever is done to one side must be done to the other. Hence the following rules:

An equation remains valid when both sides are raised to the same power, that is, when each side is multiplied by itself the same number of times as the other side.

An equation remains valid when the same root is taken of both sides.

EXAMPLE Newton's law of gravitation states that the force F between two bodies the distance r apart whose masses are m_A and m_B is given by the formula

$$F = G\frac{m_A m_B}{r^2}$$

where G is a universal constant. Solve this formula for r.

SOLUTION   We begin by transposing the r^2 to give

$$Fr^2 = Gm_A m_B$$

Next we transpose the F to give

$$r^2 = \frac{Gm_A m_B}{F}$$

and finally take the square root of both sides:

$$r = \sqrt{r^2} = \sqrt{\frac{Gm_A m_B}{F}}$$

The positive value of the root is correct here since a negative separation r between the bodies is meaningless.

Quadratic equations, which have the general form

$$ax^2 + bx + c = 0$$

are often encountered in algebraic calculations. In such an equation, a, b, and c are constants. A quadratic equation is satisfied by the values of x given by the formulas

$$x_+ = \frac{-b + \sqrt{b^2 - 4ac}}{2a}$$

$$x_- = \frac{-b - \sqrt{b^2 - 4ac}}{2a}$$

By looking at these formulas, we can see that the nature of the solution depends upon the value of the quantity $b^2 - 4ac$. When $b^2 = 4ac$, $\sqrt{b^2 - 4ac} = 0$ and the two solutions are equal to just $x = -b/2a$. When $b^2 > 4ac$, $\sqrt{b^2 - 4ac}$ is a real number and the solutions x_+ and x_- are different. When $b^2 < 4ac$, $\sqrt{b^2 - 4ac}$ is the square root of a negative number, and the solutions are different. The square root of a negative number is called an *imaginary number* because squaring a real number, whether positive or negative, always

gives a positive number: $(+2)^2 = (-2)^2 = +4$, hence $\sqrt{-4}$ cannot be either $+2$ or -2. Imaginary numbers do not occur in the sort of physical problems treated in this book.

EXAMPLE Solve the quadratic equation

$$2x^2 - 3x - 9 = 0$$

SOLUTION   Here $a = 2$, $b = -3$, and $c = -9$, so that

$$b^2 - 4ac = (-3)^2 - 4(2)(-9) = 9 + 72 = 81$$

The two solutions are

$$x_+ = \frac{-b + \sqrt{b^2 - 4ac}}{2a}$$
$$= \frac{-(-3) + \sqrt{81}}{2(2)} = \frac{3 + 9}{4} = \frac{12}{4} = 3$$

$$x_- = \frac{-b - \sqrt{b^2 - 4ac}}{2a}$$
$$= \frac{-(-3) - \sqrt{81}}{2(2)} = \frac{3 - 9}{4} = \frac{-6}{4} = -\frac{3}{2}$$

EXAMPLE Solve the quadratic equation

$$x^2 - 4x + 4 = 0$$

SOLUTION   Here $a = 1$, $b = -4$, and $c = 4$, so that

$$b^2 - 4ac = (-4)^2 - 4(1)(4) = 16 - 16 = 0$$

Hence the only solution is

$$x = \frac{-b}{2a} = \frac{-(-4)}{2} = 2$$

EXERCISES

3. Solve each of the following equations for x.

   (a) $9x^2 + 12x + 4 = 0$

   (b) $x^2 + x - 2 = 0$

   (c) $2x^2 - 5x + 2 = 0$

   (d) $x^2 + x - 20 = 0$

   (e) $4x^2 - 4x - 11 = 0$

A–7 TRIGONOMETRIC FUNCTIONS

A *right triangle* is a triangle two sides of which are perpendicular. Such triangles are frequently encountered in physics, and it is necessary to know how their sides and angles

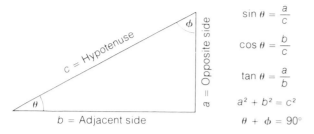

FIG. A–1. A right triangle

$$\sin \theta = \frac{a}{c}$$

$$\cos \theta = \frac{b}{c}$$

$$\tan \theta = \frac{a}{b}$$

$$a^2 + b^2 = c^2$$

$$\theta + \phi = 90°$$

are related. The *hypotenuse* of a right triangle is the side opposite the right angle, as in Fig. A–1; it is always the longest side. The three basic trigonometric functions, the *sine*, *cosine*, and *tangent* of an angle, are defined as follows:

$$\sin \theta = \frac{a}{c} = \frac{\text{opposite side}}{\text{hypotenuse}}$$

$$\cos \theta = \frac{b}{c} = \frac{\text{adjacent side}}{\text{hypotenuse}}$$

$$\tan \theta = \frac{a}{b} = \frac{\text{opposite side}}{\text{adjacent side}}$$

From these definitions we see that

$$\frac{\sin \theta}{\cos \theta} = \frac{a/c}{b/c} = \frac{a}{b} = \tan \theta$$

Numerical values of $\sin \theta$, $\cos \theta$, and $\tan \theta$ for angles form 0° to 90° are given in Appendix C. These figures may be used for angles from 90° to 360° with the help of this table:

θ	$90° + \theta$	$180° + \theta$	$270° + \theta$
sin	$\sin \theta$	$\cos \theta$	$-\sin \theta$	$-\cos \theta$
cos	$\cos \theta$	$-\sin \theta$	$-\cos \theta$	$\sin \theta$
tan	$\tan \theta$	$-\dfrac{1}{\tan \theta}$	$\tan \theta$	$-\dfrac{1}{\tan \theta}$

For example, if we require the value of $\sin 120°$, we first note that $120° = 90° + 30°$. Then, since

$$\sin (90° + \theta) = \cos \theta$$

we have

$$\sin 120° = \sin (90° + 30°) = \cos 30° = 0.866$$

Similarly, to find $\tan 342°$, we begin by noting that $342° = 270° + 72°$. Then, since

$$\tan (270° + \theta) = -\frac{1}{\tan \theta}$$

we have

$$\tan 342° = \tan (270° + 72°) = -\frac{1}{\tan 72°} = -\frac{1}{3.078}$$

$$= -0.325$$

The *inverse* of a trigonometric function is the angle whose function is given. For instance, the inverse of $\sin \theta$ is the angle θ. If $\sin \theta = x$, then the angle θ may be designated either as $\theta = \arcsin x$ or as $\theta = \sin^{-1} x$. It is important to keep in mind that an expression such as $\sin^{-1} x$ does *not* mean $1/\sin x$. The inverse trigonometric functions are as follows:

$$\sin \theta = x$$

$$\theta = \arcsin x = \sin^{-1} x = \text{angle whose sine is } x$$

$$\cos \theta = y$$

$$\theta = \arccos y = \cos^{-1} y = \text{angle whose cosine is } y$$

$$\tan \theta = z$$

$$\theta = \arctan z = \tan^{-1} z = \text{angle whose tangent is } z$$

To find an inverse function, say the angle whose cosine is 0.907, the procedure with some calculators is to enter 0.907 and then press in succession the [ARC] and [COS] keys. The result is $\theta = 25°$ to the nearest degree. With other calculators, a shift key must first be pressed that converts the [SIN], [COS], and [TAN] keys, respectively, to $\sin^{-1}$, $\cos^{-1}$, and $\tan^{-1}$, and then the [COS] key is pressed. With a table of cosines, we look for the value nearest to 0.907 in the body of the table, which is 0.906, and then read across to find the corresponding angle.

EXERCISES

4. Express the following functions in terms of angles between 0 and 90°.

    (a) $\sin 100°$

    (b) $\sin 300°$

    (c) $\cos 150°$

    (d) $\cos 350°$

    (e) $\tan 180°$

    (f) $\tan 220°$

A–8   SOLVING A RIGHT TRIANGLE

To solve a given triangle means to find the values of any unknown sides or angles in terms of the values of the known sides and angles. A triangle has three sides and three angles, and we must know the values of at least three of these six quantities, including one of the sides, to solve the triangle for the others. In a right triangle, one of the angles is always 90°, and so all we need here are the lengths of any two of its sides or the length of one side and the value of one of the other angles to find the remaining sides and angles.

Suppose we know the length of the side b and the angle θ in the right triangle of Fig. A–1. From the definitions of sine and tangent we see that

$$\tan \theta = \frac{a}{b}$$

$$a = b \tan \theta$$

$$\sin \theta = \frac{a}{c}$$

$$c = \frac{a}{\sin \theta}$$

This gives us the two unknown sides a and c. To find the unknown angle ϕ, we can use any of these formulas:

$$\phi = \sin^{-1}\frac{b}{c} \qquad \phi = \cos^{-1}\frac{a}{c} \qquad \phi = \tan^{-1}\frac{b}{a}$$

Alternatively, we can use the fact that the sum of the angles in any triangle is 180°. Because one of the angles in a right triangle is 90°, the sum of the other two must be 90°:

$$\theta + \phi = 90°$$

Hence $\phi = 90° - \theta$ here.

Another useful relationship in a right triangle is the Pythagorean theorem, which states that the sum of the squares of the sides of such a triangle adjacent to the right angle is equal to the square of its hypotenuse. For the triangle of Fig. A–1,

$$a^2 + b^2 = c^2$$

Thus we can always express the length of any of the sides of a right triangle in terms of the other sides:

$$a = \sqrt{c^2 - b^2}$$

$$b = \sqrt{c^2 - a^2}$$

$$c = \sqrt{a^2 + b^2}$$

EXAMPLE In the triangle of Fig. A–1, $a = 7$ cm and $b = 10$ cm. Find c, θ, and ϕ.

SOLUTION   From the Pythagorean theorem,

$$c = \sqrt{a^2 + b^2} = \sqrt{7^2 + 10^2}\ \text{cm}$$

$$= \sqrt{149}\ \text{cm} = 12.2\ \text{cm}$$

Since $\tan \theta = a/b$,

$$\theta = \tan^{-1}\frac{a}{b} = \tan^{-1}\frac{7\,\text{cm}}{10\,\text{cm}}$$

$$= \tan^{-1}0.7 = 35°$$

The value of the other angle ϕ is given by

$$\phi = 90° - \theta = 90° - 35° = 55°$$

EXERCISES

5. Find the values of the unknown sides and angles in the right triangles for which the following data are known:

    (a) $\theta = 45°, a = 10$

    (b) $\theta = 15°, b = 4$

    (c) $\theta = 25°, c = 5$

    (d) $a = 3, b = 4$

    (e) $a = 5, c = 13$

ANSWERS

1. (a) -1
     (b) 8
     (c) 2
     (d) 32/9
     (e) 5/3

2. (a) 4.5
     (b) 0.625
     (c) -3
     (d) -1
     (e) 6.5

3. (a) $-2/3$
     (b) $1, -2$
     (c) $2, 0.5$
     (d) $4, -5$
     (e) $0.5 \pm \sqrt{3} = 2.232, -1.232$

4. (a) $\cos 10°$
     (b) $-\cos 30°$
     (c) $-\sin 60°$
     (d) $\sin 80°$
     (e) $-1/\tan 90°$ or $\tan 0$. Since $\tan 90° = \infty$ and $\tan 0 = 0$, both expressions give the same value.
     (f) $\tan 40°$

5. (a) $b = 10, c = 14.1, \phi = 45°$
     (b) $a = 1.07, c = 4.14, \phi = 75°$
     (c) $a = 2.11, b = 4.53, \phi = 65°$
     (d) $c = 5, \theta = 37°, \phi = 53°$
     (e) $b = 12, \theta = 23°, \phi = 67°$

APPENDIX B
POWERS OF TEN
AND LOGARITHMS

B–1 POWERS OF TEN

Very small and very large numbers are common in physics. For example, the mass of an electron is 0.000, 000,000,000,000,000,000,000,000,000,910,9 kilogram, and the mass of the earth is 5,983,000,000,000, 000,000,000,000 kilograms. Such numbers in ordinary decimal form are clumsy to write and to make calculations with, and it is hard to appreciate their precise magnitudes because of the sea of zeros.

A better method for expressing numbers makes use of powers-of-ten notation. This method is based on the fact that all numbers may be represented by a number between 1 and 10 multiplied by a power of 10. In powers-of-ten notation the mass of an electron is written simply as 9.109×10^{-31} kilogram and the mass of the earth is written as 5.983×10^{24} kilogram.

Table B-1 contains powers of 10 from 10^{-10} to 10^{10}. Evidently positive powers of 10 (which cover numbers greater than 1) follow this pattern:

TABLE B-1

10^{-10}	$= 0.000,000,000,1$	10^0	$= 1$
10^{-9}	$= 0.000,000,001$	10^1	$= 10$
10^{-8}	$= 0.000,000,01$	10^2	$- 100$
10^{-7}	$= 0.000,000,1$	10^3	$= 1000$
10^{-6}	$= 0.000,001$	10^4	$= 10,000$
10^{-5}	$= 0.000,01$	10^5	$= 100,000$
10^{-4}	$= 0.000,1$	10^6	$= 1,000,000$
10^{-3}	$= 0.001$	10^7	$= 10,000,000$
10^{-2}	$= 0.01$	10^8	$= 100,000,000$
10^{-1}	$= 0.1$	10^9	$= 1,000,000,000$
10^0	$= 1$	10^{10}	$= 10,000,000,000$

$$10^0 = 1. \qquad = 1 \qquad = 1 \text{ with decimal point moved 0 places,}$$
$$10^1 = 1.0 \qquad = 10 \qquad = 1 \text{ with decimal point moved 1 place to the right,}$$
$$10^2 = 1.00 \qquad = 100 \qquad = 1 \text{ with decimal point moved 2 places to the right,}$$
$$10^3 = 1.000 \qquad = 1,000 \qquad = 1 \text{ with decimal point moved 3 places to the right,}$$
$$10^4 = 1.0000 \qquad = 10,000 \qquad = 1 \text{ with decimal point moved 4 places to the right,}$$
$$10^5 = 1.00000 \qquad = 100,000 \qquad = 1 \text{ with decimal point moved 5 places to the right,}$$
$$10^6 = 1.000000 = 1,000,000 = 1 \text{ with decimal point moved 6 places to the right, and so on.}$$

The exponent of the 10 indicates how many places the decimal point is moved *to the right* from $1.000 \cdots$

A similar pattern is followed by negative powers of 10, whose values always lie between 0 and 1:

$$10^0 \;\; = 1. \qquad\quad = 1 \qquad\quad = 1 \text{ with decimal point moved 0 places,}$$

$$10^{-1} = 0\underset{\smile}{1}. \qquad = 0.1 \qquad = 1 \text{ with decimal point moved 1 place to the left,}$$

$$10^{-2} = 0\underset{\smile}{0}1. \qquad = 0.01 \qquad = 1 \text{ with decimal point moved 2 places to the left,}$$

$$10^{-3} = 0\underset{\smile}{0}0\,1. \qquad = 0.001 \qquad = 1 \text{ with decimal point moved 3 places to the left,}$$

$$10^{-4} = 0\underset{\smile}{0}0\,01. \qquad = 0.000,1 \qquad = 1 \text{ with decimal point moved 4 places to the left,}$$

$$10^{-5} = 0\underset{\smile}{0}0\,001. \qquad = 0.000,01 \quad = 1 \text{ with decimal point moved 5 places to the left,}$$

$$10^{-6} = 0\underset{\smile}{0}0\,000\,1. = 0.000,001 = 1 \text{ with decimal point moved 6 places to the left, and so on.}$$

The exponent of the 10 now indicates how many places the decimal point is moved *to the left* from 1.

Here are a few examples of powers-of-ten notation:

$$600 = 6 \times 100 = 6 \times 10^2$$
$$7940 = 7.94 \times 1000 = 7.94 \times 10^3$$
$$93,000,000 = 9.3 \times 10,000,000 = 9.3 \times 10^7$$
$$0.023 = 2.3 \times 0.01 = 2.3 \times 10^{-2}$$
$$0.000,035 = 3.5 \times 0.000,01 = 3.5 \times 10^{-5}$$

EXERCISES

1. Express the following numbers in decimal notation.

(a) 2×10^5

(b) 8×10^{-2}

(c) 7.819×10^2

(d) 4.51×10^8

(e) 1.003×10^{-6}

(f) 10^{-10}

(g) 9.56×10^{-5}

2. Express the following numbers in powers-of-ten notation.

(a)   70

(b)   0.14

(c)   3.81

(d)   8400

(e)   1,000,000

(f)   0.007,890

(g)   351,600

B–2 USING POWERS OF TEN

Let us see how to make calculations using numbers written in powers-of-ten notation. To add or subtract numbers written in powers-of-ten notation, they must be expressed in terms of the *same* power of ten.

$$7 \times 10^4 + 2 \times 10^5 = 0.7 \times 10^5 + 2 \times 10^5$$
$$= 2.7 \times 10^5$$
$$5 \times 10^{-2} + 3 \times 10^{-4} = 5 \times 10^{-2} + 0.03 \times 10^{-2}$$
$$= 5.03 \times 10^{-2}$$
$$8 \times 10^{-3} - 7 \times 10^{-4} = 8 \times 10^{-3} - 0.7 \times 10^{-3}$$
$$= 7.3 \times 10^{-3}$$
$$4 \times 10^5 - 1 \times 10^6 = 4 \times 10^5 - 10 \times 10^5$$
$$= -6 \times 10^5$$

To multiply powers of ten together, add their exponents:

$$(10^n)(10^m) = 10^{n+m}$$

Be sure to take the sign of each exponent into account.

$$(10^2)(10^3) = 10^{2+3} = 10^5$$
$$(10^7)(10^{-3}) = 10^{7-3} = 10^4$$
$$(10^{-2})(10^{-4}) = 10^{-2-4} = 10^{-6}$$

To multiply numbers written in powers-of-ten notation, multiply the decimal parts of the numbers together and add the exponents to find the power of ten of the product:

$$(A \times 10^n)(B \times 10^m) = AB \times 10^{n+m}$$

If necessary, rewrite the result so the decimal part is a number between 1 and 10.

$$(3 \times 10^2)(2 \times 10^5) = (3 \times 2) \times 10^{2+5}$$
$$= 6 \times 10^7$$
$$(8 \times 10^{-5})(3 \times 10^7) = (8 \times 3) \times 10^{-5+7}$$
$$= 24 \times 10^2 = 2.4 \times 10^3$$
$$(1.3 \times 10^{-3})(4 \times 10^{-5}) = (1.3 \times 4) \times 10^{-3-5}$$
$$= 5.2 \times 10^{-8}$$
$$(-9 \times 10^{17})(6 \times 10^{-18}) = (-9 \times 6) \times 10^{17-18}$$
$$= -54 \times 10^{-1} = -5.4$$

To divide one power of ten by another, subtract the exponent of the denominator from the exponent of the numerator:

$$\frac{10^n}{10^m} = 10^{n-m}$$

Be sure to take the sign of each exponent into account.

$$\frac{10^5}{10^3} = 10^{5-3} = 10^2$$

$$\frac{10^{-2}}{10^4} = 10^{-2-4} = 10^{-6}$$

$$\frac{10^{-3}}{10^{-7}} = 10^{-3-(-7)} = 10^{-3+7} = 10^4$$

To divide a number written in powers-of-ten notation by another number written that way, divide the decimal parts of the numbers in the usual way and use the above rule to find the exponent of the power of ten of the quotient:

$$\frac{A \times 10^n}{B \times 10^m} = \frac{A}{B} \times 10^{n-m}$$

If necessary, rewrite the result so the decimal part is a number between 1 and 10:

$$\frac{6 \times 10^5}{3 \times 10^2} = \frac{6}{3} \times 10^{5-2} = 2 \times 10^3$$

$$\frac{2 \times 10^{-7}}{8 \times 10^4} = \frac{2}{8} \times 10^{-7-4} = \frac{1}{4} \times 10^{-11}$$

$$= 0.25 \times 10^{-11} = 2.5 \times 10^{-12}$$

$$\frac{-7 \times 10^5}{10^{-2}} = -7 \times 10^{5-(-2)} = -7 \times 10^{5+2}$$

$$= -7 \times 10^7$$

$$\frac{5 \times 10^{-2}}{-2 \times 10^{-9}} = -\frac{5}{2} \times 10^{-2-(-9)} = -2.5 \times 10^{-2+9}$$

$$= -2.5 \times 10^7$$

To find the reciprocal of a power of ten, change the sign of the exponent:

$$\frac{1}{10^n} = 10^{-n}$$

$$\frac{1}{10^{-m}} = 10^m$$

$$\frac{1}{10^5} = 10^{-5}$$

$$\frac{1}{10^{-3}} = 10^3$$

Hence the prescription for finding the reciprocal of a number written in powers-of-ten notation is

$$\frac{1}{A \times 10^n} = \frac{1}{A} \times 10^{-n}$$

For example,

$$\frac{1}{2 \times 10^3} = \frac{1}{2} \times 10^{-3} = 0.5 \times 10^{-3} = 5 \times 10^{-4}$$

$$\frac{1}{4 \times 10^{-8}} = \frac{1}{4} \times 10^8 = 0.25 \times 10^8 = 2.5 \times 10^7$$

The powers-of-ten method of writing large and small numbers makes arithmetic involving such numbers relatively easy to carry out. Here is a calculation that would be very tedious if each number were kept in decimal form.

$$\frac{(3800)(0.0054)(0.000,001)}{(430,000,000)(73)}$$

$$= \frac{(3.8 \times 10^3)(5.4 \times 10^{-3})(10^{-6})}{(4.3 \times 10^8)(7.3 \times 10^1)}$$

$$= \frac{(3.8)(5.4)}{(4.3)(7.3)} \times \frac{(10^3)(10^{-3})(10^{-6})}{(10^8)(10^1)}$$

$$= 0.65 \times 10^{(3-3-6-8-1)} = 0.65 \times 10^{-15}$$

$$= 6.5 \times 10^{-16}$$

EXERCISES

3. Perform the following additions and subtractions.

(a) $3 \times 10^2 + 4 \times 10^3$
(b) $7 \times 10^{-2} + 2 \times 10^{-3}$
(c) $4 \times 10^{-5} + 5 \times 10^{-3}$
(d) $6.32 \times 10^2 + 5$
(e) $4 \times 10^3 - 3 \times 10^2$
(f) $3.2 \times 10^{-4} - 5 \times 10^{-5}$
(g) $7 \times 10^4 - 2 \times 10^{-5}$
(h) $4.76 \times 10^{-3} - 4.81 \times 10^{-3}$

4. Evaluate the following reciprocals.

(a) $\dfrac{1}{10^2}$ (c) $\dfrac{1}{10^{-2}}$

(b) $\dfrac{1}{2 \times 10^2}$ (d) $\dfrac{1}{4 \times 10^{-4}}$

5. Perform the following calculations.

(a) $\dfrac{(500,000)(18,000)}{9,000,000}$

(b) $\dfrac{(30)(80,000,000,000)}{0.0004}$

(c) $\dfrac{(30{,}000)(0.000{,}000{,}6)}{(1000)(0.02)}$

(d) $\dfrac{(0.002)(0.000{,}000{,}05)}{0.000{,}004}$

(e) $\dfrac{(0.06)(0.0001)}{(0.000{,}03)(40{,}000)}$

(f) $\dfrac{(3 \times 10^4)(5 \times 10^{-12})}{10^3}$

(g) $\dfrac{9 \times 10^{12}}{9 \times 10^{-12}}$

(h) $\dfrac{(8 \times 10^{10})(3)}{6 \times 10^{-4}}$

(i) $\dfrac{10^{-3}}{(5 \times 10^4)(2 \times 10^2)}$

(j) $\dfrac{(5 \times 10^5)(2 \times 10^{-18})}{4 \times 10^4}$

B–3   POWERS AND ROOTS

To square a power of ten, multiply the exponent by 2; to cube a power of ten, multiply the exponent by 3:

$$(10^n)^2 = 10^{2n}$$
$$(10^n)^3 = 10^{3n}$$

In general, to raise a power of ten to the mth power, multiply the exponent by m:

$$(10^n)^m = 10^{m \times n}$$

Be sure to take the sign of each exponent into account, as in these examples:

$$(10^3)^2 = 10^{2 \times 3} = 10^6$$
$$(10^{-2})^5 = 10^{5 \times -2} = 10^{-10}$$
$$(10^{-4})^{-2} = 10^{-2 \times -4} = 10^8$$

To raise a number written in powers-of-ten notation to the mth power, multiply the decimal part of the number by itself m times and multiply the exponent of the power of ten by m:

$$(A \times 10^n)^m = A^m \times 10^{m \times n}$$

If necessary, rewrite the result so the decimal part is a number between 1 and 10:

$$(2 \times 10^5)^2 = 2^2 \times 10^{2 \times 5} = 4 \times 10^{10}$$

$$(3 \times 10^{-3})^3 = 3^3 \times 10^{3 \times -3} = 27 \times 10^{-9}$$
$$= 2.7 \times 10^{-8}$$
$$(5 \times 10^{-2})^{-4} = 5^{-4} \times 10^{-4 \times -2} = \frac{1}{5^4} \times 10^8$$
$$= \frac{1}{625} \times 10^8 = 0.0016 \times 10^8$$
$$= 1.6 \times 10^5$$

To take the mth root of a power of ten, divide the exponent by m:

$$\sqrt[m]{10^n} = (10^n)^{1/m} = 10^{n/m}$$

Thus

$$\sqrt{10^4} = (10^4)^{1/2} = 10^{4/2} = 10^2$$
$$\sqrt[3]{10^9} = (10^9)^{1/3} = 10^{9/3} = 10^3$$
$$\sqrt[3]{10^{-9}} = (10^{-9})^{1/3} = 10^{-9/3} = 10^{-3}$$

In powers-of-ten notation, the exponent of the 10 must be an integer. Hence in taking the mth root of a power of ten, the exponent should be an integral multiple of m. Instead of, for example,

$$\sqrt{10^5} = (10^5)^{1/2} = 10^{2.5}$$

which, while correct, is hardly useful, we would write

$$\sqrt{10^5} = \sqrt{10^1 \times 10^4}$$
$$= \sqrt{10} \times \sqrt{10^4}$$
$$= 3.16 \times 10^2$$

Here are two other examples:

$$\sqrt{10^{-3}} = \sqrt{10^1 \times 10^{-4}} = \sqrt{10} \times \sqrt{10^{-4}}$$
$$= 3.16 \times 10^{-2}$$
$$\sqrt[3]{10^8} = \sqrt[3]{10^2 \times 10^6} = \sqrt[3]{100} \times \sqrt[3]{10^6}$$
$$= 4.64 \times 10^2$$

To take the mth root of a number expressed in powers-of-ten notation, first write the number so the exponent of the 10 is an integral multiple of m. Then take the mth root of the decimal part of the number and divide the exponent by m to find the power of ten of the result:

$$(A \times 10^n)^{1/m} = \sqrt[m]{A} \times 10^{n/m}$$

Thus

$$\sqrt{9 \times 10^4} = \sqrt{9} \times 10^{4/2} = 3 \times 10^2$$
$$\sqrt{9 \times 10^{-6}} = \sqrt{9} \times 10^{-6/2} = 3 \times 10^{-3}$$

$$\sqrt{4 \times 10^7} = \sqrt{40 \times 10^6} = \sqrt{40} \times 10^{6/2}$$
$$= 6.32 \times 10^3$$
$$\sqrt[3]{3 \times 10^{-5}} = \sqrt[3]{30 \times 10^{-6}} \quad \sqrt[3]{30} \times 10^{-6/3}$$
$$= 3.11 \times 10^{-2}$$

EXERCISES

6. Evaluate the following powers.

(a) $(2 \times 10^7)^2$ (c) $(3 \times 10^{-8})^2$

(b) $(2 \times 10^7)^{-2}$ (d) $(5 \times 10^{-4})^{-3}$

7. Evaluate the following roots. Assume $\sqrt{4} = 2$, $\sqrt{40}$ $= 6.3$, $\sqrt[3]{4} = 1.6$, $\sqrt[3]{40} = 3.4$, and $\sqrt[3]{400} = 7.4$.

(a) $(4 \times 10^6)^{1/2}$ (f) $(4 \times 10^{13})^{1/3}$

(b) $(4 \times 10^7)^{1/2}$ (g) $(4 \times 10^{14})^{1/3}$

(c) $(4 \times 10^{-4})^{1/2}$ (h) $(4 \times 10^{-6})^{1/3}$

(d) $(4 \times 10^{-5})^{1/2}$ (i) $(4 \times 10^{-7})^{1/3}$

(e) $(4 \times 10^{-12})^{1/3}$ (j) $(4 \times 10^{-8})^{1/3}$

B–4   SIGNIFICANT FIGURES

An advantage of powers-of-ten notation is that it gives no false impression of the degree of accuracy with which a number is stated. For instance, the equatorial radius of the earth is 6378 km, but it is often taken as 6400 km for convenience in making rough calculations. To indicate the approximate character of the latter figure, all we have to do is write

$$r - 6.4 \times 10^3 \text{ km}$$

whose meaning is

$$r = (6.4 \pm 0.05) \times 10^3 \text{ km}$$

With this method, how large the number is and how accurate it is are both clear. The accurately known digits, plus one uncertain digit, are called *significant figures;* in the above case, r has two significant figures, 6 and 4. If we require greater accuracy, we would write

$$r = 6.38 \times 10^3 \text{ km}$$

which contains three significant figures, or

$$r = 6.378 \times 10^3 \text{ km}$$

which contains four significant figures.

Sometimes one or more zeros in a number are significant figures, and it is proper to retain them when expressing the number in powers-of-ten notation. There is quite a difference between 3×10^5 and 3.00×10^5:

$$3 \times 10^5 = (3 \pm 0.5) \times 10^5$$
$$3.00 \times 10^5 = (3 \pm 0.005) \times 10^5$$

When quantities are combined arithmetically, the result is no more accurate than the quantity with the largest uncertainty. Suppose a 75-kg person picks up a 0.23-kg apple. The total mass of person plus apple is still 75 kg because all we know of the person's mass is that it is somewhere between 74.5 and 75.5 kg, which means an uncertainty greater than the apple's mass. If the person's mass is instead quoted as 75.0 kg, the mass of person plus apple is 75.2 kg; if it is quoted as 75.00 kg, the mass of person plus apple is 75.23 kg. Thus

$$75 \text{ kg} + 0.23 \text{ kg} = 75 \text{ kg}$$
$$75.0 \text{ kg} + 0.23 \text{ kg} = 75.2 \text{ kg}$$
$$75.00 \text{ kg} + 0.23 \text{ kg} = 75.23 \text{ kg}$$

Significant figures must be taken into account in multiplication and division also. For example, if we divide 1.4×10^5 by 6.70×10^3, we are not justified in writing

$$\frac{1.4 \times 10^5}{6.70 \times 10^3} = 20.89552 \cdots$$

We may properly retain only two significant figures, corresponding to the two significant figures in the numerator, and so the correct answer is just 21.

In a calculation with several steps, however, it is a good idea to keep an extra digit in the intermediate steps and to round off the result only at the end. As an example,

$$\frac{5.7 \times 10^4}{3.3 \times 10^{-2}} + \sqrt{1.8 \times 10^{12}} - 1.73 \times 10^6 + 1.34 \times 10^6$$
$$= 3.07 \times 10^6 = 3.1 \times 10^6$$

If the intermediate results had been rounded off to two digits, however, the result would have been the incorrect

$$1.7 \times 10^6 + 1.3 \times 10^6 = 3.0 \times 10^6$$

B–5   LOGARITHMS

The logarithm of a number N is the exponent n to which a given base number a must be raised in order that $a^n = N$. That is, if

$$N = a^n, \qquad \text{then} \qquad n = \log_a N$$

Here are some examples using the base $a = 10$:

$1000 = 10^3$,   therefore $\log_{10} 1000 = 3$;

$5 = 10^{0.699}$,   therefore $\log_{10} 5 = 0.699$;

$0.001 = 10^{-3}$, therefore $\log_{10} 0.001 = -3$.

Since the decimal system of numbers has the base 10, this is a convenient number to use as the base of a system of logarithms. Logarithms to the base 10 are called *common logarithms* and are denoted simply as "log N." Another widely used system of logarithms uses $e = 2.718 \cdots$ as the base. Such logarithms are called *natural logarithms* (because they arise in a natural way in calculus) and are denoted "ln N." To go from one system to the other these formulas are needed:

$\log N = 0.43429 \ln N$

$\ln N = 2.3026 \log N$

Logarithms are defined only for positive numbers because the quantity a^n is positive whether n is positive, negative, or zero. Since n is the logarithm of a^n, n can describe only a positive number.

Let us consider a number N that is the product of two numbers x and y, so that $N = xy$. If $x = 10^n$ and $y = 10^m$, then

$N = xy = 10^n \times 10^m = 10^{n+m}$

$\log N = n + m$

Since $n = \log x$ and $m = \log y$, $n + m = \log x + \log y$, and so

$\log N = \log x + \log y$

Thus we have the general rule for the logarithm of a product:

$\log xy = \log x + \log y$

Similar reasoning gives the additional rules

$\log \dfrac{x}{y} = \log x - \log y$

$\log x^n = n \log x$

Before the days of electronic calculators, logarithms were widely used to simplify arithmetical work because they permit replacing multiplication and division by addition and subtraction, which are easier to do and less prone to error. Today, of course, calculators are used for such routine arithmetic. However, logarithms are still needed for finding powers and roots, with calculators replacing tables of logarithms. For example, to find $(2.13)^4$ we proceed as follows:

$\log (2.13)^4 = 4 \log 2.13 = (4)(0.3284) = 1.314$

$(2.13)^4 = \log^{-1} 1.314 = 20.6$

To obtain log 2.13 with a calculator, enter 2.13 and press the [LOG] key. To obtain $\log^{-1} 1.314$ (the *antilogarithm* of 1.314, which is the number whose logarithm is 1.314), the 1.314 is entered and the $[10^x]$ key pressed, since 10^x is the number whose logarithm is x. This key is alternatively designated [INV LOG].

The same procedure is followed for negative exponents, for instance $1/\sqrt[4]{7}$:

$$\frac{1}{\sqrt[4]{7}} = \frac{1}{7^{1/4}} = 7^{-1/4}$$

$$\log 7^{-1/4} = -\tfrac{1}{4} \log 7 = -\frac{1}{4} \times 0.8451 = -0.2113$$

$$7^{-1/4} = \log^{-1}(-0.2113) = 0.615$$

The logarithms of numbers written in powers-of-ten notation can be found with a calculator without first converting them to decimal notation. Let us consider 6.04×10^9. Since

$\log xy = \log x + \log y$

we have

$\log (6.04 \times 10^9) = \log 6.04 + \log 10^9$

But from the definition of logarithms to the base 10, $\log 10^9 = 9$. We therefore have

$\log (6.04 \times 10^9) = 0.7810 + 9 = 9.7810$

The same procedure holds for a number smaller than 1, for instance 2.4×10^{-5}:

$\log (2.4 \times 10^{-5}) = \log 2.4 + \log 10^{-5}$
$= 0.3802 - 5 = -4.6198$

EXERCISES

7. Evaluate the following with the help of logarithms.

(a) $0.0181^{1.5}$

(b) $62.2^{7.13}$

(c) $(8.15 \times 10^{14})^6$

(d) $\sqrt{156}$

(e) $(6.24 \times 10^{-4})^{1/3}$

(f) $(2.71 \times 10^5)^{1/8}$

ANSWERS

1. (a) 200,000

(b) 0.08

(c) 781.9

(d) 451,000,000

(e) 0.000,001,003

(f) 0.000,000,000,1

(g) 0.000,0956

2. (a) 7×10^1
(b) 1.4×10^{-1}
(c) 3.81
(d) 8.4×10^3
(e) 1×10^6
(f) 7.890×10^{-3}
(g) 3.516×10^5

3. (a) 4.3×10^3
(b) 7.2×10^{-2}
(c) 5.04×10^{-3}
(d) 6.37×10^2
(e) 3.7×10^3
(f) 2.7×10^{-4}
(g) -1.3×10^5
(h) -5×10^{-5}

4. (a) $10^{-2} = 0.01$
(b) $5 \times 10^{-3} = 0.005$
(c) $10^2 = 100$
(d) $2.5 \times 10^3 - 2500$

5. (a) 1×10^3
(b) 6×10^{15}
(c) 9×10^{-4}
(d) 2.5×10^{-5}
(e) 5×10^{-6}
(f) 1.5×10^{-10}
(g) 1×10^{24}
(h) 4×10^{14}
(i) 1×10^{-10}
(j) 2.5×10^{-17}

6. (a) 2×10^3
(b) 6.3×10^3
(c) 2×10^{-2}
(d) 6.3×10^{-3}
(e) 1.6×10^{-4}
(f) 3.4×10^4
(g) 7.4×10^4
(h) 1.6×10^{-2}
(i) 7.4×10^{-3}
(j) 3.4×10^{-3}

7. (a) 2.44×10^{-3}
(b) 6.16×10^{12}
(c) 2.93×10^{89}
(d) 3.53
(e) 8.55×10^{-2}
(f) 4.78

APPENDIX C
TABLES

TABLE C–1
The Elements

Atomic number	Element	Symbol	Atomic mass*	Atomic number	Element	Symbol	Atomic mass*
1	Hydrogen	H	1.008	36	Krypton	Kr	83.80
2	Helium	He	4.003	37	Rubidium	Rb	85.47
3	Lithium	Li	6.941	38	Strontium	Sr	87.62
4	Beryllium	Be	9.012	39	Yttrium	Y	88.91
5	Boron	B	10.81	40	Zirconium	Zr	91.22
6	Carbon	C	12.01	41	Niobium	Nb	92.91
7	Nitrogen	N	14.01	42	Molybdenum	Mo	95.94
8	Oxygen	O	16.00	43	Technetium	Tc	(97)
9	Fluorine	F	19.00	44	Ruthenium	Ru	101.1
10	Neon	Ne	20.18	45	Rhodium	Rh	102.9
11	Sodium	Na	22.99	46	Palladium	Pd	106.4
12	Magnesium	Mg	24.31	47	Silver	Ag	107.9
13	Aluminum	Al	26.98	48	Cadmium	Cd	112.4
14	Silicon	Si	28.09	49	Indium	In	114.8
15	Phosphorus	P	30.97	50	Tin	Sn	118.7
16	Sulfur	S	32.06	51	Antimony	Sb	121.8
17	Chlorine	Cl	35.46	52	Tellurium	Te	127.6
18	Argon	Ar	39.95	53	Iodine	I	126.9
19	Potassium	K	39.10	54	Xenon	Xe	131.3
20	Calcium	Ca	40.08	55	Cesium	Cs	132.9
21	Scandium	Sc	44.96	56	Barium	Ba	137.3
22	Titanium	Ti	47.90	57	Lanthanum	La	138.9
23	Vanadium	V	50.94	58	Cerium	Ce	140.1
24	Chromium	Cr	52.00	59	Praseodymium	Pr	140.9
25	Manganese	Mn	54.94	60	Neodymium	Nd	144.2
26	Iron	Fe	55.85	61	Promethium	Pm	(145)
27	Cobalt	Co	58.93	62	Samarium	Sm	150.4
28	Nickel	Ni	58.70	63	Europium	Eu	152.0
29	Copper	Cu	63.55	64	Gadolinium	Gd	157.3
30	Zinc	Zn	65.38	65	Terbium	Tb	158.9
31	Gallium	Ga	69.72	66	Dysprosium	Dy	162.5
32	Germanium	Ge	72.59	67	Holmium	Ho	164.9
33	Arsenic	As	74.92	68	Erbium	Er	167.3
34	Selenium	Se	78.96	69	Thulium	Tm	168.9
35	Bromine	Br	79.90	70	Ytterbium	Yb	173.0

*The unit of mass is the u. Elements whose atomic masses are given in parentheses have not been found in nature but have been produced by nuclear reactions in the laboratory. The atomic mass in such a case is the mass number of the longest-lived radioactive isotope of the element.

TABLE C-1
The Elements (continued)

Atomic number	Element	Symbol	Atomic mass*	Atomic number	Element	Symbol	Atomic mass*
71	Lutetium	Lu	175.0	89	Actinium	Ac	(227)
72	Hafnium	Hf	178.5	90	Thorium	Th	232.0
73	Tantalum	Ta	180.9	91	Protactinium	Pa	231.0
74	Tungsten	W	183.9	92	Uranium	U	238.0
75	Rhenium	Re	186.2	93	Neptunium	Np	(237)
76	Osmium	Os	190.2	94	Plutonium	Pu	(244)
77	Iridium	Ir	192.2	95	Americium	Am	(243)
78	Platinum	Pt	195.1	96	Curium	Cm	(247)
79	Gold	Au	197.0	97	Berkelium	Bk	(247)
80	Mercury	Hg	200.6	98	Californium	Cf	(251)
81	Thallium	Tl	204.4	99	Einsteinium	Es	(254)
82	Lead	Pb	207.2	100	Fermium	Fm	(257)
83	Bismuth	Bi	209.0	101	Mendelevium	Md	(258)
84	Polonium	Po	(209)	102	Nobelium	No	(255)
85	Astatine	At	(210)	103	Lawrencium	Lr	(260)
86	Radon	Rn	222	104	Rutherfordium	Rf	(257)
87	Francium	Fr	(223)	105	Hahnium	Ha	(260)
88	Radium	Ra	226.0				

TABLE C-2
Natural Trigonometric Functions

Angle Degree	Angle Radian	Sine	Cosine	Tangent	Angle Degree	Angle Radian	Sine	Cosine	Tangent
0°	.000	0.000	1.000	0.000					
1°	.017	.018	1.000	.018	16°	.279	.276	.961	.287
2°	.035	.035	0.999	.035	17°	.297	.292	.956	.306
3°	.052	.052	.999	.052	18°	.314	.309	.951	.325
4°	.070	.070	.998	.070	19°	.332	.326	.946	.344
5°	.087	.087	.996	.088	20°	.349	.342	.940	.364
6°	.105	.105	.995	.105	21°	.367	.358	.934	.384
7°	.122	.122	.993	.123	22°	.384	.375	.927	.404
8°	.140	.139	.990	.141	23°	.401	.391	.921	.425
9°	.157	.156	.988	.158	24°	.419	.407	.914	.445
10°	.175	.174	.985	.176	25°	.436	.423	.906	.466
11°	.192	.191	.982	.194	26°	.454	.438	.899	.488
12°	.209	.208	.978	.213	27°	.471	.454	.891	.510
13°	.227	.225	.974	.231	28°	.489	.470	.883	.532
14°	.244	.242	.970	.249	29°	.506	.485	.875	.554
15°	.262	.259	.966	.268	30°	.524	.500	.866	.577

TABLE C-2
Natural Trigonometric Functions (continued)

| Angle | | | | | Angle | | | | |
Degree	Radian	Sine	Cosine	Tangent	Degree	Radian	Sine	Cosine	Tangent
31°	.541	.515	.857	.601	61°	1.065	.875	.485	1.804
32°	.559	.530	.848	.625	62°	1.082	.883	.470	1.881
33°	.576	.545	.839	.649	63°	1.100	.891	.454	1.963
34°	.593	.559	.829	.675	64°	1.117	.899	.438	2.050
35°	.611	.574	.819	.700	65°	1.134	.906	.423	2.145
36°	.628	.588	.809	.727	66°	1.152	.914	.407	2.246
37°	.646	.602	.799	.754	67°	1.169	.921	.391	2.356
38°	.663	.616	.788	.781	68°	1.187	.927	.375	2.475
39°	.681	.629	.777	.810	69°	1.204	.934	.358	2.605
40°	.698	.643	.766	.839	70°	1.222	.940	.342	2.747
41°	.716	.658	.755	.869	71°	1.239	.946	.326	2.904
42°	.733	.669	.743	.900	72°	1.257	.951	.309	3.078
43°	.751	.682	.731	.933	73°	1.274	.956	.292	3.271
44°	.768	.695	.719	.966	74°	1.292	.961	.276	3.487
45°	.785	.707	.707	1.000	75°	1.309	.966	.259	3.732
46°	.803	.719	.695	1.036	76°	1.326	.970	.242	4.011
47°	.820	.731	.682	1.072	77°	1.344	.974	.225	4.331
48°	.838	.743	.669	1.111	78°	1.361	.978	.208	4.705
49°	.855	.755	.656	1.150	79°	1.379	.982	.191	5.145
50°	.873	.766	.643	1.192	80°	1.396	.985	.174	5.671
51°	.890	.777	.629	1.235	81°	1.414	.988	.156	6.314
52°	.908	.788	.616	1.280	82°	1.431	.990	.139	7.115
53°	.925	.799	.602	1.327	83°	1.449	.993	.122	8.144
54°	.942	.809	.588	1.376	84°	1.466	.995	.105	9.514
55°	.960	.819	.574	1.428	85°	1.484	.996	.087	11.43
56°	.977	.829	.559	1.483	86°	1.501	.998	.070	14.30
57°	.995	.839	.545	1.540	87°	1.518	.999	.052	19.08
58°	1.012	.848	.530	1.600	88°	1.536	.999	.035	28.64
59°	1.030	.857	.515	1.664	89°	1.553	1.000	.018	57.29
60°	1.047	.866	.500	1.732	90°	1.571	1.000	.000	∞

TABLE C-3
The Greek Alphabet

A	α	Alpha
B	β	Beta
Γ	γ	Gamma
Δ	δ	Delta
E	ε	Epsilon
Z	ζ	Zeta
H	η	Eta
Θ	θ	Theta
I	ι	Iota
K	κ	Kappa
Λ	λ	Lambda
M	μ	Mu
N	ν	Nu
Ξ	ξ	Xi
O	o	Omicron
Π	π	Pi
P	ρ	Rho
Σ	σ	Sigma
T	τ	Tau
Υ	υ	Upsilon
Φ	φ	Phi
X	χ	Chi
Ψ	ψ	Psi
Ω	ω	Omega

GLOSSARY

Absolute temperature scale   *Absolute zero* is the lowest temperature possible; in the elementary kinetic theory of matter, it is that temperature at which random molecular movement would cease, although in reality a small amount of movement would still persist. Absolute zero corresponds to $-273°C$, which is $-460°F$. The *absolute temperature scale* expresses temperature in °C above absolute zero; its unit is the *Kelvin,* denoted K. The *Rankine absolute temperature scale* expresses temperatures in °F above absolute zero of $-460°F$; temperatures in this scale are designated °R.

Acceleration   The *acceleration a* of an object is the rate at which its velocity changes with time; the change in velocity may be in magnitude or direction or both.

Acceleration of gravity   The *acceleration of gravity g* is the acceleration of a freely falling body near the earth's surface. Its value is 32 ft/s^2 in British units and 9.8 m/s^2 in SI units.

Alpha particle   An *alpha particle* is the nucleus of a helium atom. It consists of two neutrons and two protons, and it is emitted in the radioactive decay of certain nuclides.

Alternating current   The direction of an *alternating electric current* reverses itself periodically.

Ampere   The *ampere* (A) is the unit of electric current. It is equal to a flow of charge at the rate of 1 C/s.

Amplitude   The *amplitude* of an object undergoing simple harmonic motion is its maximum displacement on either side of its equilibrium position. The amplitude of a wave is the maximum value of the wave variable (for instance, displacement, pressure, electric field) regardless of sign.

Angular momentum   The *angular momentum L* of a rotating body is the product $I\omega$ of its moment of inertia and angular speed. The principle of *conservation of angular momentum* states that the total angular momentum of a system of particles remains constant when no net external torque acts upon the system.

Angular speed   The *angular speed* ω of a rotating body is the angle through which it turns per unit time. The *angular acceleration* α of a rotating body is the rate of change of its angular speed with respect to time.

Archimedes' principle   *Archimedes' principle* states that the buoyant force on a submerged object is equal to the weight of fluid it displaces.

Atom   An *atom* is the ultimate particle of an element. Every atom consists of a very small positively charged nucleus and a number of electrons at some distance from it. The nucleus contains nearly all the mass of the atom.

Atomic mass unit   Atomic and nuclear masses are expressed in *atomic mass units* (u), equal to 1.66×10^{-27} kg.

Atomic number   The *atomic number* of an element is the number of electrons in each of its atoms or, equivalently, the number of protons in each of its atomic nuclei.

Axis of rotation   The *axis of rotation* of a rigid body turning in place is that line of particles which does not move.

Back emf   A *back* (or *counter*) *emf* is induced in the rotating coils of an electric motor and is opposite in direction to the external voltage applied to them.

Beta decay   *Beta decay* is a type of radioactive decay in which a nucleus emits an electron or a positron.

Binding energy   The *binding energy* of a nucleus is the energy equivalent of the difference between its mass and the sum of the masses of its individual constituent nucleons. This amount of energy must be supplied to the nucleus if it is to be completely disintegrated. The binding energy per nucleon is least for very light and very heavy nuclei; hence the *fusion* of very light nuclei to form heavier ones and the *fission* of very heavy nuclei to form lighter ones are both processes that liberate energy.

Bohr theory of the atom   According to the *Bohr theory of the atom,* an electron can circle an atomic nucleus indefinitely without radiating energy if its orbit is an integral number of electron wavelengths in circumference. The number of wavelengths that fit into a particular permitted orbit is called the *quantum number* of that orbit. The electron energies corresponding to the various quantum numbers constitute the *energy levels* of the atom, of which the lowest is the *ground state* and the rest are *excited states*.

Boyle's law   *Boyle's law* states that, at constant temperature, the absolute pressure of a sample of a gas is inversely proportional to its volume, so that $pV = $ constant at that temperature regardless of changes in either p or V individually.

British thermal unit   The *British thermal unit* (Btu) is the unit of heat in the British system of units. It is equal to that amount of heat required to change the temperature of 1 lb of water by 1°F.

Bulk modulus The *bulk modulus B* of a material is equal to the pressure on a sample of it divided by the fractional decrease in its volume.

Capacitive reactance The *capacitive reactance X_c* of a capacitor is a measure of its effect on an alternating current; it is equal to $1/2\pi fC$, where C is the capacitance. If V_c is the effective voltage across a capacitor, the effective current that flows into and out of it is $I = V_c/X_c$. The unit of reactance is the ohm.

Capacitor A *capacitor* is a device that stores electric energy in the form of an electric field. The ratio between the charge on either plate of a capacitor and the potential difference between the plates is called its *capacitance C*. The unit of capacitance is the *farad* (F), which is equal to 1 C/V.

Carnot engine A *Carnot engine* is an idealized engine that is not subject to such practical difficulties as friction or heat losses by conduction or radiation but that obeys all physical laws. The efficiency of a Carnot engine that absorbs heat as the absolute temperature T_1 and exhausts heat at the absolute temperature T_2 is equal to $1 - T_2/T_1$; no engine operating between two given temperatures can be more efficient than a Carnot engine operating between them.

Cathode ray tube In a *cathode ray tube*, an electron beam controlled by electric or magnetic fields traces out an image on a fluorescent screen.

Celsius temperature scale In the *celsius* (centigrade) *temperature scale*, the freezing point of water is assigned the value 0°C and the boiling point of water the value of 100°C.

Center of gravity The *center of gravity* of an object is that point from which it can be suspended in any orientation without tending to rotate. The weight of an object can be considered as a downward force acting on its center of gravity.

Center of percussion The *center of percussion* of a pivoted object is that point at which it can be struck without producing a reaction force on its pivot.

Centripetal acceleration The velocity of an object in uniform circular motion continually changes in direction although its magnitude remains constant. The object's acceleration is called *centripetal acceleration*, and it points toward the center of the circular path. Centripetal acceleration is proportional to the square of the object's speed and inversely proportional to the radius of its path.

Centripetal force The inward force that provides an object in uniform circular motion with its centripetal acceleration is called *centripetal force*.

Charles's law *Charles's law* states that, at constant pressure, the volume of a sample of a gas is directly proportional to its absolute temperature, so that $V/T = $ constant at that pressure regardless of changes in either V or T individually.

Coherence Two sources of waves are *coherent* if there is a fixed phase relationship between the waves they emit during the time the waves are being observed. Interference can be observed only in waves from coherent sources.

Component of a vector A *component of a vector* is its projection in a specified direction.

Compound Two or more elements may combine chemically to form a *compound*, a new substance whose properties are different from those of the elements that compose it.

Compression When equal and opposite forces that act toward each other are applied to a body, it is said to be in *compression*.

Concave mirror A *concave mirror* curves inward toward its center and converges parallel light to a single *real focal point*. The distance from the mirror to the focal point is the *focal length* of the mirror.

Concurrent forces When the lines of action of the various forces that act on an object intersect at a common point, the forces are said to be *concurrent;* when their lines of action do not intersect, the forces are *noncurrent*.

Conduction In *conduction*, heat is transferred from one place to another by successive molecular collisions.

Convection In *convection*, heat is transferred from one place to another by the motion of a volume of hot fluid from one place to another.

Convex mirror A *convex mirror* curves outward toward its center and diverges parallel light as though the reflected light came from a single *virtual focal point* behind the mirror. The distance from the mirror to the focal point is the *focal length* of the mirror.

Coulomb The *coulomb* (C) is the unit of electric charge.

Coulomb's law *Coulomb's law* states that the force one charge exerts upon another is directly proportional to the magnitudes of the charges and inversely proportional to the square of the distance between them. The force between like charges is repulsive, and that between unlike charges is attractive.

Covalent bond In a *covalent bond* between adjacent atoms of a molecule or solid, the atoms share one or more electron pairs.

Critical point   The *critical point* is the upper limit of the vaporization curve of a substance, which cannot exist in the liquid state at a temperature above that of its critical point.

Crystalline solid   Solids whose constituent atoms or molecules are arranged in regular, repeated patterns are called *crystalline*. When only short-range order is present, the solid is *amorphous*.

De Broglie waves   A moving object behaves as though it has a wave nature. The waves representing such a particle are called *de Broglie waves*.

Density   The *density d* of a substance is its mass per unit volume. Its *weight density D* is its weight per unit volume.

Dielectric constant   The *dielectric constant K* of a particular material is a measure of how effective it is in reducing an electric field set up across a sample of the material.

Diffraction   The ability of waves to bend around the edges of obstacles in their paths is called *diffraction*. A *diffraction grating* is a series of parallel slits that produces a spectrum through the interference of light that is diffracted by them.

Diode   A *diode* is a semiconductor device that permits an electric current to flow through it in only one direction.

Dispersion   *Disperson* refers to the splitting up of a beam of light containing different frequencies by passage through a substance whose index of refraction varies with frequency.

Domain   An assembly of atoms in a ferromagnetic material whose atomic magnetic moments are aligned is called a *domain*.

Doppler effect   The *Doppler effect* refers to the change in frequency of a wave when there is relative motion between its source and an observer.

Effective value   The *effective value* of an alternating current is such that a direct current of this magnitude produces heat in a resistor at the same rate as the alternating current. The relationship between I_{eff} and I_{max} is $I_{eff} = 0.707I_{max}$. The similar relationship $V_{eff} = 0.707V_{max}$ holds for the effective and maximum voltages in an ac-circuit.

Efficiency   The *efficiency* of a machine of any kind is the ratio between the work it does and the energy supplied to it; or the ratio between its power output and its power input; or the ratio between its actual and ideal mechanical advantages.

Elastic collision   A *completely elastic collision* is one in which kinetic energy is conserved. A *completely inelastic collision* is one in which the bodies stick together upon impact, which results in the maximum possible kinetic energy loss.

Elastic limit   The *elastic limit* is the maximum stress a solid can experience without being permanently altered. Hooke's law is only valid when the elastic limit is not exceeded.

Elastic potential energy   A system under stress possesses *elastic potential energy* that is equal to the work done in deforming it.

Electric charge   *Electric charge, Q,* like rest mass, is a fundamental property of certain of the elementary particles of which all matter is composed. There are two kinds of electric charge, *positive charge* and *negative charge;* charges of like sign repel, unlike charges attract. The unit of charge is the *coulomb* (C). All charges, of either sign, occur in multiples of the fundamental *electron charge* of 1.6×10^{-19} C. The principle of *conservation of charge* states that the net electric charge in an isolated system remains constant.

Electric current   A flow of electric charge from one place to another is called an *electric current*. The unit of electric current is the *ampere* (A), which is equal to a flow of 1 C/s.

Electric field   An *electric field* **E** exists wherever an electric force acts on a charged particle. The magnitude E of an electric field at a point is defined as the force that would act on a charge of $+1$ C placed there. The unit of electric field is the volt/m, which is equal to 1 N/C.

Electrolysis   *Electrolysis* is the process by which free elements are liberated from a liquid by the passage of an electric current.

Electrolyte   A substance that separates into free ions when dissolved in water is called an *electrolyte* since the resulting solution is able to conduct electric current.

Electromagnetic induction   *Electromagnetic induction* refers to the production of an electric field wherever magnetic lines of force are in motion.

Electromagnetic waves   *Electromagnetic waves* consist of coupled electric and magnetic oscillations. The electric field of such a wave is perpendicular to its magnetic field, and both fields are perpendicular to the direction in which the wave travels. Radio waves, light waves, X rays, and gamma rays are all electromagnetic waves differing only in their wavelength. Electromagnetic waves are produced by accelerated electric charges, and in free space have the speed $c = 3.00 \times 10^8$ m/s.

Electromotive force   The *electromotive force* (emf) of a battery, generator, or other source of electrical energy is the potential difference across its terminals when no current flows. When a current is flowing, the terminal voltage is less than the emf owing to the potential drop in the *internal resistance* of the source.

Electron The *electron* is the least massive elementary particle found in matter. The electron has a charge of $-e$, where $e = 1.60 \times 10^{-19}$ coulomb.

Electron volt The *electron volt* is the energy acquired by an electron that has been accelerated by a potential difference of 1 volt. It is equal to 1.6×10^{-19} J.

Elements *Elements* are the simplest substances encountered in bulk. They cannot be decomposed or transformed into one another by ordinary chemical or physical means.

Energy *Energy E* is that which may be converted into work. When something possesses energy, it is capable of performing work or, in a general sense, of accomplishing a change in some aspect of the physical world. The unit of energy is the *joule* (J). The three broad categories of energy are *kinetic energy,* which is the energy something possesses by virtue of its motion; *potential energy,* which is the energy something possesses by virtue of its position in a force field; and *rest energy,* which is the energy something possesses by virtue of its mass. The principle of *conservation of energy* states that the total amount of energy in a system isolated from the rest of the universe always remains constant, although energy transformations from one form to another, including rest energy, may occur within the system.

Energy band The *energy bands* in a crystal are ranges of energy that correspond to energy levels in an atom. An electron in a crystal can only have an amount of energy that falls within one of its energy bands.

Equilibrium An object not acted upon by a net force is in *translational equilibrium* and may be at rest or have a constant linear velocity. An object not acted upon by a net torque is in *rotational equilibrium* and may be at rest or have a constant angular velocity.

Equivalent resistance The *equivalent resistance* of a set of interconnected resistors is the value of the single resistor that can be substituted for the entire set without affecting the currents that flow in the rest of any circuit of which it is a part.

Escape speed The minimum speed needed by an object to permanently escape from the gravitational attraction of an astronomical body such as a planet or star is called the *escape speed* of the body.

Exclusion principle According to the *exclusion principle,* no two electrons in an atom can exist in the same quantum state. Because of this principle, atoms with different numbers of electrons have different properties.

Farad The unit of capacitance is the *farad* (F), which is equal to 1 C/V.

Ferromagnetism *Ferromagnetism* refers to the ability of the atomic current loops in certain materials, notably iron and many of its alloys, to become aligned with an external magnetic field. The magnetic field produced by a current-carrying wire coil is greatly increased when the coil has a ferromagnetic core.

Force A *force F* is any influence that can cause a body to be accelerated. The unit of force is the *newton* (N); in the British system it is the *pound* (lb).

Force field A *force field* is a region of space at every point in which an appropriate test object would experience a force. Thus a *gravitational field* is a region of space in which an object by virtue of its mass is acted on by a force, and an *electric field* is a region of space in which an object by virtue of its electric charge is acted upon by a force.

Frequency The *frequency f* of something undergoing harmonic motion is the number of oscillations it makes per unit time. The frequency of a train of waves is the number of waves that pass a particular point per unit time. The unit of frequency is the *hertz* (Hz), which is equal to 1 cycle/s.

Frequency modulation In *frequency modulation,* information is contained in variations in the frequency of the carrier wave.

Friction The term *friction* refers to the resistive forces that arise to oppose the motion of a body past another with which it is in contact. *Sliding friction* is the frictional resistance a body in motion experiences, while *static friction* is the frictional resistance a stationary body must overcome in order to be set in motion. The *coefficient of friction* is the constant of proportionality for a given pair of contacting surfaces that relates the frictional force between them to the normal force with which one presses against the other. Usually the coefficient of static friction is greater than that of sliding friction. *Rolling friction* refers to the resistance a circular object experiences as it rolls over a smooth, flat surface; coefficients of rolling friction are much smaller than those of sliding friction.

Fusion, heat of The *heat of fusion* of a substance is the amount of heat that must be supplied to change a unit quantity of it at its melting point from the solid to the liquid state; the same amount of heat must be removed from a unit quantity of the substance in the liquid state at its melting point to change it to a solid.

Gamma ray A *gamma ray* is an energetic photon emitted by certain radioactive nuclei.

Gauge pressure *Gauge pressure* is the difference between true pressure and atmospheric pressure.

Generator A *generator* is a device that converts mechanical energy into electrical energy.

Gravitation Newton's *law of universal gravitation* states that every body in the universe attracts every other body with a force directly proportional to both their masses and inversely proportional to the square of the distance separating them.

Half-life The time required for half of a given sample of a radioactive substance to decay is called its *half-life*.

Harmonic motion *Simple harmonic motion* is an oscillatory motion that occurs whenever a force acts on an object in the opposite direction to its displacement from its normal position, with the magnitude of the force proportional to the magnitude of the displacement. The period of a simple harmonic oscillator is independent of its amplitude. In a *damped* harmonic oscillator, friction progressively reduces the amplitude of the vibrations.

Heat *Heat* is internal energy in transit from one body of matter to another. If a body of matter does not change state during the addition or removal of heat and neither does work nor has work done on it, the change in its internal energy results in a corresponding change in its temperature. The SI unit of heat is the joule. Another unit in common use is the *kilocalorie* (kcal), which is the amount of heat required to change the temperature of 1 kg of water by 1°C. The unit of heat in the British system is the *British thermal unit* (Btu), which is that amount of heat required to change the temperature of 1 lb of water by 1°F.

Henry The unit of inductance is the *henry* (H), which is equal to 1 V · s/A.

Hertz The unit of frequency is the *hertz* (Hz), which is equal to 1 cycle/s.

Hooke's law *Hooke's law* states that the strain (relative amount of deformation) experienced by a body under stress is proportional to the magnitude of the stress. Thus the elongation of a wire is proportional to the tension applied to it. Hooke's law is only valid when the elastic limit of the body is not exceeded.

Hydraulic press The *hydraulic press* is a machine consisting of two fluid-filled cylinders of different diameters connected by a tube. The input force is applied to a piston in one of the cylinders and the output force is exerted by a piston in the other cylinder. The theoretical mechanical advantage of a hydraulic press is equal to the inverse ratio of the cylinder diameters.

Hysteresis The permeability of a ferromagnetic material

in a given magnetizing field B_0 depends upon its past history as well as upon B_0, a phenomenon called *hysteresis*.

Ideal gas law The equation $pV/T = $ constant, a combination of Boyle's and Charles's laws, is called the *ideal gas law* and is obeyed approximately, though not exactly, by all gases.

Image A *real image* of an object is formed by light rays that pass through the image; the image would therefore appear on a properly placed screen. A *virtual image* can only be seen by the eye because the light rays that seem to come from the image actually do not pass through it.

Impedance The *impedance Z* of an alternating-current circuit is analogous to the resistance of a direct-current circuit. When the effective alternating potential difference V is applied to a circuit of impedance Z, the effective current that flows is $I = V/Z$. The unit of impedance is the ohm.

Impedance matching *Impedance matching* refers to the fact that a source of electric energy transfers maximum power to a load when the resistance (or impedance in the case of alternating current) of the load equals the internal resistance (or impedance) of the source.

Impulse The *impulse* of a force is the product of the force and the time during which it acts. Impulse is a vector quantity having the direction of the force. When a force acts on an object that is free to move, its change in momentum equals the impulse given it by the force.

Inductance The *inductance L* of a circuit is the ratio between the magnitude of the self-induced emf $\mathcal{E}$ due to a changing current in it and the rate of change $\Delta I/\Delta t$ of the current. The unit of inductance is the *henry* (H), which is equal to 1 V · s/A.

Inductive reactance The *inductive reactance X_L* of an inductor is a measure of its effect on an alternating current; it is equal to $2\pi fL$, where f is the frequency of the current and L the inductance. If V_L is the effective voltage across an inductor, the effective current that flows through it is $I = V_L/X_L$. The unit of reactance is the ohm.

Inertia The term *inertia* refers to the apparent resistance a body offers to changes in its state of motion.

Interference The interaction of different waves of the same nature is called *interference: Constructive interference* occurs when the resulting composite wave has an amplitude greater than that of either of the original waves, and *destructive interference* occurs when the resulting composite wave has an amplitude less than that of either of the original waves.

Ion An *ion* is an atom or group of atoms that carries a net electric charge. An atom or group of atoms becomes a negative ion when it picks up one or more electrons in addition to its normal number, and becomes a positive ion when it loses one or more of its usual number.

Ionic bond Electrons are transferred between the atoms of certain solids so that the resulting crystal consists of positive and negative ions rather than of neutral atoms. Such a solid is said to be held together by *ionic bonds*. The attractive forces between ions of opposite charge balance the repulsive forces between ions of like charge in an ionic solid.

Isotope The *isotopes* of an element have the same atomic number but different mass numbers. Thus the nuclei of the isotopes of an element all contain the same number of protons but have different numbers of neutrons.

Joule The *joule* (J) is the SI unit of work and energy. It is equal to $1 \text{ kg} \cdot \text{m}^2/\text{s}^2$.

Kilocalorie The *kilocalorie* (kcal) is a unit of heat equal to that amount of heat required to change the temperature of 1 kg of water by 1°C. $1 \text{ kcal} = 4185 \text{ J}$.

Kilogram The *kilogram* (kg) is the SI unit of mass. One kilogram weighs 2.21 lb at the earth's surface.

Kinetic energy *Kinetic energy* is the energy a moving object possesses by virtue of its motion. If the object has the mass m and speed v, its kinetic energy is $\frac{1}{2} mv^2$.

Kinetic theory According to the *kinetic theory of gases*, a gas consists of a great many tiny individual molecules that do not interact with one another except when collisions occur. The molecules are far apart compared with their dimensions and are in constant random motion. The ideal gas law may be derived from the kinetic theory of gases.

Kirchhoff's rules *Kirchhoff's rules* for network analysis are: (1) The sum of the currents flowing into a junction of three or more wires is equal to the sum of the currents flowing out of the junction; (2) The sum of the emf's around a closed conducting loop is equal to the sum of the *IR* potential drops around the loop.

Laminar flow In *laminar* (or *streamline*) *flow* every particle of fluid passing a particular point follows the same path, whereas in *turbulent flow* irregular whirls and eddies occur.

Laser A *laser* is a device for producing a narrow, monochromatic, coherent beam of light. The term stands for *l*ight *a*mplification by *s*timulated *e*mission of *r*adiation. Laser operation depends upon the existence of *metastable states*, which are excited atomic states that can persist for unusually long periods of time.

Lens A *lens* is a transparent object of regular form that can produce an image of an object placed before it. A *converging lens* brings parallel light to a single real focal point, while a *diverging lens* deviates parallel light outward as though it originated at a single virtual focal point.

Lenz's law *Lenz's law* states that the direction of an induced current must be such that its own magnetic field opposes the changes in flux that are inducing it.

Lines of force *Lines of force* are means for visualizing a force field. Their direction at any point is that in which a test body would move if released there, and their concentration in the neighborhood of a point is proportional to the magnitude of the force on a test particle at that point. (In the case of a magnetic field, the direction of a line of force at a point is that in which a moving charge would experience no force.)

Longitudinal waves *Longitudinal waves* occur when the individual particles of a medium vibrate back and forth in the direction in which the waves travel. Sound consists of longitudinal waves.

Magnetic field A *magnetic field* **B** exists wherever a magnetic force acts on a moving charged particle. The magnitude B of a magnetic field at a point is defined as the force that would act on a charge of $+1$ C moving at a speed of 1 m/s past that point, when the direction of the motion is such as to result in the maximum force. The unit of magnetic field is the *tesla* (T), equal to $1 \text{ N/A} \cdot \text{m}$.

Magnetic force Electric charges in motion relative to an observer appear to exert forces upon one another that are different from the forces they exert when at rest. These differences are by custom attributed to *magnetic forces*. In reality, magnetic forces represent changes in electric forces due to the motion of the charges involved.

Mass The property of matter that manifests itself as inertia is called *mass m*. The *rest mass* of a body is its mass when stationary with respect to an observer. The unit of mass is the kilogram (kg).

Mass number The *mass number* of a nucleus is the number of nucleons (protons and neutrons) it contains.

Matter waves A moving body behaves as though it has a wave character. The waves representing such a body are *matter waves*, also called *de Broglie waves*. The wave variable in a matter wave is its *wave function*, whose square is the *probability density* of the body. The value of the probability density of a particular body at a certain place and time is proportional to the probability of finding the body at that place at that time. Matter waves may thus be regarded as waves of probability.

Mechanical advantage The *mechanical advantage* of a machine is the ratio between the output force it exerts and the input force that is furnished to it. The *ideal mechanical advantage* (IMA) is its value under ideal circumstances, while the *actual mechanical advantage* (AMA) is its value when friction is taken into account.

Metallic bond The *metallic bond* that holds metal atoms together in the solid state arises from a "gas" of freely moving electrons pervading the entire metal.

Meter The *meter* (m) is the unit of length. One meter is equal to 3.28 ft.

Mirror A *mirror* is a specular reflecting surface of regular form that can produce an image of an object placed before it.

Molecule A *molecule* is a group of atoms that stick together strongly enough to act as a single particle. A molecule of a given compound always has a certain definite structure and is complete in itself with little tendency to gain or lose atoms.

Moment arm of a force The *moment arm* of a force is the perpendicular distance from its line of action to a pivot point.

Moment of inertia The *moment of inertia I* of a body about a given axis is the rotational analog of mass in linear motion. Its value depends upon the way in which the mass of the body is distributed about the axis.

Momentum, linear The momentum **p** of a body is the product of its mass and velocity. Linear momentum is a vector quantity whose direction is that of the body's velocity. The principle of *conservation of linear momentum* states that the total linear momentum of a system of particles isolated from the rest of the universe remains constant regardless of what events occur within the system.

Motion, laws of *Newton's first law of motion* states that an object at rest will remain at rest and an object in motion will continue in motion in a straight line at constant velocity in the absence of any interaction with the rest of the universe. *Newton's second law of motion* states that the net force acting on an object is equal to the rate of change of the object's linear momentum. *Newton's third law of motion* states that, when an object exerts a force on another object, the second object exerts a force on the first object of the same magnitude but in the opposite direction.

Neutron The *neutron* is an electrically neutral elementary particle, slightly heavier than the proton, that is present together with protons in atomic nuclei.

Newton The *newton* (N) is the SI unit of force. It is equal to 1 kg $\cdot$ m/s^2. One newton is equal to 0.225 pound.

Nuclear fission In *nuclear fission,* the absorption of neutrons by certain heavy nuclei causes them to split into smaller *fission fragments* with the release of energy. Because each fission also liberates several neutrons, a rapidly multiplying sequence of fissions called a *chain reaction* can occur if a sufficient amount of the proper material is assembled. A *nuclear reactor* is a device in which a chain reaction can be initiated and controlled.

Nuclear fusion In *nuclear fusion,* two light nuclei combine to form a heavier one with the evolution of energy. The sun and stars obtain their energy from fusion reactions.

Nucleon Neutrons and protons, the constituents of atomic nuclei, are jointly called *nucleons.*

Nucleus The *nucleus* of an atom is located at its center and contains all of the positive charge and most of the mass of the atom. The nucleus consists of protons and neutrons.

Nuclide A *nuclide* is a nuclear species characterized by a certain atomic number Z and mass number A. The symbol of a nuclide is $^A_Z X$, where X is the chemical symbol of the element of atomic number Z.

Ohm The *ohm* (Ω) is the unit of electrical resistance. It is equal to 1 V/A.

Ohm's law *Ohm's law* states that the current in a metallic conductor is proportional to the potential difference between its ends. Thus in such a conductor $I = V/R$.

Pascal's principle *Pascal's principle* states that an external pressure exerted on a fluid is transmitted uniformly throughout its volume.

Period The *period T* of a body undergoing simple harmonic motion is the time required for it to make one complete oscillation. The period of a wave is the time required for one complete wave to pass a particular point.

Permanent magnet A *permanent magnet* is an object composed of a ferromagnetic material whose atomic current loops have been aligned by an external magnetic field and which remain aligned after the external field is removed.

Permeability The *permeability* of a medium is a measure of its magnetic properties. *Ferromagnetic* substances have higher permeabilities than free space.

Phase relationships   The *phase relationships* between the instantaneous voltage and instantaneous current in alternating-current circuit components are as follows: The voltage across a pure resistor is in phase with the current; the voltage across a pure inductor leads the current by $\frac{1}{4}$ cycle; the voltage across a pure capacitor lags behind the current by $\frac{1}{4}$ cycle.

Phasor   A phasor is a vector rotating f times per second whose projection on a tangent line can represent either current or voltage in an alternating-current circuit whose frequency is f.

Photoelectric effect   The *photoelectric effect* refers to the emission of electrons from a metal surface when light shines on it.

Photon   Electromagnetic waves transport energy in tiny bursts called *photons* that resemble particles in a number of respects. The energy of a photon is related to the frequency f of the corresponding wave by $E = hf$, where h is Planck's constant.

Plasma   A *plasma* is a gas composed of electrically charged particles, and its behavior depends strongly upon electromagnetic forces. Most of the matter in the universe is in the plasma state.

Polarization   A *polarized* beam of transverse waves is one whose vibrations occur in only a single direction perpendicular to the direction in which the beam travels, so that the entire wave motion is confined to a plane called the *plane of polarization*. An *unpolarized* beam of transverse waves is one whose vibrations occur equally often in all directions perpendicular to the direction in which the beam travels.

Polar molecule   A *polar molecule* is one whose charge distribution is asymmetrical, so that one end is positive and the other negative even though the molecule as a whole is electrically neutral.

Pole, magnetic   The ends of a permanent magnet are called its *poles*. Magnetic lines of force leave the *north pole* of a magnet and enter its *south pole*.

Positron   A *positron* is a positively charged electron.

Potential difference   The electrical *potential difference V* between two points is the work that must be done to take a charge of 1 C from one point to the other. The unit of potential difference is the volt (V), which is equal to 1 J/C.

Potential energy   *Potential energy* is the energy an object has by virtue of its position. The gravitational potential energy of an object of mass m at a height h above a particular reference point is mgh; if its weight w is specified, its potential energy is wh. Other examples of potential energy are that of a planet with respect to the sun, that of a piece of iron with respect to a magnet, and that of a body at the end of a stretched spring with respect to its equilibrium position.

Power   The rate at which work is done is called *power*. The SI unit of power is the watt (W), which is equal to 1 J/s, and in the British system it is the ft · lb/s. One *horsepower* is equal to 746 W or 550 ft · lb/s.

Power factor   The *power factor* of an alternating-current circuit is the ratio between the power consumed in the circuit and the product of the effective current and voltage there; this ratio is equal to that between the resistance and the impedance of the circuit, and is less than 1, except at resonance. The unit of apparent power $V_{eff}I_{eff}$ is the volt-ampere, as distinct from the watt, which is the unit of consumed power.

Pressure   The *pressure* on a surface is the perpendicular force per unit area that acts upon it. *Gauge pressure* is the difference between true pressure and atmospheric pressure.

Proton   The *proton* is an elementary particle found in all atomic nuclei; its charge is $+e$ and its mass is 1836 times that of the electron.

Quantum theory of the atom   In the *quantum theory of the atom*, three quantum numbers are needed to specify the *probability cloud* that an electron occupies in an atom. The greater the probability density in a certain part of the cloud, the more likely the electron is to be found there. The electron's energy and average distance from the nucleus are chiefly determined by the quantum number n. The electrons in an atom that have the same n are said to occupy the same *shell*. An additional quantum number is needed to specify the direction of spin of each atomic electron.

Quantum theory of light   The *quantum theory of light* states that light travels in tiny bursts of energy called *quanta* or *photons*. If the frequency of the light is f, each burst has the energy hf, where h is known as *Planck's constant*. The quantum theory of light complements the wave theory of light.

Radian   The *radian* is a unit of angular measure equal to $57.30°$. If a circle is drawn whose center is at the vertex of an angle, the angle in radian measure is equal to the ratio between the arc of the circle cut by the angle and the radius of the circle. A full circle contains 2π radians.

Radiation   In *radiation*, energy is transferred from one place to another in the form of electromagnetic waves, which require no material medium for their passage.

Radioactivity *Radioactive nuclei* spontaneously transform themselves into other nuclear species by the emission of *alpha particles,* which are the nuclei of helium atoms, or *beta particles,* which are positive or negative electrons. *Electron capture* is an alternative to positron emission. The emission of *gamma rays,* which are energetic photons, enables an excited nucleus to lose its excess energy.

Reflection In *diffuse reflection* an incident beam of parallel light is spread out in many directions, while in *specular reflection* the angle of reflection is equal to the angle of incidence. In *total internal reflection,* light arriving at a medium of lower index of refraction at an angle greater than the *critical angle* of incidence is reflected back into the medium it came from.

Refraction The bending of a light beam when passing from one medium to another is called *refraction.* The quantity that governs the degree to which a light beam will be deflected in entering a medium is its *index of refraction,* defined as the ratio between the speed of light in free space and its value in the medium.

Refrigerator A *refrigerator* is a device that transfers heat from a cold reservoir to a hot one, and it must perform work in order to do this. In essence, it is a heat engine operating in reverse. The *coefficient of performance* of a refrigerator is the amount of heat removed per unit of work done. The *capacity* of a refrigerator is the rate at which it can remove heat.

Relative humidity The *relative humidity* of a volume of air is the ratio between the amount of water vapor it contains and the amount that would be present at saturation.

Relativity of mass The *relativity of mass* refers to the increase in the measured mass of an object when it is moving relative to an observer. The object's *rest mass* is its mass measured when it is at rest relative to the observer.

Resistance The *resistance R* of a body of matter is a measure of the extent to which it impedes the passage of electric current. It is defined as the ratio between the potential difference applied across the ends of the body and the resulting current. The unit of resistance is the ohm (Ω), which is equal to 1 V/A. The resistance of a conductor is proportional to its length and to the *resistivity* ρ of the material of which it is made, and inversely proportional to its cross-sectional area.

Resolution of vectors A vector can be *resolved* into two or more other vectors whose sum is equal to the original vector. The new vectors are called the *components* of the original vector and are normally chosen to be perpendicular to one another.

Resolving power The *resolving power* of an optical system refers to its ability to produce separate images of nearby objects. Resolving power is limited by diffraction; the larger the objective lens or mirror of an optical system, the greater its resolving power.

Resonance *Resonance* occurs when periodic impulses are given to an object at a frequency equal to one of its natural frequencies of oscillation.

Resonance frequency The *resonance frequency* of a series alternating-current circuit is that frequency for which the impedance is a minimum; its value is $f_0 = 1/2\pi \sqrt{LC}$.

Restitution, coefficient of The *coefficient of restitution* is the ratio between the relative speeds of two colliding objects after and before they collide. It equals 1 for a completely elastic collision and 0 for a completely inelastic collision.

Reynolds number The *Reynolds number* of a system involving fluid flow determines whether the flow will be laminar or turbulent. Systems with the same Reynolds number are *dynamically similar,* which means their patterns of fluid flow are the same.

Right-hand rule for magnetic field According to the *right-hand rule for magnetic field,* when a current-carrying wire is grasped with the right hand so that the thumb points in the direction of the current, the curled fingers of that hand then point in the direction of the magnetic field.

Right-hand rule for magnetic force Open the right hand so the fingers are together and the thumb sticks out. According to the *right-hand rule for magnetic force,* when the thumb is in the direction of motion of a positive charge and the fingers are in the direction of a magnetic field, the palm faces in the direction of the force acting on the charge. When a negative charge is involved, the force is in the opposite direction.

Scalar quantity A *scalar quantity* is one that has magnitude only.

Semiconductors *Semiconductors* are intermediate in their ability to carry electric current between conductors and insulators. An *n-type semiconductor* is one in which electric current is carried by the motion of electrons. A *p-type semiconductor* is one in which electric current is carried by the motion of *holes,* which are vacancies in the electron structure of the material that behave like positive charges.

Shear When equal and opposite forces that do not act along the same line are applied to a body, it is said to be in *shear.* The *shear modulus S* of a material is equal to the shear force per unit cross-sectional area applied to a sample of it divided by the relative distortion of the sample.

Shock wave   A *shock wave* is a shell of high pressure produced by the motion of an object whose speed is greater than that of sound.

Snell's law   *Snell's law* states that the ratio between the sine of the angle of incidence of a light ray upon a boundary between two media and the sine of the angle of refraction is equal to the ratio of the speeds of light in the two media.

Solid angle   A *solid angle* is the counterpart in three dimensions of an angle in two dimensions.

Sound   *Sound* is a longitudinal wave phenomenon that consists of successive compressions and rarefactions of the medium through which it travels. *Infrasound* and *ultrasound* respectively refer to sounds whose frequencies are too low and too high to be audible. Sound intensity level in air is measured in *decibels*.

Specific gravity   The *specific gravity* of a substance is its density relative to that of water. Specific gravity is also called *relative density.*

Specific heat capacity   The *specific heat capacity c* of a substance is the amount of heat required to change the temperature of a unit quantity of it by 1°.

Spectrum   An *absorption spectrum* results when white light that has passed through a cool gas is analyzed by a spectrometer; it is a *dark line spectrum* because it appears as a series of dark lines on a bright background, with the lines representing characteristic wavelengths absorbed by the gas. An *emission spectrum* consists of the various wavelengths of light emitted by an excited substance; it may be a *continuous spectrum,* in which all wavelengths are present, or a *bright line spectrum,* in which only a few wavelengths characteristic of the individual atoms of the substance are present.

Speed   The *average speed* of a moving object is the distance it covers in a time interval divided by the time interval. The object's *instantaneous speed* at a certain moment is the rate at which it is covering distance at that moment. Speed is a scalar quantity.

Spin   Every electron has a certain intrinsic amount of angular momentum called its *spin*. The spin of an electron is as fundamental a property as its mass or electric charge. Owing to its spin, every electron acts like a tiny bar magnet. Most other elementary particles (such as proton and neutron) also have spin associated with them.

Standing waves   *Standing waves* in a stretched string are transverse harmonic oscillations that result in a wave pattern whose amplitude varies in a sinusoidal manner along the string. Points of zero amplitude are called *nodes*. Other types of waves can also occur as standing waves.

Stress and strain   The *stress* on an object is the force (tension, compression, or shear) applied to it per unit area; the *strain* is the resulting change in a dimension of the object relative to its original value. A *modulus of elasticity* of a material is the ratio between a particular kind of applied stress and the resulting strain, providing the elastic limit is not exceeded.

Sublimation   *Sublimation* is the direct conversion of a substance from the solid to the vapor state, or vice versa, without it first becoming a liquid.

Superposition, principle of   The *principle of superposition* states that when two or more waves of the same nature travel past a given point at the same time, the amplitude at the point is the sum of the amplitudes of the individual waves.

Surface tension   The *surface tension* of a liquid refers to the tendency of its surface to contract to the minimum possible area in any situation.

Système International   The modern version of the metric system of units is the *Système International* (SI). SI units are almost universally used by scientists and are in everyday use in most of the world as well.

Temperature   The *temperature T* of a body of matter is a measure of the average kinetic energy of random translational motion of its constituent particles. When two bodies are in contact, heat flows from the one at the higher temperature to the one at the lower temperature. In the Celsius temperature scale, the unit is the °C; in the Fahrenheit scale, the unit is the °F. A *thermometer* is a device for measuring temperature.

Tension   When equal and opposite forces that act away from each other are applied to a body, it is said to be in *tension*.

Terminal speed   Because air resistance varies with speed, a falling body eventually reaches a *terminal speed* after which it ceases to be accelerated downward.

Tesla   The *tesla* (T) is the unit of magnetic field. It is equal to $1 \text{ N/A} \cdot \text{m}$.

Thermal conductivity   The *thermal conductivity* of a material is a measure of its ability to conduct heat. The *thermal resistance* of a layer of given thickness of a material is a measure of its insulating ability.

Thermal expansion   The *coefficient of linear expansion* is the ratio between the change in length of a solid rod of a particular material and its original length per 1° change in temperature. The *coefficient of volume expansion* is the ratio between the change in volume of a sample of a particular solid or liquid and its original volume per 1° change in temperature.

Thermodynamics   The *first law of thermodynamics* states that the heat input to a heat engine is equal to its work output plus any change in its internal energy. The *second law of thermodynamics* states that it is impossible to construct an engine, operating in a repeatable cycle, that does nothing besides taking energy from a source and performing an equivalent amount of work; some of the input energy must be wasted.

Thermonuclear energy   The energy liberated by nuclear fusion is called *thermonuclear energy.*

Thrust   The *thrust* of a rocket is the force that results from the expulsion of exhaust gases.

Torque   The *torque* τ of a force about a particular axis is the product of the magnitude of the force and the perpendicular distance from the line of action of the force to the axis. The latter distance is called the *moment arm* of the force. Torque plays the same role in rotational motion that force does in linear motion.

Transformer   An alternating current flowing in the primary coil of a *transformer* induces another alternating current in the secondary coil. The ratio of the emfs is proportional to the ratio of turns in the coils.

Transistor   A *transistor* is a semiconductor device that enables an electrical signal to be amplified.

Transverse waves   *Transverse waves* occur when the individual particles of a medium vibrate from side to side perpendicular to the direction in which the waves travel. The vibrations of a stretched string are transverse waves. Electromagnetic waves are transverse because the varying electric and magnetic fields of which they consist are perpendicular to the wave direction, even though nothing material is in motion.

Triple point   The *triple point* of a substance refers to the temperature and pressure at which its solid, liquid, and vapor states can exist in equilibrium with one another.

Uniform circular motion   An object traveling in a circle at constant speed is said to be undergoing uniform circular motion.

Van der Waals force   *Van der Waals forces* originate in the electric attraction between asymmetrical charge distributions in atoms and molecules. Molecular solids and liquids are held together by van der Waals forces.

Vaporization, heat of   The *heat of vaporization* of a substance is the amount of heat that must be supplied to change 1 kg of it at its boiling point from the liquid to the gaseous (or vapor) state; the same amount of heat must be removed from 1 kg of the substance at its boiling point to change it into a liquid.

Vector   A *vector* is an arrowed line whose length is proportional to the magnitude of some vector quantity and whose direction is that of the quantity. A *vector diagram* is a scale drawing of the various forces, velocities, or other vector quantities involved in the motion of a body. In *vector addition,* the tail of each successive vector is placed at the head of the previous one, with their lengths and original directions kept unchanged. The *resultant* is a vector drawn from the tail of the first vector to the head of the last. A vector can be *resolved* into two or more other vectors called the *components* of the original vector. Usually the components of a vector are chosen to be in mutually perpendicular directions.

Vector quantity   A *vector quantity* is one that has both magnitude and direction. The symbol of a vector quantity is printed in boldface type, for instance **A.**

Velocity   The *velocity* **v** of an object is a specification of both its speed and the direction in which it is moving. Velocity is a vector quantity. The *instantaneous velocity* of an object is its velocity at a specific instant of time. The *average velocity* of an object is the total displacement through which it has moved in a time interval divided by the interval.

Viscosity   The *viscosity* of a fluid is a measure of its internal friction.

Volt   The unit of electrical potential difference is the *volt* (V). It is equal to 1 J/C.

Watt   The unit of power is the *watt* (**W**), which is equal to 1 J/s.

Wavefront   A *wavefront* is an imaginary surface that joins points where all the waves from a source are in the same phase of oscillation. According to *Huygens' principle,* every point on a wavefront can be considered as a point source of secondary wavelets that spread out in all directions with the wave velocity of the medium. The wavefront at any time is the envelope of these wavelets.

Wave motion   *Wave motion* is characterized by the propagation of a change in a medium, rather than by the net motion of the medium itself. The passage of a wave across the surface of a body of water, for instance, involves the motion of a pattern of alternate crests and troughs, with the individual water molecules themselves ideally executing uniform circular motion.

Weight   The *weight w* of an object is the gravitational force exerted on it by the earth. The weight of an object is proportional to its mass.

Work   Whenever a force affects the motion of a body, the body undergoes a displacement while the force acts on it. The product of the force and the component of the displacement of the body in the direction of the force is called the *work W* done by the force on the body. Work is a measure of the change (in a general sense) a force gives rise to when it acts upon something. The unit of work is the same as that of energy, namely the joule (J).

X rays   *X rays* are high-frequency electromagnetic waves emitted when fast electrons impinge on matter.

Young's modulus   *Young's modulus Y* of a particular material is equal to the tension or compression force per unit cross-sectional area applied to a sample of that material divided by the fractional change in the length of the sample.

SOLUTIONS TO ODD-NUMBERED EXERCISES

CHAPTER 1

1. $(368 \text{ ft})(0.305 \text{ m/ft}) = 112 \text{ m} = 0.112 \text{ km}$.

3. $v = (50 \text{ km/h})(0.621 \text{ mi/km}) = 31 \text{ mi/h}$.

5. (a) $(4840 \text{ yd}^2)(3 \text{ ft/yd})^2(0.3048 \text{ m/ft})^2 = 4047 \text{ m}^2$.
(b) $(1 \text{ km}^2)(10^3 \text{ m/km})^2/(4047 \text{ m}^2/\text{acre}) = 247 \text{ acres/km}^2$.

7. (a) 1 bd ft $= (12 \text{ in})^2(1 \text{ in}) = 144 \text{ in}^3$. (b) Since $1 \text{ ft}^3 = (12 \text{ in})^3 = 1728 \text{ in}^3$, 1 bd ft $= 144 \text{ in}^3/(1728 \text{ in}^3/\text{ft}^3) = 0.0833 \text{ ft}^3$. (c) Since $1 \text{ in}^3 = (2.54 \text{ cm})^3 = 16.4 \text{ cm}^3$, 1 bd ft $= (144 \text{ in}^3)(16.4 \text{ cm}^3/\text{in}^3) = 2.36 \times 10^3 \text{ cm}^3$.

9. $v = 132 \text{ ft/s}$, $t = s/v = 0.45 \text{ s}$.

11. $t = s/v = 0.029 \text{ s}$.

13. $s = vt = 3 \times 10^{-15} \text{ m}$.

15. The snake covers 105 m in $t = s/v = 70 \text{ s}$. Therefore, your speed must be greater than $v = 100 \text{ m}/70 \text{ s} = 1.43 \text{ m/s}$.

17. $s_1 = v_1 t_1 = 1200 \text{ km}$, $s_2 = v_2 t_2 = 1000 \text{ km}$; $\bar{v} = (s_1 + s_2)/(t_1 + t_2) = 338 \text{ km/h}$.

19. If $s = $ total distance, $t_1 = s/2v_1$ and $t_2 = s/2v_2$. Hence $\bar{v} = s/(t_1 + t_2) = 2/[(1/v_1) + (1/v_2)] = 2v_1v_2/(v_1 + v_2) = 6.86 \text{ km/h}$.

21. If $D = $ total distance, $s_1 = D/3$, $s_2 = 2D/3$, $t_1 = s_1/v_1 = D/3v_1$, $t_2 = s_2v_2 = 2D/3v_2$; $\bar{v} = D/(t_1 + t_2) = 1/[(1/3v_1) + (2/3v_2)]$, $2/3v_2 = 1/\bar{v} - 1/3v_1$, $1/v_2 = 3/2\bar{v} - 1/2v_1 = 0.022857 \text{ h/mi}$, $v_2 = 43.75 \text{ mi/h}$.

23. The accelerations are the same.

25. Yes.

27. (a) $a = (v_f - v_0)/t = 1.2 \text{ m/s}^2$. (b) -2 m/s^2.

29. (a) $a = (v_f - v_0)/t = 1.1 \text{ m/s}^2$. (b) $t = (v_f - v_0)/a = 7.27 \text{ s}$.

31. The airplane's speed is 550 km/h from 4:00 P.M. to 5:41 P.M., 620 km/h from 5:41 P.M. to 7:54 P.M., and 520 km/h from 7:54 P.M. to 10:00 P.M.

33. The car had an initial speed of 21.6 mi/h and accelerated at 21.6 (mi/h)/min (1296 mi/h^2) for 3 min to 86.4 mi/h. It stayed at 86.4 mi/h for 4 min and then slowed down to a stop at 43.2 (mi/h)/min (2592 mi/h^2) in 2 min.

35. $v_f = 38.14 \text{ ft/s}$, $s = (v_0 + v_f)t/2 = 296 \text{ ft}$.

37. (a) $a = (80 \text{ m/s})/35 \text{ s}$; $t = (v_f - v_0)/a = 8.75 \text{ s}$; $s = \frac{1}{2}at^2 = 87.5 \text{ m}$. (b) $t = (v_f - v_0)/a = 8.75 \text{ s}$; $s = v_0 t + \frac{1}{2}at^2 = 612.5 \text{ m}$. (c) $s = \frac{1}{2}at^2 = 1400 \text{ m}$.

39. $s_1 = \frac{1}{2}s_1 t_1^2 = 450 \text{ m}$; $v = a_1 t_1 = 30 \text{ m/s}$, $s_2 = vt_2 = 3600 \text{ m}$; $a_2 = -v/t_3 = -2 \text{ m/s}^2$, $s_3 = vt_3 + \frac{1}{2}a_2 t_3^2 = 225 \text{ m}$; $s = s_1 + s_2 + s_3 = 4275 \text{ m}$.

41. (a) The car needs $t_1 = v_1/a = 20 \text{ s}$ to reach $v_1 = 25 \text{ m/s}$, at which time it will have gone $s_1 = v_1^2/2a = 250 \text{ m}$. Hence in the total time $t = t_1 + t_2$ it will have gone $s_{\text{car}} = s_1 + v_1 t_2 = s_1 + v_1(t - t_1)$ while the truck will have gone $s_{\text{truck}} = v_2 t$. When the car catches up with the truck, $s_{\text{car}} = s_{\text{truck}}$ and $v_2 t = s_1 + v_1(t - t_1)$, $t = (s_1 - v_1 t_1)/(v_2 - v_1) = 25 \text{ s}$. (b) $s = v_2 t = 375 \text{ m}$.

43. $s = v^2/2a = 1.02 \times 10^6 \text{ m}$.

45. $s_1 = v_0 t = 20 \text{ m}$, $s_2 = v_0^2/2a = 100 \text{ m}$; $s = s_1 + s_2 = 120 \text{ m}$.

47. $t_1 = v/a_1 = 6 \text{ s}$, $s_1 = v^2/2a_1 = 27 \text{ m}$; $t_3 = v/a_3 = 4.5 \text{ s}$, $s_3 = v^2/2a_3 = 20.25 \text{ m}$; $s_2 = 4000 \text{ m} - s_1 - s_3 = 352.75 \text{ m}$, $t_2 = s_2/v = 39.2 \text{ s}$; $t = t_1 + t_2 + t_3 = 49.7 \text{ s}$.

49. The squirrel should stay where it is, since if it lets go, it will fall with the same acceleration as the bullet and so will be struck.

51. Yes.

53. $v = \sqrt{2gh} = 26.6 \text{ m/s}$.

55. $h = \frac{1}{2}gt^2 = 78.4 \text{ m}$.

57. The time of rise and the time of fall are both $t = 1.5 \text{ s}$. Hence $v = gt = 14.7 \text{ m/s}$ is the value of both the initial and final speeds.

59. (a) $v_f = v_0 + gt = 19.8 \text{ m/s}$. (b) $v_f = 29.6 \text{ m/s}$.

61. $h = v^2/2g = 13.1 \text{ m}$.

63. (a) $v = \sqrt{2g_M h} = 10.5 \text{ m/s}$. (b) $T = 2t = 2\sqrt{2h/g_M} = 5.69 \text{ s}$.

65. (a) $h = \frac{1}{2}gt^2$, $t = \sqrt{2h/g} = 1.17 \text{ s}$. (b) $v = \sqrt{2gh} = 37.5 \text{ ft/s}$.

67. Man: $t = h/v = 25 \text{ s}$; monocle: $t = \sqrt{2h/g} = 5.5 \text{ s}$; $\Delta t = 19.5 \text{ s}$.

69. Since the coconuts met halfway down the cliff, the one thrown upward must have had the same initial speed as the speed of the dropped coconut at the place of meeting, which is $v = \sqrt{2gh} = 19.8 \text{ m/s}$.

71. (a) Calling up $+$ and down $-$, the pump falls $h = v_0 t - \frac{1}{2}gt^2 = -46.4 \text{ m}$, so the helicopter's altitude was 46.4 m when it dropped the pump. (b) $\Delta h = v_0 t = 32 \text{ m}$, so the helicopter was $h + \Delta h = 78.4 \text{ m}$ high 4 s later.

73. $t_1 = v/a = 4$ s; $s_1 = \frac{1}{2}at^2 = 12$ m; $t_3 = t_1 = 4$ s; $s_3 = s_1 = 12$ m. Hence constant-speed distance is $s_2 = 50$ m $- 24$ m $= 26$ m and $t_2 = s_2/v = 4.33$ s; $t = t_1 + t_2 + t_3 = 12.33$ s.

75. (a) $h_1 = v_0^2/2g = 5.10$ m, $t_1 = \sqrt{2h_1/g} = 1.02$ s; $h_2 = 20$ m $+ 5.1$ m $= 25.1$ m; $t_2 = \sqrt{2h_2/g} = 2.26$ s; $t = t_1 + t_2 = 3.28$ s. (b) $v = gt_2 = 22$ m/s.

CHAPTER 2

1. No. Only a net force produces an acceleration, and when a force is applied to an object, other forces may come into being that cancel it out. Thus pushing down on a book lying on a table does not accelerate the book because the table pushes back with an equal and opposite force.

3. The deceleration of a person falling onto loose earth is more gradual than if he falls onto concrete, hence the force acting on him is less.

5. $F = ma = 1.5 \times 10^{-5}$ N.

7. $F_{max} = m_0 a_0 = 2000$ N. Hence $a = F_{max}/m = 0.67$ m/s^2.

9. $a = F/m$, $v = at = Ft/m = 28.6$ m/s.

11. (a) $v_f = 55.6$ m/s; $a = (v_f - v_0)/t = 18.5$ m/s$^2 = 1.89g$. (b) $F = ma = 2.22 \times 10^5$ N.

13. $a = -v^2/2s = -4.8$ m/s^2; $F = ma = -9600$ N.

15. $a = -v_0^2/2s = -2500$ m/s^2; $F = ma = -10^4$ N (over a ton); the goat will be killed.

17. $\Delta v = 4.17$ m/s, $F = ma = m\Delta v/t = 500$ N.

19. Here $F = w = mg$, hence $a = F/m = g$.

21. Yes, by an applied downward force in addition to the downward force of gravity.

23. $m = w/g = 0.816$ kg; $a = F/m = 24.5$ m/s^2.

25. $a = (m_1 - m_2)g/(m_1 + m_2) = 1.96$ m/s^2.

27. A force equal to ma is needed in addition to the force $w = mg$ required just to lift the box without accelerating it. Hence $F = mg + ma = 59$ N.

29. $F = w - T = ma$, $T = w - ma = mg - ma = 8800$ N.

31. $v_f^2 = v_0^2 + 2as$, $a = (v_f^2 - v_0^2)/2s = 30.6$ m/s^2; $F = ma + mg = 4.04 \times 10^3$ N; $F/mg = 4.1$.

33. (a) Net upward force $= F = T - mg$, $a = F/m = (T - mg)/m = 2.7$ m/s^2. (b) 9.8 m/s^2.

35. Here $s = 0.6$ m, $h_1 = 2.0$ m, $h_2 = 1.8$ m, $m_1 = 40$ kg, $m_2 = ?$ Kangaroo by herself: $v_1^2 = 2gh_1$, $a_1 = v_1^2 2s = gh_1/s$, $F = m_1 g + m_1 a_1 = m_1 g(1 + h_1/s) = 1699$ N. With baby: $v_2^2 = 2gh_2$, $a_2 = v_2^2/2s = gh_2/s$;

$F = (m_1 + m_2)(g + a_2)$, $m_1 + m_2 = F/(g + a_2) = F/g(1 + h_2 s) = 43.3$ kg, $m_2 = 43.3$ kg $- 40$ kg $= 3.3$ kg.

37. (a) $w = mg = 48$ lb. (b) $a = F/m = 66.7$ ft/s^2.

39. $F = w + (w/g)a = 6.75$ lb.

41. $F = w_1 - w_2$, $m = (w_1 + w_2)/g$, $a = F/m = (w_1 - w_2)g/(w_1 + w_2) = 6.4$ ft/s.

43. No. Action and reaction forces act on different objects, and so a single force can certainly act on an object.

45. A propeller works by pushing backward on the air, whose reaction force in turn pushes the propeller itself forward. No air, no reaction force, so the idea is no good.

47. (a) In order to fly, the bird's wings must exert a net downward force on the air in the box equal to its weight, and this force is transmitted to the box. Hence the scale reading is the same whether the bird stands or flies. (b) In this case the forces exerted on the air by the bird's wings are not completely transmitted to the cage, and the scale reading will be smaller when the bird is flying. Because some force is nevertheless transmitted to the bottom of the cage, the reading when the bird is flying will be somewhat greater than that corresponding to the empty cage.

49. Yes; no.

51. $F = \mu N = 2$ N.

53. $F = \mu N = 72$ lb.

55. $a = F/m = \mu mg/m = \mu g$, $v_0 = \sqrt{2as} = \sqrt{2\mu gs} = 60$ ft/s $= 41$ mi/h.

57. $a = v_0^2/2s$; $\mu = F/mg = ma/mg = a/g = v_0^2/2sg = 0.041$.

59. (a) $F_1 = \mu m_1 g = 14{,}700$ N. (b) The maximum force the man can exert without slipping is $F_2 = \mu m_2 g = 784$ N. If he exerts more force than this, he will be accelerated toward the elephant by the reaction force of the rope on him. His acceleration would then be $a = (F_1 - F_2)/m = 139$ m/s^2 assuming $\mu = \mu_s$; since $\mu < \mu_s$, the acceleration would be even greater.

61. (a) $F = \mu mg = ma$, $a = \mu g = 6.86$ m/s^2. (b) $v = 27.8$ m/s, $v^2 = 2as$, $s = v^2/2a = 56$ m.

63. The truck will stop first. The coefficient of rolling friction for the truck is about nine times greater than that for the boxcar, hence the frictional force on it and its consequent deceleration are about nine times greater, and the distance it goes until it stops is about nine times smaller since $s = v^2/2a$.

65. (a) $F = m_2 g = (m_1 + m_2)a$, $a = m_2 g/(m_1 + m_2) = 4.9$ m/s^2. (b) $F = m_2 g - \mu m_1 g = (m_1 + m_2)a$, $a = (m_2 g - \mu m_1 g)/(m_1 + m_2) = 3.92$ m/s^2.

67. (a) $m_1 = 2\,\text{kg}, m_2 = 3\,\text{kg}; F = \mu(m_1 + m_2)g + \mu m_1 g = 20.58$ N. (b) $F = \mu(m_1 + m_2)g + \mu m_1 g + \mu m_1 g = 26.46$ N.

CHAPTER 3

1. A scalar quantity; a vector quantity.

3. Yes.

5. 30 N; 0.

7. $s = \sqrt{(70\,\text{m})^2 + (40\,\text{m})^2} = 80.6$ m; $\tan \theta = 40$ m/70 m $= 0.571$, $\theta = \tan^{-1} 0.571 = 30°$ above the horizontal.

9. (a) $s = \sqrt{(25\,\text{km})^2 + (35\,\text{km})^2} = 43$ km. (b) 172 km.

11. The two velocities are perpendicular, so $v = \sqrt{(7\,\text{kn})^2 + (3\,\text{kn})^2} = 7.6$ kn. If θ is the angle between **v** and NW, then $\tan \theta = 3$ kn/7 kn $= 0.429$, $\theta = 23°$. Since NW is itself 45° W of N, the direction of **v** is 45° + 23° = 68° W of N.

13. (a) She should head directly across the river. (b) $t =$ distance/speed $= 0.1875$ h $= 11$ min 15 s.

15. Alpha: $v_w = v_\alpha \cos 40° = 3.83$ km/h. Beta: $v_w = v_\beta \cos 50° = 3.86$ km/h. Beta has the higher windward component of velocity.

17. $F_y = 50$ N $+ (25$ N)$(\sin 45°) = 67.7$ N.

19. $v = v_x/\cos 37° = 87.6$ km/h.

21. (a) $v_y = v \sin 25° = 84.6$ km/h. (b) $v_x = v \cos 25° = 181$ km/h. (c) $v_{\text{south}} = v_x \sin 45° = 128$ km/h.

23. $v_{\text{rain}} = v_{\text{car}}/\tan \theta = 8.4$ m/s.

25. One way to solve this problem is to note that the angle between each towrope and the direction in which the ship moves is 15°. The force each tugboat exerts has a component in this direction of $(5.0\,\text{tons})(\cos 15°) = 4.8$ tons. The total force on the ship is 9.6 tons, because the force components perpendicular to this direction are equal and opposite and so cancel out.

27. The force **F** that accelerates each box is the component of its weight **w** parallel to the plane. Hence $F = w \sin \theta = mg \sin \theta = ma$ and $a = g \sin \theta = 3.4$ m/s^2 for each box.

29. $\mu = \tan \theta = 0.176$.

31. The friction force μN must be overcome by the horizontal component $F \cos \theta$ of the applied force **F**. Since the normal force is the sled's weight mg minus the upward vertical component $F \sin \theta$ of the force **F**, we have $F \cos \theta = \mu(mg - F \sin \theta)$, $F = \mu mg/(\mu \sin \theta + \cos \theta) = 335$ N.

33. $N = mg \cos \theta$; $F = mg \sin \theta - \mu N = mg(\sin \theta - \mu \cos \theta) = ma$; $a = g(\sin \theta - \mu \cos \theta) = 4.11$ m/s^2. Since $s = \frac{1}{2}at^2$, $t = \sqrt{2s/a} = 2.09$ s.

35. (a) From solution to Exercise 33, $a = g(\sin \theta - \mu \cos \theta) = 5.55$ m/s^2; $v = \sqrt{2as} = 23.6$ m/s. (b) $F = \mu mg = ma$, $a = \mu g = 0.98$ m/s^2, $s = v^2/2a = 283$ m.

37. The frictional force to be overcome is $F_f = \mu N = \mu mg \cos \theta = 2.144 \times 10^7$ N. The component of the ship's weight parallel to the ways is $F_w = mg \sin \theta = 2.049 \times 10^7$ N. Hence an additional force of $F_f - F_w = 9.5 \times 10^5$ N is needed.

39. Let north be the $+y$-direction and east the $+x$-direction. Then $F_{1x} = 0, F_{1y} = 10$ N; $F_{2x} = 20$ N, $F_{2y} = 0$, $F_{3x} = (10\,\text{N})(\cos 40°) = 7.66$ N, $F_{3y} = (-10\,\text{N})(\sin 40°) = -6.43$ N; $F_{4x} = (-20\,\text{N})(\sin 50°) = -11.74$ N, $F_{4y} = (-20\,\text{N})(\cos 50°) = -12.86$ N. Hence $R_x = F_{1x} + F_{2x} + F_{3x} + F_{4x} = 15.92$ N, $R_y = F_{1y} + F_{2y} + F_{3y} + F_{4y} = -9.29$ N, $R = \sqrt{R_x^2 + R_y^2} = 18.4$ N. R is south of east by the angle $\theta = \tan^{-1} R_y/R_x = 30°$.

41. Calling E the $+x$-direction and N the $+y$-direction, $s_x = 200$ km $- (100\,\text{km})(\sin 45°) = 129$ km and $s_y = -200$ km $+ (100\,\text{km})(\cos 45°) = -129$ km. Hence $s = \sqrt{s_x^2 + s_y^2} = 182$ km. The direction is NW since s_x and s_y are equal in length.

43. (a) Relative to L: $v_{Qx} = -v_L$; $v_{Qy} = v_Q$; $v_Q = \sqrt{v_{Qx}^2 + v_{Qy}^2} = 10.6$ kn. If ϕ is angle between $\mathbf{v}_Q$ (relative to L) and N, $\phi = \tan^{-1}(v_{Qx}/v_{Qy}) = -49°$, so ϕ is 49° W of N (counterclockwise from N). (b) $\mathbf{v}_L$ (relative to Q) is equal in magnitude but opposite in direction to $\mathbf{v}_Q$ (relative to L). Hence $\mathbf{v}_L$ is 10.6 kn at 229° counterclockwise from N, which is 131° clockwise from N or 41° S of E.

45. Here $A_x = A \sin 37° = 6$ cm; $A_y = A \cos 37° = 8$ cm; $B_x = B \cos 37° = 8$ cm; $B_y = -B \sin 37° = -6$ cm. (a) $R_x = A_x + B_x = 14$ cm, $R_y = A_y + B_y = 2$ cm; $R = \sqrt{R_x^2 + R_y^2} = 14.1$ cm. **R** points θ clockwise from the $+y$-direction where $\tan \theta = R_x/R_y = 7$, $\theta = 82°$. (b) $R_x = A_x - B_x = -2$ cm, $R_y = A_y - B_y = 14$ cm; $R = \sqrt{R_x^2 + R_y^2} = 14.1$ cm. **R** points θ counterclockwise from the $+y$-direction where $\tan \theta = R_x/R_y = 0.143$, $\theta = 8°$. (c) $R_x = B_x - A_x = 2$ cm, $R_y = B_y - A_y = -14$ cm; $R = \sqrt{R_x^2 + R_y^2} = 14.1$ cm. **R** points θ clockwise from the $+x$-direction where $\tan \theta = R_y/R_x = 0.143$, $\theta = 8°$.

47. Let **A** = 4 in the y-direction, so $A_y = 4$ and $A_x = 0$. Then $B_x^2 + B_y^2 = B^2 = (2)^2 = 4$ and $C_x^2 + C_y^2 = C^2 = (5)^2 = 25$. Since $C_x = B_x$ and $C_y = A_y + B_y$, $C^2 = C_x^2 + C_y^2 = B_x^2 + B_y^2 + A_y^2 + 2A_yB_y$ and $B_y = (25 - 4 - 16)/8 = 0.625$. The angle between **B** and the $+y$-direction is therefore $\theta = \sin^{-1} B_y/B = 18°$.

49. $t = s/v$, $h = \frac{1}{2}gt^2 = \frac{1}{2}gs^2/v^2 = 0.10$ m $= 10$ cm.

51. The time of fall is $t = \sqrt{2h/g} = 10.1$ s. In this time the parcel will have moved horizontally toward the ship $s_1 = v_1 t$ and the ship will have moved toward the parcel $s_2 =$

v_2t. Hence the required distance is $s_1 + s_2 = (v_1 + v_2)t = 707$ m.

53. $t = \sqrt{2h/g} = 14.3$ s, $v_x = 400$ km/h $= 111$ m/s, $v_y = gt = 140$ m/s, $v = \sqrt{v_x^2 + v_y^2} = 179$ m/s.

55. (a) $v_y = \sqrt{2gh} = 62$ ft/s; $v_x = 80$ ft/s; $v = \sqrt{v_x^2 + v_y^2} = 101$ ft/s. (b) $\theta = \tan^{-1} v_y/v_x = 38°$ below the horizontal.

57. Yes; the speed is greatest at both ends of the path and is least at the highest point.

59. At maximum range, $\theta = 45°$ and $v_0 = \sqrt{Rg} = 10.4$ m/s.

61. $R = (v_0^2 \sin 2\theta)/g$, $v_0 = \sqrt{Rg/\sin 2\theta}$. Here $R = 120$ ft and $\sin 2\theta = \sin 100° = \cos 10°$ since from Appendix A7 $\sin (90° + x) = \cos x$. Hence $v_0 = 62.4$ ft/s.

63. (a) $v_0 = \sqrt{2gh}$, $R_{max} = v_0^2/g = 2h = 200$ m. (b) $v = \sqrt{2gh} = 44.3$ m/s; 44.3 m/s.

65. Since $R_{max} = v^2/g$, $R_2/R_1 = v_2^2/v_1^2$ and $v_2 = \sqrt{R_2/R_1}v_1$. Here $R_2/R_1 = 1.2$, so $v_2 = \sqrt{1.2}v_1 = 1.095v_1$; an increase in speed of 9.5% is therefore required.

67. (a) $R = (v_0^2/g)\sin 2\theta_1 = 7953$ m; $T_1 = (2v_0/g)\sin \theta_1 = 30.6$ s; $h_1 = v_yt - \frac{1}{2}gt^2 = (v_0 \sin \theta_1)(T_1/2) - \frac{1}{2}g(T_1/2)^2 = 1148$ m. (b) $\theta_2 = 90° - \theta_1 = 60°$; $T_2 = (2v_0/g) \sin \theta_2 = 53$ s; $h_2 = (v_0 \sin \theta_2) (T_2/2) - \frac{1}{2}g (T_2/2)^2 = 3444$ m.

CHAPTER 4

1. (a) Perpendicular to the wall. (b) The force on the ground is always greater than the weight of the ladder since the force has a horizontal component equal to the reaction force of the wall on the ladder as well as a vertical component equal to the ladder's weight.

3. $T_y = T \sin \theta = w$, $T = w/\sin \theta = 156$ N.

5. The mass each rope must support is $m = 25$ kg. Since the vertical component of the tension in each rope is $F = T \cos \theta$, $T = F/\cos \theta = mg/\cos \theta = 283$ N.

7. If θ is the angle between the rope and the mast, $\sin \theta = (0.6$ m$)/(4$ m$) = 0.15$ and $\theta = 8.6°$; $F_x = mg \tan \theta = 89$ N.

9. $F_x = 80$ N, $F_y = mg = 49$ N, $F = \sqrt{F_x^2 + F_y^2} = 94$ N. $\theta - \tan^{-1} F_y/F_x = 31°$ above the horizontal.

11. (a) $F = \mu_s mg = 412$ N. (b) The first step is to find angle θ between the cable and the vertical from the force involved. $\Sigma F_x = T_x + F = -T \sin \theta + F = 0$, $\sin \theta = F/T$; $\Sigma F_y = T_y - mg = T \cos \theta - mg = 0$, $\cos \theta = mg/T$; $\tan \theta = \sin \theta/\cos \theta = F/mg = 0.210$, $\theta = 12°$. Hence $s = L \sin \theta = 1.2$ m.

13. Let T_1 and T_2 be the tensions in the two parts of the wire, where $T_1 = T_2 = T$. If θ is the angle either part makes with the horizontal, $\theta = \tan^{-1} (0.5$ m$)/(10$ m$) = 2.86°$. For the bird to be in equilibrium, $\Sigma F_y = T_{1y} + T_{2y} - mg = 2T \sin \theta - mg = 0$, $m = 0.92$ kg.

15. (a) $F_y = F \sin \theta = mg$, $F = mg/\sin \theta = 68.6$ kN. (b) $T = F_x = F \cos \theta = 59.4$ kN.

17. For each leg, $F_y = 50$ lb and $\theta = \cos^{-1} (4$ ft/10 ft$) = 66.4°$. Since $F_y = F \sin \theta$, $F = F_y/\sin \theta = 54.6$ lb.

19. From the diagram below, $\theta = \tan^{-1}(5$ cm/30 cm$) = 9.5°$ and $F_y = F \cos \theta = 0.986F$. The torques about the elbow, τ_1 exerted by the muscle, τ_2 by the forearm's weight, and τ_3 by the pail of water, are respectively $\tau_1 = F_y x_1 = (0.986F)(0.05$ m$) = 0.0493 F$ m, $\tau_2 = -w_1x_2 = -(29.4$ N$)(0.16$ m$) = -4.7$ N · m, $\tau_3 = -w_2x_3 = -(98$ N$)(0.35$ m$) = -34.3$ N · m, $F = (4.7 + 34.3)$N · m/$(0.0493$ m$) = 791$ N. This force, which is equivalent to 178 lb, is over six times the combined weights of the forearm and the pail of water.

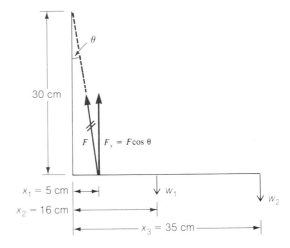

21. The location of the pivot point.

23. $F = \tau/L = 286$ N.

25. About whatever point makes the calculations easiest; the results will be the same regardless of the point chosen.

27. Let $m_1 = 15$ kg, $m_2 = 20$ kg, and $m_3 = 25$ kg. Calculating torques about the pivot with $x_1 = 2$ m and $x_3 = 2$ m gives $m_1gx_1 + m_2gx_2 = m_3gx_3$, $x_2 = 1.00$ m.

29. Calculating torques about the end of the board on the floor, $\Sigma \tau = FL - w(L/2) \cos \theta = 0$, $F = [w(L/2) \cos \theta]/L = (w/2) \cos \theta = 22.7$ lb.

31. Calculating torques about the base of the rod, $(F \sin \theta)(1$ m$) = (mg \cos \theta)(3$ m$)$, $F = 3mg/\tan \theta = 247$ N.

33. At the point of tipping over, the CG is directly above the line of the lower wheels. Hence $h = 0.9$ m/tan $30° = 1.56$ m.

35. There is no horizontal force acting on the dumbbell; hence its CG must drop vertically as the dumbbell turns. The 1-kg ball therefore moves to the left. The CG is located 20 cm from the 1-kg ball, and so the final position of this ball is 20 cm to the left of its original position.

37. Let x be the distance of the two men from the far end of the ladder and w be the ladder's weight. Calculating torques about the CG of the ladder yields $(w/3)(2$ m$) = (2w/3)(2$ m $- x)$, $x = 1$ m.

39. If w is the weight supported by the front axle, then $(24$ N $- w)$ is the weight supported by the rear axle. Calculating torques about the CG of the truck, $(w)(2.5$ m$) = (24$ kN $- w)(1.5$ m$)$, $w = 9$ kN.

41. The bear's weight is $w = mg = 1470$ N, the force shared by both feet is F_1 and that shared by both hands is F_2. Calculating torques about the feet yields $F_2L_2 = wL_1$, $F_2 = 817$ N. Hence $F_1 = w - F_2 = 653$ N. The force on each hand is $F_2/2 = 408.5$ N and that on each foot is $F_1/2 = 326.5$ N.

43. The angle between the boom and the rope is $\theta = \tan^{-1}(1$ m$)/(3$ m$) = 18.4°$ and the CG of the boom is at its midpoint. Calculating torques about the hinge pin of the boom, $(40$ kg$)(g)(1.5$ m$) = (T)(3$ m$)(\sin 18.4°)$, $T = 620$ N.

45. If x is the distance of the CG from the forward axle, calculating torques about the CG yields $m_1gx = m_2g(2.2$ m $- x)$, $x = 0.88$ m.

47. Here $F_y = F \sin 15°$, $w_1 = m_1g = 29.4$ N, $w_2 = m_2g = 9.8$ N, $L_1 = 0.13$ m, $L_2 = 0.28$ m, $L_3 = 0.60$ m. (a) Calculating torques about the shoulder joint yields $F_yL_1 = w_1L_2$, $F = w_1L_2/(L_1 \sin 15°) = 244$ N. (b) $F_y'L_1 = w_1L_2 + w_2L_2$, $F' = (w_1L_2 + w_2L_3)/(L_1 \sin 15°) = 419$ N.

49. Since the wall is frictionless, $F_y = w = mg$ and F_x is equal in magnitude to the force the ladder exerts on the wall. Calculating torques about the foot of the ladder, $(F_x)(2.4$ m$) = (mg)\frac{1}{2}\sqrt{(3\,\text{m})^2 - (2.4\,\text{m})^2} = \frac{1}{2}(mg)(1.8$ m$)$, $F_x = 55$ N.

51. (a) Let **F** be the force the upper hinge exerts on the door and $w = 200$ N be the door's weight acting from its CG. Since the door's weight is supported by the upper hinge, $F_y = w = 200$ N. To find F_x, we calculate torques about the lower hinge: $(F_x)(2.4$ m$) = (w)(0.4$ m$)$, $F_x = 33.3$ N. Hence $F = \sqrt{F_x^2 + F_y^2} = 203$ N. If θ is the angle between **F** and the vertical, $\theta = \tan^{-1} F_x/F_y = 9.5°$. The force the door exerts on the upper hinge is equal and opposite to **F**, hence 203 N at 9.5° away from vertically downward. (b) Since the

lower hinge exerts a horizontal force on the door, it must be equal in magnitude to $F_x = 33.3$ N.

53. Taking $x = 0$, $y = 0$ at the hips and calling the upper body 1, the thighs 2, and the lower legs 3, we have: $x_1 = 0$, $y_1 = 32$ cm; $x_2 = 16$ cm, $y_2 = 0$; $x_3 = 37$ cm, $y_3 = -23$ cm. Hence $X = (m_1x_1 + m_2x_2 + m_3x_3)/(m_1 + m_2 + m_3) = 7.8$ cm and similarly $Y = 18.7$ cm. The CG is 18.7 cm above the hips and 7.8 cm in front of them.

55. Let us establish $x = 0$, $y = 0$ at the lower left-hand corner of the plate. The vertical coordinate of the CG is $Y = 20$ cm since the plate is symmetric about a horizontal line at this height. To find the x-coordinate of the CG, we divide the plate into two halves by a vertical line at $x = 40$ cm. The area of the left-hand half is $A_1 = (40$ cm$)(40$ cm$) = 1600$ cm^2. The area of the right-hand half is $A_2 = A_1 - \pi r^2 = 1286$ m^2. We now imagine the left-hand half of the plate replaced by a point mass of $m_1 = 1600$ located at $x_1 = 20$ cm and the right-hand half replaced by a point mass of $m_2 = 1286$ located at $x_2 = 60$ cm. From Eq. (4.9) we have $X = (m_1x_1 + m_2x_2)/(m_1 + m_2) = 35.4$ cm.

CHAPTER 5

1. No work is done by a net force acting on a moving body when the force is perpendicular to the direction of the body's motion.

3. $h = W/mg = 13.6$ m.

5. (a) $W = Fs = 1040$ J. (b) 1040 J.

7. (a) $W = Fs = 3600$ N. (b) $F_f = \mu mg = 147$ N, $W_f = F_fs = 2940$ J. Hence the work done to accelerate the box was $W - W_f = 660$ J.

9. $W = Fs \cos \theta = 376$ kJ.

11. In both cases $h = 3000$ m and $W = mgh = 2.65$ MJ.

13. Object A travels farthest since the force on it remains **F** while the force on B decreases as v increases to keep $P = Fv$ constant.

15. $P = mgh/t = 4.04$ kW.

17. $W = Pt = (\frac{1}{3})(550$ ft $\cdot$ lb/s$)(60$ s$) = 11,000$ ft $\cdot$ lb.

19. $P = FV = mgv = 3.92$ kW.

21. $F = P/v = 23.3$ kN.

23. (a) 1 year $= 3.15 \times 10^7$ s; $P = W/t = 6.3 \times 10^{12}$ W. (b) 1.8×10^3 W $= 2.4$ hp.

25. Eff $= P_{\text{out}}/P_{\text{in}} = (mgh/t)/P = 0.78 = 78\%$.

27. The force needed to raise 200 L ($= 200$ kg) of water is $F = mg = 1960$ N. Since $t = 60$ s and $s = 1.2$ m, the power output is $P = Fs/t = 39.2$ W. Because the efficiency is $60\% = 0.60$, the power input is $P/0.60 = 65.3$ W.

29. (a) $W = Fs \cos \theta = 866$ ft · lb. (b) $P = W/t = 19.2$ ft · lb/s $= 0.035$ hp.

31. $P = (0.2)(60 \text{ kg})(6 \text{ W/kg}) = 72$ W. Since $P = mgh/t$ where $m = 72$ kg here, $h/t = P/mg = 0.102$ m/s $= 367$ m/h.

33. (a) $F_1 = P_1/v = 2698$ N. (b) $F_2 = F_1 + mg \sin \theta = 4335$ N, $P_2 = F_2v = 48.2$ kW.

35. KE $= \frac{1}{2}mv^2 = 2500$ J.

37. KE $= \frac{1}{2}(w/g)v^2 = 10^4$ ft · lb.

39. $v = s/t = 8.89$ m/s, KE $= \frac{1}{2}mv^2 = 2.77$ kJ.

41. $Fs = \frac{1}{2}mv^2$, $v = \sqrt{2Fs/m} = 7.1$ m/s.

43. $P = \frac{1}{2}m(v_2^2 - v_1^2)/t$, $t = m(v_2^2 - v_1^2)/2P = 8.8$ s.

45. $P = W/t = (\text{KE}_2 - \text{KE}_1)/t = m(v_2^2 - v_1^2)/2t = 36.2$ kW.

47. (a) $mgh_1 = 1176$ J. (b) $mgh_2 = 3528$ J.

49. (a) KE $= \frac{1}{2}mv^2$, $m = 2\text{KE}/v^2 = 0.02$ kg $= 20$ g. (b) PE $= mgh$, $h = \text{PE}/mg = 2.55$ m.

51. $Fs = mgh$, $F = mgh/s = 127$ kN.

53. $mgh_2 = 0.9 \, mgh_1$, $h_2 = 0.9 \, h_1 = 0.9$ m.

55. (a) $P = W/t = mgh/t = 2.94$ MW. (b) 29,400 bulbs.

57. KE $= \frac{1}{2}mv^2 = 4.5$ J, PE $= mgh = 98$ J, $E_0 = mc^2 = 9 \times 10^{16}$ J. Hence the largest part of the total energy is E_0 and the smallest part is KE.

59. $E = m_0c^2 = 8.19 \times 10^{-14}$ J.

61. (a) $(4m_H - m_{He})c^2 = 4.14 \times 10^{-12}$ J. (b) 2.42×10^{18} nuclei.

63. (a) $mgh = \frac{1}{2}mv^2$, $v = \sqrt{2gh} = 19.8$ m/s. (b) 12.5 m/s. (c) 19.8 m/s. (d) 0.

65. (a) $W = Fs = 1000$ ft · lb. (b) ΔPE $= wh = 800$ ft · lb. (c) ΔKE $= W - \Delta$PE $= 200$ ft · lb.

67. $\frac{1}{2}mv^2 = mg\Delta h$, $\Delta h = v^2/2g = 3.3$ m; $h = 3.3$ m $+ 1.1$ m $+ 0.6$ m $= 5.0$ m.

69. (a) KE $= W = Fs = 24$ J. (b) KE $= W - $ PE $= Fs - mgh = 9.3$ J.

71. $\Delta E = \Delta$KE $+ \Delta$PE $= \frac{1}{2}m(v_2^2 - v_1^2) + (0 - mgh) = -168$ kJ. The minus sign means that the car lost this amount of energy.

73. If L is the pendulum length, $h = L - L \cos \theta = L(1 - \cos \theta)$ and $v = \sqrt{2gh} = \sqrt{2gL(1 - \cos \theta)} = 8.65$ ft/s.

CHAPTER 6

1. Yes; yes.

3. The momentum increases by $\sqrt{2}$ since $2 \times (\frac{1}{2}mv^2) = \frac{1}{2}m(\sqrt{2}v)^2$.

5. $p = mv = wv/g = 4400$ slug · ft/s.

7. Car A has twice the KE of car B, but the two cars have the same momentum.

9. (a) $v = 242$ m/s, $mv = 3.87 \times 10^7$ kg · m/s. (b) $t = mv/F = 114$ s.

11. $F = \Delta(mv)/\Delta t$, $\Delta t = \Delta(mv)/F = 0.0102$ s.

13. (a) Newton's second law of motion: $a = v_f^2/2s$, $F = ma = mv_f^2/2s = 6.8 \times 10^5$ N. (b) Work-energy: $Fs = \frac{1}{2} mv_f^2$, $F = mv_f^2/2s = 6.8 \times 10^5$ N. (c) Impulse-momentum: $t = 2s/v_f$, $Ft = mv_f$, $F = mv_f/t = mv_f^2/2s = 6.8 \times 10^5$ N.

15. The definition of mass and the first law of motion are both included in the principle of conservation of linear momentum. The latter principle goes considerably further as well, since it can be applied to systems that consist of any number of bodies that interact with one another in any manner whatsoever, and it holds in three dimensions, not just in one.

17. (a) Yes. (b) In the opposite direction to that in which the man walks. (c) The car also comes to a stop.

19. Number of bullets $= (mv)_{\text{leopard}}/(mv)_{\text{bullet}} = 7.4$, so 8 bullets are needed.

21. $v_2 = v_1(m_1/m_2) = v_1(w_1/w_2) = 25$ ft/s.

23. (a) $v_d = m_m v_m/(m_d + m_w) = 1.9$ m/s; hence the man cannot swim back to the dinghy. (b) $\Delta v_d = m_w v_w/m_d = 2.5$ m/s; hence the dinghy now moves toward the man at 0.6 m/s and the woman can catch up with it since her own speed is 1 m/s.

25. The second piece continues in the original direction, $mv = mv_2/2 - mv_1/2$, $v_2 = 2v + v_1 = 24$ km/s.

27. From Sec. 3–7 the range of the shell is $R = v_xT$, where T is the time of flight, which is unchanged by the explosion, since it depends only on the height the shell reaches. The shell explodes at $x = R/2$. The half that continues takes all of the horizontal component of momentum, hence its speed v_x doubles and it continues for an additional distance of $2(R/2)$ for a total range of $R/2 + R = 3R/2 = 3$ km.

29. $\Delta m/\Delta t = F/v = 720$ kg/s.

31. $F = v(\Delta m/\Delta t)$, $a = (F/m) - g = [v(\Delta m/m)/\Delta t] - g = 30.2$ m/s².

33. The rocket's thrust is $F = v(\Delta m/\Delta t) = 75$ kN. Since the initial weight of the rocket is $m_0 g$, the initial force available for its acceleration is $F - m_0 g$ and so $a_0 = (F - m_0 g)/m_0 = 4$ m/s^2. The final acceleration is $a_f = (F - m_f g)/m_f = 40.2$ m/s^2.

35. $m_1 v_1 = (m_1 + m_2)V$, $V = m_1 v_1/(m_1 + m_2) = 3.3 \times 10^4$ km/s.

37. (a) $(m_1 + m_2)V = m_1 v_1 + m_2 v_2$, $V = (m_1 v_1 + m_2 v_2)/(m_1 + m_2) = 56$ km/h. (b) KE$_{\text{initial}}$ − KE$_{\text{final}}$ = $(\frac{1}{2}m_1 v_1^2 + \frac{1}{2}m_2 v_2^2) - \frac{1}{2}(m_1 + m_2)V^2 = 37.1$ kJ.

39. If V is final velocity, $V_N = m_1 v_1/(m_1 + m_2) = 0.444$ m/s; $V_w = m_2 v_2/(m_1 + m_2) = 0.889$ m/s; $V = \sqrt{V_N^2 + V_w^2} = 0.99$ m/s. If θ is angle between V and north, $\tan \theta = V_w/V_N = 2$, $\theta = 63°$ west of north.

41. (a) If V is initial velocity of wooden block, $mv = (M + m)V$ and $V = mv/(M + m)$. From PE = KE, $(M + m)gh = \frac{1}{2}(M + m)V^2 = m^2 v^2/2(M + m)$ and $v = (1 + M/m)\sqrt{2gh}$. (b) 561 m/s.

43. $m_1 \Delta v_1 = m_2 \Delta v_2$, $\Delta v_2 = \Delta v_1(m_1/m_2) = 17$ m/s, $v_f = v_1 - \Delta v = 28$ m/s.

45. From conservation of momentum, $v_{1i} = v_{1f} \cos \theta_1 + v_{2f} \cos \theta_2$ and $0 = v_{1f}\sin \theta_1 - v_{2f}\sin \theta_2$, $v_{2f} = v_{1f}(\sin \theta_1/\sin \theta_2)$. Hence $v_{1i} = v_{1f}\cos \theta_1 + v_{1f}(\sin \theta_1/\sin \theta_2)\cos \theta_2$, $v_{1f} = v_{1i}/[\cos \theta_1 + (\sin \theta_1/\sin \theta_2)\cos \theta_2] = 3.83$ m/s and $v_{2f} = v_{1f}(\sin \theta_1/\sin \theta_2) = 3.21$ m/s. Alternatively, conservation of KE, here $v_{1i} = V_{1f}^2 + v_{2f}^2$, could have been used with either of the momentum equations.

47. $h' = e^2 h = 1.88$ m.

49. (a) $m_1 v_1 = m_1 v_1' + m_2 v_2'$, $v_2' = (m_1 v_1 - m_1 v_1')/m_2 = 6.67$ m/s in the same direction as 5-kg ball. (b) $e = (v_2' - v_1')/(v_1 - v_2) = 0.78$.

51. (a) $w_1 v_1 = w_1 v_1' + w_2 v_2'$, $v_2' = (w_1 v_1 - w_1 v_1')/w_2 = 0$. (b) $e = (v_2' - v_1')/(v_1 - v_2) = 0.4$.

53. $v_2' - v_1' = e(v_1 - v_2) = 6.3$ m/s, $v_2' = v_1' + 6.3$ m/s; $m_1 v_1 + m_2 v_2 = m_1 v_1' + m_2 v_2'$, $v_2' = [(-3 \text{ kg} \cdot \text{m/s}) - v_1 \text{ kg}]/(2 \text{ kg})$; setting equal the two formulas for v_2' and solving for v_1' gives $v_1' = -5.2$ m/s and so $v_2' = 1.1$ m/s. Both balls reverse their directions as a result of the collision.

CHAPTER 7

1. Under no circumstances.

3. The equator; the poles.

5. (a) The record's circumference is $2\pi r = 3.14$ ft, so $v = (33.3 \times 3.14 \text{ ft})/60$ s $= 1.74$ ft/s. (b) $a_c = v^2/r = 6.06$ ft/s^2.

7. Since $a = v^2/r = 4g$, $r = v^2/4g = 2296$ m.

9. $v^2/r = 4g$, $v = 2\sqrt{rg}$, $f = 1/T = v/2\pi r = g/\pi\sqrt{r} = 1.18$ rev/s.

11. $mv^2/r = T$, $v = \sqrt{rT/m} = 10$ m/s.

13. $F_c = mv^2/r$, $v = \sqrt{F_c r/m} = 20$ m/s.

15. $mv^2/r = \mu mg$, $r = v^2/\mu g = 282$ ft.

17. $mv_{\max}^2/r = \mu mg$, $v_{\max} = \sqrt{\mu gr}$. Since $f = v/2\pi r$, $f_{\max} = \sqrt{\mu gr}/2\pi r = 0.863$ rev/s $= 52$ rev/min. Hence the dime will stay where it is at $33\frac{1}{3}$ rev/min but will fly off at 78 rev/min.

19. $v = 139$ m/s; $\tan \theta = v^2/gr$, $r = v^2/(g \tan 45°) = 1971$ m $= 1.97$ km.

21. $\theta = \tan^{-1} v^2/gr = 6.75°$; $h = (8 \text{ m})(\sin 6.75°) = 0.94$ min.

23. $a_c = v^2/r$ and from Eq. (7–2), $v = 2\pi r/T$; hence $T = 2\pi\sqrt{r/a_c} = 1.55 \times 10^{-3}$ s and rpm $= (60 \text{ s/min})/T = 3.86 \times 10^4$.

25. The circumference of the circle followed by the particle is $2\pi r$, so $f = v/2\pi r$. Since $r = L \sin \theta$, $\tan \theta = \sin \theta/\cos \theta = v^2/gr = 4\pi^2 f^2 r/g$ and $\cos \theta = g/4\pi^2 f^2 L$.

27. At the bottom of the circle, since here the string must support all the ball's weight of mg as well as provide the centripetal force mv^2/r. At other positions in the circle, only part of the weight is supported by the string.

29. (a) The tension in the string at the top of the circle is equal to the centripetal force of mv^2/r minus the weight mg of the stone, so that $T = mv^2/r - mg$. If the string is to be just taut, $T = 0$ and $mv^2r = mg$, $v = \sqrt{rg} = 2.80$ m/s. (b) Since the mass of the stone does not matter here, the speed is again 2.80 m/s.

31. $mv^2/r = mg$, $v = \sqrt{rg} = 10.8$ m/s.

33. $v^2/r = g$, $v = \sqrt{rg}$; $T = 2\pi r/v = 2\pi\sqrt{r/g} = 2.22$ s.

35. At top, $mv^2/r = mg$, KE(top) $= \frac{1}{2}mv^2 = \frac{1}{2}mrg$ and PE(top) $= 2mrg$. At bottom, KE(bottom) = KE(top) + PE(top) $= \frac{5}{2}mrg = \frac{1}{2}mv'^2$, $T = mv'^2/r = 5mg$.

37. (a) $F = Gm_1 m_2/r^2 = 6.67 \times 10^{-10}$ N. (b) $F = 6.67 \times 10^{-10}$ N. (c) The acceleration of the 2-kg mass is $a = F/m = 3.34 \times 10^{-10}$ m/s^2. The acceleration of the 5-kg mass is $a = F/m = 1.33 \times 10^{-10}$ m/s^2.

39. $F = Gm_A m_B/r^2 = G(w/g)^2/r^2$, $r = (w/g)\sqrt{G/F} = 24$ ft.

41. (a) $w = (r_e/r)^2 w_e = 20$ lb. (b) $m = w/g = w_e/g = 18.4$ slugs.

43. If M = sun's mass and m = earth's mass, $GMm/r^2 = mv^2/r$, $M = v^2r/G = 2.0 \times 10^{30}$ kg.

45. The earth rotates from west to east. Hence the satellite sent eastward will have its launching speed increased because of the earth's rotation, and the one sent westward will have its speed decreased. The satellite sent eastward will therefore have the larger orbit.

47. $a = (r_e/r)^2g = 6.3$ m/s^2.

49. Since v(escape) $= \sqrt{2}v_0$, KE(escape) $=$ 2KE.

51. (a) $v = \sqrt{r_e^2 g_0/r} = 7.7 \times 10^3$ m/s. (b) $T = 2\pi r/v = 5.5 \times 10^3$ s $= 92$ min.

53. From the example in Sec. 7–7, $r = (r_e^2 g_e T^2/4\pi^2)^{1/3}$ $= (190 \times 10^{20})^{1/3} = (19 \times 10^{21})^{1/3} = (19)^{1/3} (10^{21})^{1/3}$ $= 2.67 \times 10^7$ m.

CHAPTER 8

1. 2π rad/9 $= 0.698$ rad.

3. (a) $1' = 1°/60 = 0.01745$ rad/60 $= 2.91 \times 10^{-4}$ rad. (b) With such a small angle, the chord and the arc are very nearly equal. Hence $s = \theta r = 7.27 \times 10^{-3}$ cm $= 0.073$ mm.

5. (a) $\omega = 2\pi$ rad/12 hr $= 1.45 \times 10^{-4}$ rad/s. (b) $\omega = 2\pi$ rad/60 min $= 1.75 \times 10^{-3}$ rad/s. (c) $\omega = 2\pi$ rad/60 s $= 0.105$ rad/s.

7. $\omega = \theta/t = (2\pi$ rad/turn)(49,299 turns)/20,220 s $= 15.3$ rad/s.

9. (a) $v = \omega r = 94.5$ m/s. (b) $\omega = (315$ rad/s)/[0.1047 (rad/s)/rpm] $= 3009$ rpm.

11. $v = \omega r = 226$ km/h.

13. At any instant, the barrel can be regarded as rotating about its point of contact with the ground. Hence in (a) $r = 0.8$ m, $v = \omega r = 4$ m/s; in (b) $r = 0.4$ m, $v = 2$ m/s; and in (c) $r = 0$, $v = 0$. Or, we can proceed by noting that the center of the barrel is moving at $+2$ m/s. Hence the top has this velocity plus the velocity of $+2$ m/s corresponding to its motion relative to the center for a total of 4 m/s. The bottom has the velocity of the center plus its own -2 m/s velocity relative to the center for a total of $v = 0$.

15. (a) $\omega = 126$ rad/s. (b) $r = 0.0104$ ft, $v = \omega r = 1.31$ ft/s.

17. The rectangle has the smaller moment of inertia.

19. $I = mL^2/12 = 0.016$ kg $\cdot$ m^2, KE $= \frac{1}{2}I\omega^2 = 0.80$ J.

21. (a) $I = Mk^2 = \frac{2}{3}MR^2$. (b) $k = \sqrt{I/M} = \sqrt{2/5}\,R$.

23. $I = 2mR^2/5 = 0.063$ kg $\cdot$ m^2, $\omega = v/r = 33.3$ rad/s; KE $= \frac{1}{2}I\omega^2 + \frac{1}{2}mv^2 = 35$ J $+ 87.5$ J $= 122.5$ J.

25. When the rod is released, its center of gravity falls by $h = L/2$. Hence its initial potential energy is PE $= mgh = mgL/2$. The rod's KE when it has fallen is $\frac{1}{2}I\omega^2$, where $I = mL^2/3$. Since $\frac{1}{2}I\omega^2 = mgh$, $mL^2\omega^2/6 = mgL/2$, $\omega = \sqrt{3g/L}$.

27. The solid cylinder reaches the bottom first because its moment of inertia is smaller and hence less of its initial PE becomes KE of rotation.

29. The car will coast downhill faster with light tires because their moments of inertia are smaller.

31. (a) $I = mr^2$, $v = \omega r$, $h = L \sin \theta = 3.42$ m; $mgh = \frac{1}{2}I\omega^2 + \frac{1}{2}mv^2 = mr^2\omega^2$, $\omega = \sqrt{gh}/r = 11.58$ rad/s. (b) $v = \omega r = 5.8$ m/s. (c) KE(rot) $= \frac{1}{2}I\omega^2 = 50.3$ J. (d) KE $= \frac{1}{2}I\omega^2 + \frac{1}{2}mv^2 = 101$ J.

33. $\alpha = a_T/r = 0.5$ rad/s^2.

35. It is easiest to use the revolution as the angular unit here. Since $\omega_f^2 = \omega_0^2 + 2\alpha\theta$, $\theta = (\omega_f^2 - \omega_0^2)/2\alpha = 60$ revolutions.

37. It is easiest to use the revolution as the angular unit here. Since $\omega_f^2 = \omega_0^2 + 2\alpha\theta$, $\alpha = -\omega_0^2/2\theta$, and $t = -\omega_0/\alpha = 2\theta/\omega_0 = 16.7$ s.

39. The much greater I of the cue when balanced on its tip means a smaller angular acceleration when it is out of line, hence more time to rebalance it by shifting the supporting hand.

41. $I = \frac{1}{2}mr^2 = 9$ kg $\cdot$ m^2. (a) Since $\alpha = \tau/I$, $t = \omega/\alpha = \omega I/\tau = 4.5$ s. (b) KE $= \frac{1}{2}I\omega^2 = 450$ J.

43. $\alpha = (\omega_f - \omega_0)/t = 5$ rad/s^2, $I = \tau/\alpha = 100$ kg $\cdot$ m^2.

45. $\theta = \frac{1}{2}\alpha t^2$, $\alpha = 2\theta/t^2$. If L = width of door, $\tau = FL$ and $I = \tau/\alpha = FLt^2/2\theta$. Since $I = ML^2/3$, $M = 3I/L^2 = 3Ft^2/2L\theta$. Here $\theta = 90° = \pi/2$, so $M = 537$ kg.

47. $\tau = P/\omega; = 8$ N $\cdot$ m.

49. $P = 550$ ft $\cdot$ lb/s, $\omega = 126$ rad/s; $P = \tau\omega$, $\tau = P/\omega = 4.37$ lb $\cdot$ ft.

51. $\omega = 210$ rad/s; $\tau = P/\omega = 23.8$ N $\cdot$ m; $F = \tau/r = 238$ N. Each tooth exerts half this force, or 119 N.

53. The net force on the belt is $F = 180$ N $- 70$ N $= 110$ N and $\omega = (30$ rev/s$)(2\pi$ rad/rev$) = 188$ rad/s. The torque exerted by the motor is $\tau = Fr$ so $P = \tau\omega = Fr\omega = 2.07$ kW.

55. The length of the day will increase; the earth's moment of inertia will be greater when the water from the icecaps becomes uniformly distributed and hence, according to conservation of angular momentum, the angular velocity must decrease.

57. Force, MLT^{-2}; torque, ML^2T^{-2}; energy, ML^2T^{-2}; power, ML^2T^{-3}; linear momentum, MLT^{-1}; angular momentum, ML^2T^{-1}; impulse, MLT^{-1}.

59. (a) $I_1 = mR_1^2 = 0.2$ kg $\cdot$ m^2, $I_2 = mR_2^2 = 0.288$ kg $\cdot$ m^2, $\omega_1 = 31.4$ rad/s, $\omega_2 = I_1\omega_1/I_2 = 21.8$ rad/s $= 3.5$ rev/s. (b) KE$_1 = \frac{1}{2} \times \frac{1}{2}I_1\omega_1^2 = 49.3$ J, KE$_2 = \frac{1}{2} \times \frac{1}{2}I_2\omega_2^2$ $= 34.2$ J; the decrease in KE is due to the work that had to be done to stretch the string.

61. (a) $I_2 = 2I_1$, $\omega_2 = I_1\omega_1/I_2 = \omega_1/2 = 50$ rad/s. (b) KE$_1 = \frac{1}{2}I_1\omega_1^2 = 5 \times 10^3$ ft $\cdot$ lb, KE$_2 = \frac{1}{2}I_2\omega_2^2 = 2.5 \times 10^3$ ft $\cdot$ lb represents energy dissipated as heat when the moving disk was pressed against the stationary one before the two stuck together. (c) 50 rad/s. (d) No KE is lost.

CHAPTER 9

1. MA greater than 1; MA less than 1.

3. Eff $= $ (Eff$_1$)(Eff$_2$)(Eff$_3$) $= 0.26 = 26\%$.

5. $F_{\text{out}} = F_{\text{in}}L_{\text{in}}/L_{\text{out}} = 150$ N.

7. $F_{\text{out}} = w/2$; hence AMA $=$ IMA $= F_{\text{out}}/F_{\text{in}} = 4 = L_{\text{in}}/L_{\text{out}}$, $L_{\text{in}} = 4L_{\text{out}}$; $L_{\text{in}} + L_{\text{out}} = 5L_{\text{out}} = 6$ ft, $L_{\text{out}} = 1.2$ ft.

9. (a) Here $F_{\text{in}} = (6)(300$ N$) = 1800$ N. Hence $F_{\text{out}} = $ (Eff)$(F_{\text{in}}R/r) = 6120$ N. (b) $R/r = 4$, so each man had to walk 400 m.

11. With one fixed pulley and two movable pulleys, IMA $= 4$.

13. IMA $= 5$, $F_{\text{out}} = $ (IMA)$(F_{\text{in}}) = 4500$ N.

15. (a) $F_{\text{in}} = F_{\text{out}}/(\text{Eff})(\text{IMA}) = F_{\text{out}}(R - r)/(\text{Eff})(2R)$ $= 115$ lb. (b) $s_{\text{in}} = (\text{IMA})(s_{\text{out}}) = 2Rs_{\text{out}}/(R - r) = 30$ ft.

17. (a) The output RPM does not depend upon the efficiency, so RPM$_{\text{out}}$/RPM$_{\text{in}} = N_{\text{in}}/N_{\text{out}}$, RPM$_{\text{out}} = (N_{\text{in}}/N_{\text{out}})RPM_{\text{in}} = 2$ rev/min. (b) $\tau_{\text{out}} = $ (Eff)$(N_{\text{out}}/N_{\text{in}})(\tau_{\text{in}})$ $= 96$ N $\cdot$ m.

19. (a) $P = 33{,}550$ ft $\cdot$ lb/s, $\omega = 315$ rad/s, $\tau = P/\omega = 106.5$ lb $\cdot$ ft; hence $N_{\text{out}}/N_{\text{in}} = \tau_{\text{out}}/\tau_{\text{in}} = 3.29$. (b) $\omega_{\text{out}} = \omega_{\text{in}}N_{\text{in}}/N_{\text{out}} = 913$ rpm.

21. $\omega_{\text{saw}} = 3\omega_{\text{motor}} = 550$ rad/s; $v_{\text{teeth}} = \omega_{\text{saw}}r_{\text{saw}} = 115$ ft/s; $F = P/v = 2.4$ lb.

23. $F_{\text{in}} = F_{\text{out}}/\text{IMA} = mg/(L/h) = 179$ N.

25. AMA $=$ IMA $= (\text{IMA}_1)(\text{IMA}_2) = (2)(L/h) = mg/F_{\text{in}}$, $F_{\text{in}} = mgh/2L = 294$ N.

27. $F_{\text{out}} = $ (Eff)$(2\pi L/p)(F_{\text{in}}) = 31$ N.

29. $F_{\text{out}} = $ (Eff)$(F_{\text{in}})(\text{IMA}) = $ (Eff) $\times (2\pi LF_{\text{in}}/p)$, Eff $= pF_{\text{out}}/2\pi LF_{\text{in}} = 0.325 = 32.5\%$.

31. (a) $F_{\text{out}} = $ (Eff)$(F_{\text{in}})(d_{\text{out}}/d_{\text{in}})(2\pi L/p) = 2.83 \times 10^4$ N; $m = F_{\text{out}}/g = 2.9 \times 10^3$ kg. (b) $\theta_{\text{in}} = (d_{\text{out}}/d_{\text{in}})(h/p) = 375$ turns.

CHAPTER 10

1. $d = m/V = 3000$ kg/m^3, which is greater than the density of pure concrete. Hence the slab contains reinforcing rods.

3. (a) $d_{\text{water}} = 1$ kg/L, so $V = m/d = 55$ L. (b) $V = m/d = 140$ m^3.

5. $m = dV = d(\frac{4}{3}\pi r^3) = 6.03$ kg.

7. (a) $d = m/V = m/(4\pi r^3/3) = 5.52 \times 10^3$ kg/m^3. (b) The interior must consist of denser materials than those at the surface; no.

9. The weights of the liquid columns above the bottom of the oil column must be equal. Since the tube has the same cross-sectional area on both sides, $L_{\text{water}}d_{\text{water}} = L_{\text{oil}}d_{\text{oil}}$, $d_{\text{oil}} = L_{\text{water}}d_{\text{water}}/L_{\text{oil}} = 0.77$ g/cm^3.

11. $V/t = 60$ L/s $= 0.06$ m^3/s; $P = mgh/(t)(\text{Eff}) = dVgh/(t)(\text{Eff}) = 12.6$ kW.

13. The second procedure is more likely to break the string because here the tension in the string is twice as great with the reaction force exerted by the tree being equal and opposite to the pull of the two boys.

15. The perforations should have pointed ends in the direction of tearing.

17. $s = F/k = mg/k = 0.0784$ m.

19. All are under the same stress since the load is equally divided and the cross-sectional area is the same for all the legs. The strain varies as $1/Y$; hence the aluminum leg experiences the greatest strain and the steel leg the least strain.

21. $\Delta L = L_0F/YA$, $A = \pi d^2/4$, hence $\Delta L_{\text{B}} = (L_{0\text{B}}/L_{0\text{A}})(d_{\text{A}}/d_{\text{B}})^2\Delta L_{\text{A}} = 2\Delta L_{\text{A}}$.

23. Each cylinder has the same cross-sectional area A and volume V. The force corresponding to an ultimate strength of U is $F = UA$ and the mass of a cylinder is $m = dV$. Hence $F/m = UA/dV = U/dL$ where L is the cylinder length. In tension, $(F/m)/L = 7.5 \times 10^4$, 7.4×10^4, and 6.4×10^4 N $\cdot$ m/kg for bone, aluminum, and steel; in compression the figures are the same for aluminum and steel and 10.6×10^4 for bone. Bone is clearly an excellent structural material.

25. The stress due to the acrobat's weight is $F/A = mg/A = 1.47 \times 10^7$ N/m^2; hence the safety factor is $(1.7 \times 10^8)/(1.47 \times 10^7) = 11.6$.

27. $F/A = mg\pi r^2 = 1.9 \times 10^8$ N/m^2, which is just a hair less than the ultimate strength of aluminum.

29. The breaking strength F is proportional to the cross-sectional area $A = \pi d^2/4$ of the rope. Hence (a) $F = (25 \text{ kN})(6 \text{ mm}/10 \text{ mm})^2 = 9 \text{ kN}$ and (b) $F = (25 \text{ kN})(14 \text{ mm}/10 \text{ mm})^2 = 49 \text{ kN}$.

31. $A = 10^{-6} \text{ m}^2$. $\Delta L = L_0 F/YA = L_0 mg/YA = 2.94 \times 10^{-4} \text{ m}$.

33. $\Delta L = L_0 F/YA = 5.8 \times 10^{-4} \text{ ft} = 0.0070 \text{ in}$.

35. $\Delta L = L_0 F/YA = L_0 mg/YA = 2.0 \times 10^{-5} \text{ m}$.

37. $Y = (F/A)/(L_0/\Delta L) = 9.2 \times 10^{10} \text{ N/m}^2$.

39. $\Delta L = L_0 F/YA = L_0 F/Y\pi(r_{\text{outside}}^2 - r_{\text{inside}}^2) = 1.44 \times 10^{-4} \text{ m} = 0.144 \text{ mm}$.

41. The volume of lead above the lowest brick is $V = 0.9 \text{ m} \times (0.1 \text{ m})^2 = 0.009 \text{ m}^3$ and the weight of this lead is $w = Vdg = 970 \text{ N}$. Since $F = w$, $L_0 = 0.1 \text{ m}$, and $A = 0.01 \text{ m}^2$, the final height of the lowest brick is shorter by $\Delta L = L_0 F/YA = 6.06 \times 10^{-7} \text{ m}$.

43. $F_{\text{out}} = (F_{\text{in}})(2\pi L/p) = 2011 \text{ lb}$; $L = L_0 F_{\text{out}}/YA = 6.7 \times 10^{-4} \text{ in}$.

45. If the rod has the cross-sectional area A and length L, its weight is $w = mg = dgAL$ and the tensile stress it exerts on its fixed end is $w/A = dgL$. If this equals the tensile strength U, then $U = dgL$ and $L = U/dg$. For aluminum, $L = 5290 \text{ m}$, and for steel, $L = 6541 \text{ m}$.

47. $S = Fd/sA = 3333 \text{ N/m}^2$.

49. $L = 0.01 \text{ m}$, $A = 4LH$, $(F/A)_{\text{max}} = F/4Lh$, $h = F/4L(F/A)_{\text{max}} = 7.1 \times 10^{-3} \text{ m} = 7.1 \text{ mm}$.

51. $\theta = 0.436 \text{ rad}$, $R = 5 \times 10^{-4} \text{ m}$, $\tau = Fs = 3.0 \times 10^{-3} \text{ N} \cdot \text{m}$; $S = 2\tau L/\pi\theta R^4 = 4.91 \times 10^{10} \text{ N/m}^2$.

53. Experimental error; only two significant figures are quoted in the table; the theory corresponds to an ideal elastic body, and real ones may not correspond exactly to such a model.

55. $\Delta V = pV_0/B = 4 \times 10^{-8} \text{ m}^3 = 40 \text{ mm}^3$.

CHAPTER 11

1. As the steam inside the can condenses, the internal pressure falls below the external pressure of the atmosphere.

3. Atmospheric pressure.

5. 958 mb; 96.0 kPa.

7. $F = mg = 19.6 \text{ N}$, $A_1 = (0.075 \times 0.15) \text{ m}^2$, $A_2 = (0.075 \times 0.3) \text{ m}^2$, $A_3 = (0.15 \times 0.3) \text{ m}^2$. Hence $p_1 = F/A_1 = 1742 \text{ Pa}$, $p_2 = 871 \text{ Pa}$, $p_3 = 436 \text{ Pa}$.

9. $\Delta p = F/A = 90.9 \text{ lb/ft}^2$.

11. $p = F/A = F/\pi r^2 = 6.37 \text{ MPa} = 63 \text{ atm}$.

13. (a) The maximum suction equals atmospheric pressure, $14.7 \text{ lb/in}.^2$. (b) $F = pA = p\pi d^2/4 = 289 \text{ lb}$.

15. $P = 15 \text{ torr} = 2 \times 10^3 \text{ Pa}$, F/A, $F = pA = p(\pi r^2) = 0.16 \text{ N}$. The needle's diameter is irrelevant.

17. $F_{\text{out}} = mg$, $m = F_{\text{out}}/g = (F_{\text{in}})(\text{MA}_{\text{lever}})(d_{\text{out}}^2/d_{\text{in}}^2)/g = 4408 \text{ kg}$.

19. The forces are the same because the height of water is the same.

21. $p = p_{\text{atm}} + dgh = 1.10 \times 10^8 \text{ Pa}$.

23. $\Delta p = 0.1 \text{ atm} = 1.013 \times 10^4 \text{ Pa}$, $h = \Delta p/dg = 1.03 \text{ m}$.

25. $F = pA = DhA = 2.56 \times 10^4 \text{ lb}$.

27. Here $h = 0.7 \text{ m}$ so $\Delta p = dgh = 7.27 \text{ kPa} = 55 \text{ torr}$.

29. (a) $\Delta p_1 = dgh_1 = 4.16 \text{ kPa} = 31 \text{ torr}$, so $p_1 = (100 - 31) \text{ torr} = 69 \text{ torr}$. (b) $\Delta p_2 = dgh_2 = 12.5 \text{ kPa} = 94 \text{ torr}$, so $p_2 = (100 + 94 \text{ torr}) = 194 \text{ torr}$.

31. $\Delta p = dgh$, $\Delta d/d_0 = \Delta V/V_0 = \Delta p/B = dgh/B = 0.0219$; the density at 5 km depth is therefore 1.0219 times the density at the surface, or 1053 kg/m^3.

33. The water level is unchanged.

35. The heavier one, because the buoyant force is the same for both.

37. The weight of the ship does not change; hence the volume of water it displaces remains the same. Because the volume of the ship is greater, its freeboard increases.

39. The density of the atmosphere decreases with altitude, so the buoyant force on the balloon decreases as it rises. When the buoyant force equals the balloon's weight, it stops rising and floats at that altitude.

41. The volume of water whose mass is 60 kg is $60 \text{ kg}/(10^3 \text{ kg/m}^3) = 0.06 \text{ m}^3$. By Archimedes' principle the raft rises by a height h such that its submerged volume decreases by 0.06 m^3 when the woman dives off. Hence $h = (0.06 \text{ m}^3)/(3\text{m})(2 \text{ m}) = 0.01 \text{ m} = 1 \text{ cm}$.

43. $V = m/d = w/gd = 2.58 \text{ ft}^3$; $F = Vdg = 0.21 \text{ lb}$.

45. $F = Vg(d_{\text{air}} - d_{\text{hydrogen}}) - mg = 892 \text{ N}$.

47. $Vd_{\text{air}}g = (m + Vd_{\text{hydrogen}})g$, $m = V(d_{\text{air}} - d_{\text{hydrogen}}) = 2.42 \times 10^5 \text{ kg}$.

49. First we find the volume of the bracelet. The buoyant force on it in water is $F_b = (\Delta m)g = (Vd_w)g$ so $V = F_b/d_w g = \Delta m/d_w = 4 \text{ cm}^3$. Hence the bracelet's density is $d_b = m \text{ (in air)}/V = 12.5 \text{ g/cm}^3$. Since this is less than d_{gold}, the bracelet is not pure gold.

51. Mass of displaced water = 100 g, so volume of statue = 100 cm^3. Mass of displaced benzene = 88 g, so $d = m/V = 0.88 \text{ g/cm}^3 = 880 \text{ kg/m}^3$.

53. The volume of lead is $V = m/d_{lead}$. The downward force exerted by the lead is $F = mg - F_{buoyant} = mg - Vd_{water}g = mg(1 - d_{water}/d_{lead}) = 71.06$ kN. The lever arm of this force equals the horizontal displacement x of the keel's CG from being vertically below the boat, so $x = L \sin \theta = 0.513$ m and $\tau = Fx = 36.5$ kN · m.

55. $v_1 A_1 = v_2 A_2$, $A_2 = v_1 A_1/v_2$, $D_2 = \sqrt{v_1 D_1^2/v_2} = 0.447$ cm.

57. The power output of the left ventricle is $P_L = pR = 6.2$ W. Since $P_R = 0.2 P_L = 1.2$ W, the total power output is $P_L + P_R = 7.4$ W.

59. (a) $R = 10$ gal/min $= 0.0223$ ft^3/s, $A = \pi r^2 = 0.000491$ ft^2, $v = R/A = 45$ ft/s. (b) $Ft = mv$, $F = mv/t = d(V/t)v = dRv = 1.95$ lb. (c) $P_{out} = \frac{1}{2}mv^2/t = \frac{1}{2}v^2 (m/t) = \frac{1}{2}v^2 (Vd/t) = \frac{1}{2}dv^2 R = 43.8$ W; $P_{in} = P_{out}/$EFF $= 73$ W.

61. Since the heights of liquid above both openings are the same, the rates of flow will be the same.

63. $p = 120$ torr $= 1.60 \times 10^4$ Pa. This is gauge pressure, so $h = p/dg = 1.55$ m.

65. $v = \sqrt{2gh}$, $A = 2 \times 10^{-3}$ m^2; $R = vA = \sqrt{2gh}\,A = 7.14 \times 10^{-3}$ m^3/s $= 7.14$ L/s.

67. (a) $R = A_1 \sqrt{2gh/[(A_1/A_2)^2 - 1]} = 2.97 \times 10^{-4}$ m^3/s $= 0.297$ L/s. (b) 0.297 L/s.

69. (a) $h = \frac{1}{2}gt^2$, $t = \sqrt{2h/g} = 0.45$ s, $v = x/t = 4.43$ m/s. (b) $p = \frac{1}{2}dv^2 = 9.8$ kPa.

71. (a) $v_2 = v_1 A_1/A_2 = v_1 r_1^2/r_2^2 = 1.28$ m/s. (b) Here $h_1 = h_2$ so $p_2 = p_1 + \frac{1}{2}d(v_1^2 - v_2^2) = 2.31 \times 10^5$ N/m$^2 = 2.31$ bars.

73. (a) $R = v_1 A_1 = v_1 \pi r_1^2$, $v_1 = R/\pi r_1^2 = 0.884$ m/s. (b) $v_2 = v_1 A_1/A_2 = v_1(r_1/r_2)^2 = 3.54$ m/s. (c) From Bernoulli's equation, $p_2 = p_1 + \frac{1}{2}d(v_1^2 - v_2^2) + dg(h_1 - h_2) = 54.5$ kPa.

75. No. For example, water at 20° has a greater density than ethyl alcohol but a smaller viscosity.

77. $R = 16$ L/min $- 2.67 \times 10^{-4}$ m^2/s, $p/L = 8\eta R/\pi r^4 = 525$ Pa/m.

79. (a) $\Delta p = 2.66 \times 10^3$ Pa, $R = vA = v(\pi r^2)$, $v = r^2\Delta p/8\eta L = 0.665$ mm/s. (b) In each capillary, $R = vA = 8.36 \times 10^{-15}$ m^3/s $= 8.36 \times 10^{-9}$ cm^3/s, so $N = (80$ cm^3/s$)/R = 9.57 \times 10^9$ capillaries.

81. $v_a = (d_w/d_a)(\eta_a/\eta_w)V_w = 6.9$ m/s.

CHAPTER 12

1. $W = $ PE $= \frac{1}{2}ks^2$, $k = 2W/s^2 = 1$ kN/m.

3. $k = F/s = 20$ N/m; $\frac{1}{2}mv^2 = \frac{1}{2}ks^2$, $v = \sqrt{k/m}\,s = 1.4$ m/s.

5. (a) $s = 3$ in. $= 0.25$ ft; $k = F/s = 20$ lb/ft. (b) PE $= \frac{1}{2}ks^2 = 0.625$ ft · lb.

7. (a) $F = k_1 s_1 = k_2 s_2$ so $s_1 = F/k_1$, $s_2 = F/k_2$; $s = s_1 + s_2 = F/k_1 + F/k_2 = F(k_1 + k_2)/k_1 k_2$; $k = F/s = k_1 k_2/(k_1 + k_2)$. (b) 7.5 N/m.

9. No, but only if it obeys Hooke's law will the oscillations be simple harmonic in character.

11. When the object has vertical sides, so the restoring force is proportional to the displacement of the object above or below its equilibrium level.

13. $k = F/s = mg/s = 163$ N/m. (a) $T = 2\pi\sqrt{m/k} = 0.49$ s. (b) $f = 1/T = 2.0$ Hz.

15. $T = 2\pi\sqrt{m/k}$, $k = 4\pi^2 m/T^2 = 9.87$ kN/m.

17. $T = 2\pi\sqrt{m/k} = 2\pi\sqrt{w/kg}$, $k = 4\pi^2 w/T^2 g = 80.6$ N/m, $s = F/k = 0.621$ m.

19. Let $s = $ maximum sag. Then PE(gravitational) $=$ PE(elastic), $mg(h + s) = \frac{1}{2}ks^2$, $k = 2mg(h + s)/s^2 = 8575$ N/m; $T = 2\pi\sqrt{m/k} = 0.57$ s.

21. The requirement for simple harmonic motion to occur is that the acceleration of the object be proportional to its displacement from the equilibrium position and in the opposite direction. Here $a = (g/R)r$ and a is opposite to r, which fulfills this condition. From Eq. (12–3) the period is $T = 2\pi\sqrt{r/a} = 2\pi\sqrt{R/g} = 5.08 \times 10^3$ s $= 1$ hr 25 min.

23. The maximum KE and maximum PE of a harmonic oscillator are always equal.

25. $f = 1/T$, $v_{max} = 2\pi fA = 2\pi A/T = 314$ cm/s $= 3.14$ m/s.

27. $a_{max} = 4\pi^2 f^2 A$, $f = \sqrt{a_{max}/A}/2\pi = 11$ Hz.

29. (a) $v_{max} = 2\pi fA = 4.15$ m/s. (b) $a_{max} = 4\pi^2 f^2 A = 1.72 \times 10^4$ m/s$^2 = 1754g$.

31. $1/f = 2\pi\sqrt{(w/g)/k}$, $k = 4\pi^2 wf^2/g = 178$ lb/ft. (a) $E = \frac{1}{2}kA^2 = 22$ ft · lb. (b) $a_{max} = 4\pi^2 f^2 A = 178$ ft/s^2.

33. (a) $T = 2\pi\sqrt{m/k}$, $k = 4\pi^2 m/T^2 = 39.5$ N/m. (b) $v_{max} = 2\pi fA = 0.314$ m/s. (c) $a_{max} = 4\pi^2 f^2 A = 1.97$ m/s^2.

35. $f = 1/T = 5$ Hz, $v = 2\pi f\sqrt{A^2 - s^2} = 0.608$ m/s.

37. PE $= ks^2/2$; hence KE $= 3$PE $= 3ks^2/2$ and PE $+$ KE $= 2ks^2$. Since PE $+$ KE $= kA^2/2$, $s^2 = A^2/4$ and $s = A/2$.

39. (a) Since the pendulum's support is falling with the same acceleration as the pendulum bob, no oscillations occur and $f = 0$. (b) $1/T$. (c) $1/T$.

41. $T = 2\pi\sqrt{L/g} = 4.97$ s.

43. $g_M = 3.72$ m/s^2; $T = 2\pi\sqrt{L/g_M}$, $L = T^2 g_M/4\pi^2 = 0.094$ m $= 94$ mm.

45. (a) $T = 2\pi\sqrt{L/g} = 1.42$ s. (b) The motion of the pendulum bob is unaffected by motion of its support at constant velocity, hence $T = 1.42$ s. (c) The downward force on the bob here is $m(g - a)$, hence $T = 2\pi\sqrt{L/(g - a)} = 1.59$ s. (d) As in part (b), $T = 1.42$ s. (e) The downward force on the bob here is $m(g + a)$, hence $T = 2\pi\sqrt{L/(g + a)} = 1.29$ s.

47. $I = \frac{2}{5}mr^2$, $T = 2\pi\sqrt{I/K} = 2\pi r\sqrt{2m/5K} = 0.428$ s.

49. $T = 2\pi\sqrt{I/K}$, $K = 4\pi^2 I/T^2 = 95$ N $\cdot$ m/rad.

51. $K = \tau/\theta = 8.6$ lb $\cdot$ ft/rad; $T = 2\pi\sqrt{I/K}$, $I = KT^2/4\pi^2 = 5.45$ slug $\cdot$ ft^2.

53. The center of percussion should be in the axe's head so that there will be no reaction force on the hands of the user.

55. (a) $I = mL^2/3$; $h = L/2$; $T = 2\pi\sqrt{I/mgh} = 2\pi\sqrt{2L/3g} = 1.47$ s. (b) $L' = I/mh = 2L/3 = 53.3$ cm.

57. (a) $I = \frac{1}{3}mL^2$, $h = L/2$, $T = 2\pi\sqrt{I/mgh} = 2\pi\sqrt{2L/3g} = 1.6$ s. (b) Each step represents only half of a complete cycle, so such a leg swinging freely takes 0.8 s per stride. The rate of walking is therefore 1.25 strides/s and $v = (1.25)(0.8)$ m/s $= 1$ m/s $= 3.6$ km/h. To walk faster or slower than this would involve more effort.

CHAPTER 13

1. The energy becomes dissipated as heat, and the string becomes warmer as a result.

3. It increases fourfold, since the rate of flow of energy is proportional to A^2.

5. The frequency is unchanged but the wavelength is doubled.

7. (a) $f = v/\lambda = 0.857$ Hz. (b) $T = 1/f = 1.17$ s.

9. $\lambda = c/f = 3947$ km $= 2453$ mi.

11. (a) $v = \lambda/T = 28$ ft/s. (b) $v' = 2\pi r/T = 4.4$ ft/s.

13. (a) Wavelength of fundamental frequency $= \lambda = 2L = 2m$, hence $v = \lambda f = 600$ m/s. (b) $f_2 = 2f_1 = 600$ Hz; $f_3 = 3f_1 = 900$ Hz; $f_4 = 4f_1 = 1200$ Hz.

15. From Eq. (13–8), $(m/L)_B = (m/L)_A f_A{}^2/f_B{}^2 = 0.6$ g/m.

17. (a) From Eq. (13–8), $f_A/f_B = L_B/L_A$, so with $L_B = 270$ mm, $f_B = f_A L_A/L_B = 805$ Hz. $f^2 = 2f_1 = 1610$ Hz; $f_3 = 3f_1 = 2415$ Hz.

19. Sound travels fastest in solids whose constituent particles are more tightly coupled together than those of gases or liquids.

21. (a) $\lambda = v/f = 0.141$ ft. (b) 0.628 ft. (c) 2.05 ft.

23. Air: $t_1 = s/v_{air} = 1.458$ s; rail, $t_2 = s/v_{steel} = 0.100$ s; $t_1 - t_2 = 1.358$ s.

25. $\lambda = v/f = 0.29$ m $= 29$ cm, which is slightly greater than the distance between the ears.

27. $t = h/v_s$, $s = vt = vh/v_s = 1.6$ km.

29. (a) $f = 440$ Hz, $\lambda = v_{water}/f = 3.40$ m. (b) $f = 440$ Hz, $\lambda = v_{air}/f = 0.780$ m.

31. Since $v_s/v = \sin\theta$, the Mach number is $v/v_s = 1/\sin\theta = 1.15$.

33. The energy absent from locations of destructive interference is added to the wave energy at locations of constructive interference to give a greater amplitude there than if no interference has occurred. The total energy remains the same as it would be without interference, but its distribution is different.

35. The difference tone is $(396 - 264)$ Hz $= 132$ Hz, which is one octave below 264 Hz (that is, half of 264 Hz) and so is consonant.

37. (a) $f_L = f_s v/(v - v_s) = 518$ Hz. (b) $f_L = f_s v/(v + v_s) = 483$ Hz.

39. (a) $f_L = f_s(v + v_L)/(v - v_s) = (400)(343 + 16.7)/(343)$ Hz $= 419$ Hz. (b) $f_L = (400)(343 + 16.7)/(343 + 4.2)$ Hz $= 414$ Hz.

41. From Eq. (13–10) $v = c(f^2 - f_s{}^2)/(f^2 + f_s{}^2)$. Here $f = c/\lambda = 5.56 \times 10^{14}$ Hz, $f_s = c/\lambda_s = 4.84 \times 10^{14}$ Hz, hence $v = 0.138c = 4.14 \times 10^7$ m/s $= 1.49 \times 10^8$ km/h. The fine is therefore $149 million less $80.

43. Here $v_s = 0$, so $f_L = f_s(v + v_L)/v = f_s + f_s v_L/v$ and $v_L = v(f_L - f_s)/f_s = 1.34$ m/s.

45. The frequency of the waves as received by the reflector is $f_R = f(v + u)/v$. The frequency of the waves as received by the listener is $f_L = f_R v/(v - u) = f[(v + u)/v][v/(v - u)] = f(v + u)/(v - u)$. The number of beats heard by the listener is therefore $f_L - f = f[(v + u)/(v - u) - 1] = f(v + u - v + u)/(v - u) = 2fu/(v - u)$.

47. 1 atm $= 1.013 \times 10^5$ N/m^2, hence the amplitude here is 1.97×10^{-10} atm, which is $1.97 \times 10^{-8}\%$ of sea level pressure.

49. $\beta = 10 \log I/I_0 = 10 \log[(5 \times 10^6)/10^{-12}] = 10 \log (5 \times 10^6) = 67$ dB.

51. $I = I_0 \log^{-1}(\beta/10) = 10^{-12}$ W/m^2 [$\log^{-1}(120/10)$] $= 1$ W/m^2.

53. $P_{out}/P_{in} = \log^{-1}[G/10] = \log^{-1}(40/10) = \log^{-1} 4 = 10^4 = 10{,}000$.

55. Since $\log x - \log y = \log x/y$, $\beta_2 - \beta_1 = 5$ dB $= 10[\log (I_2/I_0) - \log (I_1/I_0)] = 10 \log (I_2/I_1)$, $\log (I_2/I_1) = 0.5$ and $I_2/I_1 = \log^{-1} 0.5 = 3.16$.

57. β_{20} = 70 dB = 10 log $(20I_1/I_0)$ = 10 log 20 + 10 log (I_1/I_0) = 13 dB + β_1, β_1 = (70 − 13) dB = 57 dB.

CHAPTER 14

1. $T_F = \frac{9}{5}T_C$ + 32°, hence 330°C = 626°F and 1170°C = 2138°F.

3. If $T = T_F = T_C$, $\frac{9}{5}T$ + 32° = $\frac{5}{9}(T - 32°)$, $T = -40°$.

5. $T_C = \frac{5}{9}(T_F - 32°) = -80°C$.

7. Glass expands less than copper with a rise in temperature, so the bulb would crack as it heats up. Lead-in wires must have the same coefficient of thermal expansion as glass if they are to be used successfully.

9. $\Delta L = aL_0\Delta T$ = 0.048 m = 48 mm.

11. $a = \Delta L/L_0\Delta T = 8.06 \times 10^{-6}/°C$.

13. $\Delta L = aL_0\Delta T$ = 0.0563 ft. The true width is 119.944 ft; to the nearest inch, 119 ft 11 in.

15. Glycerin: $\Delta V = bV_0\Delta T$ = 1.275 cm³; beaker: $\Delta v = bV_0\Delta T$ = 0.023 cm³; thus (1.275 − 0.023) cm³ = 1.252 cm³ overflows.

17. (a) $\Delta L = aL_0\Delta T$ = 0.0312 ft. (b) A = 7.2 in.², $F = YA\Delta L/L_0 = 3.74 \times 10^4$ lb.

19. Alcohol: V_0 = 0.5000 L, $\Delta V_1 = bV_0\Delta T$ = 0.0094 L; water: V_0 = 0.5000 L, $V_2 = bV_0\Delta T$ = 0.0026 L; $\Delta V = \Delta V_1 + \Delta V_2$ = 0.012 L so profit is ($10)($\Delta V$) = $0.12.

21. (a) For concrete $b_c = 3a_c$ is smaller than b_{water} and hence the water level rises as the temperature increases. (b) The water begins to overflow when its volume equals the pool's volume. If V_w is the water volume and V_p is the pool's total volume at 10°C, then $V_w + b_wV_w\Delta T = V_p + b_cV_p\Delta T$, $\Delta T = (V_p - V_w)/(b_wV_w - b_cV_p)$ = 13°C. The overflow temperature is 10°C + 13°C = 23°C.

23. The thermal stress is $F/A = Y\Delta L/L_0$, $\Delta L = aL_0\Delta T$, $F/A = Ya\Delta T = 9.6 \times 10^7$ N/m². This is 19% of the ultimate strength of steel.

25. $V_2 = p_1V_1/p_2$ = 2.5 m³.

27. At atmospheric pressure, using absolute values for p_1 and p_2, the volume of oxygen is $V_2 = p_1V_1/p_2$ = 218 ft³. Hence its weight is $w = DV$ = 19.6 lb.

29. The volume of air at 1 bar absolute pressure needed to fill the tire to an absolute pressure of 3 bars is $v_1 = p_2V_2/p_1$ = 15 L. Since 5 L of air are already in the tire, 10 L more are needed, hence 20 strokes.

31. Here $V_2 = 2V_1$ and $p_2 = p_a$ = atmospheric pressure. Hence $p_1V_1 = p_2V_2$, $p_1 = 2p_a$. The pressure due to the

water of depth h is $p_1 - p_2 = p_a$, so $p_a = dgh$, $h = p_a/dg$ = 10.3 m.

33. Because of the volume occupied by its molecules.

35. T_1 = 293 K, $T_2 = T_1V_2/V_1$ = 586 K = 313°C.

37. p_1 = (24 + 14.7)lb/in.² = 38.7 lb/in.², T_1 = 500°R, T_2 = 540°R; $p_2 = p_1T_2/T_1$ = 41.8 lb/in.² The corresponding gauge pressure is (41.8 − 14.7)lb/in.² = 27.1 lb/in.²

39. (a) $p_2 = p_1V_1/V_2 = 4 \times 10^5$ Pa. (b) $V_2 = p_1V_1/p_2$ = 2.67 m³. (c) $V_2 = V_1T_2/T_1$ = 2.67 m³.

41. T_1 = 293 K, $T_2 = (p_2/p_1)(V_2/V_1)T_1$ = 806 K = 533°C.

43. $p_1 = p_{atm} + dgh = 1.993 \times 10^5$ Pa, $p_2 = p_{atm}$ = 1.013×10^5 Pa, T_1 = 278 K, T_2 = 293 K; V_2/V_1 = p_1T_2/p_2T_1 = 2.07 = $D_2{}^3/D_1{}^3$, hence $D_2 = \sqrt[3]{2.07}\, D_1$ = 1.27 cm.

45. Homogeneous: carbon dioxide gas, solid carbon dioxide, iron, rust, air, oxygen, salt. Heterogeneous: blood, rock, steak, concrete, milk.

47. 6 × 12.0 u + 12 × 1.008 u + 6 × 16.00 u = 180.16 u; m ($C_6H_{12}O_6$) = (180.16 u)(1.66 × 10^{-27} kg/u) = 2.99 × 10^{-25} kg.

49. The mass of a lead atom is (207.19 u)(1.66 × 10^{-27} kg) = 3.44 × 10^{-25} kg. Hence 0.05 kg/(3.44 × 10^{-25} kg/atom) = 1.45 × 10^{23} atoms of lead.

51. Gas molecules undergo frequent collisions with one another, which considerably increases the time needed for a particular molecule to travel from one place to another.

53. No. The only temperature scale in which such comparisons make any sense is the absolute temperature scale.

55. T_0 = 27°C = 300 K; 2 (KE_0) = $\frac{3}{2}k(2T_0)$, so $T = 2T_0$ = 600 K = 327°C.

57. At the same temperature, $\frac{1}{2}m_Hv_H{}^2 = \frac{1}{2}m_0v_0{}^2$, $v_0 = v_H\sqrt{m_H/m_0}$ = 0.4 km/s.

59. (a) T = 273 K, m = (12.01 + 2 × 16.00)(1.66 × 10^{-27} kg) = 7.306 × 10^{-26} kg; $v_{av} = \sqrt{3kT/m}$ = 393 m/s. (b) $\frac{1}{2}m(2v_0)^2 = \frac{3}{2}k(4T_0)$, so $T = 4T_0$ = 1092 K = 819°C.

61. (a) $M(N_2) \cong 14M(H_2)$, hence there are 14 times more H_2 molecules. (b) The H_2 pressure is 14 times greater because there are 14 times more molecules. (c) The average molecular energy is the same for both gases since their temperatures are the same, hence the average speed of the H_2 molecules is $\sqrt{14}$ = 3.74 times greater.

63. 100%.

65. Actual vapor density = (RH)(saturated vapor density) = 17.25 g/m³, which is very nearly the saturated vapor density at 20°C.

67. RH = (saturated vapor density at 5°C)/(saturated vapor density at 20°C) = 0.39 = 39%. The dew point might actually be higher than 5°C, so this figure is a minimum.

CHAPTER 15

1. The work goes into the internal energy of the water, as manifested by an increase in its temperature.

3. $Q = mc\Delta T$ = 17 kcal.

5. $c = Q/m\Delta T$ = 1.67 kJ/kg · °C.

7. $\Delta T = Q/mc$ = 91°C, T_{final} = 209°C.

9. $Q = mc\Delta T$ = 1200 Btu.

11. $\Delta T = Q/mc$ = 67°C.

13. $Q = mc\Delta T$ = 1.2 kcal, which is 0.34% of the 350 kcal available from the digestion of the piece of pie.

15. 1 gal = 0.134 ft^3 so 50 gal = 6.7 ft^3 and $m = DV$ = 415.4 lb. $Q = mc\Delta T$ = 37,385 Btu. Since P = 0.8 $\times$ 55,000 Btu/h = 44,000 Btu/h, $t = Q/P$ = 0.85 h = 51 min.

17. $Q = Pt = mc\Delta T$, $t = mc\Delta T/P$ = 81 s.

19. (a) Q = (25,000 Btu/h)(72 h) = 1.8 $\times$ 10^6 Btu, $w = Q/c\Delta T$ = 1.9 $\times$ 10^4 lb. (b) $V = w/D$ = 306 ft^3. (c) (306 ft^3)(7.48 gal/ft^3) = 2289 gal, hence 42 drums are needed.

21. ΔT = 80°C; $t = Q/P = (m_1c_1 + m_2c_2 + m_3c_3)\Delta T/P$ = 151 s = 2.5 min.

23. Heat gained = heat lost, (70 kg)(c)(T − 26°C) = (10 kg)(c)(90°C − T), T = 34°C.

25. Heat gained = heat lost, $m_{cop}c_{cop}$ (T − 20°C) + $m_{water}c_{water}$ (T − 20°C) = $m_{iron}c_{iron}$ (120°C − T), T = 20.7°C.

27. $h = s \sin \theta$; $Mgh = Mgs \sin \theta = mc\Delta T$, ΔT = 22°C.

29. The ice, because of its heat of fusion.

31. $Q = mc_w\Delta T + mc_i\Delta T + mL_f$ = 380 kcal.

33. To heat the water to 100°C requires $Q_1 = mc\Delta T$ = 209.5 kJ. Thus Q_2 = 1000 kJ − Q_1 = 790.5 kJ is available to convert water at 100°C to steam at 100°C and m_{steam} = Q_2/L_v = 0.350 kg = 350 g.

35. (a) From Table 10–1 the mass of 1 L of seawater is 1.03 kg; hence $Q = mc\Delta T + mL_v$ = 2.68 MJ. (b) P = (2.68 MJ/L)(10^5 L)/(8.64 $\times$ 10^4 s) = 3.1 MW.

37. $mL_f = \frac{1}{2}mv^2$, $v = \sqrt{2JL_f}$ = 818 m/s.

39. $mc\Delta T + mL_f = \frac{1}{2}mv^2$, $v = \sqrt{2(c\Delta T + L_f)}$ = 327 m/s, where ΔT is the difference between the melting point of lead and the initial temperature of 100°C.

41. If M is the mass of the block of ice and m is the mass that melts and if $M \gg m$ (as it is here), then $mL_f = Fs = \mu Mgs$, $\mu = mL_f/Mgs$ = 0.854.

43. Heat gained by ice = heat lost by alcohol-water mixture, $m_{ice}c_{ice}$ [0°C − (− 10°C)] + $m_{ice}L_f$ + $m_{ice}c_{water}$ (5°C − 0°C) = $m_{water}c_{water}$ (20°C − 5°C) + $m_{alcohol}c_{alcohol}$ (20°C − 5°C), m_{ice} = 0.0263 kg = 26.3 g.

45. The boiling point of nitrogen is − 196°C, hence ΔT = 196°C; heat lost by water = heat gained by nitrogen, m_wL_{fw} + $m_wc_{ice}\Delta T = m_nL_{fn}$, m_n = 0.742 kg.

47. To establish the physical state of the mixture, we note that if all the steam is cooled to 212°F, the heat lost is Q_1 = (m_sc_s)(8°F) = 38.4 Btu, and if all the water warms to 212°F, the heat gained is $Q_2 = (m_wc_w)$(12°F) = 120 Btu. We conclude that the final temperature is 212°F and that at least some of the steam condenses. If m is the weight of steam that condenses, heat lost by steam = heat gained by water, $Q_1 + mL_v = Q_2$, $m = (Q_2 - Q_1)/L_v$ = 0.084 lb.

49. (a) No. The highest temperature the water can have while remaining liquid is its boiling point. Increasing the rate of heat supply increases the rate at which steam is produced but does not change the temperature of the boiling water. (b) Yes. In a pressure cooker, the pressure is greater than normal atmospheric pressure, which raises the boiling point and so causes the eggs to cook faster.

51. From Fig. 15–13, at p = 70 atm and T = 20°C, CO_2 is in the liquid state.

CHAPTER 16

1. The work done at constant pressure is $p(V_2 − V_1)$ and is a maximum; the final temperature would exceed the initial temperature. At constant temperature, since pV/T = constant, the pressure would drop during the expansion and less work would be done. If the expansion were adiabatic, the temperature would drop, and accordingly the final pressure would be lower than in either of the other cases. Hence the least work is done during an adiabatic expansion. (See Fig. 16–8.)

3. p = 2 $\times$ 10^5 Pa, ΔV = 10^{-3} m^3; $\Delta U = Q − W = Q − p\Delta V$ = 300 J. Since ΔU is positive, the internal energy increases.

5. p = 2160 lb/ft^2, $\Delta V = AL = (\pi d^2/4)L$; $W = p\Delta V = \pi pd^2L/4$ = 4.4 $\times$ 10^4 ft · lb.

7. (a) $a \rightarrow d$, W = 0; $d \rightarrow c$, $W = p\Delta V$ = 600 kJ (by the engine). (b) $c \rightarrow b$, W = 0; $b \rightarrow a$, $W = p\Delta V$ = − 1080 kJ (on the engine).

9. Eff $= 1 - T_2/T_1 = 44\%$.

11. Maximum efficiency is $1 - T_2/T_1 = 65\%$, hence actual efficiency is 62% of maximum.

13. $T_2/T_1 = Q_2/Q_1$, $T_2 = T_1Q_2/Q_1 = 375$ K.

15. $T_1 = 600$ K, $T_2 = 400$ K, $W = (\text{Eff})(Q_1) = (1 - T_2/T_1)(Q_1) = 3.33 \times 10^5$ J.

17. $T_1 = 960°$R, Eff $= W/Q_1 - 1 - T_2/T_1$, $T_2 = T_1 (1 - W/Q_1) = 713°$R $= 253°$F.

19. $T_1 = 773$ K, Eff $= 1 - T_2/T_1$, $T_2 = T_1(1 - \text{Eff}) = 502$ K. If T_1' is the intake temperature for an efficiency of 50%, then $T_1' = T_2/(1 - \text{Eff}) = 1005$ K $= 732°$C.

21. $mgh = mc\Delta T$, $\Delta T = gh/c = 1.1°$C; $T_1 = 11.1°$C $= 284.1$ K, $T_2 = 10°$C $= 283$ K, Eff $= 1 - T_2/T_1 = 0.0039 = 0.39\%$.

23. Eff $= P_{out}/P_{in} = 746$ W/$(0.25$ kg$)(4.6 \times 10^7$ J/kg$)(1/3600$ s$) = 0.234 = 23.4\%$.

25. $W = 1$ kW $\cdot$ h $= (10^3$ J/s$)(3600$ s/h$)(3.97$ Btu/kcal$)/4185$ J/kcal $= 3.42 \times 10^3$ Btu, $Q = (0.82$ lb$)(1.98)(10^4$ Btu/lb$) = 16.2 \times 10^3$ Btu. Eff $= W/Q_1 = 0.21 = 21\%$.

27. (a) $Q/t = (18$ kg/h$)(44.8$ MJ/kg$)(10^3$ kJ/MJ$)/(3600$ s/h$) = 224$ kW, Eff$_1 = P/(Q/t) = 0.268 = 26.8\%$. (b) Similarly Eff$_2 = 0.301 = 30.1\%$.

29. In a two-stroke engine, a power stroke occurs in each cylinder once per revolution. Since the engine has 4 cylinders, $p = P/4LAN = 6.51 \times 10^5$ Pa.

CHAPTER 17

1. Steel is a better conductor of heat than wood.

3. For the same insulating ability the ratio k/d must be the same, so here $d_2 = d_1k_2/k_1$. Hence (a) 200 cm, (b) 150 cm, (c) 32.5 cm.

5. $Q/t = kA\Delta T/d = 180$ Btu/h.

7. $d = 6 \times 10^{-3}$ m, $t = 8.64 \times 10^4$ s; $Q = kAt\Delta T/d = 5.18 \times 10^{11}$ J.

9. $A = \pi r^2 = \pi \times 10^{-2}$ m^2; $Q/t = kA\Delta T/d = 189$ kcal/s. The rate decreases as the milk gets warmer since this reduces ΔT.

11. $Q/t = 300$ Btu/h; $\Delta T = d(Q/t)/kA = 341°$F so $T_2 = T_1 + \Delta T = 541°$F.

13. $t = 1$ h $= 3600$ s; $Q = kAt\Delta T/d = 1296$ kcal; $m = 1296$ kcal/$(9500$ kcal/kg$) = 0.136$ kg.

15. We consider a 1-ft^2 area of the lake and neglect the increase in ice. The heat lost through the ice during the night

is $Q = kAt\Delta T/d = 450$ Btu. The weight of newly formed ice is $w = Q/L_f = 3.125$ lb. If h is the thickness of the newly formed ice and $A = 1$ ft^2 is its area, $w = (dg)Ah$, $h = w/(dg)A = 0.054$ ft $= 0.65$ in., and total ice thickness is 2.65 in.

17. $k = d/R = 6.25$ Btu/(ft$^2 \cdot$ h $\cdot °$F/in.).

19. $R = 2d_1/k_1 + d_2/k_2 = 0.24$ m$^2 \cdot °$C/W, $Q/tA = \Delta T/R = 12.5$ W/m^2.

21. The average temperature difference in the first case is 75°F, in the second case it is 35°F, so the time needed in the second case is 5 min $\times$ (75/35) $= 11$ min according to Eq. (17–5).

23. Under all circumstances; the higher the temperature, the shorter the predominant wavelength.

25. The rate of heat loss will decrease because polished copper has a lower emissivity.

27. The vacuum prevents heat transfer by conduction or convection, and the silvered surfaces are poor radiators.

29. $R_1/R_2 = (T_1/T_2)^4 = 0.55$.

31. $e = 1$, $T = 873$ K, $P = e\sigma AT^4 = 6.6$ W.

33. Since $R_1/R_2 = T_1^4/T_2^4$, $T_2 = T_1(R_2/R_1)^{1/4} = 800$ K $= 527°$C. Note that $X^{1/4} = \log^{-1}(\frac{1}{4}\log X)$.

35. Solar energy absorption rate $= (1.2 \times 0.5 \times 0.7)$ kW $= 0.42$ kW, so the heat loss rate is $(0.42 + 0.08)$ kW $= 0.5$ kW $= 1800$ kJ/h. Hence $(1800$ kJ/h$)/(2430$ kJ/kg$) = 0.74$ kg/h of sweat must be evaporated.

37. The kitchen will become warmer owing to the energy dissipated in the fan's operation.

39. $W = Q_1/\text{CP} = 1$ J/2.4 $= 0.417$ J.

41. $Q_1/t = (5)(12,000$ Btu/h$)(778$ ft $\cdot$ lb/Btu$)/(3600$ s/h$) = 12,967$ ft $\cdot$ lb/s, hence $P = (Q_1/t)/\text{CP} = 5187$ ft $\cdot$ lb/s $= 9.43$ hp.

43. $T_1 = 268$ K, $T_2 = 298$ K, $Q = 1$ J; $W = Q[(T_2/T_1) - 1] = 0.112$ J.

45. $T_1 = 465°$R, $T_2 = 545°$R, CP (ideal) $= T_1/(T_2 - T_1) = 5.81$.

47. $T_1 = 253$ K, $T_2 = 313$ K, $W = Q[(T_2/T_1) - 1] = 992$ J/kcal for a Carnot refrigerator, which is the most efficient possible. Hence only design C is capable of operating.

49. $Q = [150$ (Btu/h)/ft$^2](40$ ft$^2)/12,000[$(Btu/h)/ton$] = 0.5$ ton.

51. $m = 10^5$ lb, $Q_1 = mc_{water}(60°$F $- 32°$F$) = 2.8 \times 10^6$ Btu, $Q_2 = mL_f = 1.44 \times 10^7$ Btu, $Q_3 = mc_{ice}(32°$F $- 15°$F$) = 8.5 \times 10^5$ Btu. $Q = Q_1 + Q_2 + Q_3 = 1.805 \times 10^7$ Btu. Since one refrigeration ton $= 2.88 \times 10^5$ Btu/day, the required capacity is 62.7 tons.

CHAPTER 18

1. (a) No. (b) Yes.

3. An equal amount of charge of the opposite sign appears on the fur.

5. The positive charge is located in the nucleus, the negative charge in the electrons that surround the nucleus.

7. There are many reasons, but a convincing one is that all gravitational forces are observed to be attractive, whereas if they were electric in origin some would have to be repulsive.

9. $n = m/m_p$, $Q = ne = me/m_p = 95.6°C$.

11. $F_1/F_2 = (r_2/r_1)^2$, $r_2 = r_1 \sqrt{F_1/F_2} = 2$ cm.

13. $F = ke^2/r^2 = 4.18 \times 10^{-8}$ N.

15. $r = \sqrt{kQ_A Q_B/F} = 1$ cm.

17. $F = mg = ke^2/r^2$, $r = e\sqrt{k/mg} = 5.08$ m.

19. $mv^2/r = ke^2/r^2$, $v = \sqrt{ke^2/mr} = 2.19 \times 10^6$ m/s.

21. Each outer charge exerts on the middle charge a force of magnitude $F = kQ_A Q_B/r^2 = 1.35 \times 10^{-5}$ N. Since both forces are directed toward the negative charge, the total force is 2.7×10^{-5} N.

23. (a) If r_1 is the distance from the test charge Q to $Q_1 = -1 \times 10^{-8}$ C, then $r_2 = (0.4 \text{ m} - r_1)$ is the distance to $Q_2 = -3 \times 10^{-8}$ C. The forces on Q will be equal when $kQQ_1/r_1^2 = kQQ_2/r_2^2$, $r_2/r_1 = \sqrt{Q_2/Q_1} = (0.4 \text{ m} - r_1)/r_1$, $r_1 = 0.4 \text{ m}/(\sqrt{Q_2/Q_1} + 1) = 0.146$ m $= 14.6$ cm. (b) Since r_1 is independent of the magnitude and sign of Q, $r_1 = 14.6$ cm here also.

25. Each string makes an angle of $\theta = 20°$ with the vertical so the balloons are $r = 2L \sin \theta = 0.342$ m apart. If T is the tension in its string and F is the repulsive electric force on it, the conditions for the right-hand balloon to be in equilibrium are $\Sigma F_x = F - T \sin \theta = 0$ and $\Sigma F_y = T \cos \theta - mg = 0$. From $\Sigma F_x = 0$, $T = mg/\cos \theta$, and so from $\Sigma F_x = 0$ we have $F = T \sin \theta = mg \tan \theta$. Hence $mg \tan \theta = kQ^2/r^2$, $Q = r\sqrt{(mg \tan \theta)/k} = 0.37$ μC.

27. The force each of the other charges exerts has the magnitude $F = kQ^2/a^2$. The lines of action of the forces are 120° apart (since each of the interior angles of the triangle is 60°), so the vector sum of the forces is $F' = 2F \cos 60° = (2F)(0.5) = kQ^2/a^2$. The force is parallel to the side of the triangle opposite to the $+Q$ charge and points toward the $-Q$ charge.

29. $E = kQ/r^2 = 3.94 \times 10^6$ N/C.

31. $E = 0$ at the center of the square, since a charge placed there will experience no force.

33. $eE = mg$, $E = mg/e = 1.02 \times 10^{-7}$ V/m.

35. $E = 0$ at a point between the charges and the distance s from the charge $Q_1 = 1.5 \times 10^{-6}$ C. At this point $kQ_1/s^2 = kQ_2/(0.2 \text{ m} - s)^2$, $(0.2 \text{ m} - s)/s = \sqrt{Q_2/Q_1}$, $s = 0.2 \text{ m}/(\sqrt{Q_2/Q_1} + 1) = 0.828$ m.

37. No, because at a given point an appropriate test object travels along the line of force passing through that point and it can travel in only one direction at any point.

41. There is no electric field inside the box.

43. PE $= ke^2/r = 4.3 \times 10^{-18}$ J.

45. $V = W/Q = 6$ V.

47. $s = V/E = 0.005$ m $= 5$ mm.

49. $W = QV = 3.2 \times 10^8$ J.

51. $E = F/e = V/s$, $V = Fs/e = 2.5 \times 10^3$ V.

53. (a) $W = QV = 2.8 \times 10^6$ J. (b) $h = W/mg = 14.3$ km. (c) $v = \sqrt{2W/m} = 529$ m/s.

55. $E = V/s = 500$ V/m. The acceleration of the electron is $a = F/m = eE/m = 8.78 \times 10^{13}$ m/s². Since $s = 0.01$ m $= \frac{1}{2}at^2$, the electron strikes the plate $t = \sqrt{2s/a} = 1.51 \times 10^{-8}$ s later. In this time it travels $d = vt = 0.151$ m $= 15.1$ cm.

57. Since $\frac{1}{2}mv^2 = eV$, v is proportional to $\sqrt{V}$. Hence doubling V increases the electron speed by $\sqrt{2}$.

59. KE $= \frac{1}{2}mv^2 = 3.25 \times 10^{-14}$ J/1.6×10^{-19} J/eV $= 2.03 \times 10^5$ eV.

61. KE $= 8 \times 10^{-18}$ J, $\frac{1}{2}mv^2 =$ KE, $v = \sqrt{2\text{KE}/m} = 4.2 \times 10^6$ m/s.

63. $W = mc^2$, $m = W/c^2 = 1.78 \times 10^{-36}$ kg.

CHAPTER 19

1. An electric current in a metal consists of moving electrons, in an ionized gas it consists of positive and negative ions that move in opposite directions.

3. All metals are good electrical conductors, and most non-metallic solids are good electrical insulators. Liquids may be either. Gases are good insulators if not ionized. Among solids, good electrical conductors are good heat conductors, and vice versa.

5. With a single wire, charge would be permanently transferred from one end to the other, and soon so large a charge separation would occur that the electric field that produced the current would be canceled out. With two wires, charge can be sent on a round trip, so to speak, and a net flow of energy from one place to another can occur without a net transfer of charge.

7. $Q/s = 10^{20}$ e/cm $= 16$ C/cm $= 1.6 \times 10^3$ C/m, $t = s/v$, $I = Q/t = v(Q/s)$, $v = I/(Q/s) = 6.25 \times 10^{-4}$ m/s.

9. $V = IR = 2$ V.

11. $R = V/I = 15$ Ω.

13. When the bridge is balanced, no current passes through the meter. Hence the same current I_1 passes from left to right through R and C, and the same current I_2 passes from left to right through A and B. In order for the junction of R, C, and G to be at the same potential as that of A, B, and G, it must be true that $I_1R = I_2A$ and that $I_1C = I_2B$. Dividing the first of these equations by the second, $R/C = A/B$, $R = AC/B$.

15. Since R is proportional to $1/A$, the resistance per meter of the 5-mm^2 wire is $\frac{2}{5} = 40\%$ that of the 2-mm^2 wire.

17. $R = \rho L/A = 2.0$ Ω.

19. $R = \rho L/\pi r^2$, $r = \sqrt{\rho L/\pi R} = 1.427 \times 10^{-4}$ m, $d = 2r = 2.85 \times 10^{-4}$ m.

21. $\rho = RA/L = 60$ $\Omega \cdot$ m.

23. $d = 40.3$ mils, $A_{\text{cmil}} = (d_{\text{mil}})^2 = 1624$ cmil.

25. $A_{\text{cmil}} = (d_{\text{mil}})^2 = 400$ cmil, $L = RA/\rho = 192$ ft.

27. (a) $\Delta T = -20°$C, so $\Delta R = \alpha R \Delta T = -8$ Ω and $R = 100$ $\Omega - 8$ $\Omega = 92$ Ω. (b) $\Delta T = 60°$C, so $\Delta R = 24$ Ω and $R = 124$ Ω.

29. $\alpha = \Delta R/R\Delta T = 0.0051/°$C.

31. (a) The volume of a cylinder is $V = \pi d^2 L/4$; hence $d_1^2 L_1 = d_2^2 L_2$, $L_2 = d_1^2 L_1/d_2^2 = 100$ m. (b) Since the wire is 100 times longer than the rod and its cross-sectional area is 0.01 times as great, the wire's resistance is 10^4 times the rod's resistance because $R = \rho L/A$.

33. $R = V/I = 0.15$ Ω; $L = 140$ m; $A = \rho L/R = 1.59 \times 10^{-5}$ m$^2 = 15.9$ mm^2.

35. $R = V/I = 0.25$ Ω, $L = RA/\rho = 157$ ft.

37. Energy.

39. The rates of flow of charge are the same since the currents are the same. The rate of flow of energy when $V = 120$ V is twice as great as when $V = 60$ V since $P = IV$.

41. $I = P/V = 1.67$ A.

43. Eff $= (0.5$ hp$)(746$ W/hp$)/IV = 0.78 = 78\%$.

45. $P_{\text{out}} = (\text{Eff})(P_{\text{in}}) = (\text{Eff})(IV) = 1.92$ kW $= 2.57$ hp.

47. $P = W/t = \text{KE}/t = mv^2/2t = IV$, $I = mv^2/2Vt = 6.4$ A.

49. $W = I^2Rt = 4608$ J $= 1.10$ kcal; $Q = W = mc\Delta T$, $\Delta T = W/mc = 0.28°$C.

51. (a) $P = IV = 5 \times 10^3$ W. (b) 5×10^6 eV. (c) $(5 \times 10^6$ eV$)(1.6 \times 10^{-19}$ J/eV$) = 8 \times 10^{-13}$ J.

53. Water molecules are highly polar and in the presence of a charge orient themselves with their positive ends toward a negative charge and vice versa. This reduces the effective magnitude of the charge that acts on another charge nearby. Benzene molecules are much less polar and so their effect is smaller.

55. A molecular ion is a molecule with a net charge. A polar molecule is electrically neutral but has a nonsymmetrical distribution of charge.

57. 8; 8; 10.

59. (a) $Q = (1.5$ C/s$)(3600$ s$) = 5400$ C, $W = QV = 7290$ J. For $P = 0.1$ mW $= 10^{-4}$ W, $t = W/P = 7.29 \times 10^7$ s $= 844$ days $= 2.3$ years. (b) $I = PV = 7.41 \times 10^{-5}$ A $= 74.1$ μA.

CHAPTER 20

1. $I = V_1/R_1 = 0.02$ A, $V_2 = IR_2 = 5$ V.

3. A resistor in parallel is needed whose resistance is $R_2 - R_1R/(R_1 - R) = 6.67$ Ω.

5. (a) $R = (12)(5$ $\Omega) = 60$ Ω, $I = V/R = 2$ A. (b) $P = V^2/R = 240$ W.

7. Bulb 1 is the brightest because the greatest current flows in it; bulbs 3 and 4 are the dimmest because the least current flows in them.

9. (a) $1/R = 1/R_1 + 1/R_2 + 1/R_3$, $R = 10.5$ Ω. (b) $I_1 = V/R_1 = 3$ A; $I_2 = V/R_2 = 1.5$ A; $I_3 = V/R_3 = 1.2$ A.

11. (a) 5-Ω bulb: $I_1 = V/R_1 = 2.4$ A; 10-Ω bulb: $I_2 = V/R_2 = 1.2$ A. (b) 12 V. (c) $P_1 = V^2/R_1 = 28.8$ W; $P_2 = V^2/R_2 = 14.4$ W; $P = P_1 + P_2 = 43.2$ W.

13. (a) $R' = (5$ $\Omega \times 8$ $\Omega)/(5$ $\Omega + 8\Omega) = 3.08$ Ω: $R'' = 3.08$ $\Omega + 5$ $\Omega = 8.08$ Ω; $R = (8.08$ $\Omega \times 3$ $\Omega)/(8.08$ $\Omega + 3$ $\Omega) = 2.19$ Ω. (b) The current through the upper branch is $I'' = V''/R'' = 1.49$ A, hence the potential difference across the 8-Ω resistor is $V = I''r' = 4.59$ V and the current through it is $I = V/R = 0.57$ A.

15. (a) $R_1 = (5$ $\Omega \times 10$ $\Omega)/(5$ $\Omega + 10$ $\Omega) = 3.33$ Ω, $R_2 = (8$ $\Omega \times 8$ $\Omega)/(8$ $\Omega + 8$ $\Omega) = 4$ Ω, $R = R_1 + R_2 = 7.33$ Ω. (b) The resistance of the entire circuit is $R' = 7.33$ $\Omega + 1.5$ $\Omega = 8.83$ Ω, so the current is $I = V/R' = 2.72$ A. The potential difference across the first pair of resistors is $V_1 = IR_1 = 9.06$ V, hence in the 5-Ω resistor $I = V_1/5$ $\Omega = 1.81$ A and in the 10-Ω resistor $I = V_1/10$ $\Omega = 0.906$ A. The potential difference across the second pair of resistors is $V_2 = IR_2 = 10.88$ V, hence in each 8-Ω resistor $I = V_2/8 \ \Omega = 1.36$ A.

17. $R_1 = (10 \ \Omega \times 15 \ \Omega)/(10 \ \Omega + 15 \ \Omega) = 6 \ \Omega$; $R_2 = 4 \ \Omega + 6 \ \Omega + 4 \ \Omega = 14 \ \Omega$; $R_3 = (14 \ \Omega \times 14 \ \Omega)/(14 \ \Omega + 14 \ \Omega) = 7 \ \Omega$; $R_4 = 5 \ \Omega + 7 \ \Omega + 3 \ \Omega = 15 \ \Omega$. Total current is $I_4 = 60 \ V/15 \ \Omega = 4$ A. $V_3 = IR_3 = 28$ V, hence $I_2 = V_3/R_2 = 2$ A. The voltage across R_1 is $V_1 = I_2R_1 = 12$ V and so the current in the 10-Ω resistor is 12 V/10 Ω = 1.2 A.

19. $V = \mathcal{E} + Ir = 31$ V.

21. (a) From $I = \mathcal{E}/(R + r)$, $R = \mathcal{E}/I - R = 0.9 \ \Omega$. (b) $V = \mathcal{E} - Ir = 22$ V.

23. $I_1 = \mathcal{E}/(R + r) = 5.0$ A, $I_2 = 3.33$ A. Since $P = I^2R$, $P_2/P_1 = I_2{}^2/I_1{}^2 = 0.44$.

25. The equivalent resistance of the internal resistances of the batteries is $r/4$. Hence $I = \mathcal{E}/(R + r/4) = 2.9$ A.

27. $W = I^2Rt = 4608$ J; $Q = W = mc\Delta T$, $\Delta T = W/mc = 0.28°C$.

29. I_2 is downward and given by $I_2 = I_1 - I_3 = 3$A. Hence $R_2 = V_3/I_2 = 6.67 \ \Omega$. To find R_1, apply Kirchhoff's second rule to the left-hand loop, proceeding clockwise. This gives $V_1 = I_1R_1 + I_2R_2$, hence $R_1 = (V - I_2R_2)/I_1 = 6 \ \Omega$.

31. Let I_1 be the current to the right through the upper battery, I_2 be the current to the left through the lower battery, and I_3 be the current to the right through the resistor R. At the left-hand junction $I_2 = I_1 + I_3$. Proceeding clockwise in the upper loop, $\mathcal{E}_2 - \mathcal{E}_1 = I_1r_1 + I_2r_2$. Proceeding counterclockwise in the outermost loop, $\mathcal{E}_1 = I_3R - I_1r_1$. Since $I_2 = I_1 + I_3$, the first loop equation becomes $\mathcal{E}_2 - \mathcal{E}_1 = I_1r_1 + I_1r_2 + I_3r_2$. Solving this equation and the second loop equation for I_1, and setting the two expressions for I_1 equal gives $(I_3R - \mathcal{E}_1)/r_1 = (\mathcal{E}_2 - \mathcal{E}_1 - I_3r_2)/(r_1 + r_2)$, from which $I_3 = (\mathcal{E}_2r_1 + \mathcal{E}_1r_2)/(Rr_1 + Rr_2 + r_1r_2) = 0.533$ A.

33. Assume currents downward in each resistor. At the top junction, $I_1 + I_2 + I_3 = 0$. Proceeding counterclockwise in the left-hand loop, $10 \ V + 6 \ V = I_1 \times 10 \ \Omega - I_2 \times 5 \ \Omega$. Proceeding clockwise in the right-hand loop, $10 \ V + 10 \ V - I_3 \times 20 \ \Omega - I_2 \times 5 \ \Omega$. Solving yields $I_1 = 0.857$ A, $I_2 = -1.486$ A, $I_3 = 0.629$ A.

35. (a) To find the potential difference between a and b we can disregard the 8-V battery and the 6-Ω resistor. Applying Kirchhoff's second rule to the outside loop, going clockwise and assuming a clockwise current of I, yields $10 \ V - 5 \ V = I(1 \ \Omega + 1 \ \Omega + 3 \ \Omega + 9 \ \Omega)$, $I = 0.357$ A. In the upper part of the loop a potential drop occurs in both the battery

and the 3-Ω resistor, hence $V = 5 \ V + I(1 \ \Omega + 3 \ \Omega) = 6.43$ V. The lower part of the loop gives the same result. Here there is a potential increase of 10 V in the battery and a potential drop in the battery's internal resistance and in the 9-Ω resistor, so $V = 10 \ V - I(1 \ \Omega + 9 \ \Omega) = 6.43$ V. (b) Point b is at a potential of -6.43 V relative to a. No current passes through the 6-Ω resistor or the 8-V battery, so there is no IR drop in either. The 8-V battery's positive terminal is nearest to c, hence the potential of c relative to a is 8 V $-$ 6.43 V = 1.57 V.

37. (a) Let the current in the 10-Ω resistor be I_1, that in the 8-Ω resistor be I_2, and that in the 5-Ω resistor be I_3. Assume I_1 and I_2 both go from b to a and that I_3 goes from a to b; then $I_3 = I_1 + I_2$. Applying Kirchhoff's second rule to the upper loop and proceeding counterclockwise yields 8 V $-$ 5 V $= I_3(1 \ \Omega + 5 \ \Omega) + I_1(10 \ \Omega + 1 \ \Omega)$; hence $I_1 = 0.273$ A $- 0.545 \ I_3$. Proceeding clockwise around the lower loop yields 12 V $-$ 5 V $= I_3(1 \ \Omega + 5 \ \Omega) + I_2(8 \ \Omega + 1 \ \Omega)$. Since $I_2 = I_3 - I_1$, $I_1 = 1.667 \ I_3 - 0.778$ A. Setting equal the two formulas for I_1 yields $1.667 \ I_3 - 0.778$ A $= 0.273$ A $- 0.545 \ I_3$, $I_3 = 0.475$ A. (b) $V_{ab} = 5 \ V + I_3(1 \ \Omega + 5 \ \Omega) = 7.85$ V.

39. (a) $I_{coil} = 2 \times 10^{-5}$ A, $I_{shunt} = (1 \ A - I_{coil}) \approx 1$ A, $R_{shunt} = R_{coil}I_{coil}/I_{shunt} = 0.0012 \ \Omega$. (b) $R = V/I = 5 \ V/2 \times 10^{-5}$ A $= 2.5 \times 10^5 \ \Omega$, so $R_{series} = 2.5 \times 10^5 \ \Omega - R_{coil} \approx 2.5 \times 10^5 \ \Omega$.

41. The meter's resistance is $R_{coil} = 10 \ V/6 \times 10^{-5}$ A $= 1.67 \times 10^5 \ \Omega$. The required resistance is $R = 100 \ V/6 \times 10^{-5}$ A $= 1.667 \times 10^6 \ \Omega$, so $R_{series} = (1.667 - 0.167) \times 10^6 \ \Omega = 1.5 \times 10^6 \ \Omega$.

43. The equivalent resistance of the resistor and voltmeter is $R_e = V/I = 300 \ \Omega$. Since the unknown resistance R is in parallel with the resistance $R_v = 2000 \ \Omega$ of the voltmeter, $R_e = RR_v/(R + R_v)$, $R = R_vR_e/(R_v - R_e) = 353 \ \Omega$. The voltmeter resistance is significant; it is not correct to assume that the equivalent resistance V/I is equal to the unknown resistance R unless $R_v >> R$.

CHAPTER 21

1. All three fields are detected in both cases.

3. East; west.

5. $s = \mu_0I/2\pi B = 0.08$ m $= 8$ cm.

7. $I = 2Br/\mu_0 = 7.96$ A.

9. $N/L = 3000$ turns/m, $B = \mu_0I(N/L) = 0.0143$ T.

11. B $= 0$ midway between the wires because the fields of the currents are opposite in direction and of the same magnitude there.

13. $B_1 = \mu_0 I/2\pi s = 2 \times 10^{-5}$ T, $B_2 = 0.67 \times 10^{-5}$ T. Since the fields are in opposite directions, $B = B_1 - B_2 = 1.33 \times 10^{-5}$ T.

15. The fields of the two layers are in opposite directions, hence the result is the same as if the solenoid had 200 turns. Thus $B = \mu_0 NI/L = 1.26 \times 10^{-4}$ T.

17. $B_{\text{loop}} = B_{\text{solenoid}}$, $\mu_0 I_1/2r = \mu_0 I_2 (N/L)$, $I_1 = 2rI_2 (N/L) = 20$A since $(N/L) = 2 \times 10^3$ turns/m.

19. See diagram below. $B = B_1 \cos\theta + B_2 \cos\theta = 2B(s/2L) = (2)(\mu_0 I/2\pi L)(s/2L) = 2\mu_0 Is/\pi(4x^2 + s^2)$.

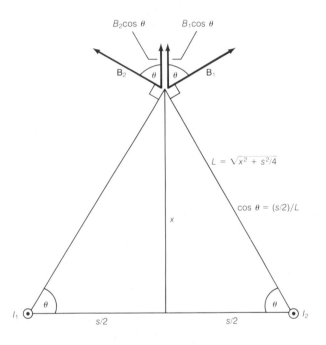

21. No. Because of hysteresis in the iron core, B is not proportional to I, and so the variations of B with time do not follow the variations of I with time.

23. $\mu/\mu_0 \approx 8900$ at $B_0 \approx 9 \times 10^{-5}$ T.

25. South.

27. $R = mv/eB = 2.84 \times 10^{-3}$ m.

29. $F_{\text{grav}} = 8.9 \times 10^{-30}$ N; if **v** is perpendicular to **B** and $B = 3 \times 10^{-5}$ T (a typical value), $F_{\text{mag}} = evB = 1.4 \times 10^{-16}$, which is over 10^{13} times greater than F_{grav}.

31. (a) The magnetic field 0.1 m from the wire is $B = \mu_0 I/2\pi s = 2 \times 10^{-4}$ T, so the force on the electron is $F = evB = 3.2 \times 10^{-16}$ N since **v** $\perp$ **B**. The force is toward the wire since the moving electron is equivalent to a current in the opposite direction. (b) Here also **v** $\perp$ **B**, so $F = 3.2 \times 10^{-15}$ N. The force is in the same direction as the current.

33. (a) $B = \mu_0 I/2\pi s - 1.6 \times 10^{-4}$ T, $F = evB = 1.54 \times 10^{-15}$ N. The force is directed away from the wire. (b) The force has the same magnitude but is directed toward the wire.

35. (a) The force has the same magnitude as in Exercise 33(a), but is in the direction of the current. (b) The force has the same magnitude but is opposite to the direction of the current.

37. If R is the orbit radius, $T = 2\pi R/v = 2\pi mv/vqB = 2\pi m/QB$, which is independent of R and v.

39. $E = V/s = vB$, $V = vBs = 2500$ V.

41. $0°$; $90°$.

43. (a) $\theta = 45°$, $F = I\Delta L/B \sin\theta = 0.283$ N; the force is downward according to the right-hand rule. (b) $F = 0.283$ N; the force is upward.

45. There the length of each turn is $2\pi r$ and $\sin\theta = 1$, so $F = 2\pi NrIB = 0.04$ N.

47. The electric force between the streams is attractive and the magnetic force is repulsive. Which force predominates depends on the speeds of the particles, since the magnetic force increases with speed whereas the electric force does not change.

49. The electron beam is deflected away from the wire.

51. A circle, because the forces that act are repulsive and a circle is the configuration in which each part of the loop is as far as possible from the other parts.

53. $P = IV$, $I = P/V = 1.67$ A $= I_1 = I_2$; $F/I = \mu_0 I_1 I_2/2\pi s = 2.78 \times 10^{-4}$ N/m; the force is repulsive, since the currents are in opposite directions.

55. There is never a net force on the loop in such a field. There is no torque on the loop when its plane is perpendicular to the field.

57. A magnetic compass is more accurate near the equator, where the horizontal component of the earth's magnetic field is strongest. In the polar regions the field has only a very small horizontal component and so can exert relatively little torque on a compass needle.

59. Two south poles.

61.

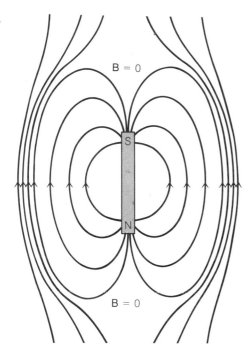

CHAPTER 22

1. The left-hand wheels have positive charges, and the right-hand wheels have negative charges. See sketch of geomagnetic field in Fig. 22–3.

3. $\mathcal{E} = B/v = 1.8 \times 10^{-3}$ v.

5. Conservation of energy.

7. Counterclockwise; zero current; clockwise.

9. (a) An emf is induced in the wire loop and a current therefore comes into being. This current interacts with the magnetic field to produce a torque that, by Lenz's law, opposes the rotation. (b) The current in a loop of aluminum wire would be smaller because of the greater electrical resistance of aluminum, and the torque opposing the rotation would be correspondingly smaller.

11. (a) $A = 5 \times 10^{-3} \text{m}^2$, $\Phi = BA = 5 \times 10^{-6}$ Wb. (b) $\mathcal{E} = \Delta\Phi/\Delta t = 1.67 \times 10^{-6}$ V.

13. $A = \pi r^2 = 7.07 \times 10^{-4} \text{ m}^2$, $\mathcal{E} = -N\Delta\Phi/\Delta t = -NA\Delta B/\Delta t = 0.017$ V.

15. $\mathcal{E} = IR = A(\Delta B/\Delta t)$, $\Delta B/\Delta t = IR/A = 10^3$ T/s.

17. When the generator is connected to an outside circuit, work must be done in turning its shaft to provide the energy of the current provided to the circuit.

19. The induced emf is zero when the plane of the coil is perpendicular to the magnetic field, which happens twice per rotation. Hence $T = 1 \text{ s}/(2 \times 100) = 0.005$ s.

21. Light bulb and electric heater.

23. (a) The emf decreases because the greater the current, the greater the IR drop in the armature and the less the potential difference across the generator. Hence the current in the magnet windings drops and, since the emf is proportional to B, it too decreases. (b) A similar argument explains the increase in the emf when the current drops.

25. $V = \mathcal{E} - IR$, $I = (\mathcal{E} - V)/R = 25$ A, $P = IV = 2875$ W.

27. Total current is $I = P/V = 25$ A, $I_f = V/R_f = 1$ A, $I_a = I - I_f = 24$ A. $P_a = I_a^2 R_a = 57.6$ W, $P_f = I_f^2 R_f = 120$ W, total power dissipated is $P' = P_a + P_f = 178$ W. Efficiency $= P(P + P') = 94\%$.

29. (a) Total current is $I = P/V = 3.5$ A, field current is $I_f = V/R_f = 0.6$ A, armature currentt is $I_a = 3.5$ A $- 0.6$ A $= 2.9$ A. Back emf is $\mathcal{E}_b = V - I_a R_a = 117$ V, power output is $\mathcal{E}_b I_a = 340$ W, efficiency is 340 W/420 W $= 81\%$. (b) $R = R_a R_f/(R_a + R_f) = 1.0$ Ω, I (start) $= V/R = 120$ A. (c) Total resistance needed is 120 V/25 A $= 4.8$ Ω, so series resistor should be 4.8 Ω $- 1.0$ Ω $= 3.8$ Ω.

31. Alternating current is necessary in order that there be a changing magnetic field to induce a current in the secondary winding.

33. The changing magnetic field produced by an alternating current in the primary winding.

35. The frequencies are the same.

37. $V_2 = V_1 N_2/N_1 = 600$ V; $I_2 = I_1 N_1/N_2 = 0.6$ A.

39. $N_1/N_2 = V_1/V_2 = 20.8$; $I_2 = P/V_2 = 41.7$ A.

41. The air gap in the recording head permits the magnetic field of its coil to reach the tape. The air gap in the playback head permits the magnetic field of the tape to enter the iron core of its coil.

CHAPTER 23

1. $E = V/s$, $w = E^2/8\pi k = V^2/8\pi k s^2 = 3.98 \times 10^{-3}$ J/m³.

3. (a) $Q = CV = 0.025$ C. (b) $W = \frac{1}{2}CV^2 = 12.5$ J.

5. $W = \frac{1}{2}CV^2$, $V = \sqrt{2W/C} = 447$ V.

7. (a) Here $K = 1$, $A = 5 \times 10^{-3}$ m², $d = 10^{-3}$ m, so $C = A/4\pi kd = 4.42 \times 10^{-11}$ F $= 44.2$ pF. (b) $Q = CV = 1.99 \times 10^{-9}$ C. (c) $W = \frac{1}{2}CV^2 = 4.48 \times 10^{-8}$ J.

9. Since for a parallel-plate capacitor $C = KA/4\pi kd$, $C' = C/2$; $W_1 = \frac{1}{2}Q^2/C$, $W_2 = \frac{1}{2}Q^2/C' = Q^2/C$, hence the work needed is $\Delta W = W_2 - W_1 = \frac{1}{2}Q^2/C$.

11. There are several ways to proceed, one of which is this. (a) $C = A/4\pi kd = 2.21 \times 10^{-11}$ F, $W = \frac{1}{2}CV^2 = 1.11 \times 10^{-7}$ J. The stored energy W is equal to the work Fd that would be done if the plates were released and flew together so $F = W/d = 1.11 \times 10^{-4}$ N. (b) The force is attractive since the plates have opposite charges. (c) $w = W/Ad = 0.044$ J/m$^3 = E^2/8\pi k$.

13. No.

15. $3W$.

17. $C = KC_0$, $K = C/C_0 = 6.25$.

19. (a) From the solution to Exercise 7, $C = 44.2$ pF; hence $C' = KC = 172$ pF. (b) $Q = C'V = 7.76 \times 10^{-9}$ C. (c) 1.75×10^{-7} J.

21. $C = C_1 + C_2 = 70$ μF.

23. $1/C = 1/C_1 + 1/C_2 + 1/C_3$, $C = 3.125$ μF, $Q = Q_1 = Q_2 = Q_3 = CV = 3.75 \times 10^{-5}$ C; $V_1 = Q_1/C_1 = 7.5$ V, $V_2 = Q_2/C_2 = 3.75$ V, $V_3 = Q_3/C_3 = 0.75$ V.

25. $V_1 = V_2 = V_3 = 12$ V; $Q_1 = C_1 = C_1V_1 = 6 \times 10^{-5}$ C, $Q_2 = C_2V_2 = 1.2 \times 10^{-4}$ C, Q_3, $= C_3V_3 = 6 \times 10^{-4}$ C.

27. $1/C' = 1/2$ μF + $1/4$ μF, $C' = 1.33$ μF; $C = C' + 6$ μF $= 7.33$ μF.

29. $C = 0$.

31. At first, $Q_{01} = C_1V_0 = 1.2 \times 10^{-3}$ C, $Q_{02} = C_2V_0 = 2.4 \times 10^{-3}$ C. Since terminals of opposite sign are joined, the total charge that remains on the combination is $Q = Q_{02} - Q_{01} = 1.2 \times 10^{-3}$ C. The charge distributes itself on the capacitors so there is the same new potential difference across each. Hence $V_1 = V_2$, $Q_1/C_1 = Q_2/C_2$, and $Q_2 = Q - Q_1$; $Q_1 = Q/(1 + C_2/C_1) = Q/3 = 4 \times 10^{-4}$ C, $Q_2 = Q - Q_1 = 8 \times 10^{-4}$ C.

33. Since $V = Q/C$, when a certain charge Q is given to a capacitor, the greater C is, the smaller V is and hence the weaker the electric field E in the capacitor is. The inductance L, on the other hand, is a measure of the magnitude of and volume occupied by the magnetic field produced by a current I in an inductor, hence the greater L is, the greater the energy of the inductor.

35. $A = \pi r^2$, $N = \sqrt{Ll/\mu_0\pi r^2} = 300$ turns.

37. (a) $L = \mu N^2A/l = 4.09$ mH. (b) $I = V/(R + r) = 0.286$ A. (c) $W = \frac{1}{2}LI^2 = 1.67 \times 10^{-4}$ J.

39. The length of the equivalent solenoid is $l = \pi D$ and $\mu = 400$ μ_0, so $L = (400 \mu_0)N^2A/\pi D = 0.32$ H.

41. $\mathscr{E} = \mathscr{E}_1 + \mathscr{E}_2 + \mathscr{E}_3 = -L_1\Delta I/\Delta t - L_2\Delta I/\Delta t - L_3\Delta I/\Delta t = -L\Delta I/\Delta t$; hence $L = L_1 + L_2 + L_3$.

43. (a) From Lenz's law, the self-induced emf will be opposite to the direction of the current. (b) For the same reason, the emf here will be in the same direction as that of the current.

45. $\mathscr{E} = L\Delta I/\Delta t = 200$ V.

47. $f = 1/(2\pi \sqrt{LC}) = 650$ Hz.

49. $C = 1/(4\pi^2Lf^2) = 0.317$ μF.

51. (a) $f = 1/(2\pi \sqrt{LC}) = 1592$ Hz. (b) $Q = CV = 10^{-5}$ C. (c) $\frac{1}{2}CV^2 = \frac{1}{2}LI^2$, $I = V\sqrt{C/L} = 0.1$ A.

53. (a) $C = 2W/V^2 = 998$ μF. (b) $T_C = RC = 0.0148$ s.

55. (a) $T_C = RC = 0.025$ s. (b) $I = V/R = 0.2$ A. (c) $I = Q/t$, $t = Q/I = CV/I = 0.025$ s.

57. $T_L = L/R$, $L = RT_L = 1.4$ H.

59. (a) $I = 0$, $\Delta I/\Delta t = \mathscr{E}/L = 4.17 \times 10^3$ A/s. (b) $\Delta I/\Delta t = (\mathscr{E} - IR)/L$, $I = (\mathscr{E} - L\Delta I/\Delta t)R = 3.25$ A. (c) $I = \mathscr{E}/R = 6.25$ A, $\Delta I/\Delta t = 0$.

61. (a) The coil is 200 mm long; hence there are 200 turns and $L = \mu N^2A/l = 1.78 \times 10^{-4}$ H. (b) Each turn requires $\pi d = 9.42$ cm of wire, hence the total length of wire is $L = 200 \pi d = 18.85$ m; $A = \pi r^2 = 7.85 \times 10^{-7}$ m^2; $R = pL/A = 0.408$ Ω. (c) $T_L = L/R = 4.36 \times 10^{-4}$ s.

CHAPTER 24

1. $T = 0.02$ s, $f = 1/T = 50$ Hz.

3. (a) $I_{max} = \sqrt{2} I_{eff} = 14.1$ A. (b) $V_{max} = \sqrt{2} V_{eff} = 84.9$ V. (c) No; they are coincident only when the frequency is the resonant frequency.

5. (a) $2\pi ft = 2\pi/8 = 45°$, $V = V_{max} \sin 2\pi ft = V_{max} \sin 45° = 70.7$ V. (b) $2\pi ft = 2\pi/4 = 90°$, $V = V_{max} \sin 90° = 100$ V. (c) $2\pi ft = 6\pi/8 = 135°$, $V = V_{max} \sin 135° = V_{max} \sin 45° = 70.7$ V. (d) $2\pi ft = 2\pi/2 = 180°$, $V = V_{max} \sin 180° = 0$.

7. The impedance increases because the coil's inductance and hence inductive reactance increase.

9. (a) $X_L = 2\pi fL = 0.314$ Ω. (b) 314 Ω.

11. $X_L = 2\pi fL$, $I = V/X_L = V/2\pi fL = 0.159$ A.

13. At $f = 200$ Hz, $L = X_L/2\pi f = 0.0955$ H. At $f = 60$ Hz, $X_L = 2\pi fL = 36$ Ω, $I = V/X_L = 6.67$ A.

15. $R = V/I = 120$ Ω. To halve the current, we require $Z = 240$ Ω. Since $Z = \sqrt{R^2 + X_L^2}$, $X_L = 2\pi fL = \sqrt{Z^2 - R^2}$ and $L = \sqrt{Z^2 - R^2}/2\pi f = 0.662$ H.

17. (a) $R = V_1/I_1 = 5 \ \Omega$, $Z = V_2/I_2 = 8 \ \Omega$; $Z = \sqrt{R^2 + X_L^2}$, $L = \sqrt{Z^2 - R^2}/2\pi f = 16.6$ mH. (b) $P_1 = I_1^2 R = 320$ W; $P_2 = I_2^2 R = 125$ W.

19. The reactance becomes $X_C/2$ because $X_C = 1/2\pi fC$.

21. (a) $X_C = 1/2\pi fC = 1.99 \times 10^5 \ \Omega$. (b) 199 Ω.

23. $C = 1/2\pi f X_C = 1.59 \times 10^{-5}$ F = 15.9 μF.

25. (a) $X_C = 1/2\pi fC = 637 \ \Omega$; $Z = \sqrt{R^2 + X_C^2} = 704 \ \Omega$; $I = V/Z = 0.170$ A. (b) $P = I^2 R = 8.7$ W.

27. (a) $Z = V/I = 240 \ \Omega$; $X_C = \sqrt{Z^2 - R^2} = 226 \ \Omega$; $C = 1/2\pi f X_C = 11.7$ μF. (b) No energy is dissipated in a capacitor. (c) $P = I^2 R = 20$ W.

29. The resistance of an ac circuit determines how much of the electrical energy that passes through it is dissipated as heat. The capacitive reactance and inductive reactance respectively determine the extent to which the capacitors and inductors in the circuit oppose the flow of current without dissipating energy. X_L and X_C vary with frequency whereas R does not, and their combined effects depend on $X_L - X_C$, so their analogy with resistance is very limited. The impedance determines the current in the circuit when a potential difference V of a certain frequency is applied according to $I = V/Z$; thus Z has the same role as R does in a dc circuit with respect to I, though the rate of energy dissipation is equal to $I^2 R$ in both cases.

31. $R = R_0$, $X_L = X_{L0}/2$, $X_C = 2C_{C0}$; in order to determine what happens to Z, the values of X_{L0} and X_{C0} must be known.

33. Since $V_C = V_L$, the circuit is in resonance, and $V_R = 60$ V.

35. (a) Since $X_L = X_C$, $X_L - X_C = 0$ and $Z = R = 20 \ \Omega$. (b) $\tan \phi = 0$, $\phi = 0$. (c) $V = IZ = 400$ V.

37. (a) $X_L = 2\pi fL = 38 \ \Omega$, $X_C = 1/2\pi fC = 265 \ \Omega$, $Z = \sqrt{R^2 + (X_L - X_C)^2} = 235 \ \Omega$, $I = V/Z = 0.51$ A. (b) $P = I^2 R = 15.6$ W. (c) $V_L = IX_L = 19.4$ V; $V_C = 135$ V; $V_R = IR = 30.6$ V.

39. Under these circumstances $X_L > X_C$, so the voltage leads the current.

41. $L = X_L/2\pi f$, $C = 1/2\pi f X_C$, $f_0 = 1/2\pi \sqrt{LC} = 1/\sqrt{X_L/f^2 X_C} = 632$ Hz.

43. (a) The power line must have a rating of at least 10 kW/(0.70 kW/kVA) = 14.3 kVA. (b) The minimum rating is now 10 kVA.

45. (a) $X_L = 2\pi fL = 113 \ \Omega$, $X_C = 1/2\pi fC = 44 \ \Omega$, $Z = \sqrt{R^2 + (X_L - X_C)^2} = 85 \ \Omega$. (b) $I = V/Z = 1.41$ A. (c) $\tan \phi = (X_L - X_C)/R = 1.38$, $\phi = 54°$, power factor $= \cos \phi = 0.588$. (d) $P = I^2 R = 99.4$ W. (e) $IV = P/\cos \phi = 169$ V·A. (f) $I = V/R = 2.4$ A.

47. (a) $\cos \phi = P/IV = 0.625$. (b) Since $\cos \phi = 0.625$, $\phi = 51°$, $\tan \phi = 1.235 = X_L/R$; $R = P/I^2 = 75 \ \Omega$, $X_L = 1.235R = 92.6 \ \Omega$; if $X_C = X_L$, $C = 1/2\pi f X_L = 28.6$ μF. (c) $I = V/R = 1.6$A. (d) $P = I^2 R = 192$ W. (e) 192 V·A since V and I are in phase at the resonance frequency.

49. The circuit is a combination of a low-pass and a high-pass filter. Output A should be connected to the tweeter and output B to the woofer.

51. High-pass filter.

53. (a) $X_L = 2\pi fL = 12.6 \ \Omega$; $I_L = V/X_L = 0.8$ A; $I_R = V/R = 1.0$ A. (b) $I = \sqrt{I_R^2 + I_L^2} = 1.3$ A. (c) $Z = V/I = 7.7 \ \Omega$. (d) $\phi = \tan^{-1}(-I_L/I_R) = -39°$. The current lags behind the voltage. (e) $P = I_R^2 R = 10$ W (alternatively, $P = IV \cos \phi = 10$ W).

55. (a) If $X_L = R$ at $f = 20$ kHz, then for the same applied voltage, $I_L > I_R$ for $f < 20$ kHz and $I_L < I_R$ for $f > 20$ kHz. Hence $X_L = R = 2\pi fL$, $L = R/2\pi f = 0.4$ mH. (b) Assume an applied voltage of 100 V for convenience. (Any other voltage could equally well be used, or we could just call the applied voltage V.) Then $X_L = 2\pi fL = 5 \ \Omega$, $I_L = V/X_L = 20$ A, $I_R = V/R = 2$ A, $I = \sqrt{I_R^2 + I_L^2} = 20.1$ A and the proportion of I that passes through the load is $I_R/I = 0.0995 = 9.95\%$. Less than 10% of the total current passes through the load. (c) When $f = 200$ kHz we proceed in exactly the same way, with the result that $I_R/I = 0.995 = 99.5\%$. Now nearly all the current passes through the load.

CHAPTER 25

1. The electric fields produced by electromagnetic induction are readily detected through the currents they produce in metal wires, but no such simple means is available for detecting the weak magnetic fields produced by changing electric fields.

3. Light waves consist of coupled electric and magnetic field fluctuations and need no material medium in which to occur. Sound waves consist of pressure fluctuations in a material medium and hence cannot occur in a vacuum.

5. The variations are always in phase.

7. $\lambda = c/f = 68.3$ m.

9. (a) $f = 1/T = 10^9$ Hz. (b) $\lambda = c/f = 0.3$ m. (c) Microwaves.

11. A real image is formed by light rays that pass through it; a screen placed at the location of the image would show the image. A virtual image is formed by the backward extension of light rays that were diverted by refraction or reflection. The light rays that seem to come from it do not actually pass through a virtual image, and so it cannot appear on a screen although it can be seen by the eye.

13. The image is always twice as far from the man as the mirror; hence it must be moving twice as fast as the mirror, namely at 60 km/h.

15. Electromagnetic waves travel more slowly in material media than they do in free space.

17. The flashes of color are the result of dispersion. In red light the flashes would be red only.

19. $i = 0$.

21. (a) $v = c/n = 2.999 \times 10^8$ m/s. (b) 1.24×10^8 m/s. (c) 1.97×10^8 m/s. (d) 2.26×10^8 m/s.

23. $\sin r = (n_1/n_2)\sin i = 0.789$, $r = 52°$.

25. $n_2 = n_1 \sin i/\sin r = 1.51$.

27. $n_1 = 1.00$. Red light: $\sin r = (n_1/n_2)\sin i = 0.4263$, $r = 25.23°$. Violet light: $\sin r = (n_1/n_2)\sin i = 0.4201$, $r = 24.84°$; $\Delta r = 0.39°$.

29. (a) By the same procedure as in the second example of Sec. 25–6, the final angle of refraction is 70°. (b) 70°.

31. $h = h'n_1/n_2 = 5.4$ cm.

33. $h' = hn_2/n_1 = 0.59$ m.

35. (a) $\sin i_c = n_2/n_1 = 0.413$, $i_c = 24°$. (b) $\sin i_c = 0.550$, $i_c = 33°$.

37. As the diagram shows, the angle of incidence must be 45°, so the critical angle of the glass must be 45° at most. Hence the minimum value of n is $1/\sin i_c = 1/\sin 45° = 1.41$.

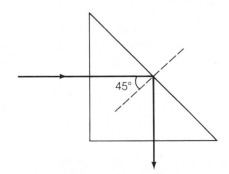

45°

39. $\Omega = A/r^2 = 3.64$ sr.

41. $F = EA = I\Omega = IA/r^2$, $I = r^2E = 3.6 \times 10^5$ cd.

43. $\Omega = A/r^2 = 0.05$ sr, $F = I'\Omega = 125$ lm, Eff = 125 lm/20 W = 6.25 lm/W.

45. $F = 4\pi I = 2513$ lm, $A = \pi d^4/4 = 0.785$ m², $E = F/A = 3200$ lx.

47. $A = 1800$ m² so $F_{total} = EA = 5.4 \times 10^5$ lm. The

useful flux emitted by each lamp is $F = \frac{2}{3}(4\pi I) = 2.09 \times 10^4$ lm/lamp so $N = F_{total}/F = 26$ lamps.

49. Below the bulb, $E = I/r^2$, $I = Er^2 = 768$ cd. At 1.0 m to one side, $\theta = \tan^{-1}(1.0$ m/0.8 m$) = 51.3°$ and $r = \sqrt{(0.8\,m)^2 + (1.0\,m)^2} = 1.28$ m, $E = I \cos \theta/r^2 = 293$ lx.

CHAPTER 26

1. No; no.

3. Crown glass has a smaller index of refraction than benzene. The immersed lens will act as a diverging lens because wavefronts passing through it are retarded less going through its center where the path length in the medium of higher light speed is greater. Thus the wavefronts are affected as in the right-hand diagram of Fig. 26–3 even though the lens is convex.

5. $R_1 = -30$ cm, $R_2 = +20$ cm, $1/f = (n - 1)(1/R_1 + 1/R_2)$, $f = 120$ cm.

7. $R_1 = -20$ cm, $R_2 = +30$ cm; $1/f = (n - 1)(1/R_1 + 1/R_2) = -120$ cm.

9. $1/R_2 = 0$, $R_1 = (n - 1)f = 4.95$ in.

11. (a) Converging. (b) The index of refraction of water is 1.33, so the relative index of refraction is $n' = 1.60/1.33 = 1.20$. From the lensmaker's equation, $f/f' = (n' - 1)/(n - 1) = \frac{1}{3}$, and $f' = 3f = 45$ cm.

13. (a) Under no circumstances. (b) When the object is between the lens and the focal point.

15. No; yes, if the object is placed between the lens and the focal point.

17. (a) Converging. (b) $p = \infty$, $q = f = 8$ cm.

19. $f = pq/(p + q) = 12$ cm; the lens is converging since its focal length is positive.

21. The object distance of the first lens is $p_1 = \infty$ and its image distance is $q_1 = f_1$. The object distance of the second lens is therefore $p_2 = -q_1 = -f_1$, and, since $1/p_2 + 1/q_2 = 1/f_2$, we obtain $1/q_2 = 1/f_1 + 1/f_2$. But q_2 is the distance from the lenses at which the original beam is brought to a focus. Hence $q_2 = f$ and we have the desired result.

23. Since the shutter is open for only $\frac{1}{4} = (\frac{1}{2})(\frac{1}{2})$ as long, the aperture should be opened two f-numbers to $f/4$.

25. (a) Inverted and larger than the object. (b) Inverted and smaller than the object. (c) Erect and smaller than the object. (d) Erect and larger than the object.

27. No image is formed when the object is at the focal point.

29. $q = pf/(p - f) = 30$ cm, $m = -q/p = -1$. The image is real, inverted, and the same size as the object.

31. $m = \frac{1}{10} = -q/p$, $q = -p/10 = -15$ ft; $1/f = 1/p + 1/q$, $f = -0.556$ m.

33. $p = 7000$ m, $f = 1$ m, $q = pf/(p - f) \approx 1$ m; $m = -q/p = -1.43 \times 10^{-4}$; $h' = mh = -1.29 \times 10^{-3}$ m $= -1.29$ mm.

35. $q = pf/(p - f) = -40$ cm, $m = -q/p = +5$, $h = h'/m = 0.2$ cm $= 2$ mm.

37. (a) $p + q = 24$ in., $p = 24$ in. $- q$; $m = -4 = -q/p$, $p = q/4 = 24$ in. $- q$, $q = 19.2$ in., $p = 4.8$ in.; $1/f = 1/p + 1/q$, $f = +3.84$ in. (b) $m = -2 = -q/p$, $p = q/2$; $1/f = 1/p + 1/q = 3/q$, $q = 3f = 11.52$ in., $p = q/2 = 5.76$ in., $p + q = 17.28$ in.

39. (a) The object distance that corresponds to $q = -25$ cm is $p = -q/m = 25$ cm/m. Hence $1/f = 1/p + 1/q = m/25$ cm $- 1/25$ cm $= (m - 1)/25$ cm and $m = (25$ cm/$f) + 1$. (b) $m = (25$ cm/5 cm$) + 1 = 6$.

41. For $p_1 = \infty$, $f_1 = q = 20$ mm. For $p_2 = 20$ cm $= 200$ mm, $f_2 = p_2q/(p_2 + q) = 18.2$ mm.

43. (a) Converging lens. (b) $p = 25$ cm, $q = -100$ cm, $f = pq/(p + q) = +33$ cm.

45. $q = -0.60$ m and $f = (1$ m$)/(-1.5) = -\frac{2}{3}$ m; hence $p = qf/(q - f) = 6$ m.

47. Here $f = 100$ cm/2.5 $= 40$ cm. The lens forms an image q away from the eye when $p = 25$ cm. Hence $q = pf/(p - f) = -67$ cm, where the minus sign means the image is in front of the eye. The near point without the lenses is therefore 67 cm.

49. (a) $q = 160$ mm, $p = qf/(q - f) = 4.1$ mm, $m = -q/p = -39$; the total magnification is $10 \times 39 = 390\times$. (b) 4.1 mm.

51. (a) $f_{eye} = f_{obj}/m_{ang} = 30$ mm. (b) $\beta = L/25$ cm, $\alpha = \beta/m_{ang} = L_0/p$, $L_0 = Lp/(25$ cm$)(m_{ang}) = 2$ m.

53. (a) Objective: $q = pf/(p - f) = 1.02$ m, $m_1 = -q/p = -0.020$. Eyepiece: $q = -25$ cm, $p = qf/(q - f) = 4.17$ m, $m_2 = -q/p = 6$. Hence $m = m_1m_2 = -0.12$, $h' = mh - -4.8$ cm, where the minus sign signifies an inverted image. (b) Here $m_1 = -0.25$, so $m = -1.5$ and $h' = -60$ cm.

55. (a) No. (b) A concave mirror forms a virtual image of any object that is between the mirror and its focal point.

57. $q = pf/(p - f) = -40$ cm, $m = -q/p = 2$; the image is therefore erect, virtual, and larger than the object.

59. $q = pf/(p - f) = 200$ cm, $m = -q/p = -4$; the image is therefore inverted, real, and larger than the object.

61. $q = pf/(p - f) = 66.7$ cm, $m = -q/p = -0.667$; the image is therefore inverted, real, and smaller than the object.

63. $q = pf/(p - f) = -6.43$ cm, $m = -q/p = 3.57$.

65. $f = -R/2 = -15$ cm, $q = -10$ cm, $p = qf/(q - f) = 30$ cm.

67. $m = \frac{1}{5}$, $p = 150$ cm, $q = -mp = -30$ cm, $f = pq/(p + q) = -37.5$ cm; $R = -2f = 75$ cm.

69. Here $m = +3$, $p = 10$ cm, $q = -mp = -30$ cm, so $f = pq/(p + q) = 15$ cm. When $m = -3$, $q = -mp = 3p$ and $1/p + 1/3p = 1/f$, $p = 4f/3 = 20$ cm. In this case the image is real and inverted.

CHAPTER 27

1. Light from incoherent sources can interfere, but the resulting interference pattern shifts continually because there is no definite phase relationship between the beams from the sources. When the sources are coherent, there is such a relationship, and the pattern is stable and therefore discernible.

3. Light waves have very short wavelengths, so their interference patterns are hard to detect. Also, natural sources of light emit waves only as short trains of random phase and so are incoherent with respect to the detection time of the eye and of most instruments.

5. The first-order spectrum contains bright lines for which $\sin \theta = \lambda/d$, where d is the spacing of the slits, and the second-order spectrum contains bright lines for which $\sin \theta = 2\lambda/d$. The second-order spectrum is deviated by more than the first-order spectrum, and is wider as well (twice as wide, in fact). A prism produces only a single spectrum.

7. (a) $d = 2.5 \times 10^{-6}$ m, $\sin \theta = \lambda/d = 0.30$, $\theta = 17°$. (b) $\sin \theta = 3\lambda/d = 0.90$, $\theta = 64°$.

9. Here $d = 2.5 \times 10^{-6}$ m. Since $\sin \theta = n\lambda/d$ and the largest value $\sin \theta$ can have is 1, the maximum number of images possible is $n = d/\lambda = 4$.

11. Here $d = 3.33 \times 10^{-6}$ m; $\sin \theta_1 = \lambda_1/d = 0.12$, $\theta_1 = 7°$; $\sin \theta_2 = \lambda_2/d = 0.21$, $\theta_2 = 12°$; $h_1 = L \tan \theta_1 = 0.246$ m, $h_2 = L \tan \theta_2 - 0.426$ m, $h_2 - h_1 - 0.18$ m.

13. The light waves reflected by the outer surface of the soap film are shifted in phase by $\lambda/2$, whereas those reflected by the inner surface of the film are not shifted in phase; hence destructive interference occurs in the reflected light where the film is extremely thin.

15. Since $n_{glass} > 1.38$, light reflected from both top and bottom of the fluoride film undergoes a phase shift of $\lambda/2$, so this factor can be ignored here. For destructive interference, the path length $2d$ in a film d thick must be an odd number of half wavelengths, the smallest of which is $2d =$

$\lambda/2$; the thinnest coating gives maximum cancellation over the greatest angle of incidence, so it is normally used. The wavelength λ in the film is λ_0/n, where λ_0 is the wavelength in air, so $d = \lambda/4 = \lambda_0/4n = 99.6$ nm.

17. The wavelengths in visible light are very small relative to the size of a building, whereas those in radio waves are more nearly comparable.

19. This is a diffraction effect.

21. Higher resolving power; greater light-gathering ability, hence ability to form images of faint objects.

23. $d_0 = 1.22 \, \lambda \, L/D = 47.1$ m. In practice, irregularities in the earth's atmosphere prevent this resolution from being attained.

25. (a) $\theta_0 = 1.22 \, \lambda/D = 3.37 \times 10^{-3}$ rad $= 0.193°$; the angular diameter is $2\theta_0 = 0.386°$. (b) The apparent angular diameter of the moon is 3476 km/3.8×10^5 km $= 0.0091$ rad $= 0.52°$.

27. (a) $L = d_0 D/1.22 \, \lambda = 1.64 \times 10^4$ m $= 16.4$ km. (b) 8.2 km.

29. For a distant object, the object distance L equals the focal length f of the lens. The ratio f/D for the required resolution, with $d_0 = 10^{-5}$ m, is therefore $d_0/1.22 \, \lambda = 15$. This is an aperture of $f/15$.

31. The individual waves in an ordinary light beam are polarized, but their planes of polarization are random so the beam itself is unpolarized.

CHAPTER 28

1. Such wave phenomena as diffraction and interference are easier to demonstrate than such quantum phenomena as the photoelectric effect.

3. $f = c/\lambda = 5 \times 10^{14}$ Hz; $n = E/hf = 3$ photons.

5. $f = E/h = 1.21 \times 10^{15}$ Hz, $\lambda = c/f = 2.49 \times 10^{-7}$ m.

7. $\lambda = c/f = ch/E = 2.22 \times 10^{-7}$ m; ultraviolet.

9. (a) $P/hf = 4.22 \times 10^{21}$ photons/m$^2 \cdot$ s. (b) The photon speed is c, hence there are $(4.22 \times 10^{21}$ photons/m$^2 \cdot$ s)/c $= 1.41 \times 10^{13}$ photons/m^3.

11. $f = eV/h = 2.41 \times 10^{18}$ Hz; X rays.

13. No.

15. Apply an electric or magnetic field and look for a deflection.

17. $\lambda = h/mv = 3.3 \times 10^{-29}$ m.

19. $\lambda = h/mv = h/\sqrt{2m\text{KE}} = h/\sqrt{2mVe}$, $V = h^2/2me\lambda^2$ $= 5.0 \, \mu$V.

21. Only if the particle's speed is also known.

23. $m = 1.01m_0$, $\sqrt{1 - v^2/c^2} = 1/1.01$, $v^2/c^2 = 0.0197$, $v = 4.2 \times 10^7$ m/s.

25. (a) $m = m_0/\sqrt{1 - v^2/c^2}$, $\lambda = h/mv = h\sqrt{1 - v^2/c^2}/m_0 v = 6.86 \times 10^{-12}$ m. (b) $\lambda = h\sqrt{1 - v^2/c^2}/m_0 v = 2.71 \times 10^{-12}$ m.

27. A negative total energy signifies that the electron is bound to the nucleus; the kinetic energy of the electron is of course a positive quantity.

29. $E = -E_1/n^2 = 0.85$ eV.

31. $\frac{3}{2}kT = E_1$, $T = 2E_1/3k = 1.05 \times 10^5$ K.

33. (a) $\lambda = h/mv = 3.68 \times 10^{-63}$ m. (b) $n\lambda = 2\pi r$, $n = 2\pi r/\lambda = 2.56 \times 10^{74}$.

35. The angular momentum is $L = mvr_n$. Since $\lambda = h/mv$, $mv = h/\lambda$, and since $n\lambda = 2\pi r_n$, $r_n = n\lambda/2\pi$. Hence $L = (h/\lambda)(n\lambda/2\pi) = nh/2\pi$.

37. A hydrogen sample contains a great many atoms, each of which can undergo a variety of possible transitions.

39. (a) Continuous emission spectrum (see Sec. 17–4). (b) Emission line spectrum. (c) Absorption line spectrum.

41. $1/\lambda = R(1 - 1/10^2)$, $\lambda = 9.21 \times 10^{-8}$ m $= 92.1$ mm; ultraviolet.

43. The shorter the wavelength, the higher the frequency, which in turn means the greater the energy change. From Fig. 28–20 the maximum energy change in the Brackett series corresponds to a transition from $n = \infty$ to $n = 4$, so that $hf = E_1 - E_f = -E_1/4^2 = 1.36 \times 10^{-19}$ J, $f = 2.06 \times 10^{14}$ Hz, and $\lambda = c/f = 1.46 \times 10^{-6}$ m.

45. The H atoms must be raised to the $n = 5$ state from the ground state, which requires $\Delta E = E_1(1/n_i^2 - 1/n_f^2) = E_1(1 - 1/5^2) = 13.06$ eV.

47. The length must be an integral number of half-wavelengths long in order that a standing wave be set up in it that causes further emissions to be exactly in step with the previous ones, thus producing a coherent beam.

CHAPTER 29

1. According to the quantum theory, an electron does not follow a definite, predictable path. At any moment, an electron may be thought of as being somewhere within a probability cloud; the denser the cloud, the greater the probability of finding the electron there.

3. In a polar covalent molecule the shared electrons spend more time on the average near one of the atoms.

5. Their ability to bond with each other as well as with other atoms.

7. In a NaCl molecule, the Na^+ and Cl^- ions interact only with each other. In an NaCl crystal, each ion of a given kind has six nearest neighbors of the other kind of ion with which it interacts strongly, plus twelve neighbors of the same kind farther away, and so on. There is no reason why the equilibrium separation in each case should be the same.

9. Van der Waals forces are too weak to hold inert gas atoms together against the forces exerted during collisions in the gaseous state.

11. A good conductor of heat is also a good conductor of electricity, because the mechanism of transport in each case involves the electron gas in the metal.

13. The exclusion principle.

15. A solid with a partly filled upper energy band is a conductor because electrons in this band can readily absorb small amounts of energy from an electric field and move through the solid. A solid with a filled upper band and a wide forbidden band above it is an insulator because electrons in the filled band require very large amounts of energy in order to jump the forbidden band to the empty band above it. A solid with a filled upper band and a narrow forbidden band is a semiconductor because a few electrons have enough thermal energy to enter the empty band above the forbidden band, and these electrons can move through the solid under the influence of an electric field.

17. p-type.

CHAPTER 30

1. The isotopes have the same atomic number and hence the same electron structures, therefore, they have the same chemical behavior. They have different numbers of neutrons, hence different atomic masses.

3. $5n, 5p;$ $12n, 10p;$ $20n, 16p;$ $50n, 38p;$ $108n, 72p.$

5. $^{10}_5B + ^1_0n \rightarrow ^4_2He + ^7_3Li;$ hence $Z = 3, A = 7$, lithium.

7. $^2_1H; ^1_0n; ^1_1H.$

9. If x is the fraction of ^{10}B present, then $(1 - x)$ is the fraction of ^{11}B present and $x(10.01$ u$) + (1 - x)(11.01$ u$) = 10.82$ u. Solving gives $x = 0.19$. Hence ordinary boron consists of 19% ^{10}B and 81% ^{11}B.

11. $2m_H + 2m_n = 4.032980$ u, $\Delta m = 0.030377$ u $= 28.3$ MeV. There are four nucleons in ^4_2He; hence the binding energy per nucleon is 7.1 MeV.

13. $E = 22.4$ MeV $= 0.02406$ u; $m_{Li} = 2m_{He} + E - m_H = 6.01516$ u.

15. Helium and radon cannot be combined chemically to form radium, nor can radium be broken down into helium and radon by chemical means.

17. $^{206}_{82}Pb.$

19. $^{233}_{92}U.$

21. (a) $^{222}_{86}Rn;$ $m_E = (4.87$ MeV$)/(931$ MeV/u$) = 0.0052$ u; hence $m(Rn) = m(Ra) - [m_E + m(He)] = 222.0176$ u. (b) The "missing" energy goes into KE of the recoiling nucleus.

23. Such particles lose energy mainly by electric interactions with the electrons of atoms in their paths. Usually the electrons are pulled away from their parent atoms to produce ion-electron pairs.

25. Photoelectric effect, Compton scattering, and pair production.

27. One-sixteenth is left, and since $\frac{1}{2} \times \frac{1}{2} \times \frac{1}{2} \times \frac{1}{2} = \frac{1}{16}$ this means four half-lives or 6400 years.

29. $3y = 1095$ d ≈ 8 half-lives, so 0.039 of the activity remains and the power output is 0.0039 W.

31. (a) The ratio between the radiocarbon and ordinary carbon contents of all living things is the same. When a plant or animal dies, its radiocarbon content decreases at a fixed rate. Hence the ratio between the radiocarbon and ordinary carbon contents of an ancient specimen of organic origin will reveal its age. (b) Determine the radioactivity per gram of a sample of carbon from the ancient wood and compare it with the radioactivity per gram of a sample from a piece of wood of recent origin. The difference between the two activities may then be converted into an age figure for the ancient wood by taking the half-life of radiocarbon into account.

33. If there is a leak in a water-cooled reactor, the coolant escapes and the fuel rods may then melt with potentially disastrous results. In a gas-cooled reactor, the heat capacity of the graphite moderator will reduce the temperature rise if the coolant gas stops circulating.

35. (a) $m = 10^{-3}$ kg, $E = mc^2 = 9 \times 10^{13}$ J. (b) 2.15×10^4 tons of TNT.

37. The density in the interior of a star is extremely high, hence there are many collisions between nuclei there; the temperature is also extremely high, hence many of the colliding nuclei have high enough energies to interact.

39. $\Delta m = 3m_a - m_c = 0.007809$ u $= 7.27$ MeV.

ACKNOWLEDGMENTS

For photographs used as chapter openers, the publisher acknowledges the following artists and sources.

1	Daniel S. Brody/Stock, Boston, Inc.	*2*
2	© Gerry Cranham/Rapho-Photo Researchers, Inc.	*34*
3	Arthur Beiser	*64*
4	Edward L. Leavitt	*90*
5	© Jack Dermid from National Audubon Society/Photo Researchers, Inc.	*118*
6	© Tim Davis 1982/Photo Researchers, Inc.	*146*
7	Natural Photo/Robert Vernon Wilson ©	*172*
8	Natural Photo/Robert Vernon Wilson ©	*200*
9	Natural Photo/Robert Vernon Wilson ©	*226*
10	American Iron & Steel Institute	*248*
11	© Dan Brinzac/Peter Arnold, Inc.	*272*
12	© Roy King	*306*
13	© Fundamental Photographs of New York	*332*
14	Peter Menzel/Stock, Boston, Inc.	*368*
15	David Powers	*398*
16	Mercedes-Benz of North America, Inc.	*418*
17	Thermo King Corp.	*448*
18	A charged Van de Graaf generator (Andrew Wilbraham). Charles H. Cox	*470*
19	© John Hendry from National Audubon Society/Photo Researchers, Inc.	*496*
20	Natural Photo/Robert Vernon Wilson ©	*528*
21	Edward L. Leavitt	*554*
22	© Roy King	*590*
23	Edward L. Leavitt	*616*
24	Natural Photo/Robert Vernon Wilson ©	*644*
25	Bell Labs	*672*
26	Kitt-Peak National Observatory	*706*
27	Fundamental Photographs of New York	*740*
28	Lawrence Livermore National Laboratory	*762*
29	Argonne National Laboratory	*794*
30	Atomic Industrial Forum, Inc.	*818*

INDEX

CONVERSION FACTORS

TIME

1 hour = 60 min = 3600 s
1 day = 1440 min = 8.64×10^4 s
1 year = 365.2 days = 3.156×10^7 s

ANGLE

1 radian (rad) = $57.30° = 57°18'$
$1° = 0.01745$ rad $(180° = \pi$ rad)
1 rev/min (rpm) = 0.1047 rad/s
1 rad/s = 9.549 rpm

LENGTH

1 meter (m) = 100 cm = 39.37 in. = 3.281 ft
1 centimeter (cm) = 10 millimeters (mm)
        = 0.3937 in.
1 kilometer (km) = 1000 m = 0.6214 mi
1 foot (ft) = 12 in. = 0.3048 m = 30.48 cm
1 inch (in.) = 2.540 cm
1 mile (mi) = 5280 ft = 1.609 km
1 nautical mile (nmi) = 6076 ft. = 1.152 mi
        = 1.852 km

AREA

$1 \text{ m}^2 = 10^4 \text{ cm}^2 = 10.76 \text{ ft}^2$
$1 \text{ cm}^2 = 10^{-4} \text{ m}^2 = 0.1550 \text{ in.}^2$
$1 \text{ ft}^2 = 144 \text{ in.}^2 = 9.290 \times 10^{-2} \text{ m}^2 = 929.0 \text{ cm}^2$
$1 \text{ in.}^2 = 6.452 \text{ cm}^2$
$1 \text{ hectare (ha)} = 10^4 \text{ m}^2 = 2.471 \text{ acres}$
$1 \text{ acre} = 43,560 \text{ ft}^2 = 0.4049 \text{ ha}$

VOLUME

$1 \text{ m}^3 = 10^3 \text{ liters} = 10^6 \text{ cm}^3 = 35.32 \text{ ft}^3$
$1 \text{ liter} = 10^3 \text{ cm}^3 = 10^{-3} \text{ m}^3 = 0.2642 \text{ gal}$
        = 1.056 quart
$1 \text{ ft}^3 = 1728 \text{ in.}^3 = 2.832 \times 10^{-2} \text{ m}^3 = 28.32 \text{ liters}$
        = 7.481 gal
$1 \text{ U.S. gallon (gal)} = 4 \text{ quarts} = 0.1337 \text{ ft}^3$
        = 3.785 liters

SPEED

1 m/s = 3.281 ft/s = 2.237 mi/h = 3.600 km/h
1 ft/s = 0.3048 m/s = 0.6818 mi/h = 1.097 km/h
1 km/h = 0.2778 m/s = 0.9113 ft/s
        = 0.6214 mi/h

1 mi/h = 1.467 ft/s = 0.4470 m/s = 1.609 km/h
1 knot = 1 nmi/h = 1.152 mi/h = 1.852 km/h
        = 1.688 ft/s = 0.5144 m/s

MASS

1 kilogram (kg) = 1000 grams (g) = 0.0685 slug
(*Note:* 1 kg corresponds to 2.21 lb in the sense that
the *weight* of 1 kg is 2.21 lb at sea level. Similarly 1
lb corresponds to 453.6 g and 1 oz to 28.35 g.)
1 slug = 14.59 kg
1 atomic mass unit (u) = 1.660×10^{-27} kg
        = 1.492×10^{-10} J = 931.5 MeV

FORCE

1 newton (N) = 0.2248 lb
1 pound (lb) = 4.448 N

PRESSURE

$1 \text{ pascal (Pa)} = 1 \text{ N/m}^2 = 1.450 \times 10^{-4} \text{ lb/in.}^2$
$1 \text{ bar} = 10^5 \text{ Pa} = 14.50 \text{ lb/in.}^2$
$1 \text{ lb/in.}^2 = 144 \text{ lb/ft}^2 = 6.895 \times 10^3 \text{ Pa}$
$1 \text{ atm} = 1.013 \times 10^5 \text{ Pa} = 1.013 \text{ bar}$
        $= 14.70 \text{ lb/in.}^2$
1 torr = 133.3 Pa

ENERGY

1 joule (J) = 0.7376 ft · lb = 2.390×10^{-4} kcal
        = 9.484×10^{-4} Btu
        = 2.778×10^{-7} kWh
1 foot-pound (ft · lb) = 1.356 J = 1.29×10^{-3} Btu
        = 3.25×10^{-4} kcal
1 kilocalorie (kcal) = 4185 J = 3.968 Btu
        = 3077 ft · lb
1 Btu = 0.252 kcal = 778 ft · lb = 1054 J
1 electron volt (eV) = 10^{-6} MeV = 10^{-9} GeV
        = 1.602×10^{-19} J
1 kilowatt-hour (kWh) = 3.600×10^6 J
        = 2.655×10^6 ft · lb
        = 860.4 kcal